IFMBE Proceedings

Volume 46

Series Editor

Ratko Magjarevic

Deputy Editors

Fatimah Binti Ibrahim
Igor Lacković
Piotr Ładyżyński
Emilio Sacristan Rock

The International Federation for Medical and Biological Engineering, IFMBE, is a federation of national and transnational organizations representing internationally the interests of medical and biological engineering and sciences. The IFMBE is a non-profit organization fostering the creation, dissemination and application of medical and biological engineering knowledge and the management of technology for improved health and quality of life. Its activities include participation in the formulation of public policy and the dissemination of information through publications and forums. Within the field of medical, clinical, and biological engineering, IFMBE's aims are to encourage research and the application of knowledge, and to disseminate information and promote collaboration. The objectives of the IFMBE are scientific, technological, literary, and educational.

The IFMBE is a WHO accredited NGO covering the full range of biomedical and clinical engineering, healthcare, healthcare technology and management. It is representing through its 60 member societies some 120.000 professionals involved in the various issues of improved health and health care delivery.

IFMBE Officers
President: Ratko Magjarevic, Vice-President: James Goh
Past-President: Herbert Voigt
Treasurer: Marc Nyssen, Secretary-General: Shankhar M. Krishnan
http://www.ifmbe.org

More information about this series at http://www.springer.com/series/7403

Vo Van Toi · Tran Ha Lien Phuong

Editors

5th International Conference on Biomedical Engineering in Vietnam

 Springer

Editors
Vo Van Toi
Biomedical Engineering Department
International University
Vietnam National Universities
Ho Chi Minh City
Vietnam

Tran Ha Lien Phuong
Biomedical Engineering Department
International University
Vietnam National Universities
Ho Chi Minh City
Vietnam

ISSN 1680-0737   ISSN 1433-9277   (electronic)
ISBN 978-3-319-11775-1   ISBN 978-3-319-11776-8   (eBook)
DOI 10.1007/978-3-319-11776-8

Library of Congress Control Number: 2014954484

Springer Cham Heidelberg New York Dordrecht London

The IFMBE Proceedings is an Official Publication of the International Federation for Medical and Biological Engineering (IFMBE)

Printed on acid-free paper

Springer is part of Springer Science+Business Media (www.springer.com)

Preface

Welcome to the Fifth International Conference on the Development of Biomedical Engineering in Vietnam.

Vietnam is a rapidly developing, socially dynamic country, where interest in biomedical engineering activities has grown considerably in recent years. The leadership of the Vietnamese government, and of research and educational institutions, are well aware of the importance of this field for the development of the country and have instituted policies to promote its development. The political, economic and social environment within the country offers unique opportunities for the international community and this conference was intended to provide a vehicle for the sharing of experiences; development of support and collaboration networks for research; and exchange of ideas on how to improve the educational and entrepreneurial environment to better address the urgent needs of Vietnam.

In **January 2004**, under the sponsorship of the U.S. National Science Foundation, Professor Vo Van Toi of the Biomedical Engineering Department of Tufts University, Medford Massachusetts USA, led a U.S. delegation that consisted of Biomedical Engineering professors from different universities in the United States, and visited several universities and research institutions in Vietnam to assess the state of development of this field. This delegation proposed a five year plan that was enthusiastically embraced by the international scientific communities to actively develop collaborations with Vietnam.

Within this framework, in July 2005, the First International Conference on the Development of Biomedical Engineering in Vietnam was held in Ho Chi Minh City. From that conference a Consortium of Vietnam- International Universities was created to advise and assist the development of Biomedical Engineering in Vietnamese universities.

In **July 2007, the Second International Conference** on the Development of Biomedical Engineering in Vietnam was held in Hanoi. During this event the Vietnamese Association of Biomedical Science and Engineering was endorsed by the Asia-Pacific International Molecular Biology Network (AIMBN), Biomedical Engineering Society Singapore (BESS), International Federation for Medical and Biological Engineering (IFMBE), Société Française de Génie Biologique et Médical (SFGBM) and IFMBE Asia-Pacific Working Group.

In March 2009, International University (IU) established its Biomedical Engineering (BME) Department and the first accredited Engineer in BME degree in Vietnam (code: 52.42.02.04). IU is a member of Vietnam National Universities – Ho Chi Minh City (VNU-HCM), one of the two elite university networks in Vietnam, and is the first public university in Vietnam that teaches all courses in English. It was created as a new model to modernize the higher education in Vietnam. The BME Department at IU has since coordinated with its national and international counterparts to promote the activities of this field in Vietnam.

In **January 2010, the Third International Conference** on the Development of Biomedical Engineering in Vietnam was organized by IU in Ho Chi Minh City. It reflected the steady growth of the activities in this field in Vietnam, and featured the contributions of researchers of 21 countries, including: Australia, Belgium, Canada, Denmark, France, India, Japan, Korea, Malaysia, New Zealand, Philippines, Poland, Russia, Singapore, Spain, Switzerland, Taiwan, Thailand, United Kingdom, the United States and Vietnam. The Conference was endorsed by the International Federation for Medical and Biological Engineering (IFMBE). It also hosted the Clinical Engineering Workshop of the IFMBE Asia Pacific Working Group. The contributed papers were published in the IFMBE Proceedings Series by Springer (ISBN 978-3-642-12019-0).

In **January 2012, the Fourth International Conference** on the Development of Biomedical Engineering in Vietnam was organized in Ho Chi Minh City as a Mega-conference. It was kicked off by the Regenerative Medicine Conference (Jan 8-10, 2012) with the theme "BUILDING A FACE" USING A REGENERATIVE MEDICINE APPROACH", endorsed mainly by the Tissue Engineering and Regenerative Medicine International Society (TERMIS) and co-organized by Professor Stephen E. Feinberg, University of Michigan Health System, USA, Professor Anh Le, University of Southern California, USA and Professor Vo Van Toi, International University-VNU HCM, Vietnam. It was followed by the Computational Medicine Conference, endorsed mainly by the Computational Surgery International Network (COSINE) and the Computational Molecular Medicine of German National Funding Agency; and the General Biomedical Engineering Conference, endorsed mainly by the International Federation for Medical and Biological Engineering (IFMBE) (Jan 10-12) and co-organized by Professor Paolo Carloni, German Research School for Simulation Sciences GmbH, Germany, Professor Marc Garbey, University of Houston,

USA and Professor Vo Van Toi, International University-VNU HCM, Vietnam. It featured the contributions of 435 scientists from 30 countries, including: Australia, Austria, Belgium, Canada, China, Finland, France, Germany, Hungary, India, Iran, Italy, Japan, Jordan, Korea, Malaysia, Netherlands, Pakistan, Poland, Russian Federation, Singapore, Spain, Switzerland, Taiwan, Turkey, Ukraine, United Kingdom, United States, Uruguay and Viet Nam. The contributed papers were published in the IFMBE Proceedings Series by Springer (ISBN 978-3-642-32182-5).

This **Fifth International Conference** on the Development of Biomedical Engineering in Vietnam (June 16-18, 2014) is a very special event. It officially opens the season for celebration of the 20th Anniversary of Vietnam National Universities – Ho Chi Minh City. It also marks the 10th Anniversary of International University and the 5th Anniversary of the Biomedical Engineering Department. This Conference features 199 papers of 508 authors and co-authors from 26 countries including Australia, Bangladesh, Belgium, Canada, China, Croatia, Czech Republic, Denmark, Finland, France, Germany, India, Israel, Italy, Japan, Korea, Malaysia, Norway, Singapore, Slovenia, Switzerland, Taiwan, Turkey, United Kingdom, USA and Vietnam. Almost all Vietnamese institutions have their delegations. Besides Vietnam the 2 countries that have the most contributors are the USA and Australia. The plenary session features the lectures on the progress in different fields of Biomedical Engineering by distinguished keynote speakers such as Prof. Ratko Magjarevic, President of IFMBE; Dr. Robert A. Lieberman, Vice-President of International Society for Optics and Photonics SPIE; Prof. Vo-Dinh Tuan, Director of Fitzpatrick Institute for Photonics, Duke University, USA; Prof. Christian Griesinger, Director of Max Planck Institute for Biophysical Chemistry, Germany; Prof. Anja Boisen, Director of VKR Centre Of Excellence 'NAMEC', Denmark; Prof. Yin Xiao, Director of Australia-China Centre for Tissue Engineering and Regenerative Medicine, Queensland University of Technology, Australia; Prof. Yukio Nagasaki, Department of Materials Science and Medical Sciences, University of Tsukuba, Japan; and Prof. Fong-Chin Su, President of Taiwanese Society of Biomedical Engineering, Director, Medical Device Innovation Center, National Cheng Kung University, Taiwan. The contributed papers will continue to be published in the IFMBE Proceedings Series by Springer.

The editors would like to thank the leadership and staff of VNU-HCM and IU, local and international sponsors, and the staff and students of Biomedical Engineering Department for their valuable support for the conference and their assistance in the editing and publication of the volume.

Editors

Vo Van Toi
Tran Ha Lien Phuong

and Editorial Board

Do Minh Thai
Nguyen Duc Thang
Nguyen Phuong Nam
Nguyen Thanh Tam
Nguyen Thi Hiep
Nguyen Van Hoa
Tran Truong Dinh Thao
Tu Thi Tuyet Nga

Organizing Institutions

Vietnam National Universities
HCM City (VNU-HCM)

International University
of VNU-HCM

Honorary Chairs

Professor Phan Thanh Binh, President of Vietnam National Universities – HCM City

Professor Ho Thanh Phong, Rector of International University VNU-HCM

Conference General Chair

Vo Van Toi, PhD
Professor and Chair of Biomedical Engineering Department
International University – Vietnam National Universities
Quarter 6, Linh Trung, Thu Duc District, Ho Chi Minh City, Vietnam
Tel: (84-8) - 37 24 42 70 Ext. 3237
Fax: (84-8) - 37 24 42 71
Email: bme2014@hcmiu.edu.vn
Website: www.hcmiu.edu.vn/bme

Organizing Committee

Dang Van Phuoc, MD, PhD
Professor, President of the School of Medicine
Vietnam National Universities
Ho Chi Minh City, Vietnam

Paolo Carloni, PhD
Professor, Head of Laboratory Computational Biophysics
German Research School for Simulation Sciences GmbH
Jülich, Germany

Hoang Dung, PhD
Professor, Co-Director of Center for Molecular and Nanoarchitectures (MANAR),
Vietnam National Universities
Ho Chi Minh City, Vietnam

Marco Santello, PhD
Professor, School of Biological and Health Systems Engineering
Arizona State University, USA

Lawrence Trong-Huan Le, PhD
Professor, Department of Radiology and Diagnostic Imaging
University of Alberta, Edmonton, Canada

Fong-Chin Su, PhD
President of Taiwanese Society of Biomedical Engineering, Director, Medical Device Innovation Center - a Global Center of Excellence, Ministry of Education, Distinguished Professor at National Cheng Kung University - Department of Biomedical Engineering, Taiwan

Nghiem Doan, PhD
Oral Surgeon, Dental Sedationist & Specialist in Public Health Dentistry, Queensland University of Technology, Australia

Vo Van Toi, PhD
Professor and Chair of Biomedical Engineering Department
International University
Vietnam National Universities Ho Chi Minh City, Vietnam

Scientific Committee

Ali Zulf	Teesside University's Graduate Research School UK
Anders Wolff	Department of Micro and Nanotechnology, Technical University of Denmark (DTU-nanotech), Denmark
Anja Boisen	Head of Nano section, Department of Micro and nanotechnology, Technical University of Denmark, Denmark
Akihiro Kishimura	Kyushu University, Japan
Asha Shekaran	Department of Biomedical Engineering, Georgia Institute of Technology and Emory University, USA
Atsushi Maruyama	Tokyo Institute of Technology, Japan
Brian Lau	Department of Biomedical Engineering, Duke University, USA
Bengang Xing	Nanyang Technological University, Singapore
Chang Young-Tae	National University of Singapore, Singapore
Christian Langton	Institute of Health & Biomedical Innovation, Queensland University of Technology, Australia
Christophe Moser	Microengineering department, EPFL, Switzerland
Christian Griesinger	Max Planck Institute for Biophysical Chemistry, Germany
Dang Duong Bang	Laboratory of Applied Micro and nanotechnology (LAMINATE), Diagnostics Engineering, Division of Food Microbiology, National Food Institute, Technical University of Denmark
David Leavesley	Queensland University of Technology, Australia
Dominique Pioletti	Director of the Laboratory of Biomechanical Orthopedics, EPFL, Switzerland
Fong-Chin Su	Taiwanese Society of Biomedical Engineering, Medical Device Innovation Center, National Cheng Kung University, Taiwan
Glen S.Kwon	University of Wisconsin, Madison, USA
Giuseppe Legname	Neurobiology Sector, SISSA, Trieste, Italy
Hoang Anh Dao	Falculty of Odontostomatology, Hue University of Medicine and Pharmacy, Vietnam
Hoang Dung	Vietnam National Universities, Ho Chi Minh City, Vietnam
Janez Plavec	Slovenian NMR Centre, National Institute of Chemistry, Slovenia
Jae-Seung Lee	Korea University, Korea
Jonathan Kaufman	CyberLogic, Inc., USA
Kang Lifeng	Department of Pharmacy, National University of Singapore, Singapore
Kazunari Akiyoshi	Kyoto University, Japan
Kazuo Nagasawa	Tokyo University of Agriculture & Technology, Japan
Keith Blackwood	Institute of Health & Biomedical Innovation, Queensland University of Technology, Australia
Le Thi Ly	International University, VNU-HCMC, Vietnam
Luiz E. Bertassoni	Faculty of Dentistry, University of Sydney, Australia
Lawrence Trong-Huan Le	University of Alberta, Canada
Marco Santello	School of Biological and Health Systems Engineering, Arizona State University, USA
Maria Laura Bolognesi	Dipartimento di Farmacia e Biotecnologie, Italy
Masayuki Yamato	Tokyo Women's Medical University, Japan
Mia Woodruff	Queensland University of Technology, Australia

Michael Canva	Directeur de Recherches au CNRS, Laboratoire Charles Fabry de l'Institut d'Optique Campus Polytechnique, France
Michinao Hashimoto	Singapore University of Technology and Design
Naoki Sugimoto	Konan University, Japan
Nguyen Vu Hieu	University of Paris, France
Nguyen Thi Hiep	International University, VNU-HCMC, Vietnam
Nguyen Thi Hue	International University, VNU-HCMC, Vietnam
Nguyen Duc Thang	International University, VNU-HCMC, Vietnam
Nguyen Hoang Khue Tu	International University, VNU-HCMC, Vietnam
Nguyen Thanh Hai	HCMC University of Technical Education, Vietnam
Nghiem Doan	Institute of Health & Biomedical Innovation, Queensland University of Technology, Australia
Nikos Stergiopulos	Laboratory of Hemodynamics and Cardiovascular Technology, Swiss Federal Institute of Technology, EFPL, Switzerland
Anh Hoang	Harvard Medical School, Massachusetts General Hospital, USA
Phan Anh Tuan	Nanyang Technological University, Singapore
Paolo Carloni	German Research School for Simulation Sciences GmbH and Institute for Advanced Simulation (IAS), Germany
Phuoc Long Truong	Nanobiotechnology Lab Department of Chemical and Biological Engineering Korea University, South Korea
Ratko Magjarevic	President, IFMBE (International Federation for Medical and Biological Engineering)
Robert A. Lieberman	President, Intelligent Optical Systems, Inc.
Sara Richter	University of Padua, Italy
Shawn Tan	Institute of Materials Research and Engineering, A*STAR, Singapore
Stefano Sensi	University of California, USA
Takao Aoyagi	MANA Coordinator, Nano-Bio Field, MANA, National Institute for Materials Science, Japan
Thanh D Nguyen	Massachusetts Institute of Technology
Tram Dang	Institute of Medical Biology, A*STAR, Singapore, School of Chemical & Biomedical Engineering, Nanyang Technology University, Singapore
Tran Quang Hung	Department of Micro and Nanotechnology, Technical University of Denmark (DTU-nanotech), Denmark
Tran Ha Lien Phuong	International University, VNU-HCMC, Vietnam
Tran Truong Dinh Thao	International University, VNU-HCMC, Vietnam
Uyen Minh Le	Department of Pharmaceutical Sciences, College of Pharmacy, Sullivan University, USA
Vincent Pizziconi	School of Biological and Health Systems Engineering, Arizona State University, USA
Vo-Dinh Tuan	Director, Fitzpatrick Institute for Photonics R. Eugene and Susie E. Goodson Distinguished Professor of Biomedical Engineering Duke University, USA
Vo Van Toi	International University, VNU-HCMC, Vietnam
Xu Chenjie	School of Chemical and Biomedical Engineering, Nanyang Technological University, Singapore

Xuewei Liu	Nanyang Technological University, Singapore
Yin Xiao	Institute of Health and Biomedical Innovation, Queensland University of Technology (QUT), Australia
Yukio Nagasaki	University of Tsukuba, Japan
Yutaka Ikeda	Department of Materials Science, University of Tsukuba, Japan
Zhang Jun	Physics and Applied Physics Division, School of Physical and Mathematical Science, Nanyang Technology University, Singapore
Zhibin Du	Institute of Health & Biomedical Innovation, Queensland University of Technology, Australia

Local Organizing Committee (LOC) & Secretary

Prof. Phan Thanh Binh	Vietnam National Universities, Ho Chi Minh City, Honorary Chair
Prof. Ho Thanh Phong	International University-VNU-HCM, Honorary Chair
Prof. Hoang Dung	Vietnam National Universities, Ho Chi Minh City, Co-Organizer
Prof. Vo Van Toi	International University-VNU-HCM, Conference General Chair
Dr. Nguyen Hong Quang	International University- VNU-HCM, LOC Co-Chair
Dr. Tran Ha Lien Phuong	International University-VNU-HCM, LOC Chair
Dr. Nguyen Trong Nam	International University-VNU-HCM, External and Public Relations Chair
Dr. Nguyen Duc Thang	International University-VNU-HCM, Poster, Sponsorship and Exhibition coordinator
Dr. Tran Truong Dinh Thao	International University-VNU-HCM, General coordinator
Dr. Nguyen Thi Hiep	International University-VNU-HCM, Volunteer coordinator
Ms. Hoang Van	International University-VNU-HCM, Facility manager
Ms. Tran Thi Thanh Lich	International University-VNU-HCM, Financial manager
Mr. Dong Quang Vinh	International University-VNU-HCM, Research and development coordinator
Ms. Nguyen Anh Dao	International University-VNU-HCM, Conference design consultant
Mr. Nguyen Thanh Tam	International University-VNU-HCM, Website manager
Mr. Nguyen Phuong Nam	International University-VNU-HCM, Poster and Exhibition coordinator
Mr. Nguyen Van Hoa	International University-VNU-HCM, Proceedings editing team
Mr. Do Minh Thai	International University-VNU-HCM, Facility coordinator
Ms. Tu Thi Tuyet Nga	International University-VNU-HCM, Conference Secretary

Acknowledgements

We would like to express our deepest appreciation to the following sponsors:

Table of Contents

Development of Individual Plasmonic Nanosensors for Clinical Diagnosis 1
  Phuoc Long Truong and Sang Jun Sim

YALES2BIO: A Computational Fluid Dynamics Software Dedicated to the Prediction of Blood Flows in Biomedical
Devices 7
  S. Mendez, C. Chnafa, E. Gibaud, J. Sigüenza, V. Moureau, and F. Nicoud

Numerical and Experimental Mixing Studies in a Split and Recombine Micromixer with Ellipse-Like Micropillars 11
  Nhut Tran-Minh, Erik Andrew Johannessen, and Frank Karlsen

Multiplex DNA Biosensor for Viral Infection Diagnosis Using SERS Molecular Sentinel-on-Chip 15
  Hoan T. Ngo, Hsin-Neng Wang, Thomas Burke, Christopher Woods, Geoffrey S. Ginsburg, and Tuan Vo-Dinh

Whispering Gallery Mode Biosensing – A Detailed Study on ZnO Microspheres 21
  Ngo Huynh Buu Trong, Paul Ching-Hang Chien, Yu-Da Chen, Shang-Hsuan Wu, Fu-Chen Hsiao,
  Linh-Nam Nguyen, and Yia-Chung Chang

Ultrasonic Assessment of the Radius 25
  J.J. Kaufman, G.M. Luo, F. Rosete, M. Bucovsky, E.M. Stein, E. Shane, and R.S. Siffert

High-Resolution Imaging of Dispersive Ultrasonic Guided Waves in Human Long Bones Using Regularized Radon
Transforms 28
  Tho N.H.T. Tran, Lawrence H. Le, and Mauricio D. Sacchi

Adaptive Noise Cancellation in the Intercept Time-Slowness Domain for Eliminating Ultrasonic Crosstalk in a
Transducer Array 32
  K.C.T. Nguyen, Lawrence H. Le, Mauricio D. Sacchi, L.Q. Huynh, and E. Lou

Simulation of Ultrasound Propagation in Long Bone with Depth-Varying Porosity 36
  Vu-Hieu Nguyen and Salah Naili

Frequency Independence of Ultrasound Transit Time Spectroscopy 39
  M.-L. Wille and C.M. Langton

In Vitro Ultrasonic Assessment of the Biomechanical Quality of the Interface Surrounding a Dental Implant 43
  R. Vayron and G. Haiat

Silicon-Based Fabrication of Biodegradable Polymer for Controlled Drug-Delivery 47
  Thanh D. Nguyen, Robert S. Langer, and Ana Jaklenec

Advancement in Gemcitabine Delivery for Cancer Treatment 51
  Uyen Minh Le

Multifunctional Drug Nanosystems: A Summary of Recent Researches at IMS/VAST 55
  X.P. Nguyen, T.T. TMai, P.T. Ha, H.N. Pham, H.N. Luu, H.M. Do, D.L. Tran, H.N. Nguyen, L.T. Nguyen, A.S. Ho,
  and T.M.N. Hoang

Preparation, Characterization and Antibacterial Curcumin Encapsulated Chitosan-PAA Silver Nanocomposite Films . . . 58
  N.V. Cuong, P.N.N. Han, N.K. Hoang, and N.N.L. Giang

Monitoring through Tissue the De-gelation of Alginate Gels by Different De-gelling Agents   62
    K.V.T. Nguyen and J.N. Anker

Use of Super Paramagnetic Iron-Oxide Nanoparticles in the Treatment of Atherosclerosis   67
    S. Chandramouli, S. Sanjana, and S. Swathi

Effect of pH on the Synthesis of Fucoidan-coated Magnetic Iron Oxide Nanoparticles for Biomedical Applications   71
    Khanh Nghia Tran, Phuong Ha-Lien Tran, Toi Van Vo, and Thao Truong-Dinh Tran

A Novel Non-invasive System for Acquiring Jugular Venous Pulse Waveforms   75
    Tam Nguyen, Anh Dinh, Francis M. Bui, and Toi Vo

Remote Monitoring of Cardiac Rhythm Management Devices in Vietnam The Role of the Biomedical Engineer   79
    Tran Thong

Inspired Sinewave Technique: A Novel Technology to Measure Cardiopulmonary Function   83
    Phi Anh Phan, Cathy Zhang, Daniel Geer, Clive Hahn, and Andrew Farmery

Wireless Sensor Network for Real Time Healthcare Monitoring: Network Design and Performance Evaluation
Simulation   87
    Minh-Thanh Vo, T.T. Thanh Nghi, Van-Su Tran, Linh Mai, and Chi-Thong Le

Estimation of Guidewire Inclination Angle for 3D Reconstruction   92
    T. Petković and S. Lončarić

Intelligent Heart Rate Variability Processing System   96
    Hoang ChuDuc, Thuan NguyenDuc, Kien NguyenPhan, and Ha NguyenThai

Analysis of Sleep Macro - and Microstructure 100
    Le Quoc Khai, Nguyen Thi Minh Huong, and Huynh Quang Linh

Design System to Remotely Monitor Patients with Parkinson's Disease 104
    Hai Tuyen Nguyen, Cao Cuong Vu, Van Quyet Phan, Viet Dung Nguyen, and The Dung Nguyen

Implementation of Telemedicine ECG System Based on Bluetooth Android Device 108
    Hung Ngoc Do, Minh-Thanh Vo, Tinh Duy Ho, and Thanh-Tam Nguyen

On Designing a System to Supervise Patients' Vital Signs through Wireless Sensor Network 113
    N.T. Dung, P.T. Nghia, D.Q. Long, N.H. Tuyen, N.V. Dung, and N.T. Ha

FPGA Implementation for Cardiac Excitation-Conduction Simulation Based on FitzHugh-Nagumo Model 117
    Nur Atiqah Adon, Mohamad Hairol Jabbar, and Farhanahani Mahmud

Determination of the Neutral Axis in Total Ankle Replacement 121
    Ha Vo, Barry Tuvel, Bich Nguyen, and Trung Le

Improving Electrospun Fibre Stacking with Direct Writing for Developing Scaffolds for Tissue Engineering for
Non-load Bearing Bone 125
    K.A. Blackwood, N. Ristovski, S. Liao, N. Bock, J. Ren, G.T.S. Kirby, M.M. Stevens, R. Steck, and M.A. Woodruff

A Comprehensive Evaluation of Flapless Dental Implant Treatment in Posterior Maxilla and the Conservative
Regeneration of Bone in Osteoporotic Rats 129
    N. Doan, Z. Du, J. Xiao, W. Xia, R. Crawford, P. Reher, S. Ivanovski, F. Yan, J. Chen, and Y. Xiao

A Measure of Clinical Outcomes in Dental Implant Surgery Flapless Surgery versus Flap Technique in Posterior
Maxilla of Post Menopause Women 133
    M. Nguyen, N. Doan, Z. Du, P. Reher, and Y. Xiao

An Analysis of Exosomes from Keratinocytes and Fibroblasts 137
    Uyen Thi Trang Than, Dominic Guanzon, Lucas Wager, Kerry J. Manton, Brett Hollier, and David Leavesley

Oral Health Problems among Adult Patients at Commune Health Centres in Central Vietnam: Prevalence and Care
Seeking Behaviour 142
    Hoang Anh Dao, Peter Hill, Ngo Hien, and Nguyen Toai

Treatment Outcomes of Periapical Lesions in Permanent Incisors Treated with Calcium Hydroxide 147
    Thai Van Nguyen and Thao Quy Le

The Primary Results of the N_2/Ar Micro-plasma Exposure on Second Degree Wound Healing 151
    Minh-Hien Ngo, Pei-Lin Shao, and Jiunn-Der Liao

Novel Approaches to Diagnosis and Therapy in Neurodegenerative Diseases 155
    Giuseppe Legname

A Reliable Semi-automatic Program to Measure the Vertebral Rotation Using the Center of Lamina for Adolescent
Idiopathic Scoliosis 159
    W. Chen, E. Lou, and Lawrence H. Le

Investigation of the Optimal Freehand Three-Dimensional Ultrasound Configuration to Image Scoliosis: An In-vitro
Study 163
    Q.N. Vo, E. Lou, Lawrence H. Le, and L.Q. Huynh

Assessment of Curve Flexibility by Ultrasonic Imaging – A Pilot Study 167
    R. Zheng, E. Lou, Lawrence H. Le, D. Hedden, J. Mahood, and M. Moreau

Prefrontal fNIRS Neuroimaging during a Sleep Induction Task Using Perception of a Red Light through Closed Eyes to
Fight Insomnia: A Pilot Study 171
    P.A. Grounauer and B. Métraux

Orthogonal Digital Radiographs - A Novel Template for a Paediatric Femur Finite Element Model Development 175
    D.S. Angadi, D.E.T. Shepherd, R. Vadivelu, and T.G. Barrett

EEG Signal Analysis and Artifact Removal by Wavelet Transform 179
    Pham Phuc Ngoc, Vu Duy Hai, Nguyen Chi Bach, and Pham Van Binh

Detection of Activities Daily Living and Falls Using Combination Accelerometer and Gyroscope 184
    Quoc T. Huynh, Uyen D. Nguyen, Kieu Trung Liem, and Binh Q. Tran

Using Near-Infrared Technique for Vein Imaging 190
    Tran Van Tien, Pham T.H. Mien, Pham T. Dung, and Huynh Quang Linh

Ribs Suppression in Chest X-Ray Images by Using ICA Method 194
    Hieu Xuan Nguyen and Tin Thanh Dang

Ultrasound Ovary Image Classification Using Kσ-Classifier 198
    B.S. Usha and S. Sandya

FPGA-in-the-Loop Simulation of Cardiac Excitation Modeling towards Real-Time Simulation 203
Norliza Othman, Mohamad Hairol Jabbar, Abd Kadir Mahamad, and Farhanahani Mahmud

G-Quadruplexes in the Human Immunodeficiency Virus-1 and Herpes Simplex Virus-1: New Targets for Antiviral Activity by Small Molecules 207
Rosalba Perrone, Sara Artusi, Elena Butovskaya, Matteo Nadai, Christophe Pannecouque, and Sara N. Richter

Cell Specific Imaging Probe Development and Biomedical Applications 211
Nam-Young Kang and Young-Tae Chang

A Case Study on Expression of Single-Chain Variable Fragment of Anti-HER2 Antibody by Using Recombinant Baculovirus in Silkworms 215
T.M.H. Nguyen, T.V.A. Nguyen, T.H. Le, T.H. La, T.T.B. Nguyen, and Q.H. Le

DNA Hypermethylation Signatures for Detection of Breast Cancer in Vietnamese Population 219
T.K. Phuong, L.D. Thuan, D.T.P. Thao, and L.H.A. Thuy

Interaction between XG and HPC in Blended-HPC/H2O/H3PO4 Tertiary System 223
S.P. Rwei, T.A. Nguyen, and H.W. Wu

The Time Based Study of Cell Morphology Using Atomic Force Microscopy 227
Wan Ibtisam Wan Omar and Chin Fhong Soon

The Effects of Enzyme to the Dissociation of Cells in Monolayer and 3D Microtissue on the Liquid Crystal Substrate 231
Kok Tung Thong, Chin Fhong Soon, and Kian Sek Tee

Microscale Tribological Response of Human Osteoarthritic Articular Cartilage under the Boundary Lubrication of Hyaluronic Acid 235
Cong-Truyen Duong and Duc-Nam Nguyen

Single Cell Traction Force Mapping Software 239
Chin Fhong Soon, Kian Sek Tee, Mansour Youseffi, and Morgan Denyer

Growth of Rutile Phased Titanium Dioxide (TiO$_2$) Nanoflowers for HeLa Cells Treatment 243
N.S. Khalid, W.S. WanZaki, and M.K. Ahmad

Cystatin C Versus Creatinine in Evaluating Glomerular Filtration Rate in Renal Transplant Recipients with Proteinuria 247
Tran Thai Thanh Tam, Hoang Khac Chuan, Du Thi Ngoc Thu, Nguyen Thi Thai Ha, Thai Minh Sam, Nguyen Thi Le, and Tran Ngoc Sinh

Genetic Mutation Types Detected in 25 Blood Samples of KHMER Patient with Beta-thalassemia in Bac Lieu Province 253
Pham Thi Ngoc Nga and Nguyen Trung Kien

Using Realtime Rt-Pcr and Sequencing Assays to Define Viral Load, Types and Subtypes of Hepatitis C Virus at Cantho Center General Hospital 257
Cao Thi Tai Nguyen, Tran Ngoc Dung, and Nguyen Thi Huynh Nga

Role of the Fifth Copper Binding Site in Prion Conversion 261
Thao Phuong Mai and Giuseppe Legname

The Electrocardiographic Characteristics of Chronic Obstructive Pulmonary Outpatients 266
   Dang Huynh Anh Thu and Le Thi Tuyet Lan

The Correlation between Peripheral Nerve Conduction Study Parametersand Level of Urinary Albumin Excretionin
Diabetic Patients 270
   Tran Vu Hoang Duong, Le Quoc Tuan, and Nguyen Thi Le

New Practical Approachs to Estimation of Glomerular Filtration Rate in Adult: A Review 274
   Le Quoc Tuan, Vo Thi Thien Huong, and Nguyen Thi Le

Testosterone – The Vital Hormone of Men: A Review 278
   Tran Ngoc Thanh and Pham Dinh Luu

Roles of Testosterone in Men with Type 2 Diabetes: A Review 281
   Tran Ngoc Thanh and Pham Dinh Luu

Investigating Chronic Obstructive Pulmonary Disease (COPD) in Vietnamese Patients Using Impulse Oscillometry
(IOS) 285
   T.X. Tan, V. Van Toi, Truong Quang Dang Khoa, H.D.H. Hanh, T.T.K. Thu, and L.T.T. Lan

Finite-Difference Time-Domain Simulations of Ultrasound Backscattered Waves in Cancellous Bone 289
   A. Hosokawa

Investigation of Solid Dispersion Methods to Improve the Dissolution Rate of Curcumin 293
   Kiet Anh Tran, Thao Truong-Dinh Tran, Toi Van Vo, Thanh Van Tran, and Phuong Ha-Lien Tran

Dissolution Enhancement of Curcumin by Solid Dispersion with Polyethylene Glycol 6000 and Hydroxypropyl
Methylcellulose 298
   Tuong Ngoc-Gia Nguyen, Phuong Ha-Lien Tran, Toi Van Vo, Thanh Van Tran, and Thao Truong-Dinh Tran

Improvement of Gliclazide Dissolution Rate Using In Situ Micronization Technique 302
   Nguyen Thanh Nhan and Tran Van Thanh

Research and Preparation of Solid Dispersion of Itraconazole in Hydroxypropyl-Beta-Cyclodextrin 306
   Tran Van-Thanh, Pham Vu Quang Vinh, and Huynh Van-Hoa

A Potential Application of Vietnamese Rice in Pharmaceutical Industry as a Sustained Release Agent 311
   Vuong Duy Ngo, Thao Truong-Dinh Tran, Toi Van Vo, and Phuong Ha-Lien Tran

Use of Microwave Method for Controlling Drug Release of Modified Sprouted Rice Starch 314
   Thinh Duc Luu, Nam Hoang Phan, Thao Truong-Dinh Tran, Toi Van Vo, and Phuong Ha-Lien Tran

Fabrication of In Situ Cross-Linking Polyvinyl Phosphonic Acid - Chitosan Hydrogel for Wound Applications 317
   Le Quoc Tuan, Dang Hoang Phuc, Vo Van Toi, and Thi-Hiep Nguyen

Development of a New Injectable PVA–Ag NPs/ Chitosan Hydrogel for Wound Dressing Application 321
   Xuan-Truong Nguyen, Vo Van Toi, and Thi-Hiep Nguyen

Investigation of the Silk Fiber Extraction Process from the Vietnam Natural Bombyx Mori Silkworm Cocoon 325
   Thu-Hien Luong, Thao-Nhi Ngoc Dang, Oanh Pham Thi Ngoc, Thanh-Ha Dinh-Thuy, Thi-Hiep Nguyen, Vo Van Toi,
   Hoang Thuy Duong, and Hoang Le Son

Fabrication of Hyaluronan – Chitosan – Polyvinyl Phosphonic Acid Hydrogel for Bioglue Applications 329
   Dang Hoang Phuc, Thi-Hiep Nguyen, Vo Van Toi, and Phan Van Tien

Investigation of the Synthetic Process of Nano-Hydroxyapatite (Hap) Using Microwave and Ultrasound   332
  Tran Thi Tuong Van, Bui Ngoc Thao Tram, Vo Van Toi, and Thi-Hiep Nguyen

Synthesis and Characterization of Hydroxyapatite Biomaterials from Bio Wastes   336
  Bui Ngoc Thao Tram, Thi-Hiep Nguyen, and Vo Van Toi

Modified DNA Extraction Method for the Detection of Aspergillus Flavus and Aspergillus Parasiticus
in Dried Foods   339
  Pham Tuong Vi, Huynh Le Thao Trinh, and Nguyen Thi Hue

DEMM: A Meta-Algorithm to Predict the pKa of Ionizable Amino Acids in Proteins   343
  T.B. Nguyen, K.P. Tan, and M.S. Madhusudhan

A Threshold Algorithm in a Fall Alert System for Elderly People   347
  Pham Ty Phu, Nguyen Thanh Hai, and Nguyen Thanh Tam

Comparative Study on Human A-Glucosidase   351
  Q. Ong and L. Le

Development of Non-Invasion Method for Prognosis and Early Diagnosis of Cervical Cancer in Vietnamese Patients
Based on DNA Methylation Specific PCR   355
  T.K. Phuong, L.D. Thuan, and L.H.A. Thuy

Antifungal Activity of Conyza Canadensis ((L.) Cronquist) Collected in Northern Viet Nam   359
  N.B. Phuong, N.T.T. Lien, and N.T.T. Hoai

Antimicrobial Activity of Senna alata (l.), Rhinacanthus nasutus and Chromolaena odorata (l.) Collected in Southern
Vietnam   362
  Thuong L.H. Pham, Trung T. Trinh, and Hoai T.T. Nguyen

Evaluation of the Optimal Multiplex PCR Method for the Detection of *Aspergillus Flavus* and *Aspergillus Parasiticus*
on Dried Peanut   367
  Nghia T. Le, Trinh Huynh, and Hue T. Nguyen

Evaluation of the Optimal Multiplex PCR Method for the Detection of Aspergillus Flavus and Aspergillus Parasiticus
on Dried Corn   371
  Tu T. Ly, Trinh Huynh, and Hue T. Nguyen

fNIRS-Based Wavelet Thresholds for Motor Area Determination   376
  Dao.V. Ha, Hai T. Nguyen, Cuong.Q. Ngo, Mai T. Tran, and Toi Van Vo

Determining the Size of a Solid Tumor   381
  Tran Thi Quynh Nhu, Nguyen Thanh Hai, Ngo Thanh Dong, and Nguyen Tan Nhu

2D Complex Shear Modulus Imaging in Gaussian Noise   385
  Nguyen Thi Anh-Dao, Tran Duc-Tan, and Nguyen Linh-Trung

Evaluation of Frontal and Visual Cortices on Mental Working Tasks Using Functional Near Infrared Spectroscopy   389
  Pham Thanh Thao, Nguyen Duc Thang, and Vo Van Toi

Differentiation of Hemodynamic Responses of the Brain with Typing and Writing   395
  Vo Nhut Tuan, Nguyen Duc Thang, Vo Van Toi, Tran Le Giang, Nguyen Huynh Minh Tam, and Dinh Dong Luong

The Relation between a Three-Day Sitting Meditation Fasting and the Participant's Psychophysiological Condition:
Through Nonlinear Chaos Analysis of Pulse Waves   399
    T. Futaba

Construction of Phantom Mimic Vessel for Study of Human Vessel Conditions in Deep Vein Thrombosis (DVT)   402
    N. Ibrahim, W.N. Wan Zakaria, N. Aziz, and M.K. Abdullah

Medical Image Contour Based Context-Aware in Contourlet Domain   405
    Nguyen Thanh Binh

Nonrigid Point Set Registration-Based 3-D Human Pose Tracking from Depth Data   409
    Dong-Luong Dinh, Nguyen Duc Thang, Sungyoung Lee, and Tae-Seong Kim

Kinematics of High-Heeled Running Gait with Consideration for Experience of Wearers   414
    Y.Q. Song, F.L. Li, J.S. Li, and Y.D. Gu

MFCC-DTW Algorithm for Speech Recognition in an Intelligent Wheelchair   417
    Le Hoang Linh, Nguyen Thanh Hai, Ngo Van Thuyen, Tran Thanh Mai, and Vo Van Toi

Lower Limb Kinematics Study on Female Latin Shoes of Different Height Heels   422
    S.R. Shao, X.X. Gao, Y. Zhang, Q.C. Mei, and Y.D. Gu

Ideal Cross-Point Regions of Prediction Errors from LOCO-I Algorithm Applied to Lossless Image Compression   426
    Tin T. Dang and Canh Xuan Huynh

Applying the Image Compression Algorithm of ICRICM to a Plugin Integrated into MIPAV Software   430
    Tin T. Dang and Khoa Anh Tran

Software Design for Training and Supporting Knowledge of Ventilators for Clinical Engineers   434
    Nguyen Nhan Thien and Huynh Quang Linh

Design of Electrotherapy Equipment Using Wireless Communication   437
    Nguyen Tuan Anh and Huynh Quang Linh

Experimental Determination of the Loss of Total Endoprosthesis Polyethylene Cup Using Holographic
Interferometry   440
    M. Houfek and L. Houfek

The Applications of Control System Approach in Biomedical Engineering Research   444
    Huynh Luong Nghia and Nguyen Van Trung

Study of Vessel Conditions for Deep Vein Thrombosis (DVT) Diagnosis According to Body Mass Index   447
    W.N. Wan Zakaria, N. Ibrahim, N. Mat Harun, Razali Tomari, and M.K. Abdullah

Tensile Stress Analyses of the Hip Joint Endoprosthesis Ceramic Head with Real Shape Deviations   450
    V. Fuis

Review and Development of Tetraplegic-Musculoskeletal FES-Elbow Joint Extension Control Strategies   454
    N.H.M. Nasir, M.K.I. Ahmad, B.S.K.K Ibrahim, and F. Sherwani

Integrated Biomedical Waste Management for Small Scale Healthcare Units in India   458
    Prasad Balachandran

Simultaneous Detection of Two Viroids Infecting Grapevines in Taiwan by Multiplex RT-PCR   461
    Nguyen Phuc Thien and Chu-Hui Chiang

Theoretical Investigation on Antioxidant Activity of Phenolic Compounds Extracted from *Artocarpus Altilis*   464
    Nguyen Minh Thong and Pham Cam Nam

An EEG-Controlled Wheelchair Using Eye Movements   470
    Hue T. Tran, Hai T. Nguyen, Hieu V. Phan, V. Van Toi, Thuyen V. Ngo, and Cao Bui-Thu

Application of Fluorescence Photography in the Evaluation of Acne   474
    Pham Thi Hai Mien, Tran Van Tien, Vu Thanh Huy, and Huynh Quang Linh

Digital Morphology Comparisons between Models of Conventional Intraoral Casting and Digital Rapid Prototyping . . .   478
    Rong-Fu Kuo, Shing-Jye Chen, Tung-Yiu Wong, Bo-Cheng Lu, and Zheng-Han Huang

Human Activity Recognition and Monitoring Using Smartphones   481
    Vu Ngoc Thanh Sang, Nguyen Duc Thang, Vo Van Toi, Nguyen Duc Hoang, and Truong Quang Dang Khoa

Evaluation of Hemodynamic Responses to Visual Tasks Using Functional Near Infrared Spectroscopy   486
    Tran Le Giang, Nguyen Duc Thang, Vo Van Toi, Nguyen Huynh Minh Tam, Dinh Dong Luong, and
    Truong Quang Dang Khoa

Investigation the Stability of Oblique Fracture Fixation of Long Bone Using Different Screw Angle   491
    Bich Nguyen, Trung Le, and Ha Van Vo

F-Scan Analysis of Prosthetic Fittings through Mercer on Mission Vietnam   495
    Emily Brett, Matthew Yin, Ha Van Vo, Edward O'Brien, Loren Sumner, and Philip McCreanor

Biomechanical Evaluation of Hybrid Bicortical and Unicortical Screw and Bone Plate Fixation in Humeral Mid-Shaft
Fractures: A Study on Cadaveric Bone   499
    Trung Le, Benjamin McDeed, and Ha Van Vo

Novel Design of a Prosthetic Foot Using Spring Mechanism   503
    Awab Umar Khan and Ha Van Vo

Development and Characterization of Porous Calcium Phosphate Cement Using α-Tricalcium Phosphate Bead   507
    Pham Trung Kien, Tsuru Kanji, and Kunio Ishikawa

Author Index   511

Keyword Index   515

Development of Individual Plasmonic Nanosensors for Clinical Diagnosis

Phuoc Long Truong and Sang Jun Sim

Department of Chemical and Biological Engineering, Korea University, Seoul 136-701, Korea

Abstract— Accurate detection of panels of protein biomarkers in serum, saliva, or tissue plays an important role in early detection of diseases. Recently, our lab has developed the localized surface plasmon resonance (LSPR)-based immunosensors with highly promising analytical approach for qualification and quantification of biochemical substances in various applications. The LSPR of noble nanoparticles shifts upon refractive index changes of the dielectric environment surrounding the nanoparticles, subsequently causing a peak shift of the LSPR; it is also very sensitive to the binding events near the nanoparticle surface. Therefore, individual nanoparticles are sufficiently sensitive to detect the molecular interactions on the nanoparticle surface induced by biological molecules at ultralow concentrations. Furthermore, LSPR λmax shifts induced by adsorbates on the nanoparticle surface are completely linear with the number of bound molecules. Owing to this reason, each nanoparticle can be used as an independent sensor. By functionalizing the nanoparticle surface with the appropriate receptors, individual nanoparticle sensors can be used as diagnostic tools for a variety of diseases. Herein, we introduce the resonant Rayleigh scattering properties of single Au nanoparticles and the use of single Au nanoparticles as the plasmonic transducers to detect the binding of protein biomarkers to specific receptors conjugated gold nanoparticles. The principle of detection based on the resonant Rayleigh scattering response of single Au nanoparticles and the LSPR λmax shift of resonant Rayleigh scattering spectrum. This approach holds great promise as a simple, label-free, ultrasensitive method for detection of protein biomarkers in clinical diagnostics.

Keywords— Au nanoparticles (AuNPs), Localized Surface Plasmon Resonance (LSPR), Rayleigh Light Scattering (RLS), Plasmonic Nanosensors.

I. INTRODUCTION

Proteins that are produced with increased amounts in disease states have emerged as important analytes or biomarkers to detect specific diseases. Biomarkers are objectively measured and evaluated as indicators of normal biological processes, pathogenic processes, or pharmacologic responses to a therapeutic intervention. Accurate detection of panels of protein biomarkers in serum, saliva, or tissue has great potential for early detection of disease and for helping direct individualized therapy. To date, protein biomarkers are already used for detection or monitoring of cancers and heart disease. For more accurate and reliable diagnosis in disease state and widespread application, detection methodologies need to be accurate, sensitive, cheap, and easy to use, to facilitate rapid diagnosis, minimize sample decomposition, and decrease patient anxiety. To accomplish the task, many studies have been devoted to development of various signal transduction methods based on optics [1], radioactivity [2, 3], fluorescence [4], electrochemistry [5], quartz crystal micro balance [6], piezoelectric cantilever [7], colorimetry [8, 9], scanometry [10, 11], surface plasmon resonance spectroscopy [12, 13], etc. However, each of them still has its shortcomings such as necessitating time-consuming procedures, complicated manipulation and low sensitivity. Therefore, the development of simple, low-cost and ultra-sensitive tools that enable real-time and label-free detection for fundamental research and clinical diagnosis is still a significant driving force in the field of biosensor research.

Nowadays, surface plasmon resonance (SPR) measurements are the simplest method for detection of protein biomarkers owing to the changes in the local refractive index upon adsorption of the target molecules to the metal surface. The great advantages of SPR-based biosensors are that bio-molecular reactions can be monitored in real time, the chip is reusable, and the detection methods enable a simple, rapid, label-free detection [14]. However, its major disadvantage is that it is difficult to determine an analyte at low concentration or with low molecular weight [15]. The advent of dark-field microscope has enabled the study of localized surface plasmon resonance (LSPR) for nanometer-sized metallic structures, which further facilitate its use in label-free detection of various biological analytes in real-time, optical sensing with ultra-high sensitivity [16, 17]. The LSPR occurs when the free conductive electrons near the metallic nanoparticle's surface undergo concerted oscillations coupled to an external optical field and, in addition, certain existence conditions are satisfied. This leads to a plasmon that oscillates locally around the nanoparticle with a frequency known as the localized surface plasmon resonance (LSPR). The excitation of free electrons of the nanoparticles by an electric field at an incident wavelength

© Springer International Publishing Switzerland 2015
V. Van Toi and T.H. Lien Phuong (eds.), *5th International Conference on Biomedical Engineering in Vietnam,*
IFMBE Proceedings 46, DOI: 10.1007/978-3-319-11776-8_1

where the resonance occurs will result in strong light scattering, or a strong UV-vis absorption band that is not present in the spectrum of the bulk metal. The unique light absorption and scattering properties of the nanoparticles are the premise of novel strategies for the development of optical biosensors and chemosensors as well optical nanodevices such as surface-enhanced spectroscopy, optical filters, plasmonic devices, and sensors [18]. The LSPR λ_{max} shift due to changes in the local refractive index upon binding of biochemicals to the nanoparticle's surface can be used to detect biomolecules such as DNA or proteins. Hence, each nanoparticle could be used as an independent sensor using a dark-field microscope in combination with a spectrophotometer. Single nanoparticle probes offer absolute detection limits due to the sensor's miniaturization down to the nanoscale level. Using single-nanoparticle measurement techniques, the detection limit can approach a few molecules, and even down to a single molecule per nanoparticle sensor [19 – 21]. In this paper, we briefly introduce the resonant Rayleigh scattering properties of single Au nanoparticles and the use of single Au nanoparticles as the plasmonic transducers for detection of protein biomarkers in clinical diagnostics, which depend on the resonant Rayleigh scattering response of single Au nanoparticles and the LSPR λ_{max} shift of resonant Rayleigh scattering spectrum of single nanoparticle.

II. RESONANT RAYLEIGH SCATTERING OF SINGLE AU NANOPARTICLES

Scientific interest in nanoscience and nanotechnology is driven by the unique and novel properties of nanometer-sized metallic materials. The distinct optical properties of nanometer-sized metallic structures are the source of various applications. Nowadays, scientists know that the characteristic hues of the noble metal nanoparticle suspensions are original from their strong interaction with incident light. This interaction induces the resonant light scattering/absorption of metallic nanoparticles, known as LSPR [18, 22, 23]. This characteristic only arises in materials such as quasi-free electron metals, Ag, Au, Cu, and Al with a negative real and a small positive imaginary dielectric constant in a given wavelength range, due to their capability to support surface plasmon resonances when submitted to electromagnetic radiation [24]. Due to its chemical stability, gold is the best studied material. For Au nanospheres, the resonance occurs in the visible spectral region at ~520 nm, resulting in the brilliant red color of the nanoparticles in solution. In comparison with spherical ones, the conductive free electron oscillation along both the long axis and short axis of the nanorods results in a stronger resonance band in the near-infrared region and a weaker

band in the visible region, respectively. As a result, the photo-physical properties of gold nanoparticles are enhanced [22]. At the resonance wavelength, Rayleigh light scattering of single nanoparticles is strongly enhanced many times larger than their geometric sizes, which enables imaging and spectroscopy of single nanoparticles [23]. Moreover, LSPR produces strong environment-dependent light scattering and absorption of nanometer-sized metallic structures, leading to their use as refractive-index nanosensors. Further studies in the field of nanoparticle optics have allowed one to deeply understand the relationship between nanomaterial properties such as composition, size, shape, and local dielectric medium. A deep understanding of the optical properties of noble nanoparticles allows one to develop practical applications [18, 25]. Recently, Truong *et al.* have analysed the relationship of the Rayleigh scattering properties of a single Au nanoparticle with its size, shape, aspect ratio and local dielectric environment by the dark-field micro-spectroscopy system [26]. They provided a detailed study on the refractive index sensitivity of three types of differently shaped Au nanoparticles, which were nanospheres, oval-shaped nanoparticles and nanorods. This study helps one differentiate the Rayleigh light scattering from individual nanoparticles of different size and/or shape and precisely obtain quantitative data as well as the correlated optical spectra of single gold nanoparticles from the inherently inhomogeneous solution of nanoparticles. For Au nanospheres, the results showed that Au nanospheres of < 30 nm diameter have a small scattering cross-section, resulting in dimly scattered green light, which is hard to visualize in a dark-field microscope. With increasing nanosphere diameter, the bulk refractive index sensitivity also increases. However, larger Au nanospheres exhibit lower surface refractive index sensitivity compared to smaller Au nanoparticles. The optimal Au nanosphere size, which is suitable for LSPR sensing with single Au nanoparticle as determined by its spectral peak position, scattering cross-section and refractive index sensitivity, was ~50 nm. The ~50 nm Au nanoparticle has green scattering color in terms of sensing application, which is the most sensitive color to the human eyes, and it enables visible colorimetric shifts from green (~560 nm) to orange (>630 nm) *via* the observation of scattered light by the naked eye under a dark-field microscope owing to chemical adsorption and plasmon coupling of the nanoparticles [26, 27]. In case of oval-shaped Au nanoparticles or Au nanorods, the resonant Rayleigh light scattering properties depend on their aspect ratios, and the plasmonic sensitivity increases proportionally as a function of the aspect ratio. In addition, their plasmonic sensitivity depends on the polarizability of the inner electrons of the metal and the refractive index of the local dielectric medium surrounding the nanorods. The optimal sensitivity can be dependent on the quality factor of an electron oscillation, which describes the number of oscillations until the oscillation is damped. As a result, a longer plasmon oscillation lifetime results in a more sensitive dependency on refractive index changes in the medium surrounding the nanorods. The sensitivity of single Au nanorods induced by changing the refractive index of the local medium surrounding the nanoparticles was highest at an aspect ratio of 3.5 (~284 nm RIU^{-1}). This value is much higher that of Au nanospheres (~70 nm RIU^{-1}). Another property of Au nanorods which ones should notice is that, if the aspect ratio of the Au nanorods is too large, the wavelength shift of the resonant Rayleigh light scattering spectrum will be higher than 800 nm after the adsorption or binding of biological materials, which is over the wavelength coverage of the micro-spectroscopy system. Therefore, the nanoparticle size and its spectral peak position must be selected with care to ensure sufficient signal intensity for LSPR assays [21, 26]. These results indicate that the shape, size and aspect ratio of single Au nanoparticles determine the resonant Rayleigh light scattering properties such as the peak position, scattering-cross-section and the refractive index sensitivity, which gives a handle for the choice of Au nanoparticles for the design and fabrication of single Au nanoparticle sensors.

III. SINGLE-NANOPARTICLE MEASUREMENT AND EXPERIMENTAL PROCEDURE

As mentioned above, the advent of dark-field microscope has enabled the study of single gold colloid particles much smaller than the light wavelength. In such a dark-field configuration, only the light scattering objects look brightly illuminated against the dark background in the image. The white-light illumination with halogen lamp allows one to differentiate the light scattering from individual nanoparticles of different size and/or shape due to different optical resonances [22]. The combination of a dark-field microscope with a spectrograph and a CCD camera helps one measure LSPR response of a single nanoparticle. Thereby a single nanoparticle can be used as a tiny signal transducer without using any electric cables, circuits, but light illumination. For measuring samples in small regions or single nanoparticles, light-scattering measurements in dark-field microscope are extremely powerful. In the dark-field microscope, white light is radiated to the sample at a high angle, and scattered light is collected at a lower angle. It means that a high-numerical aperture condenser brings light to the sample, and a low-numerical aperture microscope objective collects the scattered light at low angles. Then, the scattered light is directed to a spectrometer and detector, such as a charge-coupled device (CCD) camera. In this case, the scattered monochromatic light of a single nanoparticle, which is separated by the spectrograph, is recorded with the CCD camera and is plotted as a function of light scattering intensity versus wavelength [18, 38, 39].

For sample preparation, noble metallic nanoparticles were randomly immobilized by drop-coating a volume of the diluted Au nanoparticle solution onto a coverslip slide. Au nanoparticle concentration and incubation time were used to control inter-particle spacing. After that, the slide was washed with ultra-pure water, ethanol and dried with nitrogen gas. Immobilized nanoparticles on cover slips were inserted into a flow cell and were exposed to various dielectric environments or molecular adsorbates. The LSPR λ_{max} shift induced by the binding of adsorbates and analytes was recorded by measuring the resonant Rayleigh light scattering spectrum of single Au nanoparticle. The LSPR λ_{max} shift was calculated as follows: $\Delta\lambda_{max} = \lambda_{max}$ (after binding) - λ_{max} (before binding). Further details of the experimental method and LSPR instrumentation for single nanoparticles have been previously reported [21, 26, 27].

IV. NOVEL STRATEGIES FOR DETECTION OF PROTEIN BIOMARKERS USING INDIVIDUAL AU NANOSENSORS

Recently, several research groups have begun to develop individual nanoparticle-based sensors which enable the transduction of biological/chemical binding events into optical signals based on the wavelength shifts owing to the tunable LSPR properties of noble metallic nanoparticles [20, 28]. This application arises from the unique optical properties of certain nanomaterials such as silver nanoparticles and gold nanoparticles. It is now well-known that optical excitation of the LSPR of silver or gold nanoparticles results in absorption with large molar extinction coefficients near 3 x 10^{11} M^{-1} cm^{-1}, in case of resonant Rayleigh scattering with an efficiency equivalent to that of 10^6 fluorophors, and strong enhancement of the local electromagnetic fields near the nanoparticle surface. Moreover, the peak extinction or resonant Rayleigh scattering wavelength, intensity, and line-width of the LSPR spectra are strongly dependent on their size, shape, inter-particle spacing, and local dielectric environment [18, 29, 30]. In previous studies, Van Duyne *et al.* showed that the LSPR spectrum is ultra-sensitive to analyte-induced changes in the dielectric constant of the surrounding nanoparticle, and the resonant Rayleigh light scattering spectrum of single Ag nanoparticle can be strongly differentiated when approximately $\leq$ 100 streptavidin molecules or 60,000 molecules of 1-hexadecanthiol are absorbed on its surface.

The detection limit of a single nanosensor for absorption of small molecules is < 1,000 molecules [20]. Raschke *et al.* developed a method for detection of biomolecules using light scattering of a single biotin functionalized Au nanosphere [28]. Exposure to streptavidin led to alter the dielectric environment of the nanoparticle as well as the LSPR spectral shift of the nanoparticle. Spectral shifts as low as 2 meV could be detected. For larger molecules such as proteins, the LSPR response for one molecule can result in a greater change in the local dielectric environment. It has been reported that using single nanoparticle as a transducer captures target molecules more effectively than an ensemble of nanoparticles because the ratio of analyte to nanoparticle is increased due to the sparse distribution of nanoparticles. Another advantage of miniaturizing the sensor to a single nanoparticle is that the figure of merit is better than that of an ensemble measurement because the road resonance peak of an ensemble of nanoparticles is the result of the statistical distribution of the nanoparticle size and shape. Moreover, single nanoparticles with especially narrow bandwidths can be selected from a field of view to provide better *S/N* resolution. Thereby, the detection limit can be further improved and the sensitivity could approach the single-molecule detection limit for biopolymers [18, 20, 39]. Using single nanoparticle sensors, the detection limit can approach a few molecules, and even down to a single molecule per nanosensor.

Currently, noble nanospheres have been widely used in diagnoses. In comparison with other metallic nanostructures that are useful in various biomedical applications, Au nanospheres with various sizes, which can be either obtained from the commercial sources or conveniently produced in laboratories are still popular and promising due to their unique optical properties, simple and fast preparation, high stability and convenient surface bio-conjugation with proteins, antibodies and DNA probes by the means of covalent and non-covalent interactions [31 – 35]. In a previous study, Cao *et al.* showed a method for detection of PSA-ACT complex based on resonant Rayleigh light scattering response of single gold nanosphere with diameter of ~30 nm functionalized with PSA-ACT complex monoclonal antibody (PSA mAb); the detection limit was as low as 0.1 pg mL^{-1} (~ 1 fM) [36]. However, the shape of resonance peak was not sharp. It is well-known that the LSPR spectral shift is dependent on size, shape, composition, inter-particle spacing, and the local dielectric environment surrounding the nanoparticles [18]. Therefore, Truong *et al.* employed Au nanorods as the plasmonic transducers instead of gold nanospheres due to their higher sensitivity to refractive index changes, distinct optical properties, and clear LSPR spectral shifts. Using the

PSA-ACT complex as a protein biomarker, the lowest concentration recorded was ~ 110 aM. Compared to DNA biobarcode assays that require complicated instrumentation, multiple experimental steps, including microarrayer, complicated signal amplification, silver enhancement and light scattering measurements, single Au nanorod sensors can be fabricated with a low-cost and simple method, with label–free, attomolar detection limits [37]. In an effort to push single-nanoparticle measurement technique to its terminal detection limit, the sensitivity of Au nanorods with the various aspect ratios for single nanosensors was demonstrated and the Au nanorod with aspect ratio of ~ 3.5 were proven optimal for LSPR sensing. To reduce steric hindrance effect as well as to immobilize a large amount of ligand on the nanoparticle's surface, the various mixtures of different molar ratios of HS(CH$_2$)$_{11}$(OCH$_2$CH$_2$)$_6$OCH$_2$COOH and HS(CH$_2$)$_{11}$(OCH$_2$CH$_2$)$_3$OH were applied to form the different self-assembled monolayer (SAM) surfaces. The result showed that the best molar ratio for the antibody conjugation was observed at a molar ratio of 1:10. Using single Au nanorod sensor for PSA-ACT complex detection, the lowest concentration recorded was ~ 1 aM. These results showed that sensor miniaturization down to the nanoscale level, the reduction of steric hindrance, and optimization of size, shape, and aspect ratio of nanorods led to a significant improvement in the detection limit of sensors [21]. This result indicated that the detection platform is ultrasensitive for ultra-low concentration detection of biological analytes. Truong *et al.* also reported a strategy for attomolar level detection of small molecule-size proteins based on Rayleigh light scattering spectroscopy of single nanoplasmonic aptasensor by exploiting the outstanding characteristics of immunogold colloids to amplify the nontransparent resonant signal at ultralow analyte concentrations [27]. The fabrication method involves utilizing thiol-mediated adsorption of a DNA aptamer on the immobilized Au nanoparticle surface, the interfacial binding characteristics of the aptamer with its target molecules, and the antibody-antigen interaction through plasmonic resonance coupling of the Au nanoparticles. Using lysozyme as a model for detection of disease, the detection limit of the aptasensor was 10^2 fg mL^{-1}. Authors demonstrated up to a 380% increase in the localized resonant λ_{max} shift upon antibody binding to the analyte compared to the primary response during signal amplification using immunogold colloids. This enhancement led to a detection limit of ~7 aM which is a three orders of magnitude improvement. The results demonstrate substantial promise for developing coupled plasmonic nanostructures for ultrasensitive detection of various biological and chemical analytes.

V. CONCLUSION

This article has highlighted recent advances in using single Au nanoparticles as the plasmonic transducers for detection of protein biomarkers in clinical diagnostics, which depend on the resonant Rayleigh scattering response of single Au nanoparticles. By varying the nanomaterial, size, shape and aspect ratio of nanoparticles, the peak position, scattering-cross-section and the refractive index sensitivity, which give a handle for the choice of Au nanoparticles for the design and fabrication of single Au nanoparticle sensors, can be tuned. Moreover, the sensor miniaturization down to the nanoscale level, the reduction of steric hindrance, and optimization of size, shape, and aspect ratio of nanoparticles have led to a significant improvement in the detection limit of sensors. The single-nanoparticle spectroscopic measurement requires simple, inexpensive instrumentation and can use a variety of flexible signal transduction platforms. The LSPR sensing platforms enable the detection of the protein biomarker with ultra-sensitivity, simplicity and versatility, and the detection limit could reach a single molecule per nanoparticle sensor. Single nanoparticle sensors offer further advantages because the LSPR sensing platforms are readily implemented in multiplex detection. By controlling the size, shape and chemical functionalization of single nanoparticles, several sensing platforms can be fabricated in the same chip which each nanoparticle can be distinguished from others based on the spectral peak position, and several sensing platforms can be incorporated into one sensing device that allows for rapid, simultaneous, label-free detection of various biochemical species. Therefore, single nanoparticle-based sensors are powerful platforms for detection of biomarkers in clinical diagnostics.

REFERENCES

1. Shankaran DR, Miura N (2007) Trends in interfacial design for surface plasmon resonance based immunoassays. J Phys D: Appl Phys 40: 7187
2. Lindstedt G, Jacobsson A, Lundberg PA, Hedelin H, Pettersson S, Unsgaard B (1990) Determination of prostate-specific antigen in serum by immunoradiometric assay. Clin Chem 36: 53
3. Cuny C, Pham L, Kramp W, Sharp T, Soriano TF (1996) Evaluation of a two-site immunoradiometric assay for measuring noncomplexed (free) prostate-specific antigen. Clin Chem 42: 1243
4. Soukka T, Antonen K, Harma H, Pelkkikangas AM, Huhtinen P, Lovgren T (2003) Highly sensitive immunoassay of free prostate-specific antigen in serum using europium(III) nanoparticle label technology. Clin Chim Acta 328: 45
5. Fernandez-Sanchez C, McNeil CJ, Rawson K, Nilsson O (2004) Disposable noncompetitive immunosensor for free and total prostate-specific antigen based on capacitance measurement. Anal Chem 76: 5649
6. Henne WA, Doorneweerd DD, Lee J, Low PS, Savran C (2006) Detection of folate binding protein with enhanced sensitivity using a functionalized quartz crystal microbalance sensor. Anal Chem 78: 4880
7. Lee JH, Hwang KS, Park J, Yoon KH, Yoon DS, Kim TS (2005) Immunoassay of prostate-specific antigen (PSA) using resonant frequency shift of piezoelectric nanomechanical microcantilever. Biosens Bioelectron 20: 2157
8. Liang RQ, Tan CY, Ruan KC (2004) Colorimetric detection of protein microarrays based on nanogold probe coupled with silver enhancement. J Immunol Methods 285: 157
9. Nam JM, Wise AR, Groves JT (2005) Colorimetric bio-barcode amplification assay for cytokines. Anal Chem 77: 6985
10. Nam JM, Thaxton CS, Mirkin CA (2003) Nanoparticle-based bio-bar codes for the ultrasensitive detection of proteins. Science 301: 1884
11. Stoeva SI, Lee JS, Smith JE, Rosen ST, Mirkin CA (2006) Multiplexed detection of protein cancer markers with biobarcoded nanoparticle probes. J Am Chem Soc128: 8378
12. Cao C, Kim JP, Kim BW, Chae H, Yoon HC, Yang SS, Sim SJ (2006) A strategy for sensitivity and specificity enhancements in prostate specific antigen-α_1-antichymotrypsin detection based on surface plasmon resonance Biosens Bioelectron 21: 2106
13. Cao C, Sim SJ (2007) Signal enhancement of surface plasmon resonance immunoassay using enzyme precipitation-functionalized gold nanoparticles: A femto molar level measurement of anti-glutamic acid decarboxylase antibody. Biosens Bioelectron 22: 1874
14. Homola J (2003) Present and future of surface plasmon resonance biosensors. Anal Bioanal Chem 377: 528
15. Gomes P, Andreu D (2002) Direct kinetic assay of interactions between small peptides and immobilized antibodies using a surface plasmon resonance biosensor. J Immunol Methods 259: 217
16. Zsigmondy R, Alexander J (1909) A manual of colloid chemistry and ultramicroscopy. New York, John Wiley & Sons
17. Huang X, Jain PK, El-Sayed IH, El-Sayed MA (2007) Gold nanoparticles: interesting optical properties and recent applications in cancer diagnostics and therapy. Nanomedicine 2: 681
18. Anker JN, Hall WP, Lyandres O, Shah NC, Zhao J, Van Duyne RP (2008) Biosensing with plasmonic nanosensors. Nat Mater 7: 442
19. Haes AJ, Van Duyne RP (2002) A Nanoscale Optical Biosensor: Sensitivity and Selectivity of an Approach Based on the Localized Surface Plasmon Resonance Spectroscopy of Triangular Silver Nanoparticles . J Am Chem Soc 124: 10596
20. McFarland AD, Van Duyne RP (2003) Single silver nanoparticles as real-time optical sensors with zeptomole sensitivity. Nano Letters 3:1057
21. Truong PL, Kim BW, Sim SJ (2012) Rational aspect ratio and suitable antibody coverage of gold nanorod for ultra-sensitive detection of cancer biomarker. Lab Chip 12: 1102
22. Schatz GC, Van Duyne RP (2006) Handbook of Vibrational Spectroscopy. John Wiley & Sons
23. Lal S, Link S, Halas NJ (2007) Nano-optics from sensing to waveguiding. Nat Photonics 1: 641
24. Ringe E, Sharma B, Henry AI, Marks LD, Van Duyne RP (2013) Single nanoparticle plasmonics. Phys Chem Chem Phys 15: 4110
25. Hutter E, Fendler JH (2004) Exploitation of Localized Surface Plasmon Resonance. Adv Mater 16: 1685
26. Truong PL, Ma X, SJ Sim (2014) resonant Rayleigh light scattering of single Au nanoparticles with different size and shape. Nanoscale 6: 2307

27. Truong PL, Choi SP, Sim SJ (2013) Amplification of resonant Rayleigh light scattering response using immunogold colloids for detection of lysozyme. Small 9: 3485

28. Raschke G, Kowarik S, Franzl T, Sönnichsen C, Klar TA, Feldmann J (2003) Biomolecular recognition based on single gold nanoparticle light scattering. Nano Letters 3: 935

29. Haes AJ, Stuart DA, Nie S, Van Duyne RP (2004) Using solution-phase nanoparticles, surface-confined nanoparticle arrays and single nanoparticles as biological sensing platforms. J Fluoresc 14: 355

30. Haes AJ, Van Duyne RP (2004) A unified view of propagating and localized surface plasmon resonance biosensors. Anal Bioanal Chem 379: 920

31. Kreibig U, Vollmer M (1995) Optical Properties of Metal Clusters. Springer-Verlag, Berlin

32. Choi Y, Park Y, Kang T, Lee LP (2009) Selective and Sensitive Detection of Metal Ions by Plasmon Resonance Energy Transfer-based Nanospectroscopy. Nat Nanotechnol 4: 742

33. Liu GL, Long Y-T, Choi Y, Kang T, Lee LP (2007) Quantized plasmon quenching dips nanospectroscopy via plasmon resonance energy transfer. Nat Methods 4: 1015

34. Li H, Huang J, Lv J, An H, Zhang X, Zhang Z, Fan C, Hu J (2005) Nanoparticle PCR: nanogold-assisted PCR with enhanced specificity. Angew Chem Int Ed Engl 44: 5100

35. Frens G (1973) Controlled Nucleation for the Regulation of the Particle Size in Monodisperse Gold Suspensions. Nat Phys Sci 241: 20

36. Cao C, Sim SJ (2009) Resonant Rayleigh Light-scattering Response of Individual Au Nanoparticles to Antigen-Antibody Interaction. Lab Chip 9: 1836

37. Truong PL, Cao C, Park S, Sim SJ (2011) New method for non-labeling attomolar detection of diseases based on individual gold nanorod immunosensor. Lab Chip 11: 2591

38. Stuart DA, Haes AJ, Yonzon CR, Hicks EM, Van Duyne RP (2005) Biological applications of localised surface plasmonic phenomenae. IEE Proc-Nanobiotechnol 152: 13

39. Sannomiya T, Vörös J (2011) single plasmonic nanoparticles for biosensing. Trends Biotechnol 29: 343

YALES2BIO: A Computational Fluid Dynamics Software Dedicated to the Prediction of Blood Flows in Biomedical Devices

S. Mendez[1], C. Chnafa[1], E. Gibaud[1], J. Sigüenza[1], V. Moreau[2], and F. Nicoud[1]

[1] I3M-UMR 5149, University Montpellier II and CNRS, CC 051, 34095 Montpellier, France
[2] CORIA UMR 6614, Normandie Université, CNRS, Université and INSA de Rouen,
76801 St-Etienne-du-Rouvray, France

Abstract— **A high-fidelity computational fluid dynamics software is described. This software is designed to perform numerical simulations of blood flows and flows of red blood cells. After discussing the need for advanced flow in the context of biomedical devices, the numerical method is briefly described and examples are given to show the versatility of the flow solver.**

Keywords— **computational fluid dynamics, blood flows, complex geometries, large-eddy simulations, red blood cells.**

I. INTRODUCTION

The interest for medical devices has only been increasing over the last years. This is particularly true in the context of blood flows, where stents and flow diverters, mechanical or tissue heart valves, ventricular assist devices and artificial hearts, are few of the many implantable devices constantly developed and improved. A striking example was the first implant of a full biocompatible artificial heart, developed by the CARMAT society (http://www.carmatsa.com), in France in 2013. One can also mention external devices used either for treatment (dialyzers and extracorporeal circulation) and diagnostics (blood analyzers).

Biomedical Engineering is as fast-growing field and, as in other engineering domains, numerical simulation is expected to be an indispensable tool for the design and the optimization of biomedical devices. However, when the operating fluid is blood (full blood or isolated blood cells), computational fluid dynamics (CFD) faces tremendous challenges, both from the modeling and the numerical points of view (biological membrane modeling, hemoturbulence, flow intermittency, fluid-structure coupling, non-Newtonian rheology,...). Progresses of CFD for blood flows have been dealt with in recent reviews in the context of physiological flows [1-3] and in terms of devices optimization by Marsden [3].

Obviously, numerical approaches already exist [3] and have already provided interesting insights in the flow physics, helping the design of cardiovascular devices. However, in order to really assist the engineers in the development and optimization of such devices, numerical simulation has to improve its treatment of fluid-structure interaction, blood rheology and turbulence, the relative importance of these elements depending on the application. In the context of heart support devices for instance, rheology and turbulence are key components of both their performances and limitations (blood damage, thrombosis).

As stressed by Deutsch *et al.* [4] in 2006, challenges faced by computation of artificial blood pumps are formidable and the experimental approach remains the gold standard for researcher to study blood pumps. However, Fraser *et al.* [5] report a number of computational studies in this field a few years later. Computations use RANS (Reynolds-Averaged Navier-Stokes) modeling to account for the turbulence effects on the flow. However, Fraser *et al.* [5] state that models have rarely been compared and agreement is often qualitative. Recently the American Food & Drug Administration (FDA) started a specific program for the evaluation and the validation of biomedical CFD (https://fdacfd.nci.nih.gov/). Published results of their interlaboratory study show how turbulence modeling is problematic: as many blood flows at `high' Reynolds number have a complex regime, neither laminar nor fully turbulent, RANS models fail to provide an accurate solution of the full flow [6,7]. This is not the case for more advanced approaches like large-eddy simulations [8] or direct numerical simulations [9].

In addition of being questionable in terms of turbulence modeling, RANS approaches are used to predict thrombosis and blood damage, which are both intimately related to the shear stress history of the cells. In RANS computations, only the mean flow is predicted, which means that the instantaneous shear stress seen by the blood cells is unknown. Any prediction regarding thrombosis and hemolysis based on Lagrangian tracking of particles forced to follow the streamlines obtained by a RANS computation [5] will thus provide incomplete information.

The need for advanced CFD tools is clear at the macroscopic scale; the same statement can be made at the microscopic scale, whether it is to feed macroscopic simulation in high-fidelity data to improve models or to perform numerical simulations of the dynamics of blood cells in devices. One can cite the examples of blood analyzers [10] using the Coulter effect, where red blood cells pass through a small

© Springer International Publishing Switzerland 2015
V. Van Toi and T.H. Lien Phuong (eds.), *5th International Conference on Biomedical Engineering in Vietnam,*
IFMBE Proceedings 46, DOI: 10.1007/978-3-319-11776-8_2

aperture and perturb an electrical field, or microfluidic devices used for cells sorting [11,12]. In these applications red blood cells circulate in devices operating at high velocity, to guarantee a high throughput. The requirements are thus different from microcirculation, where a Stokes flow hy-pothesis can be made [13-15].

This introduction stresses the need for the development of advanced CFD software to perform high-fidelity numerical simulations of blood flows and flows of red blood cells in the context of the design and development of artificial devices. Here, we focus on systems operating at high speed, which generates turbulence at large scales and adds inertial effects at the scale of the red blood cells. In order to tackle these situations, our group is developing the YALES2BIO software (http://www.math.univ-montp2.fr/~yales2bio/). Its characteristics and first applications are described in the present paper.

II. NUMERICAL METHOD

This section describes the numerical framework of the YALES2BIO solver. YALES2BIO is based on another solver developed at CORIA (UMR 6614): YALES2 (http://www.coria-cfd.fr/index.php/YALES2). YALES2 is a massively parallel finite-volume research code that solves the Navier-Stokes equations on unstructured meshes. It is dedicated to computations of turbulent flows [16-18], in particular bounded domains. The incompressible Navier-Stokes equations are solved using a projection method [19]. The fluid velocity is first advanced using a 4th-order central scheme in space, and explicit low-storage four-step Runge-Kutta scheme in time [16,20]. The divergence-free velocity at the end of the time-step is obtained by solving a Poisson equation for pressure, to correct the predicted velocity. A Deflated Preconditioned Conjugate Gradient algorithm is used to solve this Poisson equation [18]. Turbulence is accounted for by direct simulations or large eddy simulations, using advanced models as the sigma model [21], which properly reproduces the turbulence damping near the walls. This description is valid for both the YALES2 and the YALES2BIO solvers.

A. YALES2BIO: Macroscopic Blood Flows

In order to predict the blood flows in large vessels, an Arbitrary Lagrangian-Eulerian (ALE) method has been developed [20]. It has specifically been used to compute flows in computational domains of known boundary velocity. This technique was used in our group to perform patient-specific simulations of the flow in arteries [22] and hearts [20,23], combining medical imaging (magnetic resonance imaging or CT scan) and CFD. This technique was first developed in the OCFIA project (http://www.ocfia.org).

The arterial movements are extracted from a medical exam, thanks to a registration algorithm [23]. Wall movements are not predicted by fluid-structure interaction, but measured directly on a patient. Once the boundary movement is known, blood flow prediction comes down to a fluid problem, solved using YALES2BIO with an ALE technique.

Valves movement in the heart could also be accounted for using this method. However, complications in terms of mesh managements become overwhelming. Even when the valves movement is supposed to be known and imposed, their opening and closure would necessitate the use of series of mesh with different connectivity. An alternative, adopted in YALES2BIO, is to treat these thin structures as immersed boundaries [24]. The numerical method is described in detail by Chnafa et al. [20].

B. YALES2BIO: Microscopic Blood Flows

In order to compute flows of red blood cells in microfluidic devices, possibly with complex geometries, an unstructured immersed boundary method has been developed [25]. It is based on the ideas by Peskin [24], adapted to be used in conjunction with unstructured grids. Red blood cells are represented thanks to a Lagrangian mesh that follows the displacement of the membrane. The membrane is displaced by the carrying fluid. When the membrane deforms, its stress state is computed using a finite-element solver [26] and membrane nodal forces are obtained. These forces are then regularized over the Eulerian fluid grid to mimic the effect of the membrane on the fluid. This technique has been used to compute vesicles, capsules and red blood cells under flow [25,27,28]. Contrary to numerous approaches to compute the dynamics of microscopic objects under flow [13-15,29], the present approach solves the Navier-Stokes equations, which enables to predict the red blood cell dy-namics in microfluidic systems using the inertial focusing principle [11,12] or more generally in devices operating at large velocity [10].

III. COMPUTATIONS

A. High-Velocity Macroscopic Blood Flows: The Example of the Flow in the Left Heart

In order to demonstrate the performances of the YALES2BIO solver to compute high-speed macroscopic blood flows, we will describe the recent large-eddy simulation of the flow in a left heart by Chnafa et al. [20,23]. By its complexity, this computation illustrates the versatility of the solver. In addition, although the case considered is not a biomedical device but a physiological case, numerous flow features encountered in the heart will also be obtained in cases of artificial hearts, for instance.

From a patient medical exam, Chnafa *et al.* [20] reconstructed the geometry and the deformations of a left heart, from the pulmonary veins to the aortic root. Imposing the endocardium and the valves movements as in the medical exam, the computation solves the flow equations and provides a characterization of the flow in the patient-specific left heart. Fig. 1 shows the flow field in the simulation at two salient instants in the simulation, both during diastole.

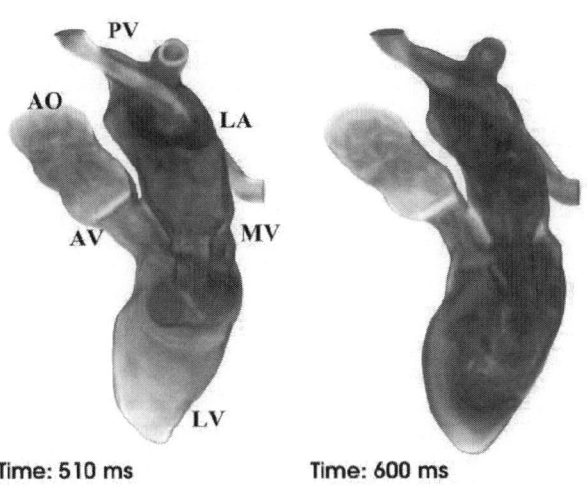

Time: 510 ms   **Time: 600 ms**

Fig. 1 Flow during diastole in the simulation of the left heart flow [20]. The simulation includes part of the pulmonary veins (PV), the left atrium (LA), left venticle (LV) and the aortic root (AO). Aortic (AV) and mitral (MV) valves are also represented. Two instants are displayed to show the mitral jet (left) and the turbulent flow during diastasis (right).

Chnafa *et al.* [20,23] extensively describe this flow, notably showing a number of non-trivial well-known physiological behaviors that are reproduced in the simulation: the strong mitral jet during the E wave (Fig. 1), the large recirculation cell during late diastole, the back flow through pulmonary veins during the atrium contraction and the large vortical movement in the atrium.

In addition, they demonstrate the presence of cycle-to-cycle variations, localized both in space and time. Varia-tions are prone to occur when jets filling cavities (pulmo-nary veins jets filling the atrium and mitral jet filling the ventricle) decelerate. There, the flow destabilizes in numer-ous small vortices (see Fig. 1, right image) and can even become turbulent. Thanks to the high-order non-dissipative numerical scheme, large-eddy simulations are able to pre-dict this intermittent behavior. The flow successively transi-tions and gets laminar depending on the flow conditions. This feature is of prime interest to study flows at moderate Reynolds numbers, characterized by an intermittent regime.

B. Red Blood Cells under Flow: The Example of the Flow in a Cytometer

In the context of flows of red blood cells, YALES2BIO is dedicated to the computation of cases where inertial effects are non-negligible [25]. Inertial effects can be used in mi-crofluidic devices for sorting, focusing, ordering and sepa-rating cells [11] with a high throughput. They can also be present in a device due to high operating speed.

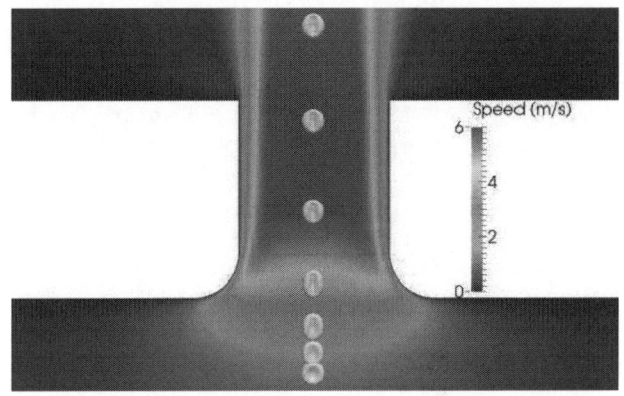

Fig. 2 Simulation of the dynamics of a red blood cell in an industrial cytometer. Sequence of shapes of one red blood cell (from bottom to top) superimposed over the carrying fluid magnitude. The aperture diameter is 50 micrometers and the main flow Reynolds number around 300.

Inside cytometers based on the Coulter effect, blood count-ing is performed at high speed (several meters per second) to obtain a short analysis time. As a consequence, red blood cells flow in a device where shear rate are extremely high.

Fig. 2 shows preliminary results of the computation of the dynamics of a red blood cell in an industrial cytometer [27]. The complex computational domain (here only a small part is presented in Fig. 2) and the operating conditions come from the actual configuration of a cytometer from Horiba Medical [10]. Red blood cells enter one by one in a small aperture (Fig. 2). Their presence will perturb an electrical field, which enables counting and sizing of the cells. In order to refine the understanding of the relationship between the red blood cells dynamics (shape, orientation, trajecto-ry,...) and the electrical pulse characteristics, YALES2BIO is used to perform numerical simulations in the industrial cytometer from Horiba Medical.

IV. CONCLUSIONS

A high-fidelity general purpose code dedicated to the computation of blood flows and flows of red cells in complex geometries and at high Reynolds number has been developed. Its performances have been illustrated in two complex flows: the flow in a patient-specific left heart, which shows intermittency and cycle-to-cycle variations and the flow in an industrial Coulter counter, where red blood cells pass through a small contraction. YALES2BIO is destined to compute flows in biomedical devices in order to help their design and optimization. Future developments will notably include the modeling of thrombosis and hemolysis and of non-Newtonian effects in the simulations.

ACKNOWLEDGMENT

The authors gratefully acknowledge Ramiro Moreno and Ghislain Lartigue for their collaboration. Damien Isèbe, Daniele Di Pietro, Marco Martins Afonso and Vanessa Lleras are also thanked. The authors thank the French Agence Nationale pour la Recherche (through the FORCE project), BPI France for the DAT@DIAG project, the CNRS, the Labex Numev (ANR-10-LABX-20) for funding and GENCI for the access to computer resources.

CONFLICT OF INTEREST

The authors declare that they have no conflict of interest.

REFERENCES

1. Sforza DM, Putman CM and Cebral JR (2009) Hemodynamics of cerebral aneurysms. Ann. Rev. Fluid Mech. 41:91-107.
2. Taylor CA and Figueroa CA (2009) Patient-specific modeling of cardiovascular mechanics. Ann. Rev. Biomed. Eng. 11:109-134.
3. Marsden AL (2014) Optimization in cardiovascular modeling. Ann. Rev. Fluid Mech. 46:519-546.
4. Deutsch S, Tarbell JM, Manning KB, Rosenberg G and Fon-taines AA (2006) Experimental fluid mechanics of pulsatile ar-tificial blood pumps. Ann. Rev. Fluid Mech. 38:65-86.
5. Fraser KH, Tskain ME, Griffith BP and Wu ZJ (2011) The use of computational fluid dynamics in the development of ventric-ular assist devices. Med. Eng. Phys. 33:263-280.
6. Stewart SFC, et al. (2012) Assessment of CFD performance in simulations of an idealized medical device - Results of FDA's first computational interlaboratory study. Cardiov. Eng. Tech. 3(2):139-160.
7. Stewart SFC, et al. (2013) Results of FDA's first interlaboratory computational study of a nozzle with a sudden contraction and conical diffuser. Cardiov. Eng. Tech. 4(4):374-391.
8. Delorme YT, Anupindi K and Frankel SH (2013) Large eddy simulation of FDA's idealized medical device. Cardiov. Eng. Tech. 4(4):392-407.
9. Passerini T, Quaini A, Villa U, Veneziani A and Canic S (2013) Validation of an open source framework for the simulation of blood flow in rigid and deformable vessels. Int. J. Numer. Meth. Biomed. Eng. 29:1192-1213.
10. Isèbe D and Nérin P (2013) Numerical simulation of particle dy-namics in an orifice-electrode system. Application to counting and sizing by impedance measurement. Int. J. Numer. Meth. Biomed. Eng., 29(4):462-475.
11. Di Carlo D (2009) Inertial microfluidics. Lab. Chip 9:3038-3046.

12. Wu Z, Willing B, Bjerketorp J, Jansson JK and Hjort K (2009) Soft inertial microfluidics for high throughput separation of bacteria from human blood cells. Lab. Chip 9:1193-1199.
13. Peng Z, Asaro RJ and Zhu Q (2011) Multiscale modelling of erythrocytes in Stokes flow. J. Fluid Mech. 686:299-337.
14. Zhao H, Isfahani AHG, Olson LN and Freund JB (2010) A spec-tral boundary integral method for flowing blood cells. J. Com-put. Phys. 229:3726-3744.
15. Freund JB (2014) Numerical simulation of flowing blood cells. Ann. Rev. Fluid Mech. 46:67-95.
16. Moreau V, Domingo P and Vervisch L (2011) Design of a massively parallel CFD code for complex geometries. Comp. Rend. Méc. 339(2-3):141-148.
17. Moreau V, Domingo P and Vervisch L (2011) From large-eddy simulation to direct numerical simulation of a lean premixed swirl flame: Filtered laminar flame-PDF modelling. Comb. Flame 158:1340-1357.
18. Malandin M, Maheu N and Moureau V (2013) Optimization of the deflated conjugate gradient algorithm for the solving of el-liptic equations on massively parallel machines. J. Comput. Phys. 238:32-47.
19. Chorin A (1968) Numerical solution of the Navier-Stokes equa-tions. Math. Comp. 22:745-762.
20. Chnafa C, Mendez S and Nicoud F (2014) Image-based large-eddy simulation in a realistic left heart. Comput. Fluids 94:173-187.
21. Nicoud F, Baya Toda H, Cabrit O, Bose S and Lee J (2011) Using singular values to build a subgrid-scale model for large eddy simulations. Phys. Fluids, 23:085106.
22. Midulla M, Moreno R, Baali A, Chau M, Negre-Salvayre A, Ni-coud F, Pruvo J-P, Haulon S and Rousseau H (2012) Haemody-namic imaging of thoracic stent-grafts by computational fluid dynamics (CFD): presentation of a patient-specific method combining magnetic resonance imaging and numerical simula-tions. Eur. Radiol. 22(10):2094-2102.
23. Chnafa C, Mendez S , Moreno R and Nicoud F (2014) Using image-based CFD to investigate the intracardiac turbulence, In `The Cardio-Circulatory System: from Modeling to Clinical Applications' Ed. A. Quarteroni. Springer MS&A, in press.
24. Peskin CS (2002) The immersed boundary method. Acta Num. 11:479-517.
25. Mendez S, Gibaud E and Nicoud F (2014) An unstructured sol-ver for simulations of deformable particles in flows at arbitrary Reynolds numbers. J. Comput. Phys. 256:465-483.
26. Charrier J-M, Shrivastava S and Wu R (1989) Free and constrai-ned inflation of elastic membranes in relation to thermoforming non-axisymmetric problems. J. St. Anal. Eng. Des. 24(2):55-74.
27. Gibaud E, Siguenza J, Mendez S and Nicoud F (2013).\ Towards numerical prediction of red blood cells dynamics within a cy-tometer. Comp. Meth. Biomech. Biomed. Eng. 16(sup1):9-10.
28. Martins Afonso M, Mendez S and Nicoud F (2014) On the dam-ped oscillations of an elastic quasicircular membrane in a two-dimensional incompressible fluid. J. Fluid Mech., in press.
29. Walter J, Salsac A-V, Barthès-Biesel D and Le Tallec P (2010) Coupling of finite element and boundary integral methods for a capsule in a Stokes flow. Int. J. Numer. Meth. Eng. 83:829-850.

Author: Simon Mendez
Institute: CNRS (I3M)
Street: UM2. I3M. CC051. Place Eugène Bataillon
City: Montpellier
Country: France
Email: smendez@um2.fr

Numerical and Experimental Mixing Studies in a Split and Recombine Micromixer with Ellipse-Like Micropillars

Nhut Tran-Minh[1,2], Erik Andrew Johannessen[1], and Frank Karlsen[1]

[1] Buskerud and Vestfold University College, Raveien 215, 3184 Borre, Norway
[2] Norchip AS, Industriveien 8, N-3490 Klokkarstua, Norway

Abstract— **A passive planar micromixer with rapid mixing has been successfully demonstrated by simulations and experiments. The structure of this micromixer contains ellipse-like micropillars in the main channel. Adding micropillars to micromixer will reduce the diffusion distance of the fluids. So, this type of design can improve mixing efficiency.**

Keywords— **Passive mixing, splitting and recombination, computational fluid dynamics, absorbance test.**

I. INTRODUCTION

Miniaturized systems, such as lab-on-a-chips system (LOCs) and micro-total analysis systems (μ-TAS) have played an increasing role in the research and commercialization because of their low sample consumption, low cost, and rapidity of the analysis. Due to mixing fluids uniform for analysis, micromixer becomes one important component in miniaturized systems. Passive micromixers require no external energy for mixing, whereas active micromixers require external energy. These external energies can be pressure, temperature, electrohydrodynamics, dielectrophoretics, electrokinetics, etc. [1]

In this study, a simple and low cost splitting and recombination micromixer with ellipse-like micropillars was investigated. The efficiency of the proposed micromixer is shown in numerical results and is verified by measurement results.

II. MICROMIXER DESIGNS

Fig.1 shows the design of the mixing unit with ellipse-like micropillars in splitting and recombination (SAR) micromixer. The term ellipse-like micropillar is an element having the shape of an ellipse [2]. When the main stream reaches the ellipse-like micropillar, the stream is then split into two separated streams on the smaller channels. Afterward, two streams within smaller channels are merged as one stream. The contact interface of fluids is increased throughout each ellipse-like micropillar so that the mixing effect is enhanced.

SAR micromixer includes 3 inlet channels (I_1, I_2, I_3), one outlet channel (O_1), and some mixing units. The geometry of SAR micromixer with ellipse-like micropillars is shown in Fig.2a.

In order to evaluate the mixing phenomena, the outlet channel of SAR micromixer is split into four sub-channels. Hence, the second design of micromixer in this study includes 3 inlet channels (I_1, I_2, I_3), four outlet channels (O_1, O_2, O_3, O_4), see Fig.2b. Absorbance test is conducted at the outlets of these sub-channels.

III. NUMERICAL SIMULATIONS

Computational fluid dynamics software (COMSOL Multiphysics) was used to solve the governing equations. The governing equations, such as the Navier-Stokes equation, the continuity equation, the diffusion-convection equation can be described in the following:

$$\frac{\partial u}{\partial t} + u \cdot \nabla u = -\frac{1}{\rho}\nabla p + \nu\nabla^2 u \qquad (1)$$

$$\nabla \cdot u = 0 \qquad (2)$$

$$\frac{\partial c}{\partial t} + (u \cdot \nabla)c = D\nabla^2 c \qquad (3)$$

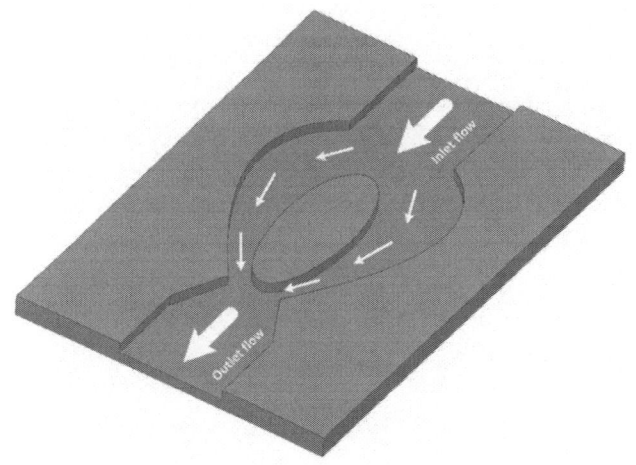

Fig. 1 Mixing unit of SAR micromixer

© Springer International Publishing Switzerland 2015
V. Van Toi and T.H. Lien Phuong (eds.), *5th International Conference on Biomedical Engineering in Vietnam*,
IFMBE Proceedings 46, DOI: 10.1007/978-3-319-11776-8_3

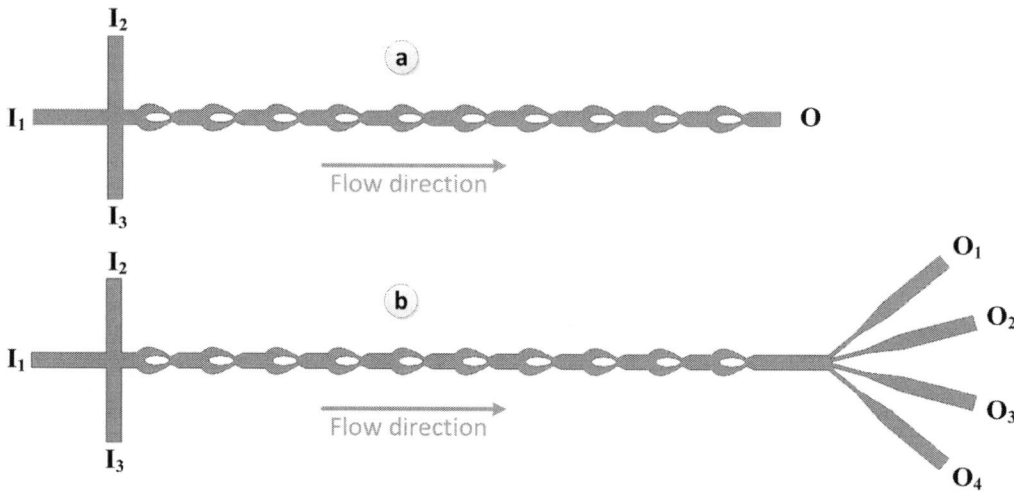

Fig. 2 SAR micromixer

In which appear velocity u, density ρ, pressure p, kinematic viscosity v of the fluid, c and D are concentration and diffusion constant of the species, respectively. In this paper, the value of ρ and v are 1000 kg.m^{-3} and 0.001 kg.s.m^{-1}, respectively. The diffusion coefficient of the water-ink mixture is 3.23×10^{-10} m^2.s^{-1} [3]. During simulation, the incompressible steady flow condition was assumed. No-slip condition is applied to the boundary on the wall. The fixed flow rate was set to the inlet of the micromixer. The fixed pressure ($p = 0$) was set to the outlet of the micromixer. The normalized molar concentration of the species was set 1 for inlet I_1, 0 for inlet I_2 and I_3. To investigate the mixing process, simulations were performed at nine flow rates as listed in Table 1. Flow rate at inlet I_2 and I_3 were kept at 0.02ml/min.

Table 1 Flow rate and *Re* value of numerical simulation

Case	Flow rate at I_1 (ml/min)	Reynolds number
i	0.020	0.952380952
ii	0.025	1.190476190
iii	0.030	1.428571429
iv	0.035	1.666666667
v	0.040	1.904761905
vi	0.045	2.142857143
vii	0.050	2.380952381
viii	0.055	2.619047619
ix	0.060	2.857142857

Assume that the flow rates of the fluids at the entrance I_1, I_2, I_3 of SAR micromixer are Fl_1, Fl_2, Fl_3, respectively. The concentration of the fluids are C_1, C_2, C_3. So, the concentration of the fluid at the outlet channel C_T can be derived in the following:

$$C_1 \times Fl_1 + C_2 \times Fl_2 + C_3 \times Fl_3 = C_T \times Fl_T \tag{4}$$

$$C_T = \frac{C_1 \times Fl_1 + C_2 \times Fl_2 + C_3 \times Fl_3}{Fl_T} \tag{5}$$

$$C_T = \frac{C_1 \times Fl_1 + C_2 \times Fl_2 + C_3 \times Fl_3}{Fl_1 + Fl_2 + Fl_3} \tag{6}$$

where Fl_T is the flow rate of the fluid at the outlet of SAR micromixer .

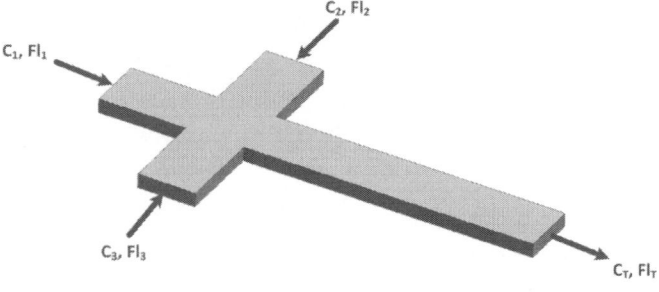

Fig. 3 Concentration and flow rates at the inlet/outlet channels

When C_1 is one, C_2 and C_3 are zero, the equation (6) can be rewritten:

$$C_T = \frac{Fl_1}{Fl_1 + Fl_2 + Fl_3} \tag{7}$$

The equation (7) will be used to verify the results from numerical simulations.

IV. EXPERIMENTAL PROCEDURES

A. Microfabrication

The negative photoresist SU-8 50 (MicroChem Corp., MA, USA) was coated on a bare silicon wafer with a thickness of 100μm. The resist must be firstly baked at 65°C for 2 minutes. Afterward, the temperature of the resist layer should be increased to 95°C and maintained at this temperature for 30 minutes.

The SU-8 thin film with thickness of 100μm was contacted with a photomask and exposed to UV light with a dose of 600 mJ/cm². The unexposed SU-8 was removed in SU-8 developer (MicroChem Corp., MA, USA) (see Fig.4a).

The rigid SU-8 mold was coated with Polydimethylsiloxane (PDMS, Sylgard 184, Dow Corning, USA) (see Fig.4b). The PDMS was degassed to remove air bubbles, then it was cured in a leveled oven at 65°C for 3 hours. The PDMS was cut and peeled off from the SU-8 mold with tweezers. Both PDMS and glass substrate were treated with O_2 plasma (see Fig.4c), and placed on a hotplate at 65°C for 30 minutes to complete the bonding process (see Fig.4d).

B. Inspection of Fluidic Mixing

The fluid flow system consists of three syringe pumps to supply two kinds of fluids into SAR micromixer. One fluid is pure water and the other is a mixture of 92 wt% pure water and 8 wt% red food dye (Idun Industri AS, Norway).

The samples at the outlet of SAR micromixer in both designs (see Fig.2), were transferred to 96-well microplate for conducting an absorbance experiment. A multidetection BioTek (Synergy 2, USA) was used to determine the absorbance spectrum of a sample solution.

V. RESULTS AND DISCUSSION

The microscopic images of the fabricated micromixers are illustrated in Fig.6. The main channels of micromixer have a width of 201μm and a depth of 92μm. The width of each sub-channel is 52μm.

According to our results (see Fig. 7), the model which was used in numerical simulations gives a good prediction of mixing performance of SAR micromixer. When the flow rate of the fluid at the inlet 1 was increased, the amount of red food dye was mixed with pure water was also increased. Hence, the concentration of the fluid at the outlet channel was also increased.

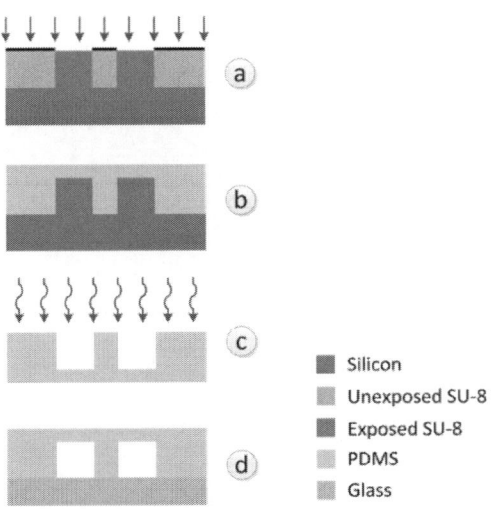

■	Silicon
■	Unexposed SU-8
■	Exposed SU-8
■	PDMS
■	Glass

Fig. 4 Process flow of microfabrication

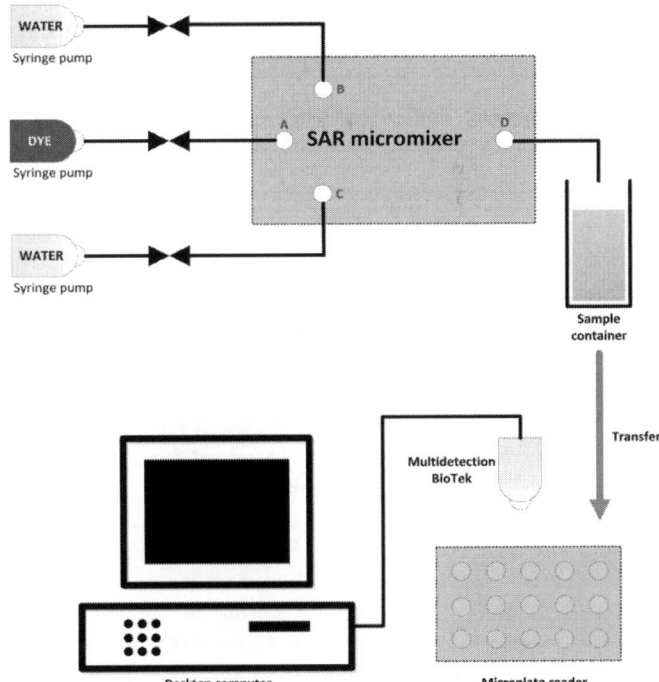

Fig. 5 Schematic of the experimental setup

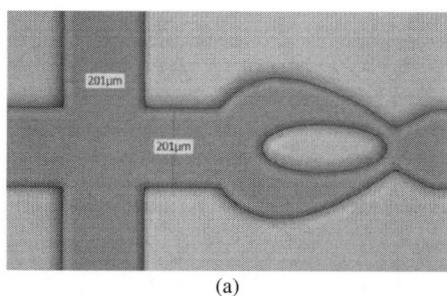

(a)

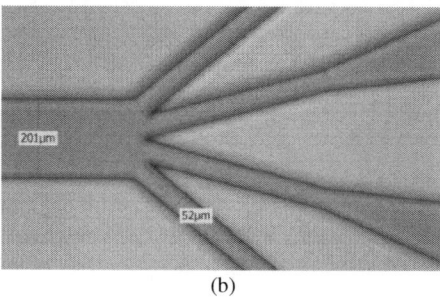

(b)

Fig. 6 Microscopic images of the SAR micromixer. (a) The junction of three inlet channels. (b) Four sub-channels.

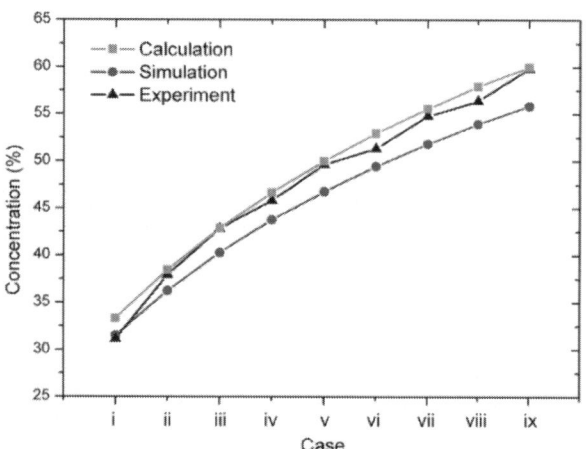

Fig. 7 Outlet concentration of SAR micromixer in different cases

Table 2 Concentration of the fluid at the outlet of each sub-channel

Fl_1 (ml/min)	O_1	O_2	O_3	O_4
0.004	18.16 %	20.65 %	19.35 %	17.57 %
0.006	11.13 %	54.26 %	49.05 %	6.56 %
0.008	3.62 %	55.15 %	49.09 %	6.67 %
0.010	6.78 %	58.72 %	54.63 %	7.75 %
0.012	4.67 %	55.41 %	51.80 %	2.81 %

VI. CONCLUSIONS

According to the numerical and experimental results, this splitting and recombination micromixer with ellipse-like micropillars shows high mixing efficiency can be achieved at low Reynold number. The homogeneity of SAR micromixer was studied by absorbance test. Due to the advantages of SAR micromixer, such as simple and low cost, proposed micromixer may be ideal for a rapid and optimal collection and mixing related sample preparation of biological fluids.

ACKNOWLEDGMENT

This research work is supported by the Research Council of Norway and Norchip AS (Norway). The Research Council of Norway is acknowledged for the support through the Norwegian Micro- and Nano-Fabrication Facility, NorFab (197411/V30).

CONFLICT OF INTEREST

The authors declare that they have no conflict of interest.

REFERENCES

1. Nguyen, N.T, Wu, Z.: Micromixers—a review. J. Micromech. Microeng. 15, R1-R16 (2005).
2. Tran-Minh, N., Dong, T., Su, Q., Yang, Z., Jakobsen, H., Karlsen, F.: Design and optimization of non-clogging counter-flow microconcentrator for enriching epidermoid cervical. Biomed. Microdevices 13, 179-190 (2011).
3. Lee, S., Lee, H-Y, Lee, I-F, Tseng, C-Y: Ink diffusion in water. Eur. J. Phys. 25, 331-336 (2004).

The address of the corresponding author:

Author: Nhut Tran-Minh
Institute: Buskerud and Vestfold University College
Street: Raveien 215
City: Borre
Country: Norway
Email: Nhut.Tran-Minh@hbv.no

In the second design (see Fig.2b), the fluids at the outlet O_1, O_2, O_3, O_4 were collected and tested separately. The results from absorbance test are listed in Table 2.

At the same flow rate of 0.004ml/min, the concentration of the outlets of four sub-channels are nearly the same. It means that the high mixing efficiency of the fluids can be achieved at this flow rate. When the Fl_1 was slightly increased, Fl_2 and Fl_3 were kept at 0.004ml/min, the concentration of outlets O_2, O_3 increases, the concentration of outlets O_1, O_4 decreases. So, the mixture of the outlet of SAR micromixer is not homogeneous. With these tests, the operation condition of SAR micromixer can be known.

Multiplex DNA Biosensor for Viral Infection Diagnosis Using SERS Molecular Sentinel-on-Chip

Hoan T. Ngo[1,2], Hsin-Neng Wang[1,2], Thomas Burke[3], Christopher Woods[4], Geoffrey S. Ginsburg[2,3], and Tuan Vo-Dinh[1,2,*]

[1] Departments of Biomedical Engineering and Chemistry, Duke University, Durham, NC 27708, USA
[2] Fitzpatrick Institute for Photonics, Duke University, Durham, NC 27708, USA
[3] Duke Institute for Genome Sciences & Policy, Duke University, Durham, NC 27708, USA
[4] Duke Global Health Institute, Duke University, Durham, NC 27708, USA
tuan.vodinh@duke.edu

Abstract— The development of sensitive and selective techniques for multiplex detection of DNA biomarkers is paramount for clinical diagnosis. Various multiplex DNA detection techniques have been reported. However, most of these techniques require multiple incubation and/or washing steps or target sequence labeling. In this work, we demonstrated a unique multiplex DNA biosensor for viral infection diagnosis using the surface-enhanced Raman scattering (SERS) "Molecular Sentinel-on-Chip" (MSC) technique. The sensing mechanism is based upon the change of SERS intensity when Raman labels tagged at 3'-ends of molecular sentinel nanoprobes are physically displaced from the Nanowave chip's surface upon target DNA hybridization. SERS measurements were performed immediately following a single hybridization reaction between the target single-stranded DNA (ssDNA) sequences and the complementary molecular sentinel nanoprobes immobilized on the Nanowave chip without requiring target labeling (i.e., label-free assay), secondary hybridization, or post-hybridization washing, thus reducing the assay time and lowering cost. Two nucleic acid transcripts, interferon alpha-inducible protein 27 (IFI27) and interferon-induced protein 44-like (IFI44L), are used as model systems for the multiplex detection concept demonstration. These two genes are well known for their critical role in host immune response to viral infections and can be used as molecular signature for viral infection diagnosis. The results indicate the effectiveness and potential of the MSC technology for multiplex DNA detection for point-of-care diagnostics and global health applications.

Keywords— Multiplex DNA detection, DNA biosensor, Surface-enhanced Raman scattering, SERS, Molecular Sentinel-on-Chip, Nanowave, Metal film over nanosphere, Infectious disease, Global health.

I. INTRODUCTION

There is currently a great interest in the development of sensitive and selective biosensing techniques for clinical diagnosis. In particular, diagnostic techniques that are capable of simultaneously detecting several biomarkers (i.e. multiplex detection) are highly needed because multiple biomarkers are usually involved in disease onset and progression. Among diagnostic techniques, molecular diagnosis based on nucleic acid biomarker (DNA or RNA) detection offers many advantages such as high specificity, low limit of detection, and the ability to quantify the level of infection [1]. Various multiplex DNA/RNA detection techniques have been reported [2-9].

Raman spectroscopy, which yields very narrow spectral vibrational features of the investigated samples, has long been considered to be a powerful tool for multiplex detection. However, Raman scattering cross-sections are extremely small, thereby limiting its practical applications. One powerful spectroscopy approach to overcome this drawback is surface-enhanced Raman scattering (SERS), which increases the Raman scattering cross-section substantially, enabling the application of this process for extremely sensitive detection of the analytes [10-12]. Large SERS enhancement factors of 10^{12} - 10^{15} have been reported, inspiring the development of new sensing materials for detection of analytes with highly sensitive detection levels [13-16]. Over the past years, we have developed various SERS-based sensing platforms, ranging from nanoparticles to nanopost arrays, nanowires and nanochips [17-21].

In this study, we show an example of a SERS-based multiplex DNA biosensor for viral infection diagnosis. The biosensor is composed of molecular sentinel (MS) probes [22-24] immobilized on a "Nanowave" SERS chip [25]. This technique is, therefore, referred to as MSC technique. In the normal configuration (i.e., in the absence of target ssDNA), the immobilized MS probes form hairpin loops, which maintain the Raman labels in close proximity of the Nanowave chip's surface, thus producing a strong plasmonic effect and inducing an intense SERS signal upon laser excitation. In the existence of complementary target ssDNA, the MS probes hybridize with the target ssDNA and open. The Raman label molecules are physically separated

---

* Corresponding author.

© Springer International Publishing Switzerland 2015
V. Van Toi and T.H. Lien Phuong (eds.), *5th International Conference on Biomedical Engineering in Vietnam*,
IFMBE Proceedings 46, DOI: 10.1007/978-3-319-11776-8_4

from the Nanowave chip's surface, thus leading to a decreased SERS signal because the plasmonic effect is a short-range process. Two nucleic acid transcripts, interferon alpha-inducible protein 27 (IFI27) and interferon-induced protein 44-like (IFI44L) were chosen as test models. These two genes have been demonstrated to play a critical role in host immune response to viral infection [26]. By profiling gene expression in peripheral blood in response to respiratory viral infection, a recent study showed that these two genes are among a set of host genes differentially expressed upon viral infection [27-29]. It is noteworthy that before our work, various SERS-based multiplex DNA detection techniques have been reported [30-36]. However, most of these techniques require multiple hybridization and/or washing steps or target sequences labeling. In contrast, our MSC requires a single hybridization step without the need of post-hybridization washing and target sequences labeling, making it simple-to-use, with short run-time and low reagent cost. The Nanowave chip's SERS enhancement factor is improved by using bi-metallic film of Ag and Au [37] instead of a single Au film for metal coating. Large area of the Nanowave chip can be fabricated with high reproducibility using self-assembly on water-air interface technique. The chip can be stored for over a month before use.

II. EXPERIMENTS

A. Materials

Premium pre-cleaned microscope slides (25 mm x 75 mm x 1 mm) were purchased from VWR (Radnor, PA). Mercaptohexanol (MCH), p-mercaptobenzoic acid (pMBA), sodium chloride (NaCl), sodium phosphate buffer pH 7.0 (SPB), and ethanol (EtOH) were purchased from Sigma Aldrich (St Louis, MO) and used as received unless noted otherwise. Polystyrene beads (PS) (5000 series, size 430 nm, $CV \leq 3\%$, 10% solids) were purchased from Fisher

Scientific (Pittsburgh, PA). Millipore Synergy ultrapure water (DI) of resistivity = 18.2 $M\Omega$ cm was used in all aqueous solutions. Au, Ag and Ti pellets were obtained from Kurt Lesker (Clairton, PA). The MS probes and ssDNA (Table 1) were synthesized by Integrated DNA Technologies (IDT, Coralville, IA).

B. Nanowave Fabrication and Characterization

Two types of Nanowave, Au film over nanospheres (Au-FON) and bimetal film over nanospheres (BMFON), were fabricated as previously described [25]. Briefly, monolayer of closely packed PS (430 nm diameter) was prepared on microscope glass slides using self-assembly at the water-air interface method. The as-prepared samples were then annealed at 80 °C for 1 hour followed by metal coating. For AuFON, PS was sequentially coated with 5 nm Ti (2 Å/s) and 200 nm of Au (10 Å/s) whereas for BMFON, the annealed substrate were sequentially coated with 5nm Ti (2 Å/s), 100 nm of Ag (5 Å/s), and 100 nm of Au (10 Å/s). All metals were deposited under 5×10^{-6} mTorr using Kurt Lesker PVD 75 electron beam evaporator. The obtained AuFON and BMFON were stored in desiccator at room temperature. The substrates were cleaned using mild oxygen plasma (10 W RF power, 200 mTorr, 10 sccm of oxygen) for 20 seconds prior use.

The fabricated Nanowaves were characterized using FEI XL30 SEM. The accelerating voltage was set at 7 kV, and working distance varied between 10 mm and 12 mm.

C. SERS Measurements

SERS enhancement of the two types of Nanowave, Au-FON and BMFON, were compared using pMBA as model molecule. First, SAM of pMBA was formed by soaking the substrates in 1 mM pMBA solutions (in EtOH) for 24h followed by EtOH rinsing and nitrogen drying. After that, SERS measurements were performed at 632.8 nm using a Renishaw InVia confocal Raman microscope with 10X objective and ~0.5 mW laser power. Each measurement has 10 seconds exposure time. Ten measurements at 10 randomly selected spots across each substrate were acquired, background subtracted, and averaged to give represent spectra for the corresponding substrate.

D. Functionalization of Nanowave with MS Probes

Before ssDNA detection, BMFON Nanowave substrates were functionalized with IFI27 and IFI44L MS probes by soaking the substrates in the MS probe solutions (1 μM MS, 0.5 M NaCl and 10 mM SPB in DI water) for 12h. Next, the substrates were gently rinsed by buffer solution (0.5 M NaCl and 10 mM SPB in DI water) to remove unbound probes. The substrates were then soaked in 1 mM MCH in

Table 1 Sequences of the MS probes, complementary target ssDNA, and non-complementary ssDNA

Name	Sequences[a]
IFI27 MS	5'-SH-AAAAA<u>GAGTA</u>CAACTGTAGCAAT **CCTGGCC**<u>GTACTC</u>-Cy5-3'
IFI44L MS	5'-SH-AAAAA<u>CCGAG</u>TCTTTAACAGAAT **ATATCCTATACCGC**<u>CTCGG</u>-ROX-3'
IFI27 Target ssDNA	5'-TTTGCCCCT**GGCCAGGATTGCTACA GTTGT**GATTGGAGGA-3'
IFI44L Target ssDNA	5'-ATAAC**CGAGCGGTATAGGATATATT CTGTTAAAGA**TGGAA-3'
Non-complementary ssDNA	5'-GTGTAGGGATTATAGAGTCGCTTTC-3'

[a]The underlined sequences indicate the complementary arms of the MS, and the bold sequences represent the sequences in the MS probes and the corresponding target ssDNA which are complementary to each other.

buffer solution for 1h to displace non-specifically adsorbed MS probes and to passivate the gold surface [38]. Finally, the substrates were gently rinsed with DI water and ready for ssDNA detection.

E. SERS-Based ssDNA Detection

Multiplex detection was demonstrated by soaking the BMFON functionalized with both IFI27 MS and IFI44L MS (IFI27 1:2 IFI44L) in following sample solutions: buffer (blank sample, no ssDNA present), target ssDNA complementary to IFI27 MS, target ssDNA complementary to IFI44L MS, and composite mixture of target ssDNA complementary to IFI27 MS and to IFI44L MS (IFI27 target 1:2 IFI44L target). In these experiments, the ssDNA solutions (complementary target and non-complementary) were prepared by diluting ssDNA stock solution to final concentration of 1 μM ssDNA, 0.5 M NaCl and 10 mM SPB in DI water. After 3h soaking in the sample solutions at 37 °C, the substrates were removed from the solutions without washing, and SERS measurements were performed. Note that the substrates must remain wet during SERS measurements to prevent MS's conformational change. Only 5% of the laser power, that is, ~0.25 mW, was used to avoid damaging the DNA sequences.

III. RESULTS AND DISCUSSIONS

A. Detection Scheme

Figure 1 schematically illustrates the detection scheme using the MSC approach with two different MS probes, each having a specific Raman label and designed to bind a specific nucleic acid target. In the absence of complementary target ssDNA, the MS probes form hairpin loop structures. The Raman labels tagged at the 3'-ends of the MS probes are in close proximity to the Nanowave chip's surface (<1 nm), resulting in a strong SERS signal due to the plasmonic enhancement effect (i.e. 'On' state). When the complementary target ssDNA sequences are added, they hybridize with the MS probes, forcing the hairpin structures to open. In this open state, the Raman labels are physically separated from the chip surface (>10 nm), resulting in decreases of SERS signals (i.e. 'Off' state). In the aforementioned detection scheme, the Nanowave substrate plays an important role in generating SERS enhancement due to its nanostructured plasmonics-active surface.

B. Nanowave Characterization

Fig. 2 shows SEM image of BMFON Nanowave chip. From the image, periodic hexagonal pattern of closely packed metal-coated nanospheres and inter-nanospheres crevices can be observed. In addition, we also notice that the metal film's surface roughness is quite substantial. This feature leads to a

strong plasmonic effect and a high SERS enhancement of BMFON Nanowave chip. Indeed, our SERS measurements show that BMFON SERS enhancement is 3.6 times higher than AuFON [39]. Therefore, BMFON is chosen as SERS substrate for ssDNA detection, which will be discussed in following section.

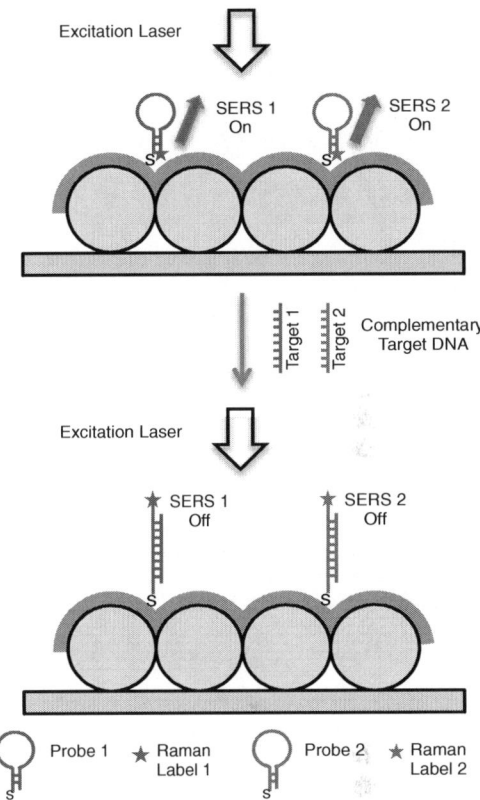

Fig. 1 Scheme for two-multiplex detection of complementary target ssDNA sequences

Fig. 2 SEM image of BMFON Nanowave chip

C. Detection of ssDNA

Fig. 3 shows the SERS spectra from the composite mixture of IFI27 and IFI44L MS probes on BMFON in the absence of any ssDNA. SERS spectra of the composite MS nanoprobes in the presence of only one of the two complementary targets are shown in Fig. 4 (only IFI27 target present) and Fig. 5 (only IFI44L target present). The result indicates that when a target ssDNA was present only the SERS peaks associated with the complementary MS probes were significantly quenched (indicated by arrows). For example, in the Fig. 4, only the SERS peaks 1 to 4 associated with the IFI27 MS nanoprobes were quenched when the IFI27 target was present, indicating that only the IFI27 MS nanoprobes were in open state. In contrast, the SERS peaks 1* and 2* associated with the second MS nanoprobes (IFI44L MS) remained high, indicating that the second nanoprobes were still in closed state due to the absence of its target ssDNA.

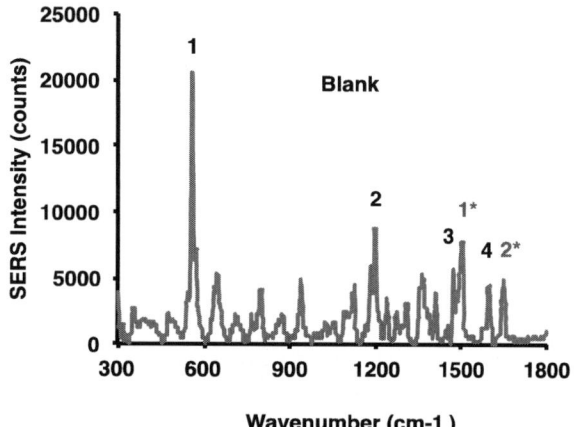

Fig. 3 SERS spectra of the composite MS probes in the absence of any target ssDNA

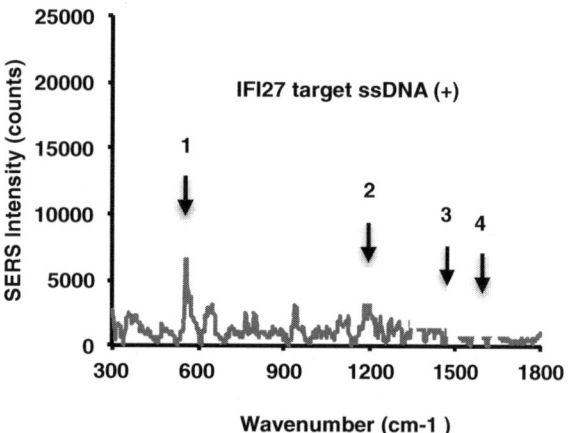

Fig. 4 SERS spectra of the composite MS probes in the presence of single target ssDNA complementary to the IFI27 MS probes

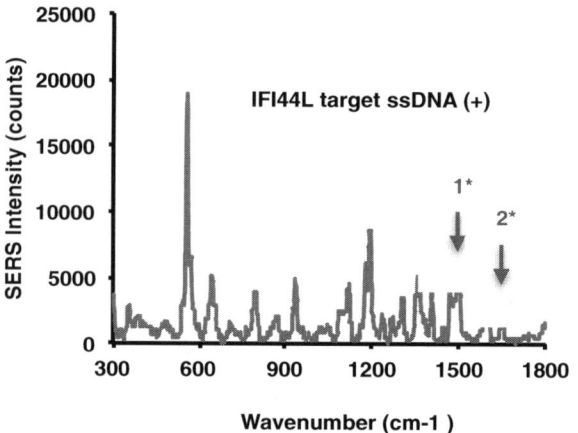

Fig. 5 SERS spectra of the composite MS probes in the presence of single target ssDNA complementary to the IFI44L MS probes

Finally, in the presence of both IFI27 and IFI44L target ssDNA (Fig. 6), all major Raman peaks associated with the two MS nanoprobes were greatly decreased. The above results demonstrate multiplex DNA detection capability of our MSC technique.

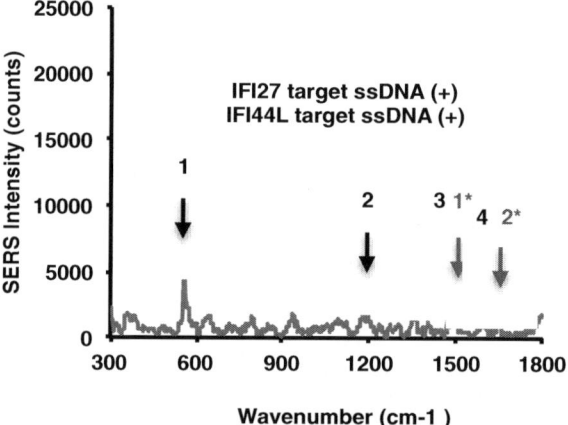

Fig. 6 SERS spectra of the composite MS probes in the presence of two target ssDNA complementary to both MS probes

IV. CONCLUSION

In conclusion, we have demonstrated the feasibility of using the MSC technique for qualitative multiplex DNA detection in a homogeneous solution. Two nucleic acid transcripts, interferon alpha-inducible protein 27 (IFI27) and interferon-induced protein 44-like (IFI44L), are used as the model system for proof of principle demonstration. The Nanowave chip fabrication is relatively simple and low-cost with high reproducibility and good storage lifetime. The bi-metallic (Ag-Au) BMFON Nanowave shows SERS enhancement higher than that of the single-metal AuFON Nanowave. The SERS measurements were performed immediately following a single hybridization reaction using a homogeneous assay without washing steps. The Nanowave chip can be fabricated in large areas with high reproducibility using self-assembly on water-air interface technique. The results of this study demonstrate that the MSC technique can provide a useful tool for multiplex DNA detection for point-of-care diagnostics and global health applications.

ACKNOWLEDGMENT

This work was sponsored by Duke Exploratory Projects and the Wallace H. Coulter Foundation Endowment. Hoan Thanh Ngo is supported by a Fellowship from the Vietnam Education Foundation.

REFERENCES

1. Cordray MS, Richards-Kortum RR (2012) Review: Emerging Nucleic Acid Based Tests for Point-of-Care Detection of Malaria. Am J Trop Med Hyg 87 (2):223-230

2. Mancuso M, Jiang L, Cesarman E, Erickson D (2013) Multiplexed colorimetric detection of Kaposi's sarcoma associated herpesvirus and Bartonella DNA using gold and silver nanoparticles. Nanoscale 5 (4):1678-1686

3. Han SH, Park LS, Lee JS (2012) Hierarchically branched silver nanostructures (HBAgNSs) as surface plasmon regulating platforms for multiplexed colorimetric DNA detection. J Mater Chem 22 (38):20223-20231

4. Stoeva SI, Lee JS, Thaxton CS, Mirkin CA (2006) Multiplexed DNA detection with biobarcoded nanoparticle probes. Angew Chem Int Ed 45 (20):3303-3306

5. Prigodich AE, Randeria PS, Briley WE, Kim NJ, Daniel WL, Giljohann DA, Mirkin CA (2012) Multiplexed Nanoflares: mRNA Detection in Live Cells. Anal Chem 84 (4):2062-2066

6. Li YG, Cu YTH, Luo D (2005) Multiplexed detection of pathogen DNA with DNA-based fluorescence nanobarcodes. Nat Biotechnol 23 (7):885-889

7. Su S, Wei XP, Zhong YL, Guo YY, Su YY, Huang Q, Lee ST, Fan CH, He Y (2012) Silicon Nanowire-Based Molecular Beacons for High-Sensitivity and Sequence-Specific DNA Multiplexed Analysis. ACS Nano 6 (3):2582-2590

8. Meade SO, Chen MY, Sailor MJ, Miskelly GM (2009) Multiplexed DNA Detection Using Spectrally Encoded Porous SiO2 Photonic Crystal Particles. Anal Chem 81 (7):2618-2625

9. Mirasoli M, Bonvicini F, Dolci LS, Zangheri M, Gallinella G, Roda A (2013) Portable chemiluminescence multiplex biosensor for quantitative detection of three B19 DNA genotypes. Anal Bioanal Chem 405 (2-3):1139-1143

10. Otto A, Mrozek I, Grabhorn H, Akemann W (1992) Surface-Enhanced Raman-Scattering. J Phys: Condens Matter 4 (5):1143-1212

11. Vo-Dinh T (1998) Surface-enhanced Raman spectroscopy using metallic nanostructures. TrAC, Trends Anal Chem 17 (8–9):557-582

12. Schatz GC (1984) Theoretical-Studies of Surface Enhanced Raman-Scattering. Acc Chem Res 17 (10):370-376

13. Nie SM, Emery SR (1997) Probing single molecules and single nanoparticles by surface-enhanced Raman scattering. Science 275 (5303):1102-1106

14. Kneipp J, Kneipp H, Kneipp K (2008) SERS - a single-molecule and nanoscale tool for bioanalytics. Chem Soc Rev 37 (5):1052-1060

15. Otto A, Bruckbauer A, Chen YX (2003) On the chloride activation in SERS and single molecule SERS. J Mol Struct 661:501-514

16. Xu HX, Bjerneld EJ, Kall M, Borjesson L (1999) Spectroscopy of single hemoglobin molecules by surface enhanced Raman scattering. Phys Rev Lett 83 (21):4357-4360

17. Vo-Dinh T, Hiromoto MYK, Begun GM, Moody RL (1984) Surface-enhanced Raman spectrometry for trace organic analysis. Anal Chem 56 (9):1667-1670

18. Meier M, Wokaun A, Vo-Dinh T (1985) Silver Particles on Stochastic Quartz Substrates Providing Tenfold Increase in Raman Enhancement. J Phys Chem 89 (10):1843-1846

19. Vo-Dinh T, Meier M, Wokaun A (1986) Surface-enhanced Raman spectrometry with silver particles on stochastic-post substrates. Anal Chim Acta 181 (0):139-148

20. Vo-Dinh T, Dhawan A, Norton SJ, Khoury CG, Wang HN, Misra V, Gerhold MD (2010) Plasmonic Nanoparticles and Nanowires: Design, Fabrication and Application in Sensing. J Phys Chem C 114 (16):7480-7488

21. Yuan H, Liu Y, Fales AM, Li YL, Liu J, Vo-Dinh T (2012) Quantitative Surface-Enhanced Resonant Raman Scattering Multiplexing of Biocompatible Gold Nanostars for in Vitro and ex Vivo Detection. Anal Chem 85 (1):208-212

22. Wabuyele MB, Vo-Dinh T (2005) Detection of human immunodeficiency virus type 1 DNA sequence using plasmonics nanoprobes. Anal Chem 77 (23):7810-7815

23. Wang H-N, Fales AM, Zaas AK, Woods CW, Burke T, Ginsburg GS, Vo-Dinh T (2013) Surface-enhanced Raman scattering molecular sentinel nanoprobes for viral infection diagnostics. Anal Chim Acta 786 (0):153-158

24. Vo-Dinh, T., et al. (2013) Plasmonic nanoprobes: from chemical sensing to medical diagnostics and therapy Nanoscale 5(21): 10127-10140

25. Ngo HT, Wang H-N, Fales AM, Vo-Dinh T (2013) Label-Free DNA Biosensor Based on SERS Molecular Sentinel on Nanowave Chip. Anal Chem 85 (13):6378–6383

26. Stetson DB, Medzhitov R (2006) Type I interferons in host defense. Immunity 25 (3):373-381

27. Zaas AK, et al. (2009) Gene Expression Signatures Diagnose Influenza and Other Symptomatic Respiratory Viral Infections in Humans. Cell Host Microbe 6 (3):207-217

28. Woods CW, et al. (2013) A Host Transcriptional Signature for Presymptomatic Detection of Infection in Humans Exposed to Influenza H1N1 or H3N2. PLoS One 8 (1)

29. Zaas AK, et al. (2013) A Host-Based RT-PCR Gene Expression Signature to Identify Acute Respiratory Viral Infection. Sci Transl Med 5 (203):203ra126

30. Li JM, Wei C, Ma WF, An Q, Guo J, Hu J, Wang CC (2012) Multiplexed SERS detection of DNA targets in a sandwich-hybridization assay using SERS-encoded core-shell nanospheres. J Mater Chem 22 (24):12100-12106

31. Kang T, Yoo SM, Yoon I, Lee SY, Kim B (2010) Patterned Multiplex Pathogen DNA Detection by Au Particle-on-Wire SERS Sensor. Nano Lett 10 (4):1189-1193

32. Cao YWC, Jin RC, Mirkin CA (2002) Nanoparticles with Raman spectroscopic fingerprints for DNA and RNA detection. Science 297 (5586):1536-1540

33. Faulds K, Jarvis R, Smith WE, Graham D, Goodacre R (2008) Multiplexed detection of six labelled oligonucleotides using surface enhanced resonance Raman scattering (SERRS). Analyst 133 (11):1505-1512

34. Faulds K, McKenzie F, Smith WE, Graham D (2007) Quantitative simultaneous multianalyte detection of DNA by dual-wavelength surface-enhanced resonance Raman scattering. Angew Chem Int Ed 46 (11):1829-1831

35. Wei X, Su S, Guo Y, Jiang X, Zhong Y, Su Y, Fan C, Lee S-T, He Y (2013) A Molecular Beacon-Based Signal-Off Surface-Enhanced Raman Scattering Strategy for Highly Sensitive, Reproducible, and Multiplexed DNA Detection. Small 9 (15):2493–2499

36. Zhang ZL, Wen YQ, Ma Y, Luo J, Jiang L, Song YL (2011) Mixed DNA-functionalized nanoparticle probes for surface-enhanced Raman scattering-based multiplex DNA detection. ChemComm 47 (26):7407-7409

37. Fu CY, Kho KW, Dinish US, Koh ZY, Malini O (2012) Enhancement in SERS intensity with hierarchical nanostructures by bimetallic deposition approach. J Raman Spectrosc 43 (8):977-985

Corresponding author: Tuan Vo-Dinh
Institute: Duke University
Street:
City: Durham, NC 27708
Country: USA
Email: tuan.vodinh@duke.edu

Whispering Gallery Mode Biosensing – A Detailed Study on ZnO Microspheres

Ngo Huynh Buu Trong[1,2,3], Paul Ching-Hang Chien[1,2,3], Yu-Da Chen[1,2,3], Shang-Hsuan Wu[1],
Fu-Chen Hsiao[1], Linh-Nam Nguyen[4], and Yia-Chung Chang[1,5,*]

[1] Research Center for Applied Sciences, Academia Sinica, Taipei, 11529, Taiwan
[2] Nano Science and Technology Program, TIGP, Academia Sinica, Taipei 11529, Taiwan
[3] Department of Engineering and System Science, National Tsing Hua University, Hsinchu 30010, Taiwan
[4] Danang College of Technology, The University of Danang, Danang 59000, Vietnam
[5] Department of Physics, National Cheng Kung University, Tainan, 701 Taiwan
buutrong85@yahoo.com,
yiachang@gate.sinica.edu.tw

Abstract— **We demonstrate the potential of using ZnO microspheres (MS) for biosensing applications. ZnO MS with smooth surface exhibiting novel whispering gallery mode (WGM) in photoluminescence and spectroscopic imaging ellipsometry behavior were synthesized by hydrothermal techniques. Sharp resonance peaks covering the entire visible range could be excited not only by UV but also visible light. Moreover, the theoretical modeling was carried out. Essential information about the position, TE and TM modes of each resonance are precisely predicted. Such information is very important for the investigation of resonance shift caused by the attachment of foreign molecules. Both experimental and theoretical results support that ZnO spherical micro cavity can be a good candidate for biosensing applications.**

Keywords— **ZnO MS, Optical resonator, WGM biosensing, Photoluminescence, Spectroscopic imaging ellipsometry.**

I. INTRODUCTION

There is a great deal of interest for using spherical microresonators (SμR) for biosensing classified as WGM biosensors [1-3]. Such sensor systems show excellent performance, thanks to their extremely high Q factors [2]. In the present paper, ZnO microspheres (MS) were investigated, aiming for applications as WGM biosensors. Because of their low toxicity [4] and high biocompatibility [5, 6], ZnO thin films and nanostructures are promising materials for biosensing applications [7, 8]. Moreover, the ultra-sensitivity can be achieved in ZnO-based sensors because of their significant enhancement of fluorescence signal compared to the one in conventional platform [9-11]. However, to our knowledge, there is no work in the literature using ZnO MS for WGM biosensing. One major problem that needs to overcome is the synthesis of ZnO MS with smooth surface. This is critical for establishing whispering gallery mode (WGM) within one SμR. Though there are few works recently dealing with WGM lasing in

ZnO MS [12-13], the possibility of exciting WGM in such system by low power and (or) white light source for sensing applications is not investigated elsewhere.

We have previously explored WGM in ZnO MS [14]. Comparing to similar investigations, we further observed the size dependence of WGM. Very uniform WGM behavior would occur as the diameter D of MS exceeds 2μm. Moreover, the number of modes and mode position are also very sensitive to D. Similar tendency was also observed in hexagonal ZnO cavity [15]. The main reason might come from the spectral dependence of refractive index. This fact was not yet considered in most of calculation, especially for ZnO resonator [16]. In biosensor systems, refractive index of the environment surrounding MS often dynamically fluctuates. The shift of resonance might, therefore, be misjudged.

Thus, the present paper would try to investigate (i) possibility of exciting WGM by using not only UV but also white light, and (ii) the inclusion of refractive index correction in the theoretical model. We report here our recent experimental results for utilizing different light sources to excite WGM in ZnO MS. WGM spectrum was observed by micro-photoluminescence (μ-PL) and spectroscopic imaging ellipsometer (SIE), based on an Optrel Multiskop system, measurements. The next part is devoted to theoretical analysis using modified-refractive-index (MRI) model. The MRI model was described in detail in [14]. This model is sufficient for determining correctly mode number, resonance position, and hence any abnormal modes.

II. MATERIALS AND CHARACTERIZATIONS

A. Synthesis and Excitation Techniques for WGM of ZnO MS

The synthesis and storage process of ZnO SμR with smooth surface was described elsewhere [14]. In short, the mixed solution of Zn(NO3)$_2$.6H$_2$O:HMT with the same

---

* Corresponding author.

© Springer International Publishing Switzerland 2015
V. Van Toi and T.H. Lien Phuong (eds.), *5th International Conference on Biomedical Engineering in Vietnam*,
IFMBE Proceedings 46, DOI: 10.1007/978-3-319-11776-8_5

molar ratio was prepared. This solution was then stirred with tri-sodium citrate of 2:1 volume ratio, respectively. The hydrothermal reaction was kept at 90°C for 90 min. The product needs to be washed, 3 times with ultrapure water and ethanol, before being kept in ethanol solution at 4°C until used.

WGM behavior is explored by using both μ-PL measurement with two laser sources (Cd - 325 nm and Ar^+ - 514 nm), and SIE measurement (λ = 450 ~ 750 nm). All optical measurements were conducted at room temperature.

B. MRI Model

MRI model was based on the fact that refractive index depends on resonator's structure properties (e.g, morphology, porosity, size of MS) and light wavelength via dispersion relation. Such complicated dependences might cause inaccurate mode spacing, missing resonance and hence failure of one-to-one assignment between theory and experiment. The MRI treatment was well described in [14]. The procedures can be summarized as follows:

(i) Firstly, we estimate the values of resonance positions $\lambda_{l,ini}$ by using Schiller's scheme [17].

(ii) Next, we utilize Newton-Raphson method to iterate λ_l via equation (1) in [14] using $\lambda_{l,ini}$ as initial values. Dispersion relation of $n_{g,bulk}(\lambda_l)$ is also taken into account during this iteration. The result is considered as analytical data.

(iii) By utilizing the group index $n_g \equiv \dfrac{\partial[n(\lambda)k]}{\partial k}$ and classical resonance condition, $2\pi R = \dfrac{l\lambda_l}{n(\lambda_l)}$, we can approximate the group index $n_g(\lambda)$ of the MS by

$$n_g(\lambda_l) \approx \frac{1}{R\Delta k_l} \equiv n_{g,Exp}(\lambda_l) \ , \qquad (1)$$

where $\Delta k_l = k_{l+1} - k_l$ is experimental mode spacing of WGM. $\Delta n = n_{g,bulk}(\lambda_l) - n_{g,Exp}(\lambda_l)$ - is then introduced as correction term for refractive index.

(iv) MRI values finally replaced to $n_{g,bulk}(\lambda_l)$ in step (ii) to give us more accurate resonance position, $\lambda_{l,fin}$.

In order to assign correctly the mode number for each resonance, we introduce an indicator function D(l, l') defined as

$$D(l, l') \equiv \frac{Re[x_l]\lambda_{Exp,l'}}{\pi}$$

$$l' = l - l_{min} + 1, \qquad (2)$$

where $l \in [l_{min}, l_{min} + N_{Exp} - 1]$, l_{min} is the minimum for the assignment, N_{Exp} is number of experimental resonances

observed. $l' \in [1, N_{Exp}]$ and $x_l \equiv \dfrac{2\pi R}{\lambda_{l,fin}}$. $\lambda_{Exp,l'}$ is the experimental peak wavelength of the resonance l'

By assigning the mode number for each resonance, the diameter of sphere is also re-optimized.

III. RESULTS AND DISCUSSIONS

A. Experimental Observations of WGM

One of the critical issues for WGM resonator is obtaining efficient and robust coupling. Free space coupling is the simplest, but least efficient one. In our current study, the efficient coupling parameter, determined by the condition $\eta < (\lambda/a)^2$ [18], is less than 0.5 % for SIE and 1.5 % for μ-PL. Yet, we can still observe WGM (Fig. 1). It is well known that defect states play an important role in photoluminescence properties of ZnO in visible range [18]. Emission of these states might be coupled to the WGM resonances in the same wavelength range (fig. 1 - left).

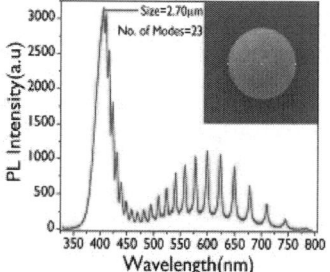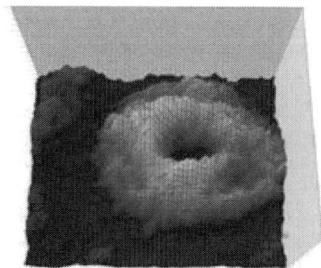

Fig. 1 WGM behavior obtained by (left) μ-PL (a = 2.7 μm, λ = 325 nm) [14] and (right) SIE (a= 7.8 μm, λ = 550 nm) (right)

Previous works [16] show that resonance peaks in PL measurement are quite stable when coating with uniform polymer layer. These peaks are also very sensitive with the size of particles [14] and their ambient environment [16]. Peak shifting did show a linear dependence on environment's refractive index. This property is crucial for detecting any fluctuation of refractive index which may be caused by foreign molecules.

More interestingly, by reconstructing Ψ and Δ images from SIE measurement, near-field on the surface of MS associated with the WGM can be detected. From such image, one can also estimate the size of MS (fig 1 - right). Ψ, Δ respectively represent for amplitude and phase difference of input/output light when it interacts with sample [18]. By employing the SIE measurement, which can focus on a single microsphere to measure WGM, we expect to detect very sensitive signal. By accompanying with more efficient light coupling, this technique could enhance the capability ZnO MS for biosensing.

B. Investigation Resonances of WGM by MRI Model

In order to determine the resonance peaks of ZnO SµR by MRI model, we first follow those steps in part II.A. The critical point in this model is to define the refractive index difference, Δn, between bulk phase and MS. This quantity show complicated size dependence (data not shown here). Strain effect in MS, comparing to its counterpart in nanostructure, do not play a significant role here. Surface and inner morphology may also affect the refractive index. In particular, by using SEM, we did see very smooth surface, but with porous structure inside the sphere. Such porosity parameter, unfortunately, is hard to determine quantitatively. Because of the nonlinear size dependence of refractive index, we might deduce that porosity parameter is arbitrary for different MS. More investigations need to be carried out to confirm the relation between porosity and refractive index. However in this stage, MRI model shows that it can predict the peak position of *MS* without knowing their inner morphology by using one fitting parameter (Fig. 2).

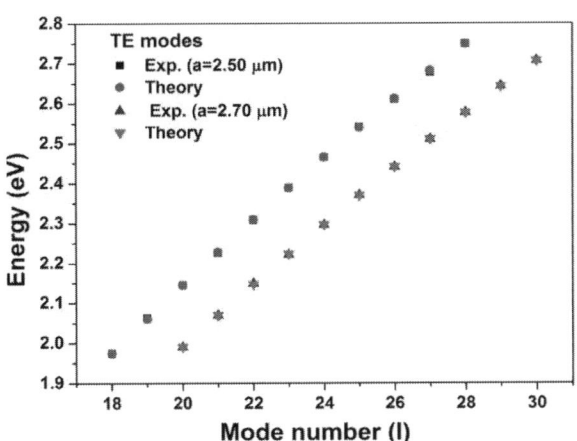

Fig. 2 The resonance peak positions of TE modes calculated by MRI model for different µR size. They are well agreement with experimental results.

From both experimental and theoretical results, we confirm the red shift of angular momentum number *l* when increasing the resonator size. Moreover, the WGM dispersion's tendency is also addressed. Both behaviors are similar to those observed for hexagonal cavity in ZnO nano-needle [15] and GaN microwires [20]. Note that even though the dispersion relation formula in [15, 20] use simple plane wave model, refractive index is considered as energy-dependent quantity. It is expected that our MRI model here is sufficient for semiconductor resonators, where the dispersion relation in the visible-UV range is not uniform.

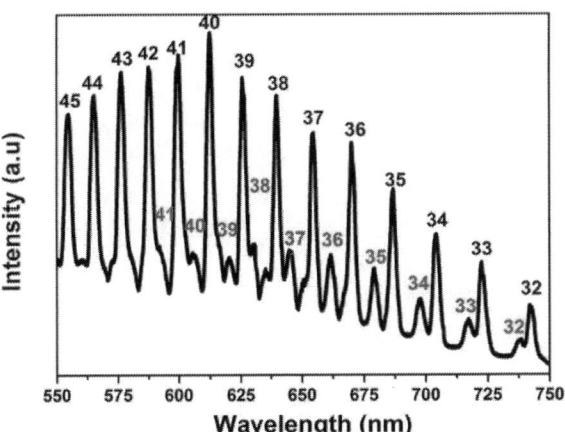

Fig. 3 Mode number of TE (black number) and TM (red number) modes in WGM spectrum of 4.78 µm MS

In order to depict the assignment of mode number we show here the result of ZnO MS of 4.78 µm. Firstly, by using MRI model, Δn was predicted as equal to 0.162. Next, x_l was calculated for every *l*. Finally, we used (2) to calculate *D (l, l')*. From (2), one can see that *D* would reveal the deviation due to a particular mode assignment when compared to the SEM diameters. Ideally, *D (l, l')* versus *l* should be converged to a constant value which is approximately equal to the measured diameter of MS. The range of *(l, l')* that give such value would also be the range of our mode number. Here, in our example, *D (l, l')* ≈ 4.85 µm. SEM measurements of the cavity diameter are performed with an error of at least 1 %. The optimized value of D here gives a difference of 0.4 % which is within the SEM error bar. Note that the WGM is very sensitive to the size, such re-optimized value is also worthy for evaluating the physical properties of our resonator.

In the next step, we explore the existence of TM mode by using Schiller's scheme for $p = 1/m^2$ (see [17]) with the modified refractive index. The final results agree well with the minor peak position in fig 3, without any further adjustment of refractive index.

IV. CONCLUSIONS

In conclusion, we have done extensive analysis of the WGM in ZnO MS. Such behavior is determined by using different measurement techniques. Adding to the recent study of interfacial refractive index sensing properties [16], ZnO MS can be considered as a good candidate for biosensing application. Moreover, to enhance the capacity of the application, MRI model was constructed. This model successfully predicted not only WGM properties (resonance position, mode number) but also physical properties (size and refractive index) of the resonator. Both experimental and theoretical analyses reinforce the possibility for using ZnO MS in biosensing applications.

ACKNOWLEDGMENT

This work was supported in part by Academia Sinica and National Science Council of the Republic of China under Contract No. NSC 101-2112-M-001-024-MY3.

REFERENCES

1. Arnold S, Teraoka I et al. (2007) Microparticle photophysis illuminates viral biosensing. Faraday Discuss 137:65-83
2. Chiasera A, Righini GC et al. (2010) Spherical whispering gallery mode microresonators. Laser & Photon Rev 4:457-482
3. Soria S, Righini GC et al. (2011) Optical microspherical resonators for biomedical sensing. Sensors 11:785-805
4. Bondarenko O, Kahru A et al. (2013) Toxicity of Ag, CuO and ZnO nanoparticles to selected environmentally relevant test organisms and mammalian cells in vitro: a critical review. Arch Toxicol 87:1181–1200
5. Zhou J, Wang ZL et al. (2006) Dissolving behavior and stability of ZnO wires in biofluids: a study on biodegradability and biocompatibility of ZnO nanostructures. Adv Mater. 18: 2432-2435
6. Li Zh, Wang ZL et al. (2008) Cellular level biocompatibility and biosafety of ZnO nanowires. J Phys Chem C 112: 20114-20117
7. Wang JX, Dong ZL et al (2006) ZnO nanocomb biosensor for glucose detection. Appl Phys Lett. DOI: 10.1063/1.2210078
8. Arya SK, Singh PS et al. (2012) Recent advances in ZnO nanostructures and thin films for biosensor applications: review. Analytica Chimica Acta 737:1-21
9. Kumar N, Hahm JI et al. (2006) Ultrasensitive DNA sequence detection using nanoscale ZnO sensor arrays. Nanotechnology 17: 2875-2881
10. Dorfman A, Hahm JI et al. (2006) Highly sensitive biomolecular fluorescence detection using nanoscale ZnO platforms. Langmuir 22: 4890-4895
11. Zhao JW, Zhi LF et al. (2008) Fabrication of micropatterned ZnO/SiO2 core/shell nanorod arrays on a nanocrystalline diamond film and their application to DNA hybridization detection. J Mater Chem 18:2459-2465
12. Okazaki K, Okada T et al. (2012) Ultraviolet whispering gallery mode lasing in ZnO micro/nano sphere crystal. Appl Phys Lett. DOI: 10.1063/1.4768696
13. Okamoto Sh, Minowa Y et al. (2012) White light lasing in ZnO microspheres fabricated by laser ablation. Proc of SPIE. DOI: 10.1117/12.907595
14. Moirangthem RS, Chang YC et al. (2013) Optical cavity modes of a single crystalline ZnO microsphere. Opt Expr 21: 3010-3020
15. Nobis T, Grundmann M et al. (2004) Whispering gallery modes in nanosized dielectric resonators with hexagonal cross section. Phys Rev Lett. DOI: 10.1103/PhysRevLett.93.103903
16. Moirangthem RS, Erbe A (2013) Interfacial refractive index using visible-excited intrinsic ZnO photoluminescence coupled to whispering gallery modes. Appl Phys Lett. DOI: 10.1063/1.4817273
17. Schiller S (1993) Asymptotic expansion of morphological resonance frequencies in Mie scattering. Appl Opt 32: 2181-2185
18. Djurisic AB, Chen XY (2010) ZnO nanostructures for optoelectronics: Material properties and device applications. Prog Quant Electron 34: 191-259
19. Fujiwara H (2007) Spectroscopic ellipsometry: principles and applications. John Wiley & Sons, West Sussex
20. Coulon PM, Perez JZ (2012) GaN microwires as optical microcavities: whispering gallery modes vs Fabry–Ferot modes. Opt Expr20: 18707-18716

Author: Yia-Chung Chang
Institute: Research Center for Applied Sciences, Academia Sinica
Street: 128 Sec. 2, Academia Road, Nankang
City: Taipei
Country: Taiwan
Email: yiachang@gate.sinica.edu.tw

Ultrasonic Assessment of the Radius

J.J. Kaufman[1,2], G.M. Luo[1], F. Rosete[3], M. Bucovsky[3], E.M. Stein[3], E. Shane[3], and R.S. Siffert[2]

[1] CyberLogic, Inc., New York, NY, USA
[2] Department of Orthopedics, The Mount Sinai School of Medicine, New York, NY, USA
[3] Department of Medicine, Columbia University College of Physicians and Surgeons, New York, NY, USA

Abstract— **The objectives of this study were to evaluate the capability of a novel ultrasound device to clinically estimate bone mineral density (BMD) at the 1/3 radius. The device rests on a desktop and is portable, and permits real-time evaluation of the radial BMD. The device measures the time delays associated with three distinct propagation pathways through the forearm, and from these time delays an ultrasound-based estimate, BMD_{US}, of the BMD is computed. A clinical IRB-approved study measured ultrasonically seventy-eight adults at the 1/3 radius. BMD was also measured ("BMD_{DXA}") at the same anatomical site and time using DXA. A linear regression of BMD_{DXA} vs BMD_{US} produced a linear correlation coefficient of 0.93 (P<0.001). In conclusion, although x-ray methods are effective in bone mass assessment, osteoporosis remains one of the largest under-diagnosed diseases in the world today. The research described here should enable significant expansion of diagnosis and monitoring of osteoporosis through a desktop device that ultrasonically assesses bone mass at the 1/3 radius.**

Keywords— **ultrasound, osteoporosis, BMD, DXA, radius.**

I. INTRODUCTION

The objective of this study is to enhance the ability of ultrasound to non-invasively assess bone. As is well known, osteoporotic fractures are a major public health problem associated with high degrees of morbidity and mortality [1, 2]. Worldwide, osteoporosis causes more than 8.9 million fractures annually, resulting in an osteoporotic fracture every 3 seconds [3]. Osteoporosis is estimated to affect 200 million women worldwide - approximately one-tenth of women aged 60, one-fifth of women aged 70, two-fifths of women aged 80 and two-thirds of women aged 90 [3]. Osteoporosis affects an estimated 75 million people in Europe, USA and Japan [3]. For the year 2000, there were an estimated 9 million new osteoporotic fractures, of which 1.6 million were at the hip, 1.7 million were at the forearm and 1.4 million were clinical vertebral fractures. Europe and the Americas accounted for 51% of all these fractures, while most of the remainder occurred in the Western Pacific region and Southeast Asia [3]. Worldwide, 1 in 3 women over 50 will experience osteoporotic fractures, as will 1 in 5 men [3]. The toll both in individual quality of life and in health care costs of osteoporotic fractures cannot be overstated. Early detection and assessment is crucial to initiating therapeutic interventions as this is the best way to prevent a fracture from occurring [2].

Presently the gold standard for bone assessment is based on x-ray densitometric techniques, such as with DXA [4]. The measurement of bone mass as represented for example by (areal) bone mineral density (BMD) is based on the principle that more mass is generally associated with a stronger bone, and a stronger bone is in turn associated with a reduced risk of fracture. Indeed, BMD is the single most important factor in estimating bone strength and fracture risk [5]. In spite of these facts, osteoporosis remains one of the world's most under-diagnosed diseases and this problem is getting worse [6, 7]. For example, in the U.S. it is estimated that only about 1 in 5 individuals at increased risk of fracture have been identified [7]. The reasons for this are three-fold. The first is that the vast majority of primary care physicians do not have bone assessment capability available; therefore patients must be referred off-site to specialists for these measurements and such referrals often do not occur. Second, the relatively high cost of such bone density tests and the high costs of the DXA machines themselves also results in less than required testing. Lastly is the issue of ionizing radiation that is associated with DXA and which provides some patients another reason to avoid the test.

Ultrasound is an alternate modality for bone assessment [8, 9, 10]. It is ionizing radiation-free, relatively inexpensive, and as an elastic wave may provide not only information on bone mass but may also contain additional information on bone strength and associated risk of fragility fracture risk [11, 12, 13]. In spite of its tremendous potential, ultrasound has had until the present relatively little impact on clinical practice. One reason for this is that the performance of present ultrasound technology—and specifically performance in terms of estimating BMD—is less than required to displace the current gold standard, DXA. As one example, presently approved thru-transmission ultrasound devices designed to measure the calcaneus provide correlations with BMD at the same anatomical site ranging from about 0.6 to a maximum of only 0.8 [14]. Therefore, this research will investigate a new method for more accurately estimating BMD. In particular, this paper will describe a new approach for ultrasonically estimating BMD at the 1/3 radius that is highly correlated ($r = 0.93$) with BMD obtained with DXA.

© Springer International Publishing Switzerland 2015
V. Van Toi and T.H. Lien Phuong (eds.), *5th International Conference on Biomedical Engineering in Vietnam,*
IFMBE Proceedings 46, DOI: 10.1007/978-3-319-11776-8_6

II. Materials and Methods

A. *Device and Signal Processing*

A new desktop device (UltraScan 650, CyberLogic, Inc., New York, NY, USA, Fig. 1) for quantitative real-time bone assessment has been constructed [15]. It processes ultrasound signals after propagating through a forearm and displays an estimate of BMD on a laptop computer that is connected via USB. Details of the device and signal processing are provided in [15]; here we report on additional clinical data that has been obtained.

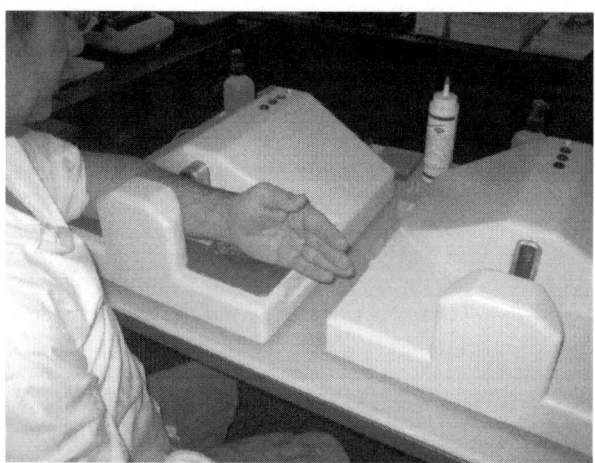

Fig. 1 UltraScan 650 Scanner (one with arm and one without.)

B. *Clinical Data*

Seventy-eight adult subjects were recruited for this study under an IRB approved protocol and informed consent was received from each participant. Pregnancy was excluded in premenopausal women before DXA scans were performed. Each subject was measured three to five times at the 1/3 radius with the *UltraScan 650*. The number of times depended on the variability of the measurements. Standard off-the-shelf isopropyl alcohol (70%) was used as an ultrasound coupling agent; it was sprayed onto the subject's forearm and the transducers' active surfaces. The *UltraScan 650* data was processed to obtain an ultrasound-based estimate, BMD_{US}, of the BMD. Each subject also had their radial bone density at the same 1/3 location measured with DXA (QDR 4500, Hologic, Inc., Bedford, MA, USA) to obtain an estimate, BMD_{DXA}, of BMD. Finally, an ultrasound reproducibility study was carried out on three additional subjects each with 15 independent measurements, and the percent coefficient of variation (%CV) was evaluated [16].

III. Results

For these clinical data, a plot of BMD_{DXA} *vs.* BMD_{US} is shown in Fig. 2. The linear correlation between BMD_{DXA} and BMD_{US} was $r = 0.93$ (P<0.001). The linear univariate regression between BMD_{DXA} and BMD_{US} produced a standard error of the prediction of 0.042 g/cm^2. Finally, the percent coefficient of variation in the reproducibility study was found to be 2.1%.

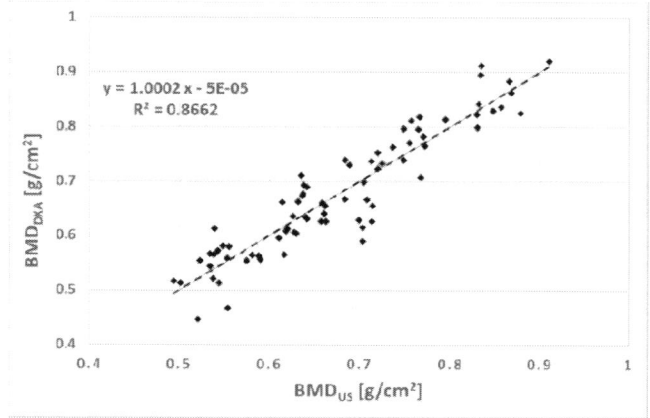

Fig. 2 BMD_{DXA} vs BMD_{US} for the 78 study subjects

IV. Conclusions

In summary a new device, the *UltraScan 650*, has been described that has the potential to enlarge the scope of ultrasound bone assessment in particular and of bone screening in general. The portability and simplicity in use of the radiation-free ultrasound scanner, combined with its high degree of accuracy and precision in estimating radial BMD, provides a basis by which to expand ultrasonic assessment to the primary care setting. This will in turn provide an opportunity to reduce the incidence of osteoporotic fractures through early and timely therapeutic interventions.

Acknowledgment

The kind support of the National Institute of Arthritis and Musculoskeletal and Skin Diseases (Grant Number AR45150) and the National Institute on Aging (Grant Number AG036879) of the National Institutes of Health, through the Small Business Innovative Research Program are gratefully acknowledged. The funders had no role in the design and conduct of the study; collection, management, analysis, and interpretation of the data; and preparation, review, or approval of the manuscript.

CONFLICT OF INTEREST

One of the authors (Jonathan J. Kaufman) is a principal and CEO of the company (CyberLogic, Inc.) which manufactures the *UltraScan 650* device, and another of the authors (Gangming Luo) is an employee of the same company. None of the other authors report any conflicts of interest.

REFERENCES

1. Kanis J (2002) Diagnosis of osteoporosis and assessment of fracture risk. The Lancet 359:1929-1936.
2. Kanis JA, Johansson H, Oden A, McCloskey EV (2009) Assessment of fracture risk. Eur J Radiol 71(3):392-7.
3. International Osteoporosis Website, www.iofbonehealth.org (Accessed March 19, 2014)
4. Ott SM, Kilcoyne RF, Chestnut III CH (1987) Ability of four different techniques of measuring bone mass to diagnose vertebral fractures in postmenopausal women. J Bone Min Res 2:201-210.
5. Lewiecki EM, Laster AJ, Miller PD, Bilezikian JP (2012) More bone density testing is needed, not less. J Bone Miner Res 27(4):739-42.
6. Zhang J, Delzell E, Zhao H et al. (2012) Central DXA utilization shifts from office-based to hospital-based settings among medicare beneficiaries in the wake of reimbursement changes. J Bone Miner Res 27(4):858-64.
7. Baim S, Leslie WD (2012) Assessment of fracture risk. Curr Osteoporos Rep 10(1):28-41.
8. Langton CM, Palmer SB, Porter RW (1984) The measurement of broadband ultrasonic attenuation in cancellous bone. Eng Med 13(2):89–91.
9. Laugier P (2008) Instrumentation for in vivo ultrasonic characterization of bone strength. IEEE Trans Ultrason Ferroelectr Freq Control 55(6):1179-96.
10. Krieg MA, Barkmann R, Gonnelli S et al. (2008) Quantitative ultrasound in the management of osteoporosis: The 2007 ISCD official positions. J Clin Dens 11(1):163-187.
11. Kaufman JJ, Einhorn TE (1993) Review - Ultrasound assessment of bone. J Bone Miner Res 8:517-525.
12. Siffert RS, Luo GM, Cowin SC, Kaufman JJ (1996) Dynamical relationships of trabecular bone density, architecture and strength in a computational model of osteopenia. Bone 18:197-206.
13. Siffert RS, Kaufman JJ (2007) Ultrasonic bone assessment: "The time has come". Bone 40:5-8.
14. Langton CM, Langton DK (2000) Comparison of bone mineral density and quantitative ultrasound of the calcaneus: site matched correlation and discrimination of axial BMD status. Brit Jour Rad 73:31-35.
15. Stein EM, Rosete F, Young P et al. Clinical assessment of the 1/3 radius using a new desktop ultrasonic bone densitometer. Ultrasound Med Biol 39(3):388–395.
16. Bonnick SL (2004) Bone densitometry in clinical practice; application and interpretation (2nd edition). Humana Press, Totowa, New Jersey.

Corresponding author:

Author: Jonathan J. Kaufman
Institute: CyberLogic, Inc.
Street: 611 Broadway, Suite 707
City: New York, NY 10012
Country: USA
Email: jjkaufman@cyberlogic.org

High-Resolution Imaging of Dispersive Ultrasonic Guided Waves in Human Long Bones Using Regularized Radon Transforms

Tho N.H.T. Tran[1], Lawrence H. Le[1,2], and Mauricio D. Sacchi[2]

[1] Department of Radiology & Diagnostic Imaging, University of Alberta, Edmonton, AB, Canada T6G 2B7
[2] Department of Physics, University of Alberta, Edmonton, AB, Canada T6G 2E1

Abstract— **Ultrasound is an indispensable imaging modality to monitor soft tissues in diagnostic radiology. Research into ultrasound has resulted in technology developments and extension of its use beyond soft tissue imaging. The use of ultrasound to probe hard tissues is not yet a common practice but in recent years modest interest has been generated to use quantitative ultrasound in bone evaluation. This is due to several advantages of ultrasound over ionizing techniques: lack of ionizing radiation, sensitivity to the mechanical elasticity of bone tissues, portability, and low cost. Recent studies using axial-transmission technique have shown that ultrasonic guided waves, which propagate within cortical bone, have great potential to characterize mechanical and structural properties of the cortical waveguide. Multi-channel dispersion analysis of ultrasonic guided waves requires a reliable mean to map the data from the time-distance domain to the frequency-phase velocity domain. In this work, linear Radon transform using various regularization strategies is considered to enhance the transform focusing power to image dispersive guide-wave energies. Four forms of linear Radon solution: adjoint, damped least-squares, Cauchy-regularized, and l_1-regularized Radon transform were applied to the simulated and *in-vivo* experimental data and the results were compared. Among the regularization strategies, the l_1-regularization renders a highly-sparse solution and images dispersion energies with the best focusing resolution. The high-resolution dispersion maps allow better wave-mode discrimination and separation. The results of this study suggest the l_1-regularized Radon transform as a valuable tool to image dispersive ultrasonic guided-wave energies propagating in long bones.**

Keywords— **Ultrasound, guided waves, dispersion, resolution, Radon transform.**

I. INTRODUCTION

Osteoporosis is a skeletal disorder that causes bones to become thin and porous, decreasing bone strength and leading to increased fracture risks. The disease can occur in both men and women at any age, but it is most common in women and elders. The bone loss occurs steadily without any symptoms until fracture occurs. Due to the serious impact of osteoporosis and its related fractures on the quality of life,

there is a huge call for reliable diagnostic approaches to diagnose osteoporosis and assess fracture risk.

Osteoporosis still remains one of the most prevalent undiagnosed diseases in the world today. Currently, bone quality is mainly assessed by measuring bone mineral density (BMD) using dual energy X-ray absorptiometry. However, BMD does not provide sufficient information for accurate diagnosis of osteoporosis because bone strength does not only depend on bone mass but also on other structural and mechanical properties. In recent years, interest to use ultrasound to study bones is increasing. Compared to X-ray based diagnostic imaging modalities, ultrasound is cost-effective, non-ionizing, and lack of hazardous bio-effects. In the context of osteoporosis diagnosis, some preliminary studies have shown the ability of ultrasonic guided waves propagating in cortical bone to assess cortical thinning [1,2]. Quantitative guided-wave ultrasonography (QGWU) is promising to evaluate skeletal quality because guided-wave generation and propagation are sensitive to the cortical thickness and bone elastic properties. Guided waves are known to be dispersive, i.e., travel with frequency-dependent velocities. Therefore determination of phase velocity is fundamental to guided wave analysis. The mapping from the time-offset (*t-x*) domain to the frequency-phase velocity (*f-c*) panel is usually performed by the conventional two-dimensional Fourier transform, followed by a replacement of wavenumber by phase velocity. However, the extracted dispersion curves lack the resolution in the transformed plane to discriminate guided-wave modes [1]. Recently, our group was the first to apply a Radon transform to image dispersive energies of the bone plate data [3,4].

In this work, we present an application of the regularized linear Radon transform to enhance the imaging resolution of the dispersive energies of ultrasonic guided waves propagating in long bones using simulated and *in-vivo* experimental data. The resolution of four Radon solutions: adjoint, damped least-squares, Cauchy-regularized, and l_1-regularized, will be discussed.

© Springer International Publishing Switzerland 2015
V. Van Toi and T.H. Lien Phuong (eds.), *5th International Conference on Biomedical Engineering in Vietnam*,
IFMBE Proceedings 46, DOI: 10.1007/978-3-319-11776-8_7

II. MATERIALS AND METHODS

A. High-Resolution Linear Radon Transform

Radon transform (RT) is an integral transform along straight lines. Consider a series of ultrasonic time signals $d(t, x_n)$ at different offsets, $x_0, x_1,..., x_{N-1}$ where t denotes time and the x-axis is not necessarily evenly sampled. We write the time signals as a superposition of Radon signals:

$$d(t, x_n) = \sum_{k=0}^{K-1} m(\tau = t - p_k x_n, p_k), \quad n = 0, ..., N-1 \quad (1)$$

where the ray parameter or slowness, p is sampled at p_0, $p_1,..., p_{K-1}$, and the intercept, τ is the arrival time at zero-offset. Taking the temporal Fourier transform of Equation (1) yields

$$D(f, x_n) = \sum_{k=0}^{K-1} M(f, p_k) e^{-i2\pi f p_k x_n} \quad . \quad (2)$$

Rewriting Equation (2) using matrix notation gives

$$\mathbf{D = LM} \text{ and } \mathbf{L} = \begin{bmatrix} e^{-i2\pi f p_0 x_0} & \cdots & e^{-i2\pi f p_{K-1} x_0} \\ \vdots & \ddots & \vdots \\ e^{-i2\pi f p_0 x_{N-1}} & \cdots & e^{-i2\pi f p_{K-1} x_{N-1}} \end{bmatrix} . \quad (3)$$

A low-resolution adjoint Radon solution $\mathbf{M}$ can be calculated by [5]

$$\mathbf{M = L}^H \mathbf{D} \quad (4)$$

where $\mathbf{L}^H$ is the adjoint operator and H denotes the complex conjugate transpose. Using the quadratic length of the model, i.e., $\|\mathbf{M}\|^2$ as the regularization, the damped least-squares (DLS) solution is:

$$\mathbf{M = (L}^H \mathbf{L} + \mu \mathbf{I)}^{-1} \mathbf{L}^H \mathbf{D} \quad (5)$$

where μ is the hyper-parameter. However, the DLS solution does not provide adequate focusing in the Radon panel and leads to moderate amplitude smearing. This problem can be alleviated by adopting non-quadratic regularization strategies based on a Cauchy- or l_1-norm for a high-resolution Radon solution

$$\mathbf{M = (L}^H \mathbf{L} + \mu \mathbf{Q(M))}^{-1} \mathbf{L}^H \mathbf{D} \quad . \quad (6)$$

The $\mathbf{Q(M)}$ is a diagonal weighting matrix with elements given by

$$Q_{ii} = \frac{1}{(1 + M_i^2 / \sigma^2)} \quad (7)$$

for the Cauchy regularization and

$$Q_{ii} = \frac{1}{|M_i|} \quad (8)$$

for the l_1-regularized Radon panel. σ^2 is the scale factor of the Cauchy distribution. Equation 6 is a non-linear system of equations which can be solved by the iterative re-weighted least-squares (IRLS) scheme for each frequency. Finally the dispersion f-c panel can be obtained by $p = 1/c$.

Following Sacchi [5], the statistical χ^2 test was used to determine the hyper-parameter μ. The μ is chosen where $\chi^2 = N$ where N is the number of observations.

B. Simulation

To simulate a linear signal arrival, $s(t)$, we calculate the signal spectrum, $S(f)$ in the Fourier space

$$S(f) = W(f) e^{\left[-i2\pi f \left(\frac{x}{c(f)} - t_0\right)\right]} \quad (9)$$

where $W(f)$ is the spectrum of the source wavelet and t_0 is a time constant. The dispersion, $c(f)$ is described by the following relationship

$$c(f) = c_{min} + (c_{max} - c_{min}) / \sqrt{1 + (f/f_c)^4} \quad (10)$$

where c_{max} is the maximum phase velocity, c_{min}, minimum phase velocity and f_c, the critical frequency. The spread, $\Delta v = c_{max} - c_{min}$, and the critical frequency, f_c determine the amount of dispersion in the data. The source spectrum is a trapezoidal wavelet. The time signal, $s(t)$ will be obtained by inverse Fourier-transforming $S(f)$.

C. In-vivo Experiment

The in-vivo experiment was performed on the right tibia of a male volunteer using axial transmission method with two 1-MHz-30° transducer-wedge systems (Figure 1a). Two rulers, laid on the relatively flat section of the tibia and held in place by medical tape, were used to guide the transducers. Similar to [6], one system acted as a stationary source and the other as a moving receiver. The transmitting transducer was taped normally to the skin surface by medical tape. Ultrasound gel was used between the transducers and the skin to ensure good coupling. The experiment was performed at room temperature (21°C). The signal detected by the receiver was digitized by and displayed on a 200-MHz digital storage oscilloscope (LeCroy 422 WaveSurfer, Chestnut Ridge, NY). A set of 40 ultrasound records was measured with 2-mm spacing interval and 46-mm closest offset. The recorded data set was further decimated for a final 0.1 μs sampling interval to form a t-x matrix, $\mathbf{d}$.

D. Theoretical Dispersion Curve Computation

To interpret the dispersion information of ultrasonic long bone data, theoretical dispersion curves of the guided-wave modes were computed using the commercial software pack age DISPERSE version 2.0.16i (Imperial College, London). The simulated model is a water-filled cylindrical model with 17-mm inner diameter and 8.5-mm cortical thickness (Figure 1b). These are the mean thicknesses of the marrow layer and the top cortex respectively. Following [7], cortical bone parameters for simulation were 4000 m/s, 1970 m/s, and 1930 kg/m^3 for compressional wave speed, shear wave speed, and density respectively. Water, whose acoustical wave speed is very similar to marrow (v_{water} = 1500 m/s versus v_{marrow} = 1435 m/s), is used as a marrow mimic.

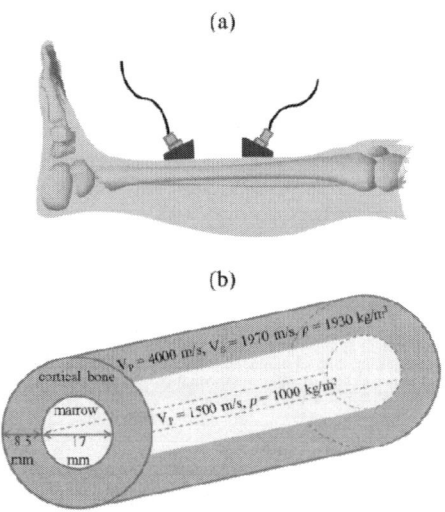

Fig. 1 (a) The *in-vivo* experimental setup; (b) Geometry of the water-filled cylindrical model.

III. RESULTS AND DISCUSSIONS

We use 64 synthetic records to validate the performance of the RTs to image the dispersion curves. The time records are 2-mm spaced and have 101 data points each with a 2-μs sampling interval. We plotted every 4 records for a total of 16 records in Figure 2a. The synthetic wave fields show the fast-travelling low-frequency components and slow-travelling high-frequency wave-trains corrupted by 14-dB white Gaussian noise. The corresponding dispersion panels (Figures 2b-e) show the dependence of phase velocity (PV) resolution on the regularization strategy used. Among the four, the adjoint Radon panel (ARP) (Figure 2b) has the worst PV resolution as the dispersion energy spreads far away from the true dispersion curve (indicated by the white curve). The DLS Radon panel (Figure 2c) has slightly better resolution than the ARP. The Cauchy-regularized RT

(Figure 2d) improved the focusing better than the DLS. The l_1-regularized RT focuses the dispersion energy far even better (Figure 2e), offering sharper image of the dispersion curve and superior resolution than the other three methods.

Thought not shown here, the PV resolution of the transform methods can be quantified by the full-width at half-maximum (FWHM) of the PV spectrum. The FWHM values for the adjoint, DLS, Cauchy, and l_1 RT are 1355 m/s, 1095 m/s, 674 m/s, and 340 m/s respectively. This indicates that the l_1-regularized RT offers 399%, 322%, and 198% better resolution than the adjoint, DLS, and Cauchy-regularized solutions respectively. The μ-values were 400, 500, and 10^3 for the DLS, Cauchy-regularized, and l_1-regularized RT respectively. For the Cauchy regularization term, $\sigma^2 = 0.8$.

The acquired *in-vivo* data shown in Figure 3a is consisted of 40 ultrasound records. The data has been self-normalized to make the small far-offset guided-wave signals visible. The *t-x* panel shows mainly two types of wave arrivals with distinct moveouts. The first type is usually the high-frequency and high-velocity bulk waves [7] and the second type is the low-frequency and low-velocity arrivals, which are usually surface or Lamb-type guided waves [6]. Between 46-70 mm, the data shows the strong presence of high frequency bulk waves. However the high frequency components decayed rapidly due to spherical divergence and preferential filtering due to absorption. After 70 mm, the low-frequency signals took over with strong build-up of guided-wave energies.

The four RT methods were successfully applied to the *in-vivo* data set to extract modal dispersion (Figures 3b-e). The results show clearly that the l_1-regularized RT offers the best resolution among the four solutions and is able to focus the energy with the least smearing. The μ-values were 1500, 1750, and 4×10^7 for DLS, Cauchy-regularized, and l_1-regularized RT respectively. With the aid of the theoretical dispersion curves, six guided modes: F(1,1), F(1,4), F(1,8), L(0,1), L(0,4), and L(0,5) were identified (Figures 3b-e).

IV. CONCLUSIONS

The results have demonstrated that the proposed linear Radon transform can reliably map the dispersion information of ultrasonic guided waves. Among the four Radon solutions, the l_1-regularized Radon transform provides the best resolution with enhanced guided-mode discrimination power. This study also confirms the presence of guided waves in axial transmission of ultrasound through human long bones. A phased array approach [4] is in progress to further study ultrasonic guided waves in long bones.

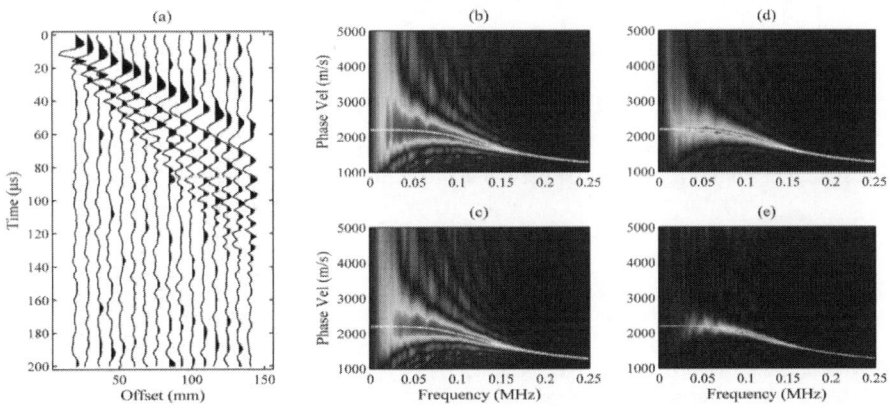

Fig. 2 The simulated dispersive signals with random noise and the corresponding (*f-c*) dispersion panels: (a) The noisy signals with 14 dB SNR, (b) The adjoint Radon panel, (c) The DLS Radon panel, (d) The Cauchy-regularized Radon panel, and (e) The l_1-regularized Radon panel. The true dispersion curve is described by the white dashed curve.

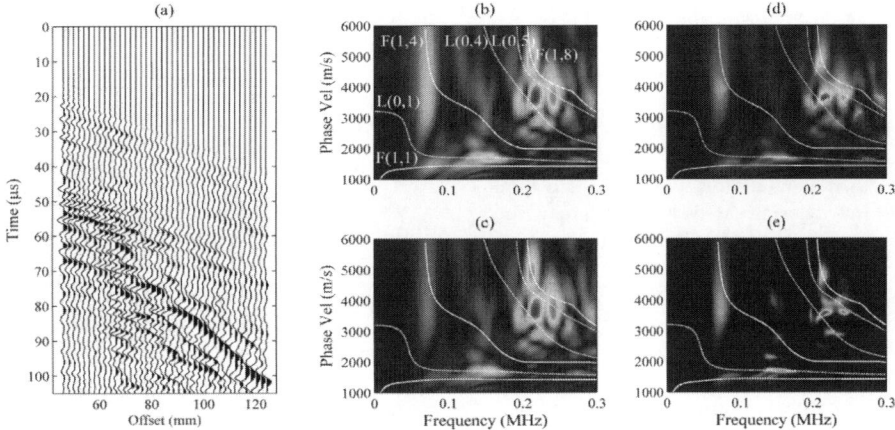

Fig. 3 Dispersion imaging of the *in-vivo* experimental data: (a) The *t-x* signals, (b) The adjoint Radon panel, (c) The DLS Radon panel, (d) The Cauchy-regularized Radon panel, and (e) The l_1-regularized Radon panel. The theoretical dispersion curves are plotted in white.

ACKNOWLEDGMENT

Tho N.H.T. Tran gratefully acknowledges the support of a PhD fellowship from the Department of Radiology & Diagnostic Imaging, University of Alberta.

REFERENCES

1. Moilanen P (2008) Ultrasonic guided waves in bone. IEEE Trans Ultrason Ferroelectr Freq Control 55:1277–1286
2. Ta D, Huang K, Wang W, Wang Y (2006) Identification and analysis of multimode guided waves in tibia cortical bone. Ultrasonics 44: e279–284
3. Le L H, Tran T N H T, Sacchi M D (2013) Radon or τ-p transform: a new tool to image dispersive guided-wave energies in long bones. Proceedings of the 5th European symposium on ultrasonic characterization of bone:52–53
4. Nguyen K C, Le L H, Tran T N T T, Sacchi M D, Lou E H (2014) Excitation of ultrasonic Lamb waves using a phased array system with two array probes: phantom and in vitro bone studies. Ultrasonics: in press (http://dx.doi.org/10.1016/j.ultras.2013.08.004)
5. Sacchi M D (1997) Reweighting strategies in seismic deconvolution. Geophys J Int 129:651–656
6. Tran T N T T, Stieglitz L, Gu Y J, Le L H (2013) Analysis of ultrasonic waves propagating in a bone plate over a water half-space with and without overlying soft tissue. Ultrasound Med Biol 39:2422–2430
7. Le L H, Gu Y J, Li Y P, Zhang C (2010) Probing long bones with ultrasonic body waves. Appl Phys Lett 96:114102

Author: Lawrence H. Le
Institute: Radiology & Diagnostic Imaging, University of Alberta
Street: 8308-114 Street
City: Edmonton, AB
Country: Canada
Email: lawrence.le@ualberta.ca

Adaptive Noise Cancellation in the Intercept Time-Slowness Domain for Eliminating Ultrasonic Crosstalk in a Transducer Array

K.C.T. Nguyen[1,2], Lawrence H. Le[2,3], Mauricio D. Sacchi[3], L.Q. Huynh[1], and E. Lou[4]

[1] Department of Biomedical Engineering, Ho Chi Minh City University of Technology, Ho Chi Minh City, Vietnam
[2] Department of Radiology and Diagnostic Imaging, University of Alberta, Edmonton, Canada
[3] Department of Physics, University of Alberta, Edmonton Canada
[4] Department of Surgery, University of Alberta, Edmonton, Canada

Abstract— **Ultrasonic waves, acquired by an array probe, are likely interfered by the crosstalk signals due to inter-communication among the elements. The crosstalk signals are not desirable and mixed with the main signals, degrading the data quality. In this work, we choose a filtering strategy to separate crosstalk and the desired signals. We design a traditional adaptive crosstalk canceller (ACC) in the Radon (intercept time, τ and slowness, p) domain. A Normalized-Least-Mean-Square strategy is used to determine the filter coefficients. Formulating the filter in the τ-p space has the merits to efficiently remove random noise and significantly enhance signal-to-noise ratio. The filtered Radon panels are then transformed back to the t-x space using the inverse Radon transform. The robustness and accuracy of the filter are studied using simulated noiseless and noisy data sets. The filtered signals tracks well with the simulated noiseless signals with an average mean-square-error (MSE) of 11%. Finally the ACC is applied to an experimental data set of a 2.4-cm thick Plexiglas plate. The result has demonstrated the ACC effectively recovered the main arrivals with reasonable coherency and continuity.**

Keywords— **ultrasonic crosstalk, phased array, adaptive noise cancellation, Radon transform, τ-p transform.**

I. INTRODUCTION

The axial transmission method is specifically designed to study the properties of long bones [1-3]. During data acquisition, the emitter and the receiver are deployed on the same side of a bone sample. The emitter is stationary while the receiver is positioned collinearly at locations on one side of the emitter. The receiver's locations are usually evenly spaced and the source-receiver distance is denoted as offset. Although a pair of transducers is still the most common means to acquire bone data, ultrasound array or phased array (UPA) systems have recently been used to study long bones [4,5]. The array system or multi-transmitter-multi-receiver system has many advantages over a single-transmitter-single-receiver system. The former has better resolution because of the smaller element footprint, fast acquisition speed, accurate coordination of the receivers, and less motion-related problems.

However, crosstalk between transducer elements inside the probe is a hindrance to the performance of the UPA system. When an element is electrically excited, its vibration creates a pressure on neighboring elements and makes them ring. This phenomenon creates delayed output signals, the so-called acoustic crosstalk or cross coupling between adjacent elements, propagate in the transfer medium. When the transducer is held in contact with the material, the crosstalk signals appear as coherent noise and are detrimental as they interfere with the signals traveling within the sample [6], which might possibly lead to inaccurate interpretation of the data. Therefore, removing or filtering the crosstalk by means of signal processing is a necessary step prior to any further data analysis.

The adaptive noise canceller (ANC) is a digital filter that predicts the noise and subtracts it from the data to obtain the desired noise-free signals. This technique has been successfully applied in many engineering and biomedical fields. For example, Wirnitzer et al. aimed to cancel interference among ultrasonic sensor arrays by using an ANC for tracking and localization [7]. Nguyen *et al.* used a least-square ANC to remove noise and artefacts from EEG data [8]. Brady et al. enhanced identification of material defects by using an ANC to eliminate backscattered noise arising from internal micro-structures [9]. The relevant noise in this study is the crosstalk and thus we call the canceller an adaptive crosstalk canceller (ACC).

In this work, we examine the use of ACC as an alternative crosstalk removal method. In section II, we present the experimental setup, Radon transform, and the Radon-based adaptive crosstalk canceller algorithm. In section III, the performance of the ACC filter is verified using simulated noiseless and noisy data. Finally we apply the ACC filer to an experimental data set.

II. MATERIALS AND METHODS

A. Experimental Setup and Acquisition

We used an Olympus TomoScan FOCUS LT™ UPA system (Olympus NDT Inc., Canada) with an array probe in this study (Fig. 1). The scanner was previously used to study scoliosis [10] and the specifications can be found in [5,10]. The probe used was a 2.25-MHz 64-element array transducer (2.25L64). The probe has an active area of 48 mm (length) by 12 mm (elevation) with a pitch of 0.75 mm.

© Springer International Publishing Switzerland 2015
V. Van Toi and T.H. Lien Phuong (eds.), *5th International Conference on Biomedical Engineering in Vietnam,*
IFMBE Proceedings 46, DOI: 10.1007/978-3-319-11776-8_8

A rexolite-matching layer covers the elements to minimize the acoustic impedance difference between the piezoelectric transducer elements and the object of investigation.

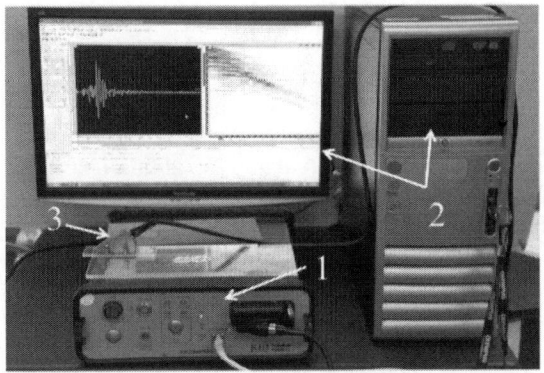

Fig. 1 The ultrasound phased array system: The TomoScan FOCUS LT™ UPA acquisition system (1), the Windows XP-based computer with the TomoView™ software to control the acquisition process (2), and the 64-element array probe (3).

We acquired axial transmission data for a 24-mm thick Plexiglas plate with a compressional wave (*P*-wave) speed of 2.7 km/s and a shear wave (*S*-wave) speed of 1.37 km/s. The plate rested on two rubber corks in air, the probe was in contact with the plate, and the two steel bars applied appropriate pressure on the probe against the plate to ensure good contact. Ultrasound gel was used between contacts to ensure good coupling in all experiments. We chose to use a single element as the emitter and receiver. The first element was electrically excited with a 40-V pulse while the other 63 elements operated as receivers. The crosstalk signals inside the transducer were recorded and measured by operating the probe in air without contact with any object. Each A-scan, which was summed 16 times and averaged, was 2000-points long with a sampling interval of 0.02 μs. The data had 63 A-scans spaced by 0.75 mm. The offset spanned from 0.75 mm to 47.25 mm with an aperture of 46.5 mm. Before analysis, data were self-normalized and filtered by a trapezoidal (0.1/0.2/2.9/3.0 MHz) bandpass filter.

B. Radon-Based Adaptive Crosstalk Canceller

1. High Resolution Radon Transform

The discrete Radon transform (RT), also known as the τ-p transform, is defined by summing the amplitudes along a line $t = \tau + px$ with moveout px where p is the ray parameter (or slowness) and τ is the arrival time at zero-offset (zero-offset time intercept) [11]. Using matrix notation in the Fourier domain, the relationship between the data, $\mathbf{d}(t, x)$ and the Radon signals, $\mathbf{m}(\tau, p)$ can be written for each frequency, f (Eq. (7) in [5]):

$$\mathbf{D} = \mathbf{LM} \tag{1}$$

where $\mathbf{D}(f, x)$ and $\mathbf{M}(f, p)$ are the corresponding temporal Fourier transform of $\mathbf{d}$ and $\mathbf{m}$. $\mathbf{L}$ is the Radon operator given by eqn. (8) in [5].

We choose to use a high-resolution least-square Radon solution with Cauchy-norm regularization to focus Radon energy [12]:

$$\mathbf{M} = \left(\mathbf{L}^H \mathbf{L} + \mu \mathbf{Q}(\mathbf{M})\right)^{-1} \mathbf{L}^H \mathbf{D} \tag{2}$$

where $\mathbf{L}^H$ is the adjoint operator, μ is the trade-off parameter, $\mathbf{Q}(\mathbf{M})$ is a diagonal weighting matrix with elements

$$Q_{kk} = \frac{1}{\left(1 + M_k^2 / \sigma^2\right)}, \tag{3}$$

and σ^2 is the scale factor of the Cauchy distribution. Equation (2) is a set of nonlinear equations and can be solved iteratively by IRLS (iterative re-weighted least-squares) scheme for each frequency. Based on our experience, 4 iterations are adequate to obtain a reasonably good result.

Once $\mathbf{M}$ is determined for all frequencies, we perform the following operations to map the *f-p* panels back to *t-x* domain: map the *f-p* panel to the *f-x* panel via Equation (1) using the Radon operator $\mathbf{L}$ and inverse Fourier-transform the *f-x* panel to the *t-x* domain.

2. Adaptive Crosstalk Canceller (ACC)

Let m_d and m_{rc} be the τ-p representation of the recorded data and reference crosstalk. The data, m_d contains the desired signal, m_s and the crosstalk, m_c. An adaptive crosstalk canceller (Fig. 2) predicts the crosstalk from the reference crosstalk and subtracts them from the data to obtain the crosstalk-free signals, which are the error signals. The error signals are fed back into the filter to adjust adaptively the filter coefficients to provide a better estimate of the signals in a least-square-sense. We adopted the normalized least-mean-square algorithm to adjust the filter coefficients [13].

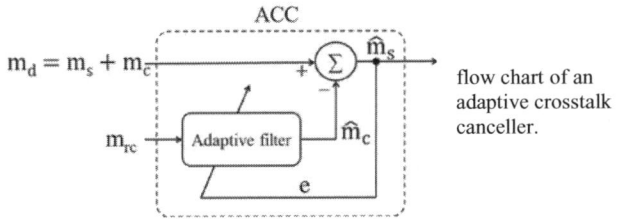

flow chart of an adaptive crosstalk canceller.

III. RESULTS AND DISCUSSION

C. Simulation Data

In order to evaluate the accuracy of the ACC to remove crosstalk, we simulated a reference crosstalk of two linear and hyperbolic events with different moveouts (Fig. 3a) by convolving the travel time matrix with a Berlage wavelet [14]. The data (Fig. 3b) contains two signal arrivals (A and B) and two crosstalk events (C and D). Being delayed by 0.6 μs and 1.6 μs, the signals, A and B, traveled at slowness of 0.29 μs/mm (3.5 km/s) and 0.77 μs/mm (1.3 km/s) respectively. The linear event C had a delay of 0.9 μs and slowness of 0.4 μs/mm (2.5 km/s). The 10-μs delayed reflection D was made up of three multiple reflections with very small delays, each of which traveled at slowness of 0.5 μs/mm (2 km/s) in a 3-mm thick layer.

small. We compared the predicted arrivals with their simulated counterparts at 17.25mm offset. The filtered signals track well with the simulated signals with a relative MSE of 13% and 9% for the mismatch of the first and second peaks respectively. For all examples presented in this study, the filter length was 4. We also used $\sigma = 0.8$. The choice of the μ was based on the L- curve to balance between the predicted error and the regulization term, which is usually around the local minimum point of the curve.

A noisy data set was simulated by adding Gaussian noise to the previous noiseless data (Fig. 4a and b). The SNR (signal-to-noise ratio) was 10 dB. Both the t-x and τ-p panels show the background noise. However the signals and their corresponding Radon energy clusters are clearly seen. The filtered signals (Fig. 4d) show that the crosstalk has been removed. As well, the random noise has also been filtered in the process. This is possible because the τ-p transform sums the amplitudes of the waveforms, thus reinforcing the coherent signals and eliminating randomness.

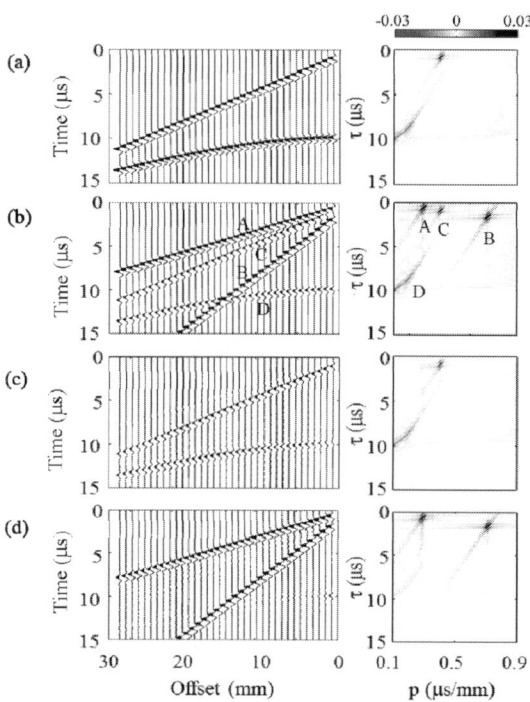

Fig. 3 A simulated noiseless example and the corresponding τ-p panels. Thirty-eight time series up to 15 μs were simulated with offset spanning from 0.75 mm to 28.5 mm. (a) The simulated reference crosstalk; (b) The simulated data consisting of signals (A and B) and crosstalk (C and D) where the amplitudes of the latter are only half of those of the reference crosstalk; (c) The predicted crosstalk; (d) The ACC-filtered signals.

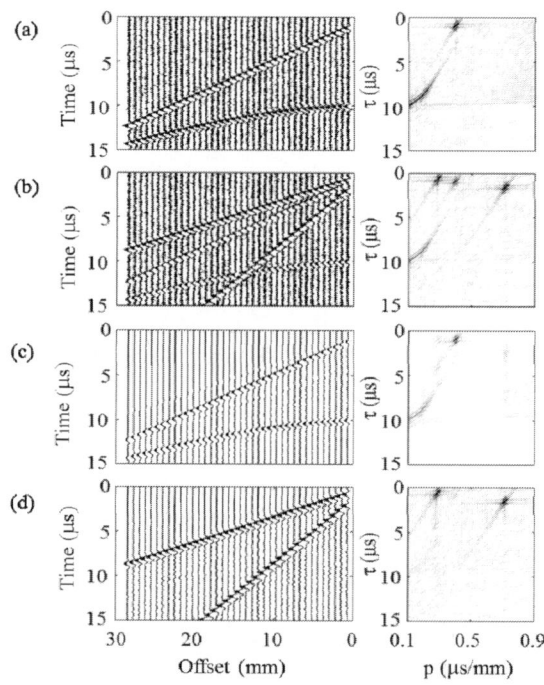

Fig. 4 A simulated noisy example and the corresponding τ-p panels. The SNR is 10 dB. (a) The simulated reference crosstalk; (b) The simulated data consisting of signals (A and B) and crosstalk (C and D); (c) The predicted crosstalk; (d) The ACC-filtered signals.

The ACC-predicted crosstalk is shown in Fig. 3c while the ACC-filtered signals are shown in Fig. 3d. The filter was not able to remove all crosstalk events and there is some residual crosstalk around 10 μs for the first few close-offset records (Fig. 3d). However, their amplitudes are very

D. Experimental Data

Fig. 5a shows the reference crosstalk in both t-x and τ-p domains. The records were self-normalized, i.e., each record was normalized by its own absolute maximum. The reference crosstalk was recorded with the array being held in air and in no contact with any material. Three main types of arrivals (A, B, and C) can be clearly identified with different phase velocities and traveling times. The recorded data (Fig. 5b) which is a mixture of crosstalk (A, B, and C) and real signals (E and F) propagating in the 24-mm Plexiglas plate. Two different Radon signal clusters appear within the 0-2 μs intercept-time window but with different slownesses, approximately 0.37 and 0.75 μs/mm. They likely correspond to the *P*-wave, *S*-wave, and Rayleigh wave traveling within the Plexiglas. After applying ACC to the data and inversing back to the *t-x* domain, the crosstalk energies were successfully removed and both modes, E and F, were recovered with high degree of coherency and continuity (Fig. 5c).

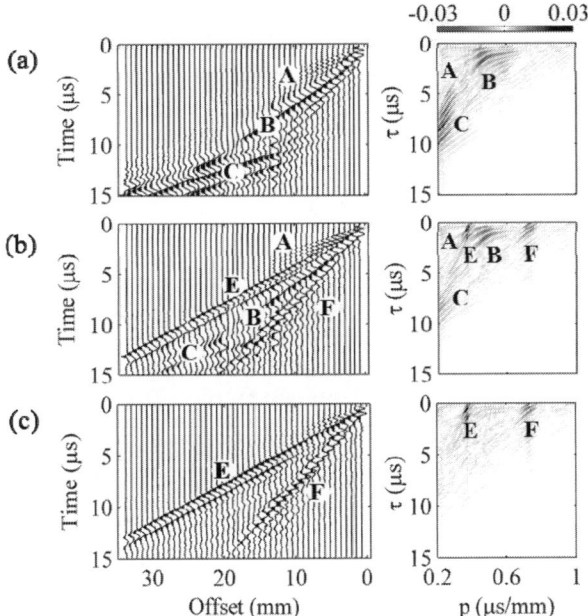

Fig. 5 The Plexiglas data set and the corresponding τ-p panels. (a) The reference crosstalk (A, B, and C); (b) the data consisting of signals (E and F) and crosstalk; (c) the ACC-filtered signals.

IV. CONCLSIONS

In this work, we applied the traditional adaptive noise cancellation scheme in the τ-p domain to minimize crosstalk artifacts in ultrasound data. The τ-p transform is able to separate wavefields based on their intercept-time and slowness. In addition, it also removes random noise, thus rendering an effective filtering technique to eliminate undesired wave modes and enhance signal-to-noise ratio of the data if necessary prior to ACC filtering. Our results using simulation and Plexiglas data have demonstrated that the Radon-based ACC is reliable in removing crosstalk and recovering the crosstalk-free signals. Our next phase of the study is to apply the filter to bone data.

ACKNOWLEDGMENT

Kim-Cuong T Nguyen would like to acknowledge Vietnam International Education Development and University of Alberta for the financial supports.

REFERENCES

1. Ta D, Huang K, Wang W, Wang Y, and Le LH (2006) Identification and analysis of multimode guided waves in tibia cortical bone. Ultrasonics, 44:279-284.
2. Le LH, Gu YJ, Li Y, and Zhang C (2010) Probing long bones with ultrasonic body waves. Appl Phys Lett, 96:114102.
3. Tran TNHT, Stieglitz L, Gu YJ, and Le LH (2013) Analysis of ultrasonic waves propagating in a bone plate over a water half-space with and without overlying soft tissue. Ultrasound Med Biol 39(12):2422-2430.
4. Minonzio JG, Talmant M, and Laugier P (2010) Guided wave phase velocity measurement using multi-emitter and multi-receiver arrays in the axial transmission configuration. J Acoust Soc Am, 127(5):2913-2919.
5. Nguyen KCT, Le LH, Tran TNHT, and Lou E (2013) Excitation of ultrasonic lamb waves using a phased array system with two array probes: Phantom and in-vitro bone studies. Ultrasonics (In press).
6. Baer RL and Kino GS (1984) Theory for cross coupling in ultrasonic transducer arrays. Appl Phys Lett, 44(10):954-956.
7. Wirnitzer B, Grimm W, Schmidt H, and Klinnert R (1998) Interference cancelation in ultrasonic sensor arrays by stochastic coding and adaptive filtering. In Proceedings of the IEEE International Conference on Intelligent Vehicles.
8. Nguyen KCT, Vo HQ, Nguyen HTM, Truong KQD, Nguyen THM, Huynh LQ, Vo TV (2010) Removing noise and artifacts from EEG using adaptive noise cancelator and blind source separation. IFMBE Proceedings Volume 27, In the 3rd International Conference on the Development of Biomed. Eng. in Vietnam, 2010, (pp. 282-286). Springer Berlin Heidelberg.
9. Brady C, Arbona J, In-Soo A, and Yufeng L (2012) Fpga-based adaptive noise cancellation for ultrasonic NDE application. In 2012 IEEE International Conference on Electro/Information Technology (EIT), pages 1-5.
10. Chen W, Le LH, Lou E (2012) Ultrasound imaging of spinal vertebrae to study scoliosis. Open J Acoustics 2:95-103. DOI 10.4236/oja.2012.23011
11. Gu YJ and Sacchi M (2009) Radon transform methods and their applications in mapping mantle reflectivity structure. Surveys in geophysics, 30(4-5):327-354.
12. Sacchi MD (1997) Reweighting strategies in seismic deconvolution. Geophys J Int, 129:651-656.
13. Sayed A (2003) Fundamentals of adaptive filtering. Hoboken, New Jersey: John Wiley& Sons.
14. Le LH (1998) An investigation of pulse-timing techniques for broadband ultrasonic velocity determination in cancellous bone: a simulation study. Phys Med Biol, 43(8):2295-2308.

Author: Kim-Cuong T Nguyen/Lawrence H. Le
Institute: Ho Chi Minh University of Technology/
           University of Alberta
City: Ho Chi Minh/Edmonton
Country: Vietnam/Canada
Email: kimcuongnguyen@hcmut.edu.vn/lawrence.le@ualberta.ca

Simulation of Ultrasound Propagation in Long Bone with Depth-Varying Porosity

Vu-Hieu Nguyen and Salah Naili

Université Paris-Est, Laboratoire Modélisation et Simulation Multi-Echelle, MSME UMR 8208 CNRS, Créteil, France

Abstract— Axial transmission quantitative ultrasound is a potential technique for assessing bone properties. In this work, to describe the link between the bone microstructure (bone matrix properties as well as porosity), Biot's anisotropic poroelastic theory model was employed for modeling the cortical/cancellous bone. A semi-analytical finite element method (SAFE) has been developed for analyzing the wave propagation problem in long bone modeled as aporoelastic medium. The developed method allows us to calculate the reflection/transmission coefficient and dispersion relation of the bone plate with a depth-varying porosity coupled with fluids. Furthermore, simulation of transient waves propagating in the coupled soft tissue/bone/marrow system can efficiently be performed by using this approach.

Keywords— Ultrasound, Bone, Poroelastic, Waveguide, Semi-analytical finite element.

I. INTRODUCTION

Evaluating bone properties is critical for diagnostics of osteoporosis. Recently, quantitative ultrasound (QUS) techniques for assessing the properties of bone have received much attention owing to its potential in estimating bone fragility and/or fracture risk. Ultrasound is non-ionizing, and moreover, ultrasonic devices are relatively inexpensive and can be made portable.

For measuring *in vivo* material properties of cortical long bones, a so-called "axial transmission" (AT) technique has been developed [1]. The axial transmission technique uses a set of ultrasonic transducers (transmitters and receivers) placed on a line in one side of the investigated skeletal site along a direction close to the long bone axis. The transmitter emits an ultrasound pulse wave (around 250 KHz 2 MHz) that propagates along the bone's longitudinal axis. The analysis of the signals received at the receivers serve for estimating the geometric parameters as well as mechanical characteristics of the cortical bone at the measured skeletal site [2,3].

Mechanical modeling of experiments on long bones using the ultrasound axial transmission technique deals with studying a coupled system of a solid waveguide (which represents the cortical bone) with fluids (which represents soft tissues and marrow). From the mechanical point of view, bone is a heterogeneous, anisotropic and porous material. There are two kinds of bone: cortical (compact) bone and cancellous (trabecular or spongy) bone. The porosity varies from 5% to 15% in cortical bone and from 50% to 90% in cancellous bone. For bone material's description, equivalent elastic medium (by neglecting the pore presence) has usually been used. However, it has been shown that the microstructural properties of the bone has strong influences on characteristics of ultrasonic wave propagation. Furthermore, in the macro-scale, the bone porosity have been shown to be very heterogeneous in the radial direction of its cross section.

In the sequel, we present a model to take into account effects of pore's presence in long bones and a numerical procedure for predicting the ultrasound response in both time and frequency domains.

II. METHODS

A. Geometrical Description

Fig. 1 presents a two-dimensional model of a bone plate surrounded by the marrow and soft tissue layers. All layers have constant thickness in the vertical direction (x_2) and infinity extents the in horizontal direction (x_1).

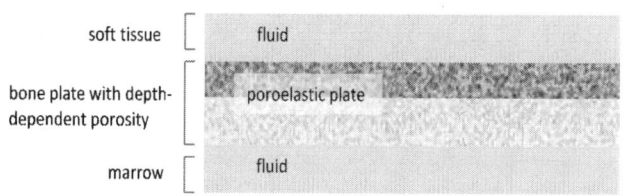

Fig. 1 Multilayer model for a soft tissue/bone/marrow system

In this study, both soft tissue and marrow domains are modeled as homogeneous acoustic fluid media. The bone is assumed to be an anisotropic poroelastic medium which is homogeneous along the longitudinal direction.

B. Poroelastic Model for Bone

The bone material may be seen as a saturated anisotropic poroelastic one. By definition, a porous material is a medium containing a solid skeleton (matrix) with pores. The pore spaces are connected and are typically filled with a fluid. A porous medium is most often characterized by its porosity (will be denoted by ϕ) which is defined by the fraction between pore volume over the total volume of material. The constitutive equations of a anisotropic porous material are given by:

© Springer International Publishing Switzerland 2015
V. Van Toi and T.H. Lien Phuong (eds.), *5th International Conference on Biomedical Engineering in Vietnam*,
IFMBE Proceedings 46, DOI: 10.1007/978-3-319-11776-8_9

$$\sigma = C{:}\epsilon - \alpha p \qquad (1)$$

$$-\frac{1}{M}p = \mathrm{div}w + \alpha{:}\epsilon \qquad (2)$$

where σ and p denote the total stress tensor and the interstitial pore pressure, respectively; ϵ is the strain tensor, $\epsilon = \frac{1}{2}(\mathrm{grad}u + \mathrm{grad}u^{T})$ with u is the solid's skeleton displacement vector; $w = \phi(u^{f} - u)$ is the fluid/solid relative displacement vector weighted by the porosity; α is the Biot effective tensor and M is the Biot modulus. By neglecting the gravity force, the linear poroelasic wave equations in the frequency domain read:

$$\mathrm{div}\sigma = -\omega^{2}\rho u - \omega^{2}\rho_{f}w, \qquad (3)$$

$$-\mathrm{grad}\,p = -\omega^{2}\rho_{f}u - \omega^{2}Aw \qquad (4)$$

where $\rho = \phi\rho_{f} + (1 - \phi)\rho_{s}$ is the mixture density (ρ_{f} and ρ_{s} are the mass densities of the solid and fluid phases, respectively); A is a frequency dependent second-order tensor which is depends on the permeability and tortuosity of the material and defined by

$$A(\omega) = \frac{\rho_{f}}{\phi}a^{\infty} + \frac{i\eta}{\omega}F\kappa^{-1} \qquad (5)$$

where a^{∞} is the static tortuosity tensor, η is the fluid viscosity, κ is the intrinsic permeability and F is a corrector factor introduced for taking into account the dynamic viscous resistance of the interstitial fluid flow at high frequencies.

In order to describe the poroelastic parameters of bone, the drained elasticity tensor C as well as Biot's effective coefficients and M used in (Eqs. 1,2) should be provided. For this study, these parameters are derived from the characteristics of the interstitial fluid and solid skeleton phases by using a continuum micromechanics model proposed by Hellmich *et al* [4]. The permeability may be derived from microstructural properties by using the Kozeny-Carman law.

C. Numerical Method

The problem presented deals with solving a system of linear partial differential equations (PDE) in which the coefficients are homogeneous in the longitudinal direction (given by x_{l}-axis) but heterogeneous in the bone's radial direction (given by x_{2}-axis). A code based on semi-analytical finite element procedure has been developed. The principal steps of this procedure may be encapsulated as follows: (i) the system of equations is firstly transformed into frequency-wavenumber domain by using a Fourier transform with respect to x_{1} combined with a Laplace transform with respect to t; as a consequence, a one-dimensional system of PDEs with respect to x1 can be established; (ii) in

the frequency-wavenumber domain, the wave equations in two fluid domains are analytically solved providing impedance boundary conditions for the solid domain; (iii) the weak and finite element formulations are then established in the bone's domain only; (iv) the space-time solution is finally obtained by performing the inverse Fourier transform (using FFT technique) and the inverse Laplace transform [5]. The formulation has been validated by comparing with conventional FE method [6]. A similar procedure, in which the Fourier transform was applied to t variable instead of using the Laplace transform, has also been developed to compute the reflection and transmission coefficients of the bone plate.

III. NUMERICAL EXAMPLES

This section will present some preliminary results in studying the effects of bone's porosity on the ultrasound response of a long bone plate. Table 1 presents the physical properties of the bone matrix and of the fluid. As mentioned before, the poroelastic in macro level will be estimated by using a homogenizing procedure.

Table 1 Physical parameters

Parameters		Unit
Mass density of bone matrix (ρ_{s})	1722	kg.m^{-3}
Elastic constants of bone matrix $[C_{11}^{m}, C_{22}^{m}, C_{12}^{m}, C_{66}^{m}, C_{16}^{m}, C_{26}^{m}]$	[28.7,23.6,9.9,7.25,0,0]	GPa
Mass density of fluid (ρ_{f})	1000	kg.m^{-3}
Bulk modulus of fluid (K_{f})	2.25	GPa
Viscosity of interstitial fluid (η)	0.001	kg.m^{-3}

Figure 2 presents a snapshot of the fluid pressure field in a bone plate (h = 4mm and $\phi = 0.05$) immersed in water due to a point impulse with a central frequency $f_{0} = 1$ MHz is emitted at $(x_{1}, x_{2}) = (0, 2)$ mm. The influence of the bone's porosity on the FAS (First Arriving Signal) velocity (denoted by V_{FAS}) has been studied.

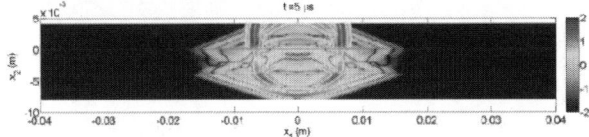

Fig. 2 Snapshot of pressure field in the coupled system

In Fig. 3, we present the variation of V_{FAS} with respect to the bone porosity obtained by studying two homogeneous bone layers with different thickness (h = 4mm and h = 0.6mm). For the both cases, V_{FAS} decreases (not linearly) with higher porosity. Moreover, the one may notice that

V_{FAS} is less influenced by bone's thickness for the higher porosity.

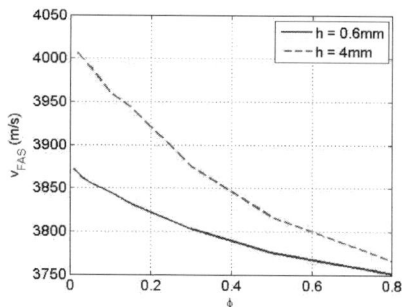

Fig. 3 Effects of the bone porosity on the FAS velocities (f_0 = 1MHz)

The effects of the frequency of emitted signal are studied in Fig. 4. A homogeneous bone plate (h = 4mm and ϕ = 0.05) is first considered. Similar to the results using viscoelastic model, V_{FAS} increases with respect to the emitted frequency and asymptotically tends to a limit value. Next, we studied a bilayer bone plate (h = 4mm) which consists of a 0.6mm thickness of cortical bone layer (ϕ = 0.05) and a 3.4mm thickness of cancellous bone layer (ϕ = 0.8). A similar tendency of frequency effects on V_{FAS} may be observed. It is interesting to notice that by adding a cancellous bone layer to the 6mm thickness cortical bone layer, the V_{FAS} is considerably changed (see Fig. 3 (ϕ = 0.05) and Fig. 4 (at 1MHz)).

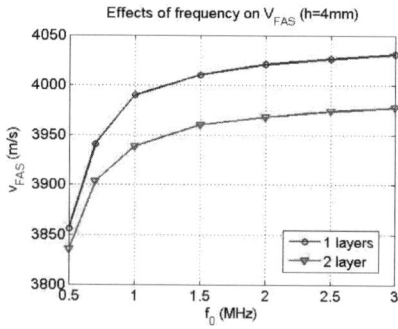

Fig. 4 Effects of the bone porosity on the FAS velocities

Fig. 5 shows effects of a linear gradient of porosity on the reflection coefficient |R| at different frequencies. While the porosity at the upper interface is fixed at ϕ_1 = 0.05, the porosity at the lower interface is varied from ϕ_2 = 0.05 to ϕ_2= 0.5. One may notice that the gradient of porosity has a

strong influence on the reflection coefficients, namely with high frequencies.

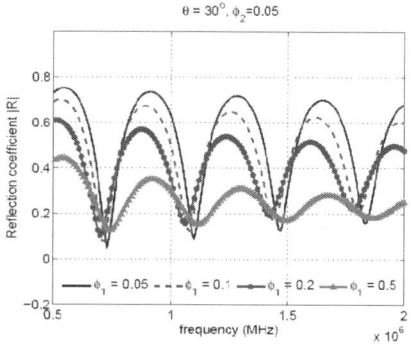

Fig. 5 Reflection coefficients of a bone with a linear gradient porosity

IV. CONCLUSIONS

An efficient numerical tool has been developed to study the ultrasonic wave propagation in a porous bone coupled with fluids. In frequency domain, the dispersion relation (phase velocities and attenuations) and the reflection/transmission coefficients. In the time domain, the wave propagation field can rapidly be calculated.

By using poroelastic model, the effects of pore presence in bone, which cannot be taken into account by using (visco)elastic model, may be studied. It has been shown that the depth-varying porosity in bone may have a ineligible effect on ultrasound response of long bone. More detail studies will be presented in a forthcoming paper.

REFERENCES

1. Lowet G, Van der Perre G. (1996) Ultrasound velocity measurements in long bones: measurement method and simulation of ultrasound wave propagation. J Biomech 29:1255–1262.
2. Nicholson P, Moilanen P, Karkkainen T, Timonen J, Cheng S. (2002) Guided ultrasonic waves in long bones: modelling, experiment and in vivo application. Physiological Measurement 23(4):755–768.
3. Moilanen P. (2008) Ultrasonic guided waves in bone. IEEE Trans. Ultrason Ferro Freq Cont, 55(6):1277–1286.
4. Hellmich C, Ulm FJ. (2005) Microporodynamics of bones: prediction of the "Frenkel–Biot" slow compressional wave. Journal of Engineering Mechanics 131(9):918–927.
5. Nguyen V-H, Naili S (2013) Ultrasonic wave propagation in viscoelastic cortical bone plate coupled with fluids: A spectral finite element study. Comput Methods Biomech Biomed Engin. 16(9):963-74
6. Nguyen VH, Naili S. (2012) Simulation of ultrasonic wave propagation in anisotropic poroelastic bone plate using hybrid spectral/FE method. Int. J. Num. Meth. Biomed. Eng. 28(8), 861–876

Frequency Independence of Ultrasound Transit Time Spectroscopy

M.-L. Wille and C.M. Langton

Biomedical Engineering & Medical Physics Discipline, Science & Engineering Faculty and Institute of Health and Biomedical Innovation, Queensland University of Technology, Brisbane, Australia

Abstract— Recent studies have shown that ultrasound transit time spectroscopy (UTTS) is an alternative method to describe ultrasound wave propagation through complex samples as an array of parallel sonic rays. This technique has the potential to characterize bone properties including volume fraction and may be implemented in clinical systems to predict osteoporotic fracture risk. In contrast to broadband ultrasound attenuation, which is highly frequency dependent, we hypothesise that UTTS is frequency independent. This study measured 1 MHz and 5 MHz broadband ultrasound signals through a set of acrylic step-wedge samples. Digital deconvolution of the signals through water and each sample was applied to derive a transit time spectrum. The resulting spectra at both 1 MHz and 5 MHz were compared to the predicted transit time values. Linear regression analysis yields agreement (R2) of 99.23% and 99.74% at 1 Mhz and 5 MHz respectively indicating frequency independence of transit time spectra.

Keywords— Deconvolution, Ultrasound, Transit Time Spectrum, Solid Volume Fraction.

I. INTRODUCTION

Osteoporosis is the systematic loss of bone leading to increased porosity, fragility and fracture risk. The disease is a significant public health burden affecting more than 200 million people worldwide. Although quantitative ultrasound (QUS) assessment of osteoporosis, in particular the measurement of broadband ultrasound attenuation (BUA), offers non-ionizing, portable, and reliable prediction of fracture risk, its widespread utilisation suffers from both a limited understanding of ultrasound wave propagation through cancellous bone and an inability to elucidate the density and structure of a cancellous bone sample.

Previous studies have shown that an ultrasound wave propagating through a complex medium such as cancellous bone may be approximated by an array of parallel sonic rays, the transit time of each determined by the proportions of bone and marrow [1], [2]. We hypothesise that the resulting transit time spectrum (TTS) has the potential to reliably estimate the solid volume fraction of a bone sample, and hence, offers for the first time using ultrasound, the application of World Health Organisation definitions of osteopenia and osteoporosis. The aim of this study was to demonstrate that the TTS is independent of ultrasound frequency.

II. MATERIALS AND METHODS

A. Experimental Ultrasound Measurements

The ultrasound experiments were performed in transmission mode utilising pairs of 1 MHz and 5 MHz broadband ultrasound transducers, all 0.75" in diameter, single element, and unfocused. The transducers were immersed in water, coaxially aligned with a fixed separation of 20.4 mm. The transmitter and receiver are connected to a high frequency pulser-receiver (Panametrics, PR5800, Austin, TX, USA). The measured ultrasound signals were acquired with 100 MHz sampling frequency by a 14-bit digitiser card and saved for further analysis. A sketch of the experimental set-up is shown in figure 1.

A range of ten different acrylic step-wedge samples, as shown in figure 2, was used. The samples have cylindrical shape with 20.4 mm height and of equal diameter to the transducer surface, varying in thickness normal to the direction of ultrasound propagation. The different number of steps results in a range of transit time inhomogeneities. Acrylic and water serve as surrogates for bone and marrow respectively with a speed of sound of v_a=2635.3 m/s for acrylic and v_w=1486.1 m/s for water respectively, measured experimentally at 21.3 °C water temperature.

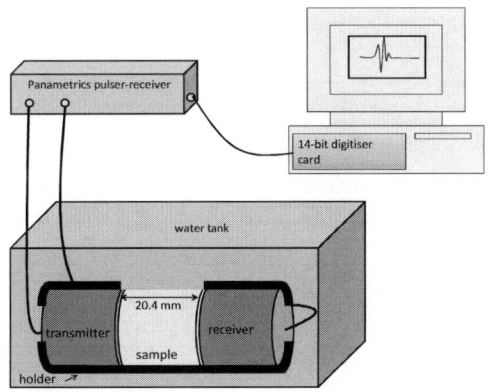

Fig. 1 Experimental set-up.

V. Van Toi and T.H. Lien Phuong (eds.), *5th International Conference on Biomedical Engineering in Vietnam*,
IFMBE Proceedings 46, DOI: 10.1007/978-3-319-11776-8_10

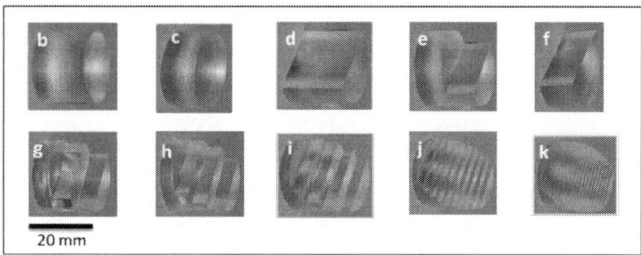

Fig. 2 Photographs of the different acrylic models. Model 'a' corresponding to 'marrow' is substituted by water and serves as a reference and is not shown in this figure.

B. Derivation of Transit Time Spectrum via Deconvolution

The transit time spectra (TTS) for each sample were derived via digital deconvolution of the measured ultrasound signals. Noting that the output signal may be described by the convolution of the sample-specific TTS and the input signal, an inverse solution for the TTS may be derived using the numerical active-set deconvolution method [2]. The 1 and 5 MHz ultrasound signals through water served as the input signal, while the measured ultrasound signal through the samples were used as the output signals for the computational deconvolution of two signals. The resulting TTS were then compared to predicted TTS values based on the sonic ray concept [3].

C. Sonic Ray Concept

Previous studies [1], [3] have shown that ultrasonic wave propagation may be described by an array of parallel sonic rays. Each ray has a unique transit time defined by the amount of material the ray is travelling through. The transit time spectrum ranges from t_{min} (transit time only through solid) to t_{max} (transit time only through liquid). The output signal measured by a phase sensitive transducer is then the superposition of all sonic rays.

III. RESULTS AND DISCUSSION

Figure 3 and 4 displays the measured 1 MHz and 5 MHz ultrasound signals solely through water (left hand side) and an example of measured ultrasound signals through a step-wedge sample, in this case model 'h' with four steps (right hand side). The experimentally derived transit time spectra (TTS) are shown in figure 5-8 along side with corresponding predicted transit time values.

It is observed that albeit the input and output signals are different for 1 MHz and 5 MHz, showing phase interference in the 1 MHz but no signal overlap in the 5 MHz output

signal, the resulting transit time spectra exhibit similar properties; for example four distinct peaks corresponding to the individual steps of model 'h'.

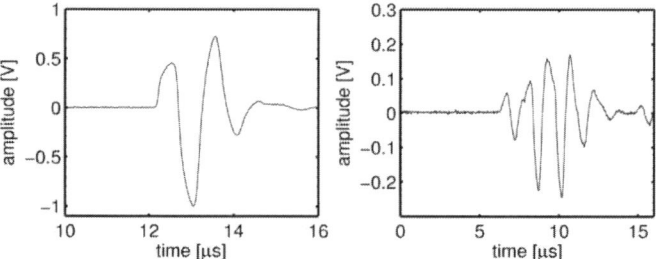

Fig. 3 *left:* 1 MHz ultrasound signal through water (input signal), *right:* 1 MHz ultrasound signal through model 'h' with 4 steps. Note the enlarged time scale for the signal through water.

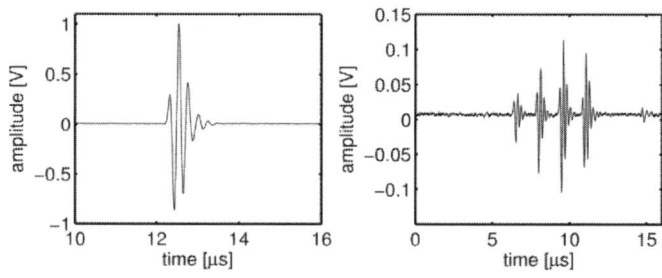

Fig. 4 *left:* 5 MHz ultrasound signal through water (input signal), *right:* 5 MHz signal through model 'h' with 4 steps. Note the enlarged time scale for the signal through water.

The black bars correspond to the experimentally derived TTS and the white bars to the predicted TTS respectively. Note that the time axis is negative with a maximum value of 0 μs = t_{max}, indicating the transit time solely through water as demonstrated for model 'a'. Consequently, all sonic rays encountering a solid portion will have shorter, i.e. negative transit times. The y-axis indicates the proportion P(t) of sonic rays with a specific transit time. The proportions within the predicted TTS were calculated by the relative sonificated area and the relative attenuation of each individual sonic ray for each step-wedge. Noting the attenuation of acrylic to be 25.3 Np/m at 1 MHz and 78.3 Np/m at 5 MHz, the proportion within the 5 MHz TTS is lower than the proportion within the 1 MHz TTS. The low amplitude peaks in the experimentally derived transit time spectra, particularly in the 1 MHz TTS (figure 5 and figure 6) are due to noise and may be avoided by applying a threshold. An interesting observation is that these deconvolution artifacts are highly suppressed in 5 MHz TTS, which is likely due to the fact that the 5 MHz output signals have less phase interference.

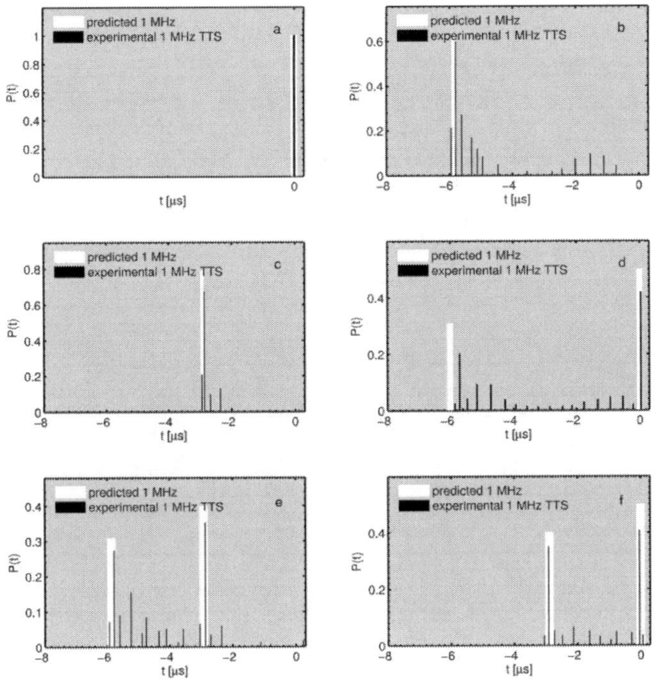

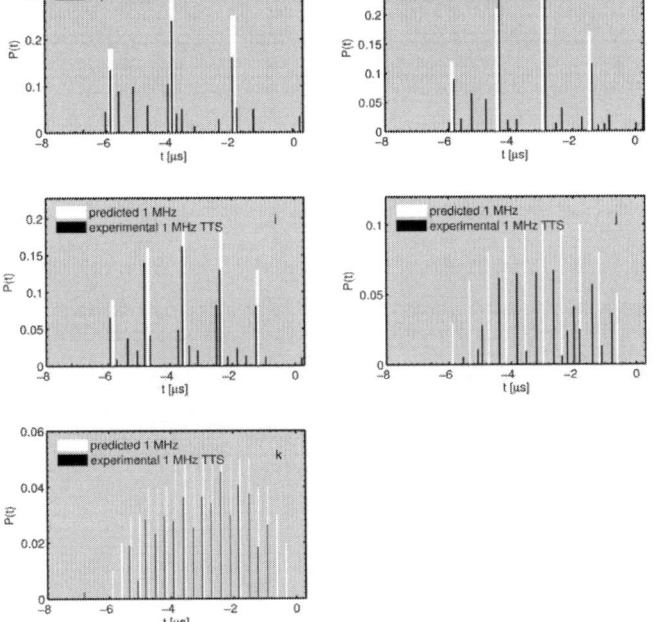

Fig. 5 Comparison of the experimental via deconvolution derived TTS with the predicted TTS for models 'a'-'f' for 1 MHz.

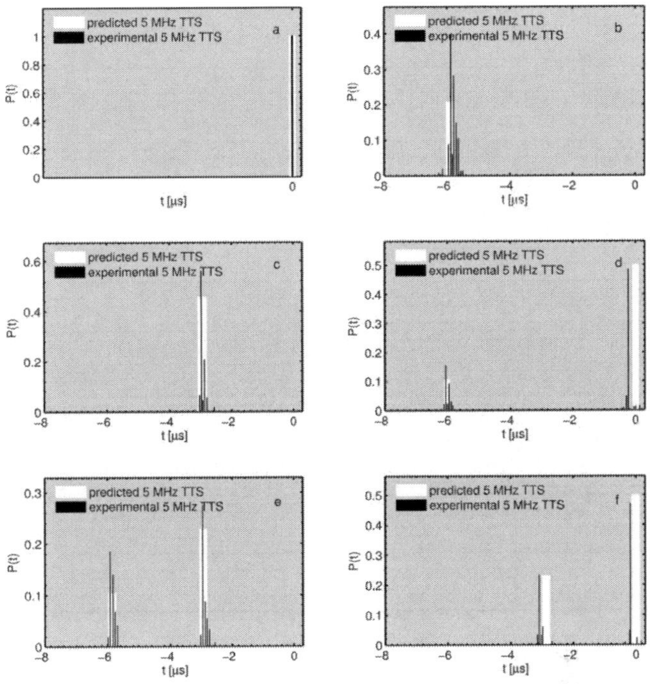

Fig. 7 Comparison of the experimental via deconvolution derived TTS with the predicted TTS for models 'a'-'f' for 5 MHz.

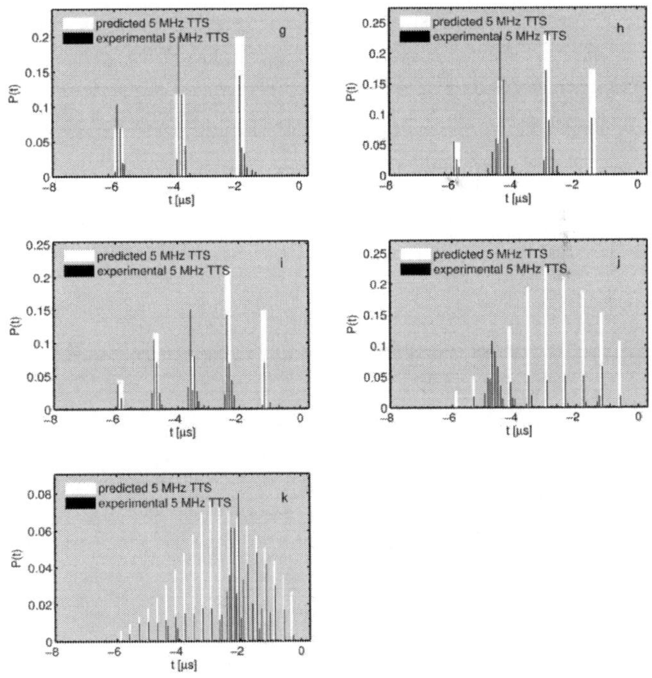

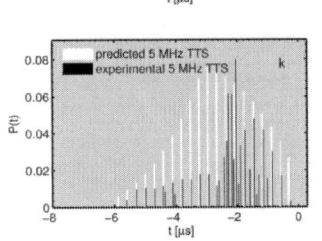

Fig. 6 Comparison of the experimental via deconvolution derived TTS with the predicted TTS for models 'g'-'k' for 1 MHz

Fig. 8 Comparison of the experimental via deconvolution derived TTS with the predicted TTS for models 'g'-'k' for 5 MHz.

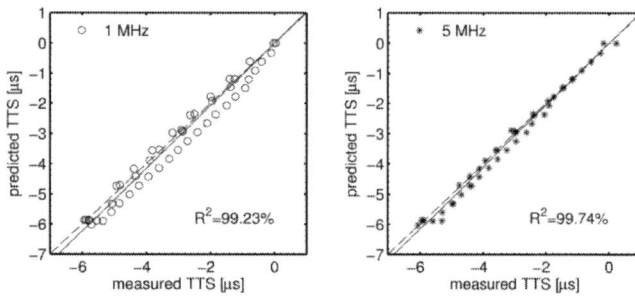

Fig. 9 Comparison of predicted TTS values (based on the parallel sonic ray model) with the experimentally derived TTS values of 1 MHz (left, circles) and 5 MHz (right, stars). The dashed line represents the line of equality, the solid line the linear regression fit.

Figure 9 shows the comparison of experimentally derived and predicted transit time values at both 1 MHz and 5 MHz. A linear regression fit yields agreements (R^2) of 99.23% (1 MHz) and 99.74% (5 MHz). The results of the linear regression analysis are listed in table 1. Frequency independence of the transit time values is given for $p_1=1$ and $p_2=0$. From our analysis with p_1 close to 1 and p_2 close to 0, we can conclude that the transit time values for 1 MHz and 5 MHz have a high agreement of more than 99% with the predicted values and hence are frequency independent.

IV. CONCLUSIONS

We have shown that ultrasound transit time spectroscopy is frequency independent. It is further envisaged that it may quantify bone morphology thereby providing both reliable estimation of WHO criteria and improved prediction of osteoporotic fracture risk.

Table 1 Linear regression analysis results

	1 MHz	5 MHz
Fit type	Linear model $f(x)=p_1 \cdot x + p_2$	Linear model $f(x)=p_1 \cdot x + p_2$
Coefficients with 95% confidence bounds	p_1=1.039 [1.027, 1.051]	p_1 = 1.028 [1.021, 1.035]
	p_2 = -1.301e-08 [-3.394e-08, 7.917e-09]	p_2= -3.209e-09 [-1.532e-08, 8.9e-09]
SSE	4.483e-12	1.495e-12
RMSE	1.434e-07	8.282e-08
R-square	**0.9923**	**0.9974**

SSE: sum of squares due to error, RMSE: root mean squared error.

REFERENCES

1. M. Langton and M.-L. Wille, "Experimental and computer simulation validation of ultrasound phase interference created by lateral inhomogeneity of transit time in replica bone:marrow composite models," *Proc. Inst. Mech. Eng. Part H-Journal Eng. Med.*, vol. 227, no. 8, pp. 888–893, 2013.
2. C. M. Langton, M.-L. Wille, and M. B. Flegg, "A Deconvolution Method for Deriving the Transit Time Spectrum for Ultrasound Propagation through Cancellous Bone Replica Models," Proc. Inst. Mech. Eng. Part H-Journal Eng. Med., in press, 2014, DOI:10.1177/0954411914523582.
3. C. M. Langton, "The 25th anniversary of BUA for the assessment of osteoporosis: time for a new paradigm?," *Proc. Inst. Mech. Eng. Part H-Journal Eng. Med.*, vol. 225, no. H2, pp. 113–125, Feb. 2011.

Author: Marie-Luise Wille
Institute: Institute of Health & Biomedical Innovation, Queensland
          University of Technology
Street: 60 Musk Avenue
City: Kelvin Grove
Country: Australia
Email: m.wille@qut.edu.au

In vitro Ultrasonic Assessment of the Biomechanical Quality of the Interface Surrounding a Dental Implant

R. Vayron and G. Haiat

CNRS, Laboratoire Modélisation et Simulation MultiEchelle, MSME UMR CNRS 8208,
61, Avenue du Général de Gaulle, 94010 Créteil, Cedex, France

Abstract— Dental implants are widely used clinically and have allowed considerable progresses in oral and maxillofacial surgery. However, implant failures, which may have dramatic consequences, still occur and remain difficult to anticipate. Accurate measurements of implants biomechanical stability are of interest since they could be used to improve the surgical strategy by adapting the choice of the healing period to each patient. Empirical methods based on palpation are still used by dental surgeons to determine when the implant should be loaded because it remains difficult to monitor bone healing in vivo. The objective of this study is to investigate the potentiality of a quantitative ultrasound method to assess the biomechanical stability of a dental implant in vitro. A 10 MHz contact transducer is located at the implant extremity. For each ultrasound measurement, a quantitative indicator I is derived based on the time variation of the amplitude of the *rf* signal. Ten implants are initially completely inserted in the proximal part of a bovine humeral bone sample. The 10 MHz ultrasonic response of the implants is then measured and an indicator I is derived based on the amplitude of the *rf* signal obtained. Then, the implants are unscrewed by 2π radians and the measurement is realized again. The procedure is repeated and the indicator I is derived after each rotation of the implants. Analysis of variance (ANOVA) ($p<10^{-5}$) tests revealed a significant effect of the amount of bone in contact with the implant on the distribution of the values of I. The results show the feasibility of QUS techniques to assess implant primary stability in vitro. This study paves the way for the development of a new biomechanical approach in oral implantology.

Keywords— Quantitative ultrasound, dental implant, primary stability, secondary stability, osseointegration.

I. INTRODUCTION

Dental implants [1] are widely used clinically and have allowed considerable progresses in oral and maxillofacial surgery, to restore one or more missing teeth caused by old age or accidents and for aesthetic purposes. However, implant failures, which may have dramatic consequences, still occur and remain difficult to anticipate.

Accurate measurements of implants biomechanical stability are of interest since they could be used to improve the surgical strategy by adapting the choice of the healing period to each patient. Empirical methods based on palpation and patient sensation are still used by dental surgeons to determine when the implant should be loaded with the prosthesis because it remains difficult to monitor bone healing in vivo [2]. Accurate noninvasive quantitative methods capable of assessing implant stability could be used to guide the surgeons and hence to reduce the risk of failure. The implant stability is determined by the quantity and biomechanical quality of bone tissue around the implant. Assessing the implant stability is a difficult multiscale problem due to the complex heterogeneous nature of bone and to remodeling phenomena [3, 4].

Different approaches have been suggested in the past to assess the implant stability in vivo: X-ray radiography, μCT, or magnetic resonance imaging (MRI), but these techniques are not entire satisfaction. As a consequence, biomechanical methods have been developed, their main advantage consisting in the absence of ionizing radiation, inexpensiveness, portability and noninvasiveness, Periotest (Bensheim, Germany) and Osstell (Gothenburg, Sweden) for example. The most commonly used biomechanical technique is the resonance frequency analysis (RFA) [5], which consists in measuring [6] the first bending resonance frequency. However, the RFA cannot be used to identify directly the bone-implant interface characteristics [7]. No correlation between the ISQ and BIC nor between ISQ and cortical thickness has been evidenced [8].

The use of quantitative ultrasound (QUS) as an alternative method to assess the implant biomechanical stability has first been suggested using an aluminum screw inserted in a metallic medium [9]. The principle of the measurement relies on the dependence of ultrasonic propagation within the implant on its boundary conditions, which are related to the bone-implant interface biomechanical properties [10]. An in vitro preliminary study was carried out by our group with a prototype titanium cylinder shaped implants inserted in bone tissue, showing the feasibility of the approach [10]. Numerical simulation [11] were also carried out to understand the experimental results, thus quantifying the sensitivity of the technique. However, the dependence of the ultrasonic response on the stability of an implant inserted in bone tissue has not yet been established.

The aim of this study is to assess potentiality of QUS to assess the biomechanical quality of the bone-implant interface surrounding a dental implant inserted in bone tissue.

© Springer International Publishing Switzerland 2015
V. Van Toi and T.H. Lien Phuong (eds.), *5th International Conference on Biomedical Engineering in Vietnam,*
IFMBE Proceedings 46, DOI: 10.1007/978-3-319-11776-8_11

II. MATERIEL AND METHODS

A. Implant and Bone Samples

Ten bone samples were cut from the proximal part of the humerus of bovine cadavers obtained at the local butcher shop. One cylindrical cavity (3.5 mm diameter and 13 mm deep) was created in each sample prior to implant insertion, similarly as what is done in the clinic. Cavities were thoroughly rinsed with isotonic saline to remove bone fragments prior to the insertion of the titanium implants. The cavity axis coincides with the axis of the resin block, which was fixed to a handy torque gauge to measure the insertion torque between the implant and ultrasonic transducer. The ten dental implants used herein are similar and manufactured by Implant Diffusion International (IDOH 1240, 12 mm of length and 4 mm of diameter).

B. Experimental Measurements

The ultrasonic device is composed of a 5 mm diameter planar ultrasonic contact transducer (Sonaxis, Besançon, France) generating a broadband ultrasonic pulse propagating perpendicularly to its active surface (monoelement transducer). The probe is used in echographic mode and its center frequency is equal to 10 MHz, with a frequency bandwidth approximately equal to 6–14 MHz. The probe is attached rigidly to a healing abutment which can be screwed into the implant so that the measurements are not influenced by positioning problems of the probe relatively to the healing abutment. The ultrasonic response of an implant embedded in air was first measured by inserting the healing abutment (with the transducer on top of it) with a torque of 0.05 N.m, e.g. see Figure 1. The healing abutment was then removed and inserted again following the same procedure in order to assess the reproducibility of the measurements. Each implant was then initially entirely inserted in the dedicated cavity. The ultrasonic transducer was screwed into the implant with a torque of 0.05 N.m and the 10 MHz ultrasonic response of the implant was measured. The healing abutment was then removed and inserted again 10 times to assess the reproducibility of the measurements. The implant was then unscrewed by 2π rad in order to reduce the surface area of the implant in contact with bone tissue. The ultrasonic response was then recorded by inserting the healing abutment with the same torque of 0.05 N.m. The procedure was repeated until the implant was spontaneously detached from the bone cavity. The number of measurements corresponding to the number of rotations before implant detachment depends on bone quality around the implant. The same experimental procedure described above was carried out for ten dental implants.

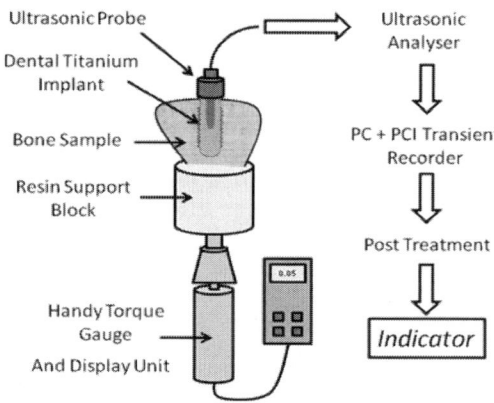

Fig. 1 Schematic description of the ultrasonic experimental set-up with the use of handy torque gauge. The dental implant is completely inserted in the bone sample.

C. Statistical Analysis

Analysis of variance (ANOVA) tests was performed to evaluate the significance of variations of I as a function of the number of rotations.

III. RESULTS

The value of the indicator I obtained when the implant was embedded in air is equal to 434.1± 1.7. The value of I significantly increases as a function of the number of rotation applied to the implant. Each rotation corresponds to 2π rad. The average of the standard deviation of the indicator (when the implant is fully inserted in bone tissue) is equal to 1.42. The average of the correlation coefficient corresponding to the linear regression analysis of the variation of I as a function of the number of rotations is equal to 0.91. The average of slope a of the same linear regression is equal to 4.47. The results show that bone quantity in contact with the implant has a significant influence on its ultrasonic response. For example, the implant of the sample #1 was detached from the surrounding bone after a total number of 5 rotations, so that six measurement points were obtained.

Table 1 summarizes the results obtained for the 10 implants (number of rotations, reproducibility Δy corresponding to the standard deviation of the indicator I obtained for ten measurements, slope a of the linear regression of the indicator I as a function of the number of rotations and the associated correlation coefficient R^2). Table 1 shows that comparable results in terms of order of magnitude are obtained for all implants. In particular, an increase of the indicator I as a function of the bone quantity in contact with the implant is obtained for all implants. ANOVA shows a significant effect of rotation on the value of I ($p<10^{-5}$, $F=29.42$).

Table 1 Results obtained for 10 implants. *a* is the slope of the linear regression of the curve, R^2 is the correlation coefficient and Δx is the incertitude of the measure.

	Number of rotations	Standard Deviation	a	R^2	Δx (in mm)
Implant #1	5	0,89	3,32	0,92	0,43
Implant #2	6	2.08	4,89	0,86	0,89
Implant #3	6	2,19	8,49	0,96	0,41
Implant #4	6	0,84	1,95	0,95	0,69
Implant #5	6	2.02	5,83	0,98	0,84
Implant #6	7	1,40	4,41	0,97	0,51
Implant #7	7	1,39	6,82	0,83	0,33
Implant #8	6	1,65	2,48	0,90	1,06
Implant #9	6	1,09	2,47	0,79	0,71
Implant #10	6	1,93	4,01	0,90	0,77
Mean ± Standard Deviation	6.1 ± 0.57	1.42 ±0.48	4.47 ± 2.10	0.91 ± 0.06	0.66 ± 0.24

IV. DISCUSSION

To the best of our knowledge, this study constitutes the first attempt to investigate the variation of the ultrasonic response of an implant as a function of bone quantity in contact with its external interface. More details on the present study can be found in [12].Previously, the 1 MHz response of a screw inserted in aluminium was measured [9] and the approach (measurements realized at 10 MHz) has been extended to cylindrical implant models inserted in bone tissue [10]. The results have been explained using finite time domain numerical simulation tools [13]. More recently, the ultrasonic response of real implants embedded in tricalcium silicate based cements and subjected to fatigue loading has been measured [14]. The originality of the present study lies in that variations of the ultrasonic response of a real implant are evidenced as a function of the insertion depth, which is related to its primary stability.

When the implant is unscrewed from the bone cavity, the amount of bone in contact with the implant decreases, due to its approximately cylindrical shape. When the bone-implant interface is debonded [15, 16], a stronger gap of mechanical properties is obtained at the implant interface, thus explaining that the transmission coefficient at the implant external interface is lower. Therefore, energy leakage of the ultrasonic wave out of the implant (which acts as a wave guide) is lower when the bone-implant is debonded, which explains the slower decrease of the ultrasonic energy recorded by the sensor. In summary, the acoustic energy recorded at the upper surface of the implant decreases faster when the implant external interface is fully bonded than when it is debonded.

The indicator *I* used herein derives from the one used in [10, 13], where cylindrical implant models were employed, which led to a possible distinction between the various echoes (see Fig. 4 of [10]). However, the implant geometry used in the present study is more complex, due to the presence of internal and external threadings, leading to possible complex multiple reflections of the ultrasonic wave within the implant. Therefore, the *rf* signals obtained with real implants are more complex and slightly longer than the one obtained with cylindrical implant model, which explains why a longer duration (75 μs compared to 60 μs in [10, 13]) was used for the indicator *I*. Despite the complex wave propagation occurring within the implant, the ultrasonic response is shown to be sensitive to the implant environment. The time window was modified in the present study because *rf* signals obtained herein are different than in [14], which is due to i) the different material surrounding the implant and ii) the integrated transducer used herein (including the healing abutment). The amplitude of the envelop of the *rf* signals obtained herein is approximately constant for all samples before 5 μs, which explains why the time window starts at 5 μs. However, using a fully integrated transducer allows to reduce positioning errors of the transducer relatively to the implant axis, which improves the reproducibility of the measurements. Despite the improvement of the reproducibility of the measurements compared to previous study, there remain different sources of errors which are discussed below.

The insertion torque of the transducer is likely to affect its ultrasonic response due to variations of contact conditions between the implant and the healing abutment. In the present study, the ultrasonic probe is screwed with a torque of 0.05 N.m before all measurements, which is around 5 times lower than torque values recommended by implant manufacturers [17]. We verified (data not shown) that changing the torque value between 0.03 and 0.08N.m does not affect the reproducibility of the measurements. A limitation of this study is that the reproducibility of the measurements was assessed only with an implant embedded in air and fully inserted in bone tissue. It was not possible to assess the reproducibility after each rotation since it may affect the implant angular position and thus modify the implant insertion. However, both situations lead to comparable results in terms of order of magnitude (1.7 and 1.4), which shows that the reproducibility weakly depends on the surrounding medium of the implant. Due to this incomplete database, it was not possible to realize ANOVA analysis of the dependence of the indicator *I* as a function of the number of rotations for one given sample.

Considering a linear variation of the indicator *I* as a function of the number of rotation, the reproducibility error Δy can be related to an error on the estimation of the number of rotation equal to $2\Delta y/a$. Due to the nature of the implant threading, each rotation of the implant corresponds to an axial displacement of around $d=0.8$ mm. Therefore, the error Δx on the implant penetration depth due to reproducibility errors is equal to $\Delta x= 2\Delta y.d/a$. Table 1 shows the corresponding value of Δx for all implants. As a consequence, the ultrasonic method leads to an error on the implant positioning equal to 0.66 ± 0.24, which shows the potentiality of the method.

V. CONCLUSION

This study paves the way for the development of a new ultrasonic tool to be used in oral implantology for the monitoring of osseointegration. The main advantage of ultrasonic based technology compared with RFA technique lies in its sensitivity to the bone-implant interface properties in terms of biomechanical quantity and quality [18].

ACKNOWLEDGMENT

This work has been supported by French National Research Agency (ANR) through EMERGENCE program (project WaveImplant n°ANR-11-EMMA-039).

REFERENCES

1. Albrektsson, T., Dahl, E., Enbom, L., Engevall, S., Engquist, B., Eriksson, A. R., Feldmann, G., Freiberg, N., Glantz, P. O., and Kjellman, O., 1988, "Osseointegrated oral implants. A Swedish multicenter study of 8139 consecutively inserted Nobelpharma implants," J Periodontol, 59(5), pp. 287-296.
2. Serra, G., Morais, L. S., Elias, C. N., Meyers, M. A., Andrade, L., Muller, C., and Muller, M., 2008, "Sequential bone healing of immediately loaded mini-implants," Am J Orthod Dentofacial Orthop, 134, pp. 44–52.
3. Frost, H. M., 2003, "Bone's mechanostat: A 2003 update," Anat Rec A Discov Mol Cell Evol Biol, 275A(2), pp. 1081-1101.
4. Wolff, J., 1892, "The law of bone remodeling."
5. Valderrama, P., Oates, T. W., Jones, A. A., Simpson, J., Schoolfield, J. D., and Cochran, D., 2007, "Evaluation of two different resonance frequency devices to detect implant stability: A clinical trial," J Periodont, 78, pp. 262–272.
6. Meredith, N., Alleyne, D., and Cawley, P., 1996, "Quantitative determination of the stability of the implant-tissue interface using resonance frequency analysis," Clin Oral Implant Res, 7, pp. 261–267.
7. Aparicio, C., Lang, N. R., and Rangert, B., 2006, "Validity and clinical significance of biomechanical testing of implant/bone interface," Clin Oral Implant Res, 17, pp. 2–7.
8. Seong, W. J., Kim, U. K., Swift, J. Q., Hodges, J. S., and Ko, C. C., 2009, "Correlations between physical properties of jawbone and dental implant initial stability," J Prosthet Dent, 101, pp. 306–318.
9. de Almeida, M. S., Maciel, C. D., and Pereira, J. C., 2007, "Proposal for an ultrasonic tool to monitor the osseointegration of dental implants," Sensors, 7, pp. 1224–1237.
10. Mathieu, V., Anagnostou, F., Soffer, E., and Haiat, G., 2011, "Ultrasonic evaluation of dental implant biomechanical stability: an in vitro study," Ultrasound Med. Biol., 37(2), pp. 262-270.
11. Mathieu, V., Anagnostou, F., Soffer, E., and Haiat, G., 2011, "Numerical simulation of ultrasonic wave propagation for the evaluation of dental implant biomechanical stability," J. Acoust. Soc. Am., 129(6), pp. 4062-4072.
12. Vayron, R., Mathieu, V., Michel, A., and Haiat, G., submitted, "Ultrasonic evaluation of dental implant biomechanical stability : an in vitro study," Ultrasound Med Biol.
13. Mathieu, V., Anagnostou, F., Soffer, E., and Haiat, G., 2011, "Numerical simulation of ultrasonic wave propagation for the evaluation of dental implant biomechanical stability," J. Acoust. Soc. Am, 129(6), pp. 4062-4072.
14. Vayron, R., Karasinski, P., Mathieu, V., Michel, A., Loriot, D., Richard, G., Lambert, G., and Haiat, G., in press, "Variation of the ultrasonic response of a dental implant embedded in tricalcium silicate-based cement under cyclic loading," J. Biomech.
15. Mathieu, V., Vayron, R., Barthel, E., Dalmas, D., Soffer, E., Anagnostou, F., and Haiat, G., 2012, "Mode III cleavage of a coin-shaped titanium implant in bone: Effect of friction and crack propagation," J Mech Behav Biomed Mater, 8, pp. 194-203.
16. Mathieu, V., Fukui, K., Matsukawa, M., Kawabe, M., Vayron, R., Soffer, E., Anagnostou, F., and Haiat, G., 2012, "Micro-Brillouin scattering measurements in mature and newly formed bone tissue surrounding an implant," J Biomech Eng-Trans ASME, 133(2).
17. Kanawati, A., Richards, M. W., Becker, J. J., and Monaco, N. E., 2009, "Measurement of clinicians' ability to hand torque dental implant components," J Oral Implant, 35(4), pp. 185-188.
18. Vayron, R., Barthel, E., Mathieu, V., Soffer, E., Anagnostou, F., and Haiat, G., 2012, "Nanoindentation measurements of biomechanical properties in mature and newly formed bone tissue surrounding an implant," J Biomech Eng, 134(2), p. 021007.

Author: Haiat Guillaume
Institute: CNRS, Laboratoire Modélisation et Simulation MultiEchelle, MSME UMR CNRS 8208
Street: 61, avenue du Général de Gaulle
City: 94010, Créteil
Country: France
Email: guillaume.haiat@univ-paris-est.fr

Silicon-Based Fabrication of Biodegradable Polymer for Controlled Drug-Delivery

Thanh D. Nguyen, Robert S. Langer, and Ana Jaklenec

Koch Institute for Integrative Cancer Research, MIT, Cambridge, 02139 MA, USA

Abstract— **Methods to fabricate biodegradable polymer microparticles with well-defined structures could offer significant impact in the field of drug delivery. To create a controlled drug-release system, drugs are usually encapsulated by a biodegradable polymer whose physical properties can be tailored to obtain desired release kinetics. Most of the current approaches to create such drug delivery systems are via emulsion methods, but they have limitations in completely entrapping the drugs inside the particle core and controlling geometry of the polymeric microstructures. Here, we present a new scheme to create microparticles of biodegradable polylactic-co-glycolic acid polymer with a precise control over the particle's structure. We use microfabrication techniques to create a Silicon micromold and use a heat-pressing process to transfer the mold's features to the polymer. Our method relies solely on common materials for fabricating electronic chips (e.g. Silicon) and doesn't require any other special synthetic materials. This method is versatile as it is applicable to many types of polymers, thereby offering a platform technology to create microcapsules with well-defined structures for controlled drug-release systems.**

Keywords— **Microfabrication, drug delivery, biodegradable polymer, heat pressing.**

I. INTRODUCTION/METHODS/RESULTS

In the last two decades, the development of drug delivery systems has garnered significant impacts due to the ability to control the release time and drug release doses. Drug-delivery systems including drug-eluting polymers, micelles, liposomes and microcapsules [1-4] have been generated for different types of drugs. They can response to stimuli such as pH, temperature, ionic strength or specific proteins. [2, 5, 6] These drugs can be used for effective treatments of cancers, diabetes and many other diseases. To systematically control the release, drugs need to be completely sealed in a well-defined microcapsule of a biodegradable polymer. This condition would allow a modification of the polymer's degradation to control the release process. However, most of current drug/polymer particles are derived from emulsion methods which have many limitations. Emulsion methods encapsulate drugs in aqueous forms with solvents and require a lyophilizing freedrying process, both of which can negatively affect drug stability. Further, the methods form very complex interwoven structures of the drug and the polymer, thereby limiting a fine control over the drug release. So far, the only system, which provides a complete encapsulation of drugs and can tailor the drug release in a systematic manner, is the microchip. [7, 8]

The device has many drug reservoirs and performs the release by a degradation of covering layers. It can be externally triggered or automatically released in multiple pulses. Yet, the microchip is bulky and requires painful and expensive implantation/removal surgeries. Thus, microparticles, which can be injected and possess well-defined structures with empty cores for containing drugs, are much more favorable. Fabrication techniques using photolithography offer an excellent approach to create such microstructures. Indeed, many nano-micro devices and materials have been created by using photolithography methods for significant biomedical applications. [9-13] The DeSimone group has reported an excellent PRINT method to create biodegradable polymeric nanoparticles. [14, 15] This method, however, needs a synthesis of a special Teflon-like fluoro-polymer. Here we report a process based solely on Silicon, the most common material in microfabrication of electronic chips, to create microparticles of biodegradable polylactic-co-glycolic acid (PLGA) polymer with a precise control over the particles' structures and dimensions. First, we create a superhydrophobic Silicon master mold and then press the mold on a melted PLGA piece to create the microparticles.

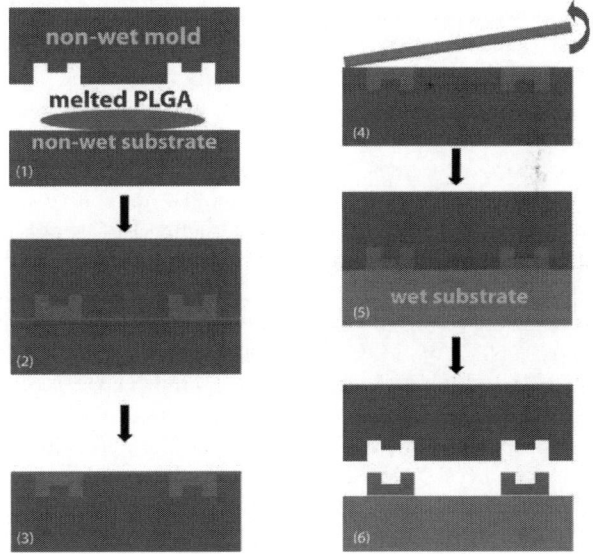

Fig. 1 Schematic of the process to create PLGA microparticles. In (1), a PLGA piece is melted before the pressing process. In (2), the Silicon mold is pressed on the PLGA piece lying on a Silicon substrate. In (3), the substrate is released and we obtain the particles with a thin flashing/scum layer of PLGA adhered with the mold. In (4), we use scotch-tape to peel off the thin layer from the mold. In (5), the particles are transferred to a different wet-sticky substrate. In (6), the mold is released and we obtain the PLGA particles on the host substrate.

© Springer International Publishing Switzerland 2015
V. Van Toi and T.H. Lien Phuong (eds.), *5th International Conference on Biomedical Engineering in Vietnam,*
IFMBE Proceedings 46, DOI: 10.1007/978-3-319-11776-8_12

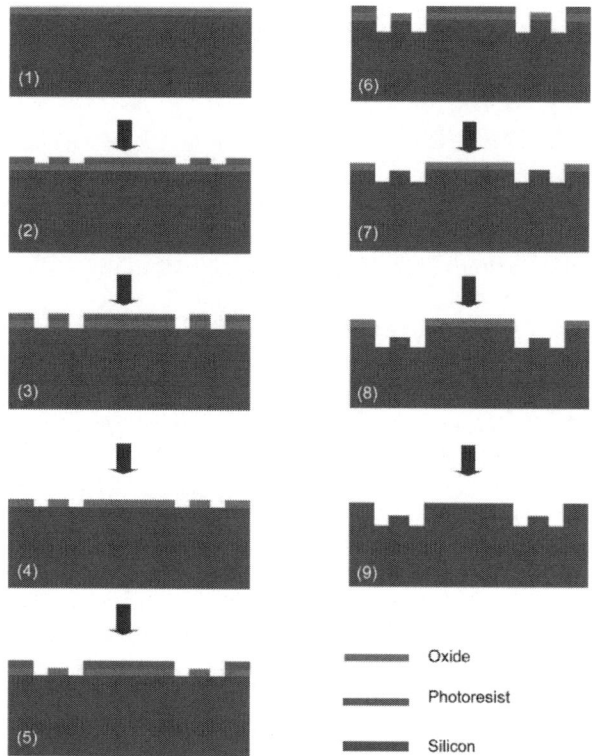

Fig. 2 Schematics of the process to create the Silicon mold used in fig.1.

We specifically design the photomask and aim to achieve the microstructure with empty cores. This microstructure can be sealed by pressing two of them face-to-face under heat. The method can be applied for any type of biodegradable polymer and offers a platform technology to create biodegradable drug microcapsules.

The process to create PLGA particles is described in the schematics of Fig. 1. First, a Silicon mold with an inverse desired feature is pressed on a piece of melted PLGA polymer with a pressure of 500 Pas at 100°C in Fig. 1.1 and 1.2. The mold and the substrate are fluorinated by a plasma of CHF$_3$ fluorine gas to be superhydrophobic and not wetable to organic materials. [16] The superhydrophobic surfaces allow most of the polymer to flow into the empty spaces of the mold and form the PLGA particles. There is a thin "flashing or scum layer" (~ 5 μm) adhered to the particles after this pressing process. Next, the substrate is removed in Fig 1.3 and we use scotch-tape to mechanically rupture and remove the flashing layer from the mold in Fig 1.4. The particles are then transferred onto a different substrate by simply bringing the mold into contact with a hydrophilic and sticky surface under heat (80°C) for 30 minutes in fig. 1.5. Finally, we obtain the microparticles with desired structure on a host substrate after removing the mold in Fig. 1.6. We can also release the particles from the host substrate by using

selective chemical etching. For example, the host substrate is made of polyethylene glycol which can be easily removed by water without affecting the PLGA particles.

Fig. 3 Silicon master mold. From the top to the bottom are the schematic of the mold, the camera image of the mold and the zoom-in microscopic images of the microfeatures from the mold, respectively. Scale bar is 500 μm in the bottom image.

In order to make the Silicon mold for the described pressing process, we use two-level photolithography and deep dry etching methods. The flow chart of this fabrication is described in Fig. 2. In Fig. 2.1, we deposit a thin layer (3 μm) of Silicon oxide on a Silicon wafer, using dielectric plasma/chemical vapor deposition (DCVD). In Fig. 2.2, the first photolithography is performed to create the photoresist mask for etching the oxide layer in Fig. 2.3. In Fig. 2.4, the photoresist is removed by using solvents and oxygen plasma. In Fig. 2.5, we perform the second aligned-photolithography to define the x, y dimensions of the empty core in the final polymeric particles. In Fig. 2.6, a deep dry-etching in plasma is performed to define the thickness of the empty core. In Fig. 2.7, we etch away the exposed oxide and remove the photoresist. In Fig. 2.8, we deep-etch the Silicon wafer to define the thickness of the bottom part of the PLGA particles. In Fig. 2.9 we remove the oxide and obtain the desired Silicon mold. Finally, the Silicon mold is fluorinated under the plasma of fluorine gas along with a non-patterned plain Silicon substrate (50 sccm CHF_3, base pressure of 60 mTorrs, power of 100 watts in 2 minutes). These processes deposit a thin layer of Teflon-like polymer on surfaces of the silicon mold and the substrate, making them very inert to the organic materials.

Fig. 3 shows the achieved Silicon master mold from the fabrication process. On one silicon mold, we can fabricate up to thousands of features, corresponding to thousands of PLGA microparticles. The dimensions of these features are easily controlled and high-aspect depth-to-width ratios of the particles can be achieved.

Fig. 4 Fabrication of single wall PLGA particles. a. Scanning electron microscope (SEM) image shows single wall PLGA microparticles with the empty cores. Scale bar is 500 µm. b. Optical microscope image shows arrays of the single wall PLGA particles were achieved on a large area in controlled positions. Scale bar is 1 mm

Fig. 4 shows the scanning electron microscopic and optical images of the achieved PLGA particles. These PLGA particle structures are well-defined with the wall thickness of 150 um and precise dimensions of the core of 100 um x 100 um x 100 um (width x length x height) as seen in Fig. 4a. The microfabrication process can easily create different

wall thicknesses by varying the photo-mask dimensions. The heights of the particles and the particle cores can be easily controlled by varying the etching times in the dry-etching processes. As shown in Fig. 4b, the method is scalable as arrays of these particles can be fabricated on a large area and the particles can be placed at exact positions on a host substrate. Our method can also create different geometries and structures of PLGA particles such as multiple-wall micro-structures by changing photomask design.

II. CONCLUSION

The presented fabrication method is significant as it allows the achievement of very well-defined biodegradable PLGA microparticles for encapsulating drugs. This drug delivery system would enable fine control over the release kinetics by allowing for a modulation of particle size, wall thickness, polymer type and polymer molecular weight.

The fabrication process is scalable, based solely on Silicon and avoids any exotic and synthetic materials. Most importantly, it provides a platform technology to generate microcapsules of biodegradable polymers with desired structures for better controlled drug-delivery.

Research on developing methods to fill the empty cores with drugs and sealing the PLGA particles are being performed to obtain the complete PLGA/drug particles.

REFERENCES

1. Cullis, P. R.; Chonn, A. Adv Drug Deliv Rev 1998, 30, (1-3), 73-83.
2. Langer, R. Acc Chem Res 2000, 33, (2), 94-101.
3. Langer, R. J Biomed Mater Res A 2013, 101, (9), 2449-55.
4. Haag, R.; Kratz, F. Angew Chem Int Ed Engl 2006, 45, (8), 1198-215.
5. Uhrich, K. E.; Cannizzaro, S. M.; Langer, R. S.; Shakesheff, K. M. Chem Rev 1999, 99, (11), 3181-98.
6. Torchilin, V. P. Nature Reviews Drug Discovery 2005, 4, (2), 145-160.
7. Richards Grayson, A. C.; Choi, I. S.; Tyler, B. M.; Wang, P. P.; Brem, H.; Cima, M. J.; Langer, R. Nat Mater 2003, 2, (11), 767-72.
8. Farra, R.; Sheppard, N. F., Jr.; McCabe, L.; Neer, R. M.; Anderson, J. M.; Santini, J. T., Jr.; Cima, M. J.; Langer, R. Sci Transl Med 2012, 4, (122), 122ra21.
9. Qi, Y.; Nguyen, T. D.; Purohit, P. K.; McAlpine, M. C., Stretchable Piezoelectric Nanoribbons for Biocompatible Energy Harvesting. In Stretchable Electronics, Wiley-VCH Verlag GmbH & Co. KGaA: 2012; pp 111-139.
10. Nguyen, T. D.; Deshmukh, N.; Nagarah, J. M.; Kramer, T.; Purohit, P. K.; Berry, M. J.; McAlpine, M. C. Nat Nanotechnol 2012, 7, (9), 587-93.
11. Nguyen, T. D.; Hogue, I. B.; Cung, K.; Purohit, P. K.; McAlpine, M. C. Lab Chip 2013, 13, (18), 3735-40.
12. Nguyen, T.; McAlpine, M., Fabrication of Nano-piezomaterials for Powering Bioelectronics and Interfaced Cellular Biomechanics. In 4th International Conference on Biomedical Engineering in Vietnam, Toi, V. V.; Toan, N. B.; Dang Khoa, T. Q.; Lien Phuong, T. H., Eds. Springer Berlin Heidelberg: 2013; Vol. 40, pp 25-29.

13. Nguyen, T. D.; Nagarah, J. M.; Qi, Y.; Nonnenmann, S. S.; Morozov, A. V.; Li, S.; Arnold, C. B.; McAlpine, M. C. Nano Lett 2010, 10, (11), 4595-9.
14. Rolland, J. P.; Maynor, B. W.; Euliss, L. E.; Exner, A. E.; Denison, G. M.; DeSimone, J. M. J Am Chem Soc 2005, 127, (28), 10096-100.
15. Enlow, E. M.; Luft, J. C.; Napier, M. E.; DeSimone, J. M. Nano Lett 2011, 11, (2), 808-13.
16. Schvartzman, M.; Mathur, A.; Hone, J.; Jahnes, C.; Wind, S. J. Appl Phys Lett 2008, 93, (15), 153105.

Corresponding author

Author: Thanh D Nguyen
Institute: Massachusetts Institute of Technology
Street: 77 Massachusetts Ave.
City: Cambridge, MA 02139
Country: USA
Email: nguyentd@mit.edu

Advancement in Gemcitabine Delivery for Cancer Treatment

Uyen Minh Le

Sullivan University College of Pharmacy, Louisvile, Kentucky, USA

Abstract— **Gemcitabine is utilized as the first-line treatment for adenocarcinoma of the pancreas that has been considered as one of the most challenging diseases. The major drawback of the gemcitabine formulations is its high hydrophilicity and short half-life. To compensate for those shortcomings, a large dose of infused gemcitabine is usually used to achieve the desire therapeutic effects. However, using this dose, could lead to a high toxicity and severe adverse effects. Hence, there has been a great of interest in the development of gemcitabine to increase the hydrophobicity, half-life, and stability of the drug. In this review, we summarize the latest approaches in drug delivery of gemcitabine to clarify the unsolved problems of drug resistance and discuss the effectiveness of the advanced delivery systems on different types of cancer.**

Keywords— **gemcitabine, advanced drug delivery, cancer treatment.**

I. INTRODUCTION

Gemcitabine, commonly known as Gemzar [®], has been approved by the FDA for the treatment of ovarian cancer, breast cancer, non-small cell lung cancer, and pancreatic cancer [1]. Gemcitabine structure was analogue to the nucleoside cytidine, in which the fluorine atoms in Gemcitabine replace the 2' carbon of deoxycytidine. The figures below illustrate the structures of Deoxycytidine and Gemcitabine.

Deoxycytidine   Gemcitabine

From its specific structure, gemcitabine works by replacing the building blocks of cytidine during the DNA replication. Hence, it prevents cells from synthesizing DNA and RNA, leading to the cell death. Gemcitabine typically used as the white powder of gemcitabine hydrochloride which is reconstituted into a solution with 0.9% sodium chloride before use. Gemcitabine hydrochloride solution, normally in large doses, is slowly infused into a vein over 30 minutes. Longer infusion time may increase toxicity. Gemcitabine can kill many cancer cells; unfortunately, it also kills the normal cells, especially for fast growth cells from the hair and skin. Typical side effects of gemcitabine are tiredness, nausea, vomiting, diarrhea, pain, flu-like symptoms, and hair loss. Some other kinds of toxicity, such as pulmonary toxicity [2], respiratory failure [3], and skin necrosis [4] were also occasionally reported. Although gemcitabine has been approved for various cancer types; however, its efficacy is still limited in prolonging the life-span for patients, especially for those with pancreatic cancers. Major possible causes for its poor effects and toxicity were thought due to its structure instability and low cell penetration; which were reasoned from the simple structure and hydrophilicity of the gemcitabine hydrochloride. Many efforts have been implemented to improve the formulation of gemcitabine. In this study, we reviewed significant literature and related research to provide the latest information in regards to the improvement of drug delivery and effectiveness of the new formulations on various types of cancers and promising direction towards gemcitabine resistance.

II. MATERIALS AND METHODS

Our main goal is to capture most of latest articles which related to gemcitabine delivery for cancer treatment. We used the data base of publications in PubMed for a period from 1993 to 2014 as our information sources. Additional papers of interest were also developed from the reference lists of selected articles. We included only articles in English which are relevant to the drug delivery of gemcitabine. When more than one articles reported similar outcomes, only the most current and inclusive one was cited. The improved delivery of gemcitabine was evaluated in terms of technology, drug formulations, characterization, release, stability, and anti-cancer effects. Furthermore, a quantitative comparison of the effectiveness in drug delivery to various types of cancer as well as to gemcitabine resistance was also comprehensively analyzed.

© Springer International Publishing Switzerland 2015
V. Van Toi and T.H. Lien Phuong (eds.), *5th International Conference on Biomedical Engineering in Vietnam*,
IFMBE Proceedings 46, DOI: 10.1007/978-3-319-11776-8_13

III. RESULTS AND DISCUSSIONS

Of the 303 titles retrieved from the search engine, 82 articles were further reviewed based on their abstracts. Subsequently, 10 articles were involved in prodrugs or modified gemcitabine, 51 articles were for nanocarriers, and 21 articles were major targeted delivery of gemcitabine.

A. Synthesis of Gemcitabine Prodrugs to Enhance the Lipophilicity of Gemcitabine

Gemcitabine modifications were mainly for an enhanced lipophilicity to improve the stability and cell penetration. The most developed derivative was 4-(N)-stearoyl gemcitabine [5], which possessed a high stability at storage, in plasma, and with the presence of enzyme cathepsins. Other successful synthesized derivatives were gemcitabine with heptamethine cyanine[6], gemcitabine-coumarin-biotin conjugates [7], gemcitabine and asymmetric bifunctional silyl ether [8], gemcitabine combining with N-[p-maleimidophenyl]-isocyanate, amphiphilic prodrug by covalently coupling gemcitabine with a derivative of squalene [9], gemcitabine phosphoramidate [10], gemcitabine and heptamethine cyanine [6], H-gemcitabine [11], and gemcitabine with azo-based polyphosphazene [12]. All of the new derivatives shown prolonged stability, improved drug uptake, and increased cytotoxicity, as compared to the free gemcitabine. In addition, some of them have specific tissue target which potentially minimize undesirable side effects of the drug. Particularly, the azo based polyphosphazene product of gemcitabine, which was stable in stimulated acidic and basic pHs of the GIT system, was active against human colorectal cancer cell lines [12] whereas H-gemcitabine was targeted to the extracellular DNA in tumors [11]. Furthermore, the heptamethine cyanine derivative was a "theranostic" agent, which facilitated the detection of drug monitoring and therapeutic effect via fluorescent imaging [6]. Moreover, the squalene[9] and phosphoramidate [10] prodrugs , respectively, demonstrated an improved efficacy on transporter-deficient resistant cancer and in the presence of transport inhibitor dypridamole. However, most of the prodrugs demonstrated a slightly improved *in vivo*, leading to additional needs for the development drug delivery. Of the search data base, nanocarriers were the most dominant and effective systems to introduce gemcitabine to tumor sites.

B. Nanocarriers for the Delivery of Gemcitabine

One of the most promising advanced drug deliveries for cancer treatment is the use of nanocarriers. Different nano-colloidal systems of gemcitabine have been developed and demonstrated highly improved efficacy *in vitro* and *in vivo*. Encapsulated into nanocarriers, typically smaller than 200 nm in diameter of size, allowed gemcitabine to prolong its circulation time in the blood, enhance the permeability and retention effect, release intra-cellular, possess the controlled release, and increase the drug uptake inside tumor cells. Generally, the most commons nanocarriers have widely been explored and shown positive effects for the delivery of gemcitabine were liposomes, nanoparticles, nanocomposites, carbon nanotubes, micelles, and polymeric conjugates.

Liposomes comprise an internal aqueous core surrounded by a lipophilic membrane of phospholipids. The use of liposomes is ideal in term of drug encapsulation, biocompatibility, and safety. The lipophilic prodrugs of gemcitabine had better encapsulation efficiency and highest stability when encapsulated in liposomes [5]. The internal aqueous core of stealth liposomes is appropriate for loading gemcitabine and allows preventing the drug from degradation in plasma before intracellularly delivered to the tumor cells[13]. The improved effects were reported on various pancreatic adenocarcinoma, colon cancer LS174-T, colorectal adenocarcinoma HT-29, multiple myeloma U266 and INA-6, and thyroid carcinoma cells. Interestingly, ultrafast temperature sensitive liposomes, which were composed of phospholipid DPPC and the surfactant BriJ78, showed the highest stability and enhanced antitumor efficacy with complete tumor regression after only a single dose [14]. Another innovative approach is the use of two-wave nanocarrier, in which polyethyleneimine (PEI)/polyethylene glycol (PEG)/coated mesoporous silica nanoparticles was simultaneously injected with a PEGylated liposome-gemcitabine, showing a strong shrinkage of tumor xenografts at the pancreatic ductal adenocarcinoma tumor sites [15].

As with liposomes, other nanocarriers, such as gold nanoparticles, polymeric particles, carbon nanotubes, micelles, and nanocomposites, demonstrated promising candidates for the delivery of gemcitabine. Most of them focused on pancreatic cancer; however the sensitivity of each type of pancreatic cancer cell line was various upon the materials, structure, and methodology of nannocarriers. Particularly, chitosan and glyceryl monnoleate nanostructure was effective for the MIA PaCa-2 and BxPC-3 cell lines [16], whereas polybutylcyanoacrylate (PBCA) nanoparticles coated with polysorbate 80 was for C6 glimo cells [17]. Importantly, the use of magnetic nanocarriers for gemcitabine demonstrated multiple applications, including treatment and diagnostics, from a single drug delivery, which lead to a new paradigm for treatment of cancer [18].

C. Targeted Delivery of Gemcitabine

Prodrugs of gemcitabine enhanced the drug stability. Additional incorporating them into nanocarriers helped the drug have better plasma half-live and intracellular trafficking. However, tumor resistance as well as inconsistent and unspecific uptake of gemcitabine is still the key challenge for the cancer treatment. As for many other chemotherapy agents, delivery of gemcitabine in a targeted manner, although more costly and difficult, is a current trend. This smart drug delivery is promising to localize and increase the concentrations of gemcitabine at the desirable sites, thus decreasing the systemic toxicity. A popular target in the targeted delivery of gemcitabine is epidermal growth factor receptor (EGFR). This receptor was over-expressed on pancreatic cancer cells [19], whose first-line treatment is gemcitabine. As a matter of fact, EGFR-targeted nanoparticles of gemcitabine all showed improved in vitro and in vivo anti-tumor effects, as compared to the non-targeted drug, on EGFR-over-expressing pancreatic cancer cell lines, such as PANC-1 and AsPC-1 [20, 21]. Similarly, other respective receptor-targeted nanocarriers of gemcitabine sensitized the cancer cells and prolong the median survival when compared to non-target. Examples for such target were liposomal gemcitabine targeted by a single-chain antibody fragment for the transferring receptor delivering wild-type human p53 [22]; gemcitabine-liposomes conjugated with hyaluronic acid (HA 12kDa) to target CD44 receptor on pancreatic adenocarcinoma [23], folate-targeted gemcitabine-loaded liposomes for MCF-7 breast cancer cells [24], gemcitabine conjugated with bisphosphonate for targeted bone-specific in breast cancer bone metastases [25], vascular targeting by endoTAG-1 for lung and pancreatic cancers [26], and targeted delivery of a peripheral benzodiazepine receptor ligand-gemcitabine conjugate to brain tumors [27].

In addition to the specific delivery, targeted delivery of gemcitabine demonstrated the capability of overcoming the tumor resistance to gemcitabine. It was shown to have the drug accumulated in the cellular membranes and delivered directly to the cell cytoplasm [9]. For instance, covalent gemcitabine-carbamate-anti-HER2/neu was effectively against chemotherapeutic-resistant SKBr-3 mammary carcinoma [28]. Similarly, nanogels conjugates of gemcitabine well penetrated through the gastrointestinal barrier in Caco-2 cells, potentially for oral against gemcitabine-resistant tumors [29].

IV. CONCLUSIONS

In conclusion, advanced delivery systems of nanocarriers have shown to be good candidates for the development of gemcitabine. For the long term, molecularly targeted and combination therapy of gemcitabine are important cancer treatment modalities.

AKNOWLEDGEMENTS

The author acknowledged Dr. Amber Cann at Sullivan University College of Pharmacy for her assistance in searching for articles. The author also acknowledged Dr. Tuan Tran from the SPeerTech LLC for his valuable proof-reading.

REFERENCES

1. NCI. [cited 2014 03/19]; Available from: http://www.cancer.gov/cancertopics/druginfo/fda-gemcitabine-hydrochloride.
2. Chi, D.C., et al., Gemcitabine-induced pulmonary toxicity. Anticancer research, 2012. 32(9): p. 4147-9.
3. Davidoff, S., R.D. Shah, and T. Arunabh, Gemcitabine-induced respiratory failure associated with elevated erythrocyte sedimentation rate (ESR). Respiratory medicine, 2006. 100(4): p. 760-3.
4. D'Epiro, S., et al., Gemcitabine-induced extensive skin necrosis. Case reports in medicine, 2012. 2012: p. 831616.
5. Immordino, M.L., et al., Preparation, characterization, cytotoxicity and pharmacokinetics of liposomes containing lipophilic gemcitabine prodrugs. Journal of controlled release : official journal of the Controlled Release Society, 2004. 100(3): p. 331-46.
6. Yang, Z., et al., Folate-based near-infrared fluorescent theranostic gemcitabine delivery. Journal of the American Chemical Society, 2013. 135(31): p. 11657-62.
7. Maiti, S., et al., Gemcitabine-coumarin-biotin conjugates: a target specific theranostic anticancer prodrug. Journal of the American Chemical Society, 2013. 135(11): p. 4567-72.
8. Parrott, M.C., et al., Incorporation and controlled release of silyl ether prodrugs from PRINT nanoparticles. Journal of the American Chemical Society, 2012. 134(18): p. 7978-82.
9. Bildstein, L., et al., Transmembrane diffusion of gemcitabine by a nanoparticulate squalenoyl prodrug: an original drug delivery pathway. Journal of controlled release : official journal of the Controlled Release Society, 2010. 147(2): p. 163-70.
10. Wu, W., et al., Synthesis and biological activity of a gemcitabine phosphoramidate prodrug. Journal of medicinal chemistry, 2007. 50(15): p. 3743-6.
11. Dasari, M., et al., H-gemcitabine: a new gemcitabine prodrug for treating cancer. Bioconjugate chemistry, 2013. 24(1): p. 4-8.
12. Sharma, R., et al., Design, synthesis and ex vivo evaluation of colon-specific azo based prodrugs of anticancer agents. Bioorganic & medicinal chemistry letters, 2013. 23(19): p. 5332-8.
13. Zhou, Q., et al., Preparation and characterization of gemcitabine liposome injections. Die Pharmazie, 2012. 67(10): p. 844-7.
14. May, J.P., et al., Thermosensitive liposomes for the delivery of gemcitabine and oxaliplatin to tumors. Molecular pharmaceutics, 2013. 10(12): p. 4499-508.
15. Meng, H., et al., Two-wave nanotherapy to target the stroma and optimize gemcitabine delivery to a human pancreatic cancer model in mice. ACS nano, 2013. 7(11): p. 10048-65.
16. Trickler, W.J., et al., Chitosan and glyceryl monooleate nanostructures containing gemcitabine: potential delivery system for pancreatic cancer treatment. AAPS PharmSciTech, 2010. 11(1): p. 392-401.
17. Wang, C.X., et al., Antitumor effects of polysorbate-80 coated gemcitabine polybutylcyanoacrylate nanoparticles in vitro and its pharmacodynamics in vivo on C6 glioma cells of a brain tumor model. Brain research, 2009. 1261: p. 91-9.
18. Viota, J.L., et al., Functionalized magnetic nanoparticles as vehicles for the delivery of the antitumor drug gemcitabine to tumor cells. Physicochemical in vitro evaluation. Materials science & engineering. C, Materials for biological applications, 2013. 33(3): p. 1183-92.
19. Xiong, H.Q. and J.L. Abbruzzese, Epidermal growth factor receptor-targeted therapy for pancreatic cancer. Seminars in oncology, 2002. 29(5 Suppl 14): p. 31-7.

20. Patra, C.R., et al., Targeted delivery of gemcitabine to pancreatic adenocarcinoma using cetuximab as a targeting agent. Cancer research, 2008. 68(6): p. 1970-8.

21. Arya, G., et al., Enhanced antiproliferative activity of Herceptin (HER2)-conjugated gemcitabine-loaded chitosan nanoparticle in pancreatic cancer therapy. Nanomedicine : nanotechnology, biology, and medicine, 2011. 7(6): p. 859-70.

22. Camp, E.R., et al., Transferrin receptor targeting nanomedicine delivering wild-type p53 gene sensitizes pancreatic cancer to gemcitabine therapy. Cancer gene therapy, 2013. 20(4): p. 222-8.

23. Dalla Pozza, E., et al., Targeting gemcitabine containing liposomes to CD44 expressing pancreatic adenocarcinoma cells causes an increase in the antitumoral activity. Biochimica et biophysica acta, 2013. 1828(5): p. 1396-404.

24. Paolino, D., et al., Folate-targeted supramolecular vesicular aggregates as a new frontier for effective anticancer treatment in in vivo model. European journal of pharmaceutics and biopharmaceutics : official journal of Arbeitsgemeinschaft fur Pharmazeutische Verfahrenstechnik e.V, 2012. 82(1): p. 94-102.

25. El-Mabhouh, A.A., et al., A conjugate of gemcitabine with bisphosphonate (Gem/BP) shows potential as a targeted bone-specific therapeutic agent in an animal model of human breast cancer bone metastases. Oncology research, 2011. 19(6): p. 287-95.

26. Eichhorn, M.E., et al., Vascular targeting by EndoTAG-1 enhances therapeutic efficacy of conventional chemotherapy in lung and pancreatic cancer. International journal of cancer. Journal international du cancer, 2010. 126(5): p. 1235-45.

27. Guo, P., et al., Targeted delivery of a peripheral benzodiazepine receptor ligand-gemcitabine conjugate to brain tumors in a xenograft model. Cancer chemotherapy and pharmacology, 2001. 48(2): p. 169-76.

28. Coyne, C.P., T. Jones, and T. Pharr, Synthesis of a covalent gemcitabine-(carbamate)-[anti-HER2/neu] immunochemotherapeutic and its cytotoxic anti-neoplastic activity against chemotherapeutic-resistant SKBr-3 mammary carcinoma. Bioorganic & medicinal chemistry, 2011. 19(1): p. 67-76.

29. Senanayake, T.H., et al., Application of activated nucleoside analogs for the treatment of drug-resistant tumors by oral delivery of nanogel-drug conjugates. Journal of controlled release : official journal of the Controlled Release Society, 2013. 167(2): p. 200-9.

Multifunctional Drug Nanosystems: A Summary of Recent Researches at IMS/VAST

X.P. Nguyen[1], T.T.TMai[1], P.T. Ha[1], H.N. Pham[1], H.N. Luu[1], H.M. Do[1], D.L. Tran[1], H.N. Nguyen[1],
L.T. Nguyen[2], A.S. Ho[2], and T.M.N. Hoang[3]

[1] Lab. of Biomedical Nanomaterials, Institute of Materials Science (IMS/VAST), 18 Hoang Q. Viet, Hanoi, Vietnam
[2] Academy of Military Medicine, Ha Dong, Hanoi, Vietnam
[3] University of Natural Sciences, Vietnam National University, 334 Nguyen Trai, Hanoi, Vietnam

Abstract— The main task of nanomedicine is to fabricate, normally by chemical engineering, nanoscale systems that can play various functions of both diagnosis and treatment. This report aims to present some researches, carried out by the Laboratory of Biomedical Nanomaterials (IMS/VAST in Hanoi), on fabrication and characterization of nanovectors for the disease of cancer. The first part deals with magnetite (Fe_3O_4) nanoparticles (MNPs) based nanoconjugates, functionalized by coating with several polymers as well as loaded with a drug of curcumin. The used MNPs were obtained by co-precipitation, exhibited spherical shape of diameter of 15-20 nm, saturation magnetization of $M_s \sim$ 65-70 emu/g. The coating polymers were acrylic acid (PAA), chitosan (CS) and Alginate (Alg) which were confirmed using the infrared (FTIR) spectra. Magnetic Inductive Heating (MIH) measurements demonstrated that the fabricated MNPs-based conjugates exhibited quite high heating performance, perspective for hyperthermia application. The application of Fe_3O_4@PAA for *in-vivo* hyperthermia treatment of cancer incubated on mice will be shown. As for imaging application, the Fe_3O_4@CS@Cur was used to demonstrate a dual possibilities, fluorescence and magnetic resonance, of monitoring cell penetration by macrophage. In the second part, we show a recent study on targeted delivery systems of paclitaxel/doxorubicin/curcumin-loaded copolymer/polymer nanoparticles, which were prepared by a modified solvent extraction/evaporation technique and decorated by folic acid. The obtained spherical nanoparticles were negatively charged with a zeta potential of about − 30 mV with the size around 50 nm and a narrow size distribution. The targeting effect of anticancer-drugs nanoparticles with folate decoration was investigated *in vitro* by the uptake in cancer cell lines and in nude mouse. The results indicate that the targeted paclitaxel/doxorubicin/curcumin-loaded copolymer/polymer nanoparticles are successful anticancer-targeted drug delivery system for effective cancer chemotherapy.

Keywords— magnetic nanoparticles, drug delivery systems, hyperthermia, magnetic resonance imaging.

I. INTRODUCTION

Recent advances in nanotechnology have driven the development of multifunctional nanoparticles which are promising for targeted delivery of both imaging and therapeutic agents in biomedical applications [1-3]. Among various materials used, magnetite nanoparticles (MNPs) have been proven to be an important lass of material thanks to their unique properties including the ablity to be guided by an external magnetic field, the ability to perturb magnetic local fields and the ability to create heat when subjected to an alternative magnetic field. Many researches have been carried out concerning the fabrication, characterization of the multifunctional nanosystems based on magnetite nanoparticles as core and polymers as shell layer with targeting moiety for selective delivery [4-7]. Besides, several polymeric nanosystems without the MNPs in the core have also paid much attention of researchers in the recent years. The micelles nanosystems which incorporate drugs such as paclitaxel, doxorubicin, curcumin in its hydrophobic domain and carry targeting and imaging moiety have tremendous potential in cancer therapy [8-10].

In this review, we will summarize the recent studies at the Laboratory of Biomedical Nanomaterials in fabrication of multifunctional nanosystems based on Fe_3O_4 magnetic nanoparicles encapsulated with different polymers as well as the targeted drug delivery systems for effective cancer chemotherapy. The preliminary application of MNPs capped with natural or synthesized polymers, namely chitosan (CS), acrylic acid (PAA) and alginate (Alg), for optical and magnetic resonance imaging (MRI) and hyperthermia will be presented. Furthermore, the cellular uptake and apoptosis of anticancer-drugs nanoparicles with folate decoration will also be investigated.

II. MATERIALS AND METHODS

A. Magnetite Nanoparticles Based Nanosystems

The Fe_3O_4 and the poly (styrene – co – acrylic acid) copolymer (PAA) were synthesized through co-precipitation and polymerization, respectively before the encapsulation process. The Fe_3O_4@CS, Fe_3O_4@PAA and Fe_3O_4@Alg nanosystems were prepared through ex-situ process by mixing the MNPs and polymer under vigorous stirring for 48 hours. Then, curcumin (Cur) was preliminarily dissolved in ethanol and absorbed on the Fe_3O_4 surface to form multifunctional nanosystem. Several nanosystems with and without Cur were prepared for further studies. The procedures are described in more detail in [11-13]. The crystal, morphological and magnetic properties of the magnetite nanoparticles based nanosystems was thoroughly characterized.

© Springer International Publishing Switzerland 2015
V. Van Toi and T.H. Lien Phuong (eds.), *5th International Conference on Biomedical Engineering in Vietnam,*
IFMBE Proceedings 46, DOI: 10.1007/978-3-319-11776-8_14

B. *Polymeric Micelle Based Nanosystems*

The copolymer PLA-TPGS or PLA-PEG was first synthesized through the ring-opening polymerization of PLA and TPGS in the presence of stannous octotate as catalyst.The targeted drug delivery systems of paclitaxel, doxorubicin, curcumin loaded copolymer nanoparticles were prepared by a modified solvent extraction/evaporation process and decorated by folic acid. More details of the synthesis procedure can be found in [14]. Surface morphology, size distribution, zeta potential and apoptosis of the polymeic nanoparticles were also investigated.

III. RESULTS AND DISCUSSIONS

As for the magnetite nanoparticles based nanosystem, the co-precipitatedFe_3O_4particles are of single phase with an average diameter of 15-20 nm and a saturation magnetization of 65-70 emu/g. The Fe_3O_4@CS, Fe_3O_4@PAA and Fe_3O_4@Alg nanoparticles are almost spherical with 50 nm in diameter and a slight decrease by 5% in magnetization. The coating layers of polymers were confirmed by the observation ofthe stretching bands of free carboxyl group (C = O) on the surface of the nanoparticles at 1702 cm^{-1} and 1750 cm^{-1} for Fe_3O_4@PAA and Fe_3O_4@Alg, correspondingly; while the characteristic vibrations of N-H group in Fe_3O_4@CS were observed at 1379 and 1604 cm^{-1}, respectively. The linking between Fe_3O_4 and the encapsulating polymers was verified by the appearance of peaks at 567, 575 and 574 cm^{-1} for Fe_3O_4@CS, Fe_3O_4@PAA and Fe_3O_4@Alg, respectively which are assigned to the stretching band of Fe – O – Fe vibration originally observed at 585 cm^{-1}. The magnetic inductive heating (MIH) experiments were carried out at a magnetic field of 236 kHz and 70 Oe (Fig. 1). It can be easily seen that both Fe_3O_4@CS,Fe_3O_4@PAA and Fe_3O_4@Alg exhibit quite high magnetic heating performance thus demonstrating a potential to be utilized as thermal nanoseeds for hyperthermia application.

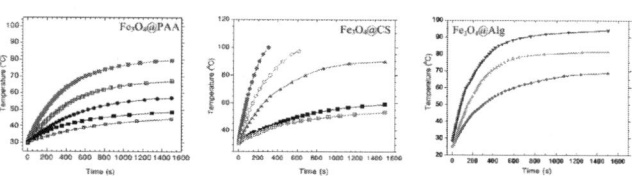

Fig. 1 Magnetic inductive heating curves ofFe_3O_4@CS, Fe_3O_4@PAA and Fe_3O_4@Alg nanosystems

In-vivo hyperthermia experiment was designed for the treatment of a Sarcoma tumor (6 × 6 mm2 size) using Fe_3O_4@PAA as a thermoseed (figure 2). As depicted by the

images, all the tumor in the control mice increased continuously with time and the mice finally died 4 weeks after the experiment started (mouse A), whereas the tumor began to shrink after the first course of irraditaion by AC magnetic field and the mice were totally recovered 3 weeks after threee courses of treatment during the first week (mouse D).

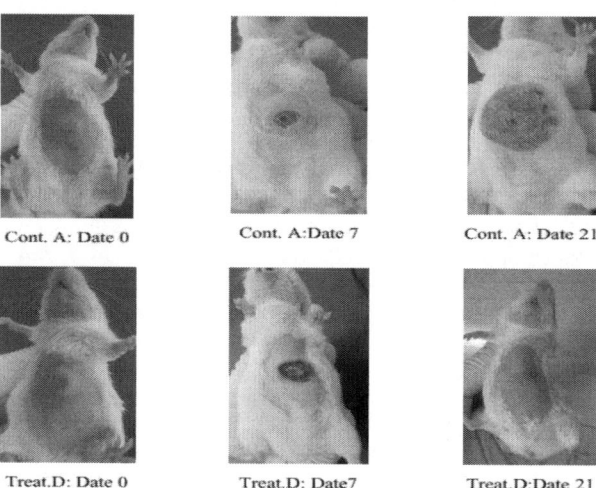

Fig. 2 Images of a control mouse (top) and the treated D mice (bottom) at three date periods

Figure 3 presents confocal microscope image of the macrophages alone and 1hour, 6 hours after its uptake with *Fe₃O₄@CS@Cur.* The conjugate evidenced by the green spots of Cur was observed to uptake in to the vacuole of the cells.From this result, one can deduce that the presence of Cur has created a wonderful enhancement in contrast of the imaging technique even at cellular level.

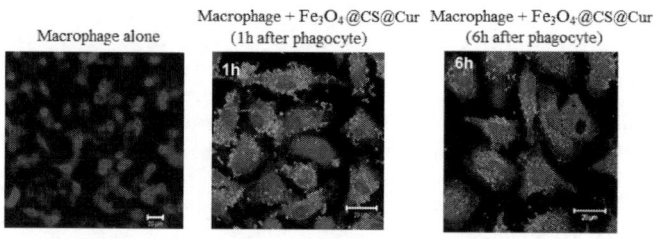

Fig. 3 Fluorescence images of macrophages alone, and macrophages with Fe_3O_4@CS@Cur 1 and 6 hours after phagocytosis

In the case of polymeric micelles based targeted drug delivery nanosystems, the obtained nanoparticles have spherical shape with an average diameter of around 50 nm. This small size makes them advantageous for the penetration to the cancer cells. The nanosystems were negatively charged with the zeta potential of about -30 mV thus proving the stability of nanosystems. Thanks to the small size, drug-loaded polymeric micelle nanoparticles can penetrate through cell membrane, interfere in the metabolite action of the cells and cause cell death at last. As can be seen in fig. 4, the effect of paclitaxel loaded PLA-TPGS with folate decoration to Hela cancer cells significantly increases compared to that of paclitaxel loaded PLA-TPGS which is much higher than free paclitaxel. Moreover, the targeting effect of drug-loaded nanoparticles with folate decoration was also investigated *in-vitro* by the uptake of the nanosystem into cancer cell lines and *in-vivo* for nude mice. The results indicate that the targeted drug delivery systems based on polymeric micelles are effective approach for chemotherapy in cancer treatment.

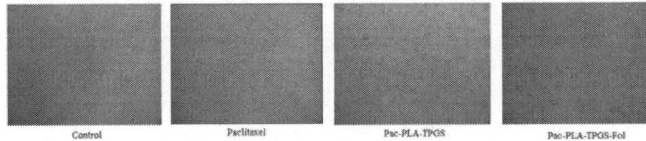

Fig. 4 Effect of Pac-PLA-TPGS and Pac-PLA-TPGS-Fol to Hela cancer cell line after 48h at concentration of 0.3 µg ml−1

IV. CONCLUSIONS

From all researches presented above, it can be summarized that: (i) several multifunctional nanosystems either with or without magnetite nanoparticles have been designed and successfully fabricated; (ii) magnetic nanoparticles based nanosystems exhibit to be a good candidate for both imaging and hyperthermia cancer treatment as well as for drug delivery systems; (iii) The targeted paclitaxel/doxorubicin/curcumin-loaded copolymer nanoparticles have been proven to be successful targeted anticancer-drug delivery systems for cancer chemotherapy.

ACKNOWLEDGMENT

The work was supported by the NAFOSTED grant coded DT.NCCB-DHUD.2012-G/08 and the the VAST grant coded VAST03.03/13-14 (HPT).

CONFLICT OF INTEREST

The authors declare that they have no conflict of interest.

REFERENCES

1. Sajja H K et al. (2009) Development of multifunctional nanoparticles for targeted drug delivery and noninvasive imaging of therapeutic effects. Curr Drug Discov Technol 6(1):43-51
2. Abeylath SC, Ganta S, Iyer AK, Amiji M (2011) Combinatorial-Designed multifunctional polymeric nanosystems for tumor-targeted therapeutic delivey. Acc Chem Res 44(10):1009-1017
3. Lee D E et al. (2012) Multifunctional nanoparticles for multimodal imaging and theragnosis. Chem. Soc. Rev. 41(7): 2656-2072
4. Yallapu M M, Othman S F, Curtis E T, Gupta B K, Jaggi M and Chauhan S C (2011) Multifunctional magnetic nanoparticles forfor magnetic resonance imaging and cancer therapy. Biomaterials 32(7):1890-1905
5. Kumar A, Jena PK, Behera S et al. (2010) Multifunctionalmagnetic nanoparticles for targeted delivery. Nanomedicine 6(1):64-69 DOI 10.1016/j.nano.2009.04.002
6. Feng C and Zhang M (2009) Multifunctional magnetic nanoparticles for medical imaging applications. J Mater Chem 19:6258-6266
7. Banobre-López M, Teijeirob A and Rivas J (2013) Magnetic nanoparticles-based hyperthermia for cancer treatment. Rep Pract Oncol Radiother 18:397-400
8. Boddu S H S, Vaishya R, Jwala J et al. (2012) Preparation and characterization of folate conjugated nanoparticles of doxorubicin using PLGA-PEG-FOL polymer. Med Chem 2:068-075
9. Zhang Z, Lee S H, Feng S S (2007) Folate decorated poly(lactide-co-glycolide)-vitamin E TPGS nanoparticles for targeted drug delivery. Biomaterials 28:1889-1899
10. Yoo H S and Park T G (2004) Folate receptor targeted biodegradable polymeric doxorubicin micelles. J Controlled Release 96(2):273-283
11. Luong T T et al. (2011) Design of carboxylated Fe3O4/poly (styrene-co-acrylic acid) ferrofluids with highly efficient magnetic heating effect. Colloids Surf., A 384:23-30
12. Mai T T T et al. (2012) Chitosan and O-carboxymethyl chitosan modified Fe_3O_4 for hyperthermic treatment. Adv Nat Sci Nano Sci Nanotechnol 3:015006-015010
13. Devkota J et al.(2014) Synthesis, inductive heating, and magnetoimpedance-based detection of multifunctional Fe_3O_4 nanoconjugates. Sensors and Actuators B 190:715-722
14. Nguyen H N et al. (2012) Apoptosis induced by paclitaxel-loade copolymer PLA-TPGS in Hep-G2 cells. Adv Nat Sci Nano Sci Nanotechnol 3:045005-045010

Author: Nguyen Xuan Phuc
Institute: Institute of Materials Science
Street: 18 Hoang Quoc Viet
City: Hanoi
Country: Vietnam
Email: phucnx@ims.vast.ac.vn

Preparation, Characterization and Antibacterial Curcumin Encapsulated Chitosan-PAA Silver Nanocomposite Films

N.V. Cuong, P.N.N. Han, N.K. Hoang, and N.N.L. Giang

Department of Chemical Engineering, Industrial University of Ho Chi Minh City,
12 Nguyen Van Bao Rd, Ho Chi Minh City, Vietnam

Abstract— **In this paper, we demonstrate the in situ fabrication of chitosan-poly (acrylic acid)-silver nanocomposite films in view of their increasing applications as antimicrobial packaging, wound dressing and antibacterial materials. The reduction of silver ions into silver nanoparticles is achieved in acidic solution of chitosan and poly (acrylic acid) (PAA) using their functional groups (-OH, -COOH, -NH2 groups). The structure of nanocomposite film was examined by Fourier Transform Infrared (FTIR) spectroscopy and Scanning Electron Microscopic (SEM). The anti-microbial activities of the chitosan-PAA silver nanoparticle films have demonstrated significant effects against Escherichia coli (E. coli) and Staphylococcus. To improve further their therapeutic efficacy as anti-microbial agents, curcumin encapsulated chitosan-PAA silver nanocomposite films are developed which improve the growth inhibition of E. coli and Staphylococcus compared to curcumin and chitosan-PAA silver nanoparticles film alone. Therefore, the present study clearly provides novel antimicrobial films which are potentially useful in preventing/treating infections.**

Keywords— **Chitosan, curcumin, anti-microbacteria.**

I. INTRODUCTION

Chitosan, a polysaccharide biopolymer derived from naturally occurring chitin, displays unique polycationic, chelating, and film-forming properties due to the presence of active amino and hydroxyl functional groups. Chitosan also exhibits a number of interesting biological activities, including antimicrobial activity, induced disease resistance in plants, and diverse stimulating or inhibiting activities toward a number of human cell types [1-2]. Moreover, chitosan can be used to prevent or treat wound and burn infections due to its intrinsic antimicrobial properties and its ability to deliver extrinsic antimicrobial agents to wounds and burns. Additionally, it can accelerate the functions of inflammatory cells, macrophages, and fibroblasts [3-4]. Chitosan polymer and its derivatives are also used to increase the stability of the drug in which the drugs are loaded in chitosan film or chitosan nanoparticles, resulting in enhancement of drug accumulation and toxicity to cancer cells [5-6]. Curcumin (diferuloylmethane, a polyphenol) is an active principle of the perennial herb Curcuma longa (commonly known as turmeric). Curcumin, a naturally occurring polyphenolic phytoconstituent which presents anticancer, antioxidant, anti-inflammatory, hyperlipidemic, antibacterial, wound healing and hepatoprotective activities.

The therapeutic efficacy of curcumin is limited due to its poor aqueous solubility and extensive first pass metabolism. Additionally, Ag and silver nanoparticles are currently used to control bacterial growth in a variety of applications, including dental work, catheters, and burn wounds. Thus, the present study focused on the preparation and evaluation of nanocomposite films made with chitosan-PAA copolymer and silver nanoparticle. Moreover, curcumin was loaded into the silver chitosan-PAA nanocomposite for improving anti-bacterial properties.

II. MATERIALS AND METHODS

A. Materials

Chitosan is extracted from crab shells which was prepared from our lab. The degree of deacetylation of the chitosan was approximately 85%. Curcumin was purchased from Vietnam Institute Dietary of Supplements (Hanoi, Vietnam). Acrylic acid and acetic acid were purchased from Sigma-Aldrich Chemical Co (USA). All chemicals used in this study were of reagent grade.

B. Preparation of Silver-chitosan-PAA Nanocomposites

Chitosan solution (1 and 2%) was prepared by adding the predetermined amount of chitosan to acrylic acid 1%. A clear solution was obtained. The temperature of solution was adjusted to 60 °C and then desired amount of initiators (ammonium persulfate) was added to the solution. The reaction mixture was allowed to react for 90 min, then curcumin 10% in acetone and silver nitrate solution were added. The yellow coloured solution started turning to red, then brown and brownish indicating the formation of silver nanoparticles. The solutions are then poured into Teflon covered glass plates and dried. The final products were named as CS1-PAA-Ag0.5-Cur5 and CS2-PAA-Ag0.5-Cur5 for chitosan 1% and chitosan 2%, repectively.

C. Characterizations

The FT-IR spectras were measured by FT-IR spectrometer (Bruker, Tensor 27) in the range from 4000 to 500 cm⁻¹. Morphology of the sponges was observed by SEM through a JEOL JSM 5500 (Tokyo, Japan).

© Springer International Publishing Switzerland 2015
V. Van Toi and T.H. Lien Phuong (eds.), *5th International Conference on Biomedical Engineering in Vietnam*,
IFMBE Proceedings 46, DOI: 10.1007/978-3-319-11776-8_15

D. Swelling Properties

The composite sponges (3×3 cm) were dispersed in 20 mL phosphate buffer solution, pH 7.4 at 37 oC. The wet weight of the swollen sponge was measured immediately after gently blotting with filter paper to remove absorbed water on surface, followed by lyophilization and reweighing. The water content of the sponge was calculated as follows: $w_t = \frac{m_t - m_o}{m_t} \times 100\%$

Where w_t is the percentage water adsorption of nanocomposite. m_t and m_0 represent the weight of the nanocomposite at time t and initial weight of the nancomposite, respectively.

E. Antimicrobial and Anti-fungal Activity

The measurements of the antimicrobial activities of individual chitosan and nanocomposite films were conducted by agar diffusion method using *Escheriachia coli* and *Staphylococus aureus*. The films are cut into a disc shape with 6 mm diameter and placed on different cultured agar plates. Petriplates were incubated for 24 h at 37°C. The zone of inhibition was determined. This was done in triplicate with each film for each organism and an average diameter of zone of inhibition was noted.

III. RESULTS AND DISCUSSIONS

A. Characterization of Nanocomposite Films

The nanocomposite films having nano-silver particle (AgNPs) and curcumin that are to be used for wound dressing purpose, these films should exhibit good water absorption abilities and anti-bacterial properties. The chitosan-PAA silver nanoparticle films are prepared via simple process. The silver nitrate is added into the chitosan-PAA copolymer solution and the solution is irradiated by sunlight thereby reducing the silver ions into silver nanoparticles. The curcumin loaded nancomposite films are developed by adding the curcumin solution in nano-silver/chitosan-PAA solution.

To examine the chemical structures of nanocomposite films prepared under different feeding ratios, the prepared films were examined by FTIR spectra. As shown in Fig.1, it can be observed the characteristic absorption peak at 3449 cm^{-1}, assigned to stretching vibration of NH_2 and OH groups. The –CH– stretching vibration of the chitosan and PAA are at 2923–2856 cm^{-1}, while peaks of chitosan at 1650 and 1560 cm^{-1} correspond to the amide carbonyl group (Amide I) and the bending frequency of the amide N-H group (Amide II), respectively (Fig.1). The FT-IR of nanocomposite film exhibited a mixture of characteristic absorptions of chitosan and the carboxylic acid groups of PAA. The C=O peaks of PAA was observed at 1712 cm^{-1}

for all composite films. In addition, the stretching vibration at 3449 corresponding to OH/NH_2 groups has shifted to 3432 cm^{-1}, indicating that the silver particles are bounded to the functional groups present both in chitosan and PAA [6].

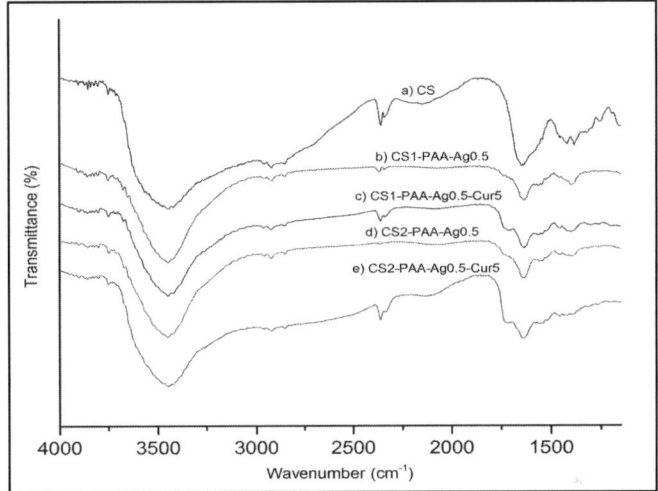

Fig.1 FT-IR spectra of prepared nancomposite films

The results of SEM for curcumin-loaded film and without curcumin-loaded film were shown in Fig.2. The results indicate that the curcumin and silver nanoparticles loaded chitosan-PAA films. The film has exhibited a dense and uniform structure. Additionally, the silver nano particle can be seen in both films. Whereas only curcumin-loaded film has shown the presence of defined curcumin particle in the film.

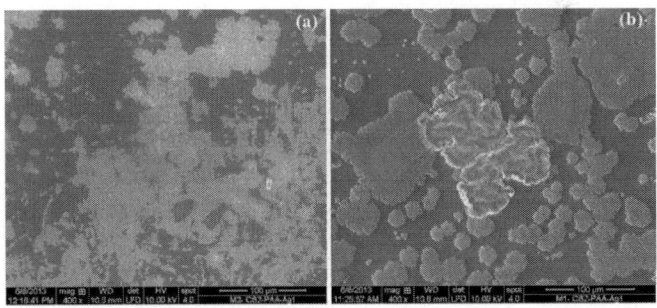

Fig. 2 SEM image of nancomposite film (a) without curcumin and (b) with curcumin

B. Swelling Properties

Water absorption capacity is an important parameter for biological applications and wound healing. It presents the capacity of films to absorb wound exudates. As the chitosan contents of prepared film were increased, the water adsorption capacities were decreased. This was explained by the swelling properties of chitosan. As a result, the water uptake capacity was 88% and 82% for ratio of chitosan 1 and 2%, respectively (Fig.3). The swelling capacity of prepared films increased with increasing of contact time and reached equilibrium at 40 min for both ratios.

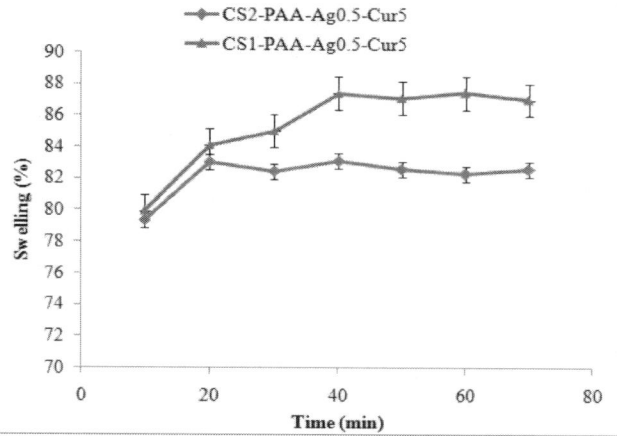

Fig. 3 The swelling capacity of prepared films

C. Curcumin Release

The *in vitro* release profile of curcumin-loaded films was conducted in phosphate buffer solution. The average percent release of curcumin was approximately 59% and 23% for chitosan 1% and 2%, respectively. The curcumin release increased with increase of chitosan content (Fig.4).

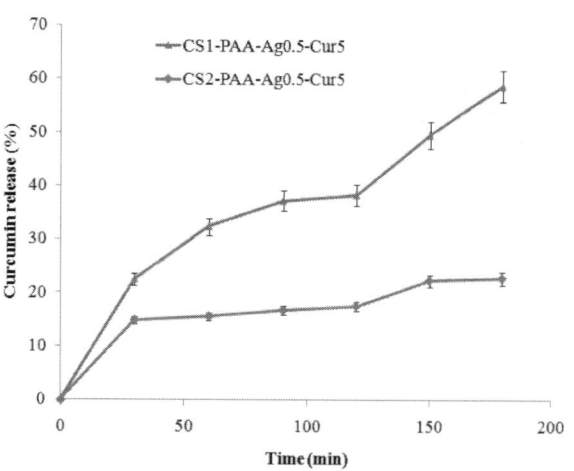

Fig. 4 The curcumin release profile of prepared films

D. Antimicrobial Activity

From the results of antibacterial activity studies of prepared films against Escherichia coli (E. coli) and Staphylococcus, it can be seen that the inhibition zone is larger in curcumin-loaded films than that of without curcumin. It can be also seen that the higher chitosan films exhibited a higher inhibition zone than those of lower chitosan content (Fig.5, 6).

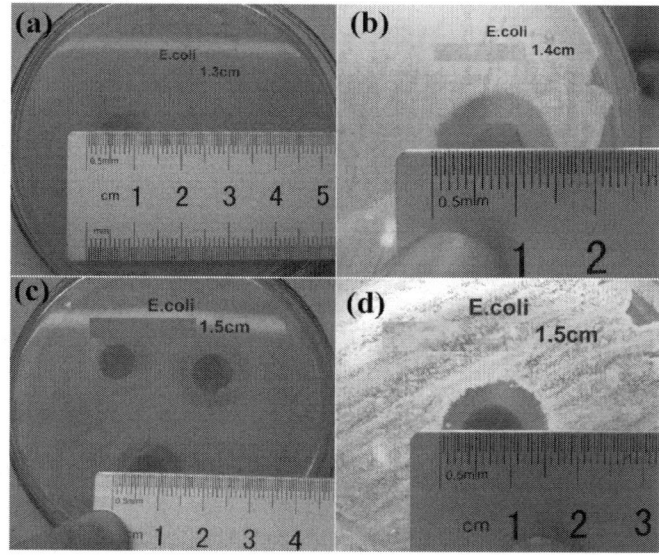

Fig. 5 Antibacterial activity of (a) CS1-PAA-Ag0.5; (b) CS1-PAA-Ag0.5-Cur5; (c) CS2-PAA-Ag0.5-Cur5 and (d) CS2-PAA-Ag0.5-Cur5 against E.Coli

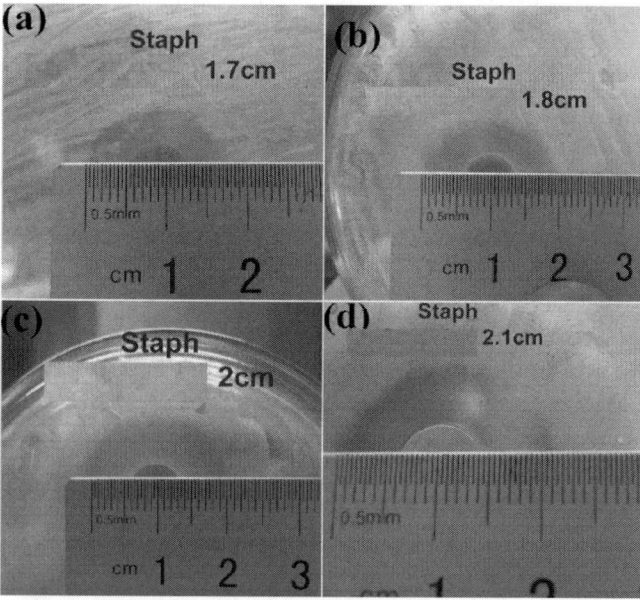

Fig. 6 Antibacterial activity of (a) CS1-PAA-Ag0.5; (b) CS1-PAA-Ag0.5-Cur5; (c) CS2-PAA-Ag0.5-Cur5 and (d) CS2-PAA-Ag0.5-Cur5 against Staphylococcus

IV. CONCLUSIONS

The present work demonstrates a simple method in preparing chitosan-PAA silver nanocomposite films. The structure and morphology were characterized using FT-IR and SEM. The prepared curcumin-loaded silver nanocomposite films have exhibited higher antimicrobial properties than those of chitosan-PAA and silver-chitosan-PAA films.

ACKNOWLEDGMENT

The author would like to thank Industrial University of Industry for facilities.

CONFLICT OF INTEREST

The authors declare that they have no conflict of interest"

REFERENCES

1. Wang B, Chen K, Jiang S, Reincke F, Tong W, Wang D, Gao C (2006) Chitosan-Mediated Synthesis of Gold Nanoparticles on Patterned Poly(dimethylsiloxane) Surfaces, Biomacromolecules, 7: 1203-1209
2. H. Yi, L.-Q. Wu, W. E. Bentley, R. Ghodssi, G. W. Rubloff, J. N. Culver, G. F. Payne, Biofabrication with Chitosan, Biomacromolecules, 6, 2881-2894 (2005).
3. Jayakumar R, Prabaharan M, Sudheesh Kumar P. T, Nair S. V, Tamura H (2011) Biomaterials based on chitin and chitosan in wound dressing applications, Biotechnology Advances, 29: 322-337
4. Muzzarelli R. A. A (2009) Chitins and chitosans for the repair of wounded skin, nerve, cartilage and bone, Carbohydrate Polymers, 76: 167-182
5. Ye Y. Q, Chen F. Y, Wu Q. A, Hu F. Q, Du Y. Z, Yuan H, Yu H. Y (2009) Enhanced cytotoxicity of core modified chitosan based polymeric micelles for doxorubicin delivery, J. Pharm. Sci., 98: 704-712
6. Vimala K et al. (2011) Fabrication of Curcumin Encapsulated Chitosan-PVA Silver Nanocomposite Films for Improved Antimicrobial Activity. : J Biomater Nanobiotechnol 2: 55-64

Author: Dr. Nguyen Van Cuong
Institute: Department of Chemical Engineering, Industrial University of Ho Chi Minh City
Street: Nguyen Van Bao
City: Ho Chi Minh
Country: Vietnam
Email: nvc@hui.edu.vn

Monitoring through Tissue the De-gelation of Alginate Gels by Different De-gelling Agents

K.V.T. Nguyen and J.N. Anker

Department of Chemistry, Center for Optical Materials Science and Engineering Technology (COMSET), SC
BioCRAFT and Environmental Toxicology Program, Clemson University, Clemson, SC 29634, USA

Abstract— **Alginate gels are widely used for drug delivery and implanted devices. The rate at which these gels break down is important for controlling drug release. However, it is challenging to monitor the gel through tissue due to optical scattering and tissue autofluorescence. Herein we describe a method to detect through tissue the gelation and de-gelation processes of alginate gel using magnetically modulated optical nanoprobes (MagMOONs). The MagMOONs are fluorescent magnetic microspheres coated with a thin layer of opaque metal on one hemisphere. The metal layer prevents excitation and emission light from passing through one side of the MagMOONs, which creates an orientation-dependent fluorescence intensity. These magnetic particles also align in an external magnetic field and give blinking signals when they rotate to follow the magnetic field modulation. The blinking signals from these MagMOONs are distinguished from background autofluorescence and can be tracked on a single particle level in the absence of tissue, or by averaging an ensemble of particles through tissue. When these MagMOONs are dispersed in alginate gel, they become sensors for the detection of either the gelation with the addition of a multivalent cation (Ca^{2+}, Cu^{2+}) or the de-gelation as de-gelling agents like ammonium chloride, sodium citrate, or alginate lyase are added. Herein the release of copper ion as an antimicrobial from the gel can be tracked. Our method also can potentially be applied to detect bacterial biofilm formation and other biosensors and drug delivery systems based on enzyme-catalyzed breakdown of gel components.**

Keywords— **MagMOON, de-gelation, tissue, alginate lyase, antimicrobial.**

I. INTRODUCTION

Alginate gels are natural polysaccharides widely used for implanted devices and especially drugs delivery [1-3] due to their unique properties such as being biodegradable, injectable and chemically modifiable [2,4]. These gels can be the localized depots in tissue for proteins, chemical drugs as well as other small molecules; and the degradation of the gels will release them [4-8]. The gels also compose in wound dressings or function as the platform for cell culture [9,10]. Recently, they have a growing application in tissue regeneration by carrying and delivering proteins and cells that promote bone, muscle, cartilage and blood vessels formation [11-14].

In situ monitoring the rate at which alginate gels carrier breaks down is important for controlling drug release *in vivo*. In this study, we developed a sensor to detect the gelation and de-gelation of alginate gels based on the modulation of magnetically modulated optical nanoprobes (MagMOONs). These MagMOONs are magnetic fluorescent particles coated with a thin layer of metal on one hemisphere and therefore are optically asymmetric. The motion of these MagMOONs can be tracked optically on a single particle level (provided that the particles can be resolved, i.e. tissue does not scatter the light) [15-19]. Measuring the motion of fluorescent particles through tissue, however, is challenging for these reasons: a) the excitation light and probe fluorescence is attenuated by the tissue; b) the tissue autofluorescence can obscure the probe signal from the probes; c) tissue scattering blurs the probe signal with a point-spread-function approximately equal to the tissue depth [20,21]; d) tissue scattering also reduces contrast between "bright" and "dark" MagMOON orientations as scattering scrambles the direction of excitation and emission light [22]. Fortunately, the difficulties can be overcome by choosing the proper wavelength of the fluorescence excitation and emission as well as particle concentration. We can measure the intensity change from an ensemble of particles that are driven to rotate and blink together. In this study, a 50 mW 514 nm excitation laser is found to be sufficient to detect MagMOONs at 2.5 to 5 x 10^4 particles/mL or ca. 10 to 100 ng/mL range concentration through 4 mm of tissue. It is expected that larger depths will be attainable for higher concentrations of red or near-infrared-exciting and emitting MagMOONs.

In brief, we aim to create a simple MagMOON-based sensor film to detect gelation and de-gelation of alginate gels. The rate of this process can indicate the activity of proteases degrading the alginate. De-gelation can also indicate the release of contents in the gel, for example, antimicrobial copper ion release in wound dressings. The technique we use in this study can also be applied in other biosensors and drug delivery systems based on enzyme-catalyzed breakdown of gel components.

© Springer International Publishing Switzerland 2015
V. Van Toi and T.H. Lien Phuong (eds.), *5th International Conference on Biomedical Engineering in Vietnam,*
IFMBE Proceedings 46, DOI: 10.1007/978-3-319-11776-8_16

II. MATERIALS AND METHODS

A. MagMOON Fabrication

MagMOONs were made by hemispherically coating 4.8 μm fluorescent Nile Red carboxyl ferromagnetic particles (Spherotech Inc., Spherotech, Lake Forest, IL) with 70 nm thickness of aluminum using thermal vapor deposition, as described in previous papers [18,23-25]. The hemispherically-coated MagMOONs were magnetized by drying the particles onto a glass slide and placing the slide in a strong magnetic field after vapor deposition, so the magnetization orientation of all MagMOONs are the same; ensuring particles are all bright or dim together during modulation (see Fig. 1a-c).

The MagMOON working principle is illustrated in Fig. 1d. The MagMOON appear to blink, "ON" when the light comes to the non-metal-coated hemisphere, and "OFF" as the light faces the metal-coated side, as they rotate in response to rotating magnetic fields, enabling separation of the particles from un-modulated backgrounds [23].

B. Chemicals and Experimental Setup

Alginate lyase from *Sphingobacterium multivorum*, powder, > 10,000 units/g solid was purchased from Sigma Aldrich (Sigma Aldrich, St. Louis, MO) and freshly made into 4 mg/mL solution in distilled water. NH_4Cl, $CaCl_2$, $CuSO_4$, and sodium citrate were purchased from Acros Organics (Acros Organics, Morris Plains, NJ) and were made into 2 M, 0.1 M, 10 mM, and 0.4 M respectively. Sodium alginate was purchased from Alfa Aesar (Ward Hill, MA) and was freshly dissolved in distilled water to make 10 mg/mL solution.

Calcium/copper alginate gels were made by the addition of 20 μL $CaCl_2$ or $CuSO_4$ solution to 10 μL of the MagMOONs – sodium alginate mixture on a 25 mm-coverslip that was previously cleaned by plasma-etching (Harrick Plasma, Ithaca, NY) The gel was left dry (about 30 minutes) before gently washing with distilled water and be ready for imaging. For the experiment on de-gelation by NH_4Cl or citrate 400 μL distilled water was added to the chamber containing the prepared coverslip with MagMOONs-trapped calcium/copper alginate gel. During the acquisition, an additional 400 μL NH_4Cl 2 M, sodium citrate 0.4 M, or alginate lyase 4 mg/mL was added by pipette to cause de-gelation. To detect the effect of alginate lyase on calcium alginate matrix through tissue, the prepared calcium alginate thin film coverslip was placed in a chamber embedded with square pieces of chicken breast, 2.5 x 2.5 inches, wrapped in clear plastic. The top slice of chicken breast was about 1 cm thick while the bottom slice was varied from 1 mm to 6 mm.

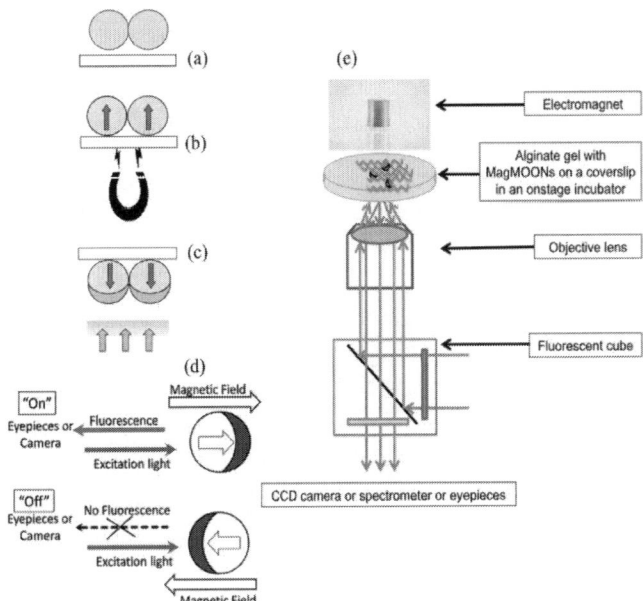

Fig. 1 MagMOON fabrication, MagMOON working principle and experimental setup. (a)-(c) MagMOON fabrication: (a) 4.8 μm fluorescent ferromagnetic microspheres deposited on a glass coverslip. (b) Microspheres magnetization. (c) Aluminum metal vapor deposited onto one hemisphere of the microspheres. (d) MagMOON working priciple: MagMOONs blink when they rotate in response to rotating magnetic field. (e) Schematic of fluorescence micoroscopy setup

To modulate the MagMOON orientation and cause them to blink, the current through an electromagnet was controlled to generate a magnetic field of ~ 0.5 mT at the center of the stage. The current flowing through the electromagnet was programed to alternate directions and switch the magnetic field polarity from North-facing to South-facing every 5 seconds (for the gelation by calcium, de-gelation by ammonium and citrate) or 2 seconds (for all other experiments).

For the single particle tracking experiment, a series of images were taken every 1 second using blue light excitation fluorescence microscopy with Leica IM6000 microscope (Leica Microsystems, Bannockburn IL) and a 10x 0.3 NA objective lens. Fig. 1e shows a microscopy scheme. The images were collected using ORCA- Flash 2.8 CMOS camera and HCImage software (Hamamatsu, Bridgewater, NJ). The fluorescent intensity was exported and analyzed using a single particle tracking script written in Matlab. For the acquisition through tissue, a 514 nm laser was used at a power of ~ 50 mW as the excitation source for a better penetration through tissue than wide field. The fluorescence emission signal was collected using DNS 300 spectrometer (DeltaNu, Laramie, WY) equipped with an Andor DU420A-BV CCD camera and Andor software

(Andor Technology, South Windsor CT) with an acquired rate of 0.2 seconds per spectrum.

III. RESULTS AND DISCUSSION

A. Single MagMOON Tracking to Monitor Calcium Alginate Gelation and De-gelation

Gelation: Alginate composes of polymer chains of vary arrangements of α-L-guluronate (G), and β-D-mannuronate (M) monomers. In sodium alginate, –COOH groups are partly replaced by –COONa [26-28]. Alginate gels form in the presence of divalent ions M^{2+} such as Ca^{2+}, Mg^{2+}, Cu^{2+}, etc. The gel structure is maintained by the coordination of M^{2+} with oxygen atoms in a cavity created by a pair of guluronate sequences along alginate chains. The gel is first formed in egg-box-liked dimer, and then laterally associated egg-box multimer [29].

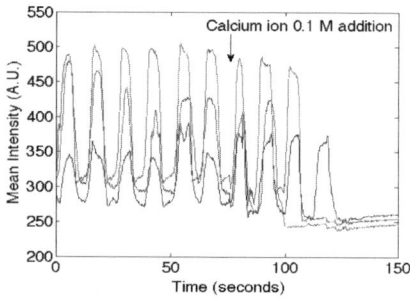

Fig. 2 Plot of three single MagMOONs intensity modulation. MagMOONs stop blinking about 30 to 50 seconds after $CaCl_2$ 0.1 M addition.

In this experiment, we added 200 μL $CaCl_2$ 0.2 M to 200 μL MagMOON mixed alginate 20 mg/mL. The MagMOONs were modulated by ~ 0.5 mT external magnetic field. Fig. 2 shows the intensity change with time of 3 single MagMOONs. These particles stopped blinking about 30 s to 50 s following the Ca^{2+} addition, and about 10 s apart from one to another. The immediate cessation of MagMOONs modulation indicated the prompt gelation and the time it took was mainly for Ca^{2+} diffusion to the view. On another note, the different time the particles stopped represented different regions that Ca^{2+} reached. Therefore, it can be used to monitor the gels property locally.

De-gelation by Ion Replacement: Unlike divalent cations, monovalent ions such as Na^+, K^+, Li^+ or NH_4^+ cannot crosslink guluronate sequences [30]. Addition of excess ammonium ions replaced the calcium ions, broke the cross-linked gels and therefore caused de-gelation. Fig. 3 shows the de-gelation of a thin film calcium alginate gel layer after 1M NH_4Cl addition. At the beginning of the experiment, when MagMOON were fixed in the gel, its fluorescent

intensity remained unchanged with time. After about 5 minutes, when the ion replacement had occurred enough to deform the gel structure, MagMOONs started to blink in response to the magnetic field as they are released from the matrix.

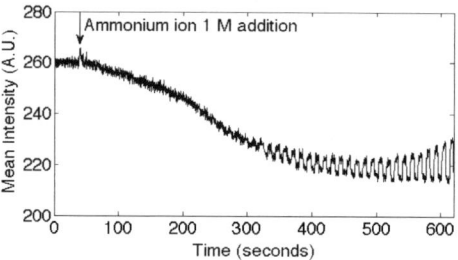

Fig. 3 Tracking MagMOON intensity to detect alginate de-gelation by ion replacement. Modulation is seen after about 300 seconds

De-gelation by Citrate Chelation: Chelating agents like EDTA, DMSA, and penicillamine have been used to remove heavy metals such as lead, mercury, arsenic and copper because they have high affinity with these metals [31,32]. Similarly, we used citrate as the chelator for the removal of copper from its alginate gel. Monitoring the release of Cu^{2+} from alginate gel can be useful as copper and silver are used as the antimicrobials in wound dressing [33,34]. Therefore, in this experiment, we added Cu^{2+} to form copper alginate gel and then use sodium citrate to cause de-gelation to demonstrate the ability to track copper release using MagMOONs. Fig. 4 presents the intensity modulation of 4 single MagMOONs after citrate 0.2 M addition. It is interesting to note that the change in viscosity of the MagMOONs surrounding presented in the shape of the modulation curve. As can be seen with all 4 particles, at first the modulation function is more round and later on it become square-like, indicating the quick drop in viscosity as more bonds around the particles are cleaved. The magnitude of the modulation also increases implying the particles become freer to rotate.

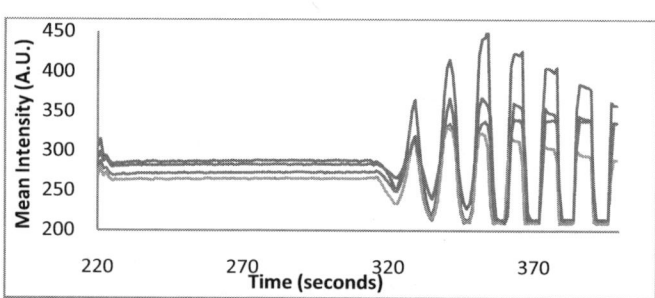

Fig. 4 Four MagMOONs start blinking after about 100 seconds following 0.2 M sodium citrate addition marked by the peaks around 220th second

The above tests demonstrate a simple way to detect the de-gelation process using MagMOON modulation. It also suggests that MagMOONs modulation can provide worthy information about the substrate properties. We moved a step further by testing the detection ability through different thicknesses of chicken breast. Although the position of individual particles can hardly be detected through thick tissue, the blinking signal can.

B. Through Tissue Detection of De-gelation by Alginate Lyase

Alginate lyase is an enzyme which catalyzes the degradation of alginate by cleaving the 1-4 O linkage between monomers [28,35]. In the next step, we used alginate lyase to cleave the MagMOON-trapped calcium alginate gel. As the chain is cleaved, MagMOONs were released and free to rotate with the magnetic field. Beside the de-gelation detection purpose, this test can also be seen as an approach for alginate lyase activity detection.

To determine if the blinking MagMOON signal could be detected through tissue, we placed a coverslip with calcium alginate film of MagMOONs in a temperature-controlled chamber between two slices of chicken breast, 1 cm thick on top and various thicknesses of 1 mm, 1.5 mm, 2.5 mm, 4 mm, and 6 mm at the bottom. During the acquisition, the temperature was maintained at 37 °C for fast and efficient enzymatic activity of alginate lyase [36]. Fig. 5 shows the set up schematic (a) and picture (b).

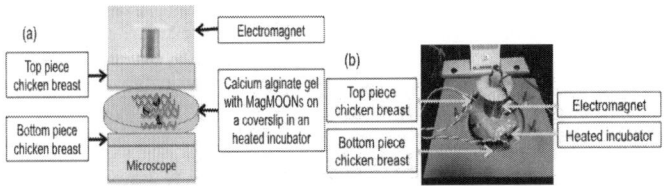

Fig. 5 Fluorescent imaging through turbid chicken breast tissue. (a) Schematic of the set up for magnetic modulation and imaging through tissue. The MagMOONs are trapped in alginate calcium gel on a coverslip in an on-stage incubator chamber. This chamber is embedded in the chicken breast tissue. An electromagnet was used to modulate the MagMOONs. (b) Photo of the experimental setup

The modulated and background-corrected fluorescent signals through 1 mm, 1.5 mm, 2.5 mm, 4 mm and 6 mm are presented in Fig. 6. Although the amplitude of modulation decreases dramatically with the increase of the tissue thickness, the modulation is clear through up to 4 mm.

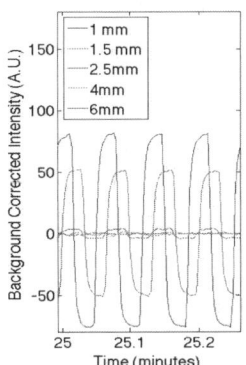

Fig. 6 Background-corrected magnetically modulated fluorescence signal through 1 mm, 1.5 mm, 2.5 mm, 4 mm and 6 mm chicken breast when alginate lyase cleaves calcium alginate gel matrix. Modulation amplitude reduces quickly as the tissue thickness increases but can be clearly seen

This part of the study confirms the modulation occurs after alginate lyase addition and MagMOONs signal can be tracked through tissue. The modulation signal reduces significantly with tissue depth and can be seen through 4 mm.

IV. CONCLUSIONS

We applied MagMOONs to detect the gelation and de-gelation using several different de-gelling approaches. We also showed MagMOONs overcame tissue scattering when imaging through tissue. This suggests a simple yet novel means to detect alginate lyase activity and copper release. Future work will be *in situ* non-invasive biophysical and chemical sensing for proteases in body. An endoscope will be used as the detector to collect fluorescent signals from MagMOONs through tissue.

ACKNOWLEDGMENT

This work was supported by NIH NIBIB grant, award number 1R15EB014560-01A1; the South Carolina Research Authority NSF grant 2002-593TO#0056, and a Vietnam Education Foundation (VEF) fellowship to KVTN. We thank Spherotech Inc. for generously donating the magnetic fluorescent particles.

CONFLICT OF INTEREST

The authors declare that they have no conflict of interest.

REFERENCES

1. Maslov, M. Y., Edelman, E. R., Wei, A. E., Pezone, M. J. & Lovich, M. A. (2013) High concentrations of drug in target tissues following local controlled release are utilized for both drug distribution and biologic effect: An example with epicardial inotropic drug delivery. Journal of Controlled Release 171: 201-207

2. Li, Y., Rodrigues, J. & Tomas, H. (2012) Injectable and biodegradable hydrogels: gelation, biodegradation and biomedical applications. Chemical Society Reviews 41: 2193-2221

3. Lee, K. Y. & Mooney, D. J. (2012) Alginate: properties and biomedical applications. Progress in polymer science 37: 106-126

4. Tønnesen, H. H. & Karlsen, J. (2002) Alginate in drug delivery systems. Drug development and industrial pharmacy 28: 621-630

5. Brulé, S., Levy, M., Wilhelm, C., Letourneur, D., Gazeau, F. et al. (2011) Doxorubicin release triggered by alginate embedded magnetic nanoheaters: a combined therapy. Advanced Materials 23: 787-790

6. Wells, L. A. & Sheardown, H. (2007) Extended release of high pI proteins from alginate microspheres via a novel encapsulation technique. European journal of pharmaceutics and biopharmaceutics 65: 329-335

7. Zhang, X., Hui, Z., Wan, D., Huang, H., Huang, J. et al. (2010) Alginate microsphere filled with carbon nanotube as drug carrier. International journal of biological macromolecules 47: 389-395

8. Chang, C.-H., Lin, Y.-H., Yeh, C.-L., Chen, Y.-C., Chiou, S.-F. et al. (2009) Nanoparticles incorporated in pH-sensitive hydrogels as amoxicillin delivery for eradication of Helicobacter pylori. Biomacromolecules 11: 133-142

9. Queen, D., Orsted, H., Sanada, H. & Sussman, G. (2004) A dressing history. International wound journal 1: 59-77

10. Rowley, J. A., Sun, Z., Goldman, D. & Mooney, D. J. (2002) Biomaterials to spatially regulate cell fate. Advanced Materials 14: 886-889

11. Krebs, M. D., Salter, E., Chen, E., Sutter, K. A. & Alsberg, E. (2010) Calcium phosphate-DNA nanoparticle gene delivery from alginate hydrogels induces in vivo osteogenesis. Journal of Biomedical Materials Research Part A 92: 1131-1138

12. Levenberg, S., Rouwkema, J., Macdonald, M., Garfein, E. S., Kohane, D. S. et al. (2005) Engineering vascularized skeletal muscle tissue. Nature biotechnology 23: 879-884

13. Thornton, A. J., Alsberg, E., Albertelli, M. & Mooney, D. J. (2004) Shape-defining scaffolds for minimally invasive tissue engineering. Transplantation 77: 1798-1803

14. Gu, F., Amsden, B. & Neufeld, R. (2004) Sustained delivery of vascular endothelial growth factor with alginate beads. Journal of Controlled Release 96: 463-472

15. Crick, F. H. C. & Hughes, A. F. W. (1950) The physical properties of cytoplasm: A study by means of the magnetic particle method Part I. Experimental. Experimental Cell Research 1: 37-80

16. McNaughton, B. H., Agayan, R. R., Clarke, R., Smith, R. G. & Kopelman, R. (2007) Single bacterial cell detection with nonlinear rotational frequency shifts of driven magnetic microspheres. Applied Physics Letters 91: 224105-224105

17. Sinn, I., Albertson, T., Kinnunen, P., Breslauer, D. N., McNaughton, B. H. et al. (2012) Asynchronous Magnetic Bead Rotation Microviscometer for Rapid, Sensitive, and Label-Free Studies of Bacterial Growth and Drug Sensitivity. Analytical chemistry 84: 5250-5256

18. Behrend, C. J., Anker, J. N., McNaughton, B. H., Brasuel, M., Philbert, M. A. et al. (2004) Metal-capped Brownian and magnetically modulated optical nanoprobes (MOONs): Micromechanics in chemical and biological microenvironments. The Journal of Physical Chemistry B 108: 10408-10414

19. Wang, G., Sun, W., Luo, Y. & Fang, N. (2010) Resolving rotational motions of nano-objects in engineered environments and live cells with gold nanorods and differential interference contrast microscopy. J Am Chem Soc 132: 16417-16422

20. Shimizu, K., Tochio, K. & Kato, Y. (2005) Improvement of transcutaneous fluorescent images with a depth-dependent point-spread function. Applied optics 44: 2154-2161

21. Ntziachristos, V. (2010) Going deeper than microscopy: the optical imaging frontier in biology. Nature methods 7: 603-614

22. Yang, Z., Nguyen, K. T., Chen, H. & Anker, J. N. (2010) Magnetically modulated fluorescent probes in turbid media. J arXiv preprint arXiv:1007.2863

23. Anker, J. N. & Kopelman, R. (2003) Magnetically modulated optical nanoprobes. Applied physics letters 82: 1102

24. Behrend, C. J., Anker, J. N. & Kopelman, R. (2004) Brownian modulated optical nanoprobes. Applied physics letters 84: 154

25. Behrend, C. J., Anker, J. N., McNaughton, B. H. & Kopelman, R. (2005) Microrheology with modulated optical nanoprobes (MOONs). Journal of magnetism and magnetic materials 293: 663-670

26. Gacesa, P. (1988) Alginates. Carbohydrate polymers 8: 161-182

27. Grasdalen, H. (1983) High-field, 1HNMR spectroscopy of alginate: sequential structure and linkage conformations. Carbohydrate Research 118: 255-260

28. Haug, A., Larsen, B. J. & Smidsrod, O. (1967) Studies on the sequence of uronic acid residues in alginic acid. Acta Chem Scand 21: 691-704

29. Fang, Y., Al-Assaf, S., Phillips, G. O., Nishinari, K., Funami, T. et al. (2007) Multiple steps and critical behaviors of the binding of calcium to alginate. The Journal of Physical Chemistry B 111: 2456-2462

30. Kevekordes, K. (1996) Using light scattering measurements to study the effects of monovalent and divalent cations on alginate aggregates. Journal of experimental botany 47: 677-682

31. Goyer, R. A., Cherian, M. G., Jones, M. M. & Reigart, J. R. (1995) Role of chelating agents for prevention, intervention, and treatment of exposures to toxic metals. Environmental health perspectives 103: 1048

32. Horowitz, S. (2014) Chelation Therapy: A Review of Applications and Empirical Evidence. Archives of Agronomy and Soil Science

33. Borkow, G., Okon-Levy, N. & Gabbay, J. (2010) Copper oxide impregnated wound dressing: biocidal and safety studies. Wounds-a compendium of clinical research and practice 22: 301-310

34. Rai, M., Yadav, A. & Gade, A. (2009) Silver nanoparticles as a new generation of antimicrobials. Biotechnology advances 27: 76-83

35. Wong, T. Y., Preston, L. A. & Schiller, N. L. (2000) Alginate lyase: review of major sources and enzyme characteristics, structure-function analysis, biological roles, and applications. Annual Reviews in Microbiology 54: 289-340

36. Takeuchi, T., Nibu, Y., Murata, K., Yoshida, S. & Kusakabe, I. (1997) Characterization of a novel alginate lyase from Flavobacterium multivolum K-11. Food Science and Technology International, Tokyo 3: 388-392

Use of Super Paramagnetic Iron-Oxide Nanoparticles in the Treatment of Atherosclerosis

S. Chandramouli, S. Sanjana, and S. Swathi

SSN College of Engineering, Anna University, Chennai, India

Abstract— *Atherosclerosis* **is a serious condition where arteries become clogged up by fatty substances, such as cholesterol. This results in the formation of plaques or atheromas, a condition where the arteries become narrowed and hardened due to an excessive build up of plaque around the artery wall. This disease disrupts the flow of blood around the body, posing serious cardiovascular complications leading to life-threatening situations. Nanoparticles have the potential to be used in many different biological and medical applications as diagnostic or therapeutic tools. Super paramagnetic iron-oxide nanoparticles (SPIONs) are widely used in various biomedical applications, such as targeted drug delivery system and magnetic resonance imaging. Utilization of biocompatible magnetic nanoparticles such as SPIONs in the removal of atherosclerotic plaque through their unique magnetic properties directing them to the site is novel. Here we hypothesize that SPIONs, can be effective in the treatment of atherosclerosis because they can be targeted in the site allowing the particles to spin backwards and forward at high speed generating heat using an external a.c magnetic field, which will reduce the hardness of the plaque, due to temporary thermal expansion. By magnetically manipulating the SPION, the plaque can be abraded**.

I. INTRODUCTION

The buildup of plaque in artery walls in atherosclerosis leads to major cardiovascular diseases. The artery is said to be damaged when low density lipoprotein accumulates in the artery wall. This initiates an inflammatory response that is initially responsible for the fatty streak. The progression of the lesion is associated with deposition of cholesterol, platelets, cellular debris and arterial calcification which changes the mechanical characteristics of the artery. This plaque clogs up the artery, disrupting the flow of blood around the body. This potentially causes blood clots that can result in life-threatening conditions such as heart attack, stroke and other cardiovascular diseases. The condition can affect the entire artery tree, but mainly affects the larger high-pressure arteries. Superparamagnetic Iron Oxide Nanoparticles (SPIONs) based on their inducible magnetization, allow them to be directed to a defined location or heated in the presence of an externally applied AC magnetic field. The properties of SPION such as the low toxicity profile, low particle dimension, large surface area, suitable magnetic properties and reactive surface [6] that can be willingly modified with biocompatible coatings, have made them highly useful for MRI contrast enhancement and

in vivo targeted imaging of the atherosclerotic plaque. In cancer treatments, these magnetic nanoparticles , with an applied field of 38 kA/m at 980 kHz, can be used to heat tumors to 60°C in 2 minutes, durably ablating them with millimeter (mm) precision, leaving surrounding tissue intact [4]. Here, we hypothesize that SPIONs, can be effective in the treatment of *atherosclerosis* because they can be targeted in the site and they can be made to spin backwards and forward at high speed, generating heat using an external magnetic field (i.e. MRI) which helps in the abrasion of plaque.

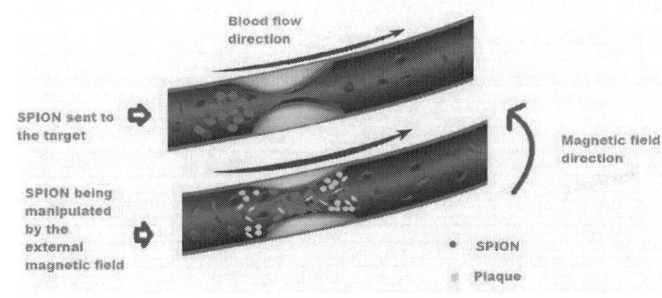

Fig. 1 Schematic representation of the use of SPIONs in the treatment of atherosclerosis.

II. BACKGROUND

A. Superparamagnetic Iron Oxide Nanoparticles

Nanoparticles may provide advanced biomedical research tools based on polymeric or inorganic formulations or a combination of both. They have the potential to be used in many different biological and medical applications as in diagnostic tests assays for early detection of diseases, to serve as tools for noninvasive imaging and drug development, and to be used as targeted drug delivery systems. Based on their widely different magnetic properties in contrast to other nanoparticles, we can allow them to be directed to a defined location or heated in the presence of an externally applied AC magnetic field. This characteristic makes them attractive for many applications, ranging from various separation techniques and contrast enhancing agents for MRI to drug delivery systems, magnetic hyperthermia. Research with SPIONs has already demonstrated that these particles have the potential to become an important tool for

© Springer International Publishing Switzerland 2015
V. Van Toi and T.H. Lien Phuong (eds.), *5th International Conference on Biomedical Engineering in Vietnam,*
IFMBE Proceedings 46, DOI: 10.1007/978-3-319-11776-8_17

enhancing magnetic resonance contrast. Due to specific adsorption they can possibly help physicians to identify dangerous arteriosclerotic plaques by MRI.

B. *Hyperthermia Treatment*

Magnetic hyperthermia is one of the suggested methods for the treatment of cancer. In this process the superparamagnetic nanoparticle is sent to the affected site. When an alternating magnetic field of particular amplitude and frequency is applied, the nanoparticles begin to heat up the surrounding tissues. Cancerous cells have less developed neurons and veins when compared to the healthy tissues. The heat dissipation capability of the cancerous tissue is very less when compared to the healthy tissues. The healthy tissues withstand the high temperature, while the cancerous tissues are destroyed by the heat produced. It has been found out that the safest value of temperature that can be produced by hyperthermia is 45degree Celsius above which irreversible cell damage happens, which leads to the destruction of the healthy cells. This method can be used to heat the atherosclerotic plaques.

III. METHODS

A. *Targeting Atherosclerotic Plaque*

Targeted delivery of diagnostic imaging agents and therapeutic drugs to atherosclerotic plaques is an ambitious goal in atherosclerotic research that is coming closer to a reality. Imaging dyes that are targeted to lesions give greater detail and information about the plaques and their morphology. Targeted delivery of currently available therapeutics such as anticoagulants to the atherosclerotic plaques would allow the drug to concentrate in the affected tissue, thereby enabling a lower dose and lowering the risk of side effects. Recent advances in imaging techniques, especially magnetic resonance technology allow imaging of vessels at much higher resolution, even at the sub millimeter level.

B. *Bi--functionality of SPION*

The SPION drug delivery system has the dual-ability of functioning both as an imaging and therapeutic delivery nanosystem. It can be locally targeted to specific sites guided by externally applied magnetic fields. The SPION-DDS enables therapeutic and diagnostic effects and thus allows for treatment follow-up by magnetic resonance imaging.

IV. MAGNETIC HYPERTHERMIA

A. *Power Loss*

In magnetic hyperthermia, temperature rise is mainly due to hysteresis loss, Brownian loss and Neels loss. In case of superparamagnetic nanoparticles, the hysteresis loss is zero. The Neels relaxation occurs due to the reorientation of the magnetic moment in the particle. The nanoparticle rotate randomly in blood, a phenomenon called as the Brownian movement. This movement causes losses when a magnetic field is applied. This is known as the Brownian loss. This power loss produced by the magnetic field is converted as the heat energy. The power loss can be written as

$$P = 2\pi^2 \mu_o \chi_o H \frac{f^2 \tau}{(1 + (2\pi f \tau)2)} \qquad (1)$$

μ_o - permeability of free space or vacuum = $4 \pi \times 10^{-7}$ Vs/Am; χ_0 - static susceptibility ; H - Magnetic field strength (A/m); f - Frequency of the a.c wave (1/s); τ – Total relaxation time (s); τ_b – relaxation time due to Brownian movement (s) ;τ_n – Neels relaxation time (s) ;

The values of the various independent variables are varied and the power loss is found out by substituting the known values which are taken from various literatures and graphs are plotted.

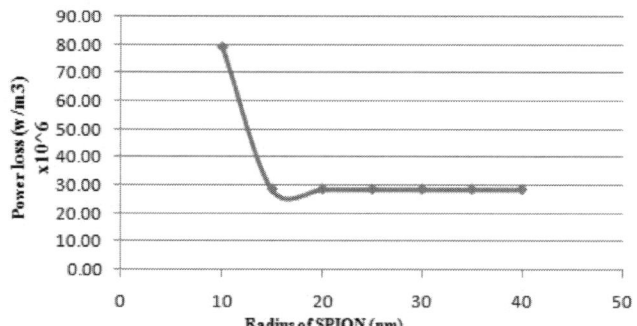

Fig. 2 Radius of SPION vs Power Loss

The graph shows the variation of the power loss with respect to the radius of the SPION, for a constant value of Frequency f=100khz, H=10kA/m, Temperature=310k.The size of the nanoparticle used is chosen in such a way that a higher value of heat is produced.

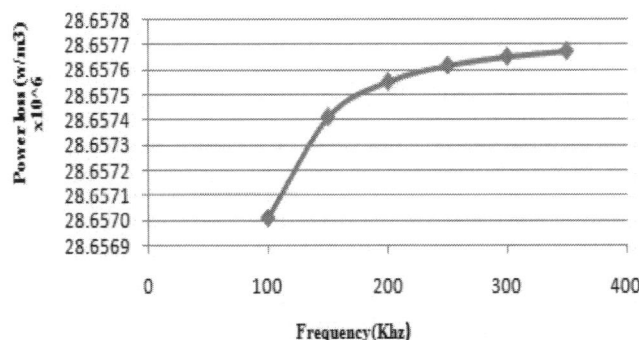

Fig. 3 Frequency of AC Voltage vs Power Loss

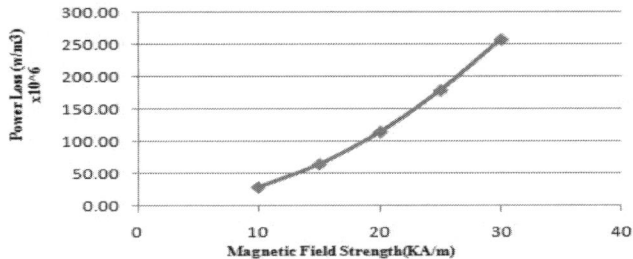

Fig. 4 Magnetic Field Strength vs Power Loss

Both the frequency and the magnetic field strength are chosen in such a way that it produces maximum heat. However the product of H-f should be less than 13.4×10^8 $Am^{-1}s^{-1}$ for safety reasons.

Once the temperature is raised from 37degree Celsius (normal human blood temperature) to 45 degree Celsius, the heating is stopped. Now the viscosity of the blood will be reduced due to the increase in temperature and hence the drag force experienced by the nanoparticle will be reduced. Due to that, the velocity of the nanoparticle can be easily increased.

B. Plaque Abrasion

The heat produced by hyperthermia will cause thermal expansion of the plaque. Since the volume of the plaque increases due to expansion, the density of the plaque and the ultimate tensile strength of the plaque is reduced. Now SPION is controlled with the help of a magnetic field. The magnetic field applied is used to control the velocity of the nanoparticle. Then either by rotation or by reciprocation, the nanoparticle is made to hit the atherosclerotic plaque. By continuously doing this process, the plaque can be removed from the body through abrasion.

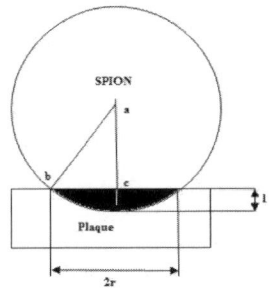

Fig. 5 SPION-Plaque modelling

Volume of the plaque removed is equal to the volume of the hemispherical impact crater of the nanoparticle.

$$V = \pi l^2 \left(\frac{d}{2} - \frac{l}{3}\right) = \frac{\pi i^2 d}{2} \tag{1}$$

$$K.E = \frac{\pi}{12} d^3 \rho_{spion} v^2 \tag{2}$$

Where ρ_{spion} is the density of SPION

On impact the plaque will be subjected to a force F which causes an indentation 'I'

Therefore,

$$F = \sigma_p \pi d I \tag{3}$$

Workdone,

$$w = \sigma_p \pi d I * I \tag{4}$$

The kinetic energy is completely used to abrade the plaque.

On Equating (4) and (2),

$$I = v d \sqrt{\frac{\rho_{spion}}{6\sigma_p}} \tag{5}$$

Where, ρ_{spion} is the density of SPION,'d'-diameter of spion.

Plaque removal rate = volume of plaque removed per cycle $*(\dfrac{Mass\ flow\ rate\ of\ spion}{Mass\ of\ the\ spion})$

On Simplification,

Plaque removal rate is,

$$\frac{M_a v^2}{(2 * \sigma_p)} \tag{6}$$

Where M_a = mass flow rate of spion(kg/s)
V= velocity of SPION.
σ_p = breaking stress of the plaque (179±56 kpa)

The velocity of the nanoparticle can be controlled by the applied magnetic field. By considering the magnetic force, the buoyant force and the Drag force acting on the SPION , the velocity of the SPION is found out as,

$$V = \frac{R^2}{9\mu_o\,\eta}(\chi_p - \chi_f)\nabla B^2 + \frac{2\,R^2}{9\eta}(\rho_{p-}\,\rho_f)\,g \qquad (7)$$

Where   R – radius of the SPION; μ_o- permeability of free space; η – Viscosity of the blood; χ_p _ volume susceptibility of SPION in blood; χ_f _ volume susceptibility of the fluid; ρ_p - Density of the spion;ρ_f- Density of the blood; g- Acceleration due to gravity; R – Reynolds number; ∇- Operator; B – External magnetic field.

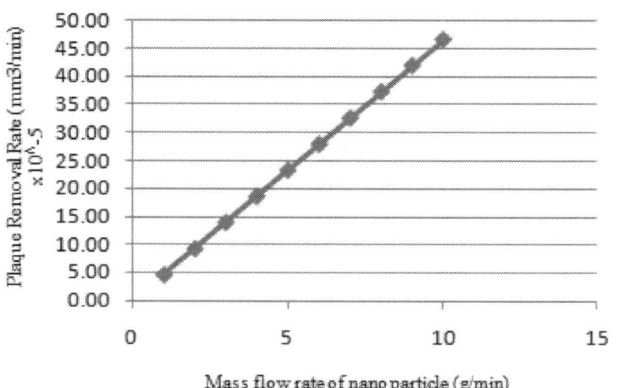

Fig. 6 Mass Flow Rate vs PlaqueRemovalRate

This graph represents the variation of the plaque removal rate due to the change in the mass flow rate of the nanoparticle for a velocity v=1mm/s.

The linear increase in mass flow rate signifies the increase in abrasion rate.

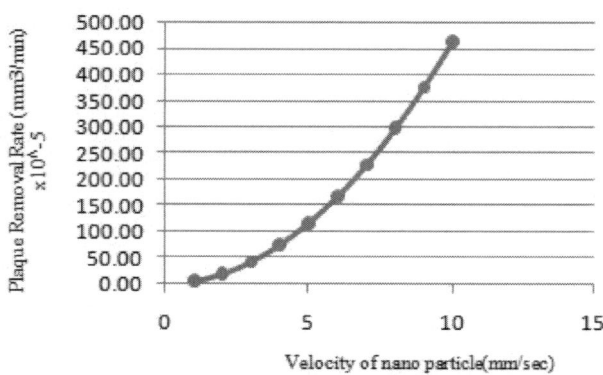

Fig. 7 Velocity of SPION vs PlaqueRemovalRate

This graph represents the variation of the plaque removal rate due to the change in the velocity for a mass flow rate of 1g/min.

V. Conclusions

The proposed method has several advantages when compared to the conventional method of treating atherosclerosis which requires catheterization and takes more recovery time. From the discussion of the above presented results, an optimum value of nanoparticle size and magnetic field strength is chosen in such a way that a large amount of heat is produced for effective reduction of the fracture strength of the plaque. The plaque removal rate increases with increase in velocity and the mass flow rate of the nanoparticle. The mass flow rate and the velocity of the SPION is controlled in such a way that an effective plaque removal rate is obtained.

Acknowledgment

The present work benefitted from the constant support and encouragement of concerned faculties of SSN College of Engineering, Anna University, Chennai, India.

References

1. Reju Thomas[1], In-Kyu Park[2] and Yong Yeon Jeong[1],* Magnetic Iron Oxide Nanoparticles for Multimodal Imaging and Therapy of Cancer ISSN 1422-0067
2. R.Ahmadi1, H.R.Madaah2, Hosseini, Department of material science and Engineering. Sharif university of Technology,Tehran ,Iran. A new approach for calculation of relaxation time and magnetic anisotropy of ferrofluids containing super   paramagnetic nanoparticles.
3. Chouly.C1, Pouliquen.D, Lucet.I, Jeune.JJ, Jallet.P Development of Superparamagnetic Nanoparticles for MRI: Effect of Particle Size, Charge and Surface Nature on Biodistribution
4. Hui S Huang1, James F Hainfeld, Nanoprobes, Yaphank, NY, USA Intravenous magnetic nanoparticle cancer hyperthermia
5. Margarethe Hofmann-Amtenbrink1, Brigitte von Rechenberg2 and Heinrich Hofmann3 Superparamagnetic Nanoparticles for Biomedical Applications ISBN: 978-81-7895-397-7
6. Rupinder K Kanwar, Rajneesh Chaudhary, Takuya Tsuzuki, Emerging Engineered Magnetic Nanoparticulate Probes for Targeted MRI of Atherosclerotic Plaque Macrophages. Nanomedicine, 2012;7(5):735-749.

Effect of pH on the Synthesis of Fucoidan-coated Magnetic Iron Oxide Nanoparticles for Biomedical Applications

Khanh Nghia Tran, Phuong Ha-Lien Tran, Toi Van Vo, and Thao Truong-Dinh Tran*

Biomedical Engineering Department, International University,
Vietnam National University, Ho Chi Minh City, Vietnam

Abstract— **Nano drug delivery system is a highlight among advanced drug delivery systems. They offer many advantages such as improved efficacy, reduced toxicity and improved patient compliance over conventional drug delivery systems. Purpose of this study was to develop new magnetic iron oxide nanoparticles containing fucoidan, a recently promising material discovered from the ocean, for diagnosis and cancer therapy. The nanoparticles were prepared using ultrasonication method. Fucoidan concentration was determined by conductometric titration. Fucoidan was investigated to be attached to iron oxide nanoparticles in the buffers of pH 6, pH 7 and pH 8. Scanning electron microscopy (SEM) was used to characterize surface morphology and particle shape. Fourier transform infrared spectroscopy (FTIR) was also used to investigate the structure of the fucoidan-coated iron oxide nanoparticles. Percentage of fucoidan, which was coated on iron oxide nanoparticles, was 87.7412%, 74.4936% and 63.0362% at pH 7, pH 6 and pH 8, respectively. The SEM image showed the spherical shape and size of fucoidan modified iron oxide nanoparticles less than 100 nm. Moreover, the FTIR spectra indicated the presence of fucoidan on iron oxide nanoparticles. The fucoidan concentration on the surface of iron oxide nanoparticles was depended on the change of pH media. The current nanoparticles can be introduced into further experiments for biomedical applications including therapeutics and diagnostics.**

Keywords— **Fucoidan, magnetic nanoparticles, cancer, biomedical, tumor.**

I. INTRODUCTION

According to estimates from the International Agency for Research on Cancer (IARC), in 2008, there were 12.7 million new cancer cases while 7.6 million People were died by cancer (about 21,000 cancer deaths a day) [1]. World Health Organization predicted that cancer would have replaced ischemic heart disease as the overall leading cause of death worldwide in 2010. Moreover, 21.4 million new cancers case with 13.2 million cancer deaths will occur in 2030[2]. Recently, drugs have been used for cancer therapy with several reports announced their toxicity which cause the kill of not only cancer cells but also normal cells and tissues.

The sulfated polysaccharides (fucoidan), which contain substantial percentages of L-fucose and sulfate ester groups, are constituents of brown seaweed, have a wide variety of biological activities: anticoagulant, antivirus, immunomodulation and antitumor activities [3-5]. A lot of researches about the antitumor activity of fucoidan have been reported: effectively inhibited proliferation and formation of cancer cells *in vitro* [6, 7], and inhibitory activity in tumors growing *in vivo* [8].

Magnetic nanoparticles (MNPs), which are particles with a size range from 10 to 100 nm in diameter [9], are conveniently tool in medicine such as targeted drug delivery[10-12], bimolecular labeling and separation[13, 14], as well as magnetic resonance imaging (MRI) [15-17]. Many different kinds of fabrication techniques to bring MNPs into biological systems such as attached surface ligands [18, 19], silica coating [20], encapsulation with amphiphilic molecules including small surfactants [21], lipids [22] and amphiphilic copolymers [23-25] have been reported.

The aim of this study was to develop a new magnetic iron oxide nanoparticle containing fucoidan for diagnosis and cancer therapy. The nanoparticles were prepared using ultrasonication method as shown in Fig.1.

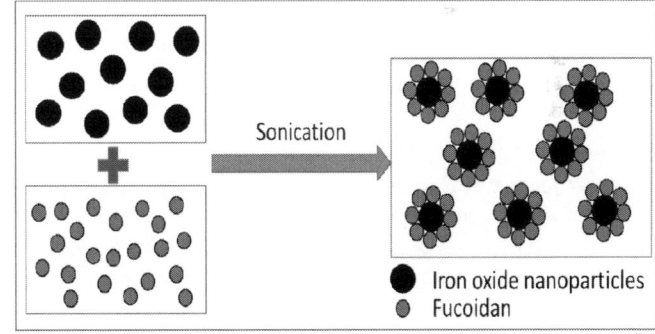

Fig. 1 Schematic illustration of fucoidan-coated iron oxide nanoparticles

II. MATERIAL AND METHODOLOGY

A. Materials

Sodium hydroxide (NaOH) were obtained from Guangzhou Jinhuada Chemical Reagent Co., Ltd. Potassium dihydrogen phosphate (KH_2PO_4) were purchased from

---

* Corresponding author.

© Springer International Publishing Switzerland 2015
V. Van Toi and T.H. Lien Phuong (eds.), *5th International Conference on Biomedical Engineering in Vietnam,*
IFMBE Proceedings 46, DOI: 10.1007/978-3-319-11776-8_18

Xilong Chemical Industry Incorporated Co., Ltd. Barium chloride dehydrate ($BaCl_2.2H_2O$) were obtained from from Xilong Chemical Industry Incorporated Co., Ltd. Fucoidan 80% were supported by Nha Trang Institute of Technology Research and Application and VietNam Fucoidan Joint Stock Company. Iron oxide nanoparticles were obtained from Sigma-Aldrich Corporation.

B. Preparation of pH Buffer Solutions

2.04g of KH_2PO_4 was dissolved in 300ml of distilled water. This solution was separated to 3 beakers with 100ml in each beaker. Solution of NaOH 1M was used to adjust the pH 6, 7 and 8.The pH meter (Jenway 3510 pH meter) was used to measure the value of pH.

C. Preparation of Fucoidan Solution and Magnetic Iron Oxide Nanoparticles Suspension

50mg of fucoidan was dissolved in 50 ml of each pH solution. Also, 100mg of iron oxide nanoparticles was dispersed in 50ml of each pH solution by ultrasonication for 5 min with amplitude of 5.

D. Coating Iron Oxide Nanoparticles with Fucoidan

Fucoidan solution was added to iron oxide nanoparticles suspension and sonicated for 120 min with amplitude of 1 by ultrasonication. Then, the nanoparticles were washed by 50ml distilled water for thrice. The nanoparticles were dried at $40^{\circ}C$ in 6 h. This process was applied to each pH solution. Washing water was used to determine the amount of fucoidan which was coated to iron oxide nanoparticles.

E. Characterization

The structure of the fucoidan-coated iron oxide nanoparticles were determined by Fourier Transfer Infrared Spectrometer (Bruker Vertex 70, Germany). The wavelength was scanned from 4000 to 500 cm^{-1} with a resolution for 2 cm^{-1}. KBr pellets were prepared by gently mixing 1mg of the sample with 200mg of KBr. The size and shape of fucoidan-coated iron oxide nanoparticles were shown by Scanning Electron Microscopy (Hitachi S-4800, Japan).

F. Determination of Fucoidan Concentration

The amount of fucoidan coated on iron oxide nanoparticles was calculated by extrapolation from the amount of fucoidan

not coated on iron oxide nanoparticles which was determined using conductometric titration with $BaCl_2$ as titrant.

III. RESULTS AND DISCUSSION

A. Characterization

FT-IR absorption spectra of fucoidan, iron oxide nanoparticles and fucoidan-coated iron oxide nanoparticles are shown in Figure 2. The peak at 1,240 cm^{-1} was attributed to the asymmetric stretching of S=O [26] and the peak at 3400 cm^{-1} was attributed to –OH groups on spectra of fucoidan. These peaks were also displayed on the spectra of fucoidan-coated iron oxide nanoparticles. These results demonstrated that fucoidan were coated on iron oxide nanoparticles.

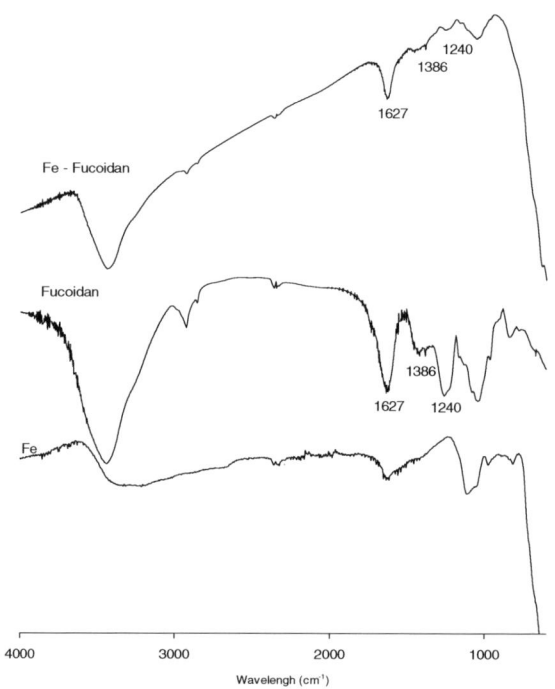

Fig. 2 FTIR spectra of fucoidan , iron oxide nanoparticles and fucoidan coated iron oxide nanoparticles

According to figure 3, SEM image of fucoidan-coated iron oxide nanoparticles demonstrates that these particles are smaller than 100nm and have spherical shape.

Fig. 3 SEM image of fucoidan modified iron oxide nanoparticles

B. Fucoidan Loading

Percentage of fucoidan, which was coated on iron oxide nanoparticles by ultrasonication, were 87.74 % at pH 7, 74.49 % at pH 6 and 63.03 % at pH 8 (Table 1). The results may be explained by the charge interaction between fucoidan and iron oxide nanoparticles.

Iron oxide nanoparticles in aqueous systems have amphoteric character through hydration [27-29]. In alkaline solution (pH 8), iron oxide has the highest negative charge (as compared to the negative charge in pH 6 and pH 7) on the surface [30]. Fucoidan also displayed the highest negative charge at pH 8 due to -OH groups, -COOH groups and -SO4$^-$ group. Therefore, the highest charge repulsion between fucoidan and iron oxide nanoparticles was observed at pH 8. Thus, the percentage of fucoidan loading is the lowest at pH 8.

Table 1 Fucoidan loading efficiency in iron oxide nanoparticles

Solution	Percentage	SD
pH 6	74.49 %	1.42
pH 7	87.74 %	1.105
pH 8	63.03 %	0.39

Meanwhile, at pH 7, a little smaller negative charge of iron oxide nanoparticles was shown as compared to the one at pH 8 [30]. However, a large amount of negative charge from -OH groups and -COOH groups of fucoidan was changed to the neutral charge at this pH. Thus, there is a significant reduction of charge repulsion and increase of fucoidan loading on iron oxide nanoparticles.

At pH 6, negative charge of fucoidan was reduced as compared to the one at pH 7 because -SO4$^-$ had a tendency to receive positive charge from acid solution. Nevertheless,

negative charge of iron oxide nanoparticles in this environment was as smaller as that at pH 7 [30]. The increase of charge repulsion may be obtained and fucoidan loading, therefore, was lower than that at pH 7.

IV. CONCLUSIONS

Fucoidan can be coated on iron oxide nanoparticles with the highest percentage of fucoidan at pH 7. These nanoparticles are potential for cancer therapy with fucoidan and can be used as a contrast agent.

ACKNOWLEDGMENT

This research is funded by Vietnam National University – Ho Chi Minh City under grant number C2014-28-09. We also would like to thank Vietnam Fucoidan Joint Stock Company and Nha Trang Institute of Technology Research and Application for provided fucoidan.

REFERENCES

1. S.H. Ferlay J, Bray F, Forman D, Mathers C, Parkin DM, GLOBOCAN 2008 v2.0, Cancer Incidence and Mortality Worldwide: IARC CancerBase No. 10 [Internet]. in: I.A.f.R.o. Cancer (Ed.), Lyon, France, 2010.
2. WHO, Ten statistical highlights in global public health. World Health Statistics 2007., World Health Organization, Geneva, 2007.
   I. Wijesekaraa, R. Pangestutia, S.-K. Kim, Biological activities and potential health benefits of sulfated polysaccharides derived from marine alga, Carbohydrate Polymers, 84 (2011) 14-21.
3. Li, F. Lu, X. Wei, R. Zhao, Fucoidan: Structure and Bioactivity, Molecules, 13 (2008) 1671-1695.
4. Y.S. Khotimchenko, Antitumor properties of nonstarch polysaccharides: Fucoidans and chitosans., Russian Journal of Marine Biology, 36 (2010) 321–330.
5. Z. Jiang, T. Okimura, T. Yokose, Y. Yamasaki, K. Yamaguchi, T. Oda, Effects of sulfated fucan, ascophyllan, from the brown Alga Ascophyllum nodosum on various cell lines: A comparative study on ascophyllan and fucoidan., Journal of Bioscience and Bioengineering, 110 (2011) 113-117.
6. S. Ermakova, R. Sokolova, S.-M. Kim, B.-H. Um, V. Isakov, T. Zvyagintseva, Fucoidans from brown seaweeds Sargassum hornery, Eclonia cava, Costaria costata: Structural characteristics and anticancer activity, Applied Biochemistry and Biotechnology, 164 (2011) 841-850.
7. H. Yea, K. Wanga, C. Zhoub, J. Liua, X. Zenga, Purification, antitumor and antioxidant activities in vitro of polysaccharides from the brown seaweed Sargassum pallidum, Food Chemistry, 111 (2008) 428-432.
8. S.R. Mudshinge, A.B. Deore, S. Patil, C.M. Bhalgat, Nanoparticles: Emerging carriers for drug delivery, Saudi Pharmaceutical Journal, 19 (2011) 129-141.
9. M.R. Zhang J, Magnetic drug-targeting carrier encapsulated with thermosensitive smart polymer: core-shell nanoparticle carrier and drug release response., Acta Biomaterialia, 3 (2007) 838-850.

10. R.R. Purushotham S, Thermoresponsive magnetic composite nanomaterials for multimodal cancer therapy, Acta Biomaterialia, 6 (2010) 502–510.

11. A.J. Durán JDG, Gallardo V, Delgado AV, Magnetic colloids as drug vehicles, Journal of Pharmaceutical Sciences, 97 (2008) 2948–2983.

12. M. Takahashi, Y. Akiyama, J. Ikezumi, T. Nagata, T. Yoshino, A. Iizuka, K. Yamaguchi, T. Matsunaga, Magnetic Separation of Melanoma-Specific Cytotoxic T Lymphocytes from a Vaccinated Melanoma Patient's Blood Using MHC/Peptide Complex-Conjugated Bacterial Magnetic Particles, Bioconjugate Chem., 20 (2009) 304–309.

13. Q.F. Gai Q-Q, Liu Z-J, Dai R-J, Zhang Y-K, Superparamagnetic lysozyme surface-imprinted polymer prepared by atom transfer radical polymerization and its application for protein separation, Journal of Chromatography A, 1217 (2010) 5035–5042.

14. X.H. Zhang L, Gao C, Carr L, Wang J, Chu B, et al, Imaging and cell targeting characteristics of magnetic nanoparticles modified by a functionalizable zwitterionic polymer with adhesive 3,4-dihydroxyphenyl-l-alanine linkages, Biomaterials, 31 (2010) 6582-6588.

15. Z.R. Kluchova K, Tucek J, Pecova M, Zajoncova L, Safarik I, et al, Superparamagnetic maghemite nanoparticles from solid-state synthesis e their functionalization towards peroral MRI contrast agent and magnetic carrier for trypsin immobilization, Biomaterials, 30 (2009) 2855-2863.

16. F.B. Hong RY, Chen LL, Liu GH, Li HZ, Zheng Y, et al, Synthesis, characterization and MRI application of dextran-coated Fe3O4 magnetic nanoparticles

17. Biochemical Engineering Journal, 42 (2008) 290-300.

18. R.J. Shultz MD, Khanna SN, Carpenter EE, Reactive nature of dopamine as a surface functionalization agent in iron oxide nanoparticles, J. Am. Chem. Soc., 129 (2007) 2482-2487.

19. A.E. Bagaria HG, Shamsuzzoha M, Nikles DE, Johnson DT, Understanding mercapto ligand exchange on the surface of FePt nanoparticles, Langmuir, 22 (2006) 7732-7737.

20. C.D. Zhang M, He X, Chen L, Zhang Y, Magnetic silica-coated submicrospheres with immobilized metal ions for the selective removal of bovine hemoglobin from bovine blood, Chemistry – An Asian Journal, 5 (2010) 1332-1340.

21. J.C. Qiu P, Charity N, Towner R, Mao C, Oil phase evaporation-induced self-assembly of hydrophobic nanoparticles into spherical clusters with controlled surface chemistry in an oil-in-water dispersion and comparison of behaviors of individual and clustered

22. iron oxide nanoparticles., J Am Chem Soc, 132 (2010) 17724-17732.

23. G.J. Pan X, Yoo J-W, Epstein AJ, Lee LJ, Lee RJ, Cationic lipid-coated magnetic nanoparticles associated with transferrin for gene delivery, International Journal of Pharmaceutics, 358 (2008) 263-270.

24. Z.J. Yang H, Tian Q, Hu H, Fang Y, Wu H, et al., One-pot synthesis of amphiphilic superparamagnetic fept nanoparticles and magnetic resonance imaging in vitro., Journal of Magnetism and Magnetic Materials, 322 (2010) 973-977.

25. M.S. Lu J, Sun J, Xia C, Liu C, Wang Z, et al, Manganese ferrite nanoparticle micellar nanocomposites as MRI contrast agent for liver imaging, Biomaterials, 30 (2009) 2919-2928.

26. D.S. Kaiser A, Schmidt AM, Kinetic studies of surface-initiated atom transfer radical polymerization in the synthesis of magnetic fluids., Journal of Polymer Science Part A: Polymer Chemistry, 47 (2009) 7012–7020.

27. M.-F. Marais, J.-P. Joseleau, A fucoidan fraction from Ascophyllum nodosum Carbohydrate Research, 336 (2001) 155-159.

28. G.A. PARKS, Aqueous Surface Chemistry of Oxides and Complex Oxide Minerals Isoelectric Point and Zero Point of Charge, in: Equilibrium Concepts in Natural Water Systems, AMERICAN CHEMICAL SOCIETY, 1967, pp. 121-160.

29. E. Illés, E. Tombácz, The effect of humic acid adsorption on pH-dependent surface charging and aggregation of magnetite nanoparticles, Journal of Colloid and Interface Science, 295 (2006) 115-123.

30. E. Illés, E. Tombácz, The role of variable surface charge and surface complexation in the adsorption of humic acid on magnetite, Colloids and Surfaces A: Physicochemical and Engineering Aspects, 230 (2003) 99-109.

31. M. Erdemoglu, M. Sarikaya, Effects of heavy metals and oxalate on the zeta potential of magnetite, J. Colloid Interface Sci., 300 (2006) 795-804.

*Corresponding Author: Thao Truong-Dinh Tran
Institute: Biomedical Engineering Department, International University, Vietnam National University - Ho Chi Minh City
City: Ho Chi Minh City
Country: Vietnam
Email: ttdthao@hcmiu.edu.vn

A Novel Non-invasive System for Acquiring Jugular Venous Pulse Waveforms

Tam Nguyen[1], Anh Dinh[1], Francis M. Bui[1], and Toi Vo[2]

[1] Division of Biomedical Engineering, University of Saskatchewan, Canada
[2] Department of Biomedical Engineering, International University, Vietnam

Abstract— **This paper lays the groundwork for the non-invasive acquisition of jugular venous pulse (JVP) waveforms, using accelerometer sensors to measure the vibrations of the jugular vein and the carotid artery. In the cardiac system, the vibrational signals quantify the pressure buildup in the blood vessels due to various hemodynamic events in the cardiac cycle. Accordingly, the accelerometer outputs are suitably amplified, filtered and digitized, so that an appropriate post-processing scheme can be applied to derive the corresponding JVP waveforms. The experimental results show that, even in a preliminary implementation, the JVP waveforms obtained using our proposed system exhibit promising consistency and accuracy, suitable for basic timing analyses of various cardiac parameters. It is anticipated that, with further development and optimization, our proposed system should offer a compelling alternative for clinical diagnoses and applications based on the JVP waveforms, potentially supplanting more complicated methods traditionally used to measure the JVP.**

Keywords— **Jugular venous pulse, accelerometer, heart activities, heart disease screening.**

I. INTRODUCTION

Based on an assessment of the jugular venous pulse (JVP) waveform, various important cardiac parameters may be characterized and quantified, including the mean venous pressure, venous pulse contour, as well as presence and type of cardiac arrhythmias [1]. For illustrative purposes, a typical JVP is depicted in Fig. 1, where several characteristic waves and corresponding labels have been designated [2]. A more detailed depiction in the context of other waveforms is shown later in Fig. 2. By convention, the **a**-wave indicates the contraction of the right atrium. The peak of this wave corresponds to the end of atrial systole. The atrial relaxation and rapid atrial filling due to low pressure result in the **x**-descent following the **a**-wave. The **c**-wave corresponds to the right ventricular contraction. The contraction causes the tricuspid valve to bulge towards the right atrium. The result of the right ventricle, pulling the tricuspid valve downward during ventricular diastole, causes the **x'**-descent following the **c**-wave. This descent can be used as a measure of right ventricle contractility [2]. The **v**-wave corresponds to the venous filling, when the tricuspid valve is closed and the venous pressure increases from venous return. Normally this occurs during and following the carotid pulse, as shown in Fig. 2. Next, the **y**-descent is due to the rapid emptying of the atrium into the ventricle following the opening of the tricuspid valve [3].

Physiologically speaking, the JVP measures the pressure in the right atrium. When the pressure is high, the JVP increases and when right atrial pressure is low the JVP will drop [2]. As such, the JVP waveform can facilitate the identification of specific heart diseases, including constrictive pericarditis, restrictive cardiomyopathy, pericardial effusion, and severe right-sided heart failure due to impaired filling of the right ventricle.

Despite its unique properties and characterization advantages, the JVP has not found widespread adoption in clinical applications, e.g., compared to the electrocardiogram (ECG). This is due, in part, to the outstanding challenges involved in the acquisition hardware design and measurement technique needed. In fact, the traditional technique of obtaining a JVP waveform is very tedious and difficult, not to mention the issues of accuracy and consistency of the waveform from one observer to another [1,4,5]. For instance, in a clinical measurement setting, the patient lies on a bed at a 30° angle, with a ruler measuring the distance between the sternal angle and the intercept of the vertical line. JVP will be calculated based on the measurement. Other techniques include ultrasonography [6] or Doppler method [7]. As a result, the accurate and consistent acquisition of JVP waveform typically requires experienced healthcare personnel.

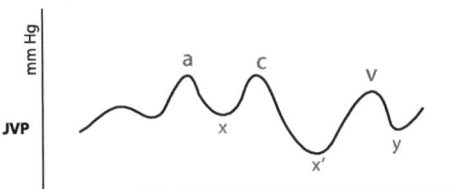

Fig. 1 A typical jugular venous pulse (JVP) waveform

Currently, the literature on effective techniques to acquire and analyze JVP waveforms is quite scant. Our work endeavors to bridge this gap, by proposing a novel method to obtain JVP waveforms, which in turn can be used to perform timing analyses for screening of certain heart diseases. The proposed non-invasive system uses accelerometer

V. Van Toi and T.H. Lien Phuong (eds.), *5th International Conference on Biomedical Engineering in Vietnam,*
IFMBE Proceedings 46, DOI: 10.1007/978-3-319-11776-8_19

sensors to receive the activities of the carotid artery and the jugular vein. From these activities, a JVP waveform can be derived, and used particularly to identify the peaks of the waveform and their timing intervals in each cardiac cycle.

II. METHODOLOGY

As illustrated in Fig. 1 and Fig. 2, it can be seen that the **a**, **c**, **v** peaks and **x**, **x'**, **y** descents of the JVP waveform correspond temporally to specific heart activities, e.g., as characterized traditionally by the more popular ECG and phonocardiogram (PCG).

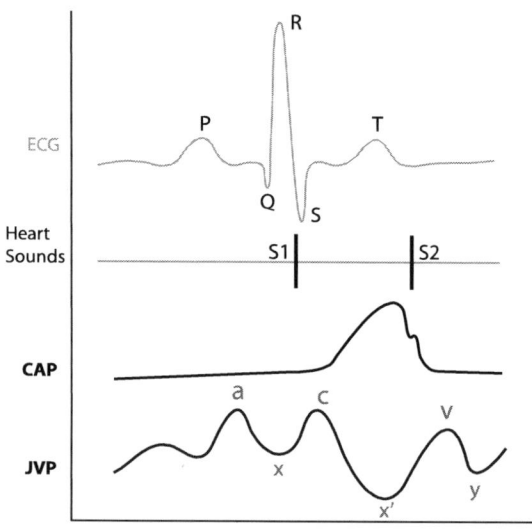

Fig. 2 Comparative overview of various heart activities, as characterized by, from top to bottom: the ECG, the heart sounds (quantified by the PCG), the carotid artery pulse (CAP), and the JVP.

In addition to ECG and PCG, recent advances in the complementary techniques of ballistocardiogram (BCG) and seismocardiogram (SCG) provide a more complete understanding the heart activities, from a mechanical perspective. For instance, the SCG is obtained using tri-axial accelerometers mounted on the chest to detect vibrations of the heart. Examples of SCG waveforms, and comparisons to other techniques for estimating hemodynamic parameters can be found in [8]. It should be noted that the peaks in an SCG waveform correspond to key events in a cardiac cycle. In particular, the mitral valve closes (MC) and the aortic valve opens (AO) during the isovolumic contraction time. The aortic valve closes (AC) and the mitral valve opens (MO) during the isovolumic relaxation time. The rapid systolic ejection (RE) can also be seen in the waveform. These activities can all be characterized suitably using the SCG [8].

Importantly, these characteristic cardiac vibrations, giving rise to the BCG and SCG waveforms, should also have an impact elsewhere in the body, albeit with varying degree of observability depending on the exact measurement site. In particular, the effects of these vibrations should be most pronounced at the carotid or other related positions. Based on this fact, MEMS accelerometers are placed at the carotid and at the jugular vein to sense the vibrations. It will be shown that the signals at these two locations contain sufficient information to construct a JVP waveform with reasonable timing accuracy. The block diagram and the picture of the prototype sensor system are shown in Fig. 3. The system will provide an ECG, a phonocardiogram, and waveforms of the carotid artery and the jugular vein activities. These two waveforms are obtained from the accelerometer axis placed perpendicular to the artery and the vein.

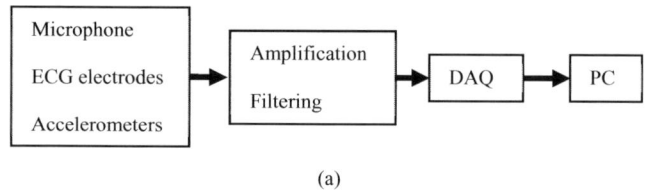

(a)

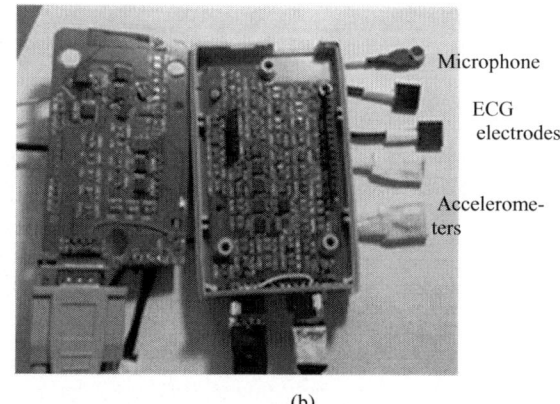

(b)

Fig. 3 (a) Block diagram and (b) The prototype of the sensing system

The dry electrodes are used as the ECG sensors while the microphone picks up the S1 and S2 heart sounds of Fig. 2. Analog circuits for the ECG and PCG signals are the traditional microelectronic circuits used to amplify and filter the signals. The two MEMS accelerometers sense the vibrations of the carotid artery and the jugular vein. The locations of the sensors are shown in Fig. 4. The signals picked up by the accelerometers need to be properly amplified and filtered, since these signals represent very small vibrations coming from the artery and the vein. These vibrations are on the order of mg (where $g=9.81 m/s^2$). The filters are used to reduce interference noise from the environment and motion artifacts, e.g., due to neck movements during measurement. Suitable post processing on the collected signals delivers the JVP waveform of interest.

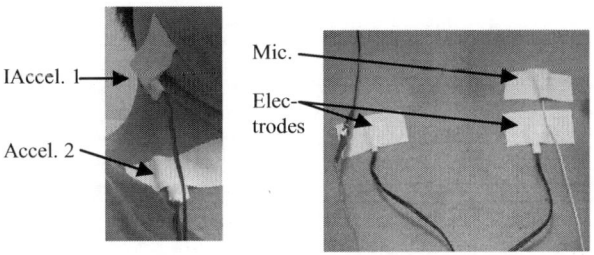

Fig. 4 Locations of various sensors

III EXPERIMENTAL RESULTS

The prototype was built and tested on a healthy male subject. The sensors were taped to the body as shown in Fig. 4. The data were collected using a DAQ system, with the sampling rate set at 1000 samples per second. The waveforms of the ECG, PCG, and the accelerometers at the top of the neck (carotid position) and at the bottom of the neck (jugular vein) are respectively depicted in Fig. 5(a).

The first waveform, from the top, in Fig. 5(a) shows the ECG signal, to provide reference timing for the P, QRS complex, and the T waves. The second waveform is the PCG, in which the two markers S1 and S2 can be identified to provide the timing check for the JVP waveform. Specifically, as illustrated in Fig. 2, S1 and S2 are in the vicinity of the two descents x and x', respectively. The 3rd waveform

from Fig. 5(a) measures the vibrations at the carotid position on the right side of the neck, as picked up by the accelerometer 1. Timings of the peaks for this waveform are comparable to those for the SCG waveform [8]. The 4th waveform is the signal collected at the bottom of the neck, as picked up by the accelerometer 2, where the carotid artery is located further insides the neck while the jugular vein is located adjacent to accelerometer 2.

The final (5th) waveform is the result of an appropriate post-processing of the 3rd and 4th waveforms, based on physiological understanding of the vibrations of the carotid artery and the jugular veins at different times during the cardiac cycle. For instance, it is known that the heart valves open or close based on the pressure buildup in the ventricles or in the atria. The pressure affects the vibrations of the jugular vein, which are picked up by the accelerometers. It is also noted that there is a difference in the distances between the two accelerometers and the heart. This difference necessitates a modest timing adjustment of the 3rd and the 4th waveforms, in order to generate the 5th signal. In particular, for the same amplification and filtering circuits, the 3rd signal should be delayed compared to the 4th signal.

Figure 5(b) depicts a more detailed view of the final signal over 2 cardiac cycles. The trace connecting the peaks of this signal is the JVP waveform, since it corresponds to the vibrations of the jugular vein caused by the increase or decrease of pressure in the vein itself. As shown, the timing is very easily obtained by observing the local maxima of the waveform. The time intervals can be easily calculated based on the sampling rate of the ADC in the DAQ system.

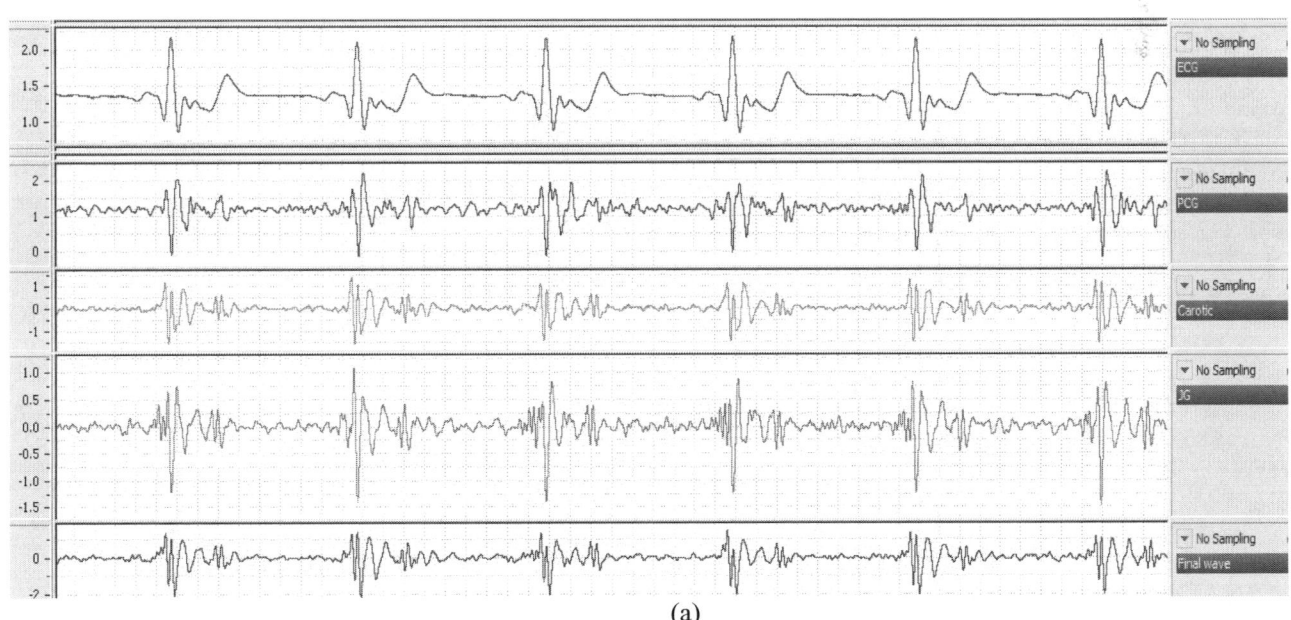

(a)

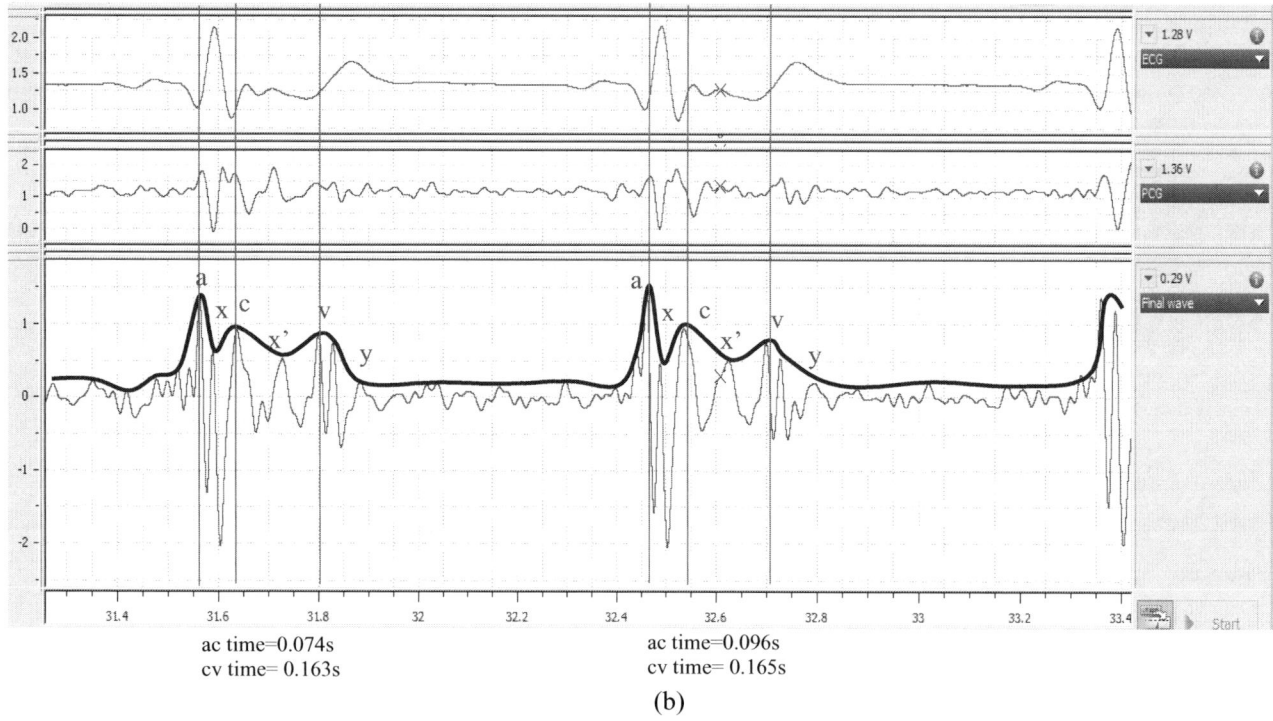

ac time=0.074s   ac time=0.096s
cv time= 0.163s   cv time= 0.165s

(b)

Fig. 5 (a) Signals obtained from the sensors and (b) A JVP waveform obtained by joining the local maxima in 2 cardiac cycles

III. CONCLUSIONS

This paper presents the preliminary bases for a system to measure JVP and derive its waveform. The proposed system is designed based on non-invasive sensors, used to detect the vibrations of the carotid artery and the jugular vein. MEMS accelerometer is used as the sensing element. With the sensed data, the shape of the JVP waveform can be obtained and timings of its peaks can be easily calculated. As future work, the calibration and calculation of the waveform amplitude need to be investigated. Together, timing and amplitude of the JVP can facilitate screening of certain heart diseases. The software-driven system should be highly user-friendly for healthcare personnel, without need for special training to measure the jugular venous pressure.

ACKNOWLEDGMENT

This work is funded in part by NSERC, and by the Grand Challenges Canada – Stars in Global Health under Grant number S6-0496-01-10.

REFERENCES

1. Mark M. Applefeld, *Clinical Methods: The History, Physical, and Laboratory Examinations*, 3rd Edition, Chapter 19: The Jugular Venous Pressure and Pulse Contour, Elsevier (1999)
2. MBBS4 Experiences, *Jugular Venous Pulse (JP)*, http://mbbs4.blogspot.ca/2012/03/jvp-waveform.html#!, accessed March 31, 2014
3. *Jugular Venous Pressure*, http://en.wikipedia.org/wiki/Jugular_venous_pressure, accessed March 31, 2014
4. David C. Wang, Roberta Klatzky, Bing Wu, Gregory Weller, Allan R. Sampson, and George D. Stetten, *Fully Automated Common Carotid Artery and Internal Jugular Vein Identification and Tracking Using B-Mode Ultrasound*, IEEE Transactions on Biomedical Engineering, Vol. 56, No. 6, pp. 1691-1699 (2009)
5. *Techniques: Jugular Venous Pressure Measurement (JVP)*, http://depts.washington.edu/physdx/neck/tech1.html, accessed March 31, 2014
6. Steven J. Socransky, et.al, *Defining normal jugular venous pressure with ultrasonography*, CJEM , 12(4):320-324 (2010)
7. M.M. Lagana, et al, *Internal Jugular Vein Blood Flow Reflux Analysis using Ultrasound Doppler Technologies and Phase Contrast Magnetic Resonance Imaging*, IEEE Computer Medical Applications (ICCMA) Conference, pp. 1-6 (2013)
8. Kouhyar Tavakolian, Andrew P. Blaber, Brandon Ngai, Bozena Kaminska, *Estimation of Hemodynamic Parameters from Seismocardiogram*, Computing in Cardiology, IEEE, pp. 1055-105 (2010)

Remote Monitoring of Cardiac Rhythm Management Devices in Vietnam The Role of the Biomedical Engineer

Tran Thong

Northwest Signal Processing, Inc., Beaverton, Oregon, USA
Systolic Medical Products, Hochiminh City, Vietnam

Abstract— Cardiac rhythm management (CRM) manufacturers of pacemakers, implantable defibrillators, offer remote monitoring of their high end products. In the West, the responsibility for monitoring devices is given to a device nurse at the cardiac center. The nurse performs daily checks of any reported issue by the manufacturer and act upon any alert reported by the device. With the patient group numbering hundreds of patients, a device nurse can be supported by reimbursement for remote monitoring.

The situation is different in Vietnam. The number of implant patients at a typical implant center is barely in the low hundreds. Patients are typically not actively followed and irregularly come back for follow-ups. Currently there is no reimbursement program for automated follow-ups.

In developing the first Remote Monitoring program for Vietnam, we had to factor in these differences and have developed an alternative approach. A biomedical engineer at the distributor assumes the duty of the device nurse and works with the patients' cardiologists. The engineer will review the daily reports and study alerts reported by the devices or the automatic analysis programs of the manufacturer. When appropriate, the cardiologist will be alerted. The cardiologist will contact the patient to come in for a follow-up. The engineer will not deal directly with the patient, except on technical matters, such as malfunctioning remote monitor.

This has enabled us to create a new model for the care of CRM patients in Vietnam. With daily report by the device, the patient does not need to visit the cardiologist, except for a yearly visit, unless contacted. Since the Home Monitoring reports had been analyzed prior to the visit, recommendations about therapy or device parameter changes will be discussed with the cardiologist before the patient arrives.

Keywords— Pacemaker, defibrillator, remote monitoring, biomedical engineer.

I. INTRODUCTION

Cardiac Rhythm Management (CRM) devices have been life savers for patient with cardiac electrical conduction problems. They consist of

- bradycardia devices: pacemakers for symptomatic patients with a slow heart rate or episodes of heart block,
- tachycardia devices: cardioverter defibrillators (ICD) for episodes of ventricular tachyarrhythmia

- Cardiac resynchronization therapy (CRT) devices for heart failure with dissynchrony between the ventricles.
- ICD and CRT patients are considered high risk patients who deserve close follow-ups.

Patients in Vietnam can now have these devices implanted at major cardiac centers throughout Vietnam. As the number of patients increases, the Vietnamese cardiologist is fast encountering the same problem that has faced his colleagues in the West. With the implanted devices reaching longevities in excess of 15 years, and requiring regular follow-ups, every 3 to 6 months, throughout their service life to maintain optimal performance, the workload of the busy cardiologist can easily reach overload conditions.

From the patient point of view, the current overload conditions at all major hospitals in Vietnam have caused a major degradation of services. A consultation with the cardiologist can only accommodate a routine device follow-up. Beyond checking the battery status of the device, the cardiologist has barely time to look over the ever increasing amount of diagnostics stored in these devices.

One approach to solving this problem has been the open CRM clinic that one hospital has adopted. On a particular day of the week, patients implantd with a CRM device of a particular manufacturer can present themselves, without appointment, at the clinic and request a device follow-up. A clinical engineer of the CRM distributor will perform the follow-up and a quick analysis of the data stored in the device. The results are summarized for and recommendations made to the cardiologist on duty. The latter will review the recommendations, discuss the results with the patient, approve any program change and if necessary write a new prescription. If further analysis of the data is needed, this is performed later at the distributor office, and if warranted, the patient is contacted to come back the following week to update the device parameter set after review and discussion with the cardiologist on duty.

This open clinic strategy can be adopted at cardiac hospitals only. The clinic at a general hospital does not yield a sufficient number of CRM patients for the CRM distributor to assign a clinical engineer for the half day of clinic.

In this paper we would like to present an alternative approach to follow-up that is aimed at the high-end devices in patient at high risks.

© Springer International Publishing Switzerland 2015
V. Van Toi and T.H. Lien Phuong (eds.), *5th International Conference on Biomedical Engineering in Vietnam*,
IFMBE Proceedings 46, DOI: 10.1007/978-3-319-11776-8_20

II. REMOTE MONITORING

A typical cardiac clinic in the West has a patient group numbering in the hundreds, if not thousands. All CRM manufacturers now offer remote monitoring (RM) programs to allow the device nurse at the clinic to follow patients remotely and the nurse will contact the patient to come in only for an actionable intervention.

The high end devices all have a short range wireless communication unit built-in. This allows the implanted device to communicate with a patient device, typically placed at the bedside. It is connected to a telephone landline or to the cellular phone network. Periodically, from 1 to 90 days, the implanted device will send a summary of its status along with supporting data to the patient device to be forwarded to a monitoring center. The center will repackage the information and make it available on-line to the device nurse and other approved staff at the cardiac clinic for review. The implanted device can also send unscheduled alerts as needed.

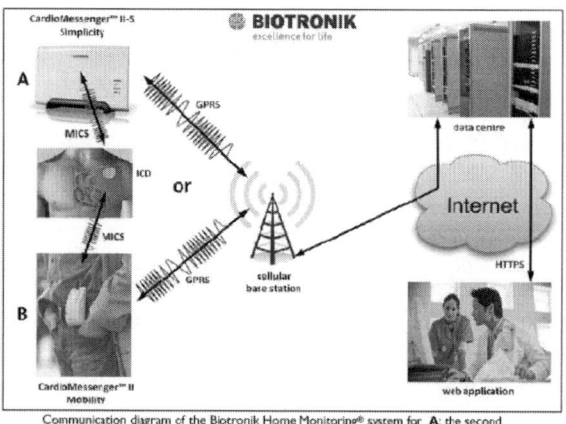

Fig. 1 Example of the Biotronik Remote Monitoring system [1]. In Vietnam, we use the device shown in B, at the bedside.

Since patients at the cardiac implant centers in Vietnam come from all over the country, and not necessarily from the local area, such a RM system should relieve the workload of the cardiologist. However, just as in the West, the busy Vietnamese cardiologist is overwhelmed by the technology (getting on line daily) and the amount of data to be reviewed. In the West, the device nurse has taken over these responsibilities. With the large population of patients, the device nurse can be supported by the reimbursement for remote monitoring. In Vietnam, there is no reimbursement for remote monitoring. As a result no device nurse is available. Thus a new operational model is needed for RM in a third world country like Vietnam.

THE VIETNAMESE REMOTE MONITORING MODEL

With the consent of the cardiologists and their patients, we have setup the following Remote Monitoring system:

- Based on Biotronik Home Monitoring® system, Berlin, Germany

  o Designed for high end pacemakers, for ICDs and CRT devices.
  o Low power transmission from the implanted device to the patient unit, CardioMessenger II® (CM II). It is in the Medical Implant Communication Service (MICS) band, 402-405 MHz [2]. In addition to measurements, multi-channel electrograms are also sent periodically or following detection and treatment of tachyarrhythmia episodes.
  o Communication between the CM II and the monitoring center in Berlin, Germany, is over the cellular GPRS (General Purpose Radio Service, so called 2.5G) network [3]. Home Monitoring® is the only remote monitoring system with GPRS as the standard method for communication with the monitoring center.
  o The GPRS message between the CM II and the monitoring center is prepaid and included in the price of the CM II. At this time, the patient does not incur any additional charge.
  o The data is analyzed and repackaged by the monitoring center. Various levels of alert are generated from analyses of the data or from therapy reports by the implanted device.
  o Authorized personnel can access the data using a protected internet account hosted by a dedicated Biotronik server in Berlin, Germany.

- An experienced Biomedical Engineer working for the Biotronik CRM distributor in Vietnam will review the data and alert the cardiologist when needed.

The Biomedical Engineer (BE), a clinical engineer with extensive experience with Biotronik CRM products, assumes the role of the device nurse in the West. The BE will review the data the Biotronik monitoring center makes available on the web. Since the number of patients is fairly modest at this time, in addition to reacting to the alarms generated by the analysis program at the monitoring center, and/or the implanted device, the BE can also be proactive and perform trend analyses and correlations of various data to arrive at preliminary diagnoses. These are then reviewed with the treating cardiologist of the particular patient. From his experience with implanted device, the BE can make recommendations for

- Adjustment of the implanted device parameter for optimal operation, like other clinical engineers.
- Recommendation on therapy adjustment. For example anti-coagulant for patients with newly discovered atrial fibrillation, shift in therapy to focus on atrial tachyarrhythmia, increased dosage of beta blocker due to increase in ventricular early systole (VES/PVC), ...
- Recommendation for referral to an Electro Physiologist for ablation of certain focal tachyarrhythmia

As noted, these are recommendations made to the treating cardiologist, never to the patient. So, effectively, the BE is working under the supervision of the cardiologist. The cardiologist is responsible for contacting the patient and making arrangement for subsequent device follow-ups. The advantage of the approach is that, by the time the patient comes to the hospital, the cardiologist knows what adjustments to the implanted device are needed and would have ready any prescription change.

The only communication the BE has with the patient is with regard to the operation of the CM II unit.

The patient no longer has to go for periodic follow-ups, which typically are routine [4, 5], and only needs to come to the clinic when contacted by the cardiologist for an actionable event, or yearly. Patients are very appreciative of this VIP treatment, especially in the overcrowded situation in Vietnam since the patient has priority due to his "requested by the doctor" appointment status.

III. CONTRIBUTIONS OF THE BIOMEDICAL ENGINEER

The BE key contribution, beyond what a device nurse would provide in response to alerts by the device or the monitoring center, is his analyses of the data. In this section we would like to present a couple of examples of this.

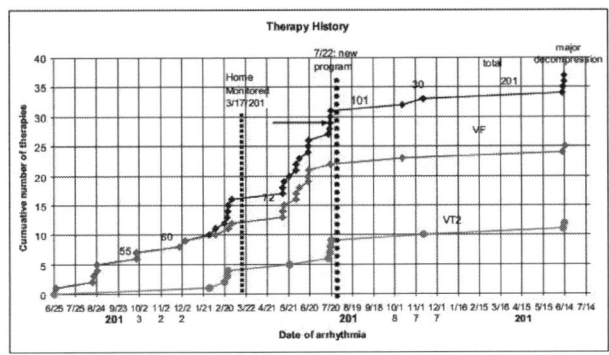

Fig. 2 Patient with arrhythmic storm.
The number on the top curve is the number of days between episodes.

A. Arrhythmic Storm

The male patient has been implanted a Biotronik Lumax 300 HF-T, a three chamber ICD with CRT for heart failure in June of year 1. For the first 7 months, he had a number of episodes of Ventricular Fibrillation (VF) and Ventricular Tachycardia (VT) that were detected and treated with burst pacing (anti-tachycardia pacing, ATP) or with shocks, most often aborted since the arrhythmia broke before charging of the shock capacitors were completed. A few episodes every other month was considered acceptable. However, starting about the 8th month, in late January of year 2, he started to have this string of tachyarrhythmia episodes, an arrhythmic storm (the strict definition of a storm is a string of episodes over a 24 hour period. We have expanded the definition here). This was unsettling. About March of year 2, we started the new RM program. The patient agreed to join. In May of the second year, he started having this long arrhythmic storm. We redoubled our effort to find a way to stop the storm using the 30 seconds of 3 channels of electrogram pre-episode we received each time an episode of arrhythmia was detected. We worked with his cardiologist trying to alter his prescription, but the storm continued unabated. After 4 month of RM, we finally uncovered the cause of his storm. The symptom in the pre-arrhythmia electrogram is delayed depolarization after a pacing pulse at normal voltage, but not at high voltage. We speculated that the right ventricular lead was implanted close to a re-entry circuit.

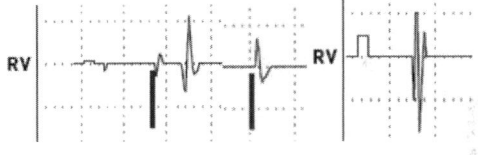

Fig. 3 Delayed depolarization; high pacing voltage; anodal stimulation

In order to achieve resynchronization of the two ventricles, it is necessary to pace both the left and right ventricles >98% of the time. The Lumax 300 HF-T allows us to pace the right ventricle anodically from the right ventricular ring (normal pacing is bipolar with the tip being the cathode, and the ring the anode) in conjunction with pacing the left ventricle (cathode) with a higher voltage (than normal) pulse. The ring being remote from the myocardium can simultaneously capture a large area of the myocardium, which we believe would include the whole arrhythmic re-entry circuit. The programming change was made on July 22 of year 2, and the arrhythmic storm stopped!

It was the ability to collect all the pre-arrhythmia electrograms that allowed us to uncover the cause of the storm.

Without RM, the patient would have to come in for a follow-up immediately after each episode, a major inconvenience for the patient. A good understanding of electrophysiology and the device allowed the solution formulation.

B. Prediction of Acute Decompensation Episode

Except for patients who need an ICD due to a genetic condition, such as Brugada or Long QT Syndrome, most patients with an ICD have heart failure, along with CRT patients. For these patients, acute decompensation [6] is a serious and constant threat. One early indication of acute decompensation is a growing pulmonary edema. It has been proposed that transthoracic impedance be used as an early warning of acute decompensation [7]. An automated alarm was proposed. The resulting high rate of false alarm has led to the failure of the DOT-HF [7] trial. It has been suggested that transthoracic impedance should require a review by an experienced operator prior to a diagnosis.

Transthoracic impedance is measured in the Biotronik Iforia 5 series of ICD. However the information is available only through RM for fluid edema diagnosis and can be used for predicting an imminent episode of decompensation. We would like to report an episode of decompensation that we were able to predict and avoid the patient a hospital stay.

The patient was implanted with a single chamber Biotronik Iforia 5 VR-T. He also signed up for the Biotronik Home Monitoring® program.

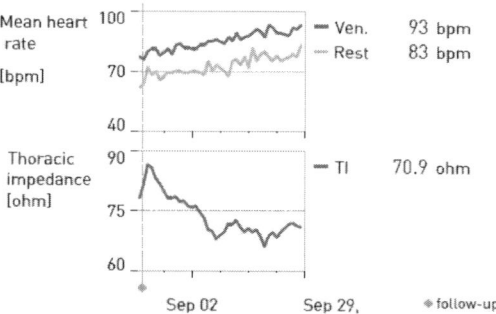

Fig. 4 Heart Rate and Thoracic Impedance report.

Since we just started monitoring this patient and this is our first patient with Thoracic Impedance measurement, we were gun-shy to the increase in fluid accumulation indicated by the decrease of the Thoracic impedance. We could have alerted the cardiologist around Sept 9, but at that time, with the rest heart rate hovering about 70 bpm, we did not feel comfortable, in view of the DOT-HF failure [7]. We waited until the daily rest heart rate, measured at 02:00, crossed the 80 bpm threshold before alerting the cardiologist. Following our Sept 27 report, the cardiologist confirmed the pulmonary edema and initiated an intensive diuretics regimen that saved the patient from a hospital stay. Since then the patient has not experienced such a low impedance concurrent with a high resting heart rate.

Because this is our first experience the transthoracic impedance, and the decompensation occurred too shortly after the implant of the device, we will need further episodes with additional patients before we can generalize this result.

IV. CONCLUSION

In this paper we have presented a model for Remote Monitoring developed for Vietnam taking into account the different operating environment. The Biomedical Engineer, an experienced clinical engineer, has assumed the role of the device nurse in the West. The BE reports his analyses and recommendations to the cardiologist, who is responsible for the interactions with the patient. Thanks to his technical training and experience, the Biomedical Engineer can do more than his device nurse counterpart. With a good understanding of electrophysiology and the devices, by analyzing trends and correlations between the parameters that are received from the remote monitoring system, the BE can be proactive in support of the cardiologist.

Thus, at this early stage of the Vietnamese Remote Monitoring program, the patient is getting first class follow-up, on par and possibly better than in the West! We are monitoring patients with pacemakers, ICD and CRT devices.

CONFLICT OF INTEREST

Northwest Signal Processing, Inc. (NSPI) and Systolic Medical Products operate as the distributor of Biotronik CRM products in Vietnam. The author is a principal of NSPI.

REFERENCES

1. Surveillance à distance des patients porteurs de défibrillateurs implantés. Evaluation de la technologie et cadre réglementaire général. KCE reports 136CB. Centre Fédéral d'expertise des soins de santé, 2010.
2. http://en.wikipedia.org/wiki/Medical_Implant_Communication_Service
3. http://en.wikipedia.org/wiki/General_Packet_Radio_Service
4. Matlock DD (2010). Big brother is watching. What do patients think about ICD Home Monitoring. Circulation 122:319-321.
5. Varma N, Epstein AE, Irimpen A, Schweikert R, et al, (2010). Efficacy and safety of automatic remote monitoring for implantable cardoverter-defibrillator follow-up. The Lumos-T Safely Reduces Routine Office Device Follow-up (TRUST) trial. Circulation 122:325-332.
6. http://en.wikipedia.org/wiki/Acute_decompensated_heart_failure
7. Van Veldhuisen DJ, Braunschweig F, Conraads V, Ford I, et al. (2011). Intrathoracic impedance monitoring, audible patient alerts, and outcomes in patients with heart failure. Circulation 124:1719-1726

Inspired Sinewave Technique:
A Novel Technology to Measure Cardiopulmonary Function

Phi Anh Phan[1], Cathy Zhang[2], Daniel Geer[3], Clive Hahn[1], and Andrew Farmery[1]

[1] Nuffield Department of Clinical Neurosciences, University of Oxford, Oxford, United Kingdom
[2] New College, University of Oxford, Oxford, United Kingdom
[3] The Queen's College, University of Oxford, Oxford, United Kingdom

Abstract— Introduction: Inspired Sinewave is a novel technique to measure dead space, alveolar volume, pulmonary blood flow, and lung inhomogeneity noninvasively. It does not require patient effort and therefore can be applied to young children, elderly, and ventilated patients with ease.

Method: In this paper, we describe a brief introduction to the principle of the technique, which involves forcing inspired concentrations to oscillate sinusoidally and measuring responding expired concentrations. Then, we give some updates to the recent developments of the device. These include comparison studies with body plethysmography for functional residual capacity measurement, and with echocardiography for cardiac output measurement.

Result and discussion: The results show that the inspired sinewave technique achieves comparable accuracy and repeatability with the "gold standard" body plethysmography method with

Conclusion: The success of these studies is a big step forward to make this novel device a useful clinical tool. The technology is patented and future work includes forming collaboration with an industry partner to bring the technology to commercialisation.

Keywords— Cardiopulmonary Function Test, Lung Function Test, Inspired Sinewave Technique, Lung Inhomogeneity, Noninvasive.

I. INTRODUCTION

Lung function testing is essential to the diagnosis of how the lung works in health and disease. Traditionally, lung function test is performed by spirometry and body plethysmography. These traditional tests however depend on patient effort and therefore are not accurate in difficult patient groups such as young children and elderly. They also cannot be used on ventilated patients.

The inspired sinewave is a novel technique to address this problem. It provides measurements of dead space, lung volume, pulmonary blood flow, and lung inhomogeneity simultaneously, non-invasively, and without patients' cooperation.

Historically, the technique was originated from Zwart's idea of using forced inspired sinusoids of halothane and acetylene in the 1970s [1], [2]. Hahn and colleagues extended this idea to the use of more patient safe gases such

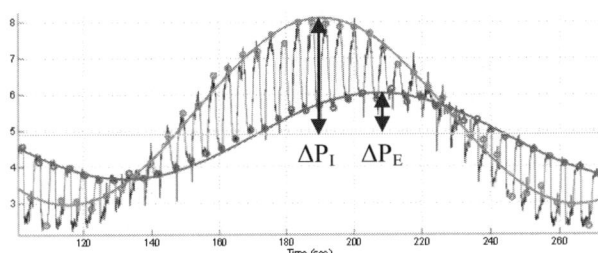

Fig. 1 Raw Data from the ISD in a healthy volunteer. Green circles and red circles are the inspired concentration and the expired concentration of each breath. Green trace and red trace are the inspired sinewave and expired sinewave respectively. cavity.) Red trace: [N2O] in expired gas. Arrows mark change in sine wave amplitude between inspired and expired gas.

as O_2 and low concentration of N_2O (3% mean) [3], [4]. They also extended the simple continuous lung model used by Zwart to more complicated models [5], [6]. Preliminary clinical studies with both animals [7]–[9] and healthy volunteers [10] showed close argeements between the inspired sinewave technique and other techniques.

The technique works based on the principle as follows. By forcing the inspired concentrations of O_2 and N_2O to oscillate sinusoidally with very low amplitudes (3-5%) and low mean for N_2O (3%), the expired concentrations of O_2 and N_2O also follow a sinewave pattern (Fig 1). Lung parameters then can be estimated from the responding amplitudes and phases of the expired concentrations. The larger the lung parameters, the greater are the attenuations of the expired oscillations:

$$\frac{\Delta P_E}{\Delta P_I} = \frac{1}{\sqrt{\left(1 + \frac{\lambda_{N_2O} \dot{Q}_P}{\dot{V}_A}\right)^2 + \left(\frac{2\pi}{T}\right)^2 \left(\frac{V_A}{\dot{V}_A}\right)^2}} \qquad (1)$$

in which:

ΔP_E : the magnitude of the expired sinewave

ΔP_I : the magnitude of the inspired sinewave

λ_{N_2O} : the solubility of nitrous oxide in the blood

© Springer International Publishing Switzerland 2015
V. Van Toi and T.H. Lien Phuong (eds.), *5th International Conference on Biomedical Engineering in Vietnam*,
IFMBE Proceedings 46, DOI: 10.1007/978-3-319-11776-8_21

$\dot{Q}_P$: the pulmonary blood flow (L/min)

$\dot{V}_A$: the volume of gas ventilated by the lung (L/min)

T : the sinewave period applied by the technique (mins)

V_A : the lung volume (L)

Based on eq (1), V_A and $\dot{Q}_P$ can be estimated by measuring $\Delta P_E / \Delta P_I$ at 2 or more periods. Interested readers can refer to more detailed publications [9], [11] for full mathematical description of the technique.

II. METHOD

In this paper, we present a recent study on the capability of the technique to estimate functional residual capacity (FRC), which is the sum of lung dead space and lung volume. Both repeatability and accuracy have been investigated in this study.

A. Study Protocols

a) Repeatability

Six measurements were made on each of 17 healthy subjects over a two-hour period. The clinical characteristics of these subjects are shown in Table 1. One subject was a light smoker and one had mild exercise-induced asthma. Each measurement consisted of 360 seconds breathing normally into the ISD with T set to be 3 mins. The mean, standard deviation (SD), coefficient of variation (CV) and values of repeatability, r, for V_A, Q_P and V_D were calculated for each volunteer to allow analysis of reproducibility of the data obtained.

Table 1 Repeatability subjects information

Characteristic	Range	Mean	StD
Age (years)	19-31	22.29	3.60
Height (m)	1.58-1.86	1.73	0.07
Weight (kg)	50-88	67.44	10.45
BMI (kg/m2)	19.48-28.09	22.47	2.73
FEV1 (L)	2.66-5.61	3.97	0.75
Forced Vital Capacity (L)	3.03-6.87	4.57	0.94
FEV1: FVC ratio (%)	79.49-93.49	87.07	4.45
Residual Volume (L)	1.15-3.04	1.91	0.42
Peak Expiratory Flow (L/min)	260-813	456	128.36
FRC (L)	2.15-4.93	3.23	0.70
Vital Capacity (L)	2.73-6.64	4.26	0.96
Total Lung Capacity (L)	4.13-8.65	6.17	1.12
ISD V_A (L)	1.53-4.21	2.50	0.66
ISD Q_P (L/min)	3.79-7.72	5.46	1.26
ISD V_D (ml)	74.6-443.4	143.10	82.76

b) Accuracy

To access the accuracy, a comparison study with body plethysmography (gold standard technique) was performed.

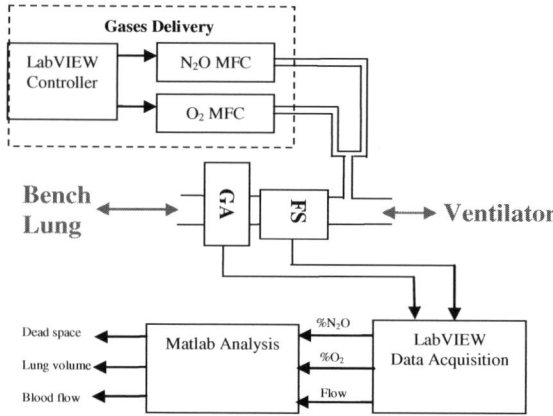

Fig. 2 Layout of the Inspired Sinewave Device. The gases delivery system employs mass flow controllers (MFCs) to inject O2 and N2O into the breathing circuit. Real-time data is read by the flow sensor (FS) and the mainstream gas analyzer (GA) and fed to LabVIEW and Matlab for estimation of dead space, lung volume and pulmonary blood flow.

44 healthy subjects with no history of cardiopulmonary diseases volunteered to participate in this study. Of which, 5 were active smokers. Participant information and relevant clinical information are given in table 2. The subjects also underwent lung volume measurements made by a variable volume body plethysmography.

Table 2 Accuracy subjects information

Characteristic	Range	Mean	StD
Age (years)	19-65	29.12	14.92
Height (m)	1.53-1.91	1.73	0.11
Weight (kg)	49.5-155.7	72.27	24.19
BMI (kg/m2)	19.8-47.0	23.88	6.25
FEV1 (L)	2.40-6.51	4.01	1.16
Forced Vital Capacity (L)	2.79-7.05	4.74	1.29
FEV1: FVC ratio (%)	73.8-94.6	84.37	6.60
Residual Volume (L)	1.23-3.13	1.98	0.52
Peak Expiratory Flow (L/min)	252-850	470	190
FRC (L)	1.86-4.76	3.30	0.85
Vital Capacity (L)	2.57-7.07	4.63	1.36
Total Lung Capacity (L)	3.96-9.15	6.62	1.51
ISD VA(L)	1.48-4.57	2.52	0.98
ISD QP (L/min)	2.90-9.34	5.78	2.04
ISD VD (ml)	51.0-164.0	114.47	31.39

B. Hardware

The layout of the inspired sinewve device (ISD) is described in Figure 2, and a photographic layout is given in appendix B. By reading the inspiration flow rate in real time, the software can decide the set-points for the mass flow controllers to inject the desired amount of O2 and N2O. The integrations of concentrations and flow signal give the inspired volume and expired volume of O2 and N2O breath-by-breath. These inspired volumes and end-expired concentration are then fed into a mathematical model of the lung in Matlab to estimate dead space, lung volume, and blood flow.

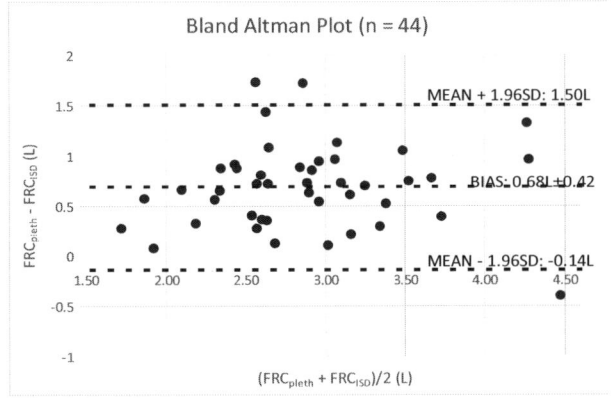

Fig. 3 Bland – Altman plot, comparing the inspired sinewave technique and the body plethysmography technique (gold standard)

The gas sensors have accuracy of 2% of reading and step response time of less than 350ms and 5ms time delay. The flow sensor is fast with updating rate of 50Hz and accuracy of 3%.

III. RESULTS

A. Repeatability

The coefficients of variance (CV) values for each individual subject were in the range 2.96% - 16.34%, and overall the mean CV values were 7.56%. Statistical analysis using ANOVA generated r values of 0.90, indicating that 10% of the variation in the data was due to differences within each individual. As VA does not change significantly over time, therefore the entirety of the variation within individuals is due to measurement variation associated with the ISD.

The CV of 7.56% obtained by the ISD is comparable to that of other techniques to measure FRC in clinical use e.g. body plethysmography (CV=5%)[12] and multiple breath helium dilution (CV=5%)[13].

B. Accuracy

The result confirmed a tendency for the ISD to underestimate FRC when compared to the reference technique. Bland – Altman plot (Figure 3) comparing the mean of FRC_{pleth} and FRC_{ISD} against the difference ($FRC_{pleth} - FRC_{ISD}$) reflected a bias of 0.76±0.36L, with a 95% CI of 0.05L to 1.47L. The standard deviation of ±0.36L suggests that the ISD has similar level of accuracy to other techniques studied in the literature (appendix A).

IV. CONCLUSIONS

This result shows that the inspired sinewave technique, at its current state of development, can be used clinically to measure functional residual capacity as an alternative to traditional "gold standard" body plethysmography. The inspired sinewave technique does not require patient and therefore can be used for young children and elderly, whom the body plethysmography is well reported to give inaccurate results[14].

In addition, the more clinical advantage of the technique is its capability to measure pulmonary blood flow and lung inhomogeneity simultaneously along with functional residual capacity. These cardiopulmonary parameters together are important to diagnose heart and lung function. However, they are not currently measured in lung function labs due to lack of techniques. If successful, the inspired sinewave technique will be an ideal technology for screening early sign of lung diseases, and for use in lung function labs, by the bedside, and at general practices.

Current efforts are to fine-tune the technology to give clinically accepted accuracy for measuring pulmonary blood flow and to validate the capability to detect early sign of lung diseases. The technology has been patented and we are working with potential industry partners to develop the technology for commercialization.

APPENDICES

Appendix A: Literature Review Comparing Methods to Measure FRC

Study	Techniques compared	Mean difference ± SD (L)
Our study	**BP** **Inspired Sinewave**	**0.68±0.42**
Cliff et al., 1999	BP Helium Dilution	0.47±0.48
	BP Mathematical Modelling	0.53±0.63
	BP Nitrogen Balance	0.08±0.48
Kauppinen – Walin et al., 1980	133 – Xenon Imaging Helium Dilution	0.72±0.58
O'Donnell et al., 2000	BP Helium Dilution	0.63±0.22
	BP CT Scan	0.87±0.21
	Helium Dilution CT Scan	0.24±0.21
Mitchell et al., 1982	Helium Dilution Oxygen Wash In	0.02±0.21

Appendix B: Photographic Layout of the Device

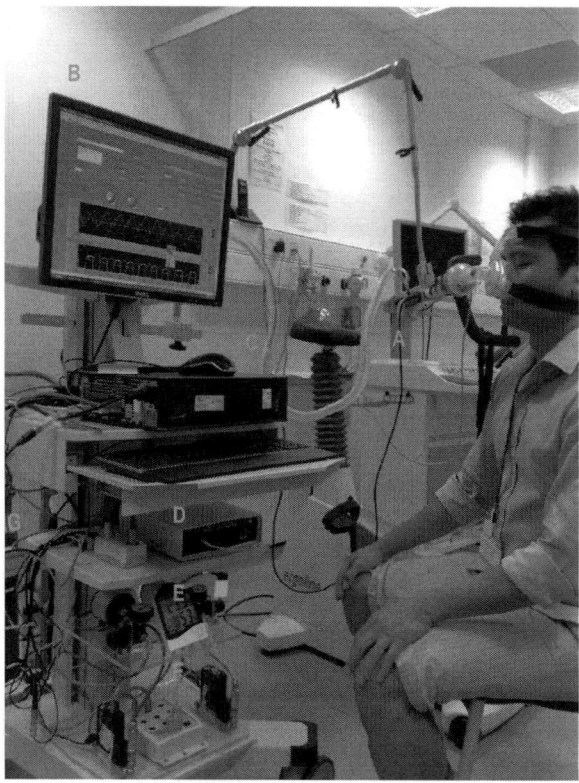

A. Flow and concentration sensing system
B. Computer and monitor (LabView Controller and
    Matlab Analysis)
C. Input from gas delivery system
D. N_2O Infra – red sensor
E. Temperature, humidity monitor
F. Gas delivery system
G. Nitrous Oxide

CONFLICT OF INTEREST

The authors declare that they have no conflict of interest.

REFERENCES

[1] A. Zwart, R. C. Seagrave, and A. V. Dieren, 'Ventilation-perfusion ratio obtained by a noninvasive frequency response technique', *J. Appl. Physiol.*, vol. 41, no. 3, pp. 419–424, Jan. 1976.

[2] A. Zwart, J. M. Bogaard, J. R. C. Jansen, and A. Versprille, 'A non-invasive determination of lung perfusion compared with the direct Fick method', *Pflüg. Arch. Eur. J. Physiol.*, vol. 375, no. 2, pp. 213–217, 1978.

[3] C. E. Hahn, A. M. Black, S. A. Barton, and I. Scott, 'Gas exchange in a three-compartment lung model analyzed by forcing sinusoids of N2O', J. Appl. Physiol., vol. 75, no. 4, pp. 1863–1876, Jan. 1993.

[4] C. E. Hahn, 'Oxygen respiratory gas analysis by sine-wave measurement: a theoretical model', J. Appl. Physiol., vol. 81, no. 2, pp. 985–997, Jan. 1996.

[5] D. J. Gavaghan and C. E. W. Hahn, 'A tidal breathing model of the forced inspired inert gas sinewave technique', Respir. Physiol., vol. 106, no. 2, pp. 209–221, Nov. 1996.

[6] J. P. Whiteley, D. J. Gavaghan, and C. E. W. Hahn, 'A tidal breathing model of the inert gas sinewave technique for inhomogeneous lungs', Respir. Physiol., vol. 124, no. 1, pp. 65–83, Dec. 2000.

[7] E. M. Williams, J. B. Aspel, S. M. Burrough, W. A. Ryder, M. C. Sainsbury, L. Sutton, L. Xiong, A. M. Black, and C. E. Hahn, 'Assessment of cardiorespiratory function using oscillating inert gas forcing signals', J. Appl. Physiol., vol. 76, no. 5, pp. 2130–2139, Jan. 1994.

[8] E. M. Williams, R. Hamilton, L. Sutton, and C. E. . Hahn, 'Measurement of respiratory parameters by using inspired oxygen sinusoidal forcing signals', J. Appl. Physiol., vol. 81, no. 2, pp. 998–1006, Jan. 1996.

[9] E. M. Williams, M. C. Sainsbury, L. Sutton, L. Xiong, A. M. S. Black, J. P. Whiteley, D. J. Gavaghan, and C. E. W. Hahn, 'Pulmonary blood flow measured by inspiratory inert gas concentration forcing oscillations', Respir. Physiol., vol. 113, no. 1, pp. 47–56, Jul. 1998.

[10] E. M. Williams, R. M. Hamilton, L. Sutton, J. P. Viale, and C. E. W. Hahn, 'Alveolar and Dead Space Volume Measured by Oscillations of Inspired Oxygen in Awake Adults', Am J Respir Crit Care Med, vol. 156, no. 6, pp. 1834–1839, Dec. 1997.

[11] L. Clifton, D. A. Clifton, C. E. W. Hahn, and A. D. Farmery, 'Assessment of lung function using a non-invasive oscillating gas-forcing technique', Respir. Physiol. Neurobiol., vol. 189, no. 1, pp. 174–182, Oct. 2013.

[12] G. Hedenstierna, P. O. Järnberg, and I. Gottlieb, 'Thoracic gas volume measured by body plethysmography during anesthesia and muscle paralysis: description and validation of a method', Anesthesiology, vol. 55, no. 4, pp. 439–443, Oct. 1981.

[13] R. Brown, D. E. Leith, and P. L. Enright, 'Multiple breath helium dilution measurement of lung volumes in adults', Eur. Respir. J., vol. 11, no. 1, pp. 246–255, Jan. 1998.

[14] D. Vilozni, O. Efrati, F. Hakim, A. Adler, G. Livnat, and L. Bentur, 'FRC measurements using body plethysmography in young children', Pediatr. Pulmonol., vol. 44, no. 9, pp. 885–891, Sep. 2009.

Wireless Sensor Network for Real Time Healthcare Monitoring: Network Design and Performance Evaluation Simulation

Minh-Thanh Vo[1], T.T. Thanh Nghi[1], Van-Su Tran[1], Linh Mai[1], and Chi-Thong Le[2]

[1] School of Electrical Engineering, International University, VNU- HCM, Vietnam
[2] Ho Chi Minh City University of Technology, Viet Nam

Abstract— Nowadays, Wireless sensor network (WSN) technologies are considered as potential solution healthcare monitoring applications. Different researches focus on network designing network for health care monitoring wireless sensor network (HCWSN), especially in the physical design of the HCWSN. However, work to evaluate the performance these network designs is largely lacking. This paper presents a HCWSN network design and simulation study to evaluate the performance in different scenarios such as network topologies, routing and media access control protocols. A practical WSN for HCWSN and a prototype SPO2 device integrated with WSN node have been designed. The testing results are also described in this paper.

Keywords— WSN, healthcare, monitoring, simulation, network design.

I. INTRODUCTION

The world population is getting older. The worldwide population over age 65 is expected to more than double from 357 million in 1990 to 761 million by 2025[1]. This results in more need for medical services to take care for this elderly population. However, the concurrent capacity of healthcare systems has not kept pace with the growing demand. The need of delivering quality care to a rapidly growing population of elderly while reducing the health-care costs is an important issue. Wireless Sensor Networks (WSN) for healthcare have emerged in the recent years as its low-cost, low-power, reliability, multi-hop networking, and scalability characteristics.

Several prototype healthcare systems based on WSN have been developed [2][3]..., These researches demonstrated the potential of WSNs for patient monitoring, the improvement of patient caring system.

This paper focuses on HCWSN network design and performance evaluation simulation. Different network designs have been studied, simulation scenarios with different network topologies, routing and media access control protocols have been simulated and evaluated before a new network design proposed. A prototype SPO2 device integrated with WSN node has been designed. A practical WSN for HCWSN has been developed and testing in this research.

The remainder of this paper is structured as follows: Section 2 presents related works, Section 3 describes the network design, Section 4 describes system implementation, Section 5 describes performance evaluation. Finally, the conclusion is drawn in Section 6.

II. RELATED WORKS

Several academic and commercial health care projects based on wireless sensor networks have been developed in the world [4].

This section provides a literature survey of some previously proposed HCWSNs. Specifically, we survey the CodeBlue[2], MEDiSN[3], and MASN[5] HCWSNs because they appear as the most cited projects in the literature. These systems are physiological monitoring applications. In these systems, low-power sensors are used to measure and report the person's vital signs (e.g. Oximetry, heart rate, temperature, etc.).

CodeBlue [2] focuses on automate patient monitoring process; this project proposed a robust wireless sensor network infrastructure, which can be implemented in different medical configurations (in the hospital, at home, etc.). CodeBlue integrates various medical sensors (EKG Electrocardiograph, pulse oximeter and EMG) and includes protocols for device discovery and a muti-hop routing model.

MEDiSN[3] is a wireless sensor network based system that aims to automate the physiological monitoring of patients in hospitals and in mass disasters. MEDiSN consists of several physiological monitoring motes for measuring physiological data of patients. MEDiSN also incorporates Relay Points that self-organize into routing trees that connect physiological monitoring motes to one or more gateways. Data routing is based on a modified version of the Collection Tree Protocol.

MASN[5]: the robust Medical Ad hoc Sensor Network (MASN) is a practical hardware and software platform designed to perform real-time collection of health care data. MASN uses a cluster-based communication scheme as its routing protocol for transmitting data. The protocol groups wireless sensor nodes in clusters to detect signals for the purpose of prolonging the lifetime of MASN, load balancing, and scalability. The clustering scheme reliably relays collected ECG data to the ECG server (sink) in the form of aggregated packets. As a result, it is able to provide fast and

© Springer International Publishing Switzerland 2015
V. Van Toi and T.H. Lien Phuong (eds.), *5th International Conference on Biomedical Engineering in Vietnam,*
IFMBE Proceedings 46, DOI: 10.1007/978-3-319-11776-8_22

accurate event detection and reliability control capabilities to the area where the event is occurring because the overhead, latency, and packet loss are reduced.

III. NETWORK DESIGN

A. Healthcare Monitoring Application Specification

In this section, we describe the specification of our WSN based medical application. We suppose our system will be implemented in a multi-storey building hospital. The aim of our system is to monitor patients' physiological parameters within the hospital. The main objective of our application is collecting the pulse oximeter parameter (SpO2) (and other vital parameters in our future development) of patients when they leave their rooms and move within the hospital. Each patient will carry a portable SPO2 device with a unique Identification (ID) code which can be called Patient ID, the real-time collection of SpO2 parameter will permit to medical caregiver to monitor the variation of patients' oxygen levels and to intervene when the collected values fall below normal ones. Our system can also be developed to support patient tracking function when they move within the hospital. The system design and arrangement of wireless sensor nodes in the hospital building should guarantee the reliability of the system. The moving nodes carried by patients can transmit data to monitoring center without interruption even though the patients are moving freely in the hospital. The system should improve the quality of health care in hospitals, this system must meet the following requirements: ease of use, mobility[3], reliable communications[2], security[3], quality of service[2], and scalability [6].

B. System Analysis and Simulation Study

To design system architecture for real time WSN for healthcare monitoring, we analyze the performance of WSN according to network topologies. In fact, wireless sensor networks can be configured to operate in three kinds of topologies including: Star, Mesh, and Cluster Tree Topology. In mesh topology, WSN nodes connect directly to each other for peer to peer communication. In Star topology, using a single PAN coordinator, each node connects directly to the central coordinator; all inter-node communications are passed through the coordinator. In cluster tree network, it consists of a number of star networks connected whose central nodes are also in direct communications with the single PAN Coordinator. Table 1 summarizes the performance of WSN with three topologies.

Table 1 Comparison WSN performance by topologies

	Star	Mesh	Cluster Tree
Scalability	No	Yes	Yes
Energy Efficiency	Yes	No	Yes
Network Synchronization	Yes	No	Yes
Redundant Paths	No	Yes	No
Node Mobility	Partial	Yes	Partial
Deterministic Routing	Yes	No	Yes
Contention-free medium access	Yes	No	Yes

From the comparison results, we can realize that star topology cannot satisfy the scalability and redundant paths criteria. Therefore, star topology is not suitable for WSNs for real time healthcare monitoring applications, especially the application in hospital multi-storey building environment. Mesh topology support scalability and node mobility, but doesn't support energy efficiency, deterministic routing and Contention-free medium access. Cluster tree support all criteria except redundant paths. To verify the performance comparison of WSN using mesh topology versus cluster tree topology we conduct simulation experiments using NS-2 simulator. The simulation parameters are summarized in table 2.

Table 2 WSN simulation parameters for mesh and cluster tree topologies

Channel	WirelessChannel
Propagation	TwoRayGround
Phy/WirelessPhy	802_15_4
MAC	802_15_4
Antenna	OmniAntenna
Max packet in interface queue	50
Number of nodes	24
Routing Protocol	AODV
Size	700m×700m
CSThresh	25m
RXThresh	25m
Traffic	CBR
Simulation Time	400s
Payload	50 bytes

The simulation results show that delay performance and packet delivery performance of cluster tree WSN is better than mesh for medium scale WSN network; MAC protocol 802.15.4 with Non-Beacon mode has better delay performance than that with Beacon mode. The explanation for this result is that the mesh topology uses redundant routing paths to enhance end-to-end reliability of data transmission but this scheme induces unpredictable end-to-end connectivity between nodes. In a cluster-tree topology, there is a single routing path between any pair of nodes, therefore, multi-hop communication is deterministic and time efficient. Fig. 1 shows one example result of the simulation.

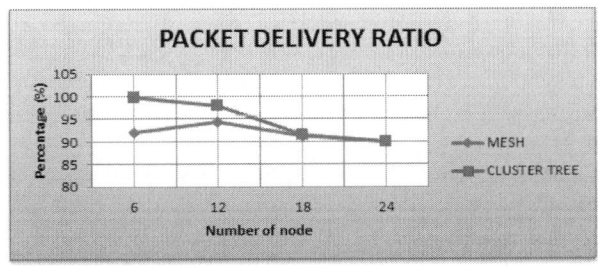

Fig. 1 Packet Delivery Ratio performance of WSN with mesh and cluster tree topologies

C. System Architecture Design

In this research, a prototype healthcare monitoring system for hospitals is designed. The wireless sensor network based healthcare monitoring system places unobtrusive wireless sensors on a person's body to form a wireless sensor network which can transmit the patient's health status with gateway node connected to the monitoring PC. The architecture of the proposed system are shown in figure 7. The system consists of four parts: (1) the sensor nodes which are responsible for collecting the physiological signals from patient, (2) the router nodes consist of a number of wireless relay nodes which is in charge of forwarding the healthcare data to the gateway node, (3) the gateway node (including coordinator node and Internet interface module which receives the relayed data and sends it to the monitoring computer through internet of direct connecting cable and (4) and the graphical user interface (GUI) which is responsible for storing, analyzing and presenting the received data in graphical and text format, this software can help the doctors and nurses in the hospitals to monitor continuously the health status of patients and take care them on time in case of emergency conditions. The designed system architecture is shown in figure 2.

To satisfy the real time requirement of health care monitoring system, we suggest selecting cluster tree topology and MAC protocol 802.15.4 with Non-Beacon mode for our WSN based system design due to their good performance in terms of delay, throughput and packet delivery ratio in our simulation results.

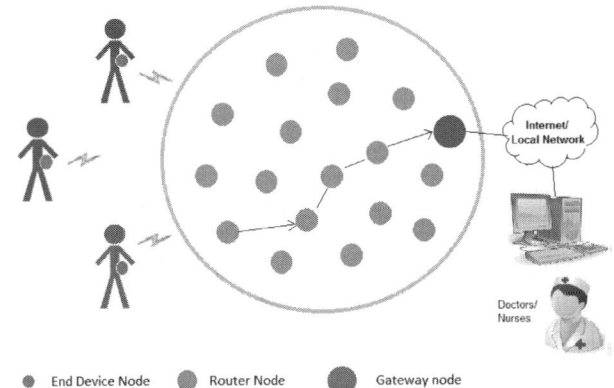

Fig. 2 Healthcare monitoring system architecture

IV. SYSTEM IMPLEMENTAION

A. Medical Sensor Node Design

To implement the healthcare monitoring system, we design the medical sensor node to collect vital sign parameters. In this research, we focus on network design, we develop medical sensor node to collect only SPO2 parameter and send to monitoring center. The block diagram of SPO2 sensor circuit is shown in Fig. 3.

In designing the SPO2 device, the sensor selected to measure the light intensity of both emitters is a digital Light to Frequency (LTF) device manufactured by Texas Advanced Optoelectronic Solutions (TAOS). It has a digital TTL output instead of a low level analog output that is used in most oximetry equipment. The advantages of such a device are better noise immunity and no analog signal conditioning required.

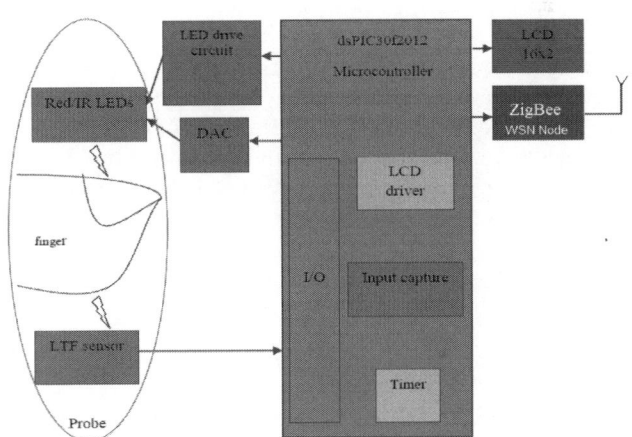

Fig. 3 SPO2 Device Block Diagram

In designing the SPO2 device, the sensor selected to measure the light intensity of both emitters is a digital Light to Frequency (LTF) device manufactured by Texas Advanced Optoelectronic Solutions (TAOS). It has a digital TTL output instead of a low level analog output that is used in most oximetry equipment. The advantages of such a device are better noise immunity and no analog signal conditioning required.

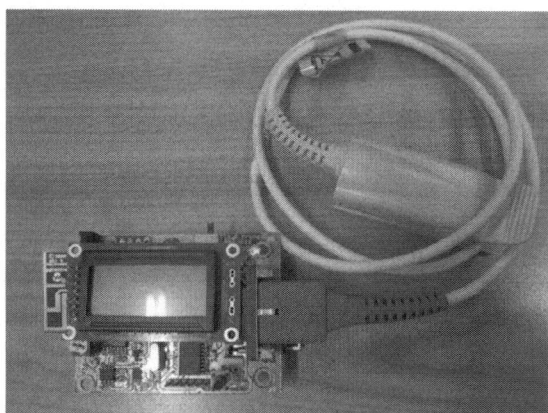

Fig. 4 Hardware Design of SPO2 sensor node

All digital signal processing techniques to process the SPO2 signal are implemented in one Microcontroller dsPIC30f2012. To transmit and receive data through WSN, 2.4GHz IEEE 802.15.4 Transceiver Module (MRF24J40MB) of Microchip is used to design the WSN node to due to their low cost, low power consumption and small size characteristics. Fig. 4 shows the hardware design of SPO2 sensor node.

V. PRACTICAL IMPLEMENTATION

In this research, a prototype system has been implemented in our university building as shown in Fig.5. The system includes one coordinator node, eight router nodes, and two end device nodes. All coordinator and router nodes are placed in fixed locations of 6th Floor of the building; the two SPO2 end device nodes are the mobile nodes which supposed to be carried by patients around the floor.

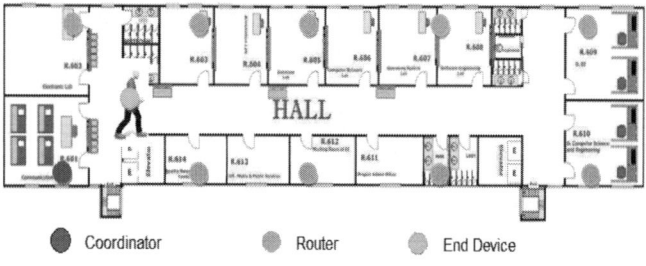

Fig. 5 Practical System Implementation

To run and test the system, the first step is to configure ZigBee wireless sensor network working with cluster tree topology. The coordinator node is preprogrammed with the PANID (Personal Area Network Identifier), although it is possible for the coordinator to dynamically scan for existing network PAN IDs in the same frequency and generate a PAN ID that does not conflict. All devices connected to the healthcare monitoring network are assigned a fixed 64 bit MAC address. At the stage of the network initialization, the coordinator assigns itself the short address 0x0000. Whenever each device moves and joins in the system, it will be assigned a dynamic 16 bit short address. After the joining stage, data can be routed from end device nodes to coordinator node through wireless connections, and vice versa.

VI. PERFORMANCE EVALUATION

Several measurements are carried out to evaluate the performance of the ZigBee wireless sensor network on healthcare monitoring system to confirm the reliability and feasibility of the system. In this paper we present Delay Performance and functional testing results.

A. Delay Performance

Average End-to-End delay indicates the length of time taken for a packet to travel from the source to the destination. It represents the average data delay an application experiences during transmission of data. In our measurement, we evaluate the quality of network by time delay corresponds with the distances. Fig. 6 shows the practical delay measurement of the WSN based health monitoring system.

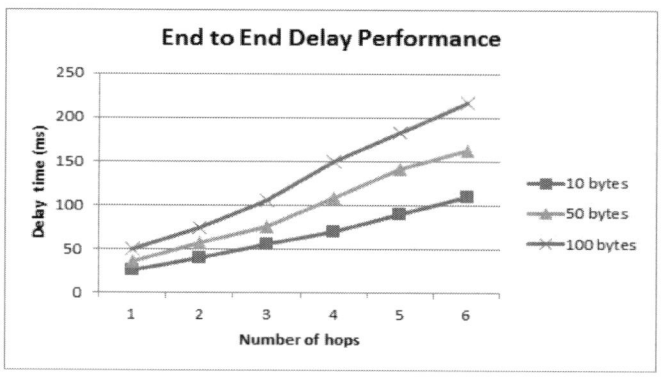

Fig. 6 Muti-hop delay performance of the system

In general, the payload size of 10, 50 and 90 bytes rises dramatically in the number of 1 to 6 hops. Three payload sizes are the similar direction but time delay of 100 bytes is the highest (from 49 to 216 ms), time delay of 50 bytes is the second (from 36 to 163 ms) and time delay of 100 bytes is the lowest (from 25 to 110 ms). In common, when the side of payload and the number of hops increase, the time delay is also increases. Actually, delay is influenced by different back-off stages due to network traffic by the number of nodes and size of payloads. In the context of healthcare monitoring application, time delay from this system is acceptable.

B. Functional Testing

Continuous SPO2 data monitoring had been tested to investigate the performance of the healthcare monitoring system in building environment, the result shows that the system work well in the building environment.

VII. Conclusion

In this paper, a simulation study of wireless sensor network for healthcare monitoring application has been carried out. Cluster tree topology, and MAC protocol 802.15.4 with Non-Beacon mode are our choice for WSN based healthcare monitoring due to their good performance in term of delay, throughput and packet delivery ratio.

A prototype WSN based health monitoring system has been designed and tested in our university building environment. The performance results of the system confirm that WSN based health monitoring system is practically applicable in building environment. However, the latency of the system is acceptable for SPO2 data monitoring, but this latency is still high for other vital signs such as ECG, EKG... it is necessary to develop better routing protocols to reduce the latency of system.

References

1. E. Egbogah and A. O. Fapojuwo, "A survey of system architecture requirements for health care-based wireless sensor networks," Sensors,vol. 11, no. 5, pp. 4875–4898, 2011.
2. V. Shnayder, B.-r. Chen, K. Lorincz, T. R. F. Jones, and M. Welsh,"Sensor networks for medical care," in SenSys, vol. 5, 2005, pp. 314–314.
3. J. Ko, J. H. Lim, Y. Chen, R. Musvaloiu-E, A. Terzis, G. M. Masson,T. Gao, W. Destler, L. Selavo, and R. P. Dutton, "Medisn: medical emergency detection in sensor networks," ACM Transactions on EmbeddedComputing Systems (TECS), vol. 10, no. 1, p. 11, 2010.
4. N. Sghaier, A. Mellouk, B. Augustin, Y. Amirat, J. Marty, M. E. A.Khoussa, A. Abid, and R. Zitouni, "Wireless sensor networks for medical care services," in Wireless Communications and Mobile ComputingConference (IWCMC), 2011 7th International. IEEE, 2011, pp. 571–576.
5. Hu, M. Jiang, L. Celentano, and Y. Xiao, "Robust medical ad hoc sensor networks (masn) with wavelet-based ecg data mining," Ad Hoc Networks, vol. 6, no. 7, pp. 986–1012, 2008.
6. K. Lorincz, B.-r. Chen, G. W. Challen, A. R. Chowdhury, S. Patel, P. Bonato, M. Welsh et al., "Mercury: a wearable sensor network platform for high-fidelity motion analysis." in SenSys, vol. 9, 2009, pp. 183–196.

Estimation of Guidewire Inclination Angle for 3D Reconstruction

T. Petković and S. Lončarić

University of Zagreb, Faculty of Electrical Engineering and Computing,
Unska 3, HR-10000 Zagreb, Croatia

Abstract— **Reconstruction of the 3D position of the guidewire is an important step toward automated navigation systems supporting minimally invasive endovascular interventions. We present a method to estimate the projected guidewire thickness for the monoplane X-ray configuration and we propose a novel method to recover the local angle between the guidewire and the X-ray path and to recover the inclination angle of the guidewire to the imaging plane, up to an orientation. We experimentally show the feasibility of the proposed approach by recovering both the local and the inclination angle of the guidewire; the average error for the recovered inclination angle is 8.6±8.2°.**

Keywords— **X-ray imaging, guidewire, inclination angle, thickness estimation.**

I. INTRODUCTION

During a minimally invasive endovascular intervention a guidewire is introduced into the vasculature and must be navigated to a point of interest, usually an aneurysm, a stenosis or an arteriovenous malformation. Accurate navigation through the vessels is a prerequisite for a successful intervention. Fluoroscopic X-ray imaging is currently used to aid in navigation, however, physicians still interpret obtained 2D fluoroscopic images themselves. Robust recovery of the 3D position from a single fluoroscopic image would represent a significant navigational improvement.

Either monoplane or biplane configuration is used for navigation. Position reconstruction for the biplane configuration is straightforward [1], but for the monoplane configuration there is an inherent ambiguity that must be overcome by using some additional knowledge. In previous works by Van Walsum et al. [2] and Petković et al. [3] an optimization method relying on known position of the blood vessels is used. Esthappan et al. [4] uses 2D/3D registration to align the model to the projection of the pre-shaped guidewire tip. Abovementioned methods only extract the guidewire position and local orientation from the 2D fluoroscopic images.

In this paper, we present a technique to estimate the projected guidewire thickness and we present a novel extension to recover the local angle between the guidewire and the X-ray path, up to an orientation, and to recover the guidewire inclination angle toward the image plane, up to an orientation, using a single fluoroscopic image. The inclination angle is significant for the reconstruction of the 3D position of the guidewire as it, together with the local 2D direction vector in the image plane, defines the local 3D direction vector of the guidewire.

The paper is structured as follows: in Section II a simple X-ray imaging model is presented; in Section III we show how to estimate the thickness and present the method to estimate the inclination angle; results are presented in Section IV; we conclude in Section V.

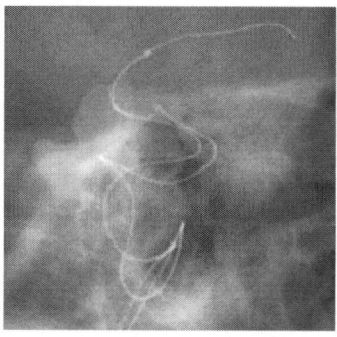

Fig. 1 A fluoroscopic image showing guidewires during a neuro-intervention.

II. IMAGING MODEL

If a homogeneous and isotropic object is placed between the X-ray source and the detector then the observed intensity can be approximated by [5,6]

$$I = I_0 \exp\left(-\sum_i \mu_i d_i\right) = I_0 \prod_i \gamma_i , \qquad (1)$$

where μ_i [cm^{-1}] is the linear attenuation coefficient and γ_i is the multiplicative attenuation factor of an i-th object on the X-ray path. The multiplicative model (1) is valid for a narrow energy band and disregards the beam diffusion. Any factor of the quantity $\prod_i \gamma_i$ may model the guidewire. Let γ_{gw} denote the multiplicative term of the guidewire. Then for pixels where the guidewire is absent we observe the intensity

$$I_1 = I_0 \prod_i \gamma_i , \qquad (2)$$

and for pixels where the guidewire is present we observe the intensity

$$I_2 = I_0 \exp\left(-\sum_i \mu_i d_i\right) \exp\left(-\mu_{gw} d_{gw}\right) = I_1 \gamma_{gw} . \quad (3)$$

The imaging model is multiplicative in intensity and is summarized by Equations (2) and (3). The multiplicative nature causes the contrast variation along the length of the guidewire; from this contrast variation the projected thickness and the inclination angle can be recovered using a single fluoroscopic image.

III. THICKNESS AND ANGLE RECOVERY

Let's assume the values (2) and (3) are known. Using division the multiplicative term γ_{gw} can be recovered,

$$\frac{I_2}{I_1} = \frac{I_1 \gamma_{gw}}{I_1} = \gamma_{gw} , \quad (1)$$

and an estimate of the projected guidewire thickness d_{gw} can be obtained as

$$d_{gw} = -\frac{1}{\mu_{gw}} \log \gamma_{gw} . \quad (2)$$

Note that d_{gw} given by (2) is the length of the path X-rays travel to pass through the guidewire and is not equal to the guidewire diameter $2r$ (Fig. 2). Also, to recover d_{gw} in real-world units the value of μ_{gw} [cm^{-1}] must be known, thus used imaging system must be calibrated and guidewire parameters must be known. However, if we are interested in recovering only local angles between the guidewire and X-rays then the value of μ_{gw} [cm^{-1}] is not needed.

Consider the geometry shown in Fig. 2. To recover the local angle α between the guidewire and an X-ray we use

$$2r = d_{gw} \cos(\alpha) , \quad (3)$$

so the diameter $2r$ must be known. However, by using (2) we obtain

$$\cos(\alpha) = \frac{2r}{d_{gw}} = \frac{\log \gamma_{gw,min}}{\log \gamma_{gw}} , \quad (4)$$

and the term μ_{gw} disappears, therefore the angle α can be recovered from the un-calibrated data. The value of $\gamma_{gw,min}$ that is tied to the diameter $2r$ can be estimated using the same data.

Once the local angle α is known the inclination angle θ can be obtained as

$$\theta = 90° + \alpha - \beta , \quad (5)$$

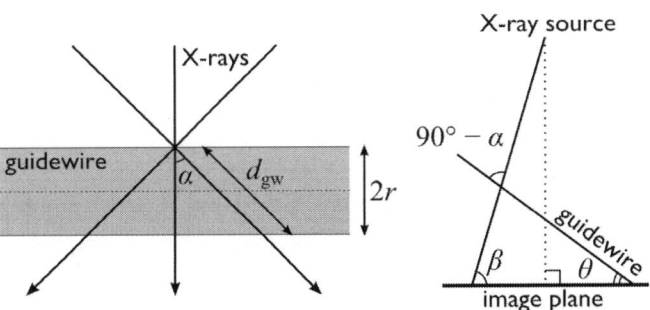

Fig. 2 Geometry of an intersection between the guidewire and an X-ray. Note that the solution is not unique, there are two rays that yield the same value of projected thickness d_{gw}.

but only up to an orientation, as the guidewire can have either falling or raising slope (Fig. 2). The angle β is the angle between X-ray and the image plane and is determined by the C-arm geometry.

The value I_2 is measured directly from the input X-ray image from pixels that are identified as the guidewire by some segmentation procedure, e.g. as described in [7,8,9, 10]. Then, using the values of adjacent background pixels the value I_1 must be estimated as it cannot be measured directly.

Therefore, the proposed method to recover the inclination angle of the guidewire from a single X-ray image is comprised of the following steps:

1. segmentation of the guidewire in the X-ray image and measurement of intensity values I_2,
2. estimation of the background values I_1 using intensity values adjacent to pixels identified in step 1,
3. computing the ratio (1) and estimating the value of $\gamma_{gw,min}$, and
4. estimation of angles α and θ using Eqs. (4) and (5).

For the step one of the implementation we use our previous work, the adaptive vesselness measure [9], which yields both the segmentation and the local guidewire direction in 2D. Combining this information with the estimated θ defines the local guidewire direction vector in 3D. Background I_1 is estimated as an arithmetic mean of adjacent pixels on both sides of the segmented guidewire.

IV. RESULTS AND DISCUSSION

To determine if the proposed estimation scheme for the inclination angle can be used a trial phantom image was acquired. Acquisition was done using Philips Allura X-ray imaging system with pixel size 0.217 mm and 1016×1016 image resolution. The guidewire was placed so it forms a large circle that is located mostly in a plane orthogonal to the imaging plane, starting with an inclination angle of about 40° sloping downward toward the image plane (Fig. 3). For this placement we expect the projected thickness to vary along the guidewire length thus indicating the local guidewire orientation relative to X-ray paths, and to cover whole range of inclination angles of interest.

The segmented guidewire in 2D is shown in Fig. 4 (computed as described in [9]). Detail of the segmentation is shown in Fig. 5: the value I_2 is measured directly from the input X-ray image at the guidewire centerline (yellow); the value I_1 is estimated as arithmetic mean of adjacent pixels on both sides of the guidewire (red). Note that if an X-ray image sequence is available then the Kalman filter approach of [10] can be used for improved temporal estimation of the background intensity I_1.

The projected thickness estimate computed directly using I_1 and I_2 must be smoothed as the initial estimate based on single pixel values is too noisy for practical use (Fig. 6, thin line). This effect is both due to present noise and due to suboptimal placement of the guidewire centerline (pixel instead of sub-pixel precision). However, applying a smoothing spline with parameter $p = 0.01$ significantly improves the result (Fig. 6, thick line). Note the thickness estimate is noisy but is varying along the length of the guidewire as expected; there are two sharp increases in the estimated thickness corresponding to the raising and the falling slopes of the circle the guidewire forms.

The value of $\gamma_{gw,min}$ must be estimated from the projected thickness estimate. For this experiment we have manually estimated $\gamma_{gw,min} = 0.9$ using the value at the central position between two peaks; this position corresponds to the segment parallel to the imaging plane. Note that due to noise and disregarded partial volume and beam diffusion effects the value of 0.9 is not the minimum value so cos of Eq. (4) is not well-defined for all data points.

For the practical use of the proposed technique the estimation procedure for $\gamma_{gw,min}$ must be automated: if the position of the blood vessels is known then it may be estimated using the known blood vessel angle and the assumption the guidewire has the same direction as blood vessels.

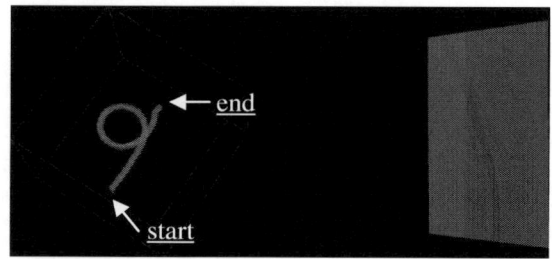

Fig. 3 Imaging geometry (3DRA reconstruction, imaging plane).

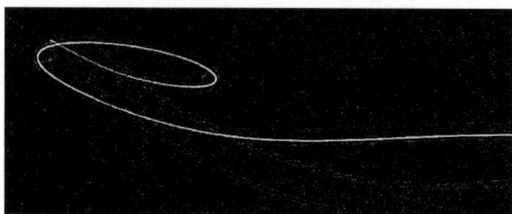

Fig. 4 Segmented guidewire (adaptive vesselness map).

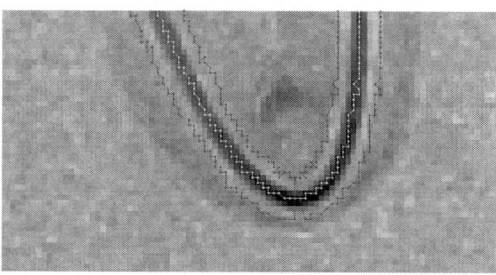

Fig. 5 Detail showing segmented guidewire centerline and adjacent pixels on both sides of the guidewire (transition area of about 3 px excluded). Note the contrast variation along the length of the guidewire and the peak at the point of maximum curvature.

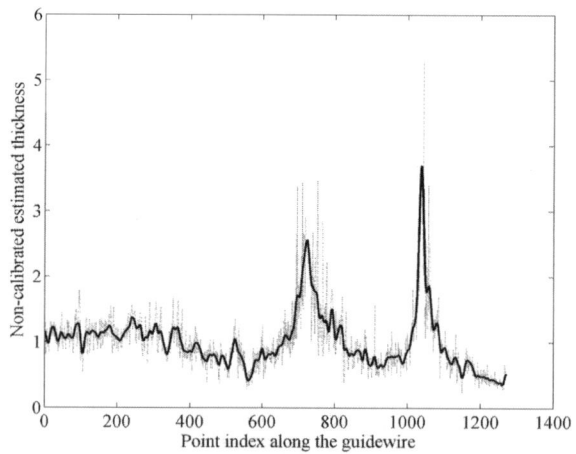

Fig. 6 Smoothed thickness estimate (initial estimate is shown as thin blue line).

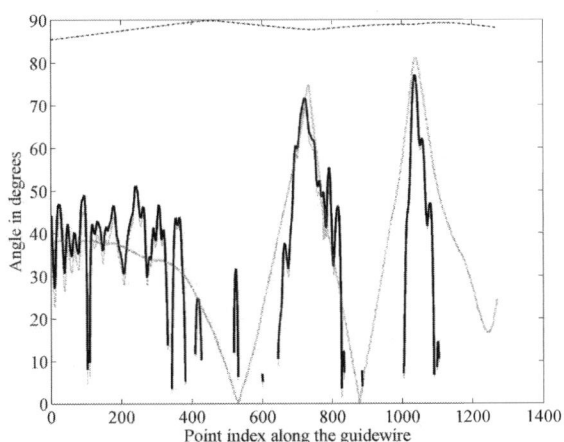

Fig. 7 Estimated angles: local angle α between the guidewire and an X-ray is represented by a thin blue line, X-ray angle β is represented by a dotted black line, and inclination angle θ is represented by a thick black line (estimation) and thin red line (true value).

Once $\gamma_{gw,min}$ is known the inclination angle can be estimated. Estimated angles are shown in Fig. 7, starting at about 40°, then dropping toward 0° as the loop starts etc. The true value of the angle recovered using 3DRA reconstruction is shown in red. A total of 1272 pixels is segmented as the guidewire; inclination angle estimation succeeded for 678 pixels and failed for 594 pixels due to Eq. (4) being larger than one. For data points where estimation succeeded the average inclination angle error is 8.6±8.2°, but the maximum error is 52.5°. The average error is a promising first result for application in 3D reconstruction, especially as further improvements are expected. However, further post-processing is needed to reduce the maximal error.

V. CONCLUSIONS

We have presented a novel estimation technique to recover the local angle between the guidewire and the X-ray path, up to an orientation, and to recover the guidewire inclination angle toward the image plane, up to an orientation, using a single fluoroscopic image. Using this information a guidewire direction in 3D can be recovered and used as an aid in the 3D position reconstruction thus further improving upon existing 3D reconstruction techniques described in the literature.

Future work should focus on improving the estimation of I_1 and I_2 using sub-pixel precision and the Poisson noise formalism. Furthermore, partial volume effects and beam diffusion must be included in the imaging model.

ACKNOWLEDGMENT

This research has been supported by the European Community's Seventh Framework Programme under grant No. 285939 (ACROSS). The authors would like to sincerely thank Robert Homan of Philips Healthcare for provided X-ray images.

CONFLICT OF INTEREST

The authors declare that they have no conflict of interest.

REFERENCES

1. Baert S, van de Kraats E, van Walsum T, Viergever M, Niessen W. Three-dimensional guide-wire reconstruction from biplane image sequences for integrated display in 3-D vasculature. IEEE Transactions on Medical Imaging 2003;22(10):1252–1258.
2. van Walsum T, Baert S, Niessen W. Guide wire reconstruction and visualization in 3DRA using monoplane fluoroscopic imaging. IEEE Transactions on Medical Imaging, 2005;24(5):612–623.
3. Petković T, Lončarić S. Real-time 3D position reconstruction of guidewire for monoplane X-ray. Comput. Med. Imaging. Graph. (2013), http://dx.doi.org/10.1016/j.compmedimag.2013.12.006
4. Esthappan J. Kupinski M, Lan L, Hoffmann K. A method for the determination of the 3D orientations and positions of catheters from single-plane X-ray images. In: Proceedings of the 22nd annual international conference of the IEEE Engineering in Medicine and Biology Society, vol. 3. 967 2000.
5. Hasegawa B. Physics of Medical X-Ray Imaging, revision of 2nd edition. Medical Physics Publishing Corporation, 1987.
6. Schram RCP. X-ray attenuation—application of X-ray imaging for density analysis, NRG, Tech. Rep. 20002/01.44395/I, Nov. 2001.
7. Bismuth V, Vancamberg L, Gorges S. A comparison of line enhancement techniques: applications to guide-wire detection and respiratory motion tracking. In Proc. SPIE, vol. 7259. SPIE, Mar. 2009.
8. Honnorat N, Vaillan R, Paragois N. Robust guidewire segmentation through boosting, clustering and linear programming. In: 2010 IEEE International Symposium on Biomedical Imaging: From Nano to Macro, Apr. 2010, pp. 924–927.
9. Petković T, Lončarić S. Using X-ray imaging model to improve guidewire detection. In: 2010 IEEE 10th International Conference on Signal Processing, Oct. 2010, pp. 805–808.
10. Petković T, Lončarić S. Guidewire tracking with projected thickness estimation. In: 2010 IEEE International Symposium on Biomedical Imaging: From Nano to Macro, Apr. 2010, pp. 1253–1256.

Author: Tomislav Petković
Institute: Faculty of Electrical Engineering and Computing
Street: Unska 3
City: Zagreb
Country: Croatia
Email: tomislav.petkovic.jr@fer.hr

Intelligent Heart Rate Variability Processing System

Hoang ChuDuc, Thuan NguyenDuc, Kien NguyenPhan, and Ha NguyenThai

School of Electronics and Telecommunications, Hanoi University of Science and Technology, Hanoi, Vietnam

Abstract— **In this research an expert system for automated detection of abnormality of heart rate variability. For this purpose, a data acquisition system from Holter ECG records, is developed by using QRS detection techniques for the future development of an standard ECG database. New and innovative medical applications based on developments in the wireless networks field are being developed in the research as well as commercial sectors. This trend has just started and we predict wireless networks are going to become an integral part of medical solutions due to its benefits in cutting down healthcare costs and increasing accessibility for patients as well as healthcare professionals. In this research we give some background on applications of wireless networks in the heart rate variability processing and discuss the issues and challenges. We have also tried to identify some of the standards in use. Another contribution due to this paper is the identification of innovative medical applications of wireless networks developed or currently being developed in the research and business sectors in Vietnam.**

Keywords— **Holter ECG records, HRV, Heart rate variability, wireless networks.**

I. INTRODUCTION

Electrocardiograms (ECGs) are used by medical professionals to monitor the heart of a patient. These devices usually operate with 12 leads connected to the patient's skin in a prescribed pattern. An ECG can be used to detect abnormal cardiovascular symptoms, measure heart rates, and monitor heart diseases. The most common non-medical application of an ECG is to measure a heart rate during a workout, however, the aim of this project is to prototype a device that could aid remote monitoring and feedback[1],[2].

A suitable procedure for cardiac surveillance should allow quasi-continuous and in most cases permanent, i.e. lifelong, and application. The used equipment should not constrain the patient in his daily routine activities and preferably not require special handling by the patient, but be forgettable if no risky situation is developing. There is increasing preliminary evidence that intramyocardial electrograms have the potential for efficient cardiac risk surveillance, especially if the system includes telemonitoring features, e.g. based on the Bluetooth and WLAN technology, cellular phone networks, intra- or internet technology or other global communication technologies which are already available, as well as advanced signal and information processing combined with powerful database systems.

The most stringent constraint until now is the limited processing capacity of the microprocessors which are used in implants, usually 16-bit devices and in many cardiac pacemakers still 8-bit devices. If more tasks have to be accomplished, e.g. pacing, then the total load for the battery may be another limiting constraint.

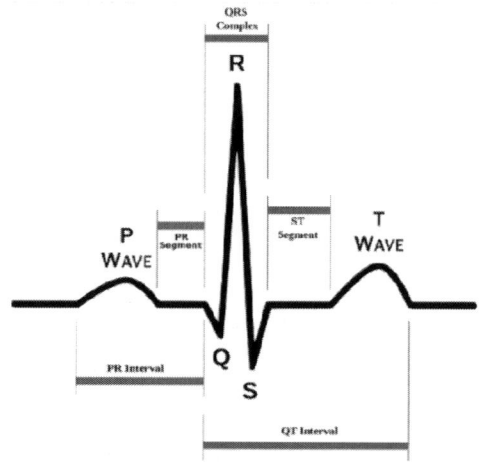

Fig. 1 ECG Parameters with P,Q,R,S,T wave

The most important components of the ECG that the device needs to report include P wave, QRS interval, and the T wave. These waveforms have the inherent issue that the measurable signals have amplitudes in the range of 0.1mV to 10mV. Another issue with these signals is that the smallest time components last as little as 50ms (The PR segment), or 80ms (The entire QRS complex). This short time means that the sample speed for the signal needs to be significantly less than 25ms to ensure adequate sampling. The polarization, depolarization, and contractions of the heart during the cardiac cycle produce signals which can be viewed on a device such as an oscilloscope [3]. The ECG signal can be visually represented by three major waves which are synchronized with the heart activity. The focus of the project relates to the basic measure that is related to the heart rhythm, the heart rate [4]. The heart rate simply describes the frequency of the cardiac cycle and is measured

V. Van Toi and T.H. Lien Phuong (eds.), *5th International Conference on Biomedical Engineering in Vietnam,*
IFMBE Proceedings 46, DOI: 10.1007/978-3-319-11776-8_24

in contractions or beats per minute (bpm). Figure 1 illustrates the wave components of the ECG signal. The PR interval is the duration of time between the beginning of the P wave, signifying atria depolarization, and the beginning of the QRS complex. It represents the time between the beginning of the contraction of the atrium and the beginning of the contraction of the ventricle. The QT interval extends from the beginning of the Q wave to the end of the T wave. It represents the time of ventricular contraction and repolarization [5]. The ST interval extends from the S wave to the end of the T wave. The standard ECG parameter values are listed in the Table 1 below:

Table 1 QRS parameter

No.	Table Column Head		
	Parameter	*Ranger*	*Unit*
1	Heart Rate	60 – 150	Beats Per Minute
2	PR Interval	0.1 - 0.25	Seconds
3	QT Interval	0.29 ± 0.14	Seconds
4	P Wave Duration	0.12 ± 0.04	Seconds
5	QRS Width	0.05 – 0.1	Seconds
6	T Wave Duration	0.08 ± 0.02	Seconds

As mentioned previously, the electric field generated by the heart is best characterized by vector quantities, however, it is generally convenient to directly measure only scalar quantities, i.e. a voltage difference (in the mV order) between given points of the body. The primary signal characteristics of an ECG signal has a useful frequency range of about 0.05Hz to 150Hz. For this reason, a good low frequency response is essential to ensure baseline stability and a good high frequency response is needed for attenuation of high frequency noise from other signals of biological origin [6],[7].

II. METHODS

A. System Block Diagram

Model ECG signal acquisition are often used as a minimum of three probes in Figure 2 below. In this model, the ECG signal acquisition, low-pass, high-pass filter, 50Hz noise filtering and variable analog-to-digital ADC. The signal from the ADC can be handled in many different ways through the MSP430 microcontroller. This microcontroller can be connected to computer systems to exchange data the two-way ECG.

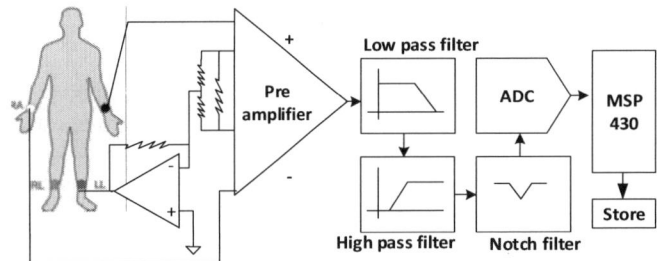

Fig. 2 Block diagram to receive ECG signals from patient

B. Intelligent Heart Rate Variability Processing System

Model ECG acquisition allows connection to many different inputs. Fig. 3. Intelligent Heart Rate Variability Processing System can be connected to the MSP430 module in Figure 2 or from the wireless block ECG, Holter, or from different sets. ECG data acquired will be synthesized and processed by a computer powerful enough configured through the software vendor or device through Matlab code, c #.

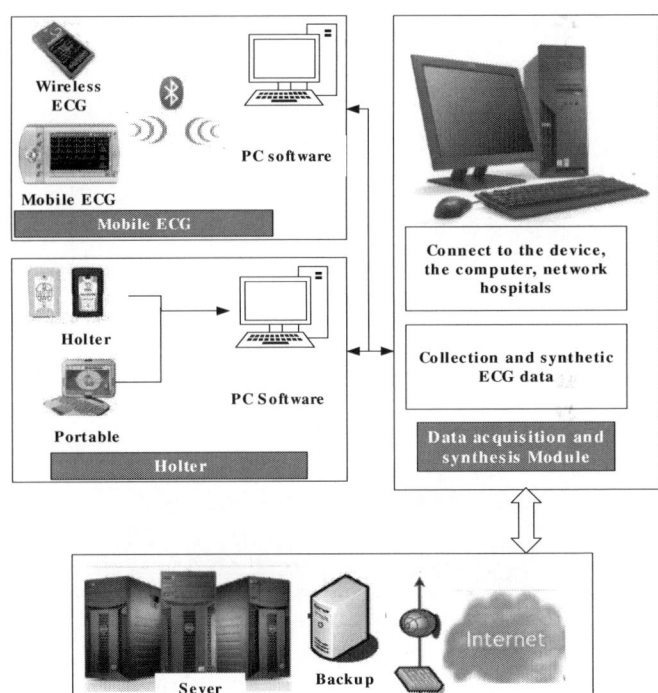

Fig. 3 Intelligent Heart Rate Variability Processing System

This Data acquisition and synthesis Module collect data and then transferr to an intermediate computer system. This machine connect to the device, the computer, network hospital, Collection and synthetic ECG data, stored in the database of the hospital network.

C. Multi Fluctualtion Detrended Analysis

Intervals between cardiac beats vary in a complex manner, presenting exponential correlations. Detrended fluctuation analysis is a method that allows the detection of long-range correlations embedded in an irregular signal, and avoids spurious detection of apparent long-range correlations that are an artifact to the object of the analysis [8]. In the context of HRV, DFA allows the distinction between complex fluctuations intrinsic to the nervous system in the command of vital actions of the human body, and those originated on the environment and that also influence the heart rate. Those fluctuations that are intrinsic to the nervous system happen to be observed throughout the signal, as opposed to the extrinsic fluctuations that present local and short term effects [9]. The main objective of DFA is to extract the extrinsic fluctuations in order to allow the analysis of the signal's variability associated exclusively with autonomic control.

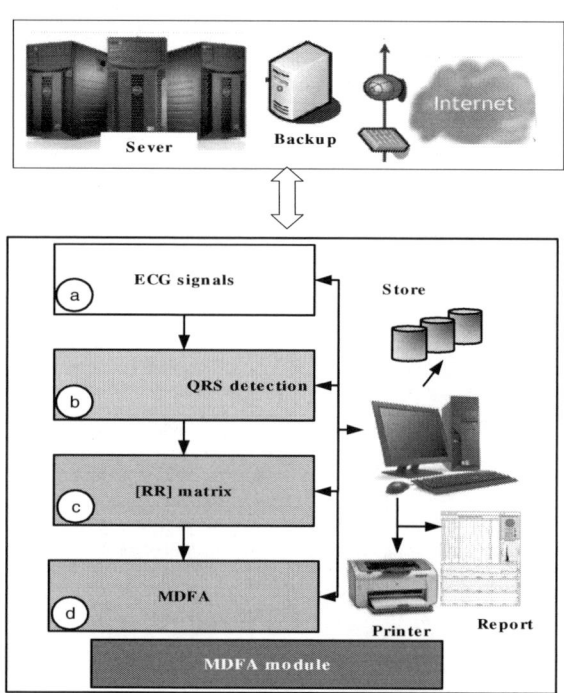

Fig 4 Multi Fluctualtion Detrended Analysis

We made some improvements to DFA transformation method to generate a sequence of values we DFA under the 20-minute interval or heart rate in 1000. This does not affect the calculation of the DFA components but create more value from 1 DFA data matrix ECG RR only.

D. Database

The MIT-BIH Arrhythmia database was used to evaluate the proposed data compression and modulation schemes. In this standard database, the ECG signals were digitized through sampling at 360 Hz with 11-bit resolution. The first 10000 samples of 10 MIT-BIH records have been tested. After data acquisition will be carried through the separation of the QRS complex to extract RR matrix [10]. RR Matrix of ECG wave is treated in the nonlinear domain method mainly MDFA (Multi Detrended Fluctuation Analysis). Results will be calculated and compared with the traditional method of DFA (Detrended Fluctuation Analysis). Filter the results will split about arrhythmias time low, keeping the amount of time the possibility of high arrhythmias.

III. RESULT

We have collected ECG signals, then transmitted wirelessly to a data processor. In the data processing, we analyzed the matrix RR domain under MDFA nonlinear method according to the data segment 20 minutes. The result of the calculation process MDFA will be classified according to three levels are stable, arrhythmias and arrhythmia medium high.

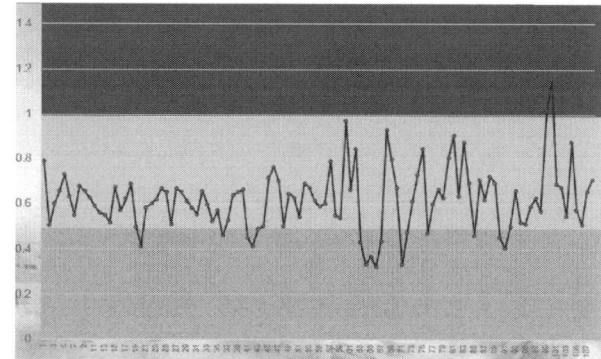

Fig. 5 Results of analysis of the degree of arrhythmia patients

Results of analysis of the degree of arrhythmia patients with follow-up time 23h53p with very high stabil-ity of 79%, the average level of arrhythmia and 3% to 18% will help doctors focus on distribution volume peri-ods occur arrhythmia respectively. This results is shown on Fig.5 and Table 2 below:

Table 2 DFA1 and DFA2 series according to the data segment 20 minutes

No	alpha	DFA1	DFA2	Meaning
1	0.5<alpha<1	84%	79%	High stability
2	0<alpha<0.5	15%	3%	Less correlated
3	1<alpha	1%	18%	Low correlation
4	alpha = 0.5	0%	0%	Noise
5	alpha = 1.0	0%	0%	Noise
6	alpha = 1.5	0%	0%	Noise

MDFA calculation process is done basically through some steps. Reduction algorithm analyzes trends on the dynamic signal analysis algorithms reducing the tendency to multi- value signal will allow limiting the downside of reducing trend analysis algorithm signal the direction of the former, while also adding some MDFA advantages:

- Track time is short enough fluctuations in the level of the heart rate associated with each other, thereby evaluate possibility of arrhythmias. Time may be short enough to allow at least 20 minutes.
- Evaluate the effectiveness and the number of beats RR minimum necessary to obtain accurate results MDFA.
- There are many values of MDFA, so evaluation is more detailed and precise.

Conclusions

We have built a model of a specific system for the wireless HRV acquisition and analysis HRV arrhythmia and provide an assessment of the level of arrhythmia, arrhythmias time. The system has been tested with sample data from the MIT-BIH with good results.

The analysis of the long-term data such as HRV with DFA along with other analysis for short-term data (up to 24 hours and overnight recording) may help in providing a tool to more accurately. More long-term data are being collected to give us an opportunity to re-test the analysis and to verify the results for different physiological states.

This system is capable of development and application in Vietnam to offload hospital, physician support for arrhythmia diagnosis quickly and accurately help patients acquire ECG signals more simple, convenient more convenient.

References

1. Peng C-K, Buldyrev SV, Havlin S, Simons M, Stanley HE, Goldberger AL. Mosaic organization of DNA nucleotides.*Phys Rev E* 1994;**49**:1685-1689.
2. Peng C-K, Havlin S, Stanley HE, Goldberger AL.Quantification of scaling exponents and crossover phenomena in nonstationary heartbeat time series. Chaos 1995;5:82-87.
3. Krishnam, R.; Nazeran, H.; Chatlapalli, S.; Haltiwanger, E.; Pamula, Y., "Detrended Fluctuation Analysis: A Suitable Long-term Measure of HRV Signals in Children with Sleep Disordered Breathing," Engineering in Medicine and Biology Society, 2005. IEEE-EMBS 2005. 27th Annual International Conference of the , vol., no., pp.1174,1177, 17-18 Jan. 2006
4. Signorini, M.G.; Sassi, R.; Cerutti, S., "Working on the Noltisalis database: measurement of nonlinear properties in heart rate variability signals," Engineering in Medicine and Biology Society, 2001. Proceedings of the 23rd Annual International Conference of the IEEE , vol.1, no., pp.547,550 vol.1, 2001
5. Leite, A.; Rocha, A.P.; Silva, M. E.; Gouveia, S.; Carvalho, J.; Costa, O., "Long-range dependence in heart rate variability data: ARFIMA modelling vs detrended fluctuation analysis," Computers in Cardiology, 2007 , vol., no., pp.21,24, Sept. 30 2007-Oct. 3 2007
6. A Goldberger, L Amaral, L Glass, J M Hausdroff, P C H Ivanov, R G Mark, J E Mietus JE, G B Moody, C K Peng, H E Stanley. Physiobank, PhysioToolkit, and PhysioNet: Components of a New Research Resource for complex Physiological Signals. Circulation 101(23):pp.215-pp.220
7. Penzel, T.; Kantelhardt, J.W.; Grote, L.; Peter, J.H.; Bunde, Armin, "Comparison of detrended fluctuation analysis and spectral analysis for heart rate variability in sleep and sleep apnea," Biomedical Engineering, IEEE Transactions on , vol.50, no.10, pp.1143,1151, Oct. 2003
8. Leite, A.; Rocha, A.P.; Silva, M.E., "Long memory and volatility in HRV: An ARFIMA-GARCH approach," Computers in Cardiology, 2009 , vol., no., pp.165,168, 13-16 Sept. 2009
9. Bojorges-Valdez, E.R.; Echeverria, J.C.; Valdes-Cristerna, R.; Pena, M.A., "Are "scaling patterns" useful tools for exploring fractality in heart rate variability data?," Computers in Cardiology, 2006 , vol., no., pp.93,96, 17-20 Sept. 2006
10. Krstacic, G.; Krstacic, A.; Martinis, M.; Vargovic, E.; Knezevic, A.; Jembrek-Gostovic, M.; Smalcelj, A.; Milicic, D.; Bergovec, M.; Gostovic, M., "Dynamic non-linear changes in heart rate variability in patients with coronary heart disease and arterial hypertension treated by amlodipine besylate," Computers in Cardiology, 2003 , vol., no., pp.485,488, 21-24 Sept. 2003

Author: Hoang ChuDuc
Institute: School of Electronics and Telecommunications
Street: No1, Dai Co Viet
City: Hanoi
Country: Vietnam
Email: hoang.chuduc@hust.edu.vn

Analysis of Sleep Macro - and Microstructure

Le Quoc Khai, Nguyen Thi Minh Huong, and Huynh Quang Linh

Faculty of Applied Science, University of Technology, Ho Chi Minh City, Viet Nam

Abstract— **Analyzing the microscopic structure of sleep is an approach which provides more useful information to evaluate the quality of the patient's sleep. Determination of arousals and events of decrease in the concentration of oxygen saturation (SpO2) which help us monitor abnormal disorders related to apnea syndrome and sleep apnea. The research focuses on developing a software with main function such as analysis of macro- and microstructure of sleep, determining the number of arousals and the number of decreased oxygen saturation of a patient. The results of mentioned software were compared with the results of the manufacturer software of polysomnography machine and analytical results of the sleep specialist, the similarity is over 85% compared with the conventional manual method of reading.**

Keywords— **micro structure, arousal, decrease SpO2, polysomnography.**

I. INTRODUCTION

Sleep was described as a succession of five repeating stages by Rechtschaffen and Kale (R&K) in 1968: the rapid eye movement (REM) stage and four nonrapid eye movement (NREM) stages, S1, S2, S3, S4 [1]. It has not been changed until 2007 the American Academy of Sleep Medicine updated part of it. S3 and S4 were grouped into one stage and signed N3 stage [2]. Thus following the AASM 2007 rules, NREM sleep has three major stages: N1, N2, N3 replace for S1, S2, S3 or S4, respectively.

Polysomnography (PSG) is a psycho-physiological method for the assessment of sleep and wake states. It is based on the concurrent recording of brain electroencephalography (EEG), chin electromyography (EMG) and electro-oculography (EOG) signals collected in human individuals using non-invasive surface electrodes. PSG allows for the description of different sleep–wake states, which may exhibit abnormal qualitative and quantitative changes with clinical conditions and environmental situations. It is the golden standard in the diagnosis of sleep disorders [3].

The scoring of sleep–wake stages is usually performed by a highly trained human expert on the basis of an epoch-by-epoch visual interpretation of the PSG signals according to a set of R&K rules, and recently updated by AASM 2007 rules. Manual scoring relies on visual extraction of specific features in EEG channels (usually C3-M2 and C4-M1 of the International 10/20 system [4]), two channels of EOG and one channel of chin EMG [3]. It is known that such manual work of sleep staging is very complex and onerous. This process is time consuming (typically 1-2 hours per subject), costly (millions of Vietnamese dong per recording) and prone to inter and intra scorer variability [5]. Although performed by trained personnel, the reliability and coherence of manual sleep scoring are usually unsatisfactory because of subjective judgment and human error [6]. In order to overcome the problems associated with manual scoring several researchers have proposed automatic sleep scoring system.

In the presented paper we consider sleep-EEG analysis based on Higuchi's fractal dimension, Df [7, 8]. We thus describe the developed staging process combine many methods to detect features of each sleep stages. These methods help to construct hypnograms (a diagrams of sleep stage vs. time).

In order to enhance the software application into clinical practice, in addition to analyzing the structure of sleep, we also combine a tool for automatic diagnosis a common slumber disorder in Vietnamese. Obstructive sleep apnea (OSA) is a common disorder in which individuals stop breathing during their sleep. These episodes last 10 seconds or more and cause oxygen levels in the blood to drop [2]. Most of sleep apnea cases are currently undiagnosed because of expenses and practicality limitations of overnight PSG at sleep labs, where an expert human observer is required. In this study, we develop and validate a tool using SpO2 measurements obtained from pulse oximetry to predict OSA. The results show that is useful as a predictive tool for OSA with a high performance and improved accuracy.

II. MATERIALS AND METHODS

A. Experimental Data

Before each experiment, subject was explained for the purpose and procedures of the study. Then they were asked to fill out a questionnaire which was kept confidential and included patient's identification. Before they sleep, we did not use any kind of medicine, stimulant or sedation. The whole night polysomnography was recorded using clinical PSG equipment (Alice 5, Respironics, Phillips [9]) at the Diagnostic Polysomnography Room of Community Health Care Center. During sessions, subjects were lying on

V. Van Toi and T.H. Lien Phuong (eds.), *5th International Conference on Biomedical Engineering in Vietnam,*
IFMBE Proceedings 46, DOI: 10.1007/978-3-319-11776-8_25

a comfortable bed within a dark and sound-isolated room. Thirty patients (17 male and 13 female) participated in this study. In this study, we use one channel of EEG were analyzed from the C3-M2 derivation, two channels of EOG include Left EOG and Right EOG signals, one channel of Chin EMG with the same sampling rate of 500 Hz .

EEG cup-electrodes were attached onto the scalp of the subjects according to the international 10–20 system for electrodes placement. EOG and EMG electrodes were attached onto the skin [3]. The patient preparation, and instrumental setup were done according to AASM guidelines [2] and Sleep Technician Guide of Alice 5 device [10].

Sleep–wake stages were scored according to conventional criteria using a fixed epoch duration of 30s. Each epoch was classified exclusively amongst six possible stages: wakefulness (W), transitional sleep (N1), shallow sleep (N2), deep sleep (N3), REM sleep (R) and movement time (MT).

B. Methods

Automatic sleep stages classification is effectuated in the hierarchical manner, presented in Figure 1. The first step in process of data analysis is apply digital signal filters for each channel of EEG, EOG and EMG signal. The filters have been used in this case include three types: 50 Hz of Notch filter to remove the noise of power source voltage fluctuations, lowpass filter and highpass filter to have a adaptive bandwidth for bio-signals.

We then used the Fast Fourier Transform (FFT) to transmute the signal from time domain to frequency domain. It is easy to detect and analyze the feature waves of EEG waves in power-frequency scale, for example Alpha, Beta, Theta, Delta waves which is define base on frequency range [11]. Beside the spectrum of EOG signal allow we to distinguish the rapid or slow movement of eyes. Applying different thresholds for many channels to observe the appearance types of wave in each epoch. We use an different threshold value of average amplitude of voltage to separate the bad signal come from electrode placement.

To separate the rapid-eye-movement s (REMs) from slow-eyes-movements (SEMs), prior to any detection, the two channels of EOG data are filtered using a fourth-order digital Butterworth bandpass filter. This effectively minimizes false detections due to SEMs as REMs as well as any high frequency noise [12]. We detect the candidate of REM in each epoch which is the characteristic of REM appearance [13]. Detection of rapid eye movement is based upon correlation coefficients between two EOG derivations and deflection between left and right EOG.

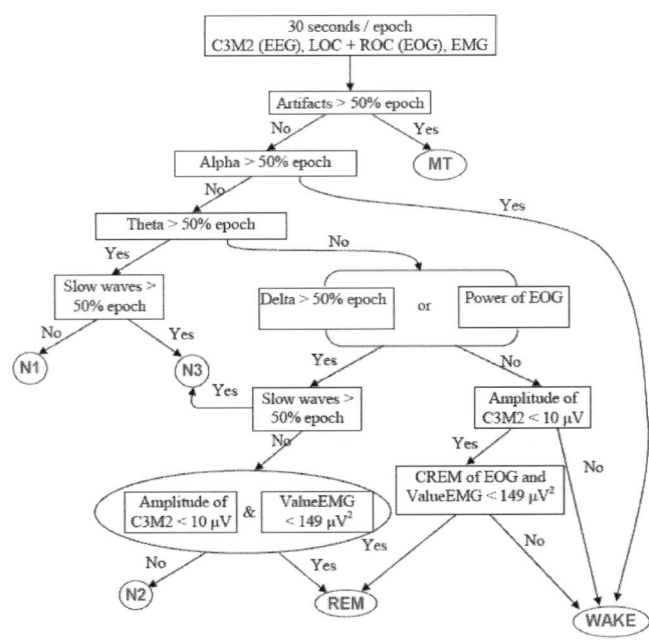

Fig. 1 A procedure of our software using to analyze and score sleep stages

- Detection of Body movements or Artifact signals

Movement time (MT) is a separate stage assigned in case of EEG, EOG or EMG signals obscured in morethan half the epoch by muscle tension or amplifier blocking artifacts associated with movement of thesubject. The values of amplitude signal are compared to thresholds, determined from the training data: ± 300 μV. The tonic EMG activity differs across the sleep stages and time of sleep. It almost always reaches its lowest level during stage REM—this fact is used in the scoring of sleep stages. Thresholds for the value of tone EMG, detecting stage REM, were determined on the basis of the data collected to train the system.

- Assigning Stages to Epochs

For each 30-s epoch is tested for muscle artifacts in EEG, EOG or EMG derivation, which can indicate the Movement Time (MT). In the next step, the algorithm detects Alpha wave, which is a feature of wakefulness by applying bandpass frequency for filtered channel C3-A2 of EEG signal form 8 to 13 Hz. Similarly, we apply frequency band to detect appearance of Beta, Theta and Delta waves. Delta waves or slow wave is a goal standard to distinguish the shallow sleep and deep sleep. Stage REM is scored in case of tone of EMG below 10 μV, in presence of stage 1.

- Detection K-complexes and Sleep Spindles in Stage N2

Applying Higuchi algorithm to calculate fractal dimension, we observe that the coefficient fractal decrease markedly at the time that EEG amplitude has strongly fluctuation. It can be characterized by a certain range of Higuchi's FD. With the appropriate value of kmax, then apply the threshold value, we easy to detect the appearance of K-complex. The method with selecting the threshold value for fractal dimension of each epoch help we get results quickly and accurately. Based on this principle and combine with the result in time-frequency domain analysis, we have made a sleep stager software that automatically.

Sleep Spindle detection characteristics by building five peaks through fractal analysis is processing in time scale, it shows a method to distinguish Sleep Spindle and alpha wave.

- Smoothing the Hypnogram

According AASM 2007, the epoch-by-epoch approach presented in the previous section does not imply that each epoch is considered in isolation. We explicitly implement the following, generally accepted smoothing rules: the "3-minute rule" for stage N2. Smoothing rules of replacing consecutive epoch of R N2 R, N2 N1N2, W R W with R RR, N2 N2N2, W WW, respectively [12]. At the end of the procedure, hypnogram is reviewed for elimination of 1-epoch stages (except MT).

- Detect Oxygen desaturation

In our work, the SpO2 signals are saved to separate files and processed by an automated system we developed to compute two of the common oximetric indices and one nonlinear metric.

Delta index: This is a common measure to detect the apneic events by measuring SpO2 variability. Levy et al. [14] calculates Δ index as the sum of the absolute variations between two successive points, divided by the number of intervals. It is usually computed for 12-sec. intervals.

Oxygen desaturation indices of 3%: This measure is obtained by calculating the number of times per hour with values of SpO2 greater than or equal to 3% from the baseline. The baseline is set initially as the mean level in the first 3 minutes of recording [15].

III. RESULTS

The system used for all the computations, including the algorithm of automatic sleep stage classification, is complete source code base on MATLAB computing language. We develop this user interface for anyone use our software easier and permit they manually adjusting the thresholds value for some specific case, for example: with human that have a sleep apnea or older people, threshold to

detect appearance of Alpha wave must be higher than another people in normal. By this way, a doctor can check correctly scored epoch following traditional method that he or she has more experiments. (Figure 2).

The detection of the K-complex and sleep spindle is an important premise for micro-structure of sleep study. It helps we build hypnogram correctly, as determined by conventional stage N2 in AASM 2007 rules [3]. One should not forget that a medical hypnogram, done by a specially trained clinician, is based not just on one channel EEG-signal but on the whole original polysomnography that contains multi-channel EEG, ECG, EOG, and others biosignals; even despite of that, the accuracy of medical hypnograms is believed to be of the order of about 70%.

The main result of all analysis process in our software is the hypnogram, the diagram of sleep stages per epoch. Processing is time latency about 2-4 minutes per subject compare with 1-2 hours per subject in classical scoring methods using visual interpretation of scorer. However, that results have been compared with the manual hypnogram based on experiment of a doctor with sleep and sleep disorder as speciality. Total concordance of the proposed automatic detection of sleep stages with hypnograms by human experts, scored for these epochs, is greater more than 75%. Moreover, the result of our software expressing that we can overcome some errors of the hypnogram is created by Alice 5 software (a software supporting combine with PSG Alice 5 device) analyses. It has a tool in the same function of our software to automatic scoring sleep stages, but results of our software and Alice 5 software have big difference.

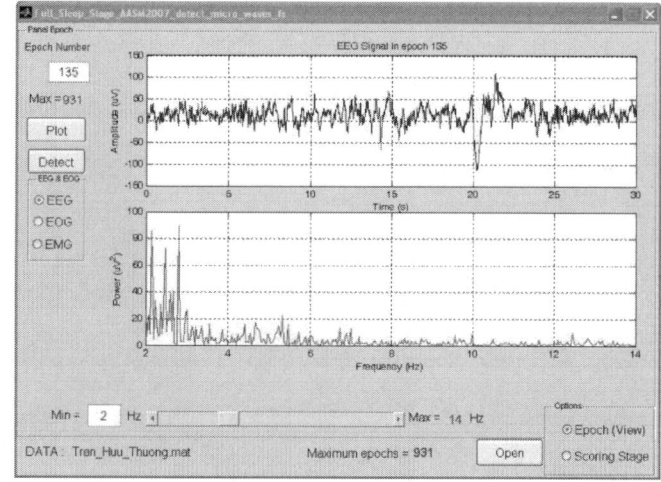

Fig. 2A Window Epoch (View)

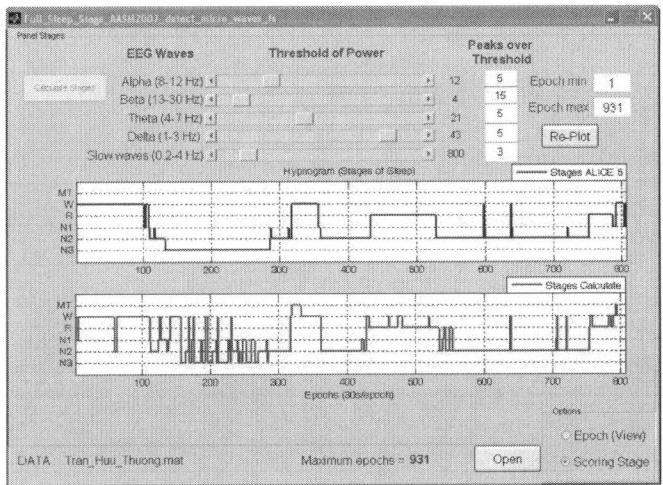

Fig. 2B Window Scoring Stage

Fig. 2 Screenshot from two main windows of our software, the full procedure built up all parts of this software is described in this paper. The window – Epoch (view) allow user easy to show the signal in both time domain and frequency domain (3 types of channel: EEG, EOG, EMG) and detect microwaves (K-complexes and sleep spindles) appeared in each epoch. The window – Scoring Stage displays two hypnograms for whole night sleep, we easy to observe and review the sleep stages of patient anytime.

IV. CONCLUSIONS

A difficult challenge is staging criteria defined in AASM 2007 leave a significant margin for subjective interpretation; therefore, hypnograms constructed for the same recording, rely on doctor experiments. Although the concordance of our software analyses and doctor scored is high, some different staged epoch show that we need more and more data for training our system to find a best threshold value.

In this paper we present a tool for automatic scoring Movement Time, Wakefulness and four stages of sleep using single channel of EEG, EMG and two channel of EOG. All major stages are plotted in hypnogram is call Macro Sleep Architecture (MSA).As a result, to reduce the dependency on complex PSG test measures, we find that software using SpO2 measurements is a practical and useful screening test to estimate whether patients have sleep apnea or not.

ACKNOWLEDGMENT

We would like to thank Thien Thanh Dang Vu, MD Ph.D, Dr. Nguyen Xuan Cam Huyen, Ho Chi Minh City Medicine and Pharmacy University, Dr Hoang Dinh HuuHanh, 115 Hospital, Dr Nguyen Xuan Bich Huyen, ChoRay Hospital for helpful suggestions on data analysis.

Finally, our deepest appreciation to all staffs of Community Health Care Center (CHAC) who helped us collect information.

REFERENCES

1. Rechtschaffen A, Kales A. A manual of standardized terminology, techniques, and scoring system for sleep stages of human subjects. Washington: Washington Public Health Service, US Government Printing Office (1968).
2. The AASM Manual for the Scoring of Sleep and Associated Events – Rules, Terminology and Technical Specifications, AmericanAcademy of Sleep Medicine (2007)
3. Aliki Minaritzoglou, Emmanouel Vagiakis. Polysomnography: Recent data on Procedure and Analysis. PNEUMON Number 4, Vol. 21 (2008).
4. Kelly A.Carden. Recording Sleep: The Electrodes, 10/20 Recording System, and Sleep System Specifications. Sleep Med Clin 4, pp.333-41 (2009)
5. Collop NA. Scoring variability between polysomnography technologists in different sleep laboratories. Sleep Med, 3(1): pp. 43-47 (2002)
6. Norman, R. G. Interobserver agreement among sleep scorers from different centers in a large dataset. Sleep Journal, 23:pp.901–908 (2000)
7. T.Higuchi. Approach to an irregular time series on the basis of the fractal theory, Physica D, vol. 31, pp. 277-283 (1988)
8. A.Accardo, M.Affinito, M.Carrozzi, and F.Bouquet. Use of the fractal dimension for the analysis of electroencephalographic time series. Biol.Cybern. vol. 77, pp. 339-350 (1997)
9. Philips, Respironics, Alice 5, Alice Sleepware v.2.8 software manual.
10. Philips, Respironics, Alice 5, Sleep Technican Guide.
11. William O. Tatum, Aatif M. Husain. Handbook of EEG Interpretation. Demos Medical Publishing (2007)
12. P. Shokrollahi. A Method for Quantifying Sleep Eye Movements That Reflects Medication Effects. IFMBE Proceedings (25) pp.1411-14 (2009)
13. Rajeev Agarwal, Tomoka Takeuchi, Suzie Laroche, JEAN Gotman. Detection of Rapid-Eye Movements in Sleep Studies. IEEE Ransactions on Biomedical Engineering. Vol. 52, No. 8 (2005)
14. P. Levy, J. Pepin,Accuracy of Oximetry for Detection of Respiratory Disturbances in Sleep Apnea Syndrome, Chest Journal, vol. 109, no. 2, pp. 395-399 (1996).
15. D. Alvarez, Nonlinear Characteristics of Blood Oxygen Saturation from Nocturnal Oximetry for Obstructive Sleep Apnoea Detection, Institute of Physics Publishing, vol. 27, no. 4, pp. 399-412, (2006).

Author: Le Quoc Khai
Institute: Faculty of Applied Science, University of Technology
Street: Ly Thuong Kiet
City: Ho Chi Minh
Country: Vietnam
Email: quockhai@hcmut.edu.vn

Author: Huynh Quang Linh
Institute: Faculty of Applied Science, University of Technology
Street: Ly Thuong Kiet
City: Ho Chi Minh
Country: Vietnam
Email: huynhqlinh@hcmut.edu.vn

Design System to Remotely Monitor Patients with Parkinson's Disease

Hai Tuyen Nguyen[1], Cao Cuong Vu[1], Van Quyet Phan[1], Viet Dung Nguyen[1], and The Dung Nguyen[2]

[1] School of Electronics and Telecommunications, Hanoi University of Science and Technology, Vietnam
[2] School of Information and Communication Technology, Thai Nguyen University, Vietnam

Abstract— **Parkinson's disease does not cause immediate danger but it should be monitored continuously because not at any time symptoms can easily observed. Aimed at saving time and costs for patients, as well as the patients cannot always go to the hospital for checking health, in this paper, we present architecture of a system to remotely monitor patients with Parkinson's disease. The system could help tracking the status of Parkinson patients at home. Movement information of patient's tremor hand is acquired then sent to central unit via GPRS. At the central unit, the data is collected, stored in 24-hour format and displayed in a tabular or continuous graph. By that mean, doctors can monitor and diagnose the status of the patients.**

Keywords— **Parkinson, Telemedicine, GPRS, socket, accelerometer.**

I. INTRODUCTION

Parkinson's disease is a condition of neural system malfunction with age that affects the state of motion, balance and control the muscles of the patient. Parkinson's disease (PD) is the most prevalent movement disorder, leading to symptoms including tremor, rigidity and bradykinesia [1].

Rest tremor is the most common and easily recognized symptom of PD [2]. However, measuring the impact of therapeutic interventions on tremor features in PD poses a challenge because the severity of tremors may vary strongly during the day.

Another important point is that most clinical rating scales are subjective. Additionally, the self-assessment of symptom severity by patients is not very reliable [3]. In previous studies, Hoffet used a commercial portable multichan-nel recorder for quantitative continuous 24-hour monitoring of tremor while the patient is at home [4]. Yanget presented a portable device for the long-term measurement and recording of tremor in which these signals are stored in a compact flash memory card [5]. More recently, Salariant developed an ambulatory system for quantification of tremor and bradykinesia in patients with PD that is based on miniature gyroscopes [6]. Meanwhile Engin described an ambulatory instrument based on two axis acceleration sensors [7], and Slack and Xianghong presented a methodology for measuring the hand tremor of surgeons [8].

There are two main methods of treatment of PD by using drug in conjunction with the supervision of a doctor and surgery. However, the surgery is not effective enough and uncontrolled. So using drug in combination with the supervision of a doctor is still the main treatment for patients. To monitor patients in the best way, it is necessary to balance between the symptoms and side effects of the drug. Patients should be monitored continuously and thus their doctors will choose the best treatment depending on the prominent symptoms of the patient. Because of the characteristics of the treatment, the patient self-care decisions at home will be difficult.

In order to overcome these difficulties, we have come up with the idea of designing a system to track Parkinson's patients at home.

II. SYSTEM ARCHITECHTURE

The system consists of two parts: mobile unit and central unit as shown in figure 1.

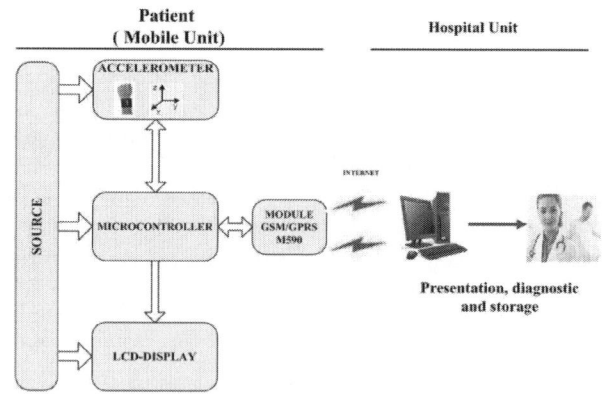

Fig. 1 Simplified block diagram of system.

The mobile unit or hardware part is attached on patient's wrist. This part composes of an accelerometer, a microcontroller and a GSM/GPRS module. The microcontroller gets movement accelerations of the patient's

V. Van Toi and T.H. Lien Phuong (eds.), *5th International Conference on Biomedical Engineering in Vietnam,*
IFMBE Proceedings 46, DOI: 10.1007/978-3-319-11776-8_26

hand from the accelerometer then sends to central unit via GSM/GPRS module.

The central unit is in fact a set of software running on PC that requires an Internet connection. At the central unit, the doctors can log in GPRS server, monitor patient's symptom remotely, continuously and provide accurate diagnosis.

III. HARDWARE DESIGN

In order to receive tremor signal from hand of Parkinson's disease, we use accelerometer MMA7455L from Freescale® [9]. It's a surface micro-machined integrated circuit accelerometer. Its operational principle is illustrated in figure 2.

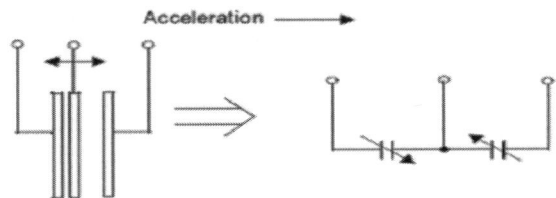

Fig. 2 Physical model of the accelerometer MMA7455L

Under the affect of acceleration, the distance between the walls change, leading to change in capacitance according to following formula

$$C = \frac{A.\varepsilon}{D} \qquad (1)$$

where C is capacitance, A is common area of the walls, ε is the dielectric constant, and D is the distance between the walls.

The capacitance in each axis is converted to voltage value through a capacitance-to-voltage conversion. Prior to measuring process, we have carried out powering and calibrating accelerometer. Due to built-in self-test mode of the accelerometer, calibration process is quite simple.

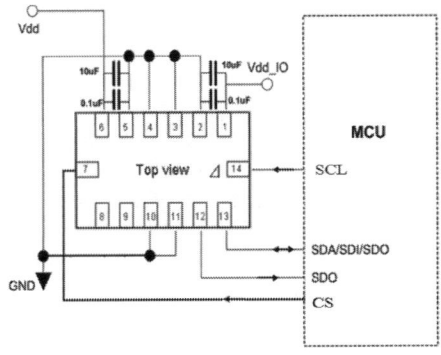

Fig. 3 Connection between the accelerometer and microcontroller.

The accelerometer can be connected to microcontroller via SPI interface. In this paper, we use a microcontroller named PIC16F877A from Microchip® [10] to acquire the data from the MMA7455L. Some of the PIC16F877A features are 256 bytes of EEPROM data memory, self-programming, 8 channels of 10-bit Analog-to-Digital (A/D) converter, the synchronous serial port can be configured as either 3-wire Serial Peripheral Interface (SPI™) or the 2-wire Inter-Integrated Circuit (I²C™) bus. They make it ideal for our application. Connections between MMA7455L and PIC16F877A is shown in figure 3.

In order to send acquired accelerations to central unit, GSM/GPRS module Neoway M590 is used [11]. It is connected to PIC16F877A through Rx and Tx pins. After getting a predefined data packet from the microcontroller, the GSM/GPSM module sends the packet to a server through GPSR network. To make it possible, the GSM/GPSM module is provided with an already registered GPRS sim.

IV. SOFTWARE DESIGN

Data acquisition software is built on the Microsoft® Visual Studio C# 2010 [12]. To make data display more visually, we utilize DotNetBar [13] to build interface and Zedgraph [14] as graphing tool. Graphic user interface (GUI) of the software is show in figure 4. It is divided into 4 separated parts. The first part is "Menu" part where the doctor can control the operations of the system. The second part is where information of the patients that are connecting to the system is shown. The third part is to show the data received. The fourth part is for display received data and movement frequency in graph forms.

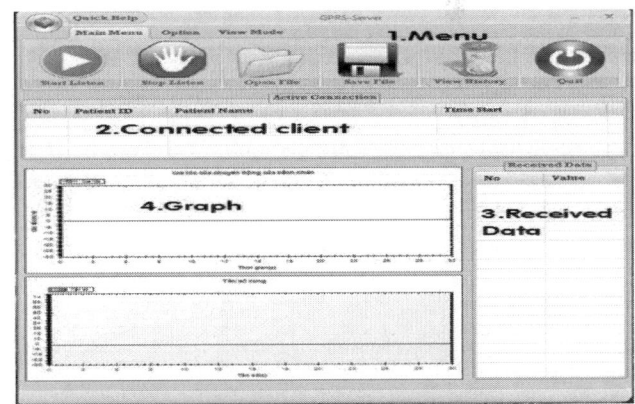

Fig. 4 The designed graphic user interface.

In "Menu" part, there are 3 tabs. The "Main Menu" tab contains all control commands of the system such as start or stop receiving data, save data to file, open a file, view history of measurements of the patients and quit the system. In "Option" tab, the users can adjust basic parameters of the system. Besides that, there are instructions to help users in using the system and authors' information. The "View Mode" tab allows the users to change data display method from "Scroll" mode to "Compact" mode and vice versa. In "Scroll" mode, new data will replace old data while they are all shown in "Compact" mode.

The most important issue for the software is to handle, peel the data sent via GPRS. To accomplish this, we use Socket method [15]. Socket is a method to establish a communication connection between a program of service request (client) and a service provider (server) on LAN, WAN or Internet, and sometimes between the processes right inside the computer. Each socket can be considered as an endpoint in a connection. In particular, we open a socket on the computer and listen to connection requests from the client or the GPRS module. When there is a connection request from the client, the link is allowed and information reading can be started. The information which is read is in the form of byte code encoded in ASCII so we must convert the information obtained to a string. After conversion into strings of characters, the data is stored as the number of posts in a row.

After getting all of accelerations in each axis, calculation of total acceleration is conducted. Because the 3-axis x, y, z are perpendicular to each other so the total acceleration is given as

$$a = \sqrt{a_x^2 + a_y^2 + a_z^2} \qquad (2)$$

where a_x, a_y and a_z is corresponding to acceleration in x y and z axis.

In addition, we can calculate movement frequency of the PD patients from received acceleration values.

V. RESULTS

The hardware unit is shown on figure 5. It is packed in a box sized 150x100x50 mm. The accelerometer is attached on the patients' wrist.

It is has been shown that the tremor frequency in Parkinson's disease patients occurs below 30 Hz. So we will use the software to calculate average frequency and maximum frequency of data.

The hardware can continuously send data to tracking software, 225 bytes of data every second. The average frequency and maximum frequency of movement of the patient are displayed in continuous graph from.

Figure 6 and 7 show the data received from normal person with unmoving hand and PD patient respectively.

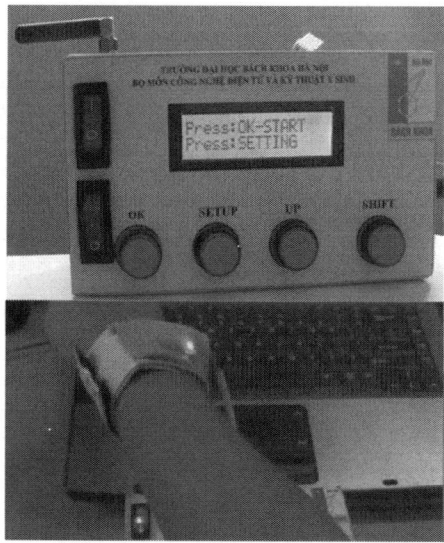

Fig. 5 The hardware unit with the accelerometer attached on patients' wrist

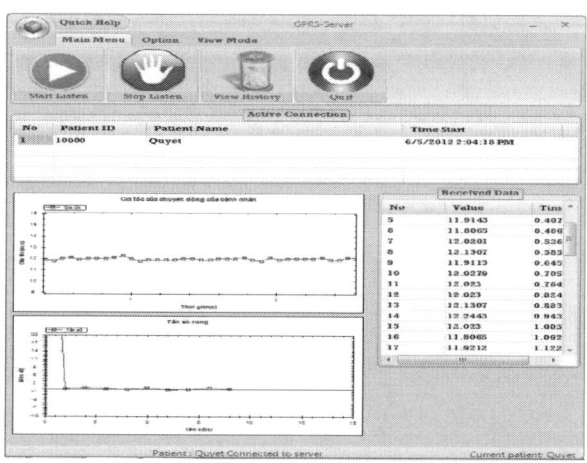

Fig. 6 Data received from normal person with unmoving hand.

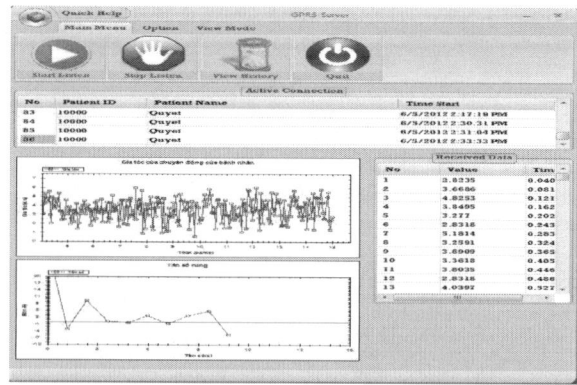

Fig. 7 Data received from patient with tremor hand.

VI. CONCLUSIONS

In this paper, we present a system that provides physicians a new method to monitor and accurately assess the Parkinson patient's status. Thereby they can offer appropriate treatment methods. Patients also do not need to spend time and costs for the hospital.

In the future, we will test this system on lot of patients to be able to offer a standard database for patients with Parkinson's disease, improve data transmission as well as display the software more visually.

REFERENCES

1. Guttman M, Kish S.J, Furukawa Y (2003) Current concepts in the diagnosis and management of Parkinson's disease. CMAJ 168:293-301.

2. Jankovic J (2008). Parkinson's disease: clinical features and diagnosis. J Neurol Neurosurg Psychiatry 79:368-376.

3. Brown R.G, MacCarthy B, Jahanshahi M, Marsden C.D (1989) Accuracy of selfreported disability in patients with parkinsonism. Arch Neurol 46:955-959.

4. Hoff J.I, Wagemans E.A, Hilten B.J (2001) Ambulatory Objective Assessment of Tremor in Parkinson's Disease. Clin Neuroph 24(5):280-283.

5. Yang M.H, Sheu Y.H, Shih Y.H, Young M.S (2003) Portable tremor monitor system for real-time full-wave monitoring and analysis. Rev Sci Instrum 74(3):1303-1309.

6. Salarian A, Russmann H, Wider C, Burkhard P.R, Vingerhoets F.J.G, Aminian K (2007) Quantification of Tremor and Bradykinesia in Parkinson's Disease Using a Novel Ambulatory Monitoring System. IEEE Trans Biom Eng 54(2):313-322.

7. Engin M (2007) A recording and analysis system for human tremor. Measurement 40:288-293.

8. Slack P.S, Ma X (2007) Tremor amplitude determination for use in clinical applications. Meas Sci Technol 18:3471-3478.

9. http://www.freescale.com/webapp/sps/site/prod_summary.jsp?code=MMA745xL

10. https://www.microchip.com/wwwproducts/Devices.aspx?dDocName=en010242

11. http://www.rfsolutions.co.uk/acatalog/info_M590.html#SID=111

12. http://www.visualstudio.com/en-us/downloads#d-2010-express

13. http://www.devcomponents.com/dotnetbar/

14. http://zedgraph.sourceforge.net/samples.html

15. http://csharp.net-informations.com/communications/csharp-multi-threaded-socket.htm

Author: Viet Dung Nguyen
Institute: Hanoi University of Science and Technology
Street: Dai Co Viet
City: Hanoi
Country: Vietnam
Email: dung.nguyenviet1@hust.edu.vn

Implementation of Telemedicine ECG System Based on Bluetooth Android Device

Hung Ngoc Do[1], Minh-Thanh Vo[1], Tinh Duy Ho[1], and Thanh-Tam Nguyen[2]

[1] School of Electrical Engineering,
International University-Vietnam National University, Hochiminh City, Vietnam
[2] Biomedical Engineering Department,
International University-Vietnam National University, Hochiminh City, Vietnam

Abstract— In recent years, mobile-based systems have become popular in variety of applications and attracted the interest of many researchers and manufacturers, especially in healthcare field. These systems help doctors assist their patients who are not at the hospitals or live in the remote areas. Although many advances in cardiovascular disease diagnose have been proposed, ECG still plays an important role. In this paper, a prototype for telemedicine application based on ECG signals is described with emphasis on the ECG measurement board design and building of Android application for mobile phone with Bluetooth wireless transmission.

Keywords— Bluetooth, ECG, Android, Telemedicine.

I. INTRODUCTION

Cardiovascular disease has been one of the most serious human health problems with high mortality [1], [2]. Although many new methods have proposed and applied in hospitals, ECG still retains its important role in diagnosis because it provides the important information about the cardiac rhythm and conduction system abnormalities. In the hospitals, the ECG of patients usually are measured, recorded and printed for monitoring and diagnosing. However, when patients stay at home or live in a remote area, the need for a compact size machine which can be used easily for ECG measurement and send data to doctors has become a necessary demand.

During the last decade, the fast development of mobile technology has brought many useful applications for human life. We can buy a smartphone equipped with a central processing unit (CPU) which can solve many complicated problems. Nowadays, there are some operation systems for mobile device like Blackberry, iOS, Symbian, etc. Among these, Android has been supported by the most manufacturers and had the largest market share. Google supports some means for programmers to build their Android applications. In this project, we implemented a prototype for ECG measurement and built an Android application for smartphone which can receive the ECG signals from the board to display into the screen and automatically send the result to the predefined telephone numbers of doctors.

The rest of the paper is divided as follows: section II introduces the overview of the system. Section III describes the details of schematics of each block of the circuit board whereas section IV shows the steps of building Android application for this project. Section V displays the flow of signal after each functional blocks and the running of application in Android smartphone. Finally, section VI concludes the paper and gives some future works.

II. SYSTEM OVERVIEW

The Fig. 1 displays the overview of our system. The hardware part includes ECG electrodes, power supply and ECG sensor board that can capture signals from human body for processing and shows result on an LCD. For the software development part, an Android application was build which can run on Android device to display the received ECG signal through Bluetooth module.

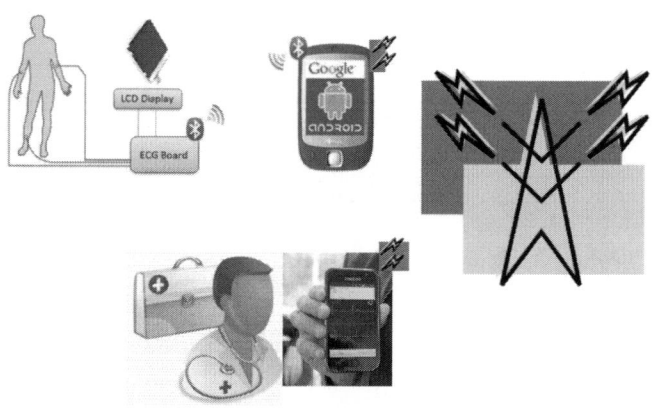

Fig. 1 System Overview

In order to capture ECG signal from human body, three electrodes are used for sensing bio electric potential. It is a transducer that converts ionic flow from the body through an electrolyte into electric current. The suitable electrodes which are used to measure ECG signals are the surface electrodes with jelly, paste, or cream [3]. When an electrode contacts to skin, a potential different up to 300mV appears known as baseline wander. This can be made worse by poor connection of electrode due to the movement of patients.

The placement of the electrodes on the body determines the view of the vector as a function of time. Fig. 2 represents the most basic form of the electrode placement

V. Van Toi and T.H. Lien Phuong (eds.), *5th International Conference on Biomedical Engineering in Vietnam,*
IFMBE Proceedings 46, DOI: 10.1007/978-3-319-11776-8_27

which is based on Eindhoven's triangle. Lead I measures the differential potential between the right and left arm, Lead II between the right arm and left leg, and Lead III between the left arm and left leg. Eindhoven's law also states that the value of any point of the triangle can be computed as long as values for the other two points are known.

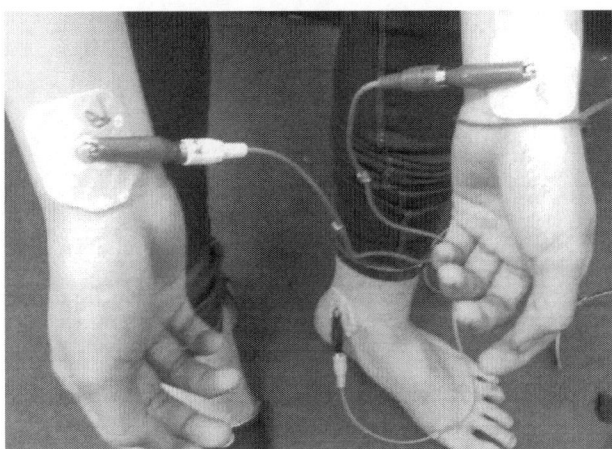

Fig. 2 Electrode Positions at the Body

The Eindhoven limb leads (standard leads) are defined in the following ways:

Lead I is the voltage between the (positive) left arm (LA) electrode and right arm (RA) electrode:

$$I = LA - RA \qquad (1)$$

Lead II is the voltage between the (positive) left leg (LL) electrode and the right arm (RA) electrode:

$$II = LL - RA \qquad (2)$$

Lead III is the voltage between the (positive) left leg (LL) electrode and the left arm (LA) electrode:

$$III = LL - LA \qquad (3)$$

The signal from electrodes will be processed by several steps in the Fig. 2. First, it is amplified by non-inverting op-amp circuit. Then, the amplified signal will be filtered to remove the unwanted parts before the MCU block. Finally, dsPic micro-controller converts analog signal to digital signal for wireless transmission via Bluetooth protocol and on-board LCD display.

III. ECG BOARD DESIGN

The front end of the ECG wireless sensor is an instrumentation amplifier. In some biomedical applications, an improvement of signal to noise ratio (SNR) may be needed since the desired signal level may be smaller than of the unwanted noise level. The main difficulty is that the

voltage level is very low, in about 0.5 - 5mV [4]. In this project, we used INA114 IC which has combination of low noise, low offset voltage, high accuracy gain, low input bias current and high linearity. This makes the INA114 is suitable for use in high resolution data acquisition system.

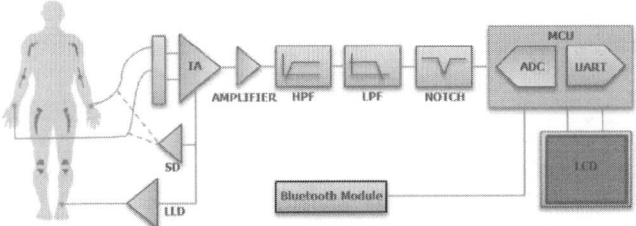

Fig. 3 Signal Processing Flow of ECG Board

Fig. 4 ECG Board

According to the AAMI (Association of the Advancement of Medical Instrumentation), there may exist an electrode offset of 300mV to 500mV at the input to the instrumentation amplifier. Being a DC voltage, this can be removed by a high-pass filter, so this filter should be designed to cut frequencies off below 0.05Hz.

In this circuit, we use low-pass filter which is designed at the cut off frequency of 100 Hz. This is because the flat frequency response for ECG signals within 0.05-100Hz. The experiment was done to see the performance of this filter using the input from signal generator and the output display at frequency spectrum analyzer. A simple fourth order low-pass Butterworth filter was selected due to this pass-band flatness. At this stage, the signal still contains considerable 50Hz noise from main electricity. A 50Hz notch or band-reject filter will attenuate signals at 50Hz, but not higher than, or lower than 50Hz.

IV. ANDOID APPLICATION DESIGN

Android is an operating system based on the Linux kernel and designed primarily for mobile devices such as smartphones and tablets. It is open source OS and Google releases the source code under the Apache License [5]. This open-source code and permissive licensing allows the software to be freely modified and distributed by device manufacturers, wireless carriers and enthusiast developers. Additionally, Android has a large community of developers writing applications that extend the functionality of devices, written primarily in the Java programming language. Because of the popularity of Android device, we decided to choose this platform for developing the software of system.

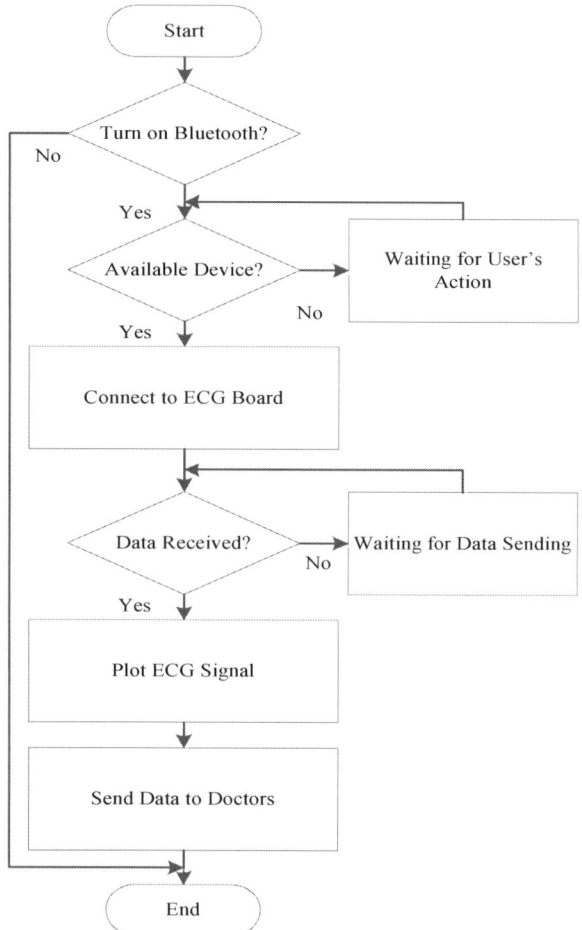

Fig. 5 Android Application Flowchart

Applications are developed in the Java language using the Android software development kit (SDK). The SDK includes a comprehensive set of development tools including

a debugger, software libraries, a handset emulator based on QEMU, documentation, sample code, and tutorials [6]. The officially supported integrated development environment (IDE) is Eclipse using the Android Development Tools (ADT) plugin. Other development tools are available, including a Native Development Kit for applications or extensions in C or C++, Google App Inventor, a visual environment for novice programmers, and various cross platform mobile web applications frameworks.

The architecture of Android OS is separated into five sections including Linux kernel, Libraries, Android runtime, Application framework, and Applications in four main layers [6]. The most important property of Android OS is open source, so it supports many features. We use the suitable parts for programming based on the requirements of application.

The main parts of the Android application in this project are connecting to other Android devices, plotting the receiving data and sending it to doctors via telecommunications networks. It starts from scanning available Bluetooth devices to connect and receiving data to process them appropriately. The whole programming steps can be summarized in the Fig. 5.

V. IMPLEMENTATION RESULTS

In this section, the results of ECG implementation are shown and discussed. The input signals can be received from ECG electrodes attached to human body and the output signals will be displayed on LCD and Android device.

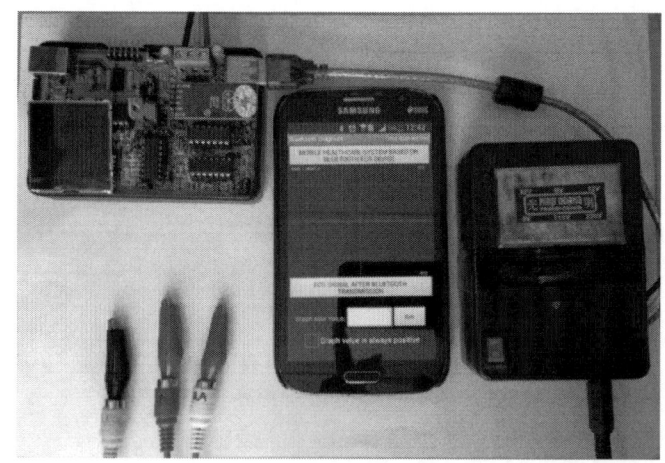

Fig. 6 Overall Picture of prototype

The Fig. 6 shows the overall of prototype with 3 main components: ECG sensor board, Android smart phone with application, and power supply. ECG wireless sensor board is constructed on a small-size printed circuit board (PCB). Three differential jacks (right arm, left arm and left leg) are used to sense signal from human body. Signals will go through analog to digital part before being sent to Android device via Bluetooth module and show on the LCD screen.

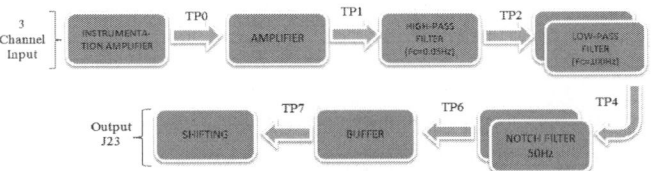

Fig. 7 Block Diagram of Signal Flow

The Fig. 2 shows the way to connect electrodes to the human body to get the input signals for the ECG board. These signals will go through some parts which shown in the block diagram as in the Fig. 7. The signals from sensors are processed after every block. In order to identify and analyze the changes of signals, after each block, we set some test points to measure the output signal by using oscilloscope.

The Fig. 8 – Fig. 14 show the measured signals at the predefined test points. It is clearly seen that the input signals from leads were processed properly step by step by some parts of circuits. Firstly, the signals were amplified by the op-amp and the result can be verified by comparing the two Fig. 9 and Fig. 10. Then, three filters removed the unwanted parts of signals shown in the Fig. 11 and Fig. 12. Finally, after summing amplifier and shifting offset parts, we have the ECG signal in the Fig. 14.

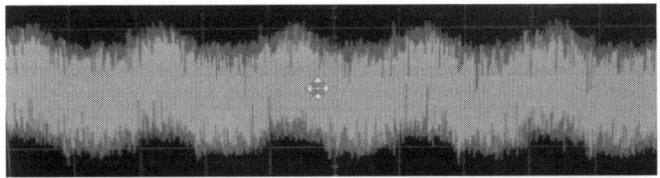

Fig. 8 Output of Instrumentation Amplifier, 20mV/Div, 10ms/Div

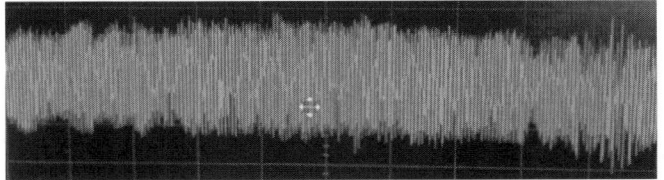

Fig. 9 Output of TL074 Amplifier, 1V/Div, 1ms/Div

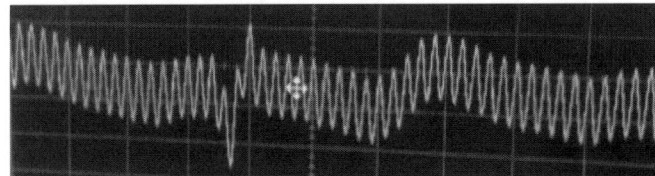

Fig. 10 Output of High-pass Filter

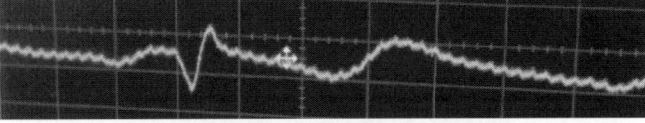

Fig. 11 Output of Notch Filter

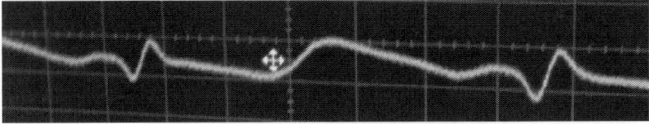

Fig. 12 Output of Low-pass Filter

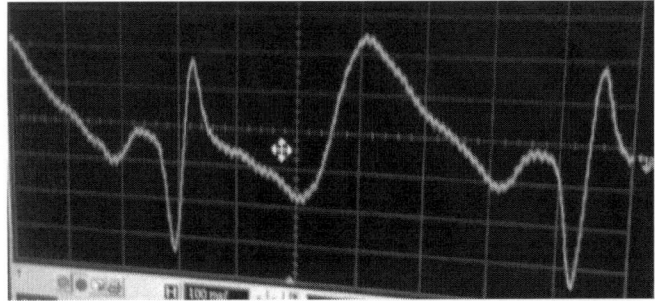

Fig. 13 Output of Summing Amplifier

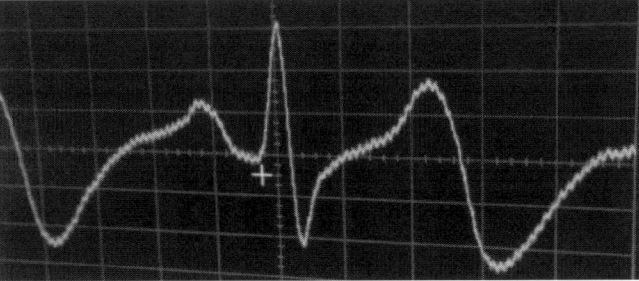

Fig. 14 Output of Offset Shifting

The Fig. 15 shows the ECG signals displayed by the screen of a smart phone. It is clear that the graph is very smooth and the same as in LCD. One more important factor is there is very small delay because of high speed in Bluetooth transmission protocol.

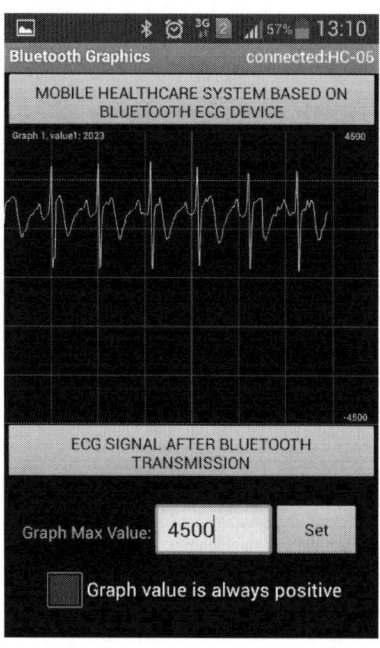

Fig. 15 Signal on Mobile Phone's Screen

VI. CONCLUSION

This paper shows the schematic design and Android application building for ECG measurement prototype. The size of this device is small and can connect easily to any Bluetooth Android device to show the ECG signal which is received from the sensors and processed by some modules. This system can be improved to have some useful functions such diagnose disease by using image and video processing, data storage and statistics in web-servers for further treatment.

REFERENCES

1. J. MacKay and G. A. Mensah, (2004) The Atlas of Heart Disease and Stroke. World Health Organization
2. A. H. Association, (2012) Heart Disease and Stroke Statistics. American Heart Association, Inc, USA
3. J. Dubovy (1978) Introduction Biomedical Electronics, McGraw-Hill, USA
4. P. O. Bobbie and C. Z. Arif, (2004) "Electrocardiogram (EKG) data acquisition and wireless transmission" in Proc of WSEAS ICOSSE 2004, pp. 2665–2672.
5. W. Jackson, (2013) Learn Android App Development. Apress
6. W. M. Lee, (2012) Beginning Android 4 Application Development. Wrox.

On Designing a System to Supervise Patients' Vital Signs through Wireless Sensor Network

N.T. Dung[1], P.T. Nghia[1], D.Q. Long[1], N.H. Tuyen[2], N.V. Dung[2], and N.T. Ha[2]

[1] Thai Nguyen University of Information and Communication Technology, Thai Nguyen University, Vietnam
[2] School of Electronics and Telecommunication, Hanoi University of Science and Technology, Vietnam

Abstract— The demand of health care of citizens has been increasing with the development of economy nowadays. In health care, the measurements of body temperature, heart rate... are the most interested. The medical specialists normally concern about these parameters when running a healthy check or evaluation of the effectiveness of a therapeutic method in general. Besides, surveillance of the parameters is important to detect bad signals of patients for timely treatment. Recently, the overcrowding in hospitals that have long become the concern of the health sectors in particular and whole society in general. With the application of wireless sensor networks (WSNs) to design and build a sensor which is capable of monitoring the heartbeat and body temperature parameters will be the solution to help take care of health for patients effectively, and will actively support for physicians in the treatment process. In this paper we introduce the design of a system that can monitor heart rate and body temperature of the patient in an effective and accurate manner through wireless sensor network.

Keywords— wireless sensor networks, body temperature, temperature sensing, heart rate, supervise patients.

I. INTRODUCTION

In recent years, wireless sensor networks (WSNs) have become increasingly important in daily life. Beginning to the development of military applications, wireless sensor networks are now also used in many other fields such as environmental monitoring, smart homes, health cares ... With the convergence of the microelectronic technology, integrated circuit technology, sensor technology and signal processing ..., scientists have created very small and multi-functional sensors with low costs increasing the applicability of wireless sensor networks .

In Vietnam, the overcrowding in hospitals that have long become the concern of the health sectors in particular and whole society in general. The application of technology in particular and wireless sensor networks in general will be effective solutions to help take care of health for patients effectively, and will actively support for physicians in the patients treatment process.

There are no units in domestic researching and designing about running equipment above. Almost similar devices have to be imported from overseas with high prices, while the demand of the treatment from patients is very large. The successful design of this device will bring enormous benefits in both science and economics terms.

II. SYSTEM DESIGN

A. System Description

The system overview is given in figure 1. It composes of sensor nodes and a base node. There heartbeat sensor (based on SpO$_2$ sensor) and a body temperature sensor (using the Microchip DS18B20) and a radio transceiver module (using the Microchip XBee) are integrated on each sensor node. The sensor nodes are worn on patient's body. A PIC 16F877a microcontroller on each sensor node receives the heartbeat and body temperature then spread them over the wireless networks. At base node, a 16F877a microcontroller receives the data and sends them to PC to display and store.

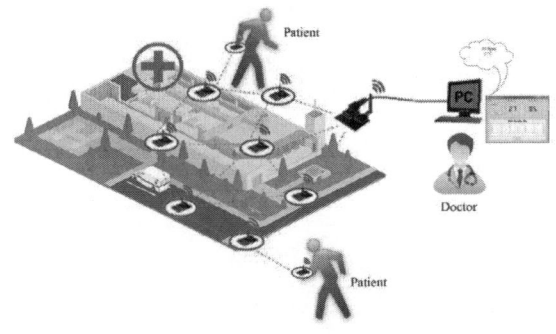

Fig. 1 The system overview.

The nodes are designed with an integrated sensors capable of monitoring heartbeat parameters and body temperature. The nodes are compact and conveniently designed so that they can be wearable and do not affect the movement of patients. To be energy saving, the sensor node on the patient's body is often different active and free statuses [1]. Normally, a node will be free more than active status. The base node is in charge of collecting data from the sensor nodes of each patient, and then displays the results on computer of doctors.

The system block diagram is given in figure 2. [2]

V. Van Toi and T.H. Lien Phuong (eds.), *5th International Conference on Biomedical Engineering in Vietnam,*
IFMBE Proceedings 46, DOI: 10.1007/978-3-319-11776-8_28

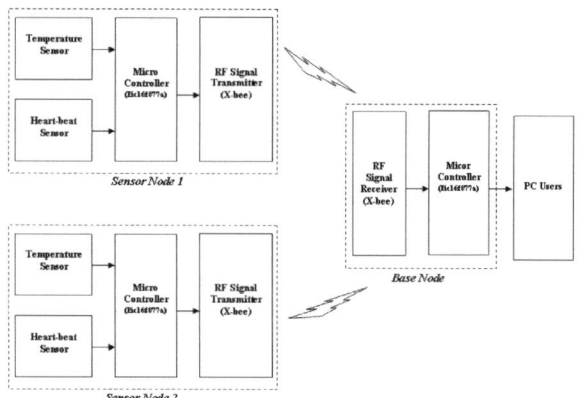

Fig. 2 The system block diagram.

B. Materials and Methods

• Microcontroller

The centre controller uses a microcontroller named PIC16F877A [3]. Currently, the microcontroller - PIC16F877A is the most popular line (be strong enough in terms of features, 40 feet, enough memory for most normal applications). The general structure of the microcontroller consists of 8KByte Flash ROM, 368 Byte RAM, 5 ports (A, B, C, D, E) in the independent signal controlling; two timers with 8 bit (Timer 0 and Timer 2), a timer with 16 bit (Timer 1) operating in the power saving mode (Sleep mode) with the external clock source; two sets of CCP (Capture / Compare / PWM); an AD transducer with 10 bit, 8 inputs, 2 analog comparators [4].

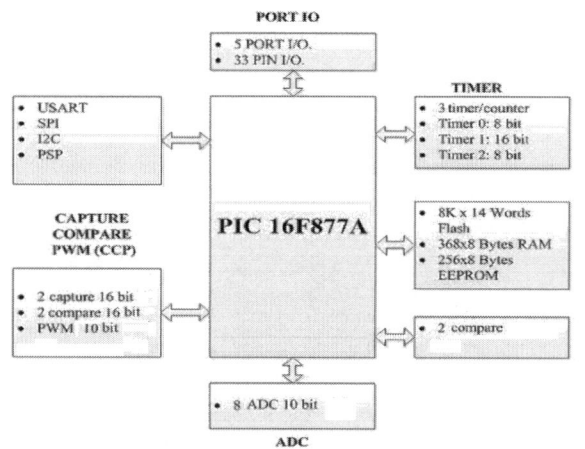

Fig. 3 Block diagram of PIC16F877A

• Temperature sensor

DS18B20 [5] is a temperature sensor, including 3 feet (operating from -55 to +125). With the temperature range

from -10 to +85, the accuracy is 0.5. The function of DS18M20 is to warn about the excess of temperature. The voltage used for the equipment from 3 to 5V, you can configure for the encryption of temperature from 9 to 12 bit because the larger the number of bits is, the higher the accuracy is. The conversion time of the temperature maximum is 750ms for 12 bit encryption.

The digital temperature sensor- DS18B20 shows data to indicate the measured temperature under the form of the binary code – 12 bit. The received information is transmitted via one wireless interface (1 - wire), therefore you need only two paths including a signal line and a ground line (GND) to connect from the microcontroller to the measured point (figure 4). The source supplying for the reading/ writing/ conversion manipulating can be obtained from the signal line, no need to add any separate lines to purvey the source voltage. Each the temperature measurement microchip of DS18B20 has a unique identification code, which is engraved by the laser in the IC fabrication process so a lot of DS18B20 microchips can be connected together in a bus having one wire without confusion. This feature makes the installation of multiple temperature sensors in many different locations become easier and with low cost. The number of sensors connected to the bus without restriction.

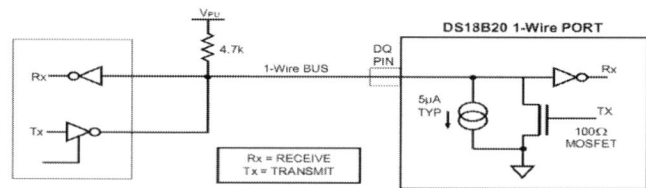

Fig. 4 Connection diagram with DS18B20

• Heartbeat counting circuit based on SpO_2 method

In the block diagram of the circuit (figure 5), IR LED is used to illuminate the user's fingers by the infrared light [6] [7]. Each heart rate, the blood will push out capillaries in the fingers making the change of infrared reflectance intensity, causing the output voltage on photo transistor.

The circuit schematic diagram is given in figure 6.

Voltages changing on the photo transistor (at point A) will be put through a high-pass filter circuit to filter the DC components on the circuit with high cut-off frequency

$$f_{C_H} = \frac{1}{2\pi R_1 C_1} \approx 0.6 \ Hz \qquad (1)$$

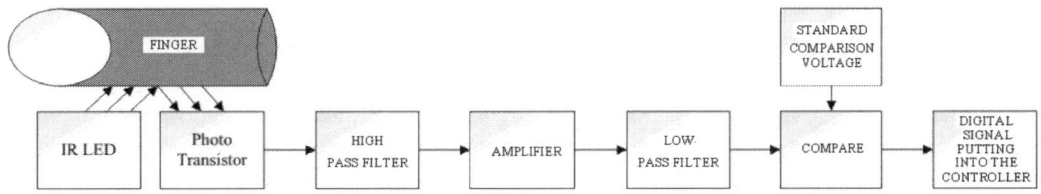

Fig. 5 Block diagram of the heartbeat counting circuit

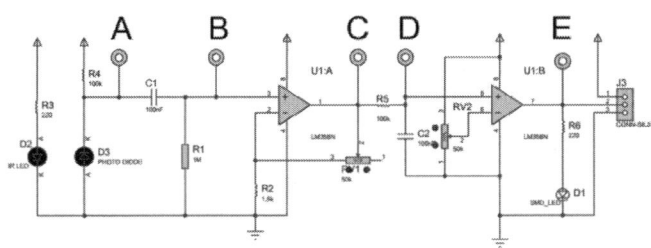

Fig. 6 Schematic diagram of the heartbeat counting circuit

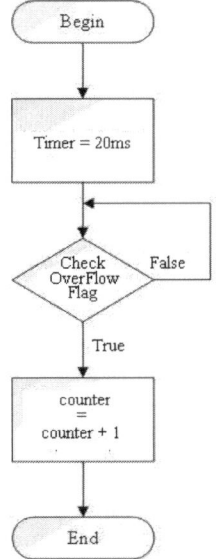

Fig. 7 The program flowchart

After high-pass filtering, the signal at point B (according to the heart rate) is amplified by $K = 1 + Rv_1 / R_2 \approx 34$ times. The amplified signal at point C is then low-pass filtered to remove high frequency noises (due to light, vibration ...). The low cut-off frequency is

$$f_{C_L} = \frac{1}{2\pi\sqrt{R_5 C_2}} \approx 15 \; Hz \qquad (2)$$

The signal at point D is compared with a standard voltage through a comparison circuit to convert from analog form to digital form that can be processed. The output signal of the circuit (at point E) has only two possible levels: "0" or "1". Whenever having a heartbeat, the output level is in "1" level.

Flowchart of counting heartbeat algorithm is illustrated in figure 7.

- Xbee communication module

The XBee RF Modules were engineered to meet IEEE 802.15.4 standards and support the unique needs of low-cost, low-power wireless sensor networks [8][9]. The XBee RF Modules interface to a host device through a logic level asynchronous serial port. Through its serial port, the module can communicate with any logic and voltage compatible UART; or through a level translator to any serial device. The Xbee module communicates with the microcontroller by DI pins transmitting data, DO receiving data, the notification flags of transmitting and receiving CTS and RTS (figure 8) [10]. Devices that have a UART interface can connect directly to the pins of the RF module as shown in the figure below.

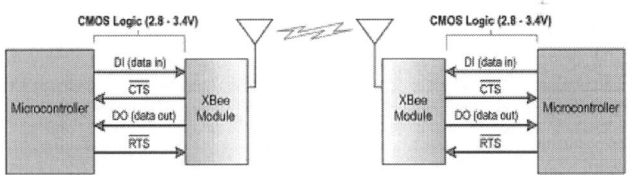

Fig. 8 System Data Flow Diagram in a UART-interfaced environment

Some parameters of the XBee ZB24 are

- ✓ Inside transmission range up to 30m
- ✓ Outside transmission range up to 90m
- ✓ Transmission Capacity: 1mW(dBm)
- ✓ Data Transmission Speed -RF: 250.000 bps
- ✓ Transmission Line: 45mA – 3.3V
- ✓ Using frequency range: 2.4 GHz
- ✓ Support Networks: Point to Point, Point to MultiPoint, peer to peer.

III. RESULTS AND DISCUSSIONS

We design a wireless sensor network for monitor heartbeat and body temperature. The network consists of two sensor nodes and a base node. An example of the data received by the base node and displayed in the PC is given in figure 9. It proves that our system allows doctors to monitor the health status of many patients at the same time and is capable of accurate warning and quick discovering bad signals on the patient's body for doctors to give handling measures timely.

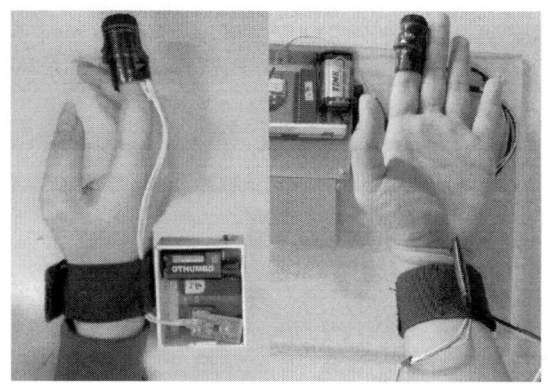

Fig. 9 The base node is in charge of collecting data from the sensor nodes of each patient

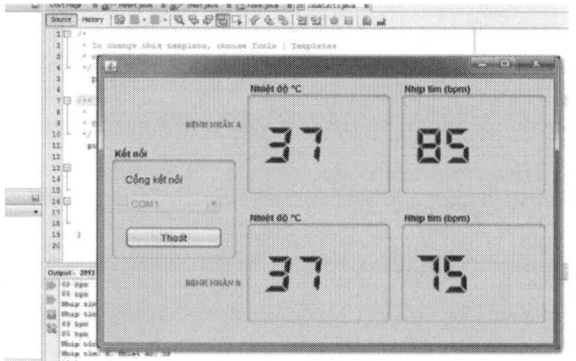

Fig. 10 Received data is displayed on PC screen

- Some features of the system:

The system is capable of monitoring temperature data, the patient's heart rate through wireless sensor networks and sending the results to the computers of the treating physicians.

✓ The system is capable of accurate warnings and quickly discovers bad signals on the patient's body for doctors to give timely remedial measures

✓ Allowing physicians can monitor the health status of many patients in the same time, thereby to facilitate the evaluation of the patient's treatment process.

✓ The implementation of the system is simple and efficient. Patients can wear the device on their body without affecting their activities.

IV. CONCLUSIONS

In this paper, a wireless sensor network for monitor heartbeat and body temperature is presented. It allows doctors to monitor the health status of patients. The system aims are to reduce the investment cost of hospitals, reduce the trade deficit contributed to implement the National Policy on medical equipment from the government. At the same time, the task will also contribute to improve the training of high qualifications human resources for the medical device sectors.

REFERENCES

1. Holger Karl and Andreas Willig, Protocols and Architectures for Wireless Sensor Networks, John Wiley & Sons, Ltd, 2005.
2. Sensor Information Networking Architecture and Applications, Chien-Chung Shen, Chavalit Srisathapornphat, and Chaiporn Jaikaeo, University of Delaware
3. Datasheet PIC16F87XA, Microchip Technology Inc, 2003.
4. Programming PIC Microcontrollers in BASIC, mikroElektronika © 1998 – 2004.
5. Datasheet DS18B20 - DS18B20 Programmable Resolution 1-Wire Digital Thermometer.
6. Design of an Infrared based Blood Oxygen Saturation and Heart Rate, Yousuf Jawahar, Electrical and Biomedical Engineering Project Report, McMaster University, 10/2009
7. Pulse Oximetry - Yun-Thai Li Department of Electronic Engineering, University of Surrey, Guildford (www.surrey.ac.uk)
8. Microchip ZigBee-2006 Residential Stack Protocol, Derrick P. Lattibeaudiere, Microchip Technology Inc.
9. Jason Lester Hill, System Architecture for Wireless Sensor Networks, page 166-177, 2003.
10. XBee®/XBee PRO® RF Modules - 802.15.4 [2009.09.23].

FPGA Implementation for Cardiac Excitation-Conduction Simulation Based on FitzHugh-Nagumo Model

Nur Atiqah Adon[1], Mohamad Hairol Jabbar[2], and Farhanahani Mahmud[1]

[1] Cardiology and Physiome Analysis Research Laboratory,
[2] Manycore System on Chip (MCSoC) Research Laboratory,
Microelectronics and Nanotechnology – Shamsudin Research Centre (MiNT-SRC),
Faculty of Electrical and Electronic Engineering,
Universiti Tun Hussein Onn Malaysia (UTHM), Batu Pahat, Johor, Malaysia
ge130013@siswa.uthm.edu.my {hairol,farhanahani}@uthm.edu.my

Abstract— The paper examines a method for development of a Field Programmable Gate Array (FPGAs)-based implementation of hardware model for the electrical excitation-conduction in cardiac tissue based on FitzHugh-Nagumo (FHN) mathematical model towards real-time simulation. The FHN model is described by a set of nonlinear Ordinary Differential Equations (ODEs) that includes two dynamic state variables for describing the excitation and the recovery states of a cardiac cell and the model is able to reproduce many characteristics of electrical excitation in cardiac tissues. In this paper, one dimensional (1D) FHN cable model is designed using MATLAB Simulink in order to simulate the conduction of cardiac excitation in coupled nonlinear systems of the heart dynamics. The designed MATLAB Simulink model is then being used for Very High Speed Integrated Circuit (VHSIC) Hardware Description Language (VHDL) code generation by using HDL Coder that will be implemented on a hardware design FPGA platform of Xilinx Virtex-6 FPGA board. In order to verify and analyze the designed algorithm on the platform, HDL Verifier is used through co-simulation with FPGA-in-the-loop (FIL) simulation and it has shown a significant result which has increased confidence that the algorithm will work in the real FPGA stand-alone application. Therefore, these approaches provide an effective FPGA design flow towards a stand-alone implementation to perform real-time simulations of the cellular excitation-conduction in a large scale cell models.

Keywords— Field Programmable Gate Array, FitzHugh-Nagumo, excitation-conduction, FPGA-in-the-loop simulation, HDL Coder.

I. INTRODUCTION

Cardiac excitation-contraction coupling mechanism is initiated by propagating electrical excitation of cardiac cell membrane in the heart tissue which controls the mechanical contractions of the cells via specialized gap junction. Nevertheless, dysfunction of electrical excitation propagation could lead disrupt the mechanical functioning of the heart underlying abnormal impulse generation or propagation and preventing the heart from supplying sufficient blood to the body [1]. Such abnormalities are often called arrhythmia and the vast majority of that disease perpetuated by reentrant mechanism [2].

In recent few decades, the action potential have been shown experimentally and clinically to reveal the underlying mechanism in electrical state of the heart [3]. Unfortunately, both of mechanisms specifically desirable but investigating the cardiac electrical behavior experimentally poses a number of challenges, such as limitation of variables for monitoring or deprivation of high-resolution data in investigating larger preparation [4]. On the other hand, mathematical modeling is useful techniques for a computer simulation of cardiac electrical behavior are not associated with such complications. Recent models of the cardiac cells starting from the electrical excitation model of the FitzHugh-Nagumo (FHN) [5], Noble [6], continued by Beeler and Reuter [7], Luo-Rudy [8-10] and many others have been developed to represent different regions of the heart.

With the progress of time, the computational techniques become more advance but complicated as parameters in the mathematical descriptions and size of the models increase which cause a drawback in the amount of computations for the dynamic simulations of the mechanism. Therefore, to overcome the computational challenge, hardware implementation appears as one of main choices recently that provides valuable tools for electrical excitation modeling [11-13]. A previous study had provided the analog-digital circuits of hardware-implemented cardiac excitation model designed by using analog circuits and a dsPIC30f4011 microcontroller [14] that could reproduce a real-time simulation of Luo-Rudy based cardiac action potential model. However, the model has shown limitations due to its power consumption and physical size.

Here, the purpose of this study is to design a hardware-implemented FHN model based cardiac excitation-conduction model in one dimensional (1D) cable model using Field Programmable Gate Array (FPGA) Xilinx Virtex-6 board towards the development of real-time

© Springer International Publishing Switzerland 2015
V. Van Toi and T.H. Lien Phuong (eds.), *5th International Conference on Biomedical Engineering in Vietnam*,
IFMBE Proceedings 46, DOI: 10.1007/978-3-319-11776-8_29

cardiac electrophysiological analysis tool. It is known that the FPGA is suitable for solving higher ordinary differential equation (ODE) with multimillion gate counts and special low-power packages [15-19]. At present, the Very High Speed Integration Circuit (VHSIC) Hardware Description Language (VHDL) code for the FPGA hardware implementation is produced by a rapid prototyping method introduced by MathWorks which are MATLAB Simulink and HDL Coder tools. The HDL Coder is applied to automate the algorithm design process by converting Simulink blocks into VHDL code for the FPGA. For verification of the design algorithm, HDL Verifier tool which is also from the MathWorks is then used to verify the designed through FPGA-in-the-loop (FIL) approaches.

The rest of the paper is organized as follows. An overview of the methodology is given in Section 2. Section 3 exposes the results and discusses the resulting outcomes. Finally, concluding remarks and further potential ideas to be explored are given in Section 4.

II. MATERIALS AND METHODS

Nowadays, model based development is familiar practice with a wide range of specialized software tools for modeling and simulation such as MATLAB and Simulink. Simulink is an extension to MATLAB which uses an icon-driven interface for the construction of a block diagram representation of a process and user-friendly due to the icon-driven interface, yet it is important to spend much time designing with its many features.

Initially, the modified FHN mathematical modeling has been built by blocks from HDL supported library in Simulink for developed design of software is made principally to solve a set of nonlinear ODEs as in Eq. (1) and Eq. (2) to simulate an action potential generation and conduction of cardiac cells [20].

$$\frac{\partial V}{\partial t} = -V(V - 0.139)(V - 1) - W + I + D\frac{\partial^2 V}{\partial x^2} \quad (1)$$

$$\frac{\partial W}{\partial t} = 0.008(V - 2.54W) \quad (2)$$

Here, V is a membrane voltage, W is a refractory period, D is a diffusion coefficient, I is a time and space dependent injected current.

Generally, the designed and verified FHN model using the MATLAB Simulink is initially represented by continuous-time modeling in a floating-point data type. However, for the FPGA hardware implementation, conversion into a fixed-point data type in discrete-time framework is needed to be done during the process of designing the modeling algorithm. Moreover, a word-length and fraction-length optimization of the fixed-point is one of challenging aspects of implementing an algorithm on a

FPGA that needs to be considered here to enhance the performance in terms of power consumption and design area.

For the rapid design, the HDL coder in the MATLAB Simulink that supports HDL code generation automatically for the FPGA implementation and the HDL Verifier that verified the design and continuously test the designed through co-simulation with the FIL approaches will be used. This design flow is as demonstrated in Fig. 1. Besides extremely flexible, another advantage offers by the HDL Coder is it has capability enables much shorter design iterations than a traditional workflow that relies on hand-written code. Moreover, through the FIL simulation from the HDL Verifier, it can increase confidence that the algorithm will work in the real world and ensure it will behave as expected when implemented in the hardware.

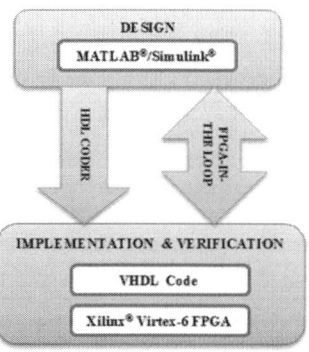

Fig. 1 FPGA Rapid design with the HDL Coder and the FIL.

III. RESULTS AND DISCUSSIONS

A. An Implementation Model Based Designed

Basically, electrophysiology of isolated cardiac cell models are coupled together to perform simulations of an action potential propagation in the cardiac tissue and it is often modeled by using significantly simplified quantitative method that can be represented by the 1D cable model. The cardiac excitations propagate diffusively in tissue, therefore cardiac electrical behavior is governed by reaction diffusion equation.

To realize the FPGA hardware implementation design in this study, the HDL Coder is used to automate the VHDL code generation from the designed MATLAB Simulink model in order to verify on the FPGA board. Thus, the designed model needs to be modified into the discrete-time framework with the conversion of floating-point to fixed-point data type. Fixed-point optimization is also being done in order to improve performance of the designed model through MATLAB Fixed-point Designer tool.

Fig. 2 shows the top level of the designed FHN 1D cable model using the MATLAB Simulink, where existing subsystems in the FHN model Subsystem block is in three layers within the model. Here, the input clock represents the timing of the stimulation and the output of the membrane voltage visualized by using the scope block.

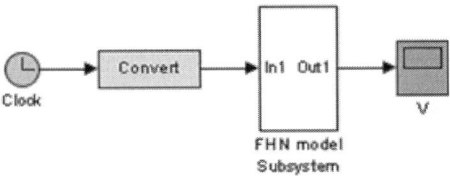

Fig. 2 The top level of FHN 1D cable model using the Simulink

The designed FHN model subsystem block which realized by coupling twenty membrane cells of the FHN model together is illustrated in detail in Fig. 3 with the diffusion coefficient of $D/\Delta x^2 = 2s^{-1}$, where $\Delta x = 0.0005cm$ is the spatial discretization at position x [20]. In this Fig. 3, the MATLAB function block is applied to control the current stimulation input.

B. FPGA-in-the-Loop Co-simulation

In this research, the FIL co-simulation is used to verify the model based design in real hardware for the generated VHDL code. The programming file is loaded onto the FPGA Xilinx Virtex-6 board via Joint Test Action Group (JTAG) connection while data are transmitted and received from Simulink to FPGA via gigabit Ethernet crossover cable. Fig. 4 demonstrates the model based designed in the Simulink connected with the FPGA Xilinx Virtex-6 board using the FIL simulation which is done through HDL Workflow Advisor.

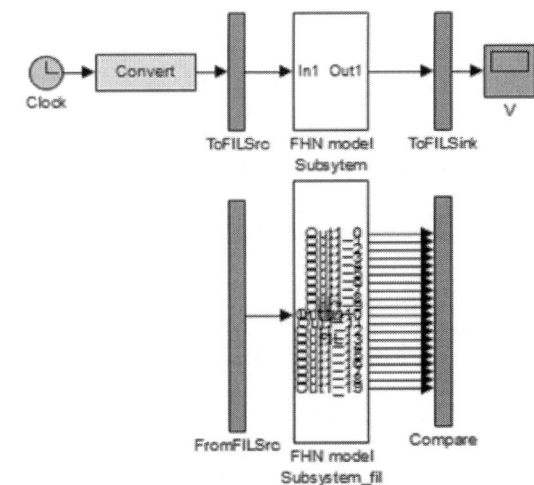

Fig. 4 FPGA-in-the-loop co-simulation

The generated blocks through the FIL workflow corresponding to error analysis between the FIL and the Simulink simulations are shown in Fig. 5.

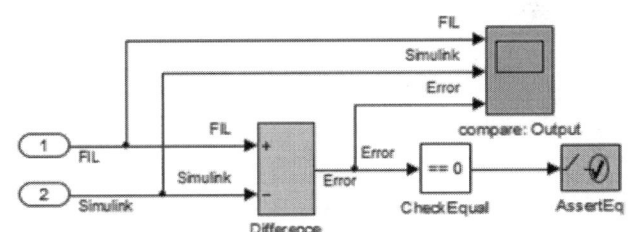

Fig. 5 Generated block for comparison of FIL simulation on FPGA hardware implementation by HDL Coder and Simulink simulation

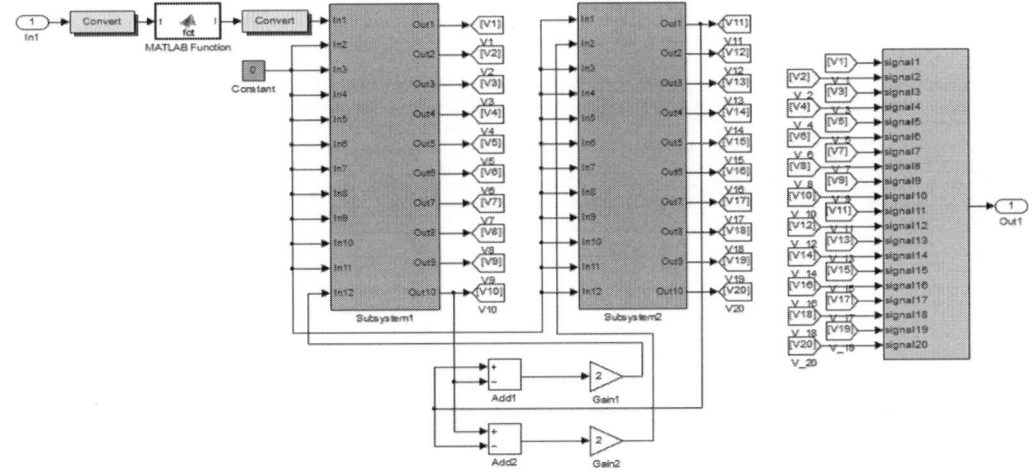

Fig. 3 Twenty membrane cells of FHN model in 1D cable model

Fig. 6 (a) and (b) show the conduction of action potential wave produced by the FIL and Simulink simulation respectively, from the first cell to the twentieth cell and the action potential is initiated at the first cell by 100ms of periodical external current stimuli starting at $t = 10$ms. Both results are noted to be comparable to the waveform in FHN model [5, 20] as both waveforms showed almost the same action potential duration around 50ms, and give the result of no difference between them as shown in Fig. 6 (c). Apparently the results in Fig. 6 (a) and (b) also show a high speed of action potential conduction which largely influenced by the value of the diffusion coefficient applied.

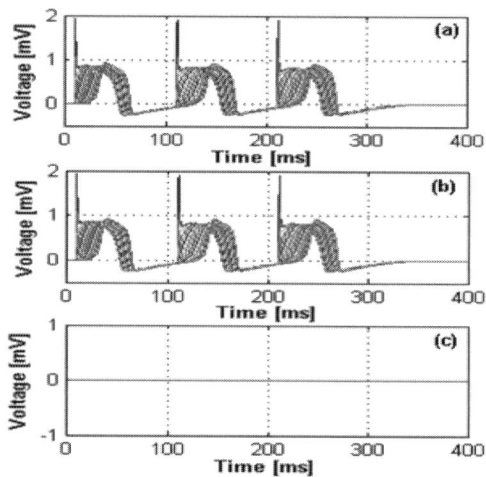

Fig. 6 Simulated 20 cells conduction of FHN model waveforms. Panel (a) and (b) represent the cardiac excitation-conduction generated by the FIL simulation and Simulink simulation, respectively. Panel (c) corresponding zero error after compare the both of simulation.

IV. CONCLUSIONS

In conclusion, the FPGA hardware implementation is successfully done from the Simulink simulation for FHN model of cardiac excitation-conduction using the HDL Coder which has been verified using the FIL simulation on the XC6VLX240T FPGA Xilinx Virtex-6 board. Comparative study for both simulations has revealed that the HDL Coder in MATLAB provides very efficient in rapid design to generate the VHDL code automatically. On-going research is focusing on the stand-alone FPGA hardware implementation to perform real-time simulations of the large scale cardiac conduction. Reentrant mechanisms will be further explored to demonstrate the capability of the proposed simulation in biomedical application systems.

ACKNOWLEDGMENT

The authors gratefully acknowledges the support by Fundamental Research Grant Scheme (FRGS) (vote no. 1053), under Ministry of Higher Education Malaysia.

REFERENCES

1. Mahmud F, Naruhiro S, Masaaki M et al. (2011) Reentrant excitation in an analog-digital hybrid circuit model of cardiac tissue. Europe PubMed Central 21:1-14
2. Mark T K, Michael C S. (2001) Molecular and Cellular Mechanisms of Cardiac Arrhythmias. Cell Press 104:569-580
3. Kerckhoffs R C P, Faris O, Bovendeerd P H M et al. (2005) Electromechanics of paced left ventrical simulated by straightforward mathematical model: Comparison with experiments. Heart Circ Phy 289:1889-1897
4. Mahmud F, Sakuhana T, Shiozaea et al. (2009) An analog-digital hybrid model of electrical excitation in a cardiac ventricular cell. Trans Jpn Soc Med Biol Eng 47(5):428-435
5. Fitzhugh R. (1960) Thresholds and plateaus in the Hodgkin-Huxley nerve equations. J Gen Phy 43(5):867-896
6. Noble D. (1960) Cardiac action potential and pacemaker potentials based on the Hodgkin-Huxley equations. Nature 188:495-497
7. Luo C H, Rudy Y A. (1991) A model of the ventricular cardiac action potential, depolarization, repolarization, and their interaction. Circ Research 68(6):1501-1526
8. Beeler G W, Reuter H. (1977) Reconstruction of the action potential of ventricular myocardial fibres. J Physiol 268:177-210
9. Luo C H, Rudy Y A. (1994) Dynamic model of the cardiac ventricular action potential. I. Simulations of ionic currents and concentration changes. Circ Res 74:1071-1096
10. Luo C H, Rudy Y A. (1994) Dynamic model of the cardiac ventricular action potential. II. After depolarizations, triggered activity, and potentiation. Circ Res 74:1097-1113
11. Harmon L D. (1961) Studies with artificial neurons, I. Properties and functions of an artificial neuron. Kybernetik 1(3):89-101
12. Nagumo J, Arimoto S, Yoshizawa S. (1962) An active pulse transmission line simulating nerve axon. Proceedings of the IRE 50(10) : 2061-2070
13. Nagumo J, Arimoto S, Yoshizawa S. (1965) Bistable transmission lines. IEEE Trans on Circ 12(3):400-412
14. Mahmud F. (2012) Real-time simulation of cardiac excitation using hardware-implemented cardiac excitation modeling. IJIE 4(3):13-18
15. Alireza F, Trong T D, Chedjou J C et al. (2012) New computational modeling for solving higher order ODE based on FPGA. IEEE Nonlinear Dynamics and Synchronization, 49-53
16. Huang C, Frank V, Tony G. (2011) A custom FPGA processor for physical model ordinary differential equation solving. IEE Embed Sys Lett 3(4):1-4
17. Dang P. (2006) VLSI architecture for real-time image and video processing systems. Journal of Real-Time Image Processing 1:57-62
18. Siwakoti P Y, Graham E T. (2013) Design of FPGA-controlled power electronics and drives using MATLAB Simulink. Macquarie University, Australia 571-577
19. Das J, Lam A, Steven J E W. (2011) An Analytical Model Relating FPGA Architecture to Logic Density and Depth. IEEE Transactions on Very Large Scale Integration (VLSI) Systems. 19(12): 2229-2242.
20. Nomura T et al. (1996) Entrainment and termination of reentrant wave propagation in a periodically stimulated ring of excitable media. The American Phy Soc 53(6):6353-6360

Determination of the Neutral Axis in Total Ankle Replacement

Ha Vo[1], Barry Tuvel[2], Bich Nguyen[1], and Trung Le[1]

[1] Department Biomedical Engineering, Mercer University, Macon, USA
[2] Ankle & Foot Care Center of Miami, Kendal, USA

Abstract— Major failures in ankle replacements include loosening, constraint, malalignment, fatigue, and impingement that are due to engineering design and implantation problems. The objective of this study is to propose a new process of implantation that includes customizing and determining specific size of ankle prosthesis using patient's specific geometric in order to maintain its natural biomechanics. The process started from a set of two dimensional (2D) Computerized Tomography (CT) data of the ankle joint, which was converted to three dimensional (3D) finite element (FE) model using Mimics 8.1. The 3D FE model was used in finite element analysis (FEA) to determine the optimal solution (the minimal stress acting on the ankle complex) using ANSYS 8.1. The new ankle design was implanted into the 3D FE model of the ankle for further analysis of the optimal solution an ankle joint to determine the specific size for this ankle joint. In this study, the 3D model was created from 180 slices of 2D CT scans of an arthritic ankle. Finite element analysis of pre and post implanted ankles are conducted to determine the solution. The result is optimized with a minimal stress of 24.365 MPa acting on the ankle in the neutral position. The 3D FE data are then translated into CNC language to generate a physiologically accurate model of the ankle from a wax block. A new ankle prosthesis designed by one of the authors is used in this study for optimization of the stress in the joint. Based on the optimized result, the proper size of the implant was determined and manufactured. A sample of the prosthesis was then pre-implanted into the cut-out wax model of the patient's ankle for pre-implantation prior to the actual operation. This process helped not only determine the proper size of the implant, but also allowed surgeons to plan the surgical process before entering the operation room.

Keywords— Ankle implant, Total Ankle Replacement, TAR, and Orthopedic Implant.

I. INTRODUCTION

Total ankle arthroplasty has become a viable treatment for patients with painful tibiotalar arthritis who have not responded to conservative treatments. Artificial ankle joint implant is a procedure that has been available since the 1970s [5]. The ankle replacement has been less successful than hip and knee replacement surgery. The main factors attributed to failure of the ankle joint implants are higher load applied to the ankle with a smaller surface area (5 times of body weight and area of 9-11 cm^2) to dissipate force, and the ankle joint's higher degrees of freedom (higher degree of movement mechanisms) [6, 7]. Over the last ten years the ankle implant has been growing in popularity as the design and the availability of the implants for replacements have improved. There are three different artificial ankle joints that are most widely used around the world. These are Alvin Agility Total Ankle (AATA), Buechel-Pappas (BP), and Scandinavian Total Ankle Replacement (STAR) or Ramses. These ankle implants reveal certain biomechanical problems and are therefore not the desired prosthesis [4, 6, 7]. Currently, the most popular of the three ankle implants is the Alvin Agility TAR [4, 5]. However, if ankle misalignment is not corrected, the replacement will have an increased chance of failure. [12] The failures that result from malalignment of the replaced ankle intraoperatively include eccentric loading of the tibial tray, progressive instability of the post-operative ankle, or increased wear of the Ultrahigh Molecular weight Polyethylene (UHMWPE) meniscus due to eccentric loading. Angular deformity is another important issue related to misalignment of the implant during surgery. Progression of this deformity will eventually cause stress fractures due to sudden increased loading on the adjacent bone and potentially result in a collapsed joint [12].

II. MATERIALS AND METHODS

To conduct the study, a CT scan of a 70-kg white female in her mid-fifties was randomly selected from the Ankle & Foot Care Centers of Miami. The patient had been treated with all invasive methods for painful arthritis of the right ankle in a continuous time of 2 years. All conservative treatments failed to alleviate the pain. Symptoms of pain and diminished range of motion of the right ankle have increased with time. Bone x-rays of the right ankle showed degenerative joint disease (DJD). Surgical treatment via total ankle arthroplasty was addressed and highly recommended to the patient. Then, CT scans were ordered for the creation of a 3D FE model to analyze stresses and determine a proper size of the ankle prosthesis. The 180 2D cross-sectional slices were obtained from the right ankle CT scan of the patient which is shown in Figure 1. The 3D FE model (Figure 2) of the ankle complex is created from the 2D images using ANSYS 8.1.

© Springer International Publishing Switzerland 2015
V. Van Toi and T.H. Lien Phuong (eds.), *5th International Conference on Biomedical Engineering in Vietnam,*
IFMBE Proceedings 46, DOI: 10.1007/978-3-319-11776-8_30

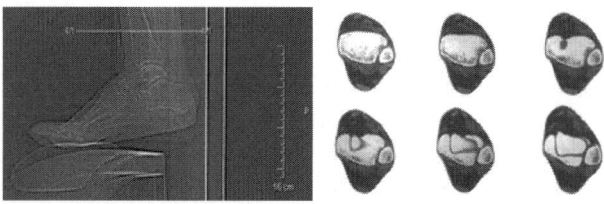

Fig. 1 The 2D images (left), generated from the CT scan of the ankle

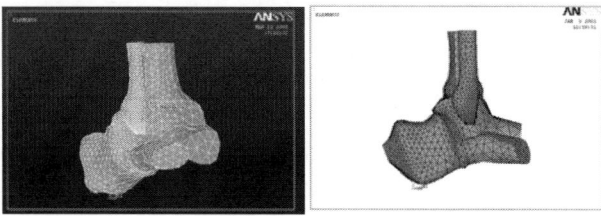

Fig. 2 Anterolateral (left) and lateral (right) views of a 3D FE model of the ankle complex

A compressive force of five times body weight, 3,430 N was applied constantly to the FE model of the ankle complex along the y-axis to analyze the contact pressure δ_{yy} of the ankle joint in neutral position. Even though the forces loading the ankle change during the cycle of gait, we decided to use a maximum load of five times body weight to conduct the study with analysis in static condition. In the FE model, thickness of meniscus increased in 2mm increments and was proportional to the same thickness of tibial plateau bone removed. Ultimate contact stresses were analyzed and compared to the ultimate contact pressures vs. bone densities. Data were collected and are shown in Table 1. First, we analyzed the effect of the thickness of UHMWPE meniscus on average contact pressure between the inferior surface of the meniscus and superior surface of the talar dome components. To investigate this goal, we used six different levels of thickness in meniscus components starting from 2 mm and ending at 12 mm with 2 mm increments. The results were graphed and exhibited by Figure 3. With increasing the meniscus thickness, the ankle joint experienced less stress on contact pressure. The contact pressure decreased dramatically with increased meniscus thickness from 2.0 mm to 6.289 mm and decreased slightly with increasing the thickness of meniscus from 6.290 mm to 12.0 mm as shown in Figure 3. Similarly, the effect of UHMWPE thickness on contact pressure was described and analyzed by Gur et al [8]. This based on the allowable contact stress for applications having compressive loading from 0 to 60 MPa applied to a typical knee design with changing the thicknesses of meniscus from 4 to 24 mm. In the Bartel et al [3] study, it was concluded that the stresses are related to the thickness of the polyethylene. It can be said that the contact stress component is a function of the meniscus thickness. Relating Gur the research to

ankle implants, it is a serious problem for ankle prostheses because the body load applied to the ankle joint is more than the load to the knee, while the ankle is much smaller in size and surface contact area. Also, Bartel et al [3] investigated the effect of thickness on stresses of polyethylene basing on an axisymetric solution of a rigid spherical indentor. In this stress condition, cortical bone of the talus exhibited mechanical properties of elastoplastic material, while cortical bone of the talus demonstrates high absorbance of load with pronounced strain softening. For ease of calculation, it was assumed that both cortical and cancellous bones are elastoplastic material models.

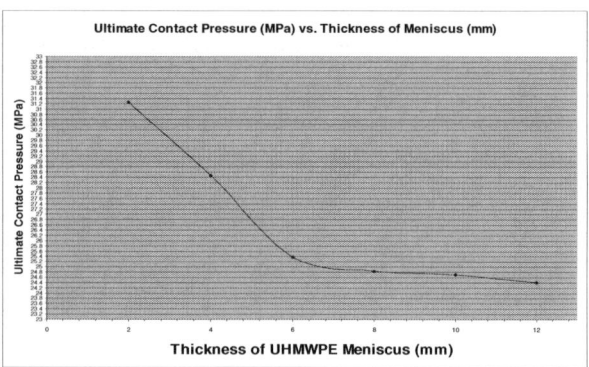

Fig. 3 Increasing the thickness of the meniscus, the ankle joint experiences less contact stress.

Table 1 Data collected from this study with employment of a load of 5 times body weight of an average person of 70kg

Applied Load (N)	Bone Mineral Density (gm/cc)	Ultimate Contact Pressure (MPa)
3430	0.6	29.673
3430	0.7	28.748
3430	0.8	26.982
3430	0.9	25.765
3430	1.0	24.912
3430	1.1	24.012
3430	1.2	23.371
3430	1.3	22.718
Applied Load (N)	Meniscus Thickness (mm)	Ultimate Contact Pressure (MPa)
3430	2	31.254
3430	4	28.489
3430	6	25.389
3430	8	24.846
3430	10	24.700
3430	12	24.410
Applied Load (N)	Tibial Plateau Bone Removal (mm)	Ultimate Contact Pressure (MPa)
3430	2	32.118
3430	4	29.012
3430	6	26.491
3430	8	24.920
3430	10	24.092
3430	12	23.874

Recently, research studies [1, 9] showed that the quantity of the bone removal around the ankle joints has seen a statistically significant decrease in bone resistance to compression and moves further from the subchondral plate. Therefore, resecting bone more than 4 - 5 mm from either side of the ankle joint may result in an increased risk of bony fracture in the edges associating with fatigue. In other words, authors of these studies suggested that the total bone removed from both sides should less than 10 mm. Related to the new ankle implant, if removed 4 – 5 mm is from both sides of medial and lateral malleoli, a cut of 8 to 10 mm to tibial plateaus will be applied; the tibial component thickness that will be added to the total talar bone removal is about 2 mm. Another study [10] from Orthoteers Company on wear rate of biomaterials determined that thin polyethylene increase wear due to increased fatigue wear if thickness is less than 6 – 8 mm. Thus the thickness level of the meniscus should be between 8 – 10 mm as concluded from previous research and studies. So the results from this study are acceptable in terms of the size of the implant for this patient. Figure 4 shows the relationship between the contact pressure and tibial plateau bone removal. More tibial bone removal (related to increase medial and lateral malleoli removal both sides of ankle bone) increases the surface contact (area of contact surface) which decreases the contact pressure. It is important to point out the region of the graph in Figure 4 showing that the ultimate compressive stress (contact pressure) slightly decreases in a domain of 8 – 12 cm of tibial bone removed.

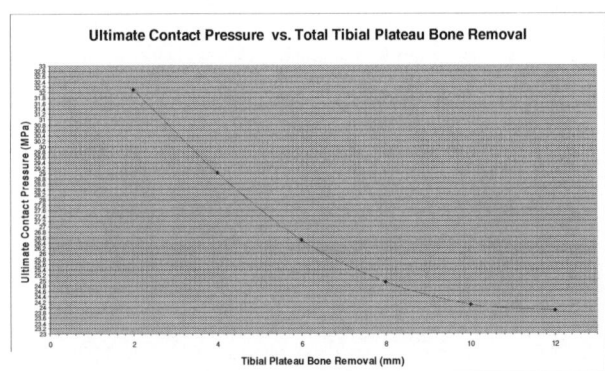

Fig. 4 The relationships between the ultimate contact pressure and the thickness bone removed of tibial plateau.

This data suggests that the tibial plateau bone removed must be determined from this domain for optimal results. In Figure 3, the graph not only exhibits the relationship between contact pressure and the thickness of the UHMWPE meniscus (size of the new implant which was dependent on the thickness of the UHMWPE meniscus) but also expresses a stable region of contact pressure with the thickness of meniscus increasing from 7 – 12 mm. In other words, the optimal size of the meniscus should be in this domain. Since the thickness of the

UHMWPE meniscus is equivalent to the total tibial bone resected, a combined graph for these two variables is shown in Figure 5. The intersection of these graphs is the optimal solution or proper size for this patient corresponding to ultimate compressive stress.

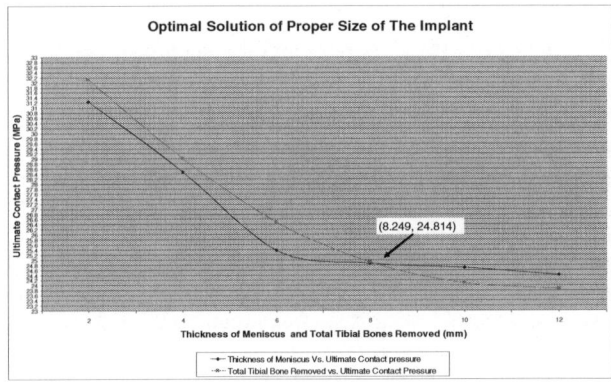

Fig. 5 The optimal solution (size of the implant) was determined from the intersection of graphs of Fig. 3 and 4.

The proper size for the implant was discussed in pervious paragraphs. In addition, bone density has a major effect on the ultimate compressive load (contact pressure). The surgeon must account for the patient's bone density in order to prevent fractures of bone or total joint collapse due to the maximum contact pressure of the implant [11]. Osteoporosis patients are not ideal candidates for any of the orthopedic implants because of the weakness of bone and these bone structures which are not strong enough to undergo the fatigue load in the interface between artificial joint components and bone. Relating this result to the optimal solution found from graphs in Figure 5, the maximum contact pressure acting on the implant that was only 24.814 MPa is much smaller than the maximum compressive stress of the actual ankle joint, 28.489 MPa. Hence, the size of this implant is the proper ankle prosthetic device for the patient under study. The implant's sizes can vary due to patient's ankle specific geometries. These FEMs of the ankle implants of the implants experience less contact pressures exerting on the joints compared to the smaller one.

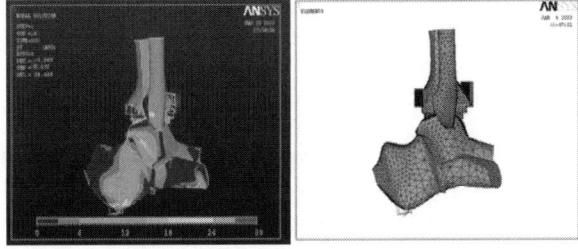

Fig. 6 FE model of the new ankle implant experiencing a constant body load of 3430N (5 times body weight of a 70 kg patient) - Lateral view

III. RESULTS AND DISCUSSIONS

The main goal of this study was to determine the proper meniscus thickness and the size the new total ankle implant (patent # US6863691) for individual patient diagnosed with painful ankle arthritis or previously fused ankle joints. As discussed earlier, the optimal solution was the best determination of the thickness and the size to the new ankle implant. Previous studies of the effect of thickness of the polyethylene on contact pressure using an axisymetric solution of a rigid spherical indentor by Bartel et al [2, 3], suggested that the thickness of the polyethylene meniscus be between 8 – 10 mm. Additionally, Beuchel – Pappas TAR [6] designed an UHMWPE meniscus with thickness ranging from 7- 10 mm. In our study, the optimal solution for the thickness of UHMWPE meniscus was 8.249 mm which belongs to the recommended domain that has been investigated by scientists and ankle implant experts in previous and current studies for the last decade.

The complete process of the this study is exhibited in Figure 7, in which the flow chart describes the step-by-step approach that guides the surgeon to determine the proper size of the implant for the total replacement. This process can also be applied to total knee and hip replacements and account for the difference in geometries of the joints. The pre-implantation of the ankle joint is displayed in Figure 7; the talar component is on the bottom and the UHMWPE meniscus is inserted into the tibial component.

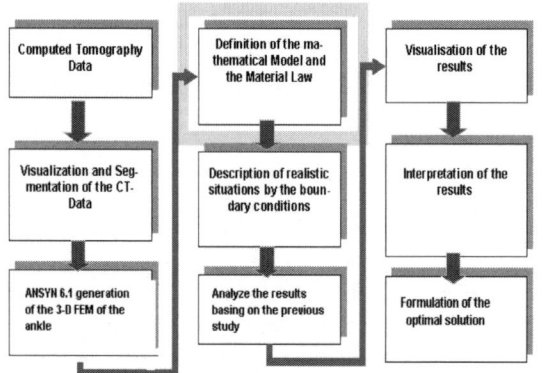

Fig. 7 The diagram shows the process starting from 2D CT to 3D FE model in order to determine the proper size for the ankle joint implant.

IV. CONCLUSIONS

The crucial goal of this study was to determine a proper size for the new ankle implant for an individual patient who is qualified for TAR due to post-traumatic ankle, painful ankle arthritis (osteoarthritis, and rheumatoid arthritis), and previous fused ankle mortise. This method also provides surgeons who perform total ankle replacements, a

pre-determination of the exact size of the implant so that the implant size does not have to be changed in the operation room (OR). A patient that chooses to have an ankle replacement will have a custom made ankle device only for his/her ankle. This not only determines precise size of the prosthesis but also guarantees stabilization and safety for the implant in the vivo.

To optimize the results of this method, more data collection is needed. In the future investigation, Mimics 8.1 software will be used to converted 2D images of the CT scans to a 3D FE model to maintain the accuracies of the development of the FE model. The studies will consider to multi-disorders of lower leg in terms of valgus, varus, cavus, and pes planus.

CONFLICT OF INTEREST

The authors declare that they have no conflict of interest.

REFERENCES

1. Aitken GK., et al., Indentation Stiffness the Cancellous Bone in the Distal Human Tibia. Clin Ortho 201: 264-270, 1985.
2. Bartel, D. L, et al., The Effect of Conformity and Plastic Thickness on Contact Stresses in Metal Backed Plastic Implants. Journal of Biomechanical Engineering, 1985. 107, p. 193-199.
3. Bartel, D. L., V. L. Bicknell, and T. M. Wright, The Effect of Conformity, Thickness, and Material on Stresses in Ultra-High Molecular Weight Components for Total Joint Replacement. Clinical Orthopaedics and Related Research, 1986. 68-A (7): p. 1041-1051.
4. Claes, L., Eberhard, O., Bauer, G., "Engineering Considerations of a New Ankle Joint", http://www.cmmaccess.com, Jan 1999.
5. Corti, Stephen F., et al., Complications of Total Ankle Replacement. 1(391), pp   105-114, Oct 2001
6. Felman, H. M., "The Total Ankle Joint Replacement", Study 98-99.
7. Felman, H. M., "This Surgical Procedure Became Available to Americans in October of 1998", wysiyg://result-doc-frame.content, 2000.
8. Hostaleu Gur, Hoechst Aktiengesellschaff, Verkauf Kunstoffe, 6230 Frankfurt am Main 80, 1982, p. 22.
9. Hvid I, Rasmusseu O., et al., Trabecular Bone Strength Profile Ankle Joint. Clin Ortho 199: 306-312, 1985.
10. Implants   &   Materials   in   Orthopedics, http://www.orthoteers.co.uk/Nrujp-ij33Lm/Orthbiomat.htm.
11. Mautalen C., et al., Ultrasound and Dual X-ray Absorptiometry Densitometry in Women with Hip Fracture. Calcif Tissue Int, 57:3, 1995 Sep, 165-168.
12. Steven M. Raikin, Mark S. Myerson, "Complications of Total Ankle Replacement"   http://www.drmyerson.com/research pubs/pressItem73.html.

Author: Dr. Ha Van Vo
Institute: Mercer University
Street: 1400 Coleman Avenue
City: Macon
Country: USA
Email: vo_hv@mercer.edu

Improving Electrospun Fibre Stacking with Direct Writing for Developing Scaffolds for Tissue Engineering for Non-load Bearing Bone

K.A. Blackwood[1], N. Ristovski[1], S. Liao[1], N. Bock[1], J. Ren[1], G.T.S. Kirby[1], M.M. Stevens[2], R. Steck[1], and M.A. Woodruff[1]

[1] Institute of Health & Biomedical Innovation, Queensland University of Technology, Brisbane, Queensland 4059, Australia
[2] Department of Materials, Imperial College, London, SW7 2AZ, United Kingdom

Abstract— Melt electrospinning can be used to produce fibres within the micro to nano scale with a deposition in a manner in-line with conventional 3D printing technology's [1]. Technical issues such as charge build up in subsequent layers lead to limitations in the precision of fibre deposition as the number of layers increases.

Polycaprolactone (PCL) is a polyester with a well-established history as a scaffold material for bone tissue engineering [2]. It is biocompatible, easy to shape and mechanically suitable for bone defects. Bioactive glasses are ceramic materials which are known to stimulate osteogenic differentiation [3]. The combination of PCL and bioactive glasses present the possibility to develop osteogenic scaffolds with a high degree of control of laydown using melt electrospinning [4].

This work develops the potential of melt electrospinning as a scaffold fabrication technique for tissue engineering non-load bearing bone defects by both developing techniques for improcing fibre laydown introducing osteogenic factors.

Keywords— Electrospinning, Polycaprolactone, Bioactive Glass.

I. INTRODUCTION

Tissue engineering is a wide field which aims to 'apply the principles of biology and engineering to the development of functional substitutes for damaged tissue' [5].

Biofabrication is a field of tissue engineering which comprises of three main disciplines, material sciences, mechanical engineering and cell & developmental biology [6]. Biofabrication offers practical design and development strategies for the manufacture of complex tissue replacement therapies.

There are numerous biofabrication technologies available, for the production of scaffolds for tissue engineering, purposes, such as salt leaching, supercritical CO_2, melt extrusion, laser sintering, microparticle formation, hydrogel casting and cell printing, and electrospinning [6–8].

Polycaprolactone (PCL) is an aliphatic polyester becoming popular in biofabrication due to it being an FDA-approved biomaterial [2]. PCL has a low melting point and easily attainable rheological and viscoelectic properties which are highly suitable for fabrication via a number of techniques, including melt electrospinning [4].

Electrospinning is a versatile technique for the production of fibers ranging from the micro- to nanoscale in diameter. Electrospinnings potential includes a large number of polymers, both natural and synthetic which can be spun [1, 9–11]. Electrospinning utilizes a static charge to drive a polymer jet towards a collector which is commonly earthed, although the use of oppositely charged collectors can be used. Electrospinning can be further classified into two separate techniques, solution electrospinning and melt electrospinning.

Solution electrospinning is the most commonly used of the two techniques, for solution electrospinning, the polymer is dissolved in a volatile solvent, which evaporates from the polymer as it travels between the tip to the collector [9]. Solution electrospinning produces fibres in the low micron to sub-micron scale.

The other technique is melt spinning [12], where the polymer is melted. Due to the increased viscosity of the melt spinning process compared to solution, the fibres produced are thicker than those produced in solution spinning [1]. A byproduct of the increased viscosity is a significantly more stable jet in melt electrospinning than that of solution. This means that melt electrospinning can be used to create structures with high degree of order by utilizing a mobile stage or spinner [1, 12].

This process makes melt electrospinning produce scaffolds with structures more like melt extrusion fused deposition modeling (FDM) [13]; but with fibres an order of magnitude smaller than the FDM technique; than classical solution electrospinning.

While PCL is a useful material in biofabrication, by itself it lacks osteoconductivity. Bioactive glasses (BGs) are a group of inorganic ion releasing glasses which have shown the ability to form a strong bond with hard and soft tissues [14]. The first commercial Bioglass; 45S5 (Bioglass®) was introduced in the 1970s. Since then, numerous variations of the 45S5 formula have been developed [15].

V. Van Toi and T.H. Lien Phuong (eds.), *5th International Conference on Biomedical Engineering in Vietnam,*
IFMBE Proceedings 46, DOI: 10.1007/978-3-319-11776-8_31

One of the newer BGs which have been developed is the strontium-substituted BG (SrBG), which have a portion of the calcium ions present in the 45S5 formula replaced with strontium [3]. It has been shown that strontium ions enhance bone regeneration [3, 16], with strontium based therapies commercially available.

Recent work from our lab has shown the potential of combining PCL and BGs together allows for additive manufacturing processing, producing scaffolds with highly controlled porosity and structure as well as enhanced bioactivity [13]. Due to melt electrospinnings ability to produce fibres significantly smaller than conventional additive manufacturing, there is the potential to develop scaffolds with high surface area for cellular attachment, high control of fibre orientation, and biological activity from BGs eluting ions. Previously we have shown it is possible to combine Strontium substituted BGs with PCL scaffolds and fabricate scaffolds via melt spinning [4], although ordered scaffolds were not achieved. This work looks at improving the manufacturing techniques in-order to develop controllable architecture for non-load bearing tissue engineering constructs.

II. MATERIALS AND METHODS

A. Materials

45S5 Bioglass® was purchased commercially. Strontium substituted bioactive glass was supplied by Prof Molly Stevens Group Imperial College London.

Both BG's had their particle size distribution determined by a Master Sorter 3000 (Marlvern).

BGs were then ground via a Micronizing mill (McCrone) using 100% ethanol and zirconia beads for 2 hours 30 min, and size distribution resampled.

PCL (*Capa 6400*; Perstorp UK Limited) was used as the structural polymer, in-order to integrate PCL with BG particles, BG particles (10% wt) were mixed with PCL in chloroform (10 ml per gram of PCL) at room temperature, followed by ethanol precipitation and air drying [13].

B. Melt Electrospinning

A bespoke melt electrospinning device with dual voltage control and an X-Y-Z stage was used for the scaffold manufacture. Setup is presented in Fig 1.

Variables for electrospinning are summarized in table 1.

Table 1 Electrospinning variables.

Material	Feed Rate	Needle Gauge	Temperature	TTC	Voltage (T/C)
PCL	40 µl/hr	G21	73°C	10 mm	7 kV/-3.5 kV
PCL/10% wt BG	40 µl/hr	G19	90°C	10 mm	7 kV/-3.5 kV

PCL alone was used as control, and PCL with 10% wt. BG (SrBG or 45S5) either as delivered, sieved through a 40 micron filter or as ground.

Fibre orientation and order was achieved by moving the stage at a rate of 750 mm per min to create a grid pattern 5 cm by 5 cm, with inter fibre spacing of 200 microns.

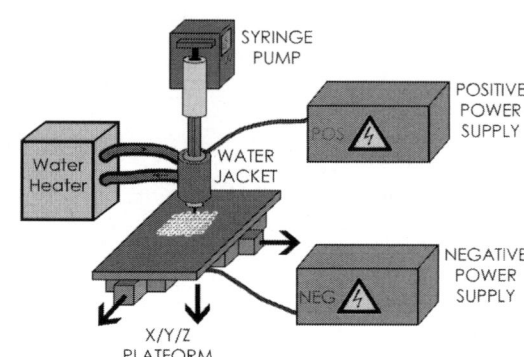

Fig. 1 Cartoon diagram setup of the melt eletcrospinning rig

C. Cell Culture

Scaffolds were first etched with 5M NaOH for 1 hour before being washed with MiliQ water until the supernatents pH reached 7.0. Scaffolds were then dried for 24 hours cultured with MC3T3s for cell attachment (~5000 cells per 30 mm^2).

D. Microscopy

Fibre size, alignment and particle distribution were assessed by light micrscopy (Ziess Axio Imager M2).

Confocal microscopy (Leica SP5 Confocal microscope) was used with DAPI/Alexa Fluor® 488 Phalloidin (Invitrogen) and Alizarin red stained BG particles as per previous studies [4].

III. RESULTS

A. Scaffold Manufacturing

Both the 45S5 and SrBGs particles were similar in size and distribtuion prior to grinding, with a median particle size of 50.4 (45S5) and 49.7 (SrBG). Both BGs were also successfully ground down with a similar profile, producing median particle sizes of 3.03 (45s5) and 2.07 (SrBG). Data is summarized in table 2.

Table 2 Bioactive glass particle size distribution.

Material	Dx (10)	Dx (50)	Dx 90
45S5 BG before grinding	13.8	50.4	110
45S5 BG after grinding	0.93	3.03	7.04
SrBG before grinding	9.61	49.7	110
SrBG after grinding	0.94	2.97	6.47

PCL was used as a control to establish baseline spin variables of voltage (spinneret and collector plate) temperature, flow rate, tip-to-collector distance and needle gauge.

The polycaprolactone scaffolds had an average fibre thickness of 30.6 μm (± 1.8 μm). Scaffolds spun with +7.5 kV at the tip and -3.5 kV exhibited a high degree of order up to 1500 μm in height (± 231 μm) (Fig 2A.), significantly higher than the +10.5 kV to ground control (500 μm ±100 μm). Past this height, undissipated charge prevented the maintaining of order during fibre laydown, resulting in a breakdown of fibre alignment.

While PCL/BG composite scaffolds could be spun with the described conditions (Fig 2. B and C), average fibre size was higher for both BGs at 46.8 μm (± 16 μm), and obtaining alignment was impossible due to instability in the jet. When attempting to spin a straight line, the jet would break and produce areas of very fine disordered fibres (Fig 2. F).

Ground PCL/BG composite scaffolds exhibited superior jet stability, and they were successfully spun in straight lines, however the fibre was unable to successfully attach to the collector plate due to excessive cooling, and ordered fibre grids were unable to be successfully spun.

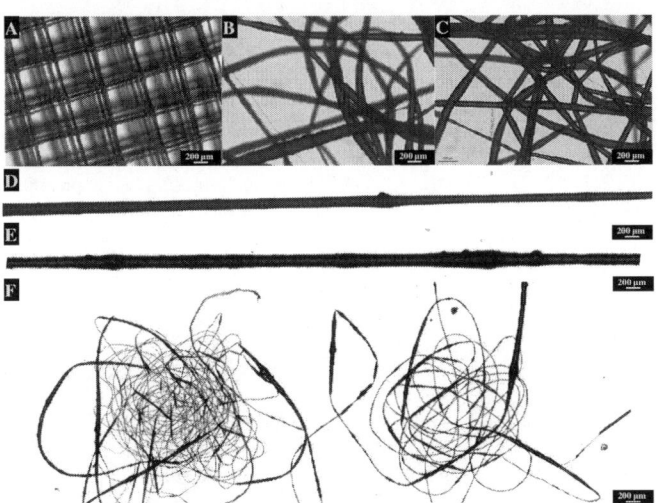

Fig. 2. Light Microscopy images of ordered melt electrospun PCL (A), PCL with 10% wt non-ground 45S5 (B), PCL with 10% wt non-ground SrBG (C), PCL with 10% wt. ground 45S5 (D), PCL with 10% wt. ground SrBG, and a representative image of attempting to align spin PCL with 10% wt unground BG (SrBG).

B. Cellular Attachment

Cells successfully penetration throughout the depth of the PCL scaffolds.

In PCL/BG scaffolds, MC3T3's exhibited preferential binding and close attachment to exposed BG particles in the scaffold (Fig 3.).

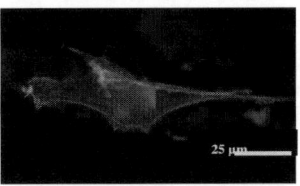

Fig 3. Representative confocal microscopy image of an MC3T3 cell (nuclei blue, and Actin green) cultured on PCL with 10% wt. SrBG (Stained red)

IV. DISCUSSION

A. Scaffold Development

Scaffolds for tissue engineering require specific features to be a viable treatment. The development of composite materials providing structural features (PCL) and bioactivity (BGs) has the potential to fulfill the required criteria.

The development of ordered fiber structure allows for greater repeatability of the scaffold manufacturing process, also allowing for the potential to develop scaffolds with defined regions of different architectures. We have shown that it is possible to maintain a high level of order by using a dual voltage system, allowing for the manufacture of ordered scaffolds up to 2 mm in height; 10× greater than previously reported [1, 12]. These results imply that the melt spinning has potential to be used for the production of tissue engineered scaffolds of high order.

We proposed that the increase in order is due to charge build up in the previous layers. Conventional melt electrospinning techniques go from a charged point towards an earthed collector [1, 4, 12]. The push-pull effect of dual charge system appears to improve the dissipation of buildup.

The addition of bioactive glass adds complexity to the melt spinning process due to the ceramics vastly different thermal and structural properties, which are largely unsuitable for melt spinning. Our initial results showed that it was possible to produce randomly orientated fibres structures of a 10% wt. strontium substituted bioactive glass, although we noted that the fibres were on average larger than PCL alone.

As the BGs do not melt, it is important that the particles are evenly distributed throughout the PCL melt, as it was observed that a conglomeration of the particles caused electrospinning jet instability. We found that the modification of a salt leaching technique [13] ensured an even distribution of BG particles.

The size of the BG particles also has a significant effect upon the stability of the jet, with large particles causing the jet to break up and 'spit' (Fig 3. F) our results have also shown that a larger distribution of BG particle size led to a larger distribution in fibre size, which is to be expected.

Both BGs tested responded similarly to grinding, with both the SrBG and 45S being ground down to approximately 3 microns average particle size, with a 90% of the particle size between 1 and 10 microns. This outcome was beneficial because we were able to eliminate particle size as a variable between the BGs. The ground particles were able to be successfully integrated into PCL via the salt leaching technique [13], and initial results have shown a much greater stability in jet flow (Fig 3. D and E) with a stabilized jet and continuous fibre being produced from the ground BG micro particles.

However, we were unable to generate scaffolds with the scaffold due to the rapid cooling effect of the BG particles causing the jet to solidify rapidly and not efficiently attach to the collector. Further work is required to facilitate attachment in order to fully realize the controlled deposition of PCL/BG composite scaffolds via melt electrospinning.

Other techniques for producing small BG particles are available [17], and the results we achieved suggest that; providing the manufacturing process can avoid conglomeration of the particles and achieve a homogenous distribution within the PCL melt; smaller particle size would be more beneficial in maintaining melt spinning jet stability.

Another phenomena observed was the potential for the BG's to be conductive, as melt spinning is an electrostatic process, too much conductivity in the material could potentially prevent the process from functioning. However it should be noted that this might allow for a higher degree of stacking due to the reduction in charge buildup on the layers.

B. Bioactivity of Scaffolds

As we have previously shown, compared with melt spun PCL scaffolds, ALP activity on PCL/SBG melt spun scaffolds were significantly higher on day 21 and day 28 in osteogenic media [4]. This result shows that the BG particles are actively eluting ions at a sufficient rate to up regulate an osteogenic response. When culturing MC3T3s on the scaffold, we observed that the cells would preferentially attach towards the scaffold regions with BG particles present on the surface. As BG particles are known to form strong bonds with tissue, it stands to reason that the cells would preferentially bind to the BG particles over the PCL fibres.

Current work has looked at integrating 10% wt BG particles to the PCL melt, as the bioactive component is the BG, the ability to incorporate a higher percentage of BG will increase the overall osteogenic potential of the scaffold,

however the previously mentioned fabrication problems will have to be overcome first.

ACKNOWLEDGEMENTS

We acknowledge the support from the Australian Research Council for Linkage Grants LP110200082 and LP100200084.

REFERENCES

1. Dalton PD, Vaquette C, Farrugia BL, Dargaville TR, Brown TD, Hutmacher DW (2013) Electrospinning and additive manufacturing: converging technologies. Biomater Sci 1:171
2. Woodruff MA, Hutmacher DW (2010) The return of a forgotten polymer—Polycaprolactone in the 21st century. Prog Polym Sci 35:1217–1256
3. Gentleman E, Fredholm YC, Jell G, Lotfibakhshaiesh N, O'Donnell MD, Hill RG, Stevens MM (2010) The effects of strontium-substituted bioactive glasses on osteoblasts and osteoclasts in vitro. Biomaterials 31:3949–3956
4. Ren J, Blackwood KA, Doustgani A, Poh PP, Steck R, Stevens MM, Woodruff MA (2013) Melt-electrospun polycaprolactone strontium-substituted bioactive glass scaffolds for bone regeneration. J Biomed Mater Res Part A. doi: 10.1002/jbm.a.34985
5. Langer R, Vacanti JP (1993) Tissue engineering. Sci 260 :920–926
6. Mironov V, Trusk T, Kasyanov V, Little S, Swaja R, Markwald R (2009) Biofabrication: a 21st century manufacturing paradigm. Biofabrication 1:022001
7. Malda J, Visser J, Melchels FP, Jüngst T, Hennink WE, Dhert WJ a, Groll J, Hutmacher DW (2013) 25th anniversary article: Engineering hydrogels for biofabrication. Adv Mater 25:5011–28
8. Ferris CJ, Gilmore KG, Wallace GG, In het Panhuis M (2013) Biofabrication: an overview of the approaches used for printing of living cells. Appl Microbiol Biotechnol 97:4243–58
9. Doshi J, Reneker DH (1995) Electrospinning process and applications of electrospun fibers. J Electrostat 35:151–160
10. Ashammakhi N, Ndreu A, Piras AM, et al (2007) Biodegradable nanomats produced by electrospinning: expanding multifunctionality and potential for tissue engineering. J Nanosci Nanotechnol 7:862–882
11. Teo W-E, Inai R, Ramakrishna S (2011) Technological advances in electrospinning of nanofibers. Sci Technol Adv Mater 12:013002
12. Brown TD, Dalton PD, Hutmacher DW (2011) Direct Writing By Way of Melt Electrospinning. Adv Mater 23:5651–5657
13. Poh PSP, Hutmacher DW, Stevens MM, Woodruff MA (2013) Fabrication and in vitro characterization of bioactive glass composite scaffolds for bone regeneration. Biofabrication 5:045005
14. Hench LL, Splinter RJ, Allen WC, Greenlee TK (1971) Bonding mechanisms at the interface of ceramic prosthetic materials. J Biomed Mater Res 5:117–141
15. Hench LL (1991) Bioceramics: From Concept to Clinic. J Am Ceram Soc 74:1487–1510
16. O'Donnell MD, Hill RG (2010) Influence of strontium and the importance of glass chemistry and structure when designing bioactive glasses for bone regeneration. Acta Biomater 6:2382–2385
17. Tsigkou O, Labbaf S, Stevens MM, Porter AE, Jones JR (2014) Monodispersed Bioactive Glass Submicron Particles and Their Effect on Bone Marrow and Adipose Tissue-Derived Stem Cells. Adv Healthc Mater 3:115–125

A Comprehensive Evaluation of Flapless Dental Implant Treatment in Posterior Maxilla and the Conservative Regeneration of Bone in Osteoporotic Rats

N. Doan[1], Z. Du[1], J. Xiao[2], W. Xia[3], R. Crawford[1], P. Reher[2], S. Ivanovski[2], F. Yan[3], J. Chen[3], and Y. Xiao[1]

[1]Institute of Health and Biomedical Innovation, Queensland University of Technology, Brisbane, Australia
[2]School of Dentistry and Oral Health, Griffith University, Gold Coast, Queensland, Australia
[3]Department of Oral Implants, Affiliated Stomatological Hospital of Fujian Medical University, Fuzhou, China

Abstract— The aim of this study is to perform a comprehensive evaluation of clinical procedures used in dental implant treatment in posterior maxilla using flapless technique; to assess osseo-integration of surface-treated implants in the posterior maxilla of osteoporotic (OP) rats, and the conservative regeneration of poor quality bone in (OP) rats with simvastatin. Materials and methods: This comprehensive consists of a systematic review of the literature on dental implant flapless technique, a retrospective study of 1241 dental implants using flapless technique, and an osseo-integration study on osteoporotic (OP) rats. The osseo-integration study used a cohort of sixty-four female Sprague-Dawley rats, aged 3 months old, split into three groups: Sham-operated (SHAM; n=20), ovariectomized (OVX; n=20) and ovariectomized treated with simvastatin (OVX+SIM; n=20). Eighty-four days following ovariectomy, screw-shaped titanium implants were immediately placed into the posterior maxilla. Simvastatin was fed orally at 5 mg/kg each day after the implant placement in the OVX+SIM group. The animals were sacrificed at either 28 or 56 days for histological analysis. The osseo-integration indices used were: bone formation rate (BFR), bone to implant contact (BIC), and bone density (BD). Results and Discussion: The systematic review showed flapless surgery had comparable, similar & high survival rate compare with flap surgery, and its cumulative 5 and 10 years implant survival rate of 97.9% and 96.5% respectively and complication rate of 6.0%. The osseo-integration indices (BFR, BIC and BD) in the three groups showed significant variations among the SHAM > OVX+SIM > OVX group, which suggested that simvastatin could encourage bone regeneration and mineralization in OVX rats. Conclusion: This study shows that implant flapless surgery has been proven to be a successful procedure, osteoporosis reduced osseo-integration, and simvastatin can positively influence on bone regeneration through the osseo-integration indices, and successfully promoted osseo-integration in the posterior maxilla in OP rats.

Keywords— Dental implants, flapless surgery, osseo-integration, osteoporosis, posterior maxilla.

I. INTRODUCTION

The review of the publications in the posterior maxilla areas shows that flapless surgery could be a viable and predictable treatment method for implant placement, indicating both efficacy and clinical effectiveness with some reservation[1].

Minimal invasive surgery has transformed modern-day surgery including dental implantology, especially flapless surgery [2].

The success of dental implants over the last twenty years, has led to a rapidly increasing number of human and animal research projects, but no research had been done using surface-treated implants in the posterior maxilla of osteoporotic rats [2].

There has been no documented study on the relationship between dental implant placement, poor quality bone, simvastatin (SIM) [a HMG-CoA reductase], and osseo-integration of surface-treated implants in the posterior maxilla of osteoporotic rats [2].

The aims of this paper is to: (1) comprehensively review the literature relating the application of dental implant flapless surgery in posterior maxillae; (2) identify the predictors of 1241 dental implants survival when using the flapless protocol in two private practices retrospectively; (3) evaluate the effects of osteoporosis on osseo-integration around titanium implants following a tooth extraction in the posterior maxilla of rats, and (4) to assess the conservative regeneration of poor quality bone in (OP) rats with simvastatin.

II. MATERIALS AND METHODS

This study was approved by the Ethic committee of the Queensland University of Technology, Brisbane, Australia and the Fujian Medical University, Fuzhou, China.

This comprehensive study consists of a systematic review of the literature on dental implant flapless technique, a retrospective study of 1241 dental implants using flapless technique, and an osseo-integration study on osteoporotic (OP) rats and the assessment on the effects of Simvastatin on osseo-intergration indices.

© Springer International Publishing Switzerland 2015
V. Van Toi and T.H. Lien Phuong (eds.), *5th International Conference on Biomedical Engineering in Vietnam*,
IFMBE Proceedings 46, DOI: 10.1007/978-3-319-11776-8_32

A. A Systematic Review of the Literature on Dental Implant Flapless Technique

Data source: The literature search was conducted using keywords: flapless, dental implants, maxilla. Hand-search and Medline search were carried out on studies published between 1971 and 2011. Study eligibility criteria: We have included study of at least 15 dental implants with a follow-up period of one year, an outcome measurement of implant survival but excluding studies involving numerous concurrent interventions and studies with missing information. The Cochrane approach for cohort studies and the criteria of the Oxford Centre for Evidence-Based Medicine were used.

B. A Retrospective Study of 1241 Dental Implants Using Flapless Technique

The gathered data was primarily computer-searched to find the patients and then later a hand-search of patient records was performed to locate 1241 flapless implants placed successively over the last 10 years. The demographic data collected or statistical predictors included: age, gender, periodontal disease and peri-implantitis status, smoking, details of implants inserted, implant locations, placement time after extraction, utilization of concurrent guided hard and soft tissue regeneration procedures, loading protocols, type of prostheses and treatment outcomes (implant survival and complications). Exclusion criteria comprised: any implants that involved flap and simultaneous guided hard and soft tissue regeneration procedures, and implants that measured less than 3.25 mm diameter.

C. An Osseo-Integration Study on Osteoporotic (OP) Rats

Forty-four 3 month old female Sprague-Dawley rats were used in this study. The rats were randomly divided into two groups: Sham-operated group (SHAM; n=22) and ovariectomized group (OVX; n=22). Surface-treated screw-shaped titanium implants were immediately inserted into the mesial extraction sites of the first molar in the posterior maxilla following tooth extraction. The animals were sacrificed at either 28 days (4 weeks) or 56 days (8 weeks) post-surgery, and undecalcified tissue sections were processed for histological analysis. Bone-to-implant contact (BIC) and bone density (BD) were evaluated.

D. The Assessment on the Effects of Simvastatin on Osseo-Intergration Indices

Sixty-four female Sprague-Dawley rats, aged 3 months old were utilized in this study, split into three groups: Sham-operated (SHAM; n=20), ovariectomized (OVX; n=20) and ovariectomized treated with simvastatin (OVX+SIM; n=20). Two rats from the SHAM and two from the OVX groups were utilized to verify osteoporosis.

Eighty-four days following ovariectomy, screw-shaped titanium implants were straight away inserted into mesial root sockets of the posterior maxilla. Simvastatin was dispensed orally at 5 mg/kg each day after the implant placement in the OVX+SIM group. The animals were slaughtered at either 28 or 56 days from the date of implant placement and the undecalcified tissue sections were processed for histological analysis. The osseo-integration indices employed were: bone formation rate (BFR), bone to implant contact (BIC), and bone density (BD).

III. RESULTS

Of the 56 published papers chosen, 14 papers on flapless technique displayed a high overall implant survival rate. The prospective studies gained 97.01% (95% CI: 90.72 to 99.0) while retrospective studies or case series illustrated 95.08% (95% CI: 91.0 to 97.93) survival. The average of intraoperative complications was 6.55% using the flapless procedure.

A life table analysis revealed a cumulative 5- and 10-year implant survival rate of 97.9% and 96.5% respectively. Most of the failed implants happened in the posterior maxilla (54%), in type 4 bone (74.0%); and 55.0% of failed implants were discovered in smokers.

The BIC and BD in the OVX group were significantly lower than those in the SHAM group at both 28 and 56 days, which showed that osteoporosis could decrease the extent of osseo-integration of dental implants in the posterior maxilla

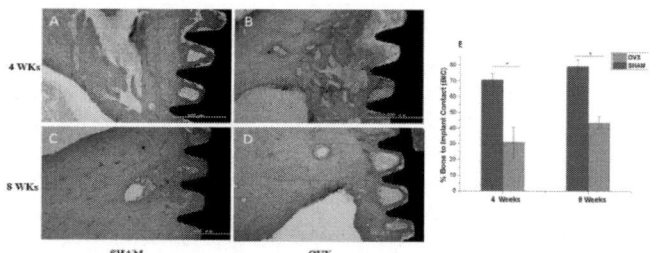

Fig. 1 Methylene blue-basic fuchsin staining (10x magnification) SHAM & OVX at 4 and 8 weeks. The newly created bone was darkly stained whereas the established bone was palely stained. More bone to implant contact (BIC) and better bone density (BD) in the SHAM group at 4 weeks (A) and 8 weeks (C) than that in the OVX group (B and D respectively). Bar= 500μm. (E) Graph of bone to implant contact (BIC) at 4 and 8 weeks which illustrated BIC in SHAM groups were superior and statistically significant as contrast to the OVX counterpart.

The osseo-integration indices (BFR, BIC and BD) in the three groups revealed significant variations among the SHAM > OVX+SIM > OVX group, which suggested that simvastatin could stimulate bone mineralization in OVX rats.

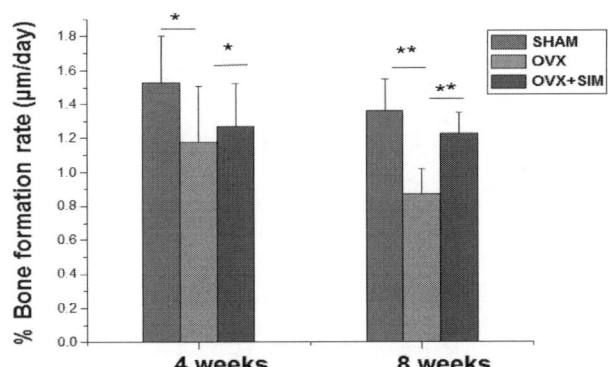

Fig. 2 Graph illustrated bone formation rate using fluorescence staining after 4 weeks and 8 weeks. The differences in BFR between OVX and SHAM and OVX and OVX+ SIM were statistically significant at 4 weeks (*p< 0.05) and 8 weeks (**p<0.005) respectively. BFR was slowed down in three groups at 8 weeks. Also at 8 weeks, the OVX+SIM group experienced the least reduction in BFR. The order of BFR was SHAM>OVX+SIM>OVX

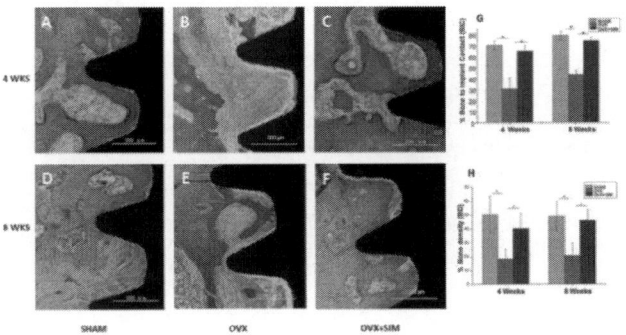

Fig. 3 Histological evaluation of bone to implant contact (BIC) and bone density (BD) at 4 weeks (A, B, and C) and 8 weeks (D, E, and F) (20x magnification using methylene blue-basic fuchsin staining). At 4 weeks: (A) SHAM, (B) OVX, and (C) OVX+SIM. At 8 weeks: (D) SHAM, (E) OVX, and (F) OVX+SIM. Contrasted with 4 weeks, there seems to be extra new bone growth at 8 weeks adjoining the implant in the three groups and the order is SHAM>OVX+SIM>OVX. Graphs demonstrate of bone to implant contact (BIC) and bone density (BD). BIC (A) and BD (B) show the subordination and statistically significant changes of the OVX groups as contrasted with the SHAM (P<0.05) and OVX+SIM groups (P<0.05).

Compared with the 4 weeks groups, the 8 weeks groups indicate BIC boosted by the three groups and BD seems to be thicker in the OVX and OVX+SIM groups but not in the SHAM group.

IV. DISSCUSSION

The obtainable short-term and long-term outcomes reported in this review elucidate that the flapless approach, initially recommended for inexperienced clinicians, necessitates additional skill and pre-surgical training than initially believed. In light of contemporary developments in digital imaging and computer-guided surgery, this technique can now be handled with more predictability. However, thoughtfulness should be used in managing flapless surgery to reduce complications and disasters. This means that implant practitioners must be prepared to acquire and familiarize to latest technology. Being thorough and thoughtful with new techniques and technology are ways that could aid to protect against nasty complications in flapless implant surgery, involving those implants inserted in the posterior maxilla.

This study has showed that flapless dental implant surgery can deliver a fairly superior survival rate in contrasted with other reports employing traditional flap methods. It has also underpinned the view that a predictable result with superior efficiency and efficacy can be attained with meticulous treatment planning and devotion to evidence-based surgical and restorative protocols.

These results steer to the supposition that the surgical engagement of an implant into the mesial root of a newly extracted maxillary first molar in rats could imitate implant placement in inferior quality bone such as is found in an osteoporotic posterior maxilla. With skillful and thorough surgical implant protocol, implant placement after tooth removal in the posterior maxilla of osteoporotic rats can yield reasonable outcomes, and implant placement ensuing to fresh extraction in low quality bone is not a total contraindication. However, in contrasted with the SHAM group, the indices of osseo-integration (BIC and BD) in the OVX group were lower. This result implies that attention should be given when inserting dental implants in osteoporotic subjects. Accordingly, there is a requirement to acquire a novel therapy modality to improve osseo-integration in osteoporosis.

There have been numerous papers studying osseo-integration of dental implants in osteoporotic subjects, but none of them concentrating on the posterior maxilla. This study is the unique of its genre that explicitly appraises the influence of simvastatin on osseo-integration of dental implants in the posterior maxilla of osteoporotic rats. Simvastatin is a HMG-CoA reductase inhibitor which hinders cholesterol biosynthesis and is commonly used as a cholesterol-reducing drug [2, 3]. Currently it has been stated that the liposoluble statin, simvastatin, could up-regulate the manifestation of BMP-2 mRNA in osteoblasts and, as a result, promote bone growth [4].

With the supplement of simvastatin not only new mineralized bone was created but it appeared to become nearer to the implant surface as revealed with calcein and alizarin staining. This entails that simvastatin had somehow up-regulated the manifestation of BMB-2 mRNA in osteoblasts to generate additional bone [4] and also through the VGF pathway [5]. The order of staining intensity of mineralized bone by the three groups was SHAM > OVX+SIM >OVX and the BFR order was analogous, namely: SHAM > OVX+SIM >OVX. Nevertheless, even in the occurrence of simvastatin, the BFR of the OVX+SIM was still lower than that of the SHAM group. Thus, it can be anticipated that osteoporotic model employed in this study functions well in the presence of simvastatin and may have assisted to minimize the effect of osteoporosis. The BFR of the OVX+SIM group was superior to the OVX group, but it was marginally lower than the SHAM group. This infers that additional research is required on the direct use of simvastatin to implant surface and medicaments necessitated to enrich osseo-integration in osteoporotic subjects.

The results obtained from Simvastatin and osseo-integration study suggested that simvastatin could stimulate bone mineralization in OVX rats. Hence, this study has revealed for the first time that simvastatin can certainly influence the osseo-integration indices, and successfully boost osseo-integration in the posterior maxilla in osteoporotic rats.

V. Conclusion

This study indicates that implant flapless surgery has been demonstrated to be an effective technique, osteoporosis diminished osseo-integration, and simvastatin can positively induce on bone rejuvenation through the osseo-integration indices, and efficaciously fostered osseo-integration in the posterior maxilla in OP rats.

Acknowledgment

This research is a collaboration effort from Queensland University of Technology (QUT), Australia and Fujian Medical University, China. The implants used in rats' studies were kindly donated by Southern Implants, South Africa. This project is also partly funded by the ITI Foundation.

Conflict of Interest

The authors declare that they have no conflict of interest.

References

1. Doan, N., et al., *Is flapless implant surgery a viable option in posterior maxilla? A review.* International journal of oral and maxillofacial surgery, 2012. **41**(9): p. 1064-71.
2. Doan N. (2014). An evaluation of clinical procedures used in dental implant treatment in posterior maxilla using flapless technique. Dissertation for the degree of Doctor of Philosophy, Queensland University of Technology.
3. Du Z, Doan N, Ivanovski S, Xiao Y. Serum bone formation marker correlation with improved osseointegration in osteoporotic rats treated with simvastatin. Clin Oral Implants Res, 2011; Nov 10.
4. Mundy, G., et al., Stimulation of bone formation in vitro and in rodents by statins. Science, 1999. 286(5446): p. 1946-9.
5. Horiuchi, N. and T. Maeda, Statins and bone metabolism. Oral diseases, 2006. 12(2): p. 85-101.

Corresponding author: Nghiem Doan, PhD
Institute: Institute of Health and Biomedical Innovation
& Science and Engineering Faculty, Queensland University of
           Technology
Street: 60 Musk Avenue, Kelvin Grove, Qld 4059
City: Brisbane
Country: Australia
Email: doanco@bigpond.com.au

A Measure of Clinical Outcomes in Dental Implant Surgery Flapless Surgery versus Flap Technique in Posterior Maxilla of Post Menopause Women

M. Nguyen[1], N. Doan[2,4], Z. Du[2], P. Reher[3], and Y. Xiao[2,4]

[1] Hue University of Medicine and Pharmacy, Vietnam
[2] Institute of Health and Biomedical Innovation, Queensland University of Technology, Brisbane, Australia
[3] School of Dentistry and Oral Health, Griffith University, Gold Coast, Queensland, Australia
[4] School of Dentistry, The University of Queensland, Brisbane, Queensland, Australia

Abstract— The aim of this study is to examine the clinical outcomes of implants inserted using split mouth study and to measure patients' satisfaction using visual analogue scale in flapless and conventional flap techniques in post menopause women age 50 years or over. Materials and methods: This study is a retrospective split mouth study of flapless vs flap technique involving the study of dental records of 16 post-menopause of patients undergoing bilateral implant surgery in the posterior maxilla. A total of 45 implants with no augmentative procedures were selected from 16 patients for the study. The patients were divided into two groups: the control group had 21 implants placed by full flap technique, and the test group consisted of 24 implants inserted using flapless procedure. Only those patients with comprehensive clinical record were included in this study. The treatment outcomes were measured using key words: implant survival, Visual Analogue Scale (VAS), Periotest, x-ray assessment. Results and Discussion: The results showed that flapless surgery had comparable, similar results as compare to flap surgery: survival rate (95.8% and 95.2%), Using visual analogue scale (VAS=0 to 10), flapless surgery revealed to have less: pain, swelling, bleeding and speech impairment and had better overall satisfaction at one day and one week than flap technique than the flap counterpart (*P<0.05). No significant difference in bone resorption at 3 months. After one year, bone change in the flap group vs the flapless group was statistically significant [-0.53 (±0.57) vs +0.08 (±0.49), **P<0.005]. No significant difference in Periotest value (PTV). Conclusion: This study showed that implant flapless surgery is a minimal invasive, effective, and novel technique that can render a significantly better early stage satisfaction outcome as compare to the traditional flap method. Flapless implantation resulted in minimal bone loss, less pain, less complications, and comparable good PTV.

Keywords— Dental implants, flapless surgery, osseo-integration, post menopause, posterior maxilla.

I. INTRODUCTION

Osseo-integration is the process that occurs immediately after placement of a dental implant into the recipient bone, involves a number of events that can be affected by many factors such as patients' health status, location selection, surgical methods, systemic and local environments, and medication used [1, 2, 3]. There are plenty of suggestions that success rates of implant procedures greatly decreased with age and specific medical conditions, such as post menopause osteoporosis [4,5]. Inferior bone quality and quantity such as those discovered in post menopause women may leave a negative outcome on osseointegration [3]. Poor bone quality are normally discovered in post menopause women [5]. Generally, in early stage of osseointegration, radiographical imaging can detect a small amount of marginal bone loss surrounding dental implants, and this is considered to be acceptable [6]. A slight loss of the interface between implant and tissue starts at the crestal region irrespective of submerged or nonsubmerged techniques, and research has revealed that during the first year of function, it is normal to have a mean bone loss between 0.9 and 1.6 mm [2,9].

The review of the publications in the posterior maxilla areas shows that flapless surgery could be a viable and predictable treatment method for implant placement, indicating both efficacy and clinical effectiveness with some reservation[1,2]. Minimal invasive surgery has transformed modern-day surgery including dental implantology, especially flapless surgery [2].

Implants survival is an important means to quantify the survival of dental implants and it was recorded as the presence of the implants at the end of the studied [2].

To measure patient satisfaction, the study employs McGill questionnaire on a visual analogue scale (VAS) [7].

The Periotest device was employed to determine the stability of implants (Periotest Values or PTV) at implant placement stage [6]

Digital x-ray assessment is the most common means for bone level or marginal bone height assessment [2,8]. Regardless of the reason of constant crestal bone loss, right use and logically thorough reading of radiographs are of ultimate significance for appraisal of dental implants [12].

The aim of this study is to examine the clinical and radiographic outcomes of implants inserted using retrospective split mouth study and measure patients' satisfaction using visual analogue scale in flapless and conventional flap techniques in post menopause women age 50 years or over.

© Springer International Publishing Switzerland 2015
V. Van Toi and T.H. Lien Phuong (eds.), *5th International Conference on Biomedical Engineering in Vietnam,*
IFMBE Proceedings 46, DOI: 10.1007/978-3-319-11776-8_33

II. MATERIALS AND METHODS

This study was approved by the Ethic committee of the Queensland University of Technology, Brisbane, Australia.

The study was carried out in a private setting using a retrospective split mouth research of flapless vs flap technique involving the review of dental records of 16 post menopause patients undergone bilateral implant surgery in the posterior maxilla.

A total of 45 implants with no augmentative procedures were selected from the records of 16 patients. The patients were divided into two groups: the control group had 21 implants placed by using a full flap technique and the test group (Fig. 4) consisted of 24 implants inserted using flapless procedure. All implants placed using non-submerged techniques (Fig.1). The inclusion criteria includes: good dental records with adequate treatment and feedback information, dental implants placed in posterior maxilla of post menopause women with one side used flap technique and the other side used flapless method.

The opposing jaw was either with fixed prosthesis or natural dentition. The implant placed on both side of the jaw was aided by the same designed surgical stent. In the flapless side no flap was raised while the contralateral side flap was used. Clinical evaluations were carried out using the following measures:

A. Implant Survival

Implants survival was recorded as the presence of the implants at the end of the studied period (1 year).

B. Visual Analogue Scale (VAS) Assessment

To measure patient satisfaction, the study employs McGill questionnaire on a visual analogue scale (VAS) ranges from 1 to 10 of which 1 as having no pain and 10 is the worst pain (Fig. 2). The patients were asked to record their overall satisfaction on sensation of discomfort on a visual-analogue-scale with 0% being totally unsatisfied and 100% being completely satisfied (Fig.3). The VAS scores were recorded for both sides at one day, one week, one month and three months follow up. The VAS scores obtained were analyzed for statistical significance.

C. Periotest Values (PTV)

The Periotest device was employed to determine the stability of implants at implant placement stage as well as at subsequent recall appointments at one month and three months. The Periotest's scale varies from -8 to +50. The lesser the Periotest value, the greater is the stability / hampering effect of the test object (tooth or implant). At these assessing visits, healing posts were connected to the implants, and the patient was positioned so that the maxilla

is in a horizontal position. The periotest probe was pushed flat upright to the implant post, and it was made to touch base as close to the alveolar crest as possible (Fig 4). The total implants involved in the study were evaluated in lateral directions. Acceptable readings were obtained only when the device registered the comparable values in three consecutive values.

D. X-ray Assessment for Bone Level

A digital periapical x-ray was carried out for each implant using identical holders to assess marginal bone height at the time of surgery, at one month, three months, and one year. The digital x-rays were calibrated to calculate the differences in bone height and bone loss.

The pertinent implant features such as: site, sizes, design, and other relevant characteristics were recorded. The x-rays were appraised by two experienced and unbiased assessors by means of a grid to determine the dimension of the implant and the proportion of bone loss in millimeters.

E. Statistical Analysis

One way analysis of variance was performed for statistical significance.

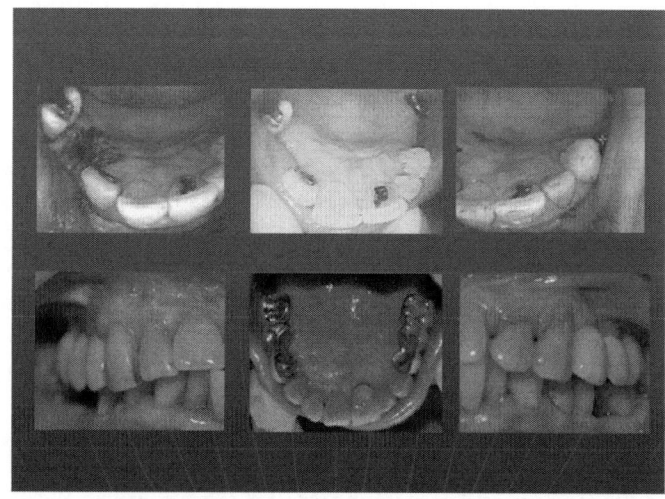

Fig. 1 Dental Implant Surgery Flapless Surgery versus Flap Technique in Posterior Maxilla of Post Menopause Women

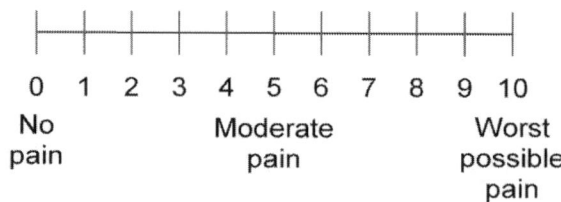

Fig. 2 Pain assessment using Visual Analogue Scale

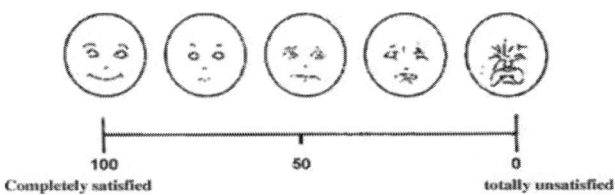

100   50
Completely satisfied   totally unsatisfied

Fig. 3 A measure of overall satisfaction

III. RESULTS

From all the dental implant records at 2 private dental practices, a total of 16 patients 45 implants were selected. The first post menopause split mouth posterior implant patient was found in 2004 and the last of these implants patient was recorded at the end of 2013. Of 45 implants inserted, 21 were flap and 24 flapless. All the placed implants were of conventional/delayed (3-4 months) loading. The restored implants consisted of 35 definitive crowns and 4 bridges.

Survival rate for the two techniques showed a 95.2% (1 lost) for flap and 95.8% (1 lost) for the flapless technique.

Using *visual analogue scale (VAS=0 to 10)*, flapless surgery revealed to have less: pain, swelling, bleeding and speech impairment and had better overall satisfaction at one day and one week than flap technique than the flap counterpart [Fig. 4, (*$P<0.05$)]. The experienced pain was significantly lesser in the flapless-group compared to the full flap group with [1.2 ($\pm$1.65) vs 4.6 ($\pm$1.94) (*$P<0.05$)]. The patients reported an overall pain of 2.9 ($\pm$1.79). Compare the flap vs flapless group, Percentage (%) of Overall Satisfactions were statistically significant at 1 day [(32.5 ($\pm$27.5) vs 93.2 ($\pm$8.70), *$P<0.05$] and 1 week [55.3 ($\pm$20.4) vs 83.2 ($\pm$9.10), *$P<0.05$] but not at 1 month [80.9 ($\pm$12.2) vs 88.1 ($\pm$8.90)] and 3 months [81.2 ($\pm$15.3) vs 87.3 ($\pm$16.2)] (Fig. 4).

No significant difference in bone resorption at 3 months. After one year, bone change in the flap group vs the flapless group was statistically significant [-0.53 ($\pm$0.57) vs +0.08 ($\pm$0.49), **$P<0.005$]. An overall marginal bone loss of 0.23 mm ($\pm$0.61) was recorded in remodeling resulted in the flapless-group to a small growth in marginal bone height of 0.08 mm ($\pm$0.49) (Fig.4). No recessions were observed. No significant difference in Periotest value (PTV).

Table 1 Overall results

		Control group flap	Test group flapless	Overall results
Number of implants placed		21	23	45
Number of implants failed		1	1	2
Loading method		Conventional/ delayed	Conventional/ delayed	
Type of prostheses		15 crowns and 2 bridges	20 crowns and 2 bridges	35 crowns 4 bridges
Survival rate (1 year)		95.2%	95.8%	95.5%
Visual	Pain	4.6 ($\pm$1.94)*	1.2 ($\pm$1.65)*	2.9 ($\pm$1.79)
Analogue Scale **(0 =**	Swelling	9.1 ($\pm$2.16)*	2.1 ($\pm$1.59)*	5.6 ($\pm$1.88)
lowest and 10=	Bleeding	7.8 ($\pm$1.75)*	1.4 ($\pm$1.65)*	4.6 ($\pm$1.70)
highest)	Speech impairment	6.9 ($\pm$1.43)*	1.1 ($\pm$1.01)*	4.0 ($\pm$1.22)
Percentage (%) of	1 day	32.5 ($\pm$27.5)*	93.2 ($\pm$8.70)*	62.9 ($\pm$18.1)
Overall Satisfaction	1 week	55.3 ($\pm$20.4)*	83.2 ($\pm$9.10)*	69.3 ($\pm$18.1)
(Visual Analogue	1 month	80.9 ($\pm$12.2)	88.1 ($\pm$8.90)	84.5 ($\pm$10.6)
Scale 0 = lowest and 100 = highest)	3 months	81.2 ($\pm$15.3)	87.3 ($\pm$16.2)	84.3 ($\pm$15.6)
Bone resorption at 3 months in mm (+= gain and -= loss)		-0.75 ($\pm$0.55)	-0.63 ($\pm$0.67)	-0.69mm
Bone changes (1 year) in mm (+= gain and -= loss)		-0.53 ($\pm$0.57)**	+0.08 ($\pm$0.49)**	-0.23mm
Periotest value [-8 (least mobile) to +20 (most mobile)]	Day 0	-3.62 ($\pm$0.81)	-3.51($\pm$ 0.87)	-3.57 ($\pm$0.84)
	1 month	-3.83 ($\pm$1.21)	-3.42 ($\pm$1.62)	-3.63 ($\pm$1.42)
	3 month	-4.01($\pm$1.37)	-4.14($\pm$1.73)	-4.08 ($\pm$1.55)

Statistical significance: *$P<0.05$ and **$P<0.005$

IV. DISSCUSSION

This study has showed that flapless dental implant surgery is a minimal invasive novel technique that can deliver a fairly good survival rate in contrasted with other reports employing traditional flap methods. It has also underpinned the view that flapless surgery can render a predictable result with superior efficiency and efficacy even in poor quality bone such as those found in post menopause women in this study. Thus, the outcome of this study also help to reinforce the notion that post menopause health status is not a relative contra indication for placement of dental in soft bone area of the posterior maxilla.

Visual analogue scales (VAS) are used widely for discomfort measurement, though it is subjective, but remain useful tool for quantifying arbitrary data, if it is used properly. In this study, it demonstrated the superior satisfaction of flapless technique to the conventional flap counterpart in the early stage of the treatment as satisfaction to dental treatment play big part in minimizing patients' complaints as well as reducing the barriers to dental implant treatment [10].

The evaluation of Periotest Value (PTV) pointed out that it is a most likely substitution for old-fashioned, unreliable dental implant strength diagnosis apparatuses. The Periotest possesses advantage of presenting consistent results by quantifying the degrees of subclinical movement utilising an ultrasonically pulsating probe. The Periotest is effective in measuring the firmness level of an implant. Though Periotest can detect terminal or failed implants, it has inherent disadvantage in identifying bone quantity in normal osseointegration. When comparing the consistency of the Periotest and Osstell Mentor in assessment of peri-implant vertical and circular bone loss it was indicated that diagnostic apparatuses for implant stability are valuable in revealing the circular bone loss [11]. However, the clinical reliability of the two gadgets for finding the fractional vertical bone loss is not so great. Therefore, digital radiography showed to be a more subtle technique of verifying peri-implant bone loss. The periapical radiographic technique used in the study is presently the favored technique for assessing implant condition based on bone loss, and digital radiographs permit straightforward calibration of the image contrast. It is proposed that reference radiographs should be captured at the stage the implant fixture penetrates the mucosal tissues and yearly afterwards, and not only average values, but also the extent of bone heights, should be given to depict the part of implants that exhibit constant crestal bone loss [12]. Though digital x-rays used in the assessment in this study did not offer the possibility of a three-dimensional evaluation. Hence, digital periapical radiographs along with Periotest apparatus were discovered to give the highest dependable appraisal of an implant's condition.

In term of overall satisfaction, patients appeared to be more satisfied in the early stage of the treatment, and not at the later stage when the implants wound were almost healed then satisfaction rate appeared to be of no difference.

V. Conclusion

This study showed that implant flapless surgery is a minimal invasive, effective, and novel technique that can render a marginally better early stage satisfaction outcome as compare to the traditional flap method in post menopause women. Flapless implantation resulted in minimal bone loss, less pain and fewer complications. Periotest is an effective alternative in measuring the firmness level of an implant, and digital periapical radiographs along with Periotest apparatus were discovered to give the highest dependable appraisal of an implant's condition.

Acknowledgment

This research is a self-funded project

Conflict of Interest

The authors declare that they have no conflict of interest.

References

1. Doan, N., et al., Is flapless implant surgery a viable option in posterior maxilla? A review. International journal of oral and maxillofacial surgery, 2012. **41**(9): p. 1064-71.
2. Doan N. (2014). An evaluation of clinical procedures used in dental implant treatment in posterior maxilla using flapless technique. Dissertation for the degree of Doctor of Philosophy, Queensland University of Technology.
3. Du Z, Doan N, Ivanovski S, Xiao Y. Serum bone formation marker correlation with improved osseointegration in osteoporotic rats treated with simvastatin. Clin Oral Implants Res, 2011; Nov 10.
4. Mundy, G., et al., Stimulation of bone formation in vitro and in rodents by statins. Science, 1999. 286(5446): p. 1946-9.
5. Horiuchi, N. and T. Maeda, Statins and bone metabolism. Oral diseases, 2006. 12(2): p. 85-101.
6. Huang HL et al. Effects of Elasticity and Structure of Trabecular Bone on the Primary Stability of Dental Implants. Journal of Medical and Biological Engineering, 30(2): 85-89.
7. Sewitch M. Measuring Differences Between Patients' and Physicians' Health Perceptions: ThePatient–Physician Discordance Scale Journal of Behavioral Medicine, Vol. 26, No. 3, June 2003.
8. Chen LC, et al. Comparison of different methods of assessing alveolar ridge dimensions prior to dental implant placement. J Periodontol 2008;79(3):401–405.
9. Blanes RJ et al. A 10-year prospective study of ITI dental implants placed in the posterior region. I: Clinical and radiographic results. Clin. Oral Impl. Res. 18, 2007 / 699–706.
10. Hof M et al. Patients' perspectives on dental implant and bone graft surgery: questionnaire-based interview survey.Clin. Oral Impl. Res. 2014 Jan;25(1):42-5. doi: 10.1111/clr.12061. Epub 2012 Oct 17.
11. Choi HH et al. Reliability of 2 implant stability measuring methods in assessment of various periimplant bone loss: an in vitro study with the Periotest and Osstell Mentor. Implant Dent. 2014 Feb;23(1):51-6.
12. De Bruyn H et al. Radiographic evaluation of modern oral implants with emphasis on crestal bone level and relevance to peri-implant health. Periodontol 2000. 2013 Jun;62(1):256-70.

Corresponding author: Nghiem Doan, PhD
Institute: Institute of Health and Biomedical Innovation
& Science and Engineering Faculty, Queensland University of Technology
Street: 60 Musk Avenue, Kelvin Grove, Qld 4059
City: Brisbane
Country: Australia
Email: doanco@bigpond.com.au

An Analysis of Exosomes from Keratinocytes and Fibroblasts

Uyen Thi Trang Than, Dominic Guanzon, Lucas Wager, Kerry J. Manton,
Brett Hollier, and David Leavesley

Institute of Health and Biomedical Innovation, Queensland University of Technology, Brisbane, Australia
Wound Management Innovation Cooperative Research Centre, Australia

Abstract— In recent years, many studies have provided evidence that exosomes secreted by cells contain various components, including microRNAs [1]. It is thought that exosomes have important roles in many biological processes. However, the role of exosomes and their components, especially miRNAs, in wound healing is poorly understood. In order to understand whether or not primary human epidermal keratinocytes and dermal fibroblasts, two important cell types contributing to wound healing process, release exosomes and what species of wound healing-associated miRNAs accumulate in these vesicles, this project will use a combination of methods to isolate and characterize exosomes, to profile exosomal cargo's, especially miRNAs in exosomes. The results showed that keratinocytes and fibroblasts released exosomes into conditioned media and these exosomes contain some target miRNAs.

Keywords— keratinocytes, fibroblasts, exosomes.

I. INTRODUCTION

Exosomes are enclosed membrane vesicles released by all cell types under physiological conditions [1-7]. They are thought to mediate cell-to-cell communication via transfer of their contents of proteins, lipids, mRNA and miRNA [1, 8, 9]. The formation and release of exosomes are tightly regulated by multiple signaling mechanisms. It is thought that exosomes have an endocytic origin; they are formed and developed via the endosome system are released by the fusion of multivesicular bodies (MVBs) with the plasma membrane via exocytosis [10]. Yet, the exact mechanisms of exosome biogenesis are not fully understood. During the formation process, cytoplasmic components, including miRNAs, solube factors and membrane complexes, are accumulated into exosomes [11].

All mammalian exosomes share common characteristics such as structure (bi-lipid layer), size, density and protein composition; however, these characteristics may be distinct depending on exosome origin. Specific molecular species are located on the surface or in the lumen of exosomes, and are considered as exosomal markers. Exosomes usually have tetraspanins, including CD9, CD24, CD63, CD81 and CD82 [12-15]. MHC class I and II molecules and heat-shock protein family such as Hsp70 and Hsp90 are also present in exosomes [2, 11, 16-19]. Cytoplasmic proteins such as tubulin, actin, actin-binding proteins, annexins and Rab protein; which are involved in intracellular membrane fusion and transport, as well as molecules responsible for signal transduction have been identified in exosomes, such as protein kinases, 14-3-3 and heterotrimeric G proteins [7, 15, 18, 20, 21]. Furthermore, exosomes possess a characteristic lipid composition, which includes cholesterol, sphingomyelin, ceramide and

Phosphatidylserine providing additional traits for identification [3, 22-24]. Depending on the cellular origin, exosomes recruit various cellular proteins that are involved in specific functions.

The recent investigation of exosomes shows that these small vesicles contain RNA molecules, including miRNAs, facilitating transfer between cells at a distance [4, 25-27]. Interestingly, miRNAs from exososmes have been identified as mediators of cell functions and can be a potential source of new cancer biomarkers [28]. Because of their functions, they represent a potential source of efficient diagnostic and disease-specific therapy [29, 30].

II. MATERIALS AND METHODS

A. Cell Culture

Donated skin discards are (Ethics number 1300000063-QUT HREC) cut into small pieces (2-3 mm^2) and placed in 0.125% trypsin overnight at 4°C (12-24 h). Keratinocytes are scraped from the underside of the epidermis and upper surface of the dermis. Fibroblasts are then isolated from the denuded dermis. Isolated primary human keratinocytes were cultured until the 3rd or 4th passage in Greens media (DMEM, Ham's F12 (3:1), 10% fetal calf serum, 1% L-glutamine, 1% penicillin-streptomycin, 0.2% adenine, 0.1% insulin, 0.1% cholera toxin, 0.1% non-essential amino acids solutions, 0.1% transferrin, 0.1% triiodothyronine, 0.08% hydrocortisone, human recombinant EGF at 37°C and 5% CO_2 [31, 32]. Fibroblasts were cultured with media containing 5% fetal calf serum, 1% L-glutamine, 1% penicillin-streptomycin at 37°C and 5% CO_2. Keratinocytes and fibroblasts were permitted to conditioned culture media (CM) (without FCS) when they reach approximately 80%

© Springer International Publishing Switzerland 2015
V. Van Toi and T.H. Lien Phuong (eds.), *5th International Conference on Biomedical Engineering in Vietnam,*
IFMBE Proceedings 46, DOI: 10.1007/978-3-319-11776-8_34

confluency [33]. After 48 hours, the CM is harvested and centrifuged at 2000g for 30 minutes to remove cells and cell debris before being subjected to extracellular vesicle isolation methods below [34] (Described below).

B. Classical Gradient Density Centrifuge

Sucrose gradients of 22.8% to 60% were prepared in Thinwall, Ultra-Clear™ Tubes (Beckman Coulter, Inc.; Fullerton, CA, USA) using the BioRad 475 Gradient Delivery system (BioRad; Hercules, CA, USA). A suitable volume (approximately 1 ml) of concentrated conditioned media was layered on the top of the sucrose gradient and centrifuged for 27 hours at 127,000 g, 4°C in an SW40Ti Rotor (Beckman Coulter, Inc., Fullerton, CA, UAS) and Optima L-90K Ultracentrifuge (Beckman Coulter Inc., Fullerton, CA, USA) [10]. A small hole was made in the bottom of the centrifuge tube, and 1 ml aliquots collected. The sucrose concentration in each fraction was determined using a PAL-10S pocket refractomater (ATAGO®; Tokyo, Japan).

C. Isolation Kit

The required volume (approximately 1 ml) of concentrated conditioned media was transferred to a new tube and 0.5 volumes (0.5 ml) of the Total Exosome Isolation reagent was added (Invitrogen, CA, USA). The solution of reagent and CM was mixed well by vortexing or pipetting up and down to create a homogenous solution. This solution was incubated at 4°C overnight. After incubation, the sample was centrifuged at 10,000 g for 1 hour at 4°C. Exosomes were contained in the pellet at the bottom of the tube. The supernatant was aspirated and discarded. Exosomes were harvested by resuspending the pellet in a convenient volume of 1X PBS (25 – 100 μl for 1 ml starting volume of CM).

D. Western Blot

Anti-CD9 antibody (ab92726), anti-CD63 antibody (ab8219), anti-Hsp70 antibody (ab47455), and anti-TSG 101 antibody (ab92726) were purchased from Abcam® (Cambridge, UK). Chicken anti-rabbit IgG secondary antibody (HAF 008, R&D System, USA) was used as secondary antibody. 11 μl sample, 4.25 μl sample buffer and 1.7 μl DTT were mixed and heated at 70 °C for 15 minutes. DTT was not used for CD9, CD63 and CD81 [35]. Extracellular vesicle preparations were separated by SDS-PAGE at 200V for 35 minutes. Separated proteins were then transferred to a nitrocellulose membrane at 200 mA for 2 hours in chilled transfer buffers (Tris Base, Glycine and Methanol). The membrane was blocked with 5% skim milk powder in Tris Buffered Saline (TBS) pH 7.4 before being probed with primary antibodies overnight at 4°C in 0.5 % skim milk powder in TBS for specific target proteins. Membranes were washed 5 times x 5 minutes before incubating with secondary antibody for 30 minutes at room temperature. Membranes were then washed 5 times x 5 minutes and incubated with detection solution (Pierce™ ECL Western Blotting Substrate, Thermo Scientific) for 5 minutes, and imaged on Curix Ultra UV-G Medical X-ray film (AFGA; Mortsel, Belgum) [35].

E. Transmission Electron Microscope (TEM)

TEM was performed to validate exosome morphology. Exosomes were mixed with 4% paraformaldehyde, deposited on Formvar-carbon coated EM grids and let stand for 20 minutes. The grid was washed with PBS, and then transferred to 1% glutaraldehyde for 5 minutes. The grids were washed with distilled water 7 times. The sample was stained with uranyl-oxalate for 5 minutes, transferred to methyl cellulose-uranyl acetate for 10 minutes on ice and dried at room temperature [35, 36]. Imaging was performed using a TEM JEOL 1400.

F. qPCR

Total RNA was extracted from exosome preparations using Trizol reagent. RNAs were reversed transcribed into cDNA and then subjected to qPCR. Oligonucleotide primers for miRNA were purchased from QIAGEN (QIAGEN, Melbourne, Australia) (Table 1). SYBR Green® Master Mix (Applied Biosystems™, Melbourne, Australia) and DNA, RNA and nuclease-free water were added to 1 μl of cDNA template in each 10 μl reaction volume in a MicroAmp® Fast optical 96 well plate (Invitrogen™, Melbourne, Australia). The cycling conditions were conserved for each gene (95°C denaturation, 55°C annealing, 70°C elongation) for 40 cycles.

III. RESULTS AND DISCUSSION

A. Exosome Characterisation

Protein Investigation

The 13 fractions from sucrose gradient density centrifuge and exosome preparation from isolation kit were subjected to SDS-PAGE and stained with Silver to visualize the proteomic component in each preparation (Figure 1). In preparations from keratinocyte cultures, the majority of proteins were observed in fractions, 2 to 7. Fractions 2 and 3 yielded the most proteins both in quantity and the broad range of protein sizes. Proteomic elements were not detected in fraction 1. It is notable that the protein contents of each fraction is unique. In terms of fibroblasts, proteomic components are observed in all fractions. While the protein distribution is similar to CM from keratinocyte culture, additional proteins are evident in fractions 2 and 3. A range of protein signals is also evident in fractions 4, 5, and 6, and in fraction 1, a strong proteomic signal can be observed, equivalent to 55 kDa and 250 kDa.

Preparations from the isolation kit yield a distinct protein profile, specific to the cell types. In the keratinocyte sample, a variety of species distributed from approximately 25 kDa and higher are evident. In the fibroblast sample, a distinct and different population of protein species is observed. These data provide evidence that exosomes produced by skin-derived keratinocytes and fibroblasts are qualitatively distinct.

Exosome Marker Identification

Western Blot was used to characterise proteins identified as extracellular vesicle biomarkers. Common surface proteins such as Hsp70, TSG 101, CD 63, CD 9, and CD 81 are considered as biomarkers of exosomes [11, 23, 37]. Western blot analysis of conditioned media (CM) from keratinocyte culture reveals that while Hsp 70 was detected in the fractions 2 and 3 (figure 2), TSG 101 was detected in 8 fractions (figure 2) and CD9, CD63 were not detected. Hsp 70 is considered as an exosomal biomarker because it has been demonstrated to be present in most discovered exosomes [9]. This suggests that exosomes may be present in fraction 2 and 3 but not elsewhere in the sucrose gradient. Fractions 2 and 3 correspond to 1.084 and 1.104 g/ml sucrose. TSG 101 is needed for the formation of multivesicular bodies (MVBs) and known to be part of their cargo [38]. It was detected in 8 fractions, from fraction 2 to 9 (figure 2). The presence of protein enriched in fraction 7, 8 and 9 was unexpected. Thus we collected fraction 7, 8, and 9 for further analysis. Notably proteins such as CD9, CD 63 and CD 81 are reported to localize to exosomes, but were not detected in our analysis of exosomes from keratinocytes and fibroblasts.

In contrast to keratinocyte CM, exosomal markers were not detected in fibroblast CM. It is not clear if this indicates that fibroblasts do not release exosomes, or that fibroblast-derived exosomes do not include CD9, CD 63, CD81, TSG 101, and Hsp70. This finding is unexpected.

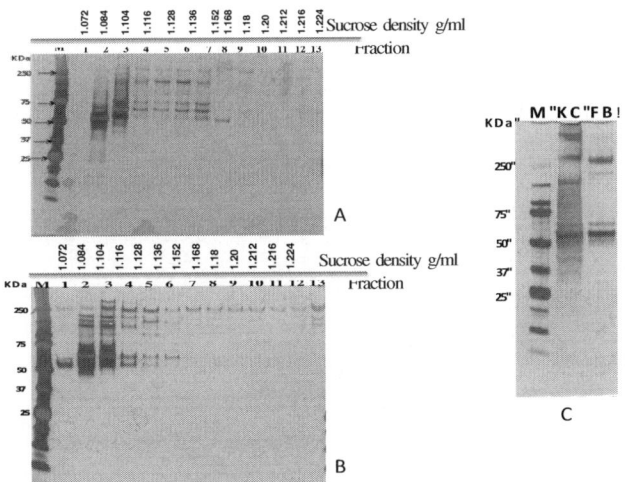

Fig. 1 Proteomic components from different isolation methods. A) and B): Protein components from keratinocyte (A) and fibroblast (B) culture isolated by classical culture isolated by classical gradient density centrifugation. C: Protein components from keratinocyte culture (KC) and fibroblast culture (FB) isolated by Isolation Kit (Invitrogen). M: Marker.

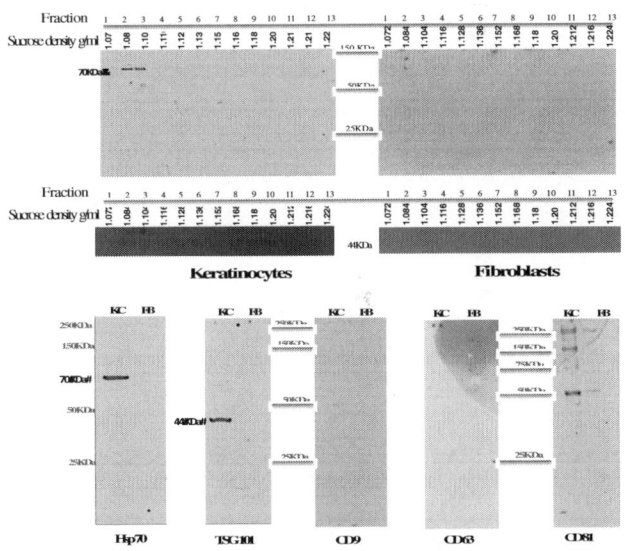

Fig. 2 Exosome marker detection. Only Hsp 70 and TSG101 were detected in keratinocyte preparation, not in fibroblast. Other protein markers, including CD9, CD63 and CD81 were not detected in any preparations. KC: keratinocytes; FB: Fibroblasts.

Transmission Electron Microscope (TEM)

TEM analysis revealed the morphology of exosomes that have cup shape and the size is around 40-100nm. This is consistent with other authors when characterising exosomes by transmission electron microscope (Figure 3).

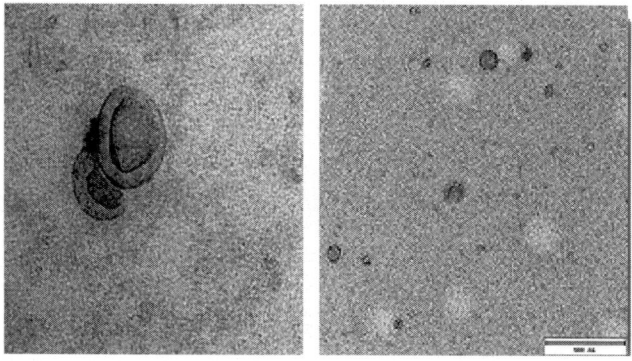

Fig. 3 Exosomes from TEM analysis have a cup shape.

B. miRNA Identification

Candidate miRNAs (hsa-miR 29a, hsa-miR 21, hsa-miR 141, hsa-miR 203) identified in preliminary studies (Dominic Guanzon, personal communication) were validated using quantitative real-time PCR (qRT-PCR). In addition, qRT-PCR was used to determine changes in the expression level of mature miRNAs isolated from culture media [39].

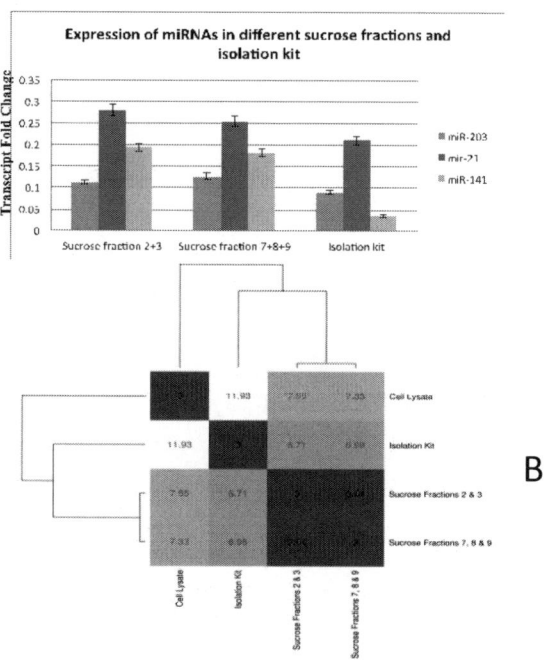

B

Fig. 4 A) Expression of targeted miRNAs from keratinocyte-derived exoxome preparation. B) The heat map based on ΔCt value shows the relationship between given samples and controls.

Analysis of exosome preparations for miRNAs revealed that selected miRNAs are expressed in preparations from both sucrose fractionation and the isolation kit. miR-21 was found to be the most abundant species compared to miR-203 and miR-141. While miR-203 and miR-21 ratio of expression changed little between the sucrose fractions and the isolation kit investigated, miR-141 was observed to decline considerably in the isolation kit (Figure 4A).

The expression level of miR21, miR203 and miR141 across 3 samples and control preparations is presented in a heat map, based on ΔCt value (Figure 4B). The colour contributed to point in the heat map grid indicates the relationship between given samples and control. Clearly, sucrose fraction 2+3 and sucrose fraction 7+8+9 exhibit few differences. However, it is noteworthy to observe a substantial difference is evident between the total RNA (from whole cell lysate) and isolation kit samples, as well as the sucrose-gradient samples.

IV. CONCLUSION

We have shown that primary epidermal keratinocytes and dermal fibroblasts secreted exosomes into the extracellular conditions continue to be optimised to demonstrate functional roles for exosome-mediated exocrine exchange of miRNAs in skin tissue. Exosomes isolated from keratinocyte culture express the exosome biomarkers Hsp70 and TSG101, but fibroblasts-derived vesicles do not. RT-PCR evidence indicates the presence of miRNAs associated with exosomes isolated fron keratinocytes. The miRNA populations are distinct between sample preparations and vesicle isolation methods. Experimental conditions continue to be optimised.

ACKNOWLEDGMENT

To do this project, we thank to Dr. Jacqui McGovern for skin collection. I acknowledge support from VIED-QUT scholarship and WMI-CRC top-up scholarship.

REFERENCES

1. Valadi, H., et al., Exosome-mediated transfer of mRNAs and microRNAs is a novel mechanism of genetic exchange between cells. Nature Cell Biology, 2007. 9(6): p. 654-659.
2. Hao, S., et al., Dendritic cell-derived exosomes stimulate stronger CD8+ CTL responses and antitumor immunity than tumor cell-derived exosomes. Cellular & molecular immunology, 2006. 3(3): p. 205.
3. Vlassov, A.V., et al., Exosomes: Current knowledge of their composition, biological functions, and diagnostic and therapeutic potentials. Biochimica Et Biophysica Acta-General Subjects, 2012. 1820(7): p. 940-948.

4. Lasser, C., et al., Human saliva, plasma and breast milk exosomes contain RNA: uptake by macrophages. Journal Of Translational Medicine, 2011. 9(1): p. 9-9.

5. Baietti, M.F., et al., Syndecan-syntenin-ALIX regulates the biogenesis of exosomes. Nature Cell Biology, 2012. 14(7): p. 677-685.

6. Meister, G., et al., Identification of Novel Argonaute-Associated Proteins. Current Biology, 2005. 15(23): p. 2149-2155.

7. Wang, Z., et al., Proteomic analysis of urine exosomes by multidimensional protein identification technology (MudPIT). Proteomics, 2012. 12(2): p. 329-338.

8. Pegtel, D.M., et al., Functional delivery of viral miRNAs via exosomes. Proceedings of the National Academy of Sciences of the United States of America, 2010. 107(14): p. 6328-6333.

9. Mathivanan, S., H. Ji, and R.J. Simpson, Exosomes: Extracellular organelles important in intercellular communication. Journal of Proteomics, 2010. 73(10): p. 1907-1920.

10. Raposo, G., et al., B lymphocytes secrete antigen-presenting vesicles. The Journal of experimental medicine, 1996. 183(3): p. 1161-1172.

11. Thery, C., Zitvogel, Laurence and S. Amigorena, Exosomes: composition, biogenesis and function. Nature Reviews Immunology, 2002. 2(8): p. 569-579.

12. Chaput, N., et al., The potential of exosomes in immunotherapy, 2005, Informa Healthcare: England. p. 737-737.

13. Escola, J.M., et al., Selective enrichment of tetraspan proteins on the internal vesicles of multivesicular endosomes and on exosomes secreted by human B-lymphocytes. The Journal of biological chemistry, 1998. 273(32): p. 20121-20127.

14. Keller, S., et al., CD24 is a marker of exosomes secreted into urine and amniotic fluid. Kidney International, 2007. 72(9): p. 1095-1102.

15. Thery, C., et al., Proteomic Analysis of Dendritic Cell-Derived Exosomes: A Secreted Subcellular Compartment Distinct from Apoptotic Vesicles. The Journal of Immunology, 2001. 166(12): p. 7309-7318.

16. Blanchard, N., et al., TCR activation of human T cells induces the production of exosomes bearing the TCR/CD3/zeta complex. Journal of immunology (Baltimore, Md. : 1950), 2002. 168(7): p. 3235-3241.

17. Wolfers, J., et al., Tumor-derived exosomes are a source of shared tumor rejection antigens for CTL cross-priming. Nature Medicine, 2001. 7(3): p. 297-303.

18. Bard, M.P., et al., Proteomic analysis of exosomes isolated from human malignant pleural effusions. American journal of respiratory cell and molecular biology, 2004. 31(1): p. 114-121.

19. Gastpar, R., et al., Heat shock protein 70 surface-positive tumor exosomes stimulate migratory and cytolytic activity of natural killer cells. Cancer research, 2005. 65(12): p. 5238-5247.

20. Skokos, D., et al., Mast Cell-Dependent B and T Lymphocyte Activation Is Mediated by the Secretion of Immunologically Active Exosomes. The Journal of Immunology, 2001. 166(2): p. 868-876.

21. Skokos, D., et al., Nonspecific B and T cell-stimulatory activity mediated by mast cells is associated with exosomes. International archives of allergy and immunology, 2001. 124(1-3): p. 133-136.

22. Keller, S., et al., Exosomes: From biogenesis and secretion to biological function. Immunology Letters, 2006. 107(2): p. 102-108.

23. Mariantonia Logozzi, A.D.M., Luana Lugini, Martina Borghi, Luana Calabrò, Massimo Spada, Maurizio Perdicchio, Maria Lucia Marino, Cristina Federici, Elisabetta Iessi, Daria Brambilla, Giulietta Venturi, Francesco Lozupone, Stefano Fais, High Levels of Exosomes Expressing CD63 and Caveolin-1 in Plasma of Melanoma Patients. PLoS ONE, 2009. 4(4): p. 10.

24. Février, B. and G. Raposo, Exosomes: endosomal-derived vesicles shipping extracellular messages. Current Opinion in Cell Biology, 2004. 16(4): p. 415-421.

25. Mittelbrunn, M., et al., Unidirectional transfer of microRNA-loaded exosomes from T cells to antigen-presenting cells. Nature communications, 2011. 2: p. 282.

26. Kosaka, N., et al., microRNA as a new immune-regulatory agent in breast milk. Silence, 2010. 1(1): p. 7-7.

27. Taylor, D.D. and C. Gercel-Taylor, MicroRNA signatures of tumor-derived exosomes as diagnostic biomarkers of ovarian cancer. Gynecologic oncology, 2008. 110(1): p. 13-21.

28. Nilsson, J., et al., Prostate cancer-derived urine exosomes: a novel approach to biomarkers for prostate cancer. British Journal of Cancer, 2009. 100(10): p. 1603-1607.

29. Gusachenko, O.N., M.A. Zenkova, and V.V. Vlassov, Nucleic acids in exosomes: disease markers and intercellular communication molecules. Biochemistry. Biokhimiiâ, 2013. 78(1): p. 1.

30. Rana, S., K. Malinowska, and M. Zöller, Exosomal tumor microRNA modulates premetastatic organ cells. Neoplasia (New York, N.Y.) 2013. 15(3): p. 281.

31. Rheinwald, J.G. and H. Green, Serial cultivation of strains of human epidermal keratinocytes: the formation of keratinizing colonies from single cells. Cell, 1975. 6(3): p. 331.

32. Staiano-Coico, L., et al., Human keratinocyte culture. Identification and staging of epidermal cell subpopulations. The Journal of clinical investigation, 1986. 77(2): p. 396-404.

33. Lässer, C., M. Eldh, and J. Lötvall, Isolation and characterization of RNA-containing exosomes. Journal of visualized experiments 2012(59): p. e3037.

34. Chiba, M., M. Kimura, and S. Asari, Exosomes secreted from human colorectal cancer cell lines contain mRNAs, microRNAs and natural antisense RNAs, that can transfer into the human hepatoma HepG2 and lung cancer A549 cell lines. Oncology Reports, 2012. 28(5): p. 1551-1558.

35. Théry, C., et al., Isolation and Characterization of Exosomes from Cell Culture Supernatants and Biological Fluids. Current Protocols in Cell Biology, 2006.

36. Tauro, B.J., et al., Comparison of ultracentrifugation, density gradient separation, and immunoaffinity capture methods for isolating human colon cancer cell line LIM1863-derived exosomes. Methods (San Diego, Calif.), 2012. 56(2): p. 293-304.

37. Lee, T.H., et al., Microvesicles as mediators of intercellular communication in cancer--the emerging science of cellular 'debris'. Seminars in immunopathology, 2011. 33(5): p. 455-467.

38. Razi, M. and C.E. Futter, Distinct roles for Tsg101 and Hrs in multivesicular body formation and inward vesiculation. Molecular biology of the cell, 2006. 17(8): p. 3469-3483.

39. Pritchard, C.C., H.H. Cheng, and M. Tewari, MicroRNA profiling: approaches and considerations. Nature Reviews Genetics, 2012. 13(5): p. 358-369.

Oral Health Problems among Adult Patients at Commune Health Centres in Central Vietnam: Prevalence and Care Seeking Behaviour

Hoang Anh Dao[1], Peter Hill[2], Ngo Hien[3], and Nguyen Toai[1]

[1] Faculty of Odontostomatology, Hue University of Medicine and Pharmacy, Vietnam
[2] Population Health School, The Univesity of Queensland, Australia
[3] School of Dentistry, The University of Queensland, Australia

Abstract— Introduction: **The burden of oral disease is an excessive issue of Vietnamese population while the accessibility of oral health services are restricted. This study were to identify the prevalence of dental caries and the oral health seeking behavior in response to oral health symptoms among adults presenting at commune health centers.** *Materials and Methods:* **A cross-sectional study was used at three commune health centres in Thua Thien Hue province, Vietnam. During the four-week study period, all patients older than 18 year old who presented to the commune health centres for either general or oral health concerns were asked to complete a social survey, and received dental examinations.** *Results and Discussion:* **There was a high proportion (92.2%) of adults had caries experience with a mean DMFT of 6.40±5.57 per person (2.98±3.04 decay teeth, 3.09±4.33 missing teeth, 0.33±0.88 filled teeth). 82.5% of adults had at least one oral health symptom in the previous year with several social and psychological impacts on quality of life. 42.5% used health facilities, 27.3% used self-medication, while 15.5% sought no treatment at all. Education level, usual source of dental care, and perceived importance of oral health were significantly associated with the use of oral health care services.** *Conclusions:* **Despite of the high use of health facilities, the high prevalence and severity of oral problems has occurred. This draws attention to large disease burden and very negligible oral health care received at such primary care settings.**

Keywords— **adult, oral health status, care seeking behaviour, Vietnam.**

I. INTRODUCTION

Poor oral health has a profound effect on general health and quality of life and represents a substantial burden for health systems worldwide [1]. The burden of oral disease is an excessive and unnecessary issue of Vietnamese population as dental caries affect more than 80% adults with a mean decayed, missing, and filled teeth score of 4.98 ±5.7, with most as untreated decay [2]. Primary health care settings in form of commune health centers (CHCs) in Vietnam, have taken the role of gate keeper, especially for special population groups with limited access to health care.Thus, they provide an ideal source of oral health care for communities in terms of prevention of oral diseases.

Many studies have been undertaken in industrialised countries to assess the nature and extent of oral health care utilization [3, 4]. Published studies in Vietnam appear to suggest a large proportion of non-utilization and low utilization of dental care. The National Oral Health survey in 1999 indicated that 55.5% of residents have never seen a dentist, of that over 60% living in rural areas [4]. Visits to dental-care facilities are mostly undertaken for symptomatic reasons rather than for preventive reasons [2, 4]. Thus, little has been published about the oral health care seeking behaviour and dental services utilization of Vietnamese population as well as impact of oral health care seeking behaviour on oral health status. The objectives of the present study were to identify the prevalence and severity of adult oral health problems presenting at CHCs and to examine the use of oral health services and self-medication in response to these problems.

II. METHODS

This was a cross-sectional study using quantitative approach. Three CHCs (one in urban and two in rural) in Thua Thien Hue province were randomly selected with the probability proportional population size to represent 150 communities of the whole province. In each CHC, participants were all eligible patients visiting for health care during the four-week period from 1st October 2011. In each day of data collection at CHC, all adult patients 18 years old or older were approached. The criteria and recording instructions used for survey form followed those stated in manual "Oral Health Surveys –Basic Methods", 4th edition, WHO 1997. A social survey using questionnaire to examine oral health seeking behavior among adult patients visited CHCs was conducted on all eligible patients.

Data were processed and analysed using the software SPSS 13.0. Calculated p values were two-tailed and statistical significant was defined as alpha of less than 0.05 (p<0.05). Logistic regression was performed in order to estimate the relative risk (Odds Ratio) of the independent variables explaining the oral health care seeking behaviour within one year.

Ethics approval was obtained from the School of Population Health Research Ethics Committee, The University of Queensland and Hue University of Medicine and Pharmacy.

© Springer International Publishing Switzerland 2015
V. Van Toi and T.H. Lien Phuong (eds.), *5th International Conference on Biomedical Engineering in Vietnam,*
IFMBE Proceedings 46, DOI: 10.1007/978-3-319-11776-8_35

III. RESULTS

B. Sampling Characteristics

The study population consists of 799 patients who met the study eligibility criteria and received the dental examination (Table 1)

Table 1 Socio-demographic characteristics of study population

Variables	n	%	Variables	n	%
Age group, years (n=799)			**Education level (n=799)**		
18-34	143	17.9	Illiterate	75	9.4
35-44	151	18.9	Primary School	231	28.9
45-60	288	36.0	Secondary School	238	29.8
>60	217	27.2	High School	185	23.1
			College University	70	8.8
Gender (n=799)			**Residence (n=799)**		
Male	187	23.4	Urban	157	19.6
Female	612	76.6	Rural	642	80.4
Occupational status (n=799)			**Monthly income * (n= 747)**		
Employee	95	11.9	<400	71	9.5
Housewife	80	10.0	400 - 799	142	19.0
Independent workers	427	53.5	800 - 1,199	152	20.3
Unemployed	16	2.0	1,200 - 1,599	132	17.7
Elderly person	154	19.3	>1,600	250	33.5
Others	27	3.4			

* in thousand VND.

C. Adult Oral Health

Evaluated Dentition Status by Clinical Examination

Of the 799 subjects selected, there was no edentulous person. In general, dental caries affected more than 90% adults with the mean DMFT of 6.40 per person (Table 2). The caries experience and prevalence as well as the severity in adults significantly statistically increased with increasing age (all $p < 0.001$).

Experience of Oral Health Problems

A striking high proportion of people had experienced at least one oral health symptom in the previous year (82.5%). Reporting of symptoms was generally similar for adults in terms of age, gender, income and education level. Among those declaring oral problems, more than half of adults reported the experience of dental or gingival problems. In addition, 57.1% of respondents stated that they more likely to avoid conversation because of oral health problems and 52.0% of adults reported sleep disturbance because of pain.

D. Oral Health Seeking Behaviour

Various predisposing factors significantly affected the probability of using oral health care facilities (Table 3). Higher education level, employee and perception that oral problems can cause other health problems or perceived positive importance of oral health significantly increased the probability of using health facilities. Regarding enabling and need factors, the prevalence of using health facilities was significantly higher for urban citizens, adults with insurance or having usual source of dental care and those who actively participate in social networks.

Table 2 Distribution on DMFT of adults by age group

Age group (years)	n	DMFT Mean±SD	DT Mean±SD	MT Mean±SD	FT Mean±SD	Caries prevalence N (%)
18-34	143	3.70±3.64[a]	2.77±3.08	0.47±0.90[a]	0.46 ±1.01[a]	116(81.1)[b]
35-44	151	4.62±3.77	2.62±2.74	1.48±1.83	0.52±1.00	136(90.1)
45-60	288	6.50±5.05	2.98±3.04	3.14±3.54	0.39 ±1.01	272(94.4)
>60	217	9.27±6.84	3.36±3.20	5.88±5.97	0.03 ±0.19	213(98.2) T
Total	**799**	**6.40±5.57**	**2.98±3.04**	**3.09± 4.33**	**0.33±0.88**	**737(92.2)**

Notes: DT = decayed teeth; MT = missing teeth; FT = filled teeth; DMFT = caries decayed, missing, and filled teeth index.
[a] Column comparison, analysis of variance $p < .001$. [b] $\chi 2$ $p <.05$.

Table 3 Logistic regression results on the predictors associated with using health facilities among adults having experienced an oral problem

	Used health facilities N (%)	Crude OR	Adjust OR	95% CI	Sig.
Education level					
Illiterate	14(24.1)	0.14	0.10	0.03-0.35	0.000
Primary School	76(45.8)	0.36	0.28	0.09-0.84	0.023
Secondary School	93(54.1)	0.50	0.36	0.12-1.05	0.061
High School	71(55.0)	0.52	0.34	0.13-0.95	0.040
College University	26(70.3)[b]	1	1		
Occupational status					
Employee	41(68.3)[a]	3.45	3.11	0.77-12.67	0.112
Housewife	36(60.0)	2.40	3.51	0.76-16.30	0.109
Independent workers	144(47.4)	1.44	1.86	0.43-7.97	0.404
Unemployed	5(41.7)	1.14	1.72	0.28-10.72	0.560
Elderly person	49(43.4)	1.23	2.62	0.56-12.27	0.222
Others	5(38.5)	1	1		
I place great value on my dental health					
Agree	220(47.9)[a]	0.67	0.62	0.39-0.98	0.040
Disagree	60(58.3)	1	1		
Residence					
Urban	56(59.6)[a]	0.62	0.64	0.37-1.10	0.106
Rural	224 (47.9)	1	1		
Insurance					
Yes	239(53.5)[b]	0.72	1.18	0.24-5.77	0.834
No	40(35.7)	1	1		
Usual source of oral health care					
Yes	117(75.0)[b]	4.50	3.88	2.48-6.07	0.000
No	163(40.1)	1	1		
Social networks					
No participation	115(47.1)	0.65	0.79	0.50-1.26	0.322
Weak participation	67(45.0)	0.59	0.87	0.52-1.46	0.596
Active participation	98(58.0)[a]	1	1		

Column comparison, Chi-square test, [a] $p<0.05$, [b] $p<0.001$.

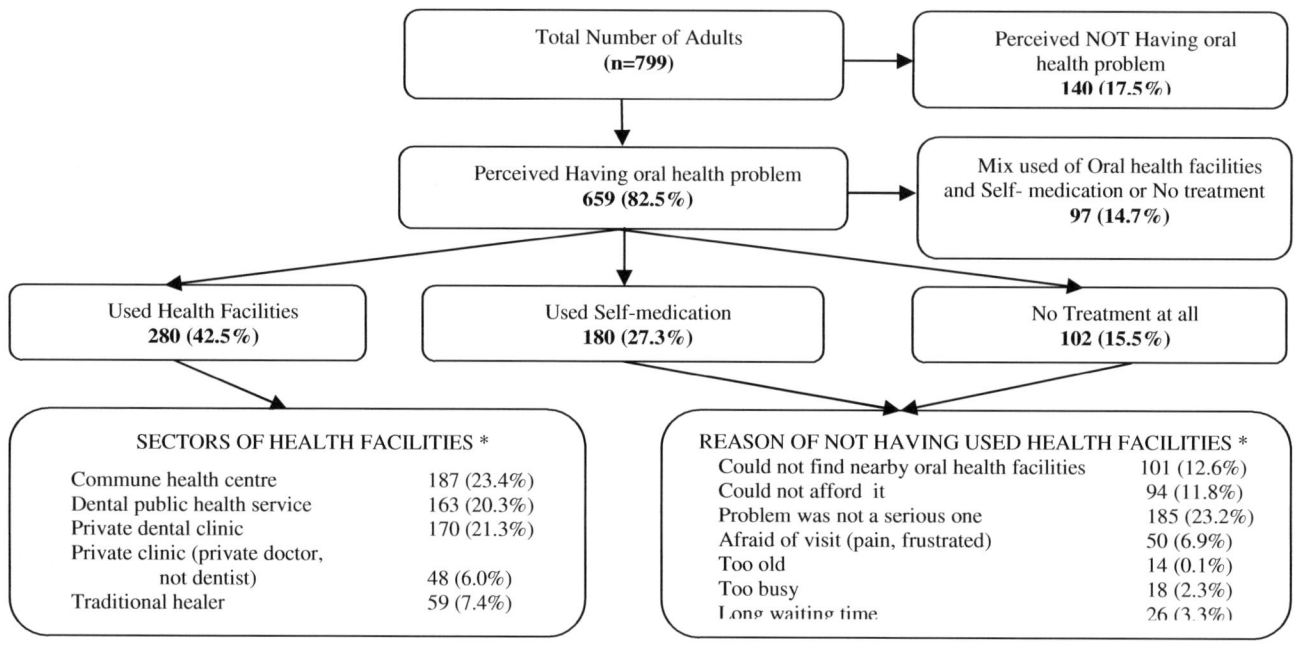

Fig. 1 Responses to an oral health problems

Among individuals who reported having experienced an oral health problem, 42.5% used health facilities, 27.3% reported self-medication, 14.7% mixed use of health facilities and self-medication and approximately 15.5% of adults chose no treatment at all (Figure 1). The nature of the problem: "problem was not a serious one" or "problem went away (by itself)" was cited by more than one-fifth of adults choosing no treatment in response to oral health problems

IV. DISCUSSION

Our results suggest an alarming situation that reflects one of the neglected high disease rates and very little oral health care service provision. 92.2% of adults had caries experience with a mean DMFT of 6.40±5.57 per person. These figures were higher than those of the National Oral Health Survey in which about 81.3% of subjects had caries experience with a mean DMFT of 4.98±5.7 [2]. In comparison with local oral health surveys, caries prevalence and severity in this study was a little higher compared to the local survey in 2000 [5], and on adults in general and adults visiting CHCs [6]. The correspondent data were 66.5% population affected with a mean DMFT of 2.24 and 85.0% patients suffered with a mean DMFT of 5.93, respectively.

The DMFT score has shown that the filled teeth were the smallest across all age groups in all centers. This pattern has been seen in other Vietnamese and developing countries' epidemiological surveys conducted in recent years [7-9]. At

health care system's level this situation might speaks for shortcomings of existing oral health promotion-programmes in controlling dental caries, thus demanding their revision.

Our data among adult patients visiting CHCs revealed profound social and psychological impacts of oral health problems on quality of life, which are not obvious from clinical measures of oral disease. Oral health symptoms reported by residence were consistent with results of a national survey in which dental and gingival problems were dominant [2]. In response to oral problems a high percentage of adults sought treatment in health facilities. Of particular interest is the similar frequency of facility choice for professional oral health care facilities (public or private dental facilities) and CHCs. This suggests evidence of using non-dentist health care providers for treatment of oral problems that was consistent with other reports [10-13].

The prevalence of self-medication for oral health problems (27.3%) in this study was much lower than the figure of 67.8% reported in a community-based research survey in Cameroon [14] and 42.0% reported in hospital-based research in Nigeria [15]. The different prevalence implies that hospital-based research might not be as valid as community-based research in quantifying the actual magnitude of oral health behaviour.

Delay or no treatment was other common practice in the present study. A high prevalence of patient's delay has also been reported in studies from other developing countries [16, 17]. Research addressing these issues is rare in Vietnam, despite its value of not only informing individuals to adopt a particular dental health action, but also potentially helping overcome obstacles in the way of accessing care [18].

Various factors significantly affected the probability of using oral health care facilities. One of the major barriers to seeking professional care was education level, as adults with lower level were less likely to use health facilities in response to oral problems. A similar pattern was reported in previous studies [19-21]. Our study also confirms the higher impact of perceived importance of oral health in response to illness episode than other health belief factors (perceived seriousness of oral disease and perceived benefits of visits) [19, 22, 23]. Having a usual source of oral health care is an indicator of an individual's linkage with the oral health care system. Such a finding corroborates various oral health care utilization studies in both developed and developing countries [21, 24, 25]. Health benefit is another aspect of the health care system that needs to be considered. In the present study, the absence of insurance was not a barrier to obtaining oral health care. This could be explained by the fact that data were collected in CHCs, and the majority of participants were health care cardholders. Thus, the comparison of insurance' effect between people with and without insurance was limited. Yet this may suggest a greater impact of having usual sources of dental care on an individual's oral health services utilization than insurance coverage.

V. Conclusion

Our study strongly confirms the prevalent oral health diseases and the unmet dental needs presented substantially in adult patients presenting at primary care practice settings. There is a strong call for a program for prevention and control of caries in adults presenting at primary care level. Furthermore, a wide range of health facilities used by the community associating with the high prevalence of self-medication reported in this study necessitate awareness creation and introduction of preventive and mitigating interventional programmes. The primary oral care component of the health care system should be strengthened to improve access to needed health care services.

Acknowledgment

The authors are grateful to the population in Thua Thien Hue province for their acceptance and to the research team for their valueable advice and great support.

Conflict of Interest

The authors declare that they have no conflict of interest.

References

1. Petersen P.E. (2005) Priorities for research for oral health in the 21st Century - the approach of the WHO Global Oral Health Programme. Community Dent Health 22:71-75.
2. Spencer A.J. et al. (2011) Oral Health Status of Vietnamese Adults: Findings From the National Oral Health Survey of Vietnam. Asia-Pacific J of Pub Health 23(2):228-236.
3. Ngo D.K. et al (1995) Vietnam oral health status.The Institute of Odonto-Stomatology: Ho Chi Minh, Vietnam.
4. National Center for Health Statistics (2002) National Oral Health Survey of Vietnam 2001. Medical Publishing House: Hanoi, Vietnam
5. Nguyen N.T.D, Nguyen T., and Tran T.P. (2010) Tinh hinh benh sau rang cua nhan dan tinh Thua Thien Hue. Y hoc Thuc Hanh,. 706+707(15):35-37
6. Bui N.L, Nguyen T.Q.H (2008) Tinh hinh benh sau rang cua nhan dan, thi tran Thuan An, tinh Thua Thien Hue.Tap chi Y DHYD Hue 4:50.
7. Marthaler T.M. (2004) Changes in Dental Caries 1953–2003. Caries Res 38(3):173-181.
8. Namal N., Vehid S., and Sheiham A. (2005) Ranking countries by dental status using the DMFT and FS-T indices. Int Dent J 55(6):373-376.
9. Khanh N. and Dong D. (2008) Report of dental care in Southern Vietnam. South Vietnam Dent Congress XI, Vietnam, 2008, pp 75-85 .
10. Anderson R., Richmond S., and Thomas D.W. (1999) Epidemiology: Patient presentation at medical practices with dental problems: an analysis of the 1996 General Practice Morbidity Database for Wales. Br Dent J 186(6):297-300.
11. Anderson J.M. (1996) Empowering patients: Issues and strategies. Soc Sci & Med 43(5):697-705.
12. Kruger E., Perera I., and Tennant M. (2010), Primary Oral Health Service Provision in Aboriginal Medical Services-based Dental Clinics in Western Australia Aus J of Pri Health 16(4):291-295.
13. Cohen L.A. et al. (2011) Comparison of patient visits to emergency departments, physician offices, and dental offices for dental problems and injuries. J of Pub Health Dent 71(1):13-22.
14. Agbor M.A. and Azodo C.C. (2011) Self medication for oral health problems in Cameroon. Int Dent J 61(4):204-209.
15. Afolabi A.O., Akinmoladun V.I., and Adebose I.J. (2010) Self-medication profile of dental patients in Ondo State, Nigeria. Niger J Med 19:96-103.
16. Jaanfar N. et al. (1992) Investigation of delay in utilisation of government dental services in Malaysia. Community Dent & Oral Epidemiol 20:144-147.
17. Develay A., Sauerborn R., and Diesfeld H.J. (1996) Utilization of health care in an African urban area: results from a household survey in Ouagadougou, Burkina-Faso. Soc Sci Med 43:1611 - 1619.
18. Freeman R. (1999) The psychology of dental patient care: Barriers to accessing and accepting dental care. Br Dent J 187(2): 81-84.
19. Chen M. et al (1997) Comparing Oral Health Care Systems: A second international collaborative study. Geneve:World Health Organisation (in press),.
20. Varenne, B., Petersen P., and Ouattara S. (2006) Oral health behaviour of children and adults in urban and rural areas of Burkina Faso, Africa. Int Dent J 56:61 - 70.

21. Davidson P.F. et al. (1999) Evaluating the effect of usual source of dental care on access to dental services: comparisons among diverse populations. Med Care Res Rev 56(1):74-93.

22. Andersen R.M. and Davidson P.L. (1997) Determinants of dental care utilization for diverse ethnic and age groups. Adv in Dent Res11(2):254.

23. Locker, D., Liddell A., and Burman D. (1991) Dental fear and anxiety in an older adult population. Community Dent Oral Epidemiol 19:120 - 124.

24. Davidson P.L. and Andersen R.M. (1997) Determinants of dental care utilization for diverse ethnic and age groups. Adv Dent Res 11(2):254-62.

25. DeVoe J.E., Petering R., and Krois L. (2008) A usual source of care: supplement or substitute for health insurance among low-income children? Med Care Res Rev 46(10):1041-8.

Treatment Outcomes of Periapical Lesions in Permanent Incisors Treated with Calcium Hydroxide

Thai Van Nguyen[1] and Thao Quy Le[2]

[1] Faculty of Odonto-Stomatology, Hue University of Medicine and Pharmacy, Hue, Vietnam
[2] Odonto-Stomatology Hospital, Hue, Vietnam

Abstract— *Introduction:* **There remains debate over the optimal initial treatment of periapical lesions that are the hallmark of chronic apical periodontitis. Calcium hydroxide may be a cost-effective initial non-surgical treatment for these lesions. The aim of this study was to evaluate the short-term treatment outcome of periapical lesions treated with calcium hydroxide both clinically and radiographically.**

Methods: **Patients with periapical lesions having Periapical Index (PAI) ≥ 3 and horizontal diameter ≤ 5mm were invited to participate in this study. A standardized treatment incorporating calcium hydroxide paste was applied followed subsequently by obturation with gutta-percha and endomethasone cement. Clinical response was determined by the presence or absence of symptoms before and at 3 and 6 months after therapy. Radiological evidence of healing was determined at both 3 and 6 months according to three predefined categories ('healed', 'healing' and 'not healed') based on changes in the PAI determined by a blinded assessor.**

Results: **Twenty-seven patients with thirty-six permanent incisors with periapical lesions were included in the study. Nine patients with ten treated lesions were lost to follow up. Clinical symptoms improved significantly after treatment with calcium hydroxide. With radiological assessments, there was a large decline in the percentage graded as 'not healed' at 6 months post therapy (from 100% at baseline to 19.2% at 6 months). The proportion of lesions rated as 'healed' and 'healing' at 6 months was 46.2% and 34.6%, respectively.**

Conclusion: **Calcium hydroxide treatment appears to be effective at facilitating the healing of periapical lesions associated with chronic periodontitis.**

Keywords— **Periapical lesions, permanent incisors, calcium hydroxide, intracanal dressing, endodontic treatment.**

I. INTRODUCTION

Chronic apical periodontitis is a relatively low-grade, long-standing response to canal bacteria and irritants mainly resulting from pulp necrosis. It is characterized by the development of asymptomatic periapical lesions which are identified by the presence of an apical radiolucency on radiograph [1]. These periapical lesions may be classified as granulomas, cysts or abscesses; although a definitive diagnosis can only be made by histological examination. While some have argued that true cysts can only be successfully treated by surgical intervention, others suggest that all periapical lesions should be initially treated with a non-surgical approach as an accurate diagnosis cannot be made through clinical assessment alone [2, 3].

The ultimate goal of endodontic treatment is to eliminate bacteria located in the root canal system that play an important role in the development and maintenance of periapical lesions [4]. Chemo-mechanical preparation accompanied by an intracanal dressing which remains in the root canal between appointments may help to eliminate bacteria entirely [5]. Calcium hydroxide, introduced by Hermann in 1920, has been considered to be a very effective intracanal dressing, especially in necrotic teeth with periapical lesions [4]. The antimicrobial activity of calcium hydroxide is related to its strongly alkaline pH and the release of hydroxyl ions. Distilled water may be the most effective vehicle to deliver the therapy as it has been shown to produce a higher pH value in an in vitro study [6]. Previous authors have suggested the calcium hydroxide dressing should remain in place for at least 15 days to achieve the best histopathological results [7].

The aim of the study was to evaluate the short-term clinical and radiological outcomes of periapical lesions in permanent incisors treated with calcium hydroxide.

II. MATERIALS AND METHODS

This prospective study was approved by the Hue University of Medicine and Pharmacy. The patients were recruited from referrals made to the Endodontic-Periodontic Department, Odonto-Stomatology Hospital.

Individuals with incisors demonstrating chronic apical periodontitis and periapical lesions demonstrating a Periapical Index (PAI) ≥ 3 and horizontal diameter ≤ 5mm (Fig. 1) were invited to participate. All lesions were verified radiographically.

Fig. 1 Horizontal diameter measurement method

The exclusion criteria were patients with periodontal disease, with contraindications to endodontic treatment.

© Springer International Publishing Switzerland 2015
V. Van Toi and T.H. Lien Phuong (eds.), *5th International Conference on Biomedical Engineering in Vietnam,*
IFMBE Proceedings 46, DOI: 10.1007/978-3-319-11776-8_36

Those whose root canal was unsuitable for obturation after 2 weeks treatment with calcium hydroxide were also excluded. This was determined according to the following selected criteria; 1) dry and clean canal; 2) canal free of odor and; 3) intact temporary restoration [8].

A. Treatment Protocol

After the initial caries excavation was performed, a standard access preparation by Endo-Access bur (Dentsply) was prepared. The working length was established radiographically. Using a step-back technique, the canals were sequentially shaped with hand K-files ISO #8-60 (Mani) depending on the canal size and the master apical file usually was # 25-30 [9]. During instrumentation, the canals were irrigated copiously with 2.5% NaOCl.

Calcium hydroxide powder (PD, Switzerland) was mixed with distilled water to form a thick paste. The canal was dried by sterile paper points, and then the paste was rotated into the canal by lentulo spirals limited within the canal space. The access was sealed temporarily with Zinc Oxide – Eugenol (ZOE) at least 3mm thick to prevent coronal leakage [10]. The paste was placed for 2 weeks.

Two weeks later, the paste was removed by irrigating with 2.5% NaOCl. After dried by sterile paper points, the canal was obturated with gutta percha points and endomethasone cement using lateral condensation technique. The tooth was restored permanently with Glass ionomer cement (GC Gold label 9).

B. Follow-Up

The symptoms and the short-term healing outcomes were clinically and raphiographically evaluated at baseline and both 3 and 6 months following treatment.

C. Clinical and Radiographic Assessments

The clinical symptoms recorded were; 1) spontaneous pain; 2) presence of a sinus tract, and; 3) sensitivity to percussion (painful, feel different, no pain).

Radiological healing was determined using the PAI which is a scoring index based on histological analysis by Brynoff. The index is a 5-point scale scored as follow:

1. Normal periapical structures
2. Small changes in bone structure
3. Changes in the bone structure with little mineral loss
4. Periodontitis with a well-defined radiolucent area
5. Severe periodontitis with exacerbating features [11]

Instructions for scoring incisors using PAI:

1. Find the reference radiograph where the periapical area most closely resembles the periapical area you are studying (Fig. 2). Assign the corresponding score to the observed root.
2. When in doubt, assign the higher score.
3. All teeth must be given a score [12].

All digital radiographic films obtained at baseline and at follow-up assessments assessed by a single blinded examiner. This approach required evaluation of reliability through a calibration process before rating radiographs.

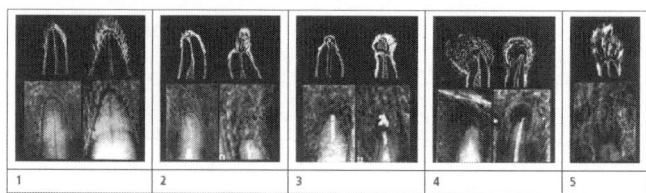

Fig. 2 The visual references of the periapical index PAI [12]

D. Outcome Classification

The outcome of the treatment was determined on the basis of radiographic and clinical evaluations.

Immediate postoperative and follow-up radiographs of studied teeth were taken, and the treatment outcome was assessed using the change in PAI over the follow-up period. Each case was categorized as follows:

1. Healed: PAI scored 1 or 2
2. Healing: PAI reduced but not yet scored 1 or 2
3. Not healed: PAI unchanged or increased

In addition to the radiographic examination, clinical symptoms were also recorded after treatment and at the recall appointments. Cases showing no pain, no sinus tract and feel different or no pain under percussion were classified as 'symptoms absent'. Otherwise, classified as 'symptoms present'.

III. RESULTS

Twenty-seven systemically healthy individuals (age range 12-59 years, 74% female) with thirty-six incisors with periapical lesions, mostly maxillary (88.9%), participated in the study.

Nine patients with 10 teeth were lost to follow-up at the 6-month review. Reasons for drop out were mainly failed to attend (n=5) and unknown reason (n=4).

Table 1 lists the frequency of clinical symptoms present at baseline and following treatment.

Table 1 Clinical symptoms

Symptoms	Baseline (n = 36)	After treatment (n = 36)	3 months (n = 36)	6 months (n = 26)
Spontaneous pain	19 (52.8%)	4 (11.1%)	0	0
Sinus tract	3 (8.3%)	3 (8.3%)	0	0
Sensitivity to percussion	29 (80.1%)	13 (36.1%)	8 (22.2%)*	2 (7.7%)*

(*) feel different.

Immediate clinical evaluation after treatment showed that 88.9% of cases were 'symptoms absent'. 6 months post-operation, all teeth were 'symptoms absent'.

Table 2 displays the radiographic assessment of healing according to the predetermined categories. The proportion assessed radiographically as 'not healed' fell from 100% at baseline to 19.2% at 6 months. Consequently the rates of 'healing and 'healed' increased considerably.

Table 2 Treatment outcome after 3 and 6 months

Outcome	Before treatment (n = 36)	3 months (n = 36)	6 months (n = 26)
Healed	0	7 (19.4%)	12 (46.2%)
Healing	0	4 (11.1%)	9 (34.6%)
Not healed	36 (100%)	25 (69.5%)	5 (19.2%)

IV. DISCUSSION

This study has demonstrated the utility of calcium hydroxide as an intracanal dressing to facilitate the healing of periapical lesions associated with chronic apical periodontitis. The treatment was associated with a high rate of both symptom resolution and radiological improvement according to the PAI. Most lesions healed completely (46.2%) or were 'healing' (34.6%) at the 6-month review. Therefore, a favorable outcome was observed in over 80% of cases. Although not confirmed in this study, healing lesions (that have demonstrated an improvement in PAI) are likely to heal over time, even though they might not have normal periodontal ligament width [13].

Our healing rate compares favorably to those in previous reports with the same period of follow-up, which range from 27.7 – 46% [14, 15]. For the longer time of follow-up, the healing rate increased up to 71 – 80.8% [16, 17]. The difference was probably because of three main reasons; (1) the period of follow-up; (2) the severity of the disease; (3) and the material and the protocol used in the study.

Elimination of microorganisms from the canals is the key to success of endodontic treatment of chronic apical periodontitis. It is, therefore, important to emphasize the combination of anti-microbial strategies employed in our protocol, in addition to the use of calcium hydroxide. They included careful mechanical preparation, copious irrigation with NaOCl, and obturating the canals in three-dimensions. This last technique serves a critical role in prevent bacteria and bacterial byproducts from invading into apex [18, 19].

As previously mentioned, chronic apical periodontitis is normally asymptomatic or produces only minimal symptoms [20]. While clinical reassessment is important, demonstrating evidence of radiological resolution is crucial. Improvement in the PAI is a reliable measure of this and has been validated for this purpose. It should be emphasized, however, that healing of periapical lesions is often slow and that while evidence of healing can be demonstrated as early as 3 months after treatment [21]. A number of professional societies recommend a minimum 12 month timeframe to determine healing status. Earlier reviews may lead to an overestimation of the rate of incomplete healing [22].

This study has a number of limitations that require comment. Firstly, the large number of participants that did not return for the 6 month review limits the strength of our conclusions. However, even if the rate of non-healing were high in this group, our healing rate would remain broadly in line with other studies in this area. Secondly, only periapical lesions ≤ 5mm were included in the study and therefore it is uncertain if this approach would be as effective in larger lesions. While Matsumato et al previously suggested that smaller lesions had a better prognosis, Siqueira et al did not find a difference in healing based on the size of the lesion [23]. This raises the possibility that even larger lesions may benefit from intracanal calcium hydroxide treatment. Finally, the conclusions that can be inferred from this observational study are limited and ideally a comparison of different treatment approaches is required in order to determine how best to manage these lesions.

V. CONCLUSION

The use of calcium hydroxide as an intracanal treatment appeared to facilitate healing in periapical lesions associated with chronic apical periodontitis. This is a cheap and simple therapy associated with a high rate of success. Further studies comparing different treatment approaches are required.

ACKNOWLEDGEMENT

I would like to express my deep gratitude to Dr. Le Quy Thao, my research supervisor, for his patient guidance, enthusiastic encouragement and useful critiques of this research work. I wish to thank Dr. Nguyen Thuc Quynh Hoa, for passing passion, knowledge and practical skill to me. I am particularly grateful for the assistance given by Dr. Pham Nu Nhu Y for her useful and constructive recommendations on this research. Besides, my special thanks are extended to everyone who supported and helped me to complete research work directly and indirectly. Finally, I wish to thank my family for their support and encouragement throughout my study.

CONFLICT OF INTEREST

The authors declare that they have no conflict of interest.

REFERENCES

1. Johnson WT (2002) Diagnosis of pulpal and periradicular pathosis. In: Johnson WT (ed) Color atlas of endodontics, 1 edn. Saunders, p 10

2. Fernandes M and de Ataide I (2010) Nonsurgical management of periapical lesions. J Conserv Dent 13(4):240-245

3. Ramachandran NPN, Pajarola G, and Schroeder HE (1996) Types and incidence of human periapical lesions obtained with extracted teeth. Oral Surgery, Oral Medicine, Oral Pathology, Oral Radiology, and Endodontology 81(1):93-102

4. Leonardo Mario R et al. (2006) Effect of a calcium hydroxide-based root canal dressing on periapical repair in dogs: a histological study. Oral Surgery, Oral Medicine, Oral Pathology, Oral Radiology, and Endodontology 102(5):680-685

5. Siqueira J. F., Jr. and Lopes H. P. (1999) Mechanisms of antimicrobial activity of calcium hydroxide: a critical review. Int Endod J 32(5):361-369

6. Pacios MG et al. (2004) Influence of different vehicles on the pH of calcium hydroxide pastes. J Oral Sci. 46(2):107-111.

7. Leonardo MR et al. (2002) Calcium hydroxide root canal dressing. Histopathological evaluation of periapical repair at different time periods. Braz Dent J 13(1):17-22

8. Bui QD (2009) Tram bit ong tuy (Canal obturation). In: Bui QD (ed) Noi nha lam sang (Clinical endodontics), edn. Medical Publisher, Ho Chi Minh City, p 119

9. Krell KV (2002) Canal preparation. In: Johnson WT (ed) Color atlas of endodontics, 1st edn. Saunders, p 70

10. Baumgartner JC (2002) Endodontic microbiology. In: Richard EW and Torabinejad M (ed) Principles and practice of endodontics, 3rd edn. WB Saunders, Philadelphia, p 291

11. Huumonen S and Ørstavik D (2002) Radiological aspects of apical periodontitis. Endodontic Topics 1(1):3-25

12. Reit C and Kirkevang L (2009) Clinical epidemiology. In: Bergenholtz G, Hørsted-Bindslev P, and Reit C (ed) Textbook of endodontology, 2nd edn. Wiley-Blackwell, UK, p 293

13. Sathorn C, Parashos P, and Messer HH (2005) Effectiveness of single- versus multiple-visit endodontic treatment of teeth with apical periodontitis: a systematic review and meta-analysis. Int Endod J 38(6):347-355

14. Cvek M, Hollender L, and Nord CE (1976) Treatment of non-vital permanent incisors with calcium hydroxide. VI. A clinical, microbiological and radiological evaluation of treatment in one sitting of teeth with mature or immature root. Odontol Revy 27(2):93-108

15. Pham NNY (2009) Nghien cuu dac diem lam sang, Xquang va ket qua dieu tri viem quanh chop man bang phuong phap noi nha (Outcome of the endodontic treatment of teeth with chronic apical periodontitis). Dissertation, Hue University of Medicine and Pharmacy

16. Caliskan MK and Sen BH (1996) Endodontic treatment of teeth with apical periodontitis using calcium hydroxide: a long-term study. Endod Dent Traumatol 12(5):215-221

17. Peters LB and Wesselink PR (2002) Periapical healing of endodontically treated teeth in one and two visits obturated in the presence or absence of detectable microorganisms. Int Endod J 35(8):660-667

18. Hammad M, Qualtrough A, and Silikas N (2009) Evaluation of root canal obturation: a three-dimensional in vitro study. J Endod 35(4):541-544

19. Siqueira JF Jr, Magalhaes KM, and Rocas IN (2007) Bacterial reduction in infected root canals treated with 2.5% NaOCl as an irrigant and calcium hydroxide/camphorated paramonochlorophenol paste as an intracanal dressing. J Endod 33(6):667-672

20. Metzger Z and Abramovitz I (2008) Periapical Lesions of Endodontic Origin. In: Ingle JI, Bakland LK, and Baumgartner JC (ed) Ingle's endodontics, 6th edn. BC Decker, p 507

21. Ørstavik D and Larheim TA (2008) Radiographic interpretation. In: Ingle JI, Bakland LK, and Baumgartner JC (ed) Ingle's Endodontics, 6th edn. BC Decker, p 607

22. Ng YL (2008) Factors affecting outcome of non-surgical root canal treatment. Dissertation, University of London

23. Siqueira JF Jr et al. (2008) Clinical outcome of the endodontic treatment of teeth with apical periodontitis. Oral Surg Oral Med Oral Pathol Oral Radiol Endod 106(5):757-762

Author: Thai Van Nguyen, DDS
Institute: Faculty of Odonto-Stomatology,
            Hue University of Medicine and Pharmacy
Street: 06 Ngo Quyen Street
City: Hue
Country: Vietnam
Email: nv_thai0106@yahoo.com.vn

The Primary Results of the N_2/Ar Micro-plasma Exposure on Second Degree Wound Healing

Minh-Hien Ngo, Pei-Lin Shao, and Jiunn-Der Liao*

Department of Materials Science and Engineering,
National Cheng Kung University, No. 1, University Road, Tainan 70101, Taiwan
jdliao@mail.ncku.edu.tw

I. INTRODUCTION

Second degree burn may reach the epidermis and partial dermis layers. Several methods and techniques have been applied to manage such burn injuries, such as different kinds of dressings, pharmacotherapies and plasma treatment. The latter has been increasingly studied. In this work, non-thermal N_2/Ar micro-plasma was applied to enhance healing on the second degree burn wound mice through the wound area reduction. Four wounds were created in the dorsal of each mouse by solid aluminum bar with 5 mm in diameter (46g) and an average temperature of $69 \pm 2°C$. The parameters for micro-plasma exposure for burn wound on mice were chosen: excitation at 13 W (< 40 °C) and addition of 0.5% N_2 in Ar, corresponding with relatively high NO peak intensity.

N_2/Ar micro-plasma was utilized to expose upon the burn wound achieved mice in these groups: continuous exposure until 2 day wound (dw) (denoted as P3), with plasma exposure and then covered by an occlusive dressing on 0, 1, 2 dw (denoted as P3+D3), continuous exposure until 4 dw (denoted as P5), with plasma exposure and then covered by an occlusive dressing on 0, 1, 2, 3, and 4 dw (denoted as P5+D5). Dressing (denoted as D3 or D5) and gas flow exposure (denoted GF3 or GF5) were also conducted as well for the references. The burn wounds were assessed every day to examine the wound size.

Until 14 dw, the mice were sacrificed for H&E staining. The wound area reduction rate was higher for the cases of N_2/Ar micro-plasma exposed wounds than those of gas flow exposed and dressing ones, while the control group exhibited the lowest wound area reduction rate. From this study, non-thermal N_2/Ar micro-plasma is presumably effective for the stimulation of newly-born cells growth and proliferation [1] and burn wound healing in mice.

---

* Corresponding author.

II. RESULTS AND DISCUSSION

A. Micro-plasma System and Plasma Plume Temperature

A custom-made micro-plasma jet source was driven by a radio-frequency power supply of 13.56 MHz (ENI ACG-3B, MKS Instruments Inc., Rochester, New York, USA) with a matching device (ACG-3B, ENI Corp., Rochester, USA).

This jet source was a capillary electrode through which the additive N_2 gas was injected. The micro-plasma device used in this work as illustrated in Figure 1, contained a quartz tube as the gas channel and a dielectric layer with an outer diameter of 2 mm.

At the center of the quartz tube, a stainless steel capillary tube (with a diameter of 0.2 mm, fixed by a perforated Teflon fitting) was used as the inner electrode as well as the N_2 feeding tube. A copper chip was used as the outer electrode, and this was connected to a generator.

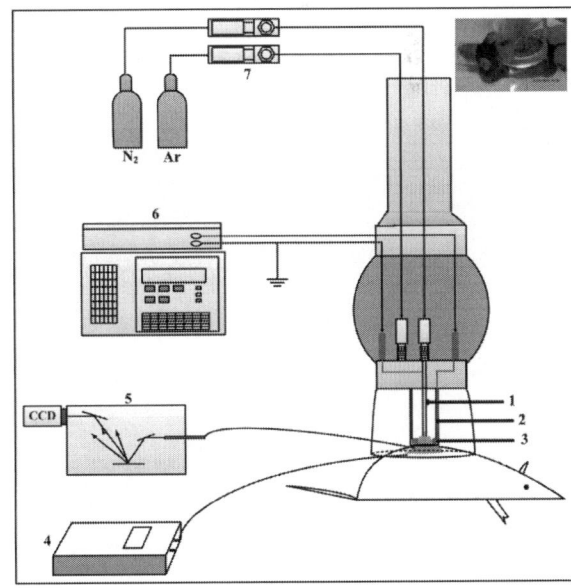

Fig. 1 Experimental set up for micro-plasma exposure to second degree burn wound on mice

© Springer International Publishing Switzerland 2015
V. Van Toi and T.H. Lien Phuong (eds.), *5th International Conference on Biomedical Engineering in Vietnam,*
IFMBE Proceedings 46, DOI: 10.1007/978-3-319-11776-8_37

B. Average Plume Temperature and Optical Emission Spectrum

The plasma plume temperature was estimated using a fiber optic thermometer (Luxtron 812, Santa Clara, USA). The fiber was placed on an XY coordinate table. The distance from the fiber to the micro-plasma jet nozzle was ≈4 mm. The temperature was measured for applied powers of 13 to 17 W and an Ar flow rate of 5 slm. To produce N- and O-containing species in an excited phase, 0.1, 0.5, and 1% N_2 additions were respectively poured into the Ar plasma.

When a relatively low power supply (e.g., 13 W) with the addition of N_2 (e.g. 0.1, 0.5, 1%) was applied to the micro-plasma, it exhibited an average temperature below 40°C (see the lines marked in the Figures 2(a)).

In these cases, the samples are presumably subjected to very minor effects due to the heat.

In Figures 2(b), taking the optical emission spectra of 0.5% N_2/Ar micro-plasma as examples, the presence of NO (237 nm), OH (306 nm), Ar I (750 nm), and O (777 nm) is obvious. Reactive plasma species (RPS) such as NO, OH, O were analyzed using Optical Emission Spectroscopy (OES, SpectraPro 2300i, Acton Research Corp., Massachusetts, USA).

The emitted light was then focused by optical fibers into the entrance slit of single monocromater (SpectraPro 2300i, Acton Ltd, MA, USA) equipped with a CCD detector (1340 × 100 pixels). The resolution of the collected spectra was 1200 grooves per millimeter with the slit width of ≈0.1 nm. Two gratings, 200~500 (1200 g.mm^{-1}) and 500~1100 nm (1200 g.mm^{-1}), were utilized to estimate the composition of RPS before interacting with the mice.

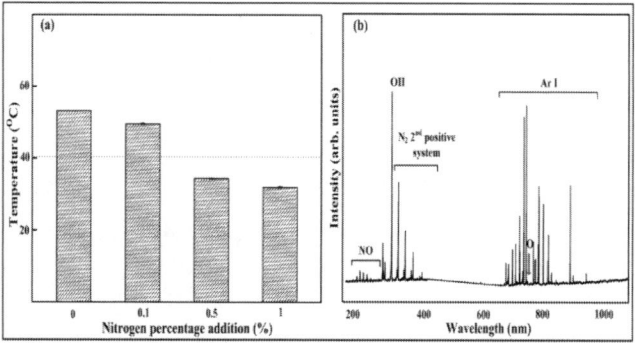

Fig. 2 Average of micro-plasma plume temperature and optical emission spectrum

C. Burn Wound Animal Model

Initially, mice were weighed and intramuscularly pre-anesthetized with 2% Isofluran inhalation (Isoflurane, USP, Baxter, Guayama, USA). Thermal injuries were made with a solid aluminum bar 5 mm in diameter, previously heated in boiling water (100 °C) until thermal equilibrium was achieved, and the Al bar surface temperature reached 69 ± 2 °C which measured with a thermometer (Figure 3a).

The bar is maintained in contact with the animal skin on the previously shaved dorsal region of the mice for 30 sec. The pressure exerted on the animal skin corresponded to the mass of 41 g of aluminum bar used in the burn induction. Four burn wound lesions were created in the dorsal skin of the mice (Figure 3b).

The second degree burn where no cellular structure and tissue structure were observed deep into the dermis was confirmed by sections stained with H&E staining on 0 and 14 dw (Figure 3c and d).

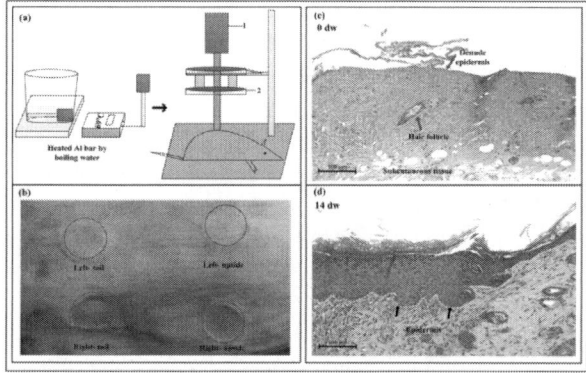

Fig. 3 Burn wound model: a) Experimental set up: Al bar temperature achievement and burn wound experimental design: 1. Aluminum bar, 2. Immobilized Al bar; b) Four burn wounds formation on mice back; c) Tissue damage of mice skin on 0 dw (H&E staining, 200X magnification), d) Tissue damage of mice skin on 14 dw (H&E staining, 200X magnification).

D. Micro-plasma Stimulated Burn Wound Closure Dynamic

Figure 4a showed the most representative photos in the wound dynamic upon the time (0, 7, 14 dw) and the healing rate was showed in Figure 4b. The measuring of burn wound open surface under P3, P5, P3+D3 and P5+D5 showed that the reducing of wound area process at a much faster than in the control group (NT, D3 and D5). For example, on 7 dw, the open surface of P5+D5 was reduced to 45.34 ± 10.45%, as compared with 87.68 ± 14.64% for NT. On 14 dw, the differences became much more significant. For example, the open surface of P5+D5 was reduced to 2.89 ± 1.4%, as compared with 17.02 ± 0.09% for NT.

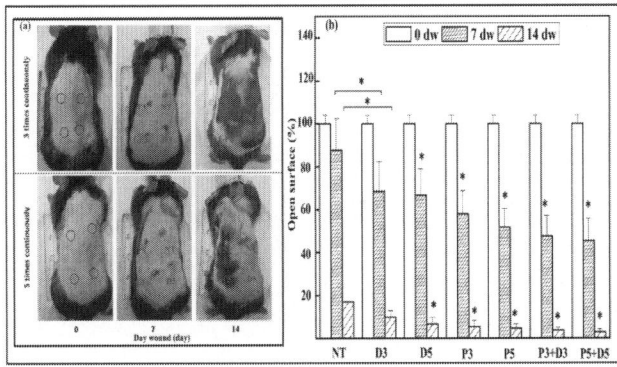

Fig. 4 Burn wound area reduction

At pre-established times for biopsy (14 dw), mice were sacrificed for collecting tissue samples, which were immediately fixed by immersion in 4% formaldehyde prepared in 1X PBS (pH 7.2), followed by routine histological processing and paraffin embedding. Paraffin sections (3 μm thick) were obtained from the wound skin sections and stained with H&E.

Morphometric images were acquired using a computer-assisted analysis system. Digital images were acquired with a microscope digital camera (DFC 450 C, Leica, Wetzlar, Germany) attached to a light microscope (DM IRB, Leica, Wetzlar, Germany).

On 14 dw, the_14_D5, _14_P5+D5 groups appeared the corrugated epidermis, with well-presented membrane spinous and corneum layers, increased the number of hair follicle in the dermis layer compared with the NT group with significant thickness in epidermis and the new epithelium formation.

Finally, the wound healing under N₂/Ar micro-plasma exposure accelerates 5 days compared to that in NT group (Figure 5).

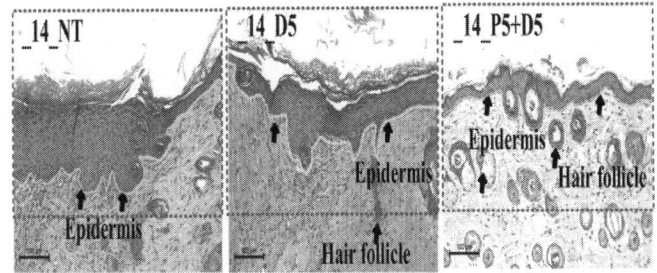

Fig. 5 H&E staining

E. Micro-plasma Stimulated the Blood Flow Cytometry

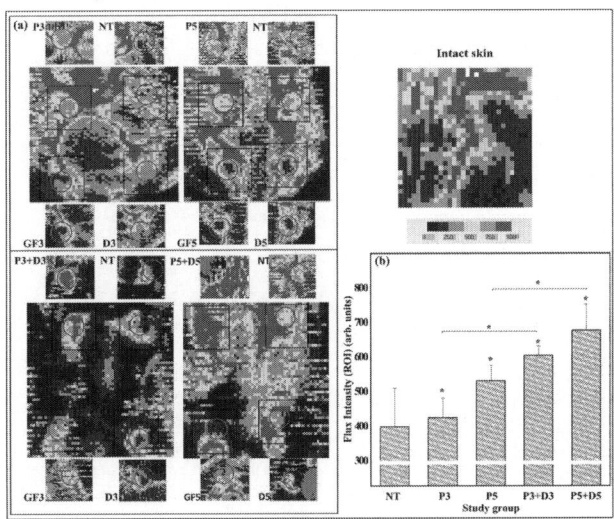

Fig. 6 Blood flow cytometry: a) The images of 7 dw for the respective study groups of NT, GF3, GF5, D3, D5, P3, P5, P3+D3, P5+D5 were taken. The parts in red represented the greatest (normal) blood flow, whereas those in blue represented the lowest (or free from) blood flow; (b) Quantification data for (a) were shown. (*$P < 0.05$).

Figure 6a showed the image of blood flow cytometry on 7 dw, N₂/Ar micro-plasma exposure (_7_P3, _7_P5, _7_P3+D3, _7_P5+D5) induced more blood flow to the burn wound area than other groups (_7_NT, _7_GF3, _7_D3, _7_GF5, _7_D5). Figure 6b showed that the blood flow under N₂/Ar micro-plasma exposure (_7_P3, _7_P5, _7_P3+D3, _7_P5+D5) were 424, 531, 604, 677 perfusions units, respectively, significantly increased compared with NT groups was only 398 perfusions units and even with others groups (_7_GF3, _7_D3, _7_GF5, _7_D5, data not showed). These results supported that laser Doppler analysis findings of increased blood flow in the wound area under N₂/Ar micro-plasma exposure, demonstrating enhanced the blood vessels growth in the wound area compared to NT group.

III. CONCLUSIONS

N₂/Ar micro-plasma with optimized parameters i.e., a supplied power of 13 W for an average temperature below 40°C, a working distance of 4 mm, and 0.5% N₂ addition to Ar plasma to applied to this burn wound model.

Finally, with the optimal exposure time 30 sec, the wound contraction and the blood flow cytometry were accelerated, the wound healing faster 5 days compared to control group. . Based on these preliminary results, a micro-plasma device with a low operating temperature and an adjustable plasma composition is a promising therapeutic alternative.

REFERENCES

1. Minh-Hien Thi Ngo, Jiunn-Der Liao, Pei-Lin Shao, Chih-Chang Weng, and Chen-Young Chang, "Increased fibroblast cells proliferation and migration using atmospheric N_2/Ar micro-plasma for the stimulated release of fibroblast growth factor-7", Plasma Processes and Polymers, Vol. 11 (1), 80-88, January 2014.

Novel Approaches to Diagnosis and Therapy in Neurodegenerative Diseases

Giuseppe Legname

SISSA - Neurobiology Sector, Italy

Abstract— Neurodegenerative diseases (NDs) are a large class of pathologies that include Alzheimer disease (AD), Parkinson disease (PD) and many other forms of senile dementias. The unique transmissible spongiform encephalopathies (TSEs), known as prion diseases, are among the rarest of these disorders.

NDs belong to age-related disorders, thus their incidence and prevalence are estimated to increase dramatically as the population ages. So far, no early diagnostic tool has been developed. NDs are usually diagnosed at very advanced stages, when neurological symptoms are evident. Conventional therapies are directed at treating the neurological symptoms, but have no effect on disease progression. Thus, the development of novel early diagnostic tools and effective therapies for NDs today is one of the major scientific challenges.

Recent evidence points at a common mechanism of pathogenesis in all NDs, which involves abnormal aggregation and deposition of misfolded proteins in the central nervous system (CNS). Furthermore, recent studies suggest that these aggregates may spread among cells in a prion-like manner, leading to cytotoxicity and cell death. These facts suggest that novel diagnostic and therapeutic approaches to different neurodegenerative disorders may share similar characteristics.

Here, we provide an overview of the most advanced and modern techniques in use to diagnose NDs, and of the various attempts at finding an effective cure for these ever increasing disorders.

Keywords— *Neurodegeneration, amyloid, protein aggregation, diagnosis, therapy.*

I. INTRODUCTION

Neurodegenerative diseases, such as AD, PD and prion diseases, are a heterogeneous group of disorders associated with gradual and progressive neuronal loss in the brain, leading to progressive memory loss, motor symptoms, cognitive impairments and death.

Age is the most important risk factor [1]. In developed countries, the increasing longevity of the population determines a significant rise in morbidity and mortality of NDs. In 2006 the worldwide prevalence of AD was 26 million, and by 2050 the prevalence will quadruple [2]. The prevalence of PD is expected to double by the year 2030 [3]. Thus, the development of early diagnostic tools as well as effective therapeutic strategies is one of the essential scientific challenges.

Despite the heterogeneity of these disorders, increasing evidence confirms that many NDs share similar pathogenic mechanisms. The common feature of NDs is the pathological aggregation of misfolded proteins, which may accumulate as amyloid deposits in different areas of the nervous system. In AD, the pathological accumulation of tau filament and amyloid β (Aβ) fibrils is implicated in the pathogenesis. Aggregated α-synuclein filaments are the main component of Lewy bodies (LB) and Lewy neurites —intracytoplasmic brain *inclusions representing a major pathological hallmark of several NDs, including PD, dementia with Lewy bodies, multiple system atrophy and other synucleinopathies [4]. In TSEs, the misfolded PrP protein (PrP^{Sc})* accumulates in CNS tissues [5].

Moreover, increasing evidence suggests that protein aggregates, fibrils and their oligomeric precursors propagate among cells in a prion-like manner leading to cytotoxicity [6]. Amyloid deposition leads to impairment of axonal transport, increased level of oxidative stress and apoptosis [7].

Common pathogenic mechanisms underlying NDs suggest that similar diagnostic and therapeutic strategies targeting different steps of the amyloidogenic pathway should be considered in current research. The development of early diagnostic tools or strategies, enabling the diagnosis before neuronal damage and other symptoms are manifest, is the highest priority in the search for novel therapies.

II. CURRENT APPROACHES TO DIAGNOSIS OF NEURODEGENERATIVE DISEASES

Although conventional diagnostic approaches for many NDs are based on the clinical evaluation of the symptoms, this method is not very reliable, especially in the early stages of the disease. NDs are therefore often diagnosed very late, and the detection of amyloid plaques in postmortem brain tissues is often necessary for a definitive diagnosis.

Traditional structural neuroimaging techniques, such as magnetic resonance imaging (MRI), may be helpful in diagnosing NDs; their clinical usage is nevertheless limited because of the relatively low diagnostic specificity. New imaging techniques, e.g. positron emission tomography (PET) and single photon emission computed tomography (SPECT), show promise in the diagnosis and monitoring of disease progression.

© Springer International Publishing Switzerland 2015
V. Van Toi and T.H. Lien Phuong (eds.), *5th International Conference on Biomedical Engineering in Vietnam,*
IFMBE Proceedings 46, DOI: 10.1007/978-3-319-11776-8_38

Currently, several studies aim at developing new reliable markers detectable in readily accessible tissues, namely blood or saliva, for disease diagnosis and monitoring. The common mechanism underlying different NDs, as well as similar clinical symptoms, raise the urgent need to identify new biomarkers or diagnostic strategies to distinguish the different forms of these disorders. Genetic markers are useful for diagnosing familiar forms of NDs and they include amyloid precursor protein and presenilin gene mutations and ApoE polymorphism in AD, gene mutations in α-synuclein protein in PD or PrP gene mutation in familiar forms in prion diseases and many others. Table 1 summarizes some potential biochemical markers in different NDs.

Table 1 Selected candidate biochemical markers for neurodegenerative diseases

Disease	Biomarker	Sources
Alzheimer disease	Aβ$_{1-42}$	plasma, CSF
	tau, phospho-tau	CSF
	BACE1	CSF
Parkinson disease	DJ-1	plasma, CSF
	α-synuclein	CSF, blood
	TNF-α	CSF
	ROS	blood
Prion diseases	14-3-14	CSF
	Tau	CSF
	PrP protein	CSF

In spite of major advances in NDs research, efficacious markers with high sensitivity and specificity are still lacking. So far, most of the identified biomarkers have shown low sensitivity and they are not available in clinical practice. The combination of some markers seems to improve their effectiveness.

Several ongoing studies focus on developing new molecular tools enabling the detection of abnormally aggregated proteins in disease-affected tissues or body fluids. Amyloid deposition may occur years before clinical symptoms, suggesting that detection of misfolded proteins is a very promising approach for early diagnosis.

Recently, new ultrasensitive assays have been developed for the detection and diagnosis of TSEs. These assays are based on the ability of the abnormal TSE-associated form of the PrP protein to seed the formation of amyloid fibrils from recombinant PrP[8]. Recombinant PrP is able to assemble in vitro into fibrils similar to those extracted from disease-affected brains. The simple addiction of biological samples containing misfolded/aggregated PrP protein into the reaction mixture accelerates fibrillization. The formation of fibrils in vitro is usually monitored by Thioflavin T (ThT), a benzothiazole dye that exhibits enhanced fluorescence upon binding to amyloid fibrils [9].

Several aggregation assays, including amyloid seeding assay (ASA) and quaking-induced conversion (QuIC) assay,

have been successfully used to detect the disease-causing isoform of the prion protein (PrP^{Sc}) in infected biological samples, including brain, cerebral spinal fluid (CSF), nasal fluid and blood plasma [10-13]. Real time-QuIC assay for diagnosing prion disease has shown 84% sensitivity and 100% specificity in CSF samples from sporadic CJD patients [12,14]. Moreover, this assay has been successfully used for detecting misfolded PrP in CSF samples from patients with various genetic forms of human prion diseases, including Gerstmann-Sträussler-Scheinker syndrome and fatal familiar insomnia [15].

The further development of the enhanced QuIC assay, employing an immunoprecipitation step before the QuIC reaction, enabled a very sensitive detection of misfolded PrP. For example, up to 10^{14}-fold dilutions of vCJD brain homogenates (BH) in human plasma, containing less than ~1 ag of PrP^{Res} have been detected [16]. This assay opens new possibilities for the detection of prions present in very low levels, hopefully allowing preclinical detection of prions in different tissues and fluids.

Furthermore, the fact that no cross-seeding between different amyloidogenic proteins occurred during the RT-QuIC reaction (results from AD and Lewy body dementia BH were similar to those obtained from normal BH) suggests that this assay may represent a promising tool to differentiate several forms of NDs [13].

As the deposition of misfolded protein is a common feature of several neurodegenerative diseases, aggregation assays may represent a powerful strategy for detecting different misfolded proteins in disease-affected samples. In fact, these assays have been used to identify Aβ aggregates in mouse AD model and misfolded huntingtin in Huntington disease [17,18]. In PD, abnormal peripheral α-synuclein aggregation occurs long before motor symptoms, suggesting a potential use of these assays in early diagnosis.

III. THERAPEUTIC STRATEGIES

Conventional therapies for NDs, for example L-dopa for PD and cholinesterase inhibitors for AD, are directed at treating the neurological symptoms, but have no or very mild effects on disease progression. Several surgical approaches for NDs have not proven to be effective so far.

Recent advances in stem cell biology suggest that stem cells transplantation may provide an attractive opportunity of therapy; this challenging issue warrants many further studies.

Recent advances in the understanding of the molecular mechanism underlying the pathogenesis of NDs have sparked several attempts to develop new strategies and drug compounds to inhibit the pathological aggregation of the

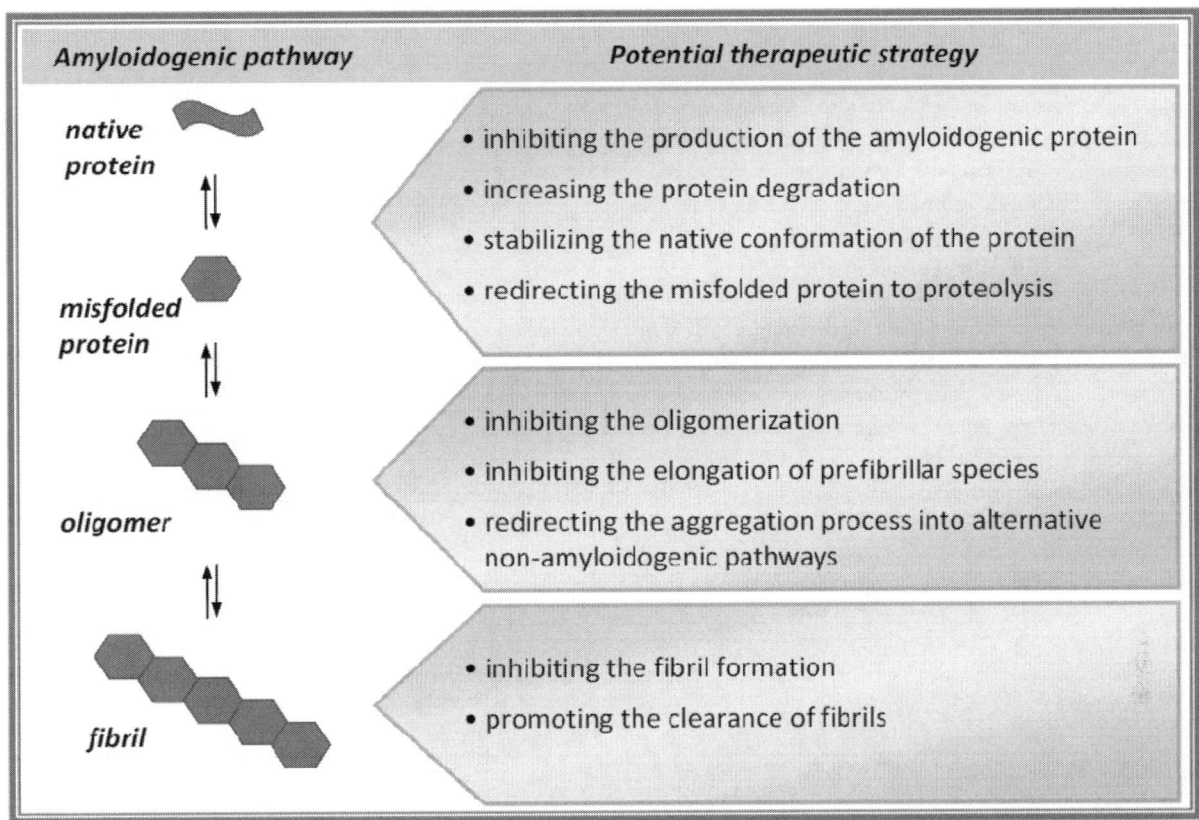

Fig. 1 Potential therapeutic strategies for NDs targeting different steps of the amyloidogenic pathway.

protein and/or promote the clearance of formed aggregates [19]. Figure 1 characterizes the potential therapeutic strategies aimed to inhibit the pathological aggregation of proteins at different steps of the amyloidogenic pathway. Some strategies have focused on preventing the production of amyloidogenic proteins, but they turned out to have several side effects [20]. Some studies have suggested that chaperone proteins, preventing abnormal protein aggregation or targeting misfolded proteins to degradation, may represent potential therapeutic agents [21].

The most common therapeutic strategy is the identification of new drug compounds inhibiting the aggregation process. In vitro aggregation assays represent fast and powerful tools for screening potential drugs inhibiting the fibrillization of recombinant proteins. In fact, several compounds, including molecular chaperones, antibodies, small chemical compounds such as polyphenols, metal chelators, tetracyclines, nanoparticles and others, have been shown to interfere with the aggregation pathway and inhibit fibril formation of different amyloidogenic proteins, e.g. Aβ, α-synuclein, PrP protein and others [22].

Several studies confirm that low molecular weight oligomers, rather than mature fibrils, are toxic species, thus the most promising therapeutic strategies should focus on preventing the formation of such oligomers or redirecting the aggregation process into alternative non-toxic pathways.

Although many potential drug compounds have proven effective in vitro and in animal models, their efficacy has not been confirmed in clinical trials. One of the central challenges in the development of new therapeutic agents lies in the crossing of the blood-brain barrier to enable their delivery to the brain. Novel nanobiotechnology-based approaches, such as the nanoparticle-mediated strategy, are very promising tools for drug delivery to target neurons.

IV. CONCLUSIONS

NDs are common disorders currently lacking early diagnostic and effective therapeutic strategies. The increasing prevalence of NDs represents a crucial challenge to the health care system worldwide. Thus, novel strategies are urgently needed to diagnose these diseases in their asymptomatic stages, when therapy would probably be most effective.

The main challenge in developing new diagnostic and therapeutic strategies lies in the fact that the biological mechanisms underlying NDs are complex and still not well understood. NDs are multifactorial disorders characterized by significant heterogeneity in clinical features of patients. Similar molecular mechanisms underlying the pathogenesis of these disorders, as well as overlapping clinical symptoms raise a need for developing novel strategies to properly diagnose and distinguish different forms of NDs. In vitro aggregation assays offer encouraging molecular tools for both early diagnosis and the development of new drug compounds.

Unfortunately, despite significant investments in the search for new therapies, to date no effective treatment strategy has been developed. The biological complexity of NDs suggests that therapeutic approaches should be directed not only to a single molecular mechanism, but a set of different strategies should be used simultaneously for successful treatment.

CONFLICT OF INTEREST

The authors declare that they have no conflict of interest.

REFERENCES

1. Levy G (2007) The relationship of Parkinson disease with aging. Arch Neurol 64:1242-12466.
2. Brookmeyer R, Johnson E, Ziegler-Graham K, Arrighi HM (2007) Forecasting the global burden of Alzheimer's disease. Alzheimers Dement 3:186-191.
3. Dorsey ER, Constantinescu R, Thompson JP et al. (2007) Projected number of people with Parkinson disease in the most populous nations, 2005 through 2030. Neurology 68:384-386.
4. Marti MJ, Tolosa E, Campdelacreu J. (2003) Clinical overview of the synucleinopathies. Mov Disord: 18 Suppl 6:S21-27.
5. Prusiner SB (2013) Biology and genetics of prions causing neurodegeneration. Annu Rev Genet 47:601-623.
6. Goedert M, Clavaguera F, Tolnay M (2010) The propagation of prion-like protein inclusions in neurodegenerative diseases. Trends Neurosci 33:317-25.
7. Glabe CG (2006) Common mechanisms of amyloid oligomer pathogenesis in degenerative disease. Neurobiol Aging 27:570-5.
8. Orru CD, Wilham JM, Vascellari S et al. (2012) New generation QuIC assays for prion seeding activity. Prion 6:147-152.
9. Nilsson MR (2004) Techniques to study amyloid fibril formation in vitro. Methods 34:151-160.
10. Colby DW, Zhang Q, Wang S et al. (2007) Prion detection by an amyloid seeding assay. Proc Natl Acad Sci U S A 104:20914-20919.
11. Wilham JM, Orru CD, Bessen RA et al. (2010) Rapid end-point quantitation of prion seeding activity with sensitivity comparable to bioassays. PLoS Pathog 6:e1001217.
12. Atarashi R, Satoh K, Sano K et al. (2011) Ultrasensitive human prion detection in cerebrospinal fluid by real-time quaking-induced conversion. Nat Med 17:175-178.
13. Peden AH, McGuire LI, Appleford NE et al. (2012) Sensitive and specific detection of sporadic Creutzfeldt-Jakob disease brain prion protein using real-time quaking-induced conversion. J Gen Virol 93:438-449.
14. Atarashi R, Sano K, Satoh K, Nishida N (2011) Real-time quaking-induced conversion: a highly sensitive assay for prion detection. Prion 5:150-153.
15. Sano K, Satoh K, Atarashi R et al. (2013) Early detection of abnormal prion protein in genetic human prion diseases now possible using real-time QUIC assay. PloS One 8:e54915.
16. Orru CD, Wilham JM, Raymond LD et al. (2011) Prion disease blood test using immunoprecipitation and improved quaking-induced conversion. MBio 2:e00078-00011.
17. Du D, Murray AN, Cohen E et al. (2011) A kinetic aggregation assay allowing selective and sensitive amyloid-beta quantification in cells and tissues. Biochemistry 50:1607-1617.
18. Gupta S, Jie S, Colby DW (2012) Protein misfolding detected early in pathogenesis of transgenic mouse model of Huntington disease using amyloid seeding assay.J Biol Chem 287:9982-9989.
19. Shaw LM, Korecka M, Clark CM et al. (2007) Biomarkers of neurodegeneration for diagnosis and monitoring therapeutics. Nat Rev Drug Discov 6:295-303.
20. Tomita T, Iwatsubo T (2004) The inhibition of gamma-secretase as a therapeutic approach to Alzheimer's disease. Drug News Perspect 17:321-325.
21. Meriin AB, Sherman MY (2005) Role of molecular chaperones in neurodegenerative disorders.Int J Hyperthermia 21:403-419.
22. Arosio P, Vendruscolo M, Dobson CM, Knowles TP (2014) Chemical kinetics for drug discovery to combat protein aggregation diseases. Trends Pharmacol Sci 35:127-135.

A Reliable Semi-automatic Program to Measure the Vertebral Rotation Using the Center of Lamina for Adolescent Idiopathic Scoliosis

W. Chen[1], E. Lou[1,2], and Lawrence H. Le[1,3]

[1] Department of Biomedical Engineering, University of Alberta, Edmonton, Canada
[2] Department of Surgery, University of Alberta, Edmonton, Canada
[3] Department of Radiology and Diagnostic Imaging, University of Alberta, Edmonton, Canada

Abstract— Adolescent idiopathic scoliosis (AIS) is a three-dimensional (3D) spinal deformity. Radiography is the standard method to diagnose AIS, but the 2D image cannot reveal the true nature of scoliosis. It also exposes patients to harmful ionizing radiation. Ultrasound (US) imaging has been reported as a non-radiation method to acquire 3D information of the spine. Vertebral rotation (VR) that can be assessed from US image is an important parameter to predict curve progression and to evaluate treatment outcomes. To improve the image quality and reduce human errors, a semi-automatic program was developed to measure the VR. The purpose of this study was to determine the intra-observer reliability and the accuracy of the measurements using the developed program. Three cadaver vertebrae, T7, L1, and L3, were scanned with 39 rotation configurations from -30° to 30° with 5° increment. The US images were randomly arranged to reduce memory bias. One observer performed the semi-automatic VR measurements twice with one-week interval. The semi-automatic VR measurements showed high intra-observer reliability (intraclass correlation coefficient (ICC): 0.999, mean absolute difference with standard deviation (MAD±SD): 0.2°±0.3°) and agreed well with the experimental setup (ICC: 0.956, MAD±SD: 2.5°±1.5°). The results demonstrated the developed program was an accurate and reliable measurement tool to assess the VR for *in-vitro* study.

Keywords— adolescent idiopathic scoliosis, center of lamina, semi-automatic program, vertebral rotation.

I. INTRODUCTION

Adolescent idiopathic scoliosis (AIS) is a three-dimensional (3D) spinal deformity with both coronal curvature and vertebral rotation. The Cobb angle [1] is the standard method to evaluate the spinal deformity. However, the 2D radiographic image does not reveal the true nature of scoliosis. The vertebral rotation (VR) is another important parameter that can be used to assess the severity of scoliosis, to indicate the risk of curve progression, and to evaluate the treatment outcomes [2-3].

Researchers have proposed different methods to measure the VR on radiographic images using the projected structures of the vertebra such as the spinous process and pedicles [1, 4-6]. However, the measurements on the radiographs are not directly performed on the transverse plane and the spinal deformity, especially the axial rotation, could change the size and orientation of the projected landmarks, which limit the accuracy of the VR measurements.

Few methods have been developed to assess the VR on the transverse images such as computed topography (CT) [7, 8] and magnetic resonance imaging (MRI) [9]. Although CT method could provide the details of the vertebra from the transverse view, it costs more time and money. Most importantly, it exposes patients to more radiation than the standard radiographs. Therefore, CT is not regularly employed to diagnose AIS in clinic. MRI is another imaging method that can show the details of the bone structure and does not introduce radiation. However it is costly and time consuming and therefore is not commonly used to diagnose AIS. In addition, both CT and the traditional MRI acquire images in supine positions, which significantly affect the VR measurements [10]. Therefore, VR is normally not measured in most of the scoliosis clinics.

Ultrasound (US) imaging has been reported as a non-radiation and cost-effective method to acquire 3D information of the spine. Suzuki [11] combined the ultrasound system with an inclinometer to measure the VR from scoliosis patients in a prone posture. This method relied on radiographs to determine the positions of the vertebrae and the measurements were performed on each vertebra in a prone position. However, the method has not been widely used. Recently, ultrasound studies [12, 13] were performed and the reports indicated that ultrasound was able to assess the coronal curvature and vertebral rotation of the AIS patients in an upright posture. In order to improve the image quality and reduce the human errors, a custom semi-automatic MATLAB program was developed to measure the VR. The intra-observer reliability and the accuracy of the semi-automatic VR measurements were investigated.

II. MATERIALS AND METHODS

A. Experimental Setup

Three cadaver vertebrae, T7, L1, and L3, were scanned using the TomoScan Focus LT phased array ultrasound equipment (Olympus NDT Inc., Canada). The acquisition system was previously described in [13]. The setup for the

© Springer International Publishing Switzerland 2015
V. Van Toi and T.H. Lien Phuong (eds.), *5th International Conference on Biomedical Engineering in Vietnam*,
IFMBE Proceedings 46, DOI: 10.1007/978-3-319-11776-8_39

vertebra phantom is as follows. A plastic platform (Fig. 1) had a plastic protractor glued on the surface. The vertebra was mounted on top of an adjustable plastic bar with a sharp pointer at the tip. The plastic bar was then secured on the platform with the pointer reading the angle of rotation from the projector. Each vertebra was rotated from -30° to 30° with 5° increment. Thirteen rotations were considered for each vertebra and therefore, a total of 39 configurations were obtained.

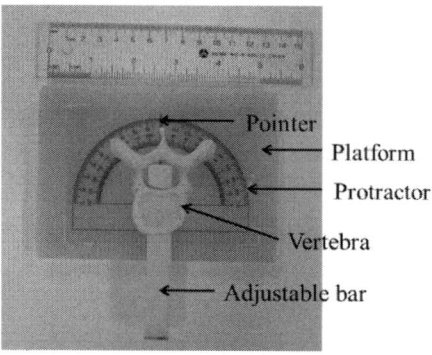

Fig. 1 The T7 vertebra phantom. The phantom includes the vertebra, the adjustable bar with a pointer, the protractor, and the platform.

The vertebra phantom was secured on the bottom of a water-filled container. The posterior side of the vertebra faced the scanning surface of the transducer. One operator scanned along the surface using a 5 MHz 128-element probe with a mini-wheel encoder, and the other one controlled the laptop to record the data. The experimental setup was similar to the setup described in [14].

B. Semi-automatic Measurement Procedure

The 3D ultrasound data including the signals from the whole vertebra was acquired for each configuration. As mentioned in the previous report [13], ultrasound signals were strongly reflected by the laminae and transverse processes (TP), which had relatively flat surfaces. The transverse image, which was acquired above the laminae, showed a 'W' shape including the laminae and transverse processes (TP). Therefore, the laminae, which were at the bottom of the 'W' shape, could be used to evaluate the VR.

The flow chart of the program is illustrated in Fig. 2. The image consisted of 3378×115 pixels. The sampling resolution was 0.0148 mm along depth. Since the sampling resolution was much smaller than the dimension of the vertebra, the data was down-sampled along the depth axis to improve the processing speed. We kept one data point for every 8 data points using the maximum intensity projection (MIP) method. Dual interpolation was performed along the lateral direction to smooth the image. After decimation and interpolation, the size of the transverse image, as shown in

Fig. 3, was 423×457. The laminae could be recognized as the red areas on the bottom of the 'W' shape.

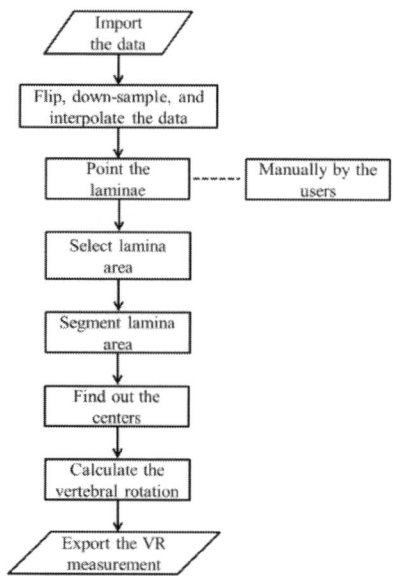

Fig. 2 The flow chart of the program.

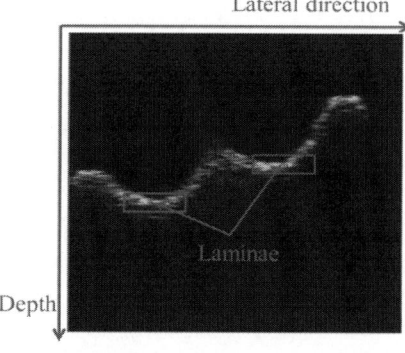

Fig. 3 The transverse image after flipping, down-sampling and interpolation. The boxes outline the regions of interest.

The user selected the laminae by using a mouse click. A region of interest (ROI) of 100×30 pixels (12.5mm×3.6mm) around the marked point was determined to ensure the laminae were covered (rectangular boxes in Fig. 3). Fig. 4a shows the magnified image of the left lamina inside the red box on Fig 3. Within the ROI, the intensities of the signals were first normalized and a median filter was applied to remove the sparse noise. A Laplacian filter, which was a second order derivative operator, was also applied to enhance the edge of the area [15]. Then the optimum global thresholding method by Otsu segmented the image into two parts [16]. The threshold value was determined by maximizing the between-class variance. The lamina is separated from the background shown in Fig. 4b. The blue area indicated the background signals below the threshold value, and the red area represented the objective signals above the threshold value. If there were more than one red area, the lamina was determined as the one including the marked point. Then the position of the geometrical center point of the lamina (red area) was calculated and displayed (Fig. 4c).

The center points of the left and right laminae were automatically connected by a line. The angle between the line and the reference horizontal line indicated the vertebral rotation (Fig. 5).

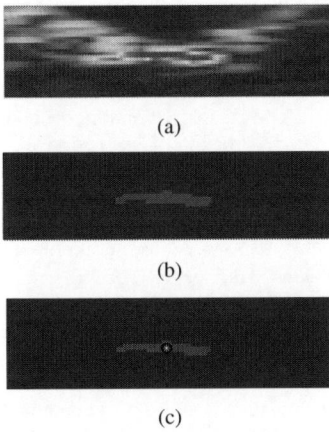

Fig. 4 (a) The image of the left lamina area, (b) the image after segmentation, and (c) the image with the marked center of the lamina.

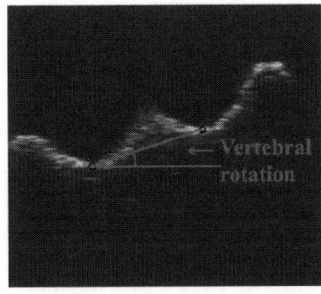

Fig. 5 Measurement of the vertebral rotation.

C. Analysis Method

One observer with 3-year ultrasound experience performed the measurements twice with one week interval to reduce the memory bias. The intraclass correlation coefficient (ICC) (2-way random and absolute agreement) [17] and mean absolute difference (MAD) with standard deviation (SD) were applied to analyze the intra-observer reliability and the agreement between the US measurements and the experimental setup. The Currier criteria for ICC

values were adopted: high reliability (0.90-1.0), good reliability (0.80-0.89), fair reliability (0.70-0.79), and poor reliability (0.69 and below) [18].

III. RESULTS

The intra-observer reliability of the semi-automatic VR measurements and the measurement comparisons between the US images and the phantoms are shown in Table 1. The ICC values (0.999 and 0.956) are large, indicating reliable measurements and significant measurement agreement between the US images and the phantoms. The MAD±SD (0.2°±0.3°) showed very small variations between the measurements from two sessions. The small difference (2.5°±1.5°) between the US measurements and experimental setup indicated the semi-automatic program provided accurate measurements.

Table 1 The results of the intra-observer reliability and the measurement comparisons between the US images and the phantoms.

	Intra-observer reliability	Comparisons between US images and phantoms
ICC	0.999	0.956
MAD±SD (°)	0.2±0.3	2.5±1.5

IV. DISCUSSIONS

The evaluation of VR measurements was reviewed and discussed on both the radiographic and CT methods [19-21]. Among these methods, the CT methods were more accurate than the radiographic methods. Ho's method measured from the CT transverse images had small standard deviation for both the intra- and inter-observer reliabilities (1.2°-3.3°) [21]. Another study [22] compared VR measurements using the CT methods with the experimental setup from 11 cadaver vertebrae. The results indicated that the Aaro's method was more accurate with a mean difference of 2.3°± 3.7° compared with the true values. The proposed COL method in this study was measured on the transverse images, which provided direct visualization of the axial rotation as the CT methods. Also, the results showed the semi-automatic measurements performed well with small variations.

The developed program made the US image smoother and the laminae could be easily recognized. Using this semi-automatic method also eliminates human errors. However, in this study, only an *in-vitro* experiment was carried out, more challenges are expected in *in-vivo* study. The soft tissue as an inhomogeneous medium will scatter US energy energies and incur speckle noise. The US signals will be attenuated when traveling through the soft tissue because the attenuation coefficient of soft tissue is much bigger than that of water. The quality of the US images would be reduced, which creates challenge to identify the laminae [13]. Therefore, prior to measurements, filtering and denoising strategies are called upon to enhance the signal-to-noise ratio of the US *in-vivo* images.

V. Conclusions

The results demonstrated the developed program provided an accurate and reliable measurement tool to assess the VR from the US phantom images. Further *in-vivo* study is necessary to confirm the performance of the program in clinical settings.

Acknowledgment

Thanks for the financial support from Chinese Scholarship Council, NSERC, WCHRI, and Alberta Health Services.

Conflict of Interest

The authors declare that they have no conflict of interest.

References

1. Cobb J (1948) Outline for the study of scoliosis. Instr Course Lect 5:261-275
2. Drerup B (1985) Improvements in measuring vertebral rotation from the projections of the pedicles. J Biomech 18:369-378
3. Kuklo TR, Potter BK, Lenke LG (2005) Vertebral rotation and thoracic torsion in adolescent idiopathic scoliosis: what is the best radiographic correlate? J Spinal Disord Tech 18:139-147. DOI 10.1097/01.bsd.0000159033.89623
4. Nash CL, Jr., Moe JH (1969) A study of vertebral rotation. J Bone Joint Surg Am 51:223-229
5. Perdriolle R, Vidal J (1985) Thoracic idiopathic scoliosis curve evaluation and prognosis. Spine 10:785-791
6. Stokes IA, Bigalow LC, Moreland MS (1986) Measurement of axial rotation of vertebrae in scoliosis. Spine 11:213-218
7. Aaro S, Dahlborn M (1981) Estimation of vertebral rotation and the spinal and rib cage deformity in scoliosis by computer tomography. Spine 6:460-467
8. Ho EK, Upadhyay SS, Chan FL, Hsu LC, Leong JC (1993) New methods of measuring vertebral rotation from computed tomographic scans. An intraobserver and interobserver study on girls with scoliosis. Spine 18:1173-1177
9. Birchall D, Hughes DG, Hindle J, Robinson L, Williamson JB (1997) Measurement of vertebral rotation in adolescent idiopathic scoliosis using three-dimensional magnetic resonance imaging. Spine 22:2403-2407. DOI 10.1097/00007632-199710150-00016
10. Yazici M, Acaroglu ER, Alanay A, Deviren V, Cila A, Surat A (2001) Measurement of vertebral rotation in standing versus supine position in adolescent idiopathic scoliosis. J Pediatr Orthop 21:252-256
11. Suzuki S, Yamamuro T, Shikata J, Shimizu K, Iida H (1989) Ultrasound Measurement of Vertebral Rotation in Idiopathic Scoliosis. J Bone Joint Surg Br 71:252-255
12. Chen W, Lou EHM, Le LH (2011) Using ultrasound imaging to identify landmarks in vertebra models to assess spinal deformity. Conf Proc IEEE Eng Med Biol Soc, Boston, USA, 2011, pp 8495-8498
13. Chen W, Le LH, Lou EHM (2012) Ultrasound imaging of spinal vertebrae to study scoliosis. Open J Acoustics 2:95-103. DOI 10.4236/oja.2012.23011
14. Chen W, Lou EHM, Zhang PQ, Le LH, Hill D (2013) Reliability of assessing the coronal curvature of children with scoliosis by using ultrasound images. J Child Orthop 7:521-529. DOI 10.1007/s11832-013-0539-y
15. Gonzalez R, Woods R, Eddins S (2009) Digital image processing using MATLAB, 2nd edn., Gatesmark, LLC, Natick
16. Otsu N (1979) A Threshold Selection Method from Gray-Level Histograms. Systems, Man and Cybernetics, IEEE Transactions on 9:62-66. DOI 10.1109/tsmc.1979.4310076
17. Shrout PE, Fleiss JL (1979) Intraclass correlations: uses in assessing rater reliability. Psychol Bull 86:420-428
18. Currier DP (1990) Elements of research in physical therapy, 3rd ed. Williams & Wilkins, Baltimore
19. Gocen S, Aksu MG, Baktiroglu L, Ozcan O (1998) Evaluation of computed tomographic methods to measure vertebral rotation in adolescent idiopathic scoliosis: an intraobserver and interobserver analysis. J Spinal Disord 11:210-214
20. Lam GC, Hill DL, Le LH, Raso JV, Lou EHM (2008) Vertebral rotation measurement: a summary and comparison of common radiographic and CT methods. Scoliosis 3:16. DOI 10.1186/1748-7161-3-16
21. Vrtovec T, Pernus F, Likar B (2009) A review of methods for quantitative evaluation of axial vertebral rotation. Eur Spine J 18:1079-1090. DOI 10.1007/s00586-009-0914-z
22. Krismer M, Chen AM, Steinlechner M, Haid C, Lener M, Wimmer C (1999) Measurement of vertebral rotation: a comparison of two methods based on CT scans. J Spinal Disord 12:126-130

Address of the corresponding author:

Author: Edmond H. M. Lou
Institute: University of Alberta
Street: 10105 - 112 Ave
City: Edmonton
Country: Canada
Email: elou@ualberta.ca

Investigation of the Optimal Freehand Three-Dimensional Ultrasound Configuration to Image Scoliosis: An In-vitro Study

Q.N. Vo[1,2], E. Lou[1,3], Lawrence H. Le[1,4], and L.Q. Huynh[2]

[1] Department of Biomedical Engineering, University of Alberta, Edmonton, Alberta, Canada
[2] Department of Biomedical Engineering, Ho Chi Minh City University of Technology, Ho Chi Minh City, Vietnam
[3] Department of Surgery, University of Alberta, Edmonton, Alberta, Canada
[4] Department of Radiology and Diagnostic Imaging, University of Alberta, Edmonton, Alberta, Canada

Abstract— Scoliosis is a three-dimensional (3D) spinal deformity. 3D ultrasound has been used to image scoliotic spines. This in-vitro study was to investigate the optimal 3D ultrasound parameters by comparing the reconstructed images and the phantom.

A medical ultrasound system, a convex and a linear ultrasound probes, both with built-in positioning sensors, were used to scan cadaveric vertebra T7 immersed in a water tank. The operating frequencies were set at 2.5 MHz, 3.3MHz, 4.0 MHz, 6.6 MHz, and 10.0 MHz. The voxel-based method was deployed for the 3D reconstruction. The minimum distances (0.1, 0.2, and 0.3 mm) between two adjacent B-scan images and the reconstruction resolutions (0.2, 0.6, and 1.0 mm) were the inputs for the reconstruction algorithm. A total of 45 configurations were investigated. Four distance parameters were measured three times on both the images and the cadaveric vertebra by one rater in one week apart to minimize memory bias. The mean absolute difference was the distance measurement difference between the images and phantom. The paired Student's t-test was used to determine the probability between the two populations. The determination of the optimal configuration was based on both accuracy and intuitive image quality.

The results showed that the convex probe with the configuration (0.2-mm minimum distance, 0.6-mm reconstruction resolution, and 4.0 MHz) provided the best reconstructed image.

Keywords— scoliosis, freehand ultrasound, mean absolute difference, voxel-based method.

I. INTRODUCTION

Scoliosis is a complicated spinal deformity associated with vertebral rotation and lateral deviation with a Cobb angle of 10° or more [1, 2]. Among different types of scoliosis, adolescent idiopathic scoliosis (AIS) is the most common type, which accounts for 80% of the cases and affects 1.5 – 3% of the population [3]. The treatment of idiopathic scoliosis depends on the severity of scoliosis and the probability of progression. Currently, the Cobb method is the gold standard recommended by the Scoliosis Research Society to assess the severity of scoliosis on posteroanterior (PA) radiographs. However, this method may underestimate the 3D severity because it only reflects the spinal curvature on the coronal plane.

As technology becomes more advanced, 3D images should be used to extract the 3D spinal deformity information. Both computed tomography (CT) and magnetic resonance imaging (MRI) can image and display 3D bone structures clearly; however, the CT modality exposes patients to more radiation than traditional X-ray imaging and requires patients in supine position, which alters the spinal curvature [4-6]. In addition, using MRI to image spine is costly and time consuming; thus, its use for imaging scoliosis is limited.

Three-dimensional images of spine can also be obtained from multi-planar standing radiography but ionizing radiation is still a concern. The EOS (EOS Imaging, France) is a medical imaging system that can perform the simultaneous acquisition of the PA and lateral radiographs in a standing position. This system generates better images and reduces radiation dosage when compared with standard radiography and CT [7, 8]. Even though the EOS exposes patients to much less radiation, the issue of radiation safety still attracts the attention of the community. In general, during the treatment or observation periods, AIS patients may take radiographs every four to twelve months; the cumulative amount of ionizing radiation may still increase the risk of developing cancer [8-12], especially for females when their incidence of scoliosis is about 80% of all cases [13].

In 2007, Purnama et al. [14] introduced a framework for human spine imaging using a freehand 3D ultrasound system with a 5 MHz curved array probe and an optical tracking system. The isotropic voxel size of 0.21 mm was used for the reconstruction. The axial vertebral rotation and the vertebral tilt of each vertebra are calculated based on the centers of mass of the transverse processes which are determined manually and semi-automatically for a comparison of accuracy. Ultrasound data from a healthy volunteer was used to build the 3D spinal model, which was then validated by 5 data sets from scoliotic patients [15]. The validation was not fully proved. Cheung et al. [16] also designed a system with a freehand 3D ultrasound system, a

© Springer International Publishing Switzerland 2015
V. Van Toi and T.H. Lien Phuong (eds.), *5th International Conference on Biomedical Engineering in Vietnam*,
IFMBE Proceedings 46, DOI: 10.1007/978-3-319-11776-8_40

wide linear probe (L53L/10-5), and an electromagnetic spatial sensing device for scoliosis applications. An in-vitro study was performed and the accuracy of the Cobb angle value was reported to be within 2 – 5° between X-ray and ultrasound measurements.

3D ultrasound has been used to image scoliotic spines, mostly on phantom and volunteers. Researchers used different ultrasound configurations including the type of transducer, the operating frequency, and the reconstruction resolution (the voxel size) to acquire images, which makes the image quality vary. This in-vitro study was to investigate the optimal 3D ultrasound parameters by comparing the reconstructed images and the phantom.

II. MATERIALS AND METHODS

A. Materials

The medical ultrasound system Ultrasonix SonixTablet (Ultrasonix Ltd., Canada), a built-in magnetic global positioning system (GPS) (SonixGPS), and a 3D Guidance device (driveBAY, Ascension Ltd., USA) (Fig. 1) were used in this study. In addition, two types of GPS-assisted transducer, which were compatible with the ultrasound equipment, were used: a flat linear transducer and a convex transducer. The flat linear transducer can operate at the ultrasound frequency range between 5 to 14 MHz, the depth range between 2 and 9 cm, and the active scanning area of 38 mm x 9 mm. Meanwhile, these parameters for the convex transducer are 2 – 5 MHz, 5 - 30 cm, and 60 mm x 15 mm, respectively. The spatial sensor (GPS receiver) embedded inside the transducer interacts with the GPS transmitter via the 3D Guidance to provide 3D orientation information of each acquired B-scan ultrasound image relative to the location of the transmitter.

B. Methods

A cadaveric vertebra T7 was scanned when immersed in a water tank. The tank was made of 4.5-mm thick acrylic to simulate the body setting. The vertebra was set with its posterior arch facing the surface of the transducer and without any vertebral rotation. The GPS transmitter was set up within the working range of 5 – 20 cm relative to the GPS receiver recommended by the manufacturer. Within this range, the GPS accuracy is ± 1.0 mm.

The operating frequencies were set at 2.5 MHz, 3.3MHz, and 4.0 MHz for the convex probe and 6.6 MHz and 10.0 MHz for the linear probe, respectively. Each frequency was used in turn to scan T7 for data acquisition. The depth was set at 9 cm and the acquisition frame rate per second was automatically set to 27, 36, 36, 8, and 10 for five frequencies, respectively. The scanning time was approximate 23 seconds for the convex probe and 46 seconds for the linear one. After the scanning process, a series of B-scans was acquired, including two data sets: intensity data (*.b8*) and GPS data (*.gps*). These two data sets were saved in the ultrasound system and then imported into an in-house developed program, which deployed the voxel-based method for the 3D reconstruction of T7.

In order to eliminate the overlap among acquired B-scans, the minimum distance (MD) between two adjacent B-scans (0.1, 0.2, and 0.3 mm) was set. If the distance between two adjacent B-scans was less than the preset minimum distance, one of the two B-scans was eliminated. In addition, the resolution of the reconstructed volume (RR) (the isotropic voxel size) (0.2, 0.6, and 1.0 mm) was also determined. The MD and RR were the inputs for the reconstruction algorithm. A total of 45 configurations (3 MD x 3 RR x 5 frequencies) were investigated. Four distance parameters (Fig. 2) were measured three times on both the images and the cadaver by one rater in one week apart to minimize memory bias. These measurements were used to evaluate the accuracy of the 3D reconstruction. The mean absolute difference (MAD) was the distance measurement difference between the images and phantom. The paired Student's t-test was used to determine the probability between the two populations (p-values). It was assumed that the two groups have unequal variance and two-tailed t-test was performed. The determination of the optimal configuration was based on a comparison among 45 configurations in terms of both accuracy and intuitive image quality.

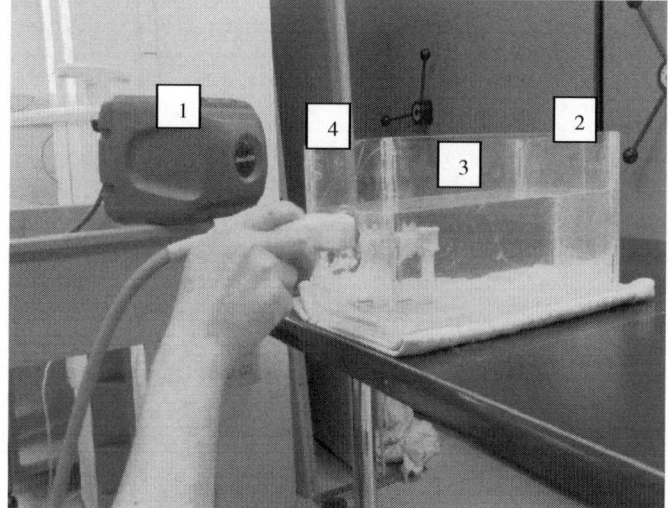

Fig. 1 The experimental setup: (1) GPS transmitter, (2) water tank, (3) cadaveric vertebra T7, and (4) GPS receiver embedded transducer.

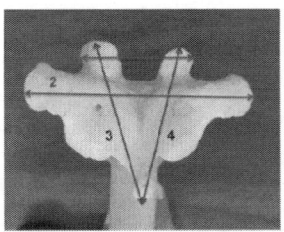

Fig. 2 The four distance parameters on the cadaveric vertebra T7: 1. superior articular process – superior articular process, 2. transverse process – transverse process, 3. Left superior articular process – spinous process, 4. right superior articular process – spinous process.

III. RESULTS

Table 1 tabulates the MAD and probability values of the 45 configurations. The best five reconstructed images based on both accuracy and intuitive image quality for each frequency are provided by the following 5 configurations (MD/RR mm, frequency MHz): (0.1/1.0, 2.5), (0.1/0.6, 3.3), (0.2/0.6, 4.0), (0.1/1.0, 6.6) and (0.1/0.6, 10.0) and shown in Fig. 3. The corresponding MAD (mm)/p-values of the measurements were $1.5\pm0.4/0.84$, $0.9\pm0.6/0.89$, $1.3\pm0.3/0.86$, $1.5\pm0.3/0.90$, and $1.7\pm0.5/0.97$ respectively. In addition, the ranges of difference between the true values and the average measurements were $0.2 - 2.4$ mm, $0 - 2.8$ mm, $0.3 - 1.8$ mm, $0.6 - 3.1$ mm, and $0.4 - 2.6$ mm for the four distances, respectively. Among the five aforementioned configurations, the best reconstructed image is given by the (0.2/0.6, 4.0) configuration.

IV. DISCUSSIONS

In this study, the probability p-values derived from the five best images are greater than 0.84, indicating a high correlation between measurements of the cadaver and those of the reconstructed images. The GPS accuracy was within ±1 mm and the MAD is not more than 1.7 mm. In order to increase the accuracy, the GPS system needs to be calibrated prior to data acquisition. Further measurements such as proxy Cobb angle and vertebral rotation will be performed on the 3D reconstructed model; therefore, the higher the accuracy, the better it is. Furthermore, the convex probe offered better images than the linear one. It was because the frame rate of the former was basically greater than that of the latter while keeping the same penetration depth and frame rate at "high" setting, resulting in a higher number of frames available for the 3D reconstruction process. Even though the scanning time of the linear probe doubled that of the convex one in order to compensate for its lower frame rates, the number of frames acquired by the linear probe was still lower. Technically, the reconstruction algorithm required an intensity value of the interpolated voxel, which was proportional to the intensities of the two nearest pixels of the two nearest frames from the voxel and inversely proportional to the distances between the voxel and the two neighboring frames. If the two frames were located far from the voxel due to the low number of acquired frames, the voxel's interpolated value was small and thus ignored, leaving a 'blank' (zero value) at the voxel's position. Increasing the scanning time is impossible because this is not reasonable in reality.

There are still some limitations due to the inherent characteristics of ultrasound. This method cannot image the vertebral body due to the acquisition configuration and the lack of ultrasound energy penetrating through bone.

V. CONCLUSIONS

The configuration with MD/RR being 0.2/0.6 mm and 4.0 MHz displayed the best image based on the in-vitro experiment. This configuration should be verified by the in-vivo data. This is an objective for our future research.

ACKNOWLEDGMENT

The authors would like to acknowledge Vietnam International Education Development and Edmonton Orthopedic Research Committee for their financial supports.

0.1/1.0, 2.5 MHz   0.1/0.6, 3.3 MHz   0.2/0.6, 4.0 MHz   0.1/1.0, 6.6 MHz   0.1/0.6, 10.0 MHz

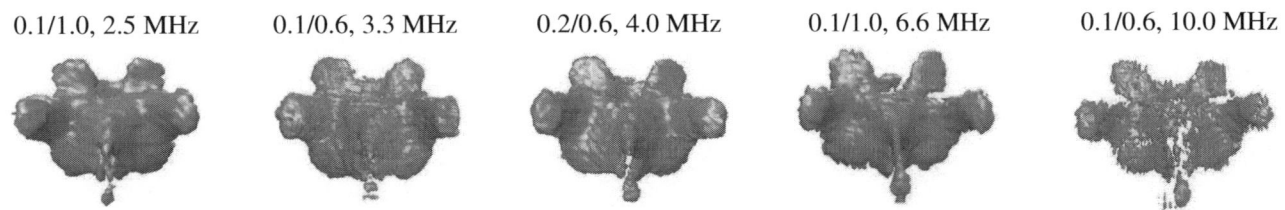

Fig. 3 The five best reconstructed images and the corresponding configurations (MD/RR mm, frequency).

Table 1 The MAD and p-values for the 45 configurations.

Frequency (MHz)		Resolutions (mm/mm)								
		0.1/0.2	0.1/0.6	0.1/1.0	0.2/0.2	0.2/0.6	0.2/1.0	0.3/0.2	0.3/0.6	0.3/1.0
2.5	MAD (mm)	2.0 ± 0.5	1.8 ± 0.4	1.5 ± 0.4	2.2 ± 0.5	1.8 ± 0.5	1.5 ± 0.4	2.1 ± 0.6	2.0 ± 0.5	1.5 ± 0.4
	p-value	0.78	0.82	0.84	0.78	0.81	0.83	0.77	0.79	0.84
3.3	MAD (mm)	1.2 ± 0.5	0.9 ± 0.6	1.2 ± 0.4	1.5 ± 0.5	1.2 ± 0.5	1.3 ± 0.5	1.3 ± 0.5	1.3 ± 0.4	1.1 ± 0.4
	p-value	0.86	0.89	0.91	0.82	0.87	0.85	0.85	0.85	0.90
4.0	MAD (mm)	2.1 ± 0.6	1.8 ± 0.5	1.9 ± 0.6	1.8 ± 0.6	1.3 ± 0.3	2.0 ± 0.6	1.7 ± 0.4	1.5 ± 0.4	1.9 ± 0.5
	p-value	0.76	0.80	0.79	0.79	0.86	0.78	0.83	0.84	0.79
6.6	MAD (mm)	2.1 ± 0.4	2.1 ± 0.6	1.5 ± 0.3	2.0 ± 0.5	1.9 ± 0.5	1.6 ± 0.3	2.3 ± 0.4	2.1 ± 0.4	2.1 ± 0.6
	p-value	0.85	0.94	0.90	0.94	0.90	0.89	0.83	0.89	0.83
10.0	MAD (mm)	1.8 ± 0.5	1.7 ± 0.5	2.1 ± 0.5	1.9 ± 0.4	1.9 ± 0.5	2.2 ± 0.5	1.9 ± 0.5	1.7 ± 0.5	1.7 ± 0.5
	p-value	0.91	0.97	0.95	0.89	0.92	0.94	0.89	0.97	0.96

REFERENCES

1. Roaf R (1958) Rotation movements of the spine with special reference to scoliosis. J Bone Joint Surg Br 40-B: 312-32
2. Graf H, Hecquet J, and Dubousset J (1983) 3-dimensional approach to spinal deformities. Application to the study of the prognosis of pediatric scoliosis. Rev Chir Orthop Reparatrice Appar Mot 69: 407-16
3. Lonstein JE (1994) Adolescent Idiopathic Scoliosis. Lancet 344:1407-1412
4. Torell G, Nachemson A, Haderspeck-Grib K, and Schultz A (1985) Standing and supine Cobb measures in girls with idiopathic scoliosis. Spine (Phila Pa 1976) 10: 425-427
5. Yazici M, Acaroglu ER, Alanay A et al (2001) Measurement of vertebral rotation in standing versus supine position in adolescent idiopathic scoliosis. J Pediatr Orthop 21: 252-256
6. Forsberg D, Lundstrom C, Andersson M et al (2013) Fully automatic measurements of axial vertebral rotation for assessment of spinal deformity in idiopathic scoliosis. Phys Med Biol 58: 1775-1787
7. Deschenes S, Charron G, Beaudoin G et al (2010) Diagnostic imaging of spinal deformities: Reducing patients radiation dose with a new slot-scanning X-ray imager. Spine 35: 989-994
8. McKenna C, Wade R, Faria R et al (2012) EOS 2D/3D X-ray imaging system: a systematic review and economic evaluation. Health Technol Assess 16
9. Miller NH (1999) Cause and natural history of adolescent idiopathic scoliosis. Orthop Clin North Am 30: 343-352
10. Doody MM, Lonstein JE, Stovall M et al (2000) Breast cancer mortality after diagnostic radiography: Findings from the U.S. scoliosis cohort study. Spine 25: 2052-2063
11. Ronckers CM, Doody MM, Lonstein JE et al (2008) Multiple diagnostic Xrays for spine deformities and risk of breast cancer. Cancer Epidemiol Biomarkers Prev 17: 605-613
12. Wade R, Yang H, McKenna C et al (2013) A systematic review of the clinical effectiveness of EOS 2D/3D X-ray imaging system. Eur Spine J 22: 296-304
13. Bunnell WP (1993) Outcome of spinal screening. Spine (Phila Pa 1976) 18: 1572-1580
14. Purnama KE, Wilkinson MH, Veldhuizen AG et al (2007) Ultrasound imaging for human spine: imaging and analysis. J CARS 2:S114-S116
15. Dewi DEO, Wilkinson MHF, Mengko TLR et al (2009) 3D ultrasound reconstruction of spinal images using an improved Olympic Hole-filling method, Proc., 2009 International Conference on Instrumentation, Communications, Information Technology, and Biomedical Engineering (ICICI-BME), Bandung, Indonesia, 2009, pp 1-5
16. Cheung CWJ, Zheng Y (2010) Development of 3-D Ultrasound System for Assessment of Adolescent Idiopathic Scoliosis (AIS), Proc., IFMBE 6th World Congress of Biomechanics (WCB 2010), Singapore, 2010, pp 584-587

The address of the corresponding author:

Author: Dr. Edmond H. M. Lou
Institute: University of Alberta
Street: 10105 – 112 Avenue
City: Edmonton
Country: Canada
Email: elou@ualberta.ca

Assessment of Curve Flexibility by Ultrasonic Imaging – A Pilot Study

R. Zheng[1], E. Lou[1,2], Lawrence H. Le[3], D. Hedden[1], J. Mahood[1], and M. Moreau[1]

[1] Department of Surgery, University of Alberta, Edmonton, Canada
[2] Glenrose Rehabilitation Hospital, Alberta Health Services, Edmonton, Canada
[3] Department of Radiology and Diagnostic Imaging, University of Alberta, Edmonton, Canada

Abstract— **Scoliosis is a three-dimensional deformity of spine and surgical treatment is recommended when the spinal curvature is severe. Supine bending radiographs are usually taken prior to scoliosis surgery to measure the curve flexibility. However the ionizing radiation is a big concern to the patients and their families. The objective of this study was to determine the accuracy of the curve flexibility measurements on the ultrasound images. In this study, two surgical cases on children who have adolescent idiopathic scoliosis were reported. Both radiographs and ultrasound scans were acquired in standing and laying down bending positions. A total of 4 measurements were compared between the ultrasonic and radiographic methods. The average difference and standard deviation of the curve flexibility measurements (correction angles) between the two methods was 3.3±0.8°. This result demonstrated that the curve flexibility estimated by the US method was comparable to the results from radiography. However, since the number of participants is small, more trials are required before making a definite conclusion.**

Keywords— **Ultrasonic imaging, Curve flexibility, Curve correction, AIS, Bending radiograph.**

I. INTRODUCTION

Adolescent idiopathic scoliosis (AIS) is a three-dimensional spinal deformity which is recognized by the lateral curvature of the spine and vertebral rotation. When the spinal curvature is severe, surgical treatment is the only option that can be offered to the patients by the orthopedic surgeons. The flexibility of the major and compensatory curves is a crucial factor to estimate the surgical treatment outcomes. The flexibility provides the information of the structural changes, helps orthopedic surgeons determine the levels that need to be instrumented and the amount of correction that can be achieved safely [1,2]. Currently, voluntarily maximal side-bending supine radiographs are the most commonly used methods to determine the curve flexibility in scoliosis clinics [2–5]. However, the ionizing radiation especially in growing children is a big concern to the patients and their families. Therefore, alternative methods have been sought to reduce or overcome the potential hazardous radiation dose caused by X-ray.

In recent years, ultrasonic (US) imaging method has been applied to both *in-vitro* and *in-vivo* studies to measure parameters and features in children with AIS, and the results showed significant consistency in comparison to the other methods including conventional radiography [6–9]. In a spine phantom study using a free hand 3D US imaging system, a strong correlation ($R^2 = 0.76$) on the spinal curvature measurements between the X-ray and US images was obtained. The intra- and inter-observer correlations are 0.99 and 0.89 respectively [6]. In an *in-vitro* study by Chen *et al.* [7], they reported that the lamina and spinous process were the strong ultrasound reflectors on each vertebra. The dimensional measurement error between the ultrasound image and the phantom was less than 4%. Instead of using spinous process and lamina, Ungi *et al.* [8] used transverse processes as vertebral landmarks to measure spinal curvature on the adult and pediatric scoliotic phantoms. Their reported MAD (mean absolute difference) and SD (standard deviation) between the US images and the radiographs were 1.27±0.84° and 0.96±0.87° respectively. Chen *et al.* [9] also did a similar *in-vivo* study on five AIS subjects and found a small 0.7±0.5° MAD. These studies illustrated that the US imaging method had the ability to accurately measure spinal curvatures of AIS, and the results were comparable to the conventional radiographic method. However all these studies were performed on patients who had mild to moderate spinal deformity, surgical candidates who are diagnosed with AIS have not been studied by using the US method.

The objectives of this study were to investigate the feasibility of using US imaging method to measure the curve flexibility on surgical candidates with AIS and to assess the accuracy of the US measurements.

II. METHODS

A. Clinical Subjects

Two surgical candidates (female, age: 16.3 and 15.9 years old) were recruited from the local scoliosis clinic. The inclusion criteria of the participants are patients: (1) diagnosed with AIS and (2) planned surgical intervention without prior surgical treatment. Ethics approval was granted from the local health ethics board. Both participants signed the consent forms before being enrolled into the study.

© Springer International Publishing Switzerland 2015
V. Van Toi and T.H. Lien Phuong (eds.), *5th International Conference on Biomedical Engineering in Vietnam,*
IFMBE Proceedings 46, DOI: 10.1007/978-3-319-11776-8_41

B. Data Acquisition

The day before surgery, three pre-op radiographs were acquired at three positions: standing, supine left bending, and supine right bending. The subjects were then directed to the research lab for the ultrasound scan within an hour. Three pre-op ultrasound scans were also performed in standing, prone left bending, and prone right bending positions. For the standing radiography, the subjects stood on a positioning chariot for the examination. Whereas they were free-standing during the ultrasound examination as shown in Fig. 1. To perform the bending radiography or US scan, the pelvis location was identified by the operators. The candidates were then asked to bend their upper body to their left or right direction for maximal correction without moving their pelvis.

All ultrasound scans were acquired by the same operator, who had more than 8-year experience in ultrasound research. The SonixTABLET system (Ultrasonix, BC Canada), equipped with 128-element C5-2/60 GPS transducer and the SonixGPS system, was used (Fig.1). The built-in GPS system can track and record the position and orientation of the ultrasound transducer relative to the transmitter for further image processing and analysis. The scanning frequency was 2.5 MHz and the imaging depth was set at 6 cm. The integral gain was tuned to 10% with linear time gain compensation (TGC) applied. The full spine ranging from vertebra level C7 to L5 was scanned, and the scan took less than 1 minute. After scanning, the data were merged and exported to a 3D image file using a custom in-house program. The duration to export the image data was approximately 3 to 5 minutes for the standing and bending images.

The post-op radiographs in standing position were also acquired one week after the surgery. All the pre-op and post-op radiographic images were also exported in JPEG format for further analysis.

C. Data Measurement

A rater, who was blinded to the clinical information, measured the US and radiographic images with 3 days apart. Both the US and radiographic images were measured using the same in-house software. The angles and vertebral levels of the spinal curvatures were first determined on the pre-op standing images, and then the corrected angles were measured on the pre-op bending images or post-op standing radiographic images based on the same vertebral levels.

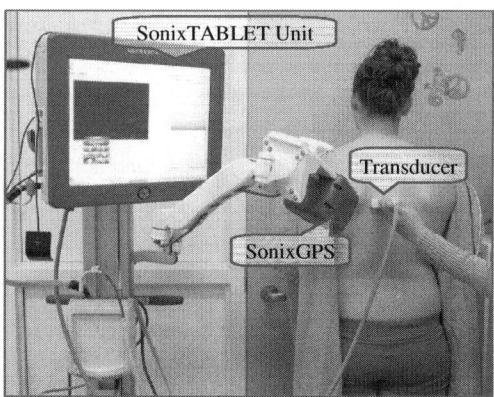

Fig. 1 The subject was scanned in standing position using SonixTABLET system (Ultrasonix, BC Canada).

To measure the spinal curvatures on the standing radiographs, the Cobb method [10] was applied, in which two lines are drawn along the end plates of the most tilted vertebra at the top and bottom end of the spinal curve. The angle between the two lines was measured and reported (the red lines in Fig. 2D). To measure the spinal curvature on the US images, the center of lamina (COL) method [7,11] was used. The COL method was started by identifying the most tilted end vertebra at the top. The centers of the left and right laminae of the selected vertebra were then highlighted by clicking on the US coronal images. Since the acquired US data was volumetric (3D) by nature, the transverse view of the measured vertebra was automatically displayed. The rater then used the transverse view to fine tune the location of the centers of laminae. A line was automatically drawn between the left and right COL. The raters would then select the bottom end vertebra and the same procedures were repeated. The angle formed by the top and bottom lines was defined as the proxy Cobb angle, which was equivalent to the spinal curvature angle from the radiograph.

Fig.2 shows the US and radiographic images from one subject. The red lines indicate the vertebral levels which were used to measure the standing or corrected Cobb angle. The corrected curve was only measured on the bending image which has the same direction as the curve, i.e., the left curve was measured on the left bending image or vice versa.

D. Data Analysis

In this study, curve flexibility was defined as the amount of curve correction (in degrees). For the pre-op radiographic or US configurations, the correction is the difference of the curve angles measured on the standing images and bending images. For the post-op radiographic configuration, the correction is the difference of the curve angles measured on

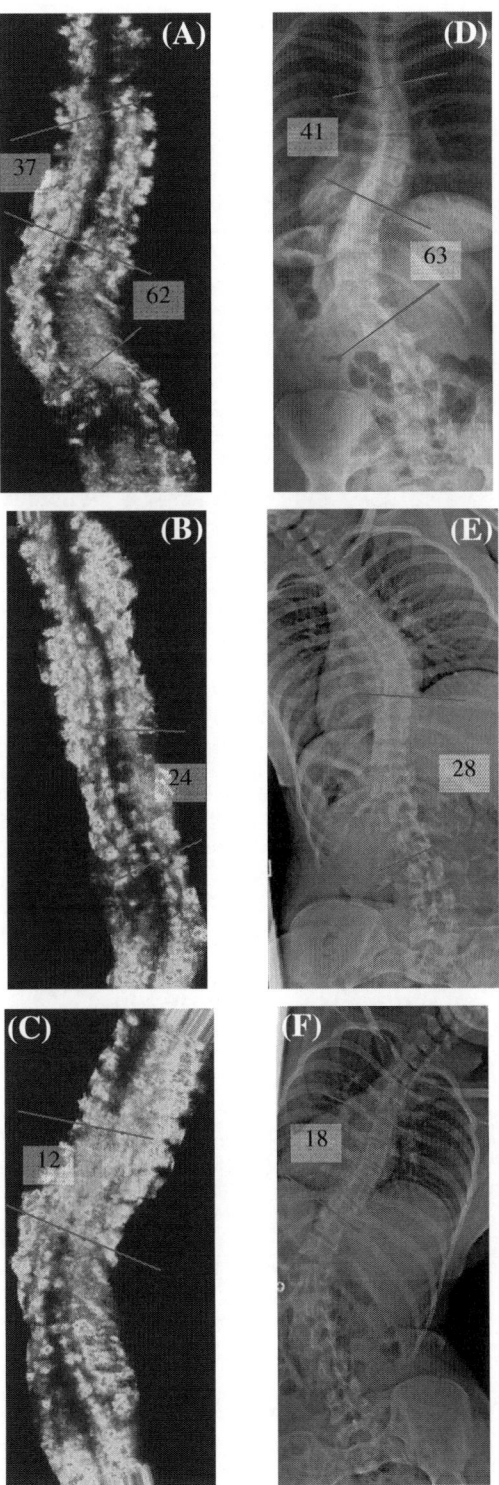

Fig. 2 The bending images of a subject: US images in (A) standing, (B) prone left, and (C) prone right positions; X-ray radiography in (D) standing, (E) supine left, and (F) supine right positions.

the pre-op and post-op standing images. The correction angles measured from the bending radiographs, bending US, and post-op radiographs were compared, and average difference and SD of the curve flexibility among the pre-op radiographic and US configurations were calculated and reported.

III. RESULTS AND DISCUSSIONS

Since both candidates had double curves, a total of 4 curves including 2 thoracic, 1 thoracolumbar, and 1 lumbar curves, were identified. The standing spinal curvatures measured from the pre-op standing radiographs ranged from 27° to 63° with an average 42.3±13.1°. Their major curves were 63° and 38°, respectively.

Table 1 listed the measurement results of the four curves and the corrections in different imaging configurations. The flexibility of all four curves using the pre-op radiographic and US configurations were calculated and compared (as shown in "Correction" column). The average curve corrections from the radiographic and US measurements were 28.3±5.0° and 31.5±5.2°, respectively. The pre-op bending corrections from the US method were larger than those of the radiographic method. The average difference of the flexibility (correction angles) between the two methods was 3.3±0.8°.

The corrected curves of two major curves measured on pre-op radiographs, pre-op US, and post-op radiographs for the two subjects (S1/S2) were 28°/7°, 24°/5° and 25°/8°, respectively. The results are comparable among three configurations.

There are several possible reasons which can cause the measurement difference between the radiographic and US methods. Firstly, the two methods applied different bending postures during the image acquisition. The radiography used the supine posture, while the US used the prone posture. The different postures can influence body flexibility of the candidates. Secondly, the US scan and the radiographs were not acquired simultaneously, but were taken within one hour apart and in different locations. The side-bending imaging is greatly dependent on candidate cooperation, and it is very difficult to maintain the same posture during the two acquisitions. On the other hand, different technicians for radiography and US imaging may give different instructions to the candidate regarding posture and positioning. Lastly, the radiographic and the US methods used different landmarks to measure the spinal curvatures. Although the COL method is highly correlated to the Cobb method [7,11], identifying the center of lamina in the COL method was less accurate than the identification of the end-plate of the vertebra.

Table 1 The curvature information of the two subjects measured by the radiographic and ultrasound methods

Subject	Curve information		Measurement method	Curvature angle		Correction[1] (°)
	Type	Direction		Standing (°)	Bending (°)	
1	Main Thoracic	Right	Pre-op radiograph	41	18	23
			Pre-op US	37	12	25
			Post-op radiograph	21	-	-
	Thoracolumbar	Left	Pre-op radiograph	63	28	35
			Pre-op US	62	24	38
			Post-op radiograph	25	-	-
2	Main Thoracic	Right	Pre-op radiograph	38	7	31
			Pre-op US	40	5	35
			Post-op radiograph	8	-	-
	Thoracolumbar	Left	Pre-op radiograph	27	3	24
			Pre-op US	30	2	28
			Post-op radiograph	15	-	-

[1] *The correction is the angle difference between curves measured on pre-op standing and bending images.*

IV. CONCLUSIONS

The ultrasound imaging method has the potential to estimate the curve flexibility for the AIS surgical candidates. Since only 2 subjects were recruited in this study, definite conclusions cannot be made. More data is required to investigate the research questions on the validation and reliability of using US to measure curve flexibility.

ACKNOWLEDGMENT

This work was supported by the Edmonton Civic Employee Research Grant.

REFERENCES

1. Kleinman RG, Csongradi JJ, Rinksy LA, Bleck EE. The radiographic assessment of spinal flexibility in scoliosis: a study of the efficacy of the prone push film. Clin Orthop. 1982 Feb;(162):47–53.
2. Hamzaoglu A, Talu U, Tezer M, Mirzanli C, Domanic U, Goksan SB. Assessment of curve flexibility in adolescent idiopathic scoliosis. Spine. 2005 Jul 15;30(14):1637–42.
3. Liu RW, Teng AL, Armstrong DG, Poe-Kochert C, Son-Hing JP, Thompson GH. Comparison of supine bending, push-prone, and traction under general anesthesia radiographs in predicting curve flexibility and postoperative correction in adolescent idiopathic scoliosis. Spine. 2010 Feb 15;35(4):416–22.
4. Klepps SJ, Lenke LG, Bridwell KH, Bassett GS, Whorton J. Prospective comparison of flexibility radiographs in adolescent idiopathic scoliosis. Spine. 2001 Mar 1;26(5):E74–79.
5. Cheung KMC, Natarajan D, Samartzis D, Wong Y-W, Cheung W-Y, Luk KDK. Predictability of the fulcrum bending radiograph in scoliosis correction with alternate-level pedicle screw fixation. J Bone Joint Surg Am. 2010 Jan;92(1):169–76.
6. Cheung C-WJ, Law S-Y, Zheng Y-P. Development of 3-D ultrasound system for assessment of adolescent idiopathic scoliosis (AIS): and system validation. Conf Proc Annu Int Conf IEEE Eng Med Biol Soc IEEE Eng Med Biol Soc Conf. 2013;2013:6474–7.
7. Chen W, Le LH, Lou EHM. Ultrasound Imaging of Spinal Vertebrae to Study Scoliosis. Open J Acoust. 2012;02(03):95–103.
8. Ungi T, King F, Kempston M, Keri Z, Lasso A, Mousavi P, et al. Spinal curvature measurement by tracked ultrasound snapshots. Ultrasound Med Biol. 2014 Feb;40(2):447–54.
9. Chen W, Lou EHM, Zhang PQ, Le LH, Hill D. Reliability of assessing the coronal curvature of children with scoliosis by using ultrasound images. J Child Orthop. 2013 Oct 22;7(6):521–9.
10. Cobb JR. Outline for the study of scoliosis. Am Acad Orthop Surg Instr Course Lect. 1948;5:261–75.
11. Chen W, Lou EHM, Le LH. Using ultrasound imaging to identify landmarks in vertebra models to assess spinal deformity. Conf Proc Annu Int Conf IEEE Eng Med Biol Soc IEEE Eng Med Biol Soc Conf. 2011;2011:8495–8.

Corresponding Authors: Edmond Lou/Rui Zheng
Institute: Department of Surgery, University of Alberta.
Street: 2D, Walter C Mackenzie Health Sciences Centre, 8440-112 St
City: Edmonton
Country: Canada
Email: elou@ualberta.ca/ruiz@ualberta.ca

Prefrontal fNIRS Neuroimaging during a Sleep Induction Task Using Perception of a Red Light through Closed Eyes to Fight Insomnia: A Pilot Study

P.A. Grounauer[1] and B. Métraux[2]

[1] University Eye Clinic, Lausanne, Switzerland
[2] CHUV, Lausanne, Switzerland

Abstract— **The use of a red light observed through closed eyes is a new CBT Cognitive Behavioral Therapy to fight insomnia. Its principles are based on the high transmission of the colour red through eyelids, the great sensitivity to light of the retina when it is adjusted to darkness and the mental distraction obtained by the perception and attentive observation of variations of intensity. The prefrontal and occipital fNIRS recordings document this method which is correlated to the sleep EEG state II.**

Keywords— **insomnia, CBT, prefrontal-occipital fNIRS, sleep EEG state II, mindfulness meditations.**

I. INTRODUCTION

The primary function of the eye is to project simultaneous two different images onto both retinas while acting as the priority sensorial captor of the density of neuro-informations transmitted to the brain centers which are highly interconnected to ensure survival of the individual and the species. This continually supplied data adapts the position of the body to its environment and provides, in 3D, the stereotactic conditions for gripping and ensures the subtlety of predation. It also allows the early recognition of the maternal face and its emotional message [16] The second function is to reset the brain clock daily by activating the retinal melanopsin--pineal gland channel to adapt the chrono-rhythm to vital social activities [21]. When this rhythm is altered and insomnia sets in, nocturnal vision, with eyes closed, becomes a powerful mental distraction capable of reducing the appearance of disturbing thoughts, their frequency and persistence. This third function, still relatively unknown, is possible thanks to the visual "giga-channel" which, with closed eyes, is always active and ready to transport visual informations, reduced here to a perception of red and black. This sensation then becomes a mental lever to lift the weight of insomnia.

II. METHOD

The aims of this pilot study are to present the objective and subjective psychophysical bases of Somnogen Visual Training SVT [24] useful for fighting insomnia which concerns almost 10% of the population. It should help prevent the use of certain potentially neurotoxic sleeping pills and avoid addiction only through the use of mental and physical resources. The first objective data is the colored spectrum transmission curve through the eyelids [1] of which red is the best transmitted wavelength. Second the ERG recordings on the closed eye (Fig. 1), adapted after 40 minutes in darkness. We observe that a very weak light 0.01 cd is sufficient to provoke an ERG response (Fig. 2). Subjectively we can observe that closed eyelids do not prevent the view of movement or the pink-colored perception of the sun. To avoid blocking the production of melatonin we must avoid a too-strong intensity, so we have chosen a minimal intensity of red which does not disturb the subject. The timing variations in the form of alternate red and black pulsations arouse the attention and maintain vigilance. This visual sensation diverts the attention and becomes a mental distraction which breaks the obsessive circle of insomnia.

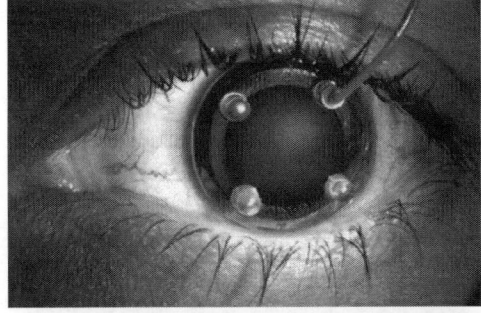
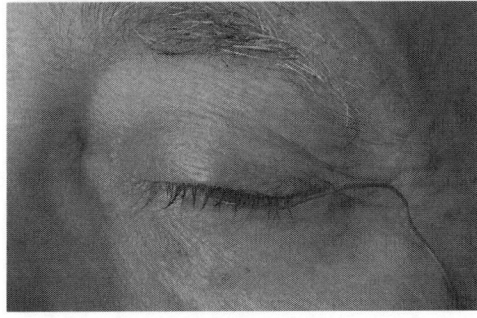

Fig. 1 ERG recording with the ERG-jet corneal contact lens electrode. Top eye open, bottom eye closed . Three pillars have been cut off [25].

© Springer International Publishing Switzerland 2015
V. Van Toi and T.H. Lien Phuong (eds.), *5th International Conference on Biomedical Engineering in Vietnam,*
IFMBE Proceedings 46, DOI: 10.1007/978-3-319-11776-8_42

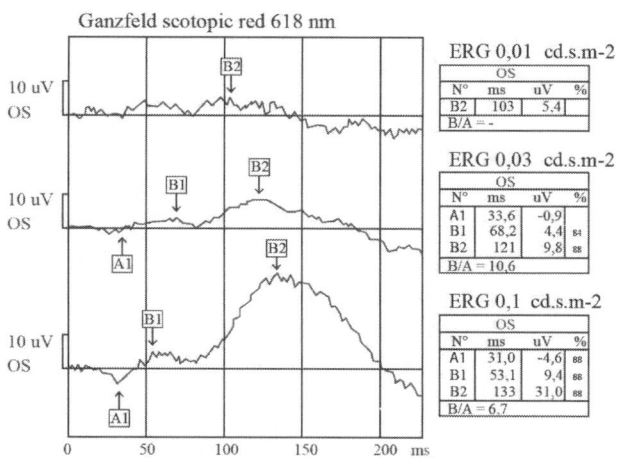

Fig. 2 Dark adapted red ERG : 0.01 cd.s.m-2 stimulation light is already effective to obtain recordings [Metrovison].

III. MATERIAL

The fNIRS recordings come from a Shimadzu FOIRE 3000 infrared spectroscope recorded in a darkened silent room. Nine optodes are placed on the prefrontal and occipital cortices 2 x 12 channels (Fig. 3). The subject tested in a sitting position is 21 years old, right-handed and in good health, with no addictions or medication. He has never previously practised SVT. The stimulator SLEAPI [24] consists of a 625 nm – 5 mcd LED placed in the centre of the forehead. The sinusoidal stimulation lasts 6 seconds, followed by a black phase lasting 11 seconds. The recording lasted 120 seconds.

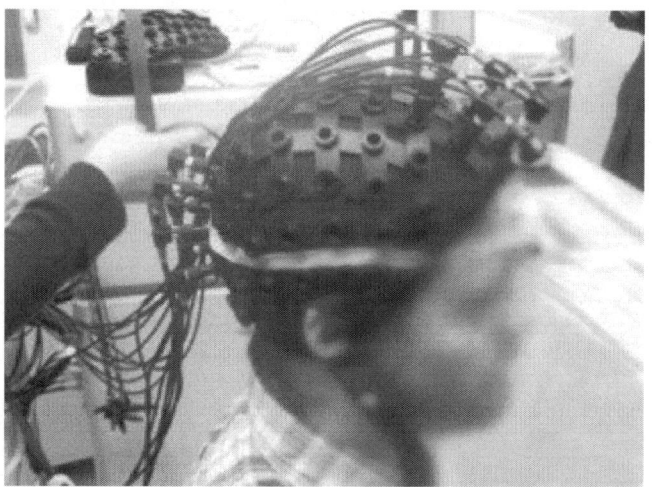

Fig. 3 Nine prefrontal and 9 occipital optodes at sites. Between the two eyes the light mental distractor SLEAPI [24] [Shimadzu].

IV. RESULTS

Fig. 4 shows after 120 seconds the oxy and deoxyhemoglobin asymmetric concentrations in prefrontal area. Interesting is the occipital visual cortex which is relatively silent.

V. DISCUSSION

Concerning the ERG on the closed eye, our results have been repeated several times on several different people using different electroretinographs. There is no equivalent in literature. Concerning the fNIRS recordings, this technique is considered to be sufficiently sensitive and specific to be valid evidence of the neocortical hemodynamic variations [2,4]. To date, it has not been applied to a similar task in a situation of minimal visual perception with closed eyes.

Our fNIRS results stand in the research of the geo-localization of intellectual activities and specifically the right and left functions of the prefrontal cortices [5-11]. If the frontal lobe is the preferred area for conceptual activities, opinions diverge on its asymmetric role [3, 12, 13, 15]. Note that in a situation of ostracism and social rejection, Peterson CK and al [14] has shown a predominance of the left prefrontal cortex. However, our study shows a hemodynamic silence of this lobe. Should we conclude that visual training SVT allows these neurons to rest and therefore facilitates sleep induction? Is this the explanation ? Note, too, that the occipital visual cortices are not aroused. This is surprising when we know that stimulation using a black and white checkerboard is the best way to obtain quality VEP visually evoked potentials. In this visual task which does not use formal vision, we must admit that these occipital neurons are resting and the essential part of the brain activity is prefrontal.

A second correlation could explain the psychophysiology of SVT by sleep EEG mapping state II (Fig. 5) where frontal predominance is obvious [22].

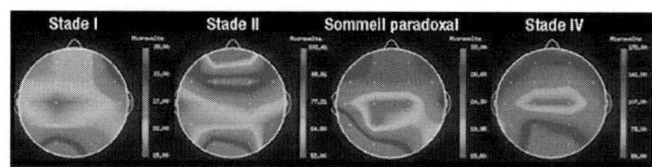

Fig 5 Sleep mapping EEG. State II shows a frontal prevalence activity. From inserm.fr/neurosciences/ le sommeil et ses troubles.

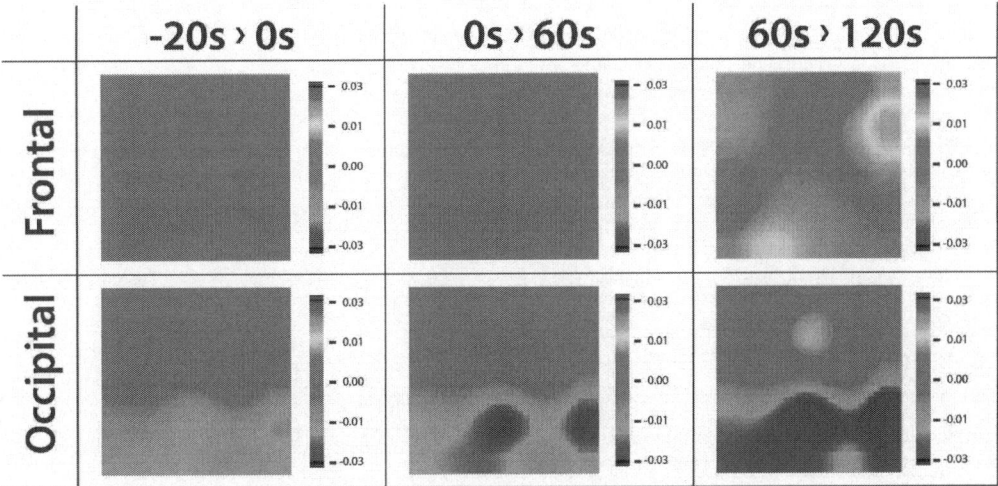

Fig. 4 fNIRS mapping. The right prefrontal is clearly more active than the left one and there is fast no activity in occipital areas.

Lastly, note that in order to increase the efficacy of SVT it is useful to associate the control of breathing and respiratory relaxation [17] together with a reduction in the muscle tonus of the lower limbs [23]. We should also note that the positioning of the red LED in the centre of the forehead is a position usually reserved for the 3rd eye, also called the serenity and diamond's, which is frequently seen on Buddhist statues. This is probably more than a methodological coincidence, possibly a hidden synergy which should not surprise those who practise meditations [18, 19, 20].

VI. CONCLUSIONS

This pilot study exposes the objective and subjective bases of a new CBT, proposes fNIRS recordings and establishes a correlation between them and stage II sleep illustrated by EEG. The use of a very weak red light allows it to be described as an intellectual distraction capable of fighting insomnia. It proposes to use its advantages and the simplicity of application encourages its daily use in order to control sleep disorders and addictions due to certain sleeping pills or thymoleptic drugs. Its innovative character is now awaiting statistical confirmation or invalidation.

ACKNOWLEDGMENT

Prof Niels Birbaumer, Prof Vo Van Toi, Prof Patrick Lemoine, Dr Thierry Faivre, Prof Claude Gronfier, Dr Jean-François Millo, Prof Michael Terman, Jacques et Nicky Fournier, Pierre-André Degoumois, David Vigo, Roland Bays, Jean-Jacques Crausaz, Gaston Schaefer, Gérard Pons, Bénédicte Wildhaber, Bruno Cholat, Charles Houriet, Jacques Charlier, Martine Crochet, Eliette Christen, Michel Onfray, Dr Bertrand Piccard, Prof. André Mazière, Prof Dan Ciulin, Prof Aki Kawasaki.

REFERENCES

1. Bierman A, Figueiro G, Rea MS, (2011) Measuring and predicting eyelid spectral transmittance. J Biomed opt 16(6): 067011.
2. Ernst LH, Plichta MM, Lutz E, et al. (2013) Prefrontal activation patterns of automatic and regulated approach - avoidance reactions - A functional near-infrared spectroscopy (fNIRS) study. Cortex 49: 131-142.
3. Herrington JD, Heller W, Mohanty A, et al (2010) Localization of Asymmetric Brain Function in Emotion and Depression. Psychophysiology 47(3): 442-454.
4. Sitaram R, Caria A, Birbaumer N (2009) Hemodynamic brain-computer interfaces for communication and rehabilitation. Neural Networks 22: 1320-1328.
5. Dao Van Ha et al.: fNIRS-Based Wavelet Thresholds for Motor Area Determination. BME5 in press (2014).
6. Pham Thanh Thao et al.: Evaluation of Frontal Visual Cortices on Mental Working Tasks Using Functional Near Infrared Spectroscopy. BME5 in press (2014).
7. Vo Nhut Tuan et al.: Differentiation of Hemodynamic Responses of the Brain with Typing and Writing. BME5 in press in press (2014).
8. Tran Le Giang et al.: Evaluation of Hemodynamic Responses to Visual Tasks Using Functional Near Infrared Spectroscopy. BME5 in press (2014).
9. Huynh Luong Nghia et al.: Investigating Physiology of Untruth in Cerebral Cortex by Functional Near-Infrared Spectroscopy (fNIRS). IFMBE Springer, vol 40.1-3 (2013).

10. Tuan Hoang et al.: High Order Moment Features for NIRS-Based Classification Problems IFMBE. Springer, vol 40.4-7 (2013) .

11. Tuan Hoang et al.: Experiments on Synchronous Nonlinear Features for 2-Class NIRS-Based Motor Imagery Problem. IFMBE. Springer, vol 40.8-12 (2013).

12. Kokubo H., Yamamoto M., Katsurugawa H., Kamada A., Kawano K., Hashizume S. & Watanabe T, Research on Brain Blood Flow during Taichi-quan by Using fNIRS, Journal of International Society of Life Information Science, 03.2008

13. Barnhofer T, Chittka T, Nightingale H, et al. (2010) State Effects of Two forms of Meditation on Prefrontal EEG Asymmetry in Previously Depressed Individuals. Mindfulness 1: 21-27.

14. Peterson CK, Gravens LC, Harmon-Jones E (2011) Asymmetric frontal cortical activity and negative responses to ostracism. Scan 6 : 227-285.

15. Travis F, Arenander A (2004) EEG asymmetry and mindfulness meditation. Letters to the Editor. Psychosomatic Medicine 66: 147-152.

16. Marendaz C, (2009) Du regard à l'émotion: la vision, le cerveau, l'affectif . Les Essais du Pommier.

17. Christen E (1996), Au centre du corps, le souffle. Edition LEP.

18. Mathieu R (2008) Lao tseu. Le Daode jing. Paris, Entrelacs 11.

19. Ricard M (2010) L'art de la Méditation, Pocket Evolution.

20. Hanson R, Mendius R (2011) Buddha's Brain, New Harbinger Publications, CA.

21. Terman M, McMahan I (2012) Chronotherapy, Avery Penguin Group Inc.

22. Münch M, Kobialka S, et al (2006) Wavelength-dependant effects of evening light exposure on sleep architecture and sleep EEG power density in men. Am J Physiol Regul Integr Comp Physiol, 290, 1421-1428.

23. Schultz JH (2005) Le Training Autogène. PUF.

24. Sleapi - Sleaping Light Emission Awaking Provider Instrument, www.sleapi.ch - Somnogen Visual Training - www.somnogenvt.ch

25. www.fabrinal.com

Author: Pierre-Alain Grounauer, MD ophthalmologist
Institute: University Eye Clinic
Street: Rue Pichard 11
City: 1003 Lausanne
Country: Switzerland
Email: pagrounauer@bluewin.ch

Orthogonal Digital Radiographs - A Novel Template for a Paediatric Femur Finite Element Model Development

D.S. Angadi[1], D.E.T. Shepherd[2], R.Vadivelu[1,3], and T.G. Barrett[1,3]

[1] School of Clinical and Experimental Medicine, University of Birmingham, Edgbaston, Birmingham, UK
[2] School of Mechanical Engineering, University of Birmingham, Edgbaston, Birmingham, UK
[3] Birmingham Childrens Hospital NHS Trust, Steelhouse Lane, Birmingham, UK

Abstract— Surgical treatment options of paediatric femur fracture include flexible or rigid intramedullary nails and submuscular plate fixation. Current computational technology has enabled virtual testing of fracture fixation implants using finite element analysis (FEA) models. Unlike FEA models based on adult femur, limited literature is available for paediatric femur. The aim of the study was to develop and validate a FEA model based on simplified geometry of composite paediatric femur using digital radiographs as a template. The model consisted of two cylinders which intersected at 130 degrees. The first hollow cylinder represented the shaft whilst the second solid cylinder represented head, neck and trochanteric region. Material properties of a composite femur (compressive strength = 157 MPa, compressive modulus = 16.7 GPa, yield strength = 93 MPa, Poisson's ratio = 0.26) were used. Simulation testing was performed in SolidWorks to estimate axial, four-point bending and torsional stiffness (k_{FEA}). An experimental study was undertaken on 4 femur specimens (k_{Exp}). FEA model predicted axial stiffness ($k_{FEA\ Axial}$ = 704.83 N/mm). In comparison mean axial stiffness of the femur specimens measured $k_{Exp\ Axial}$ = 667.39 N/mm (+/- 73.49). Four-point bending stiffness of FEA model measured ($k_{FEA\ Bending}$ = 369.1 N/mm) whereas the mean four-point bending stiffness of femur specimens was $k_{Exp\ Bending}$ = 353.49 N/mm (+/- 16.05). Torsional stiffness in external and internal rotation of FEA model was noted at $k_{FEA\ Torsion\ ER}$ = 3.49 N m/deg and $k_{FEA\ Torsion\ IR}$ = 3.49 N m/deg respectively. Mean torsional stiffness of femur specimens measured $k_{Exp\ Torsion\ ER}$ = 3.58(+/- 0.05) N m/deg and $k_{Exp\ Torsion\ IR}$ = 3.48 (+/- 0.14) N m/deg. Orthogonal digital radiographs can be used as a template to develop a simplified finite element model of paediatric femur. FEA model based on simplified geometry may be used for evaluation of routine stiffness parameters of paediatric femur.

Keywords— Paediatric, femur, finite element analysis, digital, radiographs.

I. INTRODUCTION

Paediatric femur fracture is a severe injury with a reported incidence of 20 - 33 per 100,000 a year [1]. Surgical treatment options described in the current literature include external fixation, open reduction and internal fixation with plate and screws including minimally invasive plate osteosynthesis (MIPO), flexible and rigid intramedullary nails [2]. Laboratory based biomechanical testing of implants is useful yet an expensive and time consuming process. Evolution of computational technology has enabled virtual testing of fracture fixation implants using finite element analysis (FEA) models of bones and implants.

Development of FEA bone models from computed tomography (CT) data is well documented. Some authors argue that although this approach can provide accurate geometry of the bone, it is associated with an enhanced radiation risk and is resource intensive (in terms of the additional segmentation and volume rendering software required) [3]. Biomechanical testing of paediatric fracture fixation implants has relied on the use of sawbones due to the lack of cadaveric specimens. Additionally the sawbones offer standard geometry with minimum inter-specimen variability. The standardised FEA model of sawbone (available as an open source download - http://www.biomedtown.org) is based on the geometry of an adult femur [4]. Furthermore studies have shown that this generic model may not be suitable to all research scenarios and adapting it for paediatric biomechanical research can be demanding both on time and computational resources [5]. In comparison to the adult femur FEA model, limited literature is available on the paediatric femur FEA model and only a few studies [6, 7] have addressed biomechanical testing of implants. A FEA model based on simplified geometry of paediatric femur can provide useful information to evaluate implant performance with relatively less computational resources. Hence the objectives of the current study were to i) develop a FEA model based on simplified geometry of the paediatric femur using digital radiographs of sawbone ii) validate the FEA model with the experimental data (axial, four-point bending and torsional stiffness) of the sawbones.

II. MATERIALS AND METHODS

A. Experimental model - The small fourth generation composite femur Sawbone® (model 3414, Pacific Research Laboratories Inc, Vashon, USA) was used an experimental

© Springer International Publishing Switzerland 2015
V. Van Toi and T.H. Lien Phuong (eds.), *5th International Conference on Biomedical Engineering in Vietnam*,
IFMBE Proceedings 46, DOI: 10.1007/978-3-319-11776-8_43

model. These validated composite femurs have been used extensively in paediatric biomechanical testing [8].

B. Biomechanical testing - Stiffness parameters (axial, four-point bending and torsion) of four composite femur specimens were established using custom designed jigs, polymethylmethacrylate (PMMA) split molds and Instron TT-D materials testing machine (Instron, High Wycombe, UK).

A compressive load of up to 600 N was applied at a displacement rate of 0.17 mm/s during axial loading test. Four-point bending test in the anteroposterior direction was performed with a force of 400 N applied through a top unit with two rollers separated by a distance of 70mm. The base had the first roller set perpendicular and second roller at an oblique angle to the long axis to provide a stable support for the composite femur near the lesser trochanter. Displacement rate was set at 0.17 mm/s. The resultant displacement was noted. During torsion tests proximal and distal aspects of each femur was encased in PMMA split molds and placed inside proximal and distal aluminium cylindrical units respectively. This arrangement aligned the mechanical axis of the composite femur to the mid-axis of the cylindrical units. Torque was applied in 0.2 N m increments up to 4 N m using calibrated weights suspended on an aluminium bar attached to the proximal cylindrical unit. The resultant rotation was measured using a digital inclinometer mounted on the aluminium bar.

Each of the above tests was repeated six times on each specimen. The data was entered into Microsoft Excel to determine the slope of load-displacement (or torque-rotation) curve. The slope represented the stiffness in N/mm for axial compression and bending tests and torsional stiffness in N m/degree for the torsion tests.

C. Finite Element Model - Orthogonal views (anteroposterior / lateral) of the digital radiographs of a paediatric femur (sawbone) were initially processed using Adobe Photoshop (Adobe, San Jose, California) and imported into SolidWorks™ software (Dassault Systèmes, France). Digital radiograph dimensions were adjusted in SolidWorks™ to match the geometric parameters of composite femur. Orthogonal views of the digital radiographs were used as a template for the finite element model. Using the spline and sweep functions a FEA model based on simplified geometry of the paediatric femur was developed. The model consisted of two cylinders which intersected at 130 degrees. The first hollow curved cylinder (length 350 mm, outer diameter 20 mm, inner diameter 9.5 mm, radius of curvature 1031.72 mm) represented the shaft whilst the second solid cylinder (length 96.92 mm, diameter 25 mm) represented the head, neck and trochanteric region. Fig. 1 and 2.

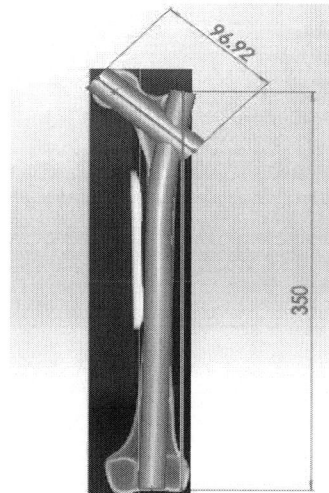

Fig. 1 Anteroposterior view

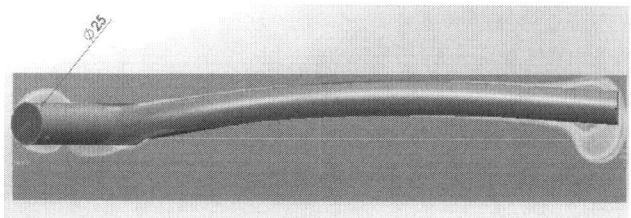

Fig. 2 Lateral view

Material properties of the composite femur (compressive strength = 157 MPa, compressive modulus = 16.7 GPa, yield strength = 93 MPa, Poisson's ratio = 0.26) were assigned to the model. The material properties of the FEA model were assumed to be isotropic and linearly elastic.

D. Mesh Selection and Convergence - The type of mesh (standard / curvature) and adaptation was selected based on the following criteria: i) Ability of the generated mesh elements to appropriately represent the geometry and shape of the model components. ii) Total solver/computational time of each type of mesh for a given simulated load. iii) Flexibility of the mesh to allow subsequent modification during iterative testing of the model.

Curvature type mesh was selected as it met the above criteria and had the least computational time. A convergence analysis was undertaken to detect the influence of element size on displacement results. The element size was sequentially decreased from the default size for a simulated load till the displacement results had plateaued. Satisfactory convergence was noted for an element size of 2 mm. The final FEA model of paediatric femur consisted of 135,788 elements in total.

E. Boundary conditions - For axial loading test the lower end of the shaft cylinder was fixed using the 'fixed geometry' option. The top end of the shaft cylinder was fixed using 'advanced fixture' option which permitted axial collapse of the shaft with incremental load but restricted mediolateral translation. Simulated axial load was applied at 60 N increments up to a maximum of 600 N. The predicted displacement value from the solver was noted for each axial loading condition and axial stiffness was estimated as described earlier.

The 'split line' command was used to create a set of four areas corresponding to the rollers (2 at the base for support and 2 at the top for application of bending load). The roller areas at the base were set at distance of 260 mm whereas the top roller areas were 70 mm apart. Simulated load was applied to top roller areas at 40 N increments up to 400 N. Displacement value from the solver was noted for each bending load condition and four-point bending stiffness was estimated.

For the torsion test the 'fixed geometry' option was applied to the distal area. The 'advanced fixture' option was used for proximal end face which allowed rotational movement but restricted excessive displacement in the anteroposterior and mediolateral directions. The long axis of the paediatric femur was marked. The femoral offset (distance from the center of rotation of the femoral head to a line bisecting the long axis of the femur) was estimated at 37 mm. Fig. 3

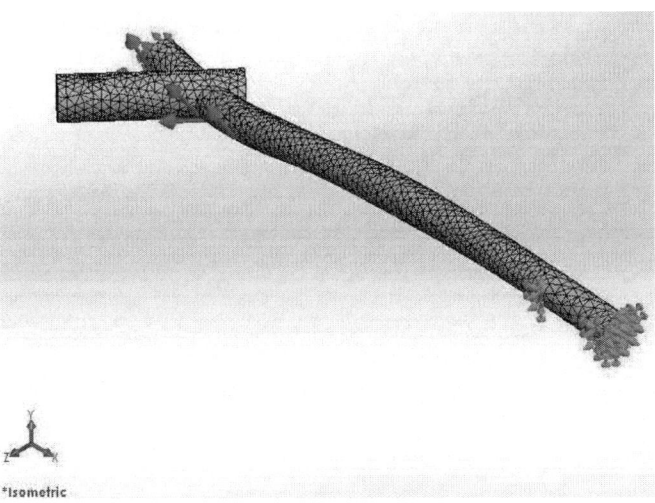

*Isometric

Fig. 3 Simulated torque (external rotation) test.

Simulated torque was applied at the proximal aspect of femoral shaft at 0.2 N m increments up to 4 N m. Positive torque produced internal rotation whereas negative torque resulted in external rotation. Displacement of the point (arc length) corresponding to the center of rotation of the femoral head was noted for each torsional loading condition. Angular displacement (in degrees) of the above point was calculated using arc of a circle principle wherein $\emptyset = $ (Arc length $\times 180$) / ($\pi \times 37$). Torsional stiffness (N m/deg) was obtained by plotting the torque against the angular displacement.

III. RESULTS

FEA model predicted a maximum axial displacement of 0.851mm at 600 N ($k_{FEA\ Axial}$ = 704.83 N/mm). In comparison mean axial stiffness of the femur specimens measured $k_{Exp\ Axial}$ = 667.39 N/mm (+/- 73.49), Table 1.

Table 1 Results of experiment and simulation tests

Stiffness Parameter	Experiment model	FEA model
Axial (N/mm)	667.39 +/- 73.49	704.83
Four-point bending (N/mm)	353.49 +/- 16.05	369.1
Torsion (Internal rotation) (N m/deg)	3.48 +/- 0.14	3.49
Torsion (External rotation) (N m/deg)	3.48 +/- 0.14	3.49

IV. DISCUSSION

An attempt was made in this study to develop a FEA model based on simplified geometry of the paediatric femur using orthogonal digital radiographs as a template. Digital radiographs enable development of FEA model that is simple yet representative of the overall dimensions and features viz. radius of curvature of the paediatric femur. This is an important requisite of femur FEA model used to assess the biomechanical parameters of intramedullary implants [9]. It has been demonstrated that omission of cancellous bone in femur FEA model does not significantly alter the overall stiffness results [10]. Therefore cancellous bone tissue was not modelled separately in the current paediatric FEA model to optimise computational time.

Results from simulation tests using the FEA model to estimate axial, four-point bending and torsional stiffness were similar to the experimental sawbone model. The above validated FEA model of paediatric femur can be used for evaluation of fracture fixation implants. Simulation data obtained using the paediatric femur FEA model can be useful in estimation of the range of loads that can be safely applied to a fracture fixation construct. This information can be helpful in planning and establishing parameters for an experimental setup [11].

Previous studies in this field [3, 12] have used specialised algorithms to develop 3 dimensional bone models from radiographs. However the lack of widespread availability of such algorithms limits their applicability. Enhanced processing capability of computers has enabled development of bone FEA models with good visual similarity through accurate geometric representation. However this approach does not necessarily guarantee the numerical accuracy of the results predicted by such models [13]. In their study Perez and colleagues [8] used the bone model available for download from the aforementioned source. The model measured 420 mm in length with a canal diameter of 9 mm which is probably representative of an adolescent femur. It has been noted that a change in the synthetic femur geometry from a large to small dimension can result in axial and torsional rigidity differences of 1.5 and 2.2 times respectively despite having the same Young's modulus for the cortical bone [14]. It will be of interest to note the predictions from their model if the overall dimensions were scaled down to be representative of a child's small femur. Krauze [9] performed biomechanical analysis comparing flexible intramedullary nails of two different materials. The femur FEA model in this study is reported to be based on a 5-7 year old child. However the details regarding development of the femur FEA model and its dimensions are not provided in the paper.

Paediatric femur has a complex shape and microarchitecture both of which contribute towards load transmission during weight bearing and activity. The simulation model in the current study was an attempt to simplify the femur anatomy using basic geometric entity like cylinder. Hence the results from this model can be used only as a general guide to predict the behaviour of the paediatric femur under different loading conditions. Accurate assessment using simulation models comes at the cost of significant computational resources. However this was not the main objective of the current study. In general simulation studies are based on a set of assumptions which are the inherent limitations and the current study is no exception to this rule.

V. CONCLUSIONS

Orthogonal digital radiographs can be used as a template to develop a simplified finite element model of a paediatric femur. FEA model based on simplified geometry may be used for evaluation of routine stiffness parameters of a paediatric femur.

ACKNOWLEDGEMENT

This study was supported by a grant from AO Foundation, UK. The authors thank Mr. C Hingley (Senior Technician) and team for their help with the experimental study.

REFERENCES

1. Hinton, R. Y., Lincoln, A. et al. (1999) Fractures of the femoral shaft in children. Incidence, mechanisms, and sociodemographic risk factors. J Bone Joint Surg Am, 81(4): 500-509.
2. Kocher, M. S., Sink, E. L. et al. (2010) American Academy of Orthopaedic Surgeons clinical practice guideline on treatment of pediatric diaphyseal femur fracture. J Bone Joint Surg Am, 92(8): 1790-1792.
3. Baudoin, A., Skalli, W. et al. (2008) Parametric subject-specific model for in vivo 3D reconstruction using bi-planar X-rays: application to the upper femoral extremity. Med Biol Eng Comput, 46(8): 799-805.
4. Viceconti, M., Casali, M. et al. (1996) The 'standardized femur program' proposal for a reference geometry to be used for the creation of finite element models of the femur. J Biomech, 29(9): 1241.
5. Greer, B., Wang, E. et al. (1999) On the appropriateness of using the standardized femur for FEA in the proximal region. ASME-PUBLICATIONS-BED 42:669–670.
6. Perez, A., Mahar, A. et al. (2008) A computational evaluation of the effect of intramedullary nail material properties on the stabilization of simulated femoral shaft fractures. Med Eng Phys, 30(6): 755-760.
7. Krauze, A., Marciniak, J. (2006) Numerical method in biomechanical analysis of intramedullary osteosynthesis in children. Journal of Achievements in Materials and Manufacturing Engineering 1-2: 120.
8. Green, J. K., Werner, F. W. et al. (2005) A biomechanical study on flexible intramedullary nails used to treat pediatric femoral fractures. J Orthop Res, 23(6): 1315-1320.
9. Egol, K. A., Chang, E. Y. et al. (2004) Mismatch of current intramedullary nails with the anterior bow of the femur. J Orthop Trauma, 18(7): 410-415.
10. Montanini, R., Filardi, V. (2010) In vitro biomechanical evaluation of antegrade femoral nailing at early and late postoperative stages. Med Eng Phys, 32(8): 889-897.
11. Zdero, R., Bougherara, H. (2010) Orthopaedic biomechanics: a practical approach to combining mechanical testing and finite element analysis. Finite Element Analysis. David Moratal: 171-194.
12. Mitton, D., Landry, C. at al (2010) 3D reconstruction method from biplanar radiography using non-stereocorresponding points and elastic deformable meshes. Med Biol Eng Comput, 38(2): 133-139.
13. Viceconti, M., McNamara, et al (1996) FEM analysis of the static stresses induced in a THR femoral component during a standardized fatigue test. Second International Symposium on Computer Methods in Biomechanics and Biomedical Engineering, London.
14. Papini, M., Zdero, R., et al (2007) The biomechanics of human femurs in axial and torsional loading: comparison of finite element analysis, human cadaveric femurs, and synthetic femurs. J Biomech Eng, 129(1): 12-19.

Address of the corresponding author:

Author: Darshan. S. Angadi
Institute: University of Birmingham
Location: Edgbaston
City: Birmingham
Country: UK
Email: DSA704@bham.ac.uk

EEG Signal Analysis and Artifact Removal by Wavelet Transform

Pham Phuc Ngoc, Vu Duy Hai, Nguyen Chi Bach, and Pham Van Binh

Hanoi University of Science and Technology/ Biomedical Engineering Department, Hanoi, Viet Nam

Abstract— **Electroencephalogram (EEG) is a non-invasive method to collect brain signals from human's scalp. EEG signals are located in low frequency range and relatively small. The amplitude of these signals are approximately 50μV with the maximum amplitude is about 100μV. Therefore, there are number of sources such as power line, EOG or ECG can extremely interfere EEG signals. Detection and elimination of artifacts plays an important role to acquire clean EEG signals to analyze and detect brain activities. Besides, the extraction of important components in recorded EEG required fast and reliable algorithm to process mix of data set. In this paper, we will demonstrate EEG acquisition from EEG Exea Ultra system of Bitmed, analyze and compare signals from volunteers in relaxation mode and contaminated EEG signals with eye blinks. We then design filters to remove powerlines and baseline noise from acquired signals. With the aim to assess the feasibility of Wavelet transform technique to identify feature in recorded EEG signals, we carried out Wavelet transform and applied threshold method to detect and remove artifacts in EEG signals with eye blinks. We achieved PSNR of original signals and wavelet filtered signal that approximately 17,7810 dB. Our preliminary results show that wavelet can be utilized as automatically detection tools for artifacts and event-related potentials and in applications require real-time processing of EEG signals.**

Keywords— **EEG, signal analysis, feature extraction, Wavelet transform, de-noise, ocular artifact, rehabilitation.**

I. INTRODUCTION

Analyzing brain wave activities plays an important role in neurosciences. Brain signals carry information allows us to identify activities of brain as well as abnormalities. In combination with other imaging techniques such as MRI, we can create an useful tools for doctors and therapeutic physicians [1,2]. Brain signals are components of recorded electroencephalogram called EEG. There are different forms of brain waves according to ages, stimulation, brain diseases or physiology conditions [3]. Components or event-related evokes can be extracted from EEG data that help doctors in diagnostic and treatment. That can be a event – related potentials happened in transient specified time or emergence in specific area in human scalp (localized in space). In our research direction, we are researching the changes of EEG signals that controlling movement statement. Frequencies range of EEG signals related to motor function are from 0,5 Hz to 30Hz [4,5]. The advantages of EEG method are non-invasive measurement and high time resolution. Therefore, this technique can be used in real-time monitoring changes of motor activities. (The time resolution is approximately 1ms). The extraction of important components in recorded EEG required fast and reliable algorithm to process mix of data set [6, 7]. However EEG signals are complex and hard to identified. Recorded signals from scalp are embedded with sources of noises that can causes problems when analyze, display and restore motor – related EEG signals. The present of noise produces spikes lead to misinterpretation of brain signals. Therefore, elimination or attenuation noises from EEG signals to improve the accuracy of diagnostic and analysis brain activities must be carried out. Eye movement is one of sources that contaminated recorded EEG. Author Senthil [8] had demonstrated the use of Wavelet transformation to canceled eye blinks without EOG references channel. V.Krishnaveni proposed method to detect eye blinks area and apply adaptive threshold algorithm based on Wavelet that allow us to keep necessary information. In this methods, author used Haar Wavelet function to detect eye blinks and eliminate them afterward using SWT with Coif3 as basic function [9,10]. Other method is use ICA independent component analysis. This is blind source separation techniques that linearly decompose EEG signals into independent sources component [11].

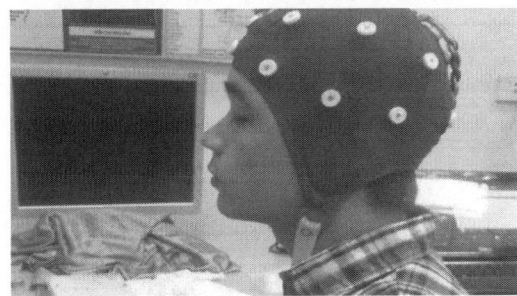

Fig. 1 EEG measurement from volunteer using EEG electrode cap following international standard 10/20.

This technique can eliminate noises by apply reconstructed algorithm without noise sources. But the

© Springer International Publishing Switzerland 2015
V. Van Toi and T.H. Lien Phuong (eds.), *5th International Conference on Biomedical Engineering in Vietnam*,
IFMBE Proceedings 46, DOI: 10.1007/978-3-319-11776-8_44

process still requires manual rejection through visual inspection on graphic interfaces [12]. In this paper, we describe EEG acquisition method from 25 channels EEG Exea Ultra of Bitmed. EEG signals are recorded from volunteers in relaxation state and with natural eye blinks to monitor and analysis. Acquired signals are preprocessed with band pass filter (FIR). Then, with the aim to assess the feasibility of Wavelet transform technique to identify feature in recorded EEG signals, we applied Wavelet transform to detect and remove eye blink by applied hard threshold and evaluate output results.

II. MEASUREMENT METHODS

EEG signals are recorded from 25 channels EXEA Ultra devices. Volunteers are sat comfortable and still to elimination movement interference to the signals. Time duration for each trial is 4 minutes. The volunteers are then required to perform followings states: Free stage with eye close and relaxation; Eye blinks state with natural eye blinks. Recorded channels are O1, O2, F8, T4, T6, Fp2, F4, C4, P4, F7, T3, T6, Fp1, F3, C3, P3, Fz, Cz, Pz. Electrode montage follows 10/20 international system for electrode placement [17]. Gels are then injected to each electrode to increase electric conduction. For good EEG recordings, electrode impedance is kept below 5KΩ. Signals are then amplified and sampled at 200Hz. In this paper, to investigate the EEG feature identification and artifacts reduction of wavelet transform technique, we used data from channel Fp1, Fp2 that located in frontal lobe. The electrodes in this area are much interfered by eye movement. Figure 1 demonstrates settings and EEG measurement from volunteers.

III. POWER LINE AND BASELINE ARTIFACT REMOVAL

Recorded signals from 19 channels are shown in Figure 2a. The power line noise can be seen relatively high, hence, covered acquired signals. Figure 2b shows power line spectrum has dominant power in 50Hz frequency. As the result, objective of this process is to minimize this line-frequency artifact. To get this goal, we use low pass filter that have cut-off frequency of 40Hz.

Still, we can observe that the signal is strongly affected by baseline noise which has the low frequency and high amplitude causing a fluctuation of amplitude in received signal. The period estimation of baseline noise is 5 to 10 second (0.1 to 0.2 Hz) and amplitude is about 150 uV.

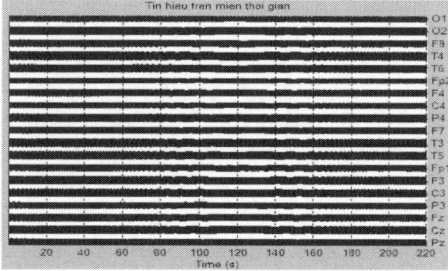

(a)

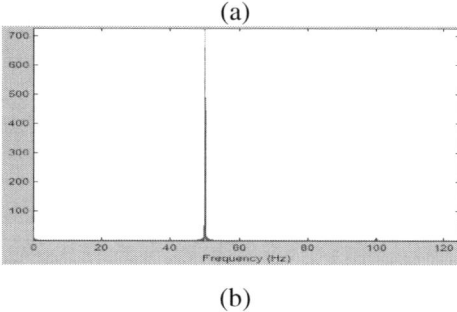

(b)

Fig. 2 Recorded signals in 19 channels (a) and spectrum magnitude of EEG with household line frequency at 50Hz (b).

We then move signals forward to high pass frequency with cut-off frequency is 0,5Hz. Figure 3 demonstrates signals before and after baseline artifact removal.

IV. EYE BLINKS REMOVAL FROM RECORDED SIGNALS USING WAVELET TRANSFORMS

After initial preprocessing stage, EEG signals are forwarded to eye blinks detection and elimination stage. As usual, conventional FFT transform can reveals spectrum component but can not identify the time when it happened. Meanwhile, EEG signals has unstable characteristic that causes frequency changes. Wavelet transform had been investigated to use for EEG signals in many researches [13,14].

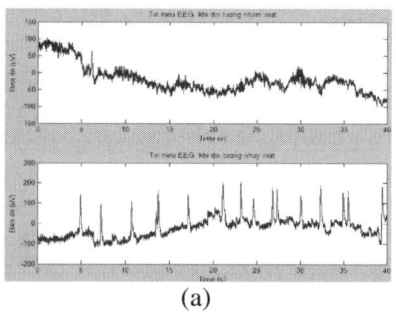

(a)

Fig. 3 (a) EEG signals when the volunteers close eyes (above) and open eyes with natural eye blinks (below) before baseline artifact removal. (b) EEG signals when the volunteers close eyes (above) and open eyes with natural eye blinks (below) before baseline artifact removal.

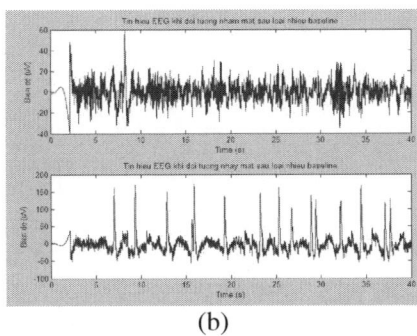

(b)

Fig. 3 (*continued*)

The most benefit of wavelet transform is that it is localized in both time and frequency. Wavelet transform are used to convert EEG signal into a series of wavelet which are shifted and scaled version of mother wavelet. Therefore, Wavelet can be suitable with hidden event that can help to detect exact frequency and location of the event in time scale. Moreover, wavelet can be suitable with abnormal event due to special shape. Wavelet transform utilizes different window sizes for each range of frequencies. Longer window for lower frequencies range and shorter window for higher frequencies range. This technique is differs from sTFT techniques as sTFT use fixed window dimension to compute transformation [15]. The fixed size window can cause drawbacks as we have the same resolution in time and frequency domain.

We can define Wavelet transform as general as following

$$C(scale, position) = \int_{-\infty}^{\infty} s(t)\Psi(scale, position, t)dt \quad (1)$$

Where C is the coefficients of Wavelet Transform, s(t) is the signal needed to be transformed, $\Psi(scale, position, t)$ is the Wavelet transform function.

Continuous Wavelet Transform formula:

$$C(a, b) = \int_R s(t)\frac{1}{\sqrt{a}}\Psi\left(\frac{t-b}{a}\right)dt \quad (2)$$

In the Discrete Wavelet Transform (DWT, SWT) the parameter scale $a = 2^{-j}$ and position $b = k2^{-j}, (j, k)\epsilon Z^2$

And the inversed function in the cases of DWT and SWT:

$$s(t) = \sum_{j\in Z}\sum_{k\in Z} C(j, k)\Psi_{j,k}(t). \quad (3)$$

Where a is scale, j is called level and b, k are shift coefficients.

DWT allows choose scale j and position k for function ψ (t) with:

$$\psi_{j,k}(t) = 2^{\frac{j}{2}}\psi(2^j t - k) \quad (4)$$

By stretching ψ (t) with the coefficients 2^j and compressing with the coefficients 2^{-j}, we can build a wavelet for any function. Compressed version of wavelet function is suitable with high frequencies components and stretched version is for low frequency components of signals. Wavelet transform can eliminate frequencies component at specific time in the EEG recordings [16]. Therefore, cancellation of bad signals and keep good ones is possible. Figure 4 presents the proposed wavelet function Sym3 [8] that have similar shape with eye blinks EEG patterns.

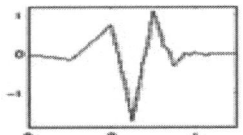

Fig. 4 Sym3 Wavelet function

It is observed that spikes (Figure 3b) have high amplitude in acquired signal because of the present of eye blinks artifact. In relaxation state, when the participant close eyes, alpha rhythm which have frequency range of 8-13Hz is prominent in EEG signal with the center frequency of approximately 10Hz. We could view it clearly on Figure 5a. In figure 5b where the spectrum of signal with eye-blinks is showed, alpha rhythm is no longer prominent.

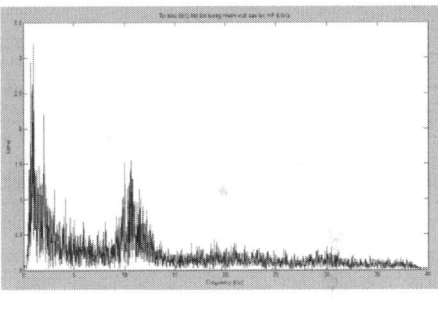

(a)

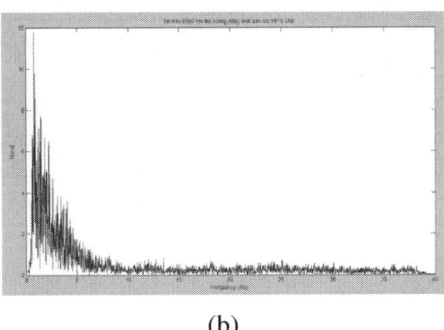

(b)

Fig. 5 EEG spectrum when volunteers close eyes (a) natural eye blinks (b) after preprocessing stage

There are other waves with frequency below 5Hz and high amplitude. When volunteers open eyes and perform natural eye blinks, these frequencies range are almost attenuation. Therefore, eye blinks has spectral components located in low frequencies range. To eliminate eye blinks components, we implemented undecimated discrete wavelet transform 8 level SWT. Wavelet Sym3 that have high correlation with eye blink artifacts is proposed to use for noise cancellation algorithm. The algorithm presents in Figure 7 has been applied for analysis recorded EEG signals. Frequencies band after decomposed by SWT has been illustrated in Table 1.

Table 1 Decomposition of acquired EEG signals after SWT 8 level wavelet transform

Frequencies band (Hz)	Decompose level	Frequencies band	Frequencies bandwidth (Hz)
62.5 – 125	D1	Noise	62.5
31.25 – 62.5	D2	Gama	31.25
15.625 – 31.25	D3	Beta	15.625
7.8 – 15.625	D4	Alpha	7.8125
3.9 – 7.8	D5	Theta	3.9
1.95 – 3.9	D6	Delta	1.95
0.98 – 1.95	D7	Delta	0.98
0.49 – 0.98	D8	Noise	0.49
0 – 0.49	A8	Noise	0.49

Figure 6 below illustrated signals with eye blink artifacts in FP1 channel and decomposition level of signals (D1-D8 and A8). Detail coefficients D1, D8 and A8 are artifact components is then removed from signals. The red pipe located in eye blinks area that has been detected in detail coefficients.

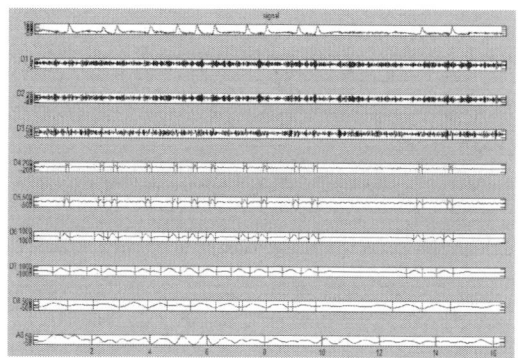

Fig. 6 EEG signal decomposition in 8 level and eye blinks area detection.

The data before and after eye blink cancellation by apply threshold level and wavelet function sym3 show in Figure 8. The received outputs are depending on wavelet function

selection that has similar shape with eye blink pattern. Higher coefficients than threshold were set to zero.

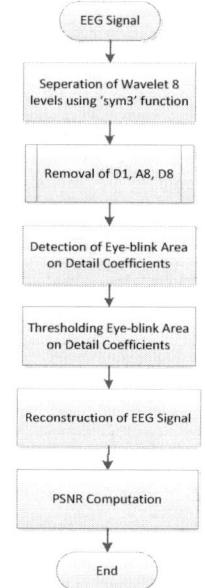

Fig. 7 Flow chart of EEG signal processing using wavelet transforms.

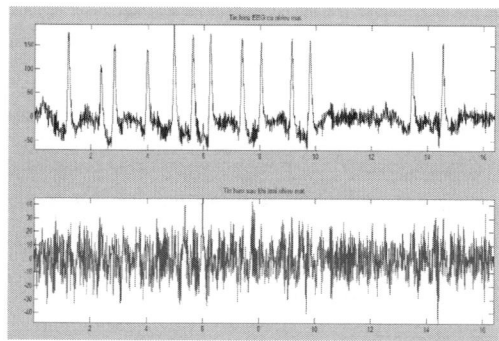

Fig. 8 EEG signal with eye blink rejection (red waves) using wavelet transform.

V. DISCUSSION

To compare and measure the quality of removing noise, the Peak Signal to Noise Ratio (PSNR) is computed between the original signal $f(k)$ and the denoised signal $f_d(k)$, given by:

$$PSNR = 10\log \left(\frac{f_{max}^2}{MSE}\right) \qquad (5)$$

Where f_{max} is the maximum value of signal given by:

$$f_{max} = \max\left(\max(f(k)), \max(f_d(k))\right) \qquad (6)$$

With the sampled EEG data from our recordings, we can achieved PSNR = 17.7810 (dB)

In this paper, we demonstrated EEG acquisition method from volunteers. The dataset is sampled at 200Hz included both clean and eye blinks contaminated EEG signals. Preprocessing signals was initially applied with bandpass filter (fc1= 0,5Hz and fc2=40hz). As the result, the filters have successfully remove line-frequency and baseline noise. However, conventional filter cannot eliminate eye blinks noise due to overlapping spectral between EEG signals and noises. Wavelet transform uses flexible window width that can deliver good signal resolution in both frequencies and time domain. Hence, using suitable wavelet can help us to identified special pattern embedded in EEG signals. The more similarities of the wavelet to the pattern of EEG signal the more possibility to analysis event of interest and the interpretation of EEG recordings. We applied threshold on details coefficients at the area of detected eye blink without distorting portions of EEG data. Application of Wavelet transforms give us an effective tool for analyzing and EEG signal processing, especially with eye blink – that have high amplitude and have spectral overlapping with EEG signals. Our preliminary results show that wavelet can be utilized as automatically detection tools for artifacts and event-related potentials and in application requires real-time processing of EEG signals.

ACKNOWLEDGMENT

We would like to give special thanks to BME staffs and other researcher in BME department, School of Electronics and Telecommunication, Hanoi University of Technology;

REFERENCES

1. Gotman J. Epileptic networks studied with EEG-fMRI. Epilepsia 49. Suppl 3:42–51. 2008.
2. Walker MC, Chaudhary UJ, Lemieux L. EEG-fMRI in adults with focal epilepsy. Springer, Heidelberg, pp. 308–331, 2010.
3. M. Teplan Fundamentals of EEG measurement – Measurement science review, Volume 2, Section 2, 2002.
4. Waldert S, Preissl H, Demandt E, Braun C, Birbaumer N, Aertsen A, Mehring C: Hand movement direction decoded from MEG and EEG. J Neurosci, 28:1000-1008, 2008.
5. Pfurtscheller G, Lopes da Silva FH: Event-related EEG/MEG synchronization and desynchronization: Basic principles. Clin Neurophysiol,110:1842-1857,1999.
6. Saeid Sanei and J. A. Chambers, EEG Signal Processing, Wiley-Interscience, 2007.
7. M. Nixon and A. Aguado, Feature Extraction & Image Processing,Elsevier, Amsterdam, 2004.
8. P. S. Kumar, 1, R. A. , 1, K. S. , 2, et al., "Removal of Ocular Artifacts in the EEG through Wavelet Transform without using an EOG Reference Channel" Int. J. Open ProblemsCompt. Math vol. 1, December, 2008.
9. V. Krishnaveni, S. Jayaraman, S. Aravind, V. Hariharasudhan, K. Ramadoss,"Automatic identification and Removal of ocular artifacts from EEG using Wavelet transform", Measurement Science Review, Vol. 6, No.4, pp.45-57,2006.
10. V. Krishnaveni, S. Jayaraman, L. Anitha, K. Ramadoss, "Removal of Ocular Artifacts from EEG using Adaptive thresholding of Wavelet coefficients", Journal of Neural Engineering, Vol.3, pp.338-346, 2006.
11. Comon P. "Independent Component Analysis, A new concept", Signal Processing 36(3), pp 287-314, 1994.
12. Delorme.A, Makeig.S & Sejnowski T. "Automatic artifact rejection for EEG data using high-order statistics and independent component analysis", Proceedings of the Third International ICA Conference, pp 9-12, 2001.
13. M. Akay, "Time Frequency and Wavelets in Biomedical Signal Processing" IEEE Press series in Biomedical Engineering, 1998.
14. R. A. G. a. H. G. C. Sidney Burrus, "Introduction to Wavelets and Wavelet Transforms" 1998.
15. R. Saab. Wavelet Based Approach for the Detection of Coupling in EEG Signals.
16. D.Lee Fugal. Conceptual Wavelets in Digital signal processing 2012.
17. C.R.Hema,M.P.Paulraj,S.Yaacob,A.H.Adom and R.Nagarajan "An Analysis of the effect of EEG Frequency Bands on the Classification of Motor Imagery Signals",Biomedical Soft Computing and Human Sciences,Vol 16,pp.121-126,1995.
18. Mattingly J. D. Elements of gas turbine propulsion. McGraw-Hill, Inc. International Editions, US, 1996.

Author contact: Phạm Phúc Ngọc
Institute: Biomedical Engineering, School of Electronics and Telecommunication, Hanoi University of Sciences and Technology
Street: No 1 Đại Cồ Việt Str.
City: Ha Noi
Country: Vietnam
Email: ngoc.phamphuc@hust.edu.vn.

Detection of Activities Daily Living and Falls Using Combination Accelerometer and Gyroscope

Quoc T. Huynh[1], Uyen D. Nguyen[1], Kieu Trung Liem[1], and Binh Q. Tran[2]

[1] School of Electrical Engineering, International University - Vietnam National University – Hochiminh City, Vietnam
[2] Department of Biomedical Engineering, School of Engineering,
Catholic University of America, Washington D.C., USA
{Quochuynhtan,kieutrungliem}@gmail.com,
nduyen@hcmiu.edu.vn, tran@cua.edu

Abstract— **This paper studied the detection of falls and activities of daily living (ADLs) with the objective: to automatically monitor health situation and prevent the elder out of injury from fallings. In this study, a wireless sensor system (WSS), based on accelerometer and gyroscope, is placed at the centre of the chest to collect real-time ADLs and fall data. The WSS contains a set of ADXL345 (3-axis digital accelerometer sensor), ITG3200 (3-axis digital gyroscope sensor), MCU LPC17680 (ARM 32-bit cortex M3), and Wi-Fi module RN131. Experiment protocols consisting of four types of falls such as forward fall, backward fall, and side way fall (left and right), and ADLs such as standing, walking, sitting down/ standing up, stepping, running along with normal gait involved 324 tests on 18 human subjects.**

The results from the experiment shows the system and algorithm could distinguish falling and ADLs with high accuracy.

Keywords— **Fall Detection, Activities of daily living, Wireless sensor system.**

I. INTRODUCTION

Activities Daily Living recognition have been receiving increasing attention in recent years. The benefit of automatically recognize different activities of humans activities makes it appealing for healthcare such as monitoring health situation, especially for elder. It provides more information on the patient's day-to-day health for the clinician. Additional, fall is the most significant causes of injury for elderly. These falls are cause many disabling fractures that could eventually lead to death due to complications, such as infection or pneumonia. More than one-third of elderly people, who is over 75 year old, have fallen at least once a year, and 24% of them has serious injuries [2] (J.Y. Hwang, J.M Kang, H.C. Kim, September 1-5, 2004). Because of that, ADLs classification and fall detection is necessary for protecting health of elder.

Most of the research on falls detection and ADLs classification in which accelerometers is used focus on determining the change in magnitude of acceleration. Base on acceleration value thresholds, ADLs or falls are distinguished. [1] (Quoc T. Huynh, Uyen D. Nguyen, Su V. Tran, Afshin Nabili, Binh Q. Tran, 2013) [3](U.Lindeann, A.Hock, M.Stuber, W.Keck, and C.Becker, 2005) [4 - 2] (M. Kangas, A. Konttila,P. Lindgren, I. Winblad, T. Jämsä, 2008 Feb 21) [5] (Ig-Jae Kim, Saemi Im, Eugene Hong, Sang Chul Ahn and Hyoung-Gon Kim, 2007). These systems successfully detect with sensitivities greater than 85% and specificities between 88-94%. However, focusing only on large acceleration can result in many false positives from fall-like activities such as sitting down quickly and running.

Furthermore, previous studies used complex algorithms like Neural Network [6] (Saisakul chernbumroong, Anthony S. Atkins, and Hongnian Yu, 08/11/2011) to classification ADLs or Support Vector Machine (SVM) [7] (Tong Zhang, Jue Wang, Liang Xu, Ping Liu, 2004) and Markov models [8] (R. K. Ganti, 2006) to detect the fall. However, accuracy of these systems has not been proven to be highly effective. They also use excessive amounts of computational resources and cannot respond in real time. In addition, activity patterns are particularly difficult to obtain for training systems.

Some fall detection algorithms also assume falls happen when the body lies prone on the floor. But they are less effective when a person's fall posture is not horizontal, e.g. fall happen on stair.

Unlike other previous researches, this project proposes using both accelerometer and gyroscope sensors to not only detect falls but also to classify ADLs with increasing in sensitivities and specificities of a system.

II. METHODOLOGY

In this study, we developed a wireless sensor system and an algorithm to classify ADLs and fall events. The system includes a Wireless Sensor System (WSS) and a detection algorithm. Figure1 shows the overall schematic of system. The WSS transmits and receives real-time accelerometer and gyro data during ADLs and fall. The detection algorithm is based on a simple threshold method.

V. Van Toi and T.H. Lien Phuong (eds.), *5th International Conference on Biomedical Engineering in Vietnam,*
IFMBE Proceedings 46, DOI: 10.1007/978-3-319-11776-8_45

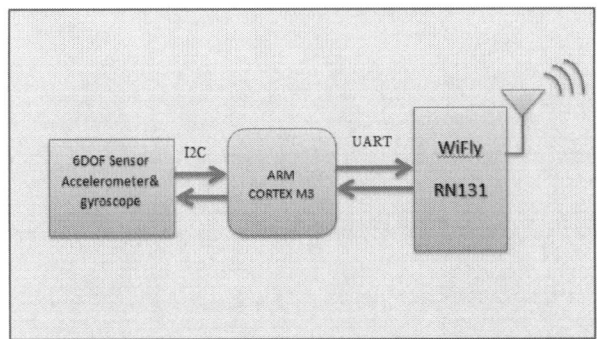

Fig. 1 The schematic of wireless sensors system

A. Wireless Sensor System (WSS)

The wireless sensors system contains a set of Sensor module, Micro Control Unit, and Wi-Fi module. The Sensor Module, the Micro Control Unit, and the Wi-Fi module are used to sense body orientation and activity data, control the flow of data, and transmit/receive data, respectively. The WSS is placed at the center of chest, see Fig 2.c.

Sensor Module

Since our system measures both acceleration and angular velocity to classify ADLs and detect falls, we chose to use the 6-DOF module with small size and power requirements. It includes a tri-axial accelerometer ADXL345 and a tri-axial gyroscope ITG3200 (Fig 2.a). The acceleration measurement range of ADXL345, with a high resolution of 13bit and 4mg/LSB, is up to $\pm$ 16g. This is an important aspect in recognizing the fall. In addition to the accelerometer, the ITG-3200 can capture the angular velocity between $\pm 2000°$/sec. The digital data output of these modules is formatted as 16 bits two complements prior to transmitting. The sensor modules are connected with MCU via I2C digital interface port.

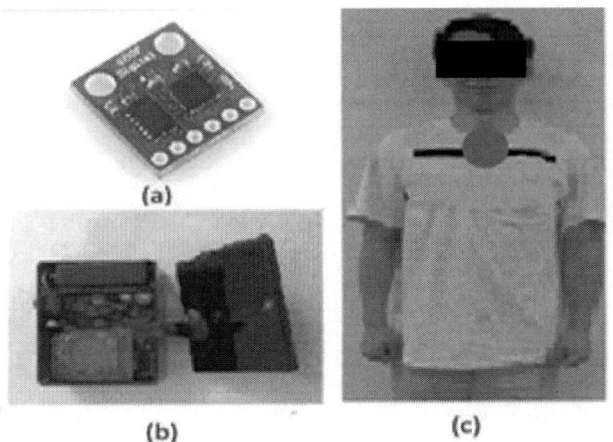

Fig. 2 ADLs classification and fall detection system: (a) sensor module (6 DOF), (b) wireless sensor system (WSS), (c) system attached on center chest.

The Micro Control Unit (MCU) module (ARM 32 bit Cortex M3)

The LPC 1786 (NXP product's) is used to develop the control system. This is an ARM Cortex-M3 32-bit based microcontroller for embedded applications requiring a high level of integration and low power dissipation. The chip can operate up to 100MHz CPU frequency. In addition, the UART interface provides the sampling frequency up to 4Mb/s.

Wireless Module

The Wifly RN131 module is a stand-alone Wi-Fi module, providing a fully integrated 2.4GHz and IP stack with IEEE 802.11 b/g standard. The RN131 can operate with the communication speed up to 11 Mbps. Due to its small form factor and extremely low power consumption; it is perfect for mobile wireless applications with portable battery operated devices. Additionally, its UART hardware interfaces for connecting with MCU can operate up to 1 Mbps data rate.

Collection Data Program

The WSS (Fig.2.b) consists of 6-DOF, MCU and Wi-Fi module. The MCU is connected to wireless module via UART port and 6DOF module via I2C port. The WSS collects the acceleration and angular velocity values. In addition, the WSS sends the real-time data to a computer via 802.11 wireless protocols to be displayed.

The collection data program is written in Matlab (Mathworks, Inc, Natick, MA). The program receives and display real-time data from the WSS. It continuously plots the acceleration and angular velocity values of each fall and saves the data for later analysis. Figure 3 (below) shows a display of the collection data program.

B. Experimental Setup

The experiment is performed on 18 young healthy subjects (age from 19 to 28 years, weight from 50 to 90 kg, and height from 154.5 to 180.0 cm). Experiments were performed at The Catholic University of America (Washington, DC) and approved by the Human Subjects/Institutional Review Board (IRB) Committee. The WSS was attached to the center of the chest (Fig. 1.c.). This is the optimum location to attach sensor for detecting as mentioned in our previous study [9] (Quoc T. Huynh, Uyen D. Nguyen, Su V. Tran, Binh Q. Tran, 2013). In this experiment, subject performs ADL such as standing, walking, sitting down/ standing up, stepping, running, lying down/ sitting up and 4 different kinds of fall tests: forward fall, backward fall, right sideway fall, and left sideway fall.

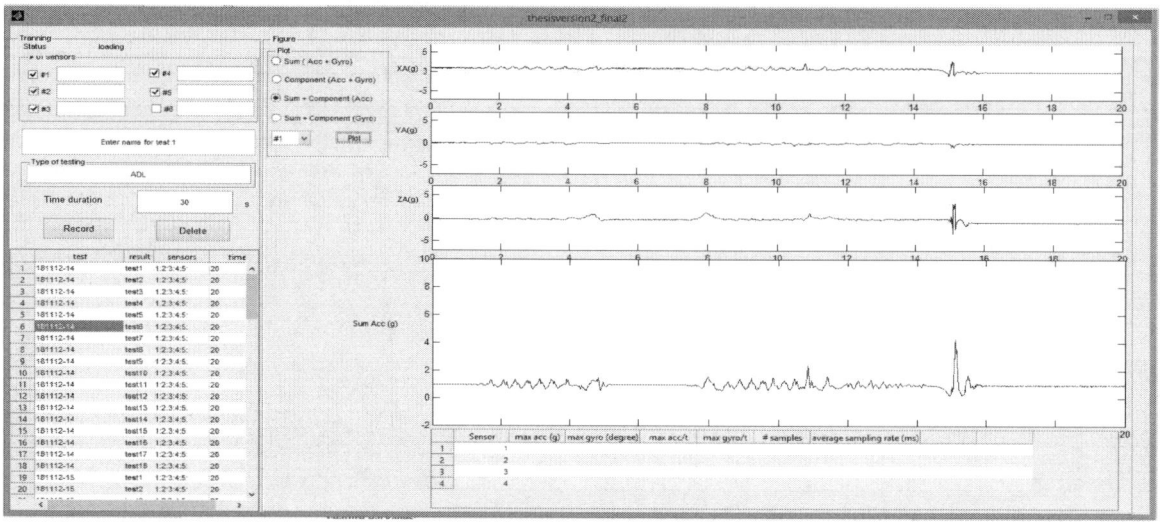

Fig. 3 The collection data program

C. Data Analysis and Algorithm

The parameters used in analyses are similar to previous studies [4](M. Kangas, A. Konttila,P. Lindgren, I. Winblad, T. Jämsä, 2008 Feb 21) [5] (Ig-Jae Kim, Saemi Im, Eugene Hong, Sang Chul Ahn and Hyoung-Gon Kim, 2007) [10](J. Klenk, C. Becker, F. Lieken, S. Nicolai, W. Maetzler, W. Alt, W. Zijlstra, 5 November 2010). The total sum acceleration vector Acc, contain both dynamic and static acceleration components, is calculated from sampled data as indicated in Eq. (1)

$$Acc = \sqrt{(A_x)^2 + (A_y)^2 + (A_z)^2} \qquad (1)$$

Where A_x, A_y, A_z is the acceleration (g) in the x, y, z axes, respectively.

Similarly to the acceleration, the angular velocity is calculated from sampled data as indicated in Eq. (2)

$$\omega = \sqrt{(_x)^2 + (_y)^2 + (_z)^2} \qquad (2)$$

Where $_{x, y, z}$ is the acceleration (g) in the x, y, z axes, respectively.

There are 324 tests collected from 18 subjects. Figure 5 shows a typical example of the acceleration (Fig.5a.) and angular velocity signals (Fig5.b) during stand, walk, sit down/stand up, step and fall.

When stationary, the acceleration, from tri-axial accelerometer is approximated +1g, and angular velocity is 0°/s. When the subject moves, the acceleration is changing and the angular velocity produces a variety of signals along activities. Over all of test, with walking activity, we found that average peak acceleration reached 1.5-1.8 g. and the average peak angular velocity reached 10-50 °/s. With sitting down/ standing up, the average peak acceleration is

same with walking. However, when human sit down, the upper torso bend down and backward, vice versa with standing up. So the average peak angular velocity of sitting down/standing up is higher walking, they reached 80-130 °/s. Similar with lying back/ sitting up (Fig.7), the average peak acceleration reached 1.4-1.8g, and the average peak angular velocity reached 130-260 °/s.

However, with high density ADLs such as stepping (Fig.5), the average peak acceleration and angular velocity is higher than walking, 1.8-2.1 g for acceleration and 40-80 °/s for angular velocity. With running, the average peak acceleration is higher stepping (2.2-2.8 g) and the average peak angular velocity reached 60-120°/s. Especially, with falling, the both acceleration and angular velocity were received high value. The average peak acceleration for falling is 2.5-5.4 g and the average peak angular velocity for falling is 200-320 °/s.

The table 1 is shown the summary of acceleration and angular velocity of ADLs and fall of 384 tests.

Table 1 Summary of acceleration and angular velocity of ADLs and fall.

Activities	Peak acceleration (g)	Peak angular velocity (°/s)
Standing	1	0
Walking	1.5-1.8	10-50
Sitting down/standing up	1.6-2.0	80-130
Lying down/ sitting up	1.4-1.8	130-260
Stepping	1.8-2.1	40-80
Running	2.2-2.8	60-120
Falling	2.5-5.4	200-320

Base on analyzing all of database, we propose a novel algorithm that can classify ADLs such as standing, walking, sitting down/standing up, stepping, running and detect the fall.

The flowchart of our algorithm is summarized in Fig.4.

The algorithm involves the following steps: Firstly, using acceleration thresholds (1g &1.8g) divides the activity to 3 groups: the first is standing, the second is normal ADLs, and the third is high density ADLs (include fall). With case normal ADLs, we can classify activity base on angular velocity thresholds (50°, 130°) such as walking, sitting down/standing up, lying down/ sitting up. With case high density ADLs, if average peak angular velocity is above 200°, the fall is detected. Otherwise, the average peak acceleration threshold (2.1g) is compared for classification between stepping and running.

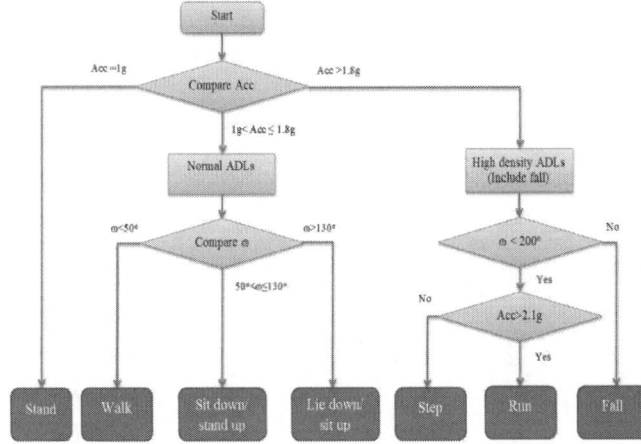

Fig. 4 Fall detection and ADLs classification algorithm schema

III. RESULTS AND DISCUSSION

The sensitive and the specificity of the system are defined as follows:

$$Sensitivity = \frac{No.TP}{No.TP+No.FN} \qquad (3)$$

$$Specificity = \frac{No.TN}{No.TN+No.FP} \qquad (4)$$

Where:

- Number of True positive (No.TP): an activity occurs, the device detects it.

- Number of False positive (No.FP): the device announces an activity, but it did not occur.
- Number of True negative (No.TN): another activity is performed; the device does not declare a activity.
- Number of False negative (No.FN): an activity occurs but the device does not detect it.

The result of 324 tests shows the sensitivity of the system and the specificity of system on table 2.

Table 2 Summary of sensitivity and specificity system.

Activities	Sensitivity (%)	Specificity (%)
Standing	100	100
Walking	97.14	95.93
Sitting down /standing up	91.35	94.67
Lying down/ sitting up	95.37	93.59
Stepping	98.75	99.07
Running	95.06	98.45
Falling	100	99.382

Compared to other algorithms, our algorithm not only classifies ADLs but also detect fall. Furthermore, it also has shown to have a higher sensitivity and specificity than previous study such as with 89.14% sensitivity and 89.97% specificity for walking [11] (Bidargaddi, N. ; E-Health Res. Centre, Brisbane ; Sarela, A. ; Klingbeil, L. ; Karunanithi, M., 2007), with 95% sensitivity for stepping [12] (Guillaume thuer and tim verwimp, 2008-2009) and with sensitivities greater than 85% and specificities between 88-94% for falling [3](U.Lindeann, A.Hock, M.Stuber, W.Keck, and C.Becker, 2005) [4] (M. Kangas, A. Konttila,P. Lindgren, I. Winblad, T. Jämsä, 2008 Feb 21) [5] (Ig-Jae Kim, Saemi Im, Eugene Hong, Sang Chul Ahn and Hyoung-Gon Kim, 2007) . The proposed method is a simple threshold method; therefore, it could be easily ported onto an electronic device worn by a person.

One of the limitations of this study is that young subjects performed the tests. Therefore, the young subjects might not be able to simulate the actual ADLs of the elderly. Therefore, the threshold values of the real condition could be significantly different than our findings. Further research is required to actually identify the correct threshold.

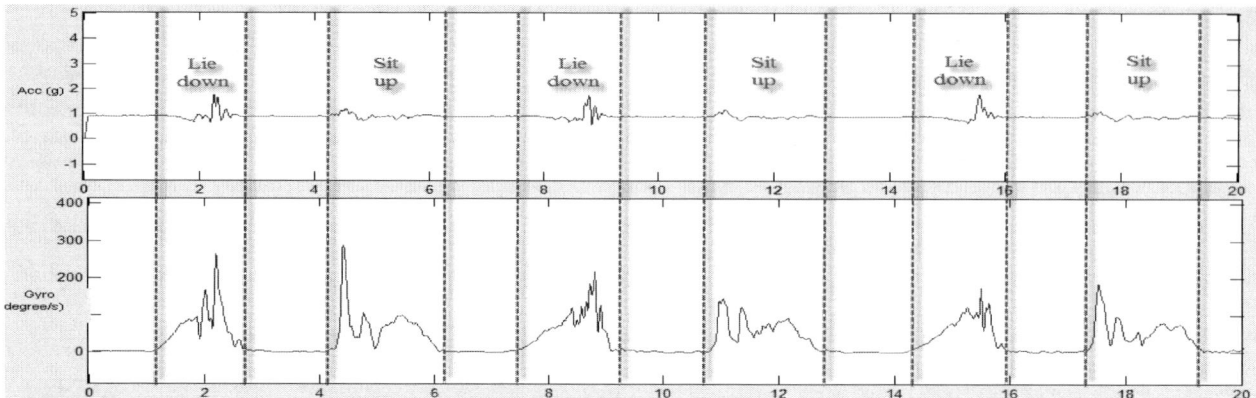

Fig. 5 The signal for standing, walking, sitting down/standing up, stepping and falling. (a)The sum of acceleration Acc.(b)The sum of angular velocity ω.

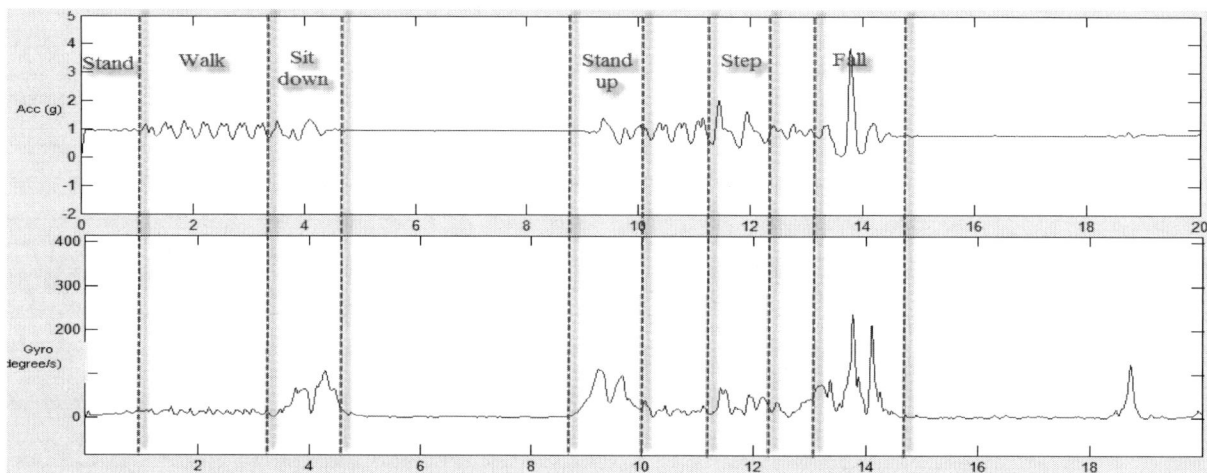

Fig. 6 The signal for running.(a)The sum of acceleration Acc.(b)The sum of angular velocity ω.

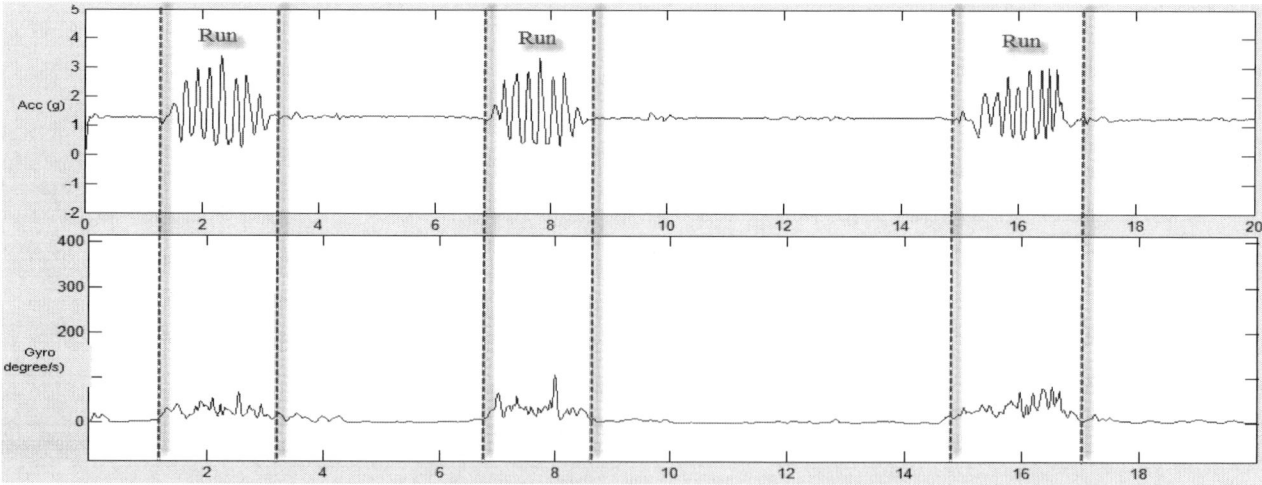

Fig. 7 The signal for lying back, sitting up.(a)The sum of acceleration Acc.(b)The sum of angular velocity ω.

IV. CONCLUSION AND FUTURE WORK

In this study, a wireless sensor system is implemented to measure the acceleration and angular velocity at center chest on the body for ADLs and four different types of fall. The collected data is used to evaluate the performance of the algorithm. There are many fall detections or ADLs classification systems investigated in previous studied. However, to increase the accuracy, we proposed a combination of accelerometer and gyroscope simultaneously. As the results, we have improved the accuracy, specifically the specificity and the sensitivity. We also have used small battery operated devices, which can be easily woven into garments.

In conclusion, fall detection and ADLs classification system has been validated to show high sensitivity and specificity results, using the combination of accelerometer and gyroscope. Future development will investigate a system to include GPS and GPRS to inform medical attention. Furthermore, an air bag is integrated the system for preventing the elder out of injury when they fall.

REFERENCES

[1] Quoc T. Huynh, Uyen D. Nguyen, Su V. Tran, Afshin Nabili, Binh Q. Tran, "Fall Detection Using Combination Accelerometer and Gyroscope," in *International Conference on Advances in Electronics Devices and Circuits-EDC 2013*, Kuala Lumpur, Malaysia, 2013, pp. 52 - 56.

[2] J.Y. Hwang, J.M Kang, H.C. Kim, "Development of novel algorithm and real-time monitoring ambulatory system using Bluetooth module for fall detection in the elderly," pp. 2204-2207, September 1-5, 2004.

[3] U.Lindeann, A.Hock, M.Stuber, W.Keck, and C.Becker, "Evauation of a fall detector based on accelerometers: A pilot study," Medical and Biological Engineering and Computing, p. 43, Oct 2005.

[4] M. Kangas, A. Konttila,P. Lindgren, I. Winblad, T. Jämsä, "Comparison of low-complexity fall detection algorithms for body attached accelerometers.," Gait & Posture, Vol.35, pp. 500-505, 2008 Feb 21.

[5] Ig-Jae Kim, Saemi Im, Eugene Hong, Sang Chul Ahn and Hyoung-Gon Kim, "ADL classification Using Triaxial Accelerometers and RFID," in International Conference on Ubiquitous Computing Convergence Technology, 2007.

[6] Saisakul chernbumroong, Anthony S. Atkins, and Hongnian Yu, "Activity classification using a single wrist-worn acceleration," in IEEE International Conference on Software, Knowledge, Information Management and Application, benevento, italy, 08/11/2011.

[7] Tong Zhang, Jue Wang, Liang Xu, Ping Liu, "Fall Detection by Wearable Sensor and One-Class SVM," Proceedings of the 26th Annual International Conference of the IEEE EMBS, pp. 2204-2207, 2004.

[8] P. Jayachandran, T.F. Abdelzaher, and J.A. Stankovic R. K. Ganti, "SATIRE: A software architecture for smart attire," Proceedings of MobiSys'06, 2006.

[9] Quoc T. Huynh, Uyen D. Nguyen, Su V. Tran, Binh Q. Tran, "Optimum Location for Sensors in Fall Detection," International Conference on Green and Human Information Technology, Feb. 2013.

[10] J. Klenk, C. Becker, F. Lieken, S. Nicolai, W. Maetzler, W. Alt, W. Zijlstra, "Comparison of acceleration signals of simulated and real-world backward falls," in Medical Engineering & Physics, VOL.33, 5 November 2010, pp. 368-373.

[11] Bidargaddi, N. ; E-Health Res. Centre, Brisbane ; Sarela, A. ; Klingbeil, L. ; Karunanithi, M., "Detecting walking activity in cardiac rehabilitation by using accelerometer," in Intelligent Sensors, Sensor Networks and Information, 2007. ISSNIP 2007. 3rd International Conference , Melbourne, 2007, pp. 555 - 560.

[12] Guillaume thuer and tim verwimp, "Step detection algorithms for accelerometers," , 2008-2009, pp. elab-master.

[13] K.M. Culhane, G.M. Lyons, D. Hilton,P.A. Grace, D. Lyons, "Long-term mobility monitoring of older adults using accelerometers in a clinical environment," in Clinical Rehabilitation, 2004, pp. 335-343.

[14] A.K. Bourkea, K.J. O'Donovan, G. ÓLaighin, "The identification of vertical velocity profiles using an inertial sensor to investigate pre-impact detection of falls," in Medical Engineering & Physics, VOL.30, September 2008, pp. 937-946.

[15] M.N. Nyan,F.E.H. Tay, A.W.Y. Tan , K.H. Seah, "Distinguishing fall activities from normal activities by angular rate characteristics and high-speed camera characterization.," in Medical Engineering & Physics, VOL.28, Epub, 2006 Jan 6, pp. 842-849.

[16] A. K. Bourke, P.van de Ven, M. Gamble, R. O'Connor, K. Murphy, E. Bogan, E. McQuade, P. Finucane, G.OLaigin, J. Nelson, "Evaluation of waist-mounted tri-axial accelerometer based fall-detection algorithms during scripted and continuos unscripted avtivities," Journal of Biomechanics, pp. 3051-3057, 2006.

Using Near-Infrared Technique for Vein Imaging

Tran Van Tien, Pham T.H. Mien, Pham T. Dung, and Huynh Quang Linh

Faculty of Applied Science, University of Technology – VNU HCM, Vietnam

Abstract— Near-infrared (NIR) imaging technique has wide applications in biomedical field, in which the NIR vein image has considerably major benefits. NIR imaging allows visualizing the veins underneath the skin of patients having problem with vein visibility, mapping the normal and abnormal veins for diagnosis and treating. In addition, finger or palm vein recognition, as a highly secure and convenient technique of personal identification, has received increasing attention recently. In this study, NIR vein images of various parts of hand, including the finger, palm, wrist and arm were described using NIR imaging techniques in various modes such as transmission mode, reflection mode, and the combination of two mentioned modes. Wavelengths in the range of 750 to 940 nm, with low absorption window, provide higher contrast of vein imaging, which can be enhanced by use of crossed polarizers and neutral density filters to remove the glare from the skin surface. Finally, analysis and evaluation of advantages and disadvantages of these methods were presented.

Keywords— NIR imaging, vein infrared, vein detection.

I. INTRODUCTION

Intravenous access is the most important phase in medical practices of daily life. However, it is sometimes difficult and time-consuming, especially when using for geriatric, pediatric or obese patients. Nurses or paramedics may fail in the first intravenous injection and have to repeat many times, which causes a lot of pains or discomfort for the patients [1]. There are few devices and techniques using X-ray and ultrasound that can be used for veins localization. However, these methods are costly, bulky, radioactive, and suitable only for deep veins localization. Near-infrared imaging is a new technology that is currently being used widely in the biomedical fields without the use of ionizing radiation. Recently more and more studies [2-5] have been initiated to define the vein imaging. Vein NIR imaging technique works on the principles of light propagation, absorption, reflection and scattering in different layers of skin. When using a NIR illuminator in 750 – 940 nm, the highest absorption of mentioned light is principally inherent to the veins. Because of the low absorption coefficient of main absorbers of the skin, like hemoglobin, oxygenated hemoglobin and water in the NIR-wavelength range, the NIR light penetrates these tissues more effectively. NIR light is absorbed or scattered in the forward direction by blood, whereas it is scattered in skin and subcutaneous fat.

Therefore, blood appears as a dark region, whereas skin and fat appear lighter [6]. In addition, the NIR light are non-ionizing radiation, which can be applied serveral times on patients without harmful effects. Arcording to different parts of patients' body, some methods were used to capture the vein networks with maximizing the contrast between skin tisssues and veins [2-8]. Firstly, by using the comfortable lighting technique such as transmission mode, reflection mode or trans– flectance mode, vein imaging will be captured [6]. Secondly, the use of multiple illuminants with wavelengths in NIR range will allow us to determine the wavelength combination that will increase the contrast of imaging [9]. Thirdly, strong glare occurring on the skin surface during illumination process is one of the reason causing contrast deterioration in the NIR vein images. The cross polarizers and the neutral density filters were used to eliminate the amount of glare light that can pass through the camera. Finally, the image processing algorithms were also used to enhance the quality images and the contrast ratio of vein imaging [2,3]. This paper presents the analysis of veins imaging using above mentioned methods, and then gives a thorough comparison of the outcome of NIR imaging.

II. METHODS

A total of 20 volunteers (10 men and 10 women) participated in this study. Experiments involve data acquisition from various parts of hand, including the finger, palm, wrist and arm are described using NIR imaging techniques. An optical system has been set up to illuminate an area of the hand with high power NIR LEDs (T.Tech LED, China) having the following wavelengths: 750 nm, 850 nm and 940 nm. An IR bandpass filter (F. IR) (Edmund optice, USA) allows IR light transmission, but blocks the visible light in the camera. Two polarizers (POL) (Edmund optice, USA) were used, one in front of the LEDs and one in front of the camera, to reduce specular reflection from the subject's skin by cross polarization. The neutral density (N.D) filter (Edmund optice, USA) was used to eliminate the amount of glare light that can pass through the camera. The contrast of vein-skin image is significantly improved and the vein networks can be clearly distinguished from the background. The NIR transmission and reflectance imaging of vein were detected by the charge couple device (CCD) camera (640 x 480 pixel) (Questek Sony, Japan), where the

V. Van Toi and T.H. Lien Phuong (eds.), *5th International Conference on Biomedical Engineering in Vietnam*,
IFMBE Proceedings 46, DOI: 10.1007/978-3-319-11776-8_46

image was converted to an electronic signal that can be displayed on a computer.

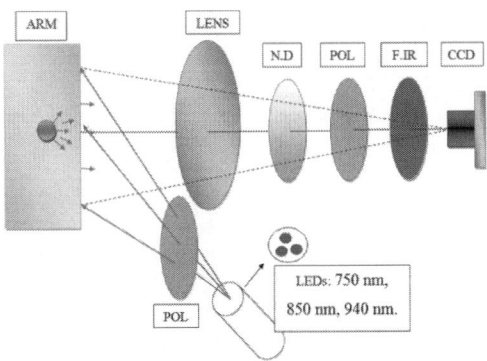

Fig. 1 Optical system model for vein imaging.

In the reflection mode, the infrared illuminator of LEDs and the camera are placed side by side in front of the hand. The camera, which is sensitive to this wavelength, captures the reflected light from the hand. In the transmission mode, the fingers, palm or wrist are placed between the camera and the infrared illuminator of LEDs. Also, in the transflectance mode is combination of two modes above. The image processing algorithms were used to enhance the quality images of vein in this study. There are various image processing techniques developed using Matlab which allow enhancing the contrast of the blood vessels. The **adapthisteq**() algorithms, which transform the values using contrast-limited adaptive histogram equalization, were used to enhance the contrast between the venous and background regions. The contrast enhancement can be limited in order to avoid amplifying the noise which might be present in the image.

To compare the veins to skin (background) contrast ratio of vein imaging, the contrast ratio, denoted by M, can be defined as below [4],

$$M = \frac{\left|U_{vein} - U_{skin}\right|}{\left(U_{vein} + U_{skin}\right)} \quad (1)$$

Where U_{vein} is the average pixel intensity of the vein area, and U_{skin} is the average pixel intensity of intact skin around the veins. In addition, the contrast ratio (C) was also quantified using Weber's law [10],

$$C = \frac{I_b - I_v}{I_b} \quad (2)$$

Where I_b is the background intensity and I_v is the intensity at the vessel region.

III. RESULTS

Firstly, the vein imaging of finger was captured using three imaging modalities: visible imaging, transmission imaging and reflectance imaging. Fig 2.a shows a visible imaging of finger which was captured by canon 550 D (18 mega pixel, Japan) and Fig 2.b, c show the NIR images of a human finger which were illuminated by high power LED 850 nm. As can be seen from the Fig 2.b (transmission mode), the blood vessels appear dark in color in contrast to the surrounding tissues that appear bright. It can be known that maximum amount of photons are absorbed by blood which is a mixture of oxygenated hemoglobin and water. The contrast ratio of transmission mode is better than of the reflectance mode (Fig 2. c). Because of vein living far from surface, the large amount of scattering and complexity of finger, the formation of good contrast becomes difficult.

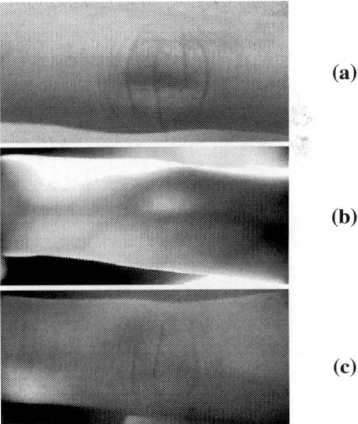

Fig. 2 Vein imaging of finger, (a) visible light, (b) NIR light transmision mode, (c) NIR light reflectance mode.

Fig 3 show the vein imaging of the palm. The palm has the same thickness as the finger and vein networks of palm are very close to the top surface. These properties make the venous appear dark in the image both transmission mode and reflectance mode. However, the vein networks in Fig 3. a so clearly than Fig 3. b. In Fig 3. c is the combination two methods above, the venous also appear dark but the contrast ratio ($M=0.40$; $C=0.60$) veins– skin imaging smaller than of the transmission mode ($M= 0.48$; $C=0.68$). Sometimes there is a need to focus on a particular part of the venous. The image enhancement algorithm increases the contrast of the region of interest from the surrounding region in order to highlight that specific area. The enhanced image is obtained by recording the image, inverting it, performing a histogram equalization to reduce the range of pixel intensities, and then taking an absolute difference of the two images. The

adapthisteq() algorithms were used to enhance the contrast between the venous and background regions. Fig 3. d illustrates the results of this algorithm (*M= 0.6 and C= 0.78*)

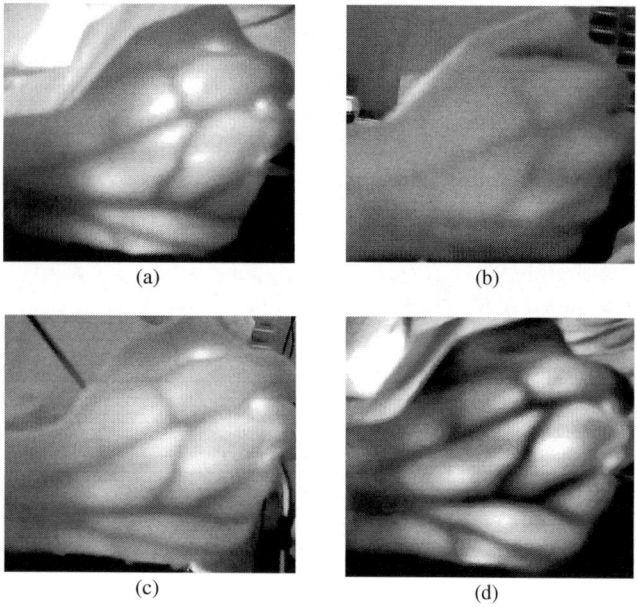

(a)   (b)

(c)   (d)

Fig. 3 Vein imaging of human palm (a) NIR light transmission mode, (b) NIR light reflectance mode, (c) NIR light trans-flectance mode, (d) Enhance image.

In the wrist (Fig 4), which being thicker than the palm, the NIR light of high power LED 850 nm also penetrates it and shows a detailed venous image.

In this part of patients' body, the transmission imaging (Fig 4. a) has the best location of vein networks, while the reflectance imaging and trans-flectance imaging (Fig 4. b, c) also allow to specify clearly vein. In the enhanced imaging (Fig 4. d), the contrast ratio of veins- background was improved (*M=0.45; C=0.62*) for a better quality of imaging.

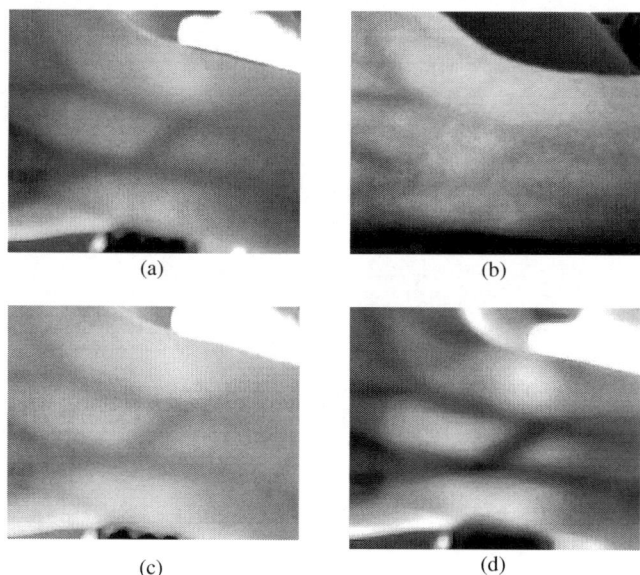

(a)   (b)

(c)   (d)

Fig. 4 Vein imaging of human wrist (a) NIR light transmision mode, (b) NIR light reflectance mode, (c) NIR light trans-flectance mode, (d) Enhance image.

For a clearer evaluation about the contrast ratio of veins to skin (background), we can see in Tab 1, where the values of contrast ratio were calculated by the formula (1,2). The NIR vein imaging of finger, palm and wrist were studied based on different methods of capture: transmission mode, reflection mode, trans– flectance mode and the enhance image.

The Tab 1 show that the contrast ratio of vein- skin changed in various tissues as well as when using different captured techniques. Firstly, in the palm the contrast ratio are the highest, following in the wrist, and lowest in the finger. Secondly, the NIR light imaging of transmission mode always has the contrast ratio higher than the reflectance mode and trans- flectance mode.

Table 1 The contrast ratio veins- skin of NIR vein imaging.

	Finger		Palm		Wrist	
	C	*M*	*C*	*M*	*C*	*M*
NIR light transmission mode	0.29	0.17	0.68	0.48	0.32	0.20
NIR light reflectance mode	0.09	0.05	0.17	0.11	0.15	0.09
T/R	3.2	3.4	4.0	4.4	2.1	2.2
Transmission enhance	0.48	0.3	0.78	0.60	0.62	0.45
Reflectance enhance	0.12	0.07	0.52	0.38	0.52	0.34
Te/Re	4	4.3	1.5	1.6	1.2	1.3

Lastly, with the image processing (in Matlab toolbox), the contrast ratio of vein- skin increases from near one to two times. The ratio of the contrast of transmission mode and reflectance mode show the index of *C* and *M* are similar. In there T/R are the ratio of transmission and reflecatance imaging contrast index, Te/Re are the ratio of transmission enhance and reflecatance enhance imaging contrast index.

When increasing the thickness of tissues such as human arm, neck or face, the NIR light of high power LED 850 nm cannot pass over all of them, so the transmission mode are less effecient in here. However with the reflectance mode, the NIR imaging also showed clealy vein. Fig 5. a shows a visible imaging of arm which was captured by canon 550 D (18 mega pixel, Japan), Fig 5. b shows a reflectance imaging of arm which was illuminated by high power LED 850 nm, and Fig 5. c is the ehanced image. We can see that the imaging in Fig 5. b (C= 0.22; M=0.12) and Fig 5. c (C= 0.36; M=0.22) have high contrast ratio when compared with Fengtao [4].

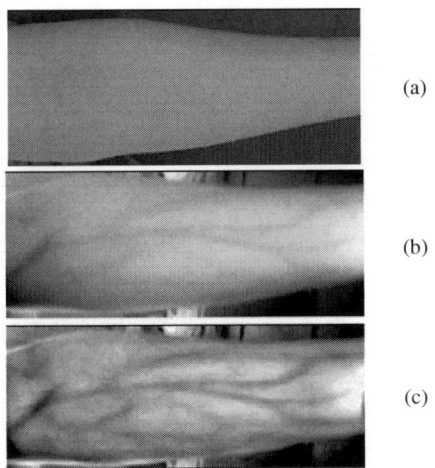

(a)

(b)

(c)

Fig. 5 Vein imaging of human arm, (a) visible light, (b) NIR light rfelectance mode, (c) Enhance image.

IV. CONCLUSIONS

The NIR imaging of vein has many advantages in biomedical field as well security field. It helps to locate vein with high contrast easily, exactly, and safely.

All of the NIR imaging techniques in this paper captured the venous in the human hand. High power LEDs, with wavelength 850 nm, give the light a deeper penetration in the thick tissues such as palm, wrist, and a part of arm. Together the cross polarizers and neutral density filters will control the light scattering in the skin surface, and allow better increase in the ratio contrast of imaging. With the use

of the transmission mode, the imaging has the clearer ratio contrast than the reflectance mode and trans-flectance mode; however, with the thickness areas such as arm the reflectance mode is more effective. The image processing helps enhance the contrast between the venous and background regions.

ACKNOWLEDGMENT

This work was supported by National Key Laboratory of Digital Control and System Engineering, and the department of Biomedical Engineering, HCMUT-VNU HCM.

REFERENCES

1. Schechter N L, Zempsky W T et al. (2007) Pain reduction duringpediatric immunizations: evidence-based review and recommendations. Pediatrics 119(5): 1184-98.
2. Vincent C P, Kenneth W T et al. (2009) 3D and Multispectral Imaging for Subcutaneous Veins Detection. Optics Express 17 (14): 11360-11365.
3. Anup P, Sam M (2011) Enhancement of Infra-Red Images of Human Hand. American Journal of Biomedical Engineering 1(1): 59-61.
4. Fengtao W, Ali B et al. (2013) High-contrast subcutaneous vein detection and localization using multispectral imaging, Journal of Biomedical Optics 18(5): 050504-1.
5. Zeman H D, Lovhoiden G et al. (2004) Prototype Vein Contrast Enhancer. Proc. SPIE, vol. 5318, 39-49.
6. Nabila B, Abdel H et al. (2010) Near-infrared image formation and processing for the extraction of hand veins. Journal of Modern Optics 57 (18): 1731–37.
7. Jinfeng Y, Ben Z, Yihua S (2012) Scattering Removal for Finger-Vein Image Restoration, Sensors 2012, 12, 3627-3640; DOI:10.3390/s120303627.
8. Tatsuhiko M, Tosiaki M et al. (2011) Qualitative near-infrared vascular imaging system with tuned aperture computed tomography. J. Biomed. Opt 16(7): 076004 DOI:10.1117/1.3595424.
9. Vincent C P, Jeffery R P et al. (2007) Combining near-infrared illuminants to optimize venous imaging. Proc. SPIE 6509, Medical Imaging 2007: Visualization and Image-Guided Procedures, 65090H (March 21, 2007); DOI:10.1117/12.712576.
10. Le V N Du, Wang Q et al. (2013) Monte Carlo modeling of light–tissue interactions in narrow band imaging. Journal of Biomedical Optics 18(1): 010504-1

Author: Tran Van Tien
Institute: Faculty of Applied Science, University of Technology
Street: Ly Thuong Kiet
City: Ho Chi Minh
Country: Vietnam
Email: tranvantien1985@gmail.com

Author: Huynh Quang Linh
Institute: Faculty of Applied Science, University of Technology
Street: Ly Thuong Kiet
City: Ho Chi Minh
Country: Vietnam
Email: huynhqlinh@hcmut.edu.vn

Ribs Suppression in Chest X-Ray Images by Using ICA Method

Hieu Xuan Nguyen and Tin Thanh Dang

Faculty of Electrical and Electronic Engineering, Ho Chi Minh City University of Technology,
Ho Chi Minh City, Vietnam

Abstract— **In many developing countries such as Vietnam, Chest X-Ray (CXR) images are still one of the common tools to diagnose the lung cancer at the early phase. Early detection of lung nodules increases the chance of survival for a patient. In this study, we apply the Independent Components Analysis (ICA) to separate the ribs and other parts in lung images. First 20 files in Japanese Society of Radiological Technology (JSRT) database are used to perform the ICA algorithm, resulting in 90% of cases that the ribs are completely and partly suppressed, 85% of cases increases the nodule visibility. Two image enhancement methods are histogram equalization and Frangi filter are used as two artificially created inputs for ICA. This research also proposes the value of free parameters of Frangi filter.**

Keywords— **Ribs suppression, Independent Component Analysis, Chest X-Ray, lung cancer, radiology.**

I. INTRODUCTION

Chest X-Ray plays an important role in diagnosis of lung cancer. According to the statistics reported by American Cancer Society for lung cancer in United State for 2013, there are about 224,210 new cases for lung cancer (116,000 in men and 108,210 in women), accounting for about 13% for all new cancers, an estimated 159,260 deaths from lung cancer (86,930 in men and 72,330 among women), accounting for about 27% of all cancer deaths. Each year, more people die of lung cancer than of colon, breast, and prostate cancers combined.

There are some advanced scanning techniques such as Computed Tomography (CT) and Magnetic Resonance Imaging (MRI) which help the radiologist have the exact diagnosis but CXR still keeps its role. The reason behind that is CT and helical CT examining the exposition of the patient to a higher dose of radiation, estimated to be about 100 times higher than that for a conventional CXR. Moreover, the widespread use of conventional chest radiographs is due to its economic feasibility.

A study showed that most lung cancer lesions that are missed on frontal chest radiographs are located behind the ribs and that the inspection of a soft tissue image can improve the detection performance of humans [1]. Automated lung cancer detection schemes also suffer from false positives caused by superposition of bony structures [2]. Including analysis of simulated soft-tissue images may improve performance of lung nodule detection systems.

As an image, the CXR is the linear combination of the ribs, the soft tissues and the noise. Noise can be removed by many of image enhancement theories. To separate the ribs from soft tissues, a specialized technique is required. In this research, we apply the method [3] that presents the usage of ICA in ribs suppression with the enhancement by using histogram equalization with some propositions on algorithm, time, and parameters. Concretely, we propose the value of three parameters of Frangi filter.

II. MATERIALS AND METHODS

A. Proposed Method

ICA needs at least two observation vectors as its input. In fact, there is only one CXR original image. To overcome this, we artificially create two inputs by using two image enhancement methods, they are histogram equalization and Frangi filter [4]. The process below is our proposal method.

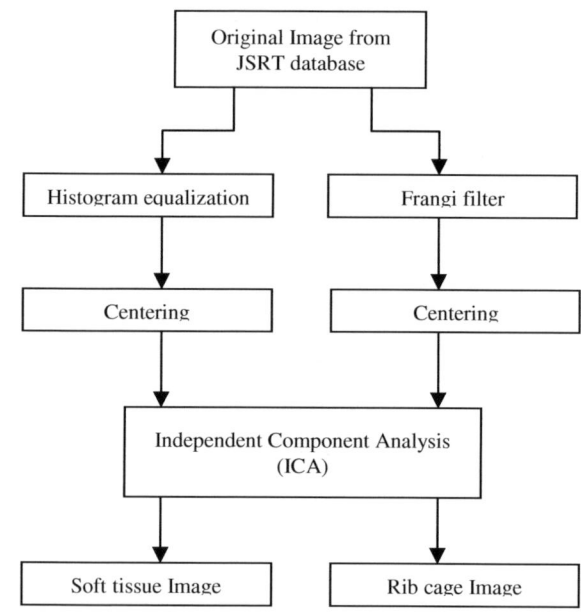

The raw file from JSRT database is pre-processed to down the gray scale to 256 levels. This will help to display the output image in most of current software. Histogram equalization method increases the contrast of image, while Frangi filter approach enhances the tubular structures in image.

V. Van Toi and T.H. Lien Phuong (eds.), *5th International Conference on Biomedical Engineering in Vietnam,*
IFMBE Proceedings 46, DOI: 10.1007/978-3-319-11776-8_47

B. JSRT Database

The standard digital image database with and without chest lung nodules (JSRT database) was created by the Japanese Society of Radiological Technology (JSRT) in cooperation with the Japanese Radiological Society (JRS) in 1998. Since then, the JSRT database has been used by a number of researchers in the world for various research purposes such as image processing, image compression, evaluation of image display, computer-aided diagnosis (CAD), picture archiving and communication system (PACS), and for training and testing.

Descriptions of the JSRT database:

- Useful for ROC (receiver operating characteristic) analysis (154 nodule and 93 non-nodule images)
- High resolution (2048 x 2048 matrix size, 0.175mm pixel size). Wide density range (12bit, 4096 gray scale)
- Universal image format (no header, big-endian raw data). Useful for diagnostic training and testing

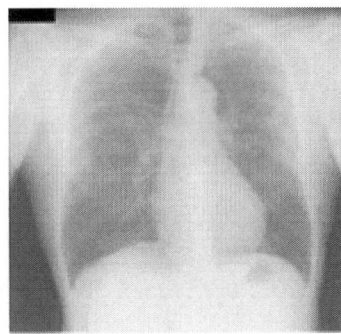

Fig. 1 JPCLN001 image from JSRT

Fig. 1 presents an example, an image in this database, the image of JPCLN001. With the image like that, it is very difficult for us to watch tissues behind ribs.

In this study, we down the gray scale from 4096 (12 bits) to 256 (8 bits). This experimental choice reduces the computational costs of the algorithm without affecting its performance. Besides that, all of current display software can display 8 bit images.

C. Histogram Equalization

This image enhancement method is a common tool in digital image processing to increase the contrast of an image. In the JSRT database, the diameters and the positions of the nodules are provided along with the images. The nodule diameters and their intensities vary from nearly invisible to very bright. The output image has uniform probability that is a non-Gaussian signal. So that, applying histogram equalization to original will get two goals: increasing the image contrast and guaranteeing the requirement of ICA method (at least one signal in non-Gaussianity).

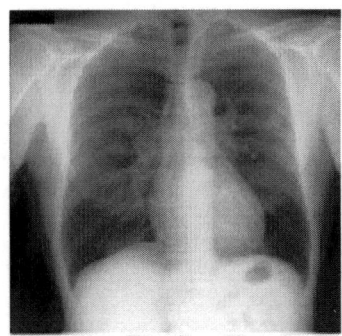

Fig. 2 Histogram equalization image of JPCLN001

Fig. 2 gives an example for the step of histogram equalization in the scheme.

D. Vessel Enhancement by Using Frangi Filter

This method was presented by Alejandro F. Frangi [4] in 1998. The vessel enhancement is a filtering process that searches for geometrical structures which can be regarded as tubular. Since vessels appear in different sizes it is important to introduce a measurement scale which varies within a certain range. A common approach to analyze the local behavior of an image, L, is to consider its Taylor expansion in the neighborhood of a point x_0,

$$L(x_0 + \delta x_{0,s}) \approx L(x_{0,s}) + \delta x_0^T \nabla_{0,s} + \delta x_0^T H_{0,s} \delta x_0 \quad (1)$$

This expansion approximates the structure of the image up to second order. $\nabla_{0,s}$ and $H_{0,s}$ are the gradient vector and Hessian matrix of the image computed in x_0 at scale s.

The third term in Equation (1) gives the second order directional derivative,

$$\delta x_0^T H_{0,s} \delta x_0 = \left(\frac{\partial}{\partial \delta x_0}\right)\left(\frac{\partial}{\partial \delta x_0}\right) L(x_{0,s}) \quad (2)$$

Base on the eigenvalue (λ_1, λ_2) of Hessian matrix, Frangi defines the parameters as 2-D patterns in Table 1 below.

Table 2 2-D Possible patterns depending on the value λ_k

2-D		Orientation pattern
λ_1	λ_2	
N	N	Noisy, no preffer direction
L	H-	Tubular structure (bright)
L	H+	Tubular structure (dark)
H-	H-	Blob-like structure (bright)
H+	H+	Blob-like structure (dark)

with N=Noisy, L=Low, H=High, +/- indicates the sign of the eigenvalue. The eigenvalues are ordered with $|\lambda_1| \le |\lambda_2|$.

In particular, the grey value of the pixel belonging to the vessel region is signaled by a small value of λ_1 (ideally zero), and a large magnitude of λ_2 (the sign is an indicator of brightness/darkness).

Frangi therefore proposed the following combination of the components to define a vesselness function,

$$V_0(s) = \begin{cases} 0 & if \ \lambda_2 > 0 \\ \exp\left(R_\beta^2\right)\left(1 - \exp\left(-\dfrac{s^2}{2c^2}\right)\right) \end{cases} \quad (3)$$

where $R_\beta = \dfrac{\lambda_1}{\lambda_2}$

The vesselness measure in Equation (3) is analyzed at different scales, s. The response of the line filter will be maximum at the scale that approximately matches the size of the vessel detected. Fig. 3 gives an example with Vessel enhancement by using Frangi filter.

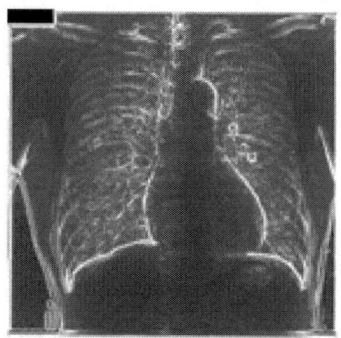

Fig. 3 Vessel enhancement of JPCLN001

E. Independent Component Analysis

Independent component analysis (ICA) is a developed method in which the goal is to find a linear representation of non-gaussian data so that the components are statistically independent, or as independent as possible. This technique was presented by Aapo Hyvärinen [5] [6] [7].

The linear ICA assumes that the signal x is a linear mixture of the independent components,

$$x_j = a_{j1}s_1 + a_{j2}s_2 + ... + a_{jn}s_n \quad \text{for all j} \quad (4)$$

Denoting by using matrix notations as the followings:

$\mathbf{x} = (x_1, x_2, ..., x_n)^T$ are observation vectors
$\mathbf{s} = (y_1, y_2, ..., y_n)^T$ are source signals
$\mathbf{A}$ is the mixing matrix with elements a_{ij}

The above mixing model is written as

$$\mathbf{x} = \mathbf{A}\,\mathbf{s} \quad (5)$$

After estimating the matrix $\mathbf{A}$, computing its inverse, $\mathbf{W}$, and obtaining the independent component simply by:

$$\mathbf{s} = \mathbf{W}\,\mathbf{x} \quad (6)$$

ICA is very closely related to the method called blind source separation (BSS) or blind signal separation. A "source" means here an original signal, i.e. independent component. "Blind" means that we know very little about the mixing matrix, and we can make little assumptions on the source signals. Fig. 4 as the following is the result, the output image after the process of ICA of the image of JPCLN001.

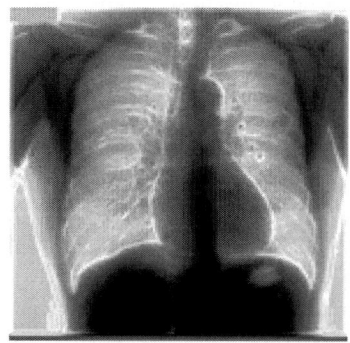

Fig. 4 Output image after ICA of JPCLN001

III. RESULTS AND DISCUSSION

First 20 files of JSRT database are used to perform the ICA estimation for evaluation. The free parameters of Frangi filter is chosen as $\beta = 0.5$, as recommended in [4], c=1, and scale s=10.

Table 1 Result of ICA for first 20 files in JSRT database

Image	Ribs suppression	Nodule visibility
JPCLN001	Partly	Increase
JPCLN002	Suppressed	Increase
JPCLN003	Suppressed	Increase
JPCLN004	Partly	Increase
JPCLN005	Partly	Same
JPCLN006	Partly	Increase
JPCLN007	Partly	Increase
JPCLN008	Partly	Increase
JPCLN009	Partly	Increase
JPCLN0010	Partly	Increase

Table 1 (continued)

JPCLN0011	Partly	Increase
JPCLN0012	Partly	Increase
JPCLN0013	Partly	Increase
JPCLN0014	Partly	Same
JPCLN0015	Partly	Increase
JPCLN0016	Partly	Increase
JPCLN0017	Partly	Increase
JPCLN0018	Not	Same
JPCLN0019	Not	Increase
JPCLN0020	Partly	Increase

The fast fixed-point algorithm (Fast ICA) using negentropy to maximize the non-gaussianity is applied to get the enhancement.

From the result, there are 2 cases (10%) where the ribs are completely removed, 16 cases (80%) where the ribs are partly suppressed and 2 cases (10%) where the ribs are not removed. From nodule visibility view, 17 cases (85%) enhance the visibility, 3 cases (15%) keep same quality of image, and there is no case that causes the image worse.

The main factor affects the result is the parameters of Frangi filter. In this work, we propose the fixed values for three parameters of Frangi filter which can make the visibility better. This is not mentioned in [4].

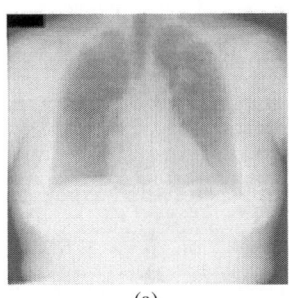 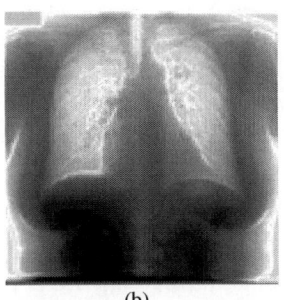

(a)   (b)

Fig. 5 JPCLN003 image
(a) Original (b) Ribs suppressed increasing the nodule visibility

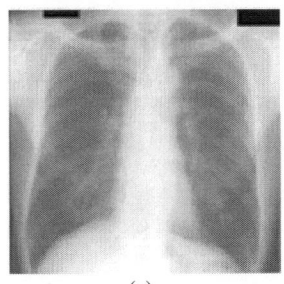 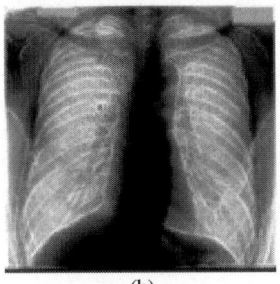

(a)   (b)

Fig. 6 JPCLN018 image.
(a) Original (b) Ribs are not suppressed with same nodule visibility

IV. CONCLUSIONS

The result from this study shows that ICA method can be applied to perform the ribs suppression, aims at improving the diagnostic result of CXR images. We propose the values of three free parameters of Frangi filter which works for first 20 cases in JSRT database. Besides that, the histogram equalization method is used to guarantee the ICA convergence. In future work, the values of Frangi filter's parameters are armed at computing based on the characterization of each image.

REFERENCES

1. Shah P.K. Missed nons mall cell lung cancer: radiographic findings of potentially resectable lesions evident only in retrospect. Radiology. 2003;226(3):235–241.
2. Loog M, Van Ginneken B, Schilham A. M. R. Filter learning: Application to suppression of bony structures from chest radiographs. Medical Image Analysis. 2006:826–840.
3. Mohammad A.U. Khan. Rib Suppression Using Blind Source Property of Independent Component Analysis. Information Technology Journal, 2008. Volume 7, pp.378-381.
4. Alejandro F. Frangi. Multiscale vessel enhancement filtering. MICCAI. Volume 1496 of Lecture Notes in Computer Science, Springer Berlin/Heidelberg, September 1998. 130–137.
5. A.Hyvarinen. Independent Component Analysis: Algorithms and Application. Neural Networks, 2000, 13(4-5): 411-430.
6. Hyvarinen. Fast and robust fixed-point algorithm for Independent Component Analysis, IEEE Transactions on Neural Networks, 2001.
7. A.Hyvarinen, J.Karhunen, and E.Oja, Independent Component Analysis, John Wiley & Sons, 2001.

Author: Hieu Xuan Nguyen
Institute: Ho Chi Minh city University of Technology
Street: Ly Thuong Kiet
City: Ho Chi Minh
Country: Viet Nam
Email: xuanhieudlm@gmail.com/hxnguyen@apm.com

Ultrasound Ovary Image Classification Using Kσ-Classifier

B.S. Usha[1] and S. Sandya[2]

[1] R N S Institute of Technology/ECE, VTU, Bengaluru, India
[2] Nitte Meenakshi Institute of Technology/ ECE, VTU, Bengaluru, India

Abstract— **Transvaginal UltraSound (TVUS) imaging is preferred imaging modality in detection of ovarian abnormalities. The ovarian parameters are measured manually by the expert and the shape of the Ovary is analyzed subjectively. There is a need for computer-assisted diagnostic support system to aid the experts in faster diagnosis as Manual measurement is time consuming. In this paper, we have extracted geometrical and shape features of the Ovary and have used Kσ-classifier to classify the Ovary as normal or abnormal. The proposed method is tested on Transvaginal ultrasound images of ovaries. The obtained experimental results are validated with the manual measurements and inferences by the medical expert and demonstrate the efficacy of the method. The algorithm could achieve a classification rate of 76.67% for Bilinear filtering-Contrast Stretched-Adaptive Thresholding (BCAT) method and 85.8% for Anisotropic filtering-CLAHE-Adaptive Thresholding (ACAT) method.**

Keywords— **Ultrasound, Ovary, Speckle filters, Feature extraction, kσ-classifier.**

I. Introduction

Ultrasound imaging has become an integral diagnostic tool to evaluate the gynecological health of a woman [1]. In this paper, we propose an algorithm for computer-assisted analysis of the ultrasound images of women's ovaries.

Ovary is most frequently scanned organ to diagnose abnormalities like infertility, irregular menstrual cycles, Polycystic Ovary Syndrome (PCOS), cysts, tumors etc. Determination of ovarian shape and size are one of the first steps in evaluation of the health of the Ovary [1, 2].The normal range of Ovary measurement is listed in Table 1 which was obtained in literature and supported by our expert Dr. Sriprada Vinekar, Gynecologist and Sonographist.

Table 1

Normal Ovary	Min	Max
Major-axis length M_1	$M_{1min} = 3cm$	$M_{1max} = 5cm$
Minor-axis length M_2	$M_{2min} = 1.5cm$	$M_{2max} = 3cm$
Thickness	0.5cm	1.5cm
Shape	Oval (Almond)	

II. Background Work

Automated measurement of ovarian parameters requires segmentation of the Ovary from the ovarian ultrasound images. The ultrasound images are of low contrast and poor quality due to the presence of speckle noise which is visible as a dense granular noise that is spread throughout the image. In computer assisted processing of ultrasound images, speckle reduction is one of the pre-processing techniques used to improve the image quality and possibly the diagnostic potential of the image. In [1] Bilateral filter proposed by Tomasi C, Manduchi R[3] was used for speckle removal, Contrast Stretching and Power Law Transformation were used for enhancement, adaptive Thresholding and Global Thresholding for segmentation, morphological operations to extract Ovary. The average Error Percentage for Major-Axis length (EM1) obtained was 11.9% and the average Error Percentage for Minor-Axis length (EM2) obtained was 17.4%. The above literature presents methods for segmentation of ovarian follicle from ovarian ultrasound image. In [2] anisotropic diffusion filter proposed by Perona and Malik [4, 5] was used for speckle reduction, CLAHE [6] and global enhancement methods followed by adaptive binary thresholding, morphological operators to extract Ovary. The algorithm could achieve an average Error Percentage of 5.27% for Major-Axis length (E_{M1}) and average Error Percentage of 6.1% for Minor-Axis length (E_{M2}). Horizontal scaling thresholding (HST) and vertical scaling thresholding (VST) techniques was proposed for extraction of follicles. Area, Ratio of major-axis length to minor-axis length, centroid, compactness and extent were the features used for classification. The classification rate of the method was 83.33%, false acceptance rate (FAR) -2.08% and false rejection rate (FRR) -16.66% [7, 8]. In this paper we have used kσ-classifier to classify Ovary as normal or abnormal based on the geometric features.

The paper is organized as follows: Section II discusses BCAT and ACAT methods where speckle reduction filters, contrast enhancement, segmentation techniques and feature extraction are explained in detail. In Section III, the experimental results are discussed along with manual measurements done by the expert and in section IV conclusion and future scope of work is discussed.

© Springer International Publishing Switzerland 2015
V. Van Toi and T.H. Lien Phuong (eds.), *5th International Conference on Biomedical Engineering in Vietnam,*
IFMBE Proceedings 46, DOI: 10.1007/978-3-319-11776-8_48

III. ALGORITHMS FOR AUTOMATIC MEASUREMENT OF OVARIAN FEATURES

This paper is focused on classifying the Ovary as normal or abnormal using the kσ-classifier. Automated measurement of ovarian size and shape parameters of ovarian ultrasound images is implemented using two algorithms, Bilinear filtering-Contrast Stretched-Adaptive Thresholding (BCAT) [1] algorithm and Anisotropic filtering-CLAHE-Adaptive Thresholding ACAT [2] algorithm are discussed in detail in Section A and B. The input is a Transvaginal ultrasound image (TVUS) of Ovary acquired from GE-LOGIQ Book XP ultrasound machine which is a high performance multipurpose color hand-carried imaging system. The images are obtained in JPEG-LS (24-bit, 532x434x3) format.

The proposed approach has two phases. In phase I Region of Interest (ROI) Ovary is extracted. Phase II is classification of Ovary. The algorithm is designed according to typical object recognition scheme. The input to Phase I, shown in Fig. 1 is a 24-bit RGB image. Since each plane is identical, only R-plane is considered as the input image. To begin with the resolution of the image i.e., the number of pixels making one cm is determined so as to measure the Ovary in cm. This can be found from the input image which has the scale marked on right hand side of the image. The position of the Ovary in all images is in the center of the fan area, so a sub-image containing only the Ovary (ROI) of size [380 410] is extracted. This image I_{ROI} is the input to the Ovary Classification System.

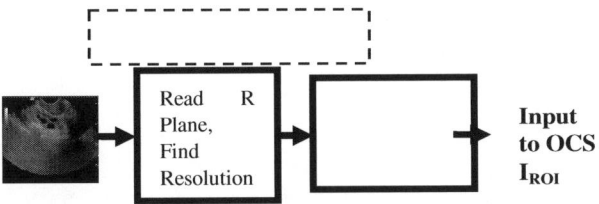

Fig. 1 Phase I :ROI extraction

A. Feature Extraction

In order to classify the Ovary as normal or abnormal it is required to extract representative features of the Ovary. In this paper we have considered geometric features of the ultrasound Ovary images. The geometric features can be classified into two feature sets: F1- the size based features and F2- the shape based features. F1 feature set includes the Major-axis length M1, Minor-axis length M2, Area A and Perimeter P. F2 feature set includes Form Factor F, Compactness Q, Eccentricity E. An additional parameter Centroid (C= Cx, Cy) is also extracted so as to select potential ovary region when multiple objects are segmented.

B. F1 Feature Set

1. Ovary is oval (ellipse) in shape and Fig 2 shows its approximate representation. Major-axis length (M_1) and Minor-axis length (M_2) define an ellipse feature. The normal range of M_1 and M_2 is given in Table 1. Fig 3 shows the expert marked values of M_1 and M_2.

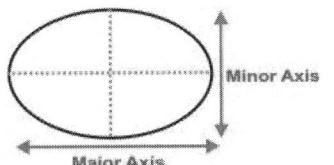

Fig 2. An ellipse

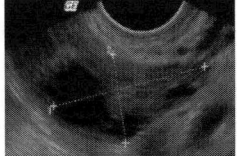

Fig 3 Ovary (1-major-axis 2-minor-axis)

2. The normal area of Ovary ranges between A_{min}=3.54cm^2 to A_{max}=11.8cm^2. Area is the number of pixels inside an Ovary which is given by (1).

$$A = \frac{\pi}{4} M1\, M2 \qquad (1)$$

3. Perimeter P is the total number of pixels around the inner boundary of the Ovary. The normal perimeter of Ovary ranges between P_{min}=7.45cm to P_{max}=12.95cm. The approximate formula to calculate perimeter of an ellipse is given by (2)

$$P \approx 2\pi \sqrt{\frac{a^2 + b^2}{2}} \qquad (2)$$

Where a=M_1/2 and b=M_2/2.

C. F2 Feature Set

1. Form Factor F is given by (3). Typical value of F of the ovary lies in the range F_{min}=0.8 to F_{max}=0.88

$$F= 4*\pi*Area/Perimeter^2 \qquad (3)$$

2. Compactness of is defined as the ratio of the square of the perimeter of a region to its area and is given by (4). Since Ovary is elliptical in shape, they have low compactness value (typical values from Q_{min}=14.2 to Q_{max}=15.6).

$$Compactness\ Q = Perimeter^2 / Area \qquad (4)$$

3. Eccentricity E is the ratio of the distance between the foci of the ellipse and its major axis length. The eccentricity equation is given in (5)

$$eccentricity = \sqrt{\left(1 - \frac{b^2}{a^2}\right)} \qquad (5)$$

Where a=M_1/2 and b=M_2/2. For an Ovary Eccentricity should be > 0.5.

4. Centriod C= Cx,Cy is the center of mass of the region. Cx and Cy are x and y coordinates of the center of mass. Objects of interest in an ultrasound image have the best visual quality when they are located in the center of the field of view. Hence we assume that Ovary lies in the center of the fan area. If segmentation leads to multiple objects then this parameter is used to choose the potential object likely to be an Ovary.

D. Bilinear filtering-Contrast Stretched-Adaptive Thresholding (BCAT)

The Ovary Classification System (OCS) is shown in Fig.4. In BCAT method, speckle noise is removed using Bilateral filter to get the filtered output I_{f1}. The Filtered output does not have good contrast between the ROI (Ovary) and its background. Hence contrast enhancement techniques-Contrast Stretching and Power Law Transformation are applied to get enhanced image I_{e1}. Adaptive thresholding (Local Region based thresholding) and Global thresholding methods are used to segment Ovary. The output of Global thresholding I_{G1} is used as an image mask to remove significant amount of background region from Adaptive thresholded image I_{A1} and the resultant segmentation output image is I_{T1}. But some spurious regions still remain in the segmented image I_{T1}. These regions are removed using morphological operators region filling, closing, erosion and dilation. The largest area corresponds to Ovary region and hence this component is chosen from the labeled components thus removing the spurious objects other than ROI (Ovary) in the image. The features of the Ovary are extracted and these features are used to classify Ovary as normal or abnormal using Kσ-classifier which is discussed in the next section.

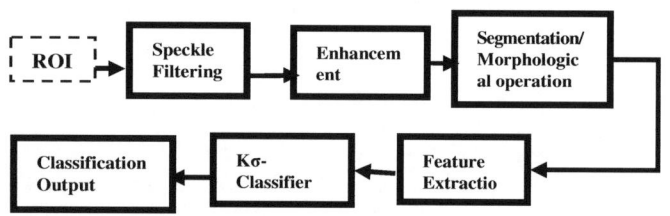

Fig. 4 Ovary Classification System

E. Anisotropic Filtering-CLAHE-Adaptive Thresholding (ACAT)

In ACAT method, Anisotropic diffusion filter is used for speckle noise removal. The filtered output I_{f2} has very good mean preservation, variance reduction, and edge localization compared to bilateral filter. As discussed in the previous section, in order to create good contrast between ROI (Ovary) and background we have used Contrast Limited Adaptive Histogram Equalization (CLAHE) [6] to get the enhanced image I_{e2}. The processing stage has Segmentation, Morphological operation and feature

extraction. The image segmentation is performed in two steps: global enhancement followed by binary thresholding.

The resulting image after binary thresholding is a binary image which contains ovary and other spurious regions. In order to remove the spurious regions, we apply morphological operations Erosion and Dilation. All regions connected to the image border are removed followed by region filling operation. But many small phantom regions still exist. A connected component labeling (CCL) algorithm finds all connected components in an image and assigns a unique label to all points in the same component. The component with the maximum area is retained which is the Region of Interest (Ovary). The geometrical properties of this region are extracted as discussed in section A. These parameters are input to the classification section.

F. Classification

The classification phase is divided into training phase and testing phase. In the training phase, the representative geometric feature set of the ovary are extracted. The eight geometric features, namely, the major-axis length M_1, minor-axis length M_2, the area A, perimeter P, Form factor F, compactness Q, Eccentricity E, Centroid C (Cx, Cy) are considered for classification of Ovary as normal or abnormal. μM_1, μM_2, μA, μP, μF, μQ, μE and μC (μCx, μCy) are the mean values and σM_1, σM_2, σA, σP, σF, σQ, σE and σC (σCx, σCx) are the standard deviation of each of these parameters. The μ and σ are listed in TABLE 2.

Table 2

Parameter	BCAT	ACAT
μM_1, σM_1	3.382, 0.774	3.389, 0.696
μM_2, σM_2	1.874, 0.499	1.838, 0.502
μA, σA	6.238, 2.599	5.014, 1.886
μP, σP	9.955, 2.712	10.28, 2.82
μF, σF	0.599, 0.08	0.59,0.1
μQ, σQ	21.16,3.01	21.62,4.99
μC_x, σC_x	189.99, 34.7	191.32, 37.65
μC_y, σC_y	109.6, 26.188	110.77, 27.277

The classification rules of the kσ-Classifier are formulated as below:

A region with major-axis length M_1, minor-axis length M_2, area A, perimeter P, form factor F, compactness Q, eccentricity E, and centroid C = (Cx, Cy), is classified as a normal Ovary, if the following conditions Ri, i=1,2,.....9 are satisfied:

R1: $\mu M_1 - k/3\ \sigma M_1 + \alpha \leq M_1 \leq \mu M_1 + k\sigma M_1 + \alpha$

R2: $\mu M_2 - k/3\ \sigma M_2 + \alpha \leq M2 \leq \mu M_2 + k\sigma M_2 + \alpha$

R3: $\mu A - k/2\ \sigma A + \alpha \leq A \leq \mu A + k\ \sigma A + 2\ \alpha$ for BCAT

$\mu A - k/2\ \sigma A + 2\alpha \leq A \leq \mu A + k^2\ \sigma A + \alpha$ for ACAT

R4: $\mu P - k/2\ \sigma P + \alpha \leq P \leq \mu P + k\ \sigma P + \alpha$

R5: $\mu F - k\ \sigma_F \leq F \leq \mu F + k\ \sigma_F$

R6: μQ - k σ_Q $\leq$ Q $\leq$ μQ + k σ_Q
R7: μE - k σ_E $\leq$ E $\leq$ μE + k σ_E
R8: μCx – k/2 σCx $\leq$ Cx $\leq$ μCx + kσCx
R9: μ Cy- k/2 σCy -γ$\leq$ Cy $\leq$ μCy + k σCy

The set of conditions Ri (R1 to R9) constitute the kσ-classifier. The value of *constant k is 2, α is 0.3 and γ is 14* in the experiment. All the values are empirically determined.

In the testing phase, Feature set F1 and F2 are computed for the segmented region of Ovary and the classification rules Ri are applied to determine whether the Ovary is normal or not. R1 to R4 classifies the Ovary based on size and are the key rules for classifying. R5 to R7 are based on shape and support the previous rules. Ovary lies in the center of the spanning area. So when multiple regions of same size are the segmentation outcome, then R8, R9 are used to eliminate regions which are not Ovary and retain Ovary.

IV. RESULTS AND DISCUSSIONS

The database consists of 75 Transvaginal ultrasound images of ovaries. The size parameter M1 and M2 are provided by the expert for all the 75 images. The classification output is validated against the expert inputs. In our experiment, 40 images are used in training phase and 35 (20 normal, 15 abnormal) in testing phase. The sample outputs of ACAT and BCAT algorithms are shown in Fig 5. The first column of images shows the expert measured values of Major-axis length M_1 and Minor axis length M_2. The second column shows the output images obtained by BCAT algorithm and the third column shows the output images obtained by ACAT algorithm. The F1 features and F2 features and classification output of the sample images shown in Table 4. It can be observed from TABLE 4 that the feature values obtained by ACAT algorithm are more nearer to the expert values than BCAT algorithm. The error percentage parameter qualifies the algorithm as to how best or how near the algorithm is to the expert measured values. It is required that ideally the error percentage is "ZERO" but a ±12% deviation is acceptable as per the expert.

The error percentage [1,2] for Major-axis length (EM1) and Minor-axis length (EM2) is calculated by using (6) and (7)

$$E_{M1} = \frac{|((length)_{obtained\ value} - (length)_{Expert\ value})|}{(length)_{Expert\ value}} X\ 100 \quad (6)$$

$$E_{M2} = \frac{|((width)_{obtained\ value} - (width)_{Expert\ value})|}{(width)_{Expert\ value}} X\ 100 \quad (7)$$

The average Error Percentage obtained is 11.9% for Major-Axis length (EM1) and 17.4% for Minor-Axis length

(EM2) using BCAT method. The average Error Percentage is 5.27% for Major-Axis length (EM1) and 6.1% for Minor-Axis length (EM2) using ACAT method. The average Error percentage for ACAT is reduced by 6.63% for Major-Axis length (E_{M1}) and by 11.3% for Minor-Axis length (E_{M2}) compared to BCAT algorithm.

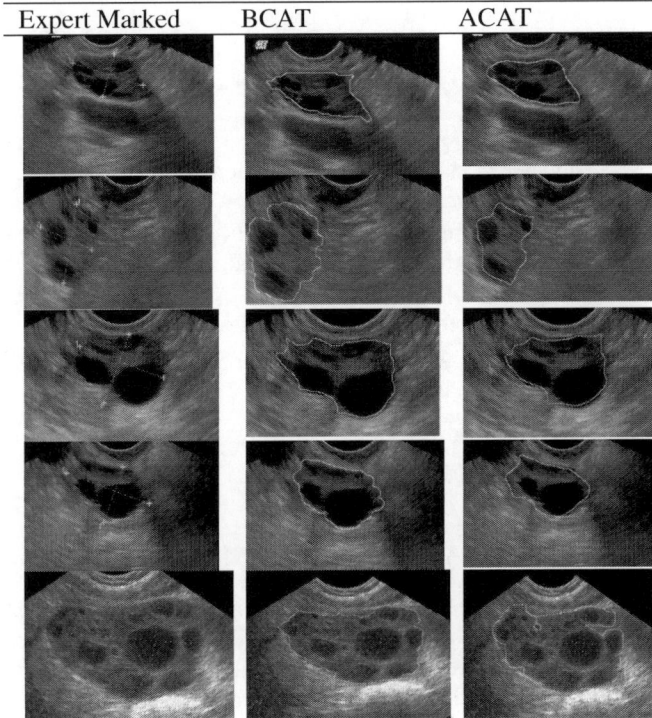

| Expert Marked | BCAT | ACAT |

Fig. 5 Sample Outputs of BCAT and ACAT methods

The K-σ classification rule discussed in section II D is applied to the feature set F1, F2 obtained after applying BCAT and ACAT algorithm to classify the Ovary as normal or abnormal. Out of 35 images used for testing, 20 images are normal and 15 images are abnormal. Classification results obtained by BCAT and ACAT is discussed below:

BCAT: Among 20 normal Ovary images 16 are classified as normal and 4 are classified as abnormal. Among 15 abnormal images, 11 are classified as abnormal and 4 images are classified as normal. The classification rate is 76.67% and false acceptance rate (FAR) is 26.66% and false rejection rate (FRR) is 20% respectively.

ACAT: Among 20 normal Ovary images 17 are classified as normal and 3 are classified as abnormal. Among 15 abnormal images, 13 are classified as abnormal and 2 images are classified as normal. The classification rate is 85.8% and false acceptance rates (FAR) is 13.33% and false rejection rate (FRR) is 15% respectively.

The Table 3 shows the classification results of BCAT and ACAT method.

Table 3

Method	BCAT	ACAT
Classification rate	76.67%	85.8
Type I error (FAR)	26.66%	13.33%
Type II error(FRR)	20%	15%

V. Conclusions

The classification rate depends on the choice of filter used, enhancement methods and segmentation method. The average Error percentage for ACAT is reduced by 6.63% for Major-Axis length (EM1) and by 11.3% for Minor-Axis length (EM2) compared to BCAT algorithm. Also from TABLE 3. It can be inferred that ACAT has a better classification rate than the BCAT method. The classification rate of ACAT is increased by 9.13% compared to BCAT method.

The algorithm can be further improved using filters that can preserve edges better and segmentation methods that can extract the Ovary.

Acknowledgment

Our sincere thanks Dr. Sriprada Vinekar, Gynaecologist and Sonographist, Ideal Nursing Home, Bangalore for supporting with the TVUS image database of ovaries, discussions and expert inputs. The authors are thankful to Prof. Prakash Hiremath, Gulbarga University for his valuable inputs and guidance.

Conflict of Interest

The authors declare that they have no conflict of interest.

References

1. Usha B S, Sandya S, Shruthi G, "Size and Shape Based Ovarian Abnormality Detection of Ultrasound Images ",ICERECT-12, LNEE, Springer, 10.1007/978-81-322-1157-0_31
2. Usha B S, Sandya S, Measurement of Ovarian Size and Shape Parameters,IEEE Xplore, 10.1109/INDCON.2013.6726079
3. Carlo Tomasi and Roberto Manduchi, "Bilateral filtering for gray and color images," in Computer Vision, 1998. Sixth International Conference on . IEEE, 1998, pp. 839– 846.
4. Pietro Perona and Jitendra Malik (July 1990). "Scale-space and edge detection using anisotropic diffusion". IEEE Transactions on Pattern Analysis and Machine Intelligence, 12 (7): 629–639. doi:10.1109/34.56205.
5. Y. Yu, S. T. Acton, and S. Member, "Speckle Reducing Anisotropic Diffusion," IEEE Transactions On Image Processing, vol. 11, no. 11, pp. 1260–1270, 2002, 88DOI 10.1109/TIP.2002.804276
6. Zuiderveld, Karel. "Contrast Limited Adaptive Histogram Equalization." Graphic Gems IV. San Diego: Academic Press Professional, 1994. 474–485.
7. P.S.Hiremath and Jyothi R. Tegnoor, Follicle Detection in Ultrasound Images of Ovaries using Scanline Thresholding Method
8. P.S.Hiremath and Jyothi R. Tegnoor, "Recognition of Follicles in Ultrasound Images of Ovaries using Geometric Features" IEEE Transactions On Medical Imaging Vol. 17, No. 6 978-1-4244-4764.

Author: Usha B S
Institute: R N S Institute of Technology
Street: Channasandra
City: Bengaluru
Country: India
Email: bsusha@gmail.com

Table 4

				BCAT					
No.	M1(Expert)	M2(Expert)	M_1(Cm)	M_2(Cm)	A(Cm^2)	P(Cm)	F	Q	E	Classification
1	3.02	1.86	3.57	1.54	4	10.17	0.485	25.85	0.9	Normal
2	2.96	2.07	3.11	2.34	5.61	10.32	0.66	18.98	0.66	Normal
3	2.38	1.88	2.8	1.68	3.44	8.5	0.59	21.02	0.8	Normal
4	3.21	2.12	2.97	1.86	4.25	8.78	0.69	18.13	0.77	Normal
5	6.24	4.27	6.28	3.97	18.43	20.06	0.57	21.83	0.77	Abnormal
				ACAT					
No.	M1(Expert)	M2(Expert)	M_1(Cm)	M_2(Cm)	A(Cm^2)	P(Cm)	F	Q	E	Classification
1	3.02	1.86	3.3	1.54	3.94	8.89	0.63	20.05	0.89	Normal
2	2.96	2.07	2.88	1.84	3.94	8.88	0.62	20.01	0.76	Normal
3	2.38	1.88	2.56	1.72	3.28	7.89	0.66	18.97	0.73	Normal
4	3.21	2.12	3.02	1.81	4.17	8.81	0.67	18.61	0.80	Normal
5	6.24	4.27	6.5	4.1	19.52	24.4	0.41	30.5	0.78	Abnormal

FPGA-in-the-Loop Simulation of Cardiac Excitation Modeling towards Real-Time Simulation

Norliza Othman[1,3], Mohamad Hairol Jabbar[2,3], Abd Kadir Mahamad[3], and Farhanahani Mahmud[1,3]

[1] Cardiology and Physiome Analysis Research Laboratory,
[2] Manycore System on Chip (MCSoC) Research Laboratory,
Microelectronics and Nanotechnology – Shamsudin Research Centre (MiNT-SRC),
[3]Faculty of Electrical and Electronic Engineering,
Universiti Tun Hussein Onn Malaysia (UTHM),
Batu Pahat, Johor, Malaysia
ge130062@siswa.uthm.edu.my, {hairol,kadir,farhanahani}@uthm.edu.my

Abstract— **Cardiac excitation is a fundamental mechanism within the heart's function. One way to understand this mechanism is by using numerical modeling techniques. However, an immense amount of computational time has been required in the simulation that generally involves a large number of parameters. In this paper a simulation study of Luo Rudy Phase I (LR-I) mathematical model by using MATLAB Simulink to solve ordinary differential equations (ODEs) using field programmable gate array (FPGA) towards a real-time simulation of cardiac excitation has been presented. The FPGA could be the best solutions because it is able to provide high performance in solving higher order ODEs for real-time hardware implementation. In fact, the FPGA hardware design can be accelerated by using MATLAB Simulink HDL Coder that automates the hardware description language (HDL) code generation from designed MATLAB Simulink blocks. Furthermore, HDL designed implementation can be verified by using HDL Verifier such as co-simulation and FPGA-in-the-Loop (FIL) approaches to simulate the generated HDL code and verify the results. In this paper, results show that the LR-I cardiac excitation modeling is successfully simulated by the MATLAB Simulink and by using the HDL Coder the designed MATLAB Simulink model is successfully converted into VHDL code and verified through the FIL. These have given a positive outlook towards the FPGA hardware implementation for real-time simulation.**

Keywords— **Ordinary Differential Equations, Luo-Rudy Phase I model, Field Programmable Gate Array, FPGA-in-the-loop simulation, cardiac, HDL Coder.**

I. INTRODUCTION

There are various methods to study the cardiac electro-physiological such as experimental, clinical and simulation by using software and hardware widely used especially for arrhythmias diagnosis. Cardiac arrhythmias are disorder in the heart rhythm that is caused by abnormalities in impulse formation or abnormalities in impulse conduction [1]. In the past few decades, the experimental studies are generally preferable [2]. Although this approach is most favored but experimental studies have the limitations such as the needs of high variables quantity for monitoring, high-resolution data in investigating larger preparations and also expensive. Meanwhile, modeling techniques for a computer simulation are not associated with such issues [3]. Computer simulation approach helps in reducing and replacing the use of animals in cardiac research. Moreover, a computer simulation also has capability to perform large-scale simulations and decreasing the time to analyze the biological behavior [4]. Therefore, many electrical and mechanical heart models have been developed such as FitzHugh-Nagumo model [5], Noble Purkinje Rudy Phase 1 (LR-I) [6], and Priebe-Beuckelmann (PB) [7] that contain 22 variables in order to model the action potentials (AP) of the heart through the simulations method.

Towards a better and a quantitative understanding of electrophysiological mechanisms of the heart, mathematical models of cardiac cells have been developed in order to simulate AP in several of conditions, where the AP provides a basis of the electrophysiological function of the heart through the cardiac excitation-contraction mechanism. A vast amount of simulation time is needed to compute the mathematical model in order to obtain the good results from the simulations [8]. Therefore, the hardware-implementation is the best solution to overcome the problems when dealing with the simulations method [11]. As an example, the hardware implementation of the cardiac mathematical model, analog-digital hybrid model had been developed by using electronic circuits and dsPIC30f4011 microcontroller [3, 8]. However, this hybrid model encounter some problems regarding noise signals, limited input and output voltage range, and gain. In addition, the designed hardware is also big in size with high power consumption [8].

Recent trend in research demand for high performance, low power consumption and small size hardware. There-fore, the aim of this study is to develop a high performance reconfigurable hardware-implemented cardiac excitation

V. Van Toi and T.H. Lien Phuong (eds.), *5th International Conference on Biomedical Engineering in Vietnam,*
IFMBE Proceedings 46, DOI: 10.1007/978-3-319-11776-8_49

using XC6VLX240T FPGA Xilinx Virtex-6 development LR-I model towards real-time performance which can provide low power consumption and low cost. Currently, Very High Speed Integration Circuit (VHSIC) Hardware Description Language (VHDL) code is needed to program the FPGA. However, the manual coding is time consuming, tedious and error prone [9]. Therefore, automatic code generation lets designers to reduce the total time of development. One way to generate automatic VHDL code is by using HDL Coder tool in MATLAB Simulink from MathWorks that is capable to automate the hardware design processes which convert Simulink blocks model design into VHDL code and verify it through the FIL approach [9]. Besides flexibility in terms of automatic code generated, another advantage offers by the HDL Coder is the capability to shorten the design iterations than a traditional workflow that relies on hand-written code.

The structure of the paper is organized as follows. An over-view of the methodology is given in Section 2. Section 3 discuss the results. Finally, concluding remarks and further potential ideas to be developed are given in Section 4.

II. MATERIALS AND METHODS

The design of software simulations of LR-I mathematical modeling by using MATLAB Simulink is made mainly to solve a set of nonlinear ODEs as in (1) to (8) to generate an action potential (AP) in a single mammalian cardiac ventricular cell. Here, the design is started by constructing blocks to represent six ion currents of sodium, potassium, calcium and chloride as in (9), where the inflow and outflow of these currents will cause changes to membrane voltage, V_m charged on membrane capacitance, C_m. The AP will be generated as an external stimulation current, I_{ext} is applied to the cell [8].

$$\frac{dV_m}{dt} = -\frac{1}{C_m}(I_{ext} + I_{ion}) \tag{1}$$

$$\frac{dm}{dt} = \alpha_m(1-m) - \beta_m(m) \tag{2}$$

$$\frac{dh}{dt} = \alpha_h(1-h) - \beta_h(h) \tag{3}$$

$$\frac{dj}{dt} = \alpha_j(1-j) - \beta_j(j) \tag{4}$$

$$\frac{dd}{dt} = \alpha_d(1-d) - \beta_d(d) \tag{5}$$

$$\frac{df}{dt} = \alpha_f(1-f) - \beta_f(f) \tag{6}$$

$$\frac{dx}{dt} = \alpha_x(1-x) - \beta_x(x) \tag{7}$$

$$\frac{d[Ca]_i}{dt} = -10^{-4} \times I_{si} + 0.07(10^{-4} - [Ca]_i) \tag{8}$$

$$I_{ion} = I_{Na} + I_{si} + I_K + I_{K1} + I_{Kp} + I_b \tag{9}$$

V_m : cardiac cell membrane voltage
C_m : membrane capacitance
I_{ion} : ionic current consist of six kind of ion currents
m : activation gate of fast sodium current, I_{Na}
h : inactivation gate of fast sodium current, I_{Na}
j : slow inactivation gate of fast sodium current, I_{Na}
d : activation gate of slow inward current, I_{si}
f : in activation gate of slow inward current, I_{si}
x : activation gate of time-dependent potassium current, I_K
Ca_i : Calcium uptake

The LR-I mathematical model is initially designed by using the MATLAB Simulink with floating-point data types. However, a fixed-point data type conversion is made to implement the algorithm to FPGA target devices. To convert the floating-point into the fixed-point data type, the word length and fraction length optimization becomes a challenging aspect of implementing an algorithm on a FPGA to achieve low power consumption and less area designed. Here, the word length and fraction length can be defined as the number of bits in the representation of the signals and the scaling or binary point location represents the number of bits of the integer part, respectively. For rapid design, the HDL coder in Simulink is used to automate the HDL code generation the designed model. Moreover, the designed model is verified using the HDL Verifier such as the FIL simulations approach as illustrated in Fig.1. The FIL increases confidence that the algorithm will behave as expected on the stand-alone implementation afterwards.

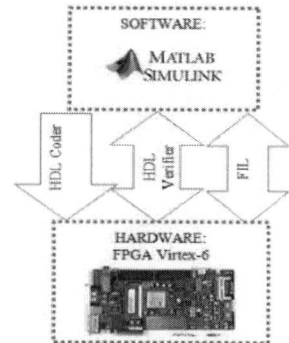

Fig. 1 FPGA designs with HDL Coder and FIL

III. RESULTS

A. MATLAB HDL Coder Implementation

Fig. 2 shows the overall design system of LR-I model using the MATLAB Simulink consists of a clock as the input LR-I subsystem and scope. Here, the input clock represents the simulation time as a time reference to control the periodic pulse of external current, I_{ext} and the result of the membrane voltage will be visualized by using the scope Simulink block. Here, the initial value of membrane voltage V_m, activation gate m, inactivation gate h, slow inactivation gate j of fast sodium current, I_{Na}, activation gate d, inactivation gate f of slow inward current, I_{si}, activation gate x of time dependent potassium current, I_K, and Calcium uptake Cai are set to -85 mV, 0.001, 0.999, 0.999, 0.001, 0.001, 0.001 and 0.0002, respectively.

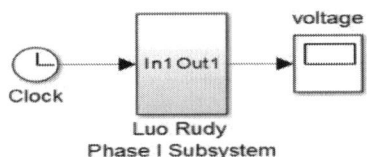

Fig. 2 The overall design of LR-I model

In the implementation of the MATLAB HDL Coder for development of the FPGA hardware design modeling, several modifications in the need to be done to match conditions required by the HDL Coder. Firstly, the continuous-time integrator block need be done by using discrete-time integrator and all the addition and multiplication blocks must have two inputs only as shown in Fig. 3. The floating-point data types need to be converted into the fixed-point data type and the suitable word length and fraction length is needed to achieve the optimum level

of design. Next, exponential and logarithmic functions must be presented in lookup tables because the output range can be very large for a relatively small input range that causes bit overflow in the fixed-point. After these several modifications, the result is comparable to the result from the initially designed MATLAB Simulink model with the floating-point data type.

Fig. 3 shows the blocks inside the LR-I subsystem. The block is designed based on the LR-I mathematical formulas as in (1) to (9) and the eight ODEs are being solved by using Simulink continuous integrator block using variable-step and *ode45* as the solver type while relative tolerance is set to 0.001. By using the MATLAB function named as Current_ext as shown in Fig. 3, the input I_{ext} is set to generate a periodic impulse current for every 500 ms starting from 100 ms, 600 ms and 1100 ms with the duration of 1 ms and magnitude of 80 μA.

Fig. 4 shows the result of action potential wave generation simulated. The resting potential is at -85 mV and action potential duration (APD) is approximately 360 ms which is taken from the starting point up to 90% of repolarization with the maximum amplitude of the AP at 60 mV. The results obtained is equivalent to the LR-I model [6].

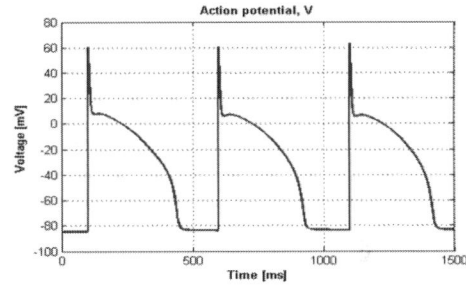

Fig. 4 Action Potential of LR-I model

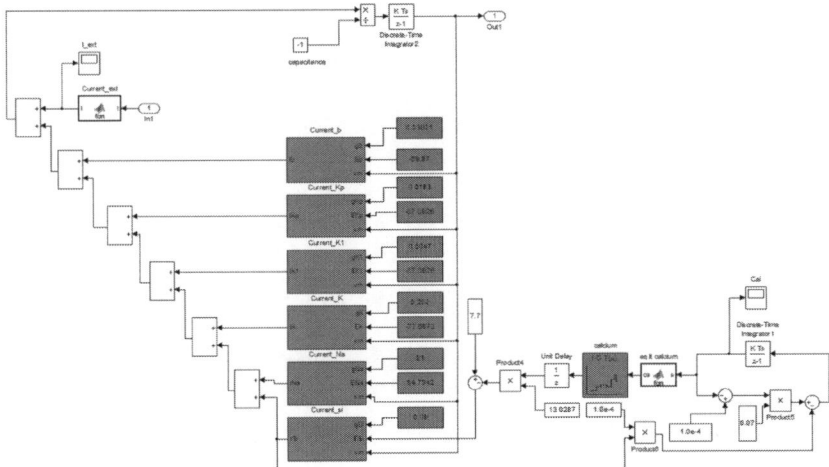

Fig .3 The LR-I model designed by MATLAB Simulink blockFPGA-in-the-Loop simulations

FPGA-in-the-Loop is one of the HDL Verifier approaches to test the behavior of the designed algorithm for FPGA hardware stand-alone implementation. The FIL is done by using HDL Workflow Advisor tool to execute the VHDL code generated from HDL Coder After the process of running the tool finished, the new model named as LR-I_fil block is generated as shown in Fig. 5. Next, the program file is loaded through the Join Test Action Group (JTAG) and Ethernet cable by clicking the generated block. Then, the program is executed to get the results. Fig. 6 illustrates the result obtained from FIL verification. The first figure in the Fig. 6 denotes as (a) shows the result of the action potential from the FIL simulation, whereas the second figure denotes as (b) shows the reference result of the action potential from the MATLAB Simulink simulation. The last figure denotes as (c) shows the error values of zero in the FIL result by referring to the result from the MATLAB Simulink simulations.

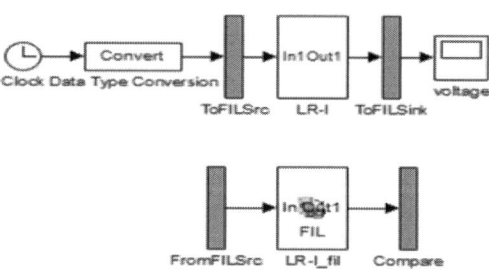

Fig. 5 New file generated for FIL simulations

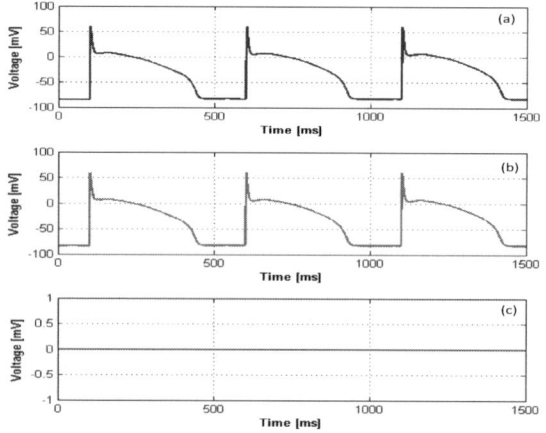

Fig. 6 FPGA-in-the-Loop results

IV. CONCLUSIONS

In conclusion, the LR-I model is successfully designed by the MATLAB Simulink and by using the HDL Coder

and successfully converted into VHDL code and verified by the FPGA-in-the-loop (FIL) on the actual FPGA board. This result gives the positive outlook for the stand-alone hardware-implementation afterwards. Here, the fixed-point optimization must be taken into consideration in order to produce reliable simulation results through the FPGA hardware implementation. For future work, the model verification using the FPGA Turnkey simulation through the HDL Workflow Advisor that enables a synchronous real-time simulation of the cardiac excitation with the FPGA board will be conducted for a stand-alone FPGA hardware implementation of the designed model. Besides, the optimization of the code can also be done manually by using ISE Design Suite or by the HDL Coder for area and speed optimization by using pipelining distribution and RAM mapping techniques to get a better performance of the design model [10].

ACKNOWLEDGMENT

The authors gratefully acknowledges the support by the Fundamental Research Grant Scheme (FRGS) (vote no. 1053), under Ministry of Higher Education Malaysia.

REFERENCES

1. S. Inada et. al, (2013) Simulation study of complex action potential conduction in atrioventricular node, 6850-6853
2. R. C. P. Kerckhoffs, et. al, (2005) Electromechanics of paced left ventrical simulated by straightforward mathematical model: Comparison with experiments, Amer. J. Physiol., Heart Circulat. Physiol. 289:1889-1897
3. F. Mahmud, (2012) Real-time simulation of cardiac excitation using hardware-implemented cardiac excitation modeling," International Journal of Integrated Engineering. 4:13-18
4. F. Martin, et. al, (2009) Mathematical models in cardiac electrophysiology research-implications for the 3Rs, 1-8.
5. R. Fitzhugh, (1960) Thresholds and plateaus in the Hodgkin-Huxley nerve equation," J. Gen. Physiol. 43:867-896.
6. C.H. Luo and Y. Rudy, (1991) A model of the ventricular cardiac action potential. Depolarization, repolarization, and their interaction, Circulation Research, 68:1501-1526.
7. F. Mahmud, S. Naruhiro, M. Masaaki, and N. Taishin, (2011) Reentrant excitation in an analog-digital hybrid circuit model of cardiac tissue, American Institute of Physics, Chaos, 21:1-14.
8. F. Mahmud, T. Sakuhana, N. Shiozawa, T. Nomura, (2009) An analog-digital hybrid model of electrical excitation in a cardiac ventricular cell," Trans. Jpn. Soc. Med. Biol. Eng. 47:428-435.
9. P. Y. Siwakoti, E. T. Graham, (2013) Design of FPGA-controlled power electronics and drives using MATLAB Simulink, Macquarie University, Australia, 571- 577.
10. K. Kintali and Y. Gu, (2012) Model-based design with Simulink, HDL Coder, and Xilinx system generator for DSP, Mathworks white paper.

G-Quadruplexes in the Human Immunodeficiency Virus-1 and Herpes Simplex Virus-1: New Targets for Antiviral Activity by Small Molecules

Rosalba Perrone[1], Sara Artusi[1], Elena Butovskaya[1], Matteo Nadai[1], Christophe Pannecouque[2], and Sara N. Richter[1]

[1] Department of Molecular Medicine, University of Padua, Padua, Italy
[2] Rega Institute for Medical Research, Katholieke Universiteit Leuven, Leuven, Belgium

Abstract— G-quadruplexes (G-4s) are G-rich non-canonical four-stranded conformations of nucleic acids that act as structural switches of cellular processes. Very little is known on the role of G-4s in viruses yet. The human immunodeficiency virus-1 (HIV-1) and the human herpes simplex virus-1 (HSV-1) are important human pathogens: HIV-1 is the etiological agent for the acquired immune deficiency syndrome (AIDS), while HSV-1 causes vesicular lesions on the mucous membranes, but it can also cause serious diseases, such as encephalitis, in immunocompromised patients and it increases sexual transmission of HIV-1. Both viruses permanently install into the human host and no cure to eradicate them has yet been developed.

We have shown that DNA G-4s arise in the integrated DNA genome, in the long terminal repeat (LTR) promoter, and inhibit viral transcription. We now show that the LTR sequence is present also in the HIV-1 RNA genome where it can fold into very stable parallel-like G-4 structures. Both DNA and RNA G-4s were stabilized by a G-4 ligand, BRACO-19, which exerted antiviral activity against a broad range of virus strains, host cells and types of infections. BRACO-19 was active both at the reverse transcription step and during post-integration events, which are compatible with BRACO-19 activity on G-4 structures.

Also HSV-1, which is characterized by a genome remarkably rich in guanines, presents clusters of repeated sequences forming very stable G-4s in key regions of the HSV-1 genome. Treatment of HSV-1 infected cells with BRACO-19 induced significant inhibition of virus production, general reduction of viral transcripts and of intracellular viral DNA. BRACO-19 was able to inhibit Taq polymerase processing at G-4 sites.

This work, besides presenting the first evidence of extended G-4 sites in key regions of the HIV-1 and HSV-1 genomes, opens up new potential antiviral therapeutic interventions based on the use of G-4 ligands.

Keywords— G-quadruplex DNA, G-quadruplex RNA, HIV-1, HSV-1, BRACO-19.

I. INTRODUCTION

G-quadruplexes (G-4s) are non-canonical four-stranded conformations of nucleic acids formed by G-rich stretches of DNA or RNA. DNA G-4s are characterized by a high conformational polymorphism which depends on the base sequence, strand orientation, loop connectivity, and available cations [1]; in contrast, RNA G-4s have been reported to fold only into parallel-stranded conformations [2]. Experimental studies and computational predictions suggested a role of DNA and RNA G-4s as cis-acting regulatory elements in gene expression and key regulatory motifs of the transcriptome, respectively [3,4]. The biological relevance of G-4 structures prompted the development of a diverse array of G-4 ligands. Their general features comprise a large flat aromatic surface and cationic groups [5]. One of the most potent reported G-4 ligands is BRACO-19, a 3,6,9-trisubstituted acridine derivative which interact with the quadruplex DNA structures formed in human telomeres [6], while displaying very low affinity towards duplex DNA [7].

Besides eukaryotic cells, our group has recently shown that DNA G-4s arise in the HIV-1 integrated DNA genome [8,9]. In particular, dynamic G-4 structures form in the LTR promoter and inhibit viral transcription, while a cluster of G-4s is present in the leading and lagging strands of *nef* impairing its expression. Viral DNA G-4s were stabilized by G-4 ligands, such as the acridine BRACO-19 and the porphyrin TMPyP4, resulting in virus inhibition with different molecular targets: while TMPyP4 activity mainly depended on the presence of the *nef* coding sequence [9], BRACO-19 activity was Nef-independent [8].

The human immunodeficiency virus-1 (HIV-1) and the human herpes simplex virus-1 (HSV-1) are important human pathogens: HIV-1 is the etiological agent for the acquired immune deficiency syndrome (AIDS) which leads to early death if not treated with a combination of drugs; HSV-1 commonly causes vesicular lesions on the mucous membranes, but it can also cause infrequent but serious diseases, such as encephalitis and disseminated neonatal infections, it may be severe in immunocompromised patients, and it increases sexual transmission of HIV-1. Importantly, both viruses install permanently into the human host and no cure that eradicates them has yet been developed.

In this work we aimed at identifying the presence and role of G-4s in HIV-1 and HSV-1 and to explore the possibility of using G-4 ligands as antiviral agents.

© Springer International Publishing Switzerland 2015
V. Van Toi and T.H. Lien Phuong (eds.), *5th International Conference on Biomedical Engineering in Vietnam*,
IFMBE Proceedings 46, DOI: 10.1007/978-3-319-11776-8_50

II. MATERIALS AND METHODS

BRACO-19 was from ENDOTERM GmbH, Saarbruecken, Germany and TMPyP2 from Livchem Logistics GmbH, Frankfurt, Germany. Oligonucleotides were purchased from Sigma-Aldrich (Milan, Italy). G-4 sequences in the viral genomes were analyzed by QGRS Mapper. Circular dichroism (CD) spectra were recorded on a JASCO-810 spectropolarimeter (Jasco, Milan, Italy) equipped with a temperature controller (Peltier PTC-4235, Jasco). UV spectra were carried out using a Lambda25 UV/Vis spectrometer (Perkin-Elmer). For electrophoretic mobility shift assay oligonucleotides were 5'-end-labelled with [γ- ^{32}P]ATP by T4 polynucleotide kinase. Samples were loaded on 16% non-denaturing polyacrylamide gels visualized by phosphorimaging (Tyhphoon FLA 7000, GE Healthcare Life Sciences, Europe). Cytotoxicity of the compounds was investigated by a 3-(4,5-dimethylthiazol-2-yl)-2,5-diphenyltetrazolium bromide (MTT) assay. Virus quantification was performed by a virus plaque assay. For real-time RT-PCR RNA was isolated at 24 h.p.i. with TRIzol reagent (Life Technologies). Reverse transcription was carried out using the Thermo Cycler Verity 96 (Applied Biosystem, Life Technologies, Monza, Italy). Q- PCR experiments were performed in ABI 7900 HT – FAST Real Time PCR System. Reverse Transcriptase Stop Assay was performed using *Recombinant HIV-1 Reverse Transcriptase* (1U/reaction, Calbiochem) at 44°C for 1h.

For a full description see references [8,10,11].

III. RESULTS AND DISCUSSION

HIV-1

The G-rich U3 region of the viral RNA genome folds into G-quadruplex, which is stabilized by BRACO-19.

We have reported that the G-rich region of the LTR promoter in the proviral genome can fold into G-4 and that this folding is stabilized by G-4 ligands, such as BRACO-19, resulting in inhibition of viral transcription [8]. Because the G-4 forming LTR DNA region derives from an identical RNA sequence present at the 3'-end of the RNA viral genome (Fig. 1), we tested if this RNA sequence could also fold into G-4.

The tested oligonucleotides showed CD signatures characteristic of RNA G-4s. T_m values of the U3 RNA G-4s were acquired by CD thermal unfolding. In the presence of K$^+$, all U3-RNAs displayed T_m above 70°C. Incubation of the oligonucleotides with BRACO-19 further greatly stabilized the U3 RNA G-4s, which could not be unfolded up to 100°C.

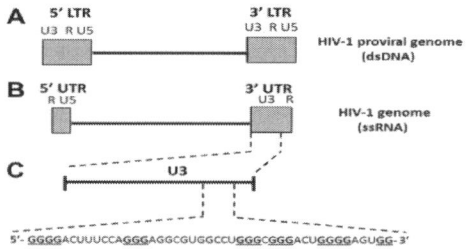

Fig. 1 Scheme of the HIV-1 proviral DNA genome (A) and of the viral RNA genome (B-C). The U3 sequence is present (A) at the 5'- and 3'-end of the proviral DNA genome, and (B) at the 3'- end of the viral RNA genome. (C) The U3 region of the viral RNA genome is enlarged. G bases involved in G-4 are in bold and underlined.

We thus asked whether G-4 formation in the G-rich region of the U3 RNA could influence RT processing. To this end, we performed a HIV-1 RT-stop assay on the U3 RNA template. As shown in Figure 2, the full-length product was obtained in the absence of K$^+$ (lane 1). In contrast, in the presence of K+, two stops corresponding to formation of G-4s were visible (lane 2). When the U3 RNA template was incubated with increasing concentrations of BRACO-19, HIV-1 RT was inhibited to a much higher extent (lanes 3-9). In contrast, only a limited aspecific inhibition was induced at the highest concentrations in a scrambled RNA template (not shown).

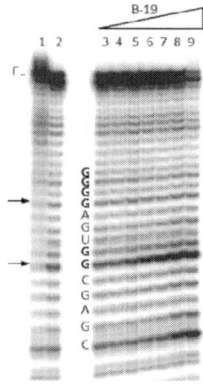

Fig. 2 RT stop assay. RT stop assay on U3 RNA template in the absence and presence of K$^+$ 100 mM (lanes 1-2) and in the presence of increasing concentration of BRACO-19 (lanes 3-9) RT pausing sites are indicated by arrows. Sequence is shown aside the gel image.

BRACO-19 is active against different viral strains, cell lines and primary cells, acting both before and after the HIV-1 integration step.

BRACO-19 was tested against different viral strains and cells: it was able to significantly inhibit viral amounts while being non-toxic against the host cells (Table1).

To determine the viral stage which BRACO-19 is active at, a time-of-addition (TOA) experiment was set up [11].

Table 1 Antiviral activity of BRACO-19

Cell type	HIV-1 strain	Detection method	Detection time post-infection	EC_{50} (µM)	CC_{50} (µM)	SI
MT-4	III_B	CPE	5 days	7.9	21.7	2.8
MT-4	III_B	p24	5 days	1.4	22.0	15.6
MT-4	III_B	p24	1 day	5.6	154.3	27.7
MT-4-LTR-eGFP	III_B	eGFP	1 day	6.2	92.9	15.0
PBMCs	III_B	p24	7 days	28.3	79.6	2.8
PBMCs	BaL	p24	7 days	17.4	79.6	4.6
MT-4/III_B	III_B	p24	5 days	5.0	28.9	5.7
MT-4/III_B + AZT	III_B	p24	5 days	3.9	32.4	8.3

The addition of BRACO-19 could be postponed up to 3 h, which corresponds to the time-frame of initial phase of the reverse transcription process (Table 2). In fact, this behavior was similar to that of the reference RT Inhibitors (AZT and nevirapine), which were able to fully block the HIV-1 replicative cycle if administered up to 3-5 h post-infection. To note that addition of BRACO-19 from 4 to 8 h post infection resulted in slight inhibition of HIV-1, since p24 levels did not reach control values. These data are compatible with a mechanism of action on G-4 structures.

Table 2 TOA indicates time how long the addition of a test compound can be postponed before it loses its antiviral activity

Drug	TOA (h)
AMD3100	0
DS	0
AZT	3
Nevirapine	5
BRACO-19	3
Ritonavir	8

HSV-1

Highly conserved G-4 forming sequences are present in key regions of the HSV-1 genome.

Following the observation that GC content of the HSV-1 genome is extremely high (68%, with peaks of 84.7% in SSRs [12]) we found nine sequences that had high propensity to fold into G-4. Six of them were mainly found in the *gp054* gene, which encodes UL36, the largest viral protein and essential viral tegument component [13]. These sequences were highly conserved among four reference HSV-1 strains. On the whole, *gp054* sequences were repeated 13-14 times and covered about 220 bps. Three more putative QGRS were found clustered at the terminal

and internal repeats (both long and short) of the HSV-1 genome. The exact role of these regions is as yet unknown and thus the QGRS in these positions were named "*un*". Each "*un*" sequence was highly repeated (5-18 times) and conserved among different HSV-1 strains. On the whole, the *un* sequences covered about 900 bps in the leading strand and 700 bps in the lagging strand. In the presence of physiological concentrations of K^+, all tested oligonucleotides displayed CD and EMSA G-4 signatures and a very high stability, with T_m around 90°C. Incubation with BRACO-19 further increased stability of these oligonucleotides up to above 95°C.

BRACO-19 Displays Anti-HSV-1 Activity

Given the ability of BRACO-19 to recognize and stabilize HSV-1 G-4s, we tested the possibility that it displayed antiviral activity. TMPyP2, a porphyrin derivative which shares chemical features, i.e. a large aromatic surface and cationic moieties, with BRACO-19, but displays a different G-4 binding mode and markedly lower G-4 binding affinity [14], was used as negative control. As shown in Fig. 3, BRACO-19 demonstrated an effect at 200 nM and its antiviral activity increased up to 70% at 25 µM. The control compound TMPyP2 was inactive up to 1 µM, while it showed 50% inhibition at the highest concentration.

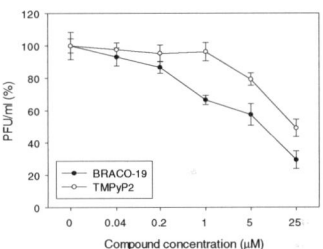

Fig. 3. Antiviral activity of BRACO-19. Plaque assay: infected cells were treated with increasing concentrations of BRACO-19 or TMPyP2; supernatants were collected 24 hpi the number of plaque forming units determined.

Stabilization of HSV-1 G-4 forming regions by BRACO-19 impairs viral DNA replication

The HSV-1 GQRS are embedded in genes encoding essential viral proteins (i.e. *gp054* for UL36 and ICP0) and are in the proximity of the transcriptional promoters of α proteins (ICP0 and ICP4), two immediate early proteins which are major regulatory proteins required to activate viral expression [15]. We thus investigated whether viral gene transcription was impaired by G-4 ligand treatment. Upon treatment of infected cells with BRACO-19, all tested viral transcripts were markedly decreased, from 30% to 50% of inhibition. In contrast, TMPyP2 reduced HSV-1 transcript levels to a much lower extent (0-20% inhibition).

The fact that all analysed transcripts were affected to a similar degree indicated that compound activity was not restricted to the transcription process of single G-4 forming genes, but it rather involved a basal level mechanism which in turn massively conditioned inhibition of viral gene expression.

Because large portions of the viral genome could fold into G-4 structures which in turn could be stabilized by BRACO-19, we investigated if compound treatment impaired polymerase processing at the viral genome. Amplified DNA corresponding to G-4 forming regions decreased in a concentration dependent manner in the presence of BRACO-19 (Fig. 4). In contrast, BRACO-19 was not able to disrupt polymerase processing in the non-G-4 forming region (Fig. 4). This data demonstrate that BRACO-19 specifically interact with G-4 forming regions in the HSV-1 genome *in vitro* and inhibit polymerase processing likely due to the steric hindrance caused by multiple tetraplex structures stabilized by the ligand.

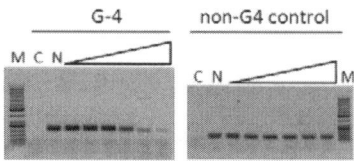

Fig. 4 Inhibition of HSV-1 DNA replication in G-4 regions *in vitro*. G-4 and non-G4 forming regions were elongated using the Taq polymerase stop assay, in the presence of increasing concentration of BRACO-19. A non-treated sample (lane N) and a sample without DNA (lane C) were used as negative controls; HSV-1 DNA was extracted from HSV-1-infected Vero cells.

IV. CONCLUSIONS

This work has provided the first evidence of 1) G-4 folding sequences in the HIV-1 RNA genome and HSV-1 DNA genome, 2) the possibility to target them with G-4 ligands and 3) G-4 ligand-mediated antiviral effects which involve basal viral mechanisms. Therefore this work paves the route to the development of selective anti-HIV-1 and HSV-1 agents with an innovative mechanism of action which could prove useful also in the treatment of viral strains resistant to current anti-viral therapies.

ACKNOWLEDGMENT

This work was supported by the Bill and Melinda Gates Foundation through the Grand Challenges Explorations Initiative (GCE grants OPP1035881 and OPP1097238 to SNR), by the Italian Ministry of University and Research (FIRB-Ideas RBID082ATK_001 to SNR). We thank Dr A. Calistri for thoughtful discussion.

CONFLICT OF INTEREST

The authors declare that they have no conflict of interest.

REFERENCES

1. Zhang DH, Fujimoto T, Saxena S, Yu HQ, Miyoshi D, Sugimoto N (2010) Monomorphic RNA G-quadruplex and polymorphic DNA G-quadruplex structures responding to cellular environmental factors. Biochemistry 49 (21):4554-4563.
2. Joachimi A, Benz A, Hartig JS (2009) A comparison of DNA and RNA quadruplex structures and stabilities. Bioorganic & medicinal chemistry 17 (19):6811-6815.
3. Huppert JL, Balasubramanian S (2007) G-quadruplexes in promoters throughout the human genome. Nucleic Acids Res 35 (2):406-413
4. Huppert JL, Bugaut A, Kumari S, Balasubramanian S (2008) G-quadruplexes: the beginning and end of UTRs. Nucleic Acids Res 36 (19):6260-6268.
5. Arora A, Dutkiewicz M, Scaria V, Hariharan M, Maiti S, Kurreck J (2008) Inhibition of translation in living eukaryotic cells by an RNA G-quadruplex motif. Rna 14 (7):1290-1296.
6. Read M, Harrison RJ, Romagnoli B, Tanious FA, Gowan SH, Reszka AP, Wilson WD, Kelland LR, Neidle S (2001) Structure-based design of selective and potent G quadruplex-mediated telomerase inhibitors. Proceedings of the National Academy of Sciences of the United States of America 98 (9):4844-4849.
7. White EW, Tanious F, Ismail MA, Reszka AP, Neidle S, Boykin DW, Wilson WD (2007) Structure-specific recognition of quadruplex DNA by organic cations: influence of shape, substituents and charge. Biophysical chemistry 126 (1-3):140-153.
8. Perrone R, Nadai M, Frasson I, Poe JA, Butovskaya E, Smithgall TE, Palumbo M, Palu G, Richter SN (2013) A dynamic G-quadruplex region regulates the HIV-1 long terminal repeat promoter. Journal of medicinal chemistry 56 (16):6521-6530.
9. Perrone R, Nadai M, Poe JA, Frasson I, Palumbo M, Palu G, Smithgall TE, Richter SN (2013) Formation of a unique cluster of G-quadruplex structures in the HIV-1 Nef coding region: implications for antiviral activity. PloS one 8 (8):e73121.
10. Pannecouque C, Daelemans D, De Clercq E (2008) Tetrazolium-based colorimetric assay for the detection of HIV replication inhibitors: revisited 20 years later. Nat Protoc 3 (3):427-434.
11. Daelemans D, Pauwels R, De Clercq E, Pannecouque C (2011) A time-of-drug addition approach to target identification of antiviral compounds. Nat Protoc 6 (6):925-933.
12. Ouyang Q, Zhao X, Feng H, Tian Y, Li D, Li M, Tan Z (2012) High GC content of simple sequence repeats in Herpes simplex virus type 1 genome. Gene 499 (1):37-40.
13. McNabb DS, Courtney RJ (1992) Analysis of the UL36 open reading frame encoding the large tegument protein (ICP1/2) of herpes simplex virus type 1. Journal of virology 66 (12):7581-7584.
14. Han H, Langley DR, Rangan A, Hurley LH (2001) Selective interactions of cationic porphyrins with G-quadruplex structures. J Am Chem Soc 123 (37):8902-8913.
15. Smith MC, Boutell C, Davido DJ (2011) HSV-1 ICP0: paving the way for viral replication. Future virology 6 (4):421-429.

Author: Sara N. Richter
Institute: University of Padua
Street: via Gabelli, 63
City: Padua
Country: Italy
Email: sara.richter@unipd.it

Cell Specific Imaging Probe Development and Biomedical Applications

Nam-Young Kang[1] and Young-Tae Chang[1,2,*]

[1] Singapore Bioimaging Consortium Agency for Science, Technology and Research, Singapore
[2] Department of Chemistry & Med Chem Program, Life Sciences Institute, National University of Singapore, Singapore

Abstract— **Bioimaging probes are reporter molecules for visualizing the cellular event. The conventional bioprobe design has been carried out by so-called hypothesis-driven approach. The basic assumption of hypothesis-driven approach is that the scientist "knows the target" in advance, and then design the recognition motif for it. An alternative approach is diversity-driven approach, in which a broad range of fluorescence molecules in a library format are constructed by combinatorial chemistry, as a tool box for unbiased screening. We have developed libraries of fluorescence small molecules by combinatorial synthesis. This Diversity Oriented Fluorescence Library Approach (DOFLA) has great advantage in terms of optical screening and target identification. The specific binding of fluorescent small molecule is readily detectable and the target protein can be tracked visibly during all the target identification processes by adding an affinity tag to the molecule. Altogether, more than 10,000 fluorescent compounds were synthesized and tested in various cell types. Using DOFLA, a broad range of colorful bioimaing probes including stem cells, microglia and pancreatic islet cells were successfully demonstrated.**

Keywords— **Fluorescence, probe, sensor, bioimaging, library.**

I. Introduction

Bioimaging probes are reporter molecules for visualizing the cellular event. Most of the colourful cell images are aided by fluorescent probes, due to their high sensitivity and visibility. While immunohistochemistry provides a powerful imaging capability, incorporation of antibody of the imaging limit the scope of study to dead cells. Small molecule probes for live cells provide tremendous new opportunities for new biological study. We have observed the revolutionary application of Fura dye for calcium imaging; this powerful small molecule calcium probe was the technical foundation for the boom of calcium signaling field during the last several decades. The original paper of this Fura dye has been cited about 20,000 times so far [1]. As there are many unexplored or unknown targets in the biological system, the need for novel bioprobes to open up new windows is ever increasing.

The conventional bioprobe design has been carried out by so-called hypothesis-driven approach. The basic assumption of hypothesis-driven approach is that the scientist "knows the target" in advance, and then design the recognition motif for it. An alternative approach is diversity-driven approach, in which a broad range of fluorescence molecules in a library format are constructed by combinatorial chemistry, as a tool box for unbiased screening. Herein, we introduce a relatively new diversity-driven approach to provide a possible solution for the facilitated bioimaging probe development platform.

II. Materials and Methods

As the tool box of DOFLA, we utilized more than 15 different fluorescent scaffolds and introduced diversity using combinatorial techniques. High quantum yield dyes with almost 100% efficiency may change its fluorescence into quenching rather further increasing upon binding target analytes. To allow the change of fluorescence in both directions, we introduced rotational flexibility to the dyes by adjusting the quantum yield in 1-10% range. Using this concept, we have synthesized extensive numbers of fluorescent compound libraries composed of more than ten thousand members by combinatorial synthesis, which is called as DOFL (Diversity Oriented Fluorescence Library) [2].

The diverse capability of DOFL has been demonstrated by selective probe generation over broad range of biological analytes, including DNA [3,4], RNA [5-9], G-quadruplex [10], heparin [11], chondroitin [12], GTP [13], human serum albumin [14-17], chymotrypsin [18], amyloid [19], immunoglobulin [20], glutathione [21], caffeine [22] and histamine [23]. This *in vitro* probe development is facilitated by fluorescence plate reader and the probes could be used for *in vitro* or *ex vivo* applications. As far as the target is known, elucidation of the possible probe candidate is straightforward.

If the goal is developing probe for live cell selective staining, there are at large two approaches. One approach is hypothesis driven, i.e. figuring out uniquely expressed protein from the target cells, then designing selective probe for the target protein (biomarker). For this approach, genomic or proteomic analysis is commonly used, but one may encounter easily hundreds of possible candidates of the target proteins. Even if it is possible to select the right target and design a perfect probe for it, the next challenge is to deliver the probe to the right location of the target protein. Predicting the behavior of the probe is extremely difficult.

---

* Corresponding author.

© Springer International Publishing Switzerland 2015
V. Van Toi and T.H. Lien Phuong (eds.), *5th International Conference on Biomedical Engineering in Vietnam,*
IFMBE Proceedings 46, DOI: 10.1007/978-3-319-11776-8_51

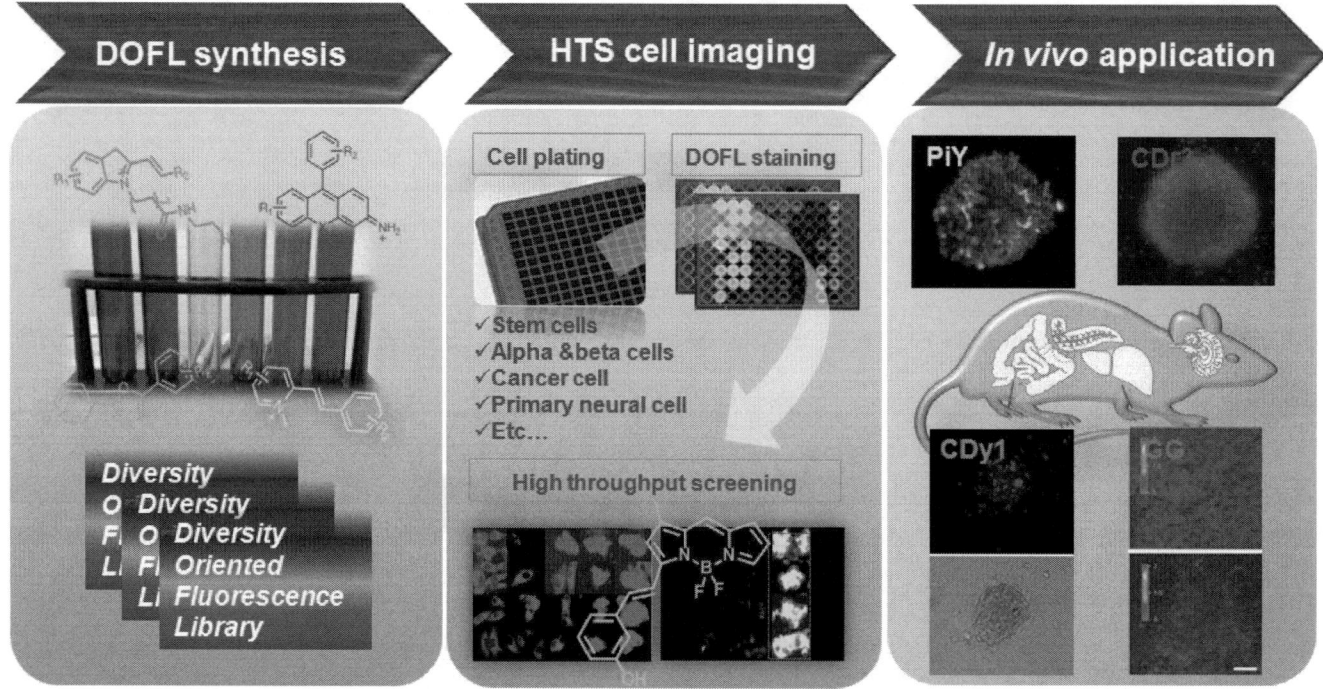

Fig. 1 Outline of bioimaging probe development using Diversity Oriented Fluorescence Library Approach (DOFLA)

Alternative approach is diversity oriented approach. We challenged this problem with DOFL and phenotypic cell screening, with a hypothesis that different cells may have thousands of reasons why they are different. They can be different protein expression level, mitochondrial membrane potential or extracellular matrix structure, etc. Therefore, our method and approach is composed of (1) DOFL synthesis, (2) high throughput screening (HTS) in cell imaging, (3) *in vivo* application of the probes in cells and animals as described in Figure 1.

In the beginning, DOFL have been screened against various stem cells. Directed differentiation of stem cells and reprogramming of somatic cells into stem cells are the key issues in stem cell research and regenerative medicine. The most demanding requisites in the basic research and clinical applications of stem cells is to develop tools and methodologies for detecting and isolating specific type of stem cells at different stage of differentiation and reprogramming.

One fluorescent dye stains one specific cell type or not is quite subjective question. Therefore, our optimized screening format in our screening is using 384 well plates with 2-4 different live cells. Same amount of fluorescence dye was added and cellular image was taken by high contents imaging machine (ImageXpress micro) and the relative staining intensity was monitored over two days. The control cells are selected among possible neighboring or co-existing cells in real situation. For example, if positive cell is stem cell, the common feeder cells were chosen as control cells.

III. RESULTS AND DISCUSSION

Using this phenotypic cell screening, we could successfully identified the first embryonic stem cell (ESC) probes, **CDy1** [24]. Following the convention of the surface antibody naming, CD (Cluster of designation), we dubbed our probe as CD (compound of designation) with their color (y: yellow, r: red, g: green, b: blue, etc) with number. **CDy1** became a popular research tool for the community with established protocol [25] and various applications [26, 27] including iPS (induced Pluripotent Stem cell) reprogramming. Different color ESC probes, **CDg4** [28] and **CDb8** [29] were also developed and multi-color staining was demonstrated.

ESC and iPS are relatively well established for their culture condition, adult stem cells are still difficult to isolate and culture. As the next goal, we challenged for neuronal stem cell (NSC) and developed **CDr3** [30] and demonstrated an isolation of NSC from adult tissue, not only from embryo tissue [31].

For specific differentiation monitoring, a muscle selective probe **CDy2** was demonstrated [32] with its molecular target as ALDH2 in live cells. Most of the DOFL compounds are bright enough to show their staining at sub-micromolar concentration, so the chance to observe their biological effect has been rare. Among more than million test points, exceptionally, one compound showed a strong muscle differentiation at hundred nM concentration [33].

For fully developed tissue test, pancreatic islet alpha cell, which make glucagon was screened. In this case, other pancreatic islet cell, insulin producing beta cells and pancreatic acinar cells were used as negative controls. From this study, glucagon targeting probe, **GY** (glucagon yellow) was successfully identified [34]. Later, beta cell selective probe **PiY** (Pancreatic islet Yellow) was also developed and visualization of islet by i.v. injection in mice was demonstrated [35].

Microglia cells are macrophage-like cells in brain. By phenotypic screening of neuronal cell culture, a unique microglia probe, **CDr10a** and **CDr10b** were developed [36]. The histamine probe discovered by in vitro screening was demonstrated its power to identify histamine containing basophiles among blood cells [23]. A new glutathione ratiometric probe **Glutathione Green** (**GG**) demonstrated its power to detect cancer area of the liver, utilizing the fact that liver cancer contains higher glutathione in comparison to neighboring tissue area [37].

The probes developed for live cell bioimaging probes are summarized in Table 1 with their structure and target cells. Once we develop the cell selective probes, a natural next question is what the mechanism of the selectivity is and what the binding target in the cells is. However, this step, so called target identification (Target ID), is known to be notoriously difficult. The conventional method for target ID is affinity matrix pull down experiment and a follow-up confirmation by genetic knock down or over expression. DOFLA provides a unique opportunity for non-conventional target ID by visual inspection. For example, **CDg4** stained extracellular structure of ESC colony, and a follow up test confirmed its target is glycogen, not a typical protein target.

IV. CONCLUSIONS

DOFLA is a unique approach for efficient and fast generation of cell specific probe. The fluorescent probes for live cells will be utilized to understand the fundamental mechanism of biological process and also expedites new drug development by providing a novel screening system. Converting the fluorescent probes to clinically relevant modality such as PET (positron emission tomography) or MRI would be the next challenge of DOFLA.

Table 1 Cell selective probes with structure and target

Probe	Structure	Target
CDy1		ESC
CDg4		ESC
CDb8		ESC
CDr3		NSC
CDy2		Muscle
GY		Alpha cell
PiY		Beta Cell
CDr10b		Microglia
Histamine Blue		Basophile
Glutathione Green (GG)		Liver Cancer

Acknowledgment

This study was supported by an intramural funding from A*STAR (Agency for Science, Technology and Research, Singapore) Biomedical Research Council and a Singapore Ministry of Education Academic Research Fund Tier 2 (MOE2010-T2-2-030).

References

1. Grynkiewicz G, Poenie M, Tsien RY (1985) A new generation of Ca^{2+} indicators with greatly improved fluorescence properties. J Biol Chem 260: 3440-3450.

2. Kang NY, Ha HH, Yun SW et al. (2011) Diversity-driven chemical probe development for biomolecules: beyond hypothesis-driven approach. Chem Soc Rev 40: 3613-3626.

3. Feng S, Kim YK, Yang Q et al. (2010) Discovery of a Green DNA Probe for Live-Cell Imaging. Chem Commun 46: 436-438.

4. Lee JW, Jung M, Rosania GR et al (2003) Development of novel cell-permeable DNA sensitive dyes using combinatorial synthesis and cell-based screening. Chem Commun 1852-1853.

5. Li Q, Kim YK, Namm J et al (2006) RNA-selective, live cell imaging probes for studying nuclear structure and function. Chem Biol 13: 615-623

6. Li Q, Chang YT (2006) A protocol for preparing, characterizing and using three RNA-specific, live cell imaging probes: E36, E144 and F22. Nat Protoc 1: 2922-2932.

7. Cervantes S, Prudhomme J, Carter D (2009) High-content live cell imaging with RNA probes: advancements in high throughput antimalarial drug discovery. BMC Cell Biol 10: 45.

8. Lee JS, Baldridge A, Feng S (2011) Fluorescence response profiling for small molecule sensors utilizing the green fluorescent protein chromophore and its derivatives. ACS Comb Sci 13: 32-38.

9. Speese SD, Ashley J, Jokhi V (2012) Nuclear envelope budding enables large ribonucleoprotein particle export during synaptic wnt signaling. Cell 149: 832-846.

10. Zhang L, Er JC, Ghosh KK et al. (2014) Discovery of a Structural-Element Specific G-Quadruplex "Light-Up" Probe. Sci Rep, in press.

11. Wang S, Chang YT (2008) Discovery of heparin chemosensors through diversity oriented fluorescence library approach. Chem Comm 1173-1175.

12. Lee SC, Zhai D, Chang YT (2013) Discovery of a chondroitin 4-sulfate fluorescent probe. Supramol Chem 25: 41-45

13. Wang S, Chang YT (2006) Combinatorial synthesis of benzimidazolium dyes and its diversity directed application toward GTP-selective fluorescent chemosensors. J Am Chem Soc 128: 10380-10381.

14. Min J, Lee JW, Ahn YH (2007) Combinatorial dapoxyl dye library and its application to site selective probe for human serum albumin. J Comb Chem 9: 1079-1083.

15. Ahn YH, Lee JS, Chang YT (2008) A selective human serum albumin sensor from the screening of a fluorescent rosamine library. J Comb Chem 10: 376-380.

16. Er JC, Tang MK, Chia CG et al. (2013) MegaStokes BODIPY-Triazoles as Environmentally-Sensitive Turn-on Fluorescent Sensors. Chem Sci 4: 2168-2176.

17. Er JC, Vendrell M, Tang MK et al. (2013) A fluorescent dye cocktail for multiplex drug-site mapping on human serum albumin. ACS Comb Sci 15: 452-457.

18. Wang S, Kim YK, Chang YT (2008) Diversity-oriented fluorescence library approach (DOFLA) to the discovery of chymotrypsin sensor. J Comb Chem 10: 460-465.

19. Li Q, Lee JS, Ha C et al. (2004) Solid phase synthesis of styryl dye library and its application to amyloid sensors. Angew Chem Int Ed Engl 43: 6331-6335.

20. Vendrell M; Krishna GG, Ghosh KK et al. (2011) Solid-Phase Synthesis of BODIPY Dyes and Development of an immunoglobulin Fluorescent Sensor. Chem Commun 47: 8424-8426.

21. Ahn YH, Lee JS, Chang YT (2007) Combinatorial rosamine library and application to in vivo glutathione probe. J Am Chem Soc 129: 4510-4511.

22. Xu W, Kim TH, Zhai D et al. (2013) Make caffeine visible: a fluorescent caffeine "traffic light" detector. Sci Rep 3: 2255.

23. Kielland N, Vendrell M, Lavilla R et al. (2012) Imaging histamine in Live Basophils and Macrophages with a Fluorescent Mesoionic Acid. Chem Commun 48: 7401-7403.

24. Im CN, Kang NY, Ha HH et al. (2010) A fluorescent rosamine compound selectively stains pluripotent stem cells. Angew Chem Int Ed Engl 49: 7497-7500.

25. Kang NY, Yun SW, Ha HH et al. (2011) Embryonic and induced pluripotent stem cell staining and sorting using live cell fluorescence imaging probe CDy1, Nat Protoc 6: 1044-1052.

26. Vendrell M, Park SJ, Chandran Y et al. (2012) A fluorescent screening platform for the rapid evaluation of chemicals in cellular reprogramming. Stem Cell Res 9: 185-191.

27. Hawley TS, Riz I, Yang W et al. (2013) Identification of an ABCB1 (P-glycoprotein)-positive carfilzomib-resistant myeloma subpopulation by the pluripotent stem cell fluorescent dye CDy1, Am J Hematol 88: 265-272.

28. Lee SC, Kang NY, Park SJ et al. Development of a fluorescent chalcone library and its application in the discovery of a mouse embryonic stem cell probe. Chem Commun 48: 6681-6683.

29. Ghosh KK, Ha HH, Kang NY et al. (2011) Solid phase combinatorial synthesis of a xanthone library using click chemistry and its application to an embryonic stem cell probe. Chem Commun 47: 7488-7490.

30. Yun SW, Leong C, Zhai D et al. (2012) Neural stem cell specific fluorescent chemical probe binding to FABP7. Proc Natl Acad Sco USA 109: 10214-10217.

31. Leong C, Zhai D, Kim B et al. (2013) Neural stem cell isolation from the whole mouse brain using the novel FABP7-binding fluorescent dye, CDr3. Stem Cell Res 11: 1314-1322.

32. Kim YK, Lee JS Bi X et al. (2011) The binding of fluorophores to proteins depends on the cellular environment. Angew Chem Int Ed Engl 50: 2761-2763.

33. Kim YK, Ha HH, Lee JS et al. (2010) Control of muscle differentiation by a mitochondria-targeted fluorophore. J Am Chem Sco 132: 576-579.

34. Lee JS, Kang NY, Kim YK et al. (2009) Synthesis of a BODIPY library and its application to the development of live cell glucagon imaging probe. J Am Chem Soc 131: 10077-10082.

35. Kang NY, Lee SC, Park SJ et al. (2013) Visualization and isolation of Langerhans islets by a fluorescent probe PiY. Angew Chem Int Ed Engl 52: 8557-8560.

36. Leong C, Lee SC, Ock J et al. (2014) Microglia specific fluorescent probes for live cell imaging. Chem Commun 50: 1089-1091.

37. Zhai D, Lee SC, Yun SW et al. (2013) A ratiometric fluorescent dye for the detection of glutathione in live cells and liver cancer tissue. Chem Commun 49: 7207-7209.

Author: Young-Tae Chang
Institute: National University of Singapore
Street: 3 Science Drive 3
City: Singapore
Country: Singapore
Email: chmcyt@nus.edu.sg

A Case Study on Expression of Single-Chain Variable Fragment of Anti-HER2 Antibody by Using Recombinant Baculovirus in Silkworms

T.M.H. Nguyen[1,*], T.V.A. Nguyen[1], T.H. Le[1], T.H. La[1], T.T.B. Nguyen[2], and Q.H. Le[1]

[1] Institute of Biotechnology/Animal Cell Biotechnology Lab., VAST, 18 Hoang Quoc Viet, Cau Giay, Hanoi, Vietnam
[2] Institute of Biotechnology/Animal Gene Technology Lab., VAST, 18 Hoang Quoc Viet, Cau Giay, Hanoi, Vietnam

Abstract— HER2 (Epidermal Human Growth Factor Receptor), is one of receptors in the epidermal growth factor receptor family. The overexpression of HER2 was determined in 25% to 35% of patients which were positive with breast cancer and in some of different cancers such as ovary cancer, uterus cancer... Due to this characteristic, HER2 has become an effective molecular marker for pre-diagnosis as well as a strike target for immune therapeutically treatment. Currently, there are different types of monoclonal antibodies (mAbs) specific to HER2 which produced based on recombinant DNA technology. These recombinant mAbs are bringing many benefits for diagnosis and making more effective in the treatment of breast cancer. In this research, the pair of primers (BacHER2F/BacHER2R) was designed to amplify antiHER2 antibody (antiHER2 Ab) fragment (Single-chain variable fragment, scFv) which had been moved into a cloning vector and confirmed by sequencing. An expression vector containing this fragment (vector Bacmid/AntiHER2) has been constructed and transfected into a cultured Sf9 insect cells to produce recombinant baculovirus (r-BV). The virus then infected to silkworm Bombyx mori and the recombinant antiHER2 Ab (r-antiHER2 Ab) was expressed successfully. The expressed protein was confirmed by Western blot analysis. We obtained r-antiHER2 Ab using Ni-NTA affinity chromatography. Our result verified that we received purified r-antiHER2 Ab with 95% purity, which can be used for next experiments.

Keywords— Baculovirus (BV), Anti-HER2 antibody, scFv, silk-worm, breast cancer.

I. INTRODUCTION

Recently, many pharmaceutical products were therapeutic proteins or mAbs (1). These products play very important role in the medicine development and they bring many benefits for human life. The advantage of protein therapeutics had described by Leader et al (2). Production of therapeutic protein or mAbs used a host cell which is various from bacterial cell (*E. coli*) to mammalian cell (CHO or HEK). Each of the host cells has advantages and disadvantage in comparison to each other (http://www.labome.com/method/Recombinant-Protein-Expression-Vector-Host-Systems.html). However, insect cell and silkworm system seem to be ideal for the production of complex proteins which require extensive post-translational modification (3,4). Among of therapeutic mAbs, antiHER2

Ab had a long development history since the first approved for use in HER2-positive metastatic breast cancer by United States Food and Drug Administration (FDA) in 1998 with trade name Trastuzumab(5). Resistant to Trastuzumab had been reported which push researcher develop a novel antiHER2 Ab or therapy (6,7) for breast cancer treatment. In Vietnam, the number of woman had breast cancer increase every years and the cost for treatment of breast cancer is very high because of patient are almost in late stage diagnose (8) (9). For developing a cheaper r-antiHER2 Ab to reduce the cost of treatment, we used silkworm as an expression system for antiHER2 antibody. This recombinant antibody is scFv form which will be more advantage than the original form of antiHER2 Ab.

II. MATERIALS AND METHODS

A. Construction of scFv Fragment of r-antiHER2 Ab for Expression in Insect Cells

Gene encoding scFv was obtained from Animal Cell Biotechnology laboratory, IBT, VAST. The DNA was cloned into pFastBac[TM]/NT-TOPO® (Invitrogen) vector using the blunt end PCR which created by Vent-polymerase with the primers BacHER2F: 5'-GATATTCAAATGACTCAGT-3' and BacHER2R: 5'-CTATGAAGATACGGTAACCA-3'. The next steps of cloning were followed the Bac-to-Bac® TOPO® cloning kit manual (Invitrogen). The recombinant plasmid was sent for sequencing to confirm the correct orientation and reading frame, then was used for generating recombinant bacmid DNA to be transfected into insect cell for generate BV which bearing scFv.

B. Recombinant BV Production

r-BV was produced by transfecting bacmid DNA into Sf9 (*Spodoptera frugiperda*) insect cell line. Briefly, Sf9 cells were cultured in Sf-900 II SFM medium (Invitrogen) supplemented with 10% FBS (Fetal Bovine Serum) and 1% Penicillin and Streptomycin (Invitrogen) at 27°C in the attached form. Bacmid DNA was transfected in Sf9 insect cells using Cellfectin® II reagent as manufacturer's instruction (Bac-to-Bac® TOPO® Expression System, Invitrogen). r-BV was collected from culture medium after three days of infection (P1).

© Springer International Publishing Switzerland 2015
V. Van Toi and T.H. Lien Phuong (eds.), *5th International Conference on Biomedical Engineering in Vietnam,*
IFMBE Proceedings 46, DOI: 10.1007/978-3-319-11776-8_52

C. Quantification of r-BV

r-BV was titered using Real-time PCR method as well as viral plaque assay method. In Real-time PCR quantification, viral DNA was extracted using phenol/chloroform purification method (10) and PCR reaction was performed in 20 µl total reaction volume with the following components: 4 µl Real-time PCR master mix, 0.5 µl BacHER2F primer, 0.5 µl BacHER2R primer, 0.2 µl Taq-man Probe, 5 µl viral DNA, 9.8 µl H$_2$O. The reaction is performed with thermal cycles: 1. Pre-incubation in 95°C for 10 seconds, 2. 45 cycles of amplification (95°C, 10 seconds; 60°C, 20 seconds; 40°C, 30 seconds).

Viral plaque assay was performed following the instruction from Bac-to-Bac® TOPO® Expression System, Invitrogen).

D. Infection of BV to Silkworm

The healthy Bombyx mori with 5th old was infected with BV (10^6 cfu/ml) in two ways, either by subcutaneously injection (injection) or by spraying of BV solution on mulberry leaves and let the silkworm eat all of these before giving the new mulberry leaves (feeding). The silkworm's blood was collected and examined for the expression of expected recombinant gene after 72 hours of infection.

III. RESULTS

A. Construction of Vector Containing the antiHER2 Gene

The gene coding for antiHER2 was amplified by PCR as described above. The PCR product has the size of about 700 bp (Fig. 1a). The PCR product was purified with PCR purification kit (Fermentas), ligated to pFastBac^{TM}/NT-TOPO® vector and transformed into E. coli DH5α. The colonies appeared were subjected to test by PCR with primers specific to pFastBac^{TM}/NT-TOPO® vector. PCR products had the size of about 1000 bp (Fig. 1b). This size is suitable with the estimation from the gene size and extra fragment of pFastBac^{TM}/NT-TOPO® vector. The correct orientation genes inserted into plasmids were confirmed by sequencing (Macrogen, Korea).

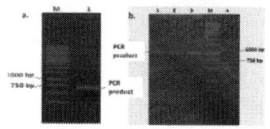

Fig. 1. a. PCR product of AntiHER2 Ab coding sequence, M. Ladder marker, 1. PCR product (size of about 720 bp). b. Colony PCR results, M. Ladder marker, 1-4. PCR product of random selected colonies (size of about 1000 bp).

B. Design Recombinant Bacmid/AntiHER2

The pFastBac^{TM}/NT-TOPO® plasmid bearing gene coding for antiHER2 Ab was transformed into MAX Efficiency® DH10Bac^{TM} Competent E. coli (Invitrogen) which contains a BV shuttle vector (bacmid) and a helper plasmid, and allow to generation of a recombinant bacmid following transposition of the pFastBac^{TM} expression construct (Bac-to-Bac® TOPO® Expression system manual). The white colonies which grown in LB medium supplemented with antibiotic were picked up randomly to purify recombinant bacmid DNA (Fig. 2a) and checked by colony PCR with specific primers for antiHER2 Ab. PCR products had a size of about 720 bp (Fig. 2b).

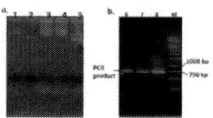

Fig. 2 a. Bacmids of AntiHER2 Ab coding sequence, 1, 3, 4. Purified recombinant bacmid DNA from white colony, 2. Purified recombinant bacmid DNA from blue colony, 5. Original bacmid DNA. b. 6, 7, 8. Colony PCR results of white colony, M. Ladder marker.

The above result was confirmed by sequencing (result not show). In summary, the recombinant bacmid containing DNA coding for antiHER2 Ab was successful designed in this experiment.

C. Generation of Baculoviral Containing AntiHER2 Gene in Sf9 Insect Cells

Sf9 insect cells derived from ovaries of Spodoptera frugiperda larvae by Smith and Cherry in 1983 from the parent cell line IPLB-SF21 AE. These cells play an important role in gene expression using BV. Advantages of using sf9 insect cells for gene expression is the potential to produce large amounts of recombinant proteins with very similar spatial structure to the protein which synthesized in mammalian cells; the ability of expression multiple genes; simple manipulation technique in the culture process; short doubling time; can be cultured in monolayer in culture flask or suspension in bioreactor system. BV is a parasite of the invertebrates and obligatory intrusions into living cell. Therefore, using BV for producing recombinant protein extremely needs insect cell culture. In this experiment, Sf9 cells were used as the host for infecting recombinant bacmid DNA which produced from last step by Cellfectin® II reagent. Figure 3 showed the different in Sf9 cell morphology after infection of recombinant bacmid DNA containing antiHER2 Ab gene. The result is summary in below table (table 1):

Table 1 The status of Sf9 cell after infection with recombinant bacmid DNA for producing r-BV

Time	Cell status observed by light microscopy
0h	Cell had round shape, shiny, adherent
24 h	Cell swelling, cell size was enlarge, adherent, nuclei could be observed obviously
48 h	Cell morphology and shape were changed, cell stopped dividing. Most of the cell detached from the surface of culture flask. Cell nuclei was observed obviously
More than 72 h	Budded virus was formed and released into the medium which caused Sf9 cell broken and lysed. Number of the Sf9 cell dramatically decreased

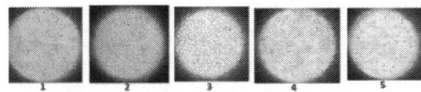

Fig. 3 Sf9 cell morphology changed by infected with recombinant bacmid DNA. 1. Normal Sf9 cells, 2. Cells before infection, 3, 4, 5. Cells after infection with bacmid DNA for 24 h, 48 h and 72h, respectively.

By observing the cell morphology in comparing to the control experiment (without transfection, result is not shown here), we could assume that transfection of recombinant bacmid DNA into Sf9 cells was successful. P1 viral stock was collected from cell culture medium and subjected to titration in the next step.

D. Viral Titration and Silkworm Infection

The viral copy number was estimated using Real-time PCR. For this purpose, the standard curve was built using the recombinant bacmid DNA which has determined the exact concentration (original from 10^8 copy number, performed serial dilution for different concentration in standard curve). Extracted viral DNA from the collected medium was estimated with concentration of about 329.52 ng/µl with purity (OD 260/280) of 1.88 by NanoDrop-1000 machine. This viral DNA was used as template for Real-time PCR in order to viral titration. Result was calculated with standard curve and estimated of about 2.59 x 10^8 copy number of virus from P1 viral stock (fig. 4). Diluted samples gave the same consequence.

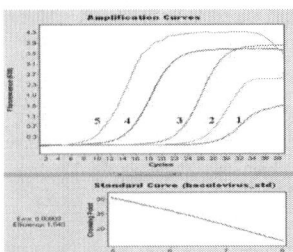

Fig. 4 Viral titration by Real-time PCR. 1. Negative control (Distilled water), 2, 3, 4. Viral DNA with 10^{-2}, $5x10^{-1}$ and no dilution, respectively, 5. Positive control

This result means that viral concentration in P1 viral stock was 2.59 x 10^8 virus particle/ml. This result matched with the result obtained from viral plaque assay which simultaneously performed with this experiment. In viral plaque assay, P1 viral stock was estimated of about 2.8 x 10^8 virus particle/ml. The small different viral particle number can be explained by the steps of experiment. In Real-time PCR, DNA must be extracted from P1 viral stock before subjected to Real-time PCR. However, viral plaque assay used serial dilution directly from P1 viral stock.

Silkworm is member of Lepidoptera family, have ability to produce silk which make it become an economic valuable insect species from very long time ago. In larval stage, silkworms have advantage as a "biofactory" to produce recombinant protein (4) by the ability to scale up, low cost and short life cycle. This make silkworm become a promising species for industrial scale of protein biopharmaceutical. In addition, recombinant product which expressed in silkworm has more bio-safety than that in bacterial system or yeast system. In comparison to Sf9, expression of a recombinant product in silkworm has more efficient (11) and lower cost. Up to now, many successful of using silkworm as an expression system to produce recombinant product (4). In our experiment, r-antiHER2 Ab was expressed in Bombyx mori by subcutaneous injection and spraying of BV solution on the mulberry leaves. Figure 5 (left) show that, different protein band appeared in lane 4, 5 and 8 which were the infected silkworm by both ways. This band has size of about 29 kDa, which suitable with estimated r-antiHER2 Ab size. It was not much clear in lane 2 which was sample with subcutaneous injection of 10µl of BV solution. The western blot with anti his-tag Ab result in figure 5 (middle) confirmed that the new appeared band was our r-antiHER2 Ab. From this experiment, infection way of BV to silkworm was not much effect to the production of recombinant protein. Since the easy of the feeding by BV containing mulberry leaves, this method will be helpful for scaling up of expression as well as a promising method in therapeutic protein production.

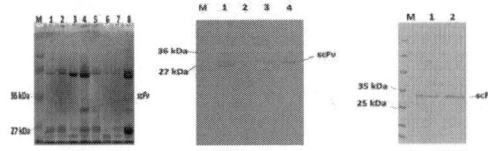

Fig. 5 SDS-PAGE (left), Western blot (middle) analysis of silkworm blood and purified scFv (right). Left figure.1, 3, 6. Injection of 10 µl, 15 µl, 20 µl of saline buffer, 2, 4, 5. Injection of 10 µl, 15 µl and 20 µl BV solution, 7, 8. Feeding without and with BV solution, Middle figure. 1, 2. Feeding with and without BV, 3, 4. Injection of 20 µl, 15 µl of BV solution, respectively, M. Marker (Fermentas), Right figure, Purified scFv by affinity column.

E. Purification of r-antiHER2 Ab

R-antiHER2 Ab was designed with C-terminal His-tag which is helpful for protein purification using affinity chromatography. The recombinant Ab was purified with Ni-NTA following Invitrogen's protocol. Result showed in fig. 5 (right). The purified recombinant protein had size of about 29 kDa. This protein size is suitable for our r-antiHER2 Ab with his-tag. The purified protein had more than 95% of purity, which can be apply for further *in vitro* experiment.

IV. DISCUSSION

Vietnam recently developed very fast in comparison with ten years ago. The effect of developing is environment change which related to the human health, results in increasing number of cancer people. Due to poor health care system, the patients are normally diagnosed in the late stage of cancer which makes treatment more costly and lower survival rate (9) (8). This circumstance drive researcher to find the cheaper products for fighting cancer. Therapeutic proteins and mAbs are new research focusing in Vietnam currently. However, it has a high potential for development of these products in Vietnam and there are several biotechnology company started to study as well as conduct their business on this discovery. Even the study requires a lot of time and effort to make such products become commercial available. Nevertheless, it is necessary for doing this research because we need to own technology for long term biopharmaceutical development. Our study is one of the beginnings of using modern technique in developing a recombinant antibody which may be useful for breast cancer treatment.

V. CONCLUSIONS

In this work, the vector containing gene coding for antiHER2 Ab sequence to express in insect system was successfully designed. BV which carried the antiHER2 gene was obtained from infected Sf9 insect cell culture. The amount of antiHER2 Ab which produced in silkworm is not depended on the method of infection of BV to the silkworm, either by subcutaneous injection or feeding. The purified antiHER2 Ab was obtained through affinity chromatography with more than 95% of purity. Using BV and silkworm system to produce recombinant protein or mAb can be a promising method in Vietnam by the available of silkworm during the year.

ACKNOWLEDGMENT

This work was financially supported by Ministry of Science and Technology of Vietnam through basic research project (code 04/2011/HD-NCCBUD) and national key project (code KC10.19/11-15).

CONFLICT OF INTEREST

The authors declare that this manuscript have no conflict of interest with any other research groups.

REFERENCES

1. Carter, P. J. *Exp Cell Res* **317**, 1261-1269
2. Leader, B., Baca, Q. J., and Golan, D. E. (2008) *Nat Rev Drug Discov* **7**, 21-39
3. Druzinec, D., Salzig, D., Brix, A., Kraume, M., Vilcinskas, A., Kollewe, C., and Czermak, P. *Adv Biochem Eng Biotechnol* **136**, 65-100
4. Kato, T., Kajikawa, M., Maenaka, K., and Park, E. Y. *Appl Microbiol Biotechnol* **85**, 459-470
5. Horton, J. (2002) *Cancer Control* **9**, 499-507
6. Shojaei, S., Gardaneh, M., and Rahimi Shamabadi, A. *Int J Breast Cancer* **2012**, 761917
7. Wong, A. L., and Lee, S. C. *Int J Breast Cancer* **2012**, 415170
8. Lan, N. H., Laohasiriwong, W., and Stewart, J. F. *Glob Health Action* **6**, 1-9
9. Hoang Lan, N., Laohasiriwong, W., Stewart, J. F., Tung, N. D., and Coyte, P. C. *Glob Health Action* **6**, 18872
10. Lo, H. R., and Chao, Y. C. (2004) *Biotechnol Prog* **20**, 354-360
11. Usami, A., Ishiyama, S., Enomoto, C., Okazaki, H., Higuchi, K., Ikeda, M., Yamamoto, T., Sugai, M., Ishikawa, Y., Hosaka, Y., Koyama, T., Tobita, Y., Ebihara, S., Mochizuki, T., Asano, Y., and Nagaya, H. *J Biochem* **149**, 219-227

Author: Thi Minh Huyen Nguyen
Institute: Institute of Biotechnology, Vietnam Academy of Science
             and Technology (VAST)
Street: 18 Hoang Quoc Viet, Cau Giay
City: Hanoi
Country: Vietnam
Email: ntminhhuyen@ibt.ac

DNA Hypermethylation Signatures for Detection of Breast Cancer in Vietnamese Population

T.K. Phuong[1], L.D. Thuan[1], D.T.P. Thao[2], and L.H.A. Thuy[1,*]

[1] Falcuty of Biotechnology, Ho Chi Minh City Open University, Vietnam
[2] Falcuty of Medicine, Ho Chi Minh City Medical University, Vietnam
lhathuy@gmail.com

Abstract— **Breast cancer is the common cause of death among women in most countries worldwide, with rapidly increases in the developing countries, including Vietnam. To establish the potential biomarker is an attempt of researchers in the world, one of the biomarkers is the disruptions of the genetic material such as the epigenetics including DNA methylation. In present study, with the aim towards using the hypermethylation at CpG islands of promoter of candidate genes as the biomarker for breast cancer in Vietnamese population, sensitive methyl specific PCR (MSP) was carried out to analyze the hypermethylation status of the panel of candidate genes including BRCA1, p16^{INK4α}, GSTP1, RASSF1A and Cyclin D2 gene in 115 samples including 95 breast cancer specimens and 20 normal breast tissues from another disease (not breast cancer) which were obtained from Ho Chi Minh City Medical Hospital, Vietnam. The results indicated that the hypermethylation of one or more genes occurred in all total of 95 tumor specimens (100% diagnostic coverage) with the frequencies for methylation of each genes reach to 82.1% (p<0.001), 62.1% (p<0.01), 49.5% (p<0.05), 43.2% (p<0.01) and 42.1% (p<0.05) for BRCA1, Cyclin D2, p16^{INK4α}, GSTP1, and RASSF1A gene, respectively. In addition, the DNA hypermethylation of the panel of candidate genes increase the possibility to be breast cancer with high incidence via calculated of odd ratio (p<0.05). In conclusion, the hypermethylation of candidate genes could be used as the promising biomarkers applying in Vietnamese breast cancer patients.**

Keywords— **BRCA1, Cyclin D2, p16^{INK4α}, GSTP1, RASSF1A, methylation.**

I. INTRODUCTION

Breast cancer was the common cause of death among women in most countries worldwide, counting for 1.6% of female deaths every year with rapidly increases in the developing countries. In Vietnam, the breast cancer incident has increased from a crude rate of 13.8 per 100,000 women in 2000 to 28.1 per 100,000 women in 2010, with an estimated 12,533 breast cancer cases [10]. One of the leading causes to tumorgenesis is thought to be the hypermethylation at promoter of tumor suppressor gene, a common change of epigenetics event [2, 7, 9]. The tumor

suppressor gene, *BRCA1, p16^{INK4α}, GSTP1, RASSF1A* and *Cyclin D2*, were silenced by the DNA hypermethylation has been defined frequently in breast cancer [1, 3, 4, 12, 15]. In the present study, using the above genes, we determined quantitatively the hypermethylation status at CpG islands of promoter belongs to *BRCA1, p16^{INK4α}, GSTP1, RASSF1A and Cyclin D2* gene in both Vietnamese breast cancer patients of all stages from premalignant to advanced metastatic breast tumor and healthy specimens by the MSP (Methylation specific PCR) method.

II. MATERIALS AND METHODS

A. Sample Collection

In present study, a total of 115 samples including 95 breast cancer specimens and 20 healthy specimens were enrolled in evaluating predictive factors including immunohistochemistry with two antibodies were HER2/neu, p53 according to the protocols of ASCO quality guideline. The healthy specimens were obtained from women who underwent a biopsy of the mammary gland because of mammographic screening and for whom histology confirmed the present of only normal tissue. All the samples were admitted to the Ho Chi Minh city Medical Hospital in Vietnam from 2010 to 2011. These tissues were obtained from the surgical specimens and then, embedded in the paraffin and stored at -20°C until the further used.

B. DNA Bisulfite Modification and Methylation Assays

DNA extractions were performed by phenol chloroform method. In addition, DNA modifications were carried out by the DNA modification kit (*Epitect Kit, Qiagen*). Methylation-Specific PCR (MSP) reaction was carried out by specific primers to evaluate the situation of methylated and unmethylated status for a given gene following our previous study [11].

C. Statistical Analysis

The status of methylation of candidate genes was calculated for each gene. Differences in the presence of methylation were determined by a two sided Fisher test and Chi squared tests for variables. Statistical analyses were performed by using Medcalc® Version 12.7.0.0. Statistical significance was assumed at two-side P value of p < 0.05.

---

* Corresponding author.

© Springer International Publishing Switzerland 2015
V. Van Toi and T.H. Lien Phuong (eds.), *5th International Conference on Biomedical Engineering in Vietnam,*
IFMBE Proceedings 46, DOI: 10.1007/978-3-319-11776-8_53

III. RESULT AND DISCUSSION

The Methylation Frequencies of BRCA1, p16^INK4a, GSTP1, RASSF1A and Cyclin D2 Gene

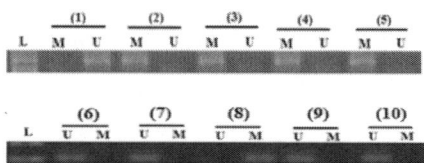

Fig.1. Methylated promoter of *RASFF1A* gene analysis on some clinical samples by MSP. The MSP products are 192 bp. U: unmethylated; M: methylated; L: 100 kp Ladder; (1), (2), (3), (4), (5) breast cancer samples; (6), (7), (8), (9), (10) non-cancer specimens

According to the result of methylation status of promoter belongs to candidate genes, the frequencies of methylation was 82.1% (*p<0.001*), 62.1% (*p<0.001*), 49.5% (*p<0.001*), 43.2% (*p<0.001*), 42.1% (*p<0.05*) for *BRCA1, Cyclin D2, p16^INK4a, GSTP1* and *RASSF1A* gene, respectively. Moreover, in the cases of non-cancer specimens, we also detected the methylation in *p16^INK4a, RASSF1A* and *Cyclin D2* counting for 15.0%, 15.0% and 10%. Twenty cases of non-cancer were fully unmethylated at *BRCA1* and *GSTP1*. Comparison with recently reports, the mean of methylation status of *GSTP1, Cyclin D2, p16^INK4a, BRCA1* in this study were also higher than the research of Yoon *et al.,* (2012) as 27.8% (*GSTP1*), 19.6% (*Cyclin D2*), Hui *et al.,* (2000) as

20% (*p16^INK4a*), and William *et al.,* (2013) as 15.1% (*BRCA1*). However, the methylation frequency of *RASSF1A* was lower than the report of Yoon *et al.,* (2012) as 85.2%. Concerning to the methylation in non-cancer samples, the similar to those studies was considered that the lower hypermethylation frequencies were found in non-cancer specimens and normal cells.

The Correlation between the Methylation Status of Panel of Genes and Several Clinicopathological Outcomes

To evaluate the correlation between the DNA methylation and protein expression, the immunohistochemical staining of HER2/neu, p53 was carried out (Table 1). Regarding to the DNA methylation and HER2/neu staining, this HER2/neu immunohistochemical characteristic was correlation with RASSF1A (p=0.01, table 1). Besides, the overexpression of mutant or overexpression p53 had been indicated that to have a considerable relationship to the tumor rate, in our report, we found out that the frequent methylation of *BRCA1, RASSF1A, p16^INK4a, GSTP1* were associated to the p53 stages. Especially, taken these two prognosis biomarkers together, the frequent methylation of individual promoter methylation in breast cancer were strongly associated to HER2/neu(-)p53(-). It meant that the DNA based marker methylation were clearly associated with the other protein based marker, in which, HER2/neu was the specific marker for breast cancer and p53 was the general marker.

Table 1 The correlation between five genes promoter methylation and HER2/neu, p53 characteristic

Characteristics	BRCA1		GSPTP1		Cyclin D2		RASSF1A		p16^INK4a	
	U (%)	M (%)	U (%)	M (%)	U (%)	M (%)	U (%)	M (%)	U (%)	M (%)
HER2/neu										
Negative	26	50	52	24	34	42	55	21	48	28
Positive	(34.2)	(65.8)	(68.4)	(31.6)	(44.7)	(55.3)	(72.4)	(27.6)	(63.2)	(36.8)
	9	30	19	20	19	20	20	19	17	22
	(23.1)	(76.9)	(48.7)	(51.3)	(48.7)	(51.3)	(51.3)	(48.7)	(43.6)	(56.4)
	p = 0.31		*p = 0.06*		*p = 0.84*		**p = 0.04**		*p = 0.07*	
p53										
Negative	16	18	30	4	20	14	29	5	25	9
Positive	(47.1)	(52.9)	(88.2)	(11.8)	(58.8)	(41.2)	(85.3)	(14.7)	(73.5)	(26.5)
	19	62	41	40	33	48	46	35	40	41
	(23.5)	(76.5)	(50.6)	(49.4)	(40.7)	(59.3)	(56.8)	(43.2)	(49.4)	(50.6)
	p = 0.02		**p = 0.0003**		*p = 0.11*		**p = 0.01**		**p = 0.03**	
HER(-)p53(-)										
No	16	66	45	37	32	50	46	36	41	41
Yes	(19.5)	(80.5)	(54.9)	(45.1)	(39.0)	(61.0)	(56.1)	(43.9)	(50.0)	(50.0)
	19	14	26	7	21	12	29	4	24	9
	(57.6)	(42.4)	(78.8)	(21.2)	(63.6)	(36.4)	(87.9)	(12.1)	(72.7)	(27.3)
	p = 0.0002		**p = 0.03**		**p = 0.03**		**p = 0.0025**		**p = 0.04**	

Table 2 Odd ratio analysis of candidate gene methylation in breast cancer.

Gene	BRCA1	Cyclin D2	$p16^{INK4\alpha}$	GSTP1	RASSF1A
OR	183.91	14.75	5.55	31.22	31.22
p value	<0.001	<0.001	0.01	0.02	0.03

Table 3 Calculation of MI value in breast cancer samples collection

The number of methylated gene	MI value	The number of breast cancer samples	Percentages* (%)	p value
None – methylated	0.0	0	0.0	
One gene	0.2	7	7.4	
Two genes	0.4	31	32.6	
Three genes	0.6	34	35.8	< 0.001
Four genes	0.8	21	22.1	
Five genes	1.0	2	2.1	
At least one gene	≥ 0.2	95	100	

In addition, to evaluate the applicability of BRCA1, Cyclin D2, $p16^{INK4\alpha}$, GSTP1, RASSF1A methylation in breast cancer patients as markers for breast cancer. Even though only 20 non-cancer specimens were compared with 95 tumor specimens, we tentatively applied the Chi^2 test to calculate the OR (Odd ratio) value, as the result shown at table 2, we found the high correlation between the breast cancer risk (at 95% confidence with the significant statistic).

Therefore, for further research, the number of non-cancer sample have to be enlarged in order to apply this phenomenon as the potential biomarkers for prognosis and early diagnosis for breast cancer. Moreover, the research would be carried out on various non-invasive samples including the milk ducts or lobules, blood, serum…

As the result shown at table 3, there was at least one gene of five candidate genes, counting for 100%, was methylated. It indicated that, in the case of screening on total five genes, the ability to detect the aberrant methylation in samples was very high. In order to evaluate the status of multiple genes methylation, the MI (methylation index) was calculated for each sample. In our sample collection, the percentage was 100%, 2.1%, 22.1%, 35.8%, 32.6% and 7.4% for MI=1 (all five genes were methylated), MI=0.8 (four of five genes were methylated), MI=0.6 (three of five genes were methylated), MI=0.4 (two of five genes were methylated), MI=0.2 (one of five genes were methylated). Interesting, no sample were found without methylation (MI=0). Regarding to the sample with low MI value, it was the useful phenomenon for the methylation drug test or hypermethylation inhibitor treatment leading the cancer cell apoptosis, which was recently mentioned such as detoxification of GSTP1 gene [6, 8]. Furthermore, the function of BRCA1 was concerned

as the DNA repaired, therefore, the hypermethylation at this gene was taken interest in considering as the indicator for evaluating the efficient of medicine which acted on the inhibitor of PARP (poly-ADP-ribose polymerase) pathway via the blocking of the unique DNA repaired pathway in several cancer including breast cancer, prostate cancer, ovarian cancer, pancreatic cancer [5, 13].

IV. CONCLUSION

95 tumors from Vietnamese breast cancer patient and 20 non-cancer specimens were carried out for BRCA1, GSTP1, Cyclin D2, $p16^{INK4\alpha}$ and RASSF1A promoter methylation analysis. The results indicated that the aberrant hypermethylation of at least one gene in total of 95 tumor specimens with the 100% diagnosis coverage was consider as to be a specific characteristic of Vietnamese breast cancer patients. The frequencies of methylation were 82.1%, 62.1%, 49.5%, 43.2%, 42.1% and the OR values were 183.91, 14.74, 5.55, 31.22 and 31.22 for BRCA1, Cyclin D2, $p16^{INK4\alpha}$, GSTP1 and RASSF1A gene, respectively. Therefore, the hypermethylation of promoter belonged to BRCA1, Cyclin D2, $p16^{INK4\alpha}$, GSTP1 and RASSF1A gene were the signatures for detection of breast cancer in Vietnamese population. For further study, the various types of cancer specimens as patient serum, invasive tissue, biopsy should be examined to have a overall vision about the methylation promoter in Vietnamese breast cancer patients.

ACKNOWLEDGEMENTS

This study was funded by The Department of Science and Technology, Ho Chi Minh City, Vietnam.

REFERENCES

1. Agathanggelou A, Honorio S, Macartney DP, Martinez A, Dallol A, Rader J, Fullwood P, Chauhan A, Walker R, Shaw JA, Hosoe S, Lerman MI, Minna JD, Maher ER, Latif F (2001) Methylation associated inactivation of RASSF1A from region 3p21.3 in lung, breast and ovarian tumours. Oncogene 20:1509–1518.

2. Donninger H, Vos MD, Clark GJ (2007) The *RASSF1A* tumor suppressor. J Cell Sci 120 (18):3163-3172.

3. Esteller M, Corn PG, Urena JM, Gabrielson E, Baylin SB, Herman JG (1998) Inactivation of glutathione S-transferase P1 gene by promoter hypermethylation in human neoplasia. Cancer Res 58:4515–4518

4. Evron E, Umbricht C, Korz D, Raman V, Loeb D, Niranjan B, Buluwela L, Weitzman S, Marks J, Sukumar S (2001) Loss of cyclin D2 expression in the majority of breast cancers is associated with promoter hypermethylation. Cancer Res 61:2782–2787.

5. Fong PC et al. (2010) Poly(ADP)-ribose polymerase inhibition: Frequent durable responses in BRCA carrier ovarian cancer correlating with platinum-free interval. J Clin Oncol 28:2512-2519

6. Fukushige S, Horii A (2013) DNA methylation in Cancer: A gene silencing mechanism and the clinical potential of its biomarkers. Tohoku J. Exp. Med 229:173-185

7. Heba HG, Hoda AH, Eman HAB (2012) RASSF1A Gene Hypermethylation in Tissue and Serum Together with Tissue Protein Expression in Breast Cancer Patients. Life Sci J 9 (3):667-675.

8. Heyn H, Esteller M (2012) DNA methylation profiling in the clinic: applications and challenges. Nat. Rev. Genet. 13:679-692

9. [9] Jia X, Priya BS, Weiwei F, Carol C, Robert CBJr, Jean-Pierre JI, Susan GH, Yinhua Y (2012) Methylation of HIN-1, RASSF1A, RIL and CDH13 in breast cancer is associated with clinical characteristics, but only RASSF1A methylation is associated with outcome. BME Cancer 12:243-249.

10. Lan NH, Wongsa L, John FS, Tung ND, Peter CC (2013) Cost of treatment for breast cancer in central Vietnam. Glob Health Action 6:18872.

11. Linh LTT, Phuong HTB, Kim TNT, Thao DTP., Thuy LHA (2011) Appraisal of potential methylation biomarkers: BRCA1, p16INK4a, cyclin D2, GSTP1, RASSF1 in breast cancer early detection. Journal of Science and Technology 49 (1A):329-337).

12. Partha MD, Rakesh S (2004) DNA Methylation and Cancer. J Clin Oncol 22:4632-4642.

13. Veeck J et al. (2010) BRCA1 CpG Island Hypermethylation Predicts Sensitivity to Poly(Adenosine Diphosphate)-Ribose Polymerase Inhibitors. J Clin Oncology 28(29):563-564

14. Rina H, Douglas M, Frances SL (2000) p16INK4a gene expression and methylation in primary breast Cancer: overexpression of p16INK4a Messenger RNA is a marker of Poor Prognosis. Clin Cancer Res 6:2777–2787.

15. Silva JM, Dominguez G, Villanueva MJ, Gonzalez R, Garcia JM, Corbacho C, Provencio M, Espana P, Bonilla F (1999) Aberrant DNA methylation of the p16INK4a gene in plasma DNA of breast cancer patients. Br J Cancer 80:1262–1264

16. William J, Simon T, Romain S, Cathy V, Anne-Claire L, Lopez-Crapez E, Frédéric B, Jean-Pierre B, Gilles R, Pierre-Jean L (2013) BRCA1 promoter hypermethylation, 53BP1 protein expression and PARP-1 activity as biomarkers of DNA repair deficit in breast cancer. BMC Cancer 13:523

17. Yoon HC, Jing S, Marilie DG, Yu-Jing Z, Qiao W, Karina G, Xinran X, Patrick TB, Susan LT, Gail G, Hanina H, Alfred IN, Jia C, Regina MS (2012) Prognostic significance of gene-specific promoter hypermethylation in breast cancer patients. Breast Cancer Res Treat 131(1):197-205

Interaction between XG and HPC in Blended-HPC/H2O/H3PO4 Tertiary System

S.P. Rwei, T.A. Nguyen*, and H.W. Wu

Institute of Organic and Polymeric Materials, National Taipei University of Technology, Taipei, Taiwan (R.O.C)

Abstract— This study is to investigate the formation of phase diagram of HPC/H$_2$O/H$_3$PO$_4$ tertiary system with adding a tiny XG to HPC by using optical methods such as polarized optical microscopy (POM) and light transmission detection. The experimental results showed that the liquid crystalline (LC) behavior of the blended system significantly enhanced in H$_2$O-rich but H$_3$PO$_4$-poor region whilst the presence of H$_3$PO$_4$ strongly suppressed the synergy between XG and HPC molecules. Accordingly, the lower critical solution temperature (LCST) of these blends was also reduced with only 3.5 % of XG in HPC. It reveals that the interaction mechanism or the anisotropic behavior of blended polysaccharide solution increased at room temperature but decreased at higher temperature to form a cloudy suspension manner and a less miscible solution. This is caused by the characteristic of the lower critical solution temperature (UCST) of XG aqueous solution.

Keywords— Hydroxypropyl cellulose (HPC), xanthan gum (XG), HPC/H$_2$O/H$_3$PO$_4$ tertiary system, lower critical solution temperature (LCST), synergistic interaction.

I. INTRODUCTION

Both xanthan gum (XG) and hydroxypropyl cellulose (HPC) are the typical lyotropic liquid crystalline polysaccharides in some specific mediums, especially in aqueous solution [1,2,4,6]. The appearance of such anisotropic phase is demonstrated to be due to the special manner of rod-like macromolecules with many hydrophilic groups, e.g., residual hydroxyl moieties, as described in Fig.1. The formation of interactions inside the blends (i.e. hydrogen bonding) simultaneously causes irregular and regular arrangements for the polysaccharide macromolecules or a semi-flexible rod manner. This can be recognized under a polarized light microscope through the textural variation since it yielded the ringed light diffraction pattern.

Furthermore, these biopolymers are found to be environmentally friend products and thus they are widely used for many commercial applications of food industry, pharmaceutics and cosmetics such as thickener, oil recovery, paper forming, coatings, fillers, dietary fiber, anti clumping agents, emulsifiers, and so on. HPC is also known to posses both lyotropic and thermotropic characteristics [3] while XG well exhibits a thermally structural stability in aqueous medium [4]. In our previously individual

publications on the phase diagrams of tertiary system in which the XG and HPC aqueous solutions in the presence of phosphoric acid (H$_3$PO$_4$), we reported that H$_3$PO$_4$ strongly suppressed the formation of LC phase according to its loadings. Besides, the increase in temperature could promote the phase transition from ordered to disordered behavior. In other words, the LC phase of both systems might be seriously shrunk in H$_3$PO$_4$-rich region. Especially, the cloudy suspension (CS) phase in HPC system was quite sensitive to high temperature. This was not seen in XG system due to the thermal stability and thus a lower critical solution temperature (LCST) was not detected afterward.

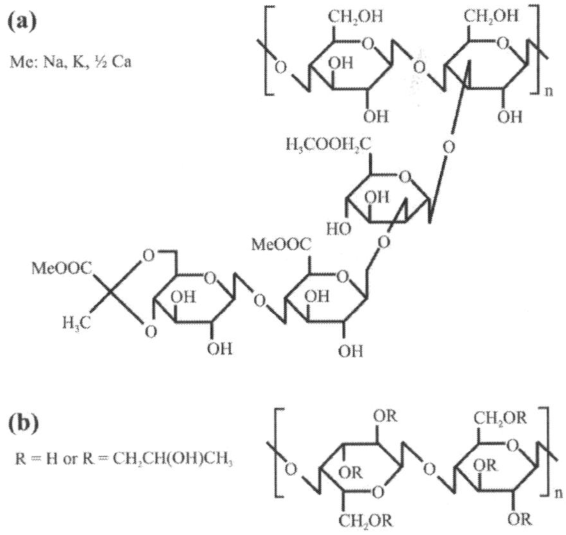

Fig.1 Structure of xanthan gum (a) and hydroxypropyl cellulose (b)

In the scope of this study, we only blended a tiny XG content with HPC before establishing a phase diagram of blended-HPC/H$_2$O/H$_3$PO$_4$ tertiary system. The further results of various blended ratios of XG and HPC will be reported in near future.

II. EXPERIMENTS

HPC and XG materials were purchased from TCI and Sigma-Aldrich. Phosphoric acid was provided by Acros Organics and deionized water was filtered by a reverse osmosis and deionizer system. Phosphoric acid was soluble

---

* Corresponding author.

© Springer International Publishing Switzerland 2015
V. Van Toi and T.H. Lien Phuong (eds.), *5th International Conference on Biomedical Engineering in Vietnam,*
IFMBE Proceedings 46, DOI: 10.1007/978-3-319-11776-8_54

in deionized water to use as a solvent. All samples were prepared by stirring the mixtures until the homogeneous solutions could be obtained. After preparing for 24 hours, the solutions were measured to determine the conformational difference at room and various temperatures.

Polarized optical microscope was used to observe the LC samples through the textural and color difference among others. Light transmission detector was connected to a laser source to evaluate the change of transmitted light intensity with temperature related to the change of phases.

The transmitted light intensity of a HPC solution behaves as a function of temperature at a specific heating rate (1°C/min). In our recent publications, the cloudy point temperature (T_C) at which a CS behavior can be defined as a steepest slope of this curve as presented in Fig.2 (Rwei et al., 2013) and the LC region of the phase diagram of $XG/H_2O/H_3PO_4$ tertiary system was displayed better than that of $HPC/H_2O/H_3PO_4$ tertiary system (Rwei et al., 2014), (Fig.3).

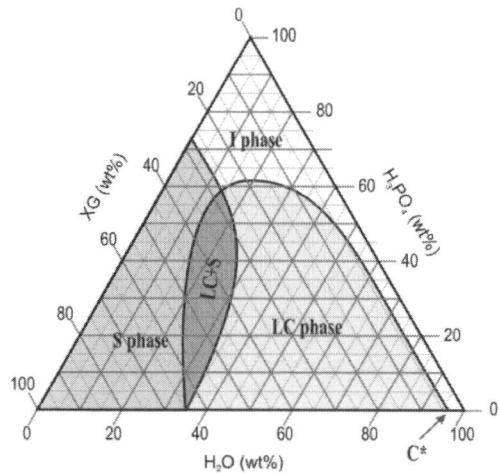

Fig. 3 Phase diagram of $XG/H_2O/H_3PO_4$ tertiary system investigated at room temperature (**Rwei et al., 2014**) [9].

III. RESULT AND DISCUSSION

A. Liquid Crystalline Phase at Room Temperature

According to our previous results obtained from both $HPC/H_2O/H_3PO_4$ and $XG/H_2O/H_3PO_4$ tertiary systems at room temperature, four distinct behaviors and a transition region – liquid crystalline (LC) miscible phase, isotropic (I) miscible phase, completely separated (S) phase and LC plus S region – were identified. Several points with various compositional concentrations of HPC, XG, water and phosphoric acid has been examined in these studies. With only 1% of XG in HPC, a newly phase diagram was developed as described in Fig.4 and there will be many interesting conclusions and discussions as follows.

By comparing between the $HPC/H_2O/H_3PO_4$ system without XG and the **blended**-$HPC/H_2O/H_3PO_4$ systems with 1% of XG as plotted in Fig.4, it can be shown that the LC region tends to expand further in H_2O-rich but H_3PO_4-poor domain. Clearly, the interaction between HPC and XG molecules was associated with the intra-molecular hydrogen bonds at which strongly occur without the presence of H_3PO_4. For more details, it was demonstrated that H_3PO_4

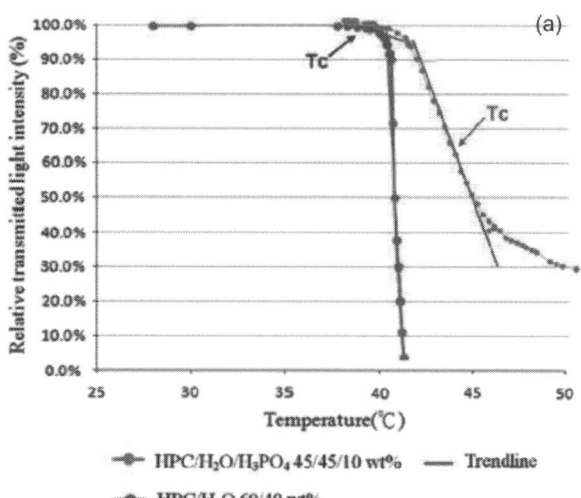

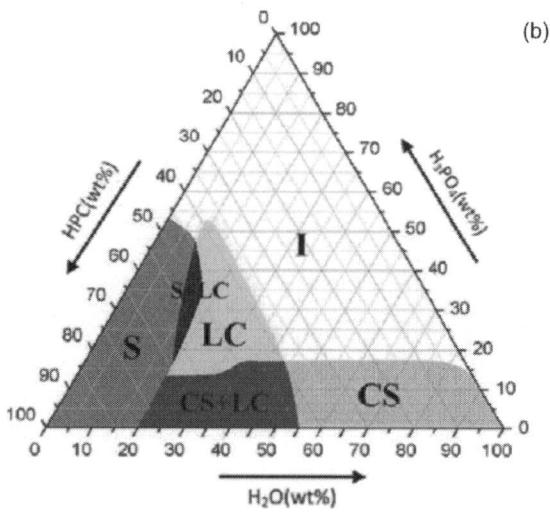

Fig. 2 Appearance of CS phase at 50°C on phase diagram of $HPC/H_2O/H_3PO_4$ tertiary system based on the cloudy point T_C at which was investigated by the steepest slope of curves of transmitted light intensity against temperature (**Rwei et al., 2013**) [7,8].

could suppress or reduce both intra-molecular and intermolecular interactions of HPC molecules to lead the degradation of ordered manner in these aqueous solutions. Accordingly, the destructed hydrogen bonding interaction by H_3PO_4 seemed to be not able enough to recover its behavior with replacing only 1% of HPC by XG.

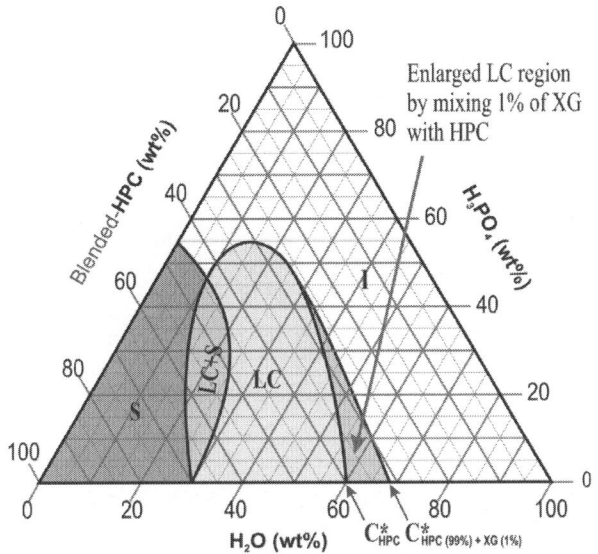

Fig. 4 Change in liquid crystalline phase as HPC was blended with 1% of XG at room temperature

Furthermore, the extension of LC phase was only mentioned at the boundary of isotropic region, not completely separated (S) region. The reason for this evolution is because the high critical concentration of HPC (about 70 wt%) at which its miscibility in H_2O could be introduced was not affected by adding a very little other solid like XG owing to almost saturated interactions.

B. Cloudy Suspension Behavior at High Temperatures

As mentioned previously, the T_C point can use for determining the shrinkage of CS phase by temperature. This phenomenon shows the coagulation limit of HPC polymer in which a rapid drop of relative transmitted light intensity is detected as temperature increases (Fig.2).

Table 1 Cloudy point temperature of HPC/H2O/H3PO4 solution as HPC is blended with 3.5% of some other chemicals.

Blended material	HPC (wt%)	T_C (°C)
None	40	46
	50	40
Xanthan gum (3.5 %)	40	36
	50	35

Values of transmitted light intensity I_S significantly reduce with temperature as well as the presence of a little XG content in blended-HPC/H2O/H3PO4 tertiary system as displayed in Table 1 and Fig.5. For the HPC solution without XG, T_C values of 40 % and 50 % HPC are about 46 and 40°C, respectively. This decrease of T_C is because the cloudy suspension manner was easier to form with more HPC content. Obviously, the significant increase of interaction amount due to the more presence of HPC in the solution was not meaningful for CS formation.

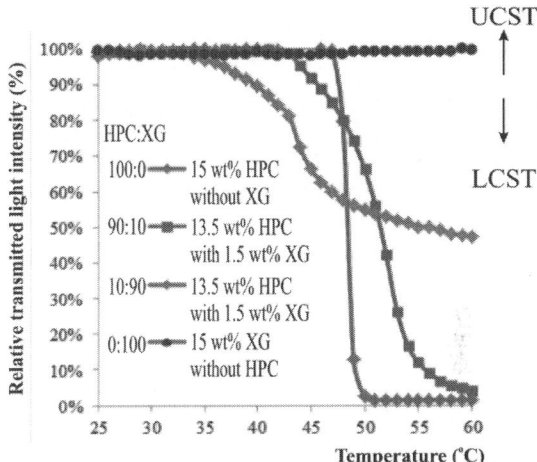

Fig. 5 Existence evidence of CS phase for HPC system, but not for XG system based on plot of light transmission intensity versus temperature

For the blended-HPC solutions (3.5 % of XG inside), the decrease of T_C values was insignificant (Fig.6). In fact, this difference for 40 and 50% of HPC (ΔT_C) was only 1°C while that for the former HPC solutions (without XG) was 6°C. Clearly, the thermal stability of these liquids was enhanced by the presence of XG. The characteristic of UCST of XG material actually affected on that of LCST of HPC material. The hydroxyl group in the structure of XG helices was useful for enforcing the cohesion among polysaccharide macromolecules (i.e. XG and HPC). The decrease in T_C clarified that the disappearance of CS regime by high temperature and the random coils of HPC molecules was able to be raised by the rigid helices of XG materials.

At the same HPC concentration, T_C value with 3.5 % of XG was much lower than that without XG. Particularly, T_C values were 46°C and 36°C at 40 % of HPC and those were 40°C and 35°C at 50 % of HPC. The decrease in T_C with XG concentration was induced by the interaction enhancement in which the coagulation formed easier. On the other hand, the addition of XG to HPC reduced the LCST characteristic of HPC as mentioned above. For more details, Fig.7 shows that the LCSTs reduced with adding XG to HPC from 39 to 34°C, being a further evidence for above explanation.

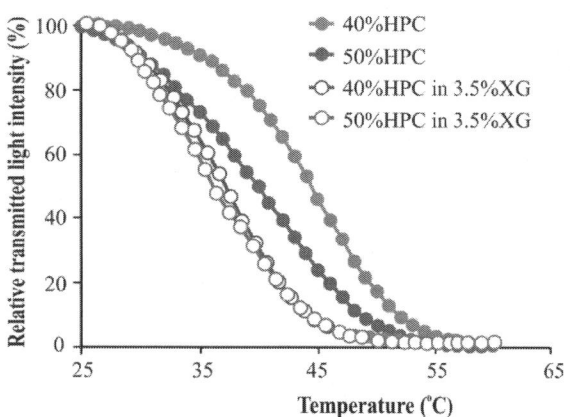

Fig. 6 Effect of XG on the light transmission as increasing temperature, demonstrating the synergistic interaction among polysaccharide molecules.

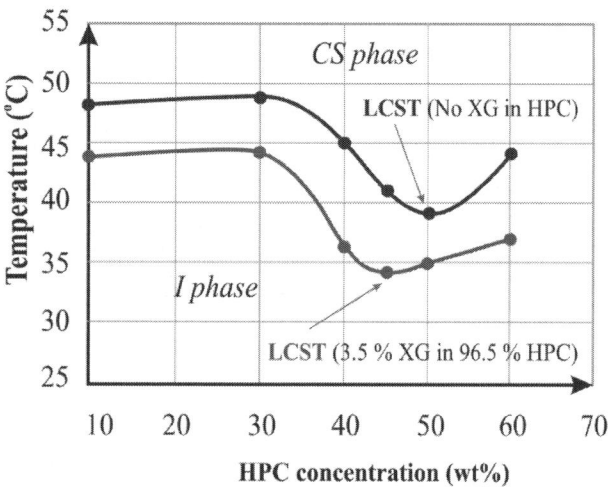

Fig. 7 Miscibility boundaries of CS and I phase for HPC solution (without XG) in dark blue color and a blended-HPC solution (with 3.5% of XG in 96.5 % of HPC) in red color

IV. CONCLUSIONS

This work is to study the formation of LC phase in HPC/H_2O/H_3PO_4 tertiary system in the presence of XG. The considerable enhancement of interactions between XG and

HPC is the main reason for enlarging the LC phase toward H_2O-rich but H_3PO_4 region. It means that the existence of H_3PO_4 in HPC aqueous solutions could reduce such incorporation owing to the strong destruction of hydrogen bonding interactions. However, the CS behavior was reduced with adding XG in HPC and increasing temperature because UCST characteristic of XG caused the decrease the LCSTs of HPC in its blended-HPC/H_2O/H_3PO_4 solutions.

ACKNOWLEDGMENT

This work was financially supported from the National Science Council of the Republic of China (Taiwan) under Contract No. NSC.

REFERENCES

1. Fischer H, Murray M, Keller A, Odell JA (1995), On the phase diagram of the hydroxypropyl cellulose – water system, J Mater Sci 30: 4623-4637
2. Fortin S, Charlet G (1989), Phase diagram of aqueous solutions of hydroxypropyl cellulose, Macromol 28: 4530-4539
3. Werbowji RS, Gray DG (1976), Liquid crystalline structure in aqueous hydroxypropyl cellulose solutions. Mol Cryst Liq Cryst 34: 97-103.
4. Buchard W (2001), Structure formation by polysaccharide in concentrated solution. Biomacromolecules 2 (2): 342-353
5. Fujiwara J, Iwanami T, Takahashi M Tanaka R, Hatakeyama T, Hatakeyama H (2000), Structural change of xanthan gum association in aqueous solutions. Thermochim Acta 352-353(3): 241-246
6. Spontak RJ, Bartolo RG, Nokaly ME, Hiler GD (1992). Enhanced anisotropic ordering and phase separation in lyotropic polysaccharide blends. Polymer 33 (24): 5343-5345
7. Rwei SP, Lyu MS (2012), 3-D phase diagram of HPC/H_2O/H_3PO_4 tertiary system, Cellulose 19 (4): 1065-1074
8. Rwei SP, Lyu MS (2013) HPC/H_2O/H_3PO_4 tertiary system: a rheological study, Cellulose 20 (1): 135-147
9. Rwei SP, Nguyen TA (2014) Phase formation and transition in XG/H_2O/H_3PO_4 tertiary system, Cellulose (online).

Tuan-Anh Nguyen
Institute of Organic and Polymeric Materials
National Taipei University of Technology
No.1, Sec.3, Chung-Hsiao E.Rd, Taipei City, Taiwan (R.O.C)
Email: nta@hcmute.edu.vn

The Time Based Study of Cell Morphology Using Atomic Force Microscopy

Wan Ibtisam Wan Omar[1] and Chin Fhong Soon[2]

[1] Faculty of Electrical and Electronic Engineering
[2] Biosensor and Bioengineering Laboratory, MiNT-SRC, Universiti Tun Hussein Onn Malaysia,
86400, Parit Raja, Batu Pahat, Johor, Malaysia

Abstract— **This paper aimed at time based study of cell morphology for human keratinocytes (HaCaTs) cultured on the cholesteryl ester liquid crystals (CELC). High resolution imaging of HaCaTs adhered on CELC substrate using Atomic Force Microscopy (AFM) will be investigated. In the AFM micrographs, the appearance of cells cultured on CELC captured in different time domain showed variation in the cell morphology including the surface roughness and thickness of cells. The AFM results revealed that soft liquid crystals substrate has triggered the remodeling of cell surface structure over time.**

Keywords— **Keratinocytes, liquid crystal, cell morphology, cell adhesion and Atomic Force Microscopy.**

I. INTRODUCTION

In the in-vitro culture, cell adhesion involves a transition from rounded cells in suspension become flattened on a substrate [1]. The patterns of the cells attachment are profoundly influenced by the physical properties of the adhesion substrate. The morphological changes of cells have close relation to the coupling of cells to a surface. Atomic Force Microscopy (AFM) is a state of the art technique used in characterizing the morphology of cells. Using a scanning tip mounted on a cantilever scanning across the surface, the AFM can do imaging at molecular resolution and can obtain topographical images of cells [2].

For biological sample application, non-contact mode is more reliable because it can avoid soft biological sample from damages [2]. Under this mode, when approaching the sample, the cantilever which is driven by a piezoelectric actuator vibrates at its resonance frequency. Van Der Waals forces from a distance are sensed by the oscillating probe, brought in close proximity to the sample. Then, the feedback keeps the amplitude at a preset value and a topographic image of the sample surface can be obtained. [3]. In addition, AFM can also provide quantitative measurement for roughness of a surface.

Recently, cholesteryl ester liquid crystals (CELC) were found able to support cell adhesion without pre-coating with ligands and they were applied in cell traction force sensing [4]. However, the cell adhesion patterns to the liquid crystals over time in close relation to the morphological changes are yet to be investigated. It was known from the literature that the cellular responses during adhesion depends on the physical characteristic of the substrate [5]. In this paper, the time dependent interaction of cells and the liquid crystals were revealed using AFM technique. AFM will be used to probe the cell surface roughness, thickness of a cell and cell morphology in relation to the adhesion of cells to the liquid crystal at different time.

II. MATERIALS AND METHODS

Liquid crystal preparation: The cholesteryl ester liquid crystal used is the formulations of Cholesteryl Chloride ($C_{27}H_{45}Cl$), cholesteryl Pelargonate ($C_{36}H_{62}O_2$) and Cholesteryl Oleyl Carbonate ($C_{46}H_{80}O_3$). All cholesteryl liquid crystal were purchased from Sigma-Aldrich. The mixtures termed CELC were formulated with 25 wt% of $C_{27}H_{45}Cl$, 38 wt% of $C_{36}H_{62}O_2$ and 38 wt% of $C_{46}H_{80}O_3$. One gram of these mixed compounds was melted to their isotropic phase (clear yellowish liquid) at a temperature range between $50^{\circ}C$ to $80^{\circ}C$ in a glass vial. Using a cell scraper (Corning Incorporation), 10 µl of CELC in isotropic phase was spread on the glass cover slips at a thickness approximately 100 µm and were cooled to their cholesteric phase at room temperature. Three glass substrates with and without the liquid crystal coatings were placed each in a petri dish.

Cell culture: Human keratinocyte cell lines (HaCaT) were used in this study and they were obtained from Cell Line Services (CLS, Germany). The cells were maintained in Dulbecco's modified Eagle medium (DMEM, Sigma Aldrich, UK) supplemented with 10% Fetal Calf Serum (FCS, Biowest, South America), L-glutamine (2mM, Sigma Aldrich, UK), fungizone (2.5 mg/l, Sigma Aldrich, UK) and penicillin (100 units/ml, Sigma Aldrich, UK). HaCaT cells were sub-cultured and cell suspensions at a density of 16 × 10^4 cells/ml were plated in each petri dish (Fig. 1). The cells were incubated for 24, 48 and 72 hours at 37 °C.

© Springer International Publishing Switzerland 2015
V. Van Toi and T.H. Lien Phuong (eds.), *5th International Conference on Biomedical Engineering in Vietnam,*
IFMBE Proceedings 46, DOI: 10.1007/978-3-319-11776-8_55

(a)   **(b)**

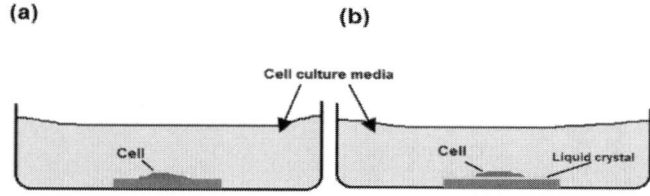

Fig. 1 Cell culture (a) without and (b) with liquid crystal coating.

Atomic force microscopy: After the cells had been cultured for 24, 48 and 72 hours, the cells on the glass substrate were removed and washed three times with Hanks Balance Salt Solution (HBSS). The cells cultured on both substrates were fixed with 1 % Formaldehyde for 5 minutes followed by serial alcohol dehydration at 35%, 50%, 75% and 90% of ethanol, each for 5 minutes. After the final dehydration, the cells were left dried at room temperature (25 °C) in a biological safety cabinet. Fixed cells on glass cover slips were transfer to a non-contact mode Atomic Force Microscopy (AFM, Park-System XE-100). A 0.2 Hz scanning rate was applied for scanning a 40 × 40 µm^2 scanning area. This measurement was performed on three randomly selected cells and the thickness of the cell membrane in the AFM images were obtained via the XE software. Cell membrane thickness was measured based on off-line scanned images obtained at different time domain and expressed as mean ± standard deviation (SD). To compare the means of cell membrane thickness, t-test analysis was performed in the Statistical Package for Social Sciences (SPSS statistic version 20) software. Differences in the means were significant at p < 0.05.

III. RESULTS

From the AFM results, HaCaT cells attached strongly to the surface with and without the presence of CELC as the substrate. Fig. 2 shows the difference of cell membrane thickness after 24, 48 and 72 hours of incubation. From the graph, the cell membrane cultured on liquid crystal indicate higher thickness value compared to cell cultured on plain glass for the three different culture period. After the cells were cultured for 24 hours, the AFM micrographs in Fig. 3a-b show that HaCaT cells cultured on plain glass and CELC substrate expressed similar rounded morphology, less spreading of membrane, with thickness of 0.32 ± 0.14 µm and 0.36 ± 0.09 µm, (N = 30) respectively (Fig. 2). However significant differences in the cell topology were seen at the cell surface after 48 hours of culture. The cell cultured on plain glass showed more extension of lamellipodia at a membrane thickness of 0.28 ± 0.07 µm, (p = 0, N = 30) (Fig. 2 and 3c) which was correlated with more expressions of focal adhesions [6]. Contrarily, cells cultured

on the liquid crystals for 48 hours showed less extension of cell membranes with a membrane thickness of 0.29 ± 0.06 µm (p = 0, N = 30) in comparison to cells cultured on plain glass as shown in Fig. 3c-d .

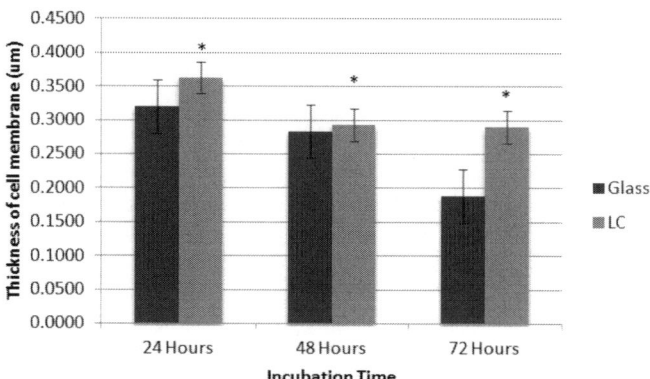

Fig. 2 Thickness of HaCaTs membrane cultured with and without the presence of liquid crystal as the substrate (N=30). The asterisk (*) indicate that thickness of cell membrane for cells cultured on glass are significantly different from cells cultured on liquid crystals for 24, 48 and 72 Hours.

Whilst after 72 hours of incubation (Fig. 3e), the lamellipodia and filopodia of the cell membrane on the glass substrate were found extended and the cells expressed more flattened morphology with a thickness of 0.19 ± 0.05 µm (p = 0, N = 30) (Fig.2), which was statistically different from the thickness of cell membrane projected on the liquid crystal substrate 0.29 ± 0.06 µm (p = 0, N = 30) (Fig. 2 and 3f).

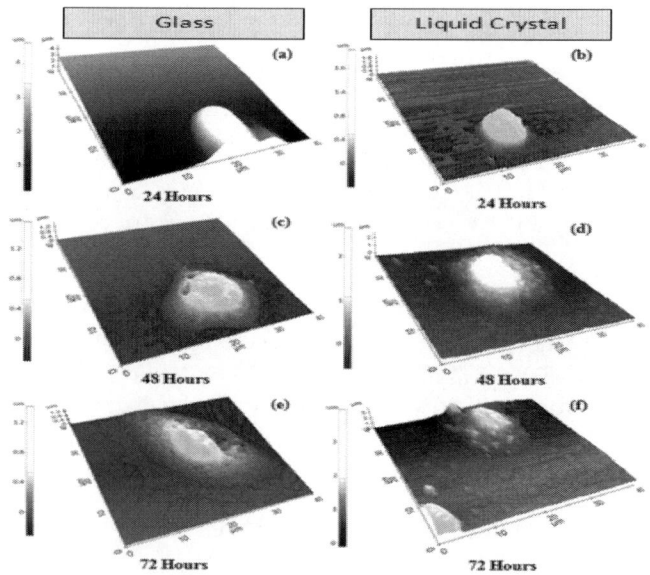

Fig. 3: Image of HaCaTs cell plated on (a, c, e) the glass substrate and (b, d, f) liquid crystal substrate after 24, 48 and 72 hours. (Scan size: 40 × 40 µm^2).

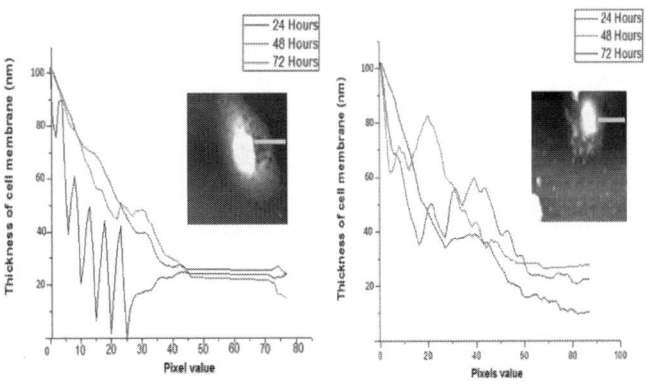

Fig. 4 Line profiles of cells cultured on plain glass and liquid crystal substrate. These line profiles were taken at the lamellipodia of the cells cultured on plain glass and liquid crystal respectively.

As indicate in the line profile of the cell adhered to substrate with and without liquid crystals (Fig. 3), lower oscillation of cell surface had been observed for the cells cultured on liquid crystals in comparison with the cell cultured on plain glass.

IV. DISCUSSION

The objective of this study is studying the morphology of cells interacted with CELC at various incubation time. As indicate in the AFM results, rounded morphology was expressed by cells during early stage of cell spreading (24 hours). At this stage, cells attached to the surface but do not required to employ to actin polymerization processes for them to crawl [7].

Once the initial attachment and secretion of extracellular matrix proteins is established, the cells began to extend their actin mesh projections that formed the lamellipodia and filopodia [6]. The significant difference of cell membrane projections can be seen when cells were grown on glass substrate for 48 and 72 hours. That is when the cells tended to sustain their bodies with broadly spread morphology. However, for cells attached to the liquid crystal for 48 and 72 hours of incubation, the cells remodeled themselves with lower adhesion area and high surface roughness. Engler et. al, 2004 demonstrated that soft substrate reduce the spreading and organization of actin into stress fibers [8]. Hence, the cells took on rounded morphology when adhered to a soft substrate. This is in good agreement with the results as shown in Fig. 4. This result shows the stiffness of a substrate influenced the morphology, cytoskeleton structure and adhesion of a cell.

Clearly, soft liquid crystal has strong influence on the surface roughness of the adherent cells and the cells membrane thickness due to the retraction of cell membrane. During adhesion, the cell surface receptors were triggered by the stiffness of the underlying substrate and sending downstream signaling to remodel the cell structure. Consequently, the cytoskeletons were regulated to project a rounded morphology leading to the reduction in lamellipodia extension. Hence, this is a result of cell mechano feedback in response to the stiffness of the substrate adhered by the cells. This report revealed the limited spreading of cells on soft liquid crystals regardless of long incubation periods.

V. CONCLUSIONS

By using AFM, time based studies of cells cultured on plain and liquid crystal coated substrate were performed. Various phases of cells formed from initial contact to firmly adhered cells topology had been shown through time based AFM studies. In different time course, the stiffness of the underlying liquid crystal substrate in which the cells adhered was found affecting the adhesion characteristic such as thickness of cell membrane, morphology, the surface roughness of the adherent cells and cells spreading.

ACKNOWLEDGMENT

The authors are grateful to the Malaysia Ministry of Higher Education for research funding support (FRGS Phase 1 Vot No. 1050) and also Universiti Tun Hussein Onn Malaysia for providing post graduate incentive research grant (GIPS Vot 1111).

REFERENCES

1. M. A. Fardin, O. M. Rossier, P. Rangamani, P. D. Avigan, N. C. Gauthier, W. Vonnegut, A. Mathur, J. Hone, R. Iyengar, and M. P. Sheetz (2010) Cell spreading as a hydrodynamic process. Soft Matter, 19:4788-4799.
2. W. I. Wan Omar, H. S. Lim, M. S. Alias, C. A. Norhidayah, N. Nayan, and C. F. Soon (2013) Application of Atomic Force Microscopy (AFM) and Field Emission Scanning Electron Microscopy (FE-SEM) in imaging HaCaT cells. Malaysian Technical Universities Conference on Engineering & Technology (MUCET), Kuantan, Pahang.
3. C. A. J. Putman, K. O. Van der Werf, B. G. De Grooth, N. F. Van Hulst, and J. Greve (1994) Tapping mode atomic force microscopy in liquid. Applied Physics Letters, 18:2454-2456.
4. C. F. Soon, M. Yousseffi, P. Twigg, N. Blagden, and M. C. T. Denyer (2012) Finite Element Quantification of the Compressive Forces Induced by Keratinocyte on a Liquid Crystal Substrate. Analysis and Design of Biological Materials and Structures: Springer, pp. 79-99.

5. T. Yeung, P. C. Georges, L. A. Flanagan, B. Marg, M. Ortiz, M. Funaki, N. Zahir, W. Ming, V. Weaver, and P. A. Janmey (2005) Effects of substrate stiffness on cell morphology, cytoskeletal structure, and adhesion. Cell motility and the cytoskeleton, 1:24-34.

6. C. F. Soon, W. I. W. Omar, R. F. Berends, N. Nayan, H. Basri, K. S. Tee, M. Youseffi, N. Blagden, and M. C. T. Denyer (2014) Biophysical characteristics of cells cultured on cholesteryl ester liquid crystals. Micron, 56:73-79.

7. J. L. McGrath (2007) Cell Spreading: The Power to Simplify. Current Biology, 10:R357-R358.

8. A. Engler, L. Bacakova, C. Newman, A. Hategan, M. Griffin, and D. Discher (2004) Substrate Compliance versus Ligand Density in Cell on Gel Responses. Biophysical journal, 1:617-628.

Address of the corresponding author:
Author: WAN IBTISAM WAN OMAR
Institute:UNIVERSITI TUN HUSSEIN ONN MALAYSIA
Street: 86400, PARIT RAJA.
City: BATU PAHAT, JOHOR
Country:MALAYSIA
Email: wanibtisamomar@gmail.com

The Effects of Enzyme to the Dissociation of Cells in Monolayer and 3D Microtissue on the Liquid Crystal Substrate

Kok Tung Thong[1], Chin Fhong Soon[2], and Kian Sek Tee[1]

[1] Faculty of Electrical and Electronic Engineering, Universiti Tun Hussein Onn Malaysia,
86400 Parit Raja, Batu Pahat, Johor, Malaysia
[2] Biosensor and Bioengineering Laboratory, MiNT-SRC Research Center,
Universiti Tun Hussein Onn Malaysia, 86400 Parit Raja, Batu Pahat, Johor, Malaysia

Abstract— **The technique first developed by our research group to culture three dimensions microtissues using liquid crystal substrate has potential to be used for cytochemical study. In order to demonstrate the differences in cell response to the cytochemical treatments, this paper applied an enzyme as a model drug to compare the enzymatic dissociation of cells grown in monolayer on a culture dish and microtissues cultured on the liquid crystal substrate. The results showed that the cells were fully dissociated layer by layer at a time course of 90 minutes. The monolayer of cells was dissociated directly by trypsinization within 6 minutes. Obviously, cells embedded deep in the microtissues were encapsulated or well protected from the treatment of EDTA-trypsin and this led to longer period of enzymatic dissociation.**

Keywords— **Liquid crystal, EDTA-trypsin, cell adhesion, keratinocytes cell, 3D microtissues.**

I. INTRODUCTION

Monolayer of cells or two-dimensional (2D) cells cultured in the culture flasks have distinct structural organization and cell-cell adhesion properties compared with three-dimensional (3D) microtissues [1]. During proliferation on a flat surface, cells persistently grow until they completely replenish the surface of the culture flasks. Essentially, cells cultured in this way could not self-regulate into multilayers of cellular structure. However, cells in 3D microtissues are able to self-regulate into spheroid or ellipsoidal population [2]. 3D microtissue is believed to have similar organization structure to the in-vivo system [3, 4]. In the biological system, hemidesmosomes and desmosomes are responsible for the binding of cell-extracellular matrix and cell-cell adhesions in arbitrary direction, respectively [5]. Comparatively, for cells cultured on stiff culture flasks, the focal adhesions are concentrated at the hemidesmosome while the function of the desmosome which supports cell-cell connection is forsaken. The difference in the cellular organization of 2D and 3D cells may affect the diffusion mechanism of a cytochemical. In our associated work, we have established a liquid crystal based 3D cell culture technique to produce 3D microtissues on the liquid crystal substrate [6]. The cells were found

self-assembling into spheroids and irregular shaped microtissues after cultured on the liquid crystal substrate for 72 hours. In this work, we are interested to find out how different structural integrity of cells affects the kinetic of drug after administration through a mechanism of distribution. EDTA-trypsin is a general enzyme commonly used in breaking the peptide bond of cells during sub-culturing of cells [7]. The effects of treating the monolayer of cells cultured on the culture flasks and microtissues formed on the liquid crystal surface with EDTA-trypsin were studied and reported in this paper.

II. MATERIALS AND METHODS

Cell culture: Human keratinocyte (HaCaT) cell lines were purchased from Cell Line Services (CLS, Germany). The cells were cultured in a $25cm^2$ culture flask with Dulbecco's Modified Eagle's medium (DMEM, Sigma-Aldrich, UK) which was supplemented with L Glutamine (2mM, Sigma-Aldrich, UK), Penicillin (100 units/ml, Sigma-Aldrich, UK), Streptomycin (100mg/ml, Sigma-Aldrich, UK), Fungizone (2.5 mg/l, Sigma-Aldrich, UK) and 10% Fetal Calf Serum (Promocell, UK). Culture flask was maintained in an incubator at 37°C with 5% CO_2. When the cells reached confluency, the existing culture media was discarded from culture flask and washed three times with Hank's Balanced Salt Solution (HBSS, Sigma-Aldrich, UK). After the wash with HBSS, 1ml of crude 0.25% EDTA-trypsin was deposited into the culture flask and incubated at 37 °C for 5 minutes to detach the cells from the culture flask. After the incubation, 5ml of DMEM was deposited into culture flask and the cells were being transferred to a tube and centrifuged at 1200 rpm for 5 minutes. After centrifugation, the supernatant was discarded and the cell pellet at the bottom of the tube was re-suspended in 6 ml of media. The cells were ready for the subsequent experiment.

Preparation of liquid crystal substrate: Liquid crystal (LC) gel will be mainly used in this research and the preparation of the liquid crystal substrates will be described. The CELC gel was prepared as described in [8]. The LC gel in a vial was heated to 80 °C on a heating stage, 5 μl of the fluid mixtures was spread at a thickness of approximately 200 μm on a petri dish using a squeegee coater developed previously [9].

3D cell culture: Cell suspension at a density of 2.6×10^6 cells/ml was gently deposited onto LC substrate coated in a Petri dish. Subsequently, the Petri dish was sealed and incubated in an incubator for 3 days in which the period allows the microtissues to form on the liquid crystal surface.

Trypsinization experiment: This experiment was performed for the cells grown in monolayer and 3D microtissues. 0.25% of EDTA-trypsin (Biowest, France) was heated up to 37 °C in a beaker. The old media used to grow the cells was discarded and replaced with 5ml of warm EDTA-trypsin. In the trypsinization experiment, the old media was discarded before an addition of the EDTA-trypsin is because the media used consists of fetal calf serum which could inhibit the effects of an enzyme such as the [7, 10]. A Nikon Eclipse TS100 inverted microscope at 10 × magnification was used to observe the trypsinization process of 2D and 3D cells. The response of the cells to trypsinization was captured using the QCapture Pro version 6.0 software at a time interval of 10 seconds. During imaging, the temperature of the Petri dish under examination was maintained at 37°C using a heating element as shown in Fig. 1. Pro's kit 303-150NCS digital multi-meter was used as a thermometer for monitoring the temperature of the EDTA-trypsin solution. Similar experiments were repeated three times.

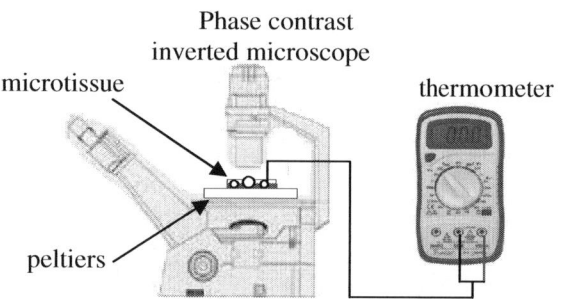

Fig. 1 Schematic diagram of a trypsinization experiment

III. RESULTS

This work was proposed to investigate the time response of HaCaT cells in monolayer and microtissues to an enzymatic dissociation buffer. The time-lapsed images of the cells response were taken every 5 seconds. After three days of culture, HaCaT cells plated in the Petri dish spread into confluency as shown in Fig. 2. Comparatively, for cells cultured on the surface of the liquid crystal substrate, aggregations of microtissues were observed as shown in Fig. 3. When the monolayer of HaCaT cells was exposed to

EDTA-trypsin, the enzyme hydrolyses cells proteins. During the first five minutes of treatment, cells morphology remained unchanged. After 5 minutes of trypsinization, the proteins of 2D cells hydrolyzed without any mechanical vibration. Cells were dissociated into single cells (Fig. 2) and suspended in the trypsin solution. In comparison, when microtissues were exposed to EDTA-trypsin for about 30 minutes, the cells at the periphery of the microtissues started to gradually dissociated. After 30 minutes of treatment, dispersion of individual cells could be observed at the adjacent area of the microtissue (Fig. 3b-c). For a microstissue with an estimated volumetric size is 9.63 μcm³, microtissue took more than 90 minutes to fully dissociate as presented in Fig. 3. The dissociation response time of the microtissues was longer than the time for the monolayer of cells.

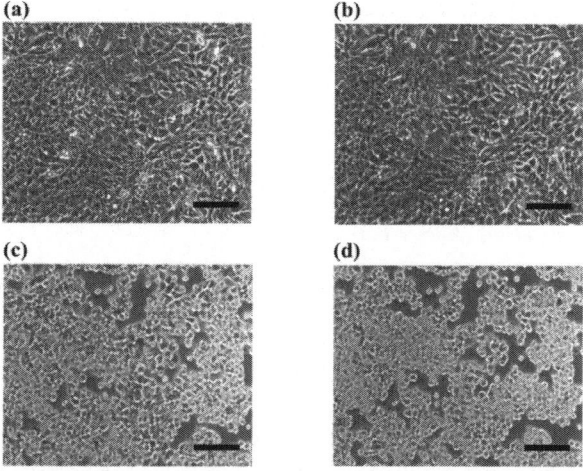

Fig. 2 The effects of trypsinization on monolayer of cells at (a) 0 minute, (b) 5 minutes, (c) 10 minutes, and (d) 15 minutes. (scale bar = 100 μm)

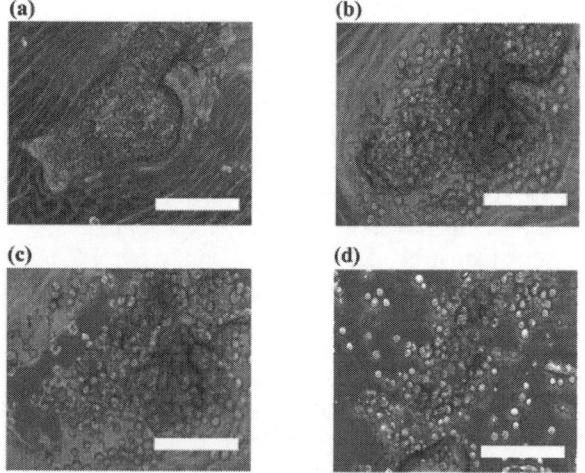

Fig. 3 The effect of trypsinization on 3D micro-tissues at (a) 0 minute, (b) 30 minutes, (c) 60 minutes, and (d) 90 minutes. (scale bar = 100 μm)

IV. DISCUSSION

The stimulation for cells to organize in 2D monolayer and 3D culture were attributed to the stiffness of the underlying substrates [10]. Stiff substrate often characterizes the adherent cells into broadly spread morphology which is contrast to the less spread cells adhering to a soft material [11, 12]. Cells tend to spread into thin monolayer on stiff substrate [13] which is due to the high exogenous tension (environment rigidity) of the substrate [4]. Hence, the cells proliferated planarly and always grown in an abnormal morphology. For microtissues, cells derive rich vinculins, α-actinins, and phosphorylated paxillins that regulate cells signaling and modelling of cytoskeleton leading to the control of cell morphology [3]. Both biophysical and biochemistry properties of a material are crucial factors that determine the behavior of the cells in-vitro and in-vivo [14, 15]. In cell biology experiment, enzymatic and non-enzymatic dissociation buffers are frequently used in cell and tissue experiment. In this study, enzymatic dissociation buffer (EDTA-trypsin) is highly recommended for dissociating microtissues due to the strong cohesion forces between the cells [16, 17]. The diffusion and distribution of enzymatic buffer occur in serial steps. First, the cells at the outer interface of the microtissues interacted with the EDTA-Trypsin. The media diffused around the cells to perform dissociate reaction. After the dissociation of the outer layer of the cells, the digestive effects of the enzyme gradually react with the next layer cells [18]. The reaction moved towards the center of microtissues and eventually, dissociated the cells in suspension. According to previous literature [2], a microtissue is divided into 3 parts histologically: the central necrotic zone, middle viable zone, and outer zone of cells with mitotic figures. The active outer and middle layer work as a protective layer to resist EDTA-trypsin penetrating the central zone of microtissue and that explained the longer time taken to dissociate microtissues in the experiment. Outer zone cells was much easier to be hydrolyzed in comparison to the central zone due to the unstable focal complexes adhesion proteins at the boundary of the microtissues [4]. Other factor such as thicker ECM proteins produced in microtissues may require longer time for the microtissues to be dissociated enzymatically in comparison to monolayer of cells. While having close encapsulation of cells, the diffusion and kinetic of solution across a semi-solid microtissue with spatially distributed adhesion proteins could literally affect the response time of a cytochemical [19]. Obviously, fast trypsinization occurred to 2D cells that have lesser adherent proteins [16]. Hence, the distribution and diffusion of enzyme to 2D cells and cells in 3D is dependent on the protein organization structure of the cells.

V. CONCLUSIONS

In this work, the microtissues took longer time to be dissociated compared with monolayer of cells using crude EDTA-trypsin. This could be due to the adherent properties and structural organization of the cells in the microtissues. The multilayer organization of cells with enriched ECM matrix and adhesion proteins are the barrier to the infusion of enzyme molecules.

ACKNOWLEDGMENT

We are thankful to Geran Insentif Penyelidik Siswazah (GIPS) for supporting Kok Tung Thong for his master degree scholarship in Universiti Tun Hussein Onn Malaysia (UTHM). We also acknowledge Malaysia Ministry of Higher Education for providing research funding support (RAGS Grant Vot No. R027).

REFERENCES

1. Cukierman, E.,R. Pankov et al. (2002) Cell interactions with three-dimensional matrices. Current opinion in cell biology 14(5): 633-640
2. Folkman, J. and M. Hochberg (1973) Self-regulation of growth in three dimensions. The Journal of experimental medicine 138(4): 745-753
3. Larsen, M.,V. V. Artym et al. (2006) The matrix reorganized: extracellular matrix remodeling and integrin signaling. Current opinion in cell biology 18(5): 463-471
4. Berrier, A. L. and K. M. Yamada (2007) Cell–matrix adhesion. Journal of cellular physiology 213(3): 565-573
5. Takahashi, Y.,D. F. Mutasim et al. (1985) The use of human pemphigoid autoantibodies to study the fate of epidermal basal cell hemidesmosomes after trypsin dissociation. Journal of investigative dermatology 85(4): 309-313
6. Soon, C. F.,M. Youseffi et al. (2009) Characterization and biocompatibility study of nematic and cholesteryl liquid crystals. Proceedings of the World Congress on Engineering
7. Huang, H.-L.,H.-W. Hsing et al. (2010) Research Trypsin-induced proteome alteration during cell subculture in mammalian cells. Journal of biomedical science 17: 36
8. Soon, C. F.,M. Youseffi et al. (2013) Development of a novel liquid crystal based cell traction force transducer system. Biosensors & Bioelectronics 39: 14-20
9. Soon, C. F.,Z. P. Goh et al. (2014) A squeegee coating appratus for producing a liquid crystal based bio-transducer. Advance Mechanics and Materials 466: 759-763
10. Sadoshima, J.-i.,L. Jahn et al. (1992) Molecular characterization of the stretch-induced adaptation of cultured cardiac cells. An in vitro model of load-induced cardiac hypertrophy. Journal of Biological Chemistry 267(15): 10551-10560
11. Yeung, T.,P. C. Georges et al. (2005) Effects of substrate stiffness on cell morphology, cytoskeletal structure, and adhesion. Cell motility and the cytoskeleton 60(1): 24-34
12. Solon, J.,I. Levental et al. (2007) Fibroblast adaptation and stiffness matching to soft elastic substrates. Biophysical journal 93(12): 4453-4461

13. Nelson, C. M. and M. J. Bissell (2006) Of extracellular matrix, scaffolds, and signaling: tissue architecture regulates development, homeostasis, and cancer. Annual review of cell and developmental biology 22: 287

14. Karamichos, D.,R. Brown et al. (2006) Complex dependence of substrate stiffness and serum concentration on cell-force generation. Journal of Biomedical Materials Research Part A 78(2): 407-415

15. Soon, C. F.,W. I. W. Omar et al. (2014) Biophysical characteristics of cells cultured on cholesteryl ester liquid crystals. Micron 56: 73-79

16. Cukierman, E.,R. Pankov et al. (2001) Taking cell-matrix adhesions to the third dimension. Science 294(5547): 1708-1712

17. Renò, F.,M. Rizzi et al. (2012) Gelatin-based anionic hydrogel as biocompatible substrate for human keratinocyte growth. Journal of Materials Science: Materials in Medicine 23(2): 565-571

18. Soon, C. F.,M. Youseffi et al. (2010) Effects of an enzyme, depolymerization and polymerization drugs to cells adhesion and contraction on lyotropic liquid crystals. Proceedings of the World Congress on Engineering 1: 556-561

19. Mehta, B. C. (2006) Optimization of enzyme dissociation process based on reaction diffusion model to predict time of tissue digestion, Ohio State University

Corresponding author:

Author   : Kok Tung, Thong
Institute   : Universiti Tun Hussein Onn Malaysia (UTHM)
Street   : 86400 Parit Raja
City   : Batu Pahat, Johor.
Country   : Malaysia
Email   : kok-tung88@hotmail.com

Microscale Tribological Response of Human Osteoarthritic Articular Cartilage under the Boundary Lubrication of Hyaluronic Acid

Cong-Truyen Duong and Duc-Nam Nguyen

Mechanical Engineering Department, Industrial University of Ho Chi Minh City, Vietnam

Abstract— Microscale frictional response by atomic force microscopy (AFM) can provide an opportunity to measure the final equilibrium of the cartilage frictional coefficient in the absence of interstitial fluid pressurization. In this study, we examined the effect of hyaluronic acid (HA) concentrations on the boundary lubrication of human osteoarthritis (OA) cartilage. Articular cartilage samples were obtained from human femoral heads with normal and advanced–stage OA cartilage. The tests samples were submerged in the lubricants of PBS and HA 1.0, 3.0, and 5 mg/ml. The microscale frictional coefficient and surface roughness of the OA cartilage was measured by AFM using triangular silicon–nitride cantilevers attached a polystyrene spherical tip. The results demonstrated HA concentrations to have ineffective boundary–lubricating ability in the normal cartilage. On the other hand, for the advanced–stage OA cartilage, HA played an important role in enhancing the boundary–lubricating ability due to the adsorption of HA molecules on the damaged cartilage surface. The microscale frictional response of the advanced–stage OA cartilage was independent on the on the HA concentrations. This suggests that some factors of the articular cartilage change during the progression of OA, which can trigger the boundary–lubricating ability of HA.

Keywords— Articular cartilage, Osteoarthritis, Atomic force microscopy, Boundary lubrication, Frictional response.

I. INTRODUCTION

Osteoarthritis (OA) is a common disease of the diarthrodial joints and is characterized by the progressive degeneration of articular cartilage that involves changes in the structure and composition of the articular cartilage [1, 2]. Many studies have suggested that a damaged cartilage surface due to OA progression results in an increase in friction and partial disruption of the biphasic lubrication mechanism by the removed cartilage surface layer [3-6]. The biphasic structure and interstitial fluid pressurization of the articular cartilage from its solid–fluid biphasic structure plays an important role in lubricating the diarthrodial joints without additional lubricants [7].

A natural synovial joint is lubricated by synovial fluid. Hyaluronic acid (HA, 3 – 4 mg/ml) [8] is the main constituents of synovial fluid and articular cartilage, which can act as the boundary lubricant [9]. HA has been reported to have effective boundary–lubricating ability in OA cartilage [10, 11]. Bell et al. reported a lower frictional coefficient in HA than that in Ringer's solution of the surfaces of both healthy and OA bovine cartilage [10]. On the contrary, Forsey et al. examined the lubricating ability of HA for human OA cartilage using a macroscale pin–on–plate tribometer and reported a decrease in frictional coefficient with HA compared to that with Ringer's solution [11]. They attributed this to the adsorption and penetration of HA molecules to the damaged cartilage layers, and reported that changes in the HA concentrations did not affect the cartilage lubricating ability [11].

A microscale frictional response by atomic force microscopy (AFM) can provide an opportunity to measure the final equilibrium of the cartilage frictional coefficient in the absence of interstitial fluid pressurization because of the extremely fast depletion of the cartilage fluid pressure on a micro-second scale caused by microscale contact between the AFM tip and cartilage surface. However, there is no data on HA concentration–dependent lubrication of human OA cartilage. Therefore, in this study, we examined the effect of HA concentrations on the boundary–lubricating ability of OA articular cartilage in human hip joints from AFM measurements of the microscale frictional coefficient and surface roughness.

II. MATERIALS AND METHODS

A. Sample Preparation

Two human femoral heads with normal and advanced–stage OA cartilage were explanted from the total hip replacement operations. The femoral heads were stored in a phosphate buffered saline (PBS) solution and frozen at −20 °C until being thawed for specimen preparation. Five cylindrical cartilage samples ($\phi = 5$ mm, $h \sim 0.5$ mm) were sectioned from each femoral head using a biopsy punch with an internal diameter of 5 mm. The test samples were stored in PBS at 4 °C for no more than 24 h before the experiments. Prior to the measurements, each OA cartilage sample was glued to a cylindrical plate ($\phi = 19$ mm, $h = 1$ mm), affixed to a 50 mm polystyrene Petri dish with cyanoacrylate adhesive, and submerged in the lubricant.

© Springer International Publishing Switzerland 2015
V. Van Toi and T.H. Lien Phuong (eds.), *5th International Conference on Biomedical Engineering in Vietnam,*
IFMBE Proceedings 46, DOI: 10.1007/978-3-319-11776-8_57

B. Lubricant Preparation

Two lubricants at different concentrations were prepared. PBS (P5493, Sigma–Aldrich, USA) was used as a control solution. HA (hyaluronic acid, 53747, Sigma–Aldrich, USA) from *Streptococcus equi* with a molecular weight of MW = 1.63 MDa at 1.0, 3.0 and 5.0 mg/ml was dissolved in PBS by stirring in an ultrasonic bath at 60 °C for 3 hours. These HA concentrations were chosen based on the physiological concentration of human synovial fluid, ranging from 3.0 to 4.0 mg/ml [8].

C. AFM Measurements

AFM (XE 70, Park Systems, South Korea) (Fig. 1) was used to measure the microscale frictional coefficient and surface roughness of human OA cartilage samples, which were submerged entirely in lubricants and measured at room temperature. An image processing program (XEI, Version 1.8.0, Park Systems, South Korea) was used to calculate the applied normal force and surface roughness. Triangular silicon–nitride (Si_3N_4) cantilevers (normal spring constant, k_N=0.35 N/m) with a polystyrene spherical tip of 4.5 μm diameter (Novascan Technologies, USA) were used for the measurements (Fig. 2). The AFM probe was submerged in the lubricants for 30 min before imaging to allow thermal equilibration at room temperature.

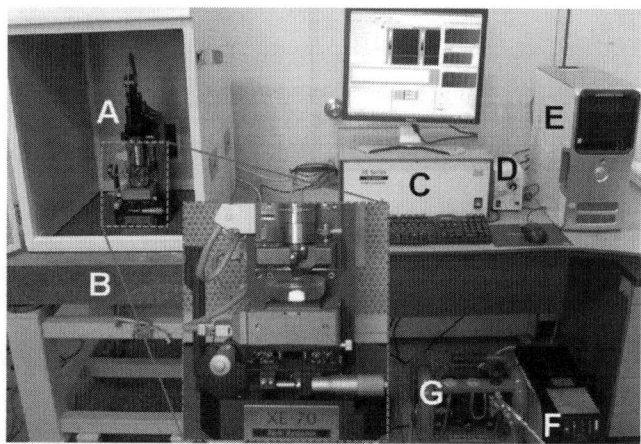

Fig. 1 AFM device was used for the measurements. (A) AFM body; (B) vibration isolation platform; (C) AFM contronller; (D) light source; (E) computer system; (G) compressor; and (F) UPS

An 40μm × 40μm area was scanned in contact mode (resolution: 256 × 256 pixels, scan rate: 1 Hz (80 μm/s)). The protocol for calculating the microscale frictional coefficient is described elsewhere [12]. Briefly, the microscale frictional coefficient (μ) was calculated from the linear slope of the plot of the frictional versus normal forces (Fig. 3). The frictional force (F_L) was calculated from the

AFM image of the 40μm × 40μm scanned area using the equation, $F_L = V_{LFM} \times S_L \times k_L$. The lateral sensitivity (S_L) was calculated analytically from the normal sensitivity (S_N = 75 nm/V) [13]. The lateral voltage signal (V_{LFM}) was obtained as one–half of the mean difference between the forward and backward voltage signals from the scanned image. The lateral spring constant (k_L) for the triangular cantilever was developed from a recent study [14] as follows

$$k_L = \frac{2L^3}{3(1+\nu)\,wh^2} \times$$
$$\left[\frac{1}{\tan\alpha} \log\left(\frac{(L-L_1)\tan\alpha + a}{a} \right) + \frac{L_1}{w} \right]^{-1} \times k_N \quad (1)$$

The triangular cantilever dimensions were measured using an optical microscope (Eclipse LV100, Nikon Instruments, Japan), and the calculated value for k_L was 110 N/m.

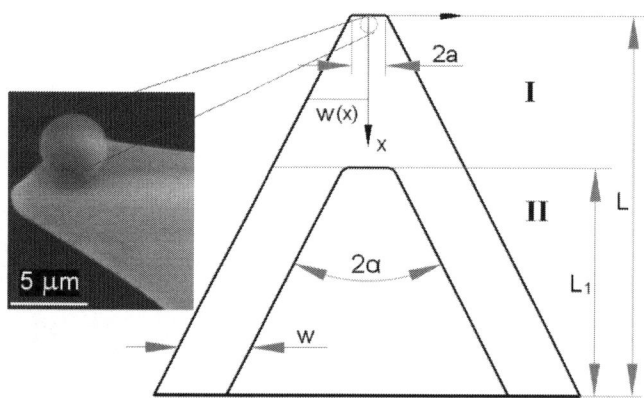

Fig. 2 SEM image of a triangular silicon-nitride cantilever with a polystyrene spherical tip (4.5 μm diameter) attached, and detailed parameters of the cantilever

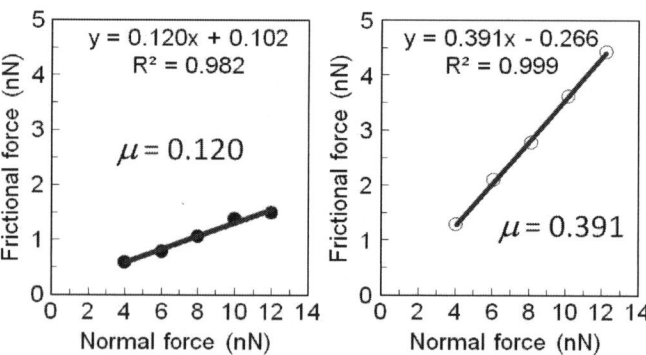

Fig. 3 Typical plots of frictional versus normal forces measured in PBS for the (a) normal cartilage and (b) advanced–stage OA cartilage

The mean value of μ was calculated from 16 different locations on the cartilage surfaces at each lubricant concentration. The value of R_q was measured simultaneously over the same scanned area and scanning velocity.

D. Statistical Analyses

A one–way analysis of variance (ANOVA) was used to detect significant differences in the surface roughness between the different OA stages and in the microscale frictional coefficient of OA cartilage between PBS and different concentrations of HA.

Table 2 lists the average cartilage frictional coefficients (μ) under different HA concentrations. For normal cartilage, all HA concentrations did not improve the cartilage frictional response (Fig. 5a). However, μ was significantly higher at a HA concentration of 5.0 mg/ml compared to the PBS and other HA concentration groups (Fig. 5(a)).

Table 1 Mean $\pm$ standard deviation (SD) values of microscale surface roughness (R_q) for the human OA cartilage measured in diverse HA concentrations ($n = 16$)

Lubricant	$R_q \pm$ SD		p value
	Normal	Advanced–stage OA	
PBS	137 ± 25	533 ± 196	< 0.0001
HA 1.0 mg/ml	173 ± 32	490 ± 227	< 0.0001
HA 3.0 mg/ml	139 ± 36	455 ± 240	< 0.0001
HA 5.0 mg/ml	202 ± 23	491 ± 115	< 0.0001

Table 2 Mean $\pm$ standard deviation (SD) values of microscale frictional coefficient (μ) and squared correlation coefficient (R^2) for the human OA cartilage measured in diverse HA concentrations ($n = 16$)

Lubricant	$\mu \pm$ SD ($R^2 \pm$ SD)		p value
	Normal	Advanced–stage OA	
PBS	0.119 ± 0.036	0.409 ± 0.119	< 0.0001
	(0.973 ± 0.031)	(0.976 ± 0.056)	
HA 1.0 mg/ml	0.126 ± 0.083	0.262 ± 0.083	< 0.0001
	(0.928 ± 0.059)	(0.959 ± 0.060)	
HA 3.0 mg/ml	0.119 ± 0.027	0.269 ± 0.119	< 0.0001
	(0.976 ± 0.017)	(0.981 ± 0.017)	
HA 5.0 mg/ml	0.181 ± 0.039	0.221 ± 0.067	$= 0.046$
	(0.971 ± 0.017)	(0.914 ± 0.074)	

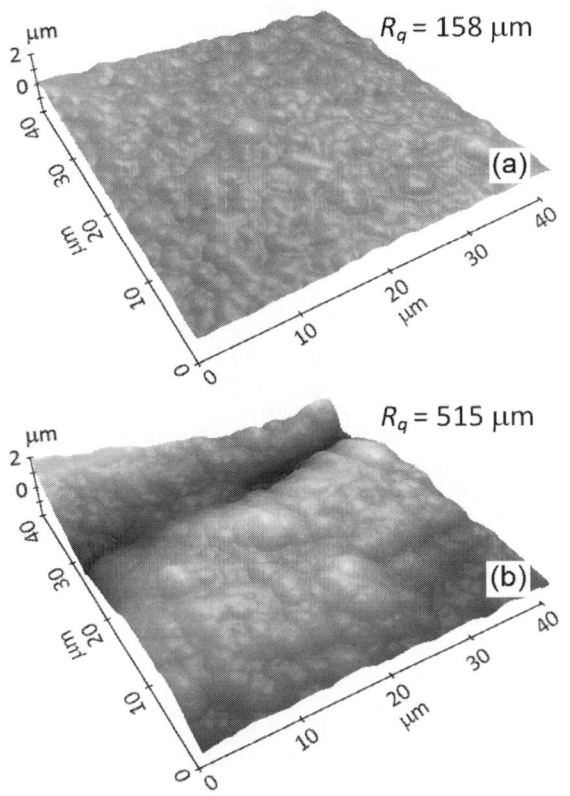

Fig. 4 Typical AFM images of cartilage surface roughness measured in PBS for (a) normal cartilage and (b) advanced–stage OA cartilage

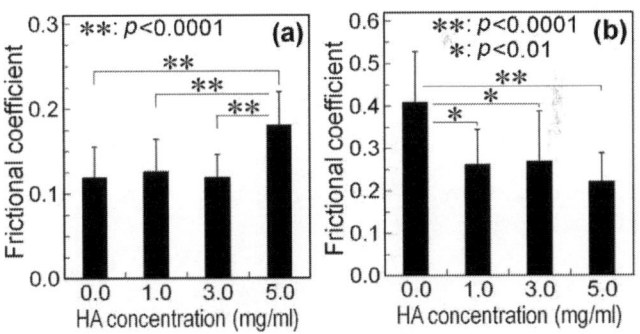

Fig. 5 Average microscale frictional coefficients of (a) normal cartilage and (b) advanced–stage OA cartilage measured in a range of HA concentrations

III. RESULTS

Fig. 4 presents typical results of the surface roughness (R_q) over the 40μm × 40μm scanned area with a 256×256 pixels resolution for all human OA cartilage measured in PBS. The mean R_q values increased significantly with increasing OA stage; In PBS, for example, $R_q = 137 \pm 25$ nm for the normal cartilage and $R_q = 533 \pm 196$ nm for the advanced–stage OA cartilage (Table 1). Significant difference in R_q were observed between the normal and advanced–stage OA cartilage ($p < 0.0001$).

In the advanced–stage OA, HA improved the cartilage frictional behavior significantly (Fig. 5(b)). Significant differences in μ showed that the improvement in cartilage lubrication by HA was concentration–independent; a significant difference in μ was observed between the PBS and all the HA concentration groups ($p < 0.01$), but not among the different HA concentration groups.

IV. Discussion

In this study, AFM was used to examine the role of HA concentrations in the boundary–lubricating ability of OA cartilage surfaces in human hip joints. The concentrations of HA were not effective in improving the boundary–lubricating ability of normal cartilage (Fig. 5(a)). For the advanced–stage OA cartilage, however, HA improved the cartilage frictional behavior significantly (Fig. 5(b)).

The uppermost superficial layer, which is covered with adsorbed molecules, consists of phospholipids, proteins and glycoproteins, and provides effective boundary lubrication [5]. In addition, proteoglycans located beneath the cartilage superficial layer maintain low friction and protect the cartilage surface [5]. In this study, the loss of cartilage superficial layer during OA progression led to an increase in R_q (~ 137 – 533 nm) and μ (~ 0.119 – 0.409) when measured in PBS (Table 1 and 2).

For the normal cartilage surface, the use of HA at a range of concentrations as lubricants for normal cartilage did not improve the frictional response, possibly because proteoglycans or SZP (superficial zone protein) in the cartilage superficial layer itself provides excellent lubricating ability; thus, HA did not give additive improvement in cartilage friction to proteoglycans or SZP.

In the advanced–stage OA, the cartilage superficial layer was removed completely, and the damaged cartilage surface (Fig. 4(b)). In this OA stage, the use of HA as boundary lubricants was quite effective due to the absence of proteoglycans, and HA molecules adsorbed on the damaged cartilage surface facilitated sliding, leading to an improved cartilage frictional response (Fig 5(b)). Previous macroscale studies reported that HA [10, 11] could help improve the frictional response of the damaged articular cartilage. Bell et al. reported the effectiveness of HA as a boundary lubricant due to the adsorption of HA molecules on the surface of damaged bovine cartilage [10].

V. Conclusions

The results demonstrated HA to have ineffective boundary–lubricating ability in the normal cartilage; but, HA played an important role in enhancing the boundary–lubricating ability in the advanced–stage OA cartilage. The microscale frictional response of the advanced–stage OA cartilage was independent on the on the HA concentrations. This suggests that some factors of the articular cartilage change during the progression of OA, which can trigger the boundary–lubricating ability of HA.

Acknowledgment

This research was funded by the Industrial University of Ho Chi Minh City, Vietnam through the University Research Foundation (No. 25/2013/HĐ-ĐHCN-KHCN).

References

1. Bonnevie E D, Burris D L et al (2011) In situ studies of cartilage microtribology: Roles of speed and contact area. Tribol Lett 41: 83-95
2. Stolz M, Gottardi R, Raiteri R et al (2009) Early detection of aging cartilage and osteoarthritis in mice and patient samples using atomic force microscopy. Nat Nanotechnol 4: 186-192
3. Basalo I M, Raj D, Krishnan R et al (2005) Effects of enzymatic degradation on the frictional response of articular cartilage in stress relaxation. J Biomech 38: 1343-1349
4. Northwood E, Fisher J (2007) A multi-directional in vitro investigation into friction, damage and wear of innovative chondroplasty materials against articular cartilage. Clin Biomech 22: 834-842
5. Kumar P, Oka M, Toguchida J et al (2001) Role of uppermost superficial surface layer of articular cartilage in the lubrication mechanism of joints. J Anat 199: 241-250
6. Krishnan R, Kopacz M., Ateshian G A (2004) Experimental verification of the role of interstitial fluid pressurization in cartilage lubrication. J Orthop Res 22: 565-570
7. Ateshian G A (2009) The role of interstitial fluid pressurization in articular cartilage lubrication. J Biomech 42: 1163-1176
8. McPherson R A, Pincus M R (2006) Henry's clinical diagnosis and management by laboratory methods. Elsevier Inc, Philadelphia
9. Katta J, Jin Z M, Ingham E (2008) Biotribology of articular cartilage-a review of the recent advances. Med Eng Phys 30: 1349-1363
10. Bell C J, Ingham E, Fisher J (2006) Influence of hyaluronic acid on the time-dependent friction response of articular cartilage under different conditions. P I Mech Eng H 220: 23-31
11. Forsey R W, Fisher J, Thompson J et al (2006) The effect of hyaluronic acid and phospholipid based lubricants on friction within a human cartilage damage model. Biomaterials 27: 4581-4590
12. Duong C T, Nam J S, Seo E M et al (2010) Tribological property of the cobalt-chromium femoral head with different regions of wear in total hip arthroplasty. P I Mech Eng H 224: 541-549
13. Liu Y H, Evans D F et al (1996) Structure and frictional properties of self-assembled surfactant monolayers. Langmuir 12: 1235-1244
14. Duong C T, Lee J H, Cho Y et al (2012) Frictional response of normal and osteoarthritic articular cartilage in human femoral head. P I Mech Eng H 227: 129-137

Authors: Cong-Truyen Duong and Duc-Nam Nguyen

Institute: Mechanical Engineering Department, Industrial University
                 of Ho Chi Minh City
Street: Nguyen Van Bao
City: Ho Chi Minh
Country: Vietnam
Email: truyendc@gmail.com; nguyennams@gmail.com

Single Cell Traction Force Mapping Software

Chin Fhong Soon[1], Kian Sek Tee[1], Mansour Youseffi[2], and Morgan Denyer[2]

[1] Biosensor and Bioengineering Laboratory, MiNT-SRC, Universiti Tun Hussein Onn Malaysia,
86400, Parit Raja, Batu Pahat, Johor, Malaysia

[2] School of Design, Engineering and Technology, School of Life Sciences, University of Bradford, BD7 3AL, Bradford,
United Kingdom

Abstract— A software was developed for the representation of cell traction forces at different levels of intensity. A cell traction force map records the distributions of forces exerted by a cell on a liquid crystal substrate which was manifested as deformation lines. The algorithm was specially developed by applying an established force-deformation coefficient. The force data points were obtained by user defined length of deformation using mouse control over the loaded image. By interpolating the discrete data points of forces in a Euclidean coordinate system, a surface in different pseudo colors representing the intensity of forces was generated. The algorithm is able to distinguish localized forces down to 5 µm of resolution. This method benefits from the fact that physical expressions of cells on the liquid crystal were integrated into a geometric surface visualization explicitly.

Keywords— force mapping, software, interpolation, cell traction force, liquid crystals.

I. INTRODUCTION

A technique to quantify cell traction forces exerted to liquid crystal surface has been developed [1]. The techniques involved with measuring the length of liquid crystal (LC) deformation line which is proportional to the bi-axial compressive forces exerted via a pair of focal adhesions at the circumference of the cell [2]. Based on the principle of Hooke's theorem, the linear relationship between the length of deformation and the forces was established and reported in our previous publication [1]. In this paper, a bespoke cell force measurement and mapping software (CTFM) will be presented. Basically, the computer program functions to map the cell traction force based on the length of deformation line which is measured by the user in the software interactively. A graphical user interface (GUI) was designed to communicate with the user, either to receive instruction from the user or to convey messages to the user. GUI allows the user to manipulate, animate and render the force map for explicit analysis [3]. The use of a GUI was to increase the speed of use, aid repeatable execution of work and provide attractiveness [4].

II. MATERIALS AND METHODS

A. Coding of Algorithm

The CTFM software was developed to be as simple as possible in the MATLAB Integrated Development Environment (IDE) with an aim to generate a cell traction force map for further analysis. The design and execution of the program was arranged sequentially and divided into three main steps: The first step involves, importing of an image and performing measurements of the deformation lengths by the user. The second step focuses on enabling the program to calculate the average traction force based on the user inputs and the generation of a visualization system that allow the display of the forces related to each LC deformation line induced by single cells in the phase contrast images. The third step then involved rendering 2D and 3D cell traction force maps into a pseudo color map, which allows representations of the force distributions across the cell.

The program initiates by executing a GUI which prompts a user to select an image of file type jpg, tiff or bmp, and reads the phase contrast image data to the program as shown in the flow chart (Fig. 1). The acquisition of the phase contrast images of keratinocytes used in this paper were obtained as reported in [5]. The GUI also contains the instructions as a guide for the user to use the software. The size of the image would be determined by the program function, *imread*, and assigned to a Euclidean coordinate system (x, y). A calibration for defining the scale of the image would be performed by the user to define the start and end point on a precluded scale bar in the image by left-clicking the computer mouse. The data of the two points would be stored in an array *[xr, yr]= ginput*, where r is the number of input. Based on the user input, the program calculates the distance between the two points in a Euclidean vector space using *norm* function in MATLAB which has included the Pythagoras theorem.

Next, a pop-up window prompts the user to input the length of the scale input through the keyboard. Next, the

© Springer International Publishing Switzerland 2015
V. Van Toi and T.H. Lien Phuong (eds.), *5th International Conference on Biomedical Engineering in Vietnam*,
IFMBE Proceedings 46, DOI: 10.1007/978-3-319-11776-8_58

user would be required to repeat similar steps in order to measure the length of the deformation lines found in the image. The length of the lines measured by the user would be multiplied by the force-deformation coefficient (20 nN/μM^2) [1] of the LC to yield the value of the traction force correlating to a deformation line. The force value would be stored in an array (z) with a counter of *meas_num,* which update the number of measurements upon completion of a measurement (Fig. 1). Once the measurement is completed, the position of the force data point would be stored in (x, y) array and then, similar measurement process can be repeated for another deformation line. To terminate the measurements, the user is required to right-click the mouse anywhere on the image and the program would take over the subsequent processing to generate the 3D traction force map and the data distribution in a coordinate system.

In the process of generating the traction force map, the x, y and z arrays are appended into a matrix and transposed into three data columns (x', y' and z') so that each data in a column can be accessed independently. A regular grid would be created using the *meshgrid* function and the x_i and y_i data are fitted on the grid. These data points can then be interpolated using the *griddata* function included with the linear and cubic interpolation functions. Then, 3D graphical visualization of the traction force distribution would be rendered by the *surf* function. The graphics produced in the *Figure* window allows rotation of the cell traction force map in 3D, zooming, panning and labeling. The software was used for analyzing the traction forces of quiescent human keratinocytes [6].

B. Determining the Spatial Resolution of CTFM Software

The resolution of the cell force mapping software was calibrated by using an image of a polydimethylsiloxane (PDMS) template containing regular line gauges. Two PDMS sheets with regular widths of 3 μm and 5 μm were prepared. These PDMS templates were fabricated by casting and curing an elastometric polymer, Sylgard 184 (Dow Corning, Midland, MI) against a photoresist micro-patterned glass template with line patterns of 3 and 5 μm. The PDMS consists of base elastomer and curing agent that were mixed in a 10:1 (v/v) ratio. After polymerization on the master template, the flexible PDMS sheets were removed from the master replica mold [7] and phase contrast microscope equipped with ImageJ software was used to capture images for each of the PDMS templates obtained.

In the CTFM software, assumptions of the deformation lines with similar magnitude around a hypothetical cell shape (oval shape) was alternately applied to the edge of the regular line patterns that were placed over the phase contrast images consisting of PDMS line patterns at a regular width

of 3 and 5 μm. The procedures used in measuring the deformation lines by using the CTFM software were as described earlier on.

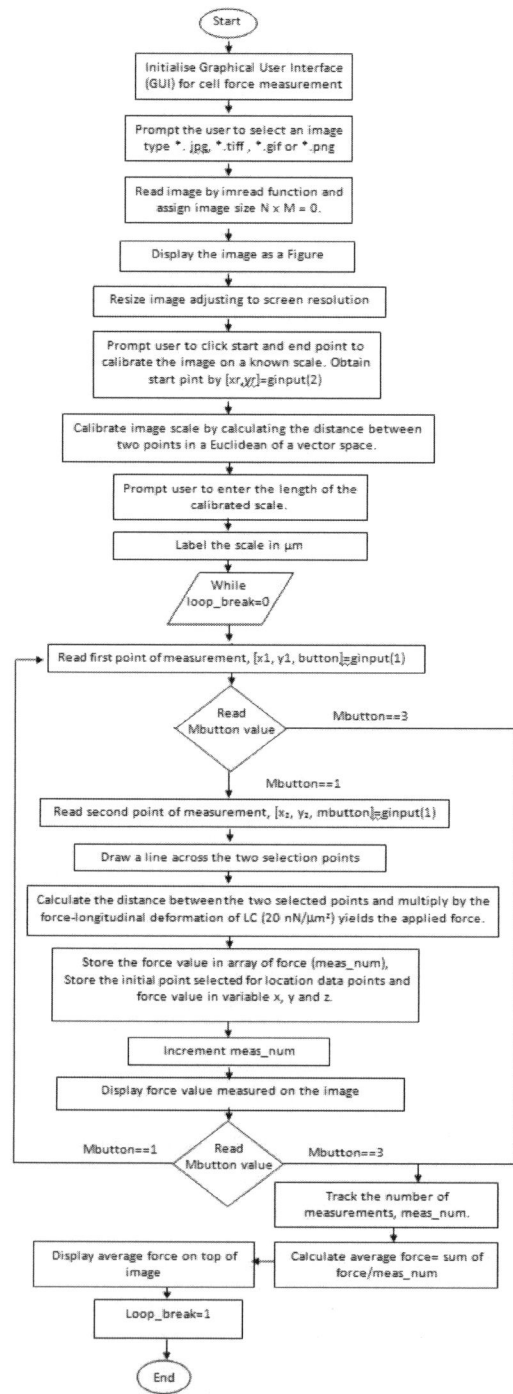

Fig. 1 Flow chart of the cell traction force mapping algorithm

III. RESULTS AND DISCUSSION

A. Cell Traction Force Measurement and Mapping Software

The main GUI of the CTFM software functions by helping the user to load image files into the program for cell force measurement. The imported image must be in a dimension (N x M) which is smaller than the pixel resolution of the computer LCD. On the main interface of the CTFM software, the user would be prompted to select an image from the computer hard disk directory, and then follows a step by step instruction in order to perform the measurements of cell traction forces. In addition, the user has the option not to do anything or to exit the program.

After loading an image of a single cell, the user calibrates the image based on the scale bar available in the image to determine the image dimension or the pixel size of the image. In the image, calibration was performed by selecting the start and end points over the scale bar as shown in Fig. 2. When the scale bar is measured, a pop-up window would request the user to enter the scale of the image (in μm).

For the calibration of the image scale, the entered value in the pop-up window was used to calculate the ratio of pixels to length of the loaded image. The example in Fig. 2 has a pixel to length ratio of 100 pixels/μm. Any subsequent measurements of the deformation lengths in pixels were multiplied by this ratio in order to calculate the length in micrometers. Subsequently, the calibrated value would be displayed over the scale bar on the image (Fig. 2). Once the value is entered, the user should select the OK button to proceed with the measurements of deformation length and force.

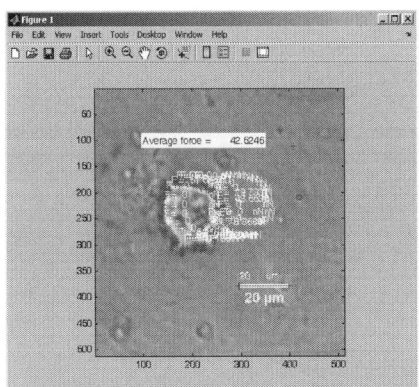

Fig. 2 The display of traction forces and the average traction force of a cell

The measurement of the deformation length in the image is again based on mouse-clicking activation. In the phase contrast image, the user would left-click the mouse to select the start and end points of a deformation line by following the same procedure used in the calibration of the image scale. After selecting the two points by using the mouse, a red line would be plotted across the two points which is over the deformation line (Fig. 2), and the computed force in nano-Newton (nN) would be indicated over the line. If there is no deformation line found in a particular area of the cell, the user would be required to double-left-click the mouse in order to select the same point, which indicates that no traction force was found for that point. In the software, the first few steps are expected to be manually performed by the user. Hence, the measurement accuracy is dependent on the user's ability to identify a deformation line. Once all measurements are completed, the average cell traction force would be computed for force points greater than 0 nN and would be displayed on top of the image (Fig. 2).

Subsequently, the rest of the computational work would be executed by the program automatically; this generates a graphical visualization of the force points over a Euclidean coordinate system, and the associated map of traction forces. Fig. 3 shows the magnitude of the force point (z) projected on a Euclidean coordinate system (x, y). In this example, the pixel size of the image is 256 × 256. From the magnitude of the force point at various positions over the coordinates, data points were fixed on the grid and then fitted by using interpolation fitting functions.

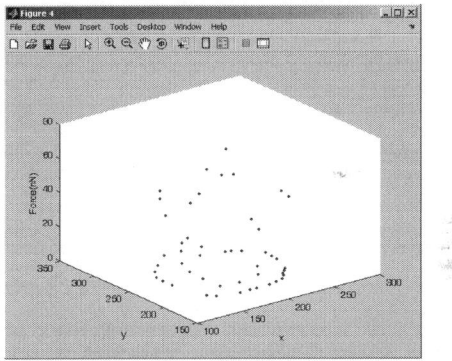

Fig. 3 A display of the force data points (z) scattering over the x and y Euclidean coordinate system.

The graphical visualization of the cell traction forces was rendered in pseudo color in which the color scale represents the intensity of the force in nano-newton. The color scale in Fig. 4 shows the force ranging between 0 to 70 nN, while the scale bar in Fig. 8b ranges between -10 to 70 nN. The negative forces found in the result of cubic interpolation are due to the minimum curvature in a curve spline creating an inflection point below 0 nN. This was found to be the flaw of cubic interpolation technique and may provide unrealistic fitting to the minimum force point or over exaggerating the curve data fitting [8].

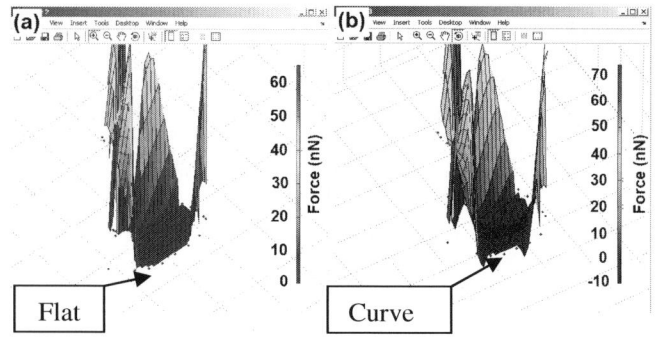

Fig. 4 The fitting of force data curves by using (a) linear and (b) cubic interpolations.

In the case of cubic interpolation, approximately 10% of measured forces were overshot at the minimum force curve. Comparatively, linear interpolation is able to terminate the transition from higher force to null force area sharply, and thus it is more suitable to the mapping of cell traction forces.

Although the 3D representation shows that cubic interpolation provides a smoother transition for the force distribution when compared with the linear interpolation, the differences in the output of the two approaches are indistinguishable in 2D force distribution maps. Nonetheless, 2D visualization is found to be more useful for the analysis of cell force distribution. Both interpolation methods showed equivalently good isolation of cell traction forces in 2D.

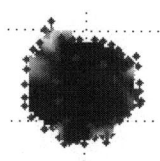

Fig. 5 2D cell traction force maps which were fitted with linear interpolation.

The resolution of the cell traction force measurement for the overall system is dependent on the CTFM software. The hypothetical simulated cell edges and deformation lines on the PDMS line patterns of 3 and 5 μm. The forces interpolated at a distance of 3 μm in the software appeared inseparable and aggregated into clusters of forces. The result from the force simulation indicates that the interpolation function in the software is able to resolve adjacent forces in clear separation within 5 μm to each other. Therefore, the resolving power of the microscope, the liquid crystals and the force mapping software collectively are crucial in determining the spatial resolution of the overall cell traction force mapping system.

IV. CONCLUSIONS

A cell traction force measurement and mapping software was custom-built to render the traction forces in computer visualization based on the length of deformation line induced by a cell in the liquid crystal surface. The 2D and 3D maps of cell traction forces rendered in pseudo color have enabled the study of traction forces for keratinocytes. The overall system has been shown to be useful for analyzing time based cell migration.

ACKNOWLEDGMENT

The author would like to thank research funding support FRGS Phase 1, Vot No. 1050 and RAGS, Vot. No. R027 awarded by Malaysia Ministry of Higher Education.

REFERENCES

1. C. F. Soon, et al., (2013) Development of a novel liquid crystal based cell traction force transducer system Biosensors & Bioelectronics vol. 39: 14-20.
2. K. Owaribe, et al., (1981) Demonstration of contractility of circumferential actin bundles and its morphogenetic significance in pigmented epithelium in vitro and in vivo The Journal of Cell Biology vol. 90: 507-514.
3. J. Chang and J. J. Zhang, "Force Mapping " presented at the Proceedings of the Theory and Practice of Computer Graphics 2004 (TPCG'04) Bournemouth, 2004.
4. C. S. Lent, (2013) Learning to program with MATLAB : building GUI tools. Wiley, .
5. C. F. Soon, et al., (2009) Interactions of cells with elastic cholesteryl liquid crystals Proceedings of IFBME vol. 25/X: 9-12.
6. C. F. Soon, et al. (2012) "Effects of Trypsin and Cytochalasin-B Treatments to Cell Traction Forces," in IEEE EMBS Conference on Biomedical Engineering and Sciences (IECBES), Langkawi, Malaysia, pp. 247 - 251.
7. H. Hassanin, et al., (2012) Fabrication of hybrid nanostructured arrays using a PDMS/PDMS replication process Lab Chip vol. 12: 4160-4167.
8. P. Wassel (2009) A general purpose Greens function-based interpolator Computers and Geosciences vol. 35: 1247-1254.

Author: Chin Fhong, Soon
Institute: Universiti Tun Hussein Onn Malaysia
Street: 86400 Parit Raja
City: Batu Pahat, Johor,
Country: Malaysia
Email: soon@uthm.edu.my

Growth of Rutile Phased Titanium Dioxide (TiO₂) Nanoflowers for HeLa Cells Treatment

N.S. Khalid[1], W.S. WanZaki[2], and M.K. Ahmad[3]

[1] Faculty of Electrical Engineering, Universiti Tun Hussein Onn Malaysia (UTHM),
86400 Parit Raja, Batu Pahat, Johor, Malaysia
[2] Medical Instrumentation Laboratory, Universiti Tun Hussein Onn Malaysia (UTHM),
86400 Parit Raja, Batu Pahat, Johor, Malaysia
[3] Solar Device Laboratory, Microelectronics & Nanotechnology - Shamsuddin Research Centre (MiNT-SRC),
Universiti Tun Hussein Onn Malaysia (UTHM), 86400 Parit Raja, Batu Pahat, Johor, Malaysia

Abstract— **Photo-catalysis process needs electron transfer to make reaction happen. In this study we want to propose a material that can make the HeLa cells lysis which is titanium dioxide (TiO₂). This paper focus on the growth of rutile phased TiO₂ nanoflowers on FTO substrate for HeLa cells treatment. The surface morphology will be characterized under FESEM and XRD while UV-vis for its optical property. The TiO₂ is fabricated by using hydrothermal method. FESEM analysis shows the size of TiO₂ nanoflowers are in range between 30 nm to 400 nm. The surface topography can be able to give data about its grain size and roughness. The TiO₂ nanoflowers sample are confirm in rutile phase mostly at lattice plane (110). By do studying on the TiO₂ characteristic, we can say that as it is important factor to do HeLa cell treatment.**

Keywords— **Titanium dioxide, hydrothermal, nanoflowers.**

I. INTRODUCTION

Titanium dioxide (TiO2) or can also be called titania is a combination of elements titanium and oxygen. This material has been widely studied in many fields such as solar cells [1], anti-reflection coating, UV protector, sensors and as environmental polution treatment [2]. Generally, TiO2 is used because it is inert and chemically stable. Besides, TiO2 is known as a good photocatalyst due to its high photosensitivity, strong oxidation and long term stability [3, 4]. The structure of TiO2 greatly affects its properties and it has low recombination rate for electron-hole pair and good optical and electrical properties. Rutile phase is stable in high temperature region, whereas anatase and brookite phases are metastable and they are transformed to rutile when annealed at high temperature. Previously, researches were more focused towards anatase phase TiO2 due to its wider bandgap [5, 6] in applications such as solar cells and other electronics. However, rutile phase is more stable and has good electron mobility [7].

TiO₂ can be fabricated by several methods such as sol-gel [3, 8], chemical vapour deposition [9, 10], spray pyrolysis [4], hydrothermal [1, 11-13] and sputtering [14-16]. Among these methods, hydrothermal method is the most practical method due to its simple preparation process, lower cost and environmental friendly nature [12]. Hydrothermal is a process that using solution-deposit method by soft chemistry giving homogenous thin film at low temperature and atmospheric pressure [1]. Generally, to produce TiO₂ rutile phase, we need to used high temperature between 700°C~900°C, however by using this method, we can produce TiO₂ rutile phase at low temperature (100°C~200°C) compared to other methods.

In this paper, we achieved to fabricate rutile phased TiO₂ nanoflowers by hydrothermal method. The hydrothermal time has affected the structural morphology of TiO₂ nanoflowers and has been characterized using field emission scanning electron microscope (FESEM), x-ray diffraction (XRD) and atomic force misroscopic (AFM). Then the TiO₂ nanoflower later will be applied to HeLa cells for cancer cell treatment.

II. METHODOLOGY

To prepare TiO₂ solution, 80ml deionized water and 80ml hydrochloric acid (35%) were stirred together on hot plate. Titanium (IV) butoxide (TBOT) from the Sigma-Aldrich was a precursor added to the mixture about 3ml by drop wise. Fluorine-doped tin oxide (FTO) substrate was washed with deionized water, ethanol and acetone with ratio 1:1:1 in ultrasonication bath for 30 minutes. The cleaned FTO then put into the autoclave (teflon lined with stainless steel). The TiO₂ solution was poured in the autoclave and put into the electric oven at 150°C for 2, 5, 10 and 24 hours. After that, the FTO substrate with TiO₂ nanoflowers was washed with deionized water and annealed at 300°C for 30 minutes.

Surface morphology of TiO₂ nanoflowers were observed with FESEM-JEOL JSM-7600F, and for surface topology including grain size and surface roughness, we used AFM-Park System 100, while for XRD analysis we used X'Pert Pro PANALYTICAL with Cu Kα radiation. The viability of the HeLa cells towards TiO₂ nanoflowers was checked by growing the cells on the TiO₂ nanoflower thin films. The cells were observed daily.

© Springer International Publishing Switzerland 2015
V. Van Toi and T.H. Lien Phuong (eds.), *5th International Conference on Biomedical Engineering in Vietnam,*
IFMBE Proceedings 46, DOI: 10.1007/978-3-319-11776-8_59

III. RESULTS AND DISCUSSIONS

A. Surface Morphology

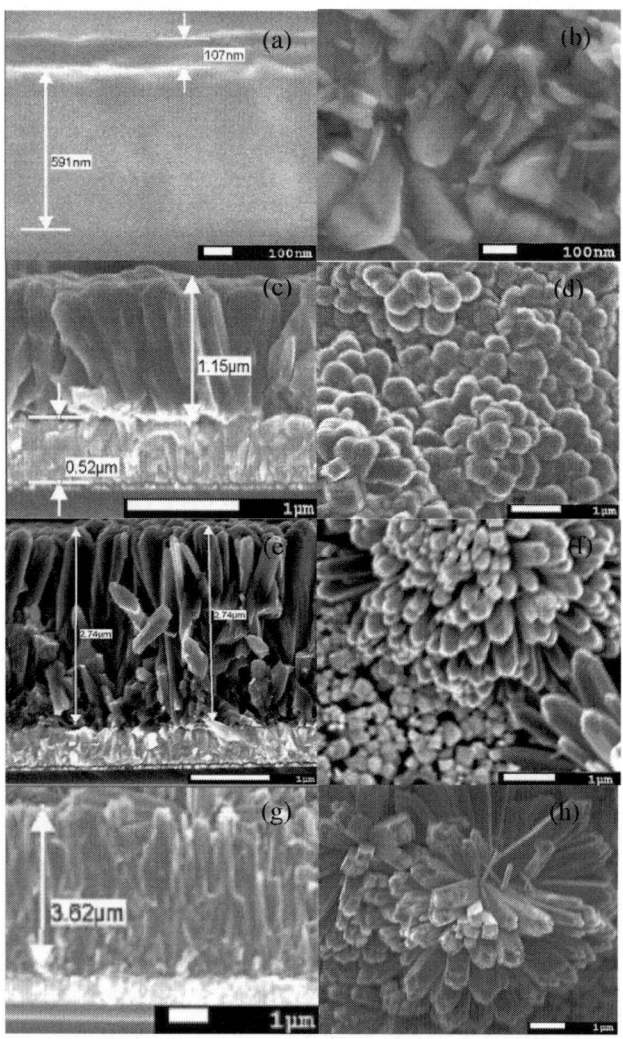

Fig. 1 The FESEM image of TiO$_2$ nanoflowers; (a, c, e, g-cross-section and b, d, f, h-surface morphology) according to hydrothermal time 2H, 5H, 10H and 24H respectively.

Fig. 1 showed the FESEM images of the prepared samples on FTO substrates. The left figures were the cross section for each sample. The right figures were the top view of TiO$_2$ nanoflowers surface. The height of TiO$_2$ nanoflowers are 102 nm, 1.15 μm, 2.74 μm and 3.62 μm for hydrothermal time deposition 2H, 5H, 10H and 24H respectively. Thus, the size of nanorods was increasing as the hydrothermal time increased. At 2H, the TiO$_2$ nanoflowers are not fully grown on the FTO substrate compare to the 5H, 10 and 24H that the intensity of TiO$_2$ is increasing gradually due to the longer synthesis time. They

grew radially on the FTO surface and eventually formed the nanoflowers on the aligned nanorods.

B. Surface Topography

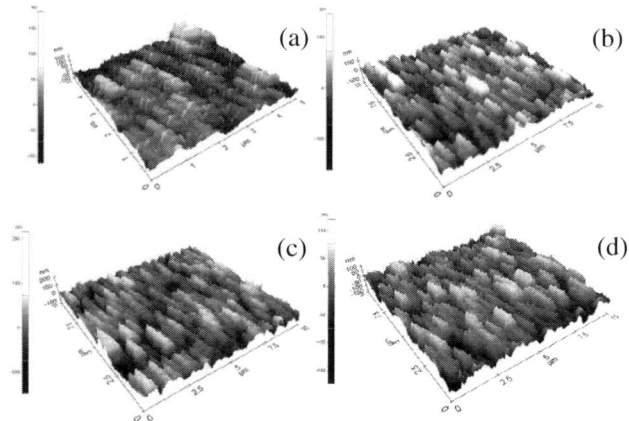

Fig.2 The surface topography images from AFM from 2H, 5H, 10H and 24H respectively.

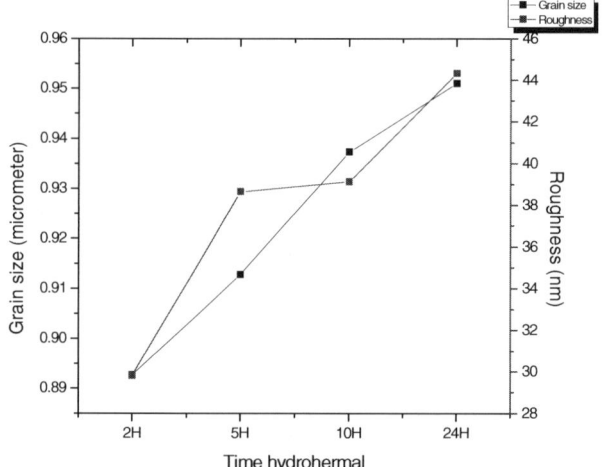

Fig. 3 The grain size and the surface roughness of TiO$_2$ nanoflower to corresponding hydrothermal time.

From Fig.2, it showed the surface topography of the samples on 10μm x 10μm area. We can see that the uniformity was achieved except for 2H hydrothermal time. Besides, from the AFM analysis also, we could get the data for grain size and the surface roughness which has been plotted as in Fig. 3. It is shown that as the hydrothermal time increased, the grain size and the roughness also increased. It is important to know the surface condition because cells most likely to attach to rough surface for best attachment.

C. XRD Analysis

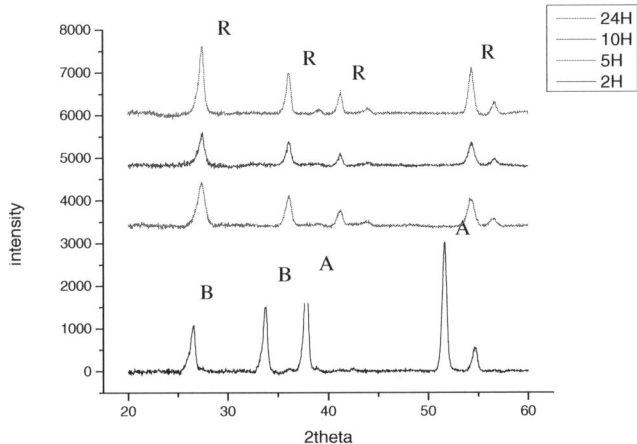

Fig. 4 The XRD patterns of TiO₂ nanoflowers sample corresponding to hydrothermal time.

Fig. 4 showed that the significant peaks for rutile phase at 2θ were at $27.35°$, $36.06°$, $41.21°$, $52.29°$ for 110, 101, 111 and 211 lattice crystal respectively for sample 5H, 10H and 24H [17, 18]. The sharp and frequent diffraction peak in these pattern showed that the synthesis of TiO₂ nanoflowers were highly crystallized. The intensity of the crystalline increased as the hydrothermal time increased. However, for sample 2H hydrothermal deposition, anatase and brookite lattice crystal were found due to insufficient time for anatase/brookite to rutile phase transformation.

D. Cells Viability

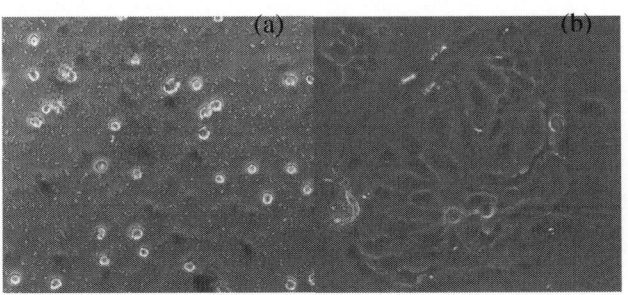

Fig. 5 The HeLa cells was grown on the TiO₂ nanoflowers (a) before and (b) after 5 days.

Fig. 5 shows the cells on the TiO₂ nanoflowers before and after 5 days. We can observe the cells can attached to the surface of TiO₂ nanoflowers. It means that TiO₂ is originally inert to the cells. This is very good property because when we applied TiO₂ nanoflower to the area that having cancer cells, the non-cancerous cells will not be affected. After this we will try to introduce UV light to excite the electrons on TiO₂ surface and photo-catalysis will occur on the cells membrane and eventually kills to HeLa cells.

IV. CONCLUSIONS

From this study, we can conclude that TiO₂ nanoflowers can be fabricated using hydrothermal method. Surface morphology showed the flower-like TiO₂ and the size was increasing as the hydrothermal time increased. The grain size and the roughness from AFM result can be used to indicate how the cell attachment to the TiO₂ nanoflowers surface. The crystalline structure was in rutile phase according to the XRD pattern for 5H, 10H and 24H except for 2H hydrothermal time since rutile structure was more stable compared to anatase and brookite. The TiO₂ nanoflower is not hazard to cells but will be hazardous to the cells if the electron excitation takes place at the cells membrane. Thus it will lead to HeLa cell treatment.

ACKNOWLEDGMENT

The author would like to acknowledge colleagues at Microelectronics and Nanotechnology-Shamsuddin Research Center (MiNT-SRC), Universiti Tun Hussein Onn Malaysia, for providing technical and moral supports. The authors are grateful to the Ministry of Education Malaysia for research funding support (Fundamental Research Grant Scheme, UTHM, Vot No. 1215).

REFERENCES

1. Mohd Khairul Ahmad, K.M., *Low temperature and normal pressure growth of rutile-phased TiO₂ nanorods/nanoflowers for DSC application prepared by hydrothermal method.* Journal of Advanced Research in Physics, 2012. 3(2): p. 1-4.
2. Chutima Srisitthiratkul, V.P., Narupol Intasata, *The potential use of nanosliver-decorated titanoum dioxide nanofibers for toxin decomposition with antimicrobial and self-cleaning properties.* Applied Surface Science, 2011. **257**: p. 8850-8856.
3. N. Venkatachalam, M.P., V. Murugesan, Sol-gel preparation and characterization of nanosized TiO2: Its potocatalytic performance. Materials Chemistry and Physics, 2007: p. 6.
4. A. Ranga Rao, V.D., Low-temperature synthesis of TiO2 *nanoparticles and preparation of TiO₂ thin films by spray* deposition. Solar Energy Materials & Solar Cells, 2007. 91: p. 1057-1080.
5. L. M. Apatiga, E.R., E.Rivera, V.M. Castano, Surface morphology of nanostructured anatase thin films prepared by pulse liquid injection MOCVD. Surface & Coating Technology, 2006. 201: p. 4136-4138.
6. Xu, M., et al., Photocatalytic Activity of Bulk TiO_{2} Anatase and Rutile Single Crystals Using Infrared Absorption Spectroscopy. Physical Review Letters, 2011. 106(13): p. 138302.
7. Cao, C., et al., UV sensor based on TiO2 nanorod arrays on FTO thin film. Sensors and Actuators B: Chemical, 2011. 156(1): p. 114-119.

8. Influence of calcination temperature on structural and optical properties of TiO2 thin films prepared by sol-gel dip coating. Mater. Lett., 2000. 57(2): p. 355.

9. Synthesis of anatase TiO2 supported on porous solids by chemical vapor deposition. Catal. Today, 2001. 68(1-3): p. 173.

10. Bae, J., S. Lee, and G. Yeom, Doped-fluorine on electrical and optical properties of tin oxide films grown by ozone-assisted thermal CVD. Journal of The Electrochemical Society, 2007. 154(1): p. D34-D37.

11. Fen, L.B., et al., Physico-chemical properties of titania nanotubes synthesized via hydrothermal and annealing treatment. Applied Surface Science, 2011. 258(1): p. 431-435.

12. Gao, M., et al., Effect of Substrate Pretreatment on Controllable Growth of TiO2 Nanorod Arrays. Journal of Materials Science & Technology, 2012. 28(7): p. 577-586.

13. Funda Sayilkan, M.A., Sadiye Sener, Sema Erdemoglu, Hydrothermal synthesis, characterization and photocatalytic activity of nanosized TiO2 based catalysts for rhodamine B degratation. Journal Chemistry of Turkey, 2007. 31: p. 211-221.

14. Boyadzhiev, S., V. Georgieva, and M. Rassovska, Characterization of reactive sputtered TiO 2 thin films for gas sensor applications. Journal of Physics: Conference Series, 2010. 253(1): p. 012040.

15. D. Kaczmarek, J.D., D. Wojcieszak, B. Gornicka, XRD and AFM studies of nanocrystalline TiO2 Thin films prepared by modified magnetron sputtering. Nanotechnology, nanomaterials and nanoelectronics, 2008. C001: p. 159-162.

16. Weinberger, B.R. and R.B. Garber, Titanium dioxide photocatalysts produced by reactive magnetron sputtering. Applied Physics Letters, 1995. 66(18): p. 2409-2411.

17. Zhu, H., et al., Growth of branched rutile TiO2 nanorod arrays on F-doped tin oxide substrate. Applied Surface Science, 2011. 257(24): p. 10494-10498.

18. He, Z., et al., Photocatalytic activity of TiO2 containing anatase nanoparticles and rutile nanoflower structure consisting of nanorods. Journal of Environmental Sciences, 2013. 25(12): p. 2460-2468.

Author: Noor Sakinah Khalid
Institute: Universiti Tun Hussein Onn Malaysia
Street: Parit Raja
City: Batu Pahat
Country: Johor, Malaysia
Email: noorsakinahkhalid@gmail.com

Cystatin C Versus Creatinine in Evaluating Glomerular Filtration Rate in Renal Transplant Recipients with Proteinuria

Tran Thai Thanh Tam[1], Hoang Khac Chuan[2], Du Thi Ngoc Thu[2], Nguyen Thi Thai Ha[2], Thai Minh Sam[2], Nguyen Thi Le[3], and Tran Ngoc Sinh[3]

[1] Can Tho University of Medicine and Pharmacy, Can Tho City, Vietnam
[2] Cho Ray Hospital, Ho Chi Minh City, Vietnam
[3] University of Medicine and Pharmacy in Ho Chi Minh City, Vietnam

Abstract— Background: **High grade proteinuria in allograft glomerular diseases is concerned with the reduced function and shortened survival of kidney allograft. Monitoring changes in glomerular filtration rate (GFR) by serum creatinine and cystatin C are the recommended methods for assessing the progression of kidney allograft function with proteinuria.**

Materials and Methods: **Sixty renal transplanted patients with allograft survival of at least 1 year, with proteinuria were included in the study. All patients had routine clinical care and underwent baseline including age, sex, weight, body mass index, serum creatinine (Scr), serum cystatin C (ScysC), urine creatinine, proteinuria at the same day. We established the correlation between creatinine clearance (mGFR) and serum creatinine and cystatin C, and the correlation between mGFR and four estimated GFR formulas (Cockcroft Gault, MDRD, CKD-EPI and Le Bricon formulas).**

Results: **The mean ScysC, Scr and mGFR were 1.49 ± 0.51 mg/L, 1.26 ± 0.25 mg/dL and 56.03 ± 20.74 ml/min/1.73m^2 respectively. There were significant correlations between mGFR and 1/ScysC (r1=0.775 (p<0.000)) and mGFR and 1/Scr (r2 =0.754 (p<0.000)). There were significant correlations between the four methods of eGFR and mGFR (r = 0.801 to 0.875, p=0.000). Among them, CKD-EPI creatinin-cystatin C 2012 formula had the strongest correlation with mGFR in both groups with mGFR≥60ml/min/1.73m^2 and mGFR<60ml/min/1.73m^2.**

Conclusion: **Serum cystatin C prove more useful in monitoring the function of kidney allograft with proteinuria compared to serum creatinine. The combination of serum creatinine and serum cystatin C (CKD-EPI creatinine cystatin C 2012) is more accurate than either marker alone for estimating GFR.**

Keywords— **serum Cystatin C, creatinine, estimated glomerular filtration rate, renal transplant recipients, proteinuria.**

I. INTRODUCTION

Since 1960s, the discovery of immunosuppressive drugs to forestall acute and chronic rejection has made the renal allograft survival increased dramatically [1]. Improving graft half-life to avoid rejection due to improvements in first year survival, whereas the attribute rates beyond the first year usually remained unchanged. Therefore, late graft loss continues to be an ongoing problem after kidney transplantation. High grade proteinuria in allograft glomerular diseases is concerned with the reduced kidney allograft survival [2]. Monitoring changes in glomerular filtration rate (GFR) is the recommended method for assessing the progression of kidney allograft function.

The gold standard to measure glomerular filtration rate (GFR) is to inject compounds such as inulin, radioisotopes (^{51}chromium-EDTA, ^{125}I-iothalamate, ^{99m}Tc-DTPA) or radiocontrast agents such as iohexol, however, these techniques are complicated, time-consuming, expensive and have potential side-effects [3][4]. Up to now, creatinine is the most widely used as a biomarker of kidney function, however, it is inaccurate at detecting mild renal impairment, and its level may be affected by muscle mass, aging, sex and protein intake [5]. Cystatin C, first proposed as a measure of GFR by Grubb and coworkers in 1985, has a low molecular weight, and is removed from the bloodstream by glomerular filtration in the kidneys [6]. In addition, cystatin C levels are less dependent on age, sex, race and muscle mass compared to creatinine. Serum levels of cystatin C are a more precise test of kidney function than serum creatinine (Scr) levels.[7]

The estimated GFR (eGFR) is the clinical standard for the assessment of kidney function. However, eGFR based on creatinine has limitations in risk prediction, particularly in patients with reduced muscle mass. Cystatin C has been considered as an alternative filtration marker with stronger and more linear risks relationships than creatinine. Several studies have suggested that the addition of cystatin C measurement to creatinine measurement significantly enhances eGFR equations in kidney transplant recipients [16]. The primary objective of this study was to determine if serum cystatin C (ScysC) and cystatin C- based GFR estimates are more accurate and could be more beneficial compared with serum creatinine and creatinine -based GFR in renal transplant population with proteinuria.

© Springer International Publishing Switzerland 2015
V. Van Toi and T.H. Lien Phuong (eds.), *5th International Conference on Biomedical Engineering in Vietnam,*
IFMBE Proceedings 46, DOI: 10.1007/978-3-319-11776-8_60

II. MATERIALS AND METHODS

This study performed using data and stored samples derived from Urology Department and Biochemistry Department of Cho Ray Hospital.

A. Patients

In this two-year long trial, sixty renal transplanted patients at Cho Ray hospital with allograft survival of at least 1 year, with proteinuria defined by excreting >1 g protein/24 h (range 1–15 g/24 h) in 3 consecutive visits, were included in the study. These patients were received immunosuppressive drug and a course of steroid in tapered doses at the time of transplant followed by prednisone tapered down to a maintenance dose of 5-10mg/day. For maintenance therapy, they received a combination of calcineurin inhibitor (cyclosporine or tacrolimus), mycophenolate mofetil or Imuran, with or without prednisone 5-10mg/day. Our exclusion criteria were hyperthyroidism, or hypothyroidism, a body mass index greater than $30kg/m^2$, pregnancy and liver cirrhosis, or administration of any medication interfering with creatinine tubular secretion.

B. Methods

All patients had routine clinical care and underwent baseline including age, sex, weight, body mass index, ScysC, Scr, urine creatinine, proteinuria at the same day. We established the correlation between creatinine clearance (mGFR) and Scr, ScysC, and the correlation between (mGFR) and creatinine and cystatin C- based GFR formulas. The study protocol conformed to the ethical guidelines of the 1975 Declaration of Helsinki.

Serum and urine creatinine was measured by kinetic Jaffe method using HITACHI 917 analyzer (traceable to National Institute of Standards and Technology Standard Reference Material): measured increase in absorbance at 505 nm with blanking at 570 nm. ScysC was analyzed by immune-turbidimetric method using MINDRAY BS300 analyzer. The manufacturer's reference interval for healthy subjects is 0.59 to 1.03 mg/L. Assay sensitivity was 0,009 – 0,01mg/L, coefficient of variation was 0,85 – 1,82%.

Table 1 Estimated glomerular filtration rate (eGFR) formula based on serum creatinine and serum cystatin C

Formula and sex	Scr (mg/dL)	Scys C (mg/L)	eGFR
Cockcroft - Gault Formula: Estimated Creatinine Clearance (eClcr)			
			$[(140 - age) \times weight]/(72 \times Scr)$ (x0.85 if female)
MDRD formula (Modification of Diet in Renal Disease Study)			
			$186 \times (Scr)^{-1.154} \times (age)^{-0.203} \times 0.742$ (if female) $\times 1.210$ (if Black)
Le Bricon formula			
			[78 x (1/Scys) + 4]
CKD-EPI creatinin 2009			
Female	$\leq 0,7$		$GFR = 144 \times (Scr/0,7)^{-0.329} \times 0,993^{Age}$

[x1,159 If Black]

Female	$> 0,7$	$GFR = 144 \times (Scr/0,7)^{-1,209} \times 0,993^{Age}$ [x1,159 If Black]
Male	$\leq 0,9$	$GFR = 141 \times (Scr/0,9)^{-0,411} \times 0,993^{Age}$ [x1,159 If Black]
Male	$> 0,9$	$GFR = 141 \times (Scr/0,9)^{-1,209} \times 0,993^{Age}$ [x1,159 If Black]

CKD-EPI cystatin C 2012

	$\leq 0,8$	$GFR = 133 \times (Scys/0,8)^{-0.499} \times 0,996^{Age}$ [x 0,932 if female]
	$> 0,8$	$GFR = 133 \times (Scys/0,8)^{-1,328} \times 0,996^{Age}$ [x 0,932 if female]

CKD-EPI creatinin-cystatin C 2012

Sex	Scr	Scys	eGFR
Female	$\leq 0,7$	$\leq 0,8$	$GFR = 130 \times (Scr/0,7)^{-0,248} \times (Scys/0,8)^{-0.375} \times 0,995^{Age}$ [x1,08 If Black]
Female	$\leq 0,7$	$> 0,8$	$GFR = 130 \times (Scr/0,7)^{-0,248} \times (Scys/0,8)^{-0.711} \times 0,995^{Age}$ [x1,08 If Black]
Female	$> 0,7$	$\leq 0,8$	$GFR = 130 \times (Scr/0,7)^{-0,601} \times (Scys/0,8)^{-0.375} \times 0,995^{Age}$ [x1,08 If Black]
Female	$> 0,7$	$> 0,8$	$GFR = 130 \times (Scr/0,7)^{-0,601} \times (Scys/0,8)^{-0.711} \times 0,995^{Age}$ [x1,08 If Black]
Male	$\leq 0,9$	$\leq 0,8$	$GFR = 135 \times (Scr/0,9)^{-0,207} \times (Scys/0,8)^{-0.375} \times 0,995^{Age}$ [x1,08 If Black]
Male	$\leq 0,9$	$> 0,8$	$GFR = 135 \times (Scr/0,9)^{-0,207} \times (Scys/0,8)^{-0.711} \times 0,995^{Age}$ [x1,08 If Black]
Male	$> 0,9$	$\leq 0,8$	$GFR = 135 \times (Scr/0,9)^{-0,601} \times (Scys/0,8)^{-0.375} \times 0,995^{Age}$ [x1,08 If Black]
Male	$> 0,9$	$> 0,8$	$GFR = 135 \times (Scr/0,9)^{-0,601} \times (Scys/0,8)^{-0.711} \times 0,995^{Age}$ [x1,08 If Black]

C. Statistical Analysis

Data were analyzed using version 16.0 of the SPSS statistical program. The normality of the distributions of variables was checked by the Kolmogrov Smirnov test. The Pearson correlation coefficient test was employed to determine the relationship and linear regression models for the relationship between sCysC, Scr and mGFR and difference methods of GFR assessment. Values of $P<0.05$ were considered statistically significant.

III. RESULTS

A. Characteristics of Recipients

The participants were 47 males and 13 females. Clinical characteristics of the patients in each data set are shown in table 2. The mean ScysC, Scr and mGFR were 1.49 ± 0.51 mg/L, 1.26 ± 0.25 mg/dL and 56.03 ± 20.74 ml/min/$1.73m^2$ respectively. There were significant differences of height, weight, BUN, ScysC and mGFR between males and females ($p<0.05$).

Table 2 The characteristics of patients.

	Total (n=60)	Males (n=47)	Females (n=13)
Age (years)	42.68±10.47	43.02±10.7	41.46±9.8
Height (cm)*	161.81±6.36	164.57±3.08	151.85±5.03
Weight (kg)*	58.15±8.22	60±6.8	51.46±9.56
BMI	22.17±2.67	22.14±2.3	22.28±3.77
BUN (mg%)*	22.85±8.51	21.53±6.87	27.61±11.99
ScysC (mg/L)*	1.49±0.51	1.39±0.4	1.84±0.7
Scr (mg%)	1.26±0.25	1.25±0.23	1.31±0.32
mGFR (ml/min/1.73m²)*	56.03±20.74	58.84±20.24	45.86±20

*: $p<0.05$.

There were significant correlations between mGFR and 1/ScysC (r1=0.775 ($p<0.000$)) and mGFR and 1/Scr (r2 =0.754 ($p<0.000$)). The linear regression were respectively:

$$1/ScysC= 0.01 \times mGFR + 0.203$$
$$1/Scr= 0.006 \times mGFR +0.488$$

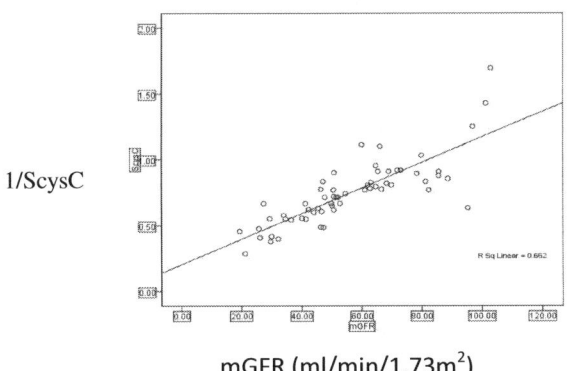

mGFR (ml/min/1.73m²)

Fig. 1 Correlation between 1/ScysC and mGFR in kidney transplant recipients with proteinuria

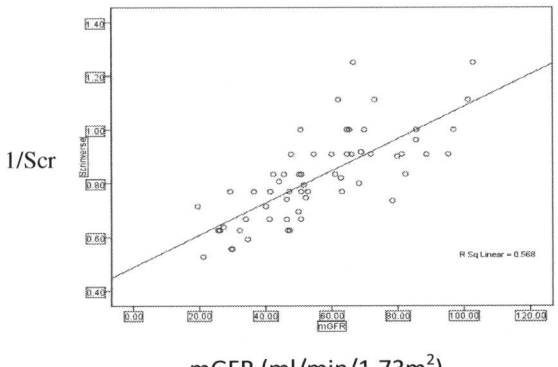

mGFR (ml/min/1.73m²)

Fig. 2 Correlation between 1/Scr and mGFR in kidney transplant recipients with proteinuria.

In both groups of proteinuria kidney transplant patient with mGFR≥60ml/min/1.73m² and mGFR<60ml/min/1.73m², the correlations between 1/ScysC and mGFR were better than between 1/Scr and mGFR. Data was shown in table 3.

Table 3. Correlation between 1/ScysC, 1/Scr and mGFR in proteinuria kidney transplant recipients classified by mGFR≥60ml/min/1.73m² and mGFR<60ml/min/1.73m²

	Correlation with mGFR	
	≥ 60ml/min/1.73m²	<60ml/min/1.73m²
1/ScysC	r3= 0.583 (p=0.002)	r5= 0.767 (p=0.000)
1/Scr	r4=0.287 (p=0.164)	r6= 0.649 (p=0.000)

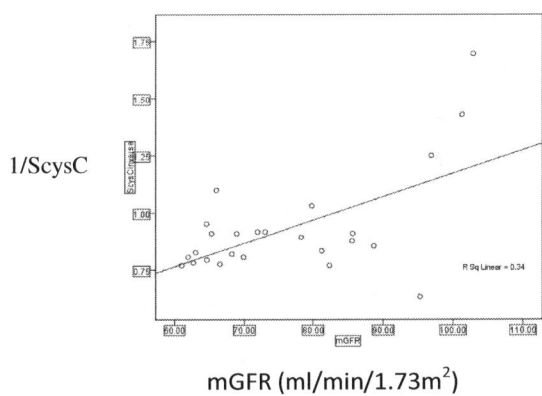

mGFR (ml/min/1.73m²)

Fig. 3 Correlation between 1/ScysC and mGFR in proteinuria kidney transplant recipients with mGFR≥60ml/min/1.73m2

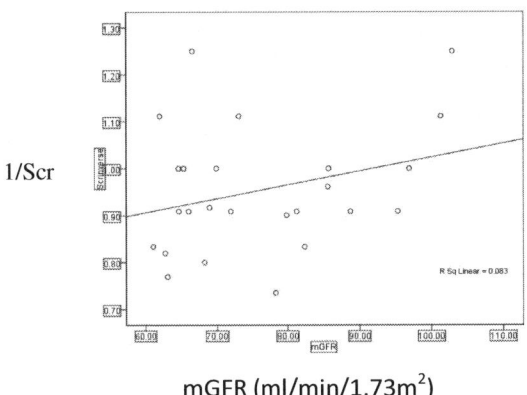

mGFR (ml/min/1.73m²)

Fig. 4 Correlation between 1/Scr and mGFR in proteinuria kidney transplant recipients with mGFR≥60ml/min/1.73m2

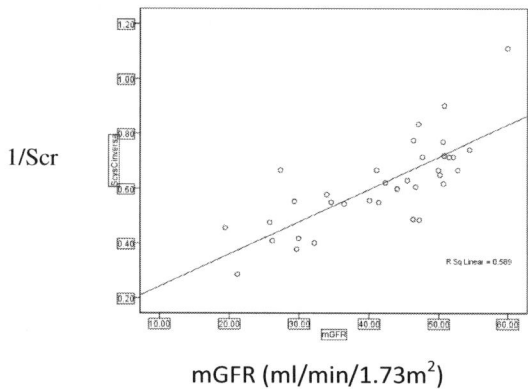

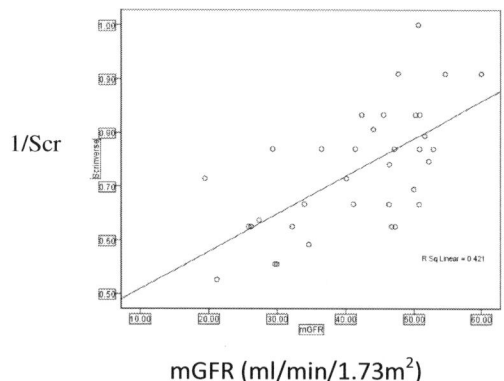

Fig. 5 Correlation between 1/ScysC and mGFR in proteinuria kidney transplant recipients with mGFR<60ml/min/1.73m2

Fig. 6 Correlation between 1/Scr and mGFR in proteinuria kidney transplant recipients with mGFR < 60ml/min/1.73m2

The mean eGFR based on Scr, ScysC alone and combined both Scr, ScysC, the correlation between difference methods of eGFR and mGFR are shown in table 4. There were significant correlations between difference methods of eGFR and mGFR (r = 0.801 to 0.875, p=0.000). Among them, CKD-EPI creatinin-cystatin C 2012 formula had the strongest correlation with mGFR.

Table 4. Mean eGFR by Scr – based and ScysC – based and combined both Scr- ScysC-based equations and the correlation between difference methods of eGFR and mGFR.

	Total	Males	Females	Correlation with mGFR
Cockcroft – Gault F (ml/min/1.73m^2)	66.91 ± 15.64	69.65± 13.74	57.23± 18.57	r= 0.801 (p=0.000)
MDRD F (ml/min/1.73m^2)	64.15 ± 1.55	66.89± 14.45	54.23± 15.7	r= 0.825 (p=0.000)
CKD-EPI creatinin 2009 F	71.33 ±	74.23± 16.94	60.84± 19.3	r=0.836 (p=0.000)

(ml/min/1.73m^2)	18.18			
CKD-EPI cystatin C 2012 F (ml/min/1.73m^2)	57.08 ± 2.35	60.94± 22.78	43.15± 21.46	r= 0.842 (p=0.000)
CKD-EPI creatinin-cystatin C 2012 F (ml/min/1.73m^2)	61.9± 2	65.45± 18.72	49.08± 19.9	r= 0,875 (p=0.000)
Le Bricon F (ml/min/1.73m^2)	62.18 ± 19.26	65.02± 19.01	51.95± 17.1	r= 0.814 (p=0.000)

In both groups of proteinuria kidney transplant patient with mGFR ≥ 60 ml/min/1.73m^2 and mGFR < 60ml/min/1.73m^2, CKD-EPI creatinin-cystatin C 2012 formula also showed more association than the other equations with mGFR. (Table 5)

Table 5. Correlation between difference methods of eGFR and mGFR in proteinuria kidney transplant recipients classified by mGFR≥60ml/min/1.73m^2 and mGFR<60ml/min/1.73m^2

	Correlation with mGFR	
	≥ 60ml/min/1.73m^2	<60ml/min/1.73m^2
Cockcroft – Gault F	r=0.511 (p=0.009)	r=0.706 (p=0.000)
MDRD F	r=0.621 (p=0.001)	r=0.671 (p=0.000)
CKD-EPI creatinin 2009 F	r=0.61 (p=0.001)	r=0.7 (p=0.000)
CKD-EPI cystatin C 2012 F	r=0.633 (p=0.001)	r=0.771 (p=0.000)
CKD-EPI creatinin-cystatin C 2012 F	r=0.647 (p=0.009)	r=0.8 (p=0.000)
Le Bricon F	r=0.583 (p=0.009)	r=0.767 (p=0.000)

IV. DISCUSSION

There is a strong association between proteinuria and reduced graft survival. 77% allografts lost within the first 5 year posttransplant had proteinuria at first year. Those studies also showed that proteinuria provided prognosis information beyond that provided by measures of kidney function, histology, and other variables associated with graft survival[8].

Accurate assessment of GFR and early kidney dysfunction recognition are essential and critical in the follow – up of kidney transplant recipients, especially with proteinuria, to interpret the symptoms, signs and laboratory abnormalities, that may indicate allograft rejection, for drug dosing, and assessing the prognosis. Therefore, a variety of markers for evaluation of glomerular dysfunction have been proposed.

Management of kidney transplantation recipients requires a simple, reliable, and accurate methods for estimating GFR. Many studies have reported that among all available markers, serum cystatin C had the best correlation with GFR [9]. In a study of 125 renal allograft recipients with stable allograft function, Christensson and colleagues showed that ScysC levels correlated significantly closer to accurate measurement of GFR and were significantly more sensitive to detect early GFR impairment than Scr [10]. In our research of sixty renal transplant patients also showed that the correlation between 1/ScysC and mGFR was better than between 1/Scr and mGFR in both groups of proteinuria kidney transplant patients with mGFR $\geq$ 60ml/min/1.73m^2 and mGFR <60ml/min/1.73m^2.

Among difference methods for cystatin C based GFR formula, the equation proposed by Le Bricon and coworkers provided a more accurate eGFR than serum creatinine and other cystatin C based GFR equations in kidney transplant recipients [11]. According to the previous studies on accuracy of cystatin C levels for measurement of GFR, we planned to compare the performance of the three creatinine based equations and two cystatin C based equations. To discuss about eGFR based on serum creatinine, the MDRD Study equation is almost the most thoroughly validated equation. The equation has been validated extensively in Caucasian and African American populations between the ages of 18 and 70 with impaired kidney function (eGFR < 60 mL/min/1.73 m^2) and has shown good performance for patients with all common causes of kidney disease. The MDRD Study equation is currently superior to other methods of approximating GFR based on creatinine such as Cockcroft-Gault. CKD-EPI cystatin C formula and CKD-EPI creatinine was first developed in 2008 and 2009 respectively, and was improved in 2011, 2012 by combining both serum creatinine and cystatin C. CKD-EPI creatinine – cystatin C 2012 was recommended to determine GFR by KDIGO 2013 [12] [13]. Our study reported that among difference methods of estimating GFR, CKD-EPI creatinine –cystatin C 2012 showed the best correlation with mGFR in proteinuria kidney transplant recipients classified by mGFR $\geq$ 60ml/min/1.73m^2 and mGFR < 60ml/min/1.73m^2. The four next equations had close relationship to mGFR were CKD-EPI cystatin C 2012, CKD-EPI creatinin 2009, MDRD and Le Bricon formulas, respectively . Cockcroft-Gault Formula had the least correlation with mGFR. Our report was in agreement with a study of Aleksandra Kukla et al that the combined serum creatinine and cystatin C estimate is more precise than either marker alone [14].

In 2002, the National Kidney Foundation published the *Kidney Disease Outcomes Quality Initiative* guideline for the diagnosis and classification of chronic kidney disease on the basis of the creatinine based formula. Several recent studies concluded that the cystatin C based equations were more precise in GFR estimation than the MDRD equations [15].

In a meta-analysis of 11 general population studies and 5 studies of Cohorts with chronic kidney disease concluded that the use of serum cystatin C alone or in combination with creatinine strengthen the association between the eGFR and the risks of death and end-stage renal diaease in diverse populations[16].

In conclusion, serum cystatin C prove more useful in monitoring the function of kidney allograft with proteinuria compared to serum creatinine. The combination of serum creatinine and serum cystatin C (CKD-EPI creatinine cystatin C 2012) is more accurate than either marker alone for estimating GFR. Besides, our study suggested that CKD-EPI cystatin C could also be used as a confirmatory test for detecting early reduction of GFR in renal transplant with proteinuria.

REFERENCES

1. J. Harold Helderman, (2011) New and Emerging Immunosuppresive Agents for kidney transplant, Dialysis and Transplatation, Volume 40, Issue 1, pp. 6.
2. First MR, Vaidya PN et al (1984), Proteinuria following transplantation. Correlation with histopathology and outcome, Transplant 38, pp 607-612.
3. Zahran A, El-Husseini A, Shoker A (2007). Can cystatin C replace creatinine to estimate glomerular filtration rate? A literature review. *Am. J. Nephrol.* 27 (2): 197–205.
4. Roos JF, Doust J, Tett SE, Kirkpatrick CM (2007).Diagnostic accuracy of cystatin C compared to serum creatinine for the estimation of renal dysfunction in adults and children--a meta-analysis. *Clin. Biochem.* 40 (5–6): 383–391.
5. King AJ, Levey AS (1993). Dietary protein and renal function. *J. Am. Soc. Nephrol.* 3 (11): 1723–37
6. Simonsen O, Grubb A, Thysell H (1985). The blood serum concentration of cystatin C (gamma-trace) as a measure of the glomerular filtration rate. *Scand. J. Clin. Lab. Invest.* 45 (2): 97–101.
7. Dharnidharka VR, Kwon C, Stevens G (2002). Serum cystatin C is superior to serum creatinine as a marker of kidney function: a meta-analysis. *Am. J. Kidney Dis.* 40 (2): 221–226.
8. Amer H, Fidler ME et al (2007). Proteinuria after kidney transplantation, relationship to allograft histology and survival. Am J transplant 7: 2748-2756.
9. Risch L, Blumber A, Huber A. Rapid and accurate assessment of glomerular filtration rate in patients with renal transplant using serum cystatin C. Nephrol dial Transplant. 1999 ; 14: 1991-6.
10. Christensson A, Ekberg J et al, Serum cystatin c is a more sensitive and more accurate maker of glomerular filtration rate than enzymatic measurements of creatinine in renal transplant. Nephrol Physiol. 2003;94:19-27.

11. White C, Akbari A, Hussain N et al. Chronic kidney disease stage in renal transplantation classification using cystatin C and creatinine – based equations. Nephrol Dial Transplant. 2007;22:3013-20.

12. Levey AS, Stevens LA, Schmid CH, Zhang YL, Castro AF, 3rd, Feldman HI, et al. A new equation to estimate glomerular filtration rate. *Ann Intern Med.* 2009;150(9):604-12.

13. Stevens LA, Schmid CH, Zhang YL, Coresh J, Manzi J, Landis R, et al. Development and validation of GFR-estimating equations using diabetes, transplant and weight. *Nephrol Dial Transplant.* 2010;25:449-57.

14. Aleksandra Kukla et al. Cystatin C enhances glomerular filtration rate estimating equation in kidney transplant recipients. Am J Nephrol 2014;39:59-65.

15. KDOQI clinical practice guidelinesfor chronic kidney disease: evaluation, classification, and stratification. Am J Kidney Dis. 2002;39:S1-266.

16. Michael G. Shlipak et al.(2013) Cystatin C versus creatinine in dertermining risk based on kidney function. N Engl J Med 369;10:932-943.

Genetic Mutation Types Detected in 25 Blood Samples of KHMER Patient with Beta-thalassemia in Bac Lieu Province

Pham Thi Ngoc Nga and Nguyen Trung Kien

Can Tho University of Medicine and Pharmacy, Vietnam

Abstract— Contributing to find out the distribution of mutations causing beta-Thalassemia in Vietnam, particularly in ethnic minority subjects, we applied the technique ARMS-PCR and sequencing in genetic molecule lab of the Ho Chi Minh University of Medicine and Pharmacy. 25 DNA samples from 25 Khmer patients with beta-Thalassemia living in Bac Lieu Province were analyzed. Beside of HbE and Fs41/42 mutations detected by ARMS-PCR technique, we found 4 other point mutations by sequencing techniques. This result contributes to the prenatal diagnosis program in the future.

Keywords— Beta – thalassemia, HbE, Khmer, ARMS-PCR, Sequencing techniques.

I. INTRODUCTION

Thalassemia is the most popular Hb disease which is inherited by single-gene Mendel model, estimatingly 7% of the world population carries the disease gene. Children with this disease will have serious consequences for the development of the body, reducing life expectancy by hemolysis and its complications. In Vietnam, Thalassemia is a leading cause of anemia, severe hemolysis in children. The percentage of distribution of the gene carriers varies in all provinces in the country according to each locality, each ethnic group. Specially, rate of gene carriers is very high among ethnic minorities such as the Muong (25%), Thai (16.6%), Nung, San Diu (14.3%), Pako (8.33%)... [2], [9]....

So far there are about 200 mutations causing β^- thalassemia described. Mutations may be classified into two groups: one causing complete loss of β globin chain number (β^0 thalassemia) and another reducing the number of β globin chain (β^+ thalassemia) [8], [10]. However, these mutations have specificity for each ethnic group. Each ethnic group has some common mutations besides a number of other less common mutations. This characteristic makes the discovery of mutations become more simple and rapid [4], [6], [10].

In the world, there are lots of laboratory applying techniques of molecular biology in the detection of β thalassemia mutations and prenatal diagnosis has been carried out extensively [2], [8]..

In Vietnam, in recent years, the molecular basis of the disease is known through a technique as ARMS-PCR, QF-PCR, ... but sequencing method is still the most popular application. Each technique has strengths and weaknesses in particular, in this study we use mainly two techniques that are ARMS-PCR and sequencing to contributingly survey the distribution of genotypes in 25 Khmer patients with beta thalassemia in Bac Lieu province.

II. SUBJECTS AND METHODS

Subjects: The Khmer ethnics who have Khmer grandparents, parents regardless of age, gender and have positive results of beta-thalassemia disease through Hemoglobin electrophoresis including:

β-thal heterozygote (10 cases): HbA2> 3.5%, 0 <HbF <10%, reduced HbA1

HbEE homozygote (3 cases): HBE + HbA2 = 100%

HbAE heterozygote (12 cases): HBE and HbA2 = 12.6 to 30.3%, rest is HbA1

Methods:
Multiplex ARMS PCR

- *DNA extraction*

Each study subject was taking 5 ml peripheral blood added EDTA anticoagulant and transferred to the molecular genetics laboratory in Medicine Pharmacy Ho Chi Minh University. These samples were extracted DNA by the Qiagen DNA extracting kits. Cells were broken, lysed by binding buffer and proteinase K, lysed fluid was filtered through filter column with high salt concentrations. Silica gel membrane in filtration column kept DNA and allowed other substances pass. Purified DNA was dissolved in the buffer solution of low salt concentrations

- *ARMS Technique*

ARMS technique was set to standard method described by John Old [5], [7]. For each mutation FS41/42, Cd17, FS71/72, IVSII-654, -28, IVSI-1 and HBE, there are 2 primer pairs were used: one pair corresponding DNA fragments carrying the normal sequence, another pair corresponding DNA mutation fragment. In addition, each PCR reaction, another primer pairs also is used to find false negative cases by out-of-work PCR. In addition, with mutations Cd95(+A), the specific mutation of the

© Springer International Publishing Switzerland 2015
V. Van Toi and T.H. Lien Phuong (eds.), *5th International Conference on Biomedical Engineering in Vietnam,*
IFMBE Proceedings 46, DOI: 10.1007/978-3-319-11776-8_61

Vietnamese, we used two pairs of normal and mutant primer according to research of Le Thi Hao [1]. Therefore, in this study to perform multiplex ARMS - PCR, we had a combination of the Old primers and specific primers by Le Thi Hao design.

- *PCR reaction*

Prepare the following 50µl reaction in a 0.5ml PCR tube on ice: 5µl 10X standard Taq reaction buffer, 3µl MgCl2, 0.4µl dNTP, 0.2µl ampli-Taq, 1µl BSA 1%, 36.4µl nuclease-free water, 1µl of each primer 1µl (10pmol/µl), 1 µl template DNA.

Each mutation have 2 separate tube: one tube with normal primer and another with mutation primer.

PCR was performed with 30 the same thermal cycle but different annealing temperature to find optimal temperature for each mutation:
3 min at 90^0C x 1 cycle
1 min at 94^0C, 1 min at $58^0/60^0$C, $62^0/65^0$C and 1 min at 72^0C x 25 cycles
3 min at 72^0C x 1 cycle

The PCR product was observed under UV irradiation after electrophoresised on 2% agarose agar in 1 hour and stained with ethidium bromide.

Agar electrophoresis results are explained as follows:

- If the PCR product is only available in the normal primers reaction, the DNA samples don't carry the mutation.

- If the PCR product is present in both reaction, the DNA samples carry the mutation in heterozygote.

- If the PCR product is only present in mutation primers reaction, the DNA samples carry the mutation in homozygote.

Sequencing technique [2], [3]
Done with negative samples in ARMS – PCR reaction .

- *DNA extraction:*

Similar to DNA extraction in Multiplex ARMS – PCR techniques

- *PCR amplification of beta globin genes:*

Using following primer pair:
Forward primer (F1) : 5' - CCA ATC TAC TCC CAG GAG CA - 3 '
Reverse primer (R3) : 5' - CCT TCA TCT GCA GCC TTC AT - 3 '

Primers have the same length of 20bp, amplified 1829bp gene fragment lenght containing the entire beta - globin gene sequence and position c.- 78.

- *Electrophoresis of PCR products:*

PCR products were detected on 1 % agarose and determined by comparison with a DNA ladder (2000bp size)

- *PCR products purification:*

After electrophoresis, PCR products had purity check to prepare for PCR sequencing using the Qiagen purification kit.

Principles:

Using the Oligonucleotide Purification Cartridge (OPC) cartridge. At high salt concentrations, DNA would be retained and other impurities would be washed away when going through the silica membrane. Then purified DNA was separated by sterile water, pH = 8 .

- *PCR sequencing:*

Beta-globin gene sequence would be run in reaction with 6 primers (3 pairs) :

+ The first primer pairs:
. Forward primer (F1): forward primer in PCR amplification beta globin gene .
. Reverse primer (R1) : 5' - GGA TCC TCA TTC AAG ACG TGG - 3 ' .

+ The second primer pairs:
. Forward primer (F2) : 5' - GGA GGA AGG TAA ACA GTA GG - 3 ' .
. Reverse primer (R2) : 5' - TGC TGG AAA GAA GAG GCATG - 3 ' .

+ The third primer pairs:
. Forward primer (F3) : 5' - TCT GGG TTA AGG CAA TAG CA - 3 ' .
. Reverse primer (R3): reverse primer in PCR amplification beta globin gene.

- *DNA sequence:*

Using Beckman Coulter CEQ8000 sequencing machine.

III. RESULT

Identifying 8 Popular Mutation Models in Vietnam by Multiplex ARMS - PCR Technique

After making techniques on control sample, 25 selected samples were carried out reactions, The obtained results are described in table 1:

Table 1 Common type of mutation by ARMS-PCR technique

Mutation types \ Name	β-thal (n=10)		HbE (n=15)		Total (n=25)	
	n	%	n	%	n	%
c.52 A → T (Cd17)	0	0	0	0	0	0
c. 79 G → A (Cd26)	0	0	15	100	15	60
c.124–127–dell TTCT (Fs41/42)	3	30	0	0	3	12
c. 214 ins A (Cd71/72)	0	0	0	0	0	0
c. – 78A → G (-28)	0	0	0	0	0	0
c. 92 + 1G → T (IVS1-1)	0	0	0	0	0	0
c. 316 – 197 C → T (IVS2 – 654)	0	0	0	0	0	0
Total	3	30	15	100	18	72
Other mutations	7	70	0	0	7	28

- 10 β-Thal patient samples:

+ 3 β-Thal patients had a common mutation in Asia (Fs 41/42 mutation)

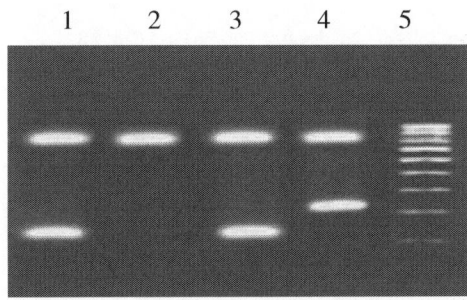

Fig. 1 Fs 41/42 mutation

1, 2: Normal and mutation primers of a control sample (helthy person).
3,4: : Normal and mutation primers of a patient sample
5: 100bp marker

+ 7 β-Thal patient mutations (28%) didn't determine through ARMS - PCR technique.

- 15 HbE patient samples:

All mutation is Cd26: 13 samples with 1 allele mutations (heterozygotes cases) and 02 samples with 2 allele mutations (homozygotes cases)

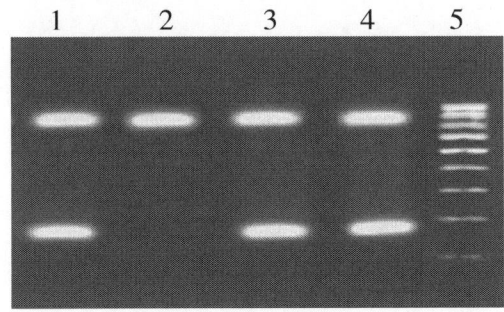

Fig. 2 Cd26 mutation

1, 2: Normal and mutation primers of a control sample (helthy person).
3,4: : Normal and mutation primers of a heterozygotes patient sample
5: 100bp marker

Identifying rare mutation models in Vietnam by sequencing technique

As many authors have described, more than 250 mutations in the HBB gene have been found to cause β-Thal. Therefore, we sequenced β-globin gene on 7/25 cases (28%) which were not detected mutation through ARMS - PCR technique. The obtained results are described in table 2.

Table 2 The results identified rare mutation by sequencing technique

Mutation types \ Name	β-thal (n=7)	
	n	%
C.9 T → C	6	85,71
IVS2 -667 G → C	3	42,86
IVS2 -185 C → T	5	71,43
IVS2 -777 T → G	2	28,57

IV. DISCUSSION

On 20 Khmer patient have determined β-Thal based on Hb electrophoresis results when conducting disease-causing mutations, we found that:

Results identify mutations through Multiplex ARMS PCR is completely consistent with Hb electrophoresis. This confirms results of PCR performed with great success. In addition, compared with ASO technique, ARMS - PCR allows rapid detection of common mutations through the steps without complex hybrid. Primers are designed for specific mutations is the important point in this technique.

With 7 carrying β-thal gene samples, sequencing results detected four point mutations occurred on a β-globin chain. We currently haven't found any research data about effect on an infected individual for this mutations, despite the fact that patients still have symptoms of iron deficiency anemia.

Also the mutant finding results by the ARMS - PCR technique once again is proven accurate through gene sequencing techniques. But the question is to study the feasibility of applying research findings widely. Besides accuracy, ARMS - PCR technique has the advantage of quickness and simpleness. Currently primers were designed and used widely around the world. Thus, this technique can easily deploy the application in the molecular biology laboratory. However, in this study we found that ARMS - PCR technique suitable for finding mutations in the community with not too diverse genotypes and having very high presence of 8 mutations (over 90 %). This observation is consistent with the author Le Thi Hao (2001). But on an ethnic with unknown types of mutations or not many common mutations as ethnic Khmer community in Bac Lieu, we recommend using the most effective method is sequence. Although its cost and time -consuming is higher than other tests but it is much cheaper than the cost of blood transfusions for patients or in cases that ARMS - PCR mutation diagnosis fail and then have to use sequencing methods.

With the advantage of more effectively detecting all mutations occur in two alleles of β-globin gene than ARMS technique which only detects 8 mutations, we sugest patients should use this method for prenatal screening because sequence just require one sampling of amniocentesis to detect mutation, therefore reducing the negative effect on the health of mother and the baby and avoid the negative psychological reaction of the mother when being tested too many times.

V. Conclusions

- 72% 8 types of mutations common in Vietnam Multilex determined by ARMS-PCR technique, in which:

+ 3/25 patients had Fs41/42 mutation in one allele of the β-globin gene

+ 13/25 patients had Cd26 mutation in one allele of β-globin gene

+ 2/25 patients carried Cd26 mutations in the two alleles of β-globin gene

- Finding 4 point mutations in β-globin gene by sequencing technique on 7 β-thalasemia patients Which was not detected mutation through ARMS - PCR technique: C.9 T → C, IVS2 -667 G → C, IVS2 -185 C → T and IVS2 - 777 T → G.

References

1. Hao Le Thi (2001) Research and application of molecular genetic techniques to detect gene mutations causing beta-thalassemia patients in Vietnam. Medicine masters thesis. Ho Chi Minh University of Medicine and Pharmacy, Ho Chi Minh City.

2. Han Nguyen Khac Hoan, Vietnamese bar discounts, Kiet Truong Dinh, Lam Thi My. (2011) Prenatal diagnosis of thalassemia in 290 cases of pregnancy. Journal of Medical Research. volume 3. pp 1-7

3. Oanh Quach Thi Hoang (2009) Applied DNA sequencing techniques to detect beta-thalassemia mutations. Medicine masters thesis, Ho Chi Minh University of Medicine and Pharmacy, Ho Chi Minh City

4. Cao A GR (2005) Beta-Thalassemia, Gene Reviews.

5. Hossein najmabadi phd (2001) Amplification Refractory Mutation System (ARMS) and Reverse Hybridization In The Detection of Beta-thalassemia Mutations. pp 165 – 170.

6. Harteveld Cl, Voskamp A, Phylipsen M et al (2005) Nine unknown rearrangements in 16p13.3 and 11p15.4 causing alpha- and beta-thalassaemia characterised by high resolution multiplex ligation-dependent probe amplification. Journal Medical Genetics 42, pp 922-931.

7. John old, Renzo Gallanello, Androulla Eleftheriou, Joanne Traeger-synodinos, Mary petrou, Michael Angastiniotis (2005) Prevention of Thal and other Haemoglobin Disoders. USA.

8. Libani IV., Guy EC. and Melchiori L (2008) Decreased differentiation of erythroid cells exacerbates ineffective erythropoiesis in beta thalassemia, Blood 112, pp 875–885.

9. M.L. Saovaros Svasti, Tran Minh Hieu, Thongperm Munkongdee, Pranee Winichagoon, Tran Van Be and Suthat Fucharoen (2002) Molecular Analysis of-Thalassemia in South Vietnam. American Journal of Hematology 71. pp 85-88.

10. Taher A., Isma'eel H. and Cappellini MD (2006) Thalassemia intermedia: revisited, Blood cells, molecules and diseases 37, pp 12-20.

Using Realtime Rt-Pcr and Sequencing Assays to Define Viral Load, Types and Subtypes of Hepatitis C Virus at Cantho Center General Hospital

Cao Thi Tai Nguyen[1], Tran Ngoc Dung[1], and Nguyen Thi Huynh Nga[2]

[1] Cantho University of Medicine and Pharmacy, Vietnam
[2] Cantho Centre General Hospital, Vietnam

Abstract— We used Realtime RT-PCR assay to count the viral load on 94 patients who having positive anti-HCV, to find the types of HCV on 85 patients with positive HCV RNA. Then, we chose 6 samples to sequence to figure out the subtypes of HCV. The first result showed that: most of patients had the viral load $\leq 2 \times 10^6$ copies/ml (92.5%). The types of HCV have been found: type 1 (47.1%), type 6 (25.9%), and type 2 (21.6%). Moreover, there was 5.8% the co-infection between type 1 and type 6 on the patients. There was the similar result between Realtime RT-PCR assay and sequencing technique on typing HCV. We have been known the subtypes of HCV by sequencing technique, they are 1b, 2c, 6a, 6e and co-infection 1b/6e.

Keywords— HCV RNA, sequencing, Realtime RT–PCR, the types and subtypes, viral load.

I. INTRODUCTION

Hepatitis C is an infectious disease affecting primarily the liver, caused by the hepatitis C virus (HCV).[6] The infection is often asymptomatic, but chronic infection can lead to scarring of the liver and ultimately to cirrhosis, which is generally apparent after many years. In some cases, those with cirrhosis will go on to develop liver failure, liver cancer, or life-threatening esophageal and gastric varices.[1]

HCV is spread primarily by blood-to-blood contact associated with intravenous drug use, poorly sterilized medical equipment, and transfusions. An estimated 150–200 million people worldwide are infected with hepatitis C.[2][3][4]

The virus persists in the liver in about 85% of those infected. Overall, 50–80% of people treated are cured. Those who develop cirrhosis or liver cancer may require a liver transplant. Hepatitis C is the leading reason for liver transplantation, though the virus usually recurs after transplantation.[7] No vaccine against hepatitis C is available.

Before giving the decision to treat this disease, determining HCV RNA concentration in the body will provide prognostic information helps to determine the manner and duration of treatment. If patients infected the types of HCV and HCV RNA concentration in the body is higher than 2×10^6 copies/ml of blood, the duration of treatment is 48 weeks instead of 24 weeks as the other case [1].

The immunoassays allow detection of most cases of chronic HCV infection but they are not entirely effective in phase "incubation". Realtime RT- PCR assay was used to find the viral load and types of HCV helping doctors prognosis, assign and treat effectively it [1].

Determining the concentration of HCV RNA in the body and the types of HCV will provide the necessary information to help doctor decide the process and time for treatment [3].

We conducted this research with the following objectives:

Determining the viral load and types of HCV on patients with hepatitis C disease by Realtime RT-PCR assay at Cantho Centre General Hospital.

Confirming the accuracy between Realtime RT – PCR assay and sequencing technique.

II. MATERIALS AND METHODS

This research is a descriptive cross-sectional study. From January 2009 to August 2010, we collected all the patients who having positive anti-HCV and totally agreeing to take part in our study. The machine, Architect i2000 SR of Abbott, was used to test anti-HCV.

After that, we checked all the positive samples with Realtime RT-PCR assay by Mx3000 Stragenatic and the series of HCV kit from Viet A corp. to identify the concentrations of HCV and the types of HCV. This machine can detect the viral load over 300 copies/ml. Realtime RT-PCR reaction using TaqMan probes specific for the HCV type 1, 2, 3 and 6 are marked with fluorescence. In [iVA]HCV tube 1-6 rPCR Mix Type, the probe of type 1 is marked with HEX color, the probe of type 6 is marked with FAM

© Springer International Publishing Switzerland 2015
V. Van Toi and T.H. Lien Phuong (eds.), *5th International Conference on Biomedical Engineering in Vietnam*,
IFMBE Proceedings 46, DOI: 10.1007/978-3-319-11776-8_62

color. In [iVA]HCV tube 2-3 rPCR Mix Type, the probe of type 3 is marked with HEX color, the probe of type 2 is marked with FAM color.

Setting up FAM and HEX colors for all samples, positive control and negative control.

Then, we just collected 6 samples having the viral load over 10^4 copies/ml to sequence to figure out the subtypes of HCV.

Statistical Analysis

Patients' data was recorded in a form and statistic analysis was performed with SPSS 13.0.

III. RESULTS

The age of study group was from 11 to 81 years. There were 60% patients living in Cantho. 40% patients were living in other provinces in Mekong Delta. By using Realtime RT-PCR assay, we found that there were 94 patients among 168 patients having positive HCV RNA (56%). 74 patients among 168 patients had negative HCV RNA (46%)

To check the viral load in 94 patients' blood, we used Realtime RT-PCR assay. The result was shown that: 92.5% patients had the viral load $\leq 2\times10^6$ copies/ml and 7.5% patients had the viral load $> 2\times10^6$ copies/ml (Table 1).

Table 1 The amount of virus in patients' blood by Realtime RT – PCR assay

The viral load (copies/ml)	Realtime RT-PCR assay	
	The number of samples	Percent
$\leq 2\times10^6$	87	92.5
$> 2\times10^6$	7	7.5
Total	94	100

85 samples were typed HCV by Realtime RT-PCR assay. The types of HCV have been found: type 1 (47.1%), type 6 (25.9%), and type 2 (21.6%), and 5.8% the co-infection between type 1 and type 6 on the patients (table 2).

Table 2 The types of HCV by Realtime RT-PCR assay on 85 patients

The types of HCV	Realtime RT-PCR assay	
	Amount of samples	Percent
1	40	47.1
2	18	21.2
6	22	25.9
Co-infection 1-6	5	5.8
Total	85	100

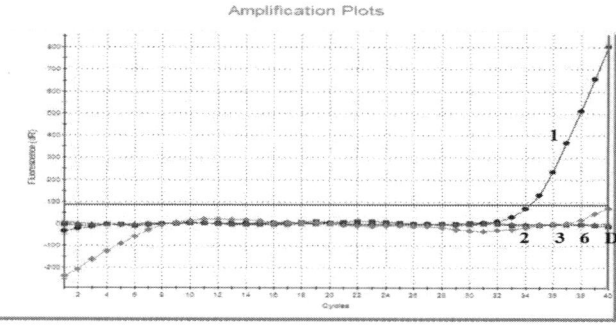

Mẫu nhiễm type 1
1, 2, 3, 6: type 1, 2, 3, 6. **D**: Chứng âm

Fig. 1 A patient infected type 1 of HCV

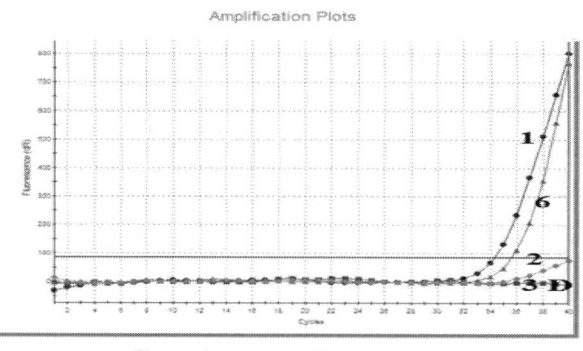

Mẫu đồng nhiễm type 1, 6
D: Chứng âm

Fig. 2 A patient co-infected type 1 and type 6 of HCV

At the end, we chose 6 samples having the viral load $>10^4$ copies/ml to sequence DNA. They are one sample of type 1, 2 samples of type 2, 2 samples of type 6 and one sample of co-type 1 and 6 (Table 3).

Table 3 The subtypes of HCV by DNA sequencing technology.

The types of HCV	DNA sequencing technology	
	Amount of samples	The subtypes of HCV
1	1	1b
2	2	2c
6	2	6a and 6e
Co-type 1 and 6	1	1b/6e

After getting the DNA sequence, we applied BLAST (Basic local alignment search tool) in NCBI (National Center for Biotechnology Information) to find the correlation and gave the conclusion of the subtypes. The answers were shown the subtypes: 1b, 2c, 6a, 6e and 1b/6e. The results of DNA sequencing of the subtypes of HCV are shown below.

This was a case of subtype 1b with 267 bases length.

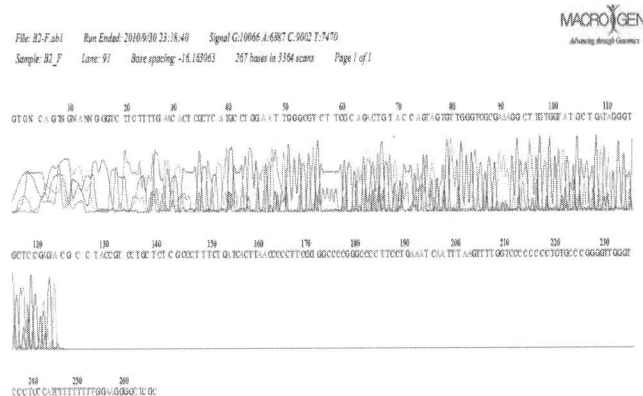

This was a case of subtype 1b/6e with 372 bases length

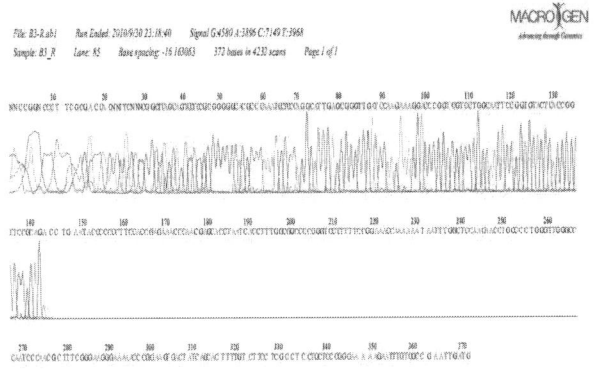

IV. DISCUSSION

By studying 168 serum samples of patients with positive anti-HCV, we recorded 94 cases, accounting for 56%, positive HCV RNA by Realtime RT-PCR asay. Our results fit with Kenny Walsh (1999), Barrett (2001) and Wiese (2000). According to Wiese (2000), studying on 917 female patients from German, there were 504 patients with positive HCV RNA accounted for 55% [7]. Our result was lower when compared with Nguyen Thanh Tong et al. (2000), the rate of positive HCV RNA accounted for 70% [9]. This can be explained by the different method used to quantify HCV; Nguyen Thanh Tong et al. used branched DNA method to quantify HCV in the blood. 54% of patients are male, so the rate of hepatitis C infected men was much more than women. It was similar to Mindie H Nguyen's research in

Vietnam (66% male) [11]. Moreover, Ho Tan Dat's study (2005) also showed that the prevalence of HCV was higher in men (57.5%). The mean age of our study was 47.4 ± 13.4 years. It was similar with Ho Tan Dat's (2005) and Mindie H Nguyen's researches. According to the studies of Ho Tan Dat et al. (2005) and Mindie H Nguyen, the mean ages were 47.9 ± 10.58 years and 49 ± 11 years [11]. The prevalence of the types of HCV in Cantho according to Huynh Van Truong et al. (2009) was type 1, 2 and 6. Our result was reported the same with Truong's study. However, there was a new thing, co-infection both type 1 and type 6, in our study. We found that patients infected with HCV type 1 accounting for the highest rate (47.1 %), following by type 6 (25.9 %) and type 2 (21.2%). This result was consistent with previous studies of HCV on patients in Vietnam was published on the website of dr.thuthuy. Ho Tan Dat's study (2005) showed that type 1 accounted for the highest percentage (58.4 %), followed by type 6 (23.9 %), type 2 (13.1%) and type 3 (0.3 %) [8] . However, when compared with studies of Huynh Van Truong et al. (2009) in Can Tho, we found a little difference in the rate of HCV types. The result of Huynh Van Truong et al. (2009) showed that 40 % of patients infected with HCV type 1; 40 % of patients infected with HCV type 6 and 20 % of patients infected with HCV type 2. This difference was probably due to the techniques and the number of samples. Huynh Van Truong used DNA sequencing to find HCV types on 30 patients. Our study was showed the subtypes 1b, 2c, 6a, 6e and 1b/6e. Huynh Van Truong et al. (2009) detected the subtypes in Can Tho 1a, 1b, 2a and 6a. It was supposed that there was a diversity of HCV subtypes in Can Tho. Subtypes of HCV in Ho Chi Minh city were 1b, 2a, 2c and 6a in Ho Tan Dat's study. Pham DA et al. did a research on type 6 of infected donors at Hanoi Hematology and Blood Transfusion Institution. Her study showed that the subtypes were 6a and 6e [12]. Therefore, comparing with above studies, our result was similar to them.

V. CONCLUSION

In summary, the prevalence of HCV was much higher in men (56%). Most of HCV patients had the viral load ≤ 2×10^6 copies/ml (92.5%). The types of HCV were prevalent in Mekong Delta: 1, 2 and 6. Moreover, the rate of co-infection between the types of HCV was recorded in our study. 5.8% patients had the co-infection of type 1 and type 6. Sequencing result was shown the subtypes of HCV: 1b, 2c, 6a, 6e and 1b/6e.

REFERENCES

1. Chau Huu Hau. 2006. Hepatitis C virus, Medical publisher.

2. Nguyen Huu Chi. 1998. Some of features of hepatitis diseases, Thanh pho Ho Chi Minh Publisher, pp.26-28

3. Pham Hoang Phiet. 2000. Hepatitis C virus: from structure to treatment, Medical publisher, pp. 69-86.

4. Adrian M Di Bisceglie. 2009. Essentials of hepatitis C infection. Current Medicine Group

5. Mindie H. Nguyen et al. 2004. Epidemiology and treatment outcomes of patients with chronic hepatitis C and genotypes 4 to 9. Reviews in Gastroenterological disorders. Vol 4, suppl 1.

6. Simmonds P. 2005. Evolution of hepatitis viruses, Viral hepatitis 3rd edition, Blackwell, pp. 65-75

7. Nguyen Thanh Tong et al. 2004. Tests to diagnosis viral hepatitis C, dr.thuthuy.com

8. Ho Tan Dat et al. 2005. Prevalence of HCV genotypes in Vietnam, dr.thuthuy.com

9. Alan Franciscus and Liz Highleyman. 2005. Hepatitis C, Hepatitis C support project

10. Mindie H. Nguyen et al. 2003. High prevalence of novel genotypes in Vietnamese patients with chronic hepatitis C, Hepatitis C support project

11. Pham DA et al. High prevalence of hepatitis C virus genotype 6 in Vietnam, Asian Pac J Allergy Immunol. 2009 Jun-Sep;27(2-3):153-60.

12. NCBI. 2005. Hepatitis C virus isolate Ra.At 5' UTR, GenBank: AY707632.1.

13. NCBI. 2007. Hepatitis C virus subtype 1b gene for polyprotein, partial cds, isolate: VT28, . GenBank: AB301760.1.

14. NCBI. 2011. Hepatitis C virus subtype 1b isolate S4 polyprotein gene, complete cds, GenBank: GU451224.1.

15. NCBI. 2007. Hepatitis C virus subtype 6e gene for polyprotein, partial cds, isolate: TV281, GenBank: AB301835.1.

16. NCBI. 2010. Hepatitis C virus subtype 6e gene for core protein, partial cds, isolate: HPA491, GenBank: AB523161.1.

17. NCBI. 2010. Hepatitis C virus subtype 6a gene for core protein, partial cds, isolate: HPA365, GenBank: AB523126.1.

18. NCBI. 1995. Hepatitis C virus type 2c (clone BE135) core protein and E1 protein mRNA, 5' end of cds, GenBank: L38321.1.

19. NCBI. 2007. Hepatitis C virus subtype 2c gene for polyprotein, partial cds, isolate: TV353, GenBank: AB301782.1.

Role of the Fifth Copper Binding Site in Prion Conversion

Thao Phuong Mai[1] and Giuseppe Legname[2]

[1] Physiology Sector, Department of Medicine, University of Medicine and Pharmacy, HoChiMinh City, Vietnam
[2] Prion Biology Laboratory, Neurobiology Sector, Scuola Internazionale Superiore di Studi Avanzati SISSA Trieste, Italy

Abstract— Prion diseases are fatal neurodegenerative disorders linked to the deposition of the abnormal prion protein isoform called PrPSc or prion. The key molecular events triggering the diseases are the conformational changes from the normal cellular α-helical prion protein PrPC to the pathological β-sheet enriched PrPSc. Therefore, understanding the mechanism and factors underlying the conversion process is essential to find possible diagnostic tools and treatments. Copper has long been known to correlate with neurodegenerative dysfunctions; PrPC is a copper binding protein via histidine residues in the highly conserved octapeptide repeats (OR) and the non-OR region located in the disordered N-terminal tail of the protein. The role of copper in facilitating protein aggregation and disease progression remains elusive. This study describes the impact of histidine residues on prion replication. By analyzing mouse PrP constructs that carry artificial mutations at histidines in the OR and non-OR, we provide cell evidence for the critical role of the non-OR copper binding site at histidine 95 in prion conversion. We also contribute to better understanding of the mechanisms and primary sites for prion conversion and replication. Our findings establish a platform for further studies aimed at elucidating the role of the H95 mutant in de novo prion diseases when expressed in transgenic mice.

Keywords— prion, copper binding, octapeptide repeats (OR), non-OR region, prion conversion.

I. INTRODUCTION

Prion diseases or transmissible spongiform encephalopathies (TSEs) are a group of rare invariably fatal neurodegenerative disorders affecting human and animals. TSEs etiologically manifest as sporadic, genetic or infectious forms. These maladies are caused by the conformational change of the physiological cellular form of the prion protein, PrP^C, into the pathological isoform denoted as prion or PrP^{Sc} [1].

The mature PrP^C encoded by the prion protein gene (*Prnp*) includes an unstructured N-terminal moiety (residues 23-126, hereafter in mouse numbering) and a C-terminal globular domain (residues 127 to 230) [2,3]. The N-terminal region features the octarepeat region (OR), an evolutionarily conserved motif of octapeptide repeats whose number varies among species [4]. In mouse (Mo) PrP^C, the OR contains four tandem copies of the eight-residue sequence (PHGGGWGQ, residues 59-90) that is able to coordinate copper ions and, to a lesser extent, zinc and manganese through histidine, glycine and tryptophan residues [5-8]. Additional histidines involved in high affinity copper binding are located in a positively charged region (residues 91-111) forming the non-OR, also known as fifth copper binding site [9-11].

PrP^C is mostly expressed in neuronal cells both in the central and peripheral nervous systems. Despite being highly conserved among mammals, the function of PrP^C has not been established with certainty [12]. Defining the physiological role of PrP^C remains essential for understanding the molecular events leading to TSEs. The ability to coordinate copper ions indicates PrP^C as a copper binding protein. A wealth of studies explored the functional aspects of this interaction, suggesting that PrP^C may act as a copper transporter, a sink of copper excess or a scavenger for free radicals [13,14]. These findings are consistent with current evidence proposing a PrP^C neuroprotective function [15-17].

Although sharing the same amino acid sequences, PrP^{Sc} bears distinct biochemical and structural features from PrP^C. In contrast to PrP^C, which is α-helical, monomeric and protease-sensitive, PrP^{Sc} is enriched in β-sheet secondary structures, amyloidogenic, infectious and partially resistant to protease-K (PK) [18]. Different structural regions are implicated in the PrP^C to PrP^{Sc} conversion. It is largely accepted that the PrP^{Sc} PK-resistant core spans approximately residues 90 to 230 [1]. The N-terminal part of this core includes the non-OR copper binding site and a hydrophobic segment containing an amyloidogenic palindromic sequence. The non-OR copper-binding site attracted special interest because it is located in a region proximal to the globular domain, involved in the conversion to PrP^{Sc}. This regional proximity raised the question whether copper bound to the non-OR site may have an impact on prion conversion.

© Springer International Publishing Switzerland 2015
V. Van Toi and T.H. Lien Phuong (eds.), *5th International Conference on Biomedical Engineering in Vietnam,*
IFMBE Proceedings 46, DOI: 10.1007/978-3-319-11776-8_63

Here, we investigated the role of OR and non-OR copper binding regions in modulating prion conversion. We used neural cell models expressing MoPrP constructs in which histidine residues, the primary amino acids involved in copper coordination in the OR and non-OR regions were substituted by tyrosines. We found that the His-mutations have different effects when transiently expressed in neuroblastoma cells chronically infected with the Rocky Mountain Laboratory (RML) prion strain (ScN2a). Interestingly, while the substitutions of each His located in the OR region with Tyr exhibited a negligible influence on PrP^{Sc} PK-resistant levels, the non-OR H95Y mutation resulted in higher PrP^{Sc} formation in ScN2a cells. Importantly, this mutant induces *de novo* prion formation in N2a cells.

II. MATERIALS AND METHODS

A. Plasmids for Cell Transfection

The 3F4 epitope tag (residues 108-111, LKHV) and the single point mutations (H60Y, H68Y, H76Y, H84Y and H95Y) were inserted into the pcDNA3.1::MoPrP(1-254). N2a and ScN2a cells were cultured in Opti-MEM (GIBCO) media supplemented with fetal bovine serum (FBS) and antibiotics, and incubated at 37°C, 5% CO_2. Transient transfections were performed using X-treme gene DNA transfection kit (Roche Biochemicals) according to the manufacturer's guidelines.

B. Biochemical Assays on PrP^{Sc} and PrP^C

Cell lysates were harvested in cold lysis buffer (10 mM Tris–HCl pH 8.0, 150 mM NaCl, 0.5% Nonidet P-40 substitute, 0.5% sodium deoxycholate). For protease-K (PK) digestion assay, quantified protein was treated with PK (Roche) at 37°C or at 25°C. PK digestions were stopped by adding 2 mM phenylmethyl-sulphonyl fluoride (PMSF). Subsequently, the samples were ultracentrifuged at 100,000 g for 1 h at 4°C (Optima TL, Beckman Coulter, Inc.). The pellets were resuspended in sample buffer. Samples were loaded onto a 10% SDS-PAGE gel and immunoblotted on Immobilion PVDF (Millipore) membranes. Membranes were blocked with 5% (w/v) non-fat milk protein in TBS-T (0.05% Tween), incubated with anti-PrP antibodies D18 (InPro Biotechnology) and 3F4 (Covance), and developed by enhanced chemiluminescence (GE Healthcare). Band intensity was acquired using the UVI Soft software (UVITEC, Cambridge).

C. De Novo Prion Formation in N2a Cells

PTA-extracted PrP^{Sc} from either N2a cells transfected with 3F4-tagged WT or H95Y MoPrP were subjected to N2a cells and regularly passaged every 7 days up to P8. The PrP^{Sc} detection was assessed by PK digestion as previously described.

D. Statistical Analysis

Two-sample *t*-tests were used for statistical analysis of the data. Differences were considered significant when $p<0.05$. Data were analyzed using the Origin 8.6 software.

III. RESULTS

A. The Non-OR H95Y Mutation Promotes Prion Conversion

To investigate the effect of His residues on prion replication, ScN2a cells were transiently transfected with 3F4-tagged wild-type (WT) MoPrP and mutant constructs in which each individual His located inside the OR and non-OR was substituted by Tyr (denoted as H60Y, H68Y, H76Y, H84Y and H95Y) (Fig. 1).

```
KKRPKPGGWN   TGGSRYPGQG   SPGGNRYPPQ   52   Mo
KKRPKPGGWN   TGGSRYPGQG   SPGGNRYPPQ   52   Hu

           60           68           76
GG- TWGQPHG  GGWGQPHGGS   WGQPHGGSWG   81   Mo
GGGGWGQPHG   GGWGQPHGGG   WGQPHGGGWG   82   Hu
    *                   *            *

84           95           110
QPHGGGWGQG   GGTHNQWNKP   SKPKTN LKHV  111  Mo
QPHGGGWGQG   GGTHNQWNKP   SKPKTNMKHM   112  Hu
                                *  *

AGAAAAGAVV   GG                        Mo
AGAAAAGAVV   GG                        Hu
```

Fig. 1 Amino acid sequences from residues 23-123 in Mouse (Mo) and Human (Hu) prion protein.

The introduction of the 3F4-epitope tag into these constructs makes it possible to discriminate between transfected and endogenous MoPrP molecules [19] (Fig. 2). The effect of the H110Y mutant was not considered in this study, as it is located inside the 3F4 tag, thus precluding its detection by 3F4 antibody. His to Tyr substitutions in MoPrP did not affect the total PrP^C expression levels; conversely, the PK digestion profiles showed remarkably different PrP^{Sc} levels among the mutants. While the H60Y, H68Y, H76Y, H84Y mutants displayed PK-resistant PrP^{Sc} levels similar to those of WT, the non-OR H95Y mutant yielded a significantly higher PK resistant PrP^{Sc} signal, providing the first evidence for the role of this mutation in prion conversion (Fig. 2).

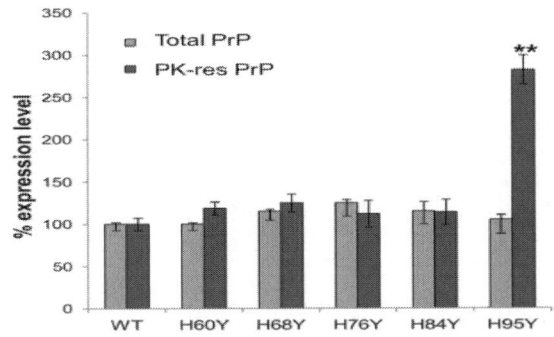

Fig. 2 H95Y mutation promotes prion conversion in ScN2a cells. Quantitative analysis of PrP expression (total PrP) and PrP^{Sc} PK-resistance (PK-res PrP) in transfected constructs (n=4, **$p<0.005$).

B. The H95Y Mutant Displayed PK-Resistance When Expressed in N2a Cells

The enhanced resistance to protease digestion is a primary feature to discriminate between PrP^{Sc} and PrP^{C} in cells chronically infected by prions. Since the transient expression of the H95Y mutant in ScN2a resulted in higher PK-resistant PrP^{Sc} signals, we reasoned that the same outcome would be observed also in non-infected N2a cells. To better characterize the PK-resistance induced by the H95Y mutation, we performed a time-dependent PK digestion at 25 °C of N2a expressing the 3F4-tagged WT or H95Y MoPrP. This experiment consistently confirmed the higher PK-resistance of the H95Y mutant over time (Fig. 3).

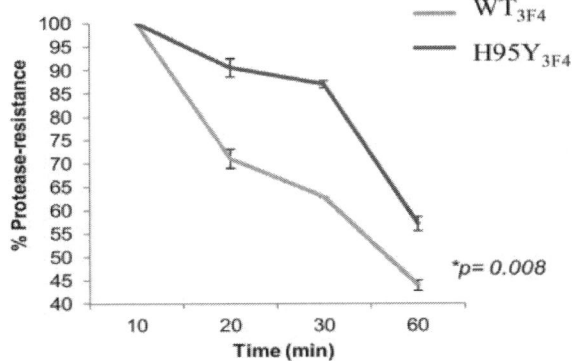

Fig. 3 H95Y mutant displays PK-resistance over time when expressed in N2a cells. Percentages of PK-resistance PrP levels at each time point (n=3, *$p<0.05$).

C. H95Y-Derived PrP^{Sc} Is Infectious in Cells

On the basis of these results, we reasoned that the H95Y mutant expressed in N2a cells acts as an infectious prion. To test this hypothesis, we isolated the H95Y seed from transiently transfected N2a cells by PTA precipitation,

subjected this seed to uninfected N2a and regularly passaged it up to P8 (Fig. 4). Interestingly, we observed an increment in PK-resistance levels over passages, thus indicating that the H95Y seed acted as a real infectious agent inducing *de novo* conformational conversion of PrP^{C} to PrP^{Sc} and propagating over passages.

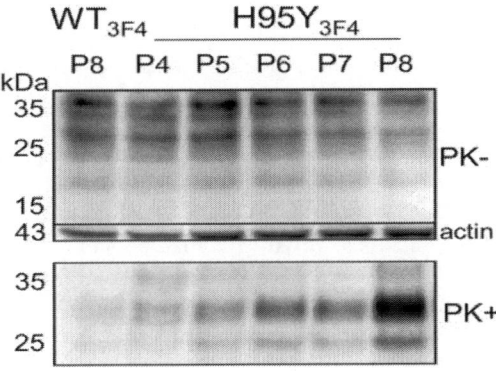

Fig. 4 H95Y mutant induces de novo prion formation. PTA-extracted PrPSc from N2a cells transfected with either 3F4-tagged WT or H95Y MoPrP were inoculated into N2a cells and regularly passaged up to passage (P) 8. PrP expressions were detected by D18 antibody.

IV. DISCUSSION

The central molecular event in prion diseases is the conversion of PrP^{C} into the disease-causing isoform PrP^{Sc}. Despite numerous investigations, the physiological function of PrP^{C} as well as its conversion mechanism leading to prion formation remain to be identified. Several lines of evidence have suggested that the essential trace metal copper may play an important role in PrP^{C} functions, although the precise contribution of copper to prion pathogenesis is still debated. Here we show that substituting all histidine residues in the OR has no effect on prion replication; nevertheless, we found that the His involved in an additional copper binding, H95, plays a critical role in prion conversion. While the substitution of each His residue in the OR region displayed a PrP^{Sc} PK-resistance profile similar to that of WT, the H95Y substitution showed a significantly high level of PrP^{Sc}. We hypothesize that there may be a compensatory effect between the four identical octapeptides of the OR, as a single His mutation in the OR is not sufficient to alter the PrP^{Sc} replication process [20]. Another possibility is that the non-OR copper-binding site at H95 is much more important for both PrP^{C} function, as shown by the higher affinity for copper binding compared to the OR region [21,22], and prion propagation and infectivity [23,24]. Residue H95 is located in a PrP^{Sc} region that is not affected by PK proteolytic cleavage; thus, copper bound to H95 may have a role during PrP^{Sc} formation even in the absence of the OR region, as shown by Cox *et al.* [25]. On the basis of our findings, we propose that the H95Y mutant may act as a pathogenic mutation located in the N-terminal domain, causing spontaneous conversion to prion. A recent molecular dynamic study proposes that N-terminal mutations may alter backbone flexibility and intra-molecular contacts in the OR and non-OR regions, thus affecting copper coordination as well as binding with other physiological interacting partners [26]. We reported the PK-resistance of the H95Y mutant and its ability to facilitate *de novo* PrP^{C} to PrP^{Sc} conversion. This is consistent with previous studies showing that a single amino acid substitution may be *per se* sufficient to generate *de novo* prions [27]. We also provide new insights into the role of histidines residues involved in copper coordination. Intriguingly, the sole substitution of H95 with tyrosine has a dramatic effect in promoting the generation of novel cell-to-cell infectious prion particles. Thus, H95 in the non-OR region has a pivotal role as molecular switch for prion conversion. We therefore argue that copper or other metal ions bound to H95 may stabilize this segment, preventing possible misfolding events that occur in the region from residue 95 to 125.

V. CONCLUSION

The data we present in this work may provide a platform for rationally designed experiments aimed at elucidating whether the H95Y mutation may cause *de novo* prion diseases when expressed in Tg mice.

ACKNOWLEDGMENT

We would like to thank Dr. Giachin for his informative comments during writing this paper.

REFERENCES

1. Colby DW, Prusiner SB (2011) Prions. Cold Spring Harb Perspect Biol 3: a006833.
2. Riek R, Hornemann S, Wider G, Glockshuber R, Wuthrich K (1997) NMR characterization of the full-length recombinant murine prion protein, mPrP(23-231). FEBS Lett 413: 282-288.
3. Zahn R, Liu A, Luhrs T, Riek R, von Schroetter C, et al. (2000) NMR solution structure of the human prion protein. Proc Natl Acad Sci U S A 97: 145-150.
4. Wopfner F, Weidenhofer G, Schneider R, von Brunn A, Gilch S, et al. (1999) Analysis of 27 mammalian and 9 avian PrPs reveals high conservation of flexible regions of the prion protein. J Mol Biol 289: 1163-1178.
5. Brazier MW, Davies P, Player E, Marken F, Viles JH, et al. (2008) Manganese binding to the prion protein. J Biol Chem 283: 12831-12839.
6. Nadal RC, Davies P, Brown DR, Viles JH (2009) Evaluation of copper2+ affinities for the prion protein. Biochemistry 48: 8929-8931.
7. Walter ED, Stevens DJ, Visconte MP, Millhauser GL (2007) The prion protein is a combined zinc and copper binding protein: Zn2+ alters the distribution of Cu2+ coordination modes. J Am Chem Soc 129: 15440-15441.
8. Burns CS, Aronoff-Spencer E, Legname G, Prusiner SB, Antholine WE, et al. (2003) Copper coordination in the full-length, recombinant prion protein. Biochemistry 42: 6794-6803.
9. Walter ED, Stevens DJ, Spevacek AR, Visconte MP, Dei Rossi A, et al. (2009) Copper binding extrinsic to the octarepeat region in the prion protein. Curr Protein Pept Sci 10: 529-535.
10. Hasnain SS, Murphy LM, Strange RW, Grossmann JG, Clarke AR, et al. (2001) XAFS study of the high-affinity copper-binding site of human PrP(91-231) and its low-resolution structure in solution. J Mol Biol 311: 467-473.
11. D'Angelo P, Della Longa S, Arcovito A, Mancini G, Zitolo A, et al. (2012) Effects of the pathological Q212P mutation on human prion protein non-octarepeat copper-binding site. Biochemistry 51: 6068-6079.
12. Aguzzi A, Sigurdson C, Heikenwaelder M (2008) Molecular mechanisms of prion pathogenesis. Annu Rev Pathol 3: 11-40.
13. Viles JH, Klewpatinond M, Nadal RC (2008) Copper and the structural biology of the prion protein. Biochem Soc Trans 36: 1288-1292.
14. Davies P, Brown DR (2008) The chemistry of copper binding to PrP: is there sufficient evidence to elucidate a role for copper in protein function? Biochem J 410: 237-244.
15. Baumann F, Tolnay M, Brabeck C, Pahnke J, Kloz U, et al. (2007) Lethal recessive myelin toxicity of prion protein lacking its central domain. EMBO J 26: 538-547.
16. Li A, Barmada SJ, Roth KA, Harris DA (2007) N-terminally deleted forms of the prion protein activate both Bax-dependent and Bax-independent neurotoxic pathways. J Neurosci 27: 852-859.
17. Caiati MD, Safiulina VF, Fattorini G, Sivakumaran S, Legname G, et al. (2013) PrPC controls via protein kinase A the direction of synaptic plasticity in the immature hippocampus. J Neurosci 33: 2973-2983.
18. Surewicz WK, Apostol MI (2011) Prion protein and its conformational conversion: a structural perspective. Top Curr Chem 305: 135-167.
19. Kaneko K, Zulianello L, Scott M, Cooper CM, Wallace AC, et al. (1997) Evidence for protein X binding to a discontinuous epitope on the cellular prion protein during scrapie prion propagation. Proc Natl Acad Sci U S A 94: 10069-10074.
20. Flechsig E, Shmerling D, Hegyi I, Raeber AJ, Fischer M, et al. (2000) Prion protein devoid of the octapeptide repeat region restores susceptibility to scrapie in PrP knockout mice. Neuron 27: 399-408.
21. Jones CE, Abdelraheim SR, Brown DR, Viles JH (2004) Preferential Cu2+ coordination by His96 and His111 induces beta-sheet formation in the unstructured amyloidogenic region of the prion protein. J Biol Chem 279: 32018-32027.
22. Millhauser GL (2007) Copper and the prion protein: methods, structures, function, and disease. Annu Rev Phys Chem 58: 299-320.
23. Quaglio E, Chiesa R, Harris DA (2001) Copper converts the cellular prion protein into a protease-resistant species that is distinct from the scrapie isoform. J Biol Chem 276: 11432-11438.
24. Qin K, Yang DS, Yang Y, Chishti MA, Meng LJ, et al. (2000) Copper(II)-induced conformational changes and protease resistance in recombinant and cellular PrP. Effect of protein age and deamidation. J Biol Chem 275: 19121-19131.

25. Cox DL, Pan J, Singh RR (2006) A mechanism for copper inhibition of infectious prion conversion. Biophys J 91: L11-13.

26. Cong X, Bongarzone S, Giachin G, Rossetti G, Carloni P, et al. (2013) Dominant-negative effects in prion diseases: insights from molecular dynamics simulations on mouse prion protein chimeras. J Biomol Struct Dyn 31: 829-840.

27. Jackson WS, Borkowski AW, Watson NE, King OD, Faas H, et al. (2013) Profoundly different prion diseases in knock-in mice carrying single PrP codon substitutions associated with human diseases. Proc Natl Acad Sci U S A 110: 14759-14764.

Author: Thao Phuong Mai
Institute: University of Medicine and Pharmacy
Street: 217 Hong Bang
City: Ho Chi Minh city
Country: Vietnam
Email: maithao292@gmail.com

The Electrocardiographic Characteristics of Chronic Obstructive Pulmonary Outpatients

Dang Huynh Anh Thu and Le Thi Tuyet Lan

Department of Physiology, University of Medical and Pharmacy, Ho Chi Minh, Vietnam

Abstract— Introduction: Chronic obstructive pulmonary disease (COPD) is the fourth leading cause of death worldwide. Cardiovascular disease is the most common comorbidities in COPD patients. ECG is the most popular and cheapest cardiovascular disease screening tool. Some research showed the high prevalence of abnomal ECG in exerbation COPD patients, but there still have no research in outpatients. The purpose of this study was to detect the characteristics of ECG in this group of patients.

Materials and Methods: The descriptive cross-sectional study was conducted in University Medical Center. A sample of 112 COPD patients attending outpatient respiratory disease center was selected randomly from June 2012 to June 2013. Patients were also evaluated electrocardiogram along with other investigations.

Results: The prevelence of abnormal ECG is 58% including rightward shiff of the QRS axis (33.9%), rightward shiff of the P wave axis (53.6%), right atrial enlargement (20.5%), right ventricular enlargement (10.7%), right atrial enlargement and right ventricular enlargement (9.8%), clockwise rotation of the heart (41.1%), low voltage QRS complexes (15.2%), arrhythmia (35.7%), right bundle branch block (7.1%), ischemic (17%). Abnormal. ECG increases with each stage, including GOLD I (33.3%), GOLD II (43.2%), GOLD III (65.1%), GOLD IV (78.3%) (p <0,05). The prevelence of abnormal ECG also increase in each group, including group A (27.3%), group B (45.7%), group C (60.0%), group D (71.4%).

Conclusions: ECG changes in COPD patients are common and correlate with the severity of disease. This suggests that COPD patients should be screened routine ECG in addition to to other clinical investigations.

Keywords— Electrocardiogram (ECG), Chronic obstructive pulmonary diseases (COPD), cardiovascular disease (CVD).

I. INTRODUCTION

COPD is among the five most prevalent diseases and causes of death worldwide [1]. COPD is characterized by chronic inflammation in the airway and lung parenchyma most commonly caused by cigarette smoking [2]. In addition to smoking, a major risk factor for cardiovascular disease (CVD) and more specifically ischemic heart disease, COPD patients share other risk factors with CVD patients due to advanced age and decrease in physical activity caused by lung disease [3]. Chronic bronchitis and lung function have also been identified as independent predictors of the occurrence of ischemic heart disease [4]. Furthermore, therapy for COPD may increase CVD [5].. Finally, once ischemic heart disease is present, CVD becomes an important comorbid factor predicting all-cause mortality in COPD patients [6]. Despite these observations, relatively minor attention has been paid to the cardiovascular status of patients with COPD. ECG is the most popular and cheapest cardiovascular disease screening tool. Some research showed the high prevalence of abnomal ECG in exerbation COPD patients, but there still have little research in COPD outpatients. Furthermore COPD patients are not usually assessed by ECG in routine medical practice particularly in developing countries like Vietnam.The purpose of this study was to detect the characteristics of ECG among COPD outpatients.

II. MATERIALS AND METHODS

Study Population

The descriptive cross-sectional study was conducted in University Medical Center. A sample of 112 COPD patients attending outpatient respiratory disease center was selected randomly from June 2012 to June 2013.

This study was approved by institutional ethical committee of University Medical and Pharmacy.

According to GOLD 2011 (Global Initiative for Chronic Obstructive lung disease) guidelines [7] of COPD, any patient who has symptoms of chronic cough, sputum production, or dyspnoea, and / or a history of exposure to risk factors for the disease are considered and included in the study, and was further confirmed by spirometry. The values of Forced Expiratory Volume in first second (FEV1) less than 80% of the expected value and ratio of forced expiratory volume in first second to the fixed vital capacity (FEV1/FVC) less than 0.7 (70%), after post bronchodilator inhalation were included in this study.

Patients with bronchial asthma, pulmonary tuberculosis, bronchectasis, known congenital or acquired heart diseases, diabetes mellitus and hypertension, were excluded.

After applying above inclusion and exclusion criteria, the 112 patients were selected.

V. Van Toi and T.H. Lien Phuong (eds.), *5th International Conference on Biomedical Engineering in Vietnam,*
IFMBE Proceedings 46, DOI: 10.1007/978-3-319-11776-8_64

Mesurements

According to GOLD 2011 criteria, the subjects with airflow limitation and FEV_1 %predicted $\geq$ 80 were identified as having mild airflow limitation (GOLD 1), $50 \leq FEV_1$ %predicted < 80 were described as moderate (GOLD 2), $30 \leq FEV_1$ %predicted < 50 were described as severe (GOLD 3) and FEV_1 %predicted < 30 were described as very severe airflow limitation.

According to combined COPD assessement, patients were divised into 4 groups A/B/C/D.

A standard 12 lead electrocardiogram (ECG) was obtained from all cases. ECG was read by cardiologists and classified according to the recommendations for measurement standards in quantitative electrocardiography of the CSE Working Party [9] and Marriott H [10]

Statistical Analysis

Comparison of various percentage among groups were performed by Chi-Squared test. Analysis was done by using Stata 10.0 computer package for statistics. Significance was inferred for differences with p < 0,05.

III. RESULTS

Characteristics of the Subjects

112 cases of COPD were studied and the following observations. Mean age was 67.0 ± 10.4. Majority of the patients were in over 60 age group (71.4%). Males were 101 and females were 11 in number. Majority of the patients were males constituting 90.2%. The male female ratio was 9:1. In the present study smoking was the major risk and 91.1% were smokers. The number of pack-years of smoking was 30 ± 19.6 packyears.

Though there are many spirometric parameters, the FEV1 and FEV /FVC ratio are often considered as indices of pulmonary function in chronic obstructive pulmonary disease. The mean values of spirometric parameters in this study were rather low. Mean FEV1was 47.8 ± 17.9. Mean FEV1/FVC was 48.4 ± 10.4.

The classification of severity of airflow limitation in COPD (based on post-bronchodilator) showed that majority cases was in severe airway obstruction (GOLD 1 (8%), GOLD 2 (33%), GOLD 3 (38.4%), GOLD 4 (20.6%)).

Similarly, according to the combined assessement of COPD (GOLD 2011), majority patients were in more symptoms and high risk group. (group A (9.8%), group B (31.3%), group C (8.9%), group D (50%)).

Electrocardiograpic Changes in COPD Outpatients

ECG changes were present in 65 (58%) patients including rightward shiff of the QRS axis (33.9%), rightward shiff of the P wave axis (53.6%), right atrial enlargement (20.5%), right ventricular enlargement (10.7%), right atrial enlargement and right ventricular enlargement (9.8%), clockwise rotation of the heart

Table 1 The characteristics of ECG changes in COPD outpatients classified by the severity of airflow limitation

ECG changes	GOLD I (n=9)		GOLD II (n=37)		GOLD III (n=43)		GOLD IV (n=23)		
	n	%	n	%	n	%	n	%	
QRS axis $\geq$ + 90°	0	0.0	9	24.3	14	32.6	15	65.2	< 0.01
P axis $\geq$ + 90°	2	22.2	15	40.5	26	60.5	17	73.9	< 0.05
Clockwise rotation	2	22.2	10	27.0	19	44.2	15	65.2	< 0.05
Low voltage QRS complexes	0	0.0	3	8.1	10	23.3	4	17.4	> 0.05
Right atrial enlargement	1	11.1	4	10.8	11	25.6	7	30.4	< 0.05
Right ventricular enlargement	0	0.0	2	5.4	5	11.6	5	21.7	< 0.05
Right atrial and ventricular enlargement	0	0.0	1	2.7	6	14.0	4	17.4	< 0.05
Left atrial enlargement	0	0.0	1	2.7	3	7.0	1	4.4	> 0.05
Left ventricular enlargement	1	11.1	1	2.7	1	2.3	1	4.4	> 0.05
Arrhythmia	1	11.1	10	27.0	16	37.2	13	56.5	> 0.05
RBBB	0	0.0	2	5.4	7	16.3	4	17.4	> 0.05
Ischemic	2	22.2	5	13.5	8	18.6	4	17.4	> 0.05

Table 2 The characteristics of ECG changes in COPD outpatients classified by the combined assessement of COPD

ECG changes	A (n=11)		B (n=35)		C (n=10)		D (n=56)		
	n	%	n	%	n	%	n	%	
QRS axis ≥ + 90°	0	0.0	9	25.7	3	30.0	26	46.4	< 0.05
P axis ≥ + 90°	1	9.1	16	45.7	6	60.0	37	66.1	< 0.01
Clockwise rotation	3	27.3	9	25.7	4	40.0	30	53.6	< 0.05
Low voltage QRS complexes	1	9.1	2	5.7	2	20.0	12	21.4	> 0.05
Right atrial enlargement	1	9.1	4	11.4	2	20.0	16	28.6	< 0.05
Right ventricular enlargement	0	0.0	2	5.7	1	10.0	9	16.1	< 0.05
Right atrial and ventricular enlargement	0	0.0	1	2.9	1	10.0	9	16.1	< 0.05
Left atrial enlargement	1	9.1	0	0.0	1	10.0	3	5.4	> 0.05
Left ventricular enlargement	1	9.1	1	2.9	1	10.0	1	1.8	> 0.05
Arrhythmia	1	9.1	10	28.6	4	40.0	25	44.6	> 0.05
RBBB	1	9.1	1	2.9	4	11.4	10	17.9	> 0.05
Ischemic	3	27.3	4	11.4	4	11.4	11	19.6	> 0.05

(41.1%), low voltage QRS complexes (15.2%), arrhythmia (35.7%), right bundle branch block (7.1%), left atrial enlargement (4.5%), left ventricular enlargement (3.6%), ischemic (17%).

Electrocardiograpic Changes in COPD Outpatients Classified by the Severity of Airflow Limitation

The prevalence of ECG changes was significantly higher in the more severe group of airflow limitation. The prevalence of ECG abnormal was 33.3% in GOLD I, 43.2% in GOLD II, 65.1% in GOLD 3 and 78.3% in GOLD IV (P < 0,005).

Table 1 shows significant difference among the incidence of rightward shiff of the QRS axis, rightward shiff of the P wave axis, clockwise rotation of the heart, right atrial enlargement, right ventricular enlargement in each stage of COPD from GOLD I to GOLD IV.

Electrocardiograpic changes in COPD outpatients classified by the combined assessement of COPD

The prevalence of ECG changes was significantly higher in the more symtomp and higher risk group classified by the combined assessement. The prevalence of ECG changes

was 27.3% in group A, 45.7% in group B, 60.0% in group C and 71.4% in group D (P < 0,005).

Table 2 shows significant difference among the incidence of rightward shiff of the QRS axis, rightward shiff of the P wave axis, clockwise rotation of the heart, right atrial enlargement, right ventricular enlargement in each group of patients from A to D

IV. DISCUSSION

COPD is a disease of late adulthood. In the present study the mean age was 67.0 ± 10.4 with 70.4% patients over 60 age. COPD is a male dominant disease, the high prevalence in males which is due to higher prevalence of smoking in this gender, and also males are more susceptible to smoking than females. In our study males accounted for 90.2%, with a male-female ratio of 9:1. In the present study 91.1% were smokers. Long-term cigarette smoking significantly reduces the vital capacity of smoker's lungs. Female patients were non-smokers but they were exposed to smoke of burnt fuels which is very common in rural. This is a strong risk factor for development of COPD [11].

In this study most common findings in ECG of patients with chronic obstructive pulmonary disease were indicating right axis deviation (rightward shiff of the QRS axis 33.9%, rightward shiff of the P wave axis 53.6%), clockwise rotation of heart (41.1%), low voltage QRS complexes (15.2%) due to the present of hyperexpanded emphysematous lungs within the chest. On the other hand, the long-term effects of hypoxic pulmonary vasoconstriction upon the right side of the heart, causing pulmonary hypertention and subsequent right atrial enlargement (20.5%) and right ventricular enlargement (10,7%), right bundle branch block (7.1%).

The present study shows significant difference among the incidence of rightward shiff of the QRS axis, rightward shiff of the P wave axis, clockwise rotation of the heart, right atrial enlargement, right ventricular enlargement in each stage of COPD from GOLD I to GOLD IV and each group of patients from A to D. According to Chou [12], the best criteria for judging the severity of COPD are an R amplitude in V6 of < 0.5 mV, R/S amplitude in V6 of 1.0 mV, and increased P wave amplitude in leads II and III. The best indicators of deteriorating pulmonary function in patients with COPD are reported to be progressive reduction of the R wave and R/S ratio in the orthogonal lead (also lead I and left precordial leads), progressive shift of the QRS axis in the superior direction and rightward shift of the P wave axis

Arrhythmia were present in 35.7%, mostly supraventricular in origin and include atrial extrasystroles atrial fibrillation.

Chronic bronchitis and lung function have also been identified as independent predictors of the ischemic heart disease [4]. The present study shows other abmormalities as ischemic 17%, left atrial enlargement 4.5%, left ventricular enlargement 3.6% although the patients had no self-reported before. There are several reasons for the association between COPD and cardiovascular disease, including a major shared risk factor (smoking) and a number of factors that may lead to increased stress on the cardiovascular system or to cardiac arrhythmia (use of β-agonist medications that may stimulate the cardiovascular system, hyperventilation leading to respiratory alkalosis, and inflammation).

V. CONCLUSIONS

ECG changes in COPD patients are common and significantly correlate with the severity of disease. Correlation was found of ECG manifestation of rightward shiff of the QRS axis, rightward shiff of the P wave axis, clockwise rotation of the heart, right atrial enlargement, right ventricular enlargement with the severity of airway obstruction. This suggests that COPD patients should be screened routine ECG in addition to to other clinical investigations.

REFERENCES

1. World Health Report (2000). Geneva: World Health Organization. Available at http://www.who.int/whr/2000/en/statistics.htm.
2. Lopez AD, Shibuya K, Rao C et al. (2006) Chronic obstructive pulmonary disease: current burden and future projections. Eur Respir J 27(2):397-412
3. Curkendall SM, DeLuise C, Jones JK et al. (2006), Cardiovascular disease in patients with chronic obstructive pulmonary disease, Saskatchewan Canada cardiovascular disease in COPD patients. Ann Epidemiol 16(1): 63-70
4. Falk JA, Kadiev S, Criner GJ et al. (2008) Cardiac disease in chronic obstructive pulmonary disease. Proc Am Thorac Soc, 5(4): 543-548
5. Sholter DE, Armstrong PW (2000) Adverse effects of corticosteroids on the cardiovascular system. Can J Cardiol 16:505-511
6. Sin DD, Man SF (2005) Chronic obstructive pulmonary disease as a risk factor for cardiovascular morbidity and mortality. Proc Am Thorac So 2(1): 8-11
7. Global Initiative for Chronic Obstructive Lung Disease (GOLD) (2011). Global strategy for the diagnosis, management and prevention of COPD. NHLBI/WHO Workshop Report, Executive Summary 2011.
8. American Thoracic Society (1991). Lung function testing: Selection of reference values and interpretative strategies. Am Rev Respir Dis, 144:1202-1217.
9. The CSE Working Party (1985) Recommendations for measurement standards in quantitative electrocardiography. The CSE Working Party". Eur Heart J, 6(10): 815-825
10. Marriott H (1998) Practical Electrocardiography. William and Wilkins, 16-32, 419-476
11. Abbey DE, Burchette RJ, Knutsen SF, et al (1998) Long-term particulate and other air pollutants and lung function in nonsmokers. Am J Respir Crit Care Med, 158(1): 289-298
12. Chou TC (1996) Electrocardiography In Clinical Practice: Adult and Pediatric. *Philadelphia, Saunders,* 4th ed, 303-304

Author: Dang Huynh Anh Thu
Institute: Department of Physiology, University of Medical and Pharmacy
Street: Hong Bang
City: Ho Chi Minh
Country: Vietnam
Email: thudanghuynhanh@gmail.com

The Correlation between Peripheral Nerve Conduction Study Parametersand Level of Urinary Albumin Excretionin Diabetic Patients

Tran Vu Hoang Duong, Le Quoc Tuan, and Nguyen Thi Le

Department of Physiology, University of Medicine and Pharmacy of Ho Chi Minh City, Vietnam

Abstract— Diabetic kidney disease (DKD) is the most common cause of end-stage renal disease (ESRD). The incidence of diabetes worldwide is increasing, hence the incidence ofDKDis also increasing.Several previous studies suggested that there may be a relationship between appearance of diabetic peripheral neuropathy (DPN) and impairment of kidney function.

The primary aim of this study was to evaluate the correlation between peripheral nerve conduction study parameters and urinary albumin excretion (UAE) level in diabetic patients. This is a cross-sectional studyperformed in 37 patients with diabetes. After the study, we identified asignificantnegative correlation between level of UAE and action potential (AP), and between level of UAE and nerve conduction velocity (NCV) of motor nerves (R = - 0.33 to R = - 0.65, $p < 0.05$). We also found asignificantnegative correlation between level of UAE and NCV ((R = - 0.36 to R = - 0.48, $p < 0.05$), and a significantpositive correlation between level of UAE and DSL (R = 0.37 to R = 0.74, $p < 0.05$) of sensory nerves.

Our results suggested that the presence of microvascular diabetic complications aresimultaneous, and the earliest presence is the DPN, especially abnormal NCSparameters. This could be a recommendation for the endocrinologist and nephrologist in early detection and treatment of DKD. We suggest that the peripheral NCS should be routinelyscreened for microvascular complications in diabetic patients.

Keywords— diabetes mellitus, diabetic peripheral neuropathy (DPN), nerve conduction study (NCS), urinary albumin excretion (UAE).

I. INTRODUCTION

Diabetic kidney disease (DKD)is the most common cause of end-stage renal disease (ESRD), causing 45% of incident cases required renal replacement therapy in the United States. The incidence of diabetes worldwide is increasing, with an estimated prevalence of 380 million by 2025, hence the incidence ofDKDis also increasing [2;15].

Several studies suggested that, there may be a relationship between appearance of DPN and impairment of kidney function [7;12;13].

As known, DPN is a common complication of diabetes [15].However, prevalence estimates vary widely, most likely depending on differences in patient selection, neuropathy definition, and methods of assessment. The routine evaluation of DPN is based on patient symptoms and a physical examination, which may include the Semmes-Weinstein monofilament and the 128-hertz tuning fork [11;[15]. However, nerve conduction studies (NCS) are the most sensitive and specific DPN detection method [6;8;10]. Their use is recommended for quantitative confirmation DPN in clinical practice. Expanded assessment to NCS has the potential for early diagnosis and improved outcomes.

The earliest presence of impairment of kidney function in diabetic patients is a urinary albumin excretion (UAE), microalbuminuria or macroalbuminuria [1;2], and the major expression of DPN in NCS is a prolongation in distal latency and a reduction in nerve conduction velocitycaused by axon damages [11;14].

The primary aim of this study was to evaluate the correlation between peripheral NCS parametersand UAElevel in diabetic patients.

II. RESEARCH DESIGN AND METHODOLOGY

The cross-sectional study was performed in 37 outpatientswith diabetes (17 men and 20 women; mean ± SD age 61.7 ± 11.4 years and duration of diabetes 7.5 ± 5.8 years)at the Nephrology Department of HCMC University Medical Center.

The UAE level is based on the albumin (μg)/creatinine (mg) ratio (ACR) measured in a random urine specimen. Microalbuminuria is defined as the ACR between 30 and 300 μg/mg; macroalbuminuriais defined as the ACR over 300 μg/mg [2].

The NCSs were performed with standard electromyography (EMG)equipment provided by major manufacturer (Neuro-MEP-Microof Neurosoft Company).Examinations were performed by trained electrophysiologists with extensive experience in recording and interpreting clinical data.Involved peripheral motor nervesare the median, ulnar, peroneal, and tibial nerveparameters evaluated includedthe compound muscle action potential (CMAP) and motor nerve conduction velocity (MNCV). Involved peripheral sensory nervesare themedian, radial, ulnar, and superficialperonealnerve -parameters

© Springer International Publishing Switzerland 2015
V. Van Toi and T.H. Lien Phuong (eds.), *5th International Conference on Biomedical Engineering in Vietnam,*
IFMBE Proceedings 46, DOI: 10.1007/978-3-319-11776-8_65

evaluated included the distal sensory latency (DSL) and sensory nerve conduction velocity (SNCV) [3].

Statistical analyses were performed with Stata software (version 10; StataCorp, College Station, TX). To evaluate the correlation between peripheral NCS parameters and UAElevel, we useSpearman rank correlation (R).

III. DATA ANALYSIS AND RESULTS

Table 1 summarizes the data of the study participants. In this study, there were 23 patients with diabetic nephropathy (62.2%). Of which, 12 subjectsare shown with microalbuminuria (32.5%) and 11 subjects areshown with macroalbuminuria (29.7%). Rate of patients with UAE in the groupwith duration of diabetes more than 5 years ishigher than the groupwith duration of diabetes less than5 years ($p < 0.05$).In addition, we identifiedasignificant moderatepositive correlation between duration of diabetes and level of UAE with correlation coefficient R = 0.35, $p < 0.05$. This result from our study was similar toresult in the study of Diep Thi Thanh Binh et al. [5], in which rate of diabetes patients with microalbuminuria is 33.7% and the correlationbetween duration of diabetes and level of urinary protein excretion is $R^2 = 0.53$ (type 1), $R^2 = 0.04$ (type 2). In this study, we not differentiate type 1 or type 2 diabetes.

Table 1 The UAEand duration of diabetesof participants.

UAE Level	Duration of diabetes			Total
	< 5 years	5 – 10 years	> 10 years	
Negative	9	3	2	14 (37.8%)
Microalbuminuria	5	2	5	12 (32.5%)
Macroalbuminuria	2	2	7	11 (29.7%)

The primary aim of this study was to evaluate the correlation between peripheral NCSparametersand UAElevel in diabetic patients. Results are showed in Table 2a and Table 2b.

Table 2a The correlation between motorNCS parameters and UAE.

Nerve	Characteristic	Side	R	p
Median	AP	L	− 0.30	0.07
		R	− 0.36	< 0.05
	NCV	L	− 0.44	< 0.01
		R	− 0.40	< 0.05
Ulnar	AP	L	− 0.02	0.93
		R	− 0.08	0.65
	NCV	L	− 0.55	< 0.001
		R	− 0.53	< 0.001
Tibial	AP	L	− 0.36	< 0.05
		R	− 0.41	< 0.05
	NCV	L	− 0.65	< 0.001
		R	− 0.48	< 0.01
Peroneal	AP	L	− 0.33	< 0.05
		R	− 0.31	0.06
	NCV	L	− 0.59	< 0.001
		R	− 0.38	< 0.05

Table 2b The correlation between sensory NCS parameters and UAE.

Nerve	Characteristic	Side	R	p
Median	DSL	L	− 0.16	0.35
		R	0.04	0.8
	NCV	L	− 0.08	0.64
		R	− 0.08	0.65
Ulnar	DSL	L	0.38	< 0.05
		R	0.56	< 0.001
	NCV	L	− 0.36	< 0.05
		R	− 0.46	< 0.01
Radial	DSL	L	0.74	< 0.0001
		R	0.65	< 0.0001
	NCV	L	− 0.46	< 0.01
		R	− 0.32	0.06
Superficial Peroneal	DSL	L	0.29	0.08
		R	0.41	< 0.05
	NCV	L	− 0.37	< 0.05
		R	− 0.48	< 0.01

Our results identified amoderatelyup to stronglysignificantnegative correlation(correlation coefficient R = -0.33 to R = - 0.65, $p < 0.05$) between level of UAE and AP,and between level of UAE and NCV of motor nerves. The strongest degree is belong tothe correlationbetween level of UAE and NCV of tibial nerve (R = - 0.65, $p < 0.001$), and peronealnerve (R = - 0.59, $p < 0.001$).

This result also found a moderatelysignificantnegative correlation (R = -0.36 to R = - 0.48, $p < 0.05$) between level of UAE and NCV of sensory nerves. The strongest degree is belong to the correlationbetween level of UAE and NCV of superficial peroneal nerve (R = - 0.48, $p < 0.01$). In addition, we identified a moderatelyup to strongly significantpositive correlation (R = 0.37 to R = 0.74, $p < 0.05$) between level of UAE and DSL of sensory nerves. The strongest degree is belong to the correlationbetween level of UAE and DSL of radial nerve (R = 0.74, $p < 0.001$).

We divided peripheral nerve conduction characteristics in diabetic patients into three groups based on level of UAE: negative, microalbuminuria, and macroalbuminuria. Results are showed in Table 3a and Table 3b. Involved upper limb, we selected characteristics of median nerve,and involved lower limb we selected characteristics of peroneal nerve. This results showed prolongation in distal latency, decrease in potential amplitude, and reduction in conduction velocity were significantly different. This result was similar to the result in study of Charles M et al. performed in 456 type 1 diabetic patients [1].

Table 3a The motor NCS parameters in three groups UAE

UAE Level	Median		Peroneal	
	AP(mV)	NCV (m/s)	AP (mV))	NCV (m/s)
Negative	10.59 ± 2.57	53.55 ± 6.38	3.39 ± 1.66	41.34 ± 2.66
Microalbu minuria	8.91 ± 2.70	51.67 ± 6.00	3.78 ± 1.43	37.91 ± 4.06
Macroalbu minuria	8.20 ± 1.85	47.29 ± 4.50	2.37 ± 1.70	34.96 ± 5.72

Table 3b The sensory NCS parameters in three groups UAE

UAE Level	Median		Superficial Peroneal	
	DSL (ms)	NCV (m/s)	DSL (ms)	NCV (m/s)
Negative	3.35 ± 1.01	44.20 ± 12.25	2.40 ± 0.72	55.77 ± 12.4
Microalbu minuria	3.08 ± 0.77	48.03 ± 9.95	2.95 ± 0.88	46.30 ± 11.51
Macroalbu minuria	3.46 ± 0.76	42.04 ± 8.29	3.12 ± 0.83	42.23 ± 11.33

IV. DISCUSSION AND CONCLUSIONS

In thiscross-sectional studyperformed in 37 patients with diabetes, we identified a significantnegative correlation between level of UAE and AP,NCV of motor nerves. We also found asignificantnegative correlation between level of UAE and NCV, and a significantpositive correlation between level of UAE and DSL of sensory nerves.Our results are similar to published researches. Results in study of Dyck et al. [6], showed that the retinopathy severity level, and mean 24-h microalbuminuria provide approximately similar correlates of DPN, R = 0.43 and R = 0.48.Because they are all thought to be markers of microvascular disease. The concept that microvascular disease implicated in causing DPN is strengthened by the finding that all of these markers are correlated with severity of DPN. Hypoxia and ischemia may result from microvascular disease, or, alternatively, the damaged endoneurial microvessels may be leaky and allow plasma constituents to enter the endoneurial environment. There is evidence that entry of plasma constituents into the endoneurial microenvironment is harmful. Similarly, Valensi P et al. [13], identified that the presence of neuropathy also correlated with presence of retinopathy,

arterial hypertension, macroangiopathy, and biological signs of nephropathy.In addition, the study found that all the electrophysiological parameters were significantly more abnormal in patients with retinopathy or nephropathy than in patients without these complications. And least one abnormal electrophysiological parameter was almost always found in patients with retinopathyor nephropathy. Abnormalities were also found in patients without these complications, but with a lesser extent. This finding supports a theory that DPN,of which the earliest presence is the abnormal in NCSs, appears earlier than other complications. Results of the EURODIAB IDDM Complications Study [12] performed in 3,250 diabetic patients identifiedthat acorrelation between severity of DPN andnot only previously known factors, such as age, duration of diabetes, and retinopathy, butalso microalbuminuria.

In summary,from this cross-sectional study, we identified asignificantcorrelation between peripheral NCS parameters and level of UAE. The results of this study suggested that the presence of microvascular diabetic complications aresimultaneous, in which the earliest presence is the DPN, especially abnormal NCSparameters. The presence of DPN may be a recommendation for the endocrinologist and nephrologistinearly detection and treatment of DKD. Clinically, there is nothe correspondencebetween symptoms of DPN and NCS parameters. Even in patients without paresthesia, abnormal NCSparameters have been presented. This is the valuableperiod for assessing proteinuria. Thus, peripheral NCS should be aroutinely screened for microvascular complications in diabetic patients.

ACKNOWLEDGMENT

Acknowledgment and References headlines without numbering.

CONFLICT OF INTEREST

The authors declare that they have no conflict of interest.

REFERENCES

1. American Diabetes Association (2010). Standards of Medical Care in Diabetes.*Diabetes Care Journals*. Vol.33(1), 2010:11-61.
2. American Journal of Kidney Disease (2007). Screening and diagnosis of diabetic kidney disease.*Clinical practice guidelines and clinical practice recommendations for Diabetes and Chronic Kidney Disease*, pp.42-61.
3. Barbara (2005).*Electromyography and neuromuscular disorders: clinical-electrophysiologic correlations* 2nd Ed, pp.03-230.

4. Charles M, et al (2010). Low Peripheral Nerve Conduction Velocities and Amplitudes Are Strongly Related to Diabetic Microvascular Complications in Type 1 Diabetes: The EURODIAB Prospective Complications Study. *Diabetes Care*, Vol.33(12), 2010:2648-2653.

5. Diep Thi Thanh Binh, et al (2001). Screening of urine microalbumin by micral test on diabetes patients. Journal of Medicine Ho Chi Minh City, Vol.5 (4):44-47.

6. Dyck PJ, O'Brien PC, Litchy WJ, Harper CM, Klein CJ (2005). Monotonicity of nerve tests in diabetes: subclinical nerve dysfunction precedes diagnosis of polyneuropathy.*Diabetes Care*, Vol.28(9), 2005:2192-2200.

7. Dyck PJ, et al (1999). Risk factors for severity of diabetic polyneuropathy: intensive longitudinal assessment of the Rochester Diabetic Neuropathy Study cohort. *Diabetes Care*, 1999:1479-1486.

8. Imada M (2007). Median-radial sensory nerve comparative studies in the detection of median neuropathy at the wrist in diabetic patients.*Clinical Neurophysiology*, Vol.118, 2007:1405-1409.

9. Partanen J (1995). Natural history of peripheral neuropathy in patients with non-insulin-dependent diabetes mellitus.*New England Journal Medicine*, Vol.333(2), 1995: 89-94.

10. Rota E, et al (2005). Electrophysiological findings of peripheral neuropathy in newly diagnosed type II diabetes mellitus.*Journal of the Peripheral Nervous System*, Vol.10, 2005:348-353.

11. Tesfaye S, Boulton AJ, Dyck PJ, Freeman R, Horowitz M, Kempler P, Lauria G, Malik RA, Spallone V, Vinik AI, Bernardi L, Valensi P (2010). Diabetic Neuropathies: Update on Definitions, Diagnostic Criteria, Estimation of Severity, and Treatments. *Diabetes Care*, Vol.33(10), 2010:2285–2293.

12. Tesfaye S, et al (1996). Prevalence of diabetic peripheral neuropathy and its relation to glycemiccontrol and potential risk factors: the EURODIAB IDDM Complications Study, *Diabetologia*, Vol.39:1377–1384.

13. Valensi P, et al (1997). Diabetic peripheral neuropathy: effects of age, duration of diabetes, glycemic control, and vascular factors. *Journal of Diabetes Complications*, Vol.11:27-34.

14. Vinik AI (1999). Diabetic neuropathy: pathogenesis and therapy. *The American Journal of Medicine*, 1999:17-26.

15. Vinik AI (2011). Complications of diabetes mellitus.*Williams textbook of Endocrinology 12ᵗʰ Ed*, Section VIII, pp.1492-1501.

Author: Tran Vu Hoang Duong
Institute: Department of Physiology, University of Medicine and Pharmacyof Ho ChiMinhCity, Vietnam
Street: 217, Hong Bang
City: Ho Chi Minh City
Country: Vietnam
Email: tranvuhoangduong@gmail.com

New Practical Approachs to Estimation of Glomerular Filtration Rate in Adult: A Review

Le Quoc Tuan, Vo Thi Thien Huong, and Nguyen Thi Le

Department of Physiology, Ho Chi Minh City Medicine and Pharmacy University, Ho Chi Minh City, Vietnam

Abstract— In nephrology, renal function is the most important indication to decide the specific treatment for patients. Accurateestimation of glomerular filtration rate (GFR) is a important test to measure the level of kidney function, to determine the stage and to manage the chronic kidney disease (CKD).GFR describes the flow rate of filtered fluid through the glomerulus in kidney. Normally GFR of a person depends on his age, sex, race, and body size. There are many different methods to estimate GFR nowadays. Some techniques are simple and not reliable. Contrastingly, some techniques are very exact but it is too difficult to perform in clinical.Therefore, we must have general sight about all of formulas for calculating GFR and choose one suitably for each patient.

Keywords— Glomerular filtration rate (GFR), Chronic kidney disease (CKD), Clearance,Creatinine, Cystatin C.

I. INTRODUCTION

One of the important functions of kidneys is to remove metabolic products and to maintain the balance of the body fluid composition[1]. The kidney is constructed by many basic units called nephrons.Glomeruli are the tiny filters which filter waste from the blood. Each of nephron composes the glomeruli and the tubules. The functions of the nephron can be divided into three groups: filtration at glomerulus, reabsorption and secretion at tubules. Blood enters the kidney through the renal artery, which then narrows into afferent arterioles for glomerulus. The clearance (C) is a measurement of the renal excretion ability, result from three processes: filtration, reabsorption and secretion; while theglomerular filtration rate (GFR) only refers the filtration at glomerulus[2]. GFR is defined as the volume of plasma that can be completely filtered by all of glomerulus per a unit of time. GFR is a clinically important indicator of renal health. Nevertheless, there is no practical way to measure GFR directly, so it must be estimated. Indeed, GFR is measured as the renal clearance of a particular substance (X), which is called the glomerular marker. The glomerular marker is freely filtered by the glomerulus, and is neither reabsorbed nor secreted by the tubules.

$$\text{GFR} = C_x = \frac{U \times V}{P} = \frac{Urine Concentration \times Urine Flow}{Plasma\ Concentration} \quad (1)$$

(C_x is the renal clearance of a glomerular marker X.)

Chronic kidney disease (CKD) is defined as abnormalities of kidney structure or function (GFR < 60 mL/min/1.73 m^2), present for > 3 months[3]. The stages of CKD are mainly based on estimated GFR (eGFR). The KDOQI stages of CKD are showed in Table 1. The stages of CKD are a excellent tool for treatment planning. Indeed, the earlier kidney disease is detected, the better its progression is controlled.The estimated GFR are not completely precise as measured GFR, but usually they can provide the reliable information.

Inulin is once considered the gold standard of exogenously administered markers of GFR. However, the scarcity and high cost of inulin eliminated its routine use[2].

Radioisotope renography is a test of kidney imaging by using radioisotope. The two most common agents used are Tc99m-MAG3 (Mercaptoacetyltriglycine) and Tc99m-DTPA (DiethyleneTriaminePentacacetic Acid)[4]. This technique is widely applied before renal transplantation to assess the perfusion of the kidney to be transplanted, but it is not recommended to check routinely the kidney function, due to the high cost and the complex procedure[5].

Table 1 GFR categories in CKD[3]

Stage	GFR (mL/min/1.73 m^2)	Term	Treatment
1	≥ 90	Normal or high	Control of blood pressure
2	60 – 89	Mildly decreased	Control of blood pressure and risk factors
3A	45 – 59	Mildly to moderately decreased	Control of blood pressure and risk factors
3B	30 – 44	Moderately to severely decreased	
4	15 – 29	Severely decreased	Planning for end stage renal failure
5	< 15	End stage renal disease (ESRD)	Renal replacement therapy

© Springer International Publishing Switzerland 2015
V. Van Toi and T.H. Lien Phuong (eds.), *5th International Conference on Biomedical Engineering in Vietnam,*
IFMBE Proceedings 46, DOI: 10.1007/978-3-319-11776-8_66

In clinical, two most common endogenous substances used for estimation of GFR are creatinine and cystatin C. They are essential center of many different studies to find the best and the most useful method in calculating eGFR.

II. SERUM CREATININE AND ESTIMATED GLOMERULAR FILTRATION RATE

Creatinine is a metabolic product of creatine phosphate in muscle. Creatinine is removed from the blood primarily by the kidneys through glomerular filtration. Beside it is also secreted a little by the proximal tubule, but it is not completely reabsorbed in renal tubules. Therefore the creatinine clearance exceeds GFR. Ketoacids, cimetidine, and trimethoprime reduce creatinine tubular secretion, so they increase the accuracy of the estimated GFR. Men tend to have higher levels of serum creatinine than women due to the greater mass of skeletal muscle.

One method of determining GFR from creatinine is using 24-hour urine collection. Creatinine clearance is calculated from the creatinine level in the 24-hour urine sample (U_{Cr}), urine flow rate (V), and the serum level (P_{Cr}) as in equation (1). In this case, the creatinine clearance is equal with GFR,

with variation in dietary intake (as vegetarian diet) or muscle mass (as amputation, pregnancy, malnutrition).

The blood creatinine level rises only with marked damage of kidneys. Therefore, the serum creatininealone cannot detect mild to moderate kidney injury, while using 24-hour urine is a complex method for patients, due to difficulty in assuring complete specimen collection. A better evaluation of renal function from the formulas for estimating glomerular filtration rate (eGFR) without a 24-hour urine collection (Table 2) is excellent way in clinical. We can easily know the creatinine-based approximation of GFR only with the serum creatinine level and some simple variables such as age, weight, sex.

A commonly equation for estimating GFR is the Cockcroft-Gault formula. It uses the serum creatinine level, the age and the weight of patient to predict the creatinine clearance [11]. The equation shows that the creatinine clearance is dependent on patient's age. This is an interesting feature compared with some other formulas.

An eGFR calculated from serum creatinine by using the Modification of Diet in Renal Disease (MDRD) Study equation is a simple and effective way for laboratories [15]. The MDRD equation is good screening test for CKD among

Table 2 Formulas for Estimating Glomerlar Filtration Rate with Serum Creatinine Level

Formula	Units	Reference
(100/Cr) − 12, if male	mL/min/1.73 m^2	Jelliffe[7, 8]
(80/Cr) − 7, if female		
W x (29.3 − 0.203 x Age) / (Cr x 14.4), if male	mL/min	Kampmann et al.[9],
W x (25.3 − 0.175 x Age) / (Cr x 14.4), if female		Mawer et al[10].
{98 − [16 x (Age -20) / 20]}/ Cr, multiply by 0.9 if female	mL/min/1.73 m^2	Jelliffe[6]
[(140 − Age) x W] / (72 x Cr), multiply by 0.85 if female	mL/mim	Cockcroft Gault[11]
[(145 − Age) / Cr] − 3, multiply by 0.85 if female	mL/min/70kg	Hull et al.[12]
[27 − (0.173 x Age)] / Cr, if male	mL/min	Bjornsson et al.[13]
[27 − (0.175 x Age)] / Cr, if female		
[7.58 / (Cr x 0.0884)] − (0.103 x Age) + (0.096 x W) − 6.66, if male	mL/min	Walser, Drew, and Guldan[14]
[6.05 / (Cr x 0.0884)] − (0.080 x Age) + (0.080 x W) − 4.81, if female		
170 x Cr$^{-0.999}$ x Age$^{-0.176}$ x 0.762 (if female) x 1.180 (if black) x SUN$^{-0.170}$ x Alb$^{0.318}$	mL/min/1.73 m^2	Levey et al.[15]
175 x Cr$^{-1.154}$ x Age$^{-0.203}$ x 0.742 (if female) x 1.212 (if black)		(MDRD)
141 x min(Cr/κ, 1)$^\alpha$ x max (Cr/κ, 1)$^{-1.209}$ x 0.993^{Age} x 1.018 (if female) x 1.59 (if black) where κ is 0.7 for females and 0.9 for males, α is -0.329 for females and -0.411 for males, *min* indicates the minimum of Cr/κ or 1, and *max* indicates the maximum of Cr/κ or 1.	mL/min/1.73 m^2	Levey et al. [16] (CKD-EPI)

* The MDRD formula uses the creatinine value obtained by using the isotope dilution mass spectrometry-traceblecreatinine assay.Alb, serum albumin level (g/dL); Cr , serum creatinine level (mg/dL); SUN, serum urea nitrogen level (mg/dL); W, body weight (kg); MDRD, Modification of Diet in Renal Disease; CKD-EPI, Chronic Kidney Disease Epidemiology Collaboration.

and the tubular secretion seems to be so little to impact on GFR result[6]. The creatinine clearance should be corrected allowing the body surface area (mL/min/1.73 m^2). The corrected result is useful to compare creatinine clearance between people of different size. This test is not widely performed in clinical, but it is useful choice for individuals

people with risk factors such as diabetes, hypertension, cardiovascular disease, or family history of kidney disease. The equation does not require weight variables because the results are reported normalized to 1.73 m^2 body surface areas. The comparison between the MDRD equation and other methods such as Cockcroft-Gault, creatinine clearance from

24-hour urine collection has showed the superiority of MDRD equation to estimate GFR [16, 17]. Although an useful way to assess renal function, eGFR calculated from MDRD equation may not be suitable for all patients. This formula should not be applied for all individuals under 18 years old. In this case, the Schwartz equation should be used to estimate GFR [18]. Additionally, the MDRD equation is not recommended for persons with variation in muscle mass and diet. Creatinine clearance is the best choice for these patients.

The Chronic Kidney Disease Epidemiology Collaboration (CKD-EPI) is a new equation for estimationg GFR by serum creatinine[16]. This equation is based on the same four variables as the MDRD equation (serum creatinine, age, sex, and race). But the equation was reported to perform better and with less bias than the MDRD equation, especially in patients with higher GFR. The eGFR values > 60 mL/min/1.73 m^2 should consider using the CKD-EPI equation[19]. However, there are also the large differences between the measured and estimated GFR in some people because of physiologic limitations of creatinine as a filtration marker [17, 20].

The problem for prevalence estimation of the various eGFRs in old age is researched. Cockcroft-Gault equation gives lower eGFR at age 70 years compared to MDRD and CKD-EPI equations, and the chronic kidney disease at age 85 years is also highest for Cockcroft-Gault (90%), lower for MDRD (55%) and CKD-EPI (68%) [19].

III. SERUM CYSTATIN C AND ESTIMATED GLOMERULAR FILTRATION RATE

The use of cystatin C as a glomerular marker for estimation of GFR was first suggested in1985 by Simonsen and colleagues [21]. Cystatin C is a low molecular weight protein

epithelial cells, with only small amounts excreted in the urine. When the kidney function decline, the plasma cystatin C levels also rise as serum creatinin. The findings from cross-sectional and longitudinal studies show a more promising result than serum creatinine[22, 23]. This is easy to understand because cystatin C concentrations are less dependent on age, sex, race, and muscle mass compared to creatinine[24, 25]. Nevertheless, there are a few researches suggesting no significant difference between serum creactinine and serum cytatin C for calculating GFR[26].In 2002, a meta-analysis using currently available data confirmed that serumcystatin C is clearly better than serum creatinine as a glomerular marker by correlation or mean ROC-plot AUC [27].

Combining serum creatinine with cystatin C in a GFR estimating equation may be a more precise technique than creatinine-based methods in both adult and pediatric patients with CKD (CKD-EPI serum cystatin C and creatinine equation, Table 3) [28]. A meta-analysis of 11 general-population studies (90750 participants) and 5 cohort studies about CKD (2960 patients) showed that the use of cystatin C in combination with creatinine strengthened the relation between the eGFR and the risks of death and end-stage renal disease [29]. However, the CKD-EPI cystatin C equation does not seem to be more precise than the CKD-EPI creatinine equation in HIV-positive patients [30].

IV. CONCLUSIONS

The endogenous substance for estimation of GFR is currently serum creatinine. The Cockcroft-Gault and Modification of Diet Renal Disease (MDRD) Study equation are two serum creatinine-based equations that are widely applied in hospitals and in laboratories to estimate

Table 3 Formulas for Estimating Glomerlar Filtration Rate with Serum Cystatin C Level

Formula	Unit	Reference
$70.69 \times (CysC)^{-0.931}$	mL/min	Schwartz et al. [7, 31]
$133 \times \min(CysC/0.8, 1)^{-0.499} \times \max(CysC/0.8, 1)^{-1.328} \times 0.996^{Age} \times 0.932$ (if female) where *min* indicates the minimum of CysC/0.8 or 1, and *max* indicates the maximum of CysC/0.8 or 1.	mL/min/1.73 m^2	Levey et al.[7, 28] (CKD-EPI)
$135 \times \min(Cr/\kappa, 1)^{\alpha} \times \max(Cr/\kappa, 1)^{-0.601} \times \min(CysC/0.8, 1)^{-0.375} \times \max(CysC/0.8, 1)^{-0.711} \times 0.995^{Age} \times 0.969$ (if female) x 1.08 (if black) where κ is 0.7 for females and 0.9 for males, α is -0.248 for females and -0.207 for males, min(Cr/κ, 1) indicates the minimum of Cr/κor 1, and max(Cr/κ, 1) indicates the maximum of Cr/κ or 1; min(CysC/0.8, 1) indicates the minimum of CysC/0.8 or 1 and max(CysC/0.8, 1) indicates the maximum of CysC/0.8 or 1.	mL/min/1.73 m^2	Inker et al.[7, 28] (CKD-EPI)

* *Cr,serum creatinine level (mg/dl); CysC, serum cystatin C level (mg/l); MDRD, Modification of Diet in Renal Disease; CKD-EPI, Chronic Kidney Disease Epidemiology Collaboration.*

(approximately 13.3 kilodaltons). In human, all cells of body seem to produce cystatin C. After filtration at the glomerulus, cystatin C is reabsorbed and catabolized by the tubular

GFR. The GFR determinations by using serum creatinine are not completely exact, due to the tubular secretion of itself creatinine. Therefore, other formulas to assess the renal function, such as MDRD and CKD-EPI equation, or othersubstances, such as cystatin C, were studied and used to improve the result of eGFR, because they are less influenced by muscle mass and diet.

REFERENCES

1. Goldman, R., *The clinical evaluation of renal function.*California Medicine, 1956.85(6): p. 376-380.
2. Maarten W. Taal, G.M.C., Philip A. Marsden, Karl Skorecki, Alan S.L. Yu, Barry M. Brenner, *Brenner and Rector's The Kidney 9th.* 2012. 1 p. 868-897.
3. *Kidney international supplements: KDIGO 2012 Clinical Practice Guideline for the Evaluation and Management of Chronic Kidney Disease.*The International Society of Nephrology.3(1).
4. Taylor A, E.D., Alazraki N, *99mTc-MAG3, a new renal imaging agent: preliminary results in patients.*Eur J Nucl med, 1987.12(10): p. 510-514.
5. Al-Nahhas AA, J.R., Britton KE, et al., *Clinical experience with 99mTc-MAG3, mercaptoacetyltriglycine, and a comparison with 99mTc-DTPA.*Eur J Nucl med, 1988.14(9-10): p. 453-462.
6. Jelliffe, *Creatinine clearance: bedside estimate.*Ann Intern Med, 1973.79: p. 604-605.
7. *K/DOQI clinical practice guidelines for chronic kidney disease: evaluation, classification and stratification.*National Kidney Foundation Kidney, 2002.39(1): p. 1-266.
8. Jelliffe R., J.S.M., *Estimation of creatinine clearance from changing serum creatinine levels.*Lancet, 1971.2: p. 710.
9. Kampmann J., S.K., Kristensen M, et al., *Rapid evaluation of creatinine clearance.*Acta Med Scand, 1974.196: p. 517-520.
10. Mawer G, K.B., Lucas SB, et al., *Computer-assisted dosing of kanamycin for patients with renal insufficiency.*Lancet, 1972.1: p. 12-14.
11. Cockcroft D, G.M., *Prediction of creatinine clearance from serum creatinine.*Nephron, 1976.16: p. 31-41.
12. Hull J, H.L., Koch GG, et al., *Influence of range of renal function and liver disease on predictability of creatinine clearance.*ClinPharmacolTher, 1981. 29: p. 516-521.
13. Bjornsson T, C.D., McGowan FX, et al., *Nomogram for estimating creatinine clearance.*ClinPharmacokinet, 1983. 8: p. 365-369.
14. Walser M, D.H., Guldan JL., *Prediction of glomerular filtration rate from serum creatinine concentration in advanced chronic renal failure.*Kidney Int, 1993.44: p. 1145-1148.
15. Levey A, B.J., Lewis JB, et al., *A more accurate method to estimate glomerular filtration rate from serum creatinine: a new prediction equation. Modification of Diet in Renal Disease Study Group.*Ann Intern Med., 1999.130: p. 461-470.
16. Levey AS, S.L., Schmid CH, et al., *A new equation to estimate glomerular filtration rate.*Ann Intern Med, 2009.150: p. 604-612.
17. Shemesh O, G.H., Kriss JP, Myers BD, *Limitations of creatinine as a filtration marker in glomerulopathic patients.*Kidney Int, 1985.28(5): p. 830-838.
18. George J. Schwartz, A.M., Susan L. Furth, *New equations to estimate GFR in children with CKD.* J Am SocNephrol, 2009. 20(3): p. 629-637.
19. Jorien M Willems, T.V., Wendy PJ den Elzen, Rudi GJ Westendorp, Ton J Rabelink, Anton JM de Craen, Gerard J Blauw, *Performance of Cockcroft-Gault, MDRD, and CKD-EPI in estimating prevalence of renal function and predicting survival in the oldest old.*BMC Geriatrics, 2013.13(113): p. 1471-2318.
20. Rule AD, B.K., Schwartz GL, Khosla S, Lieske JC, Melton LJ, *For estimating creatinine clearance measuring muscle mass gives better results than those based on demographics.*Kidney Int, 2009.75(10): p. 1071-1078.
21. Simonsen O, G.A., Thysell H, *The blood serum concentration of cystatin C (gamma-trace) as a measure of the glomerular filtration rate.*Scand J Clin Lab Invest, 1985. 45(97-101).
22. PiereDelanaye, E.C., Olivier Moranne, Laurence Lutteri, Jean-Marie Krzesinski, Olivier Bruyere, *Creatinine-or cystatin C-based equations to estimate glomerular filtration in the general population: impact on the epidemiology of chronic kidney disease.*BMC Nephrology, 2013.14(57): p. 1471-2369.
23. Hojs R, B.S., Ekart R, Gorenjak M, Puklavec L, *Serum cystatin C-based equation compared to serum creatinine-based equations for estimation of glomerular filtration rate in patients with chronic kidney disease.*ClinNephrol, 2008. 70(1): p. 10-17.
24. Premaratne E, M.R., Finch S, Panagiotopoulos S, Ekinci E, Jerums G, *Serial measurements of cystatin C are more accurate than creatinine-based methods in detecting declining renal function in type 1 diabetes.* J Am SocNephrol, 2008. 16(5): p. 1404-1412.
25. Stevens LA, C.J., Schmid CH et al, *Estimating GFR using serum cystatin C alone and in combination with serum creatinine: a pooled analysis of 3418 individuals with CKD.*Am J Kidney Dis, 2008. 51(3): p. 395-406.
26. Rule, A.D., *The CKD-EPI Equation for Estimating GFR from serum creatinine: Real Improvement or More of the same?*Clin J Am SocNephrol, 2010. 5(951-953).
27. Dharnidharka VR, K.C., Stevens G, *Serum cystatin C is superior to serum creatinine as a marker of kidney function: a meta-analysis.*Am J Kidney Dis, 2002. 40(2): p. 221-226.
28. Inker LA, S.C., Tighiouart H, Eckfeldt JH, Feldman HI, Greene T, Kusek JW, Manzi J, Van Lente F, Zhang YL, Coresh J, Levey AS, *Estimating glomerular filtration rate from serum creatinine and cystatin C.* N Eng J Med, 2012. 367(1): p. 20-29.
29. Michael G. Shlipak, K.M., Johan Arnlov, Lesley A. Inker, Ronit Katz, Kevan R. Polkinghorne, Dietrich Rothenbacher, Mark J. Sarnak, Brad C. Astor, Josef Coresh, Andrew S. Levey, Ron T. Gansevoort, *Cystatin C versus Creatinine in Determining Risk Based on Kidney Function.* N Eng J Med, 2013. 369(932-943).
30. Inker LA, W.C., Creamer R, Hellinger J, Hotta M, Leppo M, Levey AS, Okparavero A, Graham H, Savage K, Schmid CH, Tighiouart H, Wallach F, Krishnasami Z, *Performance of creatinine and cystatin C GFR estimating equations in an HIV-positive population on antiretrovirals.*J Acquir Immune DeficSyndr, 2012.61(3): p. 302-309.
31. George J. Schwartz, M.F.S., Alvaro Munoz, *Improved equations estimating GFR i children with chronic kidney disease using an immunoephelometric determination of cystatin C.* Kidney Int, 2012. 82(4): p. 445-453.

Author: Le Quoc Tuan
Institute: Ho Chi Minh City Medicine and Pharmacy University
Street: Hong Bang
City: Ho Chi Minh City
Country: Vietnam
Email: tuan_lqc@yahoo.com

Testosterone – The Vital Hormone of Men: A Review

Tran Ngoc Thanh and Pham Dinh Luu

Department of Physiology, Pathophysiology and Immunology – Pham Ngoc Thach University of Medicine,
Ho Chi Minh City, Vietnam

Abstract— **Nowsaday, the quality of life has been an important focus for health care activities. It consists of many areas such as diet, mental support … Sexual life also takes a main role in quality of male life and it is related to testosterone, a vital hormone of men. If we recognize the problems of the mechanism related to testosterone, we can fix them appropriately in time. The living quality, therefore, will become better.**

Keywords— **Testosterone, hormone of men, primary gender, secondary gender.**

I. INTRODUCTION

Nobody wonders why there is difference between male and female organism. The reason is that it is usually thought to be natural and not worth asking. Actually, the answer is quite simple; it is due to testosterone hormone. It makes the appearance of primary gender and secondary gender in men and has some other functions.

II. PRIMARY GENDER

In reality, the primitive sexual gland, a special part of the foetus, begins to produce testosterone that determines sex, which helps the foetus develop its male gender when it is at week eight. It will form male sexual organ that consists of testicles, vas deferens, seminal vesicle, penis, scrotum and dependable glands [1]. All are called primary gender, the sexual organ has been formed in the foetus.

Testosterone also makes the two testicles in the abdomen move to scrotum which is outside in the last two months of the gestation [2]. This is quite important because the only place where testicles are able to produce sperms is the scrotum where the temperature is 2 degrees lower than the body's. Some people are born with testicles in the abdomen instead of being in the scrotum, which is called undescended testicle, will be sterile because of having no sperms [3,4].

Although there are enough parts of sexual organ that develop normally for a long time, it is surprising that the sexual organ does not work. Until 15 years later, the sexual organ starts to do its role after a long sleep: seminiferous tubules of the testicles are able to produce sperms and the interstitial cells (the Leydig cells) produce testosterone, which marks a period called puberty in the sexual life.

III. SECONDARY GENDER

During the puberty, testosterone begins to cause changes of the male body which is considered as the secondary gender, changes happen after the puberty. Let consider which changes the secondary gender cause [1].

First, the sexual organs such as testicles, vas deferens, seminal vesicle, penis and scrotum develop fast and eight times as big as the primary sexual organ.

Next, the body becomes big and high rapidly due to the development of the muscular and skeleton system. The amount of muscles increase makes the body brawny and youthful; seventeen is thought to be as strong as a horse. The weight of men is usually 50% heavier than of women, women synthesize mainly fat. "Doping in sports" refers to athletes who use testosterone products before the competitions in order to make the muscles strong and increase the strength of muscles abnormally to get high results. This is considered as a deceitful action, non-sportive, which is not allowed. Testosterone increases the rate of synthesizing protein, the main material need for muscles formation. For bones, testosterone makes them long, strong and forms a firm bony frame to support the whole body. There are two cartilage passages of the two long bone link the capitulum and the diaphysis together. Testosterone also functions in ossifying the linking cartilage, the transformation of cartilaginous cells into bony cells, which separates the capitulum away from the diaphysis. Thus, the bone becomes longer, then, the cartilage is re-produced and the ossification occurs again. However, testosterone has ossified the whole cartilage of the two head of the bone, which is called closing the two head of bone at the age of over 20. The length of the bone becomes stable; it means that you are no longer higher. Meanwhile, the height is an important factor. It is a disadvantage to own a modest height [2]. For this reason, the minimum compulsory height for the Miss World or Model contest is 1.70m for men and 1.65m for women. The height depends on three factors: genetic genes, nutritious regime, and playing sport habit. The factor of genetic genes is unchangeable while the rest can be intervened in order to have a better height. It means that the intervention of good eating habit and appropriate practice have to be applied in the puberty. When you are over 20, the whole linking cartilage has been ossified, which prevents your height from

© Springer International Publishing Switzerland 2015
V. Van Toi and T.H. Lien Phuong (eds.), *5th International Conference on Biomedical Engineering in Vietnam,*
IFMBE Proceedings 46, DOI: 10.1007/978-3-319-11776-8_67

being higher. Testosterone thickens the surface of the bone that leads to increase the basis substance (protein in nature) forming bone. Calcium and phosphate will deposit on the surface to make bone hard and increase the ability of supporting the gravity of the bone. Therefore, testosterone can be used in male osteoporosis treatment.

IV. ROLES OF TESTOSTERONE

Testosterone causes a sexual desire. The higher the testosterone level is, more and more the desire will be. Actually, the testosterone level in the blood is often at the average level, it cannot be higher because of negative feedback to the anterior lobe of the pituitary gland, which declines the excretion of hormone resulting in stimulating Leydig cells produce testosterone (hormone LH – Luteinizing Hormone). The sexual desire is extremely necessary to both human and animal. People will become indifferent without the sexual desire. As a result, it leads in the risk of extinction and the animal world then will be disappear gradually [1,2,5]

Testosterone also plays a cooperative role in producing sperms. It speeds up, especially, the pre-sperm period (spermatocytogenesis) to grown-up sperms one. Under the effect of testosterone, seminal vesicle will produce fructose to raise the sperms [3].

Having an effect on the mental and nerves, testosterone determines the active, dynamic and aggressive characteristics of men. An experiment has been conducted on a grown-up cock, which is very aggressive, chases the brood of hens and fights with other cocks all day. However, when it is castrated, to remove the two testicles, it becomes virtuous and meek. It focuses on finding food during the day. Thus, it becomes fatter; its muscles are soft, which tastes delicious when being eaten. In human life, it also happens. In the feudal society, the kings were often greedy, selfish and selected thousands of beauties. However, their abilities were limited, so, they required a person who could supervise and take care of their beauties motionlessly. Therefore, they chose strong young boys who had been castrated before their puberty. When these boys reached the adolescent age, they would become eunuchs who used to be higher and fatter than ordinary people. The two heads of their bone were unable to close due to lacking of testosterone. Therefore, the bone continued to lengthen for a short passage [3].

Testosterone also increases the blood level as well as enhances the creation of red cells. It also increase the reabsorption of sodium ion and water from renal tubules, which results in raising the blood level. On the other hand, it speeds up the creation of red cells in the marrow. The amount of red cells of men is usually as much as 700.000 cells/mm^3 of blood higher than of women.

In the secondary gender, testosterone remains some other effects [2,6] such as:

- From the puberty, the system of pubic hair, armpit hair and bear begin to appear. Pubis hair can spread up near to the navel while female pubic hair appears only around the pubic bone.

- Some of the old men can be bald because of testosterone. Some are due to genetic factors.

- On the skin: make the layer thicker and increase the hardness of the tissue under the skin. Thus, it increases the excretion of sebum of the sebaceous gland, which causes comedones. In women, the estrogen of the ovary does not cause comedones but testosterone of the adrenal cortex does, it is the reticularis.

- The voice becomes bass, called broken voice due to the development of the larynx and the prolongation of the vocal cord.

The testicles of men who are over 50 will become weak. The sperm production declines, testosterone production from the Leydig cells also decreases lead to low concentration of testosterone in plasma [7]. This marks another period in sexual life, a new crisis called "sexual decline stage". What testosterone has built in the puberty seems to ruin [3,8,9,10,11] :

- Decrease or lose the sexual desire
- Muscles become flabby because the process of synthesizing protein decreases
- Skin becomes winkle because of reducing in the muscular volume, the hardness of the fat layer.
- Easy to be osteomalacia
- Easy to be tired, the active and the dynamic are declined

To resolve this situation, supplement the testosterone from outside into our body if there is a lack of testosterone is the simplest way. Although it is certainly an effective way, it may include two backwashes:

- Testosterone is entered the body originating from the outside, so, it is easy to cause negative feedback with LH hormone of the anterior lobe of the pituitary gland, a kind of the hormone that regularly stimulates Leydig cells to product testosterone.

- Easy to cause many unexpected side-effects

It is better to use herbals that can stimulate the Leydig cells directly in producing testosterone, known as endogenous testosterone. For example, there is a kind of herbal

Whose scientific name is Euricoma, has good effects and easy to use. It also has no contraindication and side-effects. If we recognize the problems of the mechanism, we can fix them appropriately in time. The living quality, therefore, will become better.

V. CONCLUSION

Testosterone is an important hormone with male life. The change in its concentration may affect the quality of life and reproductive function. Its plasma level changes should be found in time to keep the quality of life.

CONFLICT OF INTEREST

The authors declare that they have no conflict of interest.

REFERENCES

1. Ganong William F. (2010). The Gonads: Development and function of the reproductive system. Review og Medical Physiology 23th ed., 25: 915-945
2. Guyton Arthur C. (2006). Reproductive and hormonal function of the male.. Textbook of Medical Physiology 11th ed, 80: 996-1010
3. Rodney A. Rhoades, Geogre A. Tanner (2005). The male reproductive system. Medical physiology 2nd ed., 37: 649-666.
4. Stephen J. McPhee, Vishwanath R. Lingappa, William F. Ganong (2006). Disorders of the male reproductive tract. Pathophysiology of disease, 23: 653-677
5. Shlomo Melmed, Kenneth S. Polonsky, Preed Larsen, Henry M. Kronenberg (2011). Principal of endocrinology. William textbook of endocrinology 12th ed. 1: 3-12
6. Mendez T. (2005). Gender dysphobia and gender change in androgen insensivity or micropenis. Arch Sex Behav. 34: 4111
7. Allan CA, Mc Lanchlan RI (2004). Age-relaed changes in testosterone and role of replacement therapy in older men. Clin Endocrinol. 60:653
8. Traish AM, Saad F, Guay A (2009). The dark side of testosterone deficiency: II. Type 2 diabetes and insulin resistance. J Audrol., 30(1): 23 – 32
9. Shlomo Melmed, Kenneth S. Polonsky, Preed Larsen, Henry M. Kronenberg (2011). Sexual dysfunction in man and woman. William textbook of endocrinology 12th ed. 20: 778-819
10. Heaton JPW, Morales A (2003). Endocrine causes of impotence (nondiabetes). Urol Clin North Am. 30: 73
11. Roger S. Kirby, Tom F. Lue (2005). An atlas of erectile dysfunction 2nd ed., 6-15

Author: Tran Ngoc Thanh
Institute: Pham Ngoc Thach University Of Medicine
Street: 86/2 Thanh Thai District 10
City: Ho Chi Minh
Country: Vietnam
Email: dr.tranngocthanh@pnt.edu.vn (+84 (0) 903668701)

Roles of Testosterone in Men with Type 2 Diabetes: A Review

Tran Ngoc Thanh and Pham Dinh Luu

Department of Physiology, Pathophysiology and Immunology – Pham Ngoc Thach University of Medicine, Vietnam

Abstract— **Androgen deficiency has recently come to the forefront of the medical literature after being ignored for decades. The prevalence of hypogonadism is greater than previously thought. Important associations are being developed and confirmed in the literature between androgen deficiency and metabolic disorders, specially in type 2 diabetes [1]. The prevalence of type 2 diabetes is increasing on over the world and directly effects the quality of sexual life of men [2]. However, in men with type 2 diabetes, besides the effect of erectile improvement, testosterone replacement therapy also reduces insulin resistance and improves glycaemic control in hypogonadal men with type 2 diabetes. Improvements in glycaemic control, insulin resistance, cholesterol and visceral adiposity together represent an overall reduction in cardiovascular risk. The objective of this paper is reviewing the roles of testosterone in men with type 2 diabetes.**

Keywords— **Testosterone, hormone of men, type 2 diabetes, erectile dysfunction.**

I. INTRODUCTION

The prevalence of type 2 diabetes is increasing on over the world and directly effects the quality of sexual life of men. Androgen deficiency is the main cause of this circumstance because it may leads to erectile dysfunction and decreased libido, specially more in men with type 2 diabetes [3]. If we recognize the problems of the mechanism, we can fix them appropriately in time and the patient will get more benefit from appropriate treatment. The living quality, therefore, will become better.

II. TESTOSTERONE

Testosterone is one of a hormone whose nature is lipid (steroid hormone). The main place where it is produced is the interstitial cells: Leydig cells. Besides, a small amount of testosterone is produced by the adrenal cortex and ovary (androgen is produced by the reticularis of the adrenal cortex). Testosterone is transported in the blood by combining with globulin: globulin binds to sex hormone (sex hormone binding globulin: SHBG) because the lipid hormone is insoluble in plasma. More than 70% of testosterone is combined with protein; the proportion of free hormone is insignificant [4].

The main functions of testosterone are making the primary and secondary sex of male formation; taking part in protein metabolism; sperm producing and cause the sexual desire of male. In the fetus, the Y chromosome triggers the sexual germ-cells to produce a little amount of testosterone, which encourages the fetus develop its male sex (the development of male sexual organ). This is called the primary sex of male. When reaching the puberty, testicles (Leydig cells) begin to produce testosterone that forms the secondary sex of male such as: the gait, muscular development, bear-raising, armpit hair, pubic hair, bass voice … Besides, testosterone takes a role in strengthen the process of synthesizing protein, which results in muscular development. It increases the transcription of gene, DNA molecule in the cellar nucleus to create message RNA (mRNA) and increases the translation of gene on endoplasmic reticulum containing ribosomes to synthesize the protein molecules. With doping in sport, the athletes have overused the testosterone products, which makes the muscles develop rapidly and the power of muscles become stronger in order to gain high results [5,6]. Moreover, testosterone is also important in reproductive fuction. It combines with FSH hormone (Follicle Stimulating Hormone) of the anterior lobe of pituitary gland to produce sperms and cause the sexual desire of male. This is extremely important because it is a motivation for species survival of both human and animal [7,8].

III. TYPE 2 DIABETES

There are 2 types of diabetes. Type 1 diabetes or insulin dependence diabetes refers to the failure of the pancreatic endocrine glands that is unable to produce insulin, which causes the deficiency of insulin. Type 2 diabetes or insulin independence diabetes refers to the pancreatic endocrine glands whose function has not failed. It is still able to produce insulin [9]. However, why does diabetes occur? It is due to the sensitive level of the cells to insulin. When the cells want to get insulin, the membranes have to contain receptors whose nature is protein. The cells once get the insulin; they have to increase the process of synthesizing protein or exactly, synthesizing receptors. It is a pity for these people because the ability of synthesizing protein is low. As a result, there are no receptors to get insulin although the pancreas is able to produce insulin. We often use the phrase insulin resistance or lower the sensibility to insulin. Diabetes not only causes disorders in glucid metabolism but also in lipid and protein metabolism. The lipid metabolism is more important [10].

© Springer International Publishing Switzerland 2015
V. Van Toi and T.H. Lien Phuong (eds.), *5th International Conference on Biomedical Engineering in Vietnam,*
IFMBE Proceedings 46, DOI: 10.1007/978-3-319-11776-8_68

In glucid metabolic disorder, Insulin has an effect in speeding up the transportation of glucose from plasma into the cells through the membranes. The deficiency in insulin makes the process of transporting glucose into the cells lower as much as 20 times. Therefore, glucose is unable to go into the cells; it will stay in the blood, which pushes the blood sugar up. 180mg/dL is the blood sugar level that is much higher than the renal tubular absorption threshold, which causes diabetes.

In lipid metabolic disorder [11,12], insulin has an effect in inhibiting the lipase enzyme in the adipose tissue. The deficiency of insulin, thus, releases the lipase enzyme. The enzyme, then, starts its function is to digest triglyceride to form fatty acid and glycerol. The lipase enzyme digests the stored fat of the body and releases the saturated fatty acid into the blood. The liver will keep the fatty acid and increase the process of synthesizing triglyceride and cholesterol. These substances are transported into the circulation under a very low density lipoprotein (VLDL) that will transform into the low density lipoprotein (LDL) in the blood. In conclusion, the lipid blood will increase in diabetes. There is a significant increase in plasmatic level of triglyceride (fat), total cholesterol, low density lipoprotein (LDL); except for the high density lipoprotein (HDL), it decreases. LDL is the substance causing the atherosclerosis while HDL is the substance against forming the atherosclerosis. LDL brings cholesterol into the artery walls, which create many cholesterol crystal plaques on the inner walls. The fibroblasts will cover the cholesterol crystal and calcium precipitates to the cholesterol plagues to harden the walls. These factors cause many serious consequences. Firstly, the artery walls become thick as a result of some reactions: smooth muscles layer become aneurismal, the cholesterol plagues, the invading of the fibroblasts and the precipitated calcium. Then, the walls harden and loose theirs elasticity due to the fibroblasts and calcium. These cause high blood pressure. Secondly, the inner walls are narrow, which reduces the pump of the blood to the organs because of thicker arterial walls and the cholesterol plagues protruding inner arteries. Thirdly, The blood is easy to be clotted in the inner wall to form many thrombi, which cause artery obstruction throughout the body. Especially, when it happens in the heart and the brain, it causes cardiac infarction and stroke. The blood is clotted easily in the inner due to the cholesterol plagues protruding inner that help the thrombocytes stick on to trigger the process of clotting blood. Finally, the blood vessels are degenerated easily because of the sclerosis of the arterial walls, which press the blood vessels feeding the arterial wall cells. These cells that are cut off the feeding, will be degenerated and die. This easily lead to arterial aneurysm and fragile arteries. It is sometimes extremely dangerous because of breaking big blood vessels such as:

abdominal aorta. The stroke is due to breaking the cerebral blood vesels results in cerebral haemorrage. There are two reasons: high blood pressure and the arterial walls are sclerosing and degenerated. All of the above changes of the blood vessels are called atherosclerosis.

In protid metabolic disorder, insulin induces protein synthesis and inhibits protein metabolism. It increase protein synthesis by three methods. The first one is increase of amino acids transported from interstitial fluid into the cells, amino acids are ingredient for protein synthesis. In the absence of insulin, the level of amino acids transported decrease by twenty times. Without amino acids, the cells stop producing protein. The second one is increase of gene replication in the nucleus to reproduce mRNA and the last one is increase of transcription in the endoplasmic reticulum to synthesize protein. Besides, insulin also inhibits protein metabolism by inducing glucose metabolism to acetyl-CoA. Acetyl-CoA inhibits protein metabolism in the negative feedback mechanism. In the absence of insulin, glucose is not metabolized, therefore there is no acetyl-CoA to inhibit protein metabolism. In the other hand, in the lack of glucose, the cells have to metabolize protein to produce energy in the form of ATP (Adenosine Triphosphate) [13]. Therefore, without insulin, the body will decrease protein synthesis and increase protein metabolism.

Diabetes causes many complications overall the body [2,10,11,14]. They are atherosclerosis (lipid metabolism disorder increase blood LDL); cardiovascular complications (hypertension, stroke, myocardial infarction, cerebral infarction, cerebral hemorrhage); nervous system damage (both central and autonomic nervous systems are affected due to metabolism disorders), renal failure and proteinuria; severe infections caused by decreased protein synthesis and increase protein metabolism weaken the immune system and macrophage which require protein to function; limb necrosis (peripheral vessels occlusion due to thrombi); eye injury (retinal hemorrhages, or retinal vessels occlusion leading to injury to receptor cells: rods and cones, and vision loss subsequently); and erectile dysfunction, decreased libido.

IV. ERECTILE DYSFUNCTION AND DECREASED LIBIDO IN DIABETES – ROLES OF TESTOSTERONE THERAPY

Mechanism of erection is penis need a full erection to penetrate the vagina [16]. Erection is the process in which the arterioles dilate causing the erectile tissues (corpora cavernosa and corpus spongiosum) filled with blood. The veins of the corpora cavernosa are compressed restricting the egress and circulation of this blood, causing the penis erecting and firm (four times than normal). Erection regulation center is located in the spinal cord and is triggered by the stimulation from the genital organ. Erection is also caused by sexual desire stimulation from the cerebral cortex. The outlet of the signal is the pelvic nerve branches. The sympathetic stimuli constrict the arterioles, restricting blood to the erectile tissues, erection then subsides. Substances that block sympathetic receptor, like yohimbin, may cause erection. In normal person, erectile dysfunction can be affected by various factors: age, depression, exhaustion, anxiety,... In people with diabetes, beside those factors above, the main cause of erectile dysfunction [3,17,19] is the complication of diabetes, namely 4 factors below:

a – Male genital organ deficiency, where testosterone is synthesized by Leydig cells. There are two reasons of this deficiency: (1) Decreased protein synthesis cannot compensate for the normal protein loss. (2) Atherosclerosis of the supply vessels reduce blood to the organ. The male genital organ deficiency cause decreasing in testosterone production, therefore decrease sperm production.

b – Atherosclerosis narrows the lumen of arteries supply to erectile tissues, blood is not pumped enough to cause erection.

c – The decrease of testosterone level in plasma leads to libido, affects severely the quality of sexual life. They are parallel and have a closed relationship. Infact, plasma level of testosterone has never higher than physiology normal level because of negative feedback mechanism. This means whener testosterone level over the physiological level, the anterior pituitary gland will be inhibited and the secrection of LH hormone (Luteinizing Hormone, stimulating testicles secrete testosterone) from it is delayed [20].

d – The transmitted way for sexual signals from brain to sexual organ is damaged because of atherosclerosis caused by diabetes. The rate of sexual transmitance is slower and slower and finally there is no signal transmitance. There is approximately 60% of diabetes patient faced to this complication. So, the more early diabetes is diagnosed and treated, the more effective of treatment and the better of life quality are. Erectile dysfunction often appears 3 years after the patient has diabetes [1].

In men with type 2 diabetes, besides the effect of erectile improvement, testosterone replacement therapy also reduces insulin resistance and improves glycaemic control in hypogonadal men with type 2 diabetes. Improvements in glycaemic control, insulin resistance, cholesterol and visceral adiposity together represent an overall reduction in cardiovascular risk [11,22,23,24].

Testosterone is a hormone increasing the synthesis of the cell. That includes insulin receptor protein, which is a membrane receptor protein. The cause of type 2 diabetes mellitus is lacking of protein receiving insulin, not lacking of insulin. Therefore testosterone increases receptor protein, the cell is more sensitive and reduce in resistance with insulin. That mean insulin is put into the cell easily, then it will be involved in the metabolism of: protein, lipid, glucide. Then symptoms and complications of diabetes will not appear. Moreover, when insulin was present in the cell, it is also a hormone increasing the synthesis protein strongly. The effect increasing synthesis protein is double, so the effectiveness of treatment is also more powerful [3,25]. Expect from the treatment of diabetes, testosterone has two another important effect: increased libido, a factor that improve the quality of life and increased sperm production, the main role in the reproductive. Expect from directly putting testosterone into the body, there is another method that is stimulation production of testosterone male genitalia organ by herbal medicine which seems to stimulate Leydig cells to produce testosterone naturally, no need to bring from the outside, no side effects, easy to use and no contraindications. But there is still not much evidence about this.

V. Conclusion

World Health Organization (WHO) has considered diabetes such as a epidemic disease in the 21st century in the world [2]. This disease has increased rapidly, exceeding over more than the predicted rate of the health sector, causing high costs for patients in treatment, especially dangerous complications threatening their lives. The principles of diabetes treatment are early and comprehensive treatment: appropriate medication to control blood sugar levels strictly. Keep the HbA1C level below 6.5%; testosterone replacement therapy; adhere to the diet strictly; sports, exercise regularly; no alcohol and tobacco; keep healthy, comfortable and happy life.

Conflict of Interest

The authors declare that they have no conflict of interest.

References

1. M. Traish, Farid Saad, Andre Guay (2009). The Dark Side of Testosterone Deficiency: II. Type 2 Diabetes and Insulin Resistance. *Journal of Andrology*. 30(1): 23-32.
2. WHO (2003). Screening for type 2 diabetes. Report of World Health Organization and International Diabestes Federation meeting. 2-30

3. Seftel AD, Mohammed MA, Althof SE (2004). Erectile dysfunction: etiology, evaluation and treatment options. Med Clin North Am. 88: 387.

4. Shlomo Melmed, Kenneth S. Polonsky, Preed Larsen, Henry M. Kronenberg (2011). Principal of endocrinology. William textbook of endocrinology 12th ed. 1: 3-12.

5. Guyton Arthur C. (2006). Reproductive and hormonal function of the male.. Textbook of Medical Physiology 11th ed, 80: 996-1010.

6. Ganong William F. (2010). The Gonads: Development and function of the reproductive system. Review og Medical Physiology 23th ed., 25: 915-945.

7. Rodney A. Rhoades, Geogre A. Tanner (2005). The male reproductive system. Medical physiology 2nd ed., 37: 649-666.

8. Shlomo Melmed, Kenneth S. Polonsky, Preed Larsen, Henry M. Kronenberg (2011). Sexual dysfunction in man and woman. William textbook of endocrinology 12th ed. 20: 778-819.

9. Shlomo Melmed, Kenneth S. Polonsky, Preed Larsen, Henry M. Kronenberg (2011). Type 2 diabetes mellitus. William textbook of endocrinology 12th ed. 31: 1371-1435.

10. Jones TH (2007). Testosterone associations with erectile dysfunction diabetes and metabolic syndrome. Eur. Urol. Suppl., 6: 847 – 57.

11. David Preiss, Naveed Sattar (2009). Lipids, lipid modifying agents and cardiovascular risk : a review of the evidence. Clinical Endocrinology. 70: 815-828

12. Shlomo Melmed, Kenneth S. Polonsky, Preed Larsen, Henry M. Kronenberg (2011). Disorder of lipid metabolism. William textbook of endocrinology 12th ed. 37: 1633-1674.

13. Laurie Kelly (2005). The endocrine system. Essentials of human physiology for pharmacy, 10: 111-139.

14. Shlomo Melmed, Kenneth S. Polonsky, Preed Larsen, Henry M. Kronenberg (2011). Complications of diabetes. William textbook of endocrinology 12th ed. 33: 1462-1551.

15. Roger S. Kirby, Tom F. Lue (2005). An atlas of erectile dysfunction 2nd ed., 6-15.

16. Heaton JPW, Morales A (2003). Endocrine causes of impotence (nondiabetes). Urol Clin North Am. 30: 73.

17. 19. Bouvattier C, Mignot B, Lefèvre H, Morel Y, Bougnères P (2006). Impaired sexual activity in male adults with partial androgen insensitivity. J Clin Endocrinol Metab. 91:3310

18. 20. Traish AM, Saad F, Guay A (2009). The dark side of testosterone deficiency: II. Type 2 diabetes and insulin resistance. J Audrol., 30(1): 23 – 32

19. 22. Stanworth RD, Jones TH (2009). Testosterone in obesity, metabolic syndrome and type 2 diabetes. Front Horm Res., 37: 74 – 90

20. 23. D Kapoor, E Goodwin, K S Channer, T H Jones (2006). Testosterone replacement therapy improves insulin resistance, glycaemic control, visceral adiposity and hypercholesterolaemia in hypogonadal men with type 2 diabetes. Society of the European Journal of Endocrinology. http://www.eje-online.org/content/154/6/899.short

21. 24. Allan CA, Mc Lanchlan RI (2004). Age-relaed changes in testosterone and role of replacement therapy in older men. Clin Endocrinol. 60:653

22. 25. Saad F. Gooren L (2009). The role of testosterone in the metabolic syndrome: a review. J Steroid Biochem Mol Biol, 114: 1.2 - 40.3

Author: Tran Ngoc Thanh
Institute: Pham Ngoc Thach University Of Medicine
Street: 86/2 Thanh Thai District 10
City: Ho Chi Minh
Country: Vietnam
Email: dr.tranngocthanh@pnt.edu.vn (+84 (0) 903668701)

Investigating Chronic Obstructive Pulmonary Disease (COPD) in Vietnamese Patients Using Impulse Oscillometry (IOS)

T.X. Tan[1], V. Van Toi[1], Truong Quang Dang Khoa[2], H.D.H. Hanh[3], T.T.K.Thu[3], and L.T.T. Lan[3]

[1] Biomedical Engineering Department, International University, Ho Chi Minh City, Vietnam
[2] Tokyo University of Agriculture and Technology, Tokyo, Japan
[3] University Medical Center, Ho Chi Minh City, Vietnam

Abstract— Crucial measurement of Chronic Obstructive Pulmonary Disease (COPD) relies extensively on the use of spirometry, "gold standard" for diagnosis of COPD. Impulse oscillometry system (IOS) is a non-volitional way to access the mechanical structure of the respiratory system. The goals of our study were to find out the sensitivity and specificity of the IOS in diagnosis of COPD patients, and investigate which IOS parameters are related to severity and airflow obstruction in Vietnamese COPD patients. The study contain twenty-two COPD patients (stage 3 and 4) and Thirty-four healthy people, whole of them are greater than 40 years, were recruited in Community Health Care Center, Ho Chi Minh city, Vietnam. IOS measurements (R5, R20, X5, X20, AX, Fres and Delta R5-R20), and Spirometry (FEV1, FEV1/FVC) were performed. Pearson or Spearman correlation determined the relationships between IOS and Spirometry. Firstly, R5, X5, X20, AX, Fres & Delta R5-R20 were all significantly associated ($p < 0.05$) with FEV1. However, R20 were not related to FEV1. The strongest associations were observed between FEV1 and X5 ($r = 0.7737$), AX ($r = -0.7825$) and Delta R5-R20 ($r = -0.7823$). Secondly, X5, AX & Delta R5-R20 were significantly associated with FEV1/FVC. The strongest correlation was observed between Delta R5-R20 and FEV1/FVC with $r = -0.6903$. Reactance airway measurements (X5, X20, AX, Fres), peripheral airway resistance (Delta R5-R20) and total respiratory resistance (R5) are closely related to Vietnamese COPD diagnosis than central respiratory resistance (R20). The IOS measurements can be a significant value for COPD diagnosis in Vietnamese patients.

Keywords— COPD, IOS, Spirometry, CHAC.

I. INTRODUCTION

Chronic obstructive pulmonary disease (COPD) is defined by poorly reversible airflow limitation caused by the inhalation of noxious particles such as cigarette smoke [1]. COPD patients display heterogeneous pathophysiological abnormalities including small airway disease, hyperinflation and mucus hypersecretion, all of which may cause airflow obstruction [2].

Spirometry is the 'gold standard' by which airflow obstruction is assessed in COPD patients. Forced expiration is used as part of spirometry in the diagnosis and staging of COPD. This procedure can be difficult for patients to perform as it is effort dependent and can alter broncho-motor tone. Body plethysmography is an alternative pulmonary function technique, allowing assessment of airways resistance and conductance. However, it can be technically demanding for patients to perform as it requires complex 'panting' manoeuvres. Thus, there is a need for easy to perform but physiologically accurate methods to assess pulmonary mechanics in COPD patients.

The forced oscillation technique (FOT) was developed in 1956 to measure the lung impedance through application of small pressure oscillations at the mouth during normal breathing [3]. FOT systems use pseudorandom noise signals to enable the simultaneous measurement of respiratory resistance (Rrs) and reactance (Xrs). The impulse oscillometry system (IOS) [7] is a type of FOT but with 2 important differences; rectangular waveform impulses are applied instead of pseudorandom noise signals, and the IOS has a different set of data outputs. IOS shows airway resistance (R) and reactance (X) values at multiple frequencies of 5 Hz (5-20Hz).

This study describes the relationship of IOS measurements to other pulmonary function measurements in a group of Vietnamese COPD patients. There are three goals in this investigation: Multi-functional program, which was written in Matlab language, will be created to analyze IOS's data in Community Health Care Center; Investigate which IOS parameters (R5, X5, R20, X20, AX, Fres, Diff R5-R20) are related to airflow obstruction (FEV1, FEV1/FVC) in COPD patients; and Sensitivity and Specificity of the impulse oscillometry parameters in diagnosis of Vietnamese COPD.

II. METHODOLOGY

Twenty-two patients with COPD and thirty-four healthy people were recruited from Community Health Care Center, Ho Chi Minh city, Vietnam. All patients were greater than 40 years old. COPD was diagnosed according to current Global Initiative for Chronic Obstructive Lung Disease (GOLD) guidelines [1], based on a smoking history of at least 10 pack years together with typical symptoms (one or

© Springer International Publishing Switzerland 2015
V. Van Toi and T.H. Lien Phuong (eds.), *5th International Conference on Biomedical Engineering in Vietnam*,
IFMBE Proceedings 46, DOI: 10.1007/978-3-319-11776-8_69

more of productive cough, breathlessness and wheeze) and evidence of airflow obstruction. Patients with a clinical history of asthma, an exacerbation or a history of lung cancer were rejected. Written informed consent was performed.

Each patient performed pulmonary function tests in the following order is impulse oscillometry system (R5, R20, X5, X20, Fres, AX, Delta R5-R20), and spirometry (FEV1, FEV1/FVC).

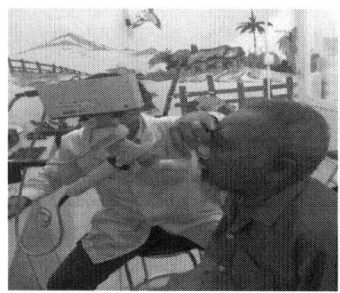 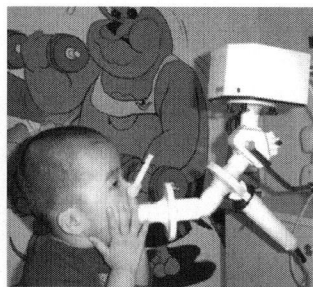

Fig. 1 IOS has been used to measure the lung function of patients

IOS measurements were performed, and the values of resistance at 5 and 20 Hz (R5 and R20, respectively), reactance at 5 and 20 Hz (X5 and X20) and resonant frequency (Fres) were recorded. The Pearson's test was analyzed the correlation between impulse oscillometry values and functional parameters. The healthy group data was used from GOLD guidelines; cut off points of impulse osillometry parameters were determined for calculating the sensitivity, specificity, negative and positive predictive values of IOS for COPD severity detection. Pearson correlation coefficient r was computed in equation (1) to express the level of association between two variables.

$$r = \frac{\sum_i (x_i - \bar{x})(y_i - \bar{y})}{\sqrt{\sum_i \sum_i (x_i - \bar{x})^2} \sqrt{\sum_i (y_i - \bar{y})^2}} \quad (1)$$

Where
r is linear coefficient
x_i and y_i are sample value
$\bar{x}$ and $\bar{y}$ are mean value
p-value was calculated by transforming the correlation to create a t-statistic having n-2 degrees of freedom. p-value < 0.05 was considered as statistically significant.

III. RESULTS

Baseline patient demographics and pulmonary function measurements at baseline are shown in below, respectively in terms of mean and standard deviation.

Table 1 Patients demographics and lung function parameters

Variable	Control	COPD
N	34	22
Age (years)	61.41 ± 11.40	64.14 ± 11.12
Female / Male	9/25	1/21
FEV1/FVC (%)	71.06 ± 17.56	45.75 ± 14.80
FEV1 (%)	82.59 ± 24.19	46.12 ± 18.99
R5 (KPa.s/L)	0.40 ± 0.10	0.51 ± 0.17
R20 (KPa.s/L)	0.31 ± 0.08	0.32 + 0.09
X5 (KPa.s/L)	-0.17± 0.11	-0.34 ± 0.23
X20 (KPa.s/L)	0.04 ± 0.06	-0.04 ± 0.08
AX	0.91 ± 1.01	2.65 ± 2.21
Fres (Hz)	15.67 ± 5.33	24.20 ± 8.03

A. Correlation between IOS and Spirometry

Firstly, R5, X5, X20, AX, Fres and Delta R5-R20 were all significantly associated (p < 0.05) with FEV1, with the resistance measurement R5 (r = -0.530), reactance measurements X5 (r = 0.774), X20 (r = 0.720), area of reactance AX (r = -0.699), resonant frequency Fres (r = -0.783) and peripheral resistance Delta R5-R20 (r = -0.782) showing the degree of correlation with the strongest relation of X5, AX and Delta R5-R20. R20 did not correlate with FEV1. Secondly, X5, Fres & Delta R5-R20 were related with *FEV1/FVC (p < 0.05)*. *Strongest association observed between Delta R5-R20 and FEV1/FVC with r = -0.690.*

Table 2 The correlation of IOS parameters with gold standard FEV1 (r: linear coefficient; NS: not statically significant)

Correlations	r
R5 vs FEV1	NS
X5 vs FEV1	0.675 (p<0.001)
X20 vs FEV1	0.418 (p<0.05)
AX vs FEV1	-0.606 (p<0.001)
Fres vs FEV1	-0.45 (p<0.05)
R5-R20 vs FEV1	-0.6259 (p<0.001)

B. Sensitivity and Specificity of IOS

Sensitivity and Specificity have done based on the diagnosis of Vietnamese Doctors. The GOLD standard and other functional pulmonary testing are used to diagnose the stages of COPD [1].

Table 3 The sensitivity and specificity of IOS and Spirometry (%FEV1) compare with Doctor Diagnosis

	Sensitivity	Specificity	PPV	NPV
IOS vs Doctor	64.30%	58.50%	54.60%	68.80%
Spiro vs Doctor	53.60%	73.00%	60.00%	67.50%

IV. DISCUSSION

In our initial assessment of 22 COPD patients, we observed total airway resistance (R5), reactance measurements (X5, X20, AX & Fres) and peripheral resistance (Delta R5-R20) to be significantly associated with measures of severity of airway obstruction (FEV1). However, R20 was not related to FEV1. The strongest associations were observed between X5, AX & Delta R5-R20 with FEV1. FEV1 is a well recognized measurement of COPD severity. Thus, the key novel finding of this study is that IOS reactance measurements (X5, X20, AX & Fres) and peripheral resistance (Delta R5-R20) are more informative than resistance measurements (R20) about the changes in pulmonary mechanics caused by airflow obstruction in Vietnamese COPD patients. R20 appears to be unrelated to the severity of airflow obstruction. Furthermore, peripheral resistance showed the strongest association with the airway obstruction (FEV1/FVC).

The strength of the correlations between FEV1 and reactance measurements (r = 0.774 for X5, r = -0.699 for AX and r = -0.783 for Fres), suggests a strong association between IOS airway reactance measurements and airflow obstruction. IOS is therefore not a replacement for FEV1, but as previously pointed out for classic FOT methods provides complimentary information on respiratory mechanics [4]-[6][23][24].

The limitation of this study is the number of subjects. Indeed, there were not many patients who could be examied in both IOS and spirometry in Community Health Care Center, Ho Chi Minh City, Vietnam. The size of two group is different which affected the computation of coefficients of correlation. Furthermore, the numbers of patietns in each stage of COPD severity differed. The average numbers should be dicussed well. The size of participants would be increased to confirm the important of IOS parameters in Vietnamese COPD patients.

Spirometry have been a difficult functional pulmonary testing for asthmatic children and COPD patients due to the lack of conveniences. The impulse oscillomerty system ,that patients just need express their normal breath during their tests, can be a crucial lung measurement in COPD patients in Vietnam.

V. CONCLUSIONS

In summary, the total airway resistance (R5), reactance measurements (X5, X20, AX & Fres) and peripheral resistance (Delta R5-R20) may be the IOS measurements more closely related with spirometry (FEV1) which can be seen as the gold standard for COPD diagnosis. These reactance measurements and Delta R5-R20 are strong associated to the degree of airflow obstruction, and confirm that they can be alternative parameters for FEV1. The ease of use of IOS and the high sensitivity, which is about 70% compare with doctor diagnosis, of this technique to obtain functional pulmonary change in COPD patients should be increased use in clinical diagnosis in Vietnam.

ACKNOWLEDGMENT

Firstly, author would like to express his gratitude to A/Prof. MD. Le Thi Tuyet Lan, Prof. PhD. Vo Van Toi, A/Prof. Truong Quang Dang Khoa for their exemplary guidance, encouragement and supports that brought him to these achievements.

Another special thank goes to MSc. Ly Ha Duy, BSc. To Nguyen Nhat Minh and BSc. Do Nguyen Trung Dung. Their technical help made these years a great learning experience. Gratitude is also expressed to the members of my reading and examination committee.

The last but not least, author would like to express his deepest gratitude to all the members in his family without them everything will not be possible.

REFERENCES

1. Global initiative for chronic obstructive lung disease (GOLD).
2. Hogg JC. Pathophysiology of airflow limitation in chronic obstructive pulmonary disease. Lancet 2004; 364(9435): 709e21.
3. Dubois AB, Brody AW, Lewis DH, Burgess Jr BF. Oscillation mechanics of lungs and chest in man. J Appl Phys 1956;8(6): 587e94.
4. Junichi Ohishi, H. K., Hiromasa Ogawa, Toshiya Irokawa, WataruHida, Masahiro Kohzuki. Application of impulse oscillometry for within-breath analysis in patients with chronic obstructive pulmonary disease: pilot study. BMJ Open 2011; 1:e000184.
5. Umme Kolsum , Zoe¨ Borrill, Kay Roy , Cerys Starkey,Jørgen Vestbo, Catherine Houghton , Dave Singh. Impulse oscillometry in COPD:Identificationof measurements related to airway obstruction,airway conductance and lung volumes. Respiratory Medicine 2009; 103, 136e143
6. Di Mango AM, Lopes AJ, Jansen JM, Melo PL. Changes in respiratory mechanics with increasing degrees of airway obstruction in COPD: detection by forced oscillation tech- nique. Respir Med 2006;100(3):399e410.
7. Goldman MD. Clinical application of forced oscillation. Pulm Pharmacol Ther 2001;14(5):341e50.

8. Singh D, Tal-Singer R, Faiferman I, Lasenby S, Henderson A, Wessels D, et al. Plethysmography and impulse oscillometry assessment of tiotropium and ipratropium bromide; a randomized, double-blind, placebo-controlled, cross-over study in healthy subjects. Br J Clin Pharmacol 2006;61(4): 398e404.

9. Houghton CM, Langley SJ, Singh SD, Holden J, Monici Preti AP, Acerbi D, et al. Comparison of bronchoprotective and bronchodilator effects of a single dose of formoterol delivered by hydrofluoroalkane and chlorofluorocarbon aerosols and dry powder in a double blind, placebo-controlled, crossover study. Br J Clin Pharmacol 2004;58(4):359e66.

10. Houghton CM, Woodcock AA, Singh D. A comparison of lung function methods for assessing dose-response effects of sal- butamol. Br J Clin Pharmacol 2004;58(2):134e41.

11. Houghton CM, Woodcock AA, Singh D. A comparison of plethysmography, spirometry and oscillometry for assessing the pulmonary effects of inhaled ipratropium bromide in healthy subjects and patients with asthma. Br J Clin Pharmacol 2005; 59(2):152e9.

12. Borrill ZL, Houghton CM, Woodcock AA, Vestbo J, Singh D. Measuring bronchodilation in COPD clinical trials. Br J Clin Pharmacol 2005;59(4):379e84.

13. Park JW, Lee YW, Jung YH, Park SE, Hong CS. Impulse oscillometry for estimation of airway obstruction and bronchodila- tion in adults with mild obstructive asthma. Ann Allergy Asthma Immunol 2007;98(6):546e52.

14. Al-Mutairi SS, Sharma PN, Al-Alawi A, Al-Deen JS. Impulse oscillometry: an alternative modality to the conventional pulmonary function test to categorise obstructive pulmonary disorders. Clin Exp Med 2007;7(2):56e64.

15. Hellinckx J, Cauberghs M, De Boeck K, Demedts M. Evaluation of impulse oscillation system: comparison with forced oscilla- tion technique and body plethysmography. Eur Respir J 2001; 18(3):564e70.

16. MacLeod D, Birch M. Respiratory input impedance measure- ment: forced oscillation methods. Med Biol Eng Comput 2001; 39(5):505e16.

17. Landser FJ, Clement J, Van de Woestijne KP. Normal values of total respiratory resistance and reactance determined by forced oscillations: influence of smoking. Chest 1982;81(5): 586e91.

18. Clement J, Landser FJ, Van de Woestijne KP. Total resistance and reactance in patients with respiratory complaints with and without airways obstruction. Chest 1983;83(2):215e20.

19. Kjeldgaard JM, Hyde RW, Speers DM, Reichert WW. Frequency dependence of total respiratory resistance in early airway disease. Am Rev Respir Dis 1976;114(3): 501e8.

20. Dellaca RL, Santus P, Aliverti A, Stevenson N, Centanni S, Macklem PT, et al. Detection of expiratory flow limitation in COPD using the forced oscillation technique. Eur Respir J 2004; 23(2):232e40.

21. Pasker HG, Schepers R, Clement J, Van de Woestijne KP. Total respiratory impedance measured by means of the forced oscillation technique in subjects with and without respiratory complaints. Eur Respir J 1996;9(1):131e9.

22. Oostveen E, MacLeod D, Lorino H, Farre R, Hantos Z, Desager K, et al. The forced oscillation technique in clinical practice: methodology, recommendations and future developments. Eur Respir J 2003;22(6):1026e41.

23. Mehdi Nikkhah, Babak Amra, Afrooz Eshaghian, Shahriar Fardad, Assadolah Asadian, Tooraj Roshanzamir, Mojtaba Akbari, Mohammad Golshan. Comparison of Impulse Osillometry System and Spirometry for Diagnosis of Obstructive Lung Disorders. *Tanaffos* (2011) 10(1), 19-25.

24. Akane Haruna, Toru Oga, Shigeo Muro, Tadashi Ohara, Susumu Sato, Satoshi Marumo, Daisuke Kinose, Kunihiko Terada, Michiyoshi Nishioka, Emiko Ogawa, Yuma Hoshino, Toyohiro Hirai, Kazuo Chin, Michiaki Mishima. Relationship between peripheral airway function and patient-reported outcomes in COPD: a cross-sectional study. *BMC Pulmonary Medicine* 2010, 10:10

Author: Tran Xuan Tan
Institute: Community Health Care Center
Street: 10 Ly Thuong Kiet
District: 10
City: Ho Chi Minh
Country: Vietnam
Email: xuantan.tran@yahoo.com.vn
Phone: (+84) 937570979

Finite-Difference Time-Domain Simulations of Ultrasound Backscattered Waves in Cancellous Bone

A. Hosokawa

Department of Electrical and Computer Engineering, Akashi National College of Technology, Akashi, Japan

Abstract— **In this study, using an elastic finite-difference time-domain (FDTD) method with numerical cancellous bone models reconstructed from microcomputed tomographic (μCT) images of bovine bone, numerical simulations of the backscattered waves in cancellous bone was performed for clarifying the backscatter characteristics. In the simulation model, an ultrasound pulse wave was transmitted toward the front surface of cancellous bone from a circular concave transmitter/receiver in water. Then, two cancellous bone models with different thicknesses, in which an artificial absorbing boundary was set at the back surface opposite to the ultrasound transmission. By calculating the difference between the simulated results of the received signals in the cases of these cancellous bone models, the reflected wave from the front surface could be canceled, and only the backscattered waves inside the bone could be extracted. Using the cancellous bone models with porosities between 0.53 and 0.86 (53% and 86%), which had main trabecular orientation parallel to the thickness direction (or the transmitted ultrasound direction), the peak-to-peak amplitude of the backscattered waves were derived as a function of the porosity. The determination coefficient between the backscattered wave amplitude and the porosity was $R^2 = 0.49$ ($P < 0.001$). The correlation was moderate, which was considered to be because the backscattered waves could be affected by not only the porosity but also the trabecular microstructure.**

Keywords— **Cancellous bone, Ultrasound backscattered waves, Numerical simulation, Finite-difference time-domain method.**

I. INTRODUCTION

Various quantitative ultrasound (QUS) techniques have been developed for clinical assessment of bone quality [1]–[4]. Among these, ultrasound backscatter measurements in cancellous bone have been attempted [4]–[7] because they are easily applicable to skeletal sites, such as vertebra, where through-transmission measurements are difficult. However, the backscatter measurements are not yet established because the complicated backscatter mechanisms are not sufficiently clarified. Several theoretical models of ultrasound scattering have been proposed [8]–[13], but all of them are simplistic and have disadvantages [5], [6]. In particular, it is difficult to detect only the backscattered waves because their intensity is low relative to that of the reflected waves at the boundaries: boundaries between soft tissues (skin, muscle, and fat) and cortical bone and between cancellous and cortical bones in

in vivo measurements, and the boundary between cancellous bone and water in *in vitro* measurements. As the backscattered waves overlapped greatly with the reflected wave, the time at which the backscattered waves appeared could not be precisely defined. To the best of the author's knowledge, an effective method for separation of the backscattered waves has not yet been proposed.

Numerical approaches can be useful because they offer no random noise and perfect reproducibility. It has been demonstrated that finite-difference time-domain (FDTD) simulations with real bone models reconstructed from three-dimensional (3-D) microcomputed tomographic (μCT) images can regarded as a surrogate for experiments of ultrasound propagation through bone [14]–[24]. In a previous study using FDTD simulations [25], only the reflected wave at the boundary between cancellous and cortical bones could be extracted by using an artificial absorbing boundary condition. In this study, using a similar method, the backscattered waves in cancellous bone were separated from the reflected wave. For an ultrasound transmission parallel to the main orientation of the trabecular network, the backscattered waves from the inside of cancellous bone were detected to investigate the backscatter characteristics.

II. SIMULATION METHOD

A. Simulation Model

3-D FDTD simulations of ultrasound backscattering in cancellous bone were performed using self-made elastic FDTD software considering medium absorptions [21]. The simulation model was constructed with consideration of *in vitro* experiments, and its cross-sectional view is shown in Fig. 1. This model had a spatial interval of 57 μm and consisted of two regions of cancellous bone and water, whose dimensions were $d \times 6.84 \times 6.84$ mm^3 and $7.98 \times 6.84 \times 6.84$ mm^3, respectively. Here, d was the cancellous bone thickness, and two values of $d_1 = 1.482$ mm and $d_2 = 2.508$ mm were adopted for each cancellous bone model. In Fig. 1, the x-direction corresponded to the direction of ultrasound transmission, and the y–z plane corresponded to the cancellous bone surface. The transmitter/receiver had a circular concave shape with a diameter of 5.7 mm, which was set at the center of the y–z plane at a distance of 7.98 mm from the cancellous bone surface. The focal

© Springer International Publishing Switzerland 2015
V. Van Toi and T.H. Lien Phuong (eds.), *5th International Conference on Biomedical Engineering in Vietnam,*
IFMBE Proceedings 46, DOI: 10.1007/978-3-319-11776-8_70

distance was 10.488 mm, which corresponded to the focal point 2.508 mm inside of the cancellous bone model. The transmitted waveform was a single sinusoid at 1.0 MHz multiplied by a Hanning window, and its time interval was 4 ns. At all boundaries surrounding the simulation region, 20 perfectly matched absorbing layers (PMLs) [26] were set to prevent artificial reflections.

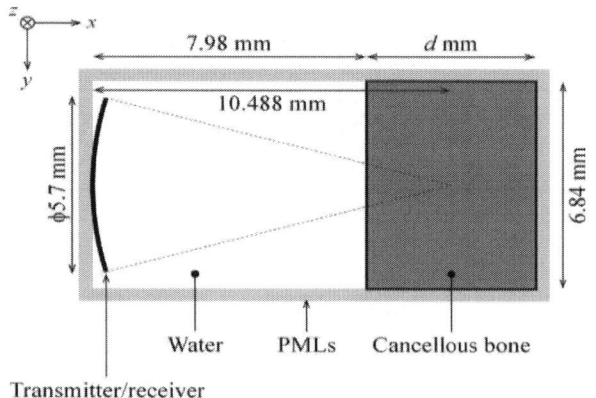

Fig. 1 Simulation model for ultrasound backscattering in cancellous bone

Eighteen 3-D cancellous bone models were reconstructed from X-ray μCT images of bovine bone, which were obtained using an X-ray μCT system (SMX-160CTS, Shimadzu Corp., Japan). The grayscale μCT images were binarized at the middle value between two peaks in a histogram calculated for all the images. The porosity was measured as the percentage of pore spaces. The porosities of the cancellous bone models used were between $\beta = 0.53$–0.86 (53%–86%). For all cancellous bone models, the thickness direction (or x-direction), which corresponded to the direction of ultrasound transmission, was set to be parallel to the main orientation of the trabecular network. The physical parameter values used in the FDTD simulations are listed in Table I. The values for the solid bone were assumed to be similar to those for cortical bone, and the first and second Lamé coefficients, $\lambda = 14.8$ GPa and $\mu = 8.3$ GPa, were calculated from the

Table 1 Physical parameter values used in FDTD simulations

	Solid bone	Water
First Lamé coefficient λ (GPa)	14.8	2.2
Second Lamé coefficient μ (Gpa)	8.3	0
Density ρ (kg/m³)	1960	1000
Normal resistance coefficient $\gamma_{\xi\xi}$ (s⁻¹)	4.5×10^5	0
Shear resistance coefficient $\gamma_{\xi\psi}$ (s⁻¹)	4.5×10^6	0

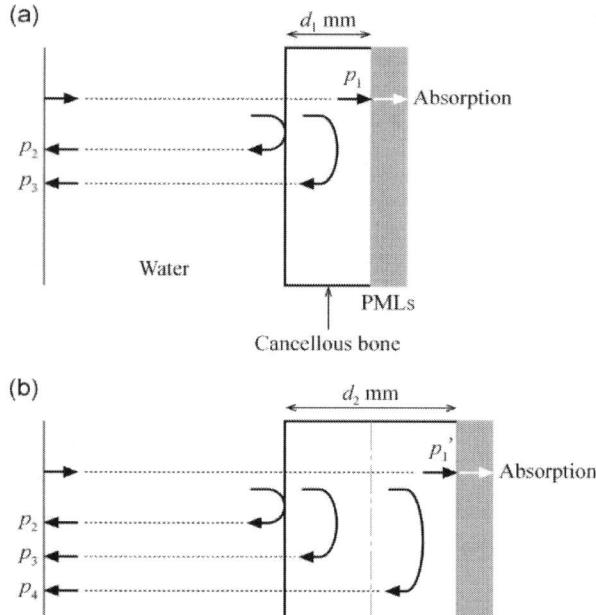

Fig. 2 Schematics of received ultrasound waves for two cancellous bone models with different thicknesses

Young's modulus and Poisson's ratio [27], and the density was $\rho = 1960$ kg/m³ [28]. The normal and shear resistance coefficients $\gamma_{\xi\xi} = 4.5 \times 10^5$ s⁻¹ and $\gamma_{\xi\psi} = 4.5 \times 10^6$ s⁻¹ were estimated by referring to the attenuation coefficients of cortical bone [29], [30]. The pore spaces were assumed to be filled with water. For water, values of $\lambda = 2.2$ GPa, $\mu = 0$ GPa, $\rho = 1000$ kg/m³, and $\gamma_{\xi\xi} = \gamma_{\xi\psi} = 0$ s⁻¹ were used.

B. Separation of Backscattered Waves

The schematics of the received ultrasound waves in the simulation model are illustrated in Fig. 2. Regardless of the cancellous bone thickness, the wave, p_1, propagating through cancellous bone cannot be reflected (or can be absorbed) at the PMLs. Therefore, in the case of the cancellous bone thickness of d_1, two waves—the reflected wave, p_2, from the cancellous bone surface and the backscattered waves, p_3, inside the cancellous bone between 0 and d_1 mm in thickness—can be received as shown in Fig. 2(a). In the case of the thickness of d_2 ($> d_1$), in addition to p_2 and p_3, the backscattered waves, p_4, inside the bone between d_1 and d_2 in thickness can be received as shown in Fig. 2(b). Owing to perfectly reproduction of numerically simulated results, only the backscattered waves, p_4, can be extracted by subtracting the simulated waveform for the thickness of d_1 from that for the thickness of d_2.

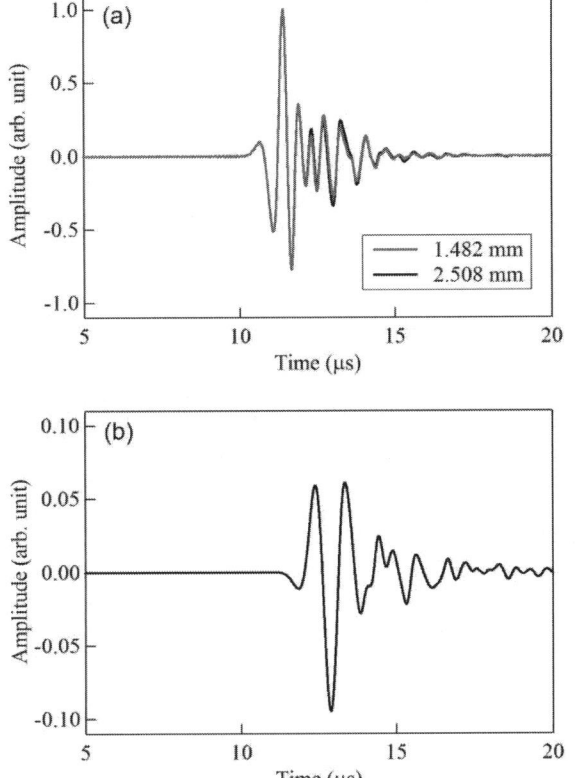

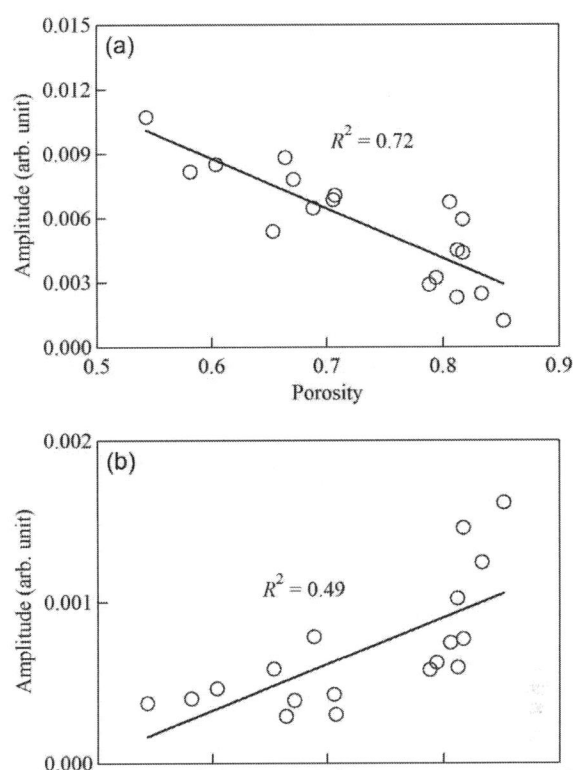

Fig. 3 Simulated waveforms; the gray and black lines in (a) illustrate the received waveforms for the cancellous bone models with thicknesses of 1.482 and 2.508 mm, and (b) shows the calculated waveform by subtracting the gray waveform from the black waveform.

Fig. 4 Variations in amplitudes of (a) reflected and (b) backscattered waves in cancellous bone with porosity.

III. SIMULATION RESULTS AND DISCUSSION

A. Observed Waveforms

The simulated results of the received signals for the cancellous bone model with a porosity of $\beta = 0.71$ are shown in Fig. 3(a). The gray and black lines illustrate the waveforms in the cases of the cancellous bone thicknesses of $d_1 = 1.482$ mm and $d_2 = 2.508$ mm, respectively. These waveforms were almost the same before the propagation time of 12 μs but slightly differed after 12 μs. The observed waveforms at 10–12 μs corresponded to the reflected wave from the cancellous bone surface and the backscattered waves inside the bone between 0 and 1.482 mm. The difference in the waveform after 12 μs is because of the backscattered waves inside the bone of 1.482–2.508 mm which can be observed only in the black waveform. The calculated result by sub- tracting the gray waveform from the black waveform is shown in Fig. 3(b). Thus, only the

backscattered waves inside the bone of 1.482–2.508 mm could be clearly observed. Comparing between Figs. 3(a) and 3(b), the amplitude of the backscattered waves was much smaller than that of the reflected wave, and the center frequency was lower.

B. Variations in Amplitudes of Reflected and Backscattered Waves with Porosity

For 18 cancellous bone models, the peak-to-peal wave amplitudes were derived. The variations in the reflected and backscattered amplitudes with the porosity are shown in Figs. 4(a) and 4(b), respectively. The determination coefficients of the reflected and backscattered wave amplitudes were $R^2 = 0.72$ ($P < 0.001$) and $R^2 = 0.49$ ($P < 0.001$), respectively. In Fig. 4(a), the reflected wave amplitude decreased with the porosity. This was considered to be because the reflection at the cancellous bone surface could be generated at the solid bone part. On the other hand, in Fig. 4(b), the backscattered wave amplitude increased with the porosity. This was considered to be because the transmission inside cancellous bone could be largely arisen in the pore part. As the determination coefficient R^2 of the backscattered wave was smaller than that of the reflected wave, it appeared that the backscattered waves could be affected by a factor other than the porosity, namely the trabecular microstructure.

IV. CONCLUSIONS

In this study, the ultrasound backscattered waves inside cancellous bone could be simulated by the FDTD method. The peak-to-peak amplitude of the backscattered waves was moderately correlated with the porosity, although the reflected wave amplitude was highly correlated. This result suggested that the backscattered waves could depend on not only the porosity but also the trabecular microstructure.

ACKNOWLEDGMENT

Part of this study was supported by the Japan Society for the Promotion of Science (JSPS) through a Grant-in-Aid for Scientific Research (B) (Grant No. 24360161).

REFERENCES

1. Laugier P (2008) Instrumentation for in vivo ultrasonic characterization of bone strength. IEEE Trans Ultrason Ferroelectr Freq Control 55:1179–1196.
2. Laugier P (2011) Quantitative ultrasound instrumentation for bone in vivo characterization. Bone Quantitative Ultrasound, Springer, Dordrecht Heidelberg London New York, 47–71.
3. Barkmann R, Glüer C C (2011) Clinical applications. Bone Quantitative Ultrasound, Springer, Dordrecht Heidelberg London New York, 73–81.
4. Karjalainen J P, Riekkinen O, Töyräs J, Jurvelin J S (2011) Linear acoustics of trabecular bone. Bone Quantitative Ultrasound, Springer, Dordrecht Heidelberg London New York , 265–289.
5. Wear K (2008) Ultrasonic scattering from cancellous bone: A review. IEEE Trans Ultrason Ferroelectr Freq Control 55:1432–1441.
6. Padilla F, Wear K (2011) Scattering by trabecular bone. Bone Quantitative Ultrasound, Springer, Dordrecht Heidelberg London New York, 123–145.
7. Hosokawa A (in press) Numerical investigation of ultrasound reflection and backscatter measurements in cancellous bone on various receiving areas. Ultrasonics.
8. Kitamura K, Pan H, Ueha S, Kimura S, Ohtomo N (1996) Ultrasonic scattering study of cancellous bone for osteoporosis diagnosis. Jpn J Appl Phys 35:3156–3162.
9. Kitamura K, Nishikouri H, Ueha S, Kimura S, Ohtomo N (1998) Estimation of trabecular bone axis for characterization of cancellous bone using scattered ultrasonic wave. Jpn J Appl Phys 37:3082–3187.
10. Strelitzki R, Nicholson P H F, Paech V (1998) A model for ultrasonic scattering in cancellous bone based on velocity fluctuations in a binary mixture. Physiol Meas 19:189–196.
11. Nicholson P H F, Strelitzki R, Cleveland R O, Bouxsein M L (2000) Scattering of ultrasound in cancellous bone: Predictions from a theoretical model. J Biomech 33:503–506.
12. Wear K A (1999) Frequency dependence of ultrasonic backscatter from human trabecular bone: Theory and experiment. J Acoust Soc Am 106:3659–3664.
13. Chaffai S, Roberjot V, Peyrin F, Berger G, Laugier P (2000) Frequency dependence of ultrasonic backscattering in cancellous bone: Autocorrelation model and experimental results. J Acoust Soc Am 108:2403–2411.
14. Bossy E, Padilla F, Peyrin F, Laugier P (2005) Three-dimensional simulation of ultrasound propagation through trabecular bone structures measured by synchrotron microtomography. Phys Med Biol 50:5545–5556.
15. Haïat G, Padilla F, Barkmann R, Gluer C -C, Laugier P (2006) Numerical simulation of the dependence of quantitative ultrasonic parameters on trabecular bone microarchitecture and elastic constants. Ultrasonics 44:e289–e294.
16. Nagatani Y, Imaizumi H, Fukuda T, Matsukawa M, Watanabe Y, Otani T (2006) Applicability of finite-difference time-domain method to simulation of wave propagation in cancellous bone. Jpn J Appl Phys 45:7186–7190.
17. Haïat G, Padilla F, Peyrin F, Laugier P (2007) Variation of ultrasonic parameters with microstructure and material properties of trabecular bone: A 3-D model simulation. J Bone Miner Res 22:665–674.
18. Nagatani Y, Mizuno K, Saeki T, Matsukawa M, Sakaguchi T, Hosoi H (2008) Numerical and experimental study on the wave attenuation in bone –FDTD simulation of ultrasound propagation in cancellous bone. Ultrasonics:607–612.
19. Haïat G, Padilla F, Peyrin F, Laugier P (2008) Fast wave ultrasonic propagation in trabecular bone: Numerical study of the influence of porosity and structural anisotropy. J Acoust Soc Am 123:1694–1705.
20. Haïat G, Padilla F, Laugier P (2008) Sensitivity of QUS parameters to controlled variations of bone strength assessed with a cellular model. IEEE Trans Ultrason Ferroelectr Freq Control 55:1488–1496.
21. Hosokawa A (2009) Numerical analysis of variability in ultrasound propagation properties induced by trabecular microstructure in cancellous bone. IEEE Trans Ultrason Ferroelectr Freq Control 56:738–747.
22. Haïat G, Padilla F, Svrcekova M, Chevalier Y, Pahr D, Peyrin F, Laugier P, Zysset P (2009) Relationship between ultrasonic parameters and apparent trabecular bone elastic modulus: A numerical approach. J Biomech 42:2033–2039.
23. Hosokawa A (2010) Effect of porosity distribution in the propagation direction on ultrasound waves through cancellous bone. IEEE Trans Ultrason Ferroelectr Freq Control 57:1320–1328.
24. Hosokawa A (2011) Numerical investigation of ultrasound refraction caused by oblique orientation of trabecular network in cancellous bone. IEEE Trans Ultrason Ferroelectr Freq Control 58:1389–1396.
25. Hosokawa A (2013) Numerical investigation of reflection properties of fast and slow longitudinal waves in cancellous bone. IEEE Trans Ferroelectr Freq Control 60:1030–1035.
26. Chew W C, Liu Q H (1996) Perfectly matched layer for elastodynamics: A new absorbing boundary condition. J Comput Acoust 4:341–359.
27. Williams J L, Johnson W J H (1989) Elastic constants of composites formed from PMMA bone cement and anisotropic bovine tibial cancellous bone. J Biomech 22:673–682.
28. Lang S B (1970) Ultrasonic method for measuring elastic coefficients of bone and results on fresh and dried bovine bones. IEEE Trans Biomed Eng BME-17:101–105.
29. Sasso M, Haïat G, Yamato Y, Naili S, Matsukawa M (2007) Frequency dependence of ultrasonic attenuation in bovine cortical bone: An in vitro study. Ultrasound Med Biol 33:1933–1942.
30. Sasso M, Haïat G, Yamato Y, Naili S, Matsukawa M (2008) Dependence of ultrasonic attenuation on bone mass and microstructure in bovine cortical bone. J Biomech 41:347–355.

Investigation of Solid Dispersion Methods to Improve the Dissolution Rate of Curcumin

Kiet Anh Tran[1], Thao Truong-Dinh Tran[1], Toi Van Vo[1], Thanh Van Tran[2], and Phuong Ha-Lien Tran[1,*]

[1] Biomedical Engineering Department, International University, Vietnam National University - Ho Chi Minh City, Vietnam
[2] School of Pharmacy, University of Medicine and Pharmacy at Ho Chi Minh City, Vietnam

Abstract— **Around current 40% of potential active agents are poorly water-soluble drugs. Solid dispersion (SD) technique is one of the most common techniques used to increase the dissolution rate of a poorly water-soluble drug. The aim of this research was to investigate a capability of SD in improving dissolution rate of curcumin, a poorly water-soluble drug in acidic and neutral media. SDs were prepared by melting and solvent methods. Hydroxypropyl methylcellulose 6, polyethylene glycol 6000, poloxamer 407 were used as carriers in the SDs. The dissolution rate of SDs was tested in simulated gastric fluid (buffer pH 1.2) and simulated intestinal fluid (buffer pH 6.8). The structural behaviors of drug were characterized by power X-ray diffraction (PXRD) and Fourier transform spectroscopy (FTIR). The presence of poloxamer 407 in SDs of curcumin improved dissolution rate of the drug. The crystalline structure of the drug was changed to amorphous form. The study indicated that the melting method with poloxamer 407 was the promising approach to enhance dissolution rate of curcumin and hence, suggesting further solutions to achieve the best product containing curcumin which could improve bioavailability of the drug.**

Keywords— **Curcumin, Solid dispersion, Polymers, Alkalizers, Dissolution test, Nano suspension.**

I. INTRODUCTION

Powdered rhizome of turmeric has been known as a preservation of food as well as traditional treatment [1] including (cardiovascular, neurological and gastrointestinal disorders, multiple sclerosis, diabetes type II, skin diseases, cystic fibrosis, cataract etc) [2]. Nowadays, the deep investigation of curcumin has been reported about their high potential in anti HIV [3], cancer and Alzheimer's disease [2], strongly proving the importance of curcumin in pharmacological actions for human healthcare. Nevertheless, bioavailability of curcumin is quite low because its solubility in water which directly affects its dissolution is very poor [4, 5]. Therefore, the most important tasks of pharmaceutical scientists are finding solutions to overcome the low bioavailability and drug stability [6]. Many tools have been shown to overcome such problems including: micronisation to increase the surface area, salt formation, use of surfactants as solubilisers, forming water soluble complexes with cyclodextrins manipulating the solid state of the drug with the aim of decreasing the drug crystallinity prodrug formation,etc [7]. Solid dispersion (SD) is one of the methods suitable in cases of poorly aqueous soluble drugs such as nimesulide, tenoxicam [8], nifedipine [9], nimodipine[10]. In this study, SD was manipulated to enhance dissolution rate of the poorly water-soluble drug curcumin. The structural behaviors of drug were characterized by power X-ray diffraction (PXRD) and Fourier transform infrared spectroscopy (FTIR).

II. MATERIALS AND METHODOLOGY

2.1 Materials

Curcumin, Acetic acid (CH_3COOH), Sodium bicarbonate ($NaHCO_3$), Sodium carbonate (Na_2CO_3), Potassium chloride (KCl), *Sodium hydroxide* (NaOH) were purchased from Guangdong Guanghua Sci-Tech company (China). Monopotassium phosphate (KH_2PO_4) was purchased from Wako Pure Chemical Industries (Japan). Hydrochloric acid, Magnesium oxide (MgO), *Sodium chloride* (NaCl) were purchased from Xilong Chemical Industry Incorporated Company (China). Hydroxypropyl methylcellulose 6 (HPMC 6) was provided by Dow Chemical Company (USA). Poloxamer 407 was purchased from the chemical company BaSF (Germany). Polyethylene glycol (PEG 6000) was purchased from Sino-Japan chemical (Taiwan). Methanol (MeOH) - HPLC grade was purchased from Thermo Fisher Scientific Inc.

---

* Corresponding author.

© Springer International Publishing Switzerland 2015
V. Van Toi and T.H. Lien Phuong (eds.), *5th International Conference on Biomedical Engineering in Vietnam,*
IFMBE Proceedings 46, DOI: 10.1007/978-3-319-11776-8_71

2.2. Solid Dispersion Preparations

Solvent method: Ethanol (C_2H_5OH) was added to the beaker containing HPMC 6 or PEG 6000 or Poloxamer 407. When polymers were dissolved in ethanol, curcumin was added in the solution and mixed again until a clear solution was obtained. Then the solution was left in the oven at 40 °C for drying. Detailed formulations are described in Table 1.

Table 1 Weight ratio of polymers and curcumin in the solvent method.

	Method	F1	F2	F3
Curcumin (mg)	Solvent	30	30	30
HPMC 6 (mg)	Solvent	90	0	0
PEG 6000 (mg)	Solvent	0	90	0
Poloxamer 407 (mg)	Solvent	0	0	90
C_2H_5OH (ml)	Solvent	12	12	12

Melting method: Carriers such as PEG 6000 or Poloxamer 407 were placed in a beaker with a magnetic stirrer and melted to a liquid using a hot plate (150 °C). When polymers and drug looked like a clear solution, curcumin was dispersed in the molten mixture with a constant stirring in 5 min. After cooling, the dispersion was grinded to powder. Detailed formulations are described in Table 2.

Table 2 Weight ratio of polymers and curcumin in the melting method

	Method	F4	F5	F6	F7	F8
Curcumin(mg)	Melting	30	30	30	30	30
PEG6000(mg)	Melting	90	0	0	0	0
Poloxamer 407(mg)	Melting	0	90	180	270	300

2.3. Dissolution Test

Dissolution rate of curcumin was tested in buffer pH 1.2 and pH 6.8 (900ml) by dissolution test machine (DT 70 Pharma Test, Germany) with the paddle at 50 rpm. In all tests, the temperature was maintained at 37 ± 0.5 ^{0}C. At regular time intervals of 10, 20, 30, 60, 90 and 120 min 1 mL of sample was withdrawn and replaced by the same volume of the fresh buffer to maintain the condition.

2.4. High Performance Liquid Chromatography (HPLC) Analysis

The quantification of curcumin was performed using HPLC system (Ultimate 3000 HPLC Thermoscientific Inc, USA). The mobile phase has the ratio MeOH: water mixture = 80:20. The aqueous phase of the mobile phase was prepared by adding 20 mL of glacial acetic acid to 980 mL of HPLC grade water and then mixing well. The flow rate was maintained at 1.2 mL/min at 25°C ± 2°C. The detection wavelength was 214 nm. 20μL of sample were injected into the HPLC system.

2.5. Fourier Transform Infrared Spectroscopy (FTIR)

Pure curcumin, Poloxamer 407, F7 and physical mixture of F7 were prepared for FTIR analysis. These spectra were recorded using a Spectrotometer (Bruker's Vertex 79 series FT-IR, Germany) which wavelength scanned from 500 to 4000 cm^{-1} and a resolution of 2 cm^{-1}. KBr were prepared by mixing 1mg of samples with 200 mg KBr.

2.6. Powder X-ray diffraction (PXRD)

Powder X-ray diffractograms of pure curcumin, Poloxamer 407, F7 and physical mixture of F7 were characterized using Bruker's D8 Advance series PXRD (Germany) equipped by a sensitive detector with scanning rate 1sec/step in increments of 0.02° from 5° to 60° (diffraction angle 2θ) using zero background sample holder.

III. RESULTS AND DISCUSSION

3.1. Dissolution Rate of Curcumin

3.1.1 Effect of Polymers

The ratios of drug to carrier (HPMC 6, PEG 6000 and Poloxamer 407) were 1:3 in the solvent method (Table 1). Among these carriers, Poloxamer 407 showed the most effective polymer in increasing the dissolution of curcumin. The dissolution rate of curcumin from poloxamer-based SD was increased about 4 folds as compared to that of pure curcumin (Figure 1).

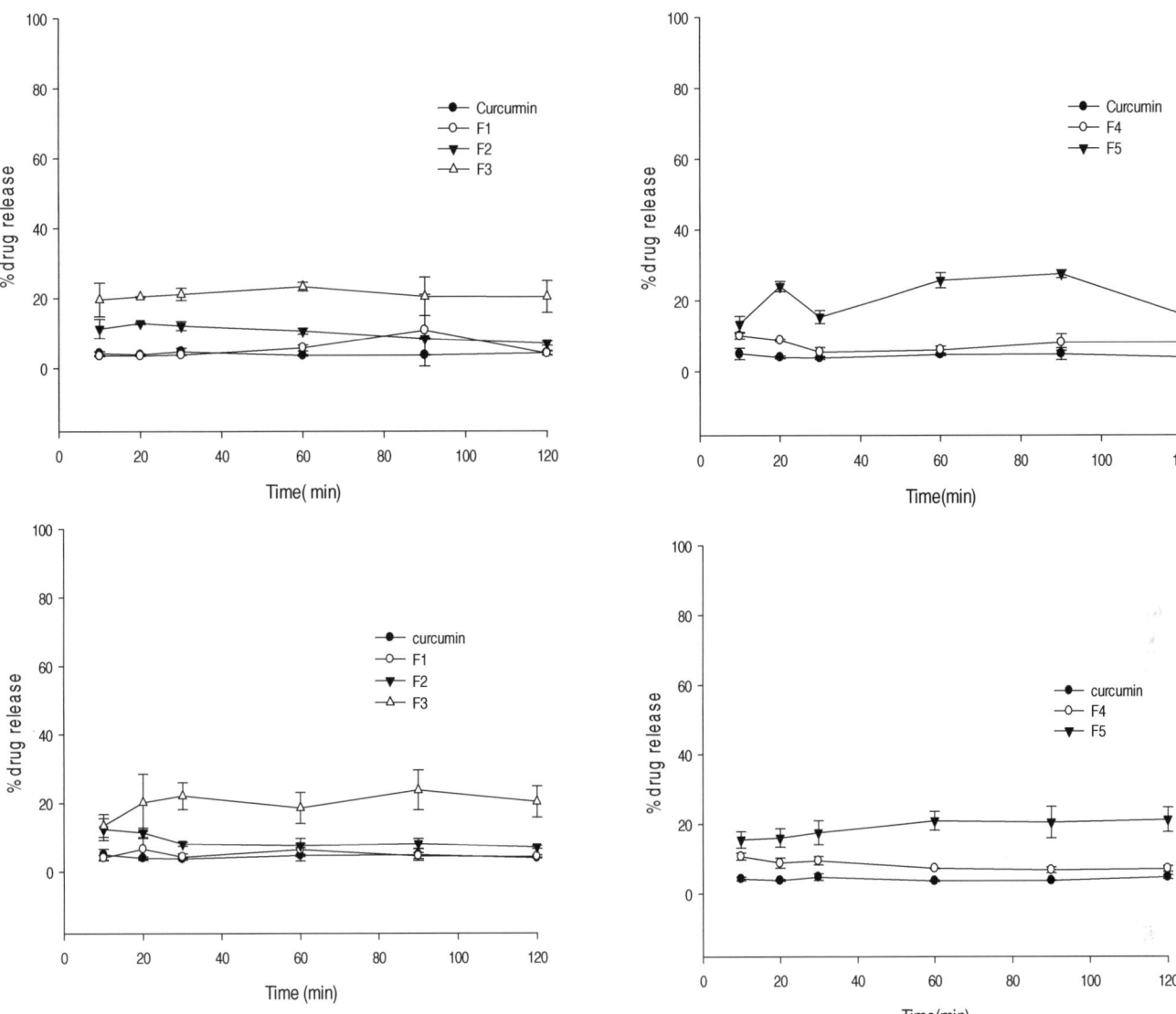

Fig. 1 Dissolution test of formulations F1, F2, F3 and curcumin at pH 6.8 (top) and pH 1.2 (bottom)

Fig. 2 Dissolution test of formulations F4, F5 and curcumin in pH 6.8 (top) and pH 1.2 (bottom)

In the melting method, carriers such as PEG 6000 and Poloxamer 407 were selected with the same ratio to drug as the one in the solvent method (Table 2). Dissolution rate of curcumin from Poloxamer 407-based SD was higher than that of PEG 6000-based SD (Figure 2). These results indicated that poloxamer is the potential polymer in increasing dissolution rate of curcumin in both of the solvent and melting method. However, the solvent method essentially requires the use of an organic solvent, and therefore, may lead to some risks of the solvent. Moreover, preparation of SDs by the melting method saves the time for solvent evaporation. The melting method hence was used in further studies.

SDs composed of drug and Poloxamer 407 at weight ratios of 1:3, 1:6, 1:9 and 1:10 were prepared by the melting method. The drug release was significantly enhanced at ratio 1:6 (Figure 3). The increasing amount of carrier resulted in the increased dissolution rate of drug. However, at the weight ratio of 1:10 between drug and carrier, the dissolution rate didn't show the best result. The addition of carrier cannot continually enhance drug release until it reaches a determined level [11].

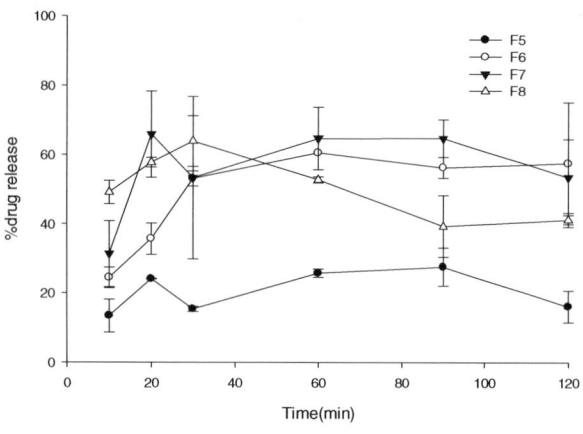

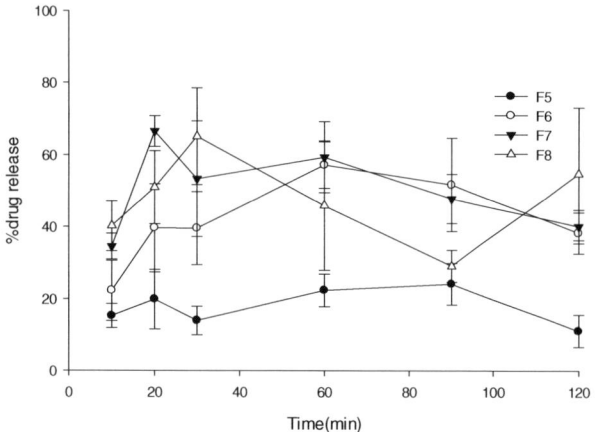

Fig. 3 Dissolution test of formulations F5, F6, F7 and F8 in pH 6.8 (top) and pH 1.2 (bottom).

3.2. PXRD and FTIR Analysis

Characteristic peaks of curcumin in physical mixture were observed at 2θ of 17.193, 21.22 and 24.63 in the PXRD diffractogram (Figure 4). However, when curcumin was introduced to F7 SD prepared by the melting method, these peaks were disappeared. This result indicated that the crystallinity of curcumin was changed to amorphous state, and hence, leading to the better dissolution rate [12].

Regarding the FTIR analysis, peaks of curcumin in the physical mixture were the same as the peaks of curcumin in F7 (Figure 5). This result indicated that there was no interaction between curcumin and poloxamer in SD.

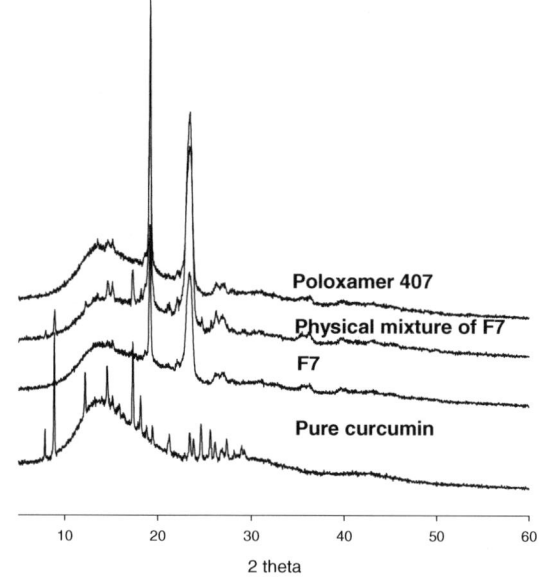

Fig.4. X-Ray diffraction spectra of pure curcumin, Poloxamer 407, F7 and physical mixture of F7.

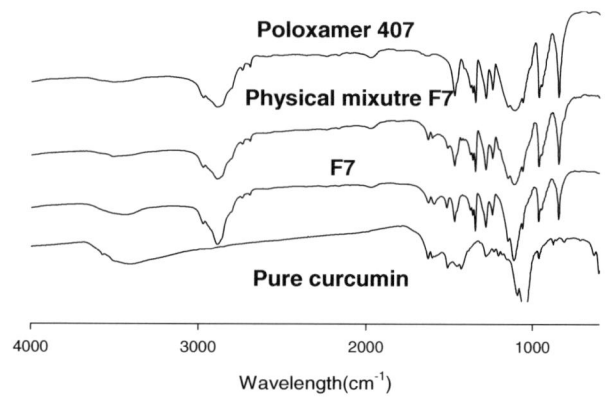

Fig. 5 FTIR spectra of pure curcumin, Poloxamer 407, F7 and physical mixture of F7

IV. CONCLUSIONS

The melting method with Poloxamer 407 as carrier was found optimal for drug dissolution release rate. Generally, as the amount of poloxamer 407 was increased, the dissolution rate of drug was increased. The best formulation to obtain the best release of curcumin was selected at the ratio 1:9 between drug and polymer by the melting method.

This study can be considered as the beginning study for further researches in attempt to enhance release of curcumin.

ACKNOWLEDGMENT

This *research* is funded by International University, Vietnam National University – Ho Chi Minh under grant number SV2013-11-BME.

REFERENCES

1. C. Moorthi, K. Kathiresan.Fabrication of highly stable sonication assisted curcumin nanocrystals by nanoprecipitation method. Drug invention today 5(2013)66-69.
2. G. Arun, P. Shweta, Kj. Upendra. Formulation and evaluation of ternary solid dispersion of curcumin. International Journal of Pharmacy and Pharmaceutical Sciences.Vol 4, Suppl 5L
3. Mazumdar A, Raghavan K, Weinstein J, Kohn K W,Pommer Y, Inhibition of human immunodeficiency virus type-1 integrase by curcumin. Biochem. Pharmacol.1995; 49;1165–1170.
4. Ravindranath V, Chandrasekhara N. Absorption and tissue distribution of Cur in rats. Toxicol. 1980, 16: 259-265
5. Chattopadhyay I. Turmeric and Cur: biological actions and clinical applications. Current Sci. 2004, 87: 44-53.
6. Ashok kumar suthar, Shailendra Singh Solank, Rakesh Kumar Dhanwani. Enhancement of dissolution of poorly water soluble raloxifene hydrochloride by preparing nanoparticles. Journal of Advanced Pharmacy Education & Research2: 189-194.
7. Aguiara. J., Zelmer A.J. And Kinkel A. W.Deaglomeration. Behaviour of relatively insoluble benzoic acid and its sodium salt, J.Pharm. Sci., 1967, 56, 1243–1252
8. Newa, M., Bhandari, K., Jong OH K,, Seob, j., Jung Ae, K., Nong Kyu, y., Jong soo,w., Han Gon, c., and Chul Soon, Y. Enhancement of Dissolution rate, Dissolution and Bioavailability of Ibuprofen in Solid Dispersion Systems, Chem. Pharm. Bull. 2008, 56: 4, 569-574
9. Ansari MJ, Ahmad S, Kholi K, Ali J, Khar RK. Stabilityindicating HPTLC determination of curcumin in bulk drug and pharmaceutical formulations. Journal of Pharmaceutical and Biomedical Analysis, 2005 Sep. 1; 39(1-2):132-138
10. Mahmoud EL-Badry, Gihan Fetih,Mohamed Fathy. Improvement of dissolution rate and dissolution rate of indomethacin by solid dispersions in Gelucire 50/13 and PEG4000. Saudi Pharmaceutical Journal (2009) 17, 217–225
11. Tran, P.H-L., Tran, T.T-D., Lee, K.-H., Kim, D-J., Lee, B-J. Dissolution-modulating mechanism of pH modifiers in solid dispersion containing weakly acidic or basic drugs with poor water solubility. Expert opinion on drug delivery, 7 (5), 647-661, 2010.
12. Wong S.M., Kellaway I.W., Murdan S., Enhancement of the dissolution rate and oral absorption of a poorly water soluble drug by formation of surfactant-containing microparticles, Int. J Pharm., (2006), 317, 61–68.

*Corresponding Author: Phuong Ha-Lien Tran
Institute: Biomedical Engineering Department, International University, Vietnam National University - Ho Chi Minh City
City: Ho Chi Minh City
Country: Vietnam
Email: thlphuong@hcmiu.edu.vn

Dissolution Enhancement of Curcumin by Solid Dispersion with Polyethylene Glycol 6000 and Hydroxypropyl Methylcellulose

Tuong Ngoc-Gia Nguyen[1], Phuong Ha-Lien Tran[1], Toi Van Vo[1],
Thanh Van Tran[2], and Thao Truong-Dinh Tran[1,*]

[1] Biomedical Engineering Department, International University,
Vietnam National University - Ho Chi Minh City, Vietnam
[2] School of Pharmacy, University of Medicine and Pharmacy at Ho Chi Minh City, Vietnam

Abstract— Curcumin, a polyphenolic compound derived from Curcuma longa, possesses diverse pharmacologic effects such as anti-inflammatory, antioxidant, antiproliferative and antiangiogenic activities. However, curcumin has poor bioavailability and low dissolution rate due to poor absorption, rapid metabolism, and rapid systemic elimination. The aim of the study was to improve the solubility and dissolution rate of curcumin by solid dispersions. Polyethylene glycol (PEG 6000) and swellable polymers such as hydroxylpropyl methylcelluloses (HPMCs) and polyethylene oxide (PEO), were applied to improve dissolution and oral absorption of curcumin. The solid dispersions (SDs) in different ratios were prepared by hot melting method. The dissolution rate of SDs was tested in simulated gastric fluid (buffer pH 1.2) and simulated intestinal fluid (buffer pH 6.8). Samples were collected at 10, 20, 30, 60, 90, 120 minutes and replaced with an equivalent amount of fresh medium to maintain a constant dissolution volume. High-performance liquid chromatography (HPLC) method was used to determine concentration of curcumin after dissolution test. The dissolution rate of curcumin was enhanced by the incorporation into hydrophilic carrier (e.g. PEG 6000, HPMC 4000, HPMC 6, and PEO). The best percentage of drug release was observed with curcumin: PEG 6000: HPMC 4000 in ratio 1:8:4. Through the selection of polymers with optimal ratio, the SDs could increase the dissolution rate of curcumin.

Keywords— Curcumin, PEG 6000, HPMC 4000, melting method.

I. INTRODUCTION

Curcumin is known as a hydrophobic polyphenol which extracted from the rhizome of Curcuma longa [1]. Curcumin contains about 77% diferuloylmethane, 17% demethoxycurcumin, and 6% bisdemethoxycurcumin. Structurally, curcumin is a diferuloylmethane; 1,7-bis(4-hydroxy-3-methoxy phenyl)-1,6-heptadiene-3,5,-dione. Curcumin contains diketon structure which is divided into two possible enol structures through intramolecular hydrogen transfer. Diketon structure exists in a tautomeric equilibrium. Indeed, the enol form is prioritized in most organic solvents [2]. Moreover, the single hydrogen atom sharing by two keto-enol structures will be delocalized when two enol tautomers are in equilibrium[3, 4]. On pharmacological evaluation, curcumin plays a role as an anti-inflammatory, antioxidant [5-8], antimicrobial, and anticarcinogenic [9-13] agent. Besides, biological activities such as the hepato- and nephro-protection [14-16], thrombosis suppression [17], myocardial infarction protection [18-20], hypoglycemia [21-24], and antirheumatics [25] have been proved. Additionally, curcumin is very safe with high dose oral administration, as proven in various animal models and studies [26, 27]. Therefore, curcumin is one of the leading agents in combating and preventing human diseases. However, curcumin has not yet been approved as a therapeutic agent because of its poor bioavailability. Curcumin is slightly absorbed in the gastrointestinal tract with the maximum solubility to be 11ng/mL in the aqueous pH 5.0 buffer. The oral bioavailability of curcumin is about 60% [28] because of its poor solubility and slow dissolution properties.

Solid dispersion (SD) is one of the most promising strategies to improve bioavailability of poor water soluble drugs by reducing drug particle size, improving wettability and forming amorphous state. Melting method was chosen because of a number of reasons such as short time process, solvent free, lower cost, and reproducible crystal form. In this study, PEG 6000 was used as a surface adsorbent to formulate SDs of curcumin. PEGs are polymers of ethylene oxide and water. It is widely used in preparing SDs because of its very high affinity towards water. HPMC was used in controlled release oral delivery as the hydrophilic matrix to improve water solubility and affinity, consequently having influences on swelling, dissolution and disintegration behavior of drug.

The aim of the study was to improve the solubility and dissolution rate of curcumin by SDs using the melting method. The effects of different parameters in SD preparation on the drug dissolution were evaluated.

II. MATERIALS AND METHODOLOGY

A. Materials

Curcumin and Sodium hydroxide (NaOH) were purchased from Guanghua Sci-Tech Company (China).

---

* Corresponding author.

V. Van Toi and T.H. Lien Phuong (eds.), *5th International Conference on Biomedical Engineering in Vietnam,*
IFMBE Proceedings 46, DOI: 10.1007/978-3-319-11776-8_72

Hydroxypropyl methyl cellulose (HPMC 4000) was provided by from Dow Chemical Company (USA). Polyethylene glycol (PEG 6000) was purchased from Sino-Japan chemical (Taiwan). Methanol (MeOH) was purchased from Fisher Scientific International, Inc (US).Hydrochloric acid (HCl) and Sodium chloride (NaCl) were purchased from Xilong Chemical Industry Incorporated Company (China). Monopotassium phosphate (KH_2PO_4) was purchased from Wako Pure Chemical Industries (Japan).

B. Solid Dispersions Preparation

SDs were prepared by the melting method to enhance dissolution rate of Curcumin using PEG 6000, HPMC 4000, HPMC 6, and PEO as carriers (Table 1). Required amount of PEG 6000 was melted in beaker at 190^0C until a molten liquid appeared. Curcumin was added into the beaker and stirred in a desired time. HPMC was dispersed in the molten mixture until a uniform mixture was obtained. These SDs then were cooled at room temperature. The obtained SDs was preserved by covering the beaker with aluminum foil and keeping in a dry place for a protection from light for further use.

Table 1 Composition and method of curcumin formulations

No.	Cur.	PEG 6000	HPMC 4000	HPMC 6	PEO	Ratio	Stirring time
F1	30	120	-	-	-	1:4	5 min
F2	30	240	-	-	-	1:8	5 min
F3	30	120	-	60	-	1:4:2	5 min
F4	30	120	60	-	-	1:4:2	5 min
F5	30	120	-	-	60	1:4:2	5 min
F6	30	120	120	-	-	1:4:4	5 min
F7	30	240	120	-	-	1:8:4	5 min
F8	30	240	180	-	-	1:8:6	5 min

C. Characterization of Solid Dispersions

a) Dissolution Test

These SDs were conducted at 37 ± 0.5 ^{0}C using paddle method. Buffer pH 1.2 and buffer pH 6.8 were used as dissolution media. 900ml of buffer pH 1.2 or pH 6.8 was added into dissolution vessel. The apparatus was set up at 50 rpm of rotation speed. 1ml of sample was collected from the media at predetermined intervals of 10, 20, 30, 60, 90, 120 min. 1ml of dissolution sample was compensated by adding 1ml of the corresponding fresh buffer. 100µl sample solutions were diluted with 900µl MeOH for the HPLC test.

b) High Performance Liquid Chromatography (HPLC)

The quantification of curcumin in microparticles was performed using Ultimate 3000 HPLC (Thermoscientific Inc., USA). The mobile phase has methanol/water mixture ratio was 4:1 with a flow rate of 1.2ml/min. The water mixture included 20ml acetic acid and 980ml H_2O. The UV/VIS detector was set to a wavelength of 425nm [29]. 20µL of sample was injected to HPLC system.

III. RESULTS AND DISCUSSION

Effect of Polymer Type on Drug Dissolution

Figure 1 and Figure 2 show the effect of polymer type on drug dissolution rate at pH 1.2 and pH 6.8. The highest drug release from F4 was 17.7 % and 18.6% after 2 h at pH 1.2 and pH 6.8, respectively. These results demonstrated that SD which was prepared with HPMC 4000 had higher dissolution rate than the SD prepared with PEO or HPMC 6.

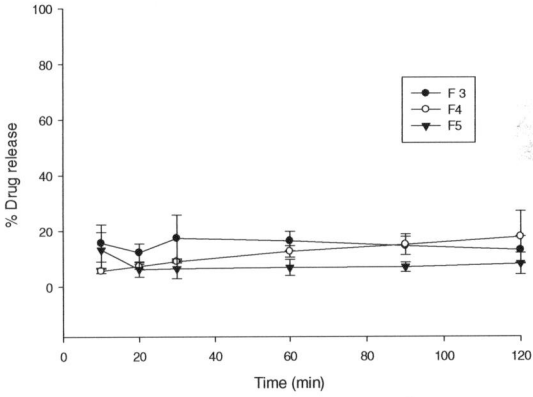

Fig. 1 Effect of polymer type on drug dissolution rate at pH 1.2

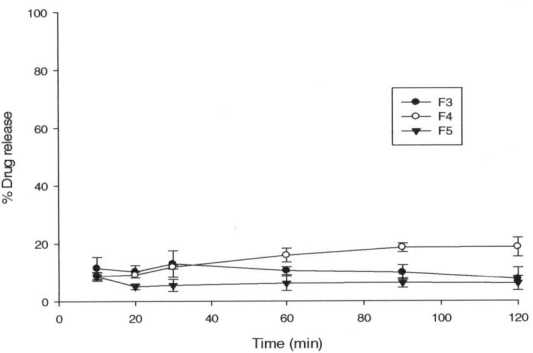

Fig. 2 Effect of polymer type on drug dissolution rate at pH 6.8

Effect of Polymer Ratio

Experiments were preceded with different ratios to figure out which ratio showed the best result. The ratios of

1:4, 1:8, 1:4:2, 1:4:4, 1:8:4 and 1:8:6 were corresponding with F1, F2, F4, F6, F7 and F8. The highest percentage drug release at pH 1.2 were 26.3%, 16.17%, 17.7%, 13.29%, 38.3% and 29.31% for F1, F2, F4, F6, F7 and F8, respectively (Figure 3). The highest percentage drug release at pH 6.8 were 18.53%, 17.07%, 18.6%, 18.93%, 42.83%, and 29.82% for F1, F2, F4, F6, F7 and F8, respectively (Figure 4).

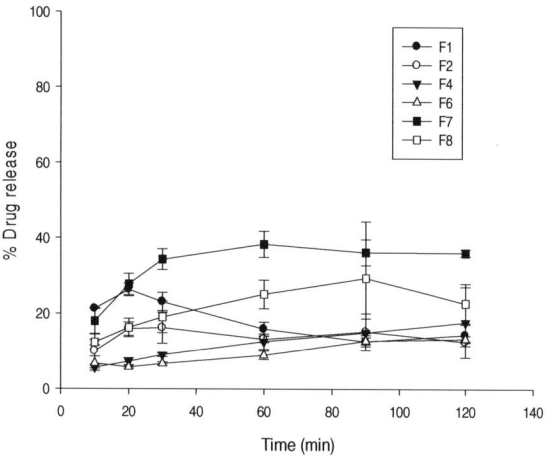

Fig. 3 Effect of polymer ratio on drug dissolution rate at pH 1.2

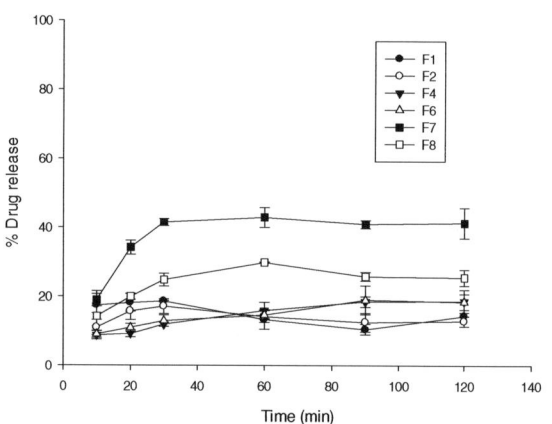

Fig. 4 Effect of polymer ratio on drug dissolution rate at pH 6.8

At pH 6.8 F7 gave the highest percentage of drug release about 42.83% (see Figure 4). It seemed that F6 did not have enough PEG 6000 to completely swell HPMC 4000. As the amount of PEG 6000 was increased two fold (F7), the higher drug dissolution rate was obtained. However, for F8 formulation, when the same amount of PEG 6000 was kept constant and the amount of HPMC 4000 was increased two fold as compared to the ones of F7, the drug release profile was fall back to be low.

IV. CONCLUSIONS

The study investigated a new SD system based on a hydrophilic carrier and a swellable polymer. The presence of a swellable polymer may induce a higher drug dissolution rate than the use of one hydrophilic carrier in SD. However, it's critical to use a suitable amount of the swellable polymer in the SD formulation. Meanwhile, hydrophilic carrier may facilitate swelling of the polymer. The study indicated that HPMC 4000 had a capability of increasing curcumin dissolution rate better than other swellable polymers when it was combined with PEG 6000 in SDs.

ACKNOWLEDGMENT

This *research* is funded by International University, Vietnam National University – Ho Chi Minh under grant number SV2013-01-BME.

REFERENCES

[1] B. B. Aggarwal, A. Kumar, and A. C. Bharti, "Anticancer potential of curcumin: preclinical and clinical studies.," *Anticancer Res.,* vol. 23, pp. 363-98, 2003.

[2] F. Payton, P. Sandusky, and W. Alworth, "NMR study of the solution structure of curcumin," *J Nat Prod,* vol. 70, 2007.

[3] K. Balasubramanian, "Molecular orbital basis for yellow curry spice curcumin's prevention of Alzheimer's disease," *J Agric Food Chem,* vol. 54, pp. 3512–3520, 2006.

[4] J. Mague, W. Alworth, and F. Payton, "Curcumin and derivatives," *Acta Crystallogr C,* vol. 60, pp. 0608-0610, 2004.

[5] O. P. Sharma, "Antioxidant activity of curcumin and related compounds.," *Biochem. Pharmacol,* vol. 25, pp. 1811–1812, 1970.

[6] A. J. Ruby, G. Kuttan, K. D. Babu, R. K. N., and R. Kuttan, " Antitumour and antioxidant activity of natural curcuminoids," *Cancer Lett,* vol. 94, pp. 79-83, 1995.

[7] Y. Sugiyama, S. Kawakishi, and T. Osawa, "Involvement of the diketone moiety in the antioxidative mechanism of tetrahydrocurcumin," *Biochem. Pharmacol.,* vol. 52, pp. 519-525, 1996.

[8] R. C. Srimal and B. N. Dhawan, "Pharmacology of diferuloyl methane (curcumin), a non-steroidal anti-inflammatory agent," *J. Pharm. Pharmacol.,* vol. 25, pp. 447-452, 1973.

[9] W. C. Jordan and C. R. Drew, "Curcumin—a natural herb with antiHIV activity.," *J. Natl. Med. Assoc.,* vol. 88, p. 333, 1996.

[10] G. B. Mahady, S. L. Pendland, G. Yun, and Z. Z. Lu, "Turmeric (Curcuma longa) and curcumin inhibit the growth of Helicobacter pylori, a group 1 carcinogen.," *Anticancer Res.,* vol. 22, pp. 4179-4181, 2002.

[11] M. K. Kim, G. J. Choi, and H. S. Lee, "Fungicidal property of Curcuma longa L. rhizome-derived curcumin against phytopathogenic fungi in a greenhouse.," *J. Agric. Food Chem.,* vol. 51, pp. 1578-1581, 2003.

[12] R. C. Reddy, P. G. Vatsala, V. G. Keshamouni, G. Padmanaban, and P. N. Rangarajan, "Curcumin for malaria therapy," *Biochem. Biophys. Res. Commun.,* vol. 326, pp. 472-474, 2005.

[13] R. Kuttan, P. Bhanumathy, K. Nirmala, and M. C. George, "Potential anticancer activity of turmeric (Curcuma longa)." *Cancer Lett,* vol. 29, pp. 197-202, 1985.

[14] Y. Kiso, Y. Suzuki, N. Watanabe, Y. Oshima, and H. Hikino, "Antihepatotoxic principles of Curcuma longa rhizomes," *Planta Med.,* vol. 49, pp. 185-187, 1983.

[15] N. Venkatesan, "Curcumin attenuation of acute adriamycin myocardial toxicity in rats," *Br. J. Pharmacol.,* vol. 124, pp. 425-427, 1998.

[16] N. Venkatesan, D. Punithavathi, and V. Arumugam, "Curcumin prevents adriamycin nephrotoxicity in rats.," *Br. J. Pharmacol.,* vol. 129, pp. 231-234, 2000.

[17] R. Srivastava, M. Dikshit, R. C. Srimal, and B. N. Dhawan, "Antithrombotic effect of curcumin.," *Thromb. Res.,* vol. 40, pp. 413-417, 1985.

[18] M. Dikshit, L. Rastogi, R. Shukla, and R. C. Srimal, "Prevention of ischaemia-induced biochemical changes by curcumin & quinidine in the cat heart. ," *Indian J. Med. Res.,* vol. 101, pp. 31-35, 1995.

[19] C. Nirmala and R. Puvanakrishnan, "Protective role of curcumin against isoproterenol induced myocardial infarction in rats," *Mol. Cell. Biochem.,* vol. 159, pp. 85-93, 1996.

[20] C. Nirmala and R. Puvanakrishnan, "Effect of curcumin on certain lysosomal hydrolases in isoproterenol-induced myocardial infarction in rats.," *Biochem. Pharmacol.,* vol. 51, pp. 47-51, 1996.

[21] M. Srinivasan, "Effect of curcumin on blood sugar as seen in a diabetic subject.," *Indian J. Med. Sci.,* vol. 26, pp. 269-270, 1972.

[22] P. S. Babu and K. Srinivasan, "Influence of dietary curcumin and cholesterol on the progression of experimentally induced diabetes in albino rat," *Mol. Cell. Biochem.,* vol. 152, pp. 13-21, 1995.

[23] P. S. Babu and K. Srinivasan, "Hypolipidemic action of curcumin, the active principle of turmeric (Curcuma longa) in streptozotocin induced diabetic rats," *Mol. Cell. Biochem.,* vol. 166, pp. 169-175, 1997.

[24] N. Arun and N. Nalini, "Efficacy of turmeric on blood sugar and polyol pathway in diabetic albino rats," *Plant Foods Hum. Nutr.,* vol. 57, pp. 41-52, 2002.

[25] S. D. Deodhar, R. Sethi, and R. C. Srimal, "Preliminary study on antirheumatic activity of curcumin (diferuloyl methane)." *Indian J. Med. Res.,* vol. 71, pp. 632-634, 1980.

[26] C. Lao, M. Demierre, and V. Sondak, "Targeting events in melanoma carcinogenesis for the prevention of melanoma.," *Expert Rev Anticancer Ther.,* vol. 6, pp. 1559-1568, 2006.

[27] D. Shoba GJoy, T. Joseph, M. Majeed, R. Rajendran, and P. Srinivas, " Influence of piperine on the pharmacokinetics of curcumin in animals and human volunteers.," *Planta Med.,* vol. 64, pp. 353-356, 1998.

[28] V. Ravindranath and N. Chandrasekhara, "Absorption and tissue distribution of curcumin in rats," *Toxicology,* vol. 16, pp. 259-265, 1980.

[29] R. M. Martin, V. P. Simone, and e. al., "Curcuminoid content and antioxidant activity in spray dried microparticles containing turmeric extract," *Food Research International,* 2011.

*Corresponding Author: Thao Truong-Dinh Tran
Institute: Biomedical Engineering Department, International University, Vietnam National University - Ho Chi Minh City
City: Ho Chi Minh City
Country: Vietnam
Email: ttdthao@hcmiu.edu.vn

Improvement of Gliclazide Dissolution Rate Using *In Situ* Micronization Technique

Nguyen Thanh Nhan and Tran Van Thanh

Faculty of Pharmacy, The University of Medicine and Pharmacy, Hochiminh-City, Vietnam

Abstract— Gliclazide (GLZ) is a second-generation sulphonylurea, widely used for the treatment of non-insulin dependent diabetes mellitus. However, due to its low solubility in water (55 mg/L) and gastric fluids, its bioavailability varied inter-individual. This research aims to enhance the dissolution rate of gliclazide using *in situ* micronization technique. The *in situ* micronization process was carried out using solvent change method by using hydroxypropyl methylcellulose E15 (HPMC E15) or tween 80 as stabilizing agents. The solution of 0.5 g or 1 g of gliclazide in 30 ml of dimethyl sulfoxide was mixed with the solution containing 0,05 g or 0,1 g of HPMC E15 or tween 80 in 100 ml of water under stirring at 1200 rpm in ice bath (about 5°C) for 30 minutes. Obtained suspensions were filtered through 0,45 µm cellulose acetate membrane. The dissolution test, scanning electron microscopy, particle size analysis, Fourier transform infrared spectroscopy (FTIR) and thermal analysis (DSC) were carried out to investigate the properties of the micronized gliclazide obtained. The *in situ* micronization technique used in this research enhanced the water-solubility and dissolution rate of gliclazide. The crystalline shape of GLZ changed from rod-shape or cube-shape to spherical shape with smaller size.

Keywords— Micronization, gliclazide, solubility, dissolution rate, stabilizing agent.

I. Introduction

Gliclazide (GLZ) is a second-generation sulphonylurea, widely used for the treatment of non-insulin dependent diabetes mellitus. However, according to the biopharmaceutics classification system (BCS), gliclazide is a class II- drug which has a low solubility in water (55 mg/l) and gastric fluids, a common characteristic of this group which determines a low dissolution rate and hence inter-individual variability on its bioavailability [1].

Various formulation strategies have been reported to improve solubility and dissolution rate of poorly water soluble drugs such as inclusion of complexation with hydroxypropyl-β-cyclodextrins, solid dispersion, use of surfactants, fluid energy mills etc [2]. However, weak points of these methods are expensiveness (for complexation with hydroxypropyl-β-cyclodextrins) and instability in chemical and physical properties (for fluid energy mills). One of the most reliable techniques to improve the dissolution rate is particle size reduction using micronization. Micronization is a term used to describe size reduction technique where micron-sized crystals of less than 10 µm are obtained during its production itself without the need for any further particle size reduction [3].

As discussed by Rasenak and Muller [4], *in situ* micronization by the presence of a stabilizing polymer covers the hydrophobic surfaces of the precipitated substances and consequently the steric hindrance caused by the polymer prevents the crystal growth. In this study an *in situ* micronization technique based on solvent change method was used for size reduction of GLZ to enhance its dissolution and bioavailability profile.

II. Materials and Methods

A. Materials

Gliclazide was purchased from Shanghai Richem International Co. Ltd (China). DMSPO, HPMC E15, tween 80 were obtained from Xilong Group (China). All the other solvents and reagents were of chemical grade.

B. In Situ Micronization Technique

The *in situ* micronization process was carried out using solvent change method by using HPMC E15 or tween 80 as stabilizing agents. Briefly, the solution of 0.5 g or 1 g of gliclazide in 30 ml of dimethyl sulfoxide (DMSO) was mixed with the solution containing 0,05 g or 0,1 g of HPMC E15 or tween 80 in 100 ml of water under stirring at 1200 rpm (STUART CB 162, England) in ice bath (about 5°C) for 30 minutes. Obtained suspensions were filtered through 0,45 µm cellulose acetate membrane and the micronization GLZ were than dried in the dry oven (Memmert ULM 500, Germany). By changing two levels of each of 3 procedure variables, eight different formulations were prepared by a full factorial design (Table 1).

C. Scanning Electron Microscopy (SEM)

Electron micrographs of crystals were obtained using a scanning electron microscope (EVOMA10, Carl Zeiss, Germany). The specimens were coated with gold to observe under vacuum.

V. Van Toi and T.H. Lien Phuong (eds.), *5th International Conference on Biomedical Engineering in Vietnam,*
IFMBE Proceedings 46, DOI: 10.1007/978-3-319-11776-8_73

Table 1 Different studied formulations for production of gliclazide microcrystals

Formulation code	Stabilizing agent	Drug concentration (g/30 ml in DMSO)	Stabilizing concentration (g/100 ml in water)
MC-GLZ1	HPMC	1	0,1
MC-GLZ2	HPMC	0,5	0,1
MC-GLZ3	HPMC	1	0,05
MC-GLZ4	HPMC	0,5	0,05
MC-GLZ5	Tween 80	1	0,1
MC-GLZ6	Tween 80	0,5	0,1
MC-GLZ7	Tween 80	1	0,05
MC-GLZ8	Tween 80	0,5	0,05

D. Particle Size Analysis

Particle size and size distribution of GLZ microcrystals and untreated GLZ was measured by micrograph analysis using software of electron microscopy (Carl Zeiss Evo MA 10, Germany) with 6 different specimens for each sample.

E. Fourier Transform Infrared Spectroscopy

Samples (about 1% w/w) were mixed with KBr powder and compressed to a disc. The spectra of untreated GLZ and MC-GLZ1 were collected on a FT-IR spectrophotometer (Shimadzu, Japan).

F. Thermal Analysis

Differential scanning calorimetry (DSC) thermograms of untreated GLZ, HPMC, MC-GLZ1 and the 1:5 HPMC: untreated GLZ physical mixture of GLZ and HPMC (5–10 mg) were recorded using a thermal analysis system (Mettler Toledo, DSC-1, Switzerland). Samples were heated at 10 °C/min in the range of 25 – 230 °C.

G. Dissolution Studies

The dissolution rates of 80 mg of GLZ microcrystals and untreated GLZ were determined according to European Pharmacopoeia by using no.4 dissolution test apparatus at 37 °C, stirred at 100 rpm. The dissolution medium was 900 ml phosphate buffer of pH 7.4. Triplicate samples (2 ml) were withdrawn from the dissolution vessels at selected time intervals and analyzed for GLZ content due to the absorbance at 226 nm after being adjusted with absorbance at 290nm on an UV spectrophotometer (Shimadzu, Japan). The results are the mean and standard deviations of three determinations.

III. RESULTS

A. SEM Morphology

The untreated GLZ and MC-GLZ1, MC-GLZ2, MC-GLZ3, MC-GLZ4 and MC-GLZ6 were observed and shown in Fig.1a, Fig.1b, c and Fig.2 a, b, c respectively. The micrographs show that MC-GLZ1, MC-GLZ2, MC-GLZ3, MC-GLZ4 are sphere-shaped while MC-GLZ6 is column-shaped and the untreated drug is rod-shaped.

B. Particle Size Distribution of Microcrystals

The size distribution of untreated GLZ and GLZ microcrystals are shown in Fig. 1 and Fig. 2. Untreated GLZ presents largest size of 26 µm with broad size distribution while other microcrystals show narrow or almost uniform of size distribution (Table 2).

Fig. 1c shows a typical size distribution of MC-GLZ2 wiht nearly uniform sphere (Fig. 1c). Statistical analysis of the particle size of microcrystals produced by different formulations showed that type of stabilizing agent had a significant effect on the particle size of the microcrystals.

Table 2 Particle size of untreated GLZ and GLZ microcrystals

Formulation code	Particle size (µm)
Untreated GLZ	26,12±3,33
MC-GLZ1	9,33±1,16
MC-GLZ2	9,57±0,64
MC-GLZ3	17,81±2,09
MC-GLZ4	9,81±0,81
MC-GLZ6	19,87±3,11

C. Fourier Transform Infrared Spectroscopy

Fourier transform infrared spectroscopy was used to analyze the microcrystals and untreated GLZ. The spectra of all samples were identical (Fig. 3) and the main absorption bands of GLZ remained unchange in all the spectra (Fig. 3): the sulphonylurea group at 1708 ± 1 cm^{1}, the amine group at 3274 ± 1 cm^{-1}, sulphonyl group bands (S=O) at 1353 ± 1 cm^{-1} and 1164 ± 1 cm^{-1}. This observation indicated that there was no difference between the internal structures and the conformation of these samples at the molecular level is intact.

D. DSC Studies

The melting point of untreated GLZ is 168.36 °C and the enthalpy of melting is 128.34 J/g. While the melting point of MC-GLZ is 156.80 °C and the enthalpy of melting is 74.6 J/g. The melting point of the physical mixture is higher than MC-GLZ's and lower than untreated GLZ's. Hence, the crystalline link formed in MC-GLZ is weaker than in untreated GLZ, resulting in higher solubility.

The appearance of the endothermic peaks and the absence of the crystallization point of MC-GLZ show that MC-GLZ is formed in crystalline state, not in amorphous state. Due to the low rate of HPMC (1:5 compared to GLZ), the endothermic peak of HPMC doesn't appear. However, in the formation of micronized particles, HPMC makes a slight interaction because the endothermic peak of MC-GLZ appears at the lower temperature compared to the endothermic peak of the physical mixture with the same rate.

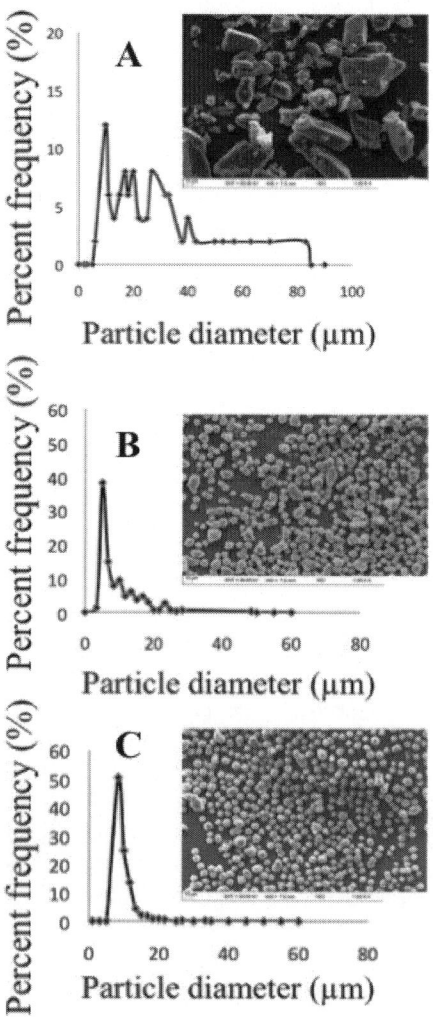

Fig. 1 Particle size distribution according to percent frequency and scanning electron micrographs of: (a) GLZ , (b) MC-GLZ1 and (c) MC-GLZ2.

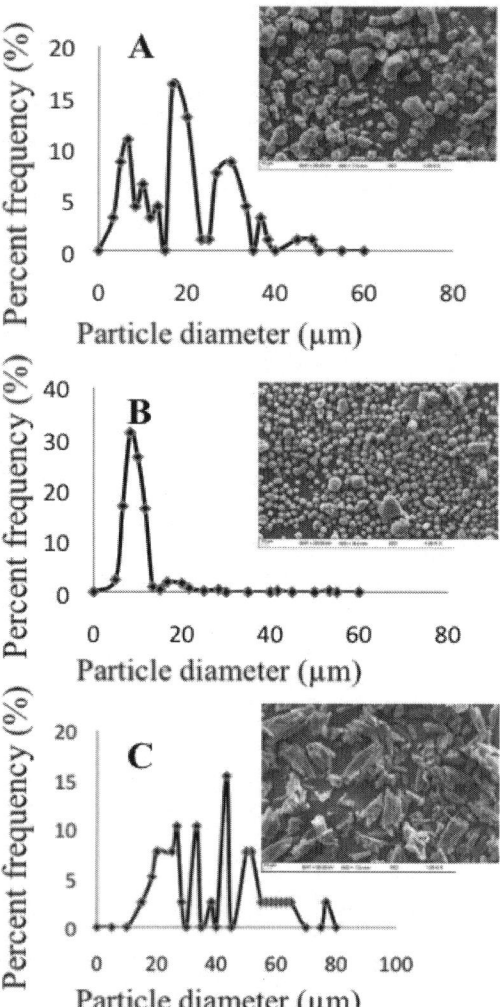

Fig. 2 Particle size distribution according to percent frequency and scanning electron micrographs of: (a) MC-GLZ3 , (b) MC-GLZ4 and (c) MC-GLZ6.

E. Dissolution Studies

As shown in Fig. 5, drug release profiles from the untreated GLZ and MC-GLZs in phosphate buffer pH 7.4 were determined. According to the results, the dissolution rate of treated samples is significantly faster than the untreated sample. It is obvious that drug release from MC-GLZs is much greater than that of the untreated powder. Statistical analysis by two way ANOVA test shows that all studied variables have effect on the dissolution rate. MC-GLZ2, the smallest formulation in term of particle size, presents the highest solubility and dissolution rate than other formulations (Fig. 5).

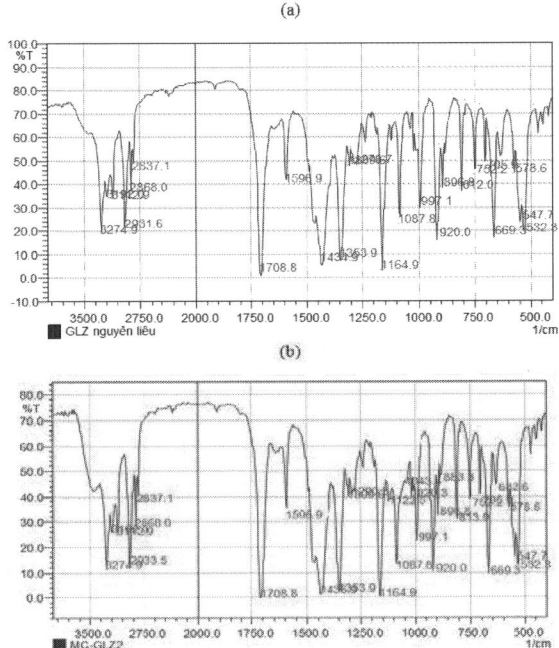

Fig. 3 FTIR spectrography of (a) untreated GLZ and (b) MC-GLZ2.

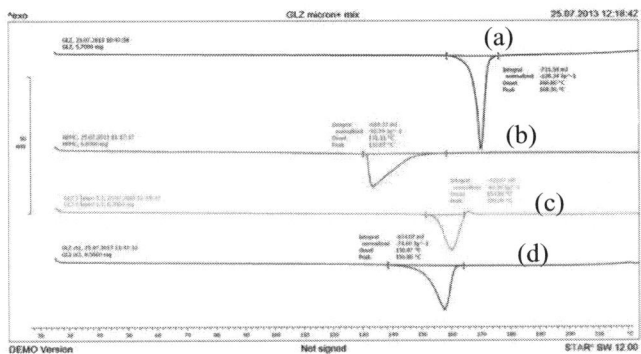

Fig. 4 DSC thermograms of (a) untreated GLZ, (b) HPMC (c) physical mixtures of gliclazide and HPMC and (d) MC-GLZ.

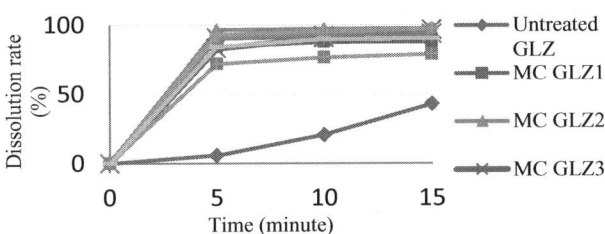

Fig. 5 Dissolution rate of untreated GLZ and MC-GLZs in phosphate buffer pH 7.4.

IV. CONCLUSIONS

The *in situ* micronization technique used in this research enhanced the water-solubility and dissolution rate of GLZ. The crystalline shape of GLZ changed from rod-shape or cube-shape to spherical shape with smaller size. The FTIR and DSC results showed no significant interaction between the drug and the stabilizers. The microcrystallization has an effect on GLZ crystal habit modification. Microcrystallization of GLZ in DMSO resulted in spherical shaped crystals whereas the untreated crystals were rod like, cube, diamond or columnar. In this process, changing stabilizing agent changed the size and shape of crystal. However, dissolution efficiency was more affected by drug and stabilizing agent concentration.

V. CONFLICT OF INTEREST

The authors declare that they have no conflict of interest.

REFERENCES

1. Neelam Seedher, Mamta Kanojia (2008), Micellar Solubilization of Some Poorly Soluble Antidiabetic Drugs: A Technical Note, AAPS Pharmaceutical Science Techniques, 9(2): 431-436.
2. Anuj Kumar, Sangram Keshri Sahoo, Kumud Padhee, Prithi Pal Singh Kochar, Ajit Satapathy and Naveen Pathak (2011), Review on solubility enhancement techniques for hydrophobic drugs, International journal of comprehensive pharmacy, 2(3):1-7
3. K.R. Vandana, Y. Prasanna Raju, V. Harini Chowdary, M. Sushma, N. Vijay Kumar (2013), An overview on in situ micronization technique – An emerging novel concept in advanced drug delivery, Saudi Pharmaceutical Journal, 5(4): 1-7
4. N. Rasenack, B.W. Muller, Dissolution rate enhancement by in situ micronization of poorly water-soluble drugs, Pharm. Res. 19 (12) (2002) :1894–1900

Author: Nguyen Thanh Nhan, Tran Van Thanh
Institute: Faculty of pharmacy, The University of Medicine and Pharmacy
Street: Dinh Tien Hoang
City: Hochiminh
Country: Vietnam
Email: *thanhpharm@gmail.com*

Research and Preparation of Solid Dispersion of Itraconazole in Hydroxypropyl-Beta-Cyclodextrin

Tran Van-Thanh, Pham Vu Quang Vinh, and Huynh Van-Hoa

Faculty of Pharmacy, University of Medicine and Pharmacy at Hochiminh City, Vietnam

Abstract— Itraconazole (ITZ) is an oral antifungal drug belongs to triazole group. However, ITZ has low and pH denpendant water solubility of about 1.1µg/ml at pH 6.8 and 6µg/ml at pH 1. Increasing solubility of ITZ is thus an urgent step for enhancement its bioavailability. Solid dispersions of ITZ and HPβCD were prepared by various methods: dry grindind, wet grinding and co-evaporation solvent with different molar rations of ITZ: HPβCD (1:1, 1:2, 1:3). Solid dispersion ITZ- HPβCD were evaluated by water solubility, dissolution profile in gastric medium. Physical characteristics of ITZ-HPβCD were assessed by DSC and FTIR spectrum. Dissolution test of solid dispersion-containing hard capsule was also experimented. Solid dispersions formulated based on dry grinding and wet grinding method showed high solubility but low dissolution profile (<20%). Solid dispersion formulation based on co-evaporation solvent showed high dissolution profile. Dissolution profile of solid dispersion-containing hard capsule showed high ITZ release after 10 mins and reach over 80% after 15 mins.

Keywords— Itraconazole, hydroxypropyl-beta-cyclodextrin (HPβCD), solid dispersion.

I. INTRODUCTION

Itraconazole (ITZ) is an oral antifungal drug belongs to triazole group. However, ITZ has low pH dependant water soluble. At neutral pH, the solubility of ITZ is about 1µg/ml When the pH decreases in acid environment, the ITZ solubility can increase up to 6µg/ml [1]. In another hand, ITZ has low oral bioavailability (of about 55 %) and is classified as high permeability and low solubility drug (according to the biopharmaceutical classification system of FDA). The solubility and absorption of ITZ thus significant change inter-individuals. The improvement of water solubility is an urgent need to increase the oral bioavailability of ITZ. There are various techniques for increasing the solubility of ITZ, where solid dispersion system with water-soluble polymers or with cyclodextrin derivated excipients have been widely used for its effectiveness [2,3].

In this research, solid dispersions of ITZ and HPβCD were prepared by various methods to find out the most suitable formulation to be encapsulated into hard capsule.

II. MATERIALS AND METHODS

A. Materials

Itraconazole was purchased from Shanghai Yung Zip Pharmaceutical Trading Co. Ltd. (China). HPβCD was synthesized at the faculty of pharmacy of the university of medicine and pharmacy at Ho Chi Minh city. PVP K30 were obtained from Xilong Group (China). All the other solvents and reagents were of chemical grade.

B. Solid Dispersion of ITZ – HPβCD Preparation

Itz and HPβCD were crushed and then sifted through 0.5 mm sieve to prepare solid dispersion system according to various methods:

Dry Grinding Method

ITZ and HPβCD were grinded in porcelain mortar for 30 minutes to obtain a homogeneous mixture. In some formulation, 10 % (w/w) of PVP is crushed and mixed in mortar with ITZ and HPβCD to create solid dispersion system. Specific components of the formulas is presented in Table 1.

Table 1 Dry grinding formulations of ITZ- HPβCD solid dispersion

Formulation code	ITZ : HPβCD (mol : mol)	Formula compositions		
		ITZ (mg)	IPβCD (mg)	PVP (mg)
TK1	1:1	705	1.936	0
TK2	1:1	705	1.936	265
TK3	1:2	705	3.872	0
TK4	1:2	705	3.872	458
TK5	1:3	705	5.808	0
TK6	1:3	705	5.808	651

© Springer International Publishing Switzerland 2015

V. Van Toi and T.H. Lien Phuong (eds.), *5th International Conference on Biomedical Engineering in Vietnam,*

IFMBE Proceedings 46, DOI: 10.1007/978-3-319-11776-8_74

Wet Grinding Method

ITZ and HPβCD were mixed in mortar porcelain for 5 minutes. Distilled water was then added to the mixture just enough to form dough texture. The mixture was grinded for 30 minutes. The obtained dough was dried at 50° C for 8 hours, and solid dispersion formulations were sifted through 0.5 mm sieve.

In some formulations, 10% (w/w) of PVP was mixed with HPβCD in porcelain mortar, ITZ was then added and grinded until homogeneous. Distilled water was added to the previous mixture to create a paste and this blocks was grinded for more than 30 minutes. Blocks forming was dried at 50° C for 8 hours, and solid dispersion formulations were sifted through 0.5 mm sieve.

Specific components of the wet grinding formula is presented in Table 2.

Table 2 Wet grinding formulations of ITZ- HPβCD solid dispersion.

Formulation codes	ITZ : HPβCD (mol : mol)	Formula compositions		
		ITZ (mg)	HPβCD (mg)	PVP (mg)
NU1	1:1	705	1.936	0
NU2	1:1	705	1.936	265
NU3	1:2	705	3.872	0
NU4	1:2	705	3.872	458
NU5	1:3	705	5.808	0
NU6	1:3	705	5.808	651

Co-evaporation Solvent

5 g of HPβCD was dissolved in 10 ml of water to create a solution (A). 1,410 g of ITZ was dissolved in 5 ml a mixture of EtOH : HCl (10:1 v/v) to obtain solution (B). Solution A was added to solution B under magnetic stirring for for 1 hour to obtain solution (C). The last was than evaporated on a hot water-bath at 55° C for 4 hours and then dried at 55° C in an oven for 8 hours more. The solid mass obtained was crushed and sifted through 0.5 mm sieve.

C. Solid Dispersion of ITZ – HPβCD Characterization

Solubility

An excess of ITZ-HPβCD corresponding to 25mg of ITZ was added in 10 ml of 0.1N HCl. This mixture was vortex in 20 minutes and let stand for 24 hours at room temperature before filtered through a 0.45 micron membrane filter. The solubility of ITZ was dertermined by means of UV - Vis spectrophotometer at a wavelength of 254 nm based on the calibration curve.

Dissolution Studies

The dissolution rates of ITZ or ITZ-HPβCD corresponding to 100mg of ITZ were determined according to European Pharmacopoeia by using no.4 dissolution test apparatus at 37°C, stirred at 100 rpm. The dissolution medium was 1000 ml solution of 0.1N HCl (pH 1.2). Triplicate samples (5 ml) were withdrawn from the dissolution vessels at selected time intervals and analyzed for ITZ content due to the absorbance at 254 nm on an UV spectrophotometer (Shimadzu, Japan). The results are the mean and standard deviations of three determinations.

Differential Thermal Analysis (DSC)

Differential scanning calorimetry (DSC) thermograms of ITZ, HPβCD, physical mixture of ITZ – HpβCD or solid dispersion of ITZ – HPβCD (5–10 mg) were recorded using a thermal analysis system (Mettler Toledo, DSC-1, Switzerland). Samples were heated at 10 °C/min in the range of 25 – 230 °C.

Fourier Transform Infrared Spectroscopy

Samples (about 1% w/w) were mixed with KBr powder and compressed to a disc. The spectra of untreated GLZ and MC-GLZ1 were collected on a FT-IR spectrophotometer (Shimadzu, Japan).

D. Formulation of Hard Capsule Containing ITZ – HPβCD

The highest soluble formulation of ITZ – HPβCD was encapsulated into hard capsule to obtain the final dosage form. ITZ – HPβCD solid dispersion was mixed with aerosil until a homogenous mixture obtained. The tapped density of this mixture was then determined to identify the size of hard capsule.

The dissolution profile of ITZ – HPβCD hard capsule was performed as described in section C.

III. RESULTS AND DISCUSSION

F. ITZ – HPβCD Solid Dispersion Formulation

Dry Grinding Method

The solubility and the dissolution profile of ITZ and ITZ – HPβCD is presented in Figure 1. The solubility of ITZ – HPβCD is higher than that of raw ITZ. Higher HPβCD ratio leads to higher solubility of ITZ – HPβCD. The solubility of ITZ in the ITZ – HPβCD solid dispersion increases 3 times, 5 times and 14 times when the the molar ratio between

HPβCD and ITZ increases from 1:1, 2:1 and 3:1 respectively. PVP does not improve the solubility of these solid dispersion systems.

Dissolution profile of ITZ – HPβCD prepared by dry grinding method is presented in Figure 1. ITZ – HPβCD has faster and higher dissolution profile than that of raw ITZ. However, the dissolution of ITZ does not reach over 20%.

the dissolution profile of all these formulations are lower than 20%. Furthermore, the ITZ – HPβCD at 1:3 (mol/mol) has too much weight to be encapsulated into oral hard capsule (data not show), so the ratio between ITZ and HPβCD was identified at 1:2 (mol/mol) for further study.

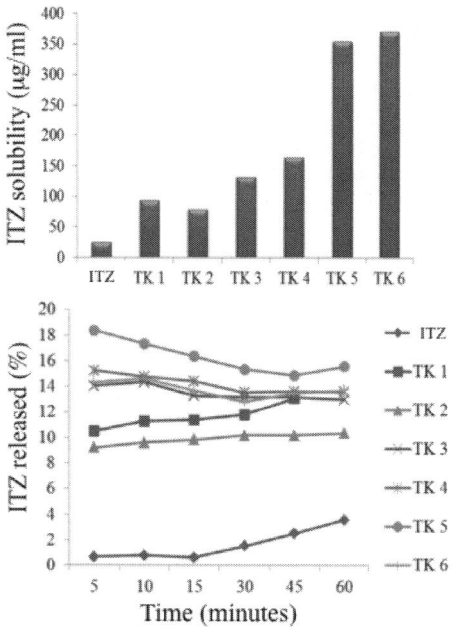

Fig. 1 Solubility and dissolution profile of ITZ and ITZ – HPβCD prepared by dry grinding method.

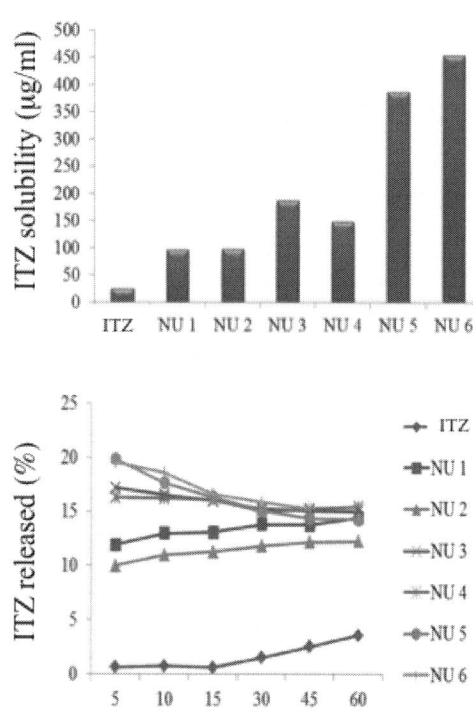

Fig. 2 Solubility and dissolution profile of ITZ and ITZ – HPβCD prepared by wet grinding method.

Wet Grinding Method

The solubility and the dissolution profile of ITZ and ITZ – HPβCD is presented in Figure 2. The solubility of the ITZ – HPβCD is higher than that of raw ITZ. Higher HPβCD ratio leads to higher solubility of ITZ – HPβCD. The solubility of ITZ in the ITZ – HPβCD solid dispersion increases 3 times, 5 times and 15 times when the the molar ratio between HPβCD and ITZ increases from 1:1, 2:1 and 3:1. For ITZ – HPβCD (1:2) formulation, PVP reduces the solubility of ITZ. For ITZ - HPβCD (1: 3) formulation, the presence of PVP increases the solubility of ITZ.

Dissolution profiles of ITZ – HPβCD are faster and higher than that of raw ITZ (Figure 2). However, the dissolution profile does not reach over 20 percent.

ITZ - HPβCD created by both dry grinding or wet grinding improves solubility of ITZ significantly. However,

Co-evaporation Solvent Method

ITZ – HPβCD solid dispersion system prepared by co-evaporation solvent method was performed at the molar ratio of itz: HPβCD 1: 2. Distilled water is selected dissolved HPβCD. ITZ solubles in water, in alcohol and has limited solubility in function of pH, thus EtOH : HCl (10:1 v/v) was selected to solubilize ITZ. This formulation was coded as DM1.

The dissolution profile of DM1 formula is much higher than dissolution profile of raw ITZ and of ITZ- HPβCD prepared by wet grinding method and reach over 80% (Figure 3).

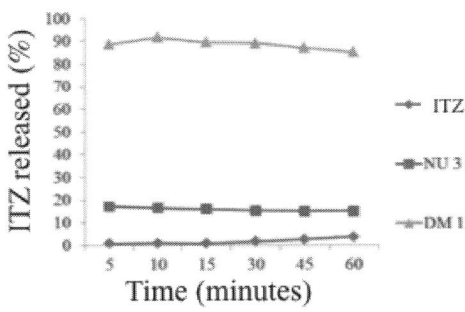

Fig. 3 Dissolution profile of raw ITZ, wet grinding formulation (NU3) and co-evaporation solvent (DM1) method.

C. Solid Dispersion System Characterization

Differential Thermal Analysis (DSC)

ITZ presents crystal structure (figure 4) with a melting point at 167.4 °C. In the physical mixture of ITZ and HPβCD, endothermic peak of ITZ still presents. The NU3 formulation has melting point at 166.15 °C that means this solid dispersion system was not formed completely. While this endo-peak does not appear in the spectrum of DM1.

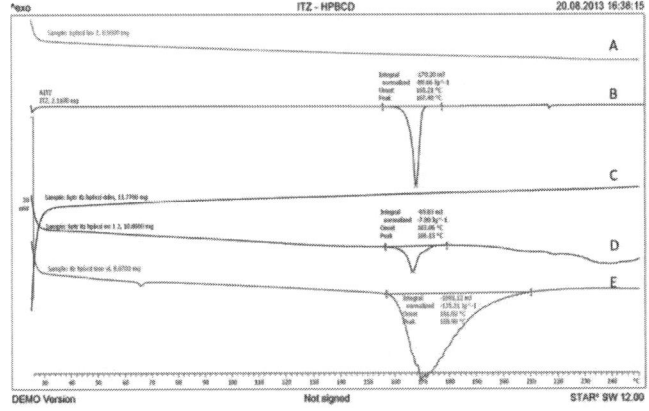

Fig. 4 Differential thermal analysis diagram of HPβCD (A), raw ITZ (B), DM1 (C), NU3 (D) and physical mixture of ITZ / HPβCD at molar ration of 1/2 (E).

Fourier Transform Infrared Spectroscopy

IR spectrum of HPβCD is characterized by: a broad peak in the range 3300-3400 cm^{-1} (3360 cm^{-1}) of the - OH group, a sharp peak at 2930 cm^{-1} of C – H, a peak at 1658 cm^{-1} due to the presence of water in cyclodextrin. The peaks in the region 1300-700 cm^{-1} corresponding C - C and 1153 cm^{-1} peak corresponding to C - O – C.

The characteristic peak ITZ form 500 cm^{-1} to 4000 cm-1 appears at a wavelength of 3325 cm^{-1} and 3111 cm^{-1} due to the absorption of NH2 group. Sharp peak at a wavelength of 1699.2 cm^{-1} is due to vibrations of C = O groups. The

characteristic peaks in the visible wavelength of 1442 cm^{-1} and 1508.2 cm^{-1} related to the links C - H.

For the Fourier transform infrared spectroscopy of DM1 (figure 5), the disappearance of the absorption at 3325 cm^{-1} and 1610.5 cm^{-1} . Moreover, the peak shift from 1442.7 cm^{-1} to 1458.2 cm^{-1} . In addition, there is the apparent disappearance of peak 1610.5 cm^{-1}. This shows ITZ and HPβCD interaction with each other.

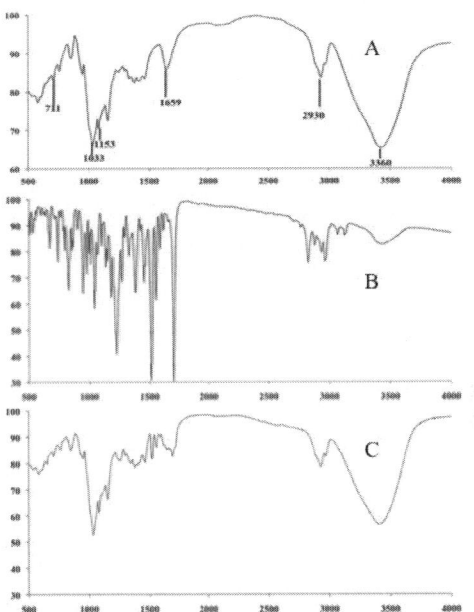

Fig. 5 Fourier spectrum of (A) HPβCD, (B) ITZ and (C) DM1.

H. Formulation of Hard Capsule Containing ITZ – HPβCD

Figure 6 shows dissolution profile of ITZ form various hard capsules contain raw ITZ, NU3 and DM1. The dissolution profile of DM1 hard capsule increase rapidly and reach over 80% after 15 minutes (84 times higher than the solubility of raw ITZ capsule and 36 times higher than NU 3 capsule).

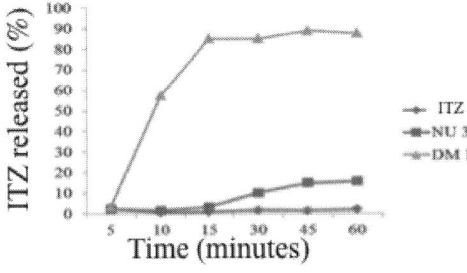

Fig. 6 Dissolution profile of ITZ form hard capsule contain ITZ, NU3 and DM1.

IV. CONCLUSIONS

Solid dispersion formulation based on co-evaporation solvent showed highest dissolution profile. Dissolution profile of solid dispersion-containing hard capsule showed high ITZ release after 10 mins and reach over 80% after 15 mins.

V. CONFLICT OF INTEREST

The authors declare that they have no conflict of interest.

ACKNOWLEDGEMENT

The authors would like to thanks Mr. Phung Duc Truyen for his kind gift of HPβCD. The production of HPβCD is one part of his PhD thesis.

REFERENCES

1. Peeters. J., Neeskens. P., Tolenaere. J. P., Van Remoortere. P., Brewster. M. E. (2002), Characterization of the interaction of 2-hydroxypropyl-β-cyclodextrin with itraconazole at pH 2, 4 and 7, J. Pharm. Sci. 92, 1414 – 1422.
2. Challa. R., Ahuja. A., Ali. J., Khar. R. (2005), *Cyclodextrin in drug delivery: an upload review.* AAPS. Pharm. Sci. Tech. 6, 329 – 35.
3. Marcus. E. Brewster., Thorsteinn Loftsson. (2007), "Cyclodextrin as pharmaceutical solubilizers", *Advanced Drug Delivery Reviews.* 59, 645 – 666.

Author: Tran Van Thanh, Pham Vu Quang Vinh, Huynh Van Hoa
Institute: Faculty of pharmacy, The University of Medicine and Pharmacy
Street: Dinh Tien Hoang
City: Hochiminh
Country: Vietnam
Email: *thanhpharm@gmail.com*

A Potential Application of Vietnamese Rice in Pharmaceutical Industry as a Sustained Release Agent

Vuong Duy Ngo, Thao Truong-Dinh Tran, Toi Van Vo, and Phuong Ha-Lien Tran*

Biomedical Engineering Department, International University, Vietnam National University - Ho Chi Minh City, Vietnam

Abstract— The aim of this study was to modify the rice starch to prolong the release of drug in gastric fluid. Various types of Vietnamese rice were introduced in the study as the matrix of sustained release dosage form. Rice was thermally modified in water for a determined time at different temperatures. The rice starch was first ground and sieved. Then the powder was physically modified by heat and finally dried in an oven. The modified rice starch appeared as an odorless fine white powder. The tablet containing modified rice starch and Paracetamol were prepared by the wet granulation method. The dissolution rate of tablets was evaluated in simulated gastric fluid without pepsin (buffer pH 1.2). HPLC method was used to determine the amount of drug release. The matrix tablet containing thermally modified starch showed sustained release as compared to the control. The thermally modified starch could be a novel pharmaceutical excipient for sustained release dosage form design.

Keywords— modified rice starch, paracetamol, sustained release.

I. INTRODUCTION

Starch like rice starch could be undergone a number of chemical or physical modifications to yield modified starches with distinctive properties for applications in food and pharmaceutical industries. A thermally modified rice starch has been expected to be a promising pharmaceutical excipient to sustain the drug release from the tablets. Without modification, native starch from any source is undesirable for pharmaceutical applications due to their inability to withstand processing conditions such as extreme temperature, diverse pH, high shear rate, and freeze thaw variation. The process of starch modification involves the destructurisation of the semi-crystalline starch granules and the effective dispersion of the component polymers. In this way, the reactive sites (hydroxyl groups) of the amylopectin polymers become accessible to electrophilic reactants [1]. Modified starch can be obtained through physical, chemical and enzymatic processes.

Chemical modification is generally achieved through etherification, esterification, crosslinking, oxidation, cationization and grafting of starch. The chemical and functional properties achieved following chemical modification of starch, depends largely on the botanical or biological source of the starch, reaction conditions (reactant concentration, reaction time, pH and the presence of catalyst), type of substituent, extent of substitution (degree of substitution, or molar substitution), and the distribution of the substituent in the starch molecule [2].

Physical modification has been used to change the granular structure and improve the ability of swelling property in cold water of some native starches. This physical modification has been studied by several methods such as extrusion, drum-drying and a controlled pregelatinization-spray-drying [3]. A method for preparing granular cold water-soluble starches by injection and nozzle-spray drying was described by Pitchon and Joseph [4]. Potato starch, corn starch, rice starch and wheat starch have been thermally modified by hot stage extrusion [5]. Waxy corn starch (100% amylopectin) [6], wheat starch [7], maize starch [8] and rice starch [9] have been thermally modified. All of them could retard the disintegration and drug release from the tablets.

Jasmine, Huong Lai Sua, Nang Hoa, Nep Bac are sticky rices, a kind of rice commonly cultivated in Viet Nam. These rices have been widely used in food industry but not have been investigated for many applications in pharmaceutical industry. They are high amylopectin starches which do not swell in cold water. Modified glutinous rice by physical modification which can swell in cold water might be an interesting candidate for controlling release of a drug from a dosage form. Hence, this study investigated the modification of rice starch to improve its properties for pharmaceutical applications including swellability and especially the sustained release of drug.

II. MATERIAL AND METHODOLOGY

1. Materials

Four varieties of sticky rice (Jasmine, Huong Lai Sua, Nang Hoa, Nep Bac) were purchased at local supermarket in Ho Chi Minh city. These rices were harvested and packed in the same year (2013). The rices were stored at room temperature (25°C) for study. Microcrystalline cellulose (Avicel PH 102) was purchased from Brenntag Group (Germany). Magnesium stearate (MgS) was purchased from

---

* Corresponding author.

Nitika Pharmaceutical Specialities Pvt. Ltd (India). Hydrochloric acid (HCl) and Sodium chloride (NaCl) were purchased from Xilong Chemical Industry Incorporated Company (China). Paracetamol (acetaminophen) was purchased from Hebei Jihengcompany (China).

2. Modification of Rice Starch

Four varieties of sticky rice were milled and sieved by 500 μm sieve. 50g of each kind was put into a 600ml beaker, treated with 100 ml distilled water and heated in an oven at different temperatures and time. These samples were dried at 40°C for 48 hours. Then, the dried samples were milled and passed through 500 μm sieve to have an odorless fine white powder. This product was named modified rice starch (MRS).

3. Preparation of Tablets

The compositions of the tablet are given in Table 1. Matrix tablets were prepared by the wet granulation method. Paracetamol and MRS were thoroughly mixed. PVP K30 was dissolved in ethanol to get a binder solution. The sample was passed through a 500 μm sieve to have granular and dried in an oven at temperature of 40°C for 60 min. Dried granules was passed through 500 μm sieve again and then mixed with Avicel PH 102 and finally, further blended with magnesium stearate. The mixtures were compressed into the tablet of diameter 13 mm using TDP 1.5 tablet press machine.

Table 1 Composition of modified starch-based tablet

Paraceltamol	Modified rice starch (MRS)	Avicel PH 102	PVP K30	Magnesium stearate	Ethanol (ml)	Weight/tablet (mg)
325 mg	100 mg	20 mg	50 mg	5 mg	120 μl	500mg

4. Dissolution Test

The tablets were conducted at 37±0.5°C in a USP specification dissolution test type II apparatus (Paddle apparatus). Each of 900ml of buffer pH 1.2 was added into dissolution vessel. The apparatus was set up at 50 rpm of rotation speed. 1ml of sample was collected from the media at predetermined intervals of 15, 30, 60 min. Each time 1ml of dissolution sample was compensated by adding 1ml of the corresponding fresh buffer. 100μl sample solutions were diluted with 900μl distilled water for HPLC test.

5. High Performance Liquid Chromatography (HPLC)

The quantification of paracetamol in tablets was performed using Ultimate 3000 HPMC ThermoscientificInc., USA. The mobile phase contained buffer solution (pH 3.5, adjusted by phosphoric acid) and acetonitrile at the ratio 3:1 with a flow rate of 1ml/min. The UV/VIS detector was set at a wavelength of 207nm. 20μL of sample was injected to the HPLC system.

III. RESULTS AND DISCUSSION

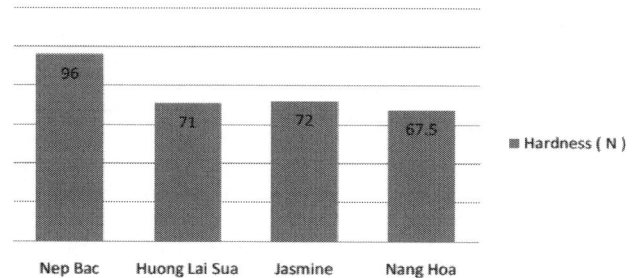

Fig. 1 Hardness of tablets from 4 types of rice

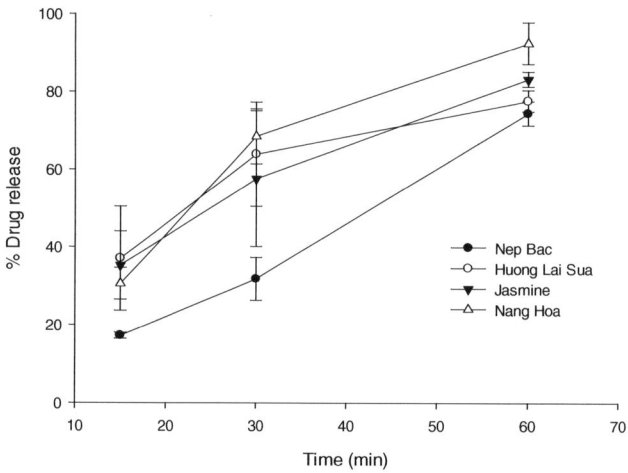

Fig. 2 Dissolution rate of tablets from 4 types of rice at pH 1.2

Figure 1 shows the hardness of tablets from modified rice starch of Nep Bac, Huong Lai Sua, Nang Hoa and Jasmine. The Nep Bac-based tablets had the highest hardness as compared to others. In adddition, the dissolution rate of tablets from Nep Bac was more sustained release than that from tablets of Huong Lai Sua, Nang Hoa and Jasmine (Figure 2). Specifically, at 15 min the drug dissolution release from Huong Lai Sua, Jasmine, Nang Hoa tablets around 30-35% while the dissolution rate of Nep Bac tablets was 17.2 %. At 30 min, the dissolution rate of tablets from Huong Lai Sua, Jasmine, Nang hoa increased to 57-68 %,; however, the dissolution rate of Nep Bac remained at 31.8 %. Therefore, Nep bac was a potential rice to prolong the release of drug and selected for further studies.

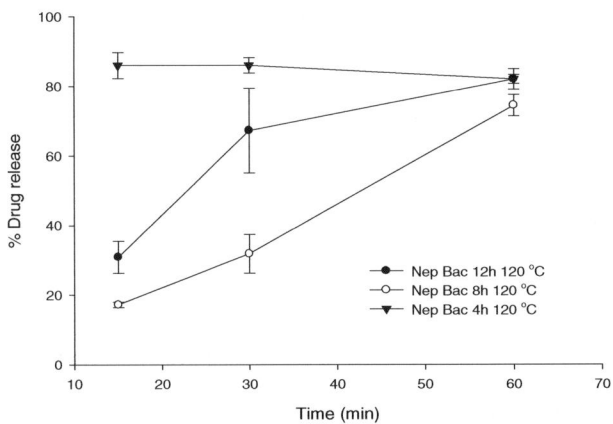

Fig. 3 Effect of modified time on dissolution rate of tablets from Nep Bac at pH 1.2

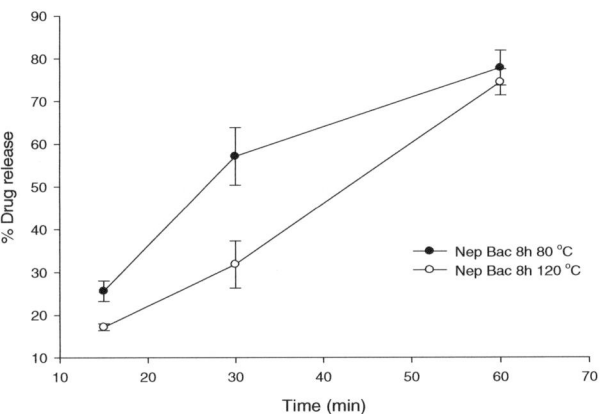

Fig. 4 Effect of modified temperature on dissolution rate of tablets from Nep Bac at pH 1.2

Figure 3 and Figure 4 show the effect of time and temperature in the modification process on the dissolution rate of tablets. By increasing the time of heating from 4 h to 8 h, the modified rice starch can retard the dissolution rate. However, by increasing to 12 h, it could reduce the sustained ability of the tablet. Therefore, the period of 8h is the maximum time for the thermal modification. On the other hand, by increasing the temperature from 80 °C to 120 °C, the dissolution rate of tablet could be retarded. These results demonstrated that factors of time and temperature critical for modifying rice starch.

IV. CONCLUSIONS

This study provides potential results of an appropriate method to modify the rice starch to prolong the release of drug. Moreover, the rice starch is really a promising candidate to apply as a hydrophilic matrix. However, further studies are required to evaluate for the applications.

ACKNOWLEDGMENT

This *research* is funded by International University, Vietnam National University – Ho Chi Minh City under grant number T2014-01-BME.

REFERENCES

1. Rajan, A., Sudha, J.D. and Abraham, T.E., (2008). Enzymatic modification of cassava starch by fungal lipase.Industrial Crops and Products, 27:50-59.
2. Singh, J., Kaur, L. and McCarthy, O.J., (2007). Factors influencing the physico-chemical, morphological, thermal and rheological properties of some chemically modified starches for food applications--A review. Food Hydrocolloids, 21:1-22.
3. J. Herman, J.P.Remon and J.De Vilder (1989). Modified starches as hydrophilic matrices for controlled oral delivery. International Journal of Pharmaceutics, 56:51-63.
4. [Pitchon. E, O.R.J.D., & Joseph T. H (1981). Process for cooking or gelatinizing materials.
5. Henrist, D., & Remon, J. P. (1999). Influence of the formulation composition on the in vitro characteristics of hot stage extrudates. International Journal of Pharmaceutics, 188:111–119.
6. Visavarungroj, N., Herman, J., & Remon, J. P. (1990). Crosslinked starch as sustained release agent. Drug Development and Industrial Pharmacy, 16: 1091–1108.
7. Sanchez, L., Torrado, S., & Lasters, J. L. (1995). Gelatinized/freeze-dried starch as excipient in sustained release tablets. International Journal of Pharmaceutics,115: 201–208.
8. A. Cristina Freire, Christiane C. Fertig, Fridrun Podczeck, Francisco Veiga, João Sousa (2009). Starch-based coatings for colon-specific drug delivery. Part I: The influence of heat treatment on the physico-chemical properties of high amylose maize starches. European Journal of Pharmaceutics and Biopharmaceutics 72:574–586.
9. Peerapattana, J., Tattawasart, A., & Srijesdaruk, V. (2004). Modified glutinous starch as a hydrophilic matrix substance.Thai Journal of Pharmaceutical Sciences, 28: 57–72.

*Corresponding author: Phuong Ha-Lien Tran
Institute: Biomedical Engineering Department, International University, Vietnam National University - Ho Chi Minh City
City: Ho Chi Minh City
Country: Vietnam
Email: thlphuong@hcmiu.edu.vn

Use of Microwave Method for Controlling Drug Release of Modified Sprouted Rice Starch

Thinh Duc Luu, Nam Hoang Phan, Thao Truong-Dinh Tran, Toi Van Vo, and Phuong Ha-Lien Tran

Biomedical Engineering Department, International University, Vietnam National University - Ho Chi Minh City,
Vietnam

Abstract— **The study aimed to investigate a new method of thermal modification which was microwave method for controlling drug release of rice starch. The Vibigaba sprouted rice starch (Vietnam) was selected as the model rice. The rice was thermally modified in water for a determined time at different power. The rice starch was first ground and sieved. Then, the powder was physically modified by heat via the microwave and finally dried in an oven. Tablets containing the modified sprouted rice starch of Paracetamol were prepared by the wet granulation method. The dissolution rate of tablets was evaluated in simulated gastric fluid without pepsin (buffer pH 1.2). The matrix tablet containing thermally modified sprouted rice starch showed a sustained release, indicating that the sprouted rice under microwave method could be a promising material for a sustained release dosage form.**

I. INTRODUCTION

Natural starch has been recognized as one of the common excipients on manufacturing of tablets. Starch is, after cellulose, the most abundantly distributed carbohydrate in the plant world. Each starch granule consists of concentric growth rings of alternating amorphous and semi-crystalline structures composed of amylase and amylopectin [1]. Amylose is a long, essentially linear polymer and only sparsely branched, consisting of 1, 4-linked-a-D-anhydroglucose units. Amylopectin is the main constituent of the starch granule (30–99%) and it appears as a highly branched polymer comprising of a large amount of short linear chains [2]. In the Pharmaceutical Industry, starch has been primarily used as a bulking agent, binder, disintegrant and thickening agent.

To control drug release, a suitable hydrophilic polymer is swellable in the presence of water, forming a substance with a gel. The release behaviors of drug depend on type of the matrix, the capability of swelling, diffusion and erosion processes [3]. One special developed product is called pregelatinized starch which has been made by mechanically processed to rupture all or part of the granules in the presence of water and subsequently dried [4]. Pregelatinized starches can improve the functional properties including swelling in cold water for gel information. Therefore, a thermally modified rice starch has been expected to be a promising pharmaceutical excipient to sustain the drug

release from the tablets. So far there has been no report using microwave method to thermally modify rice starch.

Sprouted rice named Vibigaba has been recently launched by JSC An Giang Plant Protection, Vietnam, with prominent benefits for human health. It is produced by a new technology called "sprouted grain" with many advanced features in Vietnam. The findings of HCM City Nutrition Center showed that the active ingredient in rice Vibigaba help stabilize blood sugar, an important characteristic in highly effective prevention and supportive treatment of serious diseases such as diabetes and high blood pressure.

This study investigated effects of the microwave method on thermal modification of Vibigaba rice. The modified rice has been expected to be an alternative of sustained release materials in pharmaceutical industry.

II. MATERIALS AND METHODOLOGY

A. Materials

The sprouted rice (Vibigaba) was purchased at local supermarket in Ho Chi Minh City. This rice were harvested and packed in the same year (2014). The rice was stored at room temperature (25°C) until required for experimentation. Microcrystalline cellulose (Avicel pH 102) was purchased from Brenntag Group (Germany). Magnesium stearate (MgS) was purchased from Nitika Pharmaceutical Specialties Pvt. Ltd (India), Hydrochloric acid (HCl) Sodium chloride (NaCl) was purchased from Xilong Chemical Industry Incorporated Company (China). Paracetamol (acetaminophen) was purchased from Hebei Jiheng Company (China).

B. Modification of Rice Starch

The sprouted rice was milled and sieved through 425μm sieve. 50g of sieved rice was poured into 100mL distilled water. Then, it was heated in microwave by 200, 400 and 800 W for 8.5 minutes. The sample was dried in the oven at temperature of 40 °C, then milled and passed through 500μm sieve to have a uniform particle size powder. Finally the modified rice starch (MRS) was obtained.

© Springer International Publishing Switzerland 2015
V. Van Toi and T.H. Lien Phuong (eds.), *5th International Conference on Biomedical Engineering in Vietnam,*
IFMBE Proceedings 46, DOI: 10.1007/978-3-319-11776-8_76

Fig. 1 Vibigaba sprouted rice

C. Preparation of Tablets

The compositions of the tablet are given in Table 1. Matrix tablets were prepared by wet granulation method. Paracetamol and MRS were thoroughly mixed. PVP K30 was dissolved in ethanol for a binder solution. The sample was passed through a 500 μm sieve to have granular and dried by oven at temperature of 40 °C for 60 minutes. Dried granules was sieved through 500 μm sieve again and then mixed with Avicel PH 102 and further blended with magnesium stearate. The mixtures were compressed into the tablet of diameter 13 mm using TDP 1.5 tablet press machine.

Table 1 Composition of tablet

Paracetamol	MRS	Avicel PH 102	PVP K30	MgS	Ethanol
325 mg	100 mg	20 mg	50 mg	5 mg	120 μl

D. Dissolution Test

The tablets were conducted at 37±0.5 ^{0}C on an USP specification dissolution rate test type II apparatus (Paddle apparatus). For *in vitro* dissolution test, buffer pH 1.2 was used as dissolution media. Water bath temperature was fixed at about 37 ± 0.5 ^{0}C. Each of 900ml of buffer pH 1.2 was added into dissolution vessel. The apparatus was set up at 50 rpm of rotation speed. 1ml of sample was collected from the media at predetermined intervals of 15, 30, 60 min. Each time 1ml of dissolution sample was compensated by adding 1ml of the corresponding fresh buffer. 100μl sample solutions were diluted with 900μl distilled water for HPLC test.

E. High Performance Liquid Chromatography (HPLC)

The quantification of paracetamol in tablets was performed using Ultimate 3000 HPMC Thermoscientific Inc., USA. The mobile phase contained buffer solution (pH 3.5, adjusted by phosphoric acid) and acetonitrile at the ratio 3:1 with a flow rate of 1ml/min. The UV/VIS detector was set to a wavelength of 207nm. 20μL of sample was injected to HPLC system.

III. RESULT AND DISCUSSION

Figure 2 shows hardness of the tablets which were prepared with MRS by microwave method at different powers. Hardness of the tablets prepared by MRS at 200 W was lower than that of tablets prepared by MRS at 300 W. However, as the power increased to 400 W and 800 W, hardness of the tablets decreased. Hardness of tablets prepared by MRS at 800 W was even lower than that of tablets prepared by MRS at 400 W. These results indicated an important fact that hardness of the tablets was increased as power of microwave oven increased, but only increased to a certain level of the power. If the power was over this value, the hardness would be decreased again.

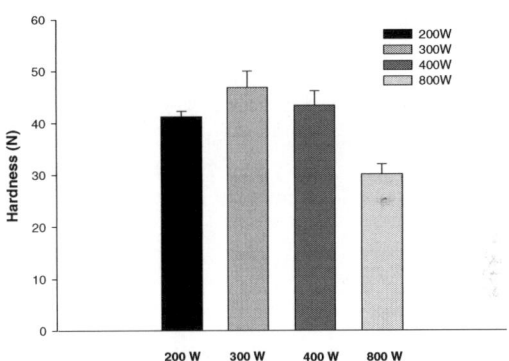

Fig. 2 Hardness of tablets prepared by various power input

The higher hardness of tablets generally leads to the longer sustained release of drug from the tablets because a longer period is needed for tablets to absorb water or disintegration of tablets in the medium. All of the dissolution profiles of tablets prepared by MRS showed their controlled drug release manner. Drug release rate from tablets prepared by MRS at 800 W reached the highest (Figure 3). This result matched with the hardness of tablets in Figure 2 where hardness of tablets prepared by MRS at 800 W was the lowest. Drug release rates from the other tablets were also corresponding to hardness of the tablets in the previous figure. The drug release from tablets prepared by MRS at 200 W, 300 W and 400 W was not different from each other. Above all, the results indicated that the microwave method is a promising way to thermally modify the Vibigaba for controlling drug release.

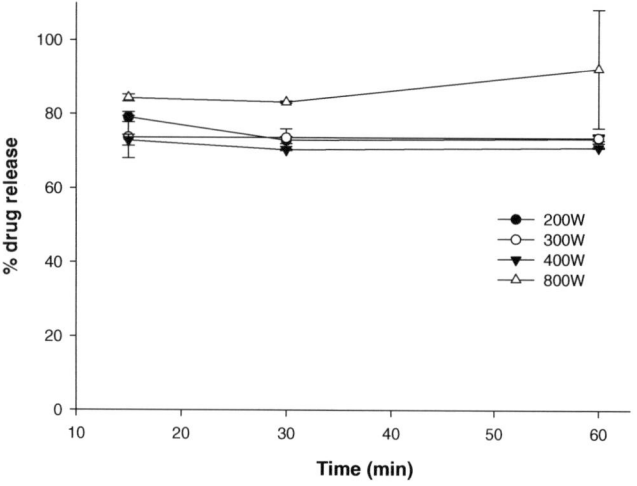

Fig. 3 Dissolution rate of tablets at pH 1.2

IV. CONCLUSION

The microwave method should be considered as an effective method for thermal modification of properties of rice starch. The modified rice starch in the study showed its potential in controlling drug release at the simulated gastric fluid. Further studies are required to evaluate the potential applications of the study in pharmaceutical applications.

ACKNOWLEGEMENT

This *research* is funded by International University, Vietnam National University – Ho Chi Minh City under grant number T2014-01-BME.

REFERENCES

1. D.J. Gallant, B. Bouchet, A. Buléon, S. Pérez, Physical characteristics of starch granules and susceptibility to enzymatic degradation, Eur. J. Clin. Nutr. 46 (1992) 3–16.
2. D.J. Gallant, B. Bouchet, P.M. Baldwin, Microscopy of starch: evidence of a new level of granule organization, Carbohydr. Polym. 32 (1997) 177–191.
3. Effionora Anwar, 2006. An Approach on Pregelatinized Cassava Starch Phosphate Ester as Hydrophilic Polymer Excipient for Controlled Release Tablet , J.Med Sci.,6 : 923 -929.
4. Colombo, P., R. Bettini, G. Massimo, P.L. Catellani, P.Sant and N.A. Peppas, 1995. Drug diffusion front movement is important in drug release control form swell able matrix tablet, J. Pharm.Sci., 8: 991-997.

*Corresponding Author: Phuong Ha-Lien Tran
Institute: Biomedical Engineering Department, International University, Vietnam National University - Ho Chi Minh City
City: Ho Chi Minh City
Country: Vietnam
Email: thlphuong@hcmiu.edu.vn

Fabrication of In Situ Cross-Linking Polyvinyl Phosphonic Acid - Chitosan Hydrogel for Wound Applications

Le Quoc Tuan, Dang Hoang Phuc, Vo Van Toi, and Thi-Hiep Nguyen*

Department of Biomedical Engineering, International University of Vietnam National Universities,
Ho Chi Minh City, Vietnam

Abstract— **The use of natural polymer as scaffold material in hydrogels has been applied for skin regeneration due to their high biocompatibility, low toxicity, and biodegradability. Hydrogel was synthesized by *in situ* crosslinking of polyvinyl phosphoric acid (PVPA) and chitosan (Cs) in this study. Fourier transform infrared spectrum (FT-IR) analysis and scanning electron microscope (SEM) were used to confirm characteristics of hydrogel. The results showed that the crosslinking was formed successfully, and the inner morphology of hydrogels was appropriate for wound application.**

Keywords— **Chitosan, PVPA, *in situ* crosslinking hydrogel.**

I. INTRODUCTION

Hydrogels are highly hydrophilic polymer networks which may absorb from 10-20% up to thousands of times their dry weight in water[1]. Hydrogel can be formed from either natural or synthetic polymers. Hydrogels from natural origin have structural similarity to the extracellular matrixes found in tissues and are considered biocompatible[2]. *In situ* crosslinking hydrogels have been widely applied in tissue engineering as well as in drug delivery and in stem cell delivery[3]. For wound healing, *in situ* crosslinking hydrogel is a good choice for scaffolds to engineer new tissue. Indeed, hydrogels have excellent mechanical characteristics to use for soft tissue, and the pore size of the gel network can be controlled to transport the biological molecules [4]and to load the stem cells.

Chitosan is a poly-(β-1,4-D-glucosamine) derived from the N-deacetylation of chitin, a naturally common component of the exoskeleton in crustaceans and insects[5]. Chitosan is polycationic at pH <6, therefore it easily interacts with negatively charged molecules, such as protein[6]. In skin regeneration, chitosan is used as component in hydrogels[7]. Chitosan is excellent for wound healing because of four main reasons. First, chitosan is non-toxic product for body environment [5, 7]. It is known for being biocompatible and has been used in many biomedical applications [3, 8] such as contact lenses in ophthalmology [5]. Moreover, chitosan is a biodegradable molecule. When chitosan is applied to the body, it is slowly metabolized by certain human enzymes, especially lysozymes and chitosanase. Second, chitosan is useful for wound treatment due to the bacteriostatic and fungi static effects. Third, chitosan structurally resembles glycosaminoglycan (GAG) [9], so it is important to maintain cell morphology, differentiation, and function. Finally, chitosan is naturally abundant, of low cost, and ecologically interesting [1].

Polyvinyl phosphonic acid (PVPA), a phosphonate-containing polymer, is hypothesized to mimic the action of bisphosphonates, a group of drugs used in the treatment of osteoporosis [10-12]. Vinylphosphonates have been known since the middle of the 20th century [13, 14]. PVPA are also the non-toxic polymers [15]. Phosphonate-containing hydrogels are applied in cartilage, dental cement, and bone engineering because of their similarity to the mineral components of hard tissues[15, 16]. Hydroxyapatite or Hap was proposed as a substituent for defective bones and teeth [17]. In addition, the PVPA hydrogel surface can support cell adhesion and proliferation through the interaction of serum protein molecules and the gel network[18], while the structure of the hydrogel scaffold allows strength over time [11].

So far, there has been no report about in situ crosslinking hydrogel PVPA/Cs. Therefore, the aim of this research is to fabricate and analyze the properties of a novel in situ crosslinking hydrogel (PVPA/Cs) for wound application. Hydrogels were formed by covalent crosslinking between –OH groups of chitosan and –P-OH groups of PVPA[6]. This reaction was observed through the presence of $-CH_2-O-P-CH_2-$ stretching peaks from FT-IR analysis. PVPA/Cs hydrogels compose of a structural similarity to GAG[9], antibacterial advantage of chitosan[15], and cell attraction of PVPA[12]. The fibroblast cell proliferated remarkably on hydrogel are basic evidences in order to approve PVPA/Cs hydrogel as a new biomaterials for skin regenerative applications.

II. MATERIALS AND METHODS

2.1. Materials

Polyvinyl phosphonic acid (PVPA, 30% solution) was purchased from Polysciences (USA). Chitosan (>75% deacetylated) was purchased from Sigma.

---

* Correspondinag author.

© Springer International Publishing Switzerland 2015
V. Van Toi and T.H. Lien Phuong (eds.), *5th International Conference on Biomedical Engineering in Vietnam,*
IFMBE Proceedings 46, DOI: 10.1007/978-3-319-11776-8_77

2.2. *Synthesis of PVPA/Chitosan Hydrogel*

Chitosan was divided into two groups of solutions: chitosan 1% and 2%. Chitosan solutions were gently stirred at room temperature to dissolve completely. PVPA30% solution in various volumes (30,200,300,400 µL) was dissolved in water. Then 100 µL chitosan solutions were taken by pipette and poured into test tubes. Hydrogels were formed by injecting PVPA solutions, and were labeled A1, A2, A3, A4, A5, A6, B, C (Table 1). The blend component were mixed by vortex, when the liquid level in the test tube stopped moving, the gelation time of each sample was recorded.

2.3. *Characterization of PVPA/Chitosan Hydrogel*

2.3.1. FT-IR Analyze

The IR spectra of chitosan (a), PVPA/Cs (b) were characterized by attenuated reflectance Fourier transform spectroscopy (Spectrum GX, PerkinElmer, USA). The infrared spectra of the samples were measured over a wavelength range of $4000 - 400$ cm^{-1} [19].

2.3.2. Scanning Electron Microscope

All of the samples were kept under vacuum conditions to remove humidity. Next, the samples were coated by platinum. The detailed morphologic structure of hydrogels was observed using scanning electron microscope (SM-65F, JEOL, Japan).

III. RESULTS AND DISCUSION

Crosslinking hydrogels possess the flexibility and the softness which are very similar to vital tissue. This property may be due to their significant water content. Water inside the polymer may bring some biological solute molecules, while the hydrogel is as the matrix to hold water together[20]. Crosslinking hydrogels are currently the good components in tissue engineering. When used as scaffolds, hydrogels can transport human cells to repair tissue. In our research, we developed an in situ crosslinking hydrogel that supports cell proliferation and growth to allow skin regeneration.

Table 1 Properties and characteristics of hydrogels

	A	B	C	CS1%	CS2%	Gelation time (seconds)
A1	200µL	-	-	-	100µL	130
A2	200µL	-	-	100µL	-	165
A3	300µL	-	-	-	100µL	150
A4	300µL	-	-	100µL	-	180
A5	400µL	-	-	-	100µL	195
A6	400µL	-	-	100µL	-	240
B	-	300µL	-	-	100µL	-
C	-	-	30µL	-	100µL	-

A: 200µL PVPA 30% + 2800µL H$_2$O   C: PVPA 30%
B: 300µL PVPA 30% +1700µL H$_2$O   CS: Chitosan

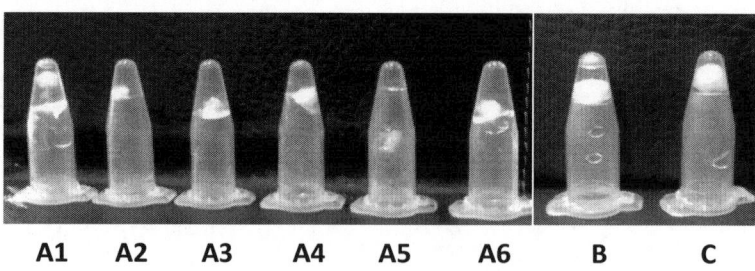

Fig. 1 Photographs of hydrogels

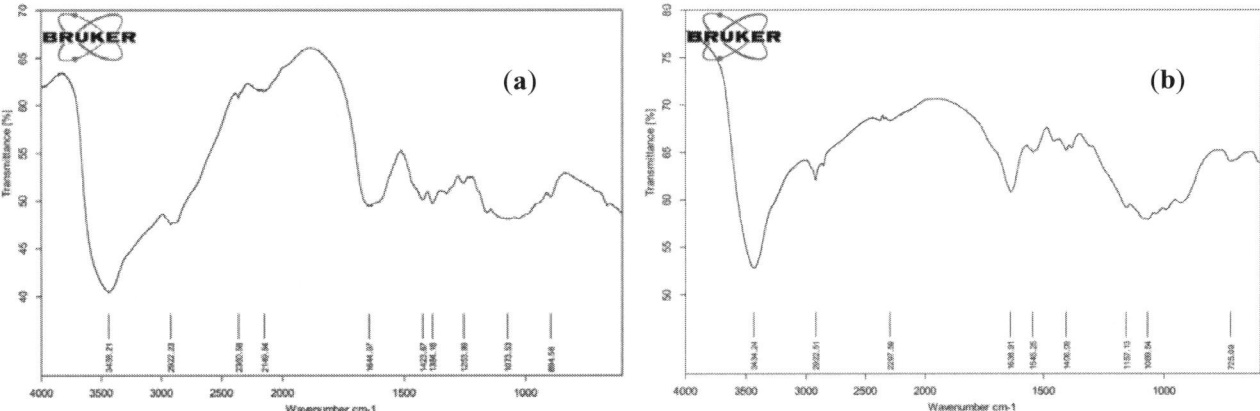

Fig. 2 FT-IR spectrum of chitosan (a) and PVPA/Cs hydrogel (sample C) (b).

Gelation time increases regularly to amount of PVPA and chitosan.

In situ crosslinking PVPA/Cs hydrogel is formed by crosslinking the -OH group of chitosan and the -P-OH group of PVPA. The FT-IR spectra of chitosan and the PVPA/Cs hydrogels were analyzed to test the formation of this crosslinking.

The FT-IR spectrum of pristine chitosan is presented in Figure 2 (a).The absorption band at 1644 cm^{-1} is corresponding to amide carboxyl group[21]. The absorption band between 1000 and 1100 cm^{-1} may be due to three differential vibrations of C-O-C, C-OH, and C-C of the ring[21]. The N-H and O-H stretching vibrations are also present at 3200 – 3500 cm^{-1}.

Figure 2 (b) shows the FT-IR spectrum of hydrogel. The FT-IR spectrum of PVPA/Cs hydrogel confirmed that chitosan and PVPA cross-linked by both covalent and ionic interaction. The significant shift in absorption band from 1600 cm^{-1} to 1545 cm^{-1} for PVPA/Cs hydrogel can be attributed to NH^{3+} bending vibration, which shows the electrostatic interaction between PVPA and chitosan [21, 22]. The covalent bonding was indicated by the presence of -CH2-O-P-O-CH2- stretching mode at 1000 – 1100 cm^{-1}[10].

For wound application, the scaffold with macroporous structure is recommended for capillaries ingrowths[9]. Thus, the morphology and pore sizes of hydrogels were observed scanning electronic microscope analysis. Result of SEM images indicated that the pore sizes of PVPA/Cs hydrogel are suitable for substitute materials in wound applications.

Fig. 3 SEM morphology of sample C.

IV. CONCLUSION

We introduced a novel in situ crosslinking hydrogel for wound application, including chitosan and polyvinyl phosphonic acid. The inner morphology and characteristics of hydrogel were evaluated by using FT-IR analysis and SEM observation. The interaction between chitosan and PVPA formed successfully in situ crosslinking through covalent bonding, which was indicated by FT-IR spectrum. The pore sizes of hydrogels seem to promise a great scaffold for skin regeneration. Therefore, PVPA/Cs in first step may be confident to become a new product for clinical wound healing.

ACKNOWLEDGMENT

This research is funded by Vietnam National University HoChiMinh City under grant number: B2013-76-03.

This research is funded by Office of Navy Research (ONR) under grant number: N62909-14-1-N011, PR No: N6290914PR00015/N00014.

REFERENCES

1. Hoffman, A.S., Hydrogels for biomedical applications Advanced Drug Delivery Reviews, 2002. 43: p. 3-12.
2. Kuen Yong Lee, D.J.M., *Hydrogels for Tisssue Enginering*. Chemical Reviews, 2001. 101: p. 1869-1879.
3. WJoshua S. Boateng, K.H.M., Howard N.E. Stevens, Gillian M. Eccleston, Wound healing dressings and drug delivery systems: A review. Journal of Pharmaceutical Sciences, 2008. 97(8): p. 2892-2923.
4. Drury JL, M.D., Hydrogels for tissue engineering: scaffold design variables and applicaions. Biomaterials, 2003. 24: p. 437-4351.
5. Pradip Kumar Dutta, J.D., V S Tripathi, Chitin and chitosan: Chemistry, properties and applications. Journal of Scientific & Industrial Research, 2004. 63: p. 20-31.
6. J. Berger, M.r., J.M. Mayer, O. Felt, N.A. Peppas, R. Gurny, Structure and interactions in covalently and ionically crosslinked chitosan hydrogels for biomedical aplications. European Journal of Pharmaceutics and Biopharmaceutics, 2004. 57: p. 19-34.
7. Willi Paul, C.P.S., Chitosan and Alginate Wound Dressings: A short review. Trends Biomater. Artif. Organs, 2004. 18: p. 18-23.
8. Gregorio Crini, P.-M.B., Application of chitosan, a natural amnopolysacchrid, for dye removal from aqueous solutions by adsorption processes using batch studies: A review of recent literature. Progress in Polymer Science, 2008. 33: p. 39-447.
9. Kelly R. Kirker, Y.L., J. Harte Nielson, Jane Shelby, Glenn D. Prestwich, Glycosaminoglycan hydrogel films as bio-interactive dressings for wound healing. Biomaterials, 2002. 23: p. 3661-3671.
10. Nguyen Thi-Hiep, D.V.H., Vo Van Toi, Injectable in situ crosslinkable hyaluronan-polyvinyl phosphonic acid hydrogels for bone engineering. Journal of Biomedical Science and Engineering, 2013. 6: p. 854-862.
11. Lavinia Macarie, G.I., Poly(vinylphosphonic acid) and its derivatives. Progress in Polymer Science 2010. 35: p. 1078-1092.
12. Z. Durmus, h.E., A. Aslan, M.S. Toprak, h. Sozeri, A. Baykal, Synthesis and characterization of poly(vinyl phosphonic acid) (PVPA)-Fe3O4 nanocomposite. Polyhedron, 2011. 30: p. 419-426.
13. Ford-Moore AH, W.J., The reaction between trialkyl phosphites and alkyl halides. J Chem Soc, 1947. 1465-1467.
14. GM, K., Isomerization of alkyl phosphites. J Am Chem Soc, 1948. 1971-1972.
15. Anbar M, N.G., St John GA, Fate and toxicity of orally administered polyethylene polyphosphonates. Food and Cosmetic Toxicology, 1973. 11(6): p. 1001-1010.
16. Wang D, e.a., Synthesis and characterization of a novel degradable phosphate-containing hydrogel. Biomaterials, 2003. 24: p. 3969-3980.
17. Wei G, M.P., Structure and properties of nano-hydroxyapatite/polymer composite scaffolds for bone tissue engineering. Biomaterials, 2004. 47: p. 49-57.
18. Jian Tan, R.A.G., Mandy Ma, W. Mark Saltzman, Improved cell adhesion and proliferation on synthetic phosphonic acid-containing hydrogels. Biomaterials, 2005. 26: p. 3663-3671.
19. J. Brugnerotto, J.L., F.M. Goycoolea, W. Arguelles-Monal, J. Desbrieres, M. Rinaudo, An infrared investigation in relation with chitin and chitosan characterization. Polymer, 2001. 42: p. 3569-3580.
20. Shakti Dwivedi, P.K., Gilhotra Ritu Mehra, Vikash Kumar, Hydrogel-A Conceptual Overview. International Journal of Pharmaceutical & Biological archives 2011. 2(6): p. 1588-1597.
21. Fadime Goktepe, S.U.C., Ayhan Bozkurt, Preparation and the proton conductivity of chitosan/poly(vinyl phosphonic acid) complex polymer electrolytes. Journal of Non-Crystalline Solids, 2008. 354: p. 3637-3642.
22. Celine Abueva, B.-T.L., Poly(vinyl phosphonic acid) immobilized on chitosan: A glycosaminoglycan-inspired matrix for bone regeneration. Intrnational Journal of Biological Macromolecules, 2014. 64: p. 294-301.

Corresponding author:

Author: Thi-Hiep Nguyen
Institute: Biomedical Engineering Department, International University, Vietnam National Universities
Street: Quarter 6, Linh Trung Ward, Thu Duc District
City: Ho Chi Minh
Country: Vietnam
Email: nthiep@hcmiu.edu.vn

Development of a New Injectable PVA–Ag NPs/ Chitosan Hydrogel for Wound Dressing Application

Xuan-Truong Nguyen, Vo Van Toi, and Thi-Hiep Nguyen*

Department of Biomedical Engineering, International University of Vietnam National University,
Ho Chi Minh City, Vietnam

Abstract— In the present time, functional and cost effective of artificial skin is an interested in study. For this aim, herein, we developed a new injectable polyvinyl alcohol–silver nanoparticles/chitosan (PVA–AgNPs/CS) hydrogel for wound dressing skin applications. Thus, silver (Ag) is an antibiotic agent, was loaded inside of hydrogel and its mechanical property, biodegradable, swelling property have been investigated. Results were showed the good flexible ability, good antibiotic, suitable the biogradable rate and swelling rate, and high mechanical property. The PVA – Ag NPs/ CS hydrogels could be used for the wound dressing skin treatment in the near future.

Keywords— Wound dressing, Polyvinyl Alcohol, Chitosan, hydrogel, Ag nanoparticles.

I. INTRODUCTION

The effects of environmental pollution and chemical have led to many strange diseases, skin cancer and severe burns [1]. Nowadays, the development of aesthetic medicine and regenerative medicine contribute to the development of research of artificial materials that improve patient healthcare. Artificial skin made from a biodegradable polymer has been the great success in the world [2]. However, the recent artificial skin production has encountered the following problems including high cost and incompatibility with patient's antibodies.

For applicationsin wound dressing, the Polyvinyl Alcohol (PVA) is a very common material [3] because of its high tensile strength, flexibility, biodegradability, biocompatible and also because of it is a water-soluble synthetic polymer [4]. Beside, Chitosan (CS) is derived from the deacetylation of chitin, the primary polysaccharide component from shrimp's shells, lobster's shells and crabs processing waste. Chitosan is an attractive material using in wound dressings because it is non-toxic, biocompatible, biodegradable and antimicrobial polymer. Furthermore, Ag^+ ions in this composition could be efficiently reduced to Ag^0[5] because of the alcohol groups contained by PVA.

In this study, five difference concentrations of Ag NO_3 were prepared before mixing with 10 wt% PVA solution

and chitosan hydrogel. These mixing solutions were then treated under microwave irradiation to obtain the PVA-Ag NPs/ CS hydrogels. Besides, it was found that the high mechanical property of PVA-Ag NPs/ CS hydrogels is useful in the artificial skin

II. MATERIALS AND METHODS

Materials

Polyvinyl Alcohol (PVA - fully hydrolyzed), Chitosan and $AgNO_3$ (99.998%) were purchased from Merck Co. (Germany), and Guangdong GuanghuaSci-Tech Co., Ltd (China) respectively.

Methods

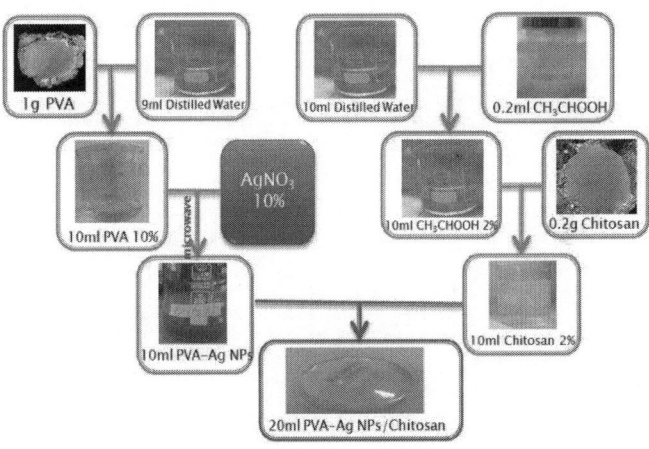

Fig. 1 Fabrication process of PVA-Ag NPs/ Chitosan hydrogels

To fabricate PVA-Ag NPs/ CS hydrogels, first, we prepared 10wt% aqueous PVA solutions by stir PVA in distilled water at 80°C for 4 hours, then mix with $AgNO_3$for 10 minutes. We divide into five set of 0%, 0.1%, 0.5%, 1% and 2% Ag NPs. Second, the solutions were irradiated for 90seconds in microwave oven. Finally, PVA-Ag NPs hydrogels and Chitosan hydrogels were well mixed together with the ratio 50:50. The hydrogels was poured into a petri disk and put into the refrigerator. After frozen, the gels were

---

* Corresponding author.

© Springer International Publishing Switzerland 2015
V. Van Toi and T.H. Lien Phuong (eds.), *5th International Conference on Biomedical Engineering in Vietnam*,
IFMBE Proceedings 46, DOI: 10.1007/978-3-319-11776-8_78

cut by the die in square shape with the width is 1cm. Then were both freeze-dried to pull out all the water inside for 8 hours.

For swelling test, five samples of each set were taken toweight and then soaked in the Phosphate Buffer Saline (BPS)solutions for 6 hours. The samples were withdrawn from the solutions at each hour and blotted with a filter paper, then re-weighted the samples and soaked in the solutions again.

For degradation test, five samples of each set were also taken to weight and soak in the BPS solutions. In the 2nd, 4th, 6th, 15th, 20th, 30th, 60th days; the samples were took out, blotted by filter paper and re-weighted. The degradation ratio was calculated using the following equation:

$$Esr(\%) = ((Ws - Wd)/Wd) \times 100$$

Where Esr is the water absorption (%wt) of the membranes, and Wd and Ws are the weights of the samples in the dry and swollen states, respectively.

The stress - strain properties were studied by gravity and tensile strength measurement methods

III. RESULTS AND DISCUSSION

Swelling and Degradation Time

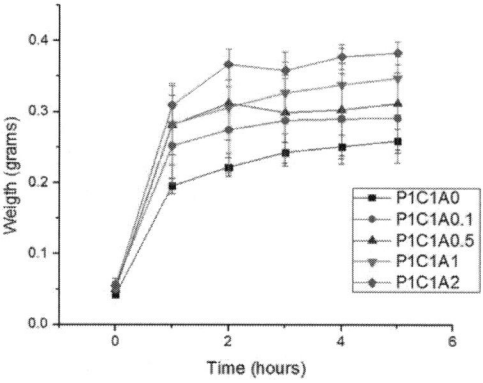

Fig. 2 The swelling behavior of hydrogel

Figure 2 shows the swelling behavior of PVA-Ag NPs/ Chitosan hydrogel after each hour. After 1 hour; P1C1A0, P1C1A0.1 and P1C1A2 increased following 4 folds, 5 folds and 7 folds, respectively; P1C1A0.5 and P1C1A1 both increased following 6 folds. After 2 hours, the all samples begin full of water with 5 folds, 6 folds following P1C1A0 and P1C1A0.1 respectively, 7 folds following P1C1A0.5 and P1C1A1, 8 folds following P1C1A2. So that, the water

absorption of the hydrogels increase if the percentage of silver increase.

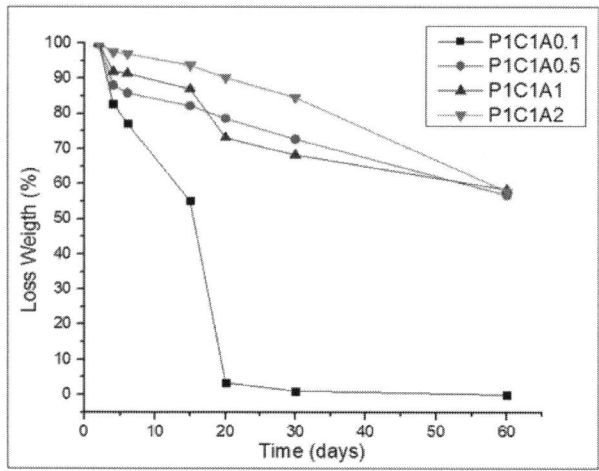

Fig. 3 The degradation behavior of hydrogel

The P1C1A0.1 was completely degraded after 30 days. On the other hand; P1C1A0.5, P1C1A1 and P1C1A2 decayed following 40% after two months. Therefore, the increase amount of silver in the hydrogels can affect the degradation behavior of the hydrogel.

Tensile Strength

The percentage of silver directly affects the stress and strain of the PVA-Ag NPs/ Chitosan hydrogel. Figure 4 shows that the stress of hydrogels increases when the percentage of silver increases. Similarly, figure 5 shows that the strain behavior of hydrogels increases when the silver percentage rises. Hence, with the increase of silver concentration, the hydrogels become more flexible and have high tensile strength.

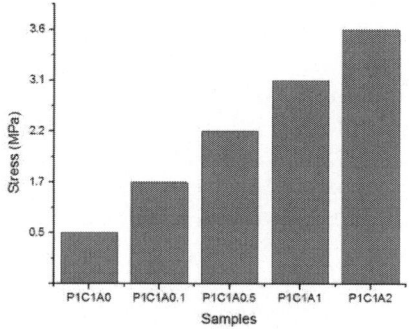

Fig. 4 The stress of the hydrogel

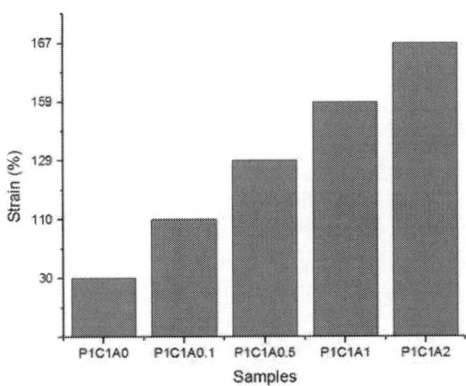

Fig. 5 The strain of the hydrogel

IV. CONCLUSIONS

The PVA-Ag NPs/ Chitosan hydrogel had the characteristics and properties depending on the percentage of silver nanoparticles. The increase amount of silver in the composition would increase the swelling, stress and strain behaviors but reduce the degradability.

ACKNOWLEDGMENT

This research is funded by Vietnam National University HoChiMinh City under grant number: B2013-76-03.

CONFLICT OF INTEREST

The authors declare that they have no conflict of interest.

REFERENCES

1. http://ykhoa.net/yhocphothong/capcuu/Bong.htm.
2. Loss, M., et al., Artificial skin, split-thickness autograft and cultured autologous keratinocytes combined to treat a severe burn injury of 93% of TBSA. Burns, 2000. 26(7): p. 644-652.
3. Gupta P, B.M., Bajpai SK Investigation of antibacterial properties of silver nanparticle- loaded poly ((acrylamide-co-itaconic acid)- grafted cotton fabric. J Cotton Sci, 2008. 12: p. 6.
4. A.R. Horrocks, S.C.A., Handbook of Technical Textiles. Woodhead Publishing Ltd. The Textiles Institute,, 2000.
5. Pal S, S.S., Devi S., Microwave-assisted synthesis of silver nanoparticles using ethanol as a reducing agent. Mater Chem Phys, 2009.
6. Supp, D.M. and S.T. Boyce, Engineered skin substitutes: practices and potentials. Clinics in Dermatology, 2005. 23(4): p. 403-412.
7. Akita, S., et al., A basic fibroblast growth factor improved the quality of skin grafting in burn patients. Burns, 2005. 31(7): p. 855-858.
8. http://www.technovelgy.com/ct/ Science-Fiction-News.asp?NewsNum=2407.
9. http://web.mit.edu/newsoffice/1996/artificialskin.html.
10. Akita, S., et al., A basic fibroblast growth factor improved the quality of skin grafting in burn patients. Burns, 2005. 31(7): p. 855-858.
11. Kim, B.-S., et al., Design of artificial extracellular matrices for tissue engineering. Progress in Polymer Science, 2011. 36(2): p. 238-268.
12. Leonardi, D., et al., Mesenchymal stem cells combined with an artificial dermal substitute improve repair in full-thickness skin wounds. Burns, (0).
13. Rustad, K.C., et al., Enhancement of mesenchymal stem cell angiogenic capacity and stemness by a biomimetic hydrogel scaffold. Biomaterials, 2012. 33(1): p. 80-90.
14. Seggewiss, R. and H. Einsele, Hematopoietic Growth Factors Including Keratinocyte Growth Factor in Allogeneic and Autologous Stem Cell Transplantation. Seminars in Hematology, 2007. 44(3): p. 203-211.
15. Gajbhiye M, K.Y., Ingle A, Gade A, Rai M., Fungus-mediated synthesis of silver nanoparticles and their activity against pathogenic fungi in combination with fluconazole. . Nanomedicine, 2009: p. 5.
16. Kokura S, H.O., Takagi T, Ishikawa T, Naito Y, Yoshikawa T., Silver nanoparticles as a safe preservative for use in cosmetics. Nanomedicine, 2010: p. 5.
17. Kamran S., M.E., Forogh M., Hamidreza D., The study of NanoSIlver (NS) antimicrobial activity and evaluation of using NS in tissue culture media. IACSIT Press, 2011. 3.
18. A. J. Kora, J.A., Assessment of antibacterial activity of silver nanoparticles on Pseudomonas aeruginosa and its mechanism of action. World J MicrobiolBiotechnol, 2010.
19. Huang, Y., et al., A randomized comparative trial between Acticoat and SD-Ag in the treatment of residual burn wounds, including safety analysis. Burns, 2007. 33: p. 5.
20. Okkyoung Choi, K.K.D., Nam-Jung Kim, Louis Ross Jr., Rao Y. Surampalli, Zhiqiang Hu, The inhibitory effects of silver nanoparticles, silver ions, and silver chloride colloids on microbial growth. Water Research 2008. 42: p. 8.
21. Foldbjerg, R., P. Olesen, M. Hougaard, D. A. Dang, H. J. Hoffmann, and H. Autrup, PVP-coated silver nanoparticles and silver ions induce reactive oxygen species, apoptosis and necrosis in THP-1 monocytes. Toxicol. Lett., 2009. 190: p. 6.
22. J, S., M.,Vlčková, B.,Pavel, I.,Šišková, K.,Šlouf,, Surface-enhanced Raman scattering from a single molecularly bridged silver nanoparticle aggregate. . J. Mol. Struct., 2009.
23. Foldbjerg, R., P. Olesen, M. Hougaard, D. A. Dang, H. J. Hoffmann, and H. Autrup, PVP-coated silver nanoparticles and silver ions induce reactive oxygen species, apoptosis and necrosis in THP-1 monocytes. Toxicol. Lett., 2009. 190: p. 6.
24. Alexandre, T., Paulino, J.I., Simionato, J.C., Garcia, J.N, , Characterization of chitosan and chitin produced from silkworm crysalides. Carbohydrate Polymers, 2006. 64: p. 5.

25. Md. Monarul Islam, S.M.M., M. Mahbubur Rahman, Md. Ashraful Islam Molla, A. A. Shaikh, S.K. Roy, Preparation of Chitosan from Shrimp Shell and Investigation of Its Properties. International Journal of Basic & Applied Sciences, 2011. 11.

26. Chunmeng Shi, Y.Z., Xinze Ran, Meng Wang, Yongping Su, and Tianmin Cheng, Therapeutic potential of chitosan and its derivatves in regenerative medicine. Surgical Research, 2005. 133: p. 7.

27. Ming Kong, X.G.C., Ke Xing, Hyun Jin Park, Antimicrobial properties of chitosan and mode of action: A state of the art review. Food Microbiology, 2010. 144: p. 12.

28. Yunli Ma, T.Z.a.C.Z., Preparation of chitosan–nylon-6 blended membranes containing silver ions as antibacterial materials. Carbohydrate Research, 2008. 343(2): p. 7.

29. Thi-Hiep Nguyen, Y.-H.K., Ho-Yeon Song, Byong-Taek Lee, Nano Ag loaded PVA nano-fibrous mats for skin applications. Applied Biomaterials, 2011. 968(2): p. 8.

Corresponding author:

Author: Thi-Hiep Nguyen
Institute: Biomedical Engineering Department, International University, Vietnam National Universities
Street: Quarter 6, Linh Trung Ward, Thu Duc District
City: Ho Chi Minh
Country: Vietnam
Email: nthiep@hcmiu.edu.vn

Investigation of the Silk Fiber Extraction Process from the Vietnam Natural Bombyx Mori Silkworm Cocoon

Thu-Hien Luong[1], Thao-Nhi Ngoc Dang[1], Oanh Pham Thi Ngoc[1], Thanh-Ha Dinh-Thuy[1], Thi-Hiep Nguyen[1,*], Vo Van Toi[1], Hoang Thuy Duong[2], and Hoang Le Son[3]

[1] Department of Biomedical Engineering, International University of Vietnam National Universities,
Ho Chi Minh City, Vietnam
[2] Research Laboratories of Saigon Hi-Tech Park, Ho Chi Minh City, Vietnam
[3] Department of Applied Chemistry, School of Biotechnology,
International University of Vietnam National Universities,
Ho Chi Minh City, Vietnam

Abstract— **Bombyx mori (B. mori) silkworms content two main protein kinds: the structural core protein, silk fibers, and the glue-like sericin coating. In contrast with silk fibers which are biocompatible, sericin is the main cause of inflammation. Hence, immunogenic sericin coating must be completely removed from silk fibers by degumming method before applying for medical uses. Several degumming processes have been applied with different chemical treatments, pH of degumming solution, time and temperature of degumming process. The purpose of this study was to investigate a new two-step degumming process that was simple, economic and suitable for B. mori silk. This process used solution with different temperatures and times of degumming. Observations under scanning electron microscope showed that there was no sign of either impurities or destruction or damage on the surface of silk fibers. It proved that the new degumming process is appropriate.**

Keywords— **B. mori silkworm, sericin, degumming, silk fiber.**

I. INTRODUCTION

The cocoons of Bombyx mori (B. mori) content two main protein kinds: the core structural fibroin protein called silk fibers and the glue-like sericin coating that covers the fibroin fibers to cement the fibroin fibers together [1].

Silk fiber is a natural polyamide fiber of repetitive protein sequences with a predominance of alanine, glycine, and serine[2]. They are generally hydrophobic. In tissue engineering and regenerative medicine, silk fiber extracted from B. mori silkworm is considered as a potential biomaterial because of its high tensile strength, good degree of toughness, elasticity, flexibility, biodegradation, and great biocompatibility that supports cell attachment and proliferation[1-6].

The second protein type is sericin that takes 20–30% of the silkworm cocoon weight. Sericin consists of 18 kinds of amino acids with strong polar side groups, unlike fibrous silk protein, sericin is hydrophilic. Its main function is to serve as a gum to bond silk fibers together. In vivo tests indicated that sericin is the main cause of allergic reactions in humans, an overreaction of the body's natural defense system that helps fight infections[1, 7]. Hence, sericin must be totally removed from the cocoon before using it in medical and clinical applications.

The thermo-chemical process that eliminating sericin and other impurities from the silk fibers called degumming is a surface modification process[1, 8]. Because of the hydrophilic of sericin, strong bases are used to weaken the non-covalent intermolecular interaction among polypeptides chains in sericin. However, strong bases affect not only sericin but also silk fibers, which lead to the decrease of their mechanical properties such as tensile strength or elasticity. The temperature and pH of degumming solution, and the applying time affect degumming efficiency as well as the silk fiber characteristics. The thermal treatment can denature the amino acid groups in the fibrous silk protein, consequently decreases the silk biocompatibility[7]. Other reported degumming processes include chemicals, auxiliaries, time for applying, temperature and pH of the silk degumming solution [6-9]. Those factors vary according to the family and bringing condition of the silkworms. The main disadvantage of these processes is that they are an one-step degumming that makes it difficult to eliminate the whole mass of sericin.

This study developed a two-step degumming process that applied on Bombyx mori silkworm. Two degumming solutions which have different pH were compared to indicate a better pH level for degumming process. The optimised solution was used in both steps of degumming process. However in the second step, temperature and time of degumming was decreased to minimize the harm to extracted silk fiber. SEM observation is used to evaluate the efficiency of this process.

---

* Corresponding author.

V. Van Toi and T.H. Lien Phuong (eds.), *5th International Conference on Biomedical Engineering in Vietnam,*
IFMBE Proceedings 46, DOI: 10.1007/978-3-319-11776-8_79

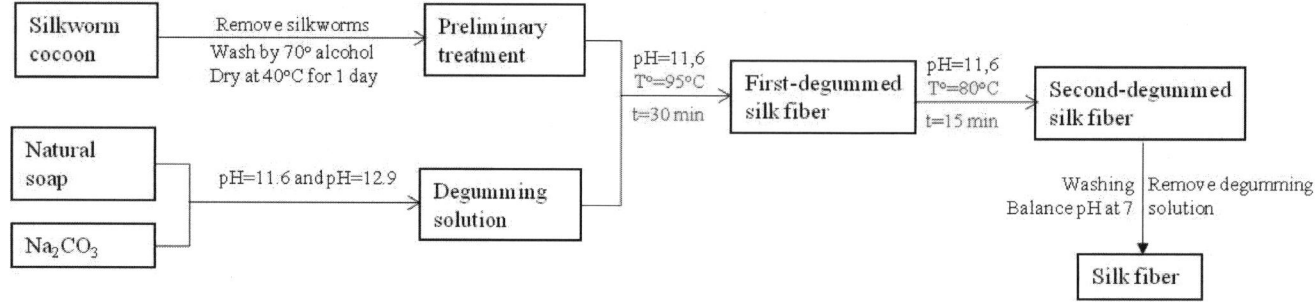

Fig. 1 Process of experiment

II. MATERIALS AND METHODS

Materials

Bombyx mori silkworm cocoons were supplied by VietSilk Limited Com., Lam Dong, Vietnam. The supplied cocoons are fresh and have not undergone any thermal, mechanical, or chemical treatment. Sodium hydroxide, sodium carbonate and citric acid were supplied by Sigma-Aldrich.

Methods

Figure 1 shows the whole procedure applied in this study which includes making degumming solution at pH=11.6 (solution A) and 12.9 (solution B), silk cocoon preliminary treatment, and the two-step degumming to get the final product which is silk fiber.

Making Degumming Solution

Pork fat is separated from pork skin, cleaned by water several time to remove dust. Pork oil was extracted by heating the pork fat at 800°C. The collected oil is then filtered by filter paper to eliminate all residues. Aqueous solution of 12.5M NaOH is prepared. NaOH solution and pork oil are mixed and stirred for 30 minutes with ratio 1:1 w/w, then, this solution is allowed to cool to room temperature to become a solid natural soap.

Solution of 0.06M sodium carbonate is boiled at 90°C. Natural soap is added to this solution to get the pH of 11.6/12.9 and the mixed solution is the degumming solution.

Preliminary Process of Silkworm Cocoon

Before the chemical treatment, B. mori silkworm cocoons undergo a preliminary process which includes: silkworms removing, sterilization and drying.

It is necessary to remove the pupa out of cocoon before any chemical treatment. After that, these cocoons are rinsed by water several times before being sterilized by 70% ethanol. Then they are dried in an oven at 40°C for 24 hours

to evaporate all water and ethanol. After this preliminary process silkworm cocoon can be stored at room temperature and in a closed environment.

Degumming the Silkworm Cocoons

In the first step, the dried cocoons are treated with boiling degumming solution at 90°C for 30 minutes. Remove degumming solution is done by washing the whole mass in 70°C distilled water until no foam appeared.

In the second step, the silk continues to be treated with degumming solution for the second time but decrease temperature at 80°C for only 15 minutes in order to totally remove the remained sericin out of silk fibers. After that, the whole mass was repeatedly washed with warm distilled water (at 70°C) until no foam is formed. The cleaned silk fibers are continuously stirred in distilled water for 10 min before 5% citric acid is used for neutralization, they were again rinsed in distilled water. Then, the silk fibers were sterilized by ethanol before drying in the oven (at 40°C) for 1 day.

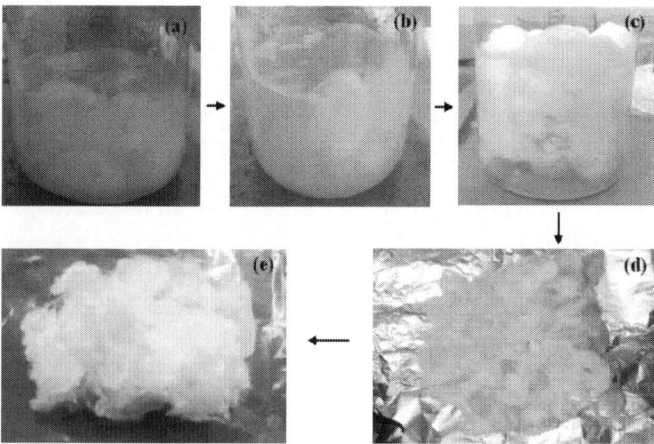

Fig. 2 Degumming process: (a) silk in first degumming solution, (b) silk in second degumming solution, (c) neutralization silk fibers, (d) degummed silk fibers before drying, (e) degummed silk fibers after drying.

Figure 2 represents the two-step degumming process using degumming solution at pH=11.6 and its result.

III. RESULTS AND DISCUSSION

Influence of pH of Degumming Solution to the Efficiency of Degumming Process (pH=11.6 and pH=12.9)

One of the most important factors that affects the degumming process is pH level. In this study, base solution was used to weaken the intermolecular interaction, hydrogen bond, among sericin molecules. However, the silk fibers are a natural structural protein type. These protein chains are also sensitive to the pH, because the strong base condition could break the peptide bond between two acid amides. Hence, high pH solution can denature the natural silk fibers, which decreases the properties of silk fibers. In this study, two levels of pH, 11.6 and 12.9, are compared in the same procedure of degumming.

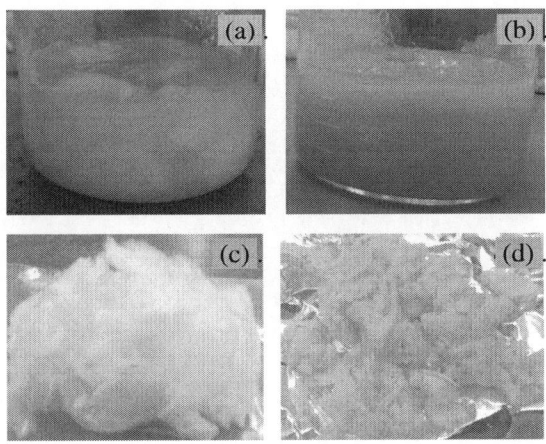

Fig. 3 Effect of pH on silk in degumming process: (a) Silk fibers in degumming solution A (pH=11.6), (b) Silk fibers in degumming solution B (pH=12.9), (c) Silk fibers after being degummed by solution A, and (d) Silk fibers after being degummed by solution B.

Figure 3 shows that high pH degumming solution denatured the silk fiber. In comparison with Fig. 3(c), the silk fibers in Fig. 3(d) had a very short length (less than 0.5 cm) or in powder. Hence, the natural structure of silk fiber was totally lost. The color of degumming solution B was also different from that of degumming solution A (Fig. 3(a) and 3(b)). It demonstrated that high pH solution which had very high concentration of OH⁻ ions could denature the fibrous silk protein and caused severe damages to the silk fiber. The test also showed that the pH at 11.6 in this study was suitable and ideal for a degumming solution.

Morphology Observation of the Natural Bombyx Mori Silk Fibers

Fig. 4 SEM of silk fibers at 3 magnifications: (a): x200, (b): x 1000, and (c): x2000.

Table 1 Statistic summary of silk fiber's diameters.

No. of counted figures	4
No. of counted fibers	35
Mean (um)	**12.084**
Std Dev	1.093
Std. Error	0.185
25%-75% (um)	11.188-12.840

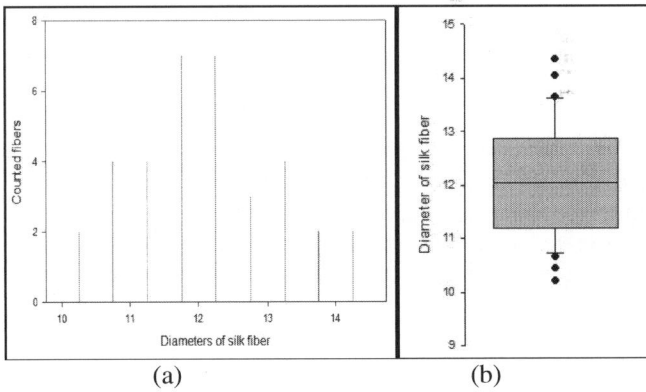

Fig. 5 (a) Histogram of silk fiber's diameter distribution, (b) 25%-75% plot box

The morphology of degummed silk fibers was observed under scanning electron microscope (SEM). Figure 4 shows SEM images of silk fibers in 3 different magnifications: 200, 1000, and 2000. Fig. 4 (b, c), show no other impurities were observed and no sign of destruction or damage on the surface of silk fibers. It demonstrated that pH and temperature of degumming solution were appropriated because there was no denatured sign and physical damage observation after degumming process. It also showed that the chemical of degumming solution was totally removed. From Figure 4(a), it is clear that each fiber was completely separated and there were no two fibers that stuck to each other. This fact showed that the 2-step degumming process could eliminate totally the sericin coating.

From figure 4(a), the uniform in diameter of silk fibers is noticeable. A closer observation in figure 4(b, c) showed that diameter of each fiber is in the range of 11 to 14 μm. The statistic summary of the fiber's diameters in 4 different figures (35 observed fibers) is described in table 1 and Fig. 5. The mean and SD diameter of the degummed silk fibers are 12.084 ±1.093 μm. The distribution of silk fibers is from 10.2 to 14.3. However, the 25%-75% range is only between 11.2 and 12.8 μm, which is a significantly small range. The natural Bombyx mori silk fiber dimension normally increased gradually when extending of the degumming period [7]. In this degumming process, the silk fibers slightly increased in diameter in comparison with the non-chemical treated silk fibers. However, the fibrous structure of silk fibers was still strongly linked to each other, and there was no separation of the fibrous structure in SEM observation.

IV. CONCLUSIONS

Degumming is an important process that must be performed before any medical and clinical application. This study investigated a 2-step degumming process that applied for Bombyx mori. These steps used the same degumming solution but with different applied time and temperature. Comparison of two products from two degumming solutions with pH 11.6 and 12.9, respectively, confirmed that high concentrated based solution (pH of 12.9) could destroy totally the natural fibrous form and denature the structural protein core of the silk while lower pH solution can get better results of degumming. SEM morphology showed that the studied process is appropriate, which could eliminate totally sericin coating without causing any damage to silk fiber. It was also proved that this process could remove all chemical, dust, and other residues out of silk fibers.

ACKNOWLEDGMENT

This research is funded by Vietnam National University HoChiMinh City under grant number: B2013-76-03.

This research is funded by International University, Vietnam National University HoChiMinh City under grant number: SV2013-06-BME.

V. CONFLICT OF INTEREST

The authors declare that they have no conflict of interest.

REFERENCES

1. V. Kearns, A.C.M., A. Crawford and P.V. Hatton, *Silk-based Biomaterials for Tissue Engineering*, R.R. N Ashammakhi, and F Chiellini, Editor. 2008.
2. Gregory H. Altman, F.D., Caroline Jakuba, Tara Calabro, Rebecca L. Horan, Jingsong Chen, Helen Lu, John Richmond, David L. Kaplan, Silk-based biomaterials. Biomaterials, 2003. 24: p. 401-416.
3. Yang Cao, B.W., Biodegradation of Silk Biomaterials. International Journal of Molecular Sciences, 2009. 10: p. 1514-1524.
4. Yongzhong Wang, H.-J.K., Gordana Vunjak-Novakovic, David L. Kaplan, Stem cell-based tissue engineering with silk biomaterials. Biomaterials, 2006. 27: p. 6064-6082.
5. O. Hakimi, F.V., A.J. Carr, Evaluation of Silk as a Scaffold for Musculoskeletal Regeneration - the Path from the Laboratory to Clinical Trials, in Comprehensive Biotechnology, S. edition, Editor. 2011.
6. Biman B. Mandal, S.C.K., Cell proliferation and migration in silk fibroin 3D scaffolds. Biomaterials, 2009. 30: p. 2956-2965.
7. M.P. Ho, K.T.L., H. Wang, Effect of Degumming on Tussah Silk Fibre, in 18th International Conference on Composite Materials.
8. Mousumi Mondal, K.T., Nirmal Kumar S, Vineet Kumar, Srinivas V. Bandlamori, Scanning Electron Microscopic Study On the Cocoon Filaments and Degummed Fibers of Two Silkmoth Hybrids of Bombyx Mori Linn. International Journal of Innovative Research and Development, 2013. 2(5): p. 1352-1362.
9. Qian Fang, D.C., Zhiming Yang, Min Li, In vitro and in vivo research on using Antheraea pernyi silk fibroin as tissue engineering tendon scaffolds. Materials Science and Engineering C, 2009. 29: p. 1527-1534.

Corresponding author:

Author: Thi-Hiep Nguyen
Institute: Biomedical Engineering Department, International University, Vietnam National Universities
Street: Quarter 6, Linh Trung Ward, Thu Duc District
City: Ho Chi Minh
Country: Vietnam
Email: nthiep@hcmiu.edu.vn

Fabrication of Hyaluronan – Chitosan – Polyvinyl Phosphonic Acid Hydrogel for Bioglue Applications

Dang Hoang Phuc[1], Thi-Hiep Nguyen[1,*], Vo Van Toi[1], and Phan Van Tien[2]

[1] Department of Biomedical Engineering, International University of Vietnam National Universities, Ho Chi Minh City, Vietnam

[2] Research Laboratories of Saigon Hi-Tech Park, Ho Chi Minh City, Vietnam

Abstract— In this study, Hyaluronan – Chitosan – Polyvinyl phosphonic acid (HA/Cs/PVPA) gel was fabricated by hydrogel method. The surface and the inner structure of the scaffold were tested by Scanning Electron Microscope (SEM). Cell proliferation on the hydrogel film was observed using light microscope. The hydrogels were expected to meet the requirement of bioglue for future application.

Keywords— Chitosan, PVPA, Hyaluronan, Fabrication.

I. INTRODUCTION

Bioglue with the properties of connecting tissue and quickly solidified had been used widely in various applications. It can be applied for blocking the blood vessel or bleeding, connecting the damage tissue, used along with or replace for surgical needle or tape in surgery hemostasis, wound healing, wound closure, and fistula [1, 2]. Bioglue can be made from synthetic polymer, natural polymer or combination of them. However using synthetic or natural polymers only cannot make an effective bioglue [3-7]. Synthetic polymer has good mechanical properties that can bear massive load during functioning with low rate of degradation. It can be used for long-term healing process [3-7]. However, after being degraded, it generates toxicity in their residues and makes it use an issue. By contrast, natural polymer has very poor mechanical properties and degrades very fast [3-7]. Highly biocompatible, natural polymer enhances cell seeding, provides proliferation better than synthetic polymer and also generates no toxicity [3-7]. Bioglue that was made only from natural polymer can be used for short term treatment, yet this bioglue can be very easily overloaded and broken down. To meet the important requirement for a successful bioglue, the useful materials are expected to be a combination of both synthetic and natural polymers.

Chitosan (Cs) – a chitin derivative – is a natural polymer that has the ability of hemostasis, bacteriostasis and wound healing and the properties of non-toxic, biodegradable and biocompatible has been widely used for skin regeneration [8-13].

Hyaluronan (HA) – An ionic water-soluble poly saccharide – is a natural polymer that commonly appears in the extracellular matrix. With the ability to enhance cell migration, cell proliferation, HA is often used as a component of scaffold for tissue engineering and wound healing [14].

Polyvinyl phosphonic acid (PVPA) – a phosphate-containing polymer – is a non-toxic synthetic polymer that is used for making hydrogel [15]. PVPA-containing hydrogel produce a strong scaffold for a long time and has the surface that can enhance cell seeding and proliferation thanks to the interaction of the gel network with the serum protein molecules [16-18]

The aim of this study is to introduce a new biomaterial by combining HA, Cs, and PVPA. In the first stage of development, the morphology of the hydrogel was tested by SEM and its biocompability was tested by cell proliferation method. This hydrogel was expected to meet the requirements of a successful bioglue for skin closure applications.

Table 1. Ratios of components of hydrogels

	HA 1%	HA 2%	HA 3%	CS1%	CS2%	PVPA 30%
H1-C1-P1	100μL	-	-	100μL	-	100μL
H2-C1-P1	-	100μL	-	100μL	-	100μL
H3-C1-P1	-	-	100μL	100μL	-	100μL
H1-C2-P1	100μL	-	-	-	100μL	100μL

* Corresponding author.

© Springer International Publishing Switzerland 2015
V. Van Toi and T.H. Lien Phuong (eds.), *5th International Conference on Biomedical Engineering in Vietnam*,
IFMBE Proceedings 46, DOI: 10.1007/978-3-319-11776-8_80

II. MATERIALS AND METHODS

Materials

Polyvinyl phosphonic acid (PVPA, 30% solution) was purchased from Polysciences (USA). Chitosan (>75% deacetylated) was purchased from Sigma. Hyaluronan was purchased from Sigma.

Fabrication of HA/Cs/PVPA Hydrogel

HA was prepared in 3 concentrations, 1%, 2% and 3%. Cs was prepared in 2 concentrations, 1%, and 2%. First, HA was mixed with Cs with the ratio showed in table 1. After stirring well using vortex, PVPA was added to the mixture with the amount showed in the table.

Characterization of HA/Cs/PVPA Hydrogel

a) Scanning Electron Microscope (SEM)

For SEM characterization, 4 samples were first freeze dried. The SEM test was performed with the magnification of 50 times by scanning electron microscope (SM-65F, JEOL, Japan).

b) Cell Proliferation

Cell proliferation characterization was performed using 12-well plate. Each sample was used in a separate plate. The test was performed with fibroblast cells in the media with the concentration of 1×10^4.

III. RESULTS AND DISCUSSION

In this study, the morphology of the samples and the ability for cell proliferation on samples were confirmed. Figure 1 showed the SEM results of 4 samples and Figure 2 showed the cell proliferation on the samples.

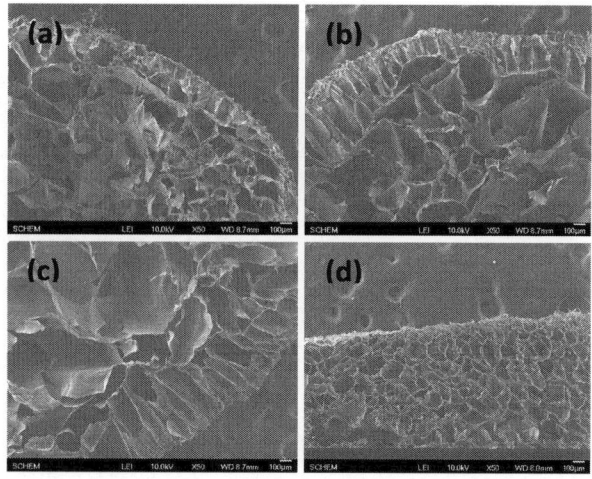

Fig. 1 SEM morphology of H1-C1-P1 (a), H2-C1-P1 (b), H3-C1-P1 (c), and H1-C2-P1 (d).

Figure 1 showed that the porous size of the sample increased when the amount of HA increased, and decreased when the amount of Cs increased. For wound application, the scaffold with macroporous structure is recommended for capillaries ingrowths [19]. In comparing between the morphology of 4 samples showed in Figure 1 from (a) to (d), it can be seen that sample H1-C1-P1, having the lowest amount of HA, had the smallest porous size, and sample H3-C1-P1, having the highest amount of HA, had the largest porous size. In comparing between the morphology of 2 samples showed in Figure 1 (a) and Figure 1 (d), having the same concentration of HA and PVPA and different concentration of Cs, we found that the higher the concentration of Cs, the smaller the porous size. Figure 2 showed the cell proliferation of 4 samples. We found that the new hydrogels supported the cell growth. The results also showed that as the amount of HA increased, the cell morphology is fully proliferated. However, the amount of chitosan increased, some cells showed the round morphology. This phenomenon is common because the chitosan has high positive charge that could prohibit cell proliferation [20].

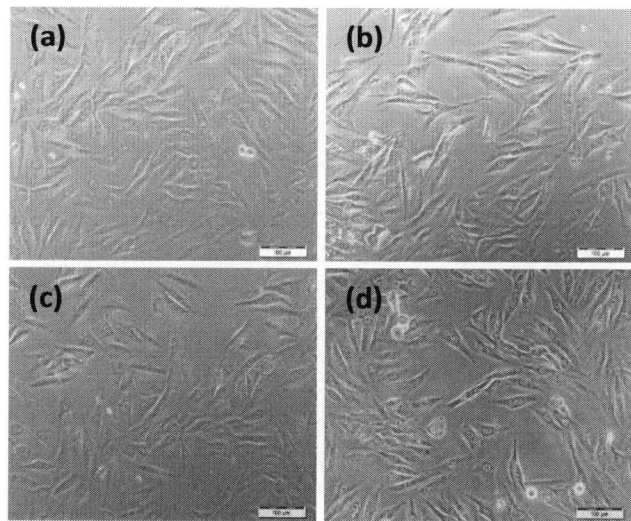

Fig. 2 Cell proliferation on H1-C1-P1 (a), H2-C1-P1 (b), H3-C1-P1 (c), and H1-C2-P1 (d).

ACKNOWLEDGMENT

This research is funded by Vietnam National University HoChiMinh City under grant number: B2013-76-03.

This research is funded by Office of Navy Research (ONR) under grant number: N62909-14-1-N011, PR No: N6290914PR00015/N00014.

REFERENCES

1. Committee AT, Bhat YM, Banerjee S, Barth BA, Chauhan SS, Gottlieb KT, Konda V, Maple JT, Murad FM, Pfau PR, et al, Tissue adhesives: cyanoacrylate glue and fibrin sealant. Gastrointestinal endoscopy 2013, 78:209-215.
2. King ME, Kinney Ay, Tissue adhesives: a new method for wound repair. Nurse Pract. 1999 Oct, 24 (10):66, 69-70, 73-4. Review.
3. Gunnatillake P.A., Adhikari R., Biodegradable synthetic polymers for tissue engineering. European Cells and Materials, 2004, 5: 1-16.
4. Drury J.L., Mooney D.J., Hydrogels for tissue engineering: scaffold design variables and applications. Biomaterials, 2003, 24(24): 4337-4351.
5. Ma P.X., Scaffolds for tissue fabrication. Materials Today, 2004, 7(5): 30-40.
6. Place E.S. et al., Synthetic polymer scaffolds for tissue engineering. Chem Soc Rev, 2009, 38(4): 1139-1151.
7. Sachlos E., Czernuszka J.T., Making tissue engineering scaffold work. Review on the application of solid freefrom fabrication technology to the production of tissue engineering scaffolds. European Cells and Materials, 2003, 5: 29-40.
8. Willi Paul, C.P.S., Chitosan and Alginate Wound Dressings: A short review. Trends Biomater. Artif. Organs, 2004. 18: 18-23.
9. Alsarra, I.A. Chitosan topical gel formulation in the management of burn wounds. Int. J. Biol. Macromol. 2009, 45, 16–21.
10. Hurler J, Škalko-Basnet N, Potentials of Chitosan-Based Delivery Systems in Wound Therapy: Bioadhesion Study. Journal of Functional Biomaterials 2012, 3:37-48.
11. WJoshua S. Boateng, K.H.M., Howard N.E. Stevens, Gillian M. Eccleston, Wound healing dressings and drug delivery systems: A review. Journal of Pharmaceutical Sciences, 2008. 97(8): p. 2892-2923.
12. Pradip Kumar Dutta, J.D., V S Tripathi, Chitin and chitosan: Chemistry, properties and applications. Journal of Scientific & Industrial Research, 2004. 63: p. 20-31.
13. Gregorio Crini, P.-M.B., Application of chitosan, a natural amnopolysacchrid, for dye removal from aqueous solutions by adsorption processes using batch studies: A review of recent literature. Progress in Polymer Science, 2008. 33: p. 39-447.
14. Chung TW, Chang YL, Silk fibroin/chitosan-hyaluronic acid versus silk fibroin scaffolds for tissue engineering: promoting cell proliferations in vitro. Journal of materials science Materials in medicine 2010, 21:1343-1351.
15. Kusunoki T., et al., Polyvinylphosphonic acid copolymer hydrogels prepared with amide and ester type crosslinkers. Journal of Applied Polymer Science 119(5): 3072-3079.
16. Lavinia Macarie, G.I., Poly(vinylphosphonic acid) and its derivatives. Progress in Polymer Science 2010. 35: p. 1078-1092.
17. Anbar M, N.G., St John GA, Fate and toxicity of orally administered polyethylene polyphosphonates. Food and Cosmetic Toxicology, 1973. 11(6): p. 1001-1010.
18. Jian Tan, R.A.G., Mandy Ma, W. Mark Saltzman, Improved cell adhesion and proliferation on synthetic phosphonic acid-containing hydrogels. Biomaterials, 2005. 26: p. 3663-3671.
19. Kelly R. Kirker, Y.L., J. Harte Nielson, Jane Shelby, Glenn D. Prestwich, Glycosaminoglycan hydrogel films as bio-interactive dressings for wound healing. Biomaterials, 2002. 23: p. 3661-3671.
20. Jiang M., Ouyang H., Ruan P., Zhao H., Pi Z., Huang S., Yi P., Crepin M., Chitosan derivatives inhibit cell proliferation and induce apoptosis in breast cancer cells. Anticancer Res. 2011, 31(4):1321-8.

Corresponding author:

Author: Thi-Hiep Nguyen
Institute: Biomedical Engineering Department, International University, Vietnam National Universities
Street: Quarter 6, Linh Trung Ward, Thu Duc District
City: Ho Chi Minh
Country: Vietnam
Email: nthiep@hcmiu.edu.vn

Investigation of the Synthetic Process of Nano-Hydroxyapatite (Hap) Using Microwave and Ultrasound

Tran Thi Tuong Van, Bui Ngoc Thao Tram, Vo Van Toi, and Thi-Hiep Nguyen[*]

Biomedical Engineering Department International University,
Vietnam National University Ho Chi Minh City, Vietnam

Abstract— In this study, nano hydroxyapatite (HAp) was synthesized using microwave and ultrasonic methods. Besides, the pH and temperature of the reaction were also the conditions of this investigation. The properties and characterizations of HA powders were analyzed by X-ray Diffraction (XRD), Fourier Transform Infrared Spectroscopyand (FT-IR). Morphology of nano-hydroxyapatite was observed by Scanning Electron Microscope (SEM) and Transmission Electron Microscope (TEM). Results showed that HAp powders were synthesized successfully by using both microwave and ultrasonic methods through FT-IR and XRD analysis. However, the SEM observation showed that the particles sizes are different.

Keywords— Hydroxyapatite, microwave, ultrasound.

I. INTRODUCTION

Calcium hydroxyapatite nanoparticles, abbreviate as HAp that has a formula is $Ca_{10}(PO_4)_6(OH)_2$, is the major mineral component of the bone, teeth and tendon[1]. HAp have a huge range of applications in a number of fields, widely used as a biocompatible ceramic in many areas of medicine[2-6], due to its resemblance to mineral bone, attractive candidates for body's hard tissues replacement [7-9].

Many diverse methods have claimed to prepare HAp nanoparticles with precise control over its microstructure. These methods involve variously chemical synthesis process [10] such as dry methods which include solid-state synthesis, and mechano-chemical method; and wet methods which include conventional chemical precipitation, sol-gel method, and hydrothermal method. These wet methods have some disadvantages, such as being time-consuming, having low quality control, chemical contamination, and powder at the sub-micron level. However, the microwave-hydrothermal and ultrasonic method was found to synthesize the ultrafine high purity nanopowder with high yield using high energy transformation in a short time [11-12].

In this project, HAp nanoparticles suspension can be prepared by chemical precipitation reaction [13] with assistance of the microwave and the ultrasonic:

$$10Ca(OH)_2 + 6H_3PO_4 \rightarrow Ca_{10}(PO_4)_6(OH)_2 + 18H_2O$$

The mophology and properties of HAp were evaluated by using X-ray diffraction (XRD), scanning electron microscopy (SEM), and transmission electron microscopy (TEM), Fourier transform infrared spectroscopy (FT-IR).

II. MATERIALS AND METHODS

Synthesis Hydroxyapatite (HAp)

Commercial-grade $Ca(OH)_2$ (98% $Ca(OH)_2$ Xilong Chemical Industry Incorporated Co.,Ltd, China) and reagent-grade H_3PO_4 (85–87.0% H_3PO_4, GuangDong GuangHua, Schitech Co., Ltd, China) were used as starting materials for the synthesis of HA nanopowders. These staring powders were weighed to coincide the stoichiometric HAp with a molar ratio of Ca/P = 1.67. The powder was mixed homogeneously by stirring it in de-ionized distilled water for 2 hours by stirring hotplate (Cimarec™ Digital Stirring Hotplates, US). The pH of the initial solution was about 4.83. Then, the pH of this solution was adjusted to 6 and 12 using a pH meter (Jenway 3510 pH meter, Camlab UK) by the addition of sodium hydroxide solution.

Then, this suspension was kept in a microwave oven (LG Electronics Co., Korea) which was operated with 2.45 GHz frequency microwave radiation while the operating power was kept at 700 W for 25 min at 300°C or allowed for ultrasonic agitation for 5 hours at 80°C. After cooling to room temperature, the precipitate of two cases was washed thoroughly with de-ionized water 10 times and dried in an oven at 90°C. Finally, the calcination was carried out at 750°C in normal condition.

---

[*] Corresponding author.

© Springer International Publishing Switzerland 2015
V. Van Toi and T.H. Lien Phuong (eds.), *5th International Conference on Biomedical Engineering in Vietnam*,
IFMBE Proceedings 46, DOI: 10.1007/978-3-319-11776-8_81

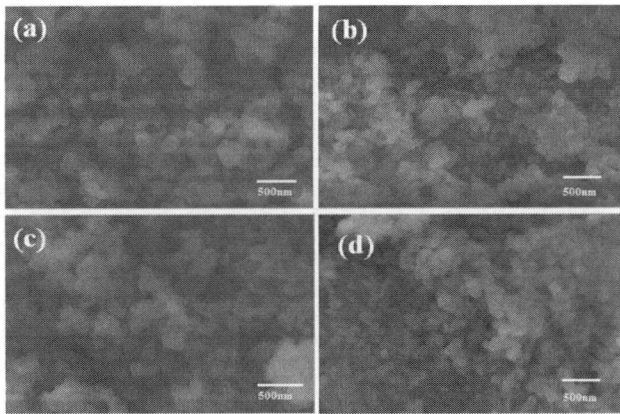

Fig. 1 SEM image of HAp using microwave with pH6 (a), pH12 (b); ultrasound with pH6 (c), and pH12 (d).

XRD Measurement

The phase analysis of HAp nanopowders depending on the pH was identified using an X-ray diffractometer. From the XRD patterns, the yields of the synthesized powders depending on the pH of the solution were calculated from the relative intensity of constituent phases. Approximately one gram of each material was ground into a fine powder using a clean agate mortar and pestle. The crystal structure of each powder sample was analyzed using an x-ray diffract meter. Data were collected over the 2θ range 20-100° at a scan speed of 2°/min.

SEM and TEM Observation

The morphology and microstructure of the synthesized powders were investigated using Scanning Electron Microscope (SEM) and Transmission Electron Microscopy (TEM).

FT-IR Spectrum Measurement

The phase composition of the powders was studied by Fourier transform infrared spectroscopy (FTIR spectrometer)

III. RESULTS AND DISCUSION

Fig.1 and Fig. 2 show the morphology of Hap particles using SEM and TEM observation. Fig. 1 shows SEM images of HAp nanoparticles which were synthesized by the microwave assisted and ultrasonic-assisted at vary pH6 (a,c) and pH12 (b,d) values. . The average particle sizes of the synthesized Hap powders decreased as the pH of the solution increased, thus, the particles sizes at pH6 and pH12 were about 50-100 and 20–50 nm in diameter, respectively.

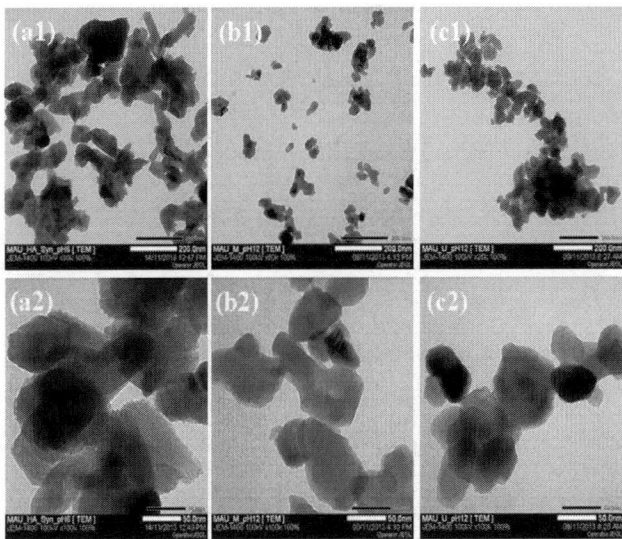

Fig. 2 TEM images of HAp using microwave with pH6 (a1),(a2); pH12 (b1),(b2); ultrasound with pH12 (c1), (c2).

At pH12, the synthesized HAp nanoparticles had a spherical shape. However, the elongated particles were observed in case of pH 6 (Fig. 2a1 and Fig. 2a2). Comparisons between two methods, the particles sizes and shape of the synthesized HAp from microwave method and ultrasonic method are similar.

All spectra were collected at room temperature at a nominal resolution of 4.00 and number of sample scans equal to 1000. The FT-IR spectra were recorded in a range of 0 - 500 cm^{-1}. The result shows in Fig.3 and Fig.4. The representative FT-IR spectrum shows all characteristic absorption peaks of HA synthesized by microwave and ultrasonic method with pH6 and pH12. The first indication for formation of HA is in the form of a strong complex broad FTIR band centered at about 900-1200 cm^{-1} due to asymmetric stretching mode of vibration for PO_4 group. By comparing four samples, as we seen there is no difference between hydroxyapatite synthesized by microwave and ultrasonic method.

Fig.5 shows the XRD profile of HAp powders by ultrasonic method at pH12 and pH6 (Fig. 5a and Fig. 5b, respectively) and microwave method at pH12 and pH6 (Fig. 5c and Fig. 5d) and the HAp nanoparticles commercial product (Fig.3e). Result XRD profiles the Bragg peaks at approximately 25, 28, 29, 31, 32, 35, 36, 37, 39, 40, 42, 43, 45, 46, 48 and 49, 51, 53, 54 (2θ) corresponds to the characteristic peaks of pure HAp phase (JCPDS File No. 9-432) [15] as shown on Fig.5(e). Based on this, it is evident that the process of thermo process in 750°C resulted in a single phase

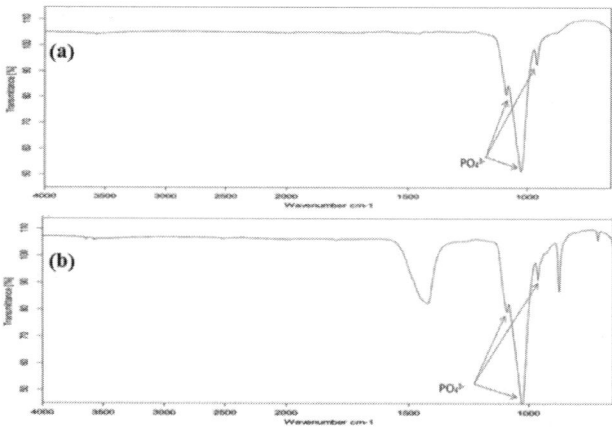

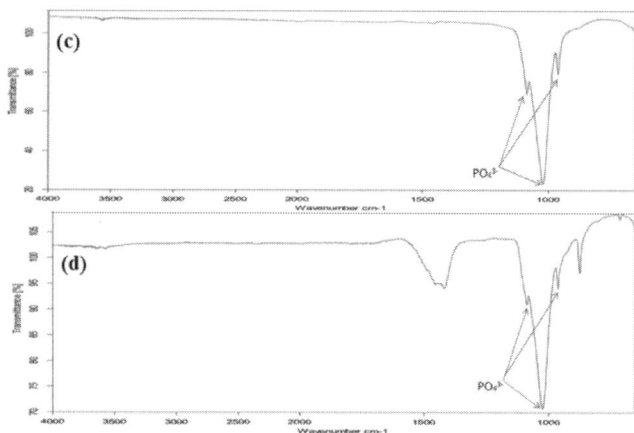

Fig. 3 FT-IR spectrum of HAp was synthesized using microwave pH6 (a) and pH12 (b).

Fig. 4 FT-IR spectrum of HAp was synthesized using ultrasonic pH6 (c) and pH12 (d).

HA peaks also exhibited a high crystallinity as indicated by the narrow diffraction peaks compare to the peaks of HA synthesized.

IV. CONCLUSION

In this study, the hydroxyapatite nanoparticles were synthesized successfully by using both microwave and ultra-sonic methods confirmed by FT-IR spectrum and XRD profiles. SEM datas observation showed that the particle sizes of hydroxyapatite at pH6 were larger than the particle-sizes of hydroxyapatite at pH12. However, the shapes of hydroxyapatite are elongated shape with pH6 and spherical shape with pH12. In addition the shaped of HAp nanoparticles of two methods were similar. Depending on the purpose of application for use, each kind of hydroxyapatite nanoparticles had the specific application.

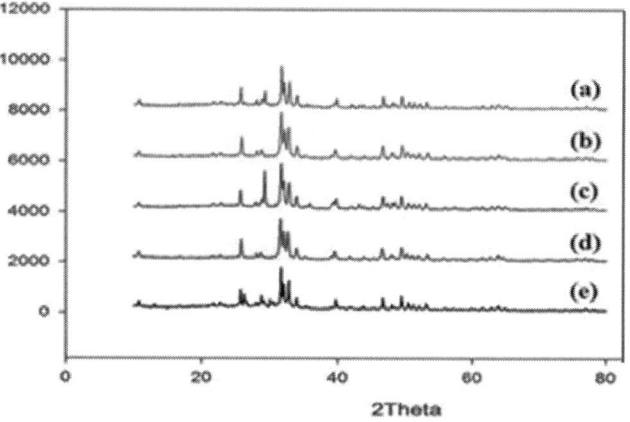

Fig. 5 XRD patterns of HAp (a); Microwave with pH6 (b), pH12 (c) and ultrasonic with pH6 (d), pH 12 (e).

ACKNOWLEDGMENT

This work was supported by International University (SV2013-07-BME), Ho Chi Minh City, Vietnam.

CONFLICT OF INTEREST

The authors declare that they have no conflict of interest.

REFERENCES

1. Li P, Ohtsuki C, Kokubo T, Nakanishi K, Soga N, Nakamura T and Yamamuro T 1992J. Am. Ceram. Soc. 1421.
2. G.X. Ni, W.W. Lu, B. Xu, K.Y. Chiu, C. Yang, Z.Y. Li, W.M. Lam, K.D.K. Luk, Interfacial behaviour of strontium-containing hydroxyapatite cement with cancellous and cortical bone, Biomaterials 27 (2006) 5127–5133.
3. A.Balamurugan,A.H.S.Rebelo,A.F.Lemos,J.H.G.Roca,J.M.G.Ventura ,J.M.F. Ferreira, Suitability evaluation of sol–gel derived Si-substituted hydroxyapatite for dental and maxillofacial applications throughin vitro osteoblasts response, Dental Materials 24 (2008) 1374–1380.
4. [W.R. Deppert, R. Lukacin, Chapter 5 hydroxyapatite chromatography, Protein Liquid Chromatography (1999) 271–299.
5. Zhou Z-H, Zhou P-L, Yang S-P, Yu X-B, Yang L-Z. Controllable synthesis of hydroxyapatite nanocry tals via a dendrimer-assisted hydrothermal process. Mater Res Bull 2007;42:1611–8.
6. G.X. Ni, W.W. Lu, B. Xu, K.Y. Chiu, C. Yang, Z.Y. Li, W.M. Lam, K.D.K. Luk, Interfacial behavior of strontium-containing hydroxyapatite cement with cancellous and cortical bone, Biomaterials 27 (2006) 5127–5133.
7. Webster TJ, Siegel RW, Bizios R. Enhanced surface and mechanical properties of nano phase ceramics to achieve orthopedic/dental implant efficacy. Key Engineering Materials 2001; 192-5: 321-4.
8. Webster TJ, Ergun C, Doremus RH, Siegel RW, Bizios R. Enhanced functions of osteoblasts on nanophase ceramics. Biomaterials 2000; 21: 1803-10.

9. Webster TJ. Specific proteins mediate enhanced osteoblast adhesion on nanophase ceramics. J Biomed Mater Res 2000; 51: 475-83.

10. Bouyer E, Gitzhofer F, Boulos MI. Morphological study of hydroxyapatite nanocrystal suspension. J Mater Sci Mater Med 2000; 11: 523-31.

11. S.K. Padmanabhan, A. Balakrishnan, M.C. Chu, Y.J. Lee, T.N. Kim, S.J. Cho, Particuology 7, 466 (2009).

12. J.K. Han, H.Y. Song, F. Saito, B.T. Lee, Mater. Chem. Phys. 99 (2006) 235.

13. W. Kim, F. Saito, Ultrason. Sonochem. 8 (2001) 85.

14. Bouyer E, Gitzhofer F, Boulos MI. Morphological study of hydroxyapatite nanocrystal suspension. J Mater Sci Mater Med 2000; 11: 523-31.

15. JCPDS File No. 9-432, International Center for Diffraction Data.

Corresponding author:

Author: Thi-Hiep Nguyen
Institute: Biomedical Engineering Department, International University, Vietnam National Universities
Street: Quarter 6, Linh Trung Ward, Thu Duc District
City: Ho Chi Minh
Country: Vietnam
Email: nthiep@hcmiu.edu.vn

Synthesis and Characterization of Hydroxyapatite Biomaterials from Bio Wastes

Bui Ngoc Thao Tram, Thi-Hiep Nguyen[*], and Vo Van Toi

Biomedical Engineering Department International University, Vietnam National University-Ho Chi Minh City (VNU-HCM), Vietnam

Abstract— In this work, hydroxyapatite HA was produced from bovine bone by the calcination at $800^{0}C$ in 3 hours. Three sources of femur were used to collect the bovine bone: waste, from cows aged 10 -12 months and 2-3 years. The purity, shape and crystal structure of HA were characterized by Scanning electron microscopy (SEM) and Transmission electron microscopy (TEM).

Keywords— hydroxyapatite, bovine bone, SEM, TEM, Nano powder, HA.

I. INTRODUCTION

Hydroxyapatite is mainly used in the medical sector such as covering prosthesis, producing bone substitutes (types of phosphocalcic ceramics and ionic cements, etc.), repair of body bony defects in defect and orthopedic application, and application there of in bone, connective tissue, fattissue and muscle tissue engineering[1]. Applications also include dental implants, periodontal treatment, maxillofacial surgery, otolaryngology, immediate tooth root replacement augmentation of alveolar ridge replacements for better denture fit, augmentation and stabilization of the jaw bone, and many more.

Most people whose related with medical agree that HA is bioactive, and it is well-known materials in regeneration for bone tissue. Thus, hydroxyapatite has been used enormous in biomedical application.

Several methods have been utilized for the synthesis of HA include precipitation technique [2-4], ultrasonic, spray drying, sold-gel approach [5], hydrothermal technique [6], multiple emulsion technique [7], biomimetic deposition technique [8-9], electrode position technique [10], etc. But these processes spend great deal of time and expense. In this study, HA was fabricated from bovine bone from different source on order to figure out the advantages of this investigation. Herein, we showed a fabricating process from starting materials (the raw bone) to collecting HA powder. Then, the morphology of hydroxyapatite particles were observed by Scanning Electron Microscopy (SEM) and Transmission electron microsopy (TEM).

---

* Corresponding author.

II. MATERIALS AND METHODOLOGY

A. Materials

There are three-source of bovine bone were used for fabricating Hydroxyapatite such as: Fresh cow/bull (2-3 years) and Buffalo calf (10-12 months), and the waste. These bone was collected from femur as shown on Figure 1.

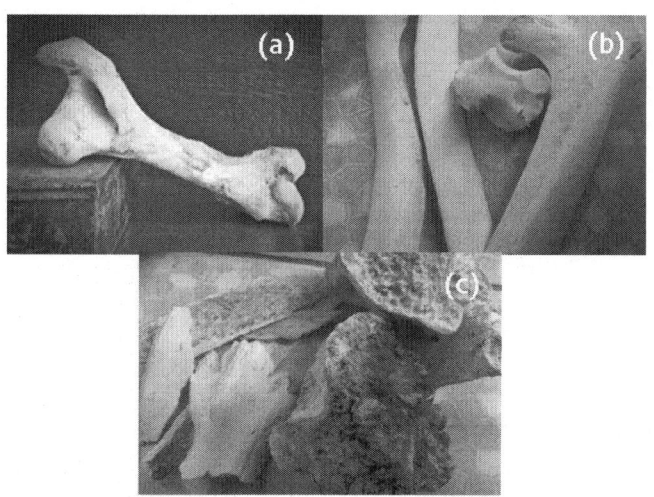

Fig. 1 Bovine bone is collected from (a) Fresh cow/bull (b) Buffalo calf (c) Waste

B. Methodology

There are three main states to fabricate the hydroxyapatite from bovine bone. First state is defatting and proteinase by boiling. The next state is burning. Final state is calcinations.

Defatting and Proteinase State

Only femur bone was collected for this investigation. Bovine bone marrows and bone tissue and tendon were cleaned by washing and boiling in the hot water for 5 -6 hours . Continue boiling in the commercial pressure cooker for 2 hours and is repeated three times for defatting and proteinase process.Figure 2 shows the bovine bone after defatting and proteinase process. The process was followed by sun drying to eliminate organic substances and to avoid soot formation in the material during the heating treatment.

© Springer International Publishing Switzerland 2015
V. Van Toi and T.H. Lien Phuong (eds.), *5th International Conference on Biomedical Engineering in Vietnam,*
IFMBE Proceedings 46, DOI: 10.1007/978-3-319-11776-8_82

Fig. 2 Bovine bone after removes bone marrows and bone tissue and tendon.

Burning State

After that, the dried bovine bone was cut by grinding to form a cube (± 1 cm³) Figure 3. Drying is continued until the color becomes yellowish white bovine bone.

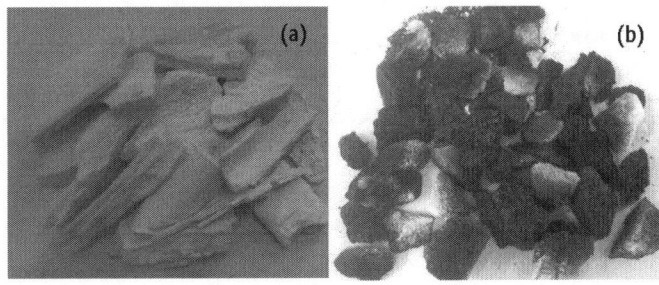

Fig. 3 Cleaned dried bovine bone (a) and burned bovine bone (b)

The cleaned dried bovine were wash and burn with ethanol (80%) to decomposed any biowaste remanding.

Calcinations State

The burned bovine bone was calcined using the electric furnace (Lenton, UK) at 800°C for 3 hours and cooled slowly to room temperature. The color of bone changed to white as shown on Figure 4

Fig. 4 The burned bovine were calcined at 800°C for 3 hours and cooled slowly to room temperature.

C. Characterization of Powder

The morphology, particle size of the fabricated HA powders were investigated using Scanningelectron microscopy (FE SEM S4800, Hitachi) and Transmission electron microsopy (TEM JEM 1400, JEOL).

III. RESULTS AND DISCUSSION

Figure 5 showed the produce of the fabricated hydroxyapatite powder. .

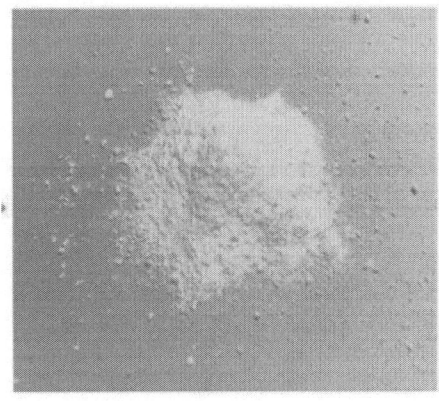

Fig. 5 Hydroxyapatite powder collected from bovine bone after calcination at 800°C in air and milling

B.N. Thao Tram, T.-H. Nguyen, and V. Van Toi

Fig. 6 FE-SEM of HA powder synthesized (a) Fresh cow/bull (b) buffalo caft (c) Waste

To observe the particles sizes of hydroxyapatite powder, SEM and TEM observation was employed. SEM micrographs showed that the fabricated HA agglomerated. The shapes of HA from different type of bone had identical spherical shape. However, the particle sizes of the nanopowders were different: about 20-200nm in diameter for the Fresh cow/bull; about 50 -100 nm for the buffalo caft and about 50-400 nm as shown on Figure 6 and Figure 7.

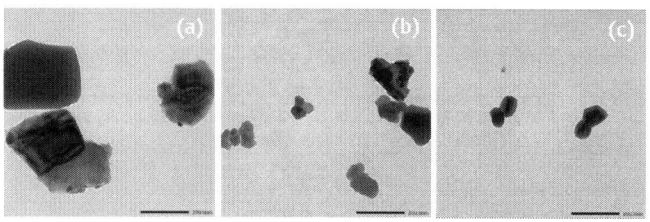

Fig. 7 TEM of HA powder synthesized (a) Fresh cow/bull (b) buffalo caft (c) Smelly cow/bull

IV. CONCLUSIONS

Hydroxyapatite was fabricated successfully by using bovine bone. The hydroxyapatite fabricated from different sources are in same shape (spherical particles) but of different size (about 20-200nm in diameter for the fresh cow/bull; about 50 -100 nm for the buffalo calf and about 50-400 nm). For the application as a bone substitute, the fabricated hydroxyapatite from bovine bone will be tested on biocompatibility, as planned in our further investigation.

ACKNOWLEDGEMENTS

This work was supported by International University (SV2013-07-BME), Ho Chi Minh City, Vietnam.

REFERENCES

1. "Biphasic calcium phosphate in periodical surgery", Chinni Suneelkumar, Krithika Datta, Manali R Srinivasan, and Sampath T Kumar J Conserv Dent. 2008 Apr-Jun; 11(2): 92–96.
2. Santos MH, de Oliveira M, de Freitas Souza P.Mansur HS, Vasconcelos WL. "Synthesis control and characterization of hydroxyapatite prepared by wet precipitation process" Mater Res. 2004; 7(4): 625-630.
3. Manuel CM, Ferraz MP, Monteiro FJ,"Nanoapatite and micro porous structures of hydroxyapatite". Proceeding of the 17th European Society of Biomaterials. Barcelona, Spain. 2002: T 153.
4. Manuel CM, Ferraz MP, Monteiro FJ. "Synthesis of hydroxyapatite and tricalcium phosphate nanoparticles." Preliminary Studies. Key EngMater. 2003; 240-242: 555-58
5. Chai CS, Ben-Nissan B. "Bioactive Nano crystalline sol-gel hydroxyapatite coatings". J Mater Sci: aterMed. 1999; 10: 465-469.
6. Manafi SA, Joughehdoust S. "Synthesis of hydroxyapatite nanostructure by hydrothermal condition for biomedical application". Iranian JPharm Sci. 2009; 5(2): 89-94.
7. Kimura I. "Synthesis of hydroxyapatite by interfacial reaction in a multiple emulsion". ResLett Mater Sci. 2007; Article ID 71284: 1-4.
8. Tas AC. "Synthesis of biomimetic Ca-hydroxyapatite powders at 37 degrees C in synthetic body fluids". Biomaterials. 2000; 21:1429-1438.
9. Thamaraiselvi TV, Prabakaran K, Rajeswari S. "Synthesis of hydroxyapatite that mimics bone mineralogy". Trends Biomater Artif Org. 2006; 19(2): 81-83.
10. Shikhanzadeh M. "Direct formation of Nano phase hydroxyapatite on catholically polarized electrodes". J Mater Sci: Mater Med. 1998; 9: 67-72.

Corresponding author:

Author: Thi-Hiep Nguyen
Institute: Biomedical Engineering Department, International University, Vietnam National Universities
Street: Quarter 6, Linh Trung Ward, Thu Duc District
City: Ho Chi Minh
Country: Vietnam
Email: nthiep@hcmiu.edu.vn

Modified DNA Extraction Method for the Detection of Aspergillus Flavus and Aspergillus Parasiticus in Dried Foods

Pham Tuong Vi, Huynh Le Thao Trinh, and Nguyen Thi Hue

School of Biotechnology, International University – VNU, Ho Chi Minh City, Vietnam

Abstract— *Aspergillus flavus* and *Aspergillus parasiticus* are two aflatoxigenic species having greatly impacts on human health and economy. About 25% of world food crops are affected and countries located in tropical and subtropical area suffer from the most risk (the United Nations Food and Agriculture Organization). To prevent the loss caused by aflatoxin contamination, they need to be detected by the most feasible way. Currently, PCR based test has been a potential door with some advantages in comparison with the morphology method. In this study, a modified protocol based on the standard DNA extraction was applied to shorten the whole procedure. For this purpose, DNA was simultaneously extracted from *A. flavus* and *A. parasiticus* artificial infected corn samples by two protocols. Standard protocol was SDS method modified from Plaza's method while the modified protocol was that standard with some modifications. After evaluation by spectrophotometry and gel electrophoresis, the extracted DNA was used for multiplex PCR to determine the fungal presence. Following, the effectiveness of both methods was evaluated via the DNA quality and PCR result. Though the DNA quality from the modified method was not good, they could still be used in multiplex PCR for the fungal detection in corn sample. 3ul of proteinase K 2mg/ml and 1 hour incubation were recommended for modified process. In conclusion, this study proved a possible method to shorten the fungal detection process. Once this method has succeeded for corn sample, the presence of *A. flavus* and *A. parasiticus* in corn could be predicted by a more effective procedure.

Keywords— **Multiplex PCR, corn, modified method.**

I. INTRODUCTION

A. flavus and *A. parasiticus* are two main aflatoxigenic species that have not only great economic impacts but also human health effects. These species are likely to produce highly toxic secondary metabolites – aflatoxin [1]. In nature, these fungi are widely distributed, especially between the 40°N and 40°S, and are capable to grow from 16°C to 35°C [2]. They can survive on various organic nutrient sources. In our project, we focus on corns, one of major commodities, since it is easily contaminated by *A. flavus* and *A. parasiticus* in tropical climates as Viet Nam [3].

About 25% of world food crops are affected and countries located in tropical and subtropical area are the most at risk according to the United Nations Food and Agriculture Organization. Aflatoxin is consider as one of the most notable group of mycotoxin due to the acute and chronic effect on human and animal health. They are known to cause chronic aflatoxicosis expressed as impaired food conversion, stunting in children [4], immune suppression, growth retardation, liver disease, and death in both humans and domestic animals [5]. Although these two fungi are both aflatoxin producers, their aflatoxigenic profile are different. *A. flavus* can produce aflatoxin B while *A. parasiticus* can produce both kind B and G [6].

Because of the potential effects of mycotoxin producing fungi and mycotoxins on human health and economy, the detection of these fungi on crops become more important. It was reported that morphological method has been known as gold standard method for fungal diagnosing in dried foods. However, it is quite inconvenient and time-consuming because they share several common points in morphology. Thus, PCR test has considered as a potential alternative method for the aflatoxigenic fungi detection because of its advantages such as sensitivity, specificity, lower cost and time-saving. Besides that, it would be better if the cost and time of this promising detection procedure is maximally reduced to fasten the whole process.

With this aim, modified and standard method were carried out on infected corn kernels for *A. flavus* and *A. parasiticus* detection. Then, these methods were evaluated by using results of electrophoresis, Nanodrop machine and PCR test. The success of this study would provide a new method for these two fungi detection rapidly.

II. MATERIALS AND METHODS

Materials

A. flavus (TCC-F-898) and *A. parasiticus* (VTCC-F-1159) were provided by Vietnam type culture collection (VTCC) from the Institute of Microbiology and Biotechnology (IMBT), National University, Hanoi, Vietnam.

The corn kernels used in this study were purchased at Ba Chieu market; they originated from Binh Thuan province.

V. Van Toi and T.H. Lien Phuong (eds.), *5th International Conference on Biomedical Engineering in Vietnam,*
IFMBE Proceedings 46, DOI: 10.1007/978-3-319-11776-8_83

Methods

a) Infected Samples Preparation

The fungal colonies were collected from the incubation in malt extract broth at 25°C, shaking 110rpm for 3 days. Then these colonies were grown on malt extract agar at 25°C in darkness for 5 days to obtain fungal spores which were inoculated on corn kernels. Before the inoculation, the corn kernels were submerged in distilled water for 4 hours; and disinfected by UV light for 30 min. And then, infected samples were used for DNA extraction.

b) DNA Extraction

• Standard method

The standard protocol was modified from Plaza's method [7, 8]. Mycelia and spores were collected from 10, 20, 30 kernels using 10ml of NaCl 0.85%. DNA extraction was performed by grinding with sterilized sand and SDS lysis buffer followed by thermal shock. Next, the DNA was collected and purified by chloroform, ice-cold isopropanol and ammonium acetate.

• Modified Method

The modified method was adopted from the standard method with some further modification and cut off steps. Distilled water was used for grinding step instead using lysis buffer. After that, proteinase K was added with 3 different volumes: 0µl, 1.5µl, 3µl before incubating at 65°C for 1 hour, 1.5 hours, and 2 hours.

c) Comparison and Analysis Extracted DNA

DNA solutions are evaluated by using 3 following methods: spectrophotometry, gel electrophoresis and multiplex PCR. First, using spectrophotometry, the isolated DNA were quantified and qualified using NanoDrop® 2000c spectrophotometer, based on the ratio A260/A280. The ideal ratio is 1.7 to 2.0. Next, the obtained DNA was used for running gel electrophoresis with 0.7% of agarose gel under 85 voltage for 45 min. Finally, multiplex PCR was carried out with two set of primers designed based on the aflatoxin biosynthesis gene cluster for amplification of two target sequences from *A. flavus* (AFLA) and *A. parasiticus* (APLA) designed in previous study [8] as follows:

APLA-F-5'-GGATTCGTGAGTGTCTTTAGGG-3',
APLA-R-5'-GGTAAATGCTCCGCACAGTC-3',
AFLA-F-5'-GGTGGTGAAGAAGTCTATCTAAGG-3',
AFLA-R-5'-AAGGCATAAAGGGTGTGGAG-3'.

25 µl PCR reaction contained 300ng of DNA template, 12.5 µl toptaq DNA polymerase, 1 µl of 10 mM each primer. PCR thermal program including 5 min initiation at 95°C, 40 cycles of 30 seconds denaturation at 95°C, 30 seconds annealing at optimal temperature, 30 seconds extension at 72°C and 3 min final extension at 72°C was used in the Eppendorf PCR system. Next, 10 µl of PCR product was analyzed on 2% ethidium bromide-containing agarose gel at 85 V for 20 min.

III. RESULTS

Spectrophotometry

The concentration of DNA extracted from various conditions was measured by Nanodrop machine and studied using statistical test. Using modified method, the result showed that the concentration of DNA extracted from 10, 20, 30 kernels were in range of 15±7.07 - 82±5.6 ng/µl, 51.5±2.1 - 89.5±0.7 ng/µl, and 107±19 - 141±14 ng/µl respectively. Among those, the condition at 3µl proteinase K and 1 hour incubation gave the highest results in all three groups, 82±5.6 ng/µl for 10 kernels, 89.5±0.7 ng/µl for 20 kernels and 507±31 ng/µl for 30 kernel (figure 1).

With $p < 0.05$ from two way ANOVA (CI 95%) result, there were significant differences in the DNA concentration of different extraction conditions for all three groups of kernels. It means that the DNA concentration of the 3µl proteinase K and 1 hour incubation was significantly higher than the other subgroups of modified method.

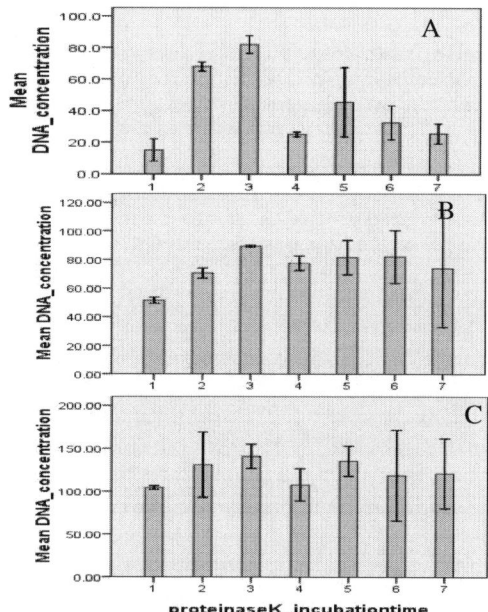

Fig. 1 DNA concentration extracted from 10 kernels (A), 20 kernels (B), 30 kernels (C) by modified method. **1**: 0 µl proteinase K (PK) + 0 hr incubation (I); **2**: 1.5 µl PK + 1 hr I; **3**: 3 µl PK + 1 hr; **4**: 1.5 µl PK + 1.5 hr; **5**: 3 µl PK + 1.5 hr; **6**: 1.5 µl PK + 2 hr; **7**: 3 µl PK + 2 hr

To compare modified method and standard method, the highest DNA concentration from modified method were used. As shown in figure 2, the concentration of DNA extracted by standard method are all higher than the DNA concentration of group 3ul proteinase K and 1 hour incubation. However, the ANOVA result showed that the difference of DNA concentration in group 10 kernels and 30 kernels were not significant (P>0.05); the concentration of group 20 kernels have difference from standard group, but it was slightly significant (P = 0.02).

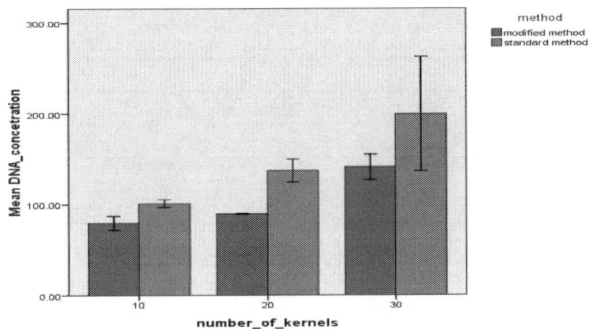

Fig. 2 DNA concentration comparison between modified method and standard method

Regarding to the purity of both methods, the A260/A280 ratio of DNA samples extracted by standard method was all in accepted range 1.7-2.0 while most of the ones obtained from modified method were 1.2-1.7, out of the acceptable range. Thus, the purity of DNA from modified extraction was not as well as standard one due to the shortening of purification steps.

Gel Electrophoresis

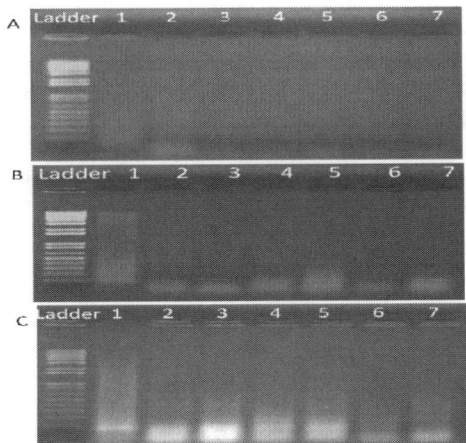

Fig. 3 Genomic DNA on gel electrophoresis. (A) from 10 kernels; (B) from 20 kernels; (C) from 30 kernels; lane 1-7 are 7 subgroup of modified method shown in figure 1

For both using methods, modified and standard one, the gel electrophoresis results all contained smeared lane, no specific band of genome DNA. However, in figure 3, the bright intensity of the lanes showed that the quantity of DNA extracted using modified method are lower than the amount of DNA from standard method (figure 4).

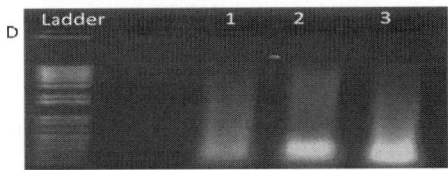

Fig. 4 Whole genome from standard method on gel electrophoresis. (1) from 10 kernels; (2) from 20 kernels; (3) from 30 kernels;

Multiplex PCR

The result of DNA quality testing in multiplex PCR assay is shown in figure 5, 6. Figure 5 shows the PCR result of 10 corn kernels. Among 7 subgroups of modified method (lane 1-7), PCR products show two clear bands of A. *flavus* and A. *parasiticus,* except lane 1 and 2. The first lane has two bands but the upper band is not quite clear. In contrast, lane 2 just appeared a band of A. *flavus*. Besides that, the PCR products of standard method appear in lane 10 with the same phenomenon as the sample in lane 1. Lane 8 and 11 represented for positive control with two clear bands that indicated that the sample was infected by two fungi; and the negative control was loaded in lane 9 and 12 with no band demonstrating that no DNA templates for PCR amplification. From lane 13 - 15, these were positive control mixed with DNA extracted from modified and standard method.

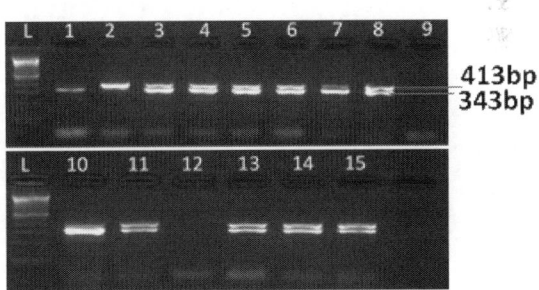

Fig. 5 Multiplex PCR result of DNA from modified method on 10 kernels (1 - 7), from standard method (10). Neg control (9, 12). Pos control (8, 11), mixture DNA (13-15)

In 20-kernels group (figure 6) two distinct and bright bands for A. *flavus* and A. *parasiticus* are observed in all bands except lane 9 and 15 of negative control. As it can be seen that the bands in lane 1-7 containing PCR product of modified method and in lane 10 of standard method are not different. Besides, there were two separately bands appeared in lane 12 to 14 revealing no inhibitors in extracted solution.

PCR assay using DNA extracted from 30 kernels was given the same result on the gel as the above groups (data not shown). Moreover, all results of both treatment methods did not show any extra products, so the PCR assay were highly specific.

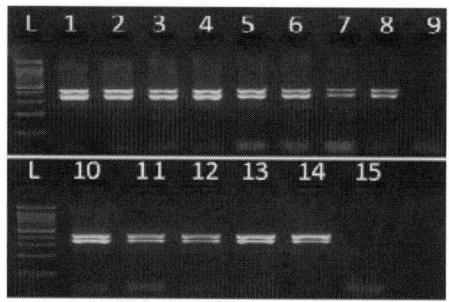

Fig. 6 Multiplex PCR result of DNA from modified method on 20 kernels (1 - 7), from standard method (10). Negative control (9, 15). Positive control (8, 11), mixture DNA (12-14)

IV. DISCUSSION

This study focused on shortening the DNA extraction without compromising the following PCR procedure. The effectiveness of a modified protocol was compared with the standard procedure through the quantity and quality of DNA in multiplex PCR. Due to the working ability of DNA in PCR assay, a new procedure could be applied to save the time of fungal detection one dried corn.

It could be realized that DNA quantity extracted by standard method was higher than modified method because of the modification DNA purification to save cost and time. The results of genome gel electrophoresis also showed that almost DNA was smeared, broken and contaminated. It could be RNA or protein from corn.

However, the statistical analysis showed that there was only slightly significant difference from two methods. Especially in group 10 and 30 kernels, the difference is not significant. So, 10 artificially infected kernels could be used in modified method. In addition, $3\mu l$ of proteinase K 2mg/ml in 1 hour incubation could likely be the most recommendable index to get highest DNA amount with modified process.

Furthermore, the multiplex PCR assay gave better result. The presence of *A. flavus* and *A. parasiticus* in corn was still detected in most cases as PCR results of standard method. Although DNA quality was slightly lower, PCR assay could still be performed and gave good result in fungal detection with DNA from modified method. It indicated that modified extraction method could replace the

standard one in isolating target DNA of *A. flavus* and *A. parasiticus* because it reduced time consuming and the cost significantly.

V. CONCLUSION

In conclusion, the modified method for DNA extraction was successfully optimized and applied in *A. flavus* and *A. parasiticus* detection on corn kernels. The best result could be generated using 3 μl of proteinase K 2mg/ml and 1 hour of incubation for 10 kernels. With the application of this new procedure, time and cost could be saved but the information derived from the test still allows to predict the fungi presence on corn kernels. For further study, the modified method should be applied in samples with unknown status of fungal infection to see whether it can work in reality or not.

ACKNOWLEDGMENT

Authors would like to thank to Vietnam National University – Ho Chi Minh City granted this study under project number B2013-28-02.

REFERENCES

1. Horn BW (2007) Biodiversity of Aspergillus section Flavi in the United States: a review. Food Addit Contam. 24(10): pp 1088-101.
2. Williams JH, et al. (2004) Human aflatoxicosis in developing countries: a review of toxicology, exposure, potential health consequences, and interventions. Am J Clin Nutr. 80(5): pp 1106-22.
3. Bennett JW, Klich M (2003) Mycotoxins. Clin Microbiol Rev. 16(3): pp 497-516.
4. Gong Y, Hounsa A, Egal S, et al. (2004) Postweaning exposure to aflatoxin results in impaired child growth: a longitudinal study in Benin, West Africa. Environ Health Perspect. 112(13): pp 1334-8.
5. Farombi EO (2006) Aflatoxin contamination of foods in the developing countries: Implications for hepatocellular carcinoma and chemopreventive strategies. African Journal of Biotechnology. 5: pp 1-14.
6. Wilson D, Mubatanhema WJurjevic Z (2002) Biology and Ecology of Mycotoxigenic Aspergillus Species as Related to Economic and Health Concerns. Mycotoxins and Food Safety, Springer, US: pp 3-17.
7. Plaza G, Upchurch R, Brigmon R, et al. (2004) Rapid DNA Extraction for Screening Soil Filamentous Fungi Using PCR Amplification. Polish Journal of Environmental Studies. 13(3).
8. Hue N, Nhien N, Thong P, et al. (2013) Detection of Toxic Aspergillus Species in Food by a Multiplex PCR, IFMBE Proc. vol 40, 4th International Conference on Biomedical Engineering, Vietnam, 2013, pp 184-189.

Corresponding Author: Dr. Nguyen Thi Hue
Institute: School of Biotechnology, International University – VNU, Ho Chi Minh City, Vietnam
Street: Block 6, Linh Trung Ward, Thu Duc District
City: Ho Chi Minh City
Country: Vietnam
Email: nthue@hcmiu.edu.vn

DEMM: A Meta-Algorithm to Predict the pKa of Ionizable Amino Acids in Proteins

T.B. Nguyen[1,2], K.P. Tan[1,3], and M.S. Madhusudhan[1,2,4]

[1] Bioinformatics Institute, 30 Biopolis Street, #07-01, Matrix, 138671, Singapore
[2] Department of Biological Sciences, National University of Singapore, 117543, Singapore
[3] School of Computer Engineering, Nanyang Technological University, 639798, Singapore
[4] Indian Institutes of Science Education and Research, Dr. Homi Bhabha Road, Pune, 4110008, India

Abstract— The protonation states of ionizable amino acid residues often have a direct influence on the functioning of a protein. The acid dissociation constant (in logarithmic scale, pKa) of these residues is hence an important determinant of protein function. To predict pKa, we integrated two complementary state of the art pKa prediction methods, DEPTH and microenvironment modulated screened Coulomb potential approximation (MM-SCP). The performance of the integrated predictor, DEMM, was benchmarked on a dataset of 47 residues with experimentally measured pKa values. DEMM has an average prediction error of < ~0.5 pH units and was statistically significantly superior to the DEPTH and MM-SCP methods. The method's utility is enhanced by its speed, accuracy and its applicability to proteins of varying sizes.

Keywords— pK$_a$ prediction, DEPTH, MM-SCP.

I. INTRODUCTION

pK$_a$ measures the tendency of protonation of a chemical group in solution. In proteins, pK$_a$ of ionizable amino acid residues determines its protonation state at different pH. Protonation states of residues modulate many protein properties, such as folding [1], stability [2,3], solubility [4], dynamics [5], interactions [6] and functions [7-10]. Estimating or predicting pK$_a$ is hence a powerful means of investigating protein functions.

The pK$_a$ value of ionizable amino acids in proteins can be measured with experiment methods such as NMR titration [11]. However, due to a myriad of technical difficulties these methods are infeasible on a large scale and computational predictions/estimations become necessary [12]. In this study, the pK$_a$s of 4 ionizable amino acids ASP, GLU, HIS and LYS were computationally predicted..

The model pK$_a$ value of an ionizable amino acid residue can be determined in solution when in isolation. In proteins, the pK$_a$ of the amino acid could differ from its model value, depending on its immediate surroundings or microenvironment. Several approaches have been proposed to predict pK$_a$ by quantifying these microenvironments, including Langevin dipoles [13,14], continuum dielectric solving the Poisson-Boltzmann equation [15-18], semi-empirical continuum dielectric based on generalized Born [19,20], empirical methods characterizing physic-chemical features of proteins [21-23], and molecular dynamics simulations involving efficient electrostatics [24,25].

Previously, we proposed an empirical method DEPTH [23] to predict pK$_a$. The DEPTH predictor is a multivariable linear regression formulation. It concisely describes the immediate environments of an ionizable amino acid by using several physical and chemical features. These features include depth [26], accessible surface area (ASA) [27], electrostatic interactions and hydrogen bonding [28].

Microenvironment modulated screened Coulomb potential approximation (MM-SCP) [29] is another state of the art method. MM-SCP attributes the shift of pK$_a$ from the model value solely to electrostatic interaction among ionizable groups in proteins. The method improved the calculation of electrostatic interactions by explicitly considering the screening of the Coulombic potential. Contributions to the Coulomb screening come from the amino acid residues in the surrounding microenvironment of the ionizable group. To model this screening effect the hydrophobicity and accessibilities of chemical groups constituting the amino acids were utilized.

Though both DEPTH and MM-SCP methods accurately predict the pK$_a$ (absolute mean error ~0.7 pH units), the consensus predictions have a modest correlation coefficient of 0.66. This suggests that the two methods complement one another. This motivated us to integrate the two methods to create a more accurate predictor, DEMM.

In the following sections we describe the formulation of the DEMM method, its benchmark performance and discuss its limitation and possible future development.

II. METHODS

Our pK$_a$ prediction procedure is a meta-algorithm integrating the two methods DEPTH and MM-SCP. The algorithm is a linear combination of the MM-SCP prediction (pK_a^{MMSCP}) [29] and the different physical/chemical features that constitute the DEPTH algorithm [23]. The predicted pK$_a$, pK_a^{pred} is computed a

V. Van Toi and T.H. Lien Phuong (eds.), *5th International Conference on Biomedical Engineering in Vietnam*,
IFMBE Proceedings 46, DOI: 10.1007/978-3-319-11776-8_84

$$pK_a^{pred} = pK_a^{model} + c_1 \cdot depth^{MC}$$
$$+ c_2 \cdot depth^{polarSC} + c_3 \cdot HB + c_4 \cdot EE_R \qquad (1)$$
$$+ c_5 \cdot ASA^{SC} + c_6 \cdot \left(pK_a^{MMSCP} - pK_a^{model}\right) + c_0$$

where, pK_a^{model} and pK_a^{MMSCP} are the model pKa and pKa predicted by MM-SCP method respectively. c_0-c_6 are coefficients of the individual features explained below. The values of the coefficients were optimized over a training set of residues (see results section).

Features

Depth: Residue (or atomic) depth measures the closest distance of the residue (or atom) to bulk solvent [26,30]. To accurately describe the solvent effect on an ionizable group, we used two complementary measures of depths in our predictor, average depth of main-chain atoms ($depth^{MC}$), and average depth of polar side-chain atoms ($depth^{polarSC}$).

Electrostatic energy (EE_R): Electrostatic energy of an ionizable residue, R, was calculated as the difference in Coulombic potential between its protonated and deprotonated states.

$$EE_R = \sum_{i \in R} \sum_{j \in R_b} \frac{Q_j}{r_{ij}} \Delta Q_i \qquad (2)$$

where ΔQ_i is the difference in partial charge of an atom i in residue R between its protonated and deprotonated forms. Q_j is the partial charge on atom j in R_b of the surrounding microenvironments (within a distance of 12 Å from atom i). r_{ij} is the distance between atom i and atom j.

The values of partial charges Q_i and Q_j were obtained from the gromos43a1 force field [31]. If R_b is an ionizable residue, the partial charge Q_j was chosen to correspond to the protonation state of residue R_b at a pH equivalent to the model pKa of residue R.

Note that this estimation of electrostatic energy described above is an improvement to the previous DEPTH model [23].

Hydrogen Bond (HB): Hydrogen bond were detected between donor-acceptor atom pairs if they were (i) within 3.5Å of one another and (ii) the donor-acceptor-acceptor antecedent angle was $100°$ or greater [28].

Solvent accessible surface area: Solvent accessible surface area of side-chain atoms (ASA^{SC}) was computed using the Shrake–Rupley algorithm [27].

Dataset

222 ionizable amino acid residues from 54 X-ray structures (resolution ranging from 1.2Å to 3.2Å) with their pKa values experimentally measured [32] were used to train (175) and test (47) our algorithm. This dataset consists of 58 ASP, 57 GLU, 71 HIS and 36 LYS residues.

III. RESULTS

The coefficients of the linear combination, c_0-c_6 (equation 1) were optimized separately for each of the residue types ASP, GLU, HIS and LYS. A conjugate gradient optimization was done over the training dataset of 175 residues for which experimentally determined pKa values were available (Table 1). Initially, the coefficient for the contribution from the MM-SCP method, c_6, was set to 0.5. The coefficients of the features contributing to original DEPTH algorithm, c_0-c_5 were taken as one half of their values from the original algorithm [23]. The optimization was performed until convergence using a tolerance value of $1e^{-8}$.

Our prediction method was tested on a set of 47 residues. The error rates for different amino acids were slightly different from each other. Our predictions for LYS were closest to the experimentally determined values (Mean error = 0.33 pH units, RMSD = 0.40 pH units), whereas predictions for HIS were the farthest (Mean error = 0.65 pH units, RMSD = 0.88 pH units). Overall, the prediction error is about 0.49 pH units (or RMSD 0.67 pH units) (Table 2). We have also shown that our method was statistically significantly superior to its individual component methods (DEPTH and MM-SCP) using a Wilcoxon paired sign rank test (Table 2).

Detailed information on the pKa predictions for the 222 ionizable residues of the training and testing set are available at http://mspc.bii.a-star.edu.sg/tankp/benchmark.html.

IV. DISSCUSSION/CONCLUSSION

DEPTH empirically describes the microenvironment using chemical and physical features. MM-SCP calculates the screened electrostatic interactions of ionizable groups by considering the hydrophobicity of the surrounding amino acid residues. The two methods complement one another as on the one hand DEPTH has a nuanced consideration of residue environments using residue depth but has a coarse treatment of the electrostatics (pairwise potential extending up to 12Å). On the other hand, MM-SCP has sophisticated electrostatics, considering the screening effect in the vicinity of ionizable group (4.25Å) but uses only solvent accessible area to describe residue environment.

This complementarity is demonstrated in the case of residue ASP 148 in *e-coli* ribonuclease H (PDB ID: 2RN2) (Fig. 1). The residue has an experimentally measured pK_a of 2. This residue has a polar side chain depth of 5.6Å and is involved in 6 hydrogen bonding interactions. As depth and hydrogen bonding effect are not considered in MM-SCP, the method overpredicted the pK_a value by 1.25 pH units. DEPTH had a smaller error in the case but underpredicted the value by -0.69 pH units. DEMM predicts a pK_a of 1.84, only -0.16 pH units from the experimental value. This is an improvement of 0.53 and 1.09 pH units over the DEPTH and MM-SCP methods respectively.

The new prediction method, DEMM, integrated the two complementary methods, DEPTH and MM-SCP. Overall, it has a mean error and RMSD of 0.49 pH units and 0.67 pH units respectively. The model improved the prediction of all 4 amino acid types ASP, GLU, HIS and LYS. Notably, the improvement in pK_a of HIS, which is usually the most difficult to predict, is at least 0.49 pH units better than the previous methods.

As compared to empirical models, quantum mechanics/molecular mechanics (QM/MM) [33] method still remains as the most accurate method in pK_a prediction (RMSD ~ 0.3 pH units). Presumably, this is because the method allows flexible and adaptive assignment of partial charge, as well as explicit consideration of protein dynamics. The QM/MM approach is however computationally expensive and not amenable to large-scale studies. In contrast, our empirical method is fast, accurate and versatile and can readily be applied to proteins/protein complexes of different sizes.

Table 2 Performance benchmark of DEPTH, MMSCP and DEMM over 47 ionizable groups in the testing. The mean absolute errors of predictions by the different methods and the corresponding RMSD (in brackets) are recorded in pH units. The improvement is the difference between DEMM and the more accurate prediction between DEPTH and MM-SCP.

Residue type (N)	DEPTH	MM-SCP	DEMM	Improvement
ASP (12)	0.72 (1.04)	0.67 (0.79)	0.53 (0.67)	0.14 (0.12)
GLU (12)	0.37 (0.48)	0.45 (0.58)	0.37 (0.5)	0.00 (-0.02)
HIS (15)	1.14 (1.45)	1.15 (1.54)	0.65 (0.88)	0.49 (0.57)
LYS (7)	0.43 (0.53)	0.48 (0.66)	0.33 (0.4)	0.10 (0.13)
Total (47)	0.71 (1.03)	0.73 (1.04)	0.49 (0.67)	0.22 (0.36)
P-value (1-tailed)	0.000*	0.004*		
P-value (2-tailed)	0.001*	0.012*		

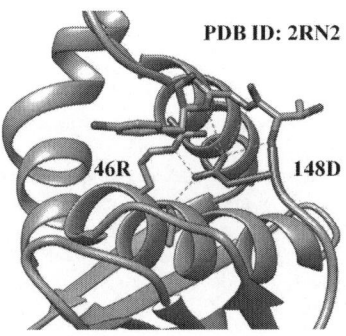

Fig. 1 A ribbon representation of the ribonuclease H (PDB ID: 2RN2). The microenvironment around residue 148D in ribonuclease H (green sticks) are shown in stick representation. The 4 residues (46, 149-151) that make hydrogen bonds (cyan lines) with 148D are also shown in stick representation. The figure was rendered using Chimera [34].

ACKNOWLEDGMENT

T.B. acknowledges Agency for Science, Technology & Research (A-STAR) for SINGA scholarship. The authors thank Prof. Mehler E.L. for providing the MM-SCP source code.

Table 1 Optimized coefficients of linear recombination for the different ionzable amino acid.

RES	Model pK_a	c_1	c_2	c_3	c_4	c_5	c_6	c_0
ASP	3.8	0.22	-0.07	-0.21	0.10	-0.02	0.66	-0.88
GLU	4.5	0.00	0.06	-0.01	0.16	0.06	0.14	-0.50
HIS	6.5	-0.20	-0.05	0.06	1.76	-1.14	0.44	1.84
LYS	10.5	-0.02	0.01	-0.01	0.83	0.66	0.69	-0.15

REFERENCES

1. Garcia-Moreno E.B., Fitch C.A. (2004) Structural interpretation of pH and salt-dependent processes in proteins with computational methods. Methods Enzymol 380: 20-51
2. Hendsch Z.S., Jonsson T., Sauer R.T. et al. (1996) Protein stabilization by removal of unsatisfied polar groups: computational approaches and experimental tests. Biochemistry 35: 7621-7625

3. Schaefer M., Sommer M., Karplus M. (1997) pH-Dependence of Protein Stability: Absolute Electrostatic Free Energy Differences between Conformations†. The Journal of Physical Chemistry B 101: 1663-1683

4. Tjong H., Zhou H.X. (2008) Prediction of protein solubility from calculation of transfer free energy. Biophys J 95: 2601-2609

5. Maciej D., Jan M.A. (2005) The impact of protonation equilibria on protein structure. Journal of Physics: Condensed Matter 17: S1607

6. Warshel A. (1981) Calculations of enzymatic reactions: calculations of pKa, proton transfer reactions, and general acid catalysis reactions in enzymes. Biochemistry 20: 3167-3177

7. Castaneda C.A., Fitch C.A., Majumdar A. et al. (2009) Molecular determinants of the pKa values of Asp and Glu residues in staphylococcal nuclease. Proteins 77: 570-588

8. Bartik K., Redfield C., Dobson C.M. (1994) Measurement of the individual pKa values of acidic residues of hen and turkey lysozymes by two-dimensional 1H NMR. Biophys J 66: 1180-1184

9. Oda Y., Yamazaki T., Nagayama K. et al. (1994) Individual ionization constants of all the carboxyl groups in ribonuclease HI from Escherichia coli determined by NMR. Biochemistry 33: 5275-5284

10. Oda Y., Yoshida M., Kanaya S. (1993) Role of histidine 124 in the catalytic function of ribonuclease HI from Escherichia coli. J Biol Chem 268: 88-92

11. Poon D.K., Schubert M., Au J. et al. (2006) Unambiguous determination of the ionization state of a glycoside hydrolase active site lysine by 1H-15N heteronuclear correlation spectroscopy. J Am Chem Soc 128: 15388-15389

12. Khandogin J., Brooks C.L., 3rd. (2006) Toward the accurate first-principles prediction of ionization equilibria in proteins. Biochemistry 45: 9363-9373

13. Sham Y.Y., Chu Z.T., Warshel A. (1997) Consistent Calculations of pKa's of Ionizable Residues in Proteins: Semi-microscopic and Microscopic Approaches. The Journal of Physical Chemistry B 101: 4458-4472

14. Sham Y.Y., Muegge I., Warshel A. (1999) Simulating proton translocations in proteins: probing proton transfer pathways in the Rhodobacter sphaeroides reaction center. Proteins 36: 484-500

15. Bashford D., Karplus M. (1990) pKa's of ionizable groups in proteins: atomic detail from a continuum electrostatic model. Biochemistry 29: 10219-10225

16. Nielsen J.E., Vriend G. (2001) Optimizing the hydrogen-bond network in Poisson-Boltzmann equation-based pK(a) calculations. Proteins 43: 403-412

17. Im W., Beglov D., Roux B. (1998) Continuum solvation model: Computation of electrostatic forces from numerical solutions to the Poisson-Boltzmann equation. Computer Physics Communications 111: 59-75

18. Warshel A., Sharma P.K., Kato M. et al. (2006) Modeling electrostatic effects in proteins. Biochim Biophys Acta 1764: 1647-1676

19. Onufriev A., Case D.A., Bashford D. (2002) Effective Born radii in the generalized Born approximation: the importance of being perfect. J Comput Chem 23: 1297-1304

20. Chen J., Im W., Brooks C.L., 3rd. (2006) Balancing solvation and intramolecular interactions: toward a consistent generalized Born force field. J Am Chem Soc 128: 3728-3736

21. Li H., Robertson A.D., Jensen J.H. (2005) Very fast empirical prediction and rationalization of protein pKa values. Proteins 61: 704-721

22 Huang R.B., Du Q.S., Wang C.H. et al. (2010) A fast and accurate method for predicting pKa of residues in proteins. Protein Eng Des Sel 23: 35-42

23. Tan K.P., Nguyen T.B., Patel S. et al. (2013) Depth: a web server to compute depth, cavity sizes, detect potential small-molecule ligand-binding cavities and predict the pKa of ionizable residues in proteins. Nucleic Acids Res 41: W314-321

24. Simonson T. (2000) Electrostatic Free Energy Calculations for Macromolecules: A Hybrid Molecular Dynamics/Continuum Electrostatics Approach. The Journal of Physical Chemistry B 104: 6509-6513

25. Lee M.S., Salsbury F.R., Jr., Brooks C.L., 3rd. (2004) Constant-pH molecular dynamics using continuous titration coordinates. Proteins 56: 738-752

26. Chakravarty S., Varadarajan R. (1999) Residue depth: a novel parameter for the analysis of protein structure and stability. Structure 7: 723-732

27. Shrake A., Rupley J.A. (1973) Environment and exposure to solvent of protein atoms. Lysozyme and insulin. J Mol Biol 79: 351-371

28. Baker E.N., Hubbard R.E. (1984) Hydrogen bonding in globular proteins. Prog Biophys Mol Biol 44: 97-179

29. Mehler E.L., Guarnieri F. (1999) A self-consistent, microenvironment modulated screened coulomb potential approximation to calculate pH-dependent electrostatic effects in proteins. Biophys J 77: 3-22

30. Tan K.P., Varadarajan R., Madhusudhan M.S. (2011) DEPTH: a web server to compute depth and predict small-molecule binding cavities in proteins. Nucleic Acids Res 39: W242-248

31. van Gunsteren W.F., Billeter S.R., Eising A.A. et al. (1996) Biomolecular Simulation: The {GROMOS96} manual and userguide. Hochschuleverlag AG an der ETH Zürich.

32. Song Y., Mao J., Gunner M.R. (2009) MCCE2: improving protein pKa calculations with extensive side chain rotamer sampling. J Comput Chem 30: 2231-2247

33. Jensen J.H., Li H., Robertson A.D. et al. (2005) Prediction and rationalization of protein pKa values using QM and QM/MM methods. J Phys Chem A 109: 6634-6643

34. Pettersen E.F., Goddard T.D., Huang C.C. et al. (2004) UCSF Chimera--a visualization system for exploratory research and analysis. J Comput Chem 25: 1605-1612

Author: Madhusudhan M.S.
Institute: *Indian Institutes of Science Education and Research*
Street: *Dr. Homi Bhabha Road*
City: *Pune*
Country: *India, 4110008*
Email: madhusudhan@iiserpune.ac.in

A Threshold Algorithm in a Fall Alert System for Elderly People

Pham Ty Phu[1], Nguyen Thanh Hai[1], and Nguyen Thanh Tam[2]

[1] Faculty of Electrical-Electronics Engineering, HCMC University of Technical Education, Vietnam
[2] Biomedical Engineering Department, International University, VNU, HCMC, Vietnam

Abstract— **Falls more likely happen with elderly people due to weak body. These falls may possibly cause serious injuries. Therefore, recognition of early falling for diagnosis and treatment is very important. In this paper, a fall alert system using an accelerometer sensor for elderly people based on a threshold algorithm is applied. A triaxial accelerometer sensor worn on the waist of people is employed to collect acceleration data. A threshold is determined by analyzing the difference between the normal daily activity and the fall. Data is collected from four daily activities and two falls: sit, stand, lie, walk, fall (walk) and fall (chair). This research potentially supports an enhanced understanding on the design and implementation of high–performance fall detections.**

Keywords— **fall detection, elderly people, threshold algorithm, human activity recognition**

I. INTRODUCTION

Due to the advancement in medical technology, the expectancy of people extends, which results in the increase elderly population. Thus, a health care mechanism for the elderly needs to be developed. According to Centers for Disease Control and Prevention (CDC) of American, unintentional fall is a common occurrence among older adults affecting approximately 30% of people aged larger than 65 years old. The fall may effect in negative psychological and physical [1]. Falls can lead to moderate to severe injuries, such as hip fractures and head traumas, and can even increase the risk of early death. The elderly people who have previously experienced a fall, fear a new fall and drop in inactivity and social isolation. The reduction in the mobility leads progressively to an increase in the risk of a fall. In addition, falls among the elderly people cost a lot of contribution from home care and social health insurance, which increases the burden to society. Therefore, a good fall detection system is important [2].

The fall detection system can be divided into three methods such as image sensing [3, 4], environmental sensing [5] and the wearable sensor [6-10]. For the image sensing method, cameras are installed in environment to detect if some of the subjects in the image fall using the image processing method. This method does not need users to wear any device, but its images are susceptible to light and noise interference resulting in low accuracy and high arithmetic difficulty. For environment sensing method, sensors such as infrared sensors or sound sensors are placed in the environment to detect fall. The method has advantages of a low price and the convenience takes for users as they have no need of wearing any device. However, the system can only detect in the environment where it is installed. In the recent years, many scholars have proposed the detection method of wearing sensors. It has advantages of low cost and no limitation environment. This method consists three phases. First, the data obtained from accelerometer are a series of time – related signal. Most researchers use single sensor, while some other use multiple sensors. The second phase discusses how to obtain useful feature form these data. There are features using commonly such as the accelerometer value of X, Y, Z axis, Signal Magnitude Vector (SMV), Signal Magnitude Area (SMA), angle of acceleration, etc. The third phase discusses how to determine whether the fall occurs. There are two determination methods based on the threshold or classifier. On the basis of the above methods, we take into account the threshold determination and SMA to carry out fall detection.

This paper presents a fall detection system from data of triaxial accelerometer sensor, which worn on waist. Data of sensor use to calculate SMA. If this value greater than the threshold, the system recognizes as a fall and then display a warning. The threshold value is determined by analyzing the difference of a normal daily activity and a fall. We collected four daily activities and two falls: sit, stand, lie, walk, fall (walk) and fall (chair).

The rest of this paper is organized as follows: Section II introduces data acquisition process and fall detection system. Section III presents the experiment result. Finally, we conclude the paper in Section IV.

II. METHODS

A. Data Collection and Data Process

We used the kit ZSTAR 3 of FreeScale. The ZSTAR3 kit accommodates an accelerometer sensor board, connected through an RF ZigBee 2.4 GHz communication to a single USB node (USB stick) connected to a PC, as shown in Fig. 1. The sensor integrated a MMA7660 digital triaxial

V. Van Toi and T.H. Lien Phuong (eds.), *5th International Conference on Biomedical Engineering in Vietnam*,
IFMBE Proceedings 46, DOI: 10.1007/978-3-319-11776-8_85

accelerometer sensor and a MC13213 microcontroller as 2.4 GHz low-power transceivers. The USB stick uses a MCHC908JW32 microcontroller to communication.

The MMA7660 is a ± 1.5g 3-axis accelerometer with digital output. It is a very low power, low profile capacitive MEMS sensor featuring a low pass filter, compensation for 0g offset and gain errors, and conversion to 6-bit digital values.

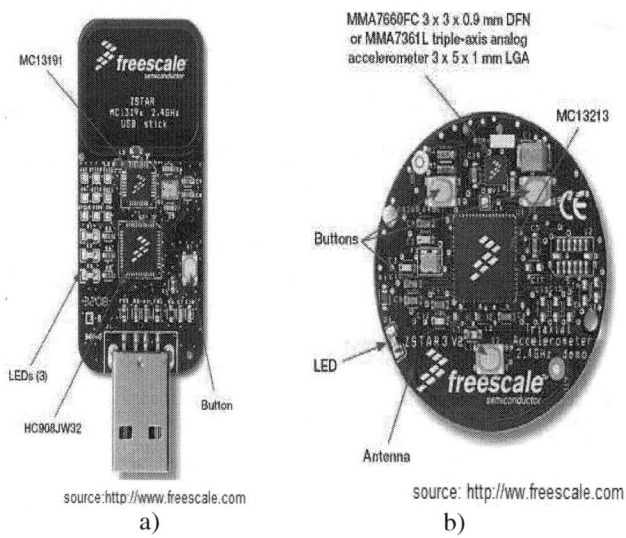

Fig. 1 Wireless Sensing Triple Axis Reference Design (ZSTAR)

a) The USB stick use MCHC908JW32 microcontroller
b) The sensor integrated a MMA7660 accelerometer

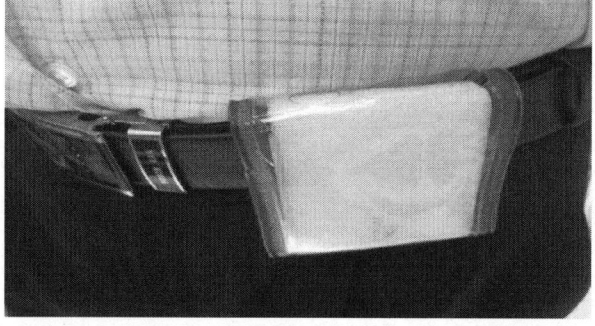

Fig. 2 The sensor board is inserted into the case, worn on the waist

The sensor board is inserted into the case and worn on the waist, as shown Fig. 2. Accelerometer data are sampled via the Analog to Digital Converter (ADC) at 30 Hz. The next step is median filtering, whereby a median filter with n=3 is applied to the raw digitized signal to remove any abnormal noise spikes produced by the accelerometers.

Fig. 3 illustrated the fall detection system with the position of wear the sensor board. Processing of the triaxial accelerometer output is to be performed by the

microcontroller embedded onto the sensor board and information regarding the movement of the user transmitted to USB stick before being forwarded to a computer for evaluation.

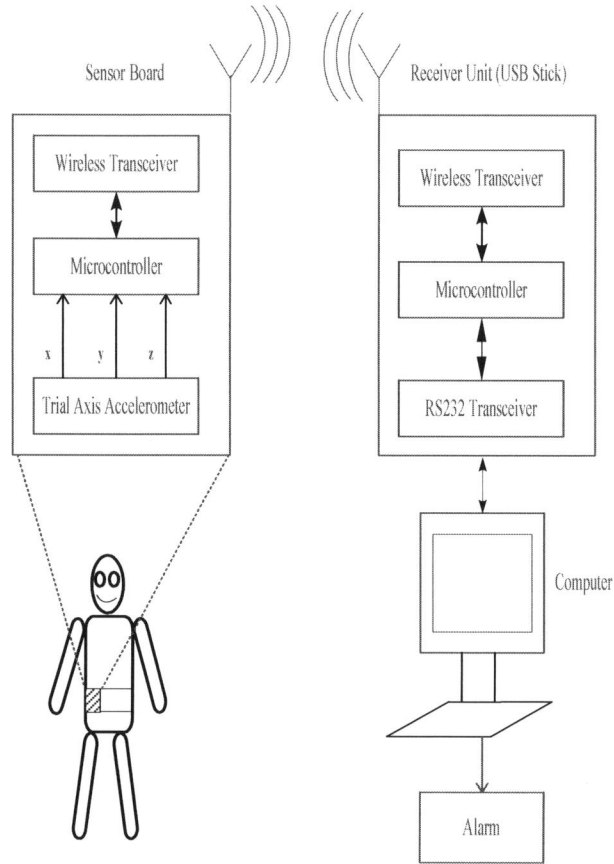

Fig. 3 Schematic diagram of the system

B. Fall Detection Algorithm

To distinguish between periods of user activity and fall, a measure that includes the effect of signal variations in all three axes is required. A suitable measure is the normalized SMA. Defined in (1), the SMA was used as the basis for identifying periods of activity:

$$SMA = \frac{1}{t}\left(\int_0^t |x(t)|dt + \int_0^t |y(t)|dt + \int_0^t |z(t)|dt\right) \quad (1)$$

where x, y and z are the acceleration signals from the tri-axial accelerometer.

In order to generate a threshold based on ADL, data were gathered from four ADL and two falls include sitting down, stand up, walking, lie down, fall when subject walks, falling when subject sit on a chair. Each of actives, 10 samples were collected. A healthy male (age: 25; height: 168 cm) was selected to perform six activities. The threshold was set at 1.5 g, after analyzing data of the ADL and the fall.

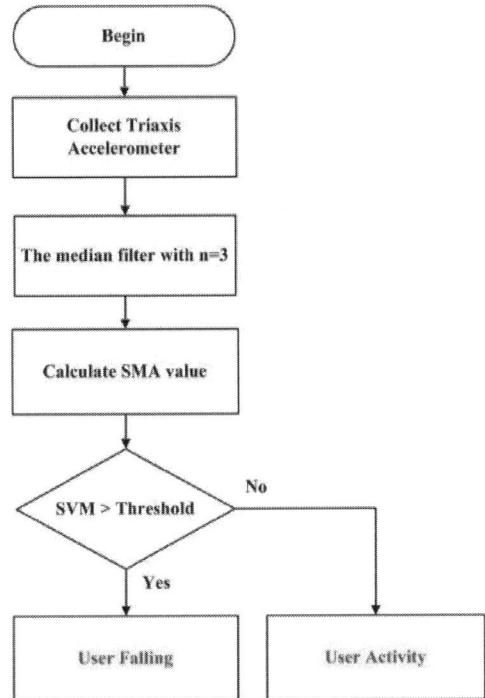

Fig. 4 Falling Detection flowchart

The flowchart (as shown Fig. 4) describes major steps of the proposed algorithm: collect data, filter data with a median filter (n=3), calculate SMA value, compare value with threshold to determine that a fall was occurred or not.

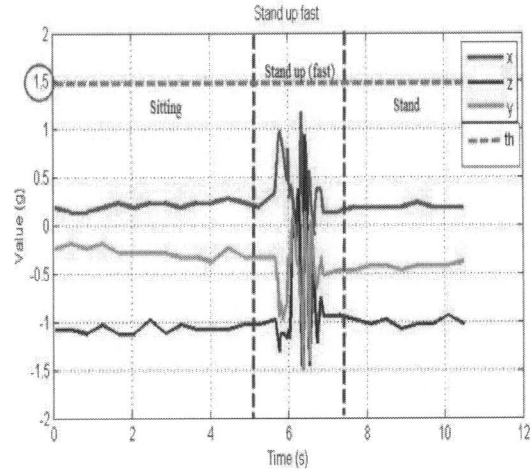

Fig. 5 The acceleration data when the subject sits down and stand up

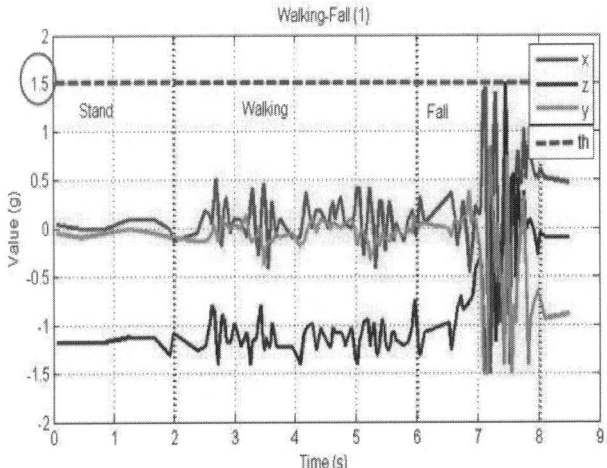

Fig. 6 The acceleration data when the subject stands, walks and falls.

Four experiments are done to analysis the difference between normal and fall status as shown in Fig. 5 and Fig. 6. The red, green, and blue lines depict acceleration data for X, Y, and Z axes, respectively. In case of normal activity as shown in Fig. 5, the acceleration value does not exceed the threshold (1.5 g). This threshold value was selected on experiment of Burchfield [10]. On the other hand, when a fall occurs, the acquired data exceeds the predefined threshold as showed in Fig. 6.

III. RESULTS

The application, which is presented with the interface shown in Fig. 6, is designed for the detection fall system. Pushing "Open" button will open file of data set. Pushing "Start" button will start to detect the fall, if the value over the threshold, a text "Fall" and a text value will be displayed. A database of random activities data are employed to test accuracy of system. One of the databases is described in Table 1. In Fig. 7, the system detected fall event and showed a caution and the value.

Table 1 Acquired acceleration data

Time (s)	X (g)	Y (g)	Z (g)
0.066	0.0469	-1.1719	-0.0469
0.466	0.0000	-1.1719	-0.0938
0.866	0.0000	-1.1719	-0.0469
1.266	0.0938	-1.1251	0.0000

Table 1 (*continued*)

1.666	0.0938	-1.1251	-0.0469
1.933	0.0000	-1.3126	-0.0938
1.999	-0.1406	-1.0782	-0.0938
...	...	...	...
8.033	0.6094	-0.0469	-0.7500
8.099	0.5157	-0.0938	-0.9376
8.499	0.4688	-0.0938	-0.8907

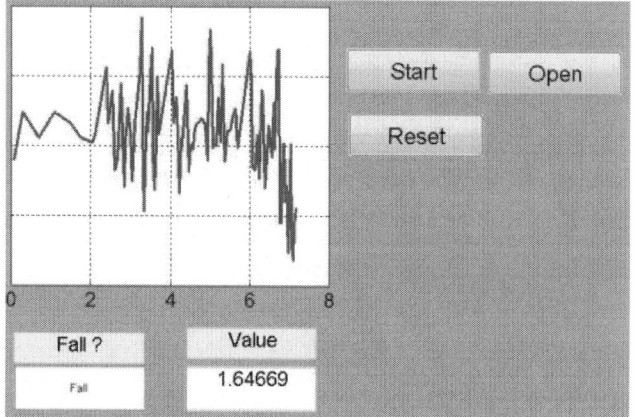

Fig. 7 The interface of application using in the fall detection system

IV. CONCLUSIONS

A fall alert system using an accelerometer sensor for elderly people was proposed in this paper. One subject was invited to perform the experiment 70 times to collect acceleration data of the X, Y, Z triaxial accelerometer worn on the waist. Experimental results showed the effectiveness of the fall detection system based on the proposed threshold algorithm. For further research, we intend to develop a real-time detection based on machine learning algorithms.

ACKNOWLEDGEMENT

Authors would like to thank Vietnam National University in Ho Chi Minh City for supporting research. Furthermore, this research was partly supported by a research fund from International University in Ho Chi Minh City. Finally, an honorable mention goes to our students and colleagues for supports on us in completing this project.

V. REFERENCES

1. Yi He YL, Chuan Yin. (2012) Falling-Incident Detection and Alarm by Smartphone with Multimedia Messaging Service. E-Health Telecommunication Systems and Networks.
2. Gonzales IJD. (2011) Fall Detection Using a Smartphone. Gjøvik University College.
3. Xiaoxiao D, Meng W, Davidson B, Mahoor M, Jun Z. (2013) Image-Based Fall Detection with Human Posture Sequence Modeling. IEEE International Conference on Healthcare Informatics (ICHI).
4. Mirmahboub B, Samavi S, Karimi N, Shirani S. (2013) Automatic monocular system for human fall detection based on variations in silhouette area. IEEE Transactions on Biomedical Engineering.
5. Litvak D, Zigel Y, Gannot I. (2008) Fall detection of elderly through floor vibrations and sound. Engineering in Medicine and Biology Society, EMBS 2008 30th Annual International Conference of the IEEE.
6. Wen-Chang C, Ding-Mao J. (2013) Triaxial Accelerometer-Based Fall Detection Method Using a Self-Constructing Cascade-AdaBoost-SVM Classifier. IEEE Journal on Biomedical and Health Informatics; 17:411-9.
7. Dean M. Karantonis MRN, Merryn Mathie, NigelH.Lovell and Branko G. Celler. (2006) Implementation of a Real-Time Human Movement Classifier Using a Triaxial Accelerometer for Ambulatory Monitoring. IEEE Transactions on Information Technology In Biomedicine.
8. Bagalà F, Becker C, Cappello A, Chiari L, Aminian K, Hausdorff JM, et al. (2012) Evaluation of accelerometer-based fall detection algorithms on real-world falls. PLoS One.
9. Farkas I, Doran E. (2011) Activity Recognition From Acceleration Data Collected With A Tri-Axial Accelerometer. Acta Technica Napocensis Electronica-Telecomunicatii.
10. Burchfield TR, Venkatesan S. (2007) Accelerometer-based human abnormal movement detection in wireless sensor networks. Proceedings of the 1st ACM Sigmobile international workshop on Systems and networking support for healthcare and assisted living environments: ACM.

Author: Pham Ty Phu
Institute: HCMC University of Technical Education
Street: Vo Van Ngan
City: Ho Chi Minh
Country: Viet Nam
Email: phupt@hcmute.edu.vn

Comparative Study on Human A-Glucosidase

Q. Ong[1] and L. Le [1,2,*]

[1] School of Biotechnology of Ho Chi Minh International University, Vietnam
[2] Life Science Laboratory of the Institute for Computational Science and Technology at Ho Chi Minh City, Vietnam
ly.le@hcmiu.edu.vn

Abstract— α-glucosidase is an enzyme in human encoded by GAA gene [1]. Normally, this enzyme is essential for the degradation of glycogen to glucose in lysosomes. However, with patients of diabetes mellitus type 2, α-glucosidase causes the unwanted elevation of blood glucose level. Therefore, α-glucosidase is now the target of anti-diabetic drugs. In this study, structure of human α-glucosidase was investigated using the approaches of bioinformatics tools. The results from phylogenetic tree also suggest the Mus musculus for animal testing. Finally, some approved α-glucosidase inhibitor drugs from Drugbank database are tested for binding affinity energy and pharmacophore features.

Keywords— α-glucosidase, diabetes mellitus type 2, multiple sequence alignment, phylogenetic tree.

I. INTRODUCTION

α-Glucosidase inhibitors are oral anti-diabetic drugs used for diabetes mellitus type 2 that work by preventing the digestion of carbohydrates (such as starch and table sugar). Carbohydrates are normally converted into simple sugars (monosaccharides), which can be absorbed through the intestine. Hence, α-glucosidase inhibitors reduce the impact of carbohydrates on blood sugar. In order to support in vivo testing of α-glucosidase inhibitor, this research of α-glucosidase will be defined on animal modeling and then observing the result before aplplying for human treatment. In addition, using computational tools including Clustal X, Treeview, VMD, Autodock Vina and other online softwares can test the drug compouds for α-glucosidase inhibitor.

II. MATERIALS AND METHODS

Multiple Sequence Alignment of Homologous α- Glucosidase Sequences

Human α-glucosidase sequence (NCBI ID: ABI53718.1) as the template protein was collected from the NCBI database [2]. It is the full-length sequence of human α-glucosidase with 952 amino acids. The homologous α-glucosidase amino acid sequences were taken from non-redundant protein database of Blastp [3] search with default

parameters (BLOSUM 62 matrix [4]). These sequences were the materials for progressive multiple sequence alignment (MSA) using Clustal X [5] with input ordered and Phylip output format. The MSA step was conducted several times to optimize by removing the most different sequences with very low identity degrees in order to gain the reliable alignment of homologies. The results from MSA were then used for refining the construction of phylogenetic tree which served to analyze the evolutionary relationship of α- glucosidase and its homologies in other species.

Obtain 3D Structure of Human α-Glucosidase

Because the crystal structure of human α- glucosidase is still a mist, 3D structure of it was obtained using homology modeling provided by Swiss-Model workspaces with default parameters [6].

Phylogenetic Tree Construction

The results of multiple sequence alignments were used as input data for constructing phylogenetic trees that would outline the interrelationships of the Homo sapiens enzyme and the other vertebrate species. To approach this, the PHYML [7] which implements the maximum likelihood method was used to estimate the phylogenetic tree using the options for amino acid data type, Jones, Taylor, and Thornton (JTT) substitution model and tree topology best search of NNI (Nearest Neighbor Interchange) and SPR (Subtree Prune and Regraft) search.

Analysis of phylogenetic tree and suggest the suitable species for testing drug on animal

Based on the results of phylogenetic study and the Three Rs principle which were first described by W.M.S. Russell and R.L. Burch in 1959 [8], one species was selected to be used as in vivo testing for α-glucosidase drug.

Prepare for Docking

Drug molecules were selected from Drugbank database (http://www.drugbank.ca).Several modifications from the original data were made using different software. VMD - Visual molecular dynamics was used to visualize and separate the receptor for docking [9]. The PRODRG Server

---

* Corresponding author.

© Springer International Publishing Switzerland 2015
V. Van Toi and T.H. Lien Phuong (eds.), *5th International Conference on Biomedical Engineering in Vietnam,*
IFMBE Proceedings 46, DOI: 10.1007/978-3-319-11776-8_86

[10] was used to convert SDF and MOL2 file formats to the pdb format. All drug molecules were modified by using this server for obtaining drug molecules in the pdb file format. ADT / Auto-dock tools — Auto-dock was used to convert pdb files to pdbqt files for the docking process. All ligands and proteins in the pdb format were converted to the pdbqt format with the correction of charges for docking [11]. Python [12] is an integrative program that facilitates ADT and AutodockVina to run on the Linux platform.

Docking

The docking procedure requires the identification of the binding box position, which is the active site of the protein. The binding energy of these bound molecules on the protein was used for control docking. Furthermore, a grid box was prepared for docking drug molecules with a box size of 98x102x97Å spacing of 0.1 Å and a level 12 of exhaustiveness.

Analyze of Docking Results

The binding conformations of drug molecules with the protein were analyzed to find the basic interactions and characterize possible residues responsible for binding with the aid of VMD and ligand-scout [13]

III. RESULTS

3D Structure Generation of Human α-Glucosidase

The 3D structure of human α-glucosidase was generated by SWISS-MODEL web server using crystal structure of the N-terminal domain of sucrase-isomaltase [14]. According to this alignment, the human α-glucosidase shares around 44.4% sequence identity and 89.0% sequence similarity with the template (Fig. 1A)

Obtain 3D Structure of Human α-Glucosidase

Because the crystal structure of human α-glucosidase is still a mist, 3D structure of it was obtained using homology modeling provided by Swiss-Model workspaces with default parameters [15].

Analysis of 3D Structure and Propose the Active Site of Human α-Glucosidase

The structure of this model has 3 main part: N-termimal domain (residues 1-346), C-terminal domain (residues 724-952) and the sub-domain (residues 347-723). The proposed active-site pocket here was composed to residues of residues W376, W402, D404, I441, D443, W481, W516, D518, M519, F525, R600, W613, D616, D645, F649, and H674 (Fig. 1B,1C). Asp-518 is predicted to be the essential carboxylate in the active site of human α-glucosidase. The functional importance of Asp-518 and other residues around the catalytic site was studied by expression of in vitro mutagenized α-glucosidase cDNA in transiently transfected COS cells [16]. The residues Trp-516 and Asp-518 are also demonstrated to be critical for catalytic function (Fig. 2A)

Phylogenetic Tree Construction

The phylogenetic tree was constructed using the result of ClustalX software. As the result, the proteins in each species share high degree of sequence similarity and common tertiary structure motif as well as the enzymatic specificity (Fig. 2B). Though all of these organisms are closely related to human beings (gorillas, gibbob, chimpanzee and monkeys) but they are limited for testing in terms of resources and animal research ethics. Therefore the Mus musculus (House mouse) is the best candidate for animal testing because they are satisfied the Three Rs principle.

Testing the Best α-Glucosidase Inhibitors Structure Using AutoDock Vina

After screening top ten approved drugs as α-glucosidase inhibitors from Drug Bank database, 3 best drugs were selected for studying pharmacophore features. They are Acarbose, Methyldopa and Voglibose. Based on docking result, Acarbose is the drug molecule had the lowest binding affinity energy among those 3 (Fig. 3).

IV. DISCUSSION

This research using the approaches of modern bioinformatics tools to give the most accurate results of Human α-glucosidase 3D structure. However, due to the rapidly increase of computational software developers, the results from updated tools may be quite different from what using in this study. Therefore, further studies should give the comparison on the results among similar softwares to give the most significant data. In addition, study on pharmacophore features of experimental drug-liked compounds as α- glucosidase inhibitors will be the next level of this research in future, which gives correlated data between computational and practical studies.

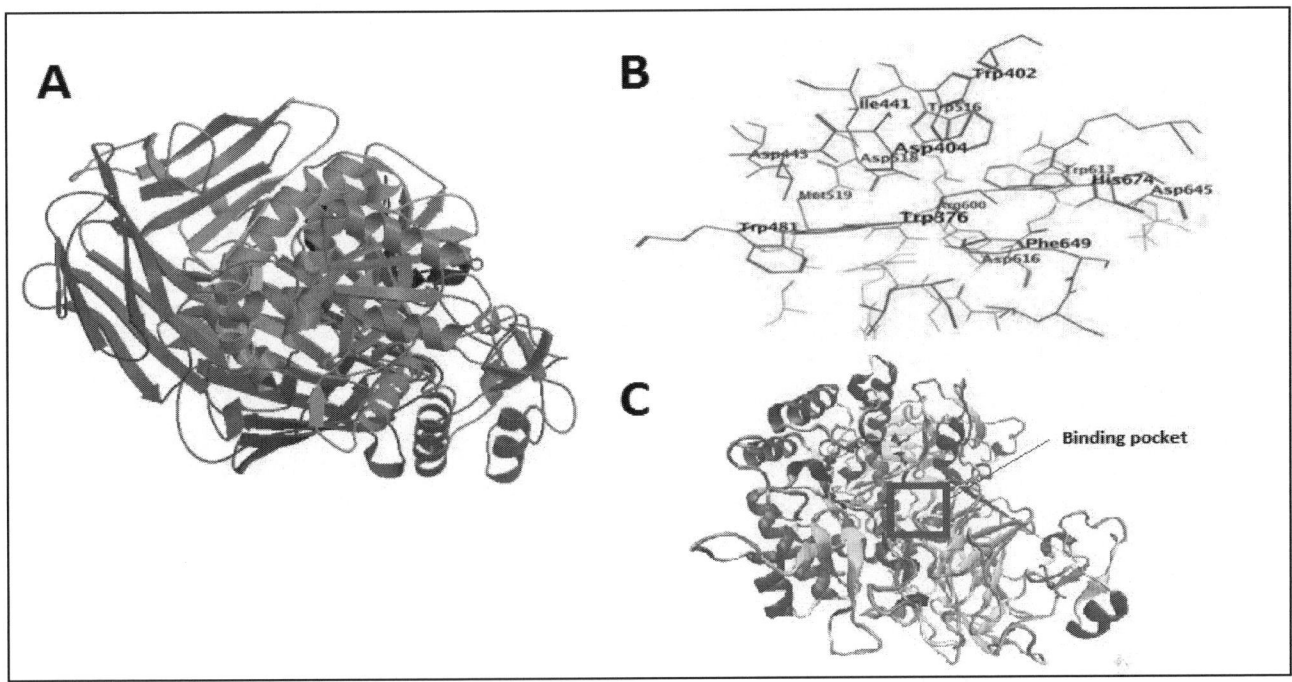

Fig. 1 A. 3D structure of human α-glucosidase generated by SWISS-MODEL. B. The proposed active-site pocket with residues W376, W402, D404, I441, D443, W481, W516, D518, M519, F525, R600, W613, D616, D645, F649, and H674. C. Binding pocket (red).

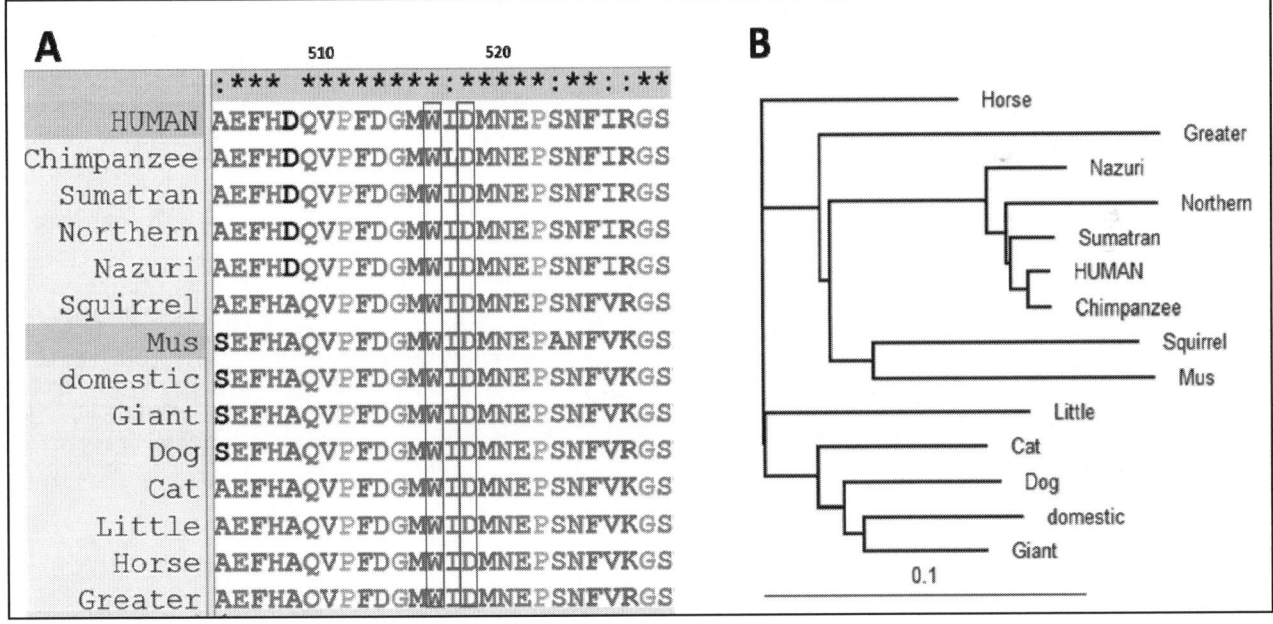

Fig. 2 A. Multiple sequences alignment using ClustalX, Trp-516 and Asp-518 (in Red column) are also demonstrated to be critical for catalytic function. B. Phylogenetic tree constructed by TreeView (version 1.6.6) in which suggests Mus Musculus for in vivo testing.

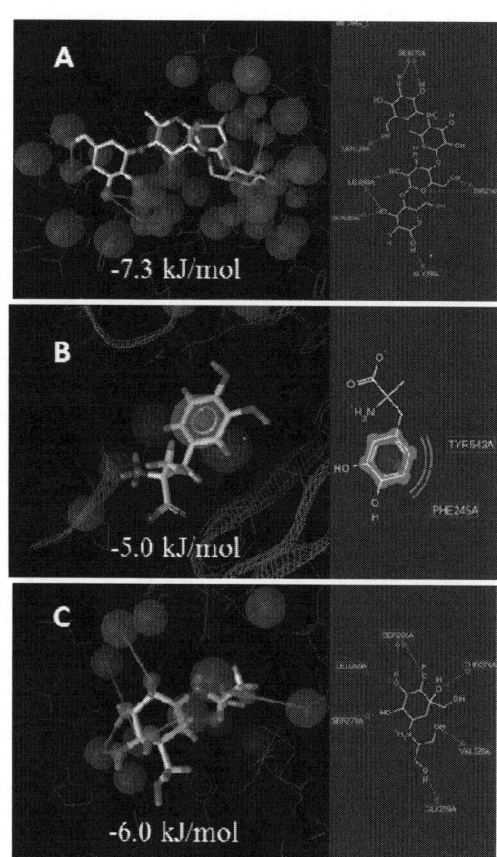

Fig. 3 A, B, C are the 3D structures of Acarbose, Methyldopa and Voglibose respectively with their binding affinity energy and 2D structures with pharmacophore features.

V. CONCLUSION

α-Glucosidase inhibitors have high potential to be used as the early treatment for diabetes mellitus type 2 by preventing the digestion of carbohydrates in our body. This research suggest Mus musculus as the animal model for α-glucosidase drug testing. In addition, Acarbose was suggested as the best drug for blocking α-glucosidase as its lowest binding affinity energy and also shown many important pharmacophore features that can be used for further study on α-glucosidase as a target treatment for diabetes mellius type 2.

REFERENCES

1. Park, Hyung-Doo, et al. "Three Patients with Glycogen Storage Disease Type II and the Mutational Spectrum of GAA in Korean Patients." Annals of Clinical & Laboratory Science 43.3 (2013): 311-316.
2. http://www.ncbi.nlm.nih.gov
3. Altschul SF, Gish W, et al. Basic local alignment search tool. Journal of Molecular Biology, 1990. 215(3): p. 403-410.
4. Henikoff S and Henikoff JG. Amino acid substitution matrices from protein blocks. Proceedings of the National Academy of Sciences, 1992. 89(22): p. 10915-10919.
5. Higgins,D.G., et al. (1992) CLUSTAL V: improved software for multiple sequence alignment. CABIOS 8,189-191.
6. Arnold K., Bordoli L., et al. (2006). The SWISS-MODEL Workspace: A web-based environment for protein structure homology modeling. Bioinformatics, 22,195-201.
7. Guindon S and Gascuel O. A Simple, Fast, and Accurate Algorithm to Estimate Large Phylogenies by Maximum Likelihood. Systematic Biology, 2003. 52(5): p. 696-704.
8. Russell, W.M.S. and Burch, R.L., (1959). The Principles of Humane Experimental Technique, Methuen, London.
9. Visual Molecular Dynamics: www.ks.uiuc.edu/Research/vmd
10. The PRODRG Server : davapc1.bioch.dundee.ac.uk/prodrg/
11. ADT/AutoDockTools-AutoDock:autodock.scripps.edu/resources/adt
12. Sanner M (1999) Python: a programming language for software integration and development. J Mol Graphics Mod 17:57–61.
13. Wolber G. Langer T (2005) LigandScout: 3-D Pharmacophores Derived from Protein-Bound Ligands and Their Use as Virtual Screening Filters. J. Chem. Inf. Model 45(1); 160-169.
14. Sim, L. et al., Structural basis for substrate selectivity in human maltase-glucoamylase and sucrase-isomaltase N-terminal domains. J.Biol.Chem (2010).
15. Arnold K., Bordoli L., Kopp J., and Schwede T. (2006). The SWISS-MODEL Workspace: A web-based environment for protein structure homology modelling. Bioinformatics, 22,195-201.
16. Hong, Yeongjin, et al. "The Lec23 Chinese hamster ovary mutant is a sensitive host for detecting mutations in α-glucosidase I that give rise to congenital disorder of glycosylation IIb (CDG IIb)." Journal of Biological Chemistry 279.48 (2004): 49894-49901.

First author: Quang D. Ong
Institute: International University
Street: Quarter 6, Linh Trung Ward, Thu Duc District
City: Ho Chi Minh
Country: Vietnam
Emai: ongquang9210@gmail.com

Corresponding author: Ly Le
Institute: Institute for Computational Science and Technology
Street: Quang Trung Software City, Tan Chanh Hiep Ward,
 District 12
City: Ho Chi Minh
Country: Vietnam
Emai: ly.le@hcmiu.edu.vn

Development of Non-Invasion Method for Prognosis and Early Diagnosis of Cervical Cancer in Vietnamese Patients Based on DNA Methylation Specific PCR

T.K. Phuong, L.D. Thuan, and L.H.A. Thuy[*]

Falculty of Biotechnology, Ho Chi Minh City Open University, Vietnam
lhathuy@gmail.com

Abstract— **Cervical cancer remains one of the common leading cause of cancer death for women worldwide, including Vietnam. Besides the infection of human papilloma virus (HPV) which is the main cause of cervical cancer, increasing evidence demonstrated that inactivation of tumor suppressor genes (TSGs) by aberrant promoter methylation is an early event during carcinogenesis. In current study, aiming to establish biomarker applied in prognosis or early diagnosis for cervical cancer, we developed a powerful assay based on Methylation Specific PCR to detect the aberrant DNA methylation of the panel of cervical cancer related genes in liquid-based Papanicolaou (Pap) tests. Total of 83 liquid-based Pap test samples which were identified whether HPV-infection or non-HPV infection, high-risk HPV oncogenetic type 16 and 18 infection or low-risk HPV infection, were carried by MSP method to evaluate the DNA aberrant methylation occurred in DAPK, RARβ and p16INK4α. According to the results, the hypermethylation reach to 78.8% for RARβ, 63.6% for DAPK and 54.5% for p16INK4α. This hypermethylation characteristic was also associated with HPV high risk genotype infection. Furthermore, the MI values with 97.0% diagnosis coverage, which meant at least one of three genes were methylated. In conclusion, these outcome suggested that the MSP assay carried out on the non-invasive samples (liqid-based pap) will lead to the potential method to prognosis and early diagnosis of cervical carcinoma, as well as allow us to have a vision for these hypermethylation of candidate genes could be a promising biomarker for cervical cancer detection in Vietnamese population.**

Keywords— **RARβ, p16INK4α, DAPK, hypermethylation, methylation specific PCR, cervical cancer.**

I. INTRODUCTION

Infection with high-risk genotypes of human papillomavirus (HPV) is the common cause which leading to the cervical carcinoma, the common cancer affecting women in developing countries, including Vietnam [3]. The most common oncogenic HPV genotype are 16 and 18, causing approximately 70% of all cervical cancer [3, 4]. Besides the high-risk HPV infection, the epigenetic alteration in regular gene, including the tumor suppressor genes (TSGs), leading to the inactivation of TSGs by promoter methylation, an early symptom in the multi-step process of carcinogenesis, including the cervical cancer [8]. To access whether aberrant methylation in the promoter of CpG islands of *DAPK* (Death-Associated Protein kinase), *RARβ* (Retinoic acid receptor beta), *p16^{INK4α}* genes which are belonged to tumor suppressor gene family, from the cervical patients using the various source of samples as Pap's smear, cervical scraping, cervical intraepithelial neoplasia (CIN), cervical squamous cell carcinomas, etc by different method including MSP (Methylation specific PCR), QMSP (Quantitative Methylation specific PCR)… has been published [23, 25]. Therefore, the frequent aberrant methylated TSGs in tumor have been proposed as the potential biomarkers for early detection of malignant cells from various clinic materials. It provides possibilities of both cancer early detection and dynamic monitoring of cancer patients after treatment [19]. In present study, based on the MSP method, the frequency of methylation of CpG belonged to promoter of *DAPK*, *RARβ*, *p16^{INK4α}* genes in Vietnamese population was evaluation by using non-invasive materials as liquid-based Pap test specimens which were defined positive or negative HPV infection and the oncogenic HPV or non-oncogenic infection. Aberrant methylation on TSG promoters is rather prevalent and tumor-specific for cervical carcinoma, which suggests its potential as an assay for cervical carcinoma diagnosis.

II. MATERIALS AND METHODS

Sample Collection

Total of 83 liquid-based Pap test samples, which pathology was identified, archived and admitted from the Au Lac clinic, Vietnam. The samples set was composed of 33 samples infecting with positive high-risk HPV infection and 50 samples which were negative HPV infection or low-risk HPV infection. The detection of HPV infection was carried out by using PCR-Reverse Dot Blot Kit (*Viet-A Corporation, Vietnam*).

DNA Extraction, Bisulfite Modification and MSP Assay

Genomic DNA from residual liquid-based Pap tests was isolated using the phenol chloroform method. In addition,

---

[*] Corresponding author.

V. Van Toi and T.H. Lien Phuong (eds.), *5th International Conference on Biomedical Engineering in Vietnam,*
IFMBE Proceedings 46, DOI: 10.1007/978-3-319-11776-8_87

Table 1 The methylation frequency of *DAPK*, *RARβ*, *p16^{INK4a}* gene

	DAPK		RARβ		p16^{INK4a}	
	M (%)	U (%)	M (%)	U (%)	M (%)	U (%)
HPV-high risk specimens	21 (63.6)	12 (36.4)	26 (78.8)	7 (21.2)	18 (54.5)	15 (45.5)
Non-HPV infection or low-risk infection	1 (2.0)	49 (98.0)	1 (2.0)	49 (98.0)	nc*	nc
p value	*< 0.001*		*< 0.001*		*nc*	
Non-HPV infection	1 (2.3)	42 (97.7)	1 (2.3)	42 (97.7)	nc	
p value	*< 0.001*		*< 0.001*		*nc*	

*Note: nc: non-calculated

Table 2 The correlation between methylation frequencies of candidate genes with HPV type infection.

Sample (n=33)	Number of methylated sample n(%)		
	DAPK	RARβ	p16^{INK4a}
HPV type 18 (n=14)	9 (64.3)	11(78.6)	8 (57.1)
HPV type 16 (n=12)	8 (66.7)	9 (75.0)	8 (66.7)
Co-infection of HPV type 18 and type 16 (n=7)	4 (57.1)	6 (85.7)	2 (28.6)
p		*< 0.0001*	
Odd Ratio	85.75	182	nc*
p		**< 0.0001**	nc

*nc: non-calculated

approximately 1 µg genomic DNA of each samples were bisulfite-modified by DNA modification kit (*Epitect Kit, Qiagen*). The MSP reaction was carried out by specific primers to determine the methylation or unmethylation status of given genes following our previous study [7, 18].

Statistical Analysis

Differences in the presence of methylation or unmethylation of candidate genes were determined by a two sided Fisher test and Chi squared tests for variables. Statistical analyses were performed by using Medcalc® Version 12.7.0.0 [9]. Statistical significance was assumed at two-side P value of $p < 0.05$.

III. RESULTS AND DISCUSSION

Methylation Frequencies of DAPK, RARβ, p16^{INK4a} Genes

According to the results, the methylation status of promoter belonged *DAPK*, *RARβ* and *p16^{INK4a}* genes were 63.6% (21 of 33), 78.8% (26 of 33) and 54.5% (18 of 33) in HPV high risk infection samples, respectively. Comparison with recently reports, in our study, the methylation status of *DAPK* was the same as the research of Niyazi *et al.*, (2012) [11] and higher than other researches of Iliopoulos et al., (2009) as 57.4% [6], Rebecca *et al.*, (2008) as 54.64% [13]. For the *RARβ*, we found out the frequency as was higher than the research of Gopeshwar *et al.*, (2003) as 43.3% [5], Tatyana *et al.*, (2002) as 40.0% [16]. Especially, there were significant differences in the frequencies of methylation

between the HPV high-risk infection specimens and non-HPV infection in both *DAPK* and *RARβ* genes counting for 2.3% (1 of 43) (*p<0.001*) (Table 1). Furthermore, concerning in the oncogenic characteristic, the HPV low-risk infection should be considered as "non-HPV" infection. We, therefore, found out another significant differences in the frequencies of methylation between the oncogenic specimens and non-oncogenic specimens in both *DAPK* and *RARβ* genes counting for 2% (1 of 50) (*p<0.001*) (Table 1). In the case of *p16^{INK4a}*, in present study, because of the limitation of time, we did not determine the status of *p16^{INK4a}* methylation in non-cancer specimens, however, the frequency of methylation on the high-risk specimens was high and nearly equal to the research of Mohammed *et al.*, (2009) counting for 59.1% [10], Arvind *et al.*, (2001) counting for 42% [2].

Statistical Analysis

According to our results in table 2, in the case of HPV type 18 infection, the hypermethylation frequencies were 64.3%, 78.6%, 57.1% for *DAPK*, *RARβ* and *p16^{INK4a}*, respectively. In total of positive HPV type 16 infection, the frequencies were 66.7%, 75.0%, 66.7% for *DAPK*, *RARβ* and *p16^{INK4a}*, respectively. With the co-infection of HPV type 16 and 18, the hypermethylation frequencies were counting for 57.1%, 85.7%, 28.6% in *DAPK*, *RARβ* and *p16^{INK4a}*, respectively. Consequently, the methylation of *DAPK*, *RARβ*, *p16^{INK4a}* genes with HPV types infection were strongly associated with each other (*p<0.0001*). It meant that the HPV type 18 and 16, which were commonly infected in Vietnamese population, were related to the methylation levels. Actually, the infection with HPV high-risk genotype acted as the carcinogens in the development of cervical cancer by integrating the viral DNA into human genome and expressing the E6 and E7 oncoproteins. According to Patti *et al.*, (2008) and the references therein, combined detection of HPV DNA intergration and methylation status may eventually be useful as an important prognostic biomarker for cervical cancer [12]. Thus, the results in our study also reinforced the above point-views. In addition, the OR value (Odd ratio) was calculated for each genes. The OR were 85.75 (95% CI, 10.46 to 702.4) (*p<0.0001*), 182 (95% CI, 21.2 to 1560.3) (*p<0.0001*) for *DAPK* and *RARβ*, respectively (Table 2). It meant that the hypermethylation characteristic in *DAPK* and *RARβ* increasing the risk up to 85.75 and 182 times, respectively, in the consideration of cervical cancer development.

As shown in the table 3, we found that at least one of three genes, counting for 97.0%, was methylated. It indicated that in the case of using these three genes for screening and detection of cancer, it brought out the high probability to detect the aberrant methylation status in samples was very high. In order to evaluate the status of multiple genes methylation, the MI (methylation index) was calculated for each sample. In our sample collection, the percentage were 3.1%, 27.3%, 39.4%, 30.3%, 97.0% for MI=0 (without methylation), MI=0.3 (one of three genes were methylated), MI=0.7 (two of three genes were methylated), MI=1.0 (all three genes were methylated) and MI≥0.3 (at least one gene was methylated), respectively. It was important to note that in our non-invasive samples with MI value, it was the useful symptom to detect the cancer development, moreover, regarding the samples with low MI value, it was used for the methylation drug test or hypermethylation inhibitor treatment leading the cancer cell apoptosis.

Due to these results, the hypermethylation of panel genes were the characteristic of the HPV oncogenic high-risk infection specimens in Vietnamese population. Furthermore, this characteristic was obtained by MSP method in non-invasive samples that could be the useful method for the clinical application in prognosis and early diagnosis of cervical cancer in future.

Table 3 Calculation of MI value in breast cancer samples collection

The number of methylated gen (n = 33)	MI value	The number of cervical cancer	Percentage (%)
None methylated gene	0	1	3.1
One gene	0.3	9	27.3
Two gene	0.7	13	39.4
Three gene	1.0	10	30.3
At least one gene	≥ 0.3	32	97.0

IV. CONCLUSION

In summary, the aberrant methylation of at least one gene of *DAPK, RARβ, p16^{INK4a}* in total of 33 cancer specimens with 97.0% diagnosis coverage was the specific characteristic of Vietnamese patients with the infection with high-risk HPV leading to cervical carcinoma. The frequencies of methylation were 63.6%, 78.8% and 54.5% for *DAPK, RARβ, p16^{INK4a}*, respectively. In the case of low-risk or non-HPV infection, there were virtually without methylation of promoter of *DAPK* and *RARβ*. The odd ratio value were 85.75, 182 for *DAPK* and *RARβ*, respectively. These finding suggested that the MSP assay carried out on the non-invasive samples (liqid-based pap) will lead to the potential method, which was easily in the clinical application, to prognosis and early diagnosis of cervical carcinoma. In further study, the methylation have to be carried out in non-cancer specimens for *p16^{INK4a}* to have a total revision of the hypermethylation symptom in *DAPK, RARβ,* and *p16^{INK4a}* genes as the biomarkers for cervical cancer in Vietnamese specimens.

REFERENCES

1. American Cancer Society. (2011). Global Cancer Facts and Figures (2nd Edition), Atlanta American Cancer Society.
2. Arvind KV, Carolyn M, Asha R (2001) Aberrant Methylation during Cervical Carcinogenesis. Clin Cancer Res 7:584-589.
3. Cutts FT, Franceschi S, Goldie S, Castellsague X, Sanjose S, Garnett G, Edmunds WJ, Claeys P, Goldenthal KL, Harper DM, Markowitz L (2007) Human papillomavirus and HPV vaccines: a review. Bulletin of the World Health Organization 85:719–726.
4. David J (2008) A review of cross-protection against oncogenic HPV by an HPV-16/18 AS04-adjuvanted cervical cancer vaccine: Importance of virological and clinical endpoints and implications for mass vaccination in cervical cancer prevention. Gynecologic Oncology 110:S18–S25.
5. Gopeshwar N, Hugo AP, Sanjay K, Hernan V, Fang FZ, Jeannine V, Achim S, Mary BT, Mahesh M, Vundavalli VM (2003) Frequent Promoter Methylation of CDH1, DAPK, RARB, and HIC1 Genes in Carcinoma of Cervix Uteri: Its Relationship to Clinical Outcome. Molecular Cancer 2:24.
6. Iliopoulos D, Oikonomou P, Messinis I, Tsezou A (2009) Correlation of promoter hypermethylation in hTERT, DAPK and MGMT genes with cervical oncogenesis progression. Oncol Rep 22(1):199-204.
7. Linh LTT, Quy LK, Phuong HTB, Thuy LHA (2011) Assessment of aberrant promoter CpG islands methylation of DcR1, DAPK1, and p16INK4α in cervical neoplasia. Journal of Biotechnology 9(4B): 1-9
8. Lu Q, Ma D, Zhao S (2012) DNA methylation changes in cervical cancers. Methods Mol Biol. 863:155-76.
9. MedCalc (2013) Statistics for biomedical research. www.medcalc.org
10. Mohammed A, Wail EH, Meriem K, Laila B, Nadia B, Abdellatif B, Mariam A, and Mohammed El Mzibri. Status of p16INK4a and E-Cadherin Gene Promoter Methylation in Moroccan Patients With Cervical Carcinoma. Oncology Research 18:1–100.
11. Niyazi M, Liu XW, Zhu KC (2012) Death-associated protein kinase promoter (DAPK) hypermethylation in uterine cervical cancer and intraepithelial neoplasia in Uyghur nationality women. Chinese journal of oncology 34(1):31-4.

12. Patti EG, Francois C, Thomas I, John WS, Wim GVQ, Cosette MW (2008) New Technologies in Cervical Cancer Screening. Vaccine 26S:K42–K52.

13. Rebecca CYL, Stephanie SL, Kelvin YKC, Kar-Fai T, Kar-Loen C, Ling-Chui W, Hextan YSN (2008) Promoter methylation of death-associated protein kinase and its role in irradiation response in cervical cancer. Oncology Reports 19:1339-1345.

14. Schulz W (2005) Qualified promise: DNA methylation assays for the detection and classification of human cancers. J Biomed Biotechnol 2005: 227–229. doi: 10.1155/JBB.2005.227.

15. Tang N, Huang S, Erickson B, Mak WB, Salituro J, Robinson J, Abravaya K (2009) High-risk HPV detection and concurrent HPV 16 and 18 typing with Abbott RealTime High Risk HPV test. J Clin Virol 1:S25-8.

16. Tatyana I, Anatolii P, Tatyana G, Svetlana V, Ernest E, Vera K, Fjodor K, Natalia K (2002) Methylation and silencing of the retinoic acid receptor-β gene in cervical cancer. BMC Cancer 2:4.

17. Thuan LD, Truong Kim Phuong TK, Anh NTQ, Neelesh S, Thuy LHA (2013) DNA methylation at the RARβ promoter: a potential biomarker for cervical cancer. Current Trends in Biotechnology and Pharmacy 7 (3):708-715.

18. Xian-Lan Z, Zhi-Ying M, Yu-Huan Q, Hui-Li Z (2008) Promoter methylation of DAPK gene in cervical carcinoma. Chinese Journal of Cancer 27(9):212-215

19. WHO/ICO Information Centre on HPV and Cervical Cancer (HPV Infor mation Centre). (2010). Human Papillomavirus and Related Cancers in Viet Nam. Summary Report. (Dateaccessed). Available at www. who. int/hpvcentre.

20. Wisman GB, Nijhuis ER, Hoque MO, Reesink-Peters N, Koning AJ, Volders HH, Buikema HJ, Boezen HM, Hollema H, Schuuring E, Sidransky D, van der Zee AG. Assessment of gene promoter hypermethylation for detection of cervical neoplasia. Int J Cancer. 2006 Oct 15;119(8):1908-14.

Antifungal Activity of Conyza canadensis ((L.) Cronquist) Collected in Northern Viet Nam

N.B. Phuong[1], N.T.T. Lien[2], and N.T.T. Hoai[1]

[1] School of Biotechnology, International University, VNU-HCM, Viet Nam
[2] Faculty of Biology, Hanoi University of Natural Sciences, VNU-HN, Viet Nam

Abstract— *Conyza canadensis* (*C. canadensis*) ((L.) Cronquist) has been used as medicinal herb in many countries. The antifungal activity of *C. canadensis* which was extracted by different solvents including ethanol, ethyl acetate and n-hexane was examined by using agar well diffusion method and minimum inhibitory concentration method. Two fungal pathogens used to determine the bioactive activity of these extracts were *Candida albicans* (*C. albicans*) and *Trichosporon insectorum* (*T. insectorum*). Among these three extracts using different polarity solvents, the antifungal activity of *C. canadensis* extracted with ethyl acetate showed the highest antifungal activity against both tested fungal pathogens. Conversely, the extracts with ethanol and n-hexane didn't show any activity towards the tested fungi in the agar well diffusion experiment. *C. canadensis* extracted with ethyl acetate showed its high effect against *T. insectorum* with 45.33 mm of inhibition zone, antifungal activity was lower to *C. albicans* which was about 25.33 mm of inhibition zone. Minimum inhibitory concentration (MIC) of ethanol, ethyl acetate and n-hexane extract against *C. albicans* were 250, 15, and 500 mg/ml, respectively. MIC values with extracts in ethanol and ethyl acetate solvents in case of *T. insectorum* were lower than which of *C. albicans*. The particular MIC values of extracts in ethanol, ethyl acetate and n-hexane against *C. albicans* were 63, 8 and 1000 mg/ml, respectively.

Keywords— *C. canadensis*, *C. albicans*, *T. insectorum*, agar well diffusion method, minimum inhibitory concentration.

I. Introduction

Conyza canadensis (*C. canadensis*) is a common medicinal herb found throughout most of North America. It is also named horseweed and belongs to Asteraceae family. In the traditional medicine of some Asian countries, *C. canadensis* functions as an effective treatment for many diseases causing by bacteria, fungi, or viruses including bronchitis, smallpox, cystitis, diarrhea [1, 2]. In China, *C. canadensis* is used as folk medicine to treat wound and pain caused by arthritis [3]. Its essential oil containing bioactive compounds can inactivate the growth of many types of microorganisms. In Viet Nam, data on the antifungal activities of this herbal plant is still limited. Furthermore, the differences in geographical origin of Vietnamese plants to that of other countries may affect their antifungal activity.

Besides, many local traditional practitioners have been using *C. canadensis* to treat fungal infections indicating its high effectiveness. Therefore, it is necessary to study thoroughly their antifungal effects in order to improve their application and usage. This study aimed to determine the antifungal activity *C. canadensis* extracting in different solvents by using agar well diffusion and minimum inhibitory concentration (MIC) method.

II. Materials and Method

A. Plant Extract Preparation

The *C. canadensis* was collected randomly in northern Viet Nam particular in Hai Duong province, in May 2013. This plant was cleaned and subjected in drying process at constant temperature of about $50^{\circ}C$ until completely dry. Then, it was grinded to fine powder by using blender (Phillips, Japan) at the Laboratory of Chemistry. Ten grams of *C. canadensis* dried powder were macerated in 100 mL of three different solvents including ethanol, ethyl acetate and n-hexane for extraction. They were kept in bottles for 3 days to be extracted according to established protocol [4]. The samples were filtrated through Whatmann filter paper by using Buchner funnel. After that, the gained solutions in the three bottles were subjected to vacuum evaporation to get crude extracts.

B. Antifungal Assay

Antifungal assay was performed with 2 fungal pathogens, *Candida albicans* (*C. albicans*) and *Trichosporon insectorum* (*T. insectorium*) by using agar well diffusion method described in previous studies [5]. The surface of SDA (Sabouroud-dextrose agar) plate was seeded by 1ml of overnight fungal culture inoculum. Three tested wells were made on the agar plate by using Pasteur pipette. Different amount of crude extract were loaded into the wells (40μL/well). Amphotericin B, a common antifungal drug, was used as positive control (40 μg/ well for *T. insectorum* and 10 μg/ well for *C. albicans*) while solvents were used as negative control. The agar plates were incubated at $37^{\circ}C$ for 48h. After 48h, diameters of inhibition zones were recorded.

V. Van Toi and T.H. Lien Phuong (eds.), *5th International Conference on Biomedical Engineering in Vietnam,*
IFMBE Proceedings 46, DOI: 10.1007/978-3-319-11776-8_88

C. Minimum Inhibitory Concentration (MIC)

The MIC values were determined by using broth dilution method [6]. The extract was diluted in 10 consecutive tubes with 1:2 ratios to 2^{11} times. Then, the fungal suspension was inoculated in each tube. Medium plus inoculum and extracted solvent plus medium and inoculums were used as positive controls while only medium or only extracted solvent plus medium were used as negative controls. All tubes were incubated at 37^0C for 48 hours. Minimum inhibitory concentrations were determined based on the turbidity of medium in compared with negative control. The concentration of the last turbid test tube was considered as MIC value of that fungi strain with *C. canadensis* extract.

III. RESULTS AND DISCUSSION

We have found a high antifungal activity of *C. canadensis* against tested fungal pathogens. Table 1 showed inhibition zones corresponding to the loading weigh of the crude extract. We observed a proportional relationship between the loading weigh and the inhibition zone, i.e. the more crude extract amount was applied, the larger inhibition zone was achieved. In details, extract in ethyl acetate solvent had highest inhibition value (25.33 mm in *C. albicans* and 45.33 mm in *T. insectorum*). However, crude extract of this plant in polar solvent (ethanol) and non-polar solvent (n-hexane) didn't show the bioactivity. The maximum inhibition zone of *C. canadensis* against *T. insectorum* was 45.33 mm in diameter (with 2 mg/ mL crude extract in ethyl acetate), against *C. albicans*, 25.33 mm in diameter (with 2 mg/ mL crude extract in ethyl acetate). This suggested that ethyl acetate – a medium polar solvent gave better bioactivity than polar solvents such as ethanol, methanol or water. Ethyl acetate solvent is often used for extraction of medium-polar substances including tertpene, coumarin, quinone,…due to the fact that it can penetrate through the cell wall of this plant and forming bonds with aldehyde or cetone group [7]. In *C. canadensis*, there are two main compound groups, Conyzolide and Conyzoflavone of *C. canadensis* with antifungal and antimicrobial activity [8]. It was known that the Conyzaflavone gave superior antifungal activity as compared to Conyzolide. The functional groups of these flavonoids contain methoxyl (-OCH$_3$) which was well dissolved in ethyl acetate. This result is in accordance with our data that *C. canadensis* extraction in ethyl acetate showed best antifungal effect.

Table 1 The inhibition zone of three types of extract were against *C. albicans* and *T. insectorum*

Fungi	Solvents	Concentration of Conyza canadesis (g/ml)	Zone of inhibition (mm)[a]
T. insectorum	Ethanol	0.5	NA
		1	NA
		1.5	NA
		2	NA
	n- hexane	0.5	NA
		1	NA
		1.5	NA
		2	NA
	Ethyl acetate	0.5	29.33 ± 1.53
		1	36 ± 2.65
		1.5	40.33 ± 1.53
		2	45.33 ± 1.16
C. albicans	Ethanol	0.5	NA
		1	NA
		1.5	NA
		2	NA
	n- hexane	0.5	NA
		1	NA
		1.5	NA
		2	NA
	Ethyl acetate	0.5	NA
		1	16.67 ± 1.53
		1.5	20 ± 1
		2	25.33 ± 1.77

[a] *Diameter of inhibition zone (mm) including diameter of well (6 mm)*
NA: not active
Three types of solvent had no effect on fungal growth.

Table 2 The inhibition zone of Amphotericin B were against *C. albicans* and *T. insectorum*

Fungi	Amount of Amphotericin B (μg)	Zone of inhibition (mm)[a]
T. insectorum	40	9.33 ± 0.58
C. albicans	10	19.67 ± 0.58

[a] *Diameter of inhibition zone (mm) including diameter of well (6 mm)*

The MIC results were similar to the agar well diffusion results described above with some slight difference (Table 3). MIC values of the crude extract ranged from 1000 mg/ml to 7.813 mg/ml. With this method, the bioactivity of the crude extracts in ethanol and n-hexane solvent was revealed (250 mg/ mL of *C. albicans* and 62.5 mg/ mL of *T. insectorum* in ethanol fraction, 500 mg/ mL of *C. albicans* and 1000 mg/ mL of *T. insectorum* in n-hexane fraction) while they were not able to be estimated by agar well diffusion method above.

Table 3 MIC results of extracted plant to *C. albicans* and *T. insectorum*

| Types of extracted solvent | MIC (mg/ml) | |
	C. albicans	*T. insectorum*
Ethanol	250 mg/mL	62.5 mg/mL
*n-h*exane	500 mg/mL	1000 mg/mL
Ethyl acetate	15.625 mg/mL	7.813 mg/mL

The lowest MIC result was with extract in ethyl acetate (15.625 mg/ mL of *C. albicans* and 7.813 mg/mL of *T. insectorum* in ethanol fraction) indicating the highest antifungal activity which was in accordance with the result done by diffusion method. However, we found that while using MIC test, extract in n- hexane showed exceptionally better activity towards *C. albicans* than to *T. insectorum*.

In general, *C. canadensis* extracts showed much higher activity towards *T. insectorum* than to *C. albicans* for both disc- test and MIC test.

IV. CONCLUSIONS

Our result on antifungal activity of *C. canadensis* revealed that this plant possessed bioactive compounds with potential to serve as antifungal agents against pathogenic fungi in this case, *C. albicans* and *T. insectorum*. In the three types of solvents which were used for extraction, the medium polar solvent such as ethyl acetate was the most appropriate solvent for the extraction process. Data also indicated the clear advantage of *C. canadensis* in inhibiting the new fungus strain *T. insectorum* compared to the commercial common antifungal drug, amphotericin B. The extraction of *C. canadensis* should be further investigated and tested in the in-vivo experiments.

ACKNOWLEDGMENT

This work was supported by the International University and the MedMic team

REFERENCES

1. Grünwald, T. Brendler, and C. Janicke, Eds., PDR for Herbal Medicine, Thompson, 2000.
2. Shinwari, M.I and M.A. Khan, 2000. Folk use of medicinal herbs of Margalla hills national park, Islamabad. Journal of Ethnopharmacology, 69(1): 45-56.
3. T.S. C. Li, Chinese and Related North American Herbs-Phytopharmacology and Therapeutic Values, CRC Press, New York, NY, USA, 2002.
4. Rauf, A., N. Muhammad, A. Khan, N. Uddin, M. Atif and Barkatullah, 2012. Antimicrobial and Phytotoxic Profile of Selected Pakistani Medicinal Plants, World Applied Sciences Journal, 20: 540-544.
5. IE Oboh, JO Akerele and O Obasuyi, 2007. Antimicrobial activity of the ethanol extract of the aerial parts of sida acuta burm.f. (malvaceae), Tropical Journal of Pharmaceutical Research, December 2007; 6 (4): 809-813.
6. Jonathan Betts, Christine Murphy, Stephen Kelly, and Stephen Haswell, 2012. Minimun inhibitory and bactericidal concentration of theaflavin and syngergistic combination with epicatechin and quercetin against clinical isolates of Stenotrophomonas Maltophilia, Journal of Microbiology, Biotechnology and Food Sciences: 1 (5) 1250-1258.
7. Nguyen Kim Phi Phung, 2007. Phuong phap co lap hop chat huu co.
8. Mohammad Shakirullah, Hanif Ahmad, Muhammad Raza Shah, Imtiaz Ahmad, Muhammad Ishaq, Nematullah Khan, Amir Badshah, Inamullah Khan. Antimicrobial activities of Conyzolide and Conyzoflavone from *Conyza canadensis*. Journal of enzyme inhibition and medicinal chemistry, Vol. 26, No. 4 , Pages 468-471.
9. Lin Jiang, 2011. Comparison of disk diffusion, agar dilution and broth micro-dilution for antimicrobial susceptibility testing of five chitosans. Fujian Agricultural and Forestry University, China.

Corresponding author:
Author: Hoai T.T. Nguyen
Institute: International University
Street: Quarter 6, Linh Trung Ward, Thu Duc District
City: Ho Chi Minh
Country: Vietnam
Email: ntthoai@hcmiu.edu.vn

Antimicrobial Activity of Senna alata (l.), Rhinacanthus nasutus and Chromolaena odorata (l.) Collected in Southern Vietnam

Thuong L.H. Pham[1], Trung T. Trinh[2], and Hoai T.T. Nguyen[1]

[1] School of Biotechnology, International University, VNU-HCMC, Vietnam
[2] Institute of Microbiology and Biotechnology, VNU-HN, Vietnam

Abstract— **This study aimed to evaluate the antimicrobial activity of *Senna alata (L.)*, *Rhinacanthus nasutus* and *Chromolaena odorata* against antibiotic resistant *Staphylococcus aureus*, *Pseudomonas aeruginosa* clinical strains; vitro – induced antibiotic resistant *S. aureus* strains; *S. aureus* ATCC 25213 and *P. aeruginosa* ATCC 9027. Plant samples were extracted with different solvents including ethanol, ethyl acetate and n-hexane. Disc diffusion method was used to evaluate the antimicrobial activity of extracts. In general, most of extracts showed good antimicrobial activity against both antibiotic resistant and sensitive *S. aureus* but no activity against *P. aeruginosa*. To clinical *S. aureus* strains, *C. odorata* and *R. nasutus* extracts using ethyl acetate showed better inhibitory zones (17.17 ± 0.29 mm ; 16.67 ± 0.58 mm) than the ones extracted using ethanol (14.17 ± 0.58mm, 11.67 ± 1.53 mm) and n- hexane (12.17 ± 0.58 mm; 13.67 ± 0.58 mm, respectively). To other *S. aureus* strains, *S. alata (L.)* extract using ethanol showed the greatest inhibition zone (22.33 ± 0.58 mm). This result suggested that these traditional herbs should be further investigated to serve as alternative treatment for multidrug- resistant *S. aureus* infections and probably also for other gram positive pathogenic bacteria.**

Keywords— **antimicrobial activity, *Senna alata (L.)*, *Rhinacanthus nasutus*, *Chromolaena odorata*, disc diffusion, inhibition zone.**

I. INTRODUCTION

Senna alata (L.), *Rhinacanthus nasutus* and *Chromolaena odorata* have been popularly used as traditional medicine for curing infectious diseases.

S. alata (L.) was found mostly in South East Asian countries [1]. This plant leaf is traditionally used for the treatment of constipation, stomach pain, ringworm and skin diseases. Yakubu claimed that the aqueous leaf extract of the plant contained saponins (1.22%), flavonoids (1.06%), cardiac glycosides (0.20%), phenolics (0.44%), alkaloids (0.52%), cardenolides and dienolides (0.18%)[2]. In 2011, Jeyaseelan and his colleagues proved that the cold extraction of *S. alata* revealed significantly higher inhibition on *S. aureus* and *P. aeruginosa* compare with hot and fresh extraction [3].

R. nasutus is widely distributed in the region of Southeast Asia and China [4]. The *R. nasutus* is cultivated particularly as a medicinal plant which has been used in treatments and preventions of diverse diseases as folklore medicine. Different parts of *R. nasutus* have been used in traditional medicine for the treatment of diseases such as eczema, several skin diseases [5]. *R. nasutus* also has potential on treatment of cancer, liver disorders, peptic ulcers, helminthiasis, scurvy, inflammation and obesity [6]. Some antimicrobial compounds such as flavonoids, steroids, terpenoids,and other components like anthraquinones, lignans, rhinacanthins were isolated and characterized from leaves and roots of *R.nasutus* plant [7][8][9]. In 2004, Stattar and his team concluded that *R.nasutus* leaves have the potential against gram positive bacteria such as *Staphylococcus aureus*, *Bacillus cereus*, *Bacillus globigill*, *Bacillus subtilis* but no inhibition activity against gram negative bacteria such as *Pseudomonas aeruginosa*, *Escherichia coli*, *Proteus morgani*, *Proteus mirabilis*, *Salmonella typhi* [10]. The ethanolic and aqueous extract of *R.nasutus* leaves inhibited well gram positive bacteria while it required double concentration to inhibit gram negative bacteria, 100 mg/mL [11].

C. odorata (L.) is a popular tropical weed [12]. *C. odorata* is found in many places from central and southern Africa to India, Sri Lanka, Bangladesh, Laos, Cambodia, Thailand, Vietnam, southern China, Taiwan, Indonesia etc.[13][14]. Traditionally, Vietnam and many other countries have been using *C. odorata* for many years for treating leech bite, soft tissue wounds, burn wounds, skin infection and dento-alveolitis [15]. In 2001, a study investigated some antimicrobial compounds in *C. odorata* such as phenolic acids, flavonol compounds in its ethanol extraction [16]. The *C. odorata* ethanol extracts inhibited against *Staphylococcus aureus*, *Proteus vulgaris* and *Escherichia coli*; however, some organisms such as *Streptococcus pyogenes*, *Pseudomonas aeruginosa* and *Klebsiella species* were found to be resistant to the extracts [17]. The methanol leaf extracts was effective against many Gram positive bacteria (*Bacillus subtilis*, *Bacillus cereus*, *Staphylococcus aereus*, *Staphylococcus epidermidis*) as well as some Gram negative bacteria such as *Escherichia coli* [18].

Staphylococcus aureus and *Pseudomonas aeruginosa* are the major cause of infections [19]. The rate of antibiotic

© Springer International Publishing Switzerland 2015
V. Van Toi and T.H. Lien Phuong (eds.), *5th International Conference on Biomedical Engineering in Vietnam,*
IFMBE Proceedings 46, DOI: 10.1007/978-3-319-11776-8_89

resistant is now considerably increasing. Particularly, in 143 human samples infected with *S. aureus* which isolated by Pasteur Insitute HCMC - Vietnam, the resistance rate of that to Penicilline G, Erythromycine, Kanamycine and Clindamycine was 93.7% , 65%, 60.8% and 58%, respectively [20]. VINARES project (studied on antibiotic resistant cooperated with 16 national hospitals in Vietnam) reported that high resistance levels were seen among 667 samples infected with *P. aeruginosa* collected in VINARES hospitals – Vietnam. The resistant rate of *P. aeruginosa* to ciprofloxacin, amikacin, gentamicin, tobramycin, ceftazidime and imipenem ranges from 29-45% [21].

For some time, these herbs were neglected in treatment because of the effectiveness, low-cost and availability of many antibiotics. However, the uncontrolled usage of antibiotics has led to the widespread of antibiotic resistant microorganisms particularly pathogenic bacteria. This study aimed to evaluate the antimicrobial activity of *S. alata (L.), R. nasutus* and *C. odorata (L.)* collected in Southern Vietnam against antibiotic resistant pathogenic *S. aureus* and *P. aeruginosa* strains by using different extraction solvents. The outcome of this study would suggest an alternative treatment for infections of antibiotic resistant bacteria.

II. MATERIALS AND METHODS

Plant Materials: Fresh leaves of three plants *Senna alata (L.), Rhinacathus nasutus* and *Chromolaena odorata* were obtained from the Ho Chi Minh City, Vietnam. Collected samples were dried at 40°C in the oven until the constant mass. After complete drying, the plant materials were milled to powder form.

Plant Extraction: Plant samples were extracted by three different solvents: ethanol, ethyl acetate and n-hexane in which 10 g powder of each plant was soaked in 100 ml of each solvent for 10 days. The mixtures were then filtered by using cellulose filter paper, the residues was re-submerged in 50 ml solvent to enhance the extraction efficacy. All the mixtures, after filtration, were concentrated under pressure and vacuum at 40°C by using rotary evaporator system. The condensed solution were dried out to paste form in Hood cabinet at room temperature and finally dissolved into 10 mL of 99% DMSO [22].

Disc Preparation for Diffusion Assay: Whatman No.1 filter paper was equally cut into 6 mm diameter discs. After sterilization, some discs were loaded with 10 mg of plant extracts; some with 10 µL of ciprofloxacin 0.5 µg/ µL as the positive control (following the amount of Nam Khoa product (Vietnam)); and some with 10 µL of DMSO 99% as the negative control.

Microorganisms: Pathogenic bacteria, *Staphylococcus aureus* and *Pseudomonas aeruginosa*, collected from infected patients in Bach Mai Hospital (Hanoi, Vietnam) and some in vitro – induced antibiotic resistant *S. aureus* strains, *S. aureus* and *P. aeruginosa* naive strains were included in the study (Table 1).

Table 2 Microorganisms used in this study

	Bacteria strains
1	*S. aureus* clinical strains (S1, S2, S3, S7, S8, S11)
2	*S. aureus* in vitro - induced antibiotic resistant (CIP,LEV)
3	*S. aureus* ATCC 25213
4	*P. aeruginosa* clinical strains (P1, P2, P3, P4, P10)
5	*P. aeruginosa* ATCC 9027

Disc Diffusion [23]: Muller Hinton agar (HiMedia, India) was prepared by following the producer's instruction. The mixture, after sterilization (121°C, 15 min), was poured to Petri dish with the depth of 4.0 ± 0.5 mm, approximately 31 mL in a 100 mm circular plate. 3-5 bacterial colonies were picked up and transferred to 4 ml of sterile tryptic soya broth (HiMedia, India) and then incubated at 37°C within 2-3 hours until OD 600nm reached 0.1 which equals approximately 1- 2x10^8 CFU/ml. Bacterial inoculums were made by diluting the culture 100 times in sodium chloride solution 0.85% to achieve bacterial concentration of 1-2x10^6 CFU/ml. A sterile cotton swab was dipped into the bacterial inoculums and stretched onto the plate. Then, prepared discs containing plant extracts and antibiotic were applied on the surface of inoculated plate within 15 min. Finally, the plate was incubated in 37°C for 24 hr. Each experiment was replicated three times.

Data Analysis: After 24 hours of incubation, the inhibition zone was measured; the average value of three replications was used to evaluate the antimicrobial activity of extracts.

III. RESULTS AND DISCUSSION

Antimicrobial activity of three plants *Senna alata (L.), Rhinacanthus nasutus* and *Chromolaena odorata* against *S. aureus* (clinical strains, ATCC 25213 strain, invitro-induced antibiotic resistant strains) and *P. aeruginosa* (clinical strains and ATCC 9027 strain) was presented in Table 2 and Table 3.

In general, most of extracts showed good inhibition activity against both antibiotic-resistant and sensitive *S. aureus* but no activity against *P. aeruginosa*.

Towards clinical *S. aureus, Senna alata* (L.) showed good antimicrobial activity when extracted by ethanol, in which the largest zone was 14.33 ± 0.58 mm. However,

Table 2 Antimicrobial activity of *Senna alata* (L.), *Rhinacanthus nasutus*, *Chromolaena odorata* extracts against different *S. aureus* strains.

S.aureus clinical strains	Senna alata (L.)(10 mg)			Rhinacanthus nasutus (10 mg)			Chromolaena odorata (10 mg)			Ciprofloxacin (5 µg)
	EtOH	EtAc	n-hx	EtOH	EtAc	n-hx	EtOH	EtAc	n-hx	
S1	11.67±0.58	-	-	13.33±1.15	13.00±1.00	10.33±0.58	10.67±1.53	14.00±1.00	11.67±0.58	-
S2	14.33±0.58	-	-	10.67±0.58	11.33±0.58	11.00±0.00	11.67±1.53	16.67±0.58	13.67±0.58	-
S3	13.67±0.58	-	-	10.33±1.53	12.33±0.58	10.00±0.00	11.50±0.87	12.67±0.58	11.00±0.00	29.33±0.58
S7	12.50±0.50	-	-	16.83±0.76	17.17±0.29	15.00±0.50	10.25±0.35	11.33±0.58	13.25±0.35	-
S8	13.33±0.58	-	-	13.17±1.04	12.33±0.58	11.17±0.29	10.33±0.58	12.67±1.53	8.33±0.29	30.00±0.00
S11	14.00±0.00	-	-	13.17±1.61	12.67±1.53	11.33±1.04	11.17±0.29	11.33±1.15	12.33±0.58	30.00±0.00
Cipro*	22.33±0.58	13.00±0.50	-	15.50±0.87	14.00±1.73	12.83±1.26	17.83±0.29	19.83±0.29	15.33±0.29	-
Levo*	18.67±0.58	9.67±1.53	-	14.17±0.58	17.67±0.58	12.17±0.58	12.67±1.04	18.67±1.04	11.50±0.50	-
ATCC 25213	11.00±0.58	9.67±1.61	-	10.67±1.04	10.67±1.26	11.17±1.44	9.33±0.58	11.50±0.50	9.00±0.50	24.00±0.00

Values are mean inhibition zone (mm) ± S.D of three replicates; (-): no inhibition zone; (*): in vitro- induced antibiotic resistant *Staphylococcus aureus*; EtOH: Ethanol; EtAc: Ethyl acetate; n-hx: n- hexane.

Table 3. Antimicrobial activity of *Senna alata* (L.), *Rhinacanthus nasutus*, *Chromolaena odorata* extracts against different *P. aeruginosa* strains.

P.aeruginosa clinical strains	Senna alata (L.)(10 mg)			Rhinacanthus nasutus (10 mg)			Chromolaena odorata (10 mg)			Ciprofloxacin (5 µg)
	EtOH	EtAc	n-hx	EtOH	EtAc	n-hx	EtOH	EtAc	n-hx	
P1	-	-	-	-	-	-	-	-	-	-
P2	-	-	-	-	-	-	-	-	-	-
P3	-	-	-	-	-	-	-	-	-	-
P4	-	-	-	-	-	-	-	-	-	-
P10	-	-	-	-	-	-	-	-	-	-
P.aeruginosa ATCC 9027	-	-	-	-	-	-	-	-	-	22.33±0.58

Values are mean inhibition zone (mm) ± S.D of three replicates; (-): no inhibition zone; EtOH: Ethanol; EtAc: Ethyl acetate; n-hx: n- hexane.

there was no activity of the plant observed when extracted with ethyl acetate and n- hexane. Towards in vitro – induced antibiotic resistant *S. aureus* strains and *S. aureus* ATCC 25213, *S. alata (L.)* ethanol extract showed the strong inhibition activity (diameter of inhibition zone: 22.33 ± 0.58 mm); while extracts in ethyl acetate and n- hexan showed lesser activity (diameter of inhibition zone: 13.00±0.50 mm). This result was in agreement with previous studies [24][25]. Phytochemical screening of *S. alata* crude extracts showed variation in the presence of bioactive compounds when using different solvents: ethanol, ethyl acetate and n-hexane [25][26][27]. This would be the reason why *S. alata* showed different antimicrobial activities when extracted in ethanol (very effective) and ethyl acetate (relatively effective) or n- hexane (low effective).

Towards clinical *S. aureus* strains, *C. odorata* extract via ethyl acetate showed better inhibitory ability (maximum 17.17 ± 0.29 mm) than *C. odorata* extracts via ethanol (maximum 16.83 ± 0.76 mm) and n- hexane (maximum 15.00 ± 0.5 mm;). Similar results were achieved with in vitro- induced antibiotic resistant *S. aureus* in which *C. odorata* extracts via ethyl acetate had better inhibitory activity (17.67 ± 0.58 mm) in comparison to the one using ethanol or n-hexane (14.17 ±0.58 and 12.17 ± 0.58 mm, respectively). This result was different to previous publish data in which ethanol was better than ethyl acetate in extracting *C. odorata* compounds with activity against *S. aureus* [28].

Ethyl acetate extracts of *R. nasutus* also exhibited better antimicrobial activity than *R. nasutus* ethanol and n-hexane extracts. The largest inhibition zone of its ethyl acetate extract was 16.67 mm ± 0.58 mm while the largest inhibition zone of its ethanol extract and n- hexane extract was 11.67 ± 1.58 mm and 13.67 ± 0.58mm, respectively. Jeyaseelan and his colleges also showed similar results to our data in the way that ethyl acetate and ethanol extracts of *R. nasutus* were able to inhibit the growth of *S. aureus*. However he pointed out that *S. aureus* was more sensitive to extracts in ethanol than to extracts in ethyl acetate [29].

Crude extract of all plants showed no inhibition activity to *P. aeruginosa* clinical strains as well as *P. aeruginosa* ATCC9027. This result was in accordance with other published data showing that Gram negative bacteria particularly *P. aeruginosa* are insensitive to these herbs [17][30][31]. It can be explained that Gram negative bacteria has an additional outer membrane which protects them from the penetration of many compounds including our bioactive compounds.

IV. CONCLUSIONS

All three plants S. *alata (L.)*, *R. nasutus* and *C. odorata* showed their potential in inhibiting multidrug resistant *S. aureus* strains. In which, *S. alata (L)* showed the higher antimicrobial properties against *S. aureus* when extracted with ethanol than with ethyl acetate and n- hexane while *C. odorata* and *R. nasutus* showed the higher antimicrobial properties against *S. aureus* when extracted with ethyl acetate than with ethanol and n- hexane.

By the way, all three plant extracts showed no activity against *P. aeruginosa*. Increasing the concentrations of these extracts might be necessary to observe antimicrobial activity against *P. aeruginosa*

This investigation suggested that these traditional herbs could be considered as good alternative treatment for diseases cause by multidrug resistant *S. aureus* and probably also for other gram positive bacteria.

REFERENCES

1. Wandee G. (2010) Ethnomedicinal plants popularly used in Thailand as laxative drugs. Ethnomedicine: A Source of Complementary Therapeutics. p295-315.
2. Yakubu MT, Olawepo OJ, Fasoranti GA.(2011) Ananas comosus: is the unripe fruit juice an abortifacient in pregnant Wistar rats? Eur J Contracept Reprod Health Care.;16(5):397–402
3. Jeyaseelan E.C, et al., (2012) Antibacterial Activity of Various Solvent Extracts of Some Selected Medicinal Plants Present in Jaffna Peninsula, International Journal of Pharmaceutical & Biological; 3(4):792-796
4. Farnsworth, N. R., and N. Bunyapraphatsara. (1992) Thai medicinal plants recommended for primary health care system. Prachachon Co., Bangkok, Thailand
5. Siripong P, Yahuafai J, Shimizu K, Ichikawa K, Yonezawa S, Asai T, et al. , (2006) Antitumor activity of liposomal naphthoquinone esters isolated from Thai medicinal plant: Rhinacanthus nasutus Kurz. Biol Pharm Bull; 29(11): 2279-83.
6. Suja SR, Lath PG, Pushpangadan P, Rajasekharan S., (2003) Evaluation of hepatoprotective effects of Rhinacanthus nasutus root extracts,Ethnomedicine and Ethnopharmacology Division, Trop. Bot. Garden and Res. Doc.; 4: 151-157
7. Wu QL, Wang SP, Du LJ, Yang JS, Xiao PG. (1998) Xanthones from Hypericum japonicum and H. Henryi. Phytochemistry; 49: 1395-1402.
8. Wu QL, Wang SP, Du LJ, Zhang SM, Yang JS, Xiao PG, (1998) Chromone glycosides and flavonoids from Hypericum japonicum. Phytochemsitry; 49: 1417-1420
9. Sendl A, Chen JL, Jolad SD, Stoddart CA, Rozhon EJ, Kernan MR, et al. (1996) Two new naphthoquinones with antiviral activity from Rhinacanthus nasutus. J Nat Prod; 59 : 808-11
10. Sattar, A.M., Abdullah, N.A., Khan, A.H., Noor, A.M., (2004) Evaluation of anti-fungal and anti-bacterial activity of a local plant Rhinacanthus nasutus (L.) J. Biol. Sci. 4, 490–500
11. Nanthakumar. R, Udhayasankar, Ashadevi.V, Arumugasamy.K and Shalimol.A (2014) In vitro antimicrobial activity of aqueous and ethanol extracts of Rrhinacanthus nasutus- a medicinal plant, international journal of pharmaceutical, chemical and biological sciences, (1), 164-166
12. Holm, L. G., Pancho, J. V., Herberger, J. P. and Plucknett, D. L, (1979) A Geographic Atlas of World Weeds. Wiley Interscience, New York
13. Bani G. (2002) Status and management of Chromolaena odorata in congo - Proceedings of the fifth international workshop on biological control and management of Chromolaena odorata, ARC-PPRI; 71-73
14. Umukoro S and Ashorobi R B. (2006) Evaluation of the anti-inflammatory and membrane stabilizing effects of Eupatorium odoratum. International Journal of Pharmacology; 2(5): 509-512
15. Le TT. (1995) The 5th European Tissue Repair Society Annual Meeting, Padova, Italy, (Abst. 30)
16. Phan T.T. et al. (2001) Phenolic Compounds of Chromolaena odorata Protect Cultured Skin Cells from Oxidative Damage: Implication for Cutaneous Wound Healing, Biol. Pharm. Bull. 24(12) 1373—1379
17. Agbabiaka T.O, Samuel T. and Sule I. O. (2010), Susceptibility of Bacterial Isolates of Wound Infections to Chromolaena odorata , Ethnobotanical Leaflets 14: 876-88.
18. Naidoo K.K et al (2011) Screening of Chromolaeana odorata (L.) king and robinson for antibacterial and antifungal properties, Journal of Medicinal Plant Research, Volume 5, Issue 19, p.4859-4862
19. Bal AM, Gould IM (2005), Antibiotic resistance in Staphylococcus aureus and its relevance in therapy, Expert Opin Pharmacother. 6(13):2257-69
20. Nguyen H.A, Tran T.T.N, Cao.H.N, Vu.L.N.L., (2013) Antibiotic resistance prevalence of Staphylococcus aureus among the specimens pathological samples in microbiological laboratory at Pasteur Institute in Ho Chi Minh city, Jounals of Prevention Medicine, XXIII, Vol 10(146), p270

21. Vinares project, Antibiotic resistance at VINARES hospitals in Vietnam during 2012-2013 , Vietnam society of infected diseases, http://hoitruyennhiem.vn/vi/nghien-cuu/du-an-vinares/34-antibiotic-resistance-at-vinares-hospitals-in-vietnam-during-2012-2013.html, reached on 09/02/2014

22. Preethy CP, Padmapriya R, Periasamy VS, Riyasdeen A, Srinag S, Krishnamurthy H, Alshatwi AA, Akbarsha MA. (2012) Antiproliferative property of n-hexane and chloroform extracts of Anisomeles malabarica (L). R. Br. in HPV16-positive human cervical cancer cells. J Pharmacol Pharmacother;13(1):26–34

23. EUCAST (2013) Disc diffusion method, European committee on antimicrobial susceptibility testing, v.3.0, 2013, <http://www.eucast.org/antimicrobial_susceptibility_testing/disc_diffusion_methodology/>

24. Alam M.T, et al. (2009) Antibacterial Activity of Different Organic Extracts of Achyranthes Aspera and Cassia Alata, J. Sci. Res 1 (2), pp 393-398

25. Salihu et al., (2012) Phytochemical, quantitative and in-vitro antimicrobial analysis of Cassia alata Linn leaves, Continental J. Biomedical Sciences 6 (2): 6 – 11

26. Usha V., Bopaiah A.K. (2012) Phytochemical investigation of the ethanol, methanol and ethyl acetate leaf extracts of six cassia species, International journal of Pharma and Bio sciences, vol 3/issue 2

27. Ugoh et al., (2013) Phytochemical Screening and Antibacterial Activity of Leaves of Senna alata on Selected Bacteria. Researcher 2013;5(9):61-65

28. Natheer S. E, et al., (2012) Evaluation of antibacterial activity of Morinda citrifolia, Vitex trifolia and Chromolaena odorata, Journal of Pharmacy and Pharmacology Vol. 6(11), pp. 783-788

29. Jeyaseelan E.C et al., (2011) In vitro evaluation of different aqueous extracts of Senna alata leaves for antibacterial activity, Clinical efficacy of Dashamoola Taila Matra Vasti …SLJIM; 01(02): 59-63

30. Wipawan P., Narumol T. and Yingmanee T. (2012) Total phenolic contents, antibacterial and antioxidant activities of some Thai medicinal plant extracts, Journal of Medicinal Plants Research Vol. 6(35), pp. 4953-4960

31. Vital P.G. et al, (2009), Antimicrobial activity and cytotoxicity of Chromolaena odorata (L. f.) King and Robinson and Uncaria perrottetii (A. Rich) Merr. Extracts, Journal of Medicinal Plants Research Vol. 3(7), pp. 511-518

Corresponding author:

Author: Hoai T.T. Nguyen
Institute: International University
Street: Quarter 6, Linh Trung Ward, Thu Duc District
City: Ho Chi Minh
Country: Vietnam
Email: ntthoai@hcmiu.edu.vn

Evaluation of the Optimal Multiplex PCR Method for the Detection of *Aspergillus Flavus* and *Aspergillus Parasiticus* on Dried Peanut

Nghia T. Le, Trinh Huynh, and Hue T. Nguyen

School of Biotechnology, International University - Vietnam National University in HCMC

Abstract— *Aspergillus flavus* and *Aspergillus parasiticus* are the common aflatoxin producing species that usually infect on foodstuff in their production line from the field to the storage place such as peanut, corn, cereal, etc., especially in tropical country like Vietnam. Aflatoxins which are considered as derived secondary metabolites assigned as a group of mycotoxins produced by several species of the *Aspergillus spp* are potent hepatotoxins, immunosuppression, carcinogen that lead to mortality or reducing the productivity of farm animals. There has been a demand for effective method to detect these two species on dried food. In previous study, a multiplex PCR method were designed to improve the detection process of *A. flavus* and *A. parasiticus* and that method showed high sensitivity and specificity by being applied on artificially infected dried peanut. In this study, that multiplex PCR method would be evaluated by testing the presence of *A. flavus* and *A. parasiticus* on natural dried peanut kernels. On this purpose, the presence of *A. flavus* and *A. parasiticus* on the collected peanut from the market was determined using two method, the conventional culturing method in Institute of Hygiene and Public Health (IHPH) and the mentioned multiplex PCR. The efficiency of multiplex PCR method would be evaluated by comparing fungi detection result of two methods using appropriate statistical tests. Next, fungal enrichment with distilled water overnight was applied to increase the detection percentage if the first analysis do not get the expected result. The result showed that 54% results from PCR method was the same with culturing method, and after fungal enrichment, this percentage increased to 76% which suggested that these two method was not significantly different with each other. Therefore, this multiplex PCR method could have more advanced points than the culturing method in detection of *A. flavus* and *A. parasiticus* on foodstuff.

Keywords— Aspergillus flavus, Aspergillus parasiticus, culturing method, dried peanut kernels, multiplex PCR.

I. INTRODUCTION

Beside rice, corn…, peanut is one of the most popular foodstuff in Vietnam. Annually, 0.46 million metric tons of peanut are produced [1]. In commercial plant, peanut seeds are stored many months before being exported or consumed domestically. Therefore, there is high chance that they get contaminated by fungi, especially in tropical area like Vietnam. Peanut is usually infected by *A. parasiticus* and *A. flavus*. Unfortunately, these fungi are the main species from section Flavi that are responsible for aflatoxins accumulation in stored peanut [2].

Aflatoxin are secondary metabolites assigned as a group of mycotoxins produced by several species of the *Aspergillus spp*. They are potent hepatotoxins, immunosuppression, carcinogen, and even mortality. Furthermore, aflatoxin infected foods have been proven to involve with a high incidence of liver cancer in human [3-5]. There are four main types of aflatoxins: B1, B2, G1, G2 [1, 6-9]. Among those, aflatoxin B1 is considered as the most popular and highly toxic metabolite.

Because of impacts on human health, the US Food and Drug Administration (FDA) has approved the safe limitation of concentration of aflatoxins B1,B2, G1, G2 in human food up to 20ng/g. There is also a need to develop a rapid, sensitive method for detection and differentiation of potential aflatoxigenic species in foods to find out any potential risk associated [10]. Many methods have been developed to detect *A. flavus* and *A. parasiticus*. Culture method (traditional method) was known as standard method currently used to detect *A. flavus* and *A. parasiticus*. However, this method has various disadvantages: time-consuming (7 - 10 days), laborious, experienced technician requiring. Recently, many methods based on molecular biology techniques for the molecular differentiation of species has resulted in substantial advances in taxonomy due to their sensitivity and specificity such as PCR-RFLP [11], real-time PCR [2], immune-sensor [12], TLC, HPLC [13]. Besides a lot of improvement on the sensitivity and specificity of fungi detection, these methods are too expensive and facility requiring to be applied into real life.

Thus, to overcome these obstacles, the project entitled "Development of a PCR method for detection of toxic/poisonous fungi in food" has been developed. In previous phase of this study, multiplex PCR procedure has been optimized to detect fungi on artificially infected samples [14]. As final step of the project, the optimized multiplex PCR method would be evaluated when being applied on natural dried peanut for detection of *A. flavus* and *A. parasiticus*. Once this project is completed, there are great developments in evaluating food quality with many considerable advantages: lower cost, time saving, more sensitive than conventional microbiological.

V. Van Toi and T.H. Lien Phuong (eds.), *5th International Conference on Biomedical Engineering in Vietnam,*
IFMBE Proceedings 46, DOI: 10.1007/978-3-319-11776-8_90

II. MATERIALS AND METHODS

A. Sample Collection

Total 50 samples of dry peanut kernel were collected at Ba Chieu market, Ho Chi Minh City (500g per sample). All of them were contained in completely-dried plastic bags to prevent contamination from surrounding environment. 400g of each was sent to Institute of Hygiene and Public Health for *A. flavus* and *A. parasiticus* detection by culture method. And another 100g would be used for PCR method.

B. DNA Extraction and Evaluation

Twenty beans of dried peanut were used for extraction of DNA using in next step. Fungal mycelia and spore were obtained by sample washing with 10ml of NaCl (0.85%). And then, 1 ml of SDS lysis buffer (10 mMTris-HCl, 20 mM EDTA, 0.7M NaCl) and grinding with glass beat were applied. Following, heat-shock was carried out with 10ul of proteinase K and 5ul of 2-mercapto-ethanol were added into the solution. Finally, chloroform, Ice-cold isopropanol, ammonium acetate, and ice-cold ethanol were used to isolate and purify fungal DNA. Isolated DNA was qualified and quantified by Nanodrop machine. The purity of DNA was based on the ratio of A260/A280.

This step could be repeated with the sample incubated overnight for fungal enrichment if the DNA extracted directly from collected sample gave different result in comparison with culture method result. 2ml autoclaved distill water was used for incubation.

C. PCR Reaction and Gel Electrophoresis

Target sequences in aflatoxin biosynthesis gene cluster of *A. parsiticus* and *A. flavus* were amplified by the optimized multiplex PCR [14]. The primers used were: for *A. flavus* AFLA-F-5'–GGTGGTGAAGAAGTCTATCTAAGG–3', AFLA-R-5'–AAGGCATAAAGGGTGTGGAG–3', which amplified 413bp fragment, and for *A. parasiticus* APLA-F-5'-GGATTCGTGAGTGTCTTTAGGG-3', APLA-R-5'-GGTAAATGCTCCGCACAGTC- 3', which amplified 343bp fragment. 25ul reaction including F-AFLA 10mM, R-AFLA 10mM, F-APLA 10mM, R-APLA 10mM, 300ng of extracted DNA and distilled water up to 25ul.The cycling conditions for PCR were: Initialization at 94°C for 5 min, 40 cycle of denaturation at 94°C for 30 sec, annealing at 63°C for 30 sec, extension at 72°C for 1 min, and final extension at 72°C for 10 min.

PCR results were analyzed by gel electrophoresis. PCR products are run on 2% ethidium bromide-containing agarose gel at 85V for 30 min. The results were observed under UV light (GelDoc-It 310, UVP). (Fig.1)

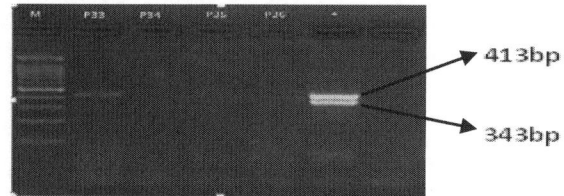

Fig. 1 An example of multiplex PCR result on agarose gel. P33 was infected by *A. flavus* with 1 band just over 400bp band of ladder (M). Positive control were infected by both with two bands.

III. RESULTS

A. Fungal Detection by Culturing Method

Institute of Hygiene and Public Health (IHPH) gave the results within 2 weeks from the sample receiving date. The results from IHPH indicated that in total 50 samples, there were 28 samples infected by *A. flavus*, 2 samples infected by *A. parasiticus*, 5 samples infected by both *A. flavus* and *A. parasiticus* and 15 samples without infection that occupied 56%, 4%, 10% and 30% respectively. The result also reflected the majority of *A. flavus* infection on foodstuff as described in previous study [1].

B. Fungal Detection of Directly Extracted Samples by PCR Method

In total 50 samples tested by PCR method without incubation, there were 10 samples infected by *A. flavus*, 2 samples infected by *A. parasiticus*, 1 samples infected by both *A. flavus* and *A. parasiticus* and 37 samples not infected, equivalent to 20%, 4%, 2% and 74% respectively. And total the number of samples that showed same results with culturing method was 27, equivalent to 54%, obviously, the unmatched results were 23, equivalent to 46%.

Table 1 Comparison between culturing method and PCR method result from directly fungal detection

Infection	PCR method		Culture method		p-value
	No.	% (n=50)	No.	%(n=50)	
A. flavus	10	20%	29	58%	0.0001
A. parasiticus	2	4%	2	4%	-
Both fungi	1	2%	5	10%	0.09296
No infection	37	74%	14	28%	0

C. Fungal Detection from Samples after Incubating by PCR Method

Screening for 50 samples, the results of PCR method after incubating peanut samples with distilled water overnight showed that there were 19 samples positive with *A. flavus*,

2 positive with *A. parasiticus*, 2 positive with both and 27 showed negative result and the respective percentages were 38%, 4%, 4% and 54%. The number of the samples that showed same results with culturing method was 38 samples, equivalent to the matching percentage of 76%, the number of unmatched results was 12, equivalent to 24%.

Table 2 Comparison between culturing method and PCR method result from fungal detection after incubation overnight

Infection	PCR method		Culture method		p-value
	No.	% (n=50)	No.	%(n=50)	
A. flavus	19	38%	29	58%	0.0455
A. parasiticus	2	4%	2	4%	0
Both fungi	2	4%	5	10%	0.238
No infection	27	54%	14	28%	0.0083

D. Data Analysis

To compare the effectiveness of PCR method with culturing method, two-sample two-tailed Z-test at $\alpha=0.05$ was conducted.

With PCR without incubating, in the case of *A. flavus* infection, the test suggested that the results from PCR method was significantly different from culturing method with p-value = 0.0001<0.05. Two methods gave the same results in cases of *A. parasiticus* infection, obviously there was no difference between them. The test also pointed out that there was no considerable difference in the case of *A. flavus* and *A. parasiticus* infection with p-value of 0.09296, and with p-value of 0 the results without infection were significantly different. (Table 1)

With PCR after incubating, the test showed that in the case of *A. flavus* infection, the p-value was 0.0455, so there was slightly difference in the results of two methods. In cases of infection by both fungi, the test proclaimed that the p-value in this case was 0.238 which is higher than significant value, indicating that the difference was not remarkable. With the p-value of 0.0083, the test also showed that there was significant difference in results of non-infection. (Table 2).

In general comparison, the Chi-Square at $\alpha=0.05$ test had implied that there was significant difference in PCR method without incubating with culturing method, and in term of PCR after incubating, there was no difference with culturing method, indicating that the fungal DNA concentration at very early infection had considerable influence on PCR reaction. (Table 3).

Table 3 Comparison between culturing method and PCR methods with and without incubation using Pearson Chi-Square test.

	X^2	Critical value
Without incubation	22.296	5.99
Incubation	7.491	5.99

IV. DISCUSSION

By comparing multiplex PCR result with the outcomes from IHPH, the Chi-Square test suggested that there were significant discrepancy between PCR method and culturing method. Multiplex PCR reaction showed 27 in total 50 cases that matched with standard results, so it could be concluded that the effectiveness of multiplex PCR was 54%. There were many factors could result in this low efficiency.

The first reason could be the contaminated of extracted DNA solution. DNA quantity just indicated total DNA that included DNA from target fungi (if any), DNA from peanut or other microorganisms. That DNA mixture could affect the sensitivity of PCR if the portion of non-target DNA was too high while target DNA concentration was lower than limitation of detection. The second reason could be the remaining chemical components of peanut kernel such as phenolic and polysaccharide considered as PCR inhibitors [15] which may bind or denature the polymerase [16]. Although DNA extraction was designed to remove those components, that method was suitable for DNA extraction of maximum of 10 kernels [17], so when being applied to larger amount of kernels, those chemical would not be removed completely out of DNA solution. Therefore, a higher ratio of chloroform such as 1:2 should be used to purify DNA solution. The other one could be low level of infection. A small amount of peanut kernels (20 kernels) were selected, unfortunately, if those kernels were infected by extremely small amount of fungi meanwhile that would lead to false negative. And additional step has been designed to overcome this problem, before DNA extraction peanut kernels were incubated in distilled water to enhance the growth of fungi.

After fungal enrichment by incubating with distilled water, there was increase in positive results by PCR method. The Chi-Square indicated that difference between PCR method after incubating and culturing method was slightly significant in general due to the significant decrease of X^2 value after incubation. It suggested that incubation step should be included in this procedure. In addition, from the comparison made by statistical tests, it could be concluded that the low fungal DNA concentration extracted from very early infected samples caused remarkable impact on the efficiency of PCR method. It could be seen that, the multiplex PCR method got 76% effectiveness with incubation step, it is slightly different culturing method. However, this difference could be caused by the very low infection level that could not be detected by PCR but could still be detected during culturing process. Moreover, PCR inhibitors from peanut is a remarkable elements should be totally eliminated to gain the effective percentage.

In exceptional cases, *A. flavus* and *A. parasiticus* were detected by PCR method while culturing method showed negative to both, which may be explained that those samples had been infected by *A. flavus* and *A. parasiticus* but they were not still alive, consequently culturing method could not detect them, while, despite dead fungi, their DNA still remained and PCR method can still detected. Those cases showed the disadvantage of culturing method that was unable to detect dead cells. And base on this weakness, the molecular based methods was developed in detecting microorganisms in foodstuff.

V. CONCLUSION

Based on 76% effectiveness in comparison with culturing method, time saving, lower cost, and simple procedure, this multiplex PCR protocol could still be considered as more advanced than culturing method in detection of both toxic fungi in dried peanut. The PCR method could be used as a confirmative or an alternative method for culturing method in *A. flavus* and *A. parasiticus* detection.

ACKNOWLEDGEMENTS

Authors would like to thank for Vietnam National University – Ho Chi Minh City granted this study under project number B2013-28-02.

REFFERENCES

1. Tran-Dinh N, Kennedy I, Bui T, et al. (2009) Survey of Vietnamese peanuts, corn and soil for the presence of Aspergillus flavus and Aspergillus parasiticus. Mycopathologia. 168(5): pp 257-268.
2. Passone MA, Rosso LC, Ciancio A, et al. (2010) Detection and quantification of Aspergillus section Flavi spp. in stored peanuts by real-time PCR of nor-1 gene, and effects of storage conditions on aflatoxin production. Int J Food Microbiol. 138(3): pp 276-281.
3. Magan NOlsen M (2004) Mycotoxins in food: Detection and control, ed. Woodhead Publishing.
4. Ventura M, Gomez A, Anaya I, et al. (2004) Determination of aflatoxins B1, G1, B2 and G2 in medicinal herbs by liquid chromatography–tandem mass spectrometry. Journal of Chromatography A. 1048(1): pp 25-29.
5. Giorni P, Magan N, Pietri A, et al. (2007) Studies on Aspergillus section Flavi isolated from maize in northern Italy. International journal of food microbiology. 113(3): pp 330-338.
6. Kabak B Prevention and Management of Mycotoxins in Food and Feed, in *Mycotoxins in Food, Feed and Bioweapons*, M. Rai and A. Varma, Editors. 2010, Springer Berlin Heidelberg. p. 201-227.
7. Bhatnagar D, Cary JW, Ehrlich K, et al. (2006) Understanding the genetics of regulation of aflatoxin production and Aspergillus flavus development. Mycopathologia. 162(3): pp 155-166.
8. Surai P, Mezes M, Fotina T, et al. Mycotoxins in Human Diet: A Hidden Danger, in *Modern Dietary Fat Intakes in Disease Promotion*. 2010, Springer. p. 275-303.
9. Valencia-Quintana R, Sánchez-Alarcón J, Tenorio MG, et al. (2012) Preventive strategies aimed at reducing the health risks of Aflatoxin B1. Toxicology and Environmental Health Sciences. 4(2): pp 71-79.
10. Valasek MARepa JJ (2005) The power of real-time PCR. Advances in physiology education. 29(3): pp 151-159.
11. El Khoury A, Atoui A, Rizk T, et al. (2011) Differentiation betweenAspergillus flavusandAspergillus parasiticusfrom Pure Culture and Aflatoxin-Contaminated Grapes Using PCR-RFLP Analysis ofaflR-aflJIntergenic Spacer. Journal of Food Science. 76(4): pp 247-253.
12. Ammida N, Micheli L, Piermarini S, et al. (2006) Detection of aflatoxin B1 in barley: comparative study of immunosensor and HPLC. Analytical letters. 39(8): pp 1559-1572.
13. Howell MVTaylor PW (1981) Determination of aflatoxins, ochratoxin a, and zearalenone in mixed feeds, with detection by thin layer chromatography or high performance liquid chromatography. Journal-Association of Official Analytical Chemists. 64(6): pp 1356-1363.
14. Hue N, Nhien N, Thong P, et al. Detection of Toxic Aspergillus Species in Food by a Multiplex PCR, in *4th International Conference on Biomedical Engineering in Vietnam*, V.V. Toi, et al., Editors. 2013, Springer Berlin Heidelberg. p. 184-189.
15. Savage GKeenan J The composition and nutritive value of groundnut kernels, in *The Groundnut Crop*. 1994, Springer. p. 173-213.
16. Wilson IG (1997) Inhibition and facilitation of nucleic acid amplification. Appl Environ Microbiol. 63(10): pp 3741-3751.
17. Huynh T, Than LNguyen H (Year) Detection of Aspergillus parasiticus on artificially infected peanut and corn kernels using PCR method, of Conference. Conference. vol National Biotechnology Conference, Location, year, pp 245-249.

Corresponding Author: Dr. Nguyen Thi Hue
Institute: School of Biotechnology, International University – VNU, Ho Chi Minh City, Vietnam
Street: Block 6, Linh Trung Ward, Thu Duc District
City: Ho Chi Minh City
Country: Vietnam
Email: nthue@hcmiu.edu.vn

Evaluation of the Optimal Multiplex PCR Method for the Detection of Aspergillus Flavus and Aspergillus Parasiticus on Dried Corn

Tu T. Ly, Trinh Huynh, and Hue T. Nguyen

School of Biotechnology, International University – Vietnam National University, HCMC, Vietnam

Abstract— **Corns, which are one of major agricultural crop in Vietnam, have high risk of being infected with A. flavus and A. parasiticus due to the high humidity of tropical climate in Vietnam which is the ideal condition for fungal development. The traditional methods for detection and identification of these mycotoxigenic fungi are very time-consuming, laborious, and facility requiring. Thus, there is a need to develop a new PCR-based method that target DNA which is considered as a good alternative for rapid diagnosis of these fungi. In this study, a multiplex PCR method which had already been optimized in previous studied was evaluated by performing the detection of A. flavus and A. parasiticus on corn kernels collected from markets. These samples were detected by both culture method and multiplex PCR method. After first time of analysis, fungal enrichment overnight was applied to increase the detection percentage. The results were then analyzed and compared with that of traditional method. The statistical outcome showed that 66% results from PCR method was the same with culture method. However, after fungal enrichment by incubating samples in moist condition overnight, this proportion increased to 82% which was suggested to be statistically the same with culture method. Therefore, The PCR method with more advantages compared to culture method in time consuming, labor demand as well as the cost, it could be considered to be the replacement for the traditional method of rapid detection of A. flavus and A. parasiticus on dried corn.**

Key words— ***Aspergillus flavus, Aspergillus parasiticus,* dried corn kernels, culture method, and multiplex PCR method.**

I. INTRODUCTION

Corn is one of major agricultural commodities in Vietnam, with 4.6 million metric tons on average produced annually [1]. The mold infected potential of this kind of crop, especially aflatoxigenic *Aspergillus parasiticus (A. parasiticus)* and *Aspergillus flavus (A. flavus)* is a well-recognized problem, particularly in tropical countries with tropical climate such as Vietnam [2].

Aflatoxin is a group of extremely toxic chemical produced by several *Aspergillus* fungi, primarily *A. flavus* and *A. parasiticus* when these molds attack and grow crops [3, 4]. This group of mycotoxins causes both acute and chronic diseases including liver diseases and certain cancers [5, 6]. Thus, they are considered as the most naturally hepatotoxic and carcinogenic secondary metabolites known [7, 8]. The United Nation's Food and Agriculture Organization has estimated that about 25% of the world's food is significantly contaminated

by these fungi [9]. Among them, the most common and most toxic compound is aflatoxin B1, which has historically been associated with the Turkey-X-Disease in the 1960s and can result in toxic hepatitis liver fibroses [9]. For instance, by the end of 2003, approximately 100 countries (covering approximately 85% of the world's inhabitants) had specific regulations or detailed guidelines for mycotoxins in food or feed [10, 11].

The traditional methods for identification and detection of these fungi in dried foods were primarily based on biological characteristics, morphological studies and culture on different media. However, this approach was very time-consuming, laborious, facility requiring and mycological expertise [12]. Therefore, other approaches using molecular biology techniques including immune-sensor (ELISA, RIA...), chemo-physical methods (TLC, HPLC), real-time PCR, etc. were applied [13-16]. However, these methods required specialized machines and were too expensive to be applied into real life. Hence, there is a need to develop a new technique for detection of mycotoxigenic fungi to overcome these obstacles.

For this purpose, a project namely "Development of a PCR method for detection of toxic/ poisonous fungi in food" was carried out. The multiplex PCR method to detect *A. flavus* and *A. parasiticus* was successfully designed in previous studies with high specificity and sensitivity by using primers that are specific to aflatoxin biosynthesis gene clusters presence in these fungi [Private communication], [17]. For further effort to commercialize the product, its efficiency and accuracy must be evaluated on real samples by comparing results between traditional and molecular method. A comprehensive assessment of this multiplex PCR method will be proposed in this study as the third step of the large project. The success development of this PCR method would be able to overcome all obstacles seen in previous ones thanks to its time efficiency, low cost, high specificity and sensitivity. These advantages bring an efficient way to quantify aflatoxins containing in commodities which directly affect to animal and human health.

II. MATERIALS AND METHODS

A. Samples Collection

50 dried corn sample was randomly collected in the markets; 500 grams of kernel per samples. The samples were

© Springer International Publishing Switzerland 2015
V. Van Toi and T.H. Lien Phuong (eds.), *5th International Conference on Biomedical Engineering in Vietnam,*
IFMBE Proceedings 46, DOI: 10.1007/978-3-319-11776-8_91

then divided into 2 parts: 400g were sent to Institute of Hygiene and Public Health-HCMC for traditional detection using culture method, 100g were tested by multiplex PCR method. The two methods were carried out simultaneously for an exact comparison.

B. DNA Extraction

Before performing PCR, all samples were extracted DNA using SDS protocol modified from Plaza's method [17, 18]. 40 kernels (10 grams) were randomly collected from each sample and washed by 10ml of sodium chloride 0.85% to collect the fungi. DNA extraction was performed using SDS lysis buffer along with grinding with sterilized sand, and then thermal shock was the following step. Next, the DNA was collected and purified using chloroform, ice-cold isopropanol and ammonium acetate.

C. Multiplex PCR Test

The multiplex PCR with optimized conditions [17] was applied to amplify target sequences within aflatoxin biosynthesis gene cluster of *A. parasiticus* and *A. flavus*. Sequences of primer sets for *A. flavus*: AFLA-F-5'–GGTGGTGAAGA AGTCTATCTAAGG–3', AFLA-R-5'–AAGGCATAAA GGGTGTGGAG–3'; and for *A. parasiticus*: APLA-F-5'-GG ATTCGTGAGTGTCTTTAGGG-3', APLA-R-5'-GGTAAA TGCTCCGCACAGTC- 3'. The total volume of 1 reaction was 25ul including 12.5µl of TopTaq Master Mix (Qiagen), 300ng of DNA template, 0.75µl of 10mM AFLA-F, 0.75µl of 10mM AFLA-R, 1µl of 10mM APLA-F, 1µl of 10mM APLA-R.

The thermal cycle for PCR were: Initialization at 94°C for 5 min, denaturation at 94°C for 30 sec, annealing at 63°C for 30 sec, extension at 72°C for 1 min, final extension at 72°C for 10 min and hold at 4°C. PCR reaction was run in 40 cycles.

PCR results were analyzed by gel electrophoresis. PCR products are run on 2% ethidium bromide-containing agarose gel at 85V for 30 min. The results were observed under UV light (GelDoc-It 310, UVP). Figure 1 is one example of results analysis based on gel electrophoresis, 15µL of PCR products were loaded into 5 lanes from 1 to 5. There was no band being seen in the first and second lane, pointed out that these samples were infected by neither *A. flavus* nor *A. parisiticus*. The third and fourth had only one band with molecular weight just over 400bp being observed, represented the presence of *A. flavus* in these samples. Besides, lane 5 had two separate bands at positions around 300bp and 400bp, implied the presences of both *A. flavus* and *A. parasiticus* in this sample. Additionally, lane 6 was positive control and lane 7 was negative control.

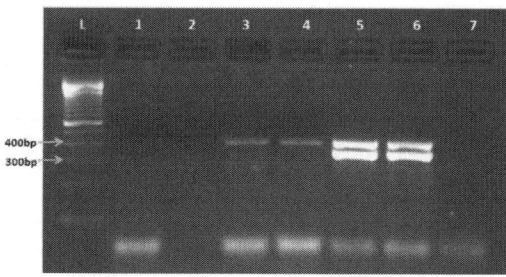

Fig. 1 Multiplex PCR and agarose gel electrophoresis result. Lane 1 – 2: sample C8 and C13, respectively. Lane 3 – 4: sample C16 and C18, respectively. Lane 5: sample C28. Lane 6: positive control. Lane 7: negative control

D. Fungal Enrichment

After performing the whole procedure for first time, regarding the samples that were negative result with the PCR test, the whole procedure was repeated with kernel incubation step with about 2ml autoclaved distill water overnight. The reason was to improve the fungal infection level, if any, resulted in increasing the fungal detectable proportion.

III. RESULTS

A. Fungal Detection Using Culture Method

The result of traditional detection from Institute of Hygiene and Public Health-HCMC would be received within 2 weeks.

The results from culture method showed that, in 50 samples, there were 43 samples infected by *A. flavus*, accounted for 86% whereas no sample was infected by *A. parasiticus*. Additionally, only one sample (2%) was infected by both *A. flavus* and *A. parasiticus* and 6 samples (12%) were not infected by both fungi. These data were summarized in the table 1 below.

Table 1 Fungal detection using culture method

Infection	Culture method	
	No.	%
A. flavus	43	86%
A. parasiticus	0	0%
AF and AP	1	2%
No infection	6	12%

B. Fungal Detection Using Multiplex PCR Method

a) Fungal Detection Directly from Collected Samples

After performing the whole procedure right after samples collection, the results showed that there were 26 samples infected by *A. flavus*, accounted for 52% whereas no sample was infected by *A. parasiticus*. Besides, 6 samples (12%) were infected by both *A. flavus* and *A. parasiticus* and 18 samples (36%) were not infected by both fungi. To be more in-depth analysis, the **two-tailed z – test at significant level α=0.05** was conducted to compare the results of each group of two methods. Regarding *A. flavus* infection, the outcome from z-test of culture method and multiplex PCR method showed that these two groups of data were statistically significantly different (p = 0.00024 << 0.05). In case of both *A. flavus* and *A. parasiticus* infection, no significant difference between results of two groups was shown with p = 0.05. As for samples without infection, the z-test result showed that the p-value was 0.00496, implied that there was a statistically significantly difference between results of two groups.

Table 2 Fungal detection directly from collected samples

infection	Culture method		PCR method		p-value
	No.	%	No.	%	
A. flavus	43	86%	26	52%	0.00024
A. parasiticus	0	0%	0	0%	-
Both fungi	1	2%	6	12%	0.05
No infection	6	12%	18	36%	0.00496

b) Fungal Detection after Incubation

To improve the detection rate, the collected samples were incubated with distill water overnight before being extracted DNA and performed PCR again. The results showed that there were 34 samples infected by *A. flavus*, accounted for 68% whereas no sample was infected by *A. parasiticus*. Besides, 6 samples (12%) were infected by both *A. flavus* and *A. parasiticus* and 10 samples (20%) were not infected by both fungi. With incubation step, there was increase in the matching results of two methods from 66% to 82%; the false negative decreased from 34% to 18%. The z-test results showed that in case of *A. flavus* infection, the detection rate increase from 52% to 68% and the p-value also greatly increased from 0.00024 to 0.03236, resulted in just a slightly difference between the results of two methods.

Table 3 Fungal detection after incubation

Infection	Culture method		PCR method		p-value
	No.	%	No.	%	
A. flavus	43	86%	34	68%	0.03236
A.parasiticus	0	0%	0	0%	-
Both fungi	1	2%	6	12%	0.05
No infection	6	12%	10	20%	0.27572

C. General Comparison between Culture and PCR Method

To compare the overall results of two methods, the **Pearson Chi-Square** test at significant level α=0.05 was carried out. The results showed that X^2 value decrease from 13.76 to 5.623 after incubation. In case of incubation, that value was smaller than critical value. Therefore, it could be concluded that there was no significant difference between the general results of two methods.

Table 4 Pearson chi-square results

	X^2	Critical value
Without incubation	13.76	5.99
Incubation	5.623	5.99

D. Exceptional Cases

Apart from those samples gave same results by both methods, there were some samples (5 samples) that are positive by PCR test whereas negative by culture test. Besides, some samples were artificially infected and sent to institute for detection. However, culture method could not detect them while PCR test could. This might be some mistakes of culture method, thus PCR test could be considered as a tool to confirm or assist the traditional test.

IV. DISCUSSION

By using statistical method to analyze data, in case of *A. flavus* infection, the results of two methods were significantly different. This might be resulted from various factors. Firstly, as being mentioned above, DNA quantification by Nanodrop machine could only estimate the total amount of DNA in solution. Thus, we could not know exactly how much fungal DNA contained in the solution, so that the amount of fungal DNA being used in PCR reaction could not be calculated. In case that target DNA was not adequate to be detectable, it might lead to false negative result. Secondly, just a small amount of corn (40 kernels) was selected to extract DNA. In case the selected kernels were not infected whereas the rest of them were, this also leaded to false negative result.

Additionally, the infection level of the collected samples were also unknown, since samples were collected randomly. Some samples were just being infected so that the amount of fungi presence on them was too small to be detected. According to previous studies, limitations for detection was 40ng of target DNA [private communication]. By comparison results between two methods using Pearson Chi Square test, the results of PCR-based method with incubation could be considered statistically the same as that of culture method. With incubation step, there was increase in the matching results of two methods from 66% to 82%. This result proved that incubation had improved the amount of fungi presence in samples, thus enhanced the detection of *A. flavus* and *A. parasiticus* by PCR method to some extent.

However, there were still 4 samples gave negative results despite of the incubation process. This might be resulted in the collected kernels were not infected by fungi whereas the remaining were. It could be suggested that before incubation, the sample must be homogenized by shaking in order to spread out the fungi presence on sample, if any.

Besides, in case that sample was positive with both *A. flavus* and *A. parasiticus* with PCR-based method while only *A. flavus* positive with culture method, the reason might because culture method could not identify dead fungi while multiplex PCR test still could detect them thanks to the presence of their genome in DNA solution.

Up to now, there are many studies about molecular method for qualifying the presence of *A. flavus* and *A. parasiticus* on dried food. However, these studies were just carried out on artificial infected samples and using singplex PCR. This study could be considered as the first study to apply multiplex PCR method on real samples collected from the markets. Based on the evaluation results and suggestions for improvement in this study, more researches should be conducted to overcome these obstacles in order to improve the efficiency of the optimal multiplex PCR protocol for fungal detection.

V. CONCLUSION

The multiplex PCR method had been evaluated with no significant difference between results of PCR method and culture method. Thanks for several advantages such as short duration (within 1 day without incubation or 1.5 day with incubation), low cost (30 000 VND compared with 250 000 VND of culture method), labor-saving and easy to perform. Besides, it could be suggested that before DNA extraction, the sample should be incubated in moist condition overnight in order to improve the fungal infection level, if any, resulted in increasing the fungal detectable proportion. With the advantage of the multiplex PCR method, it can be applied for routine diagnostics for dried food.

ACKNOWLEDGEMENTS

Authors would like to express our gratefulness to Vietnam National University – Ho Chi Minh City granted this study under project number B2013-28-02.

REFFERENCES

1. Tran-Dinh N, Kennedy I, Bui T, et al. (2009) Survey of Vietnamese peanuts, corn and soil for the presence of Aspergillus flavus and Aspergillus parasiticus. Mycopathologia. 168(5): pp 257-268.
2. Giorni P, Magan N, Pietri A, et al. (2007) Studies on Aspergillus section Flavi isolated from maize in northern Italy. Int J Food Microbiol. 113(3): pp 330-338.
3. Wrather LESaJA (2009) Alfatoxin in corn.
4. El Khoury A, Atoui A, Rizk T, et al. (2011) Differentiation between Aspergillus flavus and Aspergillus parasiticus from pure culture and aflatoxin-contaminated grapes using PCR-RFLP analysis of aflR-aflJ intergenic spacer. J Food Sci. 76(4): pp 247-253.
5. Sirot V, Fremy JM, Leblanc JC (2013) Dietary exposure to mycotoxins and health risk assessment in the second French total diet study. Food Chem Toxicol. 52: pp 1-11.
6. Klich MA (2009) Health effects of Aspergillus in food and air. Toxicol Ind Health. 25(9-10): pp 657-667.
7. Dorner JW (2008) Management and prevention of mycotoxins in peanuts. Food Addit Contam Part A Chem Anal Control Expo Risk Assess. 25(2): pp 203-208.
8. Sardinas N, Vazquez C, Gil-Serna J, et al. (2010) Specific detection of Aspergillus parasiticus in wheat flour using a highly sensitive PCR assay. Food Addit Contam Part A Chem Anal Control Expo Risk Assess. 27(6): pp 853-858.
9. Reiter E, Zentek JRazzazi E (2009) Review on sample preparation strategies and methods used for the analysis of aflatoxins in food and feed. Mol Nutr Food Res. 53(4): pp 508-524.
10. van Egmond HP, Schothorst RCJonker MA (2007) Regulations relating to mycotoxins in food: perspectives in a global and European context. Anal Bioanal Chem. 389(1): pp 147-157.
11. Dorner JW, Cole RJDiener UL (1984) The relationship of Aspergillus flavus and Aspergillus parasiticus with reference to production of aflatoxins and cyclopiazonic acid. Mycopathologia. 87(1-2): pp 13-5.
12. Wall SJEdwards DR (2002) Quantitative reverse transcription-polymerase chain reaction (RT-PCR): a comparison of primer-dropping, competitive, and real-time RT-PCRs. Anal Biochem. 300(2): pp 269-273.
13. Passone MA, Rosso LC, Ciancio A, et al. (2010) Detection and quantification of Aspergillus section Flavi spp. in stored peanuts by real-time PCR of nor-1 gene, and effects of storage conditions on aflatoxin production. Int J Food Microbiol. 138(3): pp 276-281.
14. Peters J, Bienenmann-Ploum M, de Rijk T, et al. (2011) Development of a multiplex flow cytometric microsphere immunoassay for mycotoxins and evaluation of its application in feed. Mycotoxin Res. 27(1): pp 63-72.

15. Ventura M, Gomez A, Anaya I, et al. (2004) Determination of aflatoxins B1, G1, B2 and G2 in medicinal herbs by liquid chromatography-tandem mass spectrometry. J Chromatogr A. 1048(1): pp 25-29.

16. Lopez Grio SJ, Garrido Frenich A, Martinez Vidal JL, et al. (2010) Determination of aflatoxins B1, B2, G1, G2 and ochratoxin A in animal feed by ultra high-performance liquid chromatography-tandem mass spectrometry. J Sep Sci. 33(4-5): pp 502-508.

17. Hue T.N. NHN, Thong M.P., Thanh N.C., An. V.P. (2013) Detection of Toxic Aspergillus Species in Food by a Multiplex PCR. 4th International Conference on Biomedical Engineering in Vietnam. 40: pp 184-189.

18. Płaza GA, Upchurch R, Brigmon RL, et al. (2004) Rapid DNA Extraction for Screening Soil Filamentous Fungi Using PCR Amplification. Pol. J. Environ. Stud. 13(3): pp 315-318.

Corresponding Author: Dr. Nguyen Thi Hue
Institute: School of Biotechnology, International University – VNU,
        Ho Chi Minh City, Vietnam
Street: Block 6, Linh Trung Ward, Thu Duc District
City: Ho Chi Minh City
Country: Vietnam
Email: nthue@hcmiu.edu.vn

fNIRS-Based Wavelet Thresholds for Motor Area Determination

Dao V. Ha[1], Hai T. Nguyen[1], Cuong Q. Ngo[1], Mai T. Tran[2], and Toi Van Vo[3]

[1] Faculty of Electrical-Electronics Engineering, HCMC University of Technical Education, Vietnam
[2] Faculty of Electrical-Electronics Engineering, HCMC University of Technology, Vietnam
[3] Biomedical Engineering Department, International University, VNU, HCMC, Vietnam

Abstract— **The brain is the most important part in human body. In order to determine and evaluate the activities of the body, brain signals measured using non-invasive technique - functional Near Infrared Spectroscopy (fNIRS) will be investigated. In this paper, Oxy- Hb signals in the cerebral cortex that was collected from 24 channels put into preprocessing to smooth using a Savitzky - Golay filter. Data removed noise will be transformed discrete wavelets into different frequency components. Threshold algorithms are applied to identify the channels (the mobile areas) that have Oxy - Hb changes while biting spacer and lifting object. This research is conducted experiments on five subjects with changes of biting or lifting for collection of the Oxy - Hb changes on the mobile areas. This method will help researchers save a lot of effort and archive the proper assessment of human activities through brain signals.**

Keywords— **fNIRS data, threshold algorithm, discrete wavelet transform.**

I. INTRODUCTION

In recent decades, the studies of the human brain have gained many achievements that contributed greatly to the diagnosis and treatment of human diseases. The NIRS has the advantages such as the mobility, low cost, safety, and wider studies among the data collection methods that are not invasive to the brain such as functional near-infrared spectroscopy, EEG, magnetic resonance imaging, positron emission tomography, etc. fNIRS is an analysis method that uses electromagnetic radiation in the near-infrared spectrum (around 650–950 nm). fNIRS light can travel through the skull and go to the cerebral cortex with the depth of about 3 cm [1].

When a stimulation occurs in the brain like thought, movement, etc. the amount of homodynamic changes in the brain areas corresponding to undertake such functions. So fNIRS signals reflect brain activities and related functions while we tapped hands [2]. The increase of oxy-Hb concentration in the areas near to the brain that control movement (motor cortex) and the deoxy-Hb reduction in some areas will be presented in this paper. The oxy-Hb change over time depends on the position of the channels on the cortex: operation is maintained over controlling brain areas, hyperactivity on areas somatosentory, and cumulative operation in prefrontal areas [3]. These features reflect the

functions of brain structures during operations of hand. Activities of the left brain are more powerful than those of the right brain related to experiments of lifting hands.

To identify motor areas of biting and lifting, we produce the block diagram as shown in figure 1.

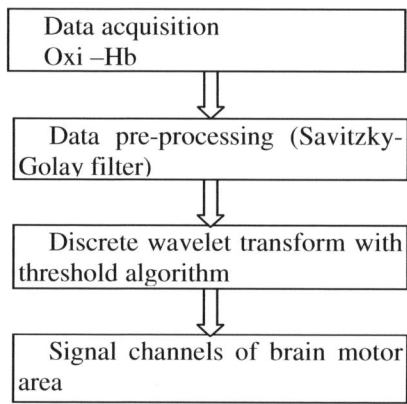

Fig. 1 Block diagram of identifying motor areas.

II. MATERIALS AND METHODS

A. Data Acquisition

Brain oxygenation signals in cortex were acquired from an FOIRE 3000 fNIRS machine (Shimadzu Corporation, Japan) at Lab 104, International University, Vietnam. Experiments were performed on 5 healthy subjects. These subjects was demanded to perform a pull down force applied on the contralateral arm while biting on a firm spacer using one side of the jaw. Data were collected with a protocol includes 20 seconds (Rest times) – 10 seconds (Task times) – 20 seconds (Rest times).

The bite and lift weights activity as described in Table 1.

Table 1 The experimental representation of bite and lift activities

Times	Spacers (mm)			Lifting weights (kg)					Experiment No
	0	2	3	0	4	5	6	7	
1	1	0	0	0	1	1	1	1	4
2	0	1	0	1	1	1	1	1	5
3	0	0	1	1	1	1	1	1	5

V. Van Toi and T.H. Lien Phuong (eds.), *5th International Conference on Biomedical Engineering in Vietnam,*
IFMBE Proceedings 46, DOI: 10.1007/978-3-319-11776-8_92

Measuring channel matrix arranged as shown in Figure 2.

This measuring matrix includes 8 light emitters and 8 light detectors used to measure on brain that controls movement area. In this study, the subjects performed lifting weights with their right hand, biting on their left jaw, so data were collected from 24 channels on the left hemisphere as shown in figure 3.

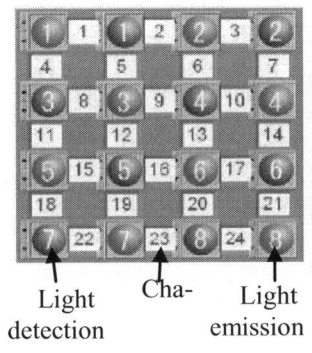 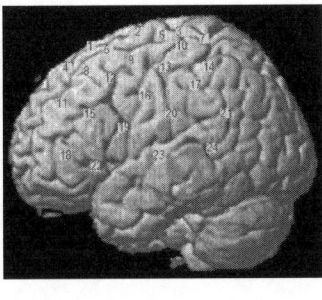

Fig. 2 Measure matrix with the 4x4 size

Fig. 3 Position of the measuring channel on area of the left hemisphere

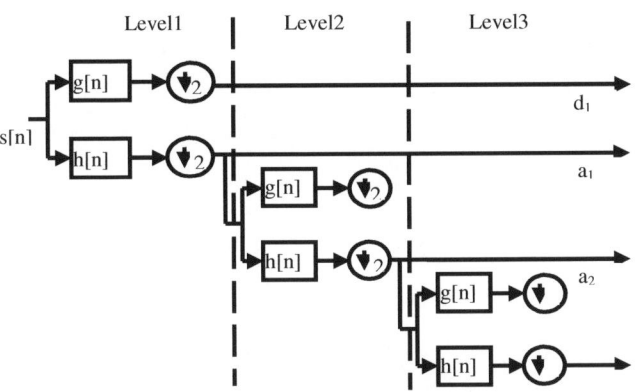

Fig. 5 Diagram of wavelet transform with level-3

Original signal sampling cycle of 0.130s has many noise components, so it will be pre-processed through a smooth filter.

B. Savitzky-Golay Filters

The Savitzky-Golay filter for smoothing fNIRS signals is given as follows:

$$g_i = \sum_{n=-n_L}^{n_R} c_n f_i \qquad (1)$$

where, g_i is output of the filter, f_i is the value of the input data, $n = -n_L \div n_R$ is the sliding window, c_n is coefficient of the filter.

The shapes of the signals before and after filtering as shown in Fig 4:

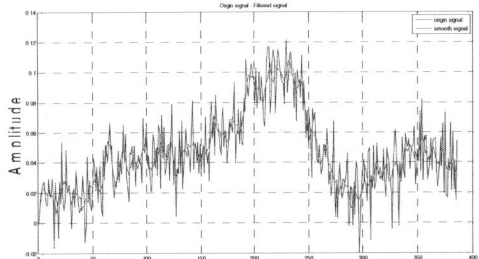

Fig. 4 Channel-1of signal before and after filtering using Savitzky – Golay filter with n = 11, M = 3.

A structured filter with a window size of 11 and a level-3 polynomial is applied to filter signals reduce noises.

C. Discrete Wavelet Transform

After flattening, smoothed signal is analyzed into smaller components through a low pass filter and a high pass filter to produce the rough approximation (a) and details (d), as shown in figure 5.

The output signal is filtered by a low-pass filter using the following equation [4]:

$$a[k] = y_{low-pass}[k] = \sum_n s[n]h[2k-n] \qquad (2)$$

And the output signal using a high-pass filter is described as follows:

$$d[k] = y_{high-pass}[k] = \sum_n s[n]g[2k-n] \qquad (3)$$

where, n is the input block size, h and g are the wavelet functions with well-chosen coefficients (filters) s(n) is the input function and $y_{low-pass}$, $y_{high-pass}$ are respectively the low-pass and the high-pass outputs, a[k] is approximation coefficien, $d_m[k]$ is detail coefficien.

Therefore, the obtained signal is expressed as [5]:

$$s[n] = a_m[k] + d_m[k] = a_{m-1}[k] + \sum_{i=1}^{m} d_i[k] \qquad (4)$$

where, $d_m[k]$ is detail coefficien at level-m, $a_m[k]$ is approximation coefficien at level-m.

Oxy-Hb signals after Bior family wavelet transform are represented in Fig 6.

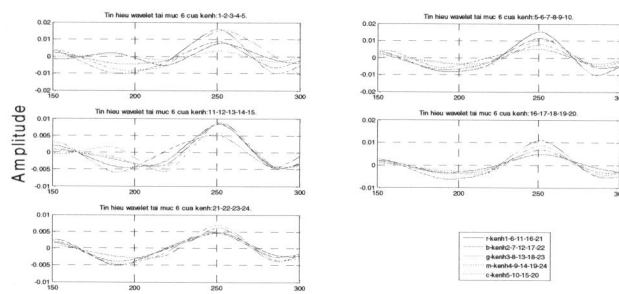

Fig. 6 Representation of wavelet signals with Bior 5.5.

D. Threshold Algorithms

* The total value for each channel of fNIRS signals is calculated as follows. [6]

$$Y_i = \sum_{m=1}^{S} y_i(n) \tag{5}$$

in which i = 1, 2, ... 24 is number of input channels; $y_i(n)$ is the ith channel signal, S = 130 is the number of signal samples, only samples are taken in the period around uptime of subjects (10 seconds) and the latency of the signal measured. Figure 7 shows at the time (t_1) started lifting weights, the oxygen-Hb begins to increase but in fact the biological signal is delayed by a period. So to get the full signal and no signal loss, the authors selected samples from 151- 280, respectively, corresponding to 19.58 - 36.31s.

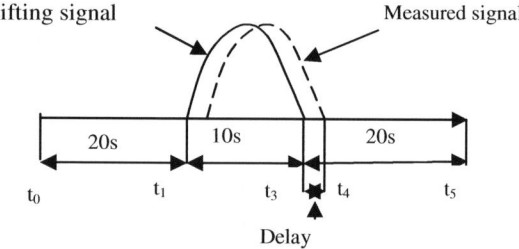

Fig. 7 Protocol of experiment

* Calculation of the average value M_i of the fNIRS signal is described as:

$$M_i = \frac{Y_i}{S} \tag{6}$$

where, Y_i is the total value of the ith channel signal, M_i is the average value of the signal in the ith channel, S is the number of signal samples is taken, i is the number of signal channels.

Therefore, the standard deviation of the signal in the ith channel is calculated by the formula:

$$SD_i = \frac{\sqrt{Y_i - M_i}}{S} \tag{7}$$

in which SD_i is the standard deviation of the ith channel, Y_i, M_i is calculated using equations 5 and 6.

The threshold algorithm is expressed as follows:

$$TH = \frac{\sum_{i=1}^{n}(M_i + a * SD_i)}{n} \tag{8}$$

where, TH is threshold, M_i, SD_i is determined from equations 6 and 7, a is respectively the standard deviation, i = 1, 2,...n.

Signal using wavetlet transform at level 6 of a subject before and after putting threshold is shown in figures 8 and 9.

Figure 8 shows the 24 channel signal formats of a subject perform no bite - lifting 6 kg activity with wavelet transform at level 6 and don't use a threshold algorithm.

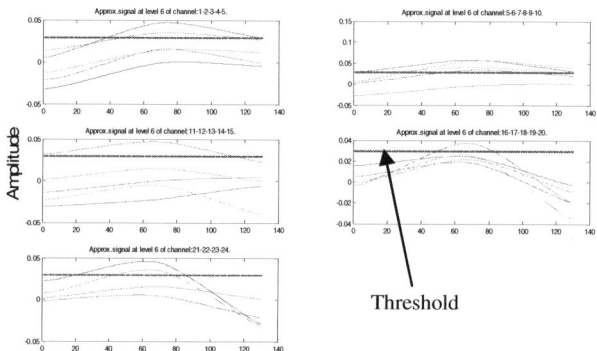

Fig. 8 The Oxy-Hb signals before using a threshold algorithm.

When we use a threshold with standard deviation a = 2, the results are represented as in Fig 9.

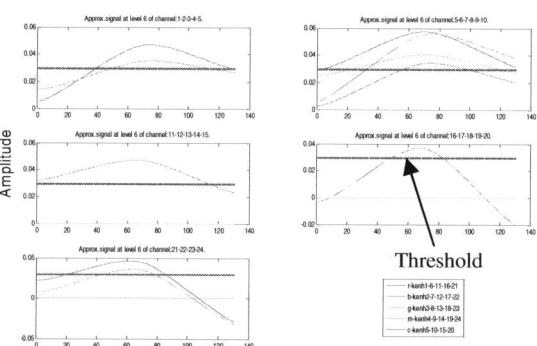

Fig. 9 The Oxy-Hb signals used a threshold algorithm.

III. RESULTS AND DISCUSSION

Some experimental results for biting - no bite and lifting weights are described in Table.2.

Table 2 Experimental results of 5 subjects with no bite and lifting 6 kg and a = 1.

Subjects	Selected channels	
	Number of channels	Channels
S_1	18	1-2-3-6-7-8-9-12-13-14-15-17-18-19-20-21-22-23
S_2	9	3-4-5-6-9-11-14-16-19
S_3	11	1-2-3-4-5-6-7-9-10-13-14
S_4	13	2-4-6-7-8-10-14-17-18-19-20-22-23
S_5	16	1-3-5-7-8-9-12-13-15-17-18-19-20-21-22-23
Result	17	**1-2-3-4-5-6-7-8-9-13-14-17-18-19-20-22-23**

Through experimental results in table 2, we see that, in three of five subjects, there is just the one common channel. This channel will be selected and the total number of channels obtained for this case is 17 channels.

Implementation of similar cases for biting-lifting 6 kg, we have the number of channels for the cases of bite 3 mm-lift 6kg including channels: 6,16,18,19,20,22,23 and channels for biting 2 mm-lifting 6 kg are: 1,6,15,18,19, 20,22, 23. Therefore, the number of channels for advocacy area activities with no bite, or bite and lifting 6 kg are channels: 1-6-18-19-20-22-23 as shown in Table 3.

Table 3 Channels of motor area when lifting 6 kg

Activities	Selected channels	
	Number of channels	Channels
No bite	17	**1-2-3-4-5-6-7-8-9-13-14-17-18-19-20-22-23**
Bite 3mm	7	**6-16-18-19-20-22-23**
Bite 2 mm	8	1-6-15-18-19-20-22-23
Result	7	**1-6-18-19-20-22-23**

The number of channels in motor area is the number general channels of each experimental case.

Similarly, the experimental results when lifting 4,5 and 7 kg including channels such as Table 4.

Through table 4 we see that when the bite and lifting operations performed on 5 subjects focused on a number of channels in bain control area were: 18-19-20-22-23.

Table 4 Channels of motor area for bite and lifting activities

Activities	Selected channels	
	Number of channels	Channels
Bite and no lift	7	15-16-18-19-20-22-23
Bite and lift 4kg	10	6-11-16-17-18-19-20-21-22-23
Bite and lift 5kg	6	18-19-20-22-23-24
Bite and lift 6kg	7	**1-6-18-19-20-22-23**
Bite and lift 7kg	6	2-18-19-20-22-23
Result	5	18-19-20-22-23

Therefore, Table 4 shows 5 channels which is considered as a typical motor area when lifting different weights and performing bite and no bite. These channels have Oxy-Hb changes related to the activities of biting on spacers and lifting weights.

In the previous article, when one needs to find out the brain activity, Oxy-Hb signals were selected based on amplitude characteristics [7]. In this research, the wavelet threshold algorithm was employed to accurately identify active channels of 24 channels. This will produce the calculable and statistic results of fNIRS data more accurate and reliable.

IV. CONCLUSION

In this paper, original brain signals of biting and lifting tasks were filtered using the Savitzky–Golay filter to produce the smooth signals. A wavelet threshold algorithm was utilized to determine the typical motor area of human brain through the activity of biting on spacers and lifting weights. A statistic method by comparing active channels was employed to determine common channels which produce reliable results. Experimental results showed to illustrate the effectiveness of this proposed method for more accurately determining active channels.

ACKNOWLEDGEMENT

The authors would like to acknowledge the support of Biomedical Engineering, IU, Vietnam. Moreover, we would send to thank students and colleagues for this research.

REFERENCES

1. F. Matthews, B.A. Pearlmutter, T. E. Ward, C. Soraghan, and C. Markham (2008), "Hemodynamics for Brain-Computer Interfaces", IEEE Signal Processing Magazine January 2008, pp. 87-94.
2. **M. I. Toshimasa Sato, Tomohiro Suto, Masaki Kameyama, Masashi Suda, Yutaka Yamagishi, Akihiko Ohshima, Toru Uehara, Masato Fukuda, Masahiko Mikuni (2007) "Time courses of brain activation and their implications for function: A multichannel near-infrared spectroscopy study during finger tapping", pp 297-304.**
3. H. Rodolphe J. Gentili, Hasan Ayaz, Patricia A. Shewokis and José L. Contreras-Vidal 2010 "Hemodynamic Correlates of Visuomotor Motor Adaptation by Functional Near Infrared Spectroscopy," presented at the 32nd Annual International Conference of the IEEE EMBS, pp 2918-2921.
4. G. A. Alonso, J. M. Gutiérrez, J. L. Marty and R. Muñoz (2011), "Implementation of the Discrete Wavelet Transform Used in the Calibration of the Enzymatic Biosensors", 2011 InTech/ Publishing Process Manager Ivana Lorkovic, pp.135-154.
5. **T. Q. D. Khoa,** Nakagawa M **(2008) "Functional Near Infrared Spectroscope for Cognition Brain Tasks by Wavelets Analysis and Neural Networks," International journal of Biological and Life science, pp 28-33.**
6. T. Nguyen, T.H. Nguyen, K.Q.D. Truong, and Toi V. Vo, "A Mean Threshold Algorithm for Human Eye Blinking Detection Using EEG" 2012, IFMBE Proceedings Vol. 40, page 275-279.
7. N. T. Hai (2013) "Temporal hemodynamic classification of two hands tapping using fNIRS", Frontiers in H. N, 2013.

The address of the corresponding author:

Author: Ha Van Dao
Institute: University of Technical Education HCMC
Street: 1 Vo Van Ngan, Thu Duc
City: HoChiMinh city
Country: Vietnam
Email: hadao77ric@gmail.com

Determining the Size of a Solid Tumor

Tran Thi Quynh Nhu, Nguyen Thanh Hai, Ngo Thanh Dong, and Nguyen Tan Nhu

Faculty of Electrical-Electronics Engineering, HCMC University of Technical Education, Vietnam

Abstract— **Determination of the size of tumors plays an important role in diagnosis and treatments of cancer disease. Two proposed methods for the calculation of total area of tumor are based on pixels of image and the division of tumor into many parts with the same distance in this paper. In comparison, simulation results are shown to determine the accurate area of each method.**

Keywords— **Tumor, Measuring, Size of Tumor, Solid Tumor.**

I. INTRODUCTION

Cancer is the second most common reason for worldwide death with over 200 kinds of one just identified. Solid tumors account for 85% of human cancers. Every year billions of dollars from both government and private funding resources is spent on Cancer researches. The most important cancer treatment is surgical removal of such tumors. The key leading to a successful cure often involves the efficient delivery of anticancer drugs to the tumor site after a surgery. The size of tumor in these cases need to be determined.

Tumor size helps doctors or medical staffs determine the stage and the appropriate treatment for the patients with tumors. Moreover, in treatment the tumor size was monitored in clinical trials to evaluate the efficacy of the treatment process and determine the next treatment. So, determination the size of a tumor plays a vital role in diagnosis and treatment of cancer disease. There are many different methods to measure the size of tumor. In this paper, the authors proposed two methods: the first one is the calculation of the total area of tumor based on pixels of image; the second is that the division of tumor into many parts with the same distance is applied.

II. MATERIALS AND METHODS

It is very difficult for doctors to determine the area of a tumor based on the original CT image. Although the accuracy of diagnosis depends partly on the qualifications of a doctor, the diagnosis will be more confident if he observed a carefully processed image. Therefore, we implement the following block diagram shown in the Figure 1. This diagram composes of 5 stages. These stages and the output of each one are shown as follow: (1) Reading the input image: The types of input image being able to read are gif, png, jpg, bmp, tiff, and pgm; (2) Pre-processing: more

easily observed image; (3) Edge detection: the edge of image; (4) ROI: the edge of region of interest in Biomedical image; (5) Measuring the size of Tumor: The area of solid Tumor in mm^2.

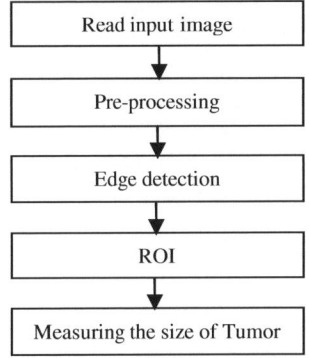

Fig. 1 Block diagram of determining the size of Tumor

After being read, the image is then pre-processed from an original image as shown in Fig. 2. At this stage, image will be filtered and enhanced in order to be more evident. The methods of enhancing image include noisy filter, histogram equalization, etc. The result of pre-processing is the better image which helps doctors or physicians determine the location of tumor more accurately.

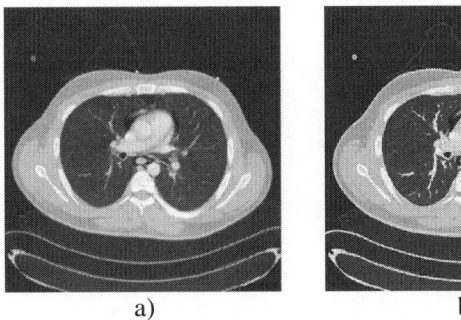

a)   b)

Fig. 2 The result of pre-processing image
a) Original image b) Pre-processed image

The second stage is the edge detection using wavelet method. The method used in this step is 2D discrete wavelet transform. The result of this step is the edge of image which is showed in Fig 3.

© Springer International Publishing Switzerland 2015
V. Van Toi and T.H. Lien Phuong (eds.), *5th International Conference on Biomedical Engineering in Vietnam,*
IFMBE Proceedings 46, DOI: 10.1007/978-3-319-11776-8_93

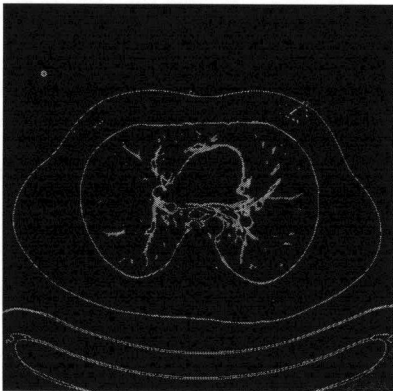

Fig. 3 Detected edge image

Therefore, we determine ROI (region of interest) in the next stage. In this stage, one can directly choose the region on the image by using PC's mouse as shown in figure 4.

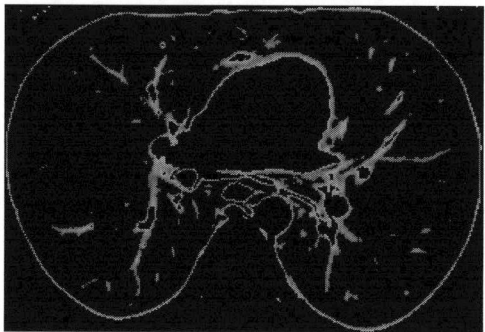

Fig. 4 Selected ROI image

The last stage is to measure the size of solid tumor. Depending on the particular image, two methods are proposed to implement this task.

The first one is the employment of the calculation of total area based on pixel of a tumor image, in which the area of the tumor is equal to the sum of all area of pixels inside it. The formula of the tumor a is described as follows:

$$S = \sum_{j=1}^{col} \sum_{i=1}^{row} a(i,j) * S_p \qquad (1)$$

where S is the size of the tumor; S_p is the area of one pixel, with $i=1,2,...,row$ and $j=1,2,...,col$.

This method has two solutions: (a) checking pixels in all rows and columns; (b) masking directly on tumor image and then counting all pixels of each line.

In the first solution, we check all pixels in rows and columns from the top to the bottom of the image, the left to the right of one and reversely. Whenever checking in row and column, if we detect bit 0, it is considered as the edge of tumor by transforming all bits inside the edge into 1. Therefore, with this ROI, an edge is shown in Fig.5.

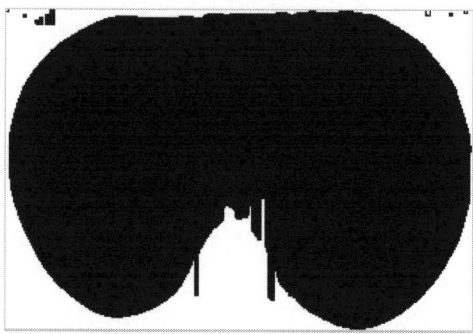

Fig. 5 The image after transforming all bits inside the edge into 1

After transforming all pixels in the edge into 1, we can count the area of all 1 pixels in the image with the size of one pixel is $S_p = 0.26 \times 0.35$ (mm^2). Based on the formula (1), one can get the area of the solid tumor.

In the second solution, doctors can directly mask the edge on a tumor image. After masking, all pixels in the masked image will be turned into 0. Therefore, the area of all 0 pixels in the image is calculated. The result is that the area of a solid tumor is obtained as shown in figure 6.

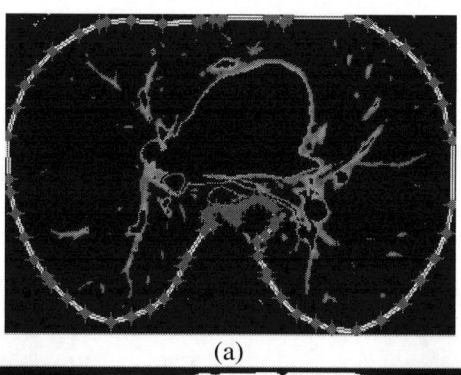

(a)

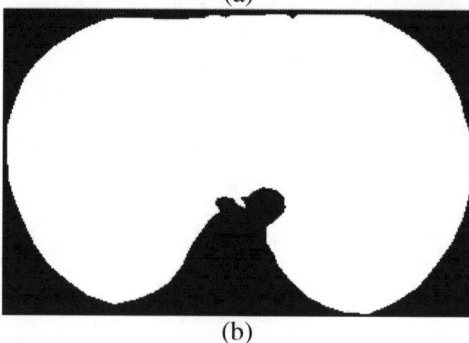

(b)

Fig. 6 Processing before measuring the size of Tumor
a) Masked image
b) The image after transforming all bits inside the edge into 0

The second method is that the tumor is divided into many discrete parts with the same distance. The size of each part is measured and then summed to produce the size of the tumor by using the following formula:

$$S = \sum_{i=1}^{n} S_i = \sum_{i=1}^{n} l_i * d \qquad (2)$$

where n is the number of slices; d is the thickness of a slice; l_i is the length of the ith slice.

After masking on the ROI image, we converse all pixels inside the mask into the value 0. For re-building the edge of a tumor, one can obtain the edge which is shown in figure 7.

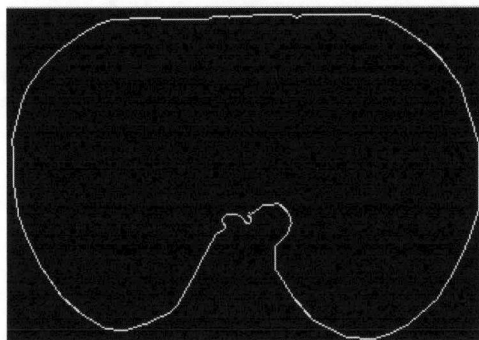

Fig. 7 Rebuilt the edge after masking

For dividing the tumor in the vertical, we have many slices with the same distance. The size of tumor is the sum of all areas of slides using formula (2). In addition, one can adjust the width of slides. This parameter affects the size error of a tumor. The smaller is parameter, the higher is the accuracy of calculating the tumor size. However, the processing speed is decreased.

III. RESULTS

This paper provided the procedure of image processing before measurement the size of tumor. In particular, an original image was filtered to reduce noise and then detected ROI for the determination the size of a solid tumor. The demonstration of feasibility using the proposed methods was shown and then compared between solutions together.

Biomedical images in this paper were processed and produced some results as follows: firstly, the original image as shown in Fig. 8a was enhanced to produce a better image (see Fig. 8b); secondly, the edge of the output image was then detected using the wavelet algorithm as shown in Fig. 9a. Therefore, the ROI of this detected was chosen by

using PC's mouse (see Figs. 9b-9c). Finally, the numeric results were calculated using two proposed methods as shown in Fig. 10, in which S1 and S2 are the two results of the first method with checking pixels and mask, while S3 is the result of the second method using dividing a tumor into many different parts.

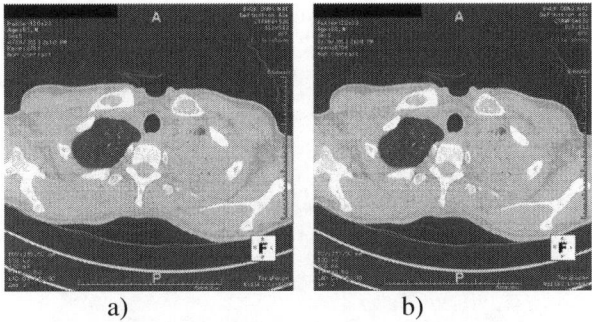

Fig. 8 The result of pre-processing image
a) Original image   b) Pre-processed image

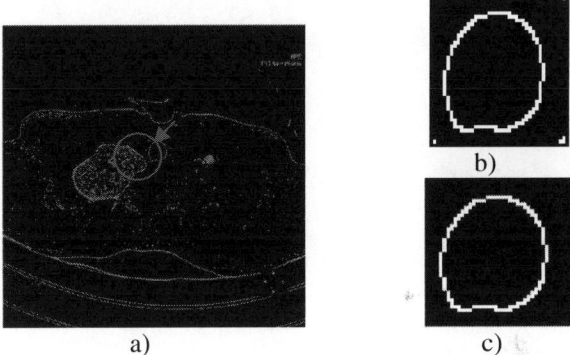

Fig. 9 The result of detecting the edge of ROI
a) Detected edge image   b) Selected ROI image   c) Processed edge

Fig. 9c describes the size of tumor using the first and third solutions and one determined S1 and S3. The result of S2 and two previous solutions are shown in Fig 10.

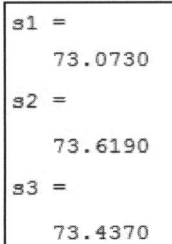

```
s1 =

      73.0730

s2 =

      73.6190

s3 =

      73.4370
```

Fig. 10 Numeric results of three solutions in the second image

IV. DISCUSSION

The comparison between the two methods of three solutions shows that the results of S1 and S3 are nearly the same. The first one is that S1 is more accurate if ROI is independent from both other parts in the image and the shape of the solid tumor. The second one is that S2 using the masking method produces a little bit different result. Because of pixel 0 in ROI, this method is considered as the edge of tumor which was shown in Fig 5. Moreover, the shape of tumor affects the result of the calculation of its size as shown in Fig 10b and 10c.

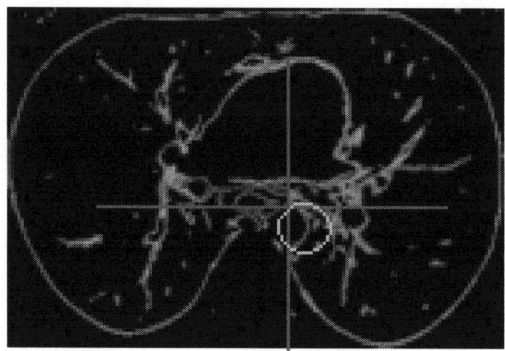

Fig. 11 The error of the first method

In these three solutions, S2 and S3 can be applied to all shapes. The error often depends on the operation of physicians in the second solution. Moreover, the error of S3 in the third solution depends on the quality of tumor edge and the parameter which can be adjusted to change the width of slides. In this paper, the width parameter of slide is chosen to be 1 for calculation.

Based on each specific case, doctors or physicians can choose the method to measure the size of tumor to give the accurate result.

V. CONCLUSION

This paper shows the pre-processing procedure before calculation of the tumor size. Two methods such as the edge detection and the division of an image into parts for calculating the size of a tumor were applied. Many shapes based on the features of image were also determined. The results showed that these methods could be used effectively for measuring the size of tumor.

ACKNOWLEDGEMENT

Authors would like to acknowledge the support of Dong Nai hospital for contribution of the success of this research.

REFERENCES

1. Alexander Herman, "*Towards a General Model for Solid Tumor Growth*", Wesleyan University, 2002.
2. M Soltani and Pu Chen, " Effect of tumor shape and size on drug delivery to solid tumors", Journal of Biological Engineering, 2012.
3. P. David Mozley , "Measurement of Tumor Volumes Improves RECIST – Based Response Assessments in Advanced Lung Cancer ", Translational Oncology, USA, 2012.
4. Soltani M, Chen P, "Numerical modeling of fluid flow in solid tumors". PLoS ONE, 2011.
5. Patrick Therasse, M.D., "New Guidelines to Evaluate the Response to Treatment in Solid Tumors", Journal of the National Cancer Institute, 1999
6. Mary Frances Dempsey, Barrie R. Condon, Donald M. Hadley, "Measurement of Tumor "Size" in Recurrent Malignant Glioma: 1D, 2D, or 3D", American Journal of reuroradiology, 2005
7. Gaia Schiavon, "Tumor Volume as an Alternative Response Measurement for Imatinib Treated GIST Patients", PLoS One, 2012
8. Lawrence H.Schwartz, Michelle S. Ginsberg, Douglas DeCorato, "Evaluation of Tumor Measurements in Oncology: Use of Film-Based and Electronic Techniques", American Society of Clinical Oncology, 2000
9. Chesnokov Yuriy, "Edge Detection in Images with Wavelet Transform", Russian Federation, 2007
10. Petrová Jana, "*Edge detection in medical images using the Wavelet Transform*", Portál pre odborné publikovanie ISSN 1338-0087, 2011

Author: Tran Thi Quynh Nhu
Institute: University of Technical Education Ho Chi Minh City
Street: Vo Van Ngan
City: Ho Chi Minh
Country: Việt Nam
Email: nhuttq@hcmute.edu.vn

2D Complex Shear Modulus Imaging in Gaussian Noise

Nguyen Thi Anh-Dao[1,2], Tran Duc-Tan[2], and Nguyen Linh-Trung[2]

[1] University of Technology and Logistics, Bac Ninh, Vietnam
[2] Electronics & Telecom., University of Engineering and Technology, Vietnam National University, Hanoi, Vietnam

Abstract— Dynamic shear-wave estimation of complex shear modulus (CSM) has demonstrated the ability to detect tumors. Ultrasound shear wave imaging is one of the methods for quantitatively estimating relevant elasticity parameters of tissues via the wave number and propagation attenuation of ultrasound waves. Maximum Likelihood Ensemble Filter (MLEF) has been efficiently applied for estimating the CSM parameters, but limited to one-dimensional (1D) scenario. This paper extends this method to detecting two-dimensional (2D) objects affected by Gaussian noise during the Doppler acquisition. A ray scanning method is used for modeling the propagation directions (lines) along each of which the MLEF is used for estimating the CSM parameters. The object 2D image is then reconstructed by transforming these estimated CSM parameters from the polar coordinates to Cartesian coordinates. it is not necessary to increase the ensemble size (which means an increase in the algorithm complexity) when the noise level is low.

Keywords— Ultrasound shear wave imaging, maximum likelihood ensemble filter (MLEF), complex shear modulus (CSM), elasticity imaging.

I. INTRODUCTION

Many pathological processes in tissues are recognized by morphological changes that reflect alterations of mechanical properties of soft tissues. Among various elasticity imaging modalities, ultrasonic shear wave imaging technique has been developed for estimating the complex shear modulus (CSM) of biphasic hydro polymers including soft biological tissues. Shear wave imaging has the potential to bridge molecular, cellular and tissue biology, and to support medical diagnoses and patient treatment.

In 2004, Chen et al. found that the propagation speed of shear waves is related to the frequency, the elasticity and viscosity of the medium [1]. Hence, they proposed a method to estimate shear elasticity and viscosity of a homogeneous medium by measuring the shear wave speed dispersion. In 2009, Orescanin et al. applied the Kelvin–Voigt model to estimate the CSM of the liver for shear wave frequencies between 50 and 300 Hz [2]. Then, the Maximum Likelihood Ensemble Filter (MLEF) was applied for CSM estimation for homogeneous medium [3, 4]. It was extended to 1D heterogeneous medium in [5]. In this paper, we extended this method to detecting two-dimensional objects[1]. In addition, we study the effect of Gaussian noise corrupting the Doppler acquisition in conjunction with the effect of the ensemble size of the MLEF.

II. MATERIALS AND METHODS

First, a needle vibrating with the frequency of f (Hz) is used for creating a shear wave whose velocities will then be measured by a Doppler scanner [6]. Second, a ray scanning method is used for modeling the propagation directions. Denote $\alpha(r)$ and $k_s(r)$ the shear wave attenuation coefficient and the wave number at the tracking location r along each ray; r is defined in the polar coordinates as: $r = \rho e^{j\theta}$. Third, the CSM of the tissue located at r is then estimated from $\alpha(r)$ and $k_s(r)$, which are the real and imaginary parts of the CSM value, based on using the Kelvin-Voigt model for a viscous medium [3]. Last, the 2D image of the object is reconstructed by transforming these estimated CSM parameters from the polar coordinates to Cartesian coordinates.

A. Shear Wave Propagation

The needle vibrates along the vertical (z) axis. Under an assumption of cylindrical shear wave propagation along the radial axis, the particle velocity of ray i is a spatio-temporal function of the radial distance r and time t, and is given by

$$v^i(r,t) = \frac{1}{\sqrt{r}} A e^{-\alpha(r)r} \cos(\omega t - k_s(r)r), \, i = 1, \dots, L \qquad (1)$$

where L is number of ray, A is the magnitude of the wave at the source location, ω is the angular shear frequency. In discrete form, we have

$$v^i_n = \frac{1}{\sqrt{r-r_0}} A e^{-\alpha(r)(r-r_0)} \cos(\omega n \Delta t - k_s(r)(r-r_0) - \phi), \qquad (2)$$

where index n denotes the discrete time, r_0 is the initial distance from the source, Δt is discrete-time step and ϕ represents the initial temporal phase.

Eq. (2) can be rewritten in a recursive form by:

---

[1] Part of this study was presented in the 2013 International Conference on Green and Human Information Technology (ICGHIT 2013) as an in-progress work.

© Springer International Publishing Switzerland 2015
V. Van Toi and T.H. Lien Phuong (eds.), *5th International Conference on Biomedical Engineering in Vietnam*,
IFMBE Proceedings 46, DOI: 10.1007/978-3-319-11776-8_94

$$v^i{}_n(r) = v^i{}_{n-1}\cos(\omega\Delta t) -$$
$$\frac{1}{\sqrt{r-r_0}}Ae^{-\alpha(r-r_0)} \times \quad\quad (3a)$$
$$\sin(\omega(n-1)\Delta t - k_s(r-r_0) - \phi)\sin(\omega\Delta t).$$

Given the effect of Gaussian noise $w^i{}_n(r)$ on the velocity at each spatial location, we have the following model:

$$v^i{}_n(r) := v^i{}_n(r) + w^i{}_n(r). \quad\quad (3b)$$

B. Attenuation Coefficient and Wave Number Estimation

In this subsection, we apply the MLEF to estimate k_s and α in each ray. The state equation can be constructed from Eq. (3a)as shown below:

$$x = \begin{bmatrix} v_n \\ \theta_n \end{bmatrix} = \begin{bmatrix} \mathcal{F}(v_{n-1}, \theta_{n-1}) \\ \theta_{n-1} \end{bmatrix} \quad\quad (4)$$

where $\theta_n = [\alpha^T, k_s^T, \phi, r, A]^T$, $\mathcal{F}$ is a nonlinear function modeling the spatial shear wave dynamics. The length of vectors v_n, α and k_s equals to the number of spatial locations. We can assume that θ_n would not be changed during the time of the experiment; hence, $\theta_n = \theta_{n-1}$ as shown in Eq. (4).

By using the Doppler acquisition, the measurements of velocities at every spatial locations are given by

$$y_n = \begin{bmatrix} \hat{v}_n \\ 0 \end{bmatrix} = [I \quad O]\begin{bmatrix} v_n \\ \theta_n \end{bmatrix} + \begin{bmatrix} w \\ 0 \end{bmatrix}, \quad\quad (5)$$

where w is the measurement noise vector.
From Eqs. (4) and (5), the shear wave attenuation coefficient $\alpha(r)$ and the wave number $k_s(r)$ of each ray are estimated by using MLEF according to the algorithm in [4].

C. CSM Estimation Using Kelvin-Voigt Rheological Model

We apply the Kelvin-Voigt model, as illustrated in Fig.1, to estimate the CSM. For a viscous elastic medium, the CSM μ is modeled by an elastic component μ_1 in parallel with the dynamic viscous component η as:

$$\mu = \bar{\mu} - i\omega\eta, \qu\quad (6)$$

The complex wave number for a viscous medium is given by

$$k'_s = \sqrt{\rho\omega^2/\mu}. \qu\quad (7)$$

Since k'_s is complex, it can be written as

$$k'_s = k_s - i\alpha, \qu\quad (8)$$

From (7) and (8), by estimating k_s and α, we can obtain μ.

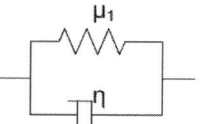

Fig. 1 Diagram of Kelvin-Voigt model

D. Detecting the Presence of 2 D Object

In this work, we verify the proposed method using a two-object simulation scenario. Each object was 'placed' at a certain spatial location $r = \rho e^{j\theta}$. We use a "ray scanning" method to cover the area of interest, in steps of a constant angle (see Fig. 2). The whole area is scanned by varying θ from $0°$ to $90°$ in step of $1°$, creating 90 rays of interest. Given the availability of 43 elements in the in-use Doppler scanner, we select 43 evenly-spaced spatial locations r along each ray. After collecting all information of these 90 rays, we estimated $\alpha(r)$ and $k_s(r)$. For a given CSM value μ corresponding to a particular material of an object under interest, we establish a detection threshold pair of (α^*, k_s^*) such that the object is detected to be present if the following conditions: $\alpha(r) > \alpha^*$ and $k_s(r) < k_s^*$ for all r. In this paper, the value of the threshold pair can be found empirically via numerical simulation.

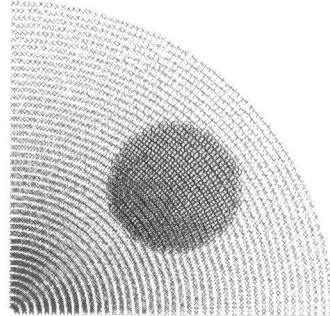

Fig. 2 Ray scanning illustration

III. NUMERICAL RESULTS AND DISCUSSIONS

In this study, we examine the proposed method for detecting 2 circular objects whose elasticity properties are different (i.e., the CSM values are different). Object 1 is placed at location (6 mm, 1.4 mm) with the radius of 1.4 mm. Object 2 is placed at (10 mm, 8 mm) with the radius of 3 mm. The CSM values of the two objects are ($\bar{\mu} = 900$ Pa, $\eta = 0.3$ Pa/s) and ($\bar{\mu} = 800$ Pa, $\eta = 0.2$ Pa/s) respectively. Accordingly, we have ($\alpha = 67.5$; $k_s = 651.7$) for Object 1 and ($\alpha = 54.3$; $k_s = 696$) for Object 2. We asssum that the first spatial location, r_0, is close to the needle: $r_0 = 0.4$ mm.

Base on empirical study, we found that the detection threshold pair for Objects 1 and 2 are ($\alpha^* = 50$; $k_s^* = 670$) and and ($\alpha^* = 44$; $k_s^* = 710$), respectively. The amplitude A and the phase ϕ are estimated using the first cycle of the particle velocity at r_0. These parameters are then used for calculating the initial state vector of the MLEF. The initial error square-root covariance matrix is Gaussianly randomly generated. After only tens of iterations, the velocity is denoised and the attenuation and wave number are estimated. Based on the results obtained from the MLEF, we apply Kelvin – Voigt rheological model to estimate CSM. Finally, we construct the 2D image. The 2D image in Cartesian coordinates of the simulation scenario is shown in Fig. 3.

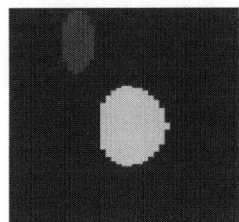

Fig. 3 Original image.

A. In Noise-Free Environment

Fig. 4 shows the estimated $\alpha(r)$ and $k_s(r)$ from rays 4 and 45, which are chosenly specifically so that ray 4 passes Object 1 and ray 45 passes Object 2, with no noise effect in the Doppler acquisition. The solid curves are the estimated attenuation and wave number, and the dashed curves are ideal ones. It is show that Object 1 is detected to be present in the interval of $r = [11, 16]$ and Object 2 in that of $r = [18, 30]$. In this study, the MLEF ensemble size is $s = 86$, which is equal to twice the number of elements of the Doppler scanner.

Then, the 2D reconstructed image and their 2D images of attenuation and wave number are shown in Fig. 5. It is obvious that the objects were detected. However, it can be seen that the wave number was better estimated then the attenuation.

B. In Gaussian Environment

Figures 6 to 9 illustrate the effect of Gaussian noise and the MLEF ensemble size on the reconstructed images and their corresponding images of attenuation and wave number, tested for the signal-to-noise ratio (SNR) of 40 and 34 dB and the ensemble size of 43 and 86. With a large value of the ensemble size ($s = 86$), we were able to detect the objects when the Doppler acquisition was corrupted by the noise.

Fig. 10 provides an insight into the effect of the ensemble size on the image reconstruction, measured by the peak-signal-to-noise ratio (PSNR), with respect to different noise levels (SNR = 20 to 40 dB). It can be seen that at a high-level of noise, a higher ensemble size offers a larger PSNR, which means a better quality. However, it is not necessary to increase the ensemble size (which means an increase in the algorithm complexity) when the noise level is low.

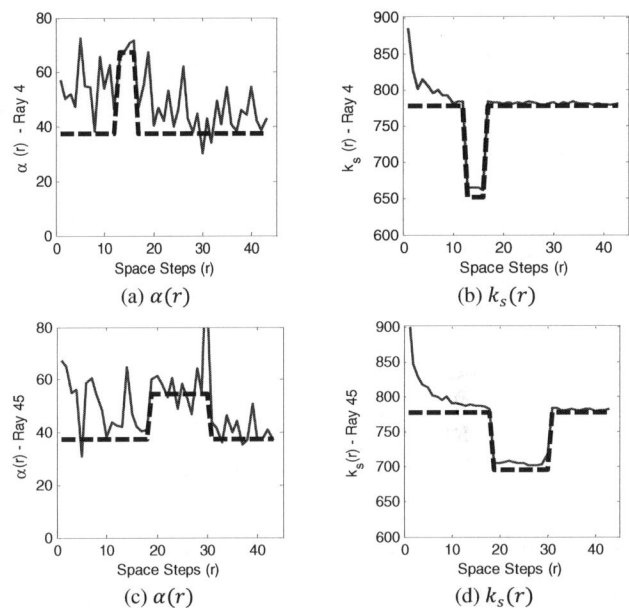

(a) $\alpha(r)$ (b) $k_s(r)$

(c) $\alpha(r)$ (d) $k_s(r)$

Fig. 4 $\alpha(r)$ and $k_s(r)$ along ray 4 (top) and ray 45 (bottom); $s = 86$.

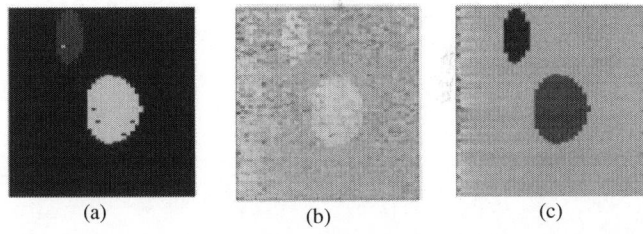

(a)   (b)   (c)

Fig. 5 Reconstructed image (a), Attenuation (b) Wave number (c) images with ensemble size of $s = 86$.

IV. CONCLUSIONS

Based on the MLEF approach, this paper has proposed a ray-tracing based method to estimate the elasticity properties of 2D objects in shear wave imaging. The experiment is quite simple when only a single vibration frequency is needed to accurately estimate the CSM in the medium. Quantitative analysis for different levels of Gaussian noise affecting the Doppler acquisition, the ensemble size were studied. In future work, it is desirable to examine further thoroughly the optimal imaging thresholds pair, (α^*, k_s^*), used for detecting the objects under various practical CSM values. In addition, better estimation of the attenuation should be investigated.

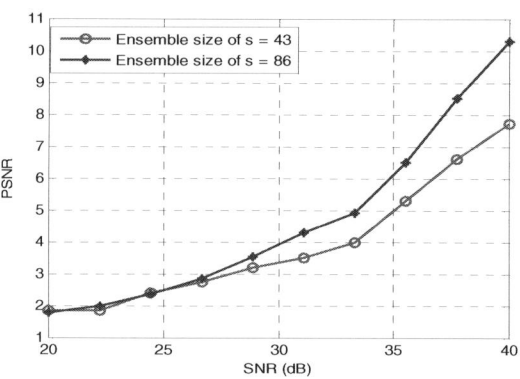

Fig. 10 Effects of noise and ensemble size.

REFERENCES

1. Chen, S et al.: Quantifying elasticity and viscosity from measurement of shear wave speed dispersion. Journal of Acoustic Soc Am. 115, 2781-2785 (2004).
2. Marko Orescanin, et al.: Complex Shear Modulus of Thermally-Damaged Liver, pp127 – 130.Ultrasonics Symposium (IUS), 2009 IEEE International.
3. Orescanin, M. et al.: Model-based complex shear modulus reconstruction: A Bayesian approach. In: IEEE Int'l Ultrasonics Symposium, pp. 61-64. IEEE Press (2010).
4. Zupanski, M.: Maximum Likelihood Ensemble Filter: Theoretical Aspects. Monthly Weather Review. 133, 1710-1726 (2005).
5. Tan Tran-Duc, et al.: Complex shear modulus estimation using the maximum likelihood ensemble filter, BME'04, 2012.
6. Orescanin et al.: Shear Modulus Estimation With Vibrating With Needle Stimulation. IEEE Trans. Ultrasonics, Ferroelectrics, and Frequency Control. 57, 1358-1367 (2010).

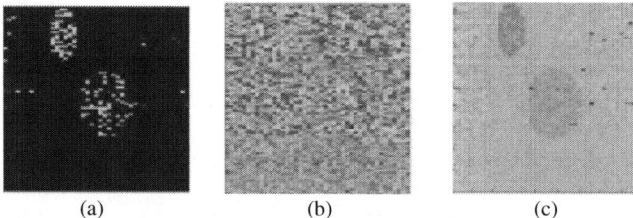

(a)   (b)   (c)

Fig. 6 Reconstructed (a), Attenuation (b), and Wave number images with $s = 43$, SNR = 40 dB. The obtained PSNR = 10.4.

Author:	Nguyen Thi Anh-Dao
Institute:	University of Technology and Logistics
Street:	Ho town, Thuan Thanh district
City:	Bac Ninh
Country:	Viet Nam
Email:	daonta81@gmail.com

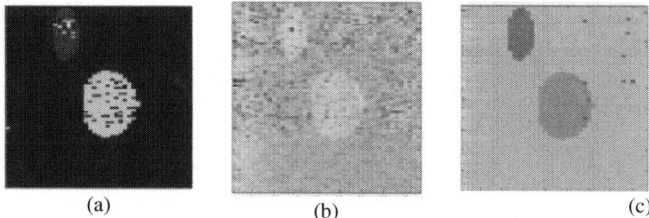

(a)   (b)   (c)

Fig. 7 Reconstructed (a), Shear attenuation (b), and Wave number images with $s = 86$, SNR = 40dB. The obtained PSNR = 10.31.

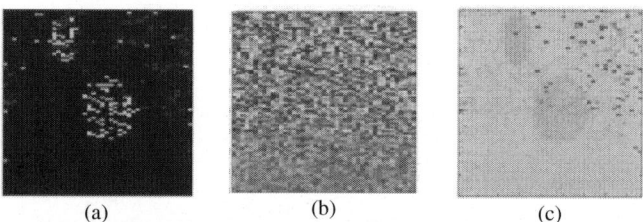

(a)   (b)   (c)

Fig. 8 Reconstructed (a), Shear attenuation (b), and Wave number images with $s = 43$, SNR = 34 dB. The obtained PSNR = 6.58.

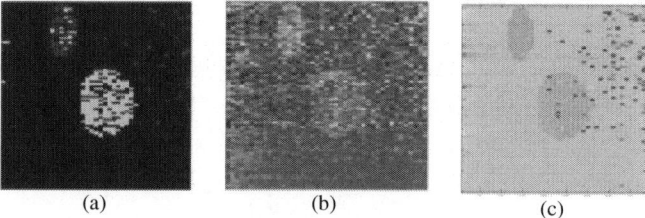

(a)   (b)   (c)

Fig. 9 (a) Reconstructed (a), Shear attenuation (b), and Wave number images with $s = 86$, SNR = 34 dB. The obtained PSNR = 8.46.

Evaluation of Frontal and Visual Cortices on Mental Working Tasks Using Functional Near Infrared Spectroscopy

Pham Thanh Thao, Nguyen Duc Thang, and Vo Van Toi

Biomedical Engineering Department, International University, Ho Chi Minh City, Vietnam

Abstract— This paper investigated the hemodynamic responses over the frontal and visual cortices in order to find out the contributions of these two brain areas to workload memory activities during a graded load-related memorizing task that involves visual perception. Factorial ANOVA was performed on the mean values of oxyHemoglobin (oxy-Hb) and deoxyHemoglobin (deoxy-Hb) concentration changes from three subjects over visual and prefrontal cortices to observe the interaction of factors like channels, subjects and workload levels. Besides, the t-maps of these two brain areas pointed out the regions of interest (channels 1, 2, 3, 4, 5, 6, 7, 8, 10, 13 and 14) are shown to reflect the significance in workloads differentiation. The results of statistical analysis and t-map investigations successfully explored the capability of multi-channel functional near infrared spectroscopy (fNIRS) to detect two neurophysiological workloads under investigation and distinguish their activation patterns over multiple cortical areas.

Keywords— fNIRS, functional Near Infrared Spectroscopy, working memory, prefrontal cortex, visual cortex.

I. INTRODUCTION

Working memory is a cognitive operation that underlies human ability to temporarily maintain representation of information in a readily accessible state for use regarding the information source have been no longer accessible [1]. While prefrontal cortex has been proved to play a crucial role in processing working load that is related to memorizing task [2, 3], the investigations on this cortical area still have limitations to explain how and where the brain maintains visual details of the stimuli during visual working memory. Over the years multiple models of working memory have been devised in an attempt to explain its properties and functions. The concept of working memory, as originally proposed by Baddeley and Hitch [4], has three dissociable components: a phonological rehearsal loop for the storage and manipulation of verbal information, a visuospatial group for visual and spatial information, and a central executive for attentional control. The visuospatial sketch pad is further divided into two subsystems: one for object based information and other for spatial information [5]. The ventral occipitotemporal pathway, known as the "what pathway", is essential for perceiving the identity of objects. And the dorsal occipitoparietal pathway, known as the "where pathway", is involved in the perception of the spatial relationships among objects. These ventral and dorsal pathways, therefore, are hypothesized to play a role in connecting the prefrontal with parietal and primary visual cortices together for the functions of reflecting memory-related maintenance processes and ensuring an active representation of visual information in working memory related to objects and locations. Therefore, recent researches have studied the relations of these neural patterns and their participation in mental workload formation and properties. J. Munneke et al. [6] employed fMRI to study the spatial working memory in early visual cortex. Piazza et al. [7] and Ritcher et al. [8] showed that parietal and especially intraparietal areas are activated in response to symbol and non-symbol arithmetic tasks by using fNIRS technique. Hong [9] introduced the use of EEG in the judgement of mental workload in visual cognition under multi-task experiments, they found that the Lempel-Ziv complexity (LZC) value of occipital lobe and parietal lobe was higher than that of other brain areas under the same task difficulty. However, to our knowledge, there have been no researches considered the differences of one mental workload task with graded levels over prefrontal and primary visual cortices at the same time. The investigation into the activation of multi-cortical regions, especially visual cortex, can give the understanding of how visual memorizing workload is generated and processed. In addition, this would give a rich information input for the mental workload discriminating system that can automatically differentiate mental workload situations and give necessary supports for the user when being overload or under pressure.

The fNIRS technology has been proved to be an effective tool in the discrimination of mental workload over prefrontal cortex [10, 11, 12]. This system measures oxygenation changes in the participants to infer information about neural activation as participants are subjected to series of stimuli. The hypothesis is that by measuring the changes in oxy-Hb and deoxy-Hb, one can infer relative changes in activation as the external demand is varied. fNIR technology is also non-ionizing, non-invasive, non-constraining and affordable.

V. Van Toi and T.H. Lien Phuong (eds.), *5th International Conference on Biomedical Engineering in Vietnam,*
IFMBE Proceedings 46, DOI: 10.1007/978-3-319-11776-8_95

In this study, fNIRS was used to record oxy-Hb and deoxy-Hb information of the participants over frontal and occipital lobes while they were performing a sequence of tasks that were randomly easy or difficult. The aim of this study included two folds. First, we identified how fNIRS was suitable in the application of detecting mental workload levels, especially mental overload. Second, we aimed to observe the activation regions over prefrontal and primary visual cortices during graded mental tasks that involves visual memorizing processes. Additionally, we considered detecting individual differences as important part of our research. In particular, the physiological measures under different workload conditions differ strongly with the individual [13, 14]. This is analogous to the differences in corresponding affective states, such as stress. This is necessary to understand and predict a user's behavior, e.g. for an estimation of how likely it is that the user will make an error.

II. Methods

A. Data Collection

Participants: Three young and healthy subjects (2 females, mean age: 20.3 years with standard deviation 1.15) participated in this study. None of them had neurological disorders. They had been acquainted prior to the experiment to ensure the motion artifact prevention. This study had been approved by the local institution review board and written informed consents were obtained from all participants. The tenets of Declaration of Helsinki were followed.

NIRS Device and Data Acquisition: FOIRE - 3000 (Shimadzu, Japan) (Figure 1(a)) was used in this study. This multi-channel system allows for measuring over large areas as well as at different cortical regions at the same time. Near infrared light with three different wavelengths of 780, 805 and 830 nm exits from each T-transmitter and passes through the skin, skull and underlying tissues, then the scattering light was detected by the surrounding R-receiver optodes. One T-R pair constitutes one channel, which provides hemodynamic information of oxy-Hb and deoxy-Hb concentration changes at the measured cortical location. The sensor has 55 milliseconds resolution and 3 cm T-R probes separation, allowing for approximately 1.5 cm penetration depth.

In our experiments, transmitters and receivers were arranged in two 2-by-3 arrays constituted 14 channels - channels 1 to 7 over frontal and 8 to 14 over occipital lobe. Figure 1(b) indicated the T and R probes and channels locations over two cognitive areas.

Experimental Procedures: During the experiment, subjects looked at one three-dimensional (3-D) graphical cube rotating on the computer screen while recording NIRS

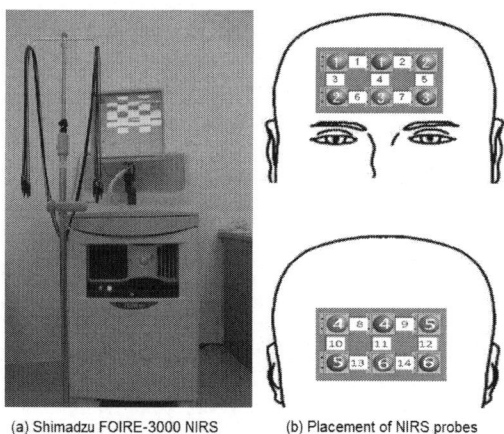

(a) Shimadzu FOIRE-3000 NIRS   (b) Placement of NIRS probes

Fig. 1 (a) Shimadzu FOIRE - 3000 fNIRS machine and (b) placement of NIRS probes, 2-by-3 arrays of T (*red*) and R (*blue*) optodes were placed over prefrontal and visual cortices

signals on prefrontal and visual cortices. This experimental protocol was designed by Tufts University researchers with the aim to discriminate mental workloads over prefrontal cortices [15]. Each cube side contained 4 small colored square sections (the colors could be red, green, blue or yellow). The cube rotated 3 times to show 5 sides (1 top and 4 lateral), allowing the subjects to view each side only once but the top side was always visible. The time for showing each lateral side was 8 seconds and the transition time for rotating to another side was 3 seconds. After showing 4 lateral sides, the cube stayed still for 4 seconds to allow the subjects to look at the top side. Therefore, one complete task lasted for 45 seconds. During each task, the subjects were asked to count and memorize the total number of square sections of each different color on the whole cube. The total color of the squares on the cube could be two or four colors randomly, corresponding to easy and difficult workload levels (i.e. named as level 2 and level 4). Then they would take another 45 seconds of rest before switching to the next task so that the NIRS signals would be shifted to baseline again. One experiment included ten rest - task cycles.

The correlation between workload difficulty levels and human mental working state could be illustrated in Figure 2, where low and high mental workloads are associated with a reduction and increment in human mental load. Workload level 2 represents a normal level of workload where subjects are not under pressure. Meanwhile, workload 4 is corresponding to the condition of overload where subjects require high concentration. In this study, the variation changes of oxy-Hb and deoxy-Hb in relative to two workload levels 2 and 4 and over prefrontal and primary visual cortices were focused.

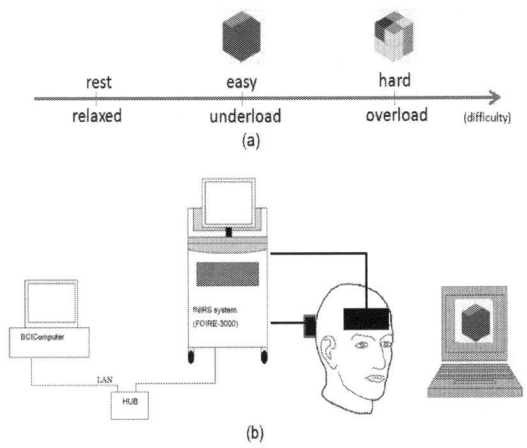

Fig. 2 (a) Correlation between two memorizing tasks with mental workload situations and (b) Experimental setups

B. Data Processing

High frequency instrument noise and low frequency drifts in oxy-Hb and deoxy-Hb signals were removed by using bandpass filter with cut off frequencies of 0.5 and 0.01 Hz. In addition, to remove spike noise due to head motion and improve signal quality and spatial specificity, we applied correlation based signal improvement (CBSI) method [16] based on assumption that true oxy-Hb and deoxy-Hb are maximally negatively correlated. By this method, the corrected activation signal was a linear combination of measured oxy-Hb and deoxy-Hb signals.

For the CBSI method, the signals were first smoothed with moving average filter. Let x_o and y_o be the recorded oxy-Hb and deoxy-Hb signals. Then the true NIRS signal recovered by CBSI method could be calculated as

$$x = \frac{1}{2}(x_o - \alpha y_o); \; y = -\frac{x_o}{\alpha} \qquad (1)$$

where $\alpha = \frac{std(x_o)}{std(y_o)}$ is the ratio of standard deviations of measured oxy-Hb and deoxy-Hb signals. The true oxy-Hb and deoxy-Hb signals were named as corrected activation signal, or denoted as CA oxy-Hb and CA deoxy-Hb.

It has been proved that CBSI method helped remove spike noise, increased contrast-to-noise ratio and improved spatial specificity of hemodynamic signals. Time series of oxy-Hb and deoxy-Hb signals preprocessed by bandpass filtered and CBSI from one subject are compared in Figure 3.

Figure 4 shows the t-map of raw, filtered and CA oxy-Hb from one participant on mean values in comparison between rest and task intervals. Compared to the uncorrected maps, the corrected map appeared to be more localized.

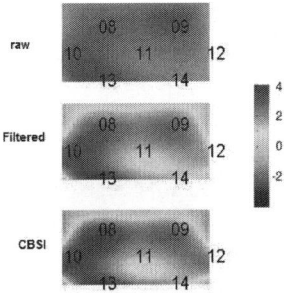

Fig. 4 t-map of raw, filtered and CA oxy-Hb from one participant that reveals the difference between rest and task mean values.

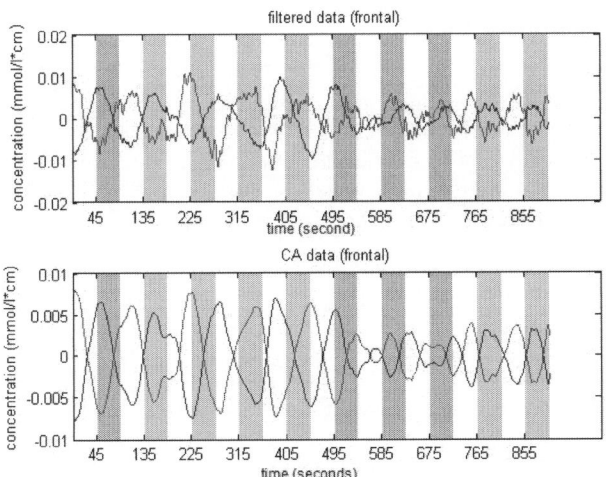

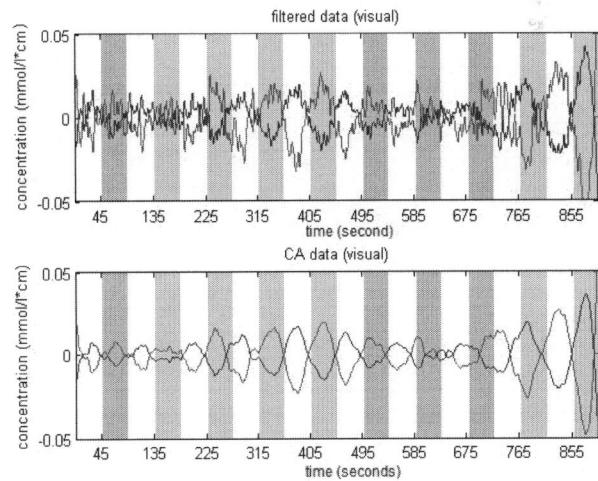

Fig. 3 Time series of filtered and CA oxy-Hb (*red lines*), deoxy-Hb (*blue lines*) signals in one experiment trial over two channels from prefrontal and visual cortices (channel 1 and channel 8). This experiment trial included level 4 (*yellow intervals*) and level 2 (*green intervals*) tasks

C. Statistical Analysis

After the signal improvement steps, we segmented the data signals into small sections initiated by 10 seconds of baseline, followed by 45 seconds of task and ended with 10 seconds of the next rest. Mean values of these sections were calculated for statistical studies. Factorial repeated measures analysis on variances (factorial ANOVA) was applied on these mean values to evaluate the statistical interference of factors like the discrimination of difficulty levels (levels 2 and 4), Channels (14), Subjects (3) and Hemoglobin types (oxy-Hb and deoxy-Hb). Difference in brain hemodynamic responses to the two memorizing workload levels 2 and 4 over each channel were also analyzed by using independent sample t-test analysis at the level of significance $\alpha = 0.05$. For visual investigation, t-map was plotted and filtered with spline method. Based on this visualization, one can observe frontal and visual cortical functions in a load-graded visual working task, as well as determine the regions of interest (ROIs) for further evaluation.

III. RESULTS

A. Interactions of factors to hemodynamic responses of NIRS signals

Factorial ANOVA (2 x 14 x 3 x 2) on mean data of raw and CA oxy-Hb and deoxy-Hb segments reported the significant interaction of factors like difficulty levels, channels, subjects and Hemoglobin types on mean values. On the raw signals, the relation of Channels and Hemoglobin types was significant ($F(13) = 1.85e3$, $p < 0.000$). In addition, the significance of the interactions Difficulty levels x Hemoglobin types ($F(1) = 39.836$, $p < 0.000$), Channels x Difficulty levels x Hemoglobin Types ($F(13) = 6.874$, $p < 0.000$) and Subjects x Levels x Hemoglobin types ($F(2) = 13.807$, $p < 0.000$) were also reported. Similarly, the hemodynamic responses on CA signals were also affected by the interactions of Channels x Hemoglobin types ($F(13) = 1.931$, $p < 0.023$), Difficulty levels x Hemoglobin types ($F(13) = 10.514$, $p < 0.001$), Channels x Difficulty levels x Hemoglobin Types

($F(13) = 1.733$, $p < 0.048$) and Subjects x Levels x Hemoglobin types ($F(13) = 3.38$, $p < 0.034$).

Figure 5 indicates the differences between total oxy-Hb and deoxy-Hb mean values of three subjects in corresponding to two workload levels. The mean values of level 4 responses for oxy-Hb appeared to be significantly higher than level 2 and vice versa for deoxy-Hb signals.

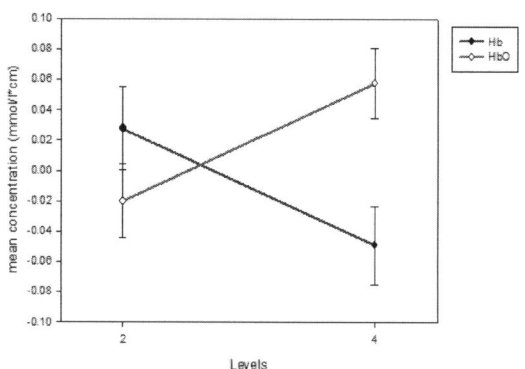

Fig. 5 Estimated marginal means and standard errors between two workload levels among three subjects in oxy-Hb (a) and deoxy-Hb (b) signals

The estimated marginal means of oxy-Hb and deoxy-Hb concentrations following the interaction of workload levels versus channels and subjects are depicted in Figure 5. There existed the difference in the discrimination between two workload levels among three subjects. For subject 2, the mean values of oxy-Hb level 4 was much higher than that of level 2. In contrast, for subject 1 the two means were nearly equal and for subject 3, the standard errors of two levels were nearly overlapped. Similarly, the deoxy-Hb signals showed the high differences between the two levels in subject 2, the nearly equal means of two levels in subject 1 and the overlapped of standard errors of two levels in subject 3. Also, Figure 6 shows the effects of channel areas on the significant difference of two workload levels. Differentiation between two workload levels was clearly indicated in some channels, which would be further implemented in the next section.

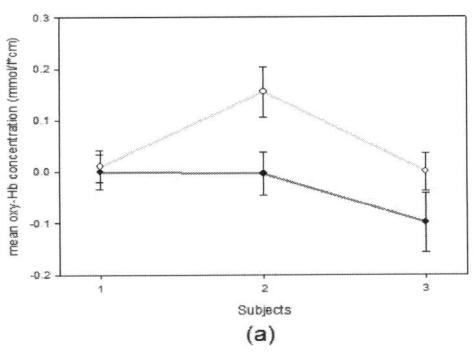 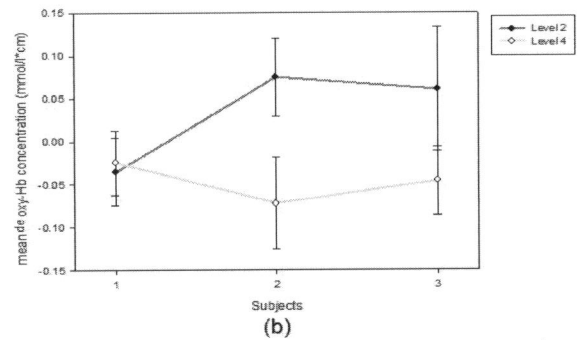

Fig. 6 Estimated marginal means and standard errors between two workload levels among three subjects in HbO (a) and Hb (b) signals

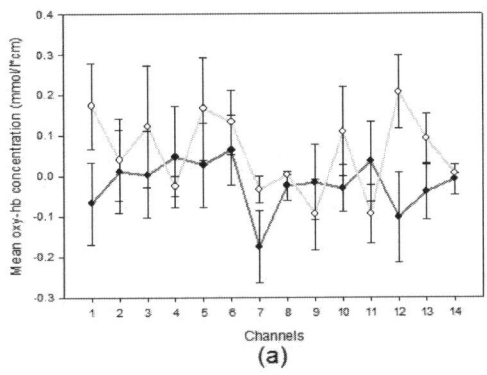 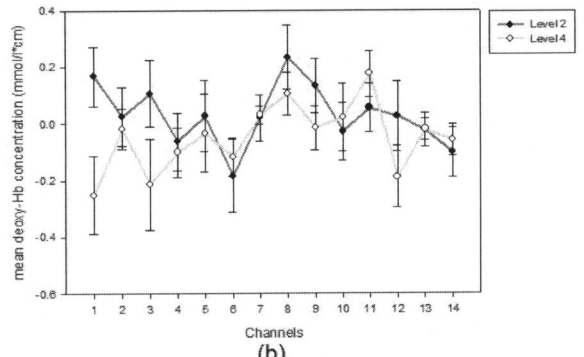

Fig. 7 Estimated marginal means and standard errors between two workload levels among 14 channels in HbO (a) and Hb (b) signals

These statistical results suggested that the oxygenation cortical dynamic responses to workload levels depended on different channels and variations of participants. For further investigation, the next part will discuss about how different brain cortical areas of three subjects responses to the two workload levels.

B. Hemodynamic responses between two difficulty levels over different channels

According to the t-maps (Figure 8), the frontal and visual cortices can reflect the discrimination between two workload levels. On the frontal cortex, the whole seven channels from subjects 2 and 3 and channels 4, 6 and 7 from subject 1 showed the significant level difference ($p < 0.05$). On the visual cortex, the significant difference were observed at channels 8, 10 and 13 with $p < 0.05$ in three subjects. When data from subject 2 and 3 showed significant level difference in channel 14 ($p < 0.001$ and $p < 0.020$ for oxy-Hb, $p < 0.001$ and $p < 0.0015$ for deoxy-Hb for the two subjects respectively), the data from subject 1 showed no difference ($p < 0.927$ for oxy-Hb and $p < 0.843$ for deoxy-Hb). Meanwhile, for channel 9, only oxy-Hb from subject 2 reflected the significant level difference

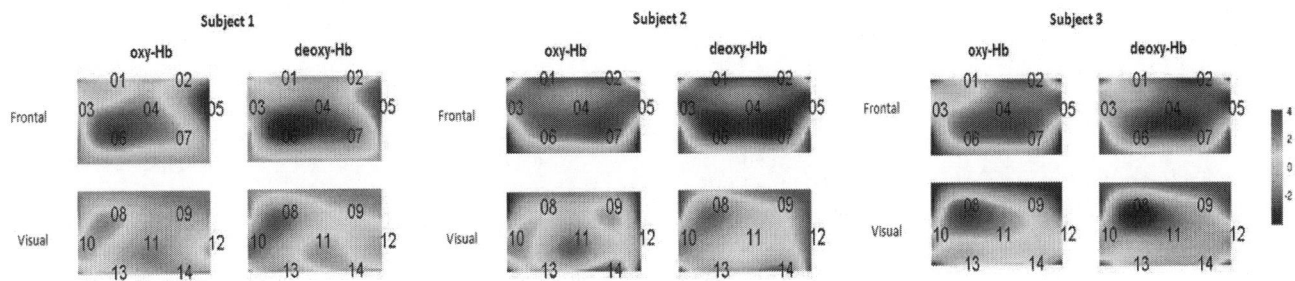

Fig. 8 t-maps of oxy-Hb and deoxy-Hb from three subjects indicating the level differences on 14 channels

(p < 0.001), whereas the deoxy-Hb showed the lower significance indication ($p < 0.081$). Also, only data from subject 3 showed significance in channels 11 and 12 with ($p < 0.009$ for oxy-Hb and deoxy-Hb in channel 11, $p < 0.036$ for oxy-Hb and $p < 0.023$ for deoxy-Hb in channel 12). The variations in the significant channels reflect the differences in brain activation according to different subjects. However, the channels of regions of interest (ROIs) can be selected as the significant regions in common between at least two subjects from either oxy-Hb or deoxy-Hb signals. Hence, the ROIs are chosen as channels 1, 2, 3, 4, 5, 6, 7, 8, 10, 13, 14.

IV. DISCUSSION AND CONCLUSION

This study investigated the hemodynamic responses of the prefrontal and visual cortices of three healthy subjects while doing a graded mental task that involved visual working memorizing. To enhance the quality of the desired fNIRS signals, CBSI method was applied to improve the NIRS signals based on the negative correlation characteristics between oxy-Hb and deoxy-Hb. The proposed method helps us to increase the spatial specificity of NIRS signal so that the localized responses could be clearly revealed.

The estimated marginal means of level 4 appeared to be greater than those of levels 2 for oxy-Hb signals and vice versa for deoxy-Hb. In addition, we found that the mean values of oxy-Hb and deoxy-Hb significantly depended on the interaction of subjects, channels and workload levels. The regions of interest ROIs selected could be useful for further investigation to localization and globalization components of hemodynamic response corresponding to workload levels. This investigation successfully confirmed that there were activation areas in prefrontal and visual cortices that had the responses to the visual stimuli following two levels of memorizing tasks. However, it appeared that for each subject, the activated channels were different. This study could not explain why there existed the variations in statistical results between different subjects. The reason would be explained by the physiological variations among individuals, which needs further investigations.

Overall, fNIRS was sensitive to hemodynamic changes induced by a mental workload task. With the benefits of non-invasive and real-time assessment, this system showed the ability to access the differentiation of various workload levels that involve memory tasks using fNIRS based on oxy-Hb and deoxy-Hb parameters. More importantly, the role of prefrontal and visual cortices in the processing of graded load-related working memory was proved. As a consequence, further investigations and discrimination between two visual workload levels should focus into the responses of these two brain regions.

REFERENCES

1. Shipstead Z, Lindsey D R B, Marshall R L et al (2014) The mechanism of working memory capacity: primary memory, secondary memory and attention control. Journal of Memory and Language 72:116–141.
2. Sato H, Yahata N, Funane T et al (2013) A NIRS – fMRI investigation of prefrontal cortex activity during a working memory task. NeuroImage 203:158–173.
3. Pu S, Yamata T, Yokoyama K et al (2012) Reduced prefrontal cortex activation during the working memory task associated with poor social functioning in late-onset depression: Multi-channel near-infrared spectroscopy study. Psychiatry research: Neuroimaging 203:222-228.
4. Baddeley A D and Hitch G J (1977) Commentary on "Working memory". G.H. Bower, New York.
5. Mammarella I C, Borella E, Pastore M, and Pazzaglia F (2013) The structure of visuospatial memory in adulthood. Learning and Individual Differences, 25:99–110.
6. Munneke J, Heslenfeld D J, and Theeuwes J et al (2010) Spatial working memory effects in early visual cortex. Brain and Cognition, 72:368–377.
7. Piazza M, Pinel P, Bihan D L, and Dehaene S (2007) A magnitude code common to numerosities and number symbols in human intraparietal cortex. Neuron, 53:293–305.
8. Richter M M, Zierhut K C, and Dresler T et al (2009) Changes in cortical blood oxygenation during arithmetical tasks measured by near-infrared spectroscopy. Journal of Neural Transmission, 116:267–273.
9. Hong J, Li X, and Xu F et al (2012) The mental workload judgment in visual cognition under multitask meter scheme. International Journal of the Physical Science, 7:787–796.
10. Durantin G, Gagnon J F, Tremblay S, and Dehais F (2014) Using near infrared spectroscopy and heart rate variability to detect mental overload. Behavioural Brain Research, 259:16–23.
11. Herff C, Heger D, and Fortmann O et al (2014) Mental workload during n-back task - quantified in the prefrontal cortex using fnirs. Frontiers in Human Neuroscience, 7:935–943.
12. Hirshfield L M, Chauncey K, and Gulotta R et al (2009) Combining electroencephalograph and functional near infrared spectroscopy to explore users' mental workload. 5th International Conference, FAC 2009 Held as Part of HCI International Proc. vol. 5638, San Diego, CA, 2009, pp 239–247.
13. Gevins A and Smith M E (2000) Neurophysiological measures of working memory and individual differences in cognitive ability and cognitive style. Cerebral Cortex, 10:829–839.
14. Riding R J, Glass A, and Douglas G (1993) Individual differences in thinking: cognitive and neurophysiological perspectives. Educational Psychology, 13:267–279.
15. Sassaroli A, Zheng F, and Hirshfield L M (2008) Discrimintaion of mental workload levels in human subjects with functional near-infrared spectroscopy. Journal of Innovative Optical Health Sciences, 1:227–237.
16. Cui X, Bray S, and Reiss A L (2010) Functional near infrared spectroscopy (nirs) signal improvement based on negative correlation between oxygenated and deoxygenated hemoglobin dynamics. NeuroImage, 49:3039–3046.

Corresponding Author: Nguyen Duc Thang
Institute: Department of Biomedical Engineering, International
        University
Street: Quarter 6, Linh Trung ward, Thu Duc district
City: Ho Chi Minh
Country: Vietnam
Email: ndthang@hcmiu.edu.vn

Differentiation of Hemodynamic Responses of the Brain with Typing and Writing

Vo Nhut Tuan[1], Nguyen Duc Thang[1], Vo Van Toi[1], Tran Le Giang[1],
Nguyen Huynh Minh Tam[2], and Dinh Dong Luong[3]

[1] Department of Biomedical Engineering, International University, Ho Chi Minh City, Viet Nam
[2] University of Saskatchewan, Canada
[3] Department of Computer Engineering, Kyung Hee University, South Korea

Abstract— Currently searchers are getting more interests in evaluating the brain functions on writing and typing activities. In order to discover the different effects of writing and typing on the brain, we establish experiments to test the oxygen consumption level in three areas of brain. We chose the Near-Infrared Spectroscopy (NIRS) to monitor hemoglobin changes. Subjects are required to type and write in the same environments while NIRS is used to record the hemoglobin dynamics. We used SPSS program as a statistical method. Image fusion was also applied to illustrate data from NIRS. The results turn out that when subjects are writing, oxygen concentration inside the brain increase more than when that inside the brain when they are typing. The experimental results show the brain has to work harder to control writing activity. By another word, writing excites brain more than typing. This fundamental information extends our knowledge of understanding on the brain behavior.

Keywords— NIRS, oxyHb, deoxyHb, typing, writing.

I. INTRODUCTION

There are clear evidences from current research have shown that writing can stimulate numerous cells of the human brain, especially reticular activating system (RAS). Stimulated RAS later sends signals back to cerebral cortex to help us to pay more attention so that we do not miss the details and see things better. Nowadays, with the popular use of computers and electronic devices, writing is gradually replaced by typing in document works. This leads to a question that whether the learning process of the human brain is affected by such changes in habits.

A research group at Paul Sabatier University has made an experiment to test the ability to remember the characters associate with the writing and typing. In this experiment, a group of adults were trained to write new characters and rewrite these words three weeks later. The result showed that handwriting has gained more advance to the brain than typing has. However, because the data is recorded after writing and typing process, this research did not cover the brain activities online [1].

Besides, a study showed that the left superior parietal lobe is responsible for writing. The functional magnetic resonance imaging (fMRI) was used to do the experiment with the parietal lobe to observe the cluster on the left superior parietal lobe and the dorsal aspects of the inferior cortex bordering the superior parietal lobe. The result indicated that the right hemisphere has no activation cluster and the writing seems to be primary organized in the hemisphere of the language-dominant. In details, the activation clusters are greater in the left superior parietal lobe than the left inferior parietal cortex [1].

Another study about the brain activity during writing was constructed with the fMRI by a research group in Japan [2]. Brain activity was recorded with twenty subjects in three conditions: written naming with the right hand, written naming with the left hand and naming silently. While the comparison of written naming with the right hand to naming silently exhibited activation only in the left frontal parietal area, the written naming with the left hand and naming silently comparison exhibited broader activation than the written naming with the right hand and naming silently comparison. In the group analysis, three areas seemed to be activated are the posterior end of left superior frontal gyrus, the anterior part of the left superior parietal lobe and the lower part of the anterior limb of the left supramarginal gyrus. In details, the interindividual discrepancy of involvement was recorded as follow: the lower part of the anterior limb of the left supramarginal gyrus involved 60%; both the right intraparietal and right frontal region involved 47% of each.

These studies have shown that writing affects the memory better than typing. However, none of them have analyzed the brain activities while the subjects doing these tasks [3-8]. Thus, in our work, we focus on the discovery of the brain process with regards to writing and typing activities, and especially on the evaluation of the hemoglobin levels in brain using near-infrared spectroscopy (NIRS). In our experiments, we record the oxygenated hemoglobin and deoxygenated hemoglobin level in brain's blood vessels. Based on the changes in hemoglobin levels, we can measure how much the brain is working. The goal of this study was to distinguish typing and writing activities on the same person in the same condition.

© Springer International Publishing Switzerland 2015
V. Van Toi and T.H. Lien Phuong (eds.), *5th International Conference on Biomedical Engineering in Vietnam*,
IFMBE Proceedings 46, DOI: 10.1007/978-3-319-11776-8_96

II. NEAR INFRARED SPECTROSCOPY

In a human brain, there are about 100 billion neurons which convert information to electrical signals. This conversion process consumes a lot of energy. During this process, hemoglobin is supplied oxygen and become oxygenated hemoglobin via the blood capillaries. While oxygen is taken from other oxygenated hemoglobin which now turns into deoxygenated hemoglobin (deoxyHb). NIRS is an imaging system which measures the relative levels of hemoglobin by using the near-infrared light with respect to changes in neural activity of the cerebral cortex, and displays these as trend graphs and images in real time. From the result, we can analyze the functional localization of the brain.

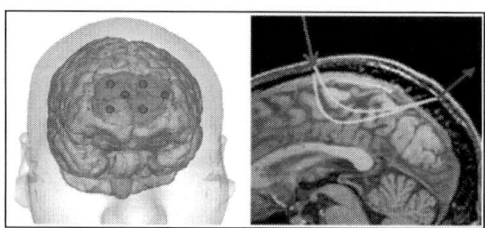

Fig. 1 Brain imaging using NIRS.

The emitted near-infrared light can penetrate into living organism harmlessly, and is absorbed by oxyHb and deoxyHb. However, the light could not penetrate easily into the human head. Therefore, the inside of the brain is irradiated from the surface of the head by near infrared light via optical fibers. In fact, after being absorbed and scattered at the cerebral cortex of the brain, the light is then condensed by optical fibers again at a distance of about 30 mm deep from the surface of the head (if subject is an adult). At this time, light reaches an area about 20 mm deep from the surface of the head, indicating that it is absorbed by hemoglobin at the cerebral cortex. Because living organisms significantly scatter light, near infrared light introduced by optical fibers is scattered by various types of tissue. Some scattered light get lost or absorbed by tissue while some of this scattered light reaches the optical fiber in the light receiving probe. Then these light are leaded to a photomultiplier where it is converted to electrical signals.

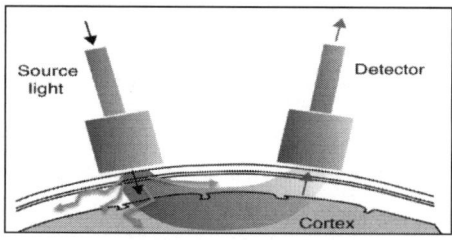

Fig. 2 Near-infrared pathway.

III. METHODOLOGY

A. Subjects

Five 20 year-old subjects participated in this study. All were right-handed, with normal history of major head injury.

The main goal of this work is to measure the NIRS signals represent writing and typing activities and to find the different appearances of the signals which correspond to these two activities. Due to the fact that typing and writing are related to the working of the cerebral cortex, the experiments are designed to measure whole three areas of the brain: the frontal cortex, the left hemisphere, and the right hemisphere. The cortex from the back (visual cortex) has visual function, thus this cortex is not considered in our experiments.

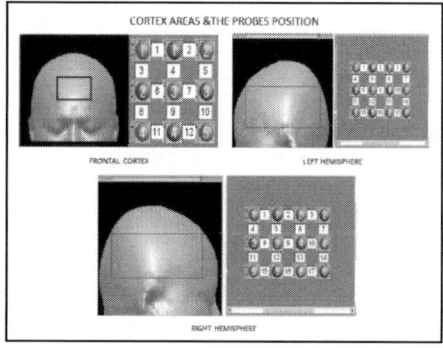

Fig. 3 Sets of probes on three areas of brain.

In NIRS program, the protocol was set: 10s – 60s – 10s. The first 10 seconds was the time for the subjects get ready to do the task. For the 60-second task (writing or typing), the subjects were asked not to do too fast, so that their brain could memorize as much as possible. And the last 10 seconds was the resting time, the subjects had to stop doing the task, and stay relax until the experiment ended. This protocol was repeated five times for each experiment.

B. Method

The raw collected data contained noise which affected significantly the final result. Hence it was band-pass filtered with the frequency range from 0.01Hz to 0.5Hz. Then the 5-set data of each subject was averaged for the next step of analysis.

The statistical method using SPSS program was then applied to analyze the mean data of both 5 subjects. After that, the univariate analysis of variance function was applied to test the differences between typing and writing. At first, we executed the test on total data (variance are sums of oxyHb and deoxyHb) to have a general result.

Then, we set up the test to oxyHb and deoxyHb levels individually. On every test, we compared the mean values in four groups: the frontal cortex, right hemisphere, left hemisphere and whole three areas. In addition of investigating the significant level, in each case, a graph was also sketched for visual illustration. The horizontal axis was the task (typing and writing) and the vertical axis was the oxyHb or deoxyHb or sum of oxyHb and deoxyHb based on the test. In addition, the significant levels show the distinction of typing and writing influences.

Finally, to strengthen the persuasion of the result, the fusion program was then applied to capture movies of the brain activities. The green color presented areas that had no change in oxyHb or deoxyHb level. If the oxygenated hemoglobin or hemoglobin levels dropped to the lowest value, the signal would turn into blue. And the red color presented the high consumption oxygenated areas.

IV. RESULT AND DISCUSSION

In the statistical method, firstly the mean values of the five subjects were taken, and then all the oxyHb and deoxyHb together in each area (frontal, right, left areas) were summed up. We applied the factorial analysis of variance (ANOVA) on Areas(3) x Activity(2) to measure the database, and the results suggest that typing and writing are different with significant level $p < 0.001$.

Then, in order to confirm that typing and writing influences are not the same, oxyHb data of the five subjects were tested. And the results turned out that the effects of typing and writing on brain was different with a significant level $p = 0.05$. This significant difference was slightly less than the total index, however it is still valuable. In short, these results are consistent with each other.

The next step was to check which area of brain is mainly responsible for the differences between typing and writing. In this step, data from the frontal and right cortex was analyzed. The ANOVA function of SPSS program was applied to set up many tests. In total hemoglobin, the results are very clear. The significant level in frontal area is $p < 0.001$ and in right hemisphere is $p = 0.025$. From these indexes, it can be inferred that typing and writing influence completely distinct in frontal cortex. Moreover, the right hemisphere also has a great change since the significant level is $p = 0.025$. When testing the oxyHb of these two areas, the results turned out that in the frontal cortex, significant level: $p < 0.06$ while in the right hemisphere: $p < 0.07$. Moreover, the data from the left hemisphere was also tested, the significant level is large hence the difference between writing and typing in the left hemisphere is not clear.

While total and oxyHb data showed very clear results, deoxyHb did not stand for changes between typing and writing. The SPSS results from deoxyHb were not significant enough, since $p > 0.1$. Moreover, when analyzing separately frontal, left and right areas, the significant levels were still large. Therefore, we were not able to make any conclusions from deoxyHb data.

Table 1. Statistical analysis with ANOVA

Data	Area	p-value
Oxy	Overall	0.050
Oxy	Frontal	0.059
Oxy	Left	0.300
Oxy	Right	0.071
Deoxy	Overall	0.352
Deoxy	Frontal	0.380
Deoxy	Left	0.773
Deoxy	Right	0.138
Total	Overall	<0.001
Total	Frontal	<0.001
Total	Left	0.235
Total	Right	0.025

The above analysis results suggest that there exist distinct influence of typing and writing on human brain. When subjects are writing, their oxyHb and deoxyHb are higher than when they are typing. Typically, the frontal cortex illustrates the most changes. It can be recommended that frontal cortex and right hemisphere take the main responsible for control activity of typing and writing. However, the difference between typing and writing is not clear in the left hemisphere.

When applying fusion program, firstly, the deoxyHb signals of the three areas were analyzed. In general, the deoxygenated signals of the two tasks are not significantly different from each other. If we observe carefully, it can be seen that the writing tasks all gave higher signals than the typing tasks. The figures below will illustrate the changes in deoxyHb in three areas of the brain.

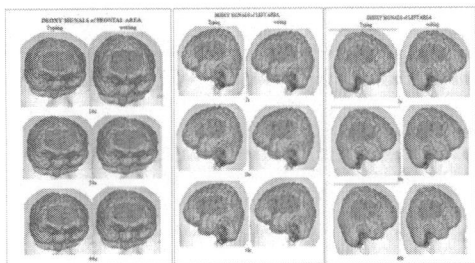

Fig. 4 Changes of deoxyHb signals in the three areas.

The next step was to analyze the oxyHb signals of the three areas. The oxygenated signals of the frontal cortex are clearly different between typing and writing tasks. In the writing tasks, the signals increase very strongly, while the typing tasks give very weak signals. Thus, the difference between writing and typing are so significant in the frontal cortex.

In contrast, the oxygenated signals in the left hemisphere are not clearly different. The signals increase very strongly and clearly in both tasks, however the difference is very difficult to distinguish. In last seconds, the signals decrease rapidly in both tasks, but the difference is still not easy to detect.

Finally, in the right hemisphere, the oxyHb signals are significantly different between typing and writing. The signals in the typing tasks decrease rapidly and continuously, while in the writing tasks signals are strongly high. Thus, the difference can be easily detected.

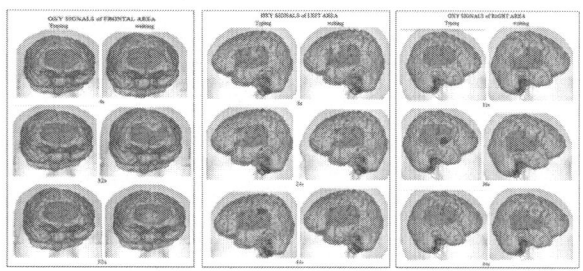

Fig. 5 Changes of oxyHb signals in the three areas.

V. CONCLUSIONS

Our study suggests that the effects on human brain activities of writing and typing are significantly different. The data from this study can also apply to BCI in the future. Although, NIRS probes are connected to the cerebral cortex, the infrared light cannot scatter deep into brain. So we do not conclude exactly what happen in deep area of brain. Therefore, a future research direction can focus on how to measure brain behavior at layer under cerebral cortex. Once the brain behaviors are discovered, people can find a way to increase the brain's effectiveness such as memorizing.

In this experiment, we have shown that functional near-infrared spectroscopy can distinguish the brain activations when a subject is writing or typing. We also demonstrated that we can differentiate typing task and writing task by using NIRS. In addition, our work support that writing helps memory more effective than typing does - the idea has been demonstrated in previous research. We believe this work to be a stepping stone to using NIRS in doing some other research like this, so that the subjects would feel more comfortable compare to using MRI, or other methods.

ACKNOWLEDGMENT

The authors would like to thank our lecturers from Biomedical Engineering department, MSc. Nguyen Huynh Minh Tam, and Dr. Nguyen Duc Thang for their guidance and encouragement. We also thank Mr. Thien and Ms. Trinh, students of Biomedical Engineering department, for their opinions and advices. Finally, we would like to thank Biomedical Engineering department IU for the NIRS equipment.

REFERENCES

1. Longcamp, M., et al. (2006). "Remembering the orientation of newly learned characters depends on the associated writing knowledge: A comparison between handwriting and typing." Human Movement Science **25**(4–5): 646-656.
2. Magrassi, L., et al. (2010). "Central and peripheral components of writing critically depend on a defined area of the dominant superior parietal gyrus." Brain Research **1346**(0): 145-154.
3. Menon, V. and Desmond J. E. (2001). "Left superior parietal cortex involvement in writing: integrating fMRI with lesion evidence." Cognitive Brain Research **12**(2): 337-340.
4. Sugihara, G., et al. (2006). "Interindividual uniformity and variety of the "Writing center": A functional MRI study." NeuroImage **32**(4): 1837-1849.
5. Hale, L. (2001). Fast typing no key to quality writing. Houston Chronicle. Houston, Tex.: 17-A.17.
6. Sugihara G., Kaminaga T., and Sugishita M. (2006), "Interindividual uniformity and variety of the "Writing center": A functional MRI study," *NeuroImage,* vol. 32, pp. 1837-1849.
7. "Writing not typing's best for remembering," in *The Post*, ed. Bristol (UK), 2013, p. 10.
8. Deardorff J. (2011) "Good penmanship has many perks; Texting, tapping and typing not enough for kids," in *Edmonton Journal*, ed. Edmonton, Alta.,.

Corresponding Author: Nguyen Duc Thang
Institute: Department of Biomedical Engineering, International University
Street: Quarter 6, Linh Trung ward, Thu Duc district
City: Ho Chi Minh
Country: Vietnam
Email: ndthang@hcmiu.edu.vn

The Relation between a Three-Day Sitting Meditation Fasting and the Participant's Psychophysiological Condition: Through Nonlinear Chaos Analysis of Pulse Waves

T. Futaba

Faculty of Intercultural Communication, Ryukoku University, Otsu, Japan

Abstract— In this research, the relationship between a three-day sitting meditation fasting called Zazen Fasting and a participant's psychophysiological conditions was examined. The subject was in a group of 50 adults in Matsumoto, Japan between 12th and 14th of January 2014. The fingertip pulse wave of the subject was measured for use as biological data before, during and after the therapy. The measured values were subjected to chaos analysis and the Lyapunov exponents calculated. The results showed that Lyapunov exponents became higher after the subject took the Zazen Fasting.

Keywords— Lyapunov exponent, fingertip pulse wave, psychophysiological condition, zazen fasting therapy, nonlinear chaos analysis.

I. INTRODUCTION

Recently, fingertip pulse waves have been discovered to be able to serve as a significant indicator of psychophysiological conditions [1], [2], [3], [4].

Dr. Koda in Osaka, Japan had created a new method of fasting and "successfully treated many patients with chronic fatigue syndrome, ulcerative colitis, collagen diseases, viral hepatitis, atopic dermatitis, and some neurodegenerative diseases" [5]. One of his patients "with cerebellospinal degeneration has successfully recovered by fasting. She has been living on a bowl of fresh green vegetable juice for more than 15 years. Her BMR is only 500 to 600 kcal/kg body weight. Her difference in calorie consumption from ordinary BMR would be accounted for by degradation and synthesis of protein. A reevaluation of fasting therapy would be useful both for dietary therapy and nutritional requirement"[6].

At the suggestion of Dr. Koda, a Zen monk, Master Noguchi, developed a three-day sitting meditation fasting called Zazen Fasting and held this Zazen Fasting in various places throughout Japan for the last 20 years. Several medical doctors followed this Zazen Fasting and "experienced good results to treat constipation and various other diseases" [7].

This study used fingertip pulse waves to measure psychophysiological conditions to gauge any difference in a subject's biological data before, during and after participating in this Zazen Fasting.

II. METHODS

A. Study Subject

The author became the subject of this study. The subject is a 56-year old male whose weight has been around 50 kilograms and whose height is 171 centimeters. Although he has been warned of being "too skinny" by annual medical checkup standards at his work, he has been healthy both mentally and physically to perform his work in the last two years.

B. Study Location and Study Period

The place of the experiment was at a typical Japanese hot spring inn called Alpine Asamaso in Matsumoto, Nagano Prefecture in Japan where only participants for the Zazen Fasting workshop stayed during the three-day period between 12th and 14th of January 2014.

C. Measurement

Fingertip pulse waves were measured by a photoplethysmography sensor (Mini MGL, Model MPULSE-01) at room temperature (about 20 degrees Celsius) and illuminated by artificial light (200 lx). Although the temperature outside during these three days dropped to minus 10 degrees Celsius one time, the room temperature had been kept around 20 degrees throughout.

D. Procedure

In this Zazen Fasting, participants gathered in a large conference room with tatami (straw) mats to do 20-minute Zazen sessions for 15 times during the three-day period. The pulse wave of the subject was recorded with the instrument, before the workshop, right after each Zazen session, and after the workshop.

E. Details of Three-Day Zazen Fasting

On day one (January 12th, 2014), each participant was advised not to eat anything for breakfast and come to the inn where they met at 6pm. They wore loose-fitting clothes to sit comfortably making a large circle facing each other.

V. Van Toi and T.H. Lien Phuong (eds.), *5th International Conference on Biomedical Engineering in Vietnam,*
IFMBE Proceedings 46, DOI: 10.1007/978-3-319-11776-8_97

Master Noguchi introduced the whole procedure including how to sit, breathe, and rest during the three days. After Master Noguchi introduced himself, each participant introduced him- or herself to everyone. At 7pm, the first 20 minute Zazen session began. During the 20 minutes, each participant sat still making a lotus flower with one's palms and holding it in front of one's body, breathed deeply with mouth closed but eyes open to count each breathing. After hearing a bell ringing to end the Zazen, they were able to relax for the next 40 minutes to prepare for the next Zazen by resting, talking to other participants, drinking tea, water or taking a pinch of natural salt. The second and third session of 20 minute Zazen started at 8pm and 9pm respectively. Sessions for the first day were over by 9:20pm. The participants took baths and went to bed.

On day 2 (January 13th), everyone woke up at 6:00am and gathered in the same room at 6:30am to start chanting a Buddhist sutra while sitting on their knees for 10 minutes. At 7am, 8am, and 9am, the same Zazen was done for the 4th, 5th and 6th time. At 9:20am, they had a longer break until they started the 7th Zazen session at 11am. After the 8th Zazen was over, 20-minute Buddhist sutra chanting started at 1pm. At 1:30pm, a glass of 150cc vegetable juice was provided to drink. At 3pm and 4pm, the same Zazen continued for the 9th and the 10th time. At 4:40pm, Master Noguchi lectured the participants about the relationship between body and mind. At 7pm and 8pm, the same Zazen was done for the 11th and 12th time. Sessions were over by 9:20pm. The participants took baths and went to bed.

On day 3 (January 14th), everyone woke up at 6:00am and gathered in the same room at 6:30am to start chanting Buddhist sutras while sitting on their knees for 10 minutes. At 6:40am, 8am and 9am, the same Zazen was done for the 13th, 14th and 15th time. After the 15th session was over, they moved to a dining room to start a meal ending the fasting. The purpose of this meal was to get rid of fecal impaction as much as possible. They were advised to do the following: 1. Drink a bowl of 300cc water, 2. Put two Umeboshi (picked plums) and 300cc warm soup of Daikon raddish in the same bowl to drink, 3. Eat vegetables such as Daikon, carrots with Miso, 4. Drink two more bowls of #2. And two more if one's body is heavier than average, 5. Wait until oneself senses the need to have a bowel movement and then go to the toilet, 6. Drink a glass of orange juice, 7. Go to the toilet, 8. Eat a cup of yoghurt, 9. Go to the toilet, 10. Eat some fruits, 11. Go to the toilet, 12. Eat some Japanese sweets, 13. Go to the toilet, 14. Eat a piece of cheese, a piece of bread, and drink a glass of milk tea.

III. RESULTS

Table 1 shows an analysis of variance of the Lyapunov exponents of the subject before the first Zazen started between November 24th of 2013 and 5:40pm of January 12th of 2014. The average of the exponents was 4.8.

Table 1 Lyapunov Exponents Before Zazen

Date and Time	Lyapunov Exponents
2013_11_24_12_42	4.0
2013_12_09_08_04	3.7
2013_12_22_18_40	5.1
2014_01_04_11_20	4.5
2014_01_12_01_59	4.0
2014_01_12_09_59	5.0
2014_01_12_12_00	6.9
2014_01_12_13_20	5.1
2014_01_12_17_40	5.0

Table 2 shows an analysis of variance of the Lyapunov exponents of the subject after the 15th Zazen and the meal between 7:40pm of January 12th of 2014 and 11:12am of the 14th. The average of the exponents was 5.7.

Table 2 Lyapunov Exponents During Zazen

Date and Time	Lyapunov Exponents
After the 1st Zazen	5.9
After the 2nd Zazen	6.5
After the 3rd Zazen	6.9
Morning of the 13th	6.9
After the Sutras chanting	7.6
After the 4th Zazen	5.3
After the 5th Zazen	5.7
After the 6th Zazen	4.5
After the 7th Zazen	6.1
After the 8th Zazen	5.2
After the Sutras chanting	4.7
After the 9th Zazen	6.2
After the 10th Zazen	5.7
After the 11th Zazen	4.6
After the 12th Zazen	8.3
Morning of the 14th	4.7
After the Sutras chanting	3.7
After the 13th Zazen	5.6
After the 14th Zazen	5.6
After the 15th Zazen	4.5
After the Meal	5.1

Table 3 shows an analysis of variance of the Lyapunov exponents of the subject after Zazen Fasting between January 14th and 27th of January, 2014. The average of the exponents was 4.5.

Table 3 Lyapunov Exponents After Zazen

Date and Time	Lyapunov Exponents
After arriving home	3.8
After dinner at home	4.4
Morning of the 17th	5.9
After breakfast on the 17th	3.9
After lunch on the 17th	3.8
After breakfast on the 19th	2.4
Before breakfast on the 20th	5.1
After dinner on the 20th	4.5
Before breakfast on the 21st	4.8
After Dinner on the 22nd	5
After breakfast on the 23rd	4.7
After breakfast on the 24th	4.9
After breakfast on the 27th	5.1

An analysis of the variance showed that Lyapunov exponents during the Zazen Fasting were higher than those before and after the Zazen Fasting. The results showed that Lyapunov exponents became higher after the subject took the Zazen Fasting.

IV. DISCUSSION

This study observed changes in largest Lyapunov exponents of finger plethysmogram. Several significant tendencies were found in the subject.

To the author's knowledge, this study is the first to use fingertip pulse waves to measure biological data before, during and after Zazen Fasting. The measured values were subjected to chaos analysis and the Lyapunov exponents calculated. It was found that there was a significant positive correlation between Zazen Fasting and the subject's psychophysiological conditions. The results showed that Lyapunov exponents became higher after the subject did the Zazen Fasting.

However, there still exist several limitations in this study. The changes in Lyapunov exponents of plethysmogram were not instrumental in the explanation. The underlying relationship between Zazen Fasting and the subject's feeling itself is unclear. What is more, to test with only one subject hardly sufficed as proof of positive results. To overcome these limitations is a subject for future research.

ACKNOWLEDGMENT

I would like to express my sincere appreciation to:

Dr. Oyama-Higa for providing me with her continuous support and encouragement in writing this paper and in conducting my research with her measurements;

Master Noguchi for allowing me to take my data during the Zazen Fasting and introducing me to Dr. Watanabe;

Dr. Watanabe for generously providing me with several issues of a medical journal to provide me with valuable information on fasting therapy in Japan.

CONFLICT OF INTEREST

The authors declare that they have no conflict of interest.

REFERENCES

1. Imanishi A, Oyama-Higa M (2006) The Relation between Observers' Psychophysiological Conditions and Human Errors During Monitoring Task. IEEE Conference on Systems, Man, and Cybernetics, Taipei, Taiwan, 2006, pp 2035-2039
2. Oyama-Higa M, Miao T (2006) Discovery and Application of New Index for Cognitive Psychology. IEEE Conference on Systems, Man, and Cybernetics, Taipei, Taiwan, 2006, pp 2040-2044
3. Oyama-Higa M, Miao T, and Mizuno-Matsumoto Y (2006) Analysis of Dementia in Aged Subjects Through Chaos Analysis of Fingertip Pulse Waves. IEEE Conference on Systems, Man, and Cybernetics, Taipei, Taiwan, 2006, pp 2863-2867
4. Hu Y, Wang W, Suzuki T, Oyama-Higa M (2011) Characteristic Extraction of Mental Disease Patients by Nonlinear Analysis of Plethysmograms. International Symposium on Computational Models for Life Sciences, AIP Conference Proceedings 1371, Toyama, Japan, 2011, pp 92-101
5. Watanabe S (2012) Japanese Longevity and Dietary Life (7). Clinical & Functional Nutriology. 4(3): 148-50.
6. Watanabe S (2012) How Much Should We Eat? Clinical & Functional Nutriology. 4(3): pp 128-31
7. Kubo A, Harada M, Watanabe S (2012) Tripartite Talk Zazen and Fasting Therapy for Health. Clinical & Functional Nutriology. 4(3): pp 148-50

Author: Terufumi Futaba
Institute: Ryukoku University
Street: 1-5 Yokotani, Oecho, Seta
City: Otsu 520-2194
Country: Japan
Email: futaba@world.ryukoku.ac.jp

Construction of Phantom Mimic Vessel for Study of Human Vessel Conditions in Deep Vein Thrombosis (DVT)

N. Ibrahim[1], W.N. Wan Zakaria[1], N. Aziz[1], and M.K. Abdullah[2]

[1] Faculty of Electrical and Electronic Engineering, Universiti Tun Hussein Onn Malaysia
[2] Faculty of Mechanical and Manufacture Engineering, Universiti Tun Hussein Onn Malaysia

Abstract— In this paper, the construction of phantom mimic vessel is presented to diagnose the Deep Vein Thrombosis (DVT). The diagnosis of DVT is commonly used by monitoring the blood velocity and present of thrombus in vessel from B-mode ultrasound image associated with the application of Doppler ultrasound. Since it is difficult to recognize the vessel condition at the early stage of DVT, this study is to assess the vessel behavior at the early stage of DVT using phantom experiment. The phantom size and elasticity is considered to be the important parameters in constructing the phantom mimic vessel. Rapid prototyping technology and solid works are used in constructing the phantom.

Keywords— Ultrasound scan, rapid prototyping technology, solid works, Deep Vein Thrombosis (DVT).

I. INTRODUCTION

Deep vein thrombosis (DVT) is a condition where a blood clot forms in deep vein. DVT usually occurs in deep leg vein that run through the muscle at the cuff and the thigh. DVT is a life threatening condition which development of pulmonary embolism (PE) gives the primary concern to society. Patients with symptomatic DVT can present with pain, swelling, and tenderness along the distribution of deep leg vein. Since the clinical diagnosis is insensitive and nonspecific, the large number of undiagnosed DVT cases provides substantial support for extensive patient screening. Although many diagnostic tools are available to evaluate the presence of DVT, the use of ultrasound for routine DVT evaluation is superior in accuracy, cost and feasibility [1]. Using ultrasound as a diagnostic tool given more advantages because it is non-invasive, accurate and fastest method of evaluating DVT at the lowest cost.

The diagnosis of Deep Vein Thrombosis (DVT) is commonly used by monitoring the blood velocity and present of thrombus in vessel from B-mode ultrasound image associated with the application of Doppler ultrasound. However, it is difficult to recognize the vessel condition at the early stage of DVT. Thus, this study intends to assess the vessel behavior at the early stage of DVT using phantom experiment. The phantom size and elasticity is the important parameters in constructing the phantom mimic vessel. To that extent, this study is to validate the *in vivo* experiment for diagnosis of early stage DVT.

II. METHOD

A. Designing Phantom Using Rapid Prototyping Technology

In considering the phantom vessel, the characteristics that need to be considered are vessel size (diameter), elasticity of vessel wall and blood pressure in the vessel. The goal of this experiment is to construct a phantom vessel using rapid prototyping technology. Rapid prototyping also known as 3D printing is the automated development of real prototypes utilizing freeform fabrication technological know-how. The process is called additive as successive layers of material are laid down in x and y directions forming different shapes build in z direction [2]. An additive manufacturing method designed for rapid prototyping uses virtual patterns from computer animation drawing tools, transforms them into slender, digital, horizontal cross sections and then also produces a series of tiers until finally the prototype is completed. In this study, the Computer Aided Design (CAD) program is employed associates with the Solid Works Software to construct the phantom.

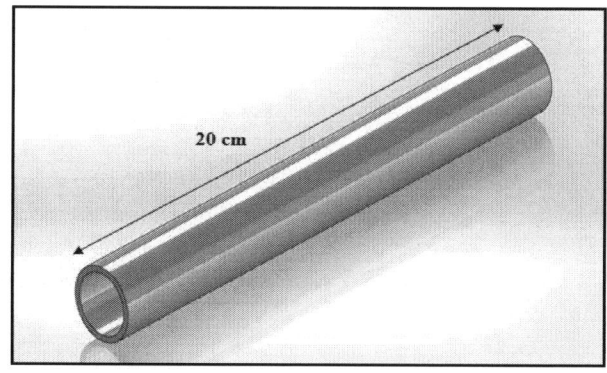

Fig. 1 Rapid prototyping vessel phantom design with dimension.

Figure 1 shows the 3D vessel phantom designed using the CAD virtual pattern and the Solid Works. The length of the phantom is constructed to be 20 cm which is longer than ultrasound transducer length. Therefore, the range of the scanning will cover the region of interest.

V. Van Toi and T.H. Lien Phuong (eds.), *5th International Conference on Biomedical Engineering in Vietnam*,
IFMBE Proceedings 46, DOI: 10.1007/978-3-319-11776-8_98

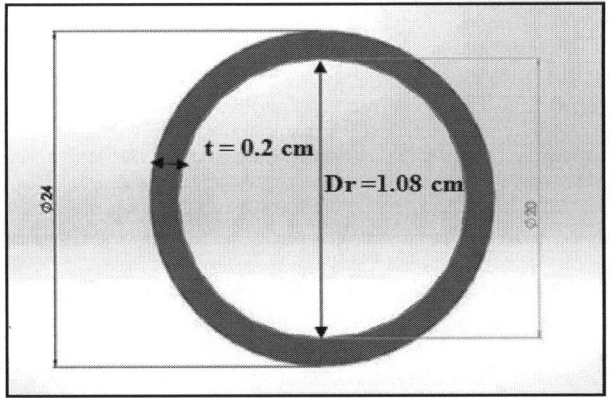

Fig. 2 Cross sectional view of phantom vessel design.

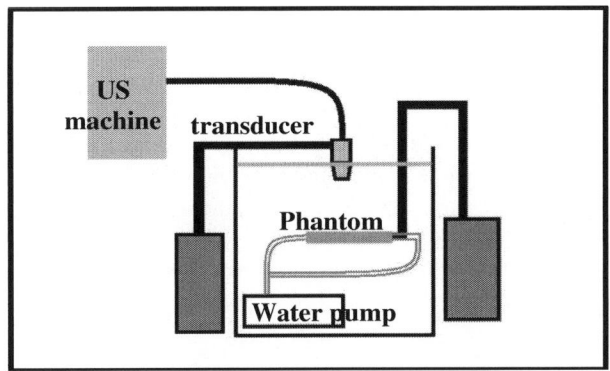

Fig. 3 The placement of phantom vessel and experimental setup.

Figure 2 shows the cross sectional of the phantom vessel. *Dr* indicates the diameter size of lumen and *t* indicates the thickness of the vein wall. The scale use in the design is 1:4 because the minimum thickness that can be used for this method is only 0.2 cm to prevent the complete design tear off. The average lumen diameter for vein is ranging from 2.7 mm to 13 mm [3] and the thickness of vein wall is between 0.5 mm and 0.9 mm [4]. The lumen diameter and wall thickness determined for this design are 1.08 cm and 0.2 cm.

The main advantage of using rapid prototyping (RP) technology in the fabrication of medical devices is easy and fast to create a complex object with convoluted feature CAD and RP system. For rapid prototyping of vessel structures with mechanical properties in the range of soft tissue, a suitable material with almost same elasticity parameters is required. The material type needs to be matched with the mechanical properties of a vein. Moreover, the stable, homogenous designs of an arbitrary, hollow and cylindrical shape are important parts in constructing the phantom. Silicon rubber compounds are used due to mechanical properties. In this project rubber material is chosen since the tensile strength is similar with the veins one.

B. Phantom Experiment

The phantom experiment setup is shown in figure 3.The phantom vessel is located inside a water tank connected with a water pump. The ultrasound transducer is used to scan the phantom and B-mode images are processed from ultrasound machine.

The phantom experiment is set up in the water because it has higher acoustic impedance mismatch. If there a large acoustic mismatch then a large proportion of energy is reflected. This will result in a strong echo, which produces a bright image on the display. Acoustic impedance mismatch is the

different of acoustic impedance between two substances. Acoustic impedance (Z) of water is 1.48 kg /m^2/s × 10^6. Eq. 1 is to calculate the intensity reflection coefficient, *R* for an interface between two substances.

$$R = \frac{I_r}{I_i} = \frac{(Z_1 - Z_2)^2}{(Z_1 + Z_2)^2} , \qquad (1)$$

where,
  I_r = Intensity of reflected echo.
  I_i = Intensity of incident beam.
  Z_1= Acoustic impedance of first medium.
  Z_2 = Acoustic impedance of second medium.

III. RESULT

In this experiment, the material type needs to match with the mechanical properties of a vein and enable the stable, homogenous design of an arbitrary, hollow and cylindrical shape. Silicon rubber compounds are used due to mechanical properties. In this project rubber-like material is chosen as the mechanical properties of the material as the structure is similar with the vein properties.

The tensile strength of vein (thigh and leg location) is 5.38 ± 0.6 MPa [5]. The rubber-like phantom which has the tensile strength of 5.8 MPa was used. In this project, an indirect fabrication method, molding process is proposed according to the desired material use to build vessel. Molding process consists of vessel shell mold and vessel arbor mold which are negative patterns of outer surface and inner surface of vessel. From the experiment, the designated phantom was being scan by the ultrasound machine.

IV. CONCLUSION

This study had described the construction of phantom mimics vessel for the evaluation of early diagnosis DVT. Using the Rapid Prototyping technology and Solid Works Software, the rubber-like phantom with tensile strength of 5.8 MPa was used which is similar to the vein leg tensile strength 5.38 MPa. In the future, the constructed phantom will be used to evaluate the vessels behavior at the early stage of DVT.

ACKNOWLEDGMENT

The authors like to thank Research Acculturation Grant Scheme (RAGS grant no: R031) for the sponsorship doing this project. They also like to thank Universiti Tun Hussein Onn Malaysia (UTHM) for having the space doing the experiment.

REFERENCES

1. Hirsh J, Hoak J. Management of deep vein thrombosis and pulmonary embolism: a statement for healthcare professionals from the Council on Thrombosis (in Consultation with the Council on Cardiovascular Radiology) Circulation. 1996;93(12):2212–2245.
2. Stephen P. Tseng and Hwa-Hsing Tang, "A Uniform Laser Energy Control for Ceramic Sintering Rapid Prototyping," in SICE Annual Conference 2010, Taiwan, August 18-21,2010.
3. Dalibor Musila, Jiri Hermanb, Julius Mazuchc, "Width of the great saphenous vein lumen in the groin and occurence of significant refluxin the sapheno-femoral junction," Biomed Pap Med Fac Univ Palacky Olomouc Czech Repub. 2008, 152(2):267–270.
4. V. Rengarajul, A.F.F da Silva, I. Sack, Ch. Kargel, "A basic study of Ultrasonic shear Wave Elastography in tissue mimicking phantom," in International Workshop on Medical Measurement and Applications, Italy, May 29-30, 2000.
5. Borhan Alhosseini Hamedani, Mahdi Navidbakash and Hossein Ahmadi Tafti, " Comparison between mechanical properties of human saphenous vein and umbilical vein," Biomedical Engineering, August 2013.

Medical Image Contour Based Context-Aware in Contourlet Domain

Nguyen Thanh Binh

Faculty of Computer Science and Engineering, Ho Chi Minh City University of Technology, Vietnam
ntbinh@cse.hcmut.edu.vn

Abstract— **Active contours are used extensively in image processing applications, including edge detection, shape modeling, medical image-analysis and detection of object boundaries. Medical images are of poor contrast, therefore boundaries of object of interest are poorly represented. This paper describes a method to represent context awareness in case of object of interest in medical images. Our proposed method uses contourlet filter bank to represent contour of object of interest present in medical images. For demonstrating the superiority of the proposed method, we have compared the results with the other recent methods such as using the simple Discrete Wavelet Transform.**

Keywords— **medical image, context-awareness, contourlet filter bank.**

I. INTRODUCTION

Medical images are generally of poor contrast and get complex types of noise due to the use of various acquisitions, transmission, storage and display devices. So it is very difficult to suggest a robust method for edge detection which works equally well for different modalities of medical images.

During the last decade, several new methods have emerged for object detection from images. There are many different methods proposed in the literature such as the Fourier Descriptors [7, 8], the moments [9], the B-Spline shape representation [11, 12, 13], the autoregressive models [7], the Hough Transform, the Fractal geometry methods [11], and the Wavelet Transform Zero-Crossing Representation [15]. Deformable templates from the introduction are limited form active contour models [3, 4, 5, 6, 7, 8, 9, 10]. The variability of the shape is limited by the prototype template parameterization and the way that we generate deformed templates. We will take a closer look at both free form and limited form of active contours.

L.D Cohen [14] proposed an external force model that significantly increases the captured range of a traditional snake. These external forces are the negative gradient of a potential function that is computed using a Euclidean (or chamfer) distance map. Therefore, distance potential forces do not solve the problem of convergence to boundary concavities.

The proposed work takes advantages of the contourlet transform. We have proposed to implement an active contour model to represent context awareness in objects of medical images. We proposed to implement it in two parts: contourlet filter bank and context-aware closed contour with boundary information. The goal of the first part is to detect the dominant edge points so that the resulting image will be composed of textures separated by edges. After that, we detect only dominant edges available in the image. We use contourlet filter bank that can act as local edge detectors. The second part is to extract a closed contour C, which covers the object.

To show the superiority of the proposed method, we have compared the results with the other recent methods such as the methods using simple Discrete Wavelet Transform. The rest of the paper is organized as follows: in section 2, we described the basic concepts of contourlet transform. Details of the proposed algorithm have been given in section 3. In section 4, the results of the proposed method for contour have been shown and compared with other methods and finally in section 5, we presented our conclusions.

II. RINCIPLES OF CONTOURLET TRANSFORM

The contourlet construction is based on the work of Do and Vetterli [1]. Contourlets constitute a new family of frames that are designed to represent smooth contours in different directions of an image. Contourlet allows for a different number of directions at each scale and aspect ratios. This feature allows an efficient contourlet-based approximation of a smooth contour at multiple resolutions. The discrete contourlet transform is a multiscale and directional decomposition using a combination of Laplacian pyramid (LP) and directional filter bank (DFB) [1].

The idea of the contourlet construction [1] is: let $a_0[n]$ be the input image, the output after the LP step is I bandpass images $b_i[n]$, i =1, 2,..., I and a lowpass image $a_I[n]$. Each bandpass image $b_i[n]$ is decomposed by an ℓ_i-level DFB into 2^{ℓ_i} bandpass directional images $c_{i,k}^{(\ell_i)}[n]$, for k = 0, 1,..., 2^{ℓ_i}-1.

III. THE PROPOSED METHOD

In this section, we will describe our proposed method. The proposed method is a two parts algorithm: contourlet coefficients computation and extraction of context-awared

© Springer International Publishing Switzerland 2015
V. Van Toi and T.H. Lien Phuong (eds.), *5th International Conference on Biomedical Engineering in Vietnam*,
IFMBE Proceedings 46, DOI: 10.1007/978-3-319-11776-8_99

closed contour boundary information. The goal of the first part is to detect the dominant edge points so that the resulting image, composed of textures be separated by edges. Now, we detect only dominant edges available in the image. We use contourlet filter bank that can act as local edge detectors. The second part is to extract a closed contour C, which covers the object. Steps of the proposed method are as follows:

Step 1: Sample Collection- First step of the proposed method is to collect sample images from hospitals at HoChiMinh city. In our approach, we have taken dataset images for testing. This data set has 1000 images.

Step 2: Preprocessing of images - The collected images are scale normalized to 256 x 256 pixel dimensions in order to reduce complexity.

Step 3: Contourlet filter bank - For contourlet filter bank computation in the proposed method, contourlet decomposition proceeds through two main steps [1]: Laplacian pyramid multiscale, directional filter bank to filter strong edges in image.

Step 4: Context- aware closed contour with boundary information - Here we extracted a closed contour C, which covers the object.

A. Contourlet Filter Bank

Contourlet decomposition proceeds through two *main* steps [1]: first, Laplacian pyramid multiscale decomposition is performed; then directional filter bank decomposition is used to link point discontinuity to linear structures. In more detail, an image is decomposed into a low pass image and bandpass images by the LP decomposition. Each bandpass output is further decomposed by the DFB step. The output of the DFB step consists of smooth contours and directional edges. Here, each directional subband at each level consists of 2^n elements, where n is a positive integer. Fig 1 shows a contourlet decomposition.

In this case, the threshold for the contourlet coefficients can be calculated using statistical properties of noise or blur. After thresholding the contourlet coefficients, the image can be reconstructed. Donoho [16] have described a threshold that depends on standard deviation of contourlet coefficients. Here, to compute the threshold value for image edge detection, we use a combination of three parameters: the contrast ratio (ratio between standard deviation and mean of contourlet coefficients), the absolute median of contourlet coefficients, and a level dependent parameter.

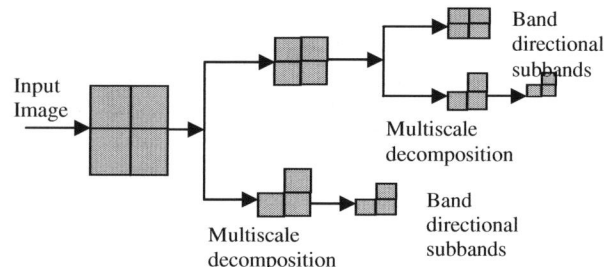

Fig. 1 Contourlet decomposition

B. Context-Aware Closed Contour with Boundary Information

The term 'context-aware', was first introduced by Schilit and Theimer [15], refer to context as location, identities of nearby people and objects, and changes to those objects.

Most of previous definitions of context are available in literature [18] that context-aware look at who's, where's, when's and what's of entities and use this information to determine why the situation is occurring. Here, our definition of context is:

"Context is any information that can be used to characterize the situation of an image such as: pixel, noise, strong edge, weak edge in medical image that is considered relevant to the interaction between pixels and pixels, including noise, weak and strong edge themselves."

In image processing [19], if a piece of information can be used to characterize the situation of a participant in an interaction, then that information is context. Contextual information can be stored in feature maps on themselves. Contextual information is collected over a large part of the image. These maps can encode high-level semantic features or low-level image features. The low-level features are image gradients, texture descriptors, shape descriptors information [17].

In image processing, contextual information can be stored in feature maps on themselves. Contextual information is collected over a large part of the image. These maps can encode high-level semantic features or low-level image features. The low-level features are image gradients, texture descriptors, shape descriptor information. A contextual feature vector can be extracted for each object in a training set [2].

Here we extracted a closed contour C, which covers the object. Firstly, we construct an edge map E. The E consists of a set of line segments that detected edges and followed by a line fitting step.

Secondly, to find an optimal closed contour C*, we identify a subset of detected segments in E and connect them together. We construct additional line segments that fill the gaps between detected segments to form closed contours (gap-filling segments).

Third, we construct a gap-filling segment between each possible pair of the endpoints of the different detected segments and using optimal closed contour C* to join contour together.

The optimal closed contour C* is defined as follows:

$$C^* = \arg\min \frac{|C_{tl}|}{\sum_{k=0}^{n} S_k(u)} \quad with \quad u \in C$$

where $|C_{tl}|$ is the total length of gaps along the contour C and $\sum_{k=0}^{n} S_k(u)$ is the total value of pixels located inside C, to join this contour together.

IV. EXPERIMENTS AND RESULTS

To demonstrate the validity of the proposed approach, the proposed method is applied on many images. We used two approaches for our experiments: the simple Discrete Wavelet Transform (DWT), and the proposed method based on context-awareness in contourlet domain. The other methods were implemented using our own program to compare the results on the same images and on similar scale. The proposed method was tested using different cases.

To test our algorithm, many images with the different images size were taken. We compared our proposed method with other methods on two cases: strong objects and weak objects. The strong objects are objects that have clear boundaries. In contrast, the weak objects are objects that have blurred boundaries. The method was tested on several images. Here, we report the results of some test images.

Fig.2 compares the proposed algorithm on the blurred medical image that comprises the object with weak edge. It is clear that the performance of the proposed method is better than DWT method in the weak objects cases.

Fig.3 compares the proposed algorithm on the medical image that comprises the object with strong edge. It is also clear that the performance of the proposed method is better than DWT method in the strong objects cases.

Other experiments also show that the proposed method works well and better than the other ones. In addition, clinicians at Hospital in Ho Chi Minh City, have evaluated

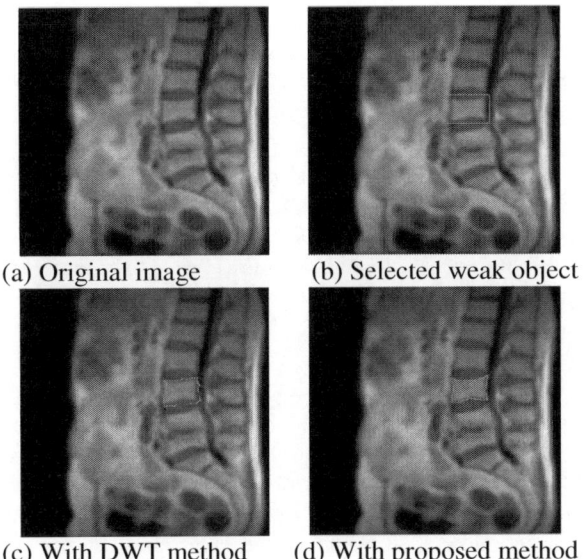

(a) Original image   (b) Selected weak object

(c) With DWT method   (d) With proposed method

Fig. 2 Performance of the proposed method with blurred medical image, compared to the other method with the weak objects.

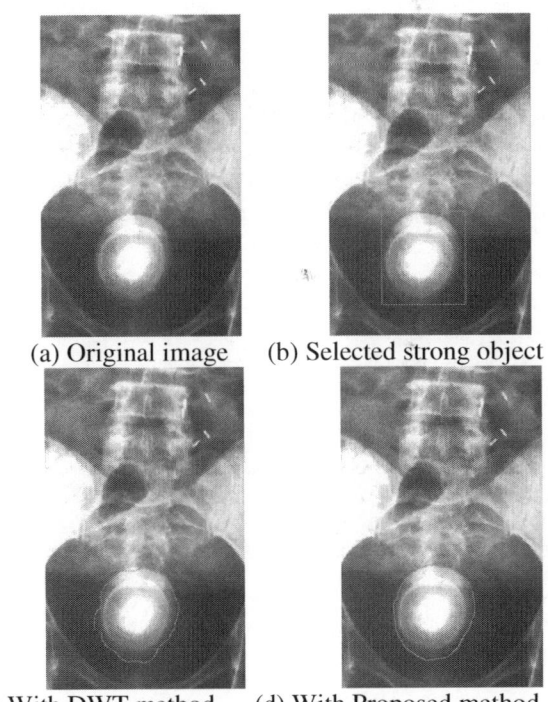

(a) Original image   (b) Selected strong object

(c) With DWT method   (d) With Proposed method

Fig. 3 Performance of the proposed method with medical image, compared to the other method with the strong objects.

these results. In their opinion, the proposed method gets more edges and information from images.

To sum up, from all the above experiments and many the other experiments, we observe that the results of the performance of the proposed method were better than other methods in the both cases: weak objects and strong objects.

As mentioned in the above, the context-aware property is one of the reasons for which our method perform better than DWT. Contourlet filter keeps the shape of the signal and carry strong edge information. The context-aware proposed method extract a closed contour. The position of point define contour is asymptote the border object than the other method.

V. CONCLUSIONS

In this paper, the contour based context awareness of medical images in contourlet domain is proposed. The proposed technique allows to estimate the contour location of a target object along an image. The contribution in the using context-awareness in form of contour coefficients for object identification was discussed. From the results of the above section, we conclude that the proposed method performs better than the other methods in both strong and weak object cases. The proposed method can be applied on any modality of images.

However, in the weak object case, the proposed method can find approximate boundaries. So, if the quality of the image is very bad such as heavy noise, heavy blur then the estimation ability is reduced. To avoid this problem, we are trying to reduce noise, blur before applying the proposed method, which is based on the characteristics of human vision.

REFERENCES

1. M. N. Do and M. Vetterli, The Contourlet Transform: An Efficient Directional Multiresolution Image Representation, IEEE Transactions on Image Processing, vol. 14, pp. 2091–2106, 2005
2. Huaizu Jiang, Jingdong Wang, Zejian Yuan, Tie Liu, Nanning Zheng, Shipeng Li. Automatic Salient Object Segmentation Based on Context and Shape Prior. British Machine Vision Conference, pp 1-12, 2011
3. T.F. Cootes, C.J. Taylor, D.H. Cooper, J. Graham Active Shape Models-Their Training and Application. Computer Vision and Image Understanding, Vol 61, pp 38–59, 1995
4. A. Blake and M. Isard, Active Contours, Springer, 1998
5. J.H. Chuang, Potential-Based Approach for Shape Matching and Recognition, Pattern Recognition, vol 29, pp 463–470, 1996.
6. A. Chakraborty et.al, Deformable boundary finding influenced by region homogeneity, Available at the web site of the Yale University, March 1994
7. K. Arbter, W. E. Snyder, H. Burkhardt, and G.Hirzinger, Application of affine-invariant Fourier descriptors to recognition of 3-D objects, IEEE Trans. PAMI, vol. 12, pp. 640-647, 1990.
8. H. Kauppinen, T. Seppanen, and M. Pietikainen, An experimental comparison of autoregressive and Fourier-based descriptors in 2D shape classification, IEEE Trans. PAMI, vol. 17, pp 201-206, 1995.
9. S. O. Belkasim, M. Shridhar, and Ahmadi, Pattern recognition with moment invariants: a comparative study and new results, Pattern Recognition, vol. 24, pp 1117-1138, 1991.
10. Z. Huang and F. S. Cohen, Affine-invariant Bspline moments for curve matching, IEEE Trans. Image Processing, vol. 5, pp. 1473-1480, 1996.
11. Y. H. Gu, and T. Tjahjadi, Efficient planar object tracking and parameter estimation using compactly represented cubic B-spline curves, IEEE Trans. SMC, vol. 29, pp. 358-367, 1999.
12. M. D. Swanson, and A. H. Tewfik, Affineinvariant multiresolution image retrieval using B-splines, Proc. of ICIP, pp 831 –834, 1997.
13. J. Y. Wang, and F. S. Cohen, Part II: 3D object recognition and shape estimation from image contours using B-splines, shape invariant matching, and neural network, IEEE Trans. PAMI, vol. 16, pp. 13-23, 1993.
14. L. D. Cohen and I. Cohen, Finite-element methods for active contour models and balloons for 2-D and 3-D images, IEEE Trans. Pattern Anal. Machine Intell., vol. 15, pp. 1131–1147, 1993.
15. Schilit, B., Theimer, M. Disseminating Active Map Information to Mobile Hosts. IEEE Network, vol 8, pp 22-32, 1994
16. D. L. Donoho and I. M. Johnstone, Ideal spatial adaptation by wavelet shrinkage, Biometrika, vol. 8, pp. 425–455, 1994
17. Huaizu Jiang, Jingdong Wang, Zejian Yuan, Tie Liu, Nanning Zheng, Shipeng Li. Automatic Salient Object Segmentation Based on Context and Shape Prior. British Machine Vision Conference, pp 1-12, 2011
18. Gregory D. Abowd, Anind K. Dey, Peter J. Brown, Nigel Davies, Mark Smith, Pete Steggles, Towards a Better Understanding of Context and Context-Awareness, Lecture Notes in Computer Science, Volume 1707, pp 304-307, 1999
19. N.T.Binh, V.T.H.Tuyet, P.C.Vinh, Ultrasound Images denoising based Context Awareness in Bandelet domain, 2nd International Conference on Context-Aware Systems and Applications, November 25–26, 2013 Phu Quoc, Vietnam.
20. N.T.Binh, L.N.Minh, Adaptive medical image edge detection in contourlet domain, Fourth International Conference on the Development of Biomedical Engineering in Vietnam, 2013 pp 97-100, Springer.

Author: Nguyen Thanh Binh
Institute: Faculty of Computer Science and Engineering
HoChiMinh City University of Technology, Vietnam.
Street: 268 Ly Thuong Kiet, Dist 10
City: Ho Chi Minh city
Country: VietNam
Email: ntbinh@cse.hcmut.edu.v

Nonrigid Point Set Registration-Based 3-D Human Pose Tracking from Depth Data

Dong-Luong Dinh[1], Nguyen Duc Thang[2], Sungyoung Lee[1], and Tae-Seong Kim[3]

[1] Department of Computer Engineering, Kyung Hee University, South Korea
[2] Department of Biomedical Engineering, International University, Vietnam
[3] Department of Biomedical Engineering, Kyung Hee University, South Korea

Abstract— **In this paper, we present a novel approach of recovering a 3-D human pose from a single human body depth silhouette using nonrigid point set registration. In our methodology, a human body depth silhouette is presented as a 3-D points set that is matched to the next 3-D points set through point correspondences between them. To recognize and maintain the body part labels, we first initialize the initial points set and their corresponding body parts, then transform them to the next points set according the point correspondences via nonrigid point set registration. Upon the point registration, we use the information of the transformed body labels of the registered pose to create a human skeleton model. Finally, a 3-D human pose is recovered by mapping the skeleton's position and orientation information to a 3-D synthetic human model. Our quantitative and qualitative evaluation on synthetic and real data show that complex poses could be tracked and recovered reliably.**

Keyword— **Human pose estimation, coherent point drift, depth image, point set registration.**

I. INTRODUCTION

Recently, 3-D human pose recovery from depth silhouettes has become an active research topic in computer vision, especially for complex human poses. This research work is triggered with an introduction of depth imaging devices which provide pixel-by-pixel distance images. Furthermore, from a sequence of depth images, a series 3-D poses, representing motion could be tracked and recovered. This research challenge is driven by many potential applications such as entertainment game, surveillance, sport science, health care technology, human computer interactions, motion tracking, and human activity recognition [1].

Many studies of this human pose recovery from depth silhouette have appeared in recent years [2]. To recover a 3-D human body pose from depth data, the techniques could be categorized into three, namely the graph-based, labeled body parts-based, or point set registration-based.

In the category of the graph-based, in [3], [4], and [5], to recover a 3-D human pose, they represented the depth data in a graph-based representation and then estimated geodesic distances of the graph to find the positions and orientations of primary human body parts such as head, hands, and feet. The computation cost of this technique is effective. However,

these methods revealed some limitations. The number of detected body parts based on primary landmarks is limited and the detected parts do not identified left or right body parts. In addition, the graph topology is sensitive to occlusions of body parts in where geodesic distance could not find a continuous path since 3-D data is disconnected or interrupted, therefore, the results of detected body parts are unstable.

In the category of body parts labeling-based, in [6], [7], and [8], they proposed the effective method to human pose recognition in body parts from a single depth silhouette inferred from a per-pixel classification via some randomized decision trees. This approach allows efficient recognition of human body parts. It could recognize up to 31 body parts from a single human depth silhouette. However, these studies required a large database for training. The training database has to be created from prerecorded motion data for automatic pixel labeling. For this reason, misrecognitions will occur if the database used for training is not properly and adequate. In addition, in some complex human poses, which contain hands or legs crossing body parts, had a low recognition accuracy of these body parts.

In the third category, the point set registration is to find point correspondences between two different point sets of rigid or nonrigid objects. For registration of 3-D shape objects, many algorithms have been proposed in [9]. Iterative Closest Point (ICP) is one of the well-known fitting or registration algorithms between two sets and it has been widely used for several applications such as 3-D model fitting, shape registration, and human motion tracking [9]. For instance, in [10] and [11], they utilized ICP algorithm to fit the 3-D human body model on the 3-D articulation data in a hierarchical manner. However, the main drawback of ICP requires that the initial position of two given point sets is adequately close. Therefore, this method may return the local optima in some complex poses of a nonrigid object like the human body.

As the approaches mentioned above, to improve robust 3-D human pose recovery from depth silhouette including complex poses, we propose a new methodology to recover 3-D human pose from depth data by tracking human body parts using nonrigid point set registration as presented in Fig. 1. Coherence Point Drift (CPD) [12-14] is used for nonrigid point set registration. This technique allows recovered human poses to

© Springer International Publishing Switzerland 2015
V. Van Toi and T.H. Lien Phuong (eds.), *5th International Conference on Biomedical Engineering in Vietnam*,
IFMBE Proceedings 46, DOI: 10.1007/978-3-319-11776-8_100

maintain their structure by preserving the motion coherence constraint during the matching process of this algorithm. In our approach, to find joint points of body parts for recovering 3-D human pose, we first initialize the initial points set and their corresponding body parts, then transform them to the next points set according the point correspondences via nonrigid point set registration. Based on the point registration, we use the body labels information of the registered pose to find the joint points and create a human skeleton model. Finally, a 3-D human pose is recovered by mapping the skeleton's joints positions and orientations to a 3-D synthetic body model.

The paper is structured as follows. In section II, we describe our proposed 3-D human pose recovery methodology. Section III presents experimental settings and obtained experimental results on both synthetic and real data. Conclusion remarks are given in sections IV.

II. THE PROPOSED 3-D HUMAN POSE RECOVERY METHOD

Fig. 1 describes the step by step of our processing framework. At initial step, a synthetic human depth map and corresponding body parts labeled map (f_o) are used to initialize the system. Given a human depth silhouette ($f_{i=1..n}$), which is extracted by removing the background, is uniformly down sampled and presented as a 3-D points cloud (f_i). The points set of this silhouette is then aligned with the points set (f_{i-1}) of previous pose to get point correspondences between them. By using the point correspondences and the body parts labeled map of the previous pose, human body parts of the given human depth silhouette are recognized. Joint position proposal is then applied on the recognized body parts. From determined joints, the known orientations and positions of the human body parts are applied on a 3-D synthetic human model [8].

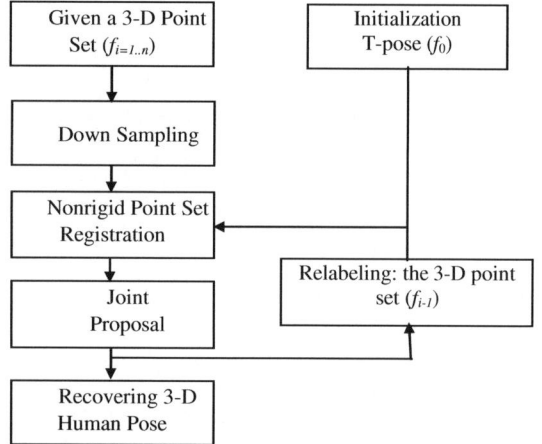

Fig. 1 The flowchart of our pose proposed 3-D human tracking framework from depth data

A. Initialization

Initialization is performed once in the beginning to help our system to identify the body part labels in the first coming depth silhouette.

B. Depth Silhouette Presentation

Let X, Y, Z be coordinates of points in 3-D space followed by x, y, and z dimensions, respectively. To convert 3-D data from depth image, the corresponding relationship between the coordinates of the scene points and these pixel of depth images are expressed as

$$X = c\frac{Z}{f}, \quad Y = v\frac{Z}{f}, \quad Z = Z, \tag{1}$$

where, the distance f is the focal length, Z the distance from camera to object (depth values), c and v are the column index and row index of the pixels in depth image.

C. Down Sampling

To decrease computation cost and improve effective point set registration, we utilize a uniform down sampling method for this purpose as presented in Fig. 2.

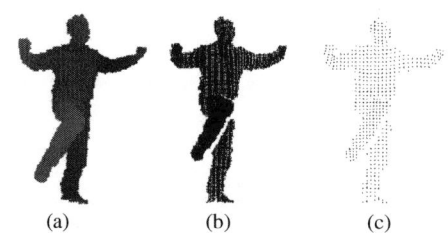

Fig. 2. Depth silhouette presentation and down sampling, (a) depth data, (b) a 3-D points set, and (c) a uniform sampled points set.

D. Nonrigid Point Set Registration

This part presents how to find point correspondences between two sets of points. We utilize the nonrigid point set registration method to find point correspondences between two complex point sets of human poses. To preserve the topological structure of human poses, we apply CPD [13], [14] to obtain the correspondences on the two point sets of human poses.

Given two 3-D points sets of human poses, the points set S_D of the previous pose and the points set S_C of the current pose. These two sets are considered the alignment as a probability density estimation problem in [13]. The CPD algorithm presents as the following:

CPD Algorithm

- Initialize parameters: β, λ
- Construct a Gaussian kernel matrix: G
- EM optimization, iterate until convergence

---

o E-step: compute the posterior probabilities of GMM components P_r

o M-step: replace current θ, δ

$$\theta, \delta \leftarrow \arg \min_{\theta', \delta'} Q(\theta', \delta' | \theta, \delta)$$

- The aligned point set is $S_C = S_{C\_init} + GW$
- The probability of correspondence is given by P_r

Where β is Gaussian smoothing filter size, λ is smoothness regularization weight, δ is standard deviation, θ is a set of the transformation parameters, G is a Gaussian kernel matrix of S_C, P_r is a posterior probability, W is a matrix of coefficients. Fig. 3 shows the result of point set registration on two poses.

F. Relabeling

To track and find the body part labels on the coming depth silhouettes, starting from the second depth silhouette, we use the labeled-parts information of the registered previous pose as target information to map on the depth silhouette of current pose using their correspondences. However, the fact that some points in the set of previous pose contain nonlabeled or mislabeled points. Therefore, these points relabel to correct their labels. In the work, we relabeled the mislabeled points based on distance from centroid points to neighbor pixels and its connectivity matrix.

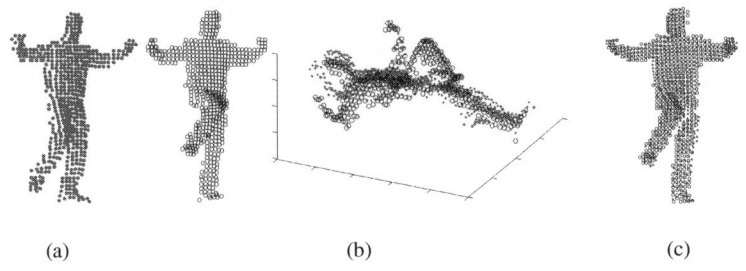

(a)   (b)   (c)

Fig. 3 Correspondences in two consecutive frames using non-rigid point set registration of two 3-D human points sets: (a) left: a previous pose, right: a current pose, (b, c) before and after pose illustrations using the point set registration, respectively.

E. 3-D Human Pose Recovery with Joints Proposals

From the point correspondences, we firstly label body parts of S_C based on the correspondences and the known label information of S_D. Secondly, the positions of joints are located by using mean shift algorithm [6] on each body part. From proposed joint positions, we create a human skeleton model. Finally, the orientation and position of each body part is determined from the skeleton model. The recovered 3-D human pose is presented in Fig. 4.

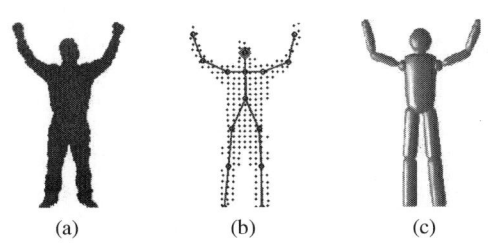

(a)   (b)   (c)

Fig. 4 Illustrational of results in our proposed system. (a) a depth silhouette, (b) a skeleton model, and (c) a 3-D recovered human pose

III. EXPERIMENTAL RESULTS

In this section, we have evaluated our proposed methodology through the quantitative and qualitative assessments using synthetic and real data.

A. Experimental Settings

In order to evaluate our proposed system quantitatively, we utilized synthetic depth silhouettes to test with the ground-truth information from the original 3-D body model. At each estimated 3-D human pose, we measured joint angles of few joints from the 3-D human body model and saved as the ground truth. Then, we derived the same joint angles from the reconstructed 3-D human pose and compared them the ground truth. In our experiment, we only focus on the evaluation of the two main joints including left-right elbows.

For qualitative assessment on real data, we utilized the coming depth silhouettes that were captured by a depth camera. Then, the human depth silhouettes were registered in our system to find point correspondences. From the point correspondences, the joints of the human body parts were determined. The orientation vectors of the body parts were estimated by joint pairs. These orientations were finally mapped on to the 3-D human body model similar to described in [8], resulting in the estimated 3-D human body pose. The testing process was run on a standard desktop PC with an Intel Pentium IV Dual-core, 2.5 GHz CPU, and 3G RAM.

B. Experimentation on Synthetic Data

We performed a quantitative evaluation using a series of 500 depth silhouettes containing various unconstrained movements. In this experiment, the evaluation results with the synthetic poses of our proposed methods are provided in Fig. 5. At each plot of Fig. 5 corresponds to an estimated joint angle by our proposed method. The solid and dashed lines indicate our estimated and its ground truth joint angles, respectively.

Table 1 The average reconstruction error of the joint angles in degree

Evaluated angles	Left elbow	Right elbow
Average reconstruction error	6.08	6.02

C. Experimentation on Real Data

For qualitative assessment of real data, we asked the subject to perform some complex pose sequences of intersected body parts. Because the ground truth joint angles are not available for real data. We only performed by visual inspection of the results of recovered poses and RGB images. Fig. 6 shows sample results of our proposed method on depth images with the occlusion of arm or leg body parts. The 1st

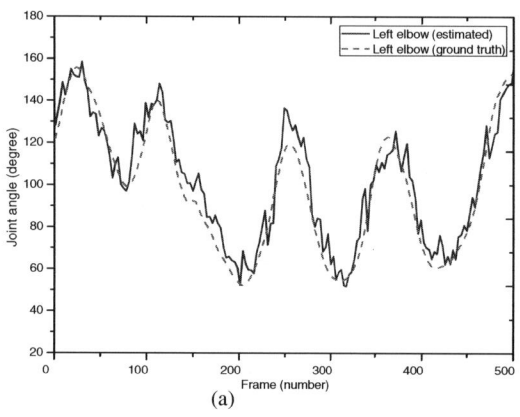

Fig. 5 A comparison between the ground-truth and the estimated joint angles in synthetic data: (a) joint angle of left elbow and (b) joint angle of right elbow.

Based on the results of estimated joint angles and the ground truth joint angles, we have computed the average reconstruction error as

$$\varepsilon_\theta = \frac{\sum_{i=1}^{n} \left| \theta_i^{est} - \theta_i^{grd} \right|}{n} \quad (3)$$

where n is the number of frames, i is the frame index, θ_i^{grd} is the ground-truth angle, and θ_i^{est} is the estimated angle. The average errors of the experiment at the two joints are given in Table 1.

and 4th column are RGB images, the 2nd and 5th human depth silhouettes, and the 3rd and 6th recovered 3-D human poses.

IV. CONCLUSION

We have presented a novel approach of recovering a 3-D human pose from a single human body depth silhouette using nonrigid point set registration. The quantitative assessments indicated the average reconstruction error of 6.06 degree. The experiments on real data show that our system reliably performs on sequences containing occlusion movements of various appearance. This approach can also

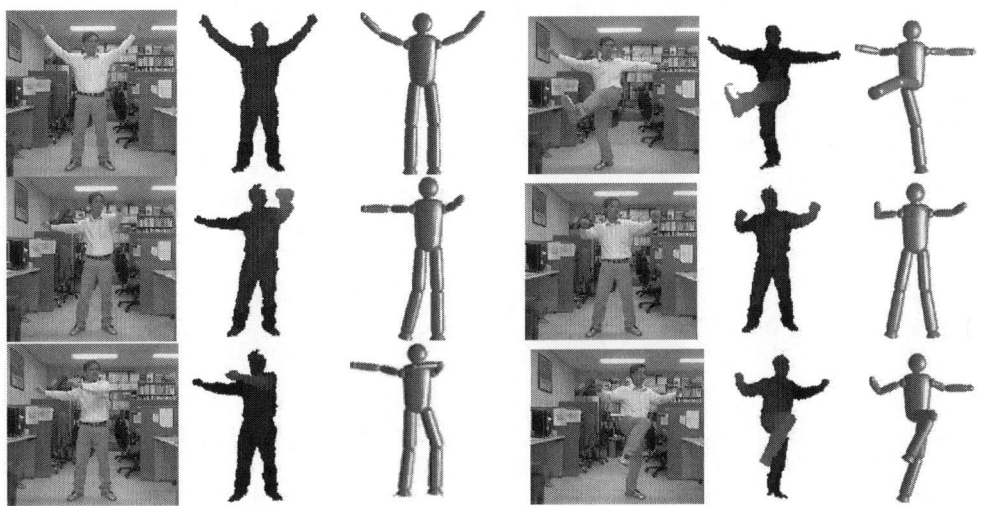

Fig. 6 Sample illustrations of our proposed 3-D human pose recovery method on depth images with the occlusion of arm or leg body parts

reconstruct some 3-D human complex pose recovery. Moreover, this method does not require any matching or training data and it is able to tracking arbitrary movements. However, the computational cost of the CPD algorithm is still high.

ACKNOWLEDGMENT

This research was supported by the MSIP (Ministry of Science, ICT & Future Planning), Korea, under the ITRC (Information Technology Research Center) support program supervised by the NIPA (National IT Industry Promotion Agency (NIPA-2013-(H0301-13-2001)).

REFERENCES

1. Poppe R (2007) Vision-based human motion analysis: An overview, Computer Vision and Image Understanding, Vol. 108, No. 1–2, pp. 4–18
2. Chen L, Wei H, and Ferryman J (2013) A survey of human motion analysis using depth imagery, Pattern Recognition Letters. ISSN 0167–8655
3. Plagemann C, Ganapathi V, Koller D, and Thrun S (2010) Real-time identification and localization of body parts from depth images, In Proc. ICRA, pp. 3108–3113
4. Schwarz L A, Mkhitaryan A, Mateus D, and Navab N (2012) Human skeleton tracking from depth data using geodesic distance and optical flow, Image and Vision Computing, Vol. 30, No. 3, pp. 217–226
5. Kalogerakis E, Hertzmann A, and Singh K (2010) Learning 3D mesh segmentation and labeling, In Proc. ACM SIGGRAPH 2010, ACM, New York, NY, 775 USA, pp. 102–114
6. Shotton J, Fitzgibbon A, Cook M, Sharp T, Finocchio M, Moore R, Kipman A, and Blake A (2011) Real-time human pose recognition in parts from single depth images, In Proc. CVPR
7. Taylor J, Shotton J, Sharp T, and Fitzgibbon A (2012) The Vitruvian Manifold: Inferring Dense Correspondences for One-Shot Human Pose Estimation, In Proc. CVPR, pp. 103–110
8. Dinh D L, Han H S, Jeon H J, Lee S Y, and Kim T S (2013) Principal direction analysis-based real-time 3D human pose reconstruction from a single depth image, In Proc SoICT'13, pp. 206–212
9. Tam G K L, Cheng Z Q, Lai Y K, Langbein F C, Liu Y H, Marshall D, Martin R R, Sun X F, and Rosin P L (2013) Registration of 3D Point Clouds and Meshes: A Survey from Rigid to Nonrigid, IEEE Transactions on Visualization and Computer Graphics, Vol. 19, No. 7, pp. 1199–1217
10. Kim D W and Kim D J (2008) A novel fitting algorithm using the ICP and the particle filters for robust 3D human body motion tracking, In Proc. ACM workshop on Vision networks for behavior analysis, pp. 69–76
11. Mundermann L, Corazza S, and Andriacchi T (2007) Accurately measuring human movement using articulated ICP with soft joint constraints and a repository of articulated models, In CVPR
12. A. L. Yuille and N. M. Grzywacz (1998) A Mathematical Analysis of the Motion Coherence Theory," Int'l J. Computer Vision, Vol. 3, No. 2, pp. 155–175
13. Myronenko A and Song X (2010) Point Set Registration: Coherent Point Drift, IEEE Trans. on PAMI, Vol. 32, No. 12, pp. 2262–2275
14. Ye M, Wang X, Yang R, Ren L, and Pollefeys M (2011) Accurate 3D pose estimation from a single depth image, In Proc. ICCV, pp. 731–738

Kinematics of High-Heeled Running Gait with Consideration for Experience of Wearers

Y.Q. Song[1], F.L. Li[1], J.S. Li[2], and Y.D. Gu[1]

[1] Faculty of Sports Science, Ningbo University, Ningbo, China
[2] Zhejiang University of Water Resources and Electric Power, Hangzhou, China

Abstract— **In today's society, many women wear high-heeled shoes. However, the effect that different experienced high-heel wearers worn same heel heights running have on the biomechanics has not been fully investigated. In the present study, the lower extremity mechanics in different experienced high-heel wearing who in same heel heights (1.5cm, 4.5cm) running were examined in 6 female subjects, three of whom could be considered experienced high-heel wearers. Kinematic data from a three-dimensional motion analysis system was collected to describe lower extremity mechanics while subjects run over ground at a natural speed. The results that in throughout the gait cycle, hip abduction was significantly decreased when experienced high-heel wearers worn low-heel shoes than in flat shoes condition, and knee flexion was significantly decreased. In addition, ankle dorsiflexion and abduction were significantly decreased when experienced high-heel wearers compared with inexperience high-heel in low-heel shoes condition. These results maybe more helpful when inexperienced high-heel wearers choose high-heel shoes.**

Keywords— **high-heel, experienced, kinematic.**

I. INTRODUCTION

Most women like wearing high-heeled shoes for the benefit of sensuous attractiveness, while foot problems are often associated. High-heeled shoes worn by women have been shown to affect gait kinematics adversely, particularly at the ankle joint, which is excessively plantar flexed. Several studies have investigated the effects of walking in high-heeled shoes. Few of these studies, however, have investigated different experienced high-heeled shoes of running in high-heeled shoes. Many more studies have been conducted on the kinematics of walking in high heels. Velocity and stride length have been observed to decrease with heel height while cadence did not change [1]. Other studies reported that, throughout the gait cycle, the ankle was more plantar flexed [2] and the range of motion of the ankle was less when wearing high heels [3]. Bih-Jen[4] reported knee flexion stance to be greater when walking in high heels than in flat shoes. In addition, these studies reported that hip flexion during the swing phase of high-heeled gait was less than in the no-heel condition, but there was no difference during the stance phase.

The biomechanical responses to human locomotion has been examined by a number of researchers. However, the majority of these studies, have focused on the active of muscles and the plantar pressure during high-heeled walking gait. Therefore, it is important to investigate and understand the effects that different experienced high-heeled shoes wearing on the biomechanics of gait. In addition, the previous studies generally compared various heel heights without considering a continuum of experienced high-heel wearing. Therefore, the purpose of the present study was, therefore, to examine the changes in low extremity mechanics while running along a continuum of experienced high-heel wearing.

II. METHODS

Six young, healthy females volunteered to take part in this study. The subjects ranged in age from21 to 25 years (23.8years±4.4),in height from 1.57 to 1.67 m (1.62m±0.71), and in body mass from 50 to 56 kg (53.2±5.2). All subjects were physically active and had no recent lower extremity injury or pain at the time of the study. Three of the subjects were classified as experienced wearers of high heels. Experienced performers were defined as individuals who wore high heels 3 or more times per week for at least 6 hours per day. The remaining three subjects were classified as inexperienced high-heel wearers; that is, individuals who wore high heels less than twice per month. Participants don't take strenuous exercise during 48 hours before the experiment.

A three-dimensional motion analysis system (Vicon MX, Vicon Motion System Ltd., Oxford, England) was used with 8 cameras (8 MX3 and 2 MX40) to capture and analyze motion of the hip , knee and ankle with a sampling frequency of 200 Hz. Participants were required to attend a single testing session at the gait laboratory. The following measurements were taken for the calculation of the joint centres: height, weight, leg length, knee and ankle width. To assess the three-dimensional motion of the lower limb, retro-reflective markers were attached in accordance with Plug In Gait (PIG). The markers were attached to the skin using double-sided tape. Each subject performed 5 running trials at natural speed in

© Springer International Publishing Switzerland 2015
V. Van Toi and T.H. Lien Phuong (eds.), *5th International Conference on Biomedical Engineering in Vietnam,*
IFMBE Proceedings 46, DOI: 10.1007/978-3-319-11776-8_101

both shoe conditions. The order of the shoe conditions was randomly assigned for each subject. Before testing, subjects were allowed to run as many trials as necessary to familiarize themselves with the testing conditions.

T-test with repeated measures were used to compare two groups. The level of 0.01 was used to test for significance. All data were analyzed using the SPSS statistical analysis software.

III. RESULTS

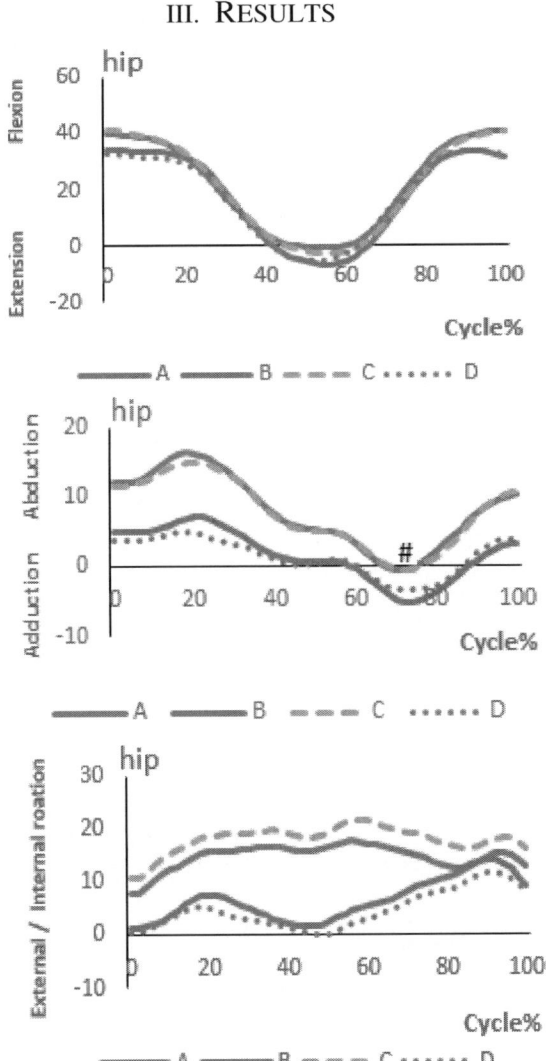

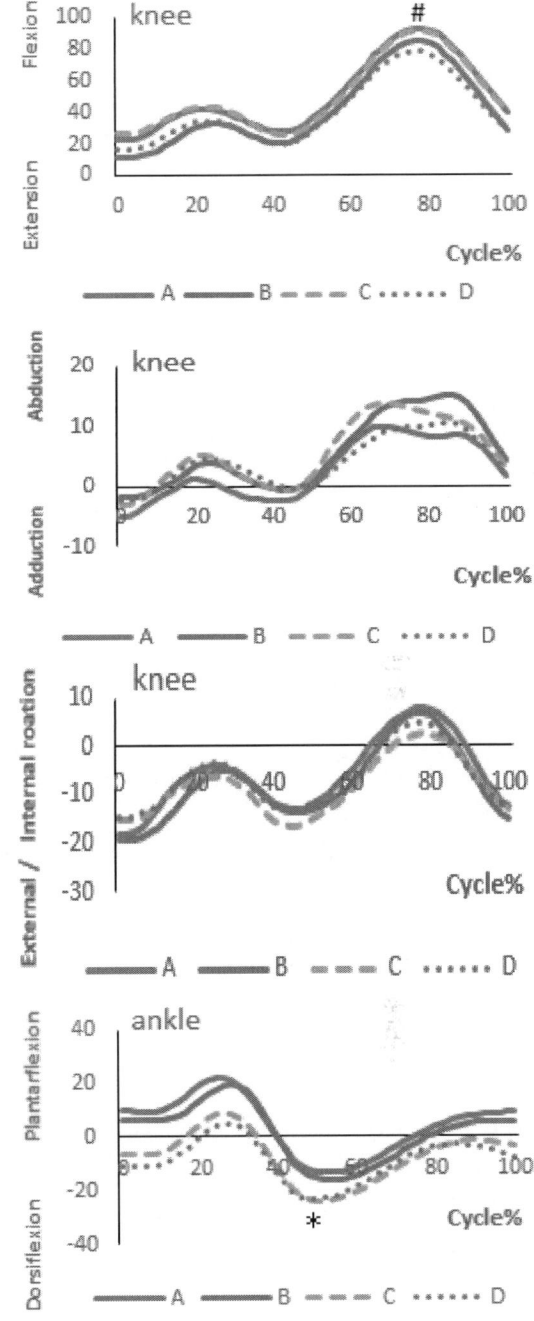

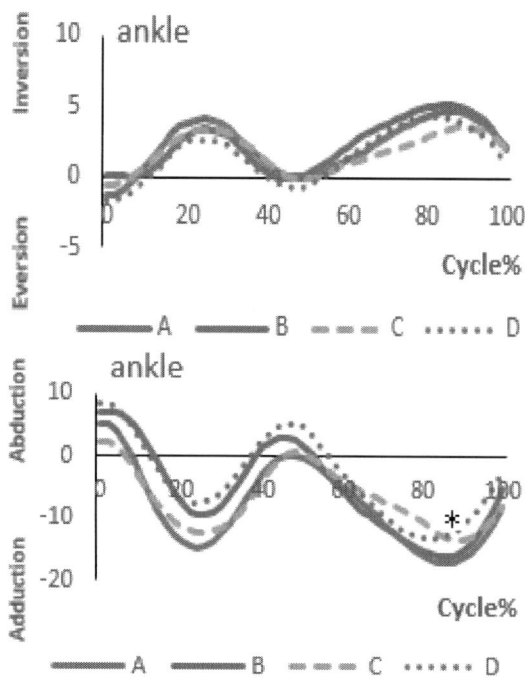

Fig. 1 A respects that experienced high-heel wearers worn flat shoes, B respects that inexperienced high-heel wearers worn flat shoes, C respects that experienced high-heel wearers worn low-heel shoes, D respects that inexperienced high-heel wearers worn low-heel shoes. * respects that there was significantly difference between two groups (p<0.01); # respects that there was significantly difference between same group (p<0.01).

Throughout the gait cycle, hip abduction was significantly decreased when experienced high-heel wearers worn low-heel shoes than in flat shoes condition, and knee flexion was significantly decreased (p=0.008, p=0.003). In addition, ankle dorsiflexion and abduction were significantly decreased when experienced high-heel wearers compared with inexperience high-heel in low-heel shoes condition (p=0.006, p=0.004).

IV. DISCUSSION

The purpose of the present study was to examine the changes in low extremity mechanics while running along a continuum of experienced high-heel wearers. According to the data from this study, it is evident that changes in different experienced high-heel wearers can greatly influence the kinematics of the lower extremity.

In this study, it was found that experienced wearers had significantly decreased knee flexion, which is contrary to Anna [5] reported. In addition, this study also found that same change in hip abduction. Perhaps that study walking was consideration, respectively, while in the current study running.

We can also know that the range of lower extremity joint angle was less when experienced wearers compared with in experienced wearers. That maybe suggested that the capability of control who experienced wearers was superior to inexperienced wearers. In fact that, it was consistent with phenomenon that inexperienced wearers would be easy to occur sprain. These results maybe more helpful when inexperienced high-heel wearers choose high-heel shoes.

ACKNOWLEDGMENT

This work was financially supported by Ningbo Natural Science Foundation (2013A610262), SRF for ROCS, SEM, and Postgraduate Key Course Development of Ningbo University (Grant NO. zdkc2012008).

REFERENCES

1. Paul H et al. (2013) High heels as supernormal stimuli: How wearing high heels affects judgements of female attractiveness. Evolution and Human Behavior 34:176-181.
2. Erik B et al. (2012) Walking on High Heels Changes Muscle Activity and the Dynamics of Human Walking Significantly. Journal of Applied Biomechanics 28:20-28.
3. Danielle et al. (2012) Heel height affects lower extremity frontal plane joint moments during walking. Gait & Posture 35:483-488.
4. Bih-Jen et al. (2009) Kinematics and kinetics of the lower extremities of young and elder women. Journal of Electromyography and Kinesiology 19:1071–1078.
5. Anna et al. (2012) The influence of heel height on lower extremity kinematics and leg muscle activity during gait in young and middle-aged women. Gait & Posture 35:677–680.

Author: Yuquan Song
Institute: Ningbo University
Street: NO.818, Fenghua Raod
City: Ningbo
Country: China
Email: songyuquan@aol.co.uk

MFCC-DTW Algorithm for Speech Recognition in an Intelligent Wheelchair

Le Hoang Linh[1], Nguyen Thanh Hai[1], Ngo Van Thuyen[1], Tran Thanh Mai[2], and Vo Van Toi[3]

[1] Faculty of Electrical-Electronics Engineering, HCMC University of Technical Education, Vietnam
[2] Faculty of Electrical-Electronics Engineering, HCMC University of Technology, Vietnam
[3] Biomedical Engineering Department, International University, VNU, HCMC, Vietnam

Abstract— This paper processed a Dynamic Time Warping (DTW) for speech recognition in an intelligent electric wheelchair. Voice is recorded for extracting features using Mel Frequency Cepstral Coefficients (MFCCs), for identifier to find the best command. Thus, the recognized signals are used control electric wheelchair. DTW identification method for produces better results and more accurate. Results of this study possibly support disabled people using the wheelchair to move easily and more convenient in everyday life.

Keywords— MFCC-DTW, wheelchair control, feature matching, electric wheelchair.

I. INTRODUCTION

Speech recognition is an active and popular method, used to translate human voice into commands. The model commonly used in identification as: Hidden Markov Model (HMM) [1], Vector Quantization (VQ), MFCC-DTW and neural networks. Identification used in wheelchair control, letter recognition [2], or number count [3]. Moreover other the HMM has high the accuracy, however it is very complex and the training time. The MFFC and DTW model is simple [4], do not take more training time, but lower accuracy than the HMM.

The voice recognition system is used to detect the attendance of speech in a background of noise. The beginning and end point of a word should be detected for processing words. The main difficulty of speech recognition is the same word spoke by different speakers depending on speaking styles, tone, regional, gender and speech patterns. In addition, noise and change of signals over time are problems considered in speech recognition.

Speech recognition plays an important role in an intelligent wheelchair system using microphone. MFCC and DTW algorithms are applied for feature extraction and identification [5]. Speech commands such as left, right, forward, backward and stop will be recognized for an electrical wheelchair control. Experiments with identified speech commands using the proposed method will be performed by users.

In the word, the number of disable people about 15% of the population. Moreover people with disabilities feel isolated and do not have access to the same opportunities as other within their own communities. Those are reasons why an intelligent wheelchair was. For more convenient in modem life, electric wheelchair is improved day by day and a smart wheelchair is inevitable need. The smart wheelchair is designed to be used for indoor environment and user can easily control it by speech commands. In fact, when the user speech to the wheelchair to move using commands [6], microphone, in which the microphone carries each command to computer through software for wheelchair control.

This paper is organized as follows: section II describes the materials and methods, section III solves results, section IV is discussion and section V provides overall conclusions.

II. MATERIALS AND METHODS

A user will perform an electronic intelligent wheelchair control by user commands such as: *Left, Right, Backward, Forward* and *Stop*. A speech signal of user is recoded in interval of 2.5 s and all signals are pre-processed with sampling frequency 16 KHz and then feature extracted for identification. The voice signal uses a combination of features based on the MFCC and voice activity detection. The DTW algorithm is used to discriminate the speech into respective classes.

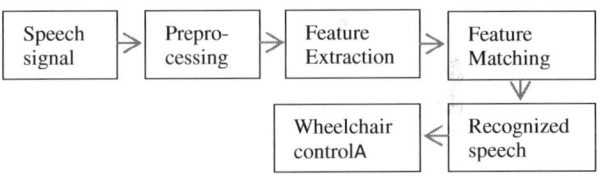

Fig. 1 Block Diagram of Methodology

A. Feature Extraction

As shown in Fig.2 of Feature extraction using the MFCC that consists of computational processes.

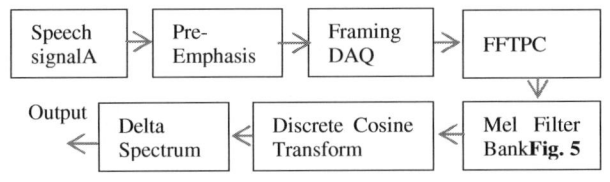

Fig. 2 Block Diagram of Feature Extraction

V. Van Toi and T.H. Lien Phuong (eds.), *5th International Conference on Biomedical Engineering in Vietnam*,
IFMBE Proceedings 46, DOI: 10.1007/978-3-319-11776-8_102

- *Pre-emphasis*

This step processes with purpose of offset high frequency components. In particular speech signal is processed using a filter which emphasizes higher frequency. This process will increase the energy of signal at the higher frequency. The output signal of the Pre-emphasis is computed the following equation:

$$H[n] = u(n) - a.u(n-1) \tag{1}$$

where, H[n] is the signal output of the pre-emphasis process, u(n) is the voice signal, typical value of a = 0.95 (>20dB gain for high frequency). The result pre-emphasis process is shown in Fig.3

- *Framing and Windowing*

The signal after Pre-emphasis is segmented due to the voice signal is continuous with time. The voice reliability can be ensured for a short time. Process frame can not wait for last sample, which split reduces the signal discontinuities at the beginning and end of each frame, in which the frame length from 10 to 30 msec. The speech signal is divided into frames of N samples. This process is important to retain short term features. Short time analysis is performed by windowing the signal. Normally Hamming window is used and its equation is given as:

If the window is defined as W(n), $0 \leq n \geq N-1$

$$Y(n) = H(n) x W(n) \tag{2}$$

where, W(n) is Hamming window, N is the number of sample in each frame, Y(n) is the output signal, H(n) is the input signal, W(n) is the Hamming window

The result Hamming function is shown below:

$$W(n) = 0.54 - 0.46 \cos\left(\frac{2\pi n}{N-1}\right) \tag{3}$$

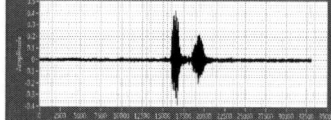 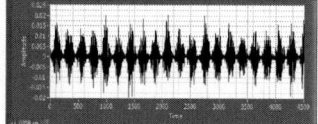

Fig. 3 Pre-emphasis of signal   Fig. 4 The signal is framed

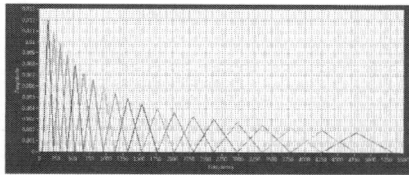

Fig. 5 Mel Filter bank with frequency band from 50 to 5400 Hz, the number of filter bank is 20

- *Mel Filter Bank*

Human hearing is not equally sensitive to all frequency bands. It is less sensitive at higher frequency, roughly greater than 1000 Hz, human perception of frequency is non-linear. The Mel spectrum is the total spectrum of the signal spectrum after discrete Fourier transform multiplied by the weight of the Mel filter. Mel filter bank is series of triangular filter of the form at the center frequency and then if decreases linearly to zero at the center frequency of two adjacent filters [7].

Each filter output is the sum of its filtered spectral components. The equation is used to compute the Mel for given frequency f in Hz:

$$f(mel) = 2595 \log_{10}(1 + \frac{f}{100}) \tag{4}$$

- *Discrete Cosine Transform*

In this final step, the log mel spectrum is converted to time. The result is called MFCC. Because the mel spectrum coefficients are real numbers, we can convert them to the time domain using the DCT. The set of coefficient is called acoustic vectors. Therefore, each input utterance is transformed into a sequence of acoustic vector.

- *Delta Spectrum and Delta Energy*

The DCT is done on Mel spectral coefficients [8] of each frame, hence obtaining the MFCC. The first 2 coefficients of the obtained MFCC are removed as they varied significantly between different utterances of the same word. Littering is done by replacing all MFCCs except the first 14 by zero. The first coefficient of the MFCC of each frame was replaced by the log energy of that frame. Delta and acceleration coefficients are found from the MFCC so as to increase the dimension of the feature vector of the frames, thereby increasing the accuracy. The energy in a frame for a signal in a window is stored in an array and the values are used to detect the threshold energy of speech signal and noise removal, the energy is represented using the equation following:

$$E = \sum_{n=0}^{N-1} x^2(n) \tag{5}$$

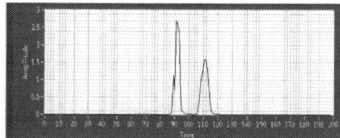 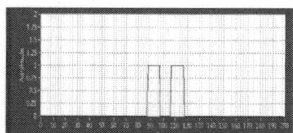

Fig. 6a Energy of signal   Fig. 6b Smooth energy of signal

The Delta coefficient are found from the following equation.

$$\delta = \frac{c(t+1) - c(t-1)}{2} \quad (6)$$

2. Feature Matching Using DTW Algorithm

In the speech identifier, the DTW algorithm is used to find the best path for the vector matrix feature which has the highest probability density. The DTW algorithm is based on dynamic programming techniques. This algorithm is for measuring similarity between two time series which may vary in time or speed.

The features of voice signal is performance from N dimensional feature vectors, the signal framed marking M vector, M vector are determined producing the M x N feature matrix. The M x N matrix is created by extracting features of speech signal for selected words. The feature of the test sequence is compared with each word in the sets using DTW and the best match in each set is outputted.

The MFCC for each of frame is computed and represented by a vector. Hence each utterance is represented by a vector sequence.

$$V = v_1, v_2, ..., v_i, ..., v_n \quad (7)$$

To align two sequences using DTW, an n by m matrix where the (i^{th}, j^{th}) element of the matrix contains the distance (v_i, p_j) between the two point v_i and p_i is constructed. Distance between individual vectors is found using the Euclidean distance formula:

$$d(v_i, p_j) = (v_i - p_j)^2 \quad (8)$$

where, $P = p_1, p_2, ..., p_j, ..., p_m$ is comparison sample vector with length m. The distance between vectors V and P is called "warped". The optimal path between the vector V and P is computed as follows:

$$O = \arg \min\{D(V, P, O)\} \quad (9)$$

Each matrix element (i, j) corresponds to the alignment between the points v_i and p_j, Key points to find the optimal path. A grid point *(i, j)* in the optimal path can have the predecessors (i-1, j), (i-1, j-1) and (i, j-1). The total distance matrix is calculated using the following formula:

$$D(i, j) = \min \begin{cases} D(i, j-1) + d(\vec{v}_i, \vec{p}_j) \\ D(i-1, j-1) + 2.d(\vec{v}_i, \vec{p}_j) \\ D(i-1, j) + d(\vec{v}_i, \vec{p}_j) \end{cases} \quad (10)$$

The accumulate distance at a point (i, j) is the distance between two sequence vectors P and V.

III. RESULT AND DISCUSSION

After running the algorithm, the result obtained. A user speaks words: left, right, backward, forward and stop, and then the system saves them. The input words will be recognized corresponding to the template with the lowest matching score. The DTW algorithm is used for distance calculation between the tested speech and the reference word bank.

The acquisition process of speech signals, using Data Acquisition (DAQ) device USB-6008 of National Instruments (NI) Company, in which voice is recorded by a microphone through the computer's soundcard. The signal is input DAQ, then signal to band-pass filter with cutoff frequency interval [80, 1200] Hz. The signal has sample rate 11025 Hz [9] and voice is recorded interval 3 sec per word.

Fig. 7a is the original input signal and Fig. 7b represents the distributed frame of the signal, in which this frame is to split the discontinuous signal at the beginning and end of each frame having the length from 10 to 30 ms. Fig. 7c shows the energy of signal smoothed using a threshold. In Fig.7d, the signal is identified. All of the identified signals of user are applied for the electrical wheelchair control to reach the desired target.

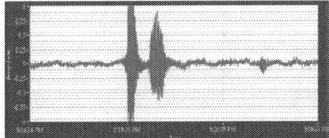

| Fig. 7a Signal input | Fig. 7b Distributed frame |

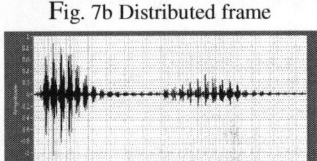

| Fig. 7c Energy signal | Fig. 7d Identified signal |

The signal recognition is the output to the wheelchair control through USB-6008 of NI Company, the output voltage level of USB-6008 shows in table 1. The voltage signal output is taken down motors of the wheelchair to control electric wheelchair movements as: left, right, backward, forward and stop. The model of the electric wheelchair is shown in Fig.9.

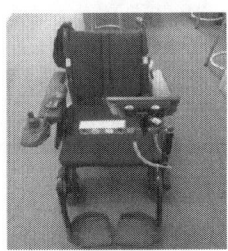

Fig.8. The model of an electric wheelchair

Table 1 Output voltage of USB-6008

Voice recognition	Output voltage	
	Speed out (V)	Steering out (V)
Left	2.5	2.8 - 3.0
Right	2.5	1.0 – 1.8
Backward	2.8 – 3.0	2.5
Forward	1.0 – 2.0	2.5
Stop	2.5	2.5

The DTW algorithm is used for speech recognition being researched and developed, because the advantages of this algorithm contribute to improve the quality and accuracy identification.

When a user says the words: left, right, backward, forward and stop, these words were recorded and processed for recognition on the LabView environment, the identification process is based on features extracted from the MFCC. The result of word recognition is shown from Fig.9 to Fig.13. The accuracy of word identification for wheelchair control is shown in table 2 in which the test result is 40 times.

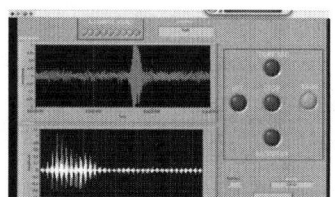

Fig. 9 The signal of "left" command

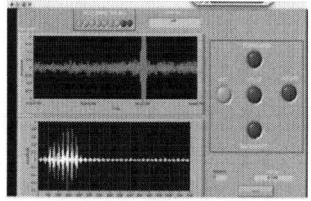

Fig.10 The signal of "right" command

Fig. 11 The signal of "backward" command

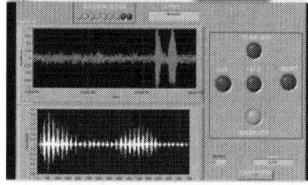

Fig. 12 The signal of "forward" command

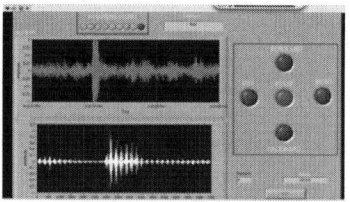

Fig. 13 The signal of "stop" command

Table 2 The accuracy of identification process

Voice recognition	Accuracy test in 40 times
Left	89%
Right	90%
Backward	91%
Forward	87%
Stop	88%

Speech recognition using the DTW algorithm, combining LabView language has been identified for controlling an electric wheelchair.

A comparison between two DTW and hidden Markov models was made to identify individual words in Arabic [10]. In this paper, the DTW method using 13 MFCC coefficients and two delta coefficients (delta, delta-delta) , in which the 256-points FFT method was used to find the energy spectrum of the signal, using Mel filter bank with 24 banks. The accuracy is about 89% in a clear environment.

IV. CONCLUSIONS

Results of identification system are using MFCC-DTW algorithms effective, the words are used to control the electric wheelchair and identification process was tested less noisy environments, with little people and move indoors. However, noisy environment is a little bit hard for people with disabilities to move. To overcome this drawback, the next will research apply neural networks for the identification process to be better and more accurate results.

ACKNOWLEDGMENT

The authors would like thank the support of BME department, IU students, Vietnam.

REFERENCES

1. Bhupinder Singh, Neha Kapur, Puneet Kaur (2012) "Speech Recognition with Hidden Markov Model: A Review", International Journal of Advanced in Computer Science and Soft Engineering, Vol. 2, pp. 400-403.
2. Dipmoy Gupta, Radha Mounima C. Navya Manjunath, Manoj PB (2012), " Isolated Word Speech Recognition Using Vector Quantization", International Journal of Advanced in Computer Science and Soft Engineering, Vol. 2, pp. 164-168.
3. Talal Bin Amin, Iftekhar Mahmood (2008) "Speech Recognition Using Dynamic Time Warping", ICAST 29th – 30th.

4. A.*Revathi*, R.Ganapathy and Y.Venkataramani (2009), *"Text Independent Speaker Recognition and Speaker Independent Speech Recognition Using Iterative Clustering Approach",* International Journal of Computer science & Information Technology, Vol. 1, No 2, pp. 30-42.

5. Lindasalwa Muda, Mumtaj Begam and I. Elamvazuthi (2010), *"Voice Recognition Algorithms using Mel Frequency Cepstral Coefficient (MFCC) and Dynamic Time Warping (DTW) Techniques"*, Journal Of Computing, Vol. 2, Issue 3, pp. 138-143.

6. Kohei Arai, Ronny Mardiyanto (2011), *Eyes Based Eletric Wheel Chair Control System,* International Journal of Advanced Computer Science and Applications, pp. 98.

7. S. M. Azam, Z.A. Mansoor, M. Shahzad Mughal, S. Mohsin (2007), *"Urdu Spoken Digits Recognition Using Classified MFCC and Backpropgation Neural Network"*, Computer Graphics, Imaging and Visualisation IEEE, Vol.1, pp. 0-7695-2928.

8. Chin Luh Tan and Adznan Jantan (2004), *"Digit Recognition Using Neural Networks"*, Malaysian Journal of Computer Science, Vol. 17 No. 2, pp. 40-54.

9. Fu-Hua Liu; Richard M. Stern; Xuedong Huang; Alejandro Acero (1993), *"Efficient cepstral normalization for robust speech recognition, human language technology"*, Proceedings of a Workshop Held at Plainsboro, New Jersey, pp. 21-24.

10. Z. Hachkar et al.(2011), *"A Comparison of DHMM and DTW for Isolated Digits Recognition System of Arabic Language"*, International J. on Computer Science and Eng.

Lower Limb Kinematics Study on Female Latin Shoes of Different Height Heels

S.R. Shao, X.X. Gao, Y. Zhang, Q.C. Mei, and Y.D. Gu

Faculty of Sports Science, Ningbo University, Ningbo, China

Abstract— Eighteen professional female Latin dancers were participated in this research. The joint angle dates were measured and analyzed using Vicon Motion System. The results showed that: With the heel increased, joint angle also increases; Latin dance shoes exist a little illogical design. We need to improve it in the future.

Keywords— female, different height of heels, latin dance shoes, joint angle.

I. INTRODUCTION

Latin dance has become an international Dance Sport now, it demonstrated the art of Latin dance style and strong athletic characteristics by passionate dancing, dashing and agile action, fast-paced dynamic music, and by the vast number of dance sports favorite. It is divided into two groups, ten species. And combining music, dance, clothing, manners, movement, body beauty in one. It is not only a rich artistic connotation, but also the sports competition form, reflecting the ornamental value and might be involved, so it is regarded as a real art. The sports dance has many functions, such as social, entertainment, arts, sports and exercise and so on. Its difficulty has been increasing follow its development, show the beauty of human body and the movement of natural beauty by the change of force, amplitude, speed and posture [1]. However, the rate of foot injury has greatly increased under the passive calf motion [2]. Variation from normal foot posture has long been thought to influence function of the foot and lower limb during gait, thereby predisposing to injury [3]. Li [4] was discovered the formation at the base of the sagittal direction force may be associated with hallux valgus through research on the gait of young women when standing and walking. Grundy et al. [5] found that the loading of forefoot was not significantly large compared with heel, but the loading time of forefoot lengthier than heel under the condition of walking barefoot. And the same time, the bearing function of forefoot is significantly weakened with the heel of the sole height increased. Sato et al. [6] found with heel height is increased, the stride will be reduced accordingly. Furthermore, the foot wearing high-heeled shoes would naturally place in plantar-flexed position, which could lead to stride length shorten [7].Through comparative study of high-heeled shoes and sports shoes, flat shoes. But H. Morris

[8] revealed that wearing high heels led to increased femininity of gait including reduced stride length and increased rotation and tilt of the hips. Anna Mika [9] observed an increase in knee flexion and decrease in ankle eversion associated with elevated heel heights. Anna Mika [10] found that the younger women exhibited an increase in pelvic range of motion in the sagittal plane during high-heeled gait compared with low-heeled gait and walking without shows. Yushin Kim [11]found that wearers of high-heeled shoes showed increased ankle range of motion on plantarflexion at 25 degrees and inversion at 10 degrees compared to flat shoe wearers (P < 0.5) but decreased dorsiflexion (about 17 degrees) and eversion (13 degrees; P <0.5). Barkema [12] found that peak internal knee abduction moment increased systematically as heel height increased, and kinetic changes at the ankle with increasing heel height may also contribute to larger medial loads at the knee.

More research on the effects of lower limb with high heels walking, but less on the changes of the joint angle in Latin dance, motion characteristics required female athletes complete a variety of complex dance under with Latin dance shoes with different heel , so the foot injury and knee injuries of the Latin dance exercise later is more. The purpose of this study was to state the Latin dance movement characteristics, revealing the movement joint pathology caused by genesis, and providing the mechanical basis for functional design of Latin dance shoes. By the research on changes of the joint angles of lower limbs when young women were wearing 1.5cm, 4.5cm, 7cm Latin dance in the Rumba square step.

II. METHODS

Eighteen professional healthy female (age:23.8±4.4years, height:165.5±7.1cm, mass:60.9±8.7kg) Latin dancers were participated in this research. The foot developed normally, without a podiatrist and motility disorders, also do not have the clinical history, further details of the foot posture screening protocol are reported by Murley et al. [13]. Participants don't take strenuous exercise during 48 hours before the experiment.

A three-dimensional motion analysis system (Vicon MX, Vicon Motion System Ltd., Oxford, England) was used with 8 cameras (8 MX3 and 2 MX40) to capture and analyse motion of the hip , knee and ankle with a sampling frequency of 200 Hz.

V. Van Toi and T.H. Lien Phuong (eds.), *5th International Conference on Biomedical Engineering in Vietnam,*
IFMBE Proceedings 46, DOI: 10.1007/978-3-319-11776-8_103

Participants were required to attend a single testing session at gait laboratory. The following measurements were taken for the calculation of the joint centers: height, weight, leg length, knee and ankle width. To assess the three-dimensional motion of the lower limb, retro-reflective markers were attached in accordance with Plug In Gait (PIG) as described by Stebbins et al. [14, 15, 16]. The markers were attached to the skin using double-sided tape. Marker trajectories were collected by an eight-camera motion capture system. According to the characteristics of Latin dance action, selecting characteristic action rumba square step for research. Most of the rumba action is completed on the basis of in square. Take an example for a step forward square in this study. Before the trial, the subjects are allowed to randomly walk barefoot in order to fit the experimental environment. Let subjects randomly selected a pair of Latin dance shoes to complete three square step, and each subject complete three times trial under each pair of shoes condition. Take the left foot for example when analyzing the data. A step forward square is divided into two phases according to the center of gravity shift, the first stage is that the center of gravity is transferred from the starting of the right to the left foot, the second stage is that the center of gravity is transferred from the left foot to the right foot again and moved back to the initial position.

After each acquisition session, 3D marker trajectories were reconstructed and the right and left step forward square phases were identified. At least three square step cycles for the left foot were selected on the basis of good quality of the marker trajectories. To satisfy the assumption of independence with statistical analysis, only measurements from a single leg were analyzed [17]. Take the left foot for example in this study. Independent samples t-tests were performed to evaluate differences between the different heel for the kinematic parameters, with p values less than 0.05 considered significant. All statistical tests were conducted using SPSS version 17 for Windows.

III. RESULTS

As shown in Fig.1, in the frontal axis 3D motion, the first stage, hip flexion amplitude reached the maximum value at the 1/4 period, compared with the heel of 4.5cm, the amplitude of hip joint flexion is more greater when wearing the heel of 7.5cm, compared with the heel of 1.5cm, the amplitude of hip joint flexion is about 1.8 times when wearing the heel of 1.5cm. At this time, the amplitude of the knee flexion reached the maximum, compared with the heel of 1.5cm, the amplitude of the knee flexion is about 2.6 times and 1.7 times when wearing the heel of 7.5cm and 4.5cm. In the beginning of the first phase compared with the heel of 1.5cm, the amplitude of the ankle planterflexion is about 3.1

times and 2 times when wearing the heel of 7.5cm and 4.5cm. In the second phase, the amplitude of hip and knee flexion reached the maximum at 4/5 period, compared with the heel of 1.5cm, the amplitude of hip and knee flexion is about 2.8 times when wearing the heel of 7.5cm. The amplitude of hip flexion is about 3.6 times and the amplitude of knee flexion is about 2.4 times when wearing the heel of 4.5cm. In the second phase, compared with the heel of 1.5cm, the amplitude of the ankle planterflexion is about 3.6 times and 2.4 times when wearing the heel of 7.5cm and 4.5cm.

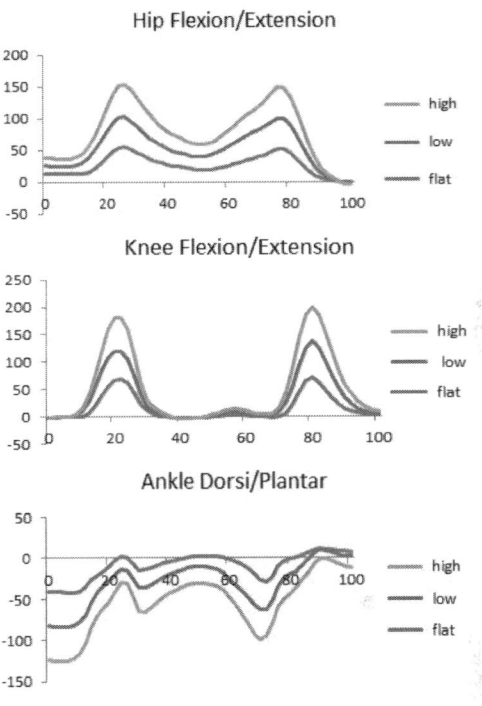

Fig. 1 Lower limb frontal axis motion variety

As shown in Fig.2, in the sagittal axis 3D motion, the first stage, the amplitude of knee adduction and ankle inversion reached the maximum value at 1/4 period, compared with the heel of 1.5cm, the amplitude of knee adduction is about 2.8 times and the amplitude of ankle inversion is about 2.5 times when wearing the heel of 7.5cm. The amplitude of knee adduction and ankle inversion is about 2.8 times and 2.5 times when wearing the heel of 4.5cm. In the end of the first stage, the amplitude of hip abduction reached a maximum, compared with the heel of 1.5cm, the amplitude of hip abduction is about 2.7 times and 1.9 times when wearing the heel of 7.5cm and 4.5cm In the second phase, the amplitude of knee and ankle inversion reached a maximum value at 4/5 period, compared with the heel of

1.5cm, the amplitude of knee inversion is about 3.8 times and the amplitude of ankle inversion is about 2.6 times when wearing the heel of 7.5cm. The amplitude of knee inversion is about 2.4 times and the amplitude of ankle inversion is about 1.9 times when wearing the heel of 4.5cm. Since then, the amplitude of hip abduction reached the maximum, compared with the heel of 1.5cm, the amplitude of hip abduction is about 3 times and 2 times when wearing the heel of 7.5cm and 4.5cm.

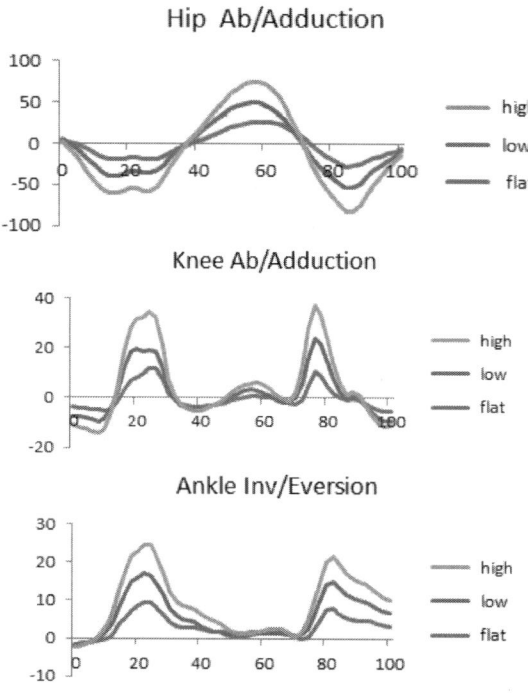

Fig. 2 Lower limb sagittal axis motion variety

As shown in Fig.3, in the vertical axis of 3D motion, the amplitude of hip rotation reached the maximum at the end of the first stage and the second stage. Compared with the heel of 1.5cm, the amplitude of hip rotation is about 2.9 times and 2 times when wearing the heel of 7.5cm and 4.5cm. Compared with the heel of 1.5cm, the amplitude of knee rotation at the end of the first stage and the second stage is about 3.1 times and 2.1 times when wearing the heel of 7.5cm and 4.5cm. The amplitude peaks of ankle pronation reached the maximum at 1/4 and 4/5 period. At the 1/4 period, compared with the heel of 1.5cm, the amplitude of ankle inversion is about 2.8 times and 1.9 times. At the 4/5 period, compared with the heel of 1.5cm, the amplitude of ankle inversion is about 2.9 times and 2 times.

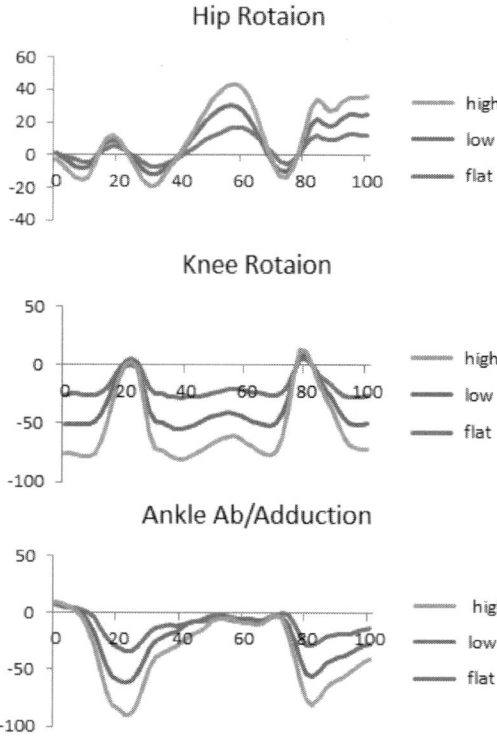

Fig.3 Lower limb vertical axis motion variety

IV. DISCUSSION

In Latin the movement, with the Latin high heel height increased, the action range also increases. This result has both advantages and disadvantages, the one hand movements increased, for the Latin dance itself, may have a better view, also in line with the Latin dance movement style; on the other hand, the action range existing is dependent on range of motion, the long term, the damaged large movement range is bound to cause the joint, especially the knee injury.

Therefore, in Latin dance movement in order to achieve better ornamental effect, also can protect the Latin dancers joint damage, athletes should choose the appropriate height of the Latin shoes.

ACKNOWLEDGMENT

This work was financially supported by Ningbo Natural Science Foundation (2013A610262), SRF for ROCS, SEM, and Postgraduate Key Course Development of Ningbo University (Grant NO. zdkc2012008).

REFERENCES

1. Li. C, (2000) Physical fitness and aerobics. Beijing: Higher education press.
2. Gu Y, Li J. (2005) The finite element analysis of full longitudinal arch stress distribution under the lift heel. Sports Science 25: 85-87.
3. McPoil T, Hunt G. (1995) Evaluation and management of foot and ankle disorders: present problems and future directions. Or hop Sports Phys Ther 21:381-388.
4. Li J. (2004) The dynamics gait research on young women wear different heel walking. Journal of Beijing Sport University 27:486-488.
5. Grundy M et al. (1975) An investigation of the centers of Pressure under the foot while walking. Bone Joint Surg 57:98-103.
6. Sato H et al. (1999) Gait Patterns of young Japanese women. Hum Ergol 20:85-88.
7. Wu J, Li J. (2003) A Research on kinematics on Young Girls' walking with High-heel Shoes. Shanghai Sport Science Research 24:9-11.
8. Paul H et al. (2012) High heels as supernormal stimuli: How wearing high heels affects judgments of female attractiveness. Evolution and Human Behavior 29:230-234.
9. Mika A et al. (2012) The influence of heel height on lower extremity kinematics and leg muscle activity during gait in young and middle-aged women. Gait & Posture 35:677-680.
10. Mika P et al. (2012) The Effect of Walking in High- and Low-Heeled Shoes on Erector Sp-nae Activity and Pelvis Kinematics During Gait. Physical Medicine & Rehabilitation 91:425-434.
11. KimY et al. (2013) Changes in Ankle Range of Motion and Muscle Strength in Habitual Wearers of High-Heeled Shoes. Foot & Ankle 34:258-262.
12. Danielle D et al. (2012) Heel height affects lower extremity frontal plane joint moments during walking. Gait & Posture 35:483-488.
13. Murley G et al. (2009) A protocol for classifying normal- and flat arched foot posture for research studies using clinical and radiographic measurements. Foot & Ankle 2:22-27.
14. Stebbins J et al. (2006) Repeatability of a model for measuring multi-segment foot kinematics in children. Gait Posture 23:401-410.
15. Ferrari A et al. (2008) Quantitative comparison of five current protocols in gait analysis. Gait & Posture 28:207-216.
16. Levinger S et al. (2010) A comparison of foot kinematics in people with normal- and flat-arched feet using the Oxford Foot Model. Gait & Posture 32:519-523.
17. Menz H. (2005) Analysis of paired data in physical therapy research: time to stop double-dipping? Orthop Sports Phys Ther 35:477-478.

Author: Shirui Shao
Institute: Ningbo University
Street: NO.818, Fenghua Raod
City: Ningbo
Country: China
Email: shaoshiruinb@sohu.com

Ideal Cross-Point Regions of Prediction Errors
from LOCO-I Algorithm Applied to Lossless Image Compression

Tin T. Dang and Canh Xuan Huynh

Ho Chi Minh city University of Technology, Ho Chi Minh City, Vietnam

Abstract— **This paper proposes the scheme of ICRICLI (Ideal Cross-point Regions for lossless Image Compression on multiple bit planes with Low complexity lossless compression for Images) that uses cross-point regions with LOCO-I algorithm for optimizing the processing of coding. The scheme ICRICLI is developed from the combination of the improved CRICM (Cross-point Regions for lossless Image Compression on Multiple bit planes) and LOCO-I, it is an algorithm for losslessly encoding and decoding images by optimizing the probability of data bits of the matrix of prediction error in cross-point regions on different bit planes. This scheme can be seen as a proposition of improvement of the standard predictive template of JPEG-LS for losslessly compressing images.**

Keywords— **cross point, entire cross-point region, ideal cross-point region, probability of bits, entropy coding.**

I. INTRODUCTION

This paper is developed from the contributions [1], [2], and [3] that presented the theory of cross-point regions and the most important is the scheme of CRICM in [3]. This work uses the idea of CRICM but using ideal cross-point regions combined with LOCO-I algorithm to build a new scheme for lossless image compression.

Cross points are neighbor points around the points of grey levels 2^n which may or may not exist in the real data, but cross points are easily specified by [4]. The original data points whose values are less than 2^n have bit states much different from those of the data points greater than or equal to 2^n [4]-[6]. The change of bit states for Gray code transformation with its characteristics has been studied and mentioned by many authors [6]-[9]; the number of bits and the distribution of these bits of Gray codes in the regions of cross points are mentioned in [1], [2]. The real data are arbitrary, so the change of bit states is systematically developed in the form of probability of data bits in cross-point regions before and after Gray coding. Being different from [3] that uses entire cross-point regions with the size fixed, the ideal cross-point regions have flexible size with grey values satisfying a certain range for optimizing the probability of data bits. Moreover, LOCO-I algorithm which is the core of JPEG-LS [9] is used to take prediction errors of the matrix of grey values to reduce interpixel redundancy between pixels, especially pixels in ideal cross-point regions. In the stage of entropy coding of the scheme, we use C.B Jones'

algorithm [10] with an improvement [11] to take the codeword of image. This scheme is suitable for big amount of data in real time processing such as telemedicine or distance learning.

This paper has six sections. After this introduction. Section II mentions the related works, the theory of cross-point regions, the characteristics of Gray codes of cross points, the change of the probability of bits in ideal cross-point regions due to Gray coding. Section III presents the property of prediction errors from LOCO-I in ideal cross-point regions. Section IV introduces the scheme ICRICLI. In Section V, some of our results obtained from using the theory above are presented. Section VI contains the conclusion and the scope for future research.

II. RELATED WORKS - GRAY CODING AND ITS EFFECT ON CROSS POINTS

Definition 1. Let the positive integer N be the bit length of data points. The region of cross points $A_o(n, p)$, with n from $(N - 1)$ to 1, p from 1 to n, and $n \geq p$, is a set of data points whose grey values are from $(2^n - 2^{n-p})$ to $(2^n + 2^{n-p} - 1)$. The point of grey value 2^n (if it exists) is called the center point of the cross-point region, and the grey value 2^n is called the central value. These regions are called the ideal cross-point regions (ICRs).

By Definition 1, data points in ICRs have grey values that satisfy the rule

$$V_A(n, p) = \{ 2^n - 2^{n-p}, ..., 2^n + 2^{n-p} - 1 \}, \tag{1}$$

where n is the exponent of grey value 2^n being estimated, and p is a value to give the number of bit plane being used. From (1) we can see the ICRs contain data points having grey values above in data of images. We will see that in the scheme for lossless image compression, the position of center points 2^n in ICRs plays an important role in determining the size and the position of ICRs because of the high correlation of grey values of pixels in images.

Proposition 1. Let n be the exponent of the central value 2^n in the ICR $A_o(n, p)$, where n is from $(N - 1)$ to 1, N is a positive integer representing bit length, p from 1 to n, and $n \geq p$. Bits of Gray codes in the ICR $A_o(n, p)$ on the bit plane $(n - p)$ are always 1 bits (if $p = 1$) or 0 bits (if otherwise).

© Springer International Publishing Switzerland 2015

V. Van Toi and T.H. Lien Phuong (eds.), *5th International Conference on Biomedical Engineering in Vietnam,*

IFMBE Proceedings 46, DOI: 10.1007/978-3-319-11776-8_104

From [3] we can see with a value of n in the interval from 1 to $(N-1)$ and $p=1$, grey values of data points in the region $A_o(n, 1)$ are from $(2^n - 2^{n-1})$ to $(2^n + 2^{n-1} - 1)$, so they may be expanded under the form of polynomials of radix 2 appropriately.

After Gray code transformation, the polynomials mentioned above always give 1 bits on the bit plane $(n-1)$. This is very good for compressing real data because the probability of 1 bit 1 in ICRs is always 1, and the probability of 0 bit in ICRs is always 0. This fact plays an important role in optimizing the probability of a data bit in the scheme.

Similarly, if p is 2, 3, ... or n, grey values of data points in the region $A_o(n, p)$ are from $(2^n - 2^{n-p})$ to $(2^n + 2^{n-p} - 1)$, so they may be also expanded under the form of polynomials of radix 2 correspondingly. After Gray coding, we always have 0 bits in ICRs $A_o(n, p)$ on the bit plane $(n-p)$. This is good for optimizing the probability of data bits of data, because the probability of 1 bits in $A_o(n, p)$s is always 0, and the probability of 0 bit in these regions is always 1.

III. IDEAL CROSS-POINT REGIONS WITH LOCO-I ALGORITM

A. Modeling with Prediction Errors

Generally, lossless image compression schemes can be divided into two main stages: modeling and coding. In the stage of modeling here we firstly use the standard of JPEG-LS where the core is LOCO-I [12] to get the matrix of prediction errors, then the theory of cross-point regions with ideal cross-point regions is applied to this matrix. Concretely, in the modeling stage the pixels are coded in a raster scan order. If x is the pixel under consideration at time i, it is referred to by assigning a conditional probability distribution of $P(.|x^{i-1})$ where $x^{i-1} = x_1 x_2 ... x_{i-1}$. Thus, the code length contributed by x_i is given by $-\log_2 P(x_i|x^{i-1})$. Ideally, this value averages to the entropy of the probabilistic model being used. The probability assignment above is usually divided into three further steps. In the first step, prediction, a value for x_i is guessed based on a subset of the past sequence x_{i-1}. The second step is context modeling, which is the determination of the context (smooth, textured, edgy, etc.) in which x_i occurs. This is also based on the past sequence of pixels, mostly pixels in the immediate neighborhood of x_i. Finally, a probabilistic model is produced for the prediction error, $e_i \triangleq (x_i - \hat{x}_i)$, based on the context of x_i. Using ideal cross-point regions over these prediction errors gives us a good way to change data in the stage of modeling to improve the compression ratio for the scheme.

The context classification process of JPEG-LS [9] and [12] is based on the simple causal template depicted in Figure 1. Here, x is the pixel to be predicted and a, b, c and d

are previously encoded pixels in the immediate neighborhood of 'x'.

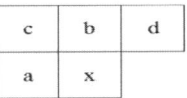

Fig. 1 JPEG-LS Predictive Template

Pixel 'x' is defined to be in a smooth region if all four neighboring pixel values, a, b, c and d are equal to each other. For non-smooth areas, JPEG-LS uses the following simple edge detection/prediction scheme, to determine the predictive value of each pixel, x.

$$\hat{x} = \begin{cases} min(a,b) & if\ c \geq max\ (a,b) \\ max(a,b) & if\ c \leq min\ (a,b) \\ a+b-c & otherwise \end{cases} \quad (2)$$

Thus, with (2) we can get the predictive values of all data pixels in images, x. Instead of using data point x themselves for the step of modeling, we compute to get prediction errors. This substep in modeling will make reduce the inter-pixel redundancy [9], and this is good for the coding step that is used to reduce coding redundancy [9].

B. Application of Ideal Cross-Point Regions to Prediction Errors

After predicting x, we can calculate the prediction residuals. It is widely accepted that the distribution of prediction residuals in continuous tone images can often be approximated by a Laplacian distribution. It makes data points in cross-point regions not change much, this leads to the fact that the probability of data bits 1/0 on bit planes $(n-p)$ increases, and as the result, we can increase the compression ratio.

In principle, a prediction residual can take on any value ε in the range $-(\alpha-1) \leq \varepsilon \leq \alpha-1$, where $\alpha = 2^\beta$ is the size of the input alphabet. It is not good because we need $(\beta+1)$ bits to store the error matrix. In order to solve this problem, we have two steps to transform the error value. First, we convert the errors into positive value by using the code: *if (err < 0) {err = - 2.err - 1} else {err = 2.err}*. Second, we subtract the errors which are larger than $\alpha-1$ by α in order to "move" these values to [0, α] and store their coordinates because we need these information in decompressing progress. After these steps, we now can store the error matrix and finish LOCO-I transform.

C. Diagonal Edge Based Prediction

Our search for diagonal edges should happen when the condition $c \geq max(a, b)$ or the condition $c \leq min(a, b)$ is satisfied. We now formulate the formal conditions for the accurate detection of diagonal edges of the nature illustrated as follows:

$$((c - max\,(a,b)) \geq T_1\; OR\; (min\,(a,b) - c) \geq T_1)\; AND\; (abs\,(a - b) \leq T_2),$$

where T_1 and T_2 are predefined positive thresholds (ideally $T_1 > T_2$) that could be used in deciding the contrast of pixel values being present across the detected diagonal edges.

We introduce the reader to a modified prediction strategy, which increases the accuracy of prediction near diagonal edges. In order to have a suitable predictor for the pixel x, in the presence of a diagonal edge of the nature, we propose to equalize the gradients, $(b - d)$ and $(a - \hat{x})$. Thus,

$$(b - d) = (a - \hat{x}) \Longrightarrow \hat{x} = a + d - b.$$

The pseudo-code for edge detection/prediction could be modified as bellow:

```
if ( c ≥ max (a,b) )
    { if ( (c − max (a, b) ) > T₁  AND  (abs (a − b) ≤ T₂ )
        x̂ = a + d − b;
    else x̂ = min (a,b) }
else if ( c ≤ min (a,b) )
    { if ( (min (a, b ) − c) > T₁ AND (abs (a − b) ≤ T₂) )
        x̂ = a + d − b;
    else x̂ = max (a,b); }
else
    x̂ = a + b − c
endif
```

Note that according to the above pseudo-code, the diagonal edges are detected from edges that would otherwise be categorized as vertical/horizontal edges and not from areas, which would otherwise be considered as non-edge detected.

So, the algorithm to detect diagonal edges based prediction can make the prediction residual at the point being estimated and its neighborhood smaller, which means the error matrix contains values that are not much different from the others. This is convenient to the ICRs which need approximate values to optimize probability of data bits after Gray coding in these regions to improve the step of coding from the scheme of entropy coding - ICRICLI. And in the next step – the step of coding, when ICRs are used we can reduce coding redundancy with the process of determining ICRs that are big enough for the process of optimizing probability of data bits. We can see if there are many more ICRs which are continuous and next to each other, we can link ICRs to get areas of ICRs whose sizes are bigger than the original ICRs.

IV. THE ICRICLI SCHEME FOR LOSSLESS IMAGE COMPRESSION

Figure 2 presents a broad overview of the ICRICLI scheme. We have 6 steps from 1 to 6. Step 1 is the LOCO-I transform in JPEG-LS [9], [12] for getting the matrix of prediction errors from the matrix of data of image. Step 2 (Cross-point regions) looks for ideal cross-point regions where we can optimize the probability of data bits; after the step we have all ICRs corresponding to the central values 2^n where we can optimize the probability of data bits in the regions. These regions are coarsely found because Step 5 (Recomputing ICRs) will identify exactly the regions where the probabilities of data bits need to be optimized in ICRs big enough, some cross-point regions satisfying step 2 but not satisfying step 5 will be canceled. Step 3 (Gray coding) carries out Gray code transformation over the matrix of prediction errors; depending on the depth of prediction errors we have the bit planes where data bits are located, these bits of the Gray codes are in ICRs being nested with the same central value 2^n but on different bit planes. Step 4 (Bit plane decomposition) gives bit planes from the matrix of Gray codes, each bit plane has the private value p that certifies the number of bit plane being used [Definition 1, Section 1] for the process. Step 5 (Recomputing ICRs) is used to cancel small ICRs determined in Step 2, in order to reduce the time and the space for the process of coding which does not affect much the compression ratio. This step needs to look up the data bit that is being encoded and the data bits that were encoded in the scheme to compare the number of data bits will be encoded, this let us get new ICRs for the process of optimizing probability of data bits. Finally, the step of Coding – Step 6, is the process of encoding using Jones' algorithm [10]; it will give us the data, the compressed image.

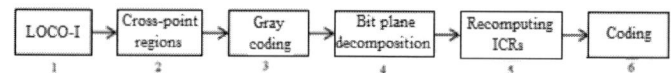

Fig. 2 The ICRICLI scheme

The decoding algorithm of the ICRICLI scheme is an inverse process of the one above. After the encoding process, we have the ICR map and all data for decoding – the codeword, the header compressed of image…, these become the input data for the decoding process.

V. EXPERIMENTAL RESULTS

The proposed scheme is lossless, so it can be used for compressing both of photographic and biomedical images. We performed experiments on some test images with 4

Table 1 Compression ratio of the scheme proposed ICRICLI and others (for whole image files)

Image	Size	ICRICLI	ICRIC [15]	CRICM [3]	AES (WinRAR 3.40) [13]	JPEG 2000 [14]
Couple	512 x 512	1.752:1	1.618:1	1.642:1	1.427:1	1.513:1
Lena	512 x 512	1.715:1	1.615:1	1.678:1	1.567:1	1.752:1 *
Mountain	640 x 480	1.441:1	1.491:1 *	1.493:1 *	1.520:1 *	1.177:1
Mandrill	512 x 512	1.280:1	1.237:1	1.242:1	1.206:1	1.291:1 *
Chest 1	512 x 480	1.946:1	1.910:1	1.966:1 *	1.673:1	1.877:1
Chest 2	512 x 480	1.883:1	1.818:1	1.900:1 *	1.457:1	1.874:1
Chest 3	512 x 480	2.403:1	2.260:1	2.363:1	2.201:1	2.268:1
Chest 4	512 x 480	2.034:1	1.932:1	2.034:1	1.728:1	1.922:1
Chest	256 x 256	1.722:1	1.703:1	1.735:1 *	1.573:1	1.243:1
Joint	512 x 400	2.165:1	2.080:1	2.151:1	1.932:1	2.225:1*

* These results are better than those of ICRICLI.

photographic (Couple, Lena, Mountain, Mandrill) and 6 medical (Chest1, Chest2, Chest3, Chest4, Chest, Joint) images.

Table 1 presents the results for compressing the bit map images (.BMP) by using ICRICLI with $T_1 = 60$, $T_2 = 8$ and $p = 2$. The compression ratio used here is the ratio between files of images, i.e. including the headers of the original image and the compressed image. The results of compression ratios from other methods are also presented here for references: CRICM [3], AES [13], lossless JPEG 2000[14], ICRIC [15] which is the proposed technique without LOCO-I algorithm. From these we can see the results of ICRICLI are mostly better than those of the others, especially with medical images.

VI. CONCLUSIONS

In this paper we have proposed the scheme of ICRICLI that uses the JPEG-LS prediction with the core of LOCO-I algorithm by introducing accurate diagonal edge detection in ideal cross-point regions. This combination gives better results over other schemes let us see that the theory of cross-point regions with ideal cross-point regions can be used as the background for LOCO-I algorithm, and furthermore for other transformations such as discrete cosine/sine transform, wavelet transform and so on in order to improve the compression ratio in lossless compression scheme. This scheme can be also used for cryptography systems when cross-point regions play an important role because of the different concentration of data bits between inside and outside these regions.

REFERENCES

1. T. Tin, V. D. Thanh, N. K. Sach, and S. Higuchi, Effect of Gray coding on the state and the distribution of data bits in cross point regions, Proceedings of ICCEA, IEEE catalog No. 04EX875, Beijing, China, pp. 553 – 556, Nov 2004.
2. Tin T. Dang, Thanh D. Vu, and Aaron Becker, A lossless coding scheme for images using cross-point regions for modeling, The proceedings of The 2007 IEEE International Conference on Electro/Information Technology , EIT2007, IEEE Catalog No. 07EX1665, Chicago, Illinois, USA, , pp. 114-119, May 17-20, 2007.
3. Tin T. Dang, Thanh D. Vu, Sach K. Nguyen, Seihaku Higuchi, Cross Point Regions on Multiple Bit Planes for Lossless Images Compression, Journal of IET, Image Processing, Vol 8, August 2011.
4. T. Tin, V. D. Thanh, and N. K. Sach, Effect of Gray codes on data areas around cross points, Vietnam Journal of science and technology, vol 42, No 3, pp. 21-26, 2004.
5. C. J. Lu, and S. C. Tsai, A note on iterating an α-ary Gray code, SIAM J. Discrete Math, vol 14, No 2, pp. 237-239, 2001.
6. N. K. Sach, Digital Images and video processing, Science and Technology Publisher, Vietnam, 1997, pp. 172-176.
7. M. C. Er, On generating the N-ary reflected Gray codes, IEEE transactions on computers, vol.33, pp. 739-741, 1984.
8. W. Gilbert, A cube - filling Hilbert curve, Math Intell 6, pp. 78, 1984.
9. R. C. Gonzalez, and R. E. Woods, Digital image processing, 2nd Editon, Prentice Hall, 2002.
10. C. B. Jones, Efficient coding system for long source sequences, IEEE Trans. Inform. Theory, vol 3, No. 27, pp. 280-291, 1981.
11. D. T. Tin, V. D. Thanh, and N. K. Sach, An improvement of Jones' method of lossless data compression, Posts and Telecommunications journal of Vietnam, No. 7, pp. 65-69, May 2002.
12. Marcelo J. Weinberger and Gadiel Seroussi, Hewlett-Packard Laboratories, Palo Alto, CA 94304, USA, Guillermo Sapiro, Department of Electrical and Computer Engineering, University of Minnesota, Minneapolis, MN 55455, USA, "The LOCO-I Lossless Image Compression Algorithm: Principles and Standardization into JPEG-LS", http://www.hpl.hp.com/loco/, Otc 2011.
13. Federal information processing standards publication 197, Announcing the Advanced Encryption Standard, National Institute of Standards and Technology (NIST), Nov 2001.
14. T. Acharya, and P. S. Tsai, JPEG2000 Standard for image compression, Wiley, 2005.
15. Dang Thanh Tin, Vu Dinh Thanh, Nguyen Kim Sach, Ideal cross point regions for lossless image compression, Vietnam Journal of Science and Technology, Vol 46, No 5, 2008.

Author: DANG Thanh Tin
Institute: Ho Chi Minh city University of Technology
Street: Ly Thuong Kiet
City: Ho Chi Minh
Country: Viet Nam
Email: dttin@hcmut.edu.vn / tindt@yahoo.com

Applying the Image Compression Algorithm of ICRICM to a Plugin Integrated into MIPAV Software

Tin T. Dang and Khoa Anh Tran

Ho Chi Minh City University of Technology, Ho Chi Minh City, Vietnam

Abstract— In this paper, we introduce MIPAV (Medical Image Processing, Analysis, and Visualization software) and its plugin architecture to build the plugin for lossless image compression by the theory of cross-point regions with ideal cross-point regions on multiple bit planes. MIPAV is an open-source software, a good tool for researching and studying operations over medical images, but it does not have the tool for image compression. To build a plugin integrated into MIPAV software we use the scheme of ICRICM (Ideal Cross-point Regions for Image Compression on Multiple bit planes). This plugin can create a new output file from the original image, and the size of this output file is smaller than before. And inversely, the output file will be used in the decompression process to get the original image without any differences.

Keywords— MIPAV, ICRICM algorithm, cross point, ideal cross-point region, lossless image compression.

I. INTRODUCTION

The purpose of this work is to build a new plugin to losslessly compress images, the scheme of ICRICM (Ideal Cross-point Regions for Image Compression on Multiple bit planes) [1] for the software of MIPAV. This paper has six sessions. The first is this introduction. We introduce MIPAV software in Section II, and its plugin architecture in Section III. In this two parts, the most information is from [2], [3] and [4]. Section IV presents the ICRICM algorithm and the way we apply it to a plugin into MIPAV. Section V gives the results we got from running the plugin in MIPAV, and the final is Section VI which gives conclusions about applying this plugin to medical imaging in future.

II. MIPAV SOFTWARE

The MIPAV (Medical Image Processing, Analysis, and Visualization) gives quantitative analysis and visualization of medical images of different medical equipments such as PET, MRI, CT.... Using MIPAV with its analysis tools, the researchers can stay at remote sites (via network/the internet) and easily obtain data, i.e. images, and analyze them. This can help us enhance the ability to research, diagnose, monitor, and treat disorders for patients.

MIPAV is a Java application and can be run on any Java-enabled platform such as Windows, UNIX, or Macintosh OS X. Fig. 1 shows the logo of MIPAV software.

Fig. 1 MIPAV logo

- Platform independence: The MIPAV application is platform independent because it is written in Java, which is an object-oriented, interpreted, programming language that was developed by Sun Microsystems. Java source code is compiled into the bytecode, which is machine-level code being compiled specifically for the Java Virtual Machine (JVM). There are versions of the Java VM for different platforms. The same program (bytecode) can run on any of those versions. If researchers run a Java program on a Windows platform, the bytecode is interpreted by the Java VM that has been specifically designed for the particular Windows platform. If the same program is run on a Solaris platform; the bytecode is then interpreted by the Java VM that was specifically designed for the Solaris platform.

- Supported image types: MIPAV supports over 20 different industry-standard image formats including: DICOM, TIFF, Analyze, and RAW. MIPAV reads and writes images in both big and little endian formats.

- Visualization of images: MIPAV allows researchers to visualize datasets using a variety of presentation formats, including lightbox, triplanar, cine, and animate. Once researchers display the image dataset, they can adjust the lookup table (LUT), apply prepackaged pseudo-color LUTs to highlight structures of interest, control the magnification level, adjust the transfer function, and more.

- Volume of interest (VOI) segmentation and analysis: MIPAV also allows researchers to perform statistical calculations on masked and contoured VOIs. Statistical

V. Van Toi and T.H. Lien Phuong (eds.), *5th International Conference on Biomedical Engineering in Vietnam,*
IFMBE Proceedings 46, DOI: 10.1007/978-3-319-11776-8_105

results can be saved to an ASCII text file and imported to another program, as needed.

- Extensibility with Java plug-ins: Many components of image dataset processing, analysis, and visualization techniques are general and can be applied to many types of data. MIPAV allows researchers, who have the programming resources, to add a customized Java plug-in to the application. To program a plug-in, researchers must have a strong understanding of the underlying structure of the application's software design.

III. MIPAV PLUGIN ARCHITECTURE

Plug-in programs, also known simply as plug-ins, are utilities or sets of instructions that add functionality to a program without changing the main purpose of the program. As a Java oriented object program, plugins in MIPAV software have the private underlying structure. Users who know how to program in Java can write a plug-in program that adds support for a new file format, creates a new view, or applies a new algorithm to an image.

1. Understanding Plug-in Programs

In MIPAV, you use Java to write and compile plug-in programs to perform specific functions. There are some types of plug-in programs supported in MIPAV:

- Algorithm: an algorithm type of plug-in performs a function on an image.
- File: A file type of plug-in allows MIPAV to support a new file format or to read/write images used in MIPAV. Because MIPAV already supports a large number of file formats, so this type is unnecessary generally.
- View: A view type of plug-in introduces a new view, or the way in which the image is displayed (i.e. animate views).
- File Transfer: This type of plug-in supports to transfer files between two file systems.
- Generic: This is generic MIPAV plugins which do not require an open image and can be run from the user interface.

After developing a plug-in program, we can then install it into MIPAV as an application and access it from the PlugIns menu of the MIPAV window. Depending on type of installed plugins they appear in exact sub-menu of Plugins MIPAV menu. For example, if our plugin type is Algorithm, it will appear under the PlugIns > Algorithm menu.

2. Developing Plug-in Programs

MIPAV provides the following classes for developing plug-in programs: PlugInAlgorithm.class, PlugInFile.class, PlugInView.class, PlugInGeneric.class, …

Plug-in programs are developed in this contribution are the Java programs integrated into MIPAV. So the following steps of building plugins need to be taken care of.

- Determining the type of plug-in program: The first step of creating a plug-in program is to determine the type writer wants to create, which depends on its purpose. MIPAV plug-in programs can be of the algorithm, file, view or other types. Each of type MIPAV provides one way to the coder in writing the source code.
- Determining which version of Java to use: it is necessary to compile the execute program, plugin must use the same version of Java that was used to build MIPAV. From menu Help > JVM Information of the MIPAV window, the coder can view the current version Java used.
- Writing the source code: Some lines of code must appear in the source code so that the plug-in program interfaces correctly with MIPAV. This includes the imported file and declares the classes of MIPAV library into each of plugin file.
- Building and compiling plug-in programs: to build a new plug-in program for MIPAV, you must first install a building environment, alter the path environment variable, and compile the plug-in files. All details are in [4].
- Creating a self-contained plug-in frame: a self-contained plug-in is a Java application that does not rely on the default MIPAV user-interface, but, instead, hides MIPAV and displays its own image(s) with the action/algorithm handling specifically to its frame.
- Installing plug-in programs: installing simple plug-in programs merely copies files into the user's home directory: c:\Documents and Settings\<user ID>\mipav\plugins (Windows system) or /user/<user ID>/mipav/plugins (UNIX system). There are two methods for copying the files: using MIPAV's plug-in installation tool, e.g. in the MIPAV window, select PlugIns > Install PlugIns, or using the operating system's tool for copying the files. Some cases need to restart MIPAV for applying the changes.

3. MIPAV Plugins Structure

To build plug-in programs, three files are typically required in its structure:

- PlugnInFoo.java: provides an interface to MIPAV and the plugin.

- PlugInDialogFoo.java: invokes the dialog to get user-supplied parameters; it can be hidden when no parameters are required.
- PlugInAlgorithmFoo.java: provides the actual algorithm to be implemented. It can be a mixture of calls to MIPAV's API, C programs, Perl, ITK, etc.

Where Foo is the name with that the coder supplies the program. For example, in this case, we call it Foo plugin.

The type of Foo plugin will be declared inside PlugInFoo.java and the running function for this type also must be overridden.

```
1       import gov.nih.mipav.plugins.PlugInAlgorithm ;
2       import gov.nih.mipav.view.*;
3       import java.awt.*;
4       public class PlugInFoo implements PlugInAlgorithm {
5       public void run(Frame parentFrame, ModelImage image){
6               // call to PlugInDialogFoo
7       }}
```

Fig. 2 PlugInFoo.java

Fig. 2 is the detail code of Foo plugin. The type of Foo plugin is Algorithm so it must implement the PlugInAlgorithm (line 4) and the class PlugInAlgorithm also need be imported (line 1). The run function in line 5 is used to call to the user interface of Foo plugin, in this case, this interface is built in PlugInDialogFoo.java.

```
1       import gov.nih.mipav.view.*;
2       import gov.nih.mipav.view.dialogs.*;
3       import gov.nih.mipav.model.algorithms.*;
4       mport gov.nih.mipav.model.structures.ModelImage;
5       import java.awt.*;
6       import java.awt.event.*;
7
8       public class PlugInDialogFoo extends JDialogBase implements AlgorithmInterface {
9       public PlugInDialogFoo(Frame theParentFrame, ModelImage im){
10              // init the user interface
11      }
12      public void actionPerformed(ActionEvent event) {
13              // call to execute Foo algorithm by PlugInAlgorithmFoo instance
14      }}
```

Fig. 3 PlugInDialogFoo.java

Fig. 3 presents the detail code in PlugInDialogFoo.java file. This file is used to generate the user interface of Foo plugin

and it also calls on the Foo algorithm by the actionPerformed function (line 13). Fig. 4 below is the detail code in PlugInAlgorithmFoo.java file where all functions of the Foo algorithm will be built and run.

```
1  import gov.nih.mipav.model.algorithms.AlgorithmBase;
2  import gov.nih.mipav.model.structures.*;
3  import gov.nih.mipav.view.*;
4  public class PlugInAlgorithmFoo extends AlgorithmBase
5  {
6  public    PlugInAlgorithmFoo(ModelImage    destImg,
   ModelImage srcImg) {
7       super(destImg, srcImg);
8  }
9  public void runAlgorithm() {
10      // run Foo algorithm
11 }}
```

Fig. 4 PlugInAlgorithmFoo.java

IV. BUILDING ICRICM PLUGIN TO MIPAV

A. ICRICM Scheme

Fig. 5 presents two processes of the ICRICM scheme. The compression process will create the output file by using the theory of cross-point regions in the step of modeling of the entropy coding. The decompression is an inverse process, with taking the output file to rebuild the same image file.

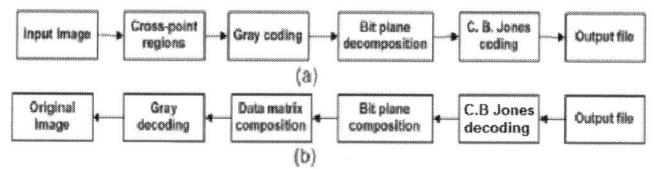

Fig. 5 The ICRICM scheme: (a) Compression process (b) Decompression process

B. ICRICM Plugin

ICRICM plugin is designed for creating the lossless image compression in MIPAV from the ICRICM scheme. Its function is to read an image file (bit map or DICOM) and to compress it to get the compressed file (i.e *.out2 extension). On the other hand, the decompression process will get the output file to obtain the image which is identical to the original. This plugin is in basic functions, so we put it in the Generic type MIPAV plugin. Therefore, the structure of ICRICM plugin has three Java files as the followings:

- PlugInICRICM.java: provides the interface to MIPAV, Fig. 6 is its source code.

- PlugInDialogICRICM.java: invokes the dialog and gets user-supplied parameters.
- PlugInAlgorithmICRICM.java: executes two methods being compress and decompress.

```
1 import gov.nih.mipav.plugins.PlugInGeneric;
2 public class PlugInICRICM implements PlugInGeneric{
3   public void run() {
4     new PlugInDialogICRICM(true);
5 }}
```

Fig. 6 PlugInICRICM.java

After compiling all Java files of ICRICM plugin with Java version 1.7.0_17 and MIPAV version 7.0.1, we can install it to MIPAV by MIPAV installation tool. The result is presented in Fig. 7. Fig. 8 is the main interface of ICRICM plugin from the MIPAV calling.

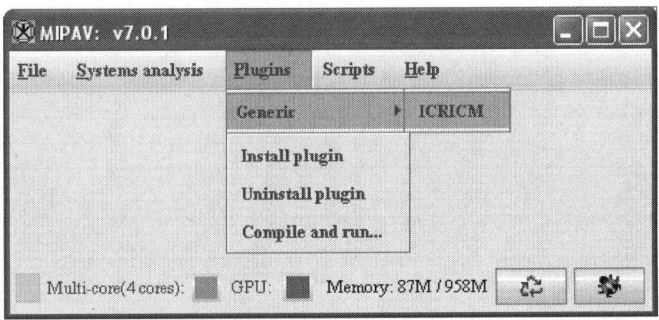

Fig. 7 ICRICM plugin in MIPAV software

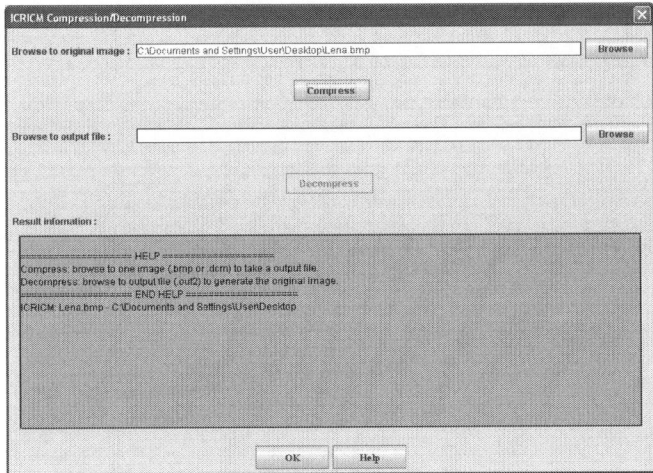

Fig. 8 ICRICM plugin user interface

V. EXPERIMENTAL RESULTS

The plugin for losslessly compressing images to MIPAV using the ICRICM scheme is completely designed, so we can use the software of MIPAV with this plugin as one of its components for losslessly compressing biomedical images. Table 1 presents some results for compressing images by MIPAV using the plugin of ICRICM. The compression ratio used here is the ratio between original images and output compressed file.

Table 1 Compression ratio of the scheme proposed ICRICM

Image	Size	ICRICM [3]
Chest	512 x 248	1.7035:1
Joint	256 x 256	2.5345:1
OT-MONO2-8-hip	512 x 512	1.3301:1

VI. CONCLUSIONS

Appling the scheme of ICRICM as a plugin integrated into MIPAV software enables us to use MIPAV more effectively. This is useful and convenient on examining and transferring images for researching and studying. Moreover, the output files from this plugin are more compact, safer because this plugin can be developed to increase the security, a private key from the user will be joined in the compression process. After this work, MIPAV can become a good tool to study and to try new ideas in the field of medical imaging for researching and teaching.

REFERENCES

1. D. T. Tin, V. D. Thanh, N. K. Sach, and S. Higuchi (2008) A lossless coding scheme for images with ideal cross point regions on multiple bit planes, PAKDD 2008. Osaka, Japan
2. MIPAV Help at http://mipav.cit.nih.gov/pubwiki/index.php/MIPAV_Help.
3. Developing Java Plugins in MIPAV at http://mipav.cit.nih.gov/documentation/presentations/plugins.pdf
4. Developing Plugin Programs at http://mipav.cit.nih.gov/pubwiki/index.php/Developing_Plugin_Programs

Author: DANG Thanh Tin
Institute: Ho Chi Minh city University of Technology
Street: Ly Thuong Kiet
City: Ho Chi Minh
Country: Viet Nam
Email: dttin@hcmut.edu.vn / tindt@yahoo.co

Software Design for Training and Supporting Knowledge of Ventilators for Clinical Engineers

Nguyen Nhan Thien and Huynh Quang Linh

Faculty of Applied Science, University of Technology – VNU HCM, Ho Chi Minh City, Viet Nam

Abstract— **Understanding ventilator principles and its operations is very essential for clinical engineers, especially in rural and remote areas in Vietnam. The purpose of paper is to present a software designed to provide theoretical and practical knowledge of ventilators, which are widely used in many departments of hospitals. This software can be used for training as well as supporting knowledge of many usual types of ventilator for physicians and clinical engineers.**

Keywords— **Ventilators, software, training, clinical engineers.**

I. INTRODUCTION

Medical ventilators nowadays play very significant role in healthcare system. They are essential devices which are used mainly in ICU departments of hospitals. Understanding ventilator principles and its operations is essential for clinical engineers who work in hospitals in order to efficiently assist physicians in clinical manipulation as well as in maintenance or repair of mentioned devices. Our purpose is to provide a flexible software used in training and supporting basic knowledge about medical ventilators and its troubleshootings for clinical engineers working in suburb hospitals where there are still a lot of lacks and difficulties in accessing actual information in mentioned domain.

II. MATERIALS AND METHODS

The software includes four main sections: Knowledge, Operation, Maintenance and Simulation.

In Knowledge section, the software has three tabs: Ventilator's mode, Ventilator's settings, and Curves-Loops-Equations of mechanical ventilation which provide knowledge of 20 modes of mechanical ventilation (Controlled Mechanical Ventilation - CMV, Assist/Control - A/C, Assisted Mechanical Ventilation - AMV, Intermittent Mandatory Ventilation - IMV, Synchronized Intermittent Mandatory Ventilation - SIMV, Continuous Positive Airway Pressure - CPAP, Bi-level Positive Airway Pressure - Bi-PAP, Bi-level Pressure Assist, Positive Control Ventilation - PCV, Pressure Controlled Inverse Ratio Ventilation - PCIRV, Pressure Support Ventilation - PSV, Volume Support - VS, Proportional Assist Ventilation - PAV, Airway Pressure Release Ventilation - APRV, Mandatory Minute Volume - MMV, Volume Assured Pressure Support - VAPS, Pressure Regulated Volume Control - PRVC, Adaptive Support Ventilation - ASV, High Frequency Ventilation - HFV, Neurally Adjusted Ventilatory Assist - NAVA) [1], 12 settings of ventilators (Volume Tidal, Flow rate, Ratio of Inspiration to Expiration - I:E Ratio, Trigger Sensitivity, Frequency - Breath rate, Positive End Expiratory Pressure - PEEP, Peak Inspiratory Pressure - PIP, End Inspiratory Plateau Pressure, Pmax, Ramp-Slope-FlowAcc, FiO2, Monitoring Gas Exchange) [2], 6 types of graph used in ventilator (Pressure vs. Time, Flow vs. Time, Volume vs. Time, Pressure - Volume Loops, Flow - Volume Loops, Trends reviewed) [3] to assess the patient's condition, and equations of mechanic ventilation [4].

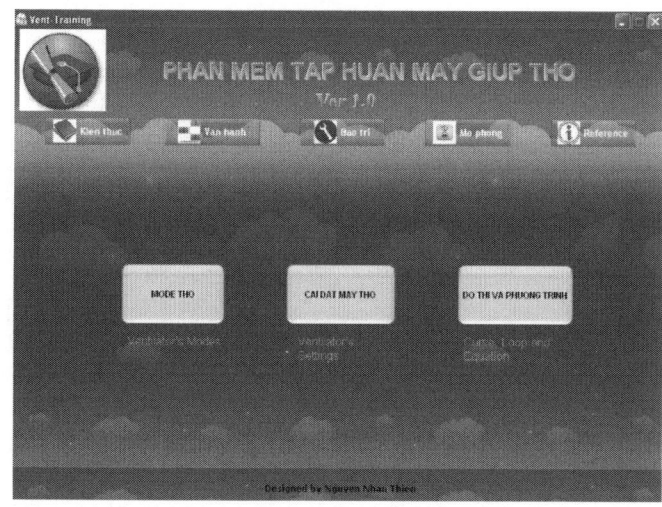

Fig. 1 Tabs in Knowledge section

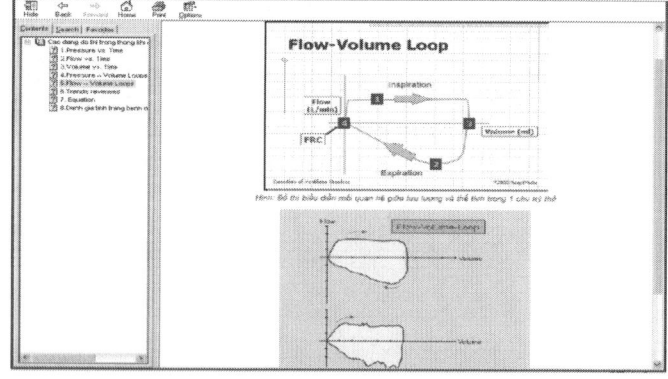

Fig. 2 Flow - Volume Loops in tab Curves-Loops- Equations

V. Van Toi and T.H. Lien Phuong (eds.), *5th International Conference on Biomedical Engineering in Vietnam,*
IFMBE Proceedings 46, DOI: 10.1007/978-3-319-11776-8_106

In Operation section, the software has four tabs: Ventilator's structure and component, Clinical application of ventilator, Ventilator's alarm and Ventilator maintenance which provide knowledge of the structure and other components of a ventilator (Power Supply, Processor, Motor and air system, Display, Breathing circuit and water trap, Expiratory Valve, Flow sensor, Oxy sensor, Humidifier, HME, Oxy source, Filter, Nebulizer, Mask), 3 basic clinical applications of ventilator in treatment ARDS - Adult Respiratory Distress Syndrome, COPD - Chronic Pulmonary Disease or Asthma patients [5], methods of preservation and inspection before operating the ventilator, ventilator's alarms and how to handle them (Low/High Tidal Volume, Low/High Minute Volume, Aiway Pressure Low/High - Paw low/high, Disconnection/Leakage, Low PEEP, High PEEP, High f, High/Low Oxygen, Apnea, Power Lost, External Power Lost, Low Battery, Continuous alarm and no signal on ventilator screen, Technical Alarm, Device Failure) [6].

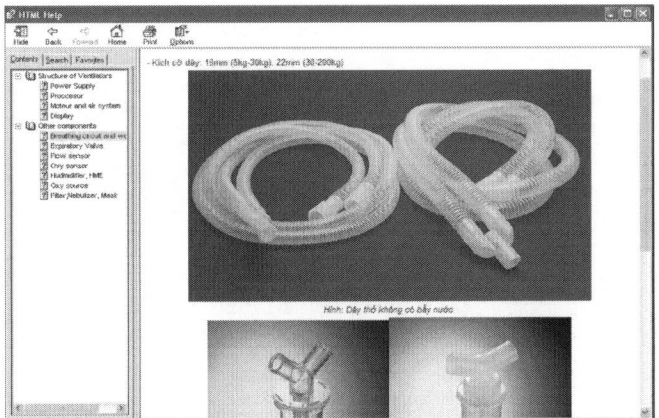

Fig. 3 Contents in tab of ventilator's structure and component

In Maintenance section, the software presents failures and troubleshooting of 7 ventilator at Cho Ray Hospital: Viasys Vela (USA), Viasys LTV1200 (USA), Hamilton Raphael Silver (Switzerland), Draeger Carina (Germany), Draeger Savina (Germany), Newport E150 (USA), Versamed iVent 201 (USA). This section also provides 16 service manual of ventilators from different vendors such as Dragger Babylog 8000, Draeger Evita 2, Draeger Evita 4, Draeger Evita XL, Draeger Savina, Hamilton C2, Hamilton Raphael, Hamilton Galileo, Maquet Servo-I, Maquet Servo-s, Newport e360, Newport HT50, Puritan Bennet 840, Respironics Esprit, Viasys Vela, Versamed iVent201 [7] which are very useful for clinical engineers in inspecting and repairing ventilators.

In Simulation section, the software provides interface guidelines and simulation of Evita 4 Edition ventilator, a Draeger Medical's ventilator (Germany). This simulation is very similar to the real ventilator, it has a full range of functions of a real ventilator that is very helpful for clinical engineers to get familiar with a virtual ventilator and have the opportunity to test theories of ventilator before using it in real.

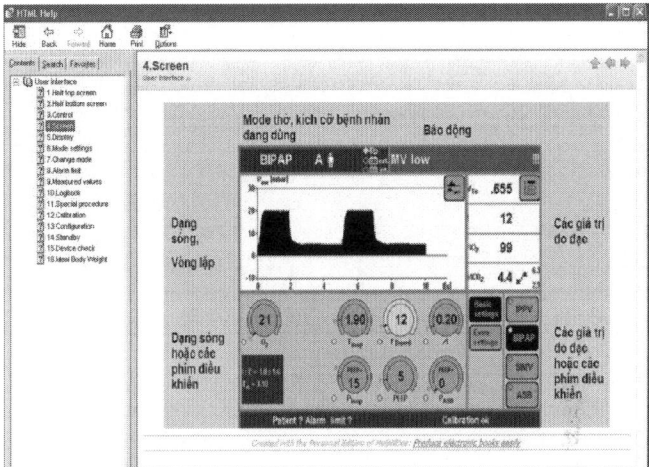

Fig. 4 Interface guidelines of Draeger Evita 4 Edition ventilator

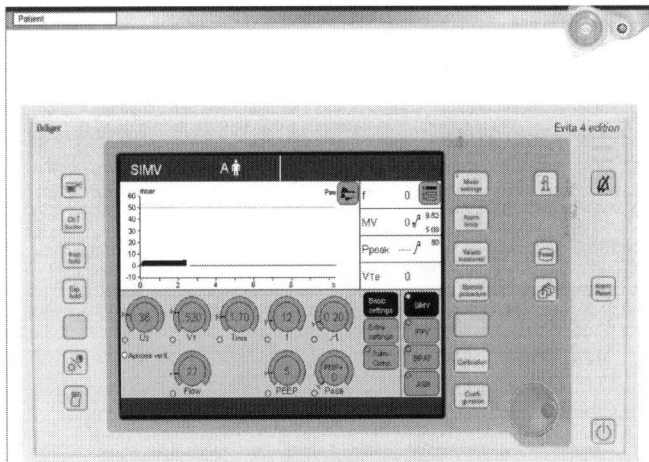

Fig. 5: Simulation of Draeger Evita 4 Edition ventilator

The software is named Vent-Training. It is written in Visual Basic Platform. Its interface and icons are created by using Autoplay Menu Builder. Databases of it are formatted as .chm which take very less space in hard drive.

This software is easy to use and compatible with all operating systems.

III. Result and Discussion

We designed and made up a user-friendly software that is helpful for beginners to become familiar with ventilator. This is a non-commercial software and its database can be updated easily. It is an incomplete software and lacks of searching's tool as well as Question-Answer tests for users. We are still working on it to make it more complete.

IV. Conclusions

With the goal of building a software used for training for clinical engineers in suburb hospitals, we designed a software which can provides and supports knowledge of 20 different modes of ventilation, 6 types of graphs, 12 settings, alarms, structures, maintenance, failures, troubleshootings of ventilators as well as its application in ARDS, COPD and Asthma patients. This software also collects 16 services manual of ventilators from different vendors and has a simulation of a real ventilators which is very helpful for clinical engineers in learning and repairing ventilators.

In the future, we will test this software in some selected hospitals to survey users opinions and comments to improve its usability and deploy it in other medical units.

References

1. Hess D R, Kacmarek R M (1996) Essentials of Mechanicals Ventilation, McGraw-Hill, USA.
2. Hasan A (2010) Understanding Mechanical Ventilation, Springer, London.
3. Deshpande V (2006) Clinical Utility of Ventilator Graphics. http://www.vsrc.org/HandoutVentGraphics.doc .
4. Rittner F, Döring M (2010) Curves and Loops in Mechanical Ventilation, Draeger Medical, Germany.
5. Papadakos P J, Lachmann B (2008) Mechanical ventilation: Clinical applications and pathophysiology, Saunders Elsevier, USA.
6. Draeger Training and Service. http://www.draeger.com/sites/enus_us/Pages/default.aspx.
7. Hamilton Training. http://www.hamilton-medical.com.

Author: Nguyen Nhan Thien
Institute: ChoRay Hospital
Street: Nguyen Chi Thanh
City: Ho Chi Minh
Country:Vietnam
Email: nguyen.nhan.thien89@gmail.com

Author:Huynh Quang Linh
Institute: Faculty of Applied Science, University of Technology
Street: Ly Thuong Kiet
City: Ho Chi Minh
Country: Vietnam
Email: huynhqlinh@hcmut.edu.vn

Design of Electrotherapy Equipment Using Wireless Communication

Nguyen Tuan Anh and Huynh Quang Linh

Faculty of Applied Science, University of Technology, Ho Chi Minh City, Viet Nam

Abstract— **The purpose of the paper is to introduce a prototype of electrotherapy equipment which can connect with the computer via wireless communication and broadcasts 12 types of pulsed waveforms such as symmetric TENS, asymmetric TENS, Galvanic, F20, F1, Trabert, Diadynamic CP, Diadynamic LP, sawtooth wave (E100, E200, E500). Patient and operational data can be registered in a database and easily managed by RFID reading system.**

Keywords— **Electrotherapy, TENS, diadynamic, galvanic, trabert, sawtooth wave, wireless, RFID system.**

I. INTRODUCTION

Electrotherapy is a popular medical treatment method with a noticeable result in pain treatment and kinesthetic stimulation. Popular electrotherapeutical equipment in the market are very diverse in types and functions, but most of them are stand-alone equipment. To improve usage flexibility an electrotherapy equipment prototype which can connect with the computer by wireless communication has been designed and made up. With common electrotherapeutical methods [1] 12 types of pulsed waveforms such as symmetric TENS, asymmetric TENS, Galvanic, F20, F1, Trabert, Diadynamic CP, Diadynamic LP, sawtooth wave (E100, E200, E500) have been designed and could be preset and controlled by the computer together with other related parameters such as frequencies, amplitudes and treatment times. Patient data as well as operational procedure can be registered in a database installed on the computer and flexibly managed by RFID reading system.

II. MATERIALS AND METHODS

Equipment prototype is composed of two main parts: host and client units. The host unit can register patient information using RFID card, exchange it with the computer database, receive preset data from the computer via wireless communication and broadcast data to client unit. The client unit receives signal from host and activates operational process according to determined parameters.

Host unit uses IC AMS1117 [2] for 3.3V power supply, central component PIC-18F4550 [3] written by CCS for connecting with computer via USB gate. Module RFID RC522 522 is used to read RFID card and to process patient

data registration. Wireless module RF24L01 is used for broadcasting to client unit.

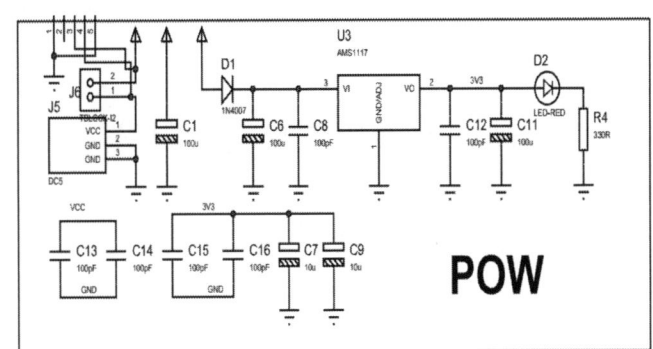

Fig. 1 Power circuit of host part

In the client unit, there are:

- RF24L01 wireless module for receiving signal from computer;
- IC LM2576 [4] for 5V power supply;
- Central component ARM7 - STM32F100C8T6 [5] for producing pulsed waveforms by using DAC of ARM – STM32F100C8T6; and
- Preamplifier IC LM358 [6] for signaling to A amplifiers. Class A power amplifier circuit is used since this device needs high power control and linear output signal.

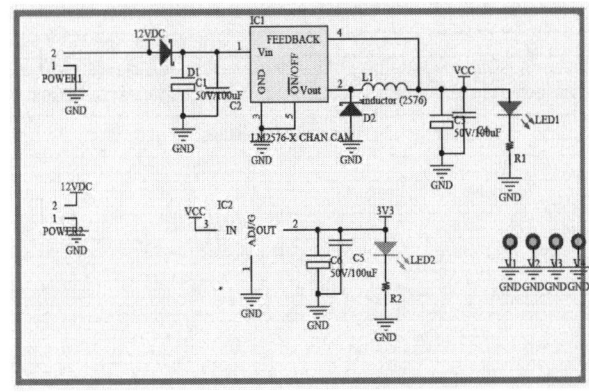

Fig. 2 Power circuit of clients part

© Springer International Publishing Switzerland 2015
V. Van Toi and T.H. Lien Phuong (eds.), *5th International Conference on Biomedical Engineering in Vietnam,*
IFMBE Proceedings 46, DOI: 10.1007/978-3-319-11776-8_107

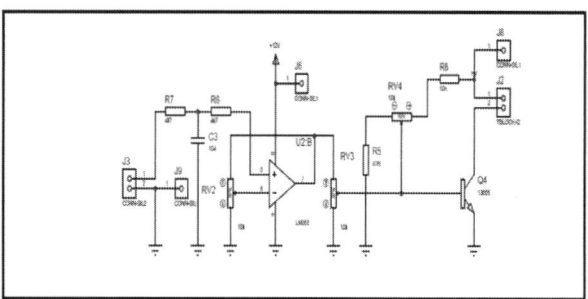

Fig. 3 Amplifiers circuit

Parameters:

- Output power U = 100V, I = 0.1A. Complex power: 10W.
- Design power 10 x √2 ≈ 15(W).
- Current transferring coefficient: ≈ 0,8. Power supply: 15 ÷ 0,8 ≈ 19(W).
- Input voltage is stable in 12V for current supply 20 ÷ 12 ≈ 2(A).
- Design current: I = 2 x √2 ≈ 3(A) . 5V source is selected.
- Transistor 13009 [7] with calculated voltage UCE= 400V and current IC= 12A.

To change 12V DC power to 100V DC, we used the Boost pulse circuit with main components such as MC34063 IC [8] of ON semiconductor.

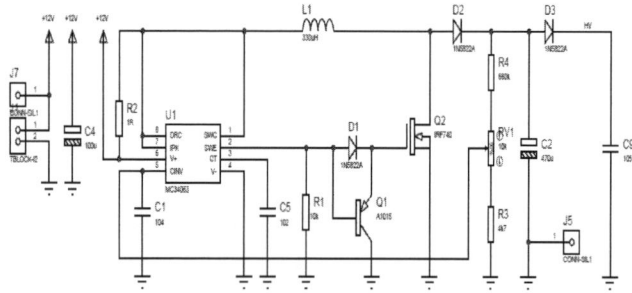

Fig. 4 Boost circuit

In the Fig. 4:

- Power source element increases voltage of Field-effect transistor FET-IRF740 [9] to decrease waste and increase suffering current.
- IC MC34063 has responsibility on creating and controlling oscillation.
- Capacitor C5 creates oscillation for IC MC34063.
- R1, D1 and Q1 create switch circuit for FET.

- Q2 is element for switching to create field variation on coil of wire.
- Spark-coil L1 with dramatically magnetic variation creates inductive current on the wire to capacitor C2 due to diode D2. R4, RV1 and R3 are elements to get samples of feedback current and control oscillation of switch element.
- Diode D3 is used in connecting serial source elements.
- Capacitor C9 filter generated voltage to avoid downgrading of the voltage and high frequency noises when applying for loads.

To increase stability of the current, system with 4 elements of power supply source has been designed.

The controlling software was written in C#.

III. RESULT AND DISCUSSION

Mentioned electrotherapeutical equipment prototype has been made up and tested with a flexible power from 0 to 100V, the amperage the range of 0-100mA and wireless distance of 20m. Results of waveforms were quite satisfactory as shown in the figures 5, 6 and 7. Control parameters were executed with considerable reliability. Utilities with RFID patient registration have increased the flexibility of the system with central computer for simultaneous control and management of multiple electrotherapeutical modules.

However there is still incomplete prototype due to lack of electrically isolating tool and controlling tool at the outgate. The database is necessarily linked and integrated with the patient management central system of health facilities. The whole system have thus to be tested with more therapeutical modules in real clinical condition.

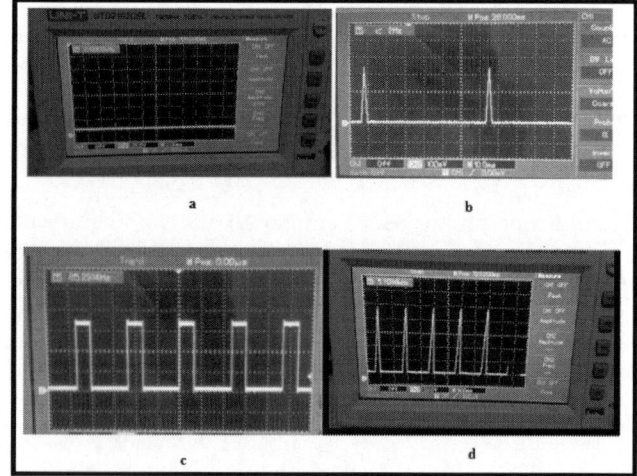

Fig. 5 Result: a) Galvanic current; b) F1 current; c) F20 current; d) sawtooth wave (E100, E200, E500)

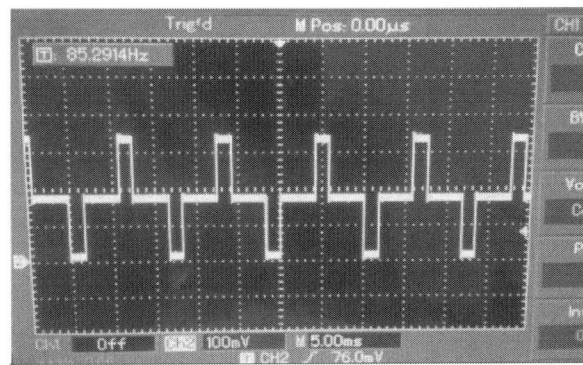

Fig. 6 Symmetric TENS

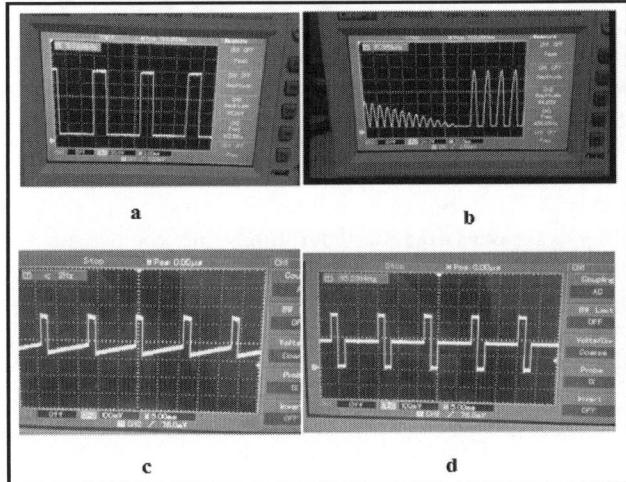

Fig. 7: a) Trabert current; b) Diadynamic LP; c) Asymmetric TENS; d) TENS current

IV. CONCLUSIONS

As a complementary note for the poster about an wireless controller solution applicable in medical instrumentation, the article presents some original features, mainly technology solutions to perform mentioned system. Basic functions of equipment prototype has been successfully tested that promise the prospect for completing the final product for practical application. That is a wireless central control system with improved flexibility such as establishing and automatic operating of predetermined treatment planning, synchronous controlling from central computer etc.

Overcoming lack of isolating tool and improving professional design of hardware and software, equipment could be clinically tested and deployed in physiotherapy facilities with different scales.

REFERENCES

1. Vu Cong Lap, Tran Cong Duyet, Do Kien Cuong, Ha Viet Hien, Nguyen Dong Son, Le Manh Hai, Nguyen Thanh Phuong, Huynh Viet Dung (2005) Physical agents using in physiotherapy, Publisher of Medicine, Vietnam.
2. Datasheet AMS1117, Alldatasheet.com 2003 – 2014, http://www.alldatasheet.com/ view.jsp? Searchword= AMS1117.
3. Datasheet PIC18F4550, Alldatasheet.com 2003 – 2014, http://www.alldatasheet.com/ view.jsp?Searchword= PIC18F4550.
4. Datasheet LM2576, Alldatasheet.com 2003 – 2014, http://www.alldatasheet.com/ view.jsp?Searchword = LM2576.
5. Datasheet STM32F100C8T6, Alldatasheet.com 2003 – 2014, http://www. alldatasheet.com/view.jsp? Searchword= STM32F100C8T6.
6. Datasheet LM358, Alldatasheet.com 2003 – 2014, http://www.alldatasheet.com/ view.jsp?Searchword= LM358.
7. Datasheet Transistor 13009, Alldatasheet.com 2003 – 2014, http://www.alldatasheet.com/view.jsp?Searchword= 13009.
8. Datasheet MC34063, Alldatasheet.com 2003 – 2014, http://www.alldatasheet.com/ view.jsp?Searchword= MC34063.
9. Datasheet IRF740, Alldatasheet.com 2003 – 2014, http://www.alldatasheet.com/ view.jsp?Searchword= IRF740.

Author: Nguyen Tuan Anh
Institute: Faculty of Applied Science, University of Technology
Street: Ly Thuong Kiet
City: Ho Chi Minh
Country: Vietnam
Email: nguyentuananh_89@yahoo.com

Author: Huynh Quang Linh
Institute: Faculty of Applied Science, University of Technology
Street: Ly Thuong Kiet
City: Ho Chi Minh
Country: Vietnam
Email: huynhqlinh@hcmut.edu.vn

Experimental Determination of the Loss of Total Endoprosthesis Polyethylene Cup Using Holographic Interferometry

M. Houfek and L. Houfek

Faculty of Mechanical Engineering, Brno University of Technology, Brno, Czech Republic

Abstract— **This article deals with determining the size of linear wear acetabulum total hip arthroplasty experimental modeling. Described in detail the creation of an experimental sample and the equipment to simulate human walking. The greatest attention is paid to the method for determining the topography of polyethylene cup and measuring the loss of polyethylene during gait simulation. In determining the loss of polyethylene cup was based on extensive analysis of selected holographic interferometry. Other parts of the article focus on first experiences with the application of this method, results and more.**

Keywords— **holographic interferometry, cup total joint replacement wear measurement.**

I. INTRODUCTION

One of the problems of contemporary total joint replacement (THR) is their limited lifespan, which is around 15 years. Modern times have seen the still higher demands on man, which is negatively reflected in the application of THA at an earlier age, or to appear as young patients with this diagnosis. Therefore, the effort of doctors, technicians and producers to design, manufacture and application of TEP with higher durability [7 – 10]. The cause of life limitation hip arthroplasty polyethylene wear it well. In [1] states that implants with polyethylene wear rate of 0.05 mm / year have very low frequency of periprosthetic osteolysis and loosening, while the wear rate of implants with greater than 0.3 mm / year have a significantly higher risk of developing and osteolysis.

All the circumstances surrounding the creation and promotion of periprosthetic osteolysis are not sufficiently illuminated, the effort to create a TEP with a minimum production of particles is therefore undeniable importance. The size of wear particles is closely related to the surgical technique, TEP structure and behavior of the patient. TEP design can affect the amount of wear particles in different ways. One is undoubtedly improve the quality of polyethylene, or other contact materials selection, and well heads. The second is to regulate the distribution of pressure between the head and the hole. Adverse effects of wear particles can be reduced by preventing or. reductions of particles entering the space between bone tissue and the hole, or bone tissue and stem TEP. This issue deals with eg paper

[6], which deals with the creation of biologically active interface between the implant and bone tissue.

Such interfaces are able to create implants with hydroxyapatite-based surfaces. Synthetic hydroxydapatit has osteoconductive properties, which means that it supports osteoprogenitorových cells grow into a suitably modified implant surface and thereby preventing the ingress of wear particles.

The change in the distribution of contact pressure has a significant effect reducing the average head. With this change, without mechanical analysis, it is difficult to say whether positive or negative. Reducing the diameter of heads even when reducing the area may mean increased contact pressure, thus increasing wear. Advertised mechanical analysis must include determining the contact pressure between the head and the cup size and the experimental determination of wear under these conditions, the method of determining the size of wear must be very sensitive.

II. DESCRIPTION OF PROBLEM SITUATIONSR

Solution deformation and stress the hip joint or hip joint with applied TEP, is a complex task of Biomechanics, which today can be solved through the development of computational techniques and numerical methods of continuum mechanics. The same applies to the contact pressure between the head and the hole TEP. In order to be able to, by mechanical quantities for this solution, predict the amount of wear particles, it is necessary experimental way – for a computed course and size of the contact pressure, determine the size of wear. Although the number of large wear particles, volumetric or mass quantities (due to their microscopic size) is small. Therefore, the aseptic loosening of arthroplasty, as a result of a polyethylene granuloma, occurs only after several years. This time is in terms of life TEP small, but the experimental determination of the size of wear is great.

Therefore, demands on the experimental method for determining the size of extreme wear. The consequence of these facts is that despite the existence of a number of methods for determining the size of wear are the subject of further research and development of new efficient methodsbased on different principles. Among the most effective are optical methods. One of the methods of optical holographic

V. Van Toi and T.H. Lien Phuong (eds.), *5th International Conference on Biomedical Engineering in Vietnam,*
IFMBE Proceedings 46, DOI: 10.1007/978-3-319-11776-8_108

interferometry, which finds application in the topography of solids, so was used to measure? Linear? Polyethylene wear hole total endoprosthesis.

III. PRINCIPLE AND EVALUATION OF THE SIZE OF MATERIAL LOSS

The method of evaluating material loss is realized by means of holographic recording. To obtain the hologram are necessary two beams - the reference and object. After the hologram recording and display in a chemical solution is necessary to place the hologram instead of the original recording. The subject is then illuminated by a beam again and performed the reconstruction. The reference beam us now serves as the reconstruction. The object beam at the holographic plate produces interference, and thereby of the wide monochrome stripes, corresponding to the surface of the feed material. Quantitatively the formation and position of the strip gets a description of the reconstruction wave according to [3]:

Condition for creating a dark strip is given by:

$$\Phi_S - \Phi_S' = \pi(2m + 1) \tag{1}$$

Where

Φ_S the original topology

Φ_S' is a topology for a given number of cycles

m is an integer

To determine the size reduction will use reflective wave interference. Path difference is equal to twice the distance between the surfaces and Δ is true:

$$\Delta = \frac{1}{2}\frac{\lambda}{2\pi}\left(\Phi_S - \Phi_S'\right) = \frac{\lambda}{2\pi}(2m + 1)\frac{\lambda}{2} \tag{2}$$

Where λ is the wavelength of the laser

IV. PRESENTATION AND ANALYSIS OF MEASUREMENT RESULTS

Analysis of loss of material was carried to a designated network of Fig. 1 in radius to areas with a radius of 2 and 6 mm.

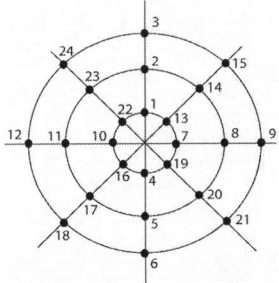

Fig. 1 Marked measuring points on the cup TEP

Table 1 Table of measured Intensities, calculated times record

Measurement of beam energy	Designation	Measured value	Unit
Image	I_O	0,609	
Reference	I_R	4,4	$\mu W/cm^2$
The measured total beam	I_C	5,01	
The ratio of beam		$\frac{I_R}{I_O} = 7,2$	-
Calculation of exposure time		$t = \frac{I_p}{I_c \cdot 2} = \frac{70}{5,01 \cdot 2} = 6,98$	s
The actual exposure time	7		

The dependence of the size of the wear on the position of points

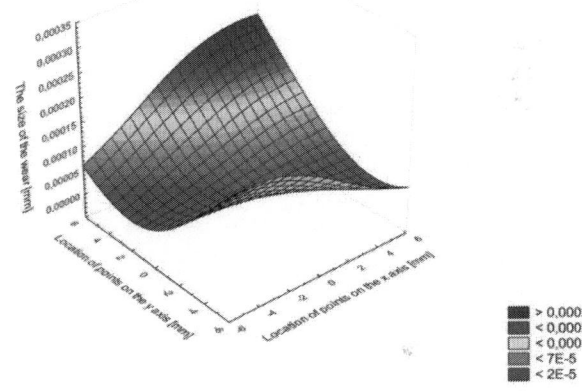

Fig. 2 Wear of polyethylene in the marked points

Table 2 Table of measured Intensities, calculated times record

Measurement of beam energy	Designation	Measured value	Unit
Image	I_O	1,2	
Reference	I_R	4,2	$\mu W/cm^2$
The measured total beam	I_C	5,4	
The ratio of beam		$\frac{I_R}{I_O} = 3,5$	-
Calculation of exposure time		$t = \frac{I_p}{I_c \cdot 2} = \frac{70}{5,01 \cdot 2} = 6,01$	s
The actual exposure time	6		

The dependence of the size of the wear on the position of points

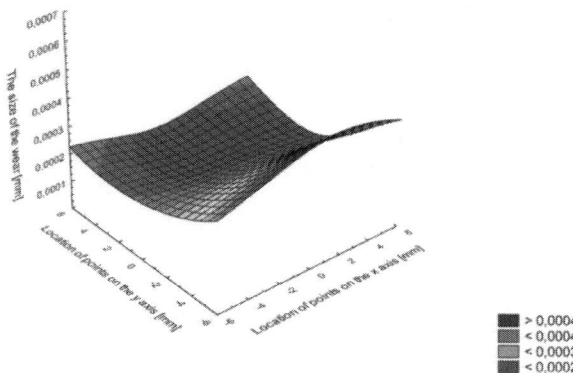

Fig. 3 Wear of polyethylene in the marked points

Table 3 Table of measured Intensities, calculated times record

Measurement of beam energy	Designation	Measured value	Unit
Image	I_O	1,63	
Reference	I_R	7,15	$\mu W/cm^2$
The measured total beam	I_C	8,9	
The ratio of beam	$\dfrac{I_R}{I_O} = 4,3$		-
Calculation of exposure time	$t = \dfrac{I_p}{I_c \cdot 2} = \dfrac{70}{5,01 \cdot 2} = 3,65$		s
The actual exposure time	3		

The dependence of the size of the wear on the position of points

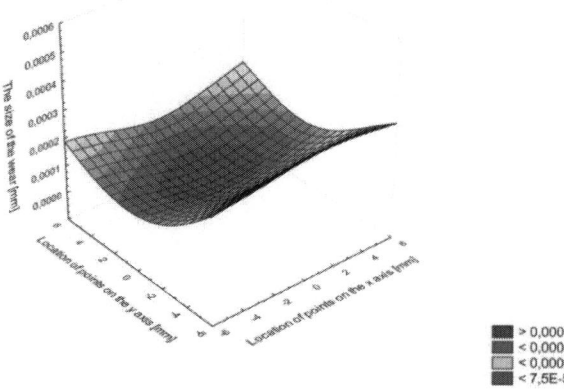

Fig. 4 Wear of polyethylene in the marked points

The dependence of the size of the total wear on the position of points

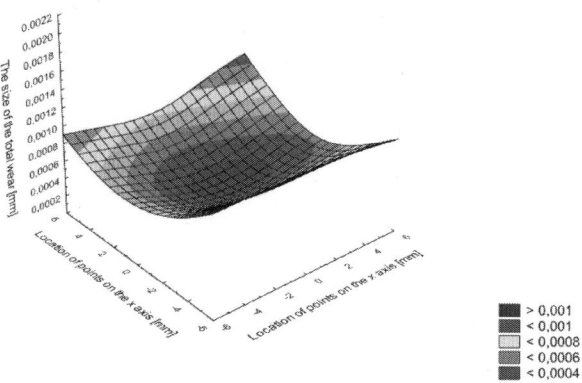

Fig. 5 Wear of polyethylene in the marked points

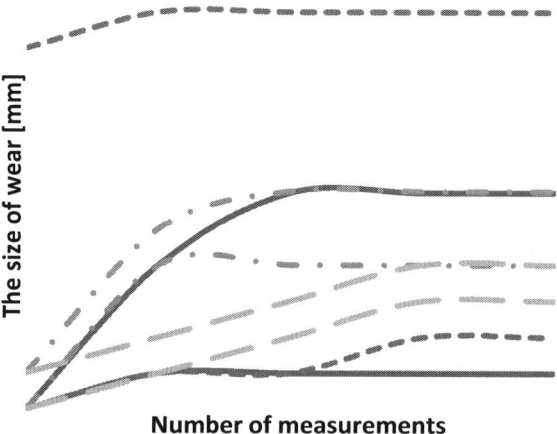

Fig. 6 The size of wear after five loaded cycles the radius R2

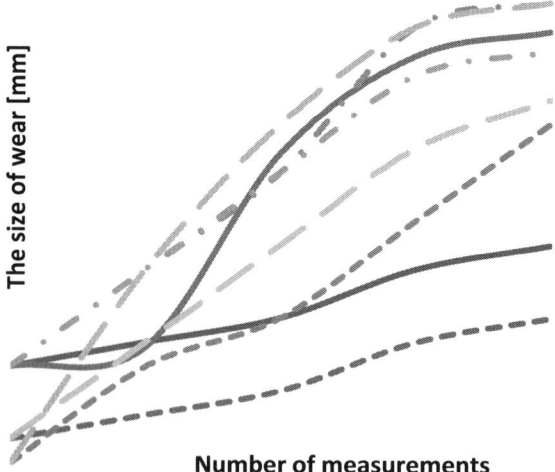

Fig. 7 The size of wear after five loaded cycles the radius R6

V. CONCLUSIONS

The wear of polyethylene described in this article is measured using the holographic optical method. The linear loss of polyethylene of the hip TEP acetabulum is calculated from the change in topology of individual surfaces (between the given measurements). A network of points was created on the acetabulum, in which the material loss and the total loss were evaluated after 500 and 2500 load cycles simulating human walking, respectively. The total loss of polyethylene in the analyzed area is in the range of <0.000079-0.001520> mm. The analysis implies that there is no significant loss of polyethylene on the cup circular areas with the radius of 2 mm after 1500 cycles, see Fig. 6. The probable cause of this phenomenon is that the femoral head and the acetabular cup were mounted with clearance, so the contact is the most frequent near the centre and the contact area increases with increasing wear. A different situation is in the radius of 6 mm, see Fig. 7, where the loss of material is increasing with an increasing number of load cycles. It may be concluded that the used method of holographic interferometry is a suitable tool for identifying small material losses that cannot be identified with a sufficient precision by another method (e.g. weighing).

ACKNOWLEDGMENT

This work is an output of cooperation between XYZ and NETME Centre, regional R&D centre built with the financial support from the Operational Programme Research and Development for Innovations within the project NETME Centre (New Technologies for Mechanical Engineering), Reg. No. CZ.1.05/2.1.00/01.0002 and, in the follow-up sustainability stage, supported through NETME CENTRE PLUS (LO1202) by financial means from the Ministry of Education, Youth and Sports under the „National Sustainability Programme I. And FSI-S-14-2311 Strength and dynamics of modern materials and designs II.

CONFLICT OF INTEREST

Every paper must contain a declaration of conflicts of interest. If there are no such conflicts write "The authors declare that they have no conflict of interest".

REFERENCES

1. Gallo. J., Havránek V., Zapletalová. J., Mandát. D.: Měření otěru polyetylenových jamek TEP kyčelního kloubu univerzálním měřicím mikroskopem, Acta Chirurgiae Orthopaedicae Et Traumatologiae Čechosl., 73, 2006, P. 28–33
2. Čihák, R.: Anatomie 1. Avicenum, Praha, 1987
3. Baláš. J., Szabó. V.: Holografickáinterferometrie v experimentálnímechanice, Bratislava 1986
4. Miler. M.:Holografie (teoretickéaexperimentálnízáklady a jejípoužití), Praha 1974.
5. Baudyš.A.: Technickáoptika, Skryptum ČVUT, Praha 1996l.
6. D´antonio. J. A., Dietrich. M.: Bioceramics and Alternative bearings in joint arthroplasty, Washington D.C., 2005.
7. Fuis V. (2009) Tensile Stress Analysis of the Ceramic Head with Micro and Macro Shape Deviations of the Contact Areas, Recent Advances in Mechatronics: 2008-2009, Proc. International Conference on Mechatronics, Brno, Czech Rep., pp. 425-430.
8. Fuis V. (2004) Stress and reliability analyses of ceramic femoral heads with 3D manufacturing inaccuracies, 11th World Congress in Mechanism and Machine Science, Tianjin, China (2004) 2197-2201.
9. Fuis V., Janicek P. (2002) Stress and reliability analyses of damaged ceramic femoral heads, Conference on Damage and Fracture Mechanics, Maui Hawaii, Structures and Materials Vol. 12, pp. 475-485.
10. Fuis V., Varga J. (2009) Stress Analyses of the Hip Joint Endoprosthesis Ceramic Head with Different Shapes of the Cone Opening. 13th International Conference on Biomedical Engineering, IFMBE Proceedings, Vol. 23, Iss. 1-3, pp. 2012-2015.

The Applications of Control System Approach in Biomedical Engineering Research

Huynh Luong Nghia and Nguyen Van Trung

Biomedical Electronic Department - Faculty of Control Technology
Le Quy Don Technical University - Hanoi – Vietnam

Abstract— **This article shows the results of applying control system approach in biomedical engineering (BME) research. Based on the advantages of the structural predetermination in comparing with traditional system approach, this methodology allows studying BME systems with more effectiveness, convenience and reality.**

Keywords— **BME research, epistemology system approach, control system approach, unilaterality, predetermination.**

I. DEFINITION OF PROBLEM

Biomedical engineering (BME) research is complex and diverse because of the following factors [1]:

a. The object of BME research varies depending on structure levels: from population → body → organs → tissue → cells → biomolecules...

b. Medical equipments include many varieties and modifications.

c. Medical equipments are based on complicated principles and different reliabilities.

How do we get these complexity and varieties?

To carry out BME research many approaches were applied such as technical (L. Zade, G.Kastler), biological (V.I. Kremenxki, K.M. Klailov, A.A. Liapunov, A.A. Malinopxki), psychological (S.Piadje, G. Oleport), linguistic (I.I.Rezvin, G.L.Melnhikov), sociobiological (L. Xorokin, V. Berkli)..., among them the system approach is outstanding [2]. However, on Epistemology, this approach has remarkable limitation that can be overcame by the control system approach below considered.

II. THE SYSTEMS APPROACH

The systems approach is defined as the methodology for scientific research and really casting the objects with complex organization, in which the top place is not the component self-analysis, but defining the object's characteristics as the whole entity in outer system and exploring the mechanisms ensuring the object's wholeness [2]. The absolute advantage of this approach is comprehensiveness anti the unilaterality in scientific research keenly illustrated in Figure 1 [4]. However, this approach has underlying weakness that does not allow to define the structure of explored object, i.e., considering it as "black box". In other words the structure and, therefore, the functions of object's components are not predeterminated.

Fig. 1 To illustrate the unilaterality in scientific researches [4]

III. THE CONTROL SYSTEMS APPROACH

The control system approach inherits all the advantages of the systems approach and overcomes underlying weakness above mentioned by next way.

Firstly, in top structure level, the control system is defined as the system at least having 2 components interactive each other. By this way, in common case, the control system is described through the scheme in Figure 2.

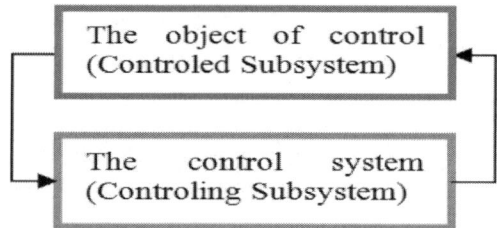

Fig. 2 The scheme of control system in top structure level

Secondly, each component of described system, in turn and by control logic, contains 3 predetermined elements:

PT1 – sensitive element
PT2 – processing element
PT3 – executive element

Therefore, the scheme of control system in second structure level is presented as in Figure 3.

Thirdly, the elements PT1, PT2, PT3, in turn and by control logic, contain again 3 predetermined elements corresponding to sensitive, processing and executive elements.

For this structure predetermination, the basic limitation of system approach is overcome, allowing its effective application in BME research.

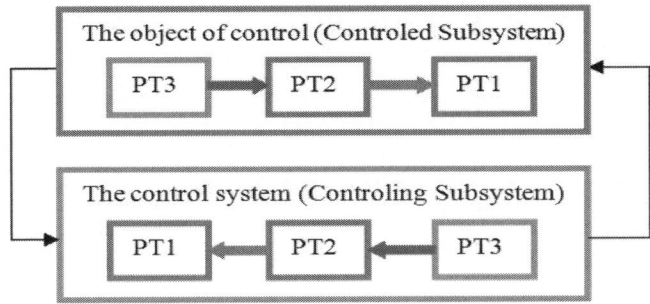

Fig. 3 The scheme of control system in second structure level

IV. THE APPLICATION OF CONTROL SYSTEMS APPROACH IN BIOMEDICAL ENGINEERING (BME) RESEARCH

For the biological object, the application of control system approach in top (body) level enables to describe the organism through the structure scheme in Figure 4 [2]:

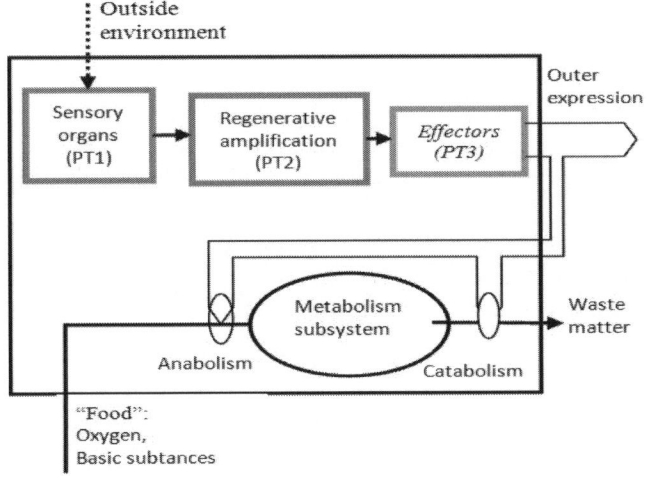

Fig. 4 The organism is described in the form of control system

Applying control system approach to medical system delivered the scheme in most common structure level is shown in Figure 5:

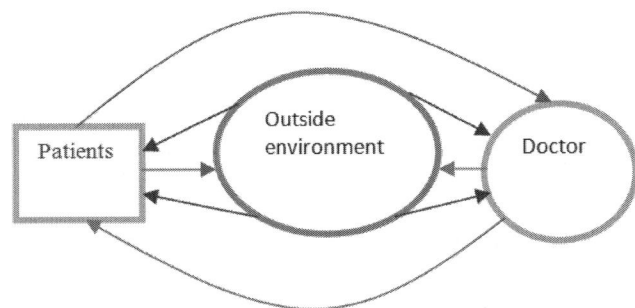

Fig. 5 The scheme of medical system in most common structure level

Applying control system approach to medical system in second structure level gives us the next scheme:

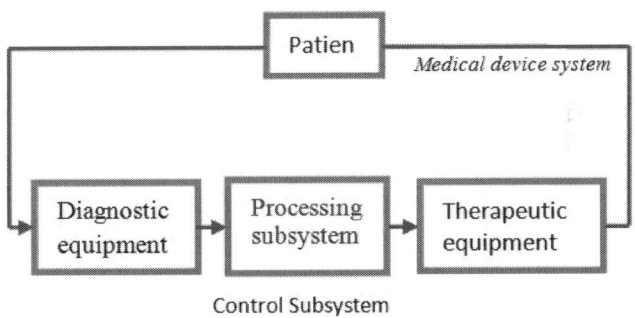

Fig. 6 The scheme of medical system in second structure level

In this scheme the diagnostic equipment play the role of sensitive element (PT1), the doctor - processing element (PT2) and therapeutic equipment - executive element (PT3). In turn these equipments have again the same structure of control system as represented in Figure 7 and Figure 8 [3]:

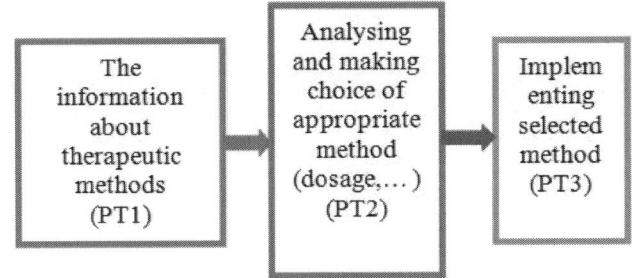

Fig. 7 The scheme of therapeutic equipment

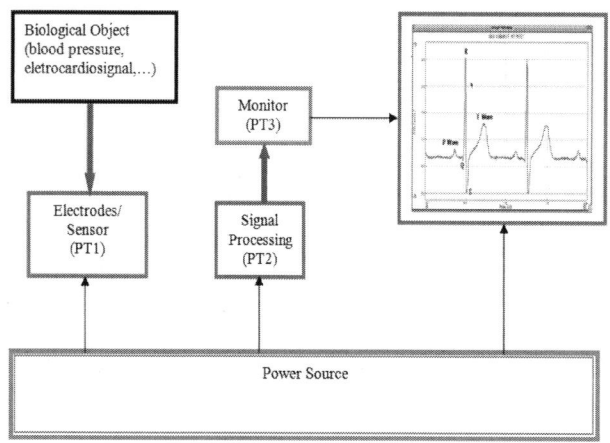

Fig. 8 The scheme of diagnostic equipment

V. Conclusions

The control system approach has the advantage over the systems approach due to structure predetermination of the object under research.

The application of control system approach in Biomedical Engineering brings special effectiveness since the objects under research are various control systems.

References

1. Huỳnh Lương Nghĩa, "*Về phương pháp luận nghiên cứu các hệ thống Điện tử - Y -Sinh*", Tạp chí khoa học kỹ thuật HVKTQS, số 104, t 111-119, năm 2003
2. Попечителев Е.П, Методы медико-биологических исследований, Учеб. Пособие, Житомир, 1997
3. Huỳnh Lương Nghĩa, Các bài giảng bổ sung môn học "Phương pháp luận nghiên cứu các hệ thống Điện tử - Y -Sinh", HVKTQS, năm 2011
4. GKA Incopareted (617) 441- 7766

Author: Huynh Luong Nghia; Nguyen Van Trung
Institute: Le Quy Don Technical University
Street: Hoang Quoc Viet
City: Ha Noi
Country: Vietnam
Email: hlnghia@yahoo.com; trung88s2@gmail.com

Study of Vessel Conditions for Deep Vein Thrombosis (DVT) Diagnosis According to Body Mass Index

W.N. Wan Zakaria[1], N. Ibrahim[1], N. Mat Harun[1], Razali Tomari[1], and M.K. Abdullah[2]

[1] Faculty of Electrical and Electronic Engineering, University Tun Hussein Onn Malaysia
[2] Faculty of Mechanical and Manufacture Engineering, University Tun Hussein Onn Malaysia

Abstract— In this paper, the clinical experiment study is presented to diagnose Deep Vein Thrombosis (DVT). The diagnosis of DVT is commonly conducted by monitoring the blood velocity and present of thrombus in vessel from B-mode ultrasound image associated with the Doppler ultrasound. Since it is difficult to recognize the vessel condition at the early stage of DVT, this study is proposed to evaluate the vein mechanism based on different BMI categories at the early stage of DVT. The wall displacement and blood flow velocity is considered to be the important parameters to construct a clinical model of DVT risk factor, thereby constitutes an important contribution for predicting probability of Deep Vein Thrombosis (DVT).

Keywords— Ultrasound scan, early diagnosis, wall displacement, blood flow velocity, Deep Vein Thrombosis (DVT).

I. INTRODUCTION

Deep vein thrombosis (DVT) is a condition in which blood clot is developed in the deep vein of the body. The clots mostly found or developed in the lower leg or thighs. However, it may also appear in the upper body such as the arms or other locations. In general, DVT is a risk for any major surgery particularly for the patient who had surgery on the legs or hips [1]. It can give a serious threat to the health of the patient.

One-way valves prevent the back-flow of blood between the contractions. Blood is squeezed up the leg against gravity and the valves prevent it from flowing back to our feet. When the circulation of the blood slows down due to illness, injury or inactivity, blood can accumulate or "pool," which provides an ideal setting for clot formation [2]. The pieces of clots can break up and travel through the bloodstream to the lungs which is called pulmonary embolism. This condition will lead to fatal soon after it occurs. There are some factors that contribute to the formation or development of clots in the veins. Stasis or stagnant blood through the veins is one of the factors [3]. It increases the contact between the blood and the vein wall irregularities. It also prevents naturally occurring anticoagulant from mixing with the blood. Moreover, prolonged the bed rest or immobility will leads to stasis. Besides, coagulation also

contributes to the clot formation [3]. Additionally, the vein wall can also be damaged during the surgery if the physician retracts soft tissues as part of the procedure [3]. To recognize the DVT, there some physical symptoms that show the person might possess DVT condition. The symptoms are swelling of the leg, a cord in a leg vein that can be felt, rapid heartbeat, slight fever, dull ache tightness, tenderness in the leg and also more visible surface veins. However, by using only clinical signs, it cannot be confirmed that the patients are having DVT.

By extrapolation, the true incidence of venous thrombosis in the general population of the United States is approximately 600000. By using the clinical sign, the percentage of people having unnecessary anticoagulation therapy is 42% of 600000 [4]. Most of the patients evaluated with ultrasonography do not have DVT. A variety of diagnostic techniques have been used to identify DVT. These include impedance plethysmography, contrast venography, ultrasonography, computed tomography, and magnetic resonance imaging. In the past, contrast venography was considered to be the gold standard. However, due to its associated expenditure of manpower resources and time, the need for specialized personnel, space and equipment, and its limited availability and associated morbidity [4], contrast venography has been replaced with other tests with more favorable risk or benefit profiles. Moreover there is D-dimer test that also being used in diagnosing the DVT. A D-dimer test is a blood test that measures a substance released as a blood clot breaks up. D-dimer levels are often higher than normal in people who have a blood clot [5]. Among these, ultrasonography is as accurate as any, with more advantages than CT, MRI, plethysmography and D-dimer test, i.e.; including low cost, portability, non-invasiveness, and simplicity[6]. Ultrasound evaluation for DVT has been shown to be successful by nonvascular specialist ranging from novice to advanced user of ultrasound and can be performed in just a few minutes at the patient bedsides [7].

Deep vein thrombosis (DVT) can give serious threat to the health condition. Therefore, it is essential to diagnose the DVT condition at the early stage. This will help in treating or improving the condition of patients with the DVT. The most common diagnoses done by the physicians are using the clinical assessment. From the result of the clinical

© Springer International Publishing Switzerland 2015
V. Van Toi and T.H. Lien Phuong (eds.), *5th International Conference on Biomedical Engineering in Vietnam,*
IFMBE Proceedings 46, DOI: 10.1007/978-3-319-11776-8_110

assessment, the condition of the patient still cannot be confirmed. Even by using the D-dimer test, there were still low probabilities in confirmation of the patient having the DVT. Therefore, as a solution for this problem, a non-invasively diagnosis method by using the ultrasound imaging is proposed. It requires no contrast medium and can be performed at the bedside condition. The important parameters, which are the wall displacement and the blood flow velocity of the popliteal vein are evaluated. Therefore, the vein mechanism can be analyzed based on these two main parameters.

II. METHOD

A. Experimental Procedure

To study the vessel condition from difference category of weight, the subjects had been categorized into three different categories of the body mass index (BMI) that are normal weight, underweight and overweight. The equipment used for the experiment is the ultrasound machine with the linear transducer with frequency range of 6MHz to 12MHz.

The patient can be placed in either a prone position, or seated on the edge of the bed with the knee flexed and the foot supported. The sonographic evaluation is performed by compressing the vein directly under the transducer while watching for complete apposition of the anterior and posterior walls. The obstructing venous thrombus is likely to be present if venous cannot be fully deformed.

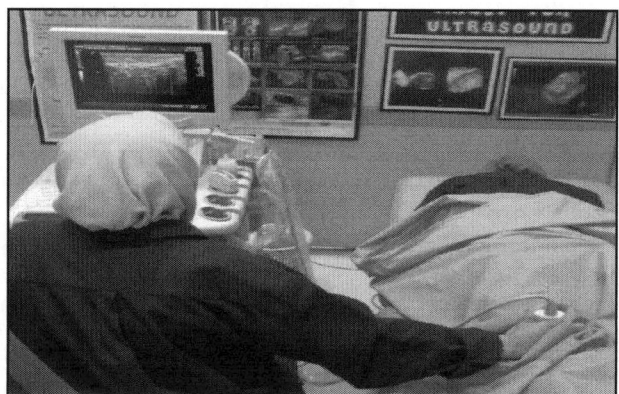

Fig. 1 Leg scanning using ultrasound.

III. RESULT

In this study, the expected result will be the elasticity of the vein obtained from the calculation. In order to obtain the elasticity, stress and strain measurements are required.

Below shows an equation analogous to Hooke's law describes the relationship between stress and strain:

$$E = \frac{STRESS}{STRAIN} = \frac{\sigma}{\varepsilon} = \frac{\frac{F}{A}}{\frac{L - L_0}{L_0}} = \frac{FL_0}{A_0 \Delta L}$$

Stress is the force per unit cross-sectional area (F/A). Strain is the fractional increase in length, that is, $(L - L_0)/L_0$ [8]. The wall displacement measurement is needed to indicate the strain value for the elasticity of the vein. In other hand, the stress value can be obtained from the blood pressure measurement over the area of the located vein.

B. Wall Displacement

The vessel-tracking algorithm is developed to extract the information of wall displacement. First, the image segmentation process is conducted to locate the region of interest. Then, the empirical threshold method is used to automatically perform clustering-based image thresholding. Finally, the morphological method is applied to filter any noise of the image. Figure 2 shows the sample of the vessel-tracking results.

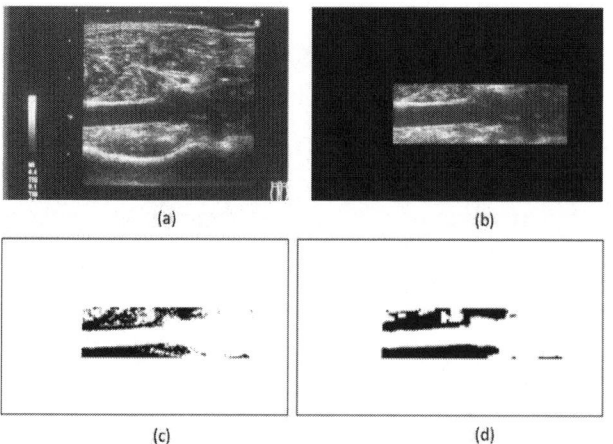

Fig. 2 Sample of Vessel-Tracking Algorithm (a) Image Segmentation, (b) Empirical threshold method, (c) Dilation process and (d) Erosion process.

C. Blood Flow Velocity

Blood flow velocity measurement is obtained by using the Doppler ultrasound. The average velocity of the blood flow will be calculated to obtain the true value of desired vein's velocity. The sample of blood flow velocity measurement is shown in Figure 3.

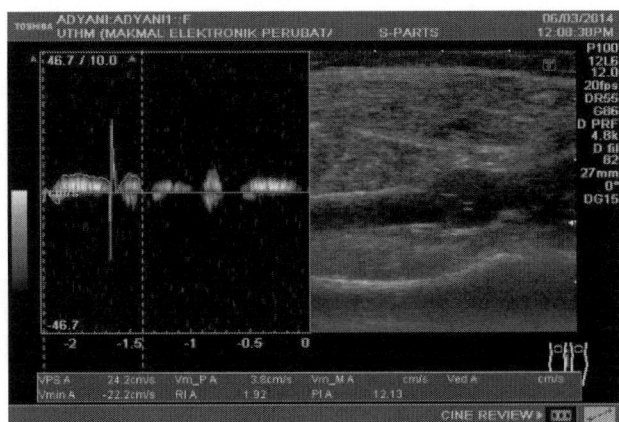

Fig. 3 Velocity measurement of the subject using pulse Doppler ultrasound.

IV. CONCLUSION

In conclusion, this study had described the experimental procedure for evaluation of early diagnosis of DVT. The main parameters which are the wall displacement and blood flow velocity of the vessel is obtained to evaluate the vessel condition based on different categories of BMI. In the future, these parameters will be used to construct a clinical model of DVT risk factor, thereby constitutes an important contribution for predicting probability of Deep Vein Thrombosis (DVT).

ACKNOWLEDGMENT

This work was supported by Universiti Tun Hussein Onn Malaysia under Multidisplinary Research Grant Vot 1311.

REFERENCES

1. University of Rochester Medical Center, 'Deep Vein Thrombosis (DVT) /Thrombophlebitis'. Retrieved from http://www.urmc.rochester.edu/encyclopedia on 25th September 2013.
2. Society of Interventional Radiology, 2013, 'Deep Vein Thrombosis Overview'. Retrieved from http://www.ctcdvt.com/what_is_dvt.html on 28th September 2013.
3. American Academy of Orthopaedic Surgeons, 2009, 'Deep Vein Thrombosis'.Retrieved from http://orthoinfo.aaos.org/topic.cfm?topic=a00219 on 25th September 2013.
4. James D. Fraser, MD, David R. Anderson, MD, 1999, 'Deep Venous Thrombosis: Recent Advances and Optimal Investigation with US', Departments of Radiology (J.D.F.) and Medicine (D.R.A.), Queen Elizabeth II Health Sciences Centre, 1278 Tower Rd, Halifax, Nova Scotia, Canada.
5. E. Gregory Thompson, MD, Jeffrey S. Ginsberg, MD, 2010, 'D-dimer test for deep vein thrombosis'. Retrieved from http://www.webmd.com/dvt/d-dimer-test-for-deep-vein-thrombosis on 20th October 2013
6. Juan N. Useche, MD, Alfredo M. Fernández de Castro, MD, Germán E Galvis, MD, Rodolfo A. Mantilla, MD and Alvaro Ariza, MD, 2008, 'Use of US in the Evaluation of Patients with Symptoms of Deep Venous Thrombosis of the Lower Extremities', Department of Radiology, Palermo Clinic, Bogotá, Colombia (J.N.U., R.A.M.); Department of Radiology, University Hospital of San Ignacio, Bogotá, Colombia (A.M.F.C., G.E.G.); and Department of Diagnostic Radiology, Clinica Nueva, Bogotá, Colombia (A.A.).
7. Stephanie Sippel, Krithika Muruganandan, Adam Levine and Sachita Shah, 2011, 'Review Article: Use Of Ultrasound In The Developing World', International Journal Of Emergency Medicine.
8. Walter F. Boron, Emile L. Boulpaep, 2008, 'Elastic Properties Of Blood Vessels', Medical Physiology, Saunders, ISBN-10: 1416031154 ISBN-13: 978-1416031154.

Tensile Stress Analyses of the Hip Joint Endoprosthesis Ceramic Head with Real Shape Deviations

V. Fuis

Institute of Thermomechanics AS CR – Branch Brno, Technicka 2, 616 69, Brno, Czech Republic

Abstract— In vivo fracture of ceramic heads of total hip joint endoprostheses has been stated in a not negligible number of patients in the Czech Republic hospitals, some time ago. The reliability of the ceramic head is based on the Weibull weakest link theory. The probability of the failure depends on the tensile stress values in the head, on the volume in which stress acts and on the material parameters of the used bioceramics. The stress analyses of the ceramic heads were realized using computational modelling - FEM system ANSYS under ISO 7206-5 loading. The maximum values of the tensile stress in the ceramic head are significantly influenced on the shape deviations of conical surfaces of the head and the stem cones. There are two types of the shape deviations – the first is macro deviations (different angle of the taper of the stem cone and the head cone), the second is micro shape deviations (the stochastic distribution of unevenness of cone contact surfaces in order micrometers). The macro and micro shape deviations of the real cones were carried out using the IMS-UMPIRE measuring equipment. By computational modelling it has been proved that the character and the size of shape deviations on a conical contact area between the stem and the head of hip joint endoprosthesis have a pronounced influence on the character and the value of stress in the head and, hence, on the probability of the head's failure. Ignorance of this influence can lead to unforeseen failures of ceramic heads and thus to cast doubt on the trust-worthiness of a certain type of total hip joint endoprosthesis and of surgical treatment.

Keywords— Total hip joint endoprosthesis, ceramics head, micro and macro shape deviations, stress, computational modelling.

I. INTRODUCTION

One of the most frequently operated joints of the human body is the hip joint [1]. In one hospital of the Czech Republic, more than 1 % of applied endoprostheses had to be reoperated - Fig. 1. That is why the state of stress and the reliability of ceramic heads of total endoprosthesis is being intensively analysed by means of computational modeling [13–14].

The analysis of the head fracture surface orientation showed that the meridionally oriented fracture surfaces prevail (Fig. 1). Since ceramics is a material characterised by the limit state of brittle strength, the cohesion failure starting in directions perpendicular to the direction of the maximum principal tensile stress, then the stress responsible for the failure of head cohesion should be the hoop stress [3]. An effective method of solving the problem of reliability is computational modelling combined with experiments. For solving the strain/stress states in the head, the finite elements method was used, namely the ANSYS program system.

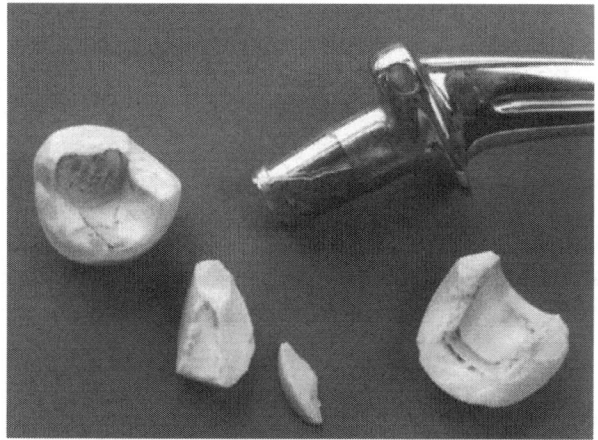

Fig. 1 *In vivo* destructed ceramic heads

II. COMPUTATIONAL MODELLING

The topology of the elements of the total hip joint endoprosthesis follows from Fig. 2 with the basic macro-dimensions of elements. Computational modelling of the strain/stress states in the head was implemented on a system whose elements were parts of the stem and the head

In order to implement computational modelling of the impact of shape deviations of the cones of stems and heads of hip joint endoprostheses, which has a stochastic character, the measurement of these deviations was carried out using the IMS-UMPIRE measuring equipment (Fig. 3). The quality of the measured contact surface of both cones can be expressed by macro-parameters (deviations) – the most important is angle of the taper of the head and the stem and straight line character), and micro-parameters (deviations) – local micro-unevenness of surface. As regards the inclination angles of the elements of ruled surfaces (degree of taper) one can say that the manufacturing documentation

© Springer International Publishing Switzerland 2015
V. Van Toi and T.H. Lien Phuong (eds.), *5th International Conference on Biomedical Engineering in Vietnam,*
IFMBE Proceedings 46, DOI: 10.1007/978-3-319-11776-8_111

permits only an alternative in which the vertex angle of the head (angle α_h) is greater than the vertex angle of the stem (angle α_s) – Fig. 2 (the value of the angle α can be found from zero to 10' and therefore the three values of angle α is assumed (0', 4' and 8')), and the circularity of the individual sections (perpendicular to the y-axis) is less than 10 μm [2 and 7].

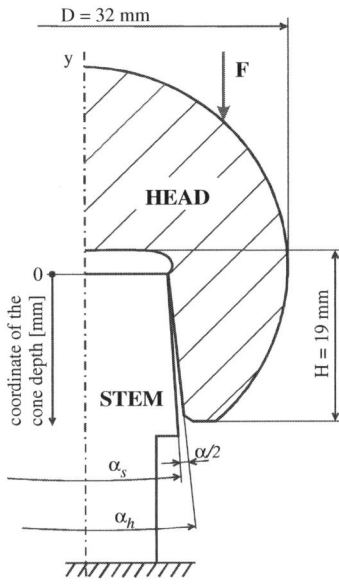

Fig. 2 Topology and macro shape deviation

Fig. 3 Measure device IMS-UMPIRE

The measured micro shape deviations from the ideal conical shape for cone of head are illustrated by the developed cones in Fig. 4a (from -1.5 to 1.5 μm). The micro shape deviations of the stem cone are show in Fig. 4b (from -2.5 to 1.5 μm). The mentioned head with macro and micro shape deviations was put and loaded on the stem cones under ISO 7206-5 loading (Fig. 2).

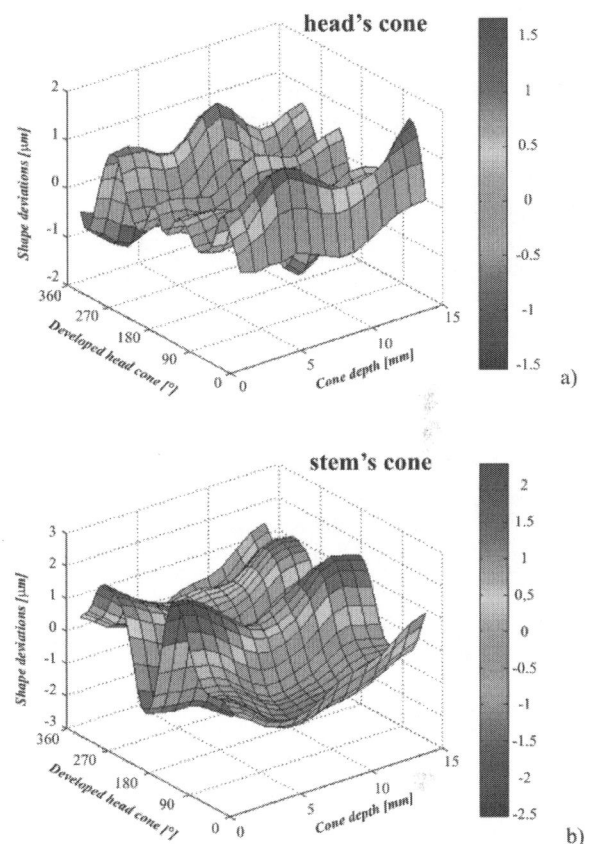

Fig. 4 Micro shape deviation of the: a) head cone and b) stem cone

The state of stress of the ceramic head of endoprosthesis can be strongly affected by the process of endoprosthesis implantation, when the surgeon fits the head on the cone of the stem. The fitting of the head on the cone of the stem is a random process, in view of the mutual position of the head to the stem (in sense of the head's slight turning around the y-axis, defined by angle β – Fig. 5) [10 and 11]. The question is, whether the local shape deviations of the cones, in case of random completing, can produce high tensile stresses that might reduce the reliability of the head. For this reason, a series of computations will be made for each pair and for various positions of the head towards the stem.

For computational analyses of the head's cohesion failure, the head was loaded with linear forces acting on a circle

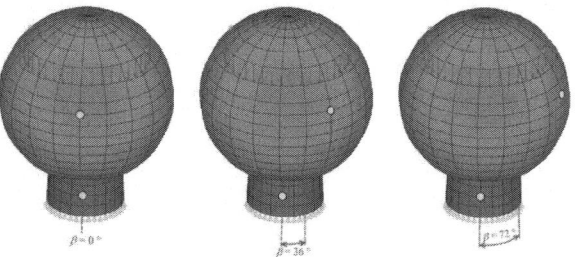

Fig. 5 Representation of various head to stem positions - angle β

the carrier of the resulting force F being identical with the axis of the stem, Fig. 2. This way of loading complies with the ISO 7206-5 standard, according to which the reliability of ceramic heads is tested for failure of cohesion.

The material of the head and the stem was considered to be a linear isotropic continuum. The stem was made of steel and the following constitutive characteristics were taken into account in the calculation: E = 210 GPa μ = 0.3. The material of the head was ceramics on the basis of aluminium oxide Al_2O_3, for which the following constitutive characteristics were considered: E = 390 GPa, μ = 0.23. Contact between the head and the stem is modelled as elastic Coulomb friction with coefficient of friction f = 0.15 [7, 8 and 12].

III. RESULTS AND DISCUSSIONS

The maximum tensile stress in the head (the most important stress in the head due to the brittleness of the ceramics [8]) is located on the contact surface near the contact region. The size of the contact region is influenced of the position of micro shape deviations of the stem and the head cone (Figs. 6 and 7). Figs. 6 and 7 show the isosurfaces of the first principal stress (left) in the head (F = 5 kN and different angles α) and for different position of the head and the stem (β = 0° and β = 240°). The position of the maximum tensile stress in the head is very different – see Figs. 6 and 7 (left side). The right figure shows the third principal stress which corresponds to the contact pressure on the contact cone with shape deviations. The contact region for α = 0' is large than for α = 8' (Figs. 6 and 7 – right side).The location of the maximum tensile stress and the shape of the isosurfaces are different for both analysed angles β due to the relative location of the micro deviations of the head's and stem's cones. The maximum tensile stress in the head grows with the increasing of the angle α (Fig. 8) due to the fact that the contact area decreases.

The distribution of maximum values of tensile stresses in the ceramic head (σ_{max}), in dependence on the value of the head's load and on value of the α, is shown in Fig. 9. The width of the curves is decresed with increasing angle α. This is due to the fact that in case the zero angle α (ideal case when the cone angle of the head and stem is the same) is in the contact local shape deviations on the entire tapered surface. Increasing of the angle α value caused the reduction of the contact area and this contact area is localized to the area of the smallest diameter of the cone.

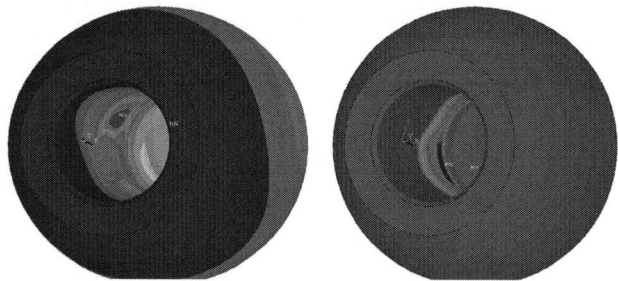

Fig. 6 Isosurfaces of the first and third principal stresses in the head, F= 5 kN, angle β = 0° and α = 0'

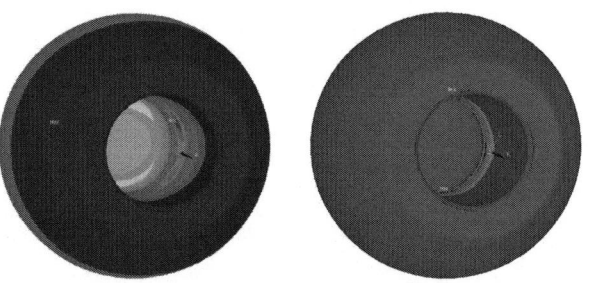

Fig. 7 Isosurfaces of the first and third principal stresses in the head, F= 5 kN, angle β = 240° and α = 8'

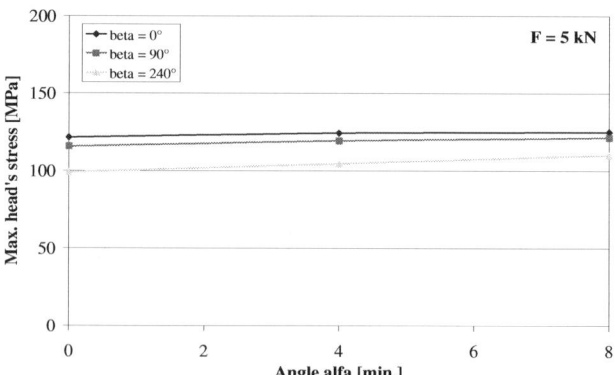

Fig. 8 Maximum tensile stress in the head for different angle α and head-stem position (angle β), F= 5 kN

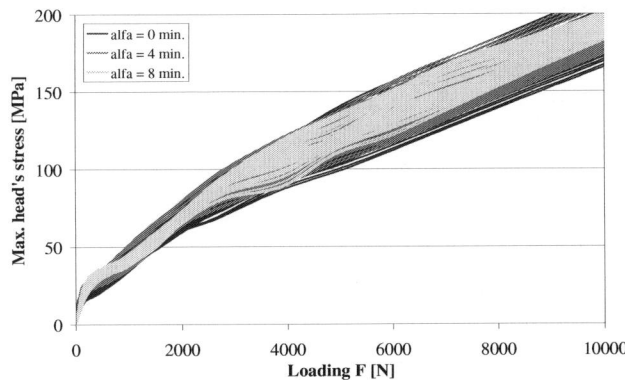

Fig. 9 Maximal tensile stress in the head during the loading for different angle α

IV. CONCLUSIONS

By computational modelling it has been proved that the character and the size of micro and macro shape deviations on a conical contact area between the stem and the head of hip joint endoprosthesis have a pronounced influence on the character and the value of stress in the head and, hence, on the probability of the head's cohesion failure. The ideal cone angles (with α = 0°) caused the higher width of the stress curves for different position head on the stem (angle β) due to the large of the contact area.

ACKNOWLEDGMENT

The research has been supported by the project of the Czech Science Foundation GA CR nr. 13-34632S.

CONFLICT OF INTEREST

The authors declare that they have no conflict of interest.

REFERENCES

1. Andrisano A. O., Dragoni E., Strozzi A. (1990) Axisymmetric mechanical analysis of ceramic heads for total hip replacement. Proc. Inst. Mech. Engrs, Part H, Vol. 204, pp. 157-167
2. Fuis V., Janicek P. (2002) Stress and reliability analyses of damaged ceramic femoral heads, Conference on Damage and Fracture Mechanics, Maui Hawaii, Structures and Materials Vol. 12, pp. 475-485
3. Fuis V., Koukal M., Florian Z. (2011) Shape Deviations of the Contact Areas of the Total Hip Replacement, 9th International Conference on Mechatronics, Warsaw, Poland pp. 203-212
4. Bush D. (1993) Designing Ceramic Components for Structural Applications, J. Mater. Eng. Perf. ASM Int., Vol. 2, pp. 851-862
5. Fuis V., Malek M., Janicek P. (2011) Probability of destruction of Ceramics using Weibull's Theory, 17th International Conference on Engineering Mechanics, Svratka, Czech Rep. (2011) 155-158
6. Teoh S. H., Chan W. H., Thampuran R. (2002) An Elasto-plastic Finite Element Model for Polyethylene Wear in Total Hip Arthroplasty, J Biomech 35: 323-330
7. Fuis V., Janicek P., Houfek L. (2008) Stress and Reliability Analyses of the Hip Joint Endoprosthesis Ceramic Head with Macro and Micro Shape Deviations, Proc. vol. 23, Issue 1-3, International Conference on Biomedical Engineering, Singapore, pp. 1580-1583
8. Fuis V. (2004) Stress and reliability analyses of ceramic femoral heads with 3D manufacturing inaccuracies, 11th World Congress in Mechanism and Machine Science, Tianjin, China pp. 2197-2201
9. McLean A. F., Hartsock D. L. (1991) Engineered materials handbook, Vol. 4, Ceramics and Glasses. ASM International, pp. 676-689
10. Fuis V., Navrat T., Hlavon P., et al. (2006) Reliability of the Ceramic Head of the Total Hip Joint Endoprosthesis Using Weibull's Weakest-link Theory, IFMBE Proc. vol. 14, World Congress on Medical Physics and Biomedical Engineering, Seoul, South Korea pp. 2941-2944
11. Fuis V. (2009) Tensile Stress Analysis of the Ceramic Head with Micro and Macro Shape Deviations of the Contact Areas, Recent Advances in Mechatronics: 2008-2009, Proc. International Conference on Mechatronics, Brno, Czech Rep., pp. 425-430
12. Fuis V., Varga J. (2009) Stress Analyses of the Hip Joint Endoprosthesis Ceramic Head with Different Shapes of the Cone Opening. 13th International Conference on Biomedical Engineering, IFMBE Proceedings, Vol. 23, Iss. 1-3, pp. 2012-2015
13. Houfek, L., Florian, Z., Brezina, T., et al. (2009) Development of Experimental Devices for Testing of the Biomechanical Systems, Conference: 13th International Conference on Biomedical Engineering (ICBME), Singapore, Vols. 1-3, Book Series: IFMBE Proceedings, Vol.: 23, Iss: 1-3, pp. 2005-2008
14. Houfek, M.; Sveda, P., Kratochvil, C. (2011) 9th International Conference on Mechatronics, Warsaw, Poland, Mechatronics: Recent Technological and Scientific Advances, pp. 293-296

Author: Vladimír Fuis
Institute: Institute of Thermomechanics AS CR v.v.i. - branch Brno
Street: Technicka 2, 616 69
City: Brno
Country: Czech Republic
Email: fuis@fme.vutbr.cz

Review and Development of Tetraplegic-Musculoskeletal FES-Elbow Joint Extension Control Strategies

N.H.M.Nasir[1], M.K.I.Ahmad[1,2], B.S.K.K. Ibrahim[1,2], and F. Sherwani[1,2]

[1] Department of Mechatronic and Robotic Engineering,
[2] Modeling and Simulation Research Laboratory,
Faculty of Electrical and Electronic Engineering,
Universiti Tun Hussein Onn Malaysia, 86400 Johor, Malaysia

Abstract— **The application of FES as a therapeutic and rehabilitation functionality is globally used especially for paralyzed person induced by SCI. The potential of this research is FES induced movement control on a significantly challenging area due to complexity and non-linearity of tetraplegic musculoskeletal system. The challenge mainly arises within FES application based on a crucial issue, in the control of musculoskeletal upper limb motor function by the artificial activation of paralyzed muscles, due to the various characteristics and parameters of the underlying physiological/biomechanical system. This research pilot study attempts to propose model theoretical framework/control strategies with computer validation simulations studies.**

Keywords— **SCI, FES, Tetraplegic, Musculoskeletal, Elbow.**

I. INTRODUCTION

Developments of control system due to Spinal Cord Injury (SCI) area nowadays are really improved and more benefits are attained by using Functional Electrical Stimulation (FES) [1,3,4,6,7]. Theoretically, FES is a treatment that uses the application of small electrical current signals to improve mobility and to restore the function of paralyzed muscles due to SCI [2,5,6,10,11,15,20]. In addition, FES is a promising method to restore mobility to individuals' paralysed muscles due to spinal cord injury (SCI) [3,6,7,9-11,16,17] . The first applications of FES were designed to restore lower limb function in patients who had experienced a stroke or a SCI. Furthermore, since 1960s, FES has been used to correct foot drop, which is a common symptom in hemiplegia, paraplegia characterized by a lack of dorsiflexion during the swing phase of gait which results in short, shuffling strides in stroke patients [3,7,8]. The concept of FES is simple, but its realization is challenging [2,5-7,14,15,17].

Therefore, numerous of articles have been published emphasizing on the importance of FES as a promising method to restore the motor functions for the patients due to spinal cord injury, brachial plexus injury, stroke, multiple sclerosis, and traumatic brain injury [3,6,9-12,13,16]. At this level, these include: muscle strengthening and cardiovascular reconditioning, endurance, standing and gait control, enhancement of limb function, wound healing, facilitation of voluntary responses, reduction of osteoporosis, improving range-of-motion (ROM), and orthotic substitution [1,10,19]. Then, rehabilitation application of FES as a therapeutic and rehabilitation modality has the potential to improve functional skill abilities, increase strength of muscle body, force production of the muscle contractions, improve the osteoporosis, balance the circulation of blood control and voluntary movement [2,3,8,13,14,21]. Past research proposed that the brain is the most complex organ in the human body and is the centre of the nervous system [2,15,17]. It produces every human thought, action, feeling and experience of the world. The spinal cord is the pathway for impulses from the brain to the body as well as from the body to the brain [11-15]. However, the bounty of this pathway could be lost due to SCI and that results in a loss of function especially mobility [2,3,11-13,15-17]. The effects of SCI depend on the type of injury and the level of the injury. For example, paraplegics are those individuals with impairment of both lower extremities but have full use of their arms and hands. Then, types and description of paralysis described detail in Table 1.

Table 1 Types and description of paralysis

Type of paralysis	Description
Quadriplegia	The loss of movement and sensation in both arms and legs, it usually occurs as a result of injury at T1 or above.
	Quadriplegia also affects the chest muscles and injuries at C4 or above, and required a mechanical breathing machine (ventilator).
Paraplegia	The loss of movement and sensation in the lower half of the body, it usually occurs as a result of injuries at T1 or below.
Triplegia	The loss of movement and sensation in one arm and both legs and usually results from incomplete SCI.
Hemiplegia	A condition in which the limbs on one side of the body have severe weakness

© Springer International Publishing Switzerland 2015
V. Van Toi and T.H. Lien Phuong (eds.), *5th International Conference on Biomedical Engineering in Vietnam*,
IFMBE Proceedings 46, DOI: 10.1007/978-3-319-11776-8_112

II. PROBLEM STATEMENT

In recent years, modeling and simulation study can greatly facilitate to test and tune various FES control strategies [10-13,15,18,21]. Applications by modeling of musculoskeletal system with spinal cord injury are challenging because of the complexity of the system. Musculoskeletal system is a complex system since it is highly non-linear and time-varying nature [10-12,15-17]. Therefore, the modeling process becomes more difficult. Thus, many researchers have developed musculoskeletal models ranging in levels of sophistication from simple to complex [2,3,14,15]. However, most musculoskeletal [2,3,6,19] models built either on experimental or physiological bases are not appropriate for control applications, since these models characterize each muscle feature alone [15,19], and sometimes there is no connection between the modeled features which may prevent from modeling the whole muscle as one model [11,13,14,18-21]. Recently, mathematical models of artificial muscle activation in healthy or tetraplegic subjects have been developed but the complexities of the system resulting mathematical representation have a large number of parameters that make the model identification process difficult [15,18]. Therefore, based on system requirement, a new approach in the way of modeling the complex system such as musculoskeletal model can be developed through this research with good performance in terms of accuracy and less time complexity [2,3,15,19-21].

III. LITERATURE REVIEW

A. Spinal Cord Injury

In previous research, SCI is defined as injury to the spinal cord that is caused by trauma and resulting in loss of function [1-,3,15]. SCI are most caused by lateral bending, dislocation, rotation, axial loading and hyperextension of the cord [2,14,15,19],. The most common cause of SCIs is motor vehicle accidents, other causes include falls, work related accidents, sports injury and gunshot [15] . Furthermore, cognitive theory of multimedia learning underscores the importance of learning when multimedia presentation is integrated with prior idea.

The symptoms experienced by a patient will vary depending on where the spine is injured. Problems that SCIs patient may face include motor uncontrollability, the muscles may contract uncontrollably, completely unresponsive and bone degeneration [15,19,20].

B. Musculoskeletal System

A Musculoskeletal system is an organ system that gives human the ability to move using the muscular and skeletal systems [2,15,19,21]. Moreover, the musculoskeletal system provides form, support, stability and movement to the body. Hence, the primary function of the musculoskeletal system includes supporting the body, allowing motion and protecting vital organs.

This paper presents an insight about models of the musculoskeletal system that are valuable tools in the study of human movement. Hence, modeling and simulation study can be facilitated to test and tune various FES control strategies [15,19,21]. In order to develop control strategies for the FES [15,21] to move the hand correctly, an accurate model of the stimulated muscle has to be used. Therefore, modeling of joint properties of upper limbs in people in SCI is significant challenging for researchers due to complexity of the system [2,3,15]. It is difficult to establish because the complexity is due to the combination of complex structural anatomy, complicated movement and dynamics as well as indeterminate muscle function [2,15,19]. The main objective of this survey is to review the upper limb model and identify its drawbacks so that the modeling approach can be improved. A schematic diagram of Fig. 3 shows the segment of upper limb.

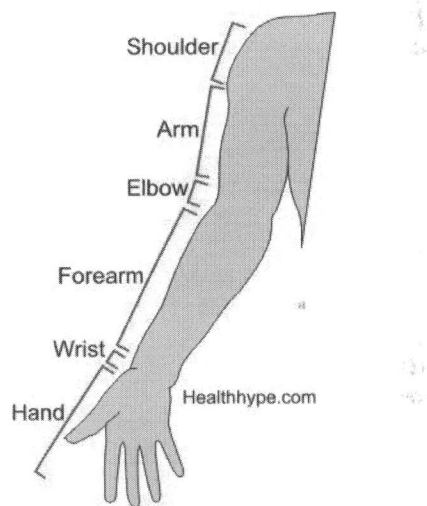

Fig. 3 The Upper Limb

C. FES-Elbow Extension Motion Control

FES induced movement control is a challenging area for researchers [2,3,14,15]. The challenge mainly arises due to many obstacles in stimulating the paralyzed neuromuscular system, such as fatigue [15,20], time varying properties and nonlinearity of paralyzed muscle [15,22]. The design of control strategies can greatly benefit from a model approach; in principle, better models mean better control [21,22]. In addition, primarily due to the complexity of the system, practically FES systems are dominantly open-loop where the controller receives no information about the actual state of the system [21]. Hence, practical success of this open loop control strategies is still, however, limited due to the fixed nature of the associated parameters. Therefore desired of accurate control of FES induced movement can be ensured with suitable closed loop control mechanism [15,26]. Such approach has several advantages over open loop, including better tracking performance and smaller sensitivity to modeling error, parameter variations, and external disturbance [1,11,15,23]. In this control strategy, fuzzy logic control (FLC) has long been known for its ability to handle a complex nonlinear system [1,3,11,12,15] without developing a mathematical model of the system. It is being used successfully in an increasing number of application areas in the control community. The control signal is computed by rule evaluation called fuzzy inference instead of by mathematical equations [1,11,12]. FLC is the preferred option to control this nonlinear multiple input-multiple output (MIMO) musculoskeletal elbow joint model [15,24,25].

IV. METHODOLOGY

A. *Flowchart Process*

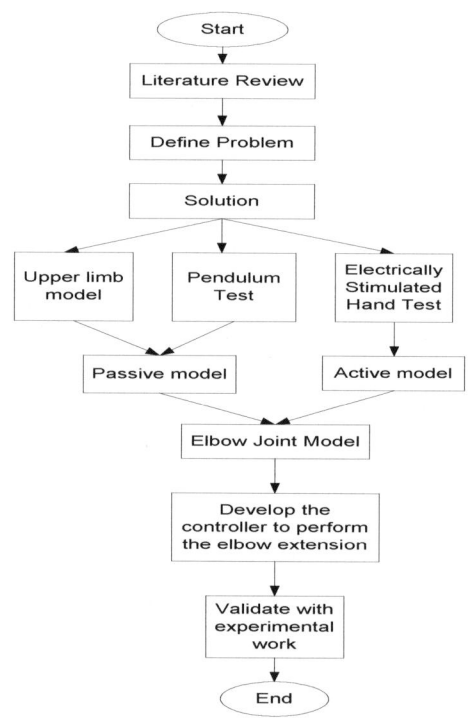

Fig. 5 Research procedure

B. *EMG Measurement*

Due to SCI problem, surface EMG measurements record the neural activity of the muscle. A passive movement is a movement of a joint without effort on the part of the examinee thus without neural activity. Consequently, the EMG record must show no significant activity of the muscles. In this study, the EMG surface will be used to measure the level of activity of the Biceps, the Brachioradialis and Triceps during experiment with the goniometer [2,3,15,17]. After full wave rectification and filtering the muscle activity [15], the activity is then normalized by the maximal activity during maximal voluntary contraction [10-13,15,25,26].

C. *Range of Motion Measurements*

Table 2 Range of motion of elbow extension

Measurement Tool	Universal Goniometer
Testing Position	Supine or sitting with the arm parallel to the midline and the forearm in the anatomical position
Stabilization	Examiner manually stabilizes the humerus
Goniometer Axis	Over the lateral epicondyle of the humerus
Stationary Arm	Parallel to the longitudinal axis of the humerus pointing towards the tip of the acromion
Moving Arm	Parallel to the longitudinal axis of the radius pointing towards the styloid process of the radius
Movement	Elbow extension
Expected ROM	0° in males; 10-15° in females is common.

V. CONCLUSIONS

The main aim of this research is the development of an accurate tetraplegic elbow joint model and the control strategy of FES-extension motion. Modeling of joint properties of upper limbs in people with SCI is significantly challenging for researchers due to the complexity of the system. A new approach of modeling the tetraplegic's elbow joint is presented in this study using FLC. On the one hand, a primary advantage to the use of a fuzzy logic-based scheme when compared to the conventional mathematical methods; fuzzy logic is simpler to implement as it eliminates the complicated mathematical modeling process by using set of fuzzy rules instead. While on the other hand, Particle swarm optimization (PSO). The great merit of a musculoskeletal model of knee joint is to help understand how the joint works and serves for control development. Again, the 'miracle' of FLC method of its ability to handle the complex nonlinear system without mathematical model will be used to develop the control strategies.

ACKNOWLEDGMENT

This research work is fully funded by Fundamental Research Grant Scheme (FRGS), Ministry of Education Malaysia, MOE and supported by the Research Management and Innovation Centre, ORICC Universiti Tun Hussein Onn Malaysia, UTHM under contract vot 0887.

REFERENCES

1. Ibrahim, B. S K K; Tokhi, M.O.; Gharooni, S. C.; Huq, M. S., "Devel-opment of fuzzy muscle contraction and activation model using multi-objective optimisation," *Systems Conference, 2010 4th Annual IEEE* , vol., no., pp.444,449, 5-8 April 2010
2. Ferrarin, M.; Pedotti, A., "The relationship between electrical stimulus and joint torque: a dynamic model," *Rehabilitation Engineering, IEEE Transactions on* , vol.8, no.3, pp.342,352, Sep 2000
3. Ahmad, M.K.I.; Ibrahim, B.S.K.K.; Rahman, K.A.A.; Masdar, A.; Nasir, N.H.M.; Mahadi Abdul Jamil, M., "Knee joint impedance hy-brid modeling and control of functional electrical stimulation (FES)-cyclingfor paraplegic: Free swinging trajectory," *Biomedical Engi-neering and Sciences (IECBES), 2012 IEEE EMBS Conference on* , vol., no., pp.984,989, 17-19 Dec. 2012
4. Elder, W., & Spetzler, R. F. (2003). Functional Electrical Stimulation (FES). In J. A. Editors-in-Chief: Michael & B. D. Robert (Eds.), Encyclopedia of the Neurological Sciences (pp. 401-403). New York: Academic Press.
5. Faghri, P. D., Garstang, S. V., & Kida, S. (2009). Functional Electrical Stimulation (pp. 407-429)
6. Ferrarin, M.; Palazzo, F.; Riener, R.; Quintern, J., "Model-based control of FES-induced single joint movements," Neural Systems and Rehabilitation Engineering, IEEE Transactions on , vol.9, no.3, pp.245,257, Sept. 2001
7. Liberson WT, Holmquest HJ, Scot D, Dow M. Functional electrotherapy: stimulation of the peroneal nerve synchronized with the swing phase of the gait of hemiplegic patients. Arch Phys Med Rehabil 1961;42:101–5
8. Taylor PN, Burridge JH, Dunkerley AL, Wood DE, Norton JA, Singleton C, et al. Clinical use of the Odstock dropped foot stimulator: its effect on the speed and effort of walking. Arch Phys Med Rehabil 1999; 80: 1577–1583
9. B.S.KSM KADER IBRAHIM, Energy-Efficient FES cycling with quadriceps stimulation Proceeding of the 13th annual conference of the IFESS 2008, 53(1): 262-264
10. B.S.K.K Ibrahim, M.O. Tokhi, M.S.Huq and S.C. Gharooni.Natural Trajectory based FES-induced Swinging Motion Control. 4th International Conference on Mechatronics (ICOM), 17-19 May 2011
11. B.S.K.K Ibrahim, M.O. Tokhi, M.S.Huq and S.C. Gharooni. Optimized Fuzzy Control For Natural Trajectory Based Fes-Swinging Motion.International Journal of Integrated Engineering, 2011, 3 (3),:17-23
12. B.S.K.K Ibrahim, M.O. Tokhi, M.S.Huq and S.C. Gharooni .Fuzzy Logic based Cycle-to-Cycle Control of FES induced Swinging Motion.International Conference on Electrical, Control and Computer Engineering,(2011)
13. Hunt, K.J.; Stone, B.; Negard, N.-O.; Schauer, T.; Fraser, M.H.; Cathcart, A.J.; Ferrario, C.; Ward, S.A.; Grant, S., "Control strategies for integration of electric motor assist and functional electrical stimulation in paraplegic cycling: utility for exercise testing and mobile cycling," Neural Systems and Rehabilitation Engineering, IEEE Transactions on , vol.12, no.1, pp.89,101, March 2004
14. Dinesh Bhatia, Gagan Bansal, R.P. Tewari, K.K. Shukla. State of art: Functional Electrical Stimulation (FES). Int. J. of Biomedical Engineering and Technology, 2011 Vol.5, No.1, pp.77 – 99
15. B.S.K.K Ibrahim (2011). Modelling and Control of Paraplegic's Knee Joint (FES-SWINGING). Doctor of Philosophy, Department of Automatic Control and Systems Engineering, University of Sheffield, UK, January 2011.
16. Masdar, A.; Ibrahim, B.S.K.K.; Jamil, M.M.A.; Hanafi, D.; Ahmad, M.K.I.; Rahman, K.A.A., "Current source with low voltage controlled for surface Electrical Stimulation," Signal Processing and its Applications (CSPA), 2013 IEEE 9th International Colloquium on , vol., no., pp.161,164, 8-10 March 2013
17. Rahman, K.A.A.; Ibrahim, B.S.K.K.; Leman, A.M.; Jamil, M.M.A., "Fundamental study on brain signal for BCI-FES system development," Biomedical Engineering and Sciences (IECBES), 2012 IEEE EMBS Conference on , vol., no., pp.195,198, 17-19 Dec. 2012
18. Massoud, R. (2007). Intelligent control techniques for spring assisted FES-cycling. PhD Thesis. The University of Sheffield, Sheffield, UK.
19. Ghafari, A.S., Meghdari, A. and Vossoughi, G.R. (2009). *Estimation of the human lower extremity musculoskeletal conditions during backpack load carrying*. Journal of Scientia Iranica, 16(5):451-462.
20. Levy, M., Mizrahi, J. and Susak, Z. (1990). Recruitment force and fatigue characteristicsof quadriceps muscles of paraplegics isometrically activated by surface functional electrical stimulation. J. Biomed. Eng.**12**:150–156.
21. Crago, P.E., Lan, N., Veltink, P. H., Abbas, J.J. and Kantor, C. (1996). *New control strategies for neuroprosthetic systems*, Journal Rehabilitation Res Device **33**(2):158-72.
22. Levy, M., Mizrahi, J. and Susak, Z. (1990). Recruitment force and fatigue characteristicsof quadriceps muscles of paraplegics isometrically activated by surface functional electrical stimulation. J. Biomed. Eng.**12**:150–156.
23. Huq M. S. (2009). *Analysis and control of hybrid orthosis in therapeutic treadmill locomotion for paraplegia*, PhD Thesis. The University of Sheffield, Sheffield, UK.
24. Giuffrida, J. P., & Crago, P. E. (2005). Functional Restoration of Elbow Extension After Spinal-Cord Injury Using a Neural Network-Based Synergistic FES Controller, *13*(2), 147–152.
25. Giuffrida, J. P., & Crago, P. E. (2004). Utilizing Remaining Voluntary Muscle Synergies to Control FES Elbow Extension after Spinal Cord Injury, 4118–4121.
26. Lynch, C. L. (2011). PHD Thesis. *Closed-Loop Control of Electrically Stimulated Skeletal Muscle Contractions. Control.* University of Toronto.
27. Wartenberg, R. (1951) Pendulousness of the leg as a diagnostic test. Neurology. **1**:18-24.

Integrated Biomedical Waste Management for Small Scale Healthcare Units in India

Prasad Balachandran

Sri Ramakrishna Engineering College/Department of Biomedical Engineering, Anna University, Coimbatore, India

Abstract— **Health care is necessary for the sustenance of human life and well-being. But the waste generated from medical activities can be hazardous, toxic and even lethal because of their high potential for diseases transmission. The wastes generated from the healthcare units are hazardous and can cause delirious biological effects. In India there are an estimate of 2.7 lacs small scale healthcare units (which includes nursing homes, pathological laboratories, etc.).The waste generated collectively nears 63% of the total biomedical waste generated in India (On an annual basis). Most of the small scale units, do not have many departments/divisions and the generation of waste is small and normally they do not have treatment facility for the bio-medical waste due to financial incapability. Thus most of the units dump the waste along the municipal wastes. This can cause adverse effects such as providing waste dumping sites as breeding grounds for the microorganisms contained in the wastes (Biological test sample wastes). This issue calls for a Common Integrated facility for treatment of these wastes.**

Keywords— **Biomedical waste management, Small scale.**

I. BIOMEDICAL WASTE

Biomedical wastes are defined as waste that is generated during the diagnosis, treatment or immunization of human beings or animals, or in research activities pertaining thereto, or in the production of biological components.

A. Quantum of Biomedical Wastes

The quantum of waste that is generated in India is estimated to be 1-2 kg per bed per day in a hospital and 600 gm per day per bed in a general practitioner's clinic. e.g. a 100 bedded hospital will generate 100 – 200 kgs of hospital waste/day. It is estimated that only 5 – 10% of this comprises of hazardous/infectious waste (5 – 10kgs/day)

B. Consequences of Poor Biomedical Waste Management

Proper disposal of biomedical waste is of paramount importance because of its infectious and hazardous characteristics. Improper disposal can result in the following: Organic portion ferments and attracts fly breeding Injuries from sharps to all categories of health care personnel and waste handlers

Increase risk of infections to medical, nursing and other hospital staff

Injuries from sharps to health workers and waste handlers · Poor infection control can lead to nosocomial infections in patients particularly HIV, Hepatitis B & C

Increase in risk associated with hazardous chemicals and drugs being handled by persons handling wastes · Poor waste management encourages unscrupulous persons to recycle disposables and disposed drugs for repacking and reselling

Development of resistant strains of microorganisms.

C. Persons Affected due to Poor Healthcare

Depending on the type of procedures, the persons at risk and mode of transmission in some common medical procedures are:

Procedure	Person at risk	Mode of Transmission
Collection of blood samples	Patient Health worker	Contaminated needle, gloves, Skin puncture by needle or container, Contamination of hands by blood
Collection of blood samples	Patient Health worker	Contaminated needle, gloves, Skin puncture by needle or container, Contamination of hands by blood
Transfer of specimens (within laboratory)	Laboratory personnel	Contamination of exterior of specimen container, Broken container,

		Splash of specimen
HIV serology and virology	Laboratory personnel	Skin puncture, splash of specimen, Broken specimen container, Perforated gloves

V. Van Toi and T.H. Lien Phuong (eds.), *5th International Conference on Biomedical Engineering in Vietnam,*
IFMBE Proceedings 46, DOI: 10.1007/978-3-319-11776-8_113

Cleaning and Main-tenance	Laboratory Personnel Sup-porting staff	Skin puncture or conta-mination, Splashes, Contaminated work sur-face
Waste Disposal	Laboratory Personnel Sup-port Staff Trans-port worker	Contact with contami-nated waste Puncture wounds and cuts
Shipment of speci-mens	Transport work-er Postal worker	Broker or leaking speci-men, containers and packages

D. Rules Regarding Biomedical Waste Management in India

The Government of India has promulgated the Biomedi-cal Waste (Management and Handling) Rules 1998. They are applicable to all persons who generate, collect, receive, store, transport, treat, dispose or handle biomedical wastes. This includes hospitals, nursing homes, clinics, dispensa-ries, veterinary institutions, animal houses, pathological laboratories and blood banks.

E. Mixing Up of Muncipal Waste and Biomedical Waste

The mixing up of municipal waste and biomedical waste takes place in India frequently because the small scale healthcare centres here do not have disposal and waste management units independently. This paves way for the epidemic outbreak as municipal waste systems are closely associated to the human communities at al hierarchies. The common epidemics are flu, rabies and malaria.

F. Need for a Common Biomedical Waste Management System

There is a dire need for an integrated biomedical system for small scale biomedical units in India. This is because of finan-cial requirement of the unit is not affordable by these units.

G. Integrated Biomedical Waste Management System

This facility consists of:

- Segregated Storage Unit
- Common Waste Transporting Unit
- Common Disposable Unit

H. Segregated Storage Unit

Collection room(s)/intermediate storage area where the waste packets/bags are collected before they are finally taken/transported to the treatment/disposal site. This is all the more important when the waste is to be taken outside the pre-mises. Two rooms - one for the general and the other for the hazardous waste are preferable. In case of shortage of rooms, the general waste (non-hazardous) can be directly stored out-side in dumper containers with lids of suitable size. Arrange-ment for separate receptacles in the storage area with promi-nent display of color code on the wall nearest to the recep-tacles has to be made. When waste carrying carts/containers arrive at this area, they have to be systematically put in the relevant receptacle/designated area.

I. Waste Transportation

The vehicle, which may be a specially designed van, should have the following specifications:

1. It should be covered and secured against accidental opening of door, leakage/spillage etc.
2. The interior of the container should be lined with smooth finish of aluminium or stainless steel, wi thout sharp edges/corners or dead spaces, which can be conveniently washed and disinfected.
3. There should be adequate arrangement for drainage and collection of any run off/leachate, which may accidentally come out of the waste bags/containers.
4. The floor should have suitable gradient, flow trap and collection container.
5. The size of the van would depend on the waste to be carried per trip.
6. In case, the waste quantity per trip is small, cov-ered container of 1-2 cubic meter mounted on 3 wheeled chassis and fitted with a tipping arrange-ment can be used.

J. Disposal Unit

Different methods have been developed for rendering bio-medical waste environmentally innocuous and aesthetically acceptable but all of them are not suitable for our condition. The 'Bio-Medical Waste (Management & Handling) Rules, 1998' has elaborately mentioned the recommended treatment and disposal options according to the 10 different categories of waste generated in health care establishments in Schedule I of the rules (Annexure 7.2). Standards for the treatment tech-nologies are given in Schedule V of the Rules, which must be complied with. The most commonly used one is incineration.

II. CONCLUSIONS

Thus the Integrated Biomedical waste management system can serve its purpose of reducing the mix-up of biomedical waste along with municipal waste initially and can prevent it in the long run thereby reducing the epidemic outbreak in India and providing a feasible solution for small scale units.

III. REFERENCES

1. Reinhardt, Peter A., and Judith G. Gordon. 1991. Infectious and medical waste management. Chelsea, Mich: Lewis Publishers

2. "Guidance on Closed Containers". Environmental Protection
3. Agency. Retrieved 15 May 2013

Author: Prasad Balachandran
Institute: Sri Ramakrishna Engineering College
Street: Rajalakshmi Street
City: Coimbatore
Country: India
Email: prasaddadmyhero@gmail.com

Simultaneous Detection of Two Viroids Infecting Grapevines in Taiwan by Multiplex RT-PCR

Nguyen Phuc Thien[*] and Chu-Hui Chiang

Department of Molecular Biotechnology, Da-Yeh University, Taiwan

Abstract— **Multiplex RT-PCR was used to detect *Hop stunt viroid* (HSVd) and *Grapevine yellow speckle viroid*-1 (GYSVd-1). Twenty five samples were collected from grapevine farms located at Changhua county of Taiwan. Specific primer pairs used for the detection of HSVd and GYSVd-1 were designed according to the conserved sequences of the viroid genomes. Multiplex one-step RT-PCR followed gel *electrophoresis analysis* showed that two DNA fragments with the sizes of 367 bp and 300 bp indicated the infection of GYSVd-1 and HSVd, respectively. The preliminary data suggests that the frequency of GYSVd-1 and HSVd infection is 80% and 90%, respectively.**

Keywords— **HSVd, multiplex RT-PCR, GYSVd-1, viroid.**

I. INTRODUCTION

Grapevine, an important crop in many areas of the Taiwan, is affected by a number of diseases.

Viroids are infectious agents with a small, single-stranded circular RNA molecule. Viroids do not code for proteins and the length of viroid genome is about 250 to 400 nucleotides. Viroids are able to replicate, move and infect plants, therefore, it frequently cause severe diseases in plants and produce symptoms of severe stunting, leaf necrosis, corky bark, leaf-roll, and fruit deformation depending on host plant and viroid species.

Up to now, viroid species are classified into two families *Pospiviroidae* and *Avsunviroidae*, which are composed of five and two genera, respectively (Owens et al., 2012).

In grapevine, some of the plants exhibited yellow leaf spots and flecks symptoms indicating of viruses or viroids infection. *Grapevine yellow speckle viroid*-1 (GYSVd-1) is the causal agent of yellow speckle disease.

A. Objective

The method was developed by using multiple RT-PCR to simultaneously detect GYSVd-1 and *Hop stunt viroid* (HSVd) from 25 grapevine samples collected from Taiwan.

II. MATERIALS AND METHODS

A. Source of Plant Materials

Young leaves of 25 grapevine samples were collected during the hotter months of summer from difference vineyards in Dacun, Taiwan during summer of 2013. Samples were from plants showing either yellow speckle or line patterns or asymptomatic plants.

Samples from four or five neighboring plants of every second row in the vineyard were analyzed for their viroid variant pattern.

Young leaves were collected and stored at -80°C until use.

B. Extraction of RNA

Total RNAs of grapevine plants were extracted according the protocol of manufacture of VIOGEN kit. The quality of the extracted RNA was measured by UV Spectrophotometer with ratio:

$$\frac{OD_{280}}{OD_{260}} = 1.0 \div 2.0$$

The full-length cDNAs of HSVd, GYSVd-1 were amplified from purified RNA extracts through RT-PCR using specific primer pairs.

C. Designed of Viroid Specific Primers

In this analysis, specific primers for HSVd and GYSVd-1 (Table 1) were determinate by sequence alignment of each viroid genomes which were available in the National Center of Biotechnology Information (NCBI, http://www.ncbi.nlm.nih.gov) with the accession numbers: HSVd (AB742225, HE575348, AB742224, JX401927, AY594202, FJ716178, JX418270, FJ716190, HQ386721, GU327606), GYSVd-1

Table 1 Design Primers

Primer	Sequence	Tm	Length of amplicon
5'HSVd	5'CTGGGGAATTCTCGAGTT	53.8	300
3'HSVd	GC-3'	51.8	
	5'AGGGGCTCRAGAGAGGM		
	TC-3'		
5'GYSVd-1	5'ACGAAGGGGTGCACTCC	60.4	367
3'GYSVd-1	GAGTG-3'	59.9	
	5'CGACGACGAGGCTCACTC		
	CC-3'		

Tm: *Melting Temperature, reference for determination the annealing temperature*

[*] Corresponding author.

© Springer International Publishing Switzerland 2015
V. Van Toi and T.H. Lien Phuong (eds.), *5th International Conference on Biomedical Engineering in Vietnam*,
IFMBE Proceedings 46, DOI: 10.1007/978-3-319-11776-8_114

(JF746188, JF746176,JF746193, JF746185, JF746182, JF746184, JF746189, JF746180, JF746183, JF746190), Specific primers for RT-PCR were designed using Clustal-W software.

D. The RT-PCR Reaction

After cDNA synthesis, PCR reaction was used to amplify the specific viroid sequences. Three steps (denaturation, primer annealing and primer extension) are carried out at discrete temperature ranges (for example, 94^0C to 98^0C, 37^0C to 65^0C, and 72^0C respectively) represent a single PCR cycle. The primer annealing and/or extension temperature and time may depend on the DNA enzyme used and on the specific sequence/length of the primers used.

For multiplex RT-PCR, primer combinations were chosen that would allow us to separate and distinguish the expected amplified fragments on a 1% agarose gel. Several concentrations of primers were tested to determine the best combinations for the amplification of the target sequences.

PCR products were analyzed by electrophoresis in 1% agarose gel, stained with ethidium bromide and visualized under UV light.

PCR products were purified using the PCR purification kit. The amplified DNA fragments were eluted from agarose gel by Micro-Elute DNA Clean extraction kit, according to the manufacturer's instructions, cloned into T&A vector and transformed into *Escherichia coli*. The recombinant cDNA clones were sequenced using automated DNA sequence.

III. RESULTS AND DISCUSSION

A. Yield and Quality of RNA Extract

Total RNA extracted from host plant using the VIOGEN kit. However, RNA extracted from grapevine contained high amounts of polysaccharides (reflected in the 260/280 ratio) making it unsuitable for RT-PCR analysis. Several other RNA extraction methods or modifications were compared with varying results in both the quantity and quality as determined by UV spectrophotometer (Table 2). 100 mg of tissue was obtained from older leaves collected during summer season. The yield was higher from younger leaves when obtained from 100 mg of tissue. The total RNA extraction method is not laborious, fast and safe as it avoids the use of organic solvents.

Table 2 Quality of RNA extraction

Tissue (100mg)	OD260/280 (Average)
Young leaves	1.95
Mature leaves	1.8

B. DNA Synthesis and PCR Amplification of Grapevine Viroids

It was efficient synthesize simultaneously. DNA from viral RNAs. Fig. 1 show typical results of an amplification of grapevine viroids from RT-PCR product.

Besides, PCR primers for these viroids were designed according to the sequence of viroid genome. Therefore, these primers would be more suitable for DNA synthesis than random primers. In amplification of these viroids, because PCR primers were designed to amplified a DNA fragment about 300bp to 367 bp. These viroids were amplified efficiently after cDNA synthesized with either primer.

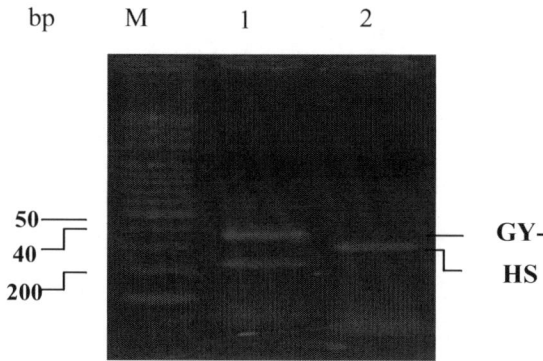

Fig 1 Agarose gel electrophoresis analysis of DNA products amplified from symptomatic grapevines by single RT-PCR using primer pairs specific for GYSVd-1 (lane1) and HSVd (lane 2), respectively

C. Development of a Multiplex RT-PCR Reaction

For multiplex RT-PCR assays, the primer pairs specific for HSVd and GYSVd-1 were added to the same RT-PCR mixture. In all cases, a final concentration of 0.5 µM for all primers resulted in a clear DNA fragment of HSVd, however, a weak DNA band or or undetectable of GYSVd-1 was found. Therefore, the concentration of GYSVd-1 and HSVd primers were varied to determine which combinations of primer concentrations gave best amplification of the expected viroid-specific fragments and finally, a . Combination of 0.4 µM of the primer pairs for HSVd and 0.5 µM of the primer pairs for GYSVd-1 was able to effectively in detection of HSVd and GYSVd-1 in grapevine extracts (Fig 2). From the agarose gel assays, in addition to the DNA fragments amplified from the target viroid genomes, another not specific DNA band was observed. A minor adjustment of the RT-PCR condition may need to improve the specificity of the detection.

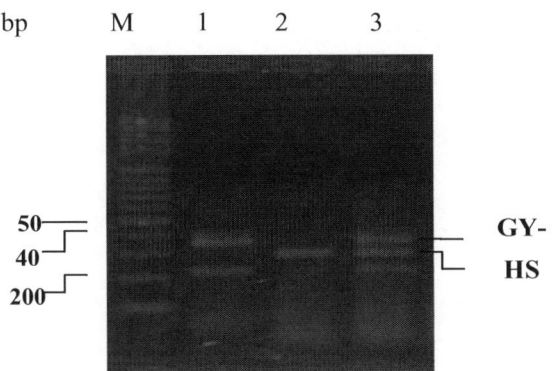

Fig 2: Agarose gel analysis of multiplex RT-PCR (lane3) assays with the primers for the detection of HSVd and GYSVd-1on extracts from grape-vines. Lane M molecular weight markers. Lane 1 and 2 were control, each been amplified using GYSVd-1 and HSVd primers.

IV. CONCLUSIONS

The list of the two sets of primers and multiplex RT-PCR and gel electrophoresis procedure described here provides a fast, convenient, low-cost and feasible tool for diagnostic of HSVd and GYSVd-1 at grapvines.

ACKNOWLEDGEMENT

This work has been supported by Prof. Chu Hui Chiang and her staffs at the Molecular Biotechnology Department, Da-Yeh University, Taiwan for providing samples and specific primer pairs.

REFERENCES

1. Hajizadeh, M., et al. (2012). "Development and validation of a multiplex RT-PCR method for the simultaneous detection of five grapevine viroids." J Virol Methods 179: 62-69.
2. Nakaune, R. and M. Nakano (2006). "Efficient methods for sample processing and cDNA synthesis by RT-PCR for the detection of grapevine viruses and viroids." J Virol Methods 134: 244-249.
3. Staub, U., et al. (1995). "Two rapid microscale procedures for isolation of total RNA from leaves rich in polyphenols and polysaccharides: application for sensitive detection of grapevine viroids." J Virol Methods 52: 209-218.
4. Jiang, D., et al. (2012). "Comprehensive diversity analysis of viroids infecting grapevine in China and Japan." Virus Res 169: 237-245.
5. Kolunow, A. M., et al. (1989). "Two related viroids cause grapevine yellow speckle disease independently." J Gen Virol 70: 3411-3419.
6. Rezaian, M. A. (1990). "Australian grapevine viroid--evidence for extensive recombination between viroids." Nucleic Acids Res 18: 1813-1818.
7. Rezaian, M. A., et al. (1992). "Common identity of grapevine viroids from USA and Australia revealed by PCR analysis." Intervirology 34: 38-43.
8. Staub, U., et al. (1995). "Two rapid microscale procedures for isolation of total RNA from leaves rich in polyphenols and polysaccharides: application for sensitive detection of grapevine viroids." J Virol Methods 52: 209-218.
9. Wan Chow Wah, Y. F. and R. H. Symons (1997). "A high sensitivity RT-PCR assay for the diagnosis of grapevine viroids in field and tissue culture samples." J Virol Methods 63: 57-69.

Theoretical Investigation on Antioxidant Activity of Phenolic Compounds Extracted from *Artocarpus Altilis*

Nguyen Minh Thong[1,2] and Pham Cam Nam[3]

[1] Department of Chemistry – Hue University of Sciences, Hue City, Viet Nam
[2] The University of Da Nang - Campus in Kon Tum, Kon Tum City, Viet Nam
[3] Department of Chemistry, Danang University of Science and Technology, The University of Da Nang, Viet Nam

Abstract— **Theoretical calculations have been performed to predict antioxidant property for phenolic compounds extracted from *Artocarpus Altilis*. The chosen ONIOM model contains only two atoms of the breaking bond as the core zone and is able to provide reliable evaluation for BDE(O–H) for phenolic compunds. Important characteristics of antioxidants such as the homolytic O-H bond dissociation enthalpy (BDE) and the adiabatic ionization energy (IE) were determined both in gas phase and in solvents. The BDE(O–H) values of compounds 1, 2, 3 and 4 are predicted to be 80.8, 80.3, 79.3, and 77.3 kcal/mol, respectively.**

Keywords— **Antioxidant, bond dissociation enthalpy, ionization potential, phenolic compounds.**

I. INTRODUCTION

Numerous physiological and biochemical processes in the human body may produce oxygen-centered free radicals and other reactive oxygen species as byproducts [1]. Some of radicals that are most abundantly produced in biochemical reactions are the so-called reactive oxygen species (ROS). The over-production of ROS and free radicals can cause oxidative damage to biomolecules, which lead to cellular damage including lipid peroxidation, DNA adduct formation, protein oxidation, and enzyme inactivation, which can all ultimately lead to cell death [2]. All human cells protect themselves by multiple mechanisms especially enzymatic and non-enzymatic antioxidant systems against free radical damage. However, these protective mechanisms may not be enough for severe or continued oxidative stress. Hence, certain amounts of antioxidant supplements are constantly required to maintain an adequate level of antioxidants in order to balance the reactive oxygen species in human body [3]. Many natural organic compounds extracted from leaves, seeds, and other parts of plants are considered as the potential antioxidants [4,5]. The great advantages of these compounds are in their high antioxidant activity, non-toxic effects on human-beings and safety to the environment [6]. In nature, *Artocarpus altilis* is known as a good source of phenolic compounds [7-9] whose beneficial role in the digestive process in humans is well established. The leaves, roots, and root bark of this plant are used as traditional medicines for the treatment of antioxidant, gout, hepatitis, hypertension, fever, and liver disorders and diabetes [10,11].

Recently, theoretical methods have been successfully used to evaluate chemical properties, such as bond dissociation enthalpy (BDE) and the adiabatic ionization energy (IE) of polyphenol compounds and to elucidate the structure–activity relationship (SAR) for phenolic antioxidants [12,13]. Furthermore, the study of the electronic and molecular properties is of great importance that helps to understand the mechanism of the antioxidant activity of these compounds.

Despite these advances in experimental and theoretical works, an understanding of the relationship between the structure and antioxidant activity and the antioxidant mechanism is still lacking. Therefore, theoretical investigation on antioxidant activity of phenolic compounds extracted from *Artocarpus altilis* (Fig. 1) is a matter of great interest to researchers in different fields.

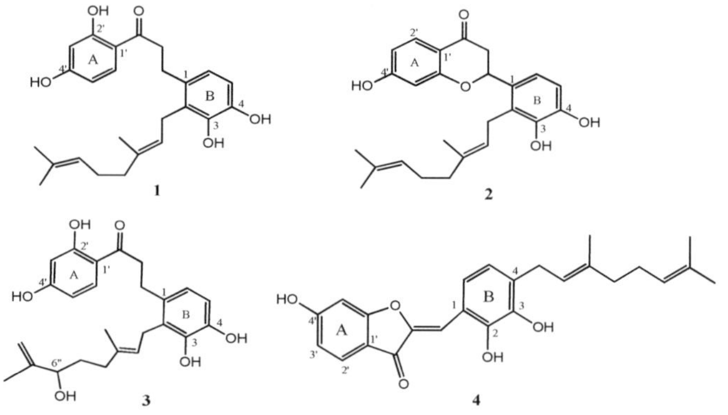

Fig. 1 Structures of investigated compounds

In the literature, two main mechanisms by which antioxidants can play their protective role were proposed and widely analyzed [14]. The first one is refered as Hydrogen Atomic Transfer (HAT, Eq.(1)) from the antioxidant ArOH that becomes itself a radical and is governed by the O–H bond dissociation energy (BDE) of ArOH. The second one is refered as Electron Transfer-Proton Transfer (ET-PT, Eq. (2)). This mechanism is governed by the electron transfer capacity, in other words, the ionization energy (IE), in

© Springer International Publishing Switzerland 2015
V. Van Toi and T.H. Lien Phuong (eds.), *5th International Conference on Biomedical Engineering in Vietnam*,
IFMBE Proceedings 46, DOI: 10.1007/978-3-319-11776-8_115

which the antioxidant gives an electron to the free radical becoming a radical cation.

$$\text{ROO}^\bullet + \text{ArOH} \longrightarrow \text{ROOH} + \text{ArO}^\bullet \quad (1)$$

$$\text{ROO}^\bullet + \text{ArOH} \longrightarrow \text{ROO}^- + \text{ArOH}^{+\bullet} \longrightarrow \text{ROOH} + \text{ArO}^\bullet \quad (2)$$

Nam and coworkers studied the thiophenol, phenylphosphine, toluene, phenolic and their derivatives [15-18] using DFT with the (RO)B3LYP method and two-layer ONIOM method [19] to accurately determine the BDE and IE values. In the present work, we propose to use the DFT restricted open-shell (RO)B3LYP/6-311++G(2df,2p) for a high layer and the semi-empirical PM6 method for the lower layer with the aim to further shed light on the structural and electronic properties of phenolic extracted from *Artocarpus altilis* and their radicals.

II. COMPUTATION METHODS

All computations were performed using the Gaussian 09 (version A.02) suite of programs [20]. Geometry optimizations and vibrational frequency calculations were conducted using the semi-empirical PM6 method. Vibrational frequencies obtained at the PM6 level were subsequently scaled by a factor of 1.078 for estimating the zero-point vibrational energies (ZPE) [21]. The enthalpy values at higher level were evaluated from the calculated single-point electronic energy based on PM6 optimized structures.

We thus propose a new ONIOM scheme integrating the (RO)B3LYP/6-311++G(2df,2p) level with the semi-empirical PM6 method for evaluating the BDE(O–H)'s of all phenolic compounds. The restricted open-shell (RO) formalism was applied in this scheme for the radical species using the same basis set. The use of an RO method improves the energies of the radicals and thereby the BDE values. In Scheme 1, we describe the three ways of choosing the layer in our proposed ONIOM scheme.

In this proposed ONIOM treatment, each molecule is divided into two layers, the atoms in the circle is treated as a high layer; whereas the leftover atoms of the molecule in the rectangle belong to the second layer which is treated as a lower layer. The (RO)B3LYP/6-311++G(2df,2p) method is thus applied for the atoms in the high layer, whereas the PM6 procedure is applied for the low layer. We consider three different ways of selecting the ONIOM model denoted as 1A and 3A. For the 1A, the model has only one oxygen atom and one hydrogen atom related to the target bond for estimating BDE at the high level (the circle in Scheme 1). The 3A ONIOM model has three heavy atoms including oxygen and two carbon atoms at the ipso- and ortho-positions involving hydrogen atoms (as shown in the circle) at the high level. The rest are defined as the low layer.

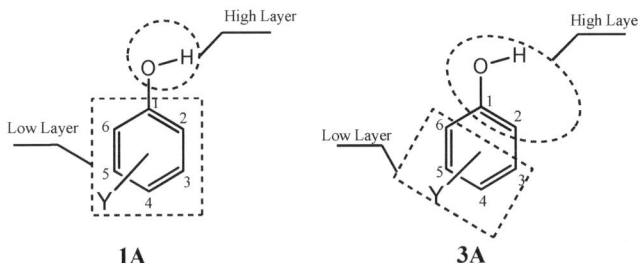

Scheme 1. Schematic description of two-layer proposed ONIOM models: (a) 1A and (b) 3A

The homolytic BDE(O-H) value in gas phase at 298.15 K and 1.00 atm for the polyphenol compound (ArOH) was calculated from the expression:

$$BDE(O\text{-}H) = H(ArO^\bullet) + H(H^\bullet) - H(ArOH)$$

where H's are the enthalpies of different species at 298.15 K. The enthalpies were estimated from the usual expression: $H(T) = E_0 + ZPE + H_{trans} + H_{rot} + H_{vib} + RT$. The H_{trans}, H_{rot}, and H_{vib} are the translational, rotational, and vibrational contributions to enthalpy, respectively; E_0 is the total energy at 0 K and ZPE is the zero-point vibrational energy. The enthalpy value for the hydrogen atom in the gas phase was taken at its exact energy of -0.5 hartree.

The IE values were obtained according to the formula $IE = E_0(ArOH^{+\bullet}) - E_0(ArOH)$, in which $E_0(ArOH)$ is the total energy of the parent molecule whereas $E_0(ArOH^+)$ denotes the corresponding total energy of the cation radical generated after the electron transfer. ZPE was added to the electronic energy to obtain E_0 at 0 K. In all computations, cation radicals from the optimized neutral compounds in the global energy minimum were generated and further fully optimized.

III. RESULTS AND DISCUSSION

A. Validating the Suitable Methods for BDEs Calculation

Since the two ONIOM models are newly proposed, first it is necessary to check their efficiency in estimating reliable BDE values. If the proposed procedures provide accurate BDE values for model compounds whose BDEs are accurately known from experiment, then it can be applied to unknown target compounds. To this end, we have calibrated our proposed ONIOM methods (1A and 3A) in calculating the BDE(O-H) for a series of substituted phenolic compounds. Table 1 presents our calculated BDE(O–H) values of phenolic compounds along with available experimental results.

For the model 1A, the largest deviation between our calculated BDE(O–H) and experimental values of the corresponding molecules is of 2.1 kcal/mol. For the other substituted compounds of phenol, the calculated BDE(O–H)s are underestimated in a range of 1–2 kcal/mol. The estimated BDE(O–H) of 87.2 kcal/mol for phenol from this work is in quite good agreement to the value of 87.5 kcal/mol obtained by Chandra and Uchimaru [22] using ROB3LYP/6-311++G(2df,2p) //B3LYP/6-311G(d,p), as well as with the recommended value of 88.0 ± 1.5 kcal/mol [23].

However when we apply the model 3A with three heavy atoms (one oxygen and two carbons), the difference between our calculated and the experimental BDE(O–H) becomes even larger, ranging between 4 and 6 kcal/mol. The largest deviation of -6.4 kcal/ mol is seen form *m*-OH-C$_6$H$_4$OH. The reference values listed in Table 1 show that the BDE(O–H) values obtained from our model 1A are

reasonably accurate and comparable with the best experimental data. Hence we choose mode 1A as our method of choice for further study. However, the reliability of the model 1A needs to be checked further for the larger molecules, like tocopherols. The tocopherols exist in alpha, beta, gamma and delta forms, in which alpha-tocopherol is known as a strong antioxidant. Figure 2 depicts the optimized structure of tocopherol and the chosen model 1A. The higher layer is displayed in bond and bond type format and the lower layer is in wire frame format. The calculated BDE(O–H) of tocopherols are given in Table 2. It can be observed that our calculated BDE(O–H) values are in very good agreement with the corresponding experimental values for tocopherols, with the deviation of only ±1.0 kcal/mol.

Table 1 Comparison of calculated BDE(O–H) values of phenolic systems using two ONIOM models: 1A and 3A.

Molecule	ONIOM		BDE(O–H) exptl [a] kcal/mol	DBDE(O–H) [b] in kcal/mol	
	Model 1A	Model 3A		Model 1A	Model 3A
C$_6$H$_5$OH	87.2	83.2	88.0 ± 1.5 (87.5 ± 1.5)	0.8 (0.3)	4.8(4.5)
o-OH-C$_6$H$_4$OH	80.3	77.3	81.2	0.9	3.9
m-OH-C$_6$H$_4$OH	88.8	84.5	90.9 (88.2)	2.1 (+0.6)	6.4 (3.7)
p-OH-C$_6$H$_4$OH	83.6	79.7	84.1 (82.2)	0.5 (1.4)	4.4 (2.5)
o-CH$_3$-C$_6$H$_4$OH	84.6	80.9	86.1	1.5	5.2
m-CH$_3$-C$_6$H$_4$OH	87.4	83.4	87.6	0.2	4.2
p-CH$_3$-C$_6$H$_4$OH	85.0	81.2	86.1 ± 0.5	1.1	4.9
o-OCH$_3$-C$_6$H$_4$OH	84.6	82.7	86.0	1.5	3.4
m-OCH$_3$-C$_6$H$_4$OH	88.2	84.1	88.4	0.2	4.3
p-OCH$_3$-C$_6$H$_4$OH	82.8	79.0	82.8	0.0	3.8
o-Ph-C$_6$H$_4$OH	84.1	80.5	86.5	2.4	6.0
p-Ph-C$_6$H$_4$OH	85.5	81.7	87.6	2.1	5.9

[a] Experimental data taken from Ref. [23].
[b] DBDE(O–H) = BDE(O–H)$_{calc}$ -BDE(O–H)$_{exptl}$.

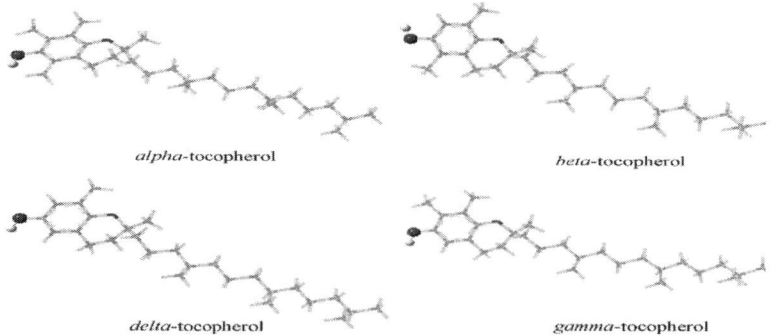

alpha-tocopherol   *beta*-tocopherol

delta-tocopherol   *gamma*-tocopherol

Fig. 2 Four forms of the tocopherol (high layer in ball and bond type format; low layer in wire frame format)

Table 2 Calculated BDE(O–H) of tocopherols. Data are in kcal/mol

Form	BDE(O–H)	Experimental value[a]	DBDE(O–H)[b]
alpha-Tocopherol	78.2	77.3	+0.9
beta-Tocopherol	79.4	80.2	-0.8
delta-Tocopherol	81.1	80.2	+0.9
gamma-Tocopherol	79.6	80.1	-0.5

[a]Experimental values taken from Ref. [23].
[b]DBDE(O–H) = BDE(O–H)$_{Calc}$ - BDE(O–H)$_{exptl}$.

B. Finding the Position of the Weakest O-H Bond

For a compound possessing more than one phenolic hydroxyl, its radical-scavenging activity is determined by the one with the lowest O–H BDE. To reduce computation time, we used the PM6 method to calculate preliminary the O–H BDE at any position. Then, the weakest bond was continued to calculate at higher levels (ONIOM 1A model). The calculated results are given in Table 3. The lowest BDE(O–H) is at position 3 of ring B for compounds 1 and 3, which are estimated to be about 69.0 and 68.4 kcal/mol, respectively. In compound 2, 4-OH had the lowest BDE value, 67.7 kcal/mol. Similarly, the BDE value of 2-OH for compound 4 was lower than other positions.

Table 3 The O-H bond dissociation enthalpies using PM6 method

Site of broken O–H bond [a]	BDE(O–H), kcal/mol			
	(1)	**(2)**	**(3)**	**(4)**
2'(ring A)	83.8	-	83.8	-
4'(ring A)	82.8	80.2	82.7	79.9
2(ringB)	-	-	-	**65.0**
3(ring B)	**69.0**	71.9	**68.4**	71.2
4(ring B)	83.8	**67.7**	70.5	-
6''	-	-	100.0	-

[a] See Figure 1 for definition of atom numbering.

C. BDE Computation – The Influence of Solvents

The hydrogen donating ability of the wide class of phenolic and the ability of these compounds to form the radical forms are characterized by BDE. The BDE corresponds to the O-H bond breaking (hydrogen abstraction), thus this parameter describes the stability of the hydroxyl bonds. The molecules with lower values of BDE are endowed with higher antioxidant activity. Table 4 presents the calculated BDE values in gas phase and solvents (methanol and water) using ONIOM 1A model. On the basis of the calculated O–H BDEs (Table 4), the hydrogen donating ability of phenolic follows the order: (4)>(3)>(2)>(1). Moreover, among the phenolic hydroxyls at different positions, the hydroxyl

at position 2 in compound 4 has the lowest O–H BDE, 77.3, 79.0 and 78.3 kcal/mol in gas phase, methanol and water, respectively.

Table 4: ONIOM(ROB3LYP/6-311++G(2df,2p):PM6)-computed BDE (O-H) and IE of phenolic

Compounds	BDE (kcal/mol)			IE (kcal/mol)
	Gas phase	Methanol	Water	
1	80.8	82.2	82.4	177.6
2	80.3	83.1	83.9	179.6
3	79.3	80.5	81.3	178.8
4	77.3	78.3	79.0	176.4
Phenol	87.2	-	88.6	190.7
α-Tocopherol	78.2	-	79.8	154.2
β-Tocopherol	79.4	-	80.8	-
δ-Tocopherol	81.1	-	82.3	-
γ-Tocopherol	79.6	-	81.3	-

From Table 4, it can be seen that the BDE values of each of the OH groups present in all radicals of phenolic are smaller than those of phenol calculated at the same level of theory. This indicates that most of the phenolic hydroxyls have stronger hydrogen donating ability than phenol. It can also be seen from Table 4 that the BDE of 2-OH group in compound 4 (77.3 kcal/mol in gas phase) is similar to that of the α-tocopherol (78.2 kcal/mol).

D. IE Computation

As stated above, scavenging of free radicals by phenolic may also be achieved via donation of a single electron. In this case, IE is an important physical factor indicating the range of electron donation. The molecules with lower values of IE are endowed with higher antioxidant activity. As shown in Table 4, the lowest IE value is found for compound 4 (176.4 kcal/mol), which is largely lower than that of phenol (190.7 kcal/mol), followed by compound 1 (177.6 kcal/mol), compound 3 (178.8 kcal/mol), compound 2 (179.6 kcal/mol). By comparison, we found that the trend for calculated IE values is different from that of BDE values. This discrepancy can be attributed to the fact that BDE is affected by the local phenomena induced by the substituents, whereas IE value is affected by the structure of the whole molecule [24]. In other words, within the mechanism of the electron transfer, the main factors affecting the value of IE are the extended delocalization and conjugation of the π-electrons, rather than the presence of particular functional groups such as additional hydroxyls. A close look at the IE values of compounds 1, 2, 3 and 4 shows that they are about

10–15 kcal/mol lower than that of phenol (190.7 kcal/mol). In addition, as seen from Table 4, the IE values of phenolic are are higher than that of α-tocopherol. The antioxidant mechanism of α-tocopherol was testified to be H-atom transfer because it is difficult to donate electrons [25], thus these compounds act as the α-tocopherol, and they are also difficult to donate electrons. Therefore, the dominant antioxidant mechanism of the phenolic should correspond to the H-atom transfer (HAT).

E. Spin Density

The spin density is another important parameter to characterize the stability of free radicals, because the energy of a free radical can be efficiently decreased if the unpaired electron is highly delocalized through the conjugated system [26]. Therefore, the spin densities on the various radical forms of compounds 1, 2, 3 and 4 are also analyzed to help understand the differences in reactivity of the various OH groups, and consequently the differences in BDE values. It should be pointed out that the more delocalized the spin density in the radical form is, the easier the radical is formed, and thus the lower the BDE value is [27]. As observed in Figure 3, the spin density of O-atom after H-removal in 6 radical is lower than other compounds. Consistently, the more delocalized the spin density in the radical, the easier is the radical formed and thus the lower is the BDE. Nevertheless, differences in the BDE cannot be explained only on the basis of the spin density value on the O-atom where H abstraction occurred. For example, the BDE of compound 2 is higher than that of compound 4 whereas the spin density on the O-atom of the radical 2 is less than that on the O-atom of radical 4. That difference is related to effect of substituents on the stability of radicals.

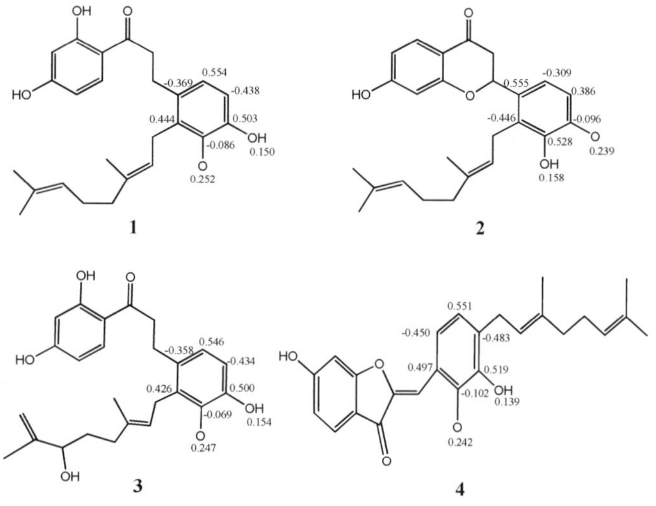

IV. CONCLUSIONS

In this article, we have applied the integration of the DFT (RO)B3LYP/6-311++G(2df,2p) and the semi-empirical PM6 methods into the ONIOM method to the study of the structure–activity relationship for a series of phenolic compounds. The two primary indicators of antioxidant activity, the O–H BDE and the adiabatic IE have been computed both in the gas phase and in solutions. In addition, the electronic feature such as spin density of radical species has also been presented. On the basis of our investigation, we can outline the following conclusions: the ONIOM(ROB3LYP/6-311++G(2df,2p): PM6) results for BDE(O–H) values of several substituted phenols are in good agreement with experimental values, within the deviation of ±1–2 kcal/mol, but the computer time cost is reduced substantially. The BDE(O–H) of alpha-, beta-, delta-, and gamma-tocopherol are predicted to be 78.2, 79.4, 81.1, and 79.6 kcal/mol respectively, in which alpha-tocopherol is evaluated as a strong antioxidant. The BDE(O–H)s of 1, 2, 3 and 4 amount to 80.8, 80.3, 79.3, and 77.3 kcal/mol, respectively. To sum up, this study will contributes to the ongoing interest on the antioxidant activity of phenolic compounds and their future exploitation for food or pharmaceutical applications.

ACKNOWLEDGEMENT

This research is funded by Vietnam National Foundation for Science and Technology Development (NAFOSTED) under grant number 104.06-2013.21.

REFERENCES

1. Harold Corke , Life Sciences 74 (2004) 2157–2184
2. Hileman, E. O.; Liu, J.; Albitar, M.; Keating, M. J.; Huang, P.Cancer Chemother. Pharmacol.53 (2004) 209–219
3. Anagnostopoulou MA, Kefalas P, Papageorgiou VP, Assimepoulou AN, Boskou D., Food Chem. 9 (2006) 419–25
4. M. Richelle, I. Tavazzi, E. Offord, J. Agric. Food Chem. 49 (7) (2001) 3438
5. N. Balasundram, K. Sundram, S. Samman, Food Chem. 99 (1) (2006) 191.
6. M.S. Brewer, Compr. Rev. Food Sci. Food Safety 10 (4) (2011) 221
7. Yuanjiang Pan, Xiaoxiang Zheng , Phytochemistry 68 (2007) 1300–1306.
8. Horng-Huey Ko, Phytochemistry 89 (2013) 78–88
9. Nguyen Trung Nhan, Phytochemistry Letters 5 (2012) 647–650
10. Do, T.L., Vietnamese Medicinal Plant, 13th ed. Medicine Publisher, Hanoi, 2005, pp 936
11. Adewole, S.O., Ojewole, J.O., Cardiovasc. J. Afr. 18 (2007) 221–227.
12. J.S. Wright, E.R. Johnson, G.A. DiLabio, J. Am. Chem. Soc. 123 (2001) 1173–1183
13. M. Reis, B. Lobato, J. Lameira, A.S. Santos, C.N. Alves, Eur. J. Med. Chem. 42 (2007) 440–446.

Fig. 3 Spin density distribution in radicals of 1–4 in gas phase calculated at ONIOM(ROB3LYP/6-311++G(2df,2p):PM6) level.

14. M. Leopoldini, N. Russo, M. Toscano, Food Chem. 125 (2011) 288

15. A.K. Chandra, P.C. Nam, M.T. Nguyen, J. Phys. Chem. A 107 (2003) 9182.

16. A.K. Chandra, P.C. Nam, M.T. Nguyen, J. Phys. Chem. A 107 (2003) 9182.

17. P.C. Nam, M.T. Nguyen, A.K. Chandra, J. Phys. Chem. A 108 (2004) 11362.

18. P.C. Nam, M.T. Nguyen, A.K. Chandra, J. Phys. Chem. A 109 (2005) 10342

19. P.C. Nam, M.T. Nguyen, Chemical Physics Letters 555 (2013) 44–50

20. M.J. Frisch et al., Gaussian 09, Revision A.2, Gaussian, Inc., Wallingford, CT, 2009.

21. I.M. Alecu, J. Zheng, Y. Zhao, D.G. Truhlar, J. Chem. Theory Comput. 6 (2010) 2872

22. A.K. Chandra, T. Uchimaru, Int. J. Mol. Sci. 3 (2002) 407

23. Y.-R. Luo, Handbook of Dissociation Energies in Organic Compounds, CRC Press LLC, 2003.

24. J.S. Wright, E.R. Johnson, G.A. DiLabio, J. Am. Chem. Soc. 123 (2001) 1173–1183

25. M. Leopoldini, T. Marino, N. Russo, M. Toscano, J. Phys.Chem. A 108 (2004) 4916–4922

26. W. Chen, P. Guo, J. Song, W. Cao, J. Bian, Bioorg. Med. Chem. Lett. 16 (2006) 3582.

27. M. Leopoldini, I.P. Pitarch, N. Russo, M. Toscano, J. Phys. Chem. A 108 (2004) 92–96

Author: Nguyen Minh Thong
Institute: The University of Da Nang-Campus in Kon Tum
Street: 704 Phan Dinh Phung
City: Kon Tum City
Country: Viet Nam
Email: thongsphoa@gmail.com

An EEG-Controlled Wheelchair Using Eye Movements

Hue T. Tran[1], Hai T. Nguyen[2], Hieu V. Phan[3], V. Van Toi[1], Thuyen V. Ngo[2], and Cao Bui-Thu[3]

[1] Biomedical Engineering Department, International University, VNU-HCMC, Vietnam
[2] Faculty of Electrical-Electronics Engineering, UTE-HCMC, Vietnam
[3] Electronics Technology Department, Industrial University of Ho Chi Minh City, Vietnam

Abstract— **This paper proposes a threshold algorithm to extract features and then create standard thresholds of EEG online signals using eye movements. Based on the standard thresholds, an electrical wheelchair can be controlled to reach the desired target. In this project, a Hamming band-pass filter is applied to detect noise of EEG signal. Moreover, calculation of blink-times and detection of confusion between left glance and right glance signals program are built and added to increase the accuracy of the wheelchair control. Additionally, a quality testing program is also built in Labview environment to improve effort of data collection. Finally, experimental results of controlling the electrical wheelchair in indoor environments demonstrated the effectiveness of the proposed method.**

Keywords— **Threshold algorithms, EEG technology, Eye movements and Electric wheelchair.**

I. INTRODUCTION

According to disability statistic, 10% of the total population of the world has handicaps, people is usually isolated to society. To improve integration to society of handicaps, many researchers have developed electrical wheelchair such as an intelligent wheelchairs using the user's face direction [1], voice command control wheelchair [2] or wheelchair control using a special camera that mounted on glasses to detect the movement of pupil and direction control [3].

Currently, a direct communication pathway between the brain and an external device, is an interesting topic in society. However, human brain is a complex and important structure in body, many methods have been applied into brain researches such as comparison concentrations of blood oxygenation, HbO, Hbb and etc in brain when subjects do different tasks using fNIRS [4], or using fMRI to find specific areas of the brain that respond to some tasks or stimulus [5] and evaluate neural network (figure out what areas of the brain are talking to eah other) [6]. In another method, electroencephalogram (EEG) provides a unique chance in terms of temporal resolution and cost compared to other neuro-imaging techniques; therefore, EEG is used in Brain Computer Interface researches such as Motor Imagery, Steady State Visual Evoked Potentials and what's known as P300 for control of computer software based on the real-time analysis of EEG signals [7].

On the background, wheelchair researches have applied EEG technology into wheelchair control such as EEG-controlled wheelchair for ALS patients [8], EEG command wheelchair with cognitive data, eyes movements and head gestures [9].

In the project, threshold algorithms are used to analyze, classify and create standards thresholds of EEG signals using eyes movements. Moreover, a wheelchair control program is built in Labview environment, the program not only connects to EEG device, subject and electric wheelchair, but also processes EEG online signals to control wheelchair in real time. Furthermore, a calculation of blink-times and a detection of confusion between left-glance and right-glance signals are supplemented to increase the accuracy of control. Additionally, a testing program is built to improve the effectiveness of data.

II. MATERIALS AND METHODS

A. Data Collection

EEG data are collected on 10 subjects with 137 samples at Fp1, F7, F8, CMS and DRL channels by using a Biosemi Active Two system show as figure 1 and 2.

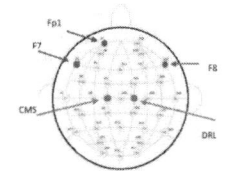

Fig. 1 Positions of electrodes for wheelchair control.

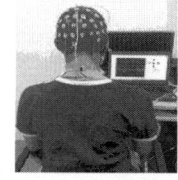

Fig. 2 Subject is doing experiments.

The subject took 4 independent tasks such as opening, blink, left-glance, and right-glance in 8 seconds. The eyes movements are performed at 2 moments: from 2nd to 4th second and from 6th to 8th second show in figure 3.

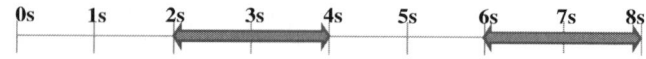

Fig. 3 Protocol of data acquisition

V. Van Toi and T.H. Lien Phuong (eds.), *5th International Conference on Biomedical Engineering in Vietnam,*
IFMBE Proceedings 46, DOI: 10.1007/978-3-319-11776-8_116

B. Signal Pre-processing

EEG signal has much noise, so a Hamming band-pass filter (from 0.5Hz to 4Hz) is used. The impulse response of the Hamming band-pass filter $H(n)$ is calculated as follows:

$$H(n) = h_d(n) \times W(n) \tag{1}$$

where $h_d(n)$ is the impulse response an ideal filter and $W(n)$ is the Hamming window.

The theory output of Hamming band-pass filter $y'(n)$ is convolution operation of the original EEG signal $x(n)$ and the impulse response of designed filter is described as follows:

$$y'(n) = x(n) * H(n) = \sum_{k=1}^{N} x(k) * H(n-k) \tag{2}$$

where, N is the order of the filter, n=1,2,3,…,N.

To reject influence by voltage drift, the real output of the designed filter $y(n)$ is calculated as follows:

$$y(n) = y'(n) - \frac{\sum_{n=1}^{N} y'(n)}{N} \tag{3}$$

The signal is divided into 8 milliseconds(t_s), the sample rate (F_s) is 128 Hz. The number of the EEG signal samples is:

$$S = F_s * t_s = 128 * 8 = 1024 \ samples \tag{4}$$

C. Feature Extraction

Threshold algorithm is applied to extract features of EEG signal using eyes movements. The values of a threshold are determined by statistics on the obtained EEG data. The threshold value using standard deviation is expressed follow as:

$$THR = M \pm a * SD \tag{5}$$

where M is the mean value, SD is the standard deviation, and a is the coefficient of standard deviation of EEG signal. These parameters are described in equations (6), (7), and (8).

$$M = \frac{\sum_{n=1}^{S} y(n)}{S} \tag{6}$$

$$SD = \sqrt{\frac{\sum_{n=1}^{S} (Y_n - M)^2}{S}} \tag{7}$$

$$a = \frac{SD}{M} \tag{8}$$

where $y(n)$ is the EEG filtered signal, S is the number of samples. Because the eyes movements are worked out in 2 seconds (from 2nd to 4th second), the samples S for calculating thresholds are:

$$S = F_s * t_s = 128 * 2 = 256 \ samples \tag{9}$$

Based on Equations (6), (7), and (9), the mean values of EEG signal for 4 eyes movements are calculated as shown in table 1.

Table 1 Mean values of signals in 4 tasks of 10 subjects

Channel	Open		Blink		Glance to left		Glance to right	
	M_o	SD_o	M_b	SD_b	M_L	SD_L	M_R	SD_R
Fp1	1	0.03	0.7	0.14	1	0.03	1	0.03
F7	2	0.05	2	0.05	1.9	0.27	2.1	0.29
F8	3	0.05	3.2	0.09	3.1	0.31	2.9	0.30

where, M_o, M_b, M_L, M_R are the mean values of opening, blinking, left-glance, and right-glance signals. SD is the standard deviation of the corresponding mean values.

From table 1, relationships between eyes movements and channels are recognized. Concretely, blink affects the mean value of Fp1 and F8 channels, left-right glance affects the mean values of F7 and F8 channels. From that point, thresholds of eyes movements are calculated at corresponding channels. Then, EEG signals of blink, left-glance, and right-glance are classified.

D. Classification of EEG Online Signal

Eyes movements are classified in 3 types: blink, left glance, and right-glance. However, controlling wheelchair requires at the least 5 commands: forward, backward, right, left, and stop. Therefore, movements of eyes should be divided into 5 types which corresponds to 5 commands of the wheelchair control as shown table 2.

Table 2 Eyes movements and commands for wheelchair control.

EYES MOVEMENTS	COMMAND
Two-times blink	STOP
Three-times blink	FORWARD
Four-times blink	BACKWARD
Left-glance	TURN LEFT
Right-glance	TURN RIGHT

a) Calculation of Blink-Times

To count the times of blink, a blink-signal program is built as shown figure 4. When a blink-signal is classified, this program starts to count it from 2 to 4 in 2 seconds. Then, the result of this program is executed to corresponding commands.

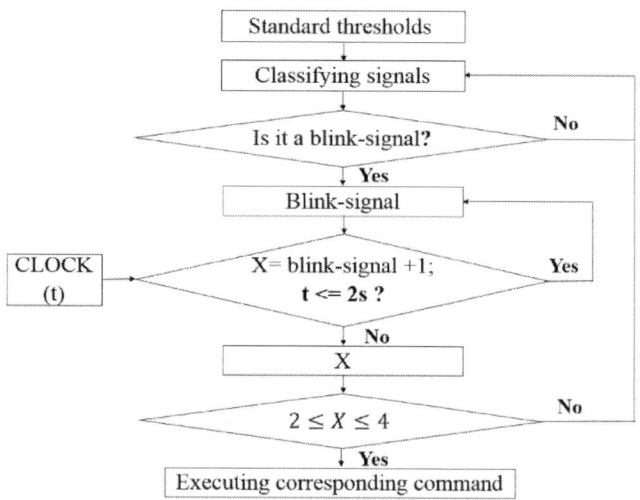

Fig. 4 Flow chart of blink-signals.

b) Detection of Left-Glance and Right-Glance Signal

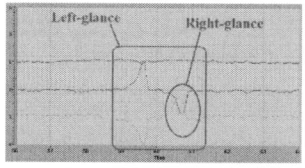 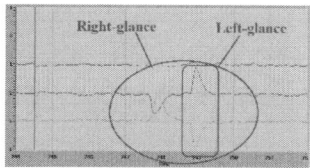

Fig. 5 Left-glance movements   Fig. 6 Right-glance movements

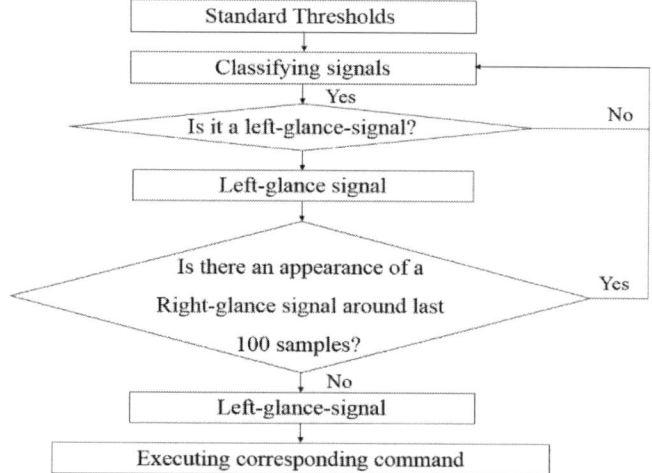

Fig. 7 Flow chart of left-glance signal detection

In some experiments, a confusion between left-glance and right-glance is recognized. Concretely, when subject glances to left, his/her eyes restoration is confused with right-glance, and otherwise as shown fig.5 and fig. 6.

Therefore, a detection of left-glance and right-glance signals is built as shown fig. 7. When a left-glance signal is classified, the detection program examines the appearance of right-glance signal around last 100 samples; if right-glance signal is not found, this is left-glance movement; conversely, the left-glance signal is a confused signal. Similarly, detecting right-glance signal is also processed in the same way.

A wheelchair control program is built in Labview environment. Firstly, the program collected EEG online signals using eyes movements from Biosemi Active Two System. Next, the EEG online signal is filtered, extracted and classified into commands to control the electric wheelchair. If the signal of eye movement is too noise or an unsuitable command, the wheelchair stops immediately.

III. RESULTS AND DISCUSSION

A. Quality Signal Testing

Firstly, the standard thresholds of EEG good signal is analyzed and reckoned as shown in table 3.

Table 3 Standard thresholds of EEG signal in Labview environment

Channel	MIN		MAX	
	Mean	*Standard Deviation*	*Mean*	*Standard Deviation*
Fp1	0.9	0.03	1.1	0.04
F7	1.9	0.04	2.1	0.03
F8	2.9	0.03	3.1	0.03

Based on the thresholds, a signal quality testing program is built in Labview environment as shown figure 8 and figure 9. Three LEDs represent the quality of three corresponding channels; if a channel has a good connection, a corresponding LED turns on; conversely, if a channel is lost the connection, the corresponding LED turns off.

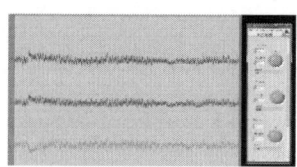 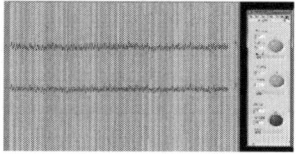

Fig. 8 Three signals are good   Fig. 9 One signal is noised

B. Feature Extraction of Real Signal

Statistical values and thresholds of eyes movements are calculated and reckoned on 10 subjects as shown table 4.

Table 4 Thresholds of eye movements.

Channel	Opening	Blink	Left-Glance	Right-Glance
Fp1	0.9 – 1.1	<0.8		
F7	1.9 – 3.1		<1.7	>2.5
F8	2.9 – 3.1	>3.2	>3.2	<2.7

C. Wheelchair Control

A program an EEG-controlled wheelchair using eye movements is built, when subject moves his/her eyes, the LED which represents a corresponding command, turns on as shown figure 10. If no command or an unavailable movement is recognized, the wheelchair stops to avoid collision.

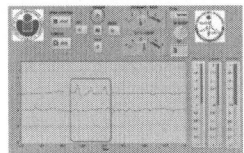 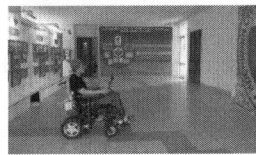

Fig. 10 Experiment of wheelchair control program.

Moreover, a subprogram detecting confusion between left-glance and right-glance signal is supplemented to improve the accuracy of wheelchair control. Table 5 show eyes movements, corresponding commands, accuracy of the wheelchair control without detection subprogram (accuracy 1), and the accuracy with the detection subprogram (accuracy 2).

Table 5 Accuracy of commands of wheelchair control after 10 users control 50 times in indoor environments

Eyes movements	Commands	Accuracy 1	Accuracy 2
Three-times blink	FORWARD	86%	87%
Four-times blink	BACK	85%	84%
Left-glance	LEFT	65%	77%
Right-glance	RIGHT	68%	75%
Two-times blink	STOP	92%	90%
Total		**79.2%**	**82.6%**

By using the detection subprogram, the accuracy of left-glance is increased from 65% to 77%, the accuracy of right-glance is improved from 68% to 75%. The great differences demonstrate the effectiveness of this subprogram.

Although the standard threshold in table 4 is reckoned on 137 samples, however, some special cases can not apply them. Moreover, continuous movements of eyes can cause eyestrain and stresses for user. Besides, Biosemi Active Two System is wordy, so it may be a barrier in movements of an electric wheelchair.

In future, a neural network algorithm is suggested to apply into similar research; expectably, the accuracy of wheelchair control will increase clearly. Furthermore, the EEG-controlled wheelchair using thoughts may be developed from results of this paper. In addition, combining an additional camera for obstacle avoidance can be proposed to avoid collision on the electric wheelchair.

IV. CONCLUSION

This paper proposed threshold algorithms in an EEG-controlled wheelchair for disabled people. The original EEG signals are filtered by a Hamming band-pass filter. After that, extracting features, and creating standard thresholds of EEG signal using eyes movements are performed by surveys and statistics on 10 subjects with 137 samples. Then, the signals are classified into commands for wheelchair control. Moreover, a wheelchair control program, using eyes movements, is built in Labview environment; the program not only connects an EEG device and an electric wheelchair to subject but also processes EEG online signals. Therefore, a blink-signals program and a detecting confusion of left-glance and right-glance signals are supplemented to improve the accuracy of wheelchair control. Finally, experimental results showed an electric wheelchair using blink, left-glance and right-glance with accuracy 82.6% in indoor environment.

REFERENCES

1. S. Nakanishi, Y. Kuno, N. Shimada, and Y. Shirai, "Robotic wheelchair based on observations of both user and environment," *the Inter. Conference on Intelligent Robots and Systems*, pp. 912-917.
2. S. Hemachandra, T. Kollar, N. Roy, and S. Teller, "Following and interpreting narrated guided tours," the *International Conference on Robotics and Automation*, 2011, pp. 2574-2579.
3. K. Arai and R. Mardiyanto, "A Prototype of Electric Wheelchair Controlled by Eye-Only for Paralyzed User," Journal of Robotics and Mechatronics vol. Vol.23, No.1, pp. 66-74, 2011.
4. Unlu, H. Bolay, and A. Akin, "Hemodynamic correlates of mental arithmetic task in migraine," in *Biomed. Eng. Meeting*, 2009, pp. 1-4.
5. T. Yanmei, R. O. Suarez, S. Whalen, I. H. Norton, and A. J. Golby, "Identification of Essential Language Areas by Combination of fMRI from Different Tasks using Probabilistic Independent Component Analysis," the *2nd International Conf. on Bioinformatics and Biomedical Engineering*, 2008, pp. 2060-2063.
6. Z. Shu, L. Jinglei, L. Xiang, J. Xi, G. Lei, and L. Tianming, "Activated cliques: Network-based activation detection in task-based FMRI," the *10th Inter. Symp. on Bio-Imaging*, 2013, pp. 274-277.
7. H. Al-Negheimish, L. Al-Andas, L. Al-Mofeez, A. Al-Abdullatif, N. Al-Khalifa, and A. Al-Wabil, "Brainwave Typing: Comparative Study of P300 and Motor Imagery for Typing Using Dry-Electrode EEG Devices," in *HCI International*, 2013, pp. 569-573.
8. Kodi, D. Kumar, D. Kodali, and I. A. Pasha, "EEG-controlled Wheelchair for ALS Patients," in the *International Conference on Communication Sys. and Network Technologies*, 2013, pp. 879-883.
9. Ben Taher, N. Ben Amor, and M. Jallouli, "EEG control of an electric wheelchair for disabled persons," in the *International Conference on Individual and Collective Behaviors*, 2013, pp. 27-32

Application of Fluorescence Photography in the Evaluation of Acne

Pham Thi Hai Mien, Tran Van Tien, Vu Thanh Huy, and Huynh Quang Linh

Ho Chi Minh City University of Technology, Ho Chi Minh City, Vietnam

Abstract— **This study introduces applying a simple optical system using UVA light for stimulating and obtaining *in vivo* facial fluorescence images. The red fluorescence was observed in all volunteers and it's intensity was higher at the T-zone than at the U-zone in all cases regardless of their skin type. It's known that the sebum amount of the T-zone is higher than of the U-zone. This result leads to the correlation between facial red fluorescence with sebum level, not with Propionibacterium acnes so far as is known [1, 2]. In addition, the blue fluorescence was found in inflammatory acne area. To the best of our knowledge, the nature of blue fluorescence has not been reported so far. Analysis of the skin fluorescence imaging can give information about the disease state of skin and be used for designing non-invasive optical devices for skin diagnosis.**

Keywords— **facial, fluorescence, acne, sebum.**

I. INTRODUCTION

In recent years the number of people suffering from skin diseases in general and acne in particular has extremely increased due to the environmental pollution, the widely applying chemicals in agriculture and food industry, etc, especially in Vietnam with the tropical climate and the major population working in the agricultural sector. Some traditional devices have usually been used in the diagnosis of skin diseases, such as lamp or microscope for skin surface examinations but not always effectively, especially in early-stage disease. Recently, some new tools have been introduced such diagnostic systems into standard clinical practice, such as fiber-based fluorimeter – SkinScan system (Jobin Yvon, France), where fluorescence of endogenous aminoacids is used for cutaneous lesions' s investigations, or DYADERM system (Biocam GmbH, Germany), which is applied for photodynamic diagnosis with exogenous photosensitizers [3]. However, up to our days there is no such universal clinical device, based on autofluorescence detection of skin surface, which could be used as a general tool for early-stage detection of skin diseases in general and acne in particular.

This investigation is a part of clinical trial for introduction of skin diagnostic device based on fluorescence imaging with some advantages such as safety, rapid test time, mobility and ability to detect early-stage disease with high accuracy. The facial skin stimulated by ultraviolet A (UVA) light 365 nm emitted the red fluorescence in both normal skin and skin with non-inflammatory acne, while the inflammatory acne emitted the blue fluorescence. The red fluorescence of facial ultraviolet photographs has been thought to originate exclusively from Propionibacterium acnes (P. acnes) [1-2, 4-5]. However, the association between P. acnes and the red facial fluorescence is still controversial [4-6]. A recent studies reported that the red fluorescence was associated with facial sebum secretion [6-8], while the nature of blue facial fluorescence of inflammatory acne has not been reported so far.

This study was designed to investigate two aspects of facial fluorescence: 1) the origination of facial fluorescence and 2) correlation between fluorescence and grades of acne.

II. MATERIALS AND METHODS

A. Subjects

Twenty volunteers of both sexes and various ages (age range 15-30 years, 10 females and 10 males) were included in this study. Among them, there are 5 volunteers with normal skin and 15 others with mild to moderate acne. The skin surface was wiped with alcohol before photography.

B. Ultraviolet Photography

Equipment for the ultraviolet photographs included a Canon DSLR camera 550D (Fig. 1); The light source was a 3w-power LED emitting radiation at 365 nm at and placed 5 cm from the skin surface; one UV bandpass filter (UG-1, Edmund) passes only UV light and blocks all visible light (from 400 nm) from the LED. Two polarized mirrors located before camera were designed for blocking unpolarized light. A 10x-magnification multiple lens system was used for magnifying fluorescence images. The subject's face was positioned for photographs under UVA conditions.

Fig. 1 Experimental instrumentation for fluorescence photography: 1 - DSLR camera, 2 & 3 - polarized mirrors, 4 - 10x-magnification multiple lens system, 5 – LED & UV bandpass filter.

V. Van Toi and T.H. Lien Phuong (eds.), *5th International Conference on Biomedical Engineering in Vietnam,*
IFMBE Proceedings 46, DOI: 10.1007/978-3-319-11776-8_117

Photograph was performed in the dark. The measured areas classified into two zones: T-zone (the forehead, nose and chin) and the U-zone (both cheeks) stimulated by UVA LED after wiping with alcohol.

III. RESULTS

The face's area was classified into T-zone and U-zone by the level of *sebum* secretion. It's known that the T-zone typically possesses a greater number of sebaceous glands, which can result in an extra production of sebum in those areas. For all volunteers, T- and U-zones were stimulated by 365-nm radiation in the same conditions and the fluorescence imaging is illustrated in Fig. 2.

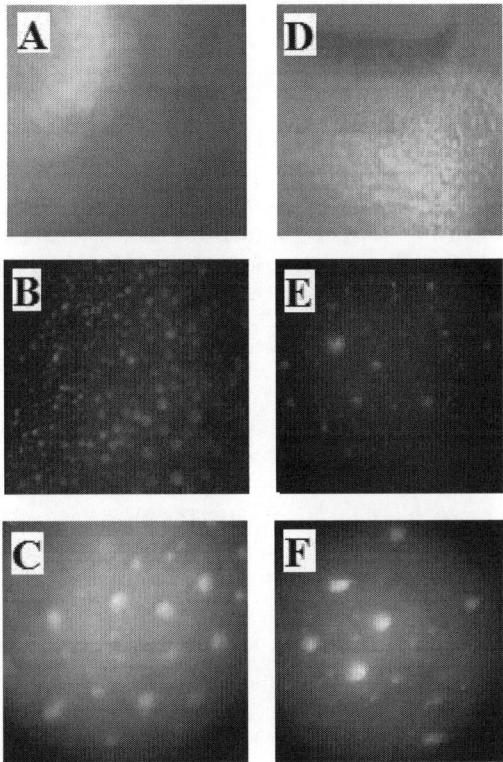

Fig. 2 Fluorescence photographs at T-zone (left) and U-zone (right) under white light (A, D), UVA 365 nm (B, E) and 10x magnification (C, F)

As can be seen in Fig.2, the density of red color spots is higher at T-zone than at U-zone in all cases regardless of their skin type. Besides, comparing the fluorescence photography of volunteers with normal skin and with acnes showed the exciting result: in some volunteers with normal skin the number of red color spots is even more than in others with acne.

In this research, the red fluorescence was found in all skin types including normal skin, skin with non-inflammatory and inflammatory acnes, while the blue fluorescence was observed only in skin with inflammatory acnes (Fig. 3).

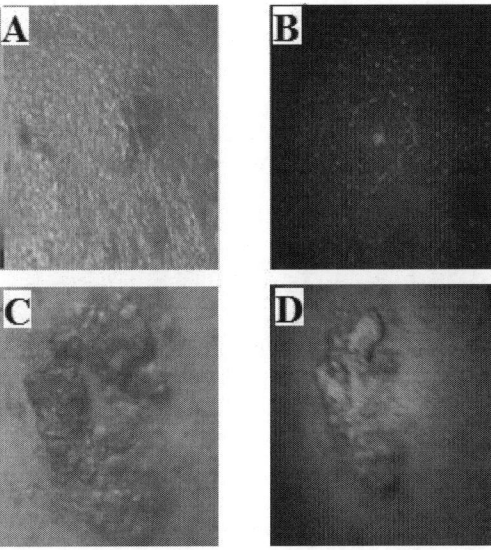

Fig. 3 Examples of fluorescence photographs at inflammatory acnes under white light (A, C) and 365 nm (B, D)

IV. DISCUSSION

A. The Origination of Facial Red Fluorescence

Acne is a chromic inflammatory disorder of the pilosebaceous follicles with a multifactorial etiology and pathogenesis. The follicular impactions develop into initially invisible lesions (microcomedones) and then into clinically evident comedones. Microcomedones and comedones are a suitable microenvironment for colonization by cutaneous bacteria, especially P. acnes. These bacteria produce proinflammatory mediators and free fatty acids, which are responsible for the appearance of inflamed acne lesions [9]. Porphyrins are further endogenous metabolic products of Propionbacteria, which might additionally contribute to the perifollicular inflammatory reaction. Porphyrins are native fluorophores and strongly fluorescent. Their presence can be demonstrated by orange-red fluorescence in the follicle openings by examining facial skin under appropriate UVA light.

The intensity of red fluorescence was considered proportional to the quantity of P. acnes and directly correlated with their reduction and clinical improvement after the treatment applied [1, 5, 10]. However, recent studies did not confirm the direct correlation of red fluorescence with P. acnes. Han et al. [7] reported that the fluorescence distribution was related to sebum spots. According to Youn et al. [6], the red fluorescence was more closely related to the amount of sebum excreted than to the presence of P. acnes. Hristo Dobrev [9] found that sebum level and intensity of fluorescence were higher at the T-zone than at the U-zone in all patients regardless of their skin type. Sebum is secreted from the sebocytes, terminally differentiated epithelial cells. Human sebum is composed of triglycerides, squalene, wax esters and cholesterol. The regulation of sebum production is not fully understood. However, androgens, peroxisome proliferator-activated receptors and growth hormones have been postulated to regulate sebum production. Increased sebum secretion is thought to be one of the major causes of acne [11]. The sebum amount is higher at the T-zone than at the U-zone.

In this study, we found the positive correlation between the red fluorescence and the sebum level: the red fluorescence intensity was proportional to sebum amount, while there was no significant correlation between the porphyrin fluorescence and the P. acne amount. However, the sebum is favorable to the growth of Propionibacteria, which is responsible for inflamed acne breakouts. Therefore, there is a indirect correlation between the red fluorescence and P. acnes.

B. The Origination of Facial Blue Fluorescence. Correlation between Fluorescence and Grades of Acne

In many years the facial red fluorescence has been observed and studied by different research groups, but the other fluorescence types, to the best of our knowledge, has not yet been.

Dermatologists classify types of acne into four grades. Determining acne grade is done by a simple visual inspection of the skin. Specific criteria are used to classify acne symptoms, including: grade 1 - mild and non- inflammatory acne; grade 2 - moderate and slight inflammatory acne; grade 3 - considered severe acne with more amount of inflammation; grade 4 - pronounced amount of inflammation and breakouts are severe.

Inflammation is a process by which the body's white blood cells and chemicals protect us from infection with foreign substances, such as bacteria and viruses. White blood cells, or leukocytes, are the cells of the immune system that are involved in defending the body against both infectious disease and foreign materials. Five different and diverse types of leukocytes exist, and several types (including monocytes and neutrophils) are phagocytic.

When inflammation occurs, chemicals from the body's neutrophils are released into the blood or affected tissues to protect your body from foreign substances. Monici M. et al. [12] studied the fluorescence property of neutrophil granulocytes and reported that neutrophil excited by 366 nm and 436 nm wavelength emits wide fluorescence spectrum in visible range, with high intensity at 450 nm (blue light) and 520 nm (green light), respectively. According to the study of Kim et al. [13] the inflammatory lesions showed the green fluorescence at 400-420 nm excitation.

In this research, inflammatory acnes were excited by 365 nm and emitted the blue fluorescence. We suppose the nature of the blue fluorescence is neutrophil. Neutrophils appear only in the area of inflammatory acnes and that's why the blue fluorescence was obtained only in inflammatory, not in non-inflammatory acnes. In addition, note that comparing with inflammatory acne in skin with non-inflammatory acne wasn't observed blue fluorescence, but was obtained the higher red fluorescence. This factor gives support to above mentioned conclusion about the origination of red fluorescence and the research results of Hristo [10]. Hristo reported that mild non-inflammatory lesions are associated with higher sebum amount and higher red fluorescence intensity, while severe inflammatory lesions have lower sebum amount and lower fluorescence intensity.

V. CONCLUSION

We have determined the origination of facial fluorescence and correlation between fluorescence and grades of acne. The results give support to some recent studies that the sebum amount is responsible for the red fluorescence at the early grade of acne (non-inflammatory). In contrast, the blue fluorescence was observed only at the late grade - inflammatory acne. We suggest that the blue fluorescence is caused by neutrophils appearing in the area of inflammatory acne. The nature of facial fluorescence and its association with grades of acne requires further investigation for future application of this non-invasive optical technique for skin disease diagnosis.

ACKNOWLEDGMENT

The present research was supported by the university project of the National Key Laboratory of Digital Control and System Engineering, HCMUT-VNU HCM. We would like to thank the laboratory leadership and other people who greatly help us in project fulfillment.

REFERENCES

1. Leslie C, Nikiforos K et al. (1996) Fluorescence photography in the evaluation of acne. J Am Acad Dermatol 35:58–63
2. Cornelius C E, Ludwig G D (1967) Red fluorescence of comedones: production of porphyrins by Corynebacterium acnes. J Invest Dermatol 49:368–370
3. De Leeuw J, Van de Beek N et al. (2009) Fluorescence detection and diagnosis of nonmelanoma skin cancer at an early stage. Las Surg Med 41:96–103
4. McGinley K J, Webster G F et al. (1980) Facial follicular porphyrin fluorescence: correlation with age and density of Propionibacterium acnes. Br J Dermatol 102:437–441
5. Pagnoni A, Kligman A M et al. (1999) Digital fluorescence photography can assess the suppressive effect of benzoyl peroxide on Propionibacterium acnes. J Am Acad Dermatol 41:710–716
6. Youn S W, Kim J H et al. (2009) The facial red fluorescence of ultraviolet photography: is this color due to Propionibacterium acnes or the unknown content of secreted sebum? Skin Res Technol 15:230–236
7. Han B, Jung B et al. (2007) Analysis of facial sebum distribution using a digital fluorescent imaging system. J Biomed Opt 12:014006–014011
8. Son T, Han B et al. (2008) Fluorescent image analysis for evaluating the condition of facial sebaceous follicles. Skin Res Technol 14:201–207
9. Hristo D (2010) Fluorescence diagnostic imaging in patients with acne. Photodermatology, Photoimmunology & Photomedicine 26:285–289
10. Borelli C, Merk K et al. (2006) In vivo porphyrin production by P. acnes in untreated acne patients and its modulation by acne treatment. Acta Derm Venereol 86:316–319
11. Choi C W, Choi J W et al. (2012) Ultraviolet-induced red fluorescence of patients with acne reflects regional casual sebum level and acne lesion distribution: quanlitative and quantitative analyses of facial fluorescence. British Association of Dermatologists 166:59–66
12. Monici M, Pratesi R et al. (1995) Natural fluorescence of white blood cells" spectroscopic and imaging study. Journal of Photochemistry and Photobiology B: Biology 30:29–37
13. Kim Y, Choi T et al. (2008) The clinical usefulness of portable digital skin fluorescence equipment in acne patients. Korean J Dermatol

Author: Pham Thi Hai Mien
Institute: Ho Chi Minh City University of Technology
Street: Ly Thuong Kiet 268
City: Ho Chi Minh City
Country: Vietnam
Email: haimiennt@gmail.com

Digital Morphology Comparisons
between Models of Conventional Intraoral Casting and Digital Rapid Prototyping

Rong-Fu Kuo[1], Shing-Jye Chen[1,*], Tung-Yiu Wong[2], Bo-Cheng Lu[3], and Zheng-Han Huang[4]

[1] National Cheng Kung University - Medical Device Innovation Center
[2] National Cheng Kung University Hospital-Stomotology Department
[3] Taiwan Care Technology Corporation
[4] National University of Tainan, Taiwan

Abstract— **Rapid prototyping (RP) technology used in digital design of dental industry has been rapidly developed. Most users in dental industry have concerned accuracy of the dental model printed by an RP device in comparison to that of a conventional intraoral impression. Thus, the purpose of the study was to compare digital morphology of a subject's teeth between a dental impression and a printed dental RP model of the same subject whose digital dental morphology was created. In this study, one subject's entire oral teeth was impressed and casted; a plastic RP model was also produced after the subject's teeth were scanned using an intraoral scanner (3Shape TRIOS) to produce a whole dental images in digits. In order to digitally compare morphological dimensions between the impression and RP models, each model's shape and dimension were scanned using a reverse engineering based optical scanner in producing digital dental images in 3D, a STL based format; meanwhile, a quality control industrial software (ATOS GOM Inspect V7.5 SR1) was used to compare the STL format based digital images of each model. The preliminary findings of the study show that an increased diverse visual coloring based error distribution of the scanned images between the RP plastic and dental impression casting models was found as well as a maximum average morphological error (0.1409mm) over eleven selected spots. The maximum relative errors could be due to a precision of the devices used in general during scanning and printing of an intraoral scanner and RP device, respectively, in addition to the errors from hardware and software operated during image reconstruction. In the future, an increased sample size of comparisons among impaired morphology in teeth, various usages of intraoral scanners or optical based scanners, specifically, will gain a better statistical meaningful finding of the study.**

Keywords— **Digitalization, 3D digital model, dental morphology, rapid prototyping, intraoral scanning.**

I. INTRODUCTION

Rapid prototyping (RP) technology used in digital design of dental industry has been rapidly developed. A trend of saving time applied from a conventional dental impression technique transformed into a digital based RP printed dental plastic model has been widely observed. The digitalization time-saving based dental model design seen in dentistry will definitely create a great impetus on to the current time consumed impression based dental model done in the most traditional dental laboratory for dental clinics. Thus, a digitalization based technology in dentistry will be further advanced in a fashion being necessarily state-of- the-art technique in modern world.

However, an accuracy of the printed digital based dental model has been greatly concerned in comparison to that of a conventional dental impression or casting [1, 2]. Without verifying the accuracy of a printed RP model prior to patient's dental treatment, development in accuracy of a digital based RP model used for clinics and industrial dentistry become an essential step for further improved technique seen in dentistry. Thus, the purpose of the pilot study was to compare digital morphology of the physical models between a cast based and a RP based dental model taken from one subject's lower jaw.

II. MATERIALS AND METHODS

In this study, a dental morphology of one healthy subject's entire lower jaw teeth was conventionally impressed and casted by an experienced dental laboratory technician; a digital intraoral scanner (3Shape TRIOS) was also used to scan the subject's teeth to produce its digital image (Image A) that was directly exported in a STL file format used for printing out a plastic RP model. The plastic model was produced by a means of an OBJECT 3D (Connex350) digital printer. Additionally, to digitally compare morphological dimensions between the models of the RP and casting, each model was physically scanned using a reverse engineering based optical scanner (smartSCAN ^{3D}, Breuckmann) in producing digital dental images of each model in 3D. The image was a STL based format (a digital RP model-Image A′ & a casting model-Image B); meanwhile, a digital image quality control software (ATOS GOM Inspect V7.5 SR1) was used to import each STL based dental image of the models to compare the digital images by superimposing the images of each scanned dental model. At the end, three pairs of the image comparisons were produced and evaluated (see Fig. 1 a, b and

---

* Corresponding author.

V. Van Toi and T.H. Lien Phuong (eds.), *5th International Conference on Biomedical Engineering in Vietnam,*
IFMBE Proceedings 46, DOI: 10.1007/978-3-319-11776-8_118

c). Furthermore, eleven self-selected spots of the superimposed images were also calculated in showing morphological relative differences in a range of visualization color based scaling bar (±0.3mm; see Fig. 2): dark red (+0.3) light red、dark/light yellow、light green(±0) 、light blue to dark blue(-0.3) 。

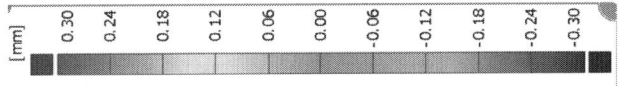

Fig. 2 A visualization color based scaling bar (±0.3mm)

*The self-s*elected eleven spots were 0.140, 0.062 and 0.0900 (mm) in average, respectively for models of (a). Casting (Image B) in comparison with digital RP (Image A′), (b). Digital RP (Image A′) in comparison with Intraoral Scanned (Image A), (c). Intraoral scanned (Image A) in comparison with casting (Image B).

(a) Casting (Image B) vs. digital RP (Image A′)
Selected spots in average (mm): 0.1409

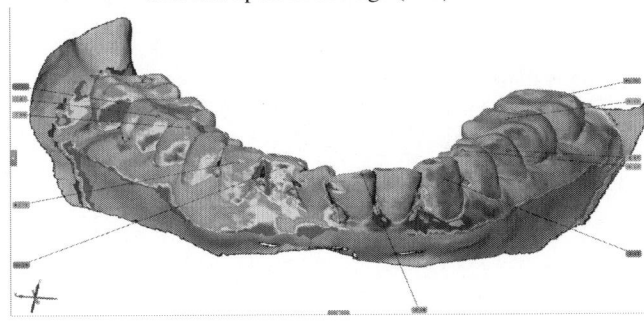

(b) Digital RP (Image A′) vs. Intraoral Scanned (Image A; Selected spots in average (mm): 0.0627

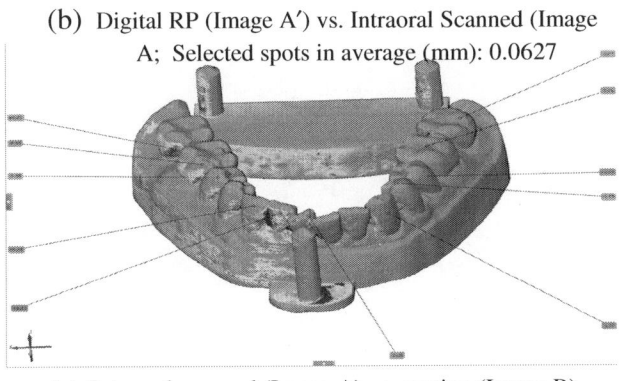

(c) Intraoral scanned (Image A) vs. casting (Image B)
Selected spots in average (mm): 0.0900

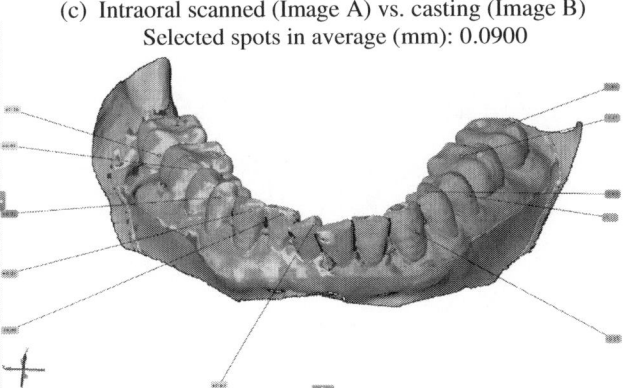

Fig. 1 (a). Casting (Image B) vs. digital RP (Image A′), (b). Digital RP (Image A′) vs. Intraoral Scanned (Image A), (c). Intraoral scanned (Image A) vs. casting (Image B)

III. RESULTS AND DISCUSSION

The preliminary findings of the digital image comparisons from an overall visualization based of each model show that the image pair between casting Image B and digital RP Image A′ reveals a greater color variation than the other pairs of the images seen between digital RP Image A′ and intraoral Scanned Image A as well as between intraoral scanned Image A and casting Image B. Specifically, the largest diverse visual color ranges and the average relative morphological differences of the self-selected spots were 0.1409mm seen for (a) Image A′ vs. Image B. On the contrary, the least relative color changes and differences of the self-selected spot values (0.0627mm) and (0.0900mm), respectively, were found in (b) Image A′ vs. Image A and (c) Image A vs. Image B (see Fig. 1).

IV. CONCLUSIONS

The relative morphological differences shown between the dental models of RP and casting impressed from one subject's lower jaw are relatively acceptable (<0.3mm) [2,3] even though one subject data were collected and analyzed in this preliminary study. The maximum relative errors of the digital dental models created between by the conventional impression-casting technique and by the digital rapid prototyping device could be a result of the digital devices used in general while being utilized to scan and print via an intraoral scanner and a RP device, respectively. In addition, possible errors for the image differences could be suspected during manual casting and impression as well as the hardware device and software operated during image reconstruction and superimposing. In the future, an increased subject sample size for more paired digital image comparisons taken among various intraoral scanners and optical based scanners, specifically, will gain a better significant finding and reliable comparison in digitalization in dentistry.

ACKNOWLEDGMENT

We would like to thank our funding agent: Southern Taiwan Science Park (STSP, Tainan, Taiwan) - A Science-Based Industrial Park cooperation program between developed advanced industry-university: Numbered 102-CP02.

CONFLICT OF INTEREST

The authors declare that they have no conflict of interest.

REFERENCES

1. Aletta H.,James J.R., Huddleston S., Yijin R. (2014). Accuracy and reproducibility of dental replica models reconstructed by different rapid prototyping techniques. American Journal of Orthodontics and Dentofacial Orthopedics, 145(1):108–115. http://dx.doi.org/10.1016/j.ajodo.2013.05.011
2. Brian R, P. Lionel S., Andre F., et al. (2005) An Evaluation of the Use of Digital Study Models in Orthodontic Diagnosis and Treatment Planning. The Angle Orthod, 75 (3):300-304.
3. Leifert MF, Leifert MM, Efstratiadis SS, & Cangialosi TJ. (2009). Comparison of space analysis evaluations with digital models and plaster dental casts. Am J Orthod Dentofacial Orthop. 136(1):16.e1-e4.
4. Wiranto M, et.al. (2013). Validity, reliability, and reproducibility of linear measurements on digital models obtained from intraoral and cone-beam computed tomography scans of alginate impressions. Am J Orthod Dentofacial Orthop, 14(1):140-147.

Human Activity Recognition and Monitoring Using Smartphones

Vu Ngoc Thanh Sang[1], Nguyen Duc Thang[1], Vo Van Toi[1], Nguyen Duc Hoang[2],
and Truong Quang Dang Khoa[3]

[1] Department of Biomedical Engineering, International University - VNU-HCMC, Vietnam
[2] Posts and Telecommunications Institute of Technology-HCM, Vietnam
[3] Tokyo University of Agriculture and Technology, Japan

Abstract— **The sedentary lifestyle is becoming popular especially for intellectual work. Although physical inactivity lifestyle may cause many unexpected illnesses, it is complicated to build up a positive lifestyle due to the lacks of reminder systems to manage and monitor physical activities of people. This research represents an effective way for daily activity monitoring using accelerator and gyroscope sensors embedded in a smartphone. Signals were recorded from accelerator and gyroscope sensors while a user wearing the smartphone performs different activities (going downstairs, going upstairs, sitting with the phone in a pocket, driving and putting the phone on the table). The classification algorithms with k-nearest-neighbor (kNN) and artificial neural network (ANN) were applied to recognize user's activities. The overall accuracy of recognizing five activities is 74% for kNN and 75.3% for ANN respectively. Based on the activities recognized during the day, users are able to manage their daily activities for a better life.**

Keywords— **Smartphone, activity recognition, activity monitoring.**

I. INTRODUCTION

As reported by the World Health Organization, 3.2 million annual deaths were associated with physical inactivity lifestyle[1]. In order to have a better health, people at all ages should practice physical activities at least thirty minutes per day according to United States Department of Health and Human Services [2]. These activities may include gardening, bicycling, fast dancing, two-mile walking, stair walking, and playing sports, etc. Such regular physical activities can decrease premature death by preventing heart disease, reducing diabetes risk and controlling high blood pressure. However, despite the great benefits that physical activities offer, it is complicated to build up a good habit to achieve the desired healthy lifestyle. One of the major obstacles is that most people are too busy to spend at least thirty minutes per day to perform physical exercises. The best they can do is to practice any activities whenever it is possible, leading to a problem that they cannot manage duration as well as the intensity of the physical activities.

Different approaches have been proposed to monitor user activities by using accelerometers and gyroscopes. They can be divided into two major approaches: wearable sensor based-approaches and smartphone-based approaches. In the wearable sensor-based approaches, three gyroscope and accelerometer sensors were used to measure joint angles including hip, knee and ankle joints of monitor subjects for walking, sit-to-stand, and stand-to-sit activities[3]. Another sensor-based approach was developed by Electronic Cervical Range of Motion Measurement System (ECROM)[4]: The three-axis accelerometer has been used as a tilt sensor to measure both the frontal and extension of the cervical spine while the resonator gyroscope played a role as a rotation sensor to measure the transverse of the cervical spine. Alternatively, with the widespread smartphone nowadays, many researchers have currently focused on existing sensors embedded in a smartphone for the task of recognizing users' activities, e.g. the implementation of real time activity recognizer[5]. The achieved accuracies ranging from 70% to 90% for different activities consisting of walking, climbing downstairs, climbing upstairs, sitting down, standing up, falling and so forth. In another aspect, a smartphone was used to detect falls in human [6]: Subjects were asked to repeat both falls as well as normal actions such as walking, sitting down and jumping to generate training dataset. Observing differences of the two activities, researchers investigated adaptive thresholds to distinguish them in a feature space. Utilizing a smart phone as a reminder to ask people to take regular breaks from prolonged sitting seems promising since it motivates a mobile-phone user to become more active. In an application[7], a smartphone generated acoustic warnings and vibration with virtual persuasive messages to remind a user to take a break.

In this paper, we propose a new method using a smartphone to recognize daily activities of a user. The recognized activities include going downstairs, going upstairs, sitting, driving and putting the phone on the table. For going downstairs, going upstairs, sitting and driving activities, the smartphone is simply considered to be in the pocket. Obviously such activities are mainly found by an individual during a working day, thus our approach plays an important role to evaluate how much energy a user has spent during a day from which he or she can adjust their lifestyle accordingly.

© Springer International Publishing Switzerland 2015
V. Van Toi and T.H. Lien Phuong (eds.), *5th International Conference on Biomedical Engineering in Vietnam,*
IFMBE Proceedings 46, DOI: 10.1007/978-3-319-11776-8_119

II. Methods

The overall system is described in Fig. 1 where the signals from the accelerometers and gyroscopes of a mobile phone is preprocessed and input into Feature Extraction module to extract main features. Certain characteristics of signals such as auto regressive (AR) coefficients, fractal dimension, mean and standard deviation are calculated. The features are later classified with k-nearest neighbor (k-NN) and artificial intelligence network (ANN) to recognize different activities.

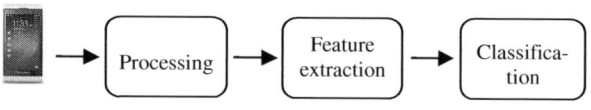

Fig. 1 The process of classifying signals.

A. Device

a) Hardware

The system was implemented on the infrastructure of Blackberry Z10, STL100-2 with a dual-core central processing unit (CPU) running at 1.5 GHz, 2 GB RAM and 16GB storage. The phone is 130 x 65.6 x 9 mm in size and 137.5 grams in weight. The touch screen has a 768 x 1280 resolution. Most importantly, the phone supports accelerometers, gyroscopes, proximities and compass sensors. However, this paper used only accelerometers and gyroscope sensors to record motions from users. Electronic accelerometer and gyroscope have been effective for physical activity detection in living subjects since accelerometers respond to changing velocities of a subject's movements and gyroscope sensors respond to changing orientation of a subject's movements.

b) Software

An application was developed on the Blackberry 10 Native Platform. Blackberry 10 Native is an open source framework designed for mobile devices with a powerful Software Development Kit (SDK) based on the combination of C++ and Qt/QML framework[8]. Blackberry 10 Native SDK provides necessary drivers needed to interact with the hardware and libraries needed to deploy a smartphone application. Furthermore, Blackberry 10 Native allows programmers to access accelerometer and gyroscope sensors and to save their values into a log for further analyses.

B. Processing

Our approach requires two recorded datasets for training and testing the classifier. The activity labels of both training and testing are stored in a database, hence the accuracy of our approach is evaluated. We have recorded accelerometer and gyroscope sensors' values and the corresponding time index. The accelerometer is designed to detect changes due to movements and vibrations of the phone. The gyroscope sensor is used to measure orientation and rotation of the phone. In the processing stage, total six variables are shown on the screen and sent to the storage at the same time. The taken time for recording those variables is a period of 60 seconds. Different activities are saved into corresponding text files with their labels. There are five labeled activities including going downstairs, going upstairs, driving, sitting, and putting the phone on the table.

C. Feature Extraction

a) Auto Regressive Coefficient (AR)

The most common model used to approach spectrum estimation of random signals is AR. In AR model, the data output is assumed to be generated by its previous values by the following input – output equation:

$$y[n] = \sum_{k=1}^{p} a_k y[n-k] + \varepsilon[n] \tag{1}$$

where $y[n]$ is the sum of the relation constant of $y[n-k]$ as the previous values, a_k is the coefficients at the lag k, $\varepsilon[n]$ is a zero mean white noise process with unknown variance and p is the order of the system.

b) Fractal Dimension

Fractal dimension presents the irregular characteristic of a time series based on the power law index[9]. Although the fractal dimension can be considered similar to Euclidian dimension, there were significant differences between them. The main difference lied in the fact that fractal dimension may be a non-integer. The Euclidian dimension represented the mathematical system of a set, while the fractal dimension demonstrated how a set in micro-scale is similar to its counterpart in the macro-scale. The closest integer value of a fractal dimension shows its equivalence to the Euclidian dimension. For example, if the fractal dimension is close to zero, it describes a point (0-dimension in Euclidian). It describes lines when it approaches one (1-dimension in Euclidian) and surfaces when it approaches to two (2-dimension in Euclidian), and so on.

Higuchi's algorithm is an effective method to calculate the fractal dimension in time domain[10]. It is based on the calculation of the length of the curve in time series. The length $L_m(k)$ of the curve has k-segments was calculated by the following formula

$$L_m(k) = \frac{1}{k} \left\{ \left(\left[\sum_{i=1}^{\left[\frac{N-m}{k}\right]} \left| X\left(m+ik\right) - X\left(m+(i-1)k\right) \right| \right) \frac{N-1}{\left[\frac{N-m}{k}\right]k} \right\} \quad (2)$$

where N is the number of data samples, m is an integer ($m = 1,2,...,k$). The relationship of the length $L_m(k)$ and the fractal dimension D is shown as the following formula:

$$\langle L(k) \rangle \sim k^{-D} \quad (3)$$

where $\langle L(k) \rangle$ is the average value over k set of $L_m(k)$.

c) Mean

The use of mean is usually for measuring central tendency. Without weights assigned on data, mean is referred to the arithmetic mean. The arithmetic mean $\bar{x}$ is equal to the sum of all variables x_n divided by the number of variables N and calculated as follow.

$$\bar{x} = \frac{1}{N} \sum_{n=0}^{N} x_n \quad (4)$$

d) Standard Deviation

The standard deviation is the measurement of variability evaluating the range of the data set and the relationship of the mean with the rest of the data. The standard deviation will be small if the data points are close to the mean. On the other hand, if many data points are far from the mean, it means that the variance of the responses is wide and the standard deviation is large. If data values are equal at all points, then the standard deviation is zero. In addition, the standard deviation is used to predict the power of the signal and calculated using the following formula,

$$\sigma = \sqrt{\frac{1}{N} \sum_{i=1}^{N} \left(x_i - \bar{x}\right)^2} \quad (5)$$

where $\bar{x}$ is the mean of data and N is the number of data points.

D. Classification

a) k-Nearest Neighbor (kNN)

Firstly, the non-parametric kNN algorithm is used for classification. In this research, the neighbors were taken from five classes (going downstairs, going upstairs, sitting, driving, walking and putting the phone on the table). Recorded data with class labels were divided into two sets: training set (classified) and testing set (unclassified). For the recognition of the test set, the Euclidian distance[11] was computed. Here, the Euclidean distance is the distance between points $x = (x_1, x_2, ..., x_n)$ and $y = (y_1, y_2, ..., y_n)$ in a feature space and calculated as follows

$$d(x,y) = \sqrt{\sum_{i=1}^{n} \left(y_i - x_i\right)^2} \cdot \quad (6)$$

In the next step, the distance from each sample in the testing set to the training set is estimated and the minimum value is chosen. The classified label is therefore based on the nearest distance of each testing point to the corresponding class.

b) Artificial Neuron Network (ANN)

ANN was defined as a computational model based on biological neural networks[12]. Fig.2 shows the model of a neuron which forms an ANN. There are three basic elements of a neural model: weight, summation and activation function. An input signal x_j connected to neuron k is multiplied with its own weight w_{kj}. The summation module adds all the input signals together and combines the result to the bias b_k. The activation potential v_k was the summation of the inputs with weights and bias. The activation function was the function that limits the amplitude of the output of a neuron. Common activation functions were threshold function, piecewise and sigmoid function.

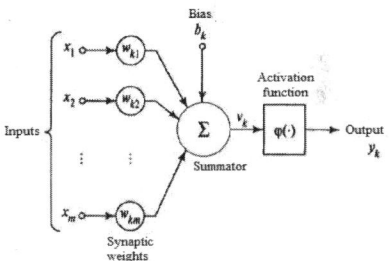

Fig. 2 The neural model.

An ANN had three layers described as input, hidden and output. The input layer received signals from the external environment or other networks. The hidden layer intervened the input and the output in several useful manners. The more added hidden layers, the higher order statistic the network was enabled to extract. The output layer constituted the overall response of the network to the activation pattern.

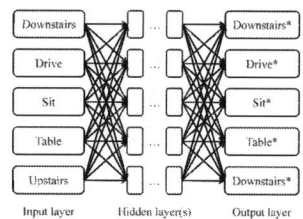

Fig. 3 The artificial neural network layer.

III. RESULTS

A. Downstairs and Upstairs Walking – Quantitative Evaluation

The raw signals of the accelerometer and gyroscope sensors were shown in the Fig. 4. For each axis, although they were different in gyroscope sensor, the accelerometer data of going downstairs and upstairs were similar. Quantitative calculations were based on AR coefficient, Higuchi, mean, and standard deviation. The result showed confusions between those activities (Table 1 and Table 2).

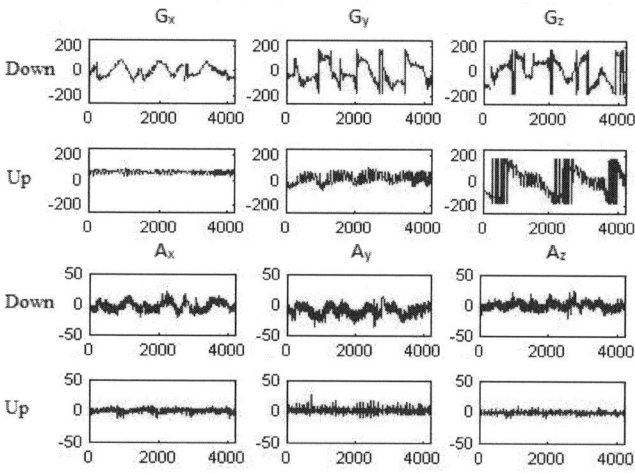

Fig. 4 The raw signal of going downstairs and upstairs.

B. Driving and Sitting – Qualitative Evaluation

Although the raw data of going downstairs and upstairs were confused, driving and sitting action were clearly differentiated by observation. The further calculation confirmed that it was possible to distingue those two activities. The result of feature extraction and classification showed that the accuracy of both activities was high. For example, the accuracy of driving and sitting were 80% and 100% when the kNN was applied. In addition, with the ANN method, the driving and sitting activities were 97% and 76% respectively.

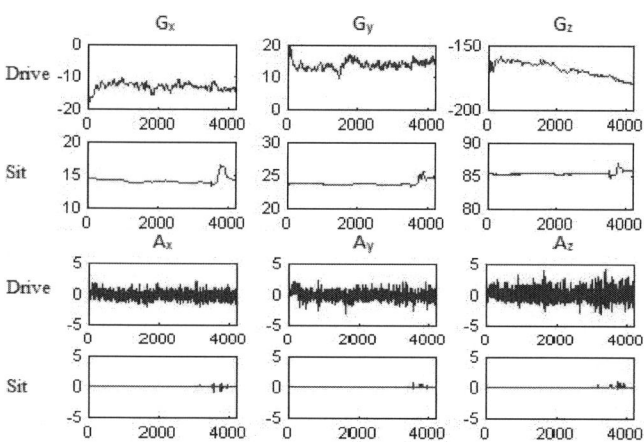

Fig. 5 The raw signal of driving and sitting.

C. General Result

As can be seen in Table 1, sitting with the phone in a pocket and putting it in a table had the highest degree of accuracy (100%), while the downstairs and upstairs walking were less accurate and easily confused with each other. The action of driving a motorbike had 80% accuracy and 20% false detection is due to the confusion of driving with the resting and sitting action.

Table 1 The accuracy of kNN method (%)

	Downstairs	Driving	Sitting	On-table	Upstairs
Down-stairs	30	30	0	0	40
Driving	0	80	20	0	0
Sitting	0	0	100	0	0
On-table	0	0	0	100	0
Upstairs	40	0	0	0	60

The result shown in Table 2 represented the level of accuracy of classification using ANN method. The accuracy of downstairs, driving, sitting, on-table and upstairs actions were 64%, 97%, 76%, 90.5%, 49% respectively. The overall accuracy was 75.3%.

Table 2 The accuracy of ANN method (%)

	Downstairs	Driving	Sitting	On-table	Upstairs
Down-stairs	64	15	4	0	17
Driving	1.5	97	1.5	0	0
Sitting	2	15	76	7	0
On-table	5	3.5	1	90.5	0
Upstairs	49	2	0	0	49

IV. CONCLUSION AND DISCUSSION

This paper has focused on the daily activity recognition by analyzing the data of accelerometer and gyroscope sensors of a smartphone. The results showed that by using just a smartphone, we could recognize various basic activities such as driving motorbike, going stairs, sitting with a smartphone in pocket and putting the smartphone on a table.

In both methods of classification including kNN and ANN, the results of recognizing driving, sitting and putting the phone on a table are considerably accurate: It ranged from 76 to 100 percentages. On the other hand, poor recognition results were presented for going downstairs and upstairs activities due to the ambiguity of accelerometer and gyroscope data. Therefore, those two activities should be concerned as a single activity for better recognition results in future.

ACKNOWLEDGEMENT

This research is funded by International University, VNU-HCM under grant number T2014-04-BME.

REFERENCES

1. Department of Health Statistics and Informatics in the Information "Global health risks: mortality and burden of disease attributable to selected maor risks", Geneva, World Health Organization, 2009, pp 11.
2. US Department of Health and Human Services, Centers for Disease Control and Prevention, National Center for Chronic Disease Prevention and Health Promotion, "Physical activity and health: a report of the Surgeon General.Atlanta", 1996. pp.9-14.
3. Qi A., Ishikawa Y., Nakagawa, Kuroda A., Oka H., and Yamakawa H., *et al.*, "Evaluation of wearable gyroscope and accelerometer sensor (PocketIMU2) during walking and sit-to-stand motions," in Roman, 2012, pp. 731-736.
4. Lian E., Hachadorian J., Hoan N., and Toi V., "A Novel Electronic Cervical Range of Motion Measurement System," in *The Third International Conference on the Development of Biomedical Engineering in Vietnam.* vol. 27, V. Toi and T. Khoa, Eds., ed: Springer Berlin Heidelberg, 2010, pp. 140-143.
5. Brezmes T., Gorricho , and J. Cotrina, "Activity Recognition from Accelerometer Data on a Mobile Phone," in *Distributed Computing, Artificial Intelligence, Bioinformatics, Soft Computing, and Ambient Assisted Living.* vol. 5518, 2009, pp. 796-799.
6. Yabo C., Yujiu Y., and Wenhuang L., "E-FallD: A fall detection system using android-based smartphone," in *Fuzzy Systems and Knowledge Discovery (FSKD), 9th International Conference on,* 2012, pp. 1509-1513.
7. Dantzig S., Geleijnse G., and Halteren A., "Toward a persuasive mobile application to reduce sedentary behavior," *Personal and Ubiquitous Computing,* vol. 17,2013, pp. 1237-1246.
8. Open source framework Native SDK for Blackbery 10. [Online]. Available: http://developer.blackberry.com/native/
9. Higuchi, T., Relationship between the fractal dimension and the power law index for a time series: a numerical investigation,Physica D, 46, 1990, pp.254 -264.
10. Higuchi T., "Approach to an irregular time series on the basis of the fractal theory," *Physica D: Nonlinear Phenomena,* vol. 31, 1988, pp. 277-283.
11. Jones P.W., Osipov A., and Rokhlin V., "A randomized approximate nearest neighbors algorithm," *Applied and Computational Harmonic Analysis,* vol. 34, 2013, pp. 415-444.
12. Sammut C. and Webb G. "Artificial Neural Networks," in *Encyclopedia of Machine Learning,* , Eds., ed: Springer US, 2010, pp. 44-44.

Corresponding Author: Nguyen Duc Thang
Institute: Department of Biomedical Engineering, International University
Street: Quarter 6, Linh Trung ward, Thu Duc district
City: Ho Chi Minh
Country: Vietnam
Email: ndthang@hcmiu.edu.vn

Evaluation of Hemodynamic Responses to Visual Tasks Using Functional Near Infrared Spectroscopy

Tran Le Giang[1], Nguyen Duc Thang[1], Vo Van Toi[1], Nguyen Huynh Minh Tam[2],
Dinh Dong Luong[3], and Truong Quang Dang Khoa[4]

[1] Department of Biomedical Engineering, International University, Ho Chi Minh City, Viet Nam
[2] University of Saskatchewan, Canada
[3] Department of Computer Engineering, Kyung Hee University, South Korea
[4] Tokyo University of Agriculture and Technology, Japan

Abstract— **Functional near-infrared spectroscopy (fNIRS) is becoming widely applied in many practical researches, especially in vivo researches on human. This study focuses on the hemodynamic responses of visual cortex when the human eye is excited by different conditions of flickering light stimulus. In our experiment, visual cortex is activated by flickering light at various spatial frequencies and modulation depths while the hemodynamic responses of the visual cortex are simultaneously monitored by fNIRS. Our experimental results suggest that flickering light can activate hemodynamic responses of the visual cortex but those changes are not significantly distinct among different stimulus conditions.**

Keywords— **Visual cortex, flickering light, fNIRS, Papillometre, hemodynamic response.**

I. INTRODUCTION

In this paper, we aim at investigating the behavior of visual cortex under different stimuli. Despite of the complexity of ophthalmic system, we can partially measure and analyze its hemodynamic responses using functional near infrared spectroscopy (fNIRS). Recently, there have been researches investigating on monitoring the visual cortex using fNIRS [1-3]. Our goal is to extend the current studies to conduct a mathematical model of the ophthalmic system which can be used in practice to explain causes of diseases and disorders related to visual system.

In our study, we concerned the ophthalmic system as a black box that would send out a response whenever it received an input. Thus if the eye is activated by a sinusoidal flickering light, it would generate a biological signal corresponding to the target stimulus through optic nerve to the visual cortex. This process may induce the changes in oxygenated and deoxygenated hemoglobin (HbO and Hb) concentration in blood. Measuring those behaviors brings a hope for the understanding of ophthalmic physiology and pathology.

Recent studies suggested that ophthalmic pathologies with a a normal person [4] and even with the baby on the early developing stage of the brain [5, 6] can be diagnosed non-invasively and objectively by fNIRS. Therein, one of the important aspects to understand of the visual system is to examine the principle and behavior of visual cortex with regards to sinusoidal light. In literature, the sinusoidal light played an important role to form a basic stimulus with controlled parameters. By varying the stimulus parameters, it is feasible to analyze effects of those parameters to the ophthalmic system. Here, we utilized Papillometre [7] to study occipital lope's responses with respect to the variation of stimulation. This device has not only been experienced on human [3, 8-10] but also on animal [1]. However, there are lacks of researches covering the flickering light with Papillometre to understand the visual cortex with fNIRS. Many researchers elsewhere have just conducted experiments with other visual patterns to stimulate the visual cortex when using fNIRS [2,5,11]. This work investigates on the hemodynamic responses of the visual cortex with respect to the flickering light generated by Papillometre.

Among different functional imaging modalities, we select functional near-infrared spectroscopy (fNIRS) as a tool to monitor hemodynamic behavior of brain. This technique allows us to capture behavioral and cortical activities simultaneously. fNIRS can monitor concentration change of HbO, Hb and total hemoglobin (THC) deep inside brain by emitting a known near- infrared light with the wavelength from 700 to 900 nm and detecting the intensity of reflected beams. Usually, fNIRS releases multi-wavelength lights that are refracted into the brain. A part of the light is absorbed by water, HbO and Hb while the rest is reflected back to head skin and captured by the detectors. Input and output intensity of the light reveal us how much HbO, Hb and THC concentration changed inside the monitored regions [12].

In comparison with MRI and EEG, fNIRS allows subjects to perform physical activities during measuring cognitive task [13] and is harmless to both participants and operators [14-16]. Furthermore, fNIRS signal is less influenced by electrical noises. Therefore, fNIRS had been utilized in many current researches including comparing active and passive observations[17], investigating on the effect of spatial and temporal frequency on brain [11], and evaluating on the absolute quantification of fNIRS [18].

© Springer International Publishing Switzerland 2015
V. Van Toi and T.H. Lien Phuong (eds.), *5th International Conference on Biomedical Engineering in Vietnam,*
IFMBE Proceedings 46, DOI: 10.1007/978-3-319-11776-8_120

The aim of this study is to evaluate the principle of visual cortex through investigating the effects of flickering light at different conditions to the visual cortex. We conducted experiments with fNIRS and Papillometre on real subjects. The data obtained were further preprocessed to remove noises and analyzed by statistic toolbox to check whether there exists a mathematical model to explain physiology principle of the ophthalmic system including both the eyes and visual cortex.

The rest of paper is organized as follows. Section 2 describes material and methods. We present experimental results with discussion in Section 3. Finally, we conclude the paper with Section 4.

II. MATERIAL AND METHODS

A. fNIRSI Instrument

We used FOIRE 3000 (Shimadzu Corporation, Japan) using three wavelengths (780±5nm, 805±5nm and 830±5nm) of the infrared light to monitoring hemodynamic responses of visual cortex (Fig.1.a). This system has totally 16 probes (8 light sources (S) and 8 detectors (D)) that forms up to maximum 24 channels. The device can acquire hemodynamic changes in vital organism with three parameters including HbO, Hb and THC.

In this study, a total 3 source-detector pairs (3 × 2 block) were used to induce totally 7 channels with the time resolution of 0.1s (Fig.1.a). The inter-sensor distance was 3cm. Sensors were plugged into a custom-build head cap constructed from plastic which was comfortably fitted to participants. For each participant, the fNIRS head cap was positioned where lower edge of the cap was just above the nape.

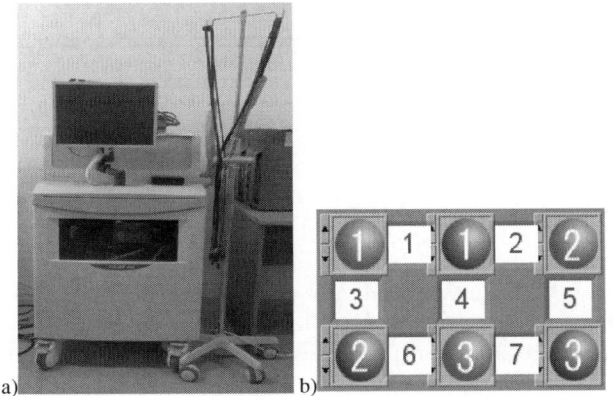

Fig. 1 a) fNIRS machine. b) Pattern of sources and detectors. The red and blue buttons present emitter and receiver respectively. Each adjacent S-D pair forms a channel.

B. Stimulus

In our experiment, Papilometre was used to generate sinusoidal flickering light which respect to formula:

$$L(t) = L_o(1 + m\cos 2\pi f t) \qquad (1)$$

with $L(t)$ is instantaneous luminance (amount of light per unit of area), L_0 is mean luminance (varied from 0 to 100 Trolands), which is kept constant in this study, m is modulation depth (varied from 0 to 100%), f is spatial frequency (varied from 0 to 100Hz).

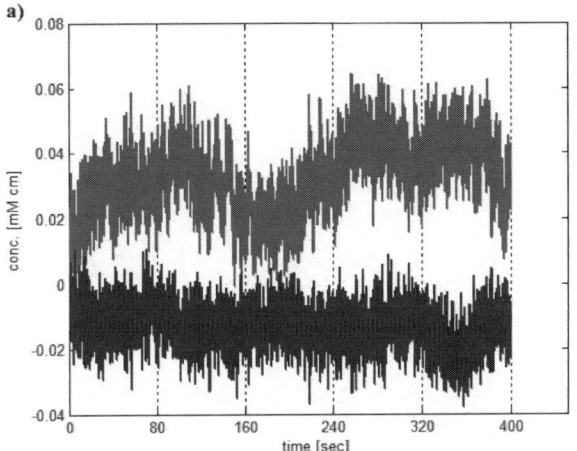

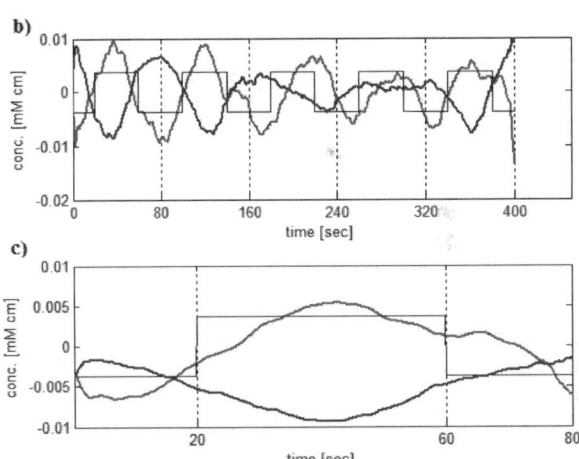

Fig. 2 Sample fNIRS signal of hemodynamic response (HbO: red, Hb: blue). (a) The raw illustration of a single measurement without any reprocessing. (b) Ploy of data has been filtered with bandpass filter (bandwidth 0.01 – 0.5 Hz), smoothing and cICA. (c) The plot of mean of five trials in a measurement after filtered.

We varied spatial frequencies and modulation depths to check whether those factors have alternative effects on the visual cortex. Stimulus was flickered at frequencies: 0Hz, 5Hz, 10Hz and 15Hz while modulation depth presented at

50% and 100% level. Those spatial frequency levels were proved to cause significant effects on human perception [19-23]. Inconsequence, we had 8 pairs of "spatial frequency-modulation depth" and each pair was measured twice per a participant.

C. Participants

This study accessed a total eight healthy participants (two females and six males with an average age 20.8). All had no

was simulated while the other eye was covered by a black bandage.

E. Preprocessing and Statistical Analysis

Data preprocessing consisted of a band-pass filter, smoothing and constrained independent component analysis (cICA) [25]. cICA was used to extract the basic components presenting the NIRS signals over all channels. Finally, statistical analysis with t-test and analysis of variance (ANOVA) was used to draw analysis results. All statistical

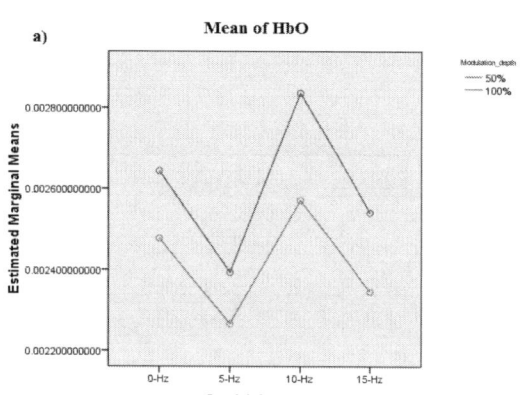

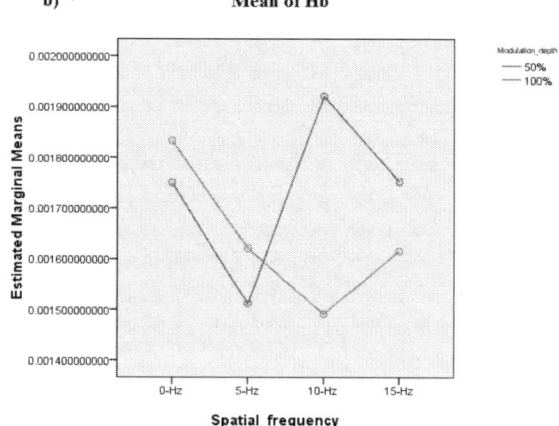

Fig. 3 Plot for illustration of mean of HbO (a) and Hb (b) in all measurements (50% modulation depth: blue, 100% modulation depth: green).

history of mental damage or neurological illness. Participants provided informed consent paper after being fully explained about the experiment. The tenets of the Declaration of Helsinki were followed.

D. Experimental Design

Flickering light with adjustable spatial frequencies and modulation depths was used as the stimulus to excite the subjects' eyes, meanwhile hemodynamic response was monitored at the visual cortex using fNIRS. Each participant performed 16 measures respect to eight pairs of stimulus. One measure consists of five experimental blocks repeated continuously. The protocol for a single block was designed as 20sec (rest) – 40sec (task) – 20sec (rest).

Before joining an experiment, participants get instructed about the protocol and experimental procedure in details. When entered the room, participants sit on a chair in front of Papillometre and they were allowed to freely adjust their seat and Papillomete to fix their position. Then, an operator placed fNIRS probes on the participant's head according to the arrangement depicted in Fig.1.b. During experiment, participant required to stay still to mitigate unnecessary motions. One eye

calculations were performed on HbO and Hb measurement. The samples of data before and after processing are illustrated in Fig.2.

III. RESULTS AND DISCUSSION

The experimental t-test results reveal significant changes ($p < 0.0001$) of the hemodynamic responses of the visual cortex from the baseline hemoglobin levels when the flickering stimulus is activated. This result suggests that flickering light can excite neuron activity thus induces an increase in HbO concentration in visual cortex.

There is no statistically significant difference between the effects of two factors spatial frequency and modulation depth in channels HbO and Hb. ANOVA analysis on HbO responses with two factors of modulation depth and spatial frequency do not show significant effect ($p>0.990$). No significant differences in with spatial frequency ($p>0.211$) or modulation depth ($p>0.336$) are reported. Besides, the interaction between spatial frequency and modulation depth is also not significant ($p>0.299$). The same are shown for the effect of spatial frequency ($p<0.362$) as well as modulation depth ($p<0.495$).

To investigate the interaction of modulation depth and spatial frequency on visual cortex, the mean of HbO and Hb are plotted on Fig. 3. It is noticed that HbO when stimulus presented at full modulation depth induces a stronger excitation than that is caused by the stimulus at a half modulation depth. Moreover, at 0Hz and 5Hz frequency, Hb level becomes higher when the modulation depth becomes higher. Meanwhile at 10Hz and 15Hz, low modulation depth induces a stronger Hb response in comparison with full modulation depth.

Table 1 ANOVA result on HbO and Hb measures

Channel	Factor	df	F	P
HbO	Spatial frequency	3.632	1.131	0.336
	Modulation depth	1.632	1.566	0.211
	Modulation depth x spatial frequency	3.632	0.038	0.990
Hb	Spatial frequency	3.632	0.799	0.495
	Modulation depth	1.632	0.833	0.362
	Modulation depth x spatial frequency	3.632	1.227	0.299

IV. CONCLUSION

The t-test and ANOVA analysis demonstrate that flickering light can activate the hemodynamic responses of the visual cortex but those changes are not distinct among different stimulus condition. This can be due to the small sample size of testing data and the limitations of the NIRS system to monitor deep inside the brain.

ACKNOWLEDGEMENT

This research was supported by Vietnam National University-Ho Chi Minh City Research grant B2011-28-01. Besides, we gratefully thanks to the Department of Biomedical Engineering-International University as well as staffs and laboratories, who had given us helps and supports to complete the project.

REFERENCES

1. Vo Van Toi Fau - Riva CE, Riva CE: Variations of blood flow at optic nerve head induced by sinusoidal flicker stimulation in cats.
2. Garhöfer G, Zawinka C, Resch H, Huemer KH, Dorner GT, Schmetterer L: Diffuse luminance flicker increases blood flow in major retinal arteries and veins. *Vision Research* 2004, 44:833-838.
3. Yamasaki T, Goto Y, Tobimatsu S: Can we estimate the activating effects of visual channels in primary visual cortex by flicker VEPs? *International Congress Series* 2005, 1278:73-76.
4. Miki A, Nakajima T, Takagi M, Usui T, Abe H, Liu C-SJ, Liu GT: Near-infrared spectroscopy of the visual cortex in unilateral optic neuritis. *American Journal of Ophthalmology* 2005, 139:353-356.
5. Taga G, Asakawa K, Hirasawa K, Konishi Y: Hemodynamic responses to visual stimulation in occipital and frontal cortex of newborn infants: a near-infrared optical topography study. *Pathophysiology* 2004, 10:277-281.
6. Liao SM, Gregg NM, White BR, Zeff BW, Inder TE, Culver JP, Bjerkaas KA: Neonatal hemodynamic response to visual cortex activity: high-density near-infrared spectroscopy study. *BIOMEDO* 2010, 15:026010-026010-026019.
7. Van Toi V, Grounauer PA: Visual stimulator. *Review of Scientific Instruments* 1978, 49:1403-1406.
8. Allefeld C, Pütz P, Kastner K, Wackermann J: Flicker-light induced visual phenomena: Frequency dependence and specificity of whole percepts and percept features. *Consciousness and Cognition* 2011, 20:1344-1362.
9. Knapen T, Paffen C, Kanai R, van Ee R: Stimulus flicker alters interocular grouping during binocular rivalry. *Vision Research* 2007, 47:1-7.
10. Carmel D, Lavie N, Rees G: Conscious Awareness of Flicker in Humans Involves Frontal and Parietal Cortex. *Current Biology* 2006, 16:907-911.
11. Bridge T: Measuring the haemodynamic responses elicited in the visual cortex from various spatial and temporal frequencies using NIRS *The Plymouth Student Scientist* 2012, 5:94-118.
12. Jobsis FF: Noninvasive, infrared monitoring of cerebral and myocardial oxygen sufficiency and circulatory parameters. *Science* 1977, 198:1264-1267.
13. Cui X, Bray S, Bryant DM, Glover GH, Reiss AL: A quantitative comparison of NIRS and fMRI across multiple cognitive tasks. *Neuroimage* 2011, 54:2808-2821.
14. P K Kaiser JARS, Non-U.S. Gov't: Color fusion and flicker fusion frequencies using tritanopic pairs y nakano *Vision research* 2003, 32.
15. S.M. Coyle TEW, and C.M. Markham: Brain-computer interface using a simplified functional near-infrared spectroscopy system. *J Neural Eng* 2007, 4:219-226.
16. Gilles Vandewalle PMaD-JD: LIGHT AS A MODULATOR OF COGNITIVE BRAIN FUNCTION, , Volume 13, Issue 10, Pages 429-438, October 2009. *Trends in Cognitive Sciences* 2009, 13:429-438.
17. Kojima H, Suzuki T: Hemodynamic change in occipital lobe during visual search: Visual attention allocation measured with NIRS. *Neuropsychologia* 2010, 48:349-352.
18. McIntosh Ma Fau - Shahani U, Shahani U Fau - Boulton RG, Boulton Rg Fau - McCulloch DL, McCulloch DL: Absolute quantification of oxygenated hemoglobin within the visual cortex with functional near infrared spectroscopy (fNIRS).
19. C. Guignard VVT, C. W. Burckhardt and J. L. Schelling: Sensitivity to the flickering light in digitalized patients. *British Journal in Clinical Pharmacology* 1983, 15:189-196.
20. Toi VV: Loss of flicker perception due to artificial ocular hypertension. *Investigative Ophthalmology and Visual Science* 1987, 28:218.
21. Vo Van Toi CWBaPAG: Flicker-fusion perception investigation: Design, Modeling and Applications *Medical Engineering and Mecha-Optoelectronics, International Journal of Precision Machinery* 1987, 1:355-376.
22. Toi VV: Derivation of a unified transfer function in the theory of flicker. *Optics Letters* 1989, 14:907-909.

23. Vo Van Toi PAGaCWB: Artificially increasing intraocular perssure cause flicker sensitivity losses. *Investigative Ophthalmology and Visual Science*, 31.

24. Lloyd-Fox S, Blasi A, Elwell CE: Illuminating the developing brain: The past, present and future of functional near infrared spectroscopy. *Neuroscience & Biobehavioral Reviews* 2010, 34:269-284.

25. Z. Wang, "Fixed-point Algorithms for Constrained ICA and their Applications in fMRI Data Analysis", Magnetic Resonance Imaging. 29, 1288-1303 (2011).

Corresponding Author: Nguyen Duc Thang
Institute: Department of Biomedical Engineering, International University
Street: Quarter 6, Linh Trung ward, Thu Duc district
City: Ho Chi Minh
Country: Vietnam
Email: ndthang@hcmiu.edu.vn

Investigation the Stability of Oblique Fracture Fixation of Long Bone Using Different Screw Angle

Bich Nguyen[1], Trung Le[1], and Ha Van Vo[2]

[1] Department Biomedical Engineering, Mercer University, Macon, USA
[2] Department of Orthopedic Surgery,
School of Medicine, Mercer University, Macon, USA

Abstract— Orthopedic surgery still hosts a number of known problems. Screw loosening and implant failure are common and pose a major issue for successful bone fracture fixation. Research shows that optimal screw fixation depends on the number and placement of screws or using bone cement as an adjunct. There is a lack of focus and attention to the biomechanical effects of screw angle placement into the bone. The purpose of this study is to investigate the biomechanical effects of placing screws at different angles with respect to the fracture plane on the stability of the fixation of oblique fractures of long bone. The 45° angle oblique fracture was investigated. Twenty one porcine femurs were harvested, cleaned, measured for dimensions, and randomly divided into three groups of seven (N=7 per group). The fracture model was 1 mm oblique osteonomy at midshaft. The method was screw fixation only using 2 cortical screws at different angles. The three different screw angle fixations of interest were perpendicular (45° group), 45° angle (T group) and 70° angle (70° group) with respect to the fracture plane. Compression testing to failure was performed at a constant rate of 3 mm/min using a servo-hydraulic MTS 858 Bionix testing machine. The load and displacement were collected through LabVIEW then being processed to obtain average maximum load and stresses. One-way unstacked ANOVA comparison test was conducted to determine the significant difference between the group data. The T group demonstrated superiority in both maximum compressive load and maximum normal stress as compared to the other groups. Statistical analysis shows a significant different between the T group and 45° group ($p < 0.05$) but no significant different between the 90° group and 70° group for the maximum normal stress.

Keywords— Oblique Fracture, Screw Angle, Screw Fixation, and Orthopedic Implant.

I. INTRODUCTION

Oblique fracture of long bone is a common type trauma where the fracture lies on a plane at an angle with the axis of the bone. The injury can be highly unstable, especially for an oblique fracture at a high angle of obliquity. [1] The fixation methods for this unstable fracture are screw fixation with bone plate or screw fixation alone.[2] Many research has been done to determine the best method for screw fixation like using different number or placement of screws or using bone cement as an adjunct. However, little has paid attention to the screw angle of placement into the bone. The purpose of this study is to investigate the effects of placing screws at different angles with respect to the fracture plane on the stability of the fixation of oblique fractures of long bone. The AO technique recommended that screw fixation through an oblique fracture should be place perpendicular to the fracture plane. [3] However, there has been no biomechanical testing done in literature to evaluate the effect of this technique. Therefore, in this study the three different screw configurations investigated for 45 degree oblique fracture from superior_lateral to inferior_medial are: screws are placed perpendicular to the fracture line; screws are placed perpendicular to the axis of the bone (which is easier performed in clinical scenario); screws are placed at the angle 70 degree to the fracture line which is the angle between the previous two configuration.

II. MATERIALS AND METHODS

Biomechanical testing was performed on fresh frozen porcine femur, which has similar mechanical properties as compared to human bone. [4] Twenty one porcine femurs were harvested, cleaned, measured for dimensions, and randomly divided into three groups of seven (N=7). An osteotomy gap of 1 mm at the midshaft was cut with a reciprocating saw to create a 45° oblique fracture. The bone marrow was removed and the cortical thickness was measured after the cut. The major and minor semi-axes of the elliptical cross section of the bone at the fracture plane were measured to obtain the cross section area for the stress calculation. The two bone halved was then fixed to a special fixture to keep the axis alignment before being fixated. Two stainless steel cortical screws were then inserted at different angle with respect to the fracture plane. The three different screw angle fixations of interest were perpendicular (90° group), 45° angle (where the screws are placed transverse – group T) and at angle in between the previous two configuration which was chosen at 70° angle (70° group). The screws were fixated applying the lag screw principle which compresses the two bone fragments in order to obtain the best fixation for bone healing in clinical scenario. The three different configuration of fixation are show in Figure 1 below.

© Springer International Publishing Switzerland 2015
V. Van Toi and T.H. Lien Phuong (eds.), *5th International Conference on Biomedical Engineering in Vietnam,*
IFMBE Proceedings 46, DOI: 10.1007/978-3-319-11776-8_121

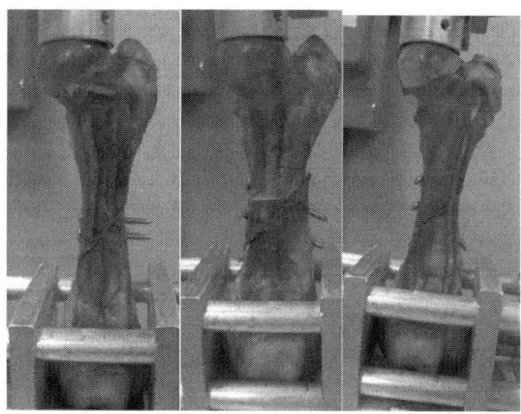

Fig. 1 The three different angle screw fixation configurations

Once the two bone halves were fixated, the specimen was loaded into a servo-hydraulic 858 Mini Bionix MTS testing system. A vice was attached to the bottom load cell transducer, and a second vice with an aluminum socket chucked to it was attached to the MTS piston, as seen in Figure 2.

Fig. 2 MTS testing setup

The free body diagram below demonstrate the force and Compression to failure test was then performed on each sample to at a rate of 3 mm/min and the data was collected at a sample rate of 25 ms using LabView software. The data collected were in terms of time in ms, axial displacement in mm, axial load in lbs. The data was then analyzed to get the results in terms of maximum load, maximum normal stress and maximum shear stress.

For statistical analysis, Minitab was used in order to conduct a one-way unstacked ANOVA comparison test for the

normal stress, as well as the shear stress. The unstacked ANOVA allows for a comparison between the three separate groups. The confidence interval was set to 95%, and a Tukey's comparison was used for the groups. Results from this analysis would determine whether or not the three types of fixation have statistically significant differences in normal stress and shear stress.

III. RESULTS AND DISCUSSIONS

The test setup with force applied on the femoral head which create a moment about the axis of the screws. This gave rise to much instability at the fixation site that resulted as different modes of implant failure observed. The free body diagram analysis for the test set up was demonstrated in Figure 3 and 4.

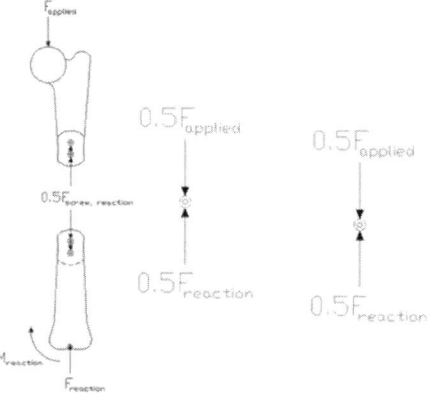

Fig. 3 Anterior View Free body Diagram

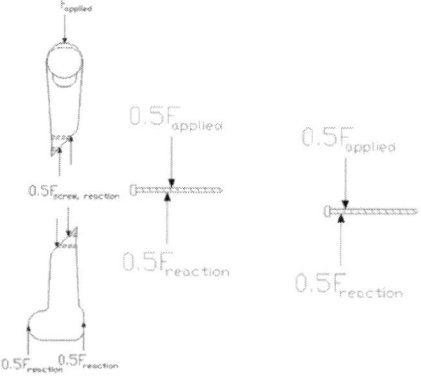

Fig. 4 Medial view free body diagram

Due to the fixation of the bone using screws alone, the configuration was highly unstable. Therefore the mode of failure was fairly inconsistent. The common modes of failure observed for the 70° group and the 90° group was that the proximal bone half shear off the axis of alignment and the bone failed along the longitudinal crack which propagated through the screw head. The common mode of failure for the T group was posterior or anterior bending of the proximal bone half due to bending moment applied along the axis of transverse screws.

The result for the maximum load, maximum normal stress and maximum shear stress for the three group were documented in the Table 1 below.

Table 1 Test Result

Group	Compressive Load (N)	σ (MPa)	τ (MPa)
45°	2288.286 ± 318.883	118.781 ± 16.638	0.0539 ± 0.0056
70°	2115.429 ± 652.708	146.057 ± 44.483	0.081 ± 0.0388
T	2824.857 ± 914.406	206.971 ± 67.105	NA

Since there is fairly large standard deviation, the result was compared in term of median instead of the mean. Among the three groups, the T group demonstrated superior characteristic with maximum force (33% greater than the 70° group and 23% greater than the 45° group). However, due to a small number of samples, there was no significant level detected between groups for the maximum load result. The T group also the median is skewed, and the most of the data lies on the lower quartile as shown in Figure 5 and 6.

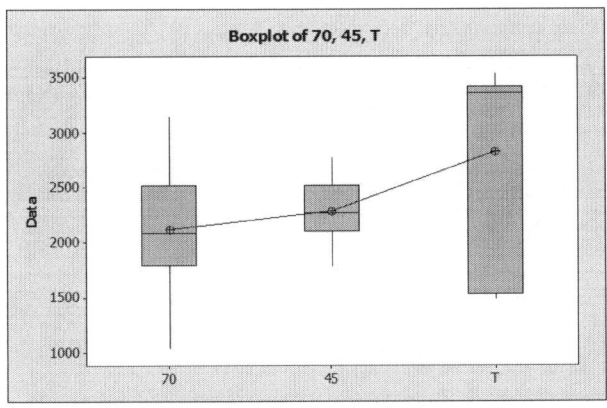

Fig. 5 Box plot of maximum load

For the normal stress, the T group also is the highest (29 % greater than the 70° group and 42% than 45° group) and is actually significantly different from the group 45°. No significant difference between the 70° and 45° group detected.

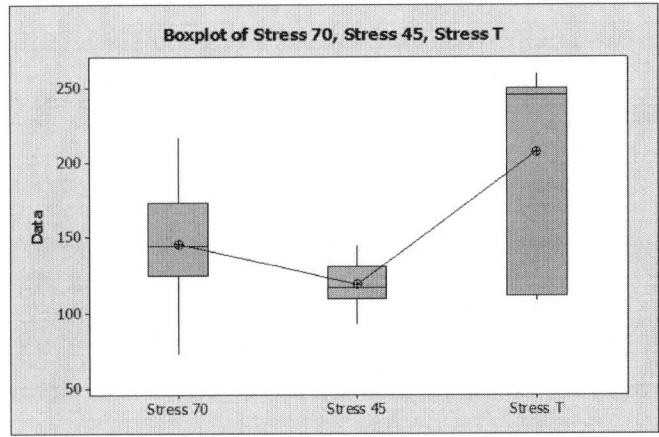

Fig. 6 Box plot of maximum normal stress

In this particular test setup, the 45° group has the most consistent in term of the test result. The standard deviation is low and the quartile and the extreme point is relatively small. This is the group where the screws were placed perpendicular to the fracture plane. The 70° group and the 45° group have relatively similar result. The result of the T group is largely different as compared to the rest of the group. The T group seems to have superior characteristic where it demonstrates highly load. However its mode of failure is more catastrophic.

IV. CONCLUSIONS

The crucial goal of this study was to determine if the recommended technique (placing the screws at a perpendicular angle to the oblique fracture) by the AO foundation is valid. It appeared that this technique provide a more stable fixation demonstrated as less shear stress observed and more consistent test result. The T group where the screws are placed perpendicular to the bone axis was the most superior. However, this result is not conclusive as the amount of instability presented in the data of this group. Future studies should increase the number of sample to increase the power of the study as well as investigate different mode of testing like cyclic, bending and torsion.

CONFLICT OF INTEREST

The authors declare that they have no conflict of interest.

REFERENCES

1. Metcalfe, A. J., Saleh M. & Yang, L. Asymetrical fracture fixation: stability of oblique fractures is influenced by orientation. Clinical Bio-mechanics (2005) 91-96.
2. Stiffler, K. S. Internal Fracture Fixation. Clinical Techniques in Small Animal Practice (2004).
3. Muller, M.E. & Allgower, M. Manual of Internal Fixation: Techniques Recommended by the AO-ASIF Group (1991).

4. An, Y. H. & Draughn, R. A. Mechanical Testing of Bone and Bone-Implant Interface (2000).

Author: Bich Ngoc Nguyen, MSc. Cad
Institute: Mercer University
Street: 1400 Coleman Avenue
City: Macon
Country: USA
Email: bn.nguyen09@gmail.com

F-Scan Analysis of Prosthetic Fittings through Mercer on Mission Vietnam

Emily Brett[1], Matthew Yin[1], Ha Van Vo[2], Edward O'Brien[2], Loren Sumner[3], and Philip McCreanor[4]

[1] Biomedical Master's Candidate, Mercer University, Macon, USA
[2] Department of Biomedical Engineering, Mercer University, Macon, USA
[3] Department of Mechanical Engineering, Mercer University, Macon, USA
[4] Director of Engineering Honors, Department of Environmental Engineering, Mercer University, Macon, USA

Abstract— **Mercer on Mission Vietnam is an annual service learning course which students spend time fitting prosthetic limbs on amputees. The prosthetic utilized in these trips includes a universal socket designed at Mercer University. The design is a low cost prosthesis that accommodates as many impoverished amputees as possible. During the summer of 2013, 18 students and 3 professors fitted 272 amputees. Fifty-three sets of Tekscan F-Scan Plantar Pressure data were collected. Students analyzed the plantar pressure data and tested for statistical trends between age, weight, height, amputation date, and gender to seek trends that could lead to improvements for the next trip. None of the five factors tested correlated with the quality of prosthetic fit. It was determined that a more systematic method of F-Scan data collection was needed in order to increase consistency and sample size for more in depth trend analysis.**

Keywords— **F-Scan, Analysis, prosthetic, trends, gait.**

I. INTRODUCTION

Mercer on Mission Vietnam is an annual service learning course in which students spend time fitting prosthetic limbs on amputees. The prosthetic utilized is a universal socket designed at Mercer University. It is low cost, adjustable, and designed to fit a wide range of amputees. During the summer of 2013, 18 students and 3 professors fitted a total of 272 amputees. In order to find areas of improvement for the following trips, F-Scan data was collected while amputees walked.

F-Scan is a pressure data collection system from the company Tekscan. It utilizes thin sensors that are placed inside the shoes of the test subject. The system reveals information about the interaction between foot and footwear. F-Scan measures force, timing, and contact pressure distribution through plantar pressure mapping. F-Scan is mobile and allowed for data collection in the remote regions that the Mercer on Mission Vietnam team fit in.

Data collected from F-Scan can be analyzed through Tekscan software provided. In order to do so however, a thorough understanding of the human gait cycle is required. The human gait cycle is composed of two phases: the stance phase and the swing phase, Figure 1. By understanding the

forces the foot undergoes each phase it is possible to determine the quality of gait when a patient walks.

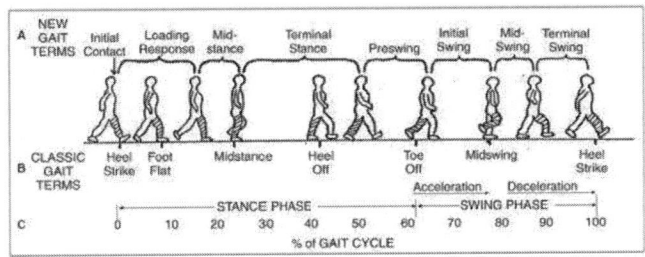

Fig. 1 Human Gait Cycle.

II. METHODS AND MATERIALS

A. Data Collection

Every amputee that comes to Mercer on Mission Vietnam must go through check in. During check in, relevant medical information is collected; every patient had to fill out a patient information sheet. Basic information was collected for each patient such as name, age, gender, height, weight, medical history, illnesses, blood pressure, and pulse. Medical information was recorded and stored.

Then a prosthetics team will be assigned to fit the amputee with a prosthetic. First, the patient's range of motion (ROM) and muscle strength were measured. Then, the length, proximal circumference, medial circumference, and distal circumference of the residual limb were measured as this information helps with further organization of the various types of amputations. Once the prosthetic was fit and approved by a certified prosthetist or orthopedic surgeon present, the F-scan system was equipped, Figure 2.

F-Scan pressure data was then collected for the patient. Using the weight data collected above, the F-Scan was calibrated to each patient's weight. Two sets of eight second F-Scan data was collected for each patient. Data collected for each patient via paper forms was transferred to Excel for analysis.

V. Van Toi and T.H. Lien Phuong (eds.), *5th International Conference on Biomedical Engineering in Vietnam,*
IFMBE Proceedings 46, DOI: 10.1007/978-3-319-11776-8_122

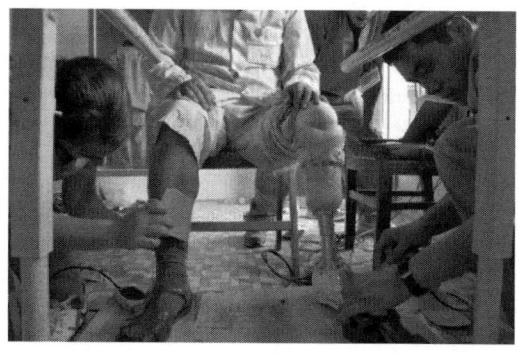

Fig. 2 Two Mercer on Mission Vietnam team members setting up a patient for F-Scan.

B. Data Organization

Data was organized and streamlined for analysis through Microsoft Excel. Each row is reserved for the medical information and analysis results for an individual patient, figure 3.

Patient Code	Gender	Age	Height (cm)	Weight (kg)	Years as Amputee	Amputee Classification
1	Female	47.00	149.00	50.00	30.00	BKL
2	Female	55.00	164.00	58.00	42.00	BKL
3	Female	82.00	152.00	27.00	45.00	BKL
4	Female	20.00	106.00	39.00	14.00	BKR

Fig. 3 Example of Excel data sheet

C. F-Scan Analysis

Analysis was completed through graphical comparison in Tekscan's F-Scan software. Two graphs are most effective for determining a proper fit with F-scan analysis. The first is a Force vs. Time graph. In this graph, the y-axis is the force applied (kilograms) and the x-axis represents time (seconds). There are two sets of data represented on each graph: one for the force applied on the left foot (red color), and one representing the force applied to the right foot (green color). In the Force vs. Time graph, the swing phase discrepancies were calculated. Swing phases are characteristically defined as the time that each foot has no force shown on the sensor. Because the foot is swinging through the air as opposed to contacting the ground, 0 kg of force occurs in this phase.

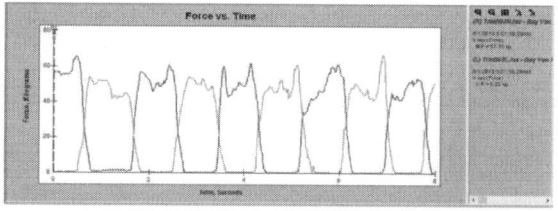

Fig. 4 Force vs. Time Graph for F-Scan Analysis

In a properly fit prosthetic, the swing phases between both legs are equal [1]. Total stride time is the amount of time it takes for one foot to go from stance phase to swing phase and back to the beginning of the next stance phase again. For analysis, a percent swing phase difference was used. By taking into account the total stride time of a patient, this ensures that analysis stays consistent between patients with differing gait speeds. In order to calculate the percent swing phase, the equation below was used:

$$Percent\ Swing\ Phase\ difference = \frac{|T(left) - T(right)|}{Ttotal}$$

The second type of graph, the Force vs. Percentage graph, effectively outlines how the weight is distributed across the foot, Figure 5. The y-axis represents the average force applied while the x-axis divides the foot length into specified percentages; 0% represents the heel of the foot and 100% represents the toes.

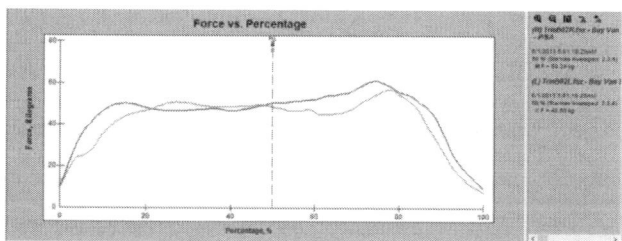

Fig. 5 Force vs. Percentage Graph

The Force vs. Percentage graph qualitatively displays a symmetrical distribution of forces amongst both feet, which gives evidence of a smooth, natural gait [2]. The force differences between the two feet were estimated using the percent difference of average force between the two legs was calculated. By using the percent difference as opposed to absolute, analysis is standard between patients with different weights. The following equation describes the percent difference of force:

$$Force\ Percent\ Difference = \frac{|F(left) - F(right)|}{Body\ Weight}$$

D. Quality of Fit Rating System

F-Scan analysis as outlined above was done for each of the patients. Calculated data was used to rate each prosthetic fit based on differences between the left and right foot. A graphical representation of the scale used is shown below in figure 6: X represents the percent difference for either swing phase or average force. A score of six is the best prosthetic fit, one that shows less than 5% difference in both the swing and force distributions between the right and left foot.

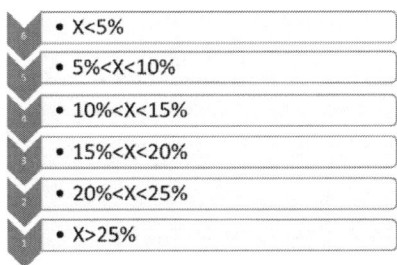

- X<5%
- 5%<X<10%
- 10%<X<15%
- 15%<X<20%
- 20%<X<25%
- X>25%

Fig. 6 Quality of fit rating according to percentage differences

E. Trend Analysis

Five factors were tested for trends: time as amputee, weight, height, age, and body mass index (BMI). Each factor was graphed. The quality of fit rating is represented along the y-axis and the x-axis is defined by the independent factors. Through regression analysis, a determination could be made in regards to how much of an impact each factor seemed to make on the quality of fit.

III. RESULTS AND DISCUSSION

In Figure 7, the y-axis represents the fitting rating and the x-axis outlines the amount of time the patient has been as an amputee. The regression analysis of the graph shows 2%, which supports no correlation. It was expected that patients that have been amputees for a longer amount of time would have better ratings for their prosthetic fits. Patients that have been amputated longer will have had time to adjust to a new center of gravity as well as increase their chance of already having a prosthetic. If a patient already has had a prosthetic, then he/she will have a better idea of what kind of prosthetic they can walk with most naturally. Those patients will have better input when the students fitting them ask for preferences concerning how they want their prosthetic to fit.

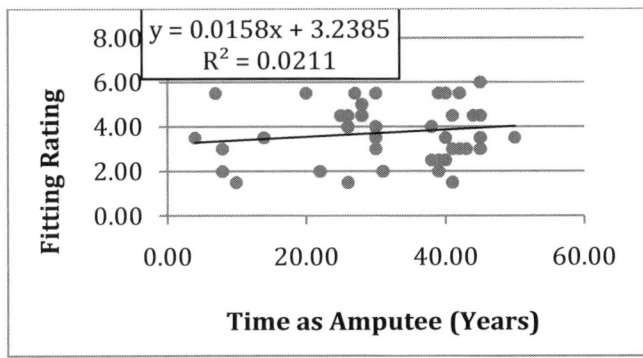

Fig. 7 Rating of Fitting vs. Time as Amputees

Figure 8 presents the relationship between fitting rating and the weight of the patient. The regression showed 2%

and supported no trend. It was hypothesized that those within a certain weight range would have the best fits. A patient that is too heavy would have more trouble walking with the prosthetic because it is specifically designed for weights under 70 kg. At weights much greater than that, the prosthetic may not be as supportive. A patient that is too light however, may not have the muscle mass needed to move around in a prosthetic. Therefore it was hypothesized that a patient between 40 kg and 70 kg would, on average, have the better prosthetic fit.

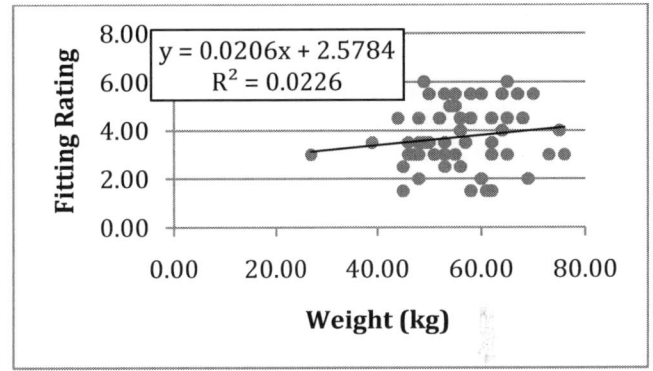

Fig. 8 Fitting Rating vs. Weight

In the Fitting Rating vs. Height analysis, Figure 9, regression analysis yielded 0% supporting no correlation. This is a clear indicator that the height of the patient does not negatively affect the quality of fitting. It was hypothesized that height would not play as much of an impact on prosthetic fit because the device is designed to support a wide range of heights.

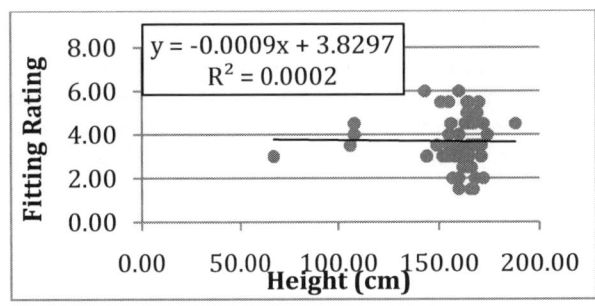

Fig. 9 Fitting Rating vs. Height

In the Fitting Rating vs. Age analysis, Figure 10, no evidence of correlation could be determined; regression analysis yielded 0%. This indicates that age of the amputee does not play a role in the prosthetic fit.

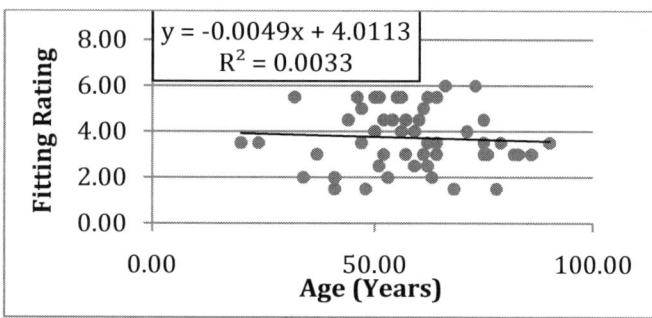

Fig. 10 Fitting Rating vs. Age

Fitting rating vs body mass index yielded a regression of 2%, indicating no correlation between BMI and quality of fit. Individually height and weight did not affect prosthetic fit and therefore BMI was not predicted to be correlated either.

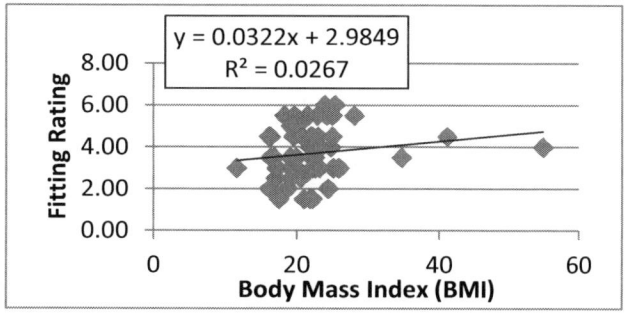

Fig. 11 Fitting Rating vs. BMI

IV. CONCLUSIONS

Of the five statistical tests done for age, weight, height, time as amputee, and BMI, none of them produced results that would suggest that the Mercer University Universal Prosthetic was better suited for any of those differences. This was with a limited sample of amputees however. Only a total of 53 patient data was cross-referenced with their F-Scan data analysis for this study. This was due to a number of limiting factors:

F-Scan data was not collect for all amputees that were fitted. Of the 272 patients that were fitted, only 150 sets of data were collected due to situational restraints. Of the 150 sets of data collected in Vietnam some of the data collection was not done correctly and made analysis difficult. Issues like incorrect calibration and misaligned sensors did not allow for proper analysis of a significant amount of data. This resulted in only 80 sets of data being fit for analysis.

Roughly 30 of the usable F-Scan data sets did not have a corresponding patient medical sheet that went with it. This data could not be used for trend analysis and left the study with a sample size of 53 patients. Current analysis yielded no correlations however a larger sample size may yield better results.

ACKNOWLEDGMENT

We would like to acknowledge the ongoing work of Dr. Vo, Dr. O'Brien, and Dr. Sumner for their continued work fitting prosthetics on impoverished amputees in Vietnam. We would also like to thank the students of the 2013 Mercer on Mission trip for their work fitting amputees.

CONFLICT OF INTEREST

The authors declare that they have no conflict of interest.

REFERENCES

1. Mueller, MJ. Strube, MJ. (1996) Generalizability of in-shoe peak pressure measures using the F-scan system. Clinical Biomechanics. Volume 11. Issue 3.
2. Catalfamo, Paola. Moser, David. Detection of gait events using an F-Scan in-shoe pressure measurement system. Gait and Psoture. Volume 28. Issue 3.

Author: Matthew Yin
Institute: Mercer University
Street: 1400 Coleman Avenue
City: Macon
Country: United States of America
Email: 10783131@live.mercer.edu

Biomechanical Evaluation of Hybrid Bicortical and Unicortical Screw and Bone Plate Fixation in Humeral Mid-Shaft Fractures: A Study on Cadaveric Bone

Trung Le[1], Benjamin McDeed[1], and Ha Van Vo[2]

[1] BME's Master Candidate, Department of Biomedical Engineering, Mercer University, Macon, USA
[2] Department of Biomedical Engineering, Mercer University, Macon, USA
Department of Orthopedic Surgery, School of Medicine, Mercer University, Macon, USA

Abstract— Internal bone plate fixation is currently considered the optimal bone fixation method for open fractures of long bone. This study aims to investigate the effects of non-locking bicortical and unicortical orthopedic bone screws and bone plate configuration on an open fracture of a long bone, the humerus. Eight cadaver humerus bones (n=8) were collected and an open gap, 1.5 cm mid-shaft transverse fracture, was modeled in each one. They were fixated by a 4.5mm non-locking (D.C) plate with two different configurations of 4.5mm non-locking bone screws: Group 1 and Group 2. The orthopedic non-locking bone plate had a total of 7 holes with 3 holes on either side of the fracture site. Therefore, 2 bicortical screws (30 mm long) and 4 unicortical screws (14mm long) were implemented into the fixation in both groups. Group 1 was defined as the bicortical screws being placed close to the fracture gap while Group 2 had the bicortical screws placed further away from fracture gap. Unicortical screws were placed in the other four holes in each setup. In order to complete mechanical evaluation, fatigue and strength of the bone plate construct was assessed by using the Material Testing System (858 Mini Bionix) to simulate cyclic loading and axial failure tests. The cyclic failure test was conducted on the two group bone specimens for 4000 cycles with the cyclic loading being approximately 240N. All bone plate and bone screw setups passed cyclic loading without any problem or failures. The axial failure test was completed by conducting an axial compression force with MTS's run rate of 0.1mm/sec and was performed on the two bone groups. The average maximum axial failure force was observed at 1064 ± 188 (N) for Group 1 and 1446 ± 467 (N) for Group 2. A two-tail t-test revealed that the strength of Group 2 was statistically significant. Thus, it was shown that Group 2's configuration was the better option for long bone fixation.

Keywords— Humeral long bone transverse fracture, bi-cortical and uni-cortical non-locking bone screws, Non-invasive fixation.

I. INTRODUCTION

Orthopedic bone plate fixation ideally minimizes damage to the infra-neurovascular structure or bone marrow inside the bone, thus the method significantly improves healing time of the fracture. Bone plates and screws can either be a locking or a non-locking mechanism; they are often made of Surgical Stainless Steel 316L, or Titanium alloy metals.

Bone cement (PMMA) can be added to non-locking and locking setups to provide even more stability and bonding strength to the screws and bone plate. The bone plates and screws fixation strives to provide the most stability at the fracture site by eliminating as much micro-motion as it can while maintaining high fixation stiffness and low strain at the fracture site [2,4]. The screws can be cortical screws, cancellous crews, or locking screws; they vary in material, diameter, and length depending on the bone and circumstance in which they will hold the bone plates to the bone [1]. The best system to implement for simple transverse fractures of the humeral shaft would be using cortical screws with a concave, compression bone plate. The concave bone plate ensures that the compressive forces from the plate and screws occur both near and far from the cortices of the fracture site. The best, most stable configuration would be using all bi-cortical screws, or screws long enough to go through both sides of the bone [4]. However, the more bone used to fixate, the more invasive the surgery and longer the bone will take to heal. Therefore, to make a less invasive, cheaper, and quicker healing time, doctors often will use non-locking screws with a combination of bicortical and unicortical screws (which only go through one side of the bone). The technique is less invasive and allows for faster healing with less bone damage while still providing adequate compression and stability at the fracture site [5]. However, the position of the bicortical and unicortical screws relative to the fracture site is left up to the doctors' decision. Therefore, a question arises: is it better to place the bicortical screws further from or closer to the fracture site?

II. MATERIALS AND METHODS

To conduct the study, eight total cadaveric arms were harvested at Mercer University, School of Medicine. The specimens were then processed through tissue dissection to obtain humeral bone specimens for the experiment. The eight specimens were paired in groups of two specimens which were actually harvested from the same cadaver in order to have uniform bone length and density to compare between both groups. Thus, the eight samples were separated into two groups so that one of the bones from each

© Springer International Publishing Switzerland 2015
V. Van Toi and T.H. Lien Phuong (eds.), *5th International Conference on Biomedical Engineering in Vietnam,*
IFMBE Proceedings 46, DOI: 10.1007/978-3-319-11776-8_123

matched pair was placed in group 1, and the other humeral bone of the pair was placed in group 2. As a result, four bones were placed in group 1, and the other four placed in group 2. This study utilized the non-locking orthopedic bone plate and bone screw system for bone fracture fixation. The orthopedic bone plates and screws were ordered from Ortho Image Company in India. The orthopedic bone plates were each made of a 316L stainless steel and were D.C. Plates (NERROW). Each bone plate is a straight, non-locking, orthopedic compression bone plate with seven holes with a diameter of 4.5 mm for each hole. To be consistent with the bone plates, orthopedic cortical bone screws were also ordered with diameters of 4.5mm but in two different length dimensions, 14 mm and 30 mm in length to serve as the uni-cortical and bi-cortical screws respectively. All of the bones of the two groups were fixated by bone plates and screws with two different configurations as it is shown in Figure 1. The bones of Group 1 were fixated with the bicortical screws close to the fracture gap while the bones of Group 2 were fixated with the bicortical screws further away from fracture gap.

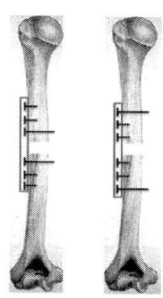

Fig. 1 Bi-cortical and unicortical bone plates and screws configuration in Group 1 (left) and Group 2 (right)

After finishing the bone fixation, the bones needed to be cut to model a simple, transverse open gap fracture with a size of approximately 1.5 cm. The reason for creating a simple, transverse gap fracture in this study is to allow the maximum possible applied force to be able to be generated on the bone construct in order to assess the stability and stiffness of the bone plate with different bone screw configurations. The transverse humeral fracture is also the quickest and most efficient to create. The reference 1cm fracture gap is exactly replicated according to *Toole, 2008.*

In order to evaluate the mechanical stability, strength and stiffness of the bone plates and screws construct, the Material Testing System (858 Mini Bionix) was utilized to simulate axial cyclic loading and axial compression which is shown in figure 2. These two main tests were carried out to analyze the fatigue failure and axial compression failure of the bone plate constructs. The cyclic loading test was performed by using a sine wave function. The frequency of the sine wave was set at 2 Hz with the span or amplitude set to ±5mm. The experiments were conducted for 4000 cycles on each specimen. The recommended force to conduct the cyclic loading ranged from 0-240N to represent anticipated physiologic loading of one-half body weight for an average (50 kg) patient [2].

The axial failure test was carried out after the cyclic loading test by using a ramp function. All the specimens passed the cyclic failure test, and each then went through the axial failure test with a compression rate of about 0.1 mm/sec. The reason for choosing that small of a displacement rate is due to the micro-motion between the bone plate construct and the bone itself under weight-bearing conditions. The microstructure failure of the bone plate constructs is analyzed in term of maximum bending stress and displacement.

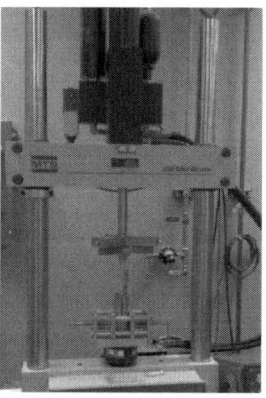

Fig. 2 MTS (858 Mini Bionix) set up for axial cyclic loading and axial compressive test

III. RESULTS AND DISCUSSIONS

The eight humeral bone specimens having two different screw configurations passed the cyclic compression test without any destruction or damage. This study had 1 set of bones that had osteoporosis, which caused one bone of the set to be severely damaged at the humeral neck. However, the setup of axial cyclic loading test only allowed the bone to be compressed and decompressed within 5mm. As a result, the integrity and structure of the bones were maintained. Since all of the specimens achieved their merit in the cyclic failure test. They were processed into the second phase with the axial compressive test to analyze the strength of the bone plate constructs.

The axial compression test was performed by a ramp function. The MTS's run rate was set at 0.1 mm/sec in axial compression mode to imitate exact micro-motion scenario between the bone plate construct and the bone itself. The maximum failure forces were recorded for each individual bone of the two groups as well as the failure mode shapes of the bone plate constructs. Most of the failure shape modes indicate the severe bending of the bone plate at the fracture site. The maximum axial failure force was generally observed to achieve maximum peak force in the group 2 which has the bicortical screws configuration set far away from the fracture gap. Also, the higher the bone mass and bone density, the more force the bone construct can withstand under an axial compressive load test. It is easy to interpret that high bone density usually occurs with thick cortical bone which is an important factor to provide a strong frame and strong support under high impact forces. By comparing the maximum failure force between two different bone groups, it is interesting to report that the bicortical screw constructs in group 2 placed far away from the fracture gap achieved higher merits in holding higher amount of axial compressive force than the bone specimens in group 1 (Figure 3). The results also displayed a consistency among the set of bones; (i.e two bones of the same sets with two different bicortical screw configurations will express two different results according to the group characteristics.)

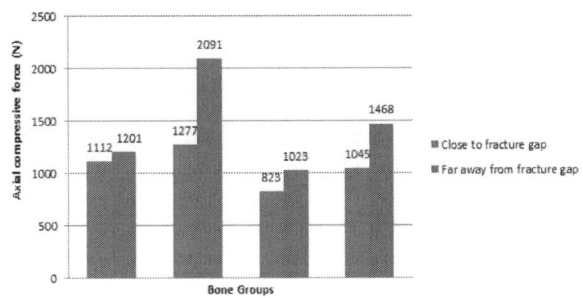

Fig. 3 Comparison of maximum axial failure force between group 1 and group 2

The changing between the groups is now analyzed on whether or not this change is significantly different statistically. By using a statistical analysis method, the two sample t-test was carried out on the two sets of data from the two different groups. Using this test, the level of confidence of the result is confirmed to be significantly different or not. The statistical t-test was analyzed by using the two tailed t-test with a 5% probability. This test basically defines a 95% level of confidence. The null hypothesis is likely to be rejected if the result of the t-test's probability is less than 5% or p < 0.05. The average means and standard deviations of the maximum failure force of the two groups were found to be 1064 ± 188 (N) and 1446 ± 467 (N) for group 1 and

group 2 respectively (figure 4). The null hypothesis (H_o) of the two tail t-test was assigned such that the two groups are statistically different since their average means are not approximately equal.

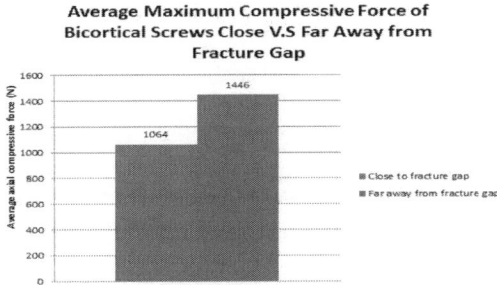

Fig. 4 Comparison of average maximum axial failure force between group 1 and group 2

The results of two tail t-test indicated that the probability of the t-test is approximately 0.21 or P_t=21% > 2.5% (Table 1), so the null hypothesis is accepted. As a result, the results of maximum compressive force between two groups are significantly different. This also means that group 2 with the bicortical screw configuration far away from the fracture gap shows its merits as an effective bone fixation method and the better of the two options to fixate a simple transverse humeral shaft fracture.

Table 1 Results of the Student's two tail t-test

Statistical Factors	Group 1	Group 2
Mean	1064.2	1445.7
Variance	35331	218313
Observations	4	4
Hypothesized Mean Difference	0	
Df	4	
t Statistic	-1.5	
P(T<=t) two-tail	0.21	
t Critical two-tail	2.8	

In order to further explain why the bicortical screw configuration far away from fracture gap is the optimal configuration, engineering analysis is used to explain the differences. The applied axial compression force exerts pressure on the bone and the bone plate off loads the force via the bone and bone screws. The force is transferred through the fracture plane to the other side, and is displaced back into the bone and throughout the multiple screws. All the force and energy is therefore eventually absorbed by the bone screw and bone plate. By investigating the force exerted on the bone via the screws, the detailed free body diagram cross section of the bone of the two groups with different unicortical and bicortical screw configurations is shown in figure 5.

The unicortical screw only penetrates one cortical section of the humeral bone while the bicortical screw penetrates both sections of cortical bone. The effect of the bicortical with double penetration will increase the reaction force from the screw at that area, thus it helps to absorb more energy under axial applied force.

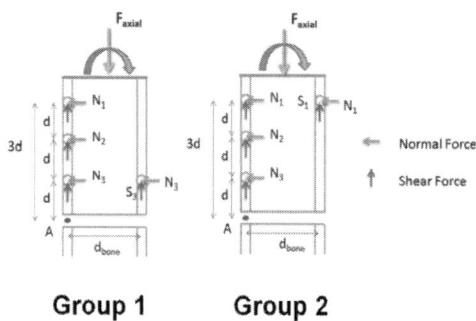

Group 1   Group 2

Fig. 5 Free-body-diagram of bone cross sections

The bending moment is determined by force times distance ($M = F \times d$). From the free body diagram, the bending moment was assumed to occur at point A due to the force applied and the weak points observed from the axial compression failure tests. In group 1, the bicortical screw asserts a uniform 2d distance away from the applied force. This causes higher moment arm or bending moment which caused the bone plate construct in group 1 to bend at a lower applied force. In group 2, the distance between the bicortical screw and applied force decreases which significantly reduces the bending moment. Thus it needs a sufficient amount of axial compressive force to fail the bone plate construct which explains the higher axial failure force that was seen in group 2.

IV. CONCLUSIONS

The results of the completed research experiment revealed that the configuration in which the bicortical screws are located at the farthest position on both sides of the fracture site with unicortical screws in all of the remaining positions in between is the stronger and sturdier configuration. It had a statistically significant increase in the amount of force the fixture and fractured bone could take before any part of the fixture or bone failed relative to the other configuration with the bicortical screws closest to the fracture site and the unicortical screws in the other holes. Therefore, based on the data collected and analyzed in this research, a combination bicortical and unicortical screw setup with the bicortical screws placed at the farthest holes in the bone plate and

unicortical screws at all of the other holes on either side of the transverse fracture site is the better of the less invasive configurations to use to stabilize a simple transverse humeral shaft fracture while it heals.

To optimize the results of this study, a much larger scale experiment with a larger sample size would need to be conducted. A future study with more samples and a much larger amount of cyclic loading cycles completed on each bone, bone plate, and bone screw configuration would be needed to determine if the results of this smaller scale study hold true and the position of the bicortical screws does affect the effectiveness of the humeral bone fixation.

ACKNOWLEDGEMENT

We would like to thank those people who donated part of their bodies to science, making this study possible. Without this great contribution, this study could not have been achieved. Finally, we also want to give a special admiration and thank you to Dr. Ha Van Vo for his work and aid in guiding us in the completion of this research study.

CONFLICT OF INTEREST

The authors declare that they have no conflict of interest.

REFERENCES

1. Dunlap, J., George, L., & Alexander, C. (2011). Biomechanical evaluation of locking plate fixation with hybrid screw constructs in analogue humeri. The American Journal of Orthopedics, 40(2), 20-25.
2. Toole, R., Anderson, C., & Vesnovsky, O. (2008). Are locking screws advantageous with plate fixation of humeral shaft fractures? A biomechanical analysis of synthetic and cadaveric bone. Journal of Orthopedic Trauma, 22(10), 709-715.
3. Rubel IF, Kloen P, Campbell D, et al. Open reduction and internal fixation of humeral nonunions: a biomechanical and clinical study. J Bone Joint Surg Am. 2002; 84:1315–1322.
4. Zimmerman MC, Waite AM, Deehan M, et al. A biomechanical analysis of four humeral fracture fixation systems. J Orthop Trauma. 1994; 8:233–239.
5. Gardner MJ, Griffith MH, Demetrakopoulos D, et al. Hybrid locked plating of osteoporotic fractures of the humerus. J Bone Joint Surg Am. 2006; 88:1962–1967.
6. Gardner MJ, Brophy RH, Campbell D, et al. The mechanical behavior of locking compression plates compared with dynamic compression plates in a cadaver radius model. J Orthop Trauma. 2005; 19:597–603.

Author: Trung Le
Institute: Mercer University
Street: 1400 Coleman Avenue
City: Macon
Country: USA
Email: trungle.bme@gmail.com

Novel Design of a Prosthetic Foot Using Spring Mechanism

Awab Umar Khan[1] and Ha Van Vo[2]

[1] Master's Biomedical Engineering, Mercer University, Macon, USA
[2] Department of Biomedical Engineering, Mercer University, Macon, USA

Abstract— **Mercer on Mission was the spark for construction of a prosthetic that provides added force during ambulation for amputee's utilizing a prosthetic. This study attempts to design a novel prosthetic foot that has a spring mechanism in the ankle region of the device that allows for added force that is provided back to the patient as he or she ambulates. The design hopes to provide the patient operating on it a reasonable amount of range of motion with respect to the movements at the ankle including dorsiflexion, plantarflexion, inversion, and eversion. Proper testing of the prosthetic was done through Tekscan F-scan®, which is a plantar pressure analysis software. The analysis of the data showed that when the patient ambulated with the Spring Ankle Prosthetic (SAP), the force vs. percentage graph showed a reduced stance time, indicating that the springs within the ankle region of the prosthetic are functioning properly and helping aid the patient in ambulation.**

Keywords— **Prosthetic foot, springs, F-scan, atrophy, gait.**

I. INTRODUCTION

The Biomedical Engineering department at Mercer University has been manufacturing and refining above and below knee prosthetics for use during Mercer on Mission for several years. Using Dr. Ha Van Vo's universal socket design all amputees can be fit; however, in fitting these amputees there is a specific criterion that has to be met by the patient in order to receive a prosthetic. Inevitably every year patients are turned away due to several complications. Muscle strength plays a prominent role in the decision process; the amputee must have less than 40% muscle atrophy in their residual stump. This is a regrettable reality to turn away amputees, therefore the goal of this study was to build a novel design of a below knee prosthetic that has an assistive mechanism that may help provide amputees with some added force during ambulation. The aspiration of this new prosthetic design is to accommodate a greater range of amputees and aid in their aptitude to achieve mobility once.

It is difficult to fit patients with muscle atrophy, Figure 1. Amputees that do not own a prosthetic or do not exercise their muscles develop muscle atrophy over time and it is a serious condition. This condition makes it problematic to fit these patients with a prosthetic knowing that it will not be utilized to its full potential. The loss of muscle mass makes it difficult to operate a prosthetic, which puts added weight on the patient. For this reason, this study attempted to design a novel prosthetic foot that has a spring mechanism feature near the ankle portion that gives patients an added force during the gait cycle. This prosthetic does not take over the actions of the muscles, but merely involves an assistive feature that helps the patient ambulate. In addition, it does not claim to make ambulation for patients with muscle atrophy possible. It merely provides a small amount of added force to test how significantly it affects the patient's gait. The testing of the prosthetic was done subjectively at first and then through the utilization of the Tekscan F-Scan®.

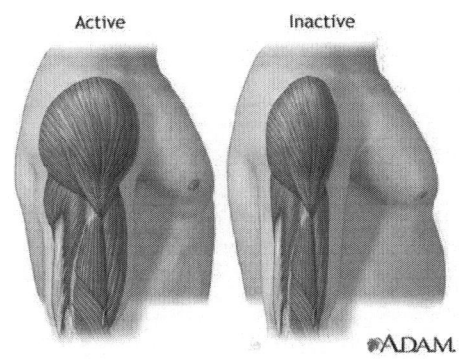

Fig. 1 Healthy Muscle (Left); Atrophied Muscle (Right) [1]

II. METHODS AND MATERIALS

The construction method for the SAP was long and involved the utilization of numerous tools. The machine shop and the prosthetics lab resources were expired to make a successful and operating prototype of the SAP. The procedure of the final prototype construction that was tested through F-scan® is described henceforth. It should be cautioned that proper handling of the materials and machinery should be used. Safety is the number one priority when building the device. Supervision of experienced personnel is advised!

The socket portion of the prosthetic is adapted from Dr. Ha Van Vo's Universal Prosthetic design and its construction method is not discussed. In the construction of the ankle portion of the prosthetic, it is easy to forget the overall design when so much detail is involved with each

V. Van Toi and T.H. Lien Phuong (eds.), *5th International Conference on Biomedical Engineering in Vietnam*,
IFMBE Proceedings 46, DOI: 10.1007/978-3-319-11776-8_124

component that has to be manufactured. For this reason, a simple overview of the SAP is shown in Figure 2. This design roughly portrays the shape of the top plate that is desired, the bottom plate and how it has a sliding mechanism, and how a revolute joint attaches the top and bottom plates in that region. The middle plate that rests above the springs and below the flat portion of the top plate is not shown in this image.

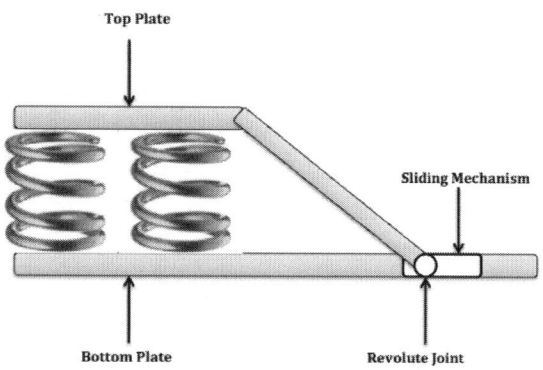

Fig. 2 Simplified SAP Mechanism Design

The prosthetic foot was constructed utilizing the band-saw, milling machine, drill machine, vice grips, aluminum mold, oven, springs, and various other tools. Due to the constraints of the paper the details of the materials and methods cannot be discussed extensively. The bottom plate was made first and bent into the shape desired. The sliding mechanism was milled and so were the slits for the revolute joint. After this the top plate was made using the aluminum mold to take its specific shape. After cutting and sanding the top plate it was ready to be attached with the bottom plate at the sliding mechanism region. The bottom plate had circles milled into it for the spring housings. The springs were placed onto the bottom plate in a diamond formation. Two different types of springs were used, the LHC 148 0CS, which has a spring rate constant of 147.4 lb/in and the LHC 148M 0CM, which has a spring rate constant of 169.6 lb/in. The lower rate constant springs were placed on the anterior and posterior of the diamond formation, and the higher rate constant springs were placed on the lateral sides. This particular setup was used to allow for easier range of motion with dorsiflexion and plantarflexion. The higher rate constant springs on the lateral sides controls the inversion and eversion motions morel, which occur less in the ankle during gait. These springs were chosen based on the patient that would be fit with the prosthetic for testing. All the other dimensions of the springs are identical besides the material they are made of and their rate constants.

After the bottom and top plates are finished the middle plate is made and placed on the superior aspect of the springs with circle housings that match perfectly with the bottom plate. The middle plate is also secured to the top plate through a screw. The finished prosthetic foot with the socket portion attached is shown in Figure 3.

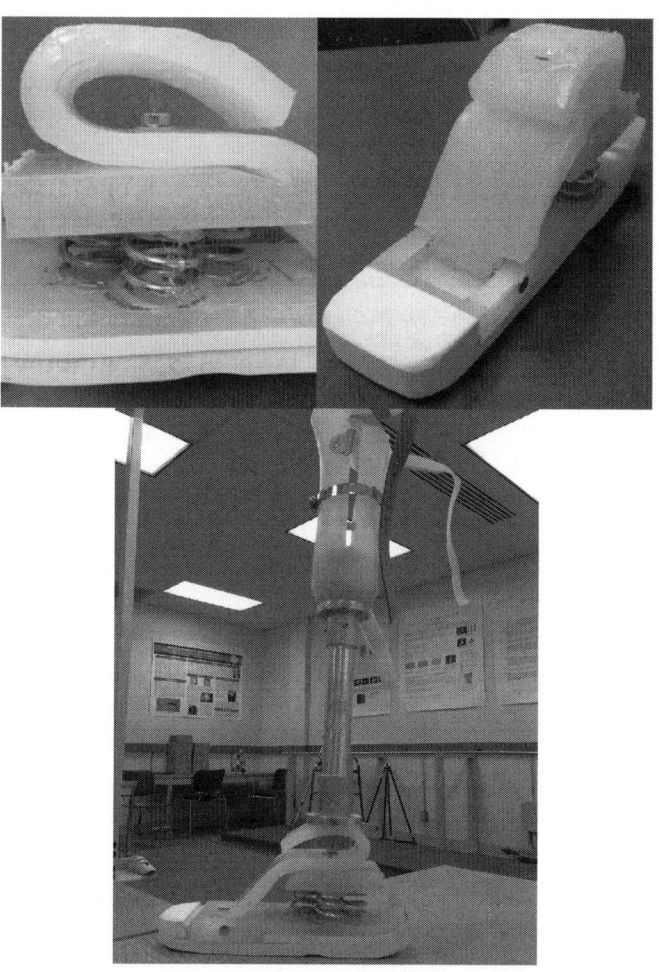

Fig. 3 Complete Assembly of SAP

III. RESULTS AND DISCUSSION

The patient to be fit has a trans-tibial amputation of the left leg. The prosthetic fittings are for her left leg. F-scan is a plantar pressure collection software that allows for analysis of gait to determine pathologies, compare prosthetics, and various other purposes. F-scan data was collected for the patient ambulating with her personal prosthetic, which is a college park prosthetic with a Trustep® foot, and the SAP. The data for the personal prosthetic is illustrated first, with the SAP data following. The prosthetic data follows with the figure captions including either '(Personal)' or '(SAP)' to indicate which prosthetic the data is for.

The following Figure 4 shows the Force vs. Time. The x-axis is time in seconds and the y-axis is force in pounds. The graph shows a green line representing the right foot and red line representing the left foot.

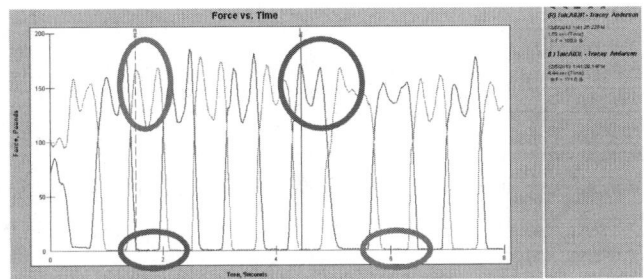

Fig. 4 Force vs. Time Graph (Personal)

The first peak indicates heel strike and the second peak indicates toe-off, green and red circles highlight these peaks. The green circle corresponds with the heel strike and toe-off of the right leg, and the red circle corresponds with the heel strike and toe-off of the prosthetic leg. The alternation in the peaks for the left and right foot are present because as one foot heel strikes, the other foot goes into the swing phase, then as that foot goes from mid-stance to toe-off, the other foot braces for impact at heel strike. When one leg is in the swing phase there is zero force being exerted on the pressure sensors that are inserted inside the shoes. The bottom portion of the graph that has green and red lines almost on the x-axis are regions where the respective feet are in swing phase. Had there been a gap between the bottom of these green and red lines and the x-axis, the region that is circled on the graph, this would have meant that the data was offset. This offset usually occurs due to preloading on the pressure sensors. However, this was not the case and the pressure sensors worked properly. Forces at heel strike for the left foot is 171 lbs and the right foot is 167 lbs. Forces are toe-off for the left foot is 168.4 lbs and the right foot is 168.5 lbs.

Another graph that is generated through F-Scan® is the Force vs. Percentage graph (Figure 5). This graph has a y-axis of force in pounds and an x-axis of percentage meaning the percentage of the gait cycle that has been completed as time continues. This graph averages the forces recorded through the stance phase of the gait cycle with the gait cycle presented as a percentage.

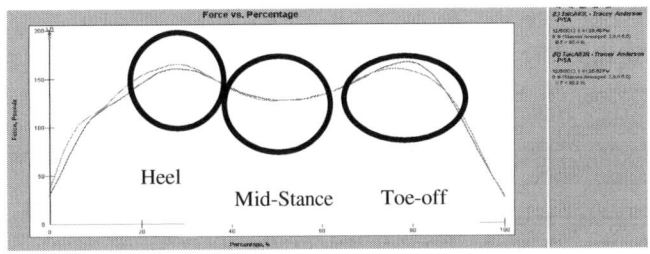

Fig. 5 Force vs. Percentage Graph (Personal)

This graph helps prove that the patient's personal prosthetic closely mimics the actual leg. The first peak in this graph is the heel strike, the dip represents the mid-stance, and the second peak represents the toe-off portion of gait. The force at heel strike for both feet when an average is taken is around 160.3 lbs for the left foot and 164.8 lbs for the right foot. At toe-off however, the left foot force is higher at 167 lbs and the right foot is lower at 159.8 lbs. This graph helps prove that the patient's gait with her prosthetic is fair with exception of a couple of discrepancies. Critical review of the graph shows that there are actually 3 peaks for both feet. Before the prominent heel strike peaks that are circled there is a slight deviation that is seen in both lines. This deviation leads to the lines not being as smooth indicating that the gait is not as smooth. The deviations are due to a one aspect of the patient's gait being poor and that is during heel strike for either leg the patient is hesitant and does not trust her leg or her prosthetic. As the patient braces for heel strike, she almost heel strikes twice, due to the fumbling nature of the heel strike as can be seen from the graph. The average forces at heel strike are higher for the right leg as compared to her prosthetic leg. This leads to the conclusion that the prosthetic leg is slightly shorter. To compensate for this, the force at toe-off for the prosthetic leg is slightly higher. The biomechanics of the gait try and balance each other as to reduce any risk of injury to the patient.

The force vs. time graph for the SAP is shown in Figure 6.

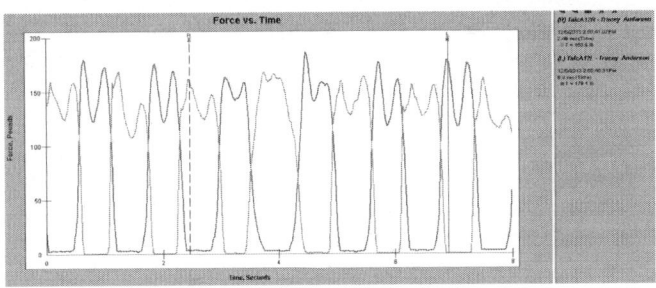

Fig. 6 Force vs. Time Graph (SAP)

The force at heel strike for the left foot is 179.1 lbs and the right foot is 160.9 lbs. The force at toe-off for the left foot is 176.2 lbs and the right foot is 164 lbs. The smoothness of the red line (SAP leg) shows that the patient trusted the prosthetic fully and was able to ambulate on it without fumbling. Initially a large difference can be seen between heel strike and toe-off forces for the right leg, but as the patient continues to ambulate and gets used to the SAP, her gait gets better. The patient is trying to compensate due to the new prosthetic, as the patient continues to walk and becomes more comfortable with the prosthetic, the right leg gait gets better.

The force vs. percentage graph is in Figure 7.

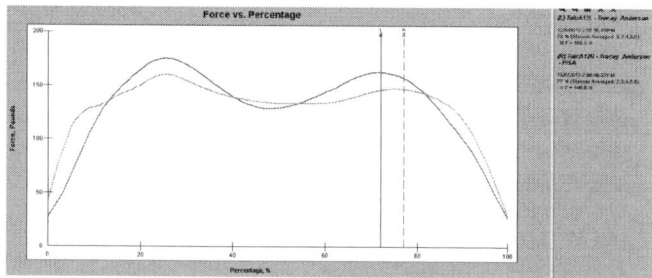

Fig. 7 Force vs. Percentage Graph (SAP)

As the patient walked the average amount of force at heel strike for the left foot was 175.2 lbs and the right foot was 160.2 lbs. This is approximately a 15-pound difference, which is significant. The average toe-off force for the left foot was 162.8 lbs and the right was 146.9 lbs. The force difference for the toe-off stance is slightly over 15 lbs, similar to the heel strike force difference.

The x-axis is labeled as the percentage of the stance phase of the gait cycle that has been complete. This does not directly convert time, but it does prove that the stance phase for the left foot completes faster than the right foot because the toe-off peak for the left foot comes before the right foot toe-off peak. This was not the case with the patient's personal prosthetic and this difference in the graph is attributed to the SAP. Since the novel design utilizes springs that are implemented to provide added force during the gait cycle the extra force shortens the amount of time that the patient spends in the stance phase for the left foot. This is a significant observation as it is a slight difference in the gait cycle that can be attributed to the SAP to help prove that it has an effect on the patient's gait [2].

IV. CONCLUSIONS

The purpose of this research study was to design and construct a prosthetic foot. The prosthetic was to implement an assistive feature that would help the patient ambulate. The design and construction of the Spring Ankle Prosthetic (SAP) was a success. A four-spring design in a diamond formation was placed in the ankle portion of the prosthetic with a sliding mechanism in the foot to allow for the springs to compress linearly. The utilization of four springs in a diamond pattern allowed for motions including plantarflexion, dorsiflexion, inversion, and eversion. The patient fitting was a success and the patient was able to ambulate on the prosthetic without any parts failing. The patient ambulated on the SAP just as well as she did on her own prosthetic; this signified that the prosthetic was sturdy and it fit the patient well and she trusted it enough to do this.

F-Scan® was used to analyze the plantar pressure data for the patient's gait as she ambulated. The results were promising and showed that there was a difference when the patient ambulated with her prosthetic versus the SAP. The Force vs. Percentage graph did well in showing how the swing phase on the patient's left foot was faster with the SAP due to the spring mechanism. The stance phase was also shorter as when the patient progressed from heel strike to mid-stance to toe-off the springs provided some form of energy back to the patient.

ACKNOWLEDGMENT

A special thanks goes out to Dr. Ha Van Vo for all the support he has provided me through this research and to Mrs. Tracey Anderson for being an understanding and committed participant.

CONFLICT OF INTEREST

The authors declare that they have no conflict of interest.

REFERENCES

1. Muscle Atrophy. Digital image. Scripps Health. N.p., n.d. Web. 13 Sept. 2013. <http://www.scripps.org/articles/3558-muscle-atrophy>.
2. Mechanics of the Ankle. N.p.: Dr. Ha Van Vo, n.d. PowerPoint.

Author: Awab Umar Khan
Institute: Mercer University
Street: 1400 Coleman Avenue
City: Macon
Country: United States of America
Email: 10702918@live.mercer.ed

Development and Characterization of Porous Calcium Phosphate Cement Using α-Tricalcium Phosphate Bead

Pham Trung Kien[1], Tsuru Kanji[2], and Kunio Ishikawa[2]

[1] Department of Ceramic Materials, Ho Chi Minh University of Technology (HCMUT), Ho Chi Minh City, Vietnam
[2] Department of Biomaterials, Kyushu University, Fukuoka City, Japan

Abstract— Interconnected set porous-calcium phosphate cement (set porous-CPC), which has fully interconnected pores could be an ideal bone substitute. In this study, set porous-CPC consisting of α-tricalcium phosphate (α-TCP) beads and acidic calcium phosphate solution was developed and its basic setting reaction was studied. When α-TCP beads were exposed to the acidic calcium phosphate solution, brushite (CaHPO4·2H2O) was formed on the surface of the α-TCP beads. The formed brushite crystals interlocked with each other, resulting in a setting reaction of α-TCP microspheres within 10 min at 37°C. As a result of this setting reaction, a fully-interconnected calcium phosphate macroporous structure was obtained. The set porous-CPC were immersed into simulated body fluid (SBF) at 37oC for different immersion time, The result clearly showed that brushite surface layer transformed to apatite layer after 3 days immersion into SBF solution by dissolution-precipitation reaction. Since apatite layer show excellent tissue response, the set porous-CPC would be a promising biomaterial used as artificial bone substitutes, and as the scaffold for tissue engineering.

Keywords— α-TCP, brushite, apatite, porous cement, setting reaction.

I. INTRODUCTION

Calcium phosphate cement (CPC) sets to form hydroxyapatite (HA; $Ca_{10}(PO_4)_6(OH)_2$) or brushite ($CaHPO_4·2H_2O$) when its powder phase is mixed with its liquid phase [1-4]. Due to the compositional transformation to HA or brushite, set CPC shows excellent tissue response and good osteoconductivity. Self-setting ability allows ideal intimate filling of bone defect and minimum invasive surgery. Also, it is reported that set CPC could be replaced by new bone. Set CPC shows a microporous structure of interlocking HAp or brushite crystals formed by a dissolution-precipitation reaction. For example, the setting reaction of β-Tricalcium Phosphate (TCP; $Ca_3(PO_4)_2$) powder with Monocalcium phosphaphate monohydrate (MCPM; $Ca(H_2PO_4)_2·H_2O$) is shown in equation (1):

$$Ca_3(PO_4)_2 + Ca(H_2PO_4)_2·H_2O + 7H_2O \rightarrow 4CaHPO_4·2H_2O \quad (1)$$

However, the size of the micropores is less than 1μm, and thus, too small to allow cell penetration or tissue ingrowth [5-6]. Many attempts have been made to fabricate macroporous CPC but not succeed to obtain the interconnected porous structure. In order to overcome this drawback, this research developed new method to fabricate set porous calcium phosphate cement with interconnected porous structure by using α-TCP beads instead of TCP powder.

II. MATERIALS AND METHODS

A. Preparation of α-TCP Beads

The preparation of α-TCP beads are described in previous publications [7-8]. In brief, commercially obtained α-TCP powder (type B; Taihei, Osaka, Japan) was placed in a stainless steel mould and pressed uniaxially with an oil-pressure machine (Riken Power, Tokyo, Japan). The cylindrical compacts of 3 mm in diameter and 3 mm in height were calcined at 1000°C for 5 hours in an electric furnace (Super burn SC 1500D, Motoyama, Osaka, Japan) and then cooled inside the furnace. The calcined α-TCP compacts were ground between abrasive papers pasted on glass plates. Alumina balls with 2.0 mm diameter (Sogorikagaku, Kyoto, Japan) were placed at the four corners of the glass plates to regulate the final diameter of the α-TCP beads. The ground α-TCP beads were sintered again at 1400°C for 5 hours in the electric furnace and cooled down inside the furnace. The diameter of the resultant α-TCP beads were 1.3 mm.

B. Preparation of the Acidic Calcium Phosphate Solution

Acidic calcium phosphate aqueous solution was used as a liquid phase of porous CPC so that the precipitate formed by the setting reaction of α-TCP beads and liquid phase would be brushite. Monocalcium phosphate monohydrate (MCPM; $Ca(H_2PO_4)_2·H_2O$) powder obtained commercial (Sigma-Aldrich, Saint Louis, USA) was dissolved into 0.1 mol/L H_3PO_4 solution (Wako, Osaka,Japan) so that the concentration of MCPM became 0.2 mol/L. This solution is almost saturated with respect to MCPM.

C. Setting Reaction of Porous Calcium Phosphate Cement

The α-TCP beads were exposed to MCPM-H_3PO_4 solution so that the ratio of liquid to α-TCP microspheres would be 2 μL per microsphere. The mixture was placed into a split stainless steel mould (6 mm in diameter x 6 mm in

© Springer International Publishing Switzerland 2015
V. Van Toi and T.H. Lien Phuong (eds.), *5th International Conference on Biomedical Engineering in Vietnam,*
IFMBE Proceedings 46, DOI: 10.1007/978-3-319-11776-8_125

height) and kept in an incubator (Eyela SLI-220, Tokyo, Japan) at 37°C for 1, 5 and 10 min. After incubation for prescribed time, the set porous CPC was rinsed with distilled water 3 times, followed by immersion in acetone to stop further the reaction.

D. Immersion Set Porous-CPC with Brushite Layer in SBF

SBF solution with ion concentrations nearly equal to those of human blood plasma was prepared as described previously [9]. Subsequently, the set porous-CPC specimens with brushite layer were immersed into 20 mL SBF solution at 37°C for 1, 3 and 7 days. After immersion, the specimens were rinsed with distilled water 3 times, followed by immersion in acetone to stop the reaction.

E. Characterization of Set Porous CPC

Surface morphology of the interface between the α-TCP beads were observed by scanning electron microscope (SEM; JSM 5400LV, Tokyo, Japan) at an acceleration voltage of 20 kV after gold sputtering.

Composition of set CPC before and after the setting reaction was evaluated by means of powder and thin layer X-ray diffraction (XRD; D8 Advance-Bruker AXS, Leipzig, Germany) using CuKα (λ=1.5406 Å) radiation generated at 40 kV and 40 mA. For the powder XRD, whole specimen was ground to fine powder whereas set porous CPC was used in the case of thin layer XRD. The irradiation angle was fixed at 1 degree for thin layer XDR measurement. The specimens were scanned from 5° to 40° in 2θ under a continuous mode at a scanning rate of 0.02°/sec. Similarly, the composition of set porous-CPC before and after immersed into SBF was analysed by powder XRD.

Interconnectivity of set porous CPC was evaluated using micro-computed tomography (μ-CT; Skyscan 1075 KHS, Kontich, Belgium). Based on 2D μ-CT slices, a 3D structure of the set porous CPC was reconstructed.

Porosity of the set porous CPC were estimated using the quantitative evaluation program included with the μ-CT software package. The result of porosity was confirmed using pycnometer method, as shown in equation (2):

$$Porosity = 100 \times (d_1 - d_2) / d_1 \qquad (2)$$

where d_1 is true density of set porous CPC, which was measured by pycnometer flask; d_2 is the bulk density of the set porous CPC, which was calculated as the weight to volume ratio.

For statistical analysis, one-way factorial ANOVA and Fisher's LSD method as a post-hoc test, were performed using KaleidaGraph 4.0. Values are expressed as mean±SD. A p-value of < 0.05 was considered statistically significant.

III. RESULTS

Fig. 1 shows typical photographs of (a) α-TCP beads and (b) μ-CT images of set porous CPC. As shown in Fig. 1, α-TCP beads set when exposed to MCPM-H₃PO₄ solution at 37°C and 100% relative humidity for 10 min. The pore is confirmed interconnected structure with pore size is > 100μm diameter of. Porosity of the set porous CPC calculated based on μ-CT method and pycnometer method were 45.8 ± 2.4 % and 49.7 ± 2.5 %, respectively.

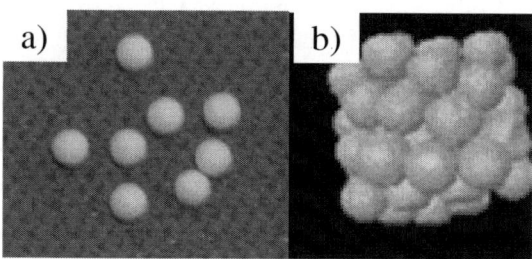

Fig. 1 Set CPC fabricated by setting reaction of α-TCP bead. The pore is confirmed interconnected structure with pore size is > 100μm. a) α-TCP beads as raw materials and b) Set CPC fabricated by α-TCP beads. The setting time is 10 minutes

Fig. 2 shows typical SEM photographs at an interface of α-TCP beads before and after being treated with acidic phosphate solution at 37°C and 100% humidity for 1, 5, and 10 min. Before treatment, the α-TCP beads show the typical smooth surface due to sintering effect. Naturally, no bonding was observed between the microspheres (Fig. 2(a)). In contrast, plate-like precipitates were observed on the surface of the α-TCP beads after being exposed to MCPM-H₃PO₄ solution (Fig. 2(b-d)). When set porous CPC was kept in an incubator for 1 min, surface of α-TCP beads was covered with the plate-like precipitate partially (Fig. 2(b)) and the length of plate-like precipitates was 20-30 μm. In contrast, whole surface of α-TCP bead was covered with the precipitates and the length of the precipitates increased to 30-40 μm when kept in an incubator for 5 and 10 min (Fig. 2(c-d)). Also, interlocking of plate-like crystals precipitated on the surface of α-TCP beads were observed as indicated by white arrow.

Fig. 3 summarizes the powder XRD patterns of the set porous-CPC when kept in an incubator at 37°C and 100% relative humidity for 1 min (Fig. 3b), 5 min (Fig. 3c), and 10 min (Fig. 3d). Peaks corresponding to α-TCP were observed on all specimens. In addition, a small peak assigned to brushite was found at 2θ=11.6° when the set porous-CPC was kept in an incubator for 5 and 10 min (Fig. 3(c-d)). The peak intensity of brushite was larger for set porous-CPC kept in an incubator for 10 min (Fig. 3d) when compared to that for 5 min (Fig. 3c).

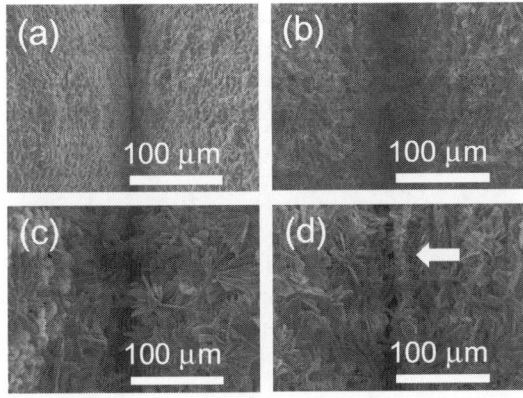

Fig. 2 SEM photographs of set porous CPC at the interface between two α-TCP beads (a) before and after being treated with MCPM-H₃PO₄ solution at 37°C for (b) 1 min; (c) 5 mins and (d) 10 mins. White arrow indicates interlocking between 2 α-TCP beads

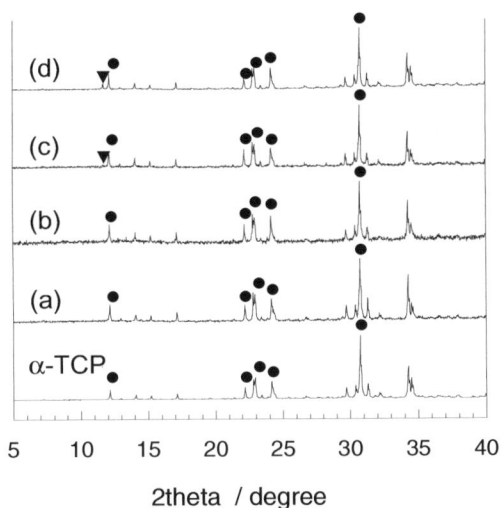

Fig. 3 Powder XRD patterns of α-TCP beads (a) before and after being reacted with MCPM-H₃PO₄ solution for (b) 1 min; (c) 5 min and (d) 10 mins. Standard α-TCP powder is also shown for comparison (●: α-TCP crystals; ▼: brushite crystals).

Fig. 4 summarizes the powder XRD patterns of set porous-CPC before (a) and after being immersed into SBF solution at 37°C for 1 day (b); 3 days (b); and 7 days (b). Before immersion into SBF solution, the XRD pattern of set porous-CPC shows the peak corresponding to α-TCP ($2\theta = 30.7°$) and brushite ($2\theta = 11.6°$) (Fig 4(a)). The peak of brushite appears after 1 day immersion into SBF (Fig. 4(b)) and disappears after 3 and 7 days immersion (Fig 4(c, d)). On the other hand, the broaden diffraction peaks are detected around $26°$ and $32°$ in 2θ which corresponding to apatite peak when the specimens are immersed into SBF solution for 3 and 7 days. This simply indicates that brushite layer is completely transformed to apatite layer after 3 days immersion.

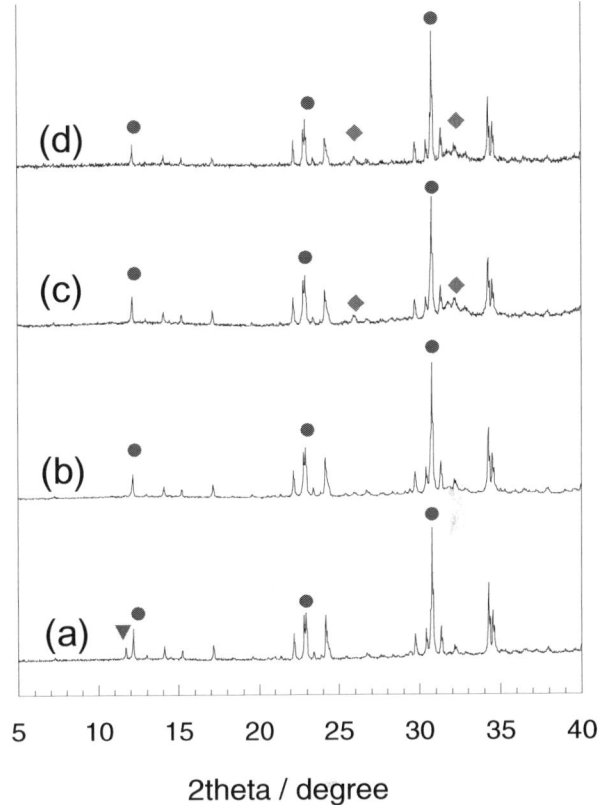

Fig. 4 Powder XRD patterns of set porous-CPC (a) before and after immersion into SBF at 37°C for (b) 1 day; (c) 3 dayd and (d) 7 days. (●: α-TCP crystals; ▼: brushite crystals; ◆ apatite crystals).

IV. DISCUSSION

The results obtained in the present study demonstrated clearly that the set porous-CPC can be developed based on the combination of α-TCP beads and MCPM-H₃PO₄ solution. Although no optimization was done in the present study, the current set porous-CPC can be formed in 10 min at 37°C and 100% relative humidity. The setting reaction of set porous-CPC is thought to be similar to that of the brushite cement invented by Mirtchi et al [3-4] When the α-TCP beads are exposed to MCPM-H₃PO₄ solution, the α-TCP beads dissolve to supply Ca^{2+} and PO_4^{3-} as shown in equation (3). The dissolution of α-TCP and resulting supply of Ca^{2+} and PO_4^{3-} lead to the MCPM-H₃PO₄ solution supersaturated with respect to brushite and thus Ca^{2+} and PO_4^{3-} are precipitated as brushite on the surface of α-TCP micro-spheres as shown in equation (4).

$$Ca_3(PO_4)_2 \rightarrow 3Ca^{2+} + 2PO_4^{3-} \qquad (3)$$
$$Ca^{2+} + H^+ + PO_4^{3-} + 2H_2O \rightarrow CaHPO_4 \cdot 2H_2O \qquad (4)$$

Based on this dissolution-precipitation reaction, the brushite crystals are precipitated on the surface of α-TCP beads and prolong with exposure time. Then brushite crystals on the surface of α-TCP beads interlock with each other to set. As a result of setting reaction a fully-interconnected porous structure would be formed due to hexagonally closed packing of α-TCP beads.

Key difference between set porous-CPC and brushite cement is the use of α-TCP beads instead of β-TCP powder. Of course, α-TCP beads are employed for the fabrication of a fully-interconnected porous structure. It should be noted that specific surface area of α-TCP beads is much smaller than that of β-TCP powder. The smaller specific surface area resulted in a smaller amount dissolution of α-TCP beads and smaller supply of Ca^{2+} and PO_4^{3-} to MCPM-H_3PO_4 solution. As a result, set porous-CPC took 10 mins to set even though brushite cement set in 30 sec when free from retarder.

The pore size of set porous-CPC is in the range of macropore, and has advantage to allow cell penetration and tissue ingrow in compare with micropore of CPC prepared by TCP powder. The macroporous structure encourages brushite layer transform to apatite layer quickly, within 3 days when porous-CPC immersed into SBF. Since the chemical composition of SBF is similar with human blood, set-porous CPC expose the feasibility to transform to apatile layer when implant into the bone defect. It is well know that apatite layer can bond directly with host tissue, thus promote the advabtage for set macroporous CPC. This set porous-CPC is expected to be used for the application in bone reconstruction and bone regeneration.

V. Conclusion

Set porous-CPC consisting of α-TCP beads and MCPM-H_3PO_4 solution was developed in this study. When α-TCP beads were exposed to MCPM-H_3PO_4 solution, brushite crystals were formed at the interface between α-TCP beads and the precipitated brushite crystals interlocked with each other to set. Due to the hexagonal closest packing of the α-TCP beads, set porous-CPC showed a fully-interconnected porous structure. Set porous-CPC is thought to have great potential value to be used as artificial bone substitutes and the scaffold for tissue engineering. Further assessment on animal study will be investigated to confirm the usefulness of set porous-CPC.

Acknowledgment

The authors gratefully acknowledge the AUN-SEED/Net - JICA for their financial support. This study was also supported in part by Ho Chi Minh University of Technology and Kyushu University.

Conflict of Interest

The authors declare that they have no conflict of interest.

References

1. Monma H, Kanazawa T (1976) The hydration of α-Tricalcium phosphate. J Ceram Soc Jpn 84(4):209-213.
2. Brown WE, Chow LC (1986) A new calcium phosphate, water-setting cement. J Am Ceram Soc 352-379.
3. Mirtchi AA, Lemaitre J, Terao N (1989) Calcium phosphate cements: Study of the β-tricalcium phosphate-monocalcium phosphate system. Biomaterials 10:475-480.
4. Mirtchi AA, Lemaitre J, Munting E (1989) Calcium phosphate cements: Action of setting regulators on the properties of the β-tricalcium phosphate-monocalcium phosphate cements. Biomaterials 10(9):634-638.
5. Burguera EF, Xu HHK, Weir MD (2006) Injectable and rapid-setting calcium phosphate bone cement with dicalcium phosphate dihydrate J Biomed Mater Res B 77(1):126-134.
6. Real RP, Wolke JGC, Vallet-Regi M, Jansen JA (2002) A new method to produce macropores in calcium phosphate cements. Biomaterials 23:3673-3680.
7. Pham TK, Michito M, Kanji T, Shigeki M, Kunio I (2010) Effect of phosphate solution on setting reaction of α-TCP spheres. J Aust Ceram Soc 46(2):63-67.
8. Kunio I, Kanji T, Trung KP, Michito M, Shigeki M (2012) Fully-interconnected Pore Forming Calcium Phosphate Cement. Key Engineering Materials 493-494:832-835.
9. Kokubo T, Takadama H (2006) How useful is SBF in predicting in vivo bone bioactivity? Biomaterials 27(15): 2907-2915.

Author Index

A

Abdullah, M.K.   402, 447
Adon, Nur Atiqah   117
Ahmad, M.K.I.   243, 454
Angadi, D.S.   175
Anh, Nguyen Tuan   437
Anh-Dao, Nguyen Thi   385
Anker, J.N.   62
Artusi, Sara   207
Aziz, N.   402

B

Bach, Nguyen Chi   179
Balachandran, Prasad   458
Barrett, T.G.   175
Binh, Nguyen Thanh   405
Blackwood, K.A.   125
Bock, N.   125
Brett, Emily   495
Bucovsky, M.   25
Bui, Francis M.   75
Bui-Thu, Cao   470
Burke, Thomas   15
Butovskaya, Elena   207

C

Chandramouli, S.   67
Chang, Yia-Chung   21
Chang, Young-Tae   211
Chen, J.   129
Chen, Shing-Jye   478
Chen, W.   159
Chen, Yu-Da   21
Chiang, Chu-Hui   461
Chien, Paul Ching-Hang   21
Chnafa, C.   7
Chuan, Hoang Khac   247
ChuDuc, Hoang   96
Crawford, R.   129
Cuong, N.V.   58

D

Dang, Thao-Nhi Ngoc   325
Dang, Tin Thanh   194, 426, 430
Dao, Hoang Anh   142

Denyer, Morgan   239
Dinh, Anh   75
Dinh, Dong-Luong   409
Dinh-Thuy, Thanh-Ha   325
Do, H.M.   55
Do, Hung Ngoc   108
Doan, N.   129, 133
Dong, Ngo Thanh   381
Du, Z.   129, 133
Duc-Tan, Tran   385
Dung, N.T.   113
Dung, N.V.   113
Dung, Pham T.   190
Dung, Tran Ngoc   257
Duong, Cong-Truyen   235
Duong, Hoang Thuy   325
Duong, Tran Vu Hoang   270

F

Farmery, Andrew   83
Fuis, V.   450
Futaba, T.   399

G

Gao, X.X.   422
Geer, Daniel   83
Giang, N.N.L.   58
Giang, Tran Le   395, 486
Gibaud, E.   7
Ginsburg, Geoffrey S.   15
Grounauer, P.A.   171
Gu, Y.D.   414, 422
Guanzon, Dominic   137

H

Ha, Dao.V.   376
Ha, N.T.   113
Ha, Nguyen Thi Thai   247
Ha, P.T.   55
Hahn, Clive   83
Hai, Nguyen Thanh   347, 381, 417
Hai, Vu Duy   179
Haiat, G.   43
Han, P.N.N.   58
Hanh, H.D.H.   285
Harun, N. Mat   447

Hedden, D.   167
Hien, Ngo   142
Hill, Peter   142
Ho, A.S.   55
Ho, Tinh Duy   108
Hoai, N.T.T.   359
Hoang, Nguyen Duc   481
Hoang, N.K.   58
Hoang, T.M.N.   55
Hollier, Brett   137
Hosokawa, A.   289
Houfek, L.   440
Houfek, M.   440
Hsiao, Fu-Chen   21
Huang, Zheng-Han   478
Hue, Nguyen Thi   339
Huong, Nguyen Thi Minh   100
Huong, Vo Thi Thien   274
Huy, Vu Thanh   474
Huynh, Canh Xuan   426
Huynh, L.Q.   32, 163
Huynh, Quoc T.   184
Huynh, Trinh   367, 371

I

Ibrahim, B.S.K.K   454
Ibrahim, N.   402, 447
Ishikawa, Kunio   507
Ivanovski, S.   129

J

Jabbar, Mohamad Hairol   117, 203
Jaklenec, Ana   47
Johannessen, Erik Andrew   11

K

Kang, Nam-Young   211
Kanji, Tsuru   507
Karlsen, Frank   11
Kaufman, J.J.   25
Khai, Le Quoc   100
Khalid, N.S.   243
Khan, Awab Umar   503
Khoa, Truong Quang Dang   285, 481, 486

Kien, Nguyen Trung   253
Kien, Pham Trung   507
Kim, Tae-Seong   409
Kirby, G.T.S.   125
Kuo, Rong-Fu   478

L

La, T.H.   215
Lan, Le Thi Tuyet   266
Lan, L.T.T.   285
Langer, Robert S.   47
Langton, C.M.   39
Le, Chi-Thong   87
Le, L.   351
Le, Lawrence H.   28, 32, 159, 163, 167
Le, Nghia T.   367
Le, Nguyen Thi   247, 270, 274
Le, Q.H.   215
Le, T.H.   215
Le, Thao Quy   147
Le, Trung   121, 491, 499
Le, Uyen Minh   51
Leavesley, David   137
Lee, Sungyoung   409
Legname, Giuseppe   155, 261
Li, F.L.   414
Li, J.S.   414
Liao, Jiunn-Der   151
Liao, S.   125
Liem, Kieu Trung   184
Lien, N.T.T.   359
Linh, Huynh Quang   100, 190, 434,
    437, 474
Linh, Le Hoang   417
Linh-Trung, Nguyen   385
Lončarić, S.   92
Long, D.Q.   113
Lou, E.   32, 159, 163, 167
Lu, Bo-Cheng   478
Luo, G.M.   25
Luong, Dinh Dong   395, 486
Luong, Thu-Hien   325
Luu, H.N.   55
Luu, Pham Dinh   278, 281
Luu, Thinh Duc   314
Ly, Tu T.   371

M

Madhusudhan, M.S.   343
Mahamad, Abd Kadir   203
Mahmud, Farhanahani   117, 203
Mahood, J.   167
Mai, Linh   87

Mai, Thao Phuong   261
Mai, Tran Thanh   417
Manton, Kerry J.   137
McCreanor, Philip   495
McDeed, Benjamin   499
Mei, Q.C.   422
Mendez, S.   7
Métraux, B.   171
Mien, Pham Thi Hai   474
Mien, Pham T.H.   190
Moreau, M.   167
Moureau, V.   7

N

Nadai, Matteo   207
Naili, Salah   36
Nam, Pham Cam   464
Nasir, N.H.M.   454
Nga, Nguyen Thi Huynh   257
Nga, Pham Thi Ngoc   253
Nghia, Huynh Luong   444
Nghia, P.T.   113
Ngo, Cuong.Q.   376
Ngo, Hoan T.   15
Ngo, Minh-Hien   151
Ngo, Thuyen V.   470
Ngo, Vuong Duy   311
Ngoc, Oanh Pham Thi   325
Ngoc, Pham Phuc   179
Nguyen, Bich   121, 491
Nguyen, Cao Thi Tai   257
Nguyen, Duc-Nam   235
Nguyen, Hai T.   376, 470
Nguyen, Hai Tuyen   104
Nguyen, Hieu Xuan   194
Nguyen, H.N.   55
Nguyen, Hoai T.T.   362
Nguyen, Hue T.   367, 371
Nguyen, K.C.T.   32
Nguyen, K.V.T.   62
Nguyen, Linh-Nam   21
Nguyen, L.T.   55
Nguyen, M.   133
Nguyen, T.A.   223
Nguyen, Tam   75
Nguyen, T.B.   343
Nguyen, Thai Van   147
Nguyen, Thanh D.   47
Nguyen, Thanh-Tam   108
Nguyen, The Dung   104
Nguyen, Thi-Hiep   317, 321, 325, 329,
    332, 336
Nguyen, T.M.H.   215
Nguyen, T.T.B.   215

Nguyen, Tuong Ngoc-Gia   298
Nguyen, T.V.A.   215
Nguyen, Uyen D.   184
Nguyen, Viet Dung   104
Nguyen, Vu-Hieu   36
Nguyen, X.P.   55
Nguyen, Xuan-Truong   321
NguyenDuc, Thuan   96
NguyenPhan, Kien   96
NguyenThai, Ha   96
Nhan, Nguyen Thanh   302
Nhu, Nguyen Tan   381
Nhu, Tran Thi Quynh   381
Nicoud, F.   7

O

O'Brien, Edward   495
Omar, Wan Ibtisam Wan   227
Ong, Q.   351
Othman, Norliza   203

P

Pannecouque, Christophe   207
Perrone, Rosalba   207
Petković, T.   92
Pham, H.N.   55
Pham, Thuong L.H.   362
Phan, Hieu V.   470
Phan, Nam Hoang   314
Phan, Phi Anh   83
Phan, Van Quyet   104
Phu, Pham Ty   347
Phuc, Dang Hoang   317, 329
Phuong, N.B.   359
Phuong, T.K.   219, 355

R

Reher, P.   129, 133
Ren, J.   125
Richter, Sara N.   207
Ristovski, N.   125
Rosete, F.   25
Rwei, S.P.   223

S

Sacchi, Mauricio D.   28, 32
Sam, Thai Minh   247
Sandya, S.   198
Sang, Vu Ngoc Thanh   481
Sanjana, S.   67
Shane, E.   25

Shao, Pei-Lin   151
Shao, S.R.   422
Shepherd, D.E.T.   175
Sherwani, F.   454
Siffert, R.S.   25
Sigüenza, J.   7
Sim, Sang Jun   1
Sinh, Tran Ngoc   247
Son, Hoang Le   325
Song, Y.Q.   414
Soon, Chin Fhong   227, 231, 239
Steck, R.   125
Stein, E.M.   25
Stevens, M.M.   125
Sumner, Loren   495
Swathi, S.   67

T

Tam, Nguyen Huynh Minh   395, 486
Tam, Nguyen Thanh   347
Tam, Tran Thai Thanh   247
Tan, K.P.   343
Tan, T.X.   285
Tee, Kian Sek   231, 239
Than, Uyen Thi Trang   137
Thang, Nguyen Duc   389, 395, 409, 481, 486
Thanh, Tran Ngoc   278, 281
Thanh, Tran Van   302
Thanh Nghi, T.T.   87
Thao, D.T.P.   219
Thao, Pham Thanh   389
Thien, Nguyen Nhan   434
Thien, Nguyen Phuc   461
Thong, Kok Tung   231
Thong, Nguyen Minh   464
Thong, Tran   79
Thu, Dang Huynh Anh   266
Thu, Du Thi Ngoc   247
Thu, T.T.K.   285
Thuan, L.D.   219, 355
Thuy, L.H.A.   219, 355
Tien, Phan Van   329

TMai, T.T.   55
Toai, Nguyen   142
Tomari, Razali   447
Tram, Bui Ngoc Thao   332, 336
Tran, Binh Q.   184
Tran, D.L.   55
Tran, Hue T.   470
Tran, Khanh Nghia   71
Tran, Khoa Anh   430
Tran, Kiet Anh   293
Tran, Mai T.   376
Tran, Phuong Ha-Lien   71, 293, 298, 311, 314
Tran, Thanh Van   293, 298
Tran, Thao Truong-Dinh   71, 293, 298, 311, 314
Tran, Tho N.H.T.   28
Tran, Van-Su   87
Tran-Minh, Nhut   11
Trinh, Huynh Le Thao   339
Trinh, Trung T.   362
Trong, Ngo Huynh Buu   21
Truong, Phuoc Long   1
Tuan, Le Quoc   270, 274, 317
Tuan, Vo Nhut   395
Tuvel, Barry   121
Tuyen, N.H.   113

U

Usha, B.S.   198

V

Vadivelu, R.   175
Van, Tran Thi Tuong   332
Van Binh, Pham   179
Van-Hoa, Huynh   306
Van-Thanh, Tran   306
Van Thuyen, Ngo   417
Van Tien, Tran   190, 474
Van Toi, V.   285, 470
Van Toi, Vo   317, 321, 325, 329, 332, 336, 389, 395, 417, 481, 486

Van Trung, Nguyen   444
Van Vo, Ha   491, 495, 499, 503
Van Vo, Toi   71, 293, 298, 311, 314, 376
Vayron, R.   43
Vi, Pham Tuong   339
Vinh, Pham Vu Quang   306
Vo, Ha   121
Vo, Minh-Thanh   87, 108
Vo, Q.N.   163
Vo, Toi   75
Vo-Dinh, Tuan   15
Vu, Cao Cuong   104

W

Wager, Lucas   137
Wang, Hsin-Neng   15
WanZaki, W.S.   243
Wille, M.-L.   39
Wong, Tung-Yiu   478
Woodruff, M.A.   125
Woods, Christopher   15
Wu, H.W.   223
Wu, Shang-Hsuan   21

X

Xia, W.   129
Xiao, J.   129
Xiao, Y.   129, 133

Y

Yan, F.   129
Yin, Matthew   495
Youseffi, Mansour   239

Z

Zakaria, W.N. Wan   402, 447
Zhang, Cathy   83
Zhang, Y.   422
Zheng, R.   167

Keyword Index

3D digital model 478
3D microtissues 231
α-glucosidase 351
α-TCP 507
τ-p transform 32

A

Absorbance test 11
Accelerometer 75, 104
Acne 474
Activities of daily living 184
Activity monitoring 481
Activity recognition 481
Adaptive noise cancellation 32
Adolescent idiopathic scoliosis 159
Adult 142
Advanced drug delivery 51
Ag nanoparticles 321
Agar well diffusion method 359
AIS 167
Alginate lyase 62
Alkalizers 293
Amyloid 155
Analysis 495
Android 108
Ankle implant 121
Anti-HER2 antibody 215
Anti-microbacteria 58
Antimicrobial 62
Antimicrobial activity 362
Antioxidant 464
Apatite 507
ARMS-PCR 253
Arousal 100
Articular cartilage 235
Aspergillus flavus 367, 371
Aspergillus parasiticus 367, 371
Atomic force microscopy 235
Atomic Force Microscopy 227

Atrophy 503
Au nanoparticles (AuNPs) 1

B

B. mori silkworm 325
Baculovirus (BV) 215
Bending radiograph 167
Beta – thalassemia 253
Bi-cortical non-locking bone screws 499
Bioactive Glass 125
Biodegradable polymer 47
Bioimaging 211
Biomedical 71
Biomedical engineer 79
Biomedical waste management 458
Blood flow velocity 447
Blood flows 7
Bluetooth 108
BMD 25
BME research 444
Body temperature 113
Bond dissociation enthalpy 464
Bone 36
Boundary lubrication 235
Bovine bone 336
BRACO-19 207
BRCA1 219
Brushite 507

C

C. albicans 359
C. canadensis 359
Calcium hydroxide 147
Cancellous bone 289
Cancer 71
Cancer treatment 51
Cardiac 203
Cardiopulmonary Function Test 83

Cardiovascular disease (CVD) 266
Care seeking behaviour 142
CBT 171
Cell adhesion 227 231
Cell morphology 227
Cell traction force 239
Center of lamina 159
Ceramics head 450
Cervical cancer 355
CHAC 285
Chest X-Ray 194
Chitosan 58, 317, 321, 329
Chromolaena odorata 362
Chronic kidney disease (CKD) 274
Chronic obstructive pulmonary diseases (COPD) 266
Clearance 274
Clinical engineers 434
Coherent point drift 409
Complex geometries 7
Complex shear modulus (CSM) 385
Computational fluid dynamics 7 11
Computational modelling 450
Context-awareness 405
Contourlet filter bank. 405
Control system approach 444
COPD 285
Copper binding 261
Corn 339
Creatinine 247 274
Cross point 426 430
Culture method 371
Culturing method 367
Cup total joint replacement wear measurement 440
Curcumin 58 293 298
Curve correction 167
Curve flexibility 167
Cyclin D2 219
Cystatin C 274

D

DAPK   355
Deconvolution   39
Decrease SpO$_2$ 100
Deep Vein Thrombosis (DVT)
    402, 447
Defibrillator   79
De-gelation   62
Degumming   325
De-noise   179
Dental implant   43
Dental implants   129, 133
Dental morphology   478
DeoxyHb   395
DEPTH   343
Depth image   409
Diabetes mellitus   270
Diabetes mellitus type 2   351
Diabetic peripheral neuropathy (DPN)
    270
Diadynamic   437
Diagnosis   therapy   155
Different height of heels   422
Digital   175
Digitalization   478
Disc diffusion   362
Discrete wavelet transform   376
Dispersion   28
Dissolution rate   302
Dissolution test   293
DNA biosensor   15
Dried corn kernels   371
Dried peanut kernels   367
Drug delivery   47
Drug delivery systems   55
DXA   25

E

Early diagnosis   447
ECG   108
EDTA-trypsin   231
EEG   179
EEG technology   470
Elasticity imaging   385
Elbow   454
Elderly people   347
Electric wheelchair   417, 470
Electrocardiogram (ECG)   266
Electrospinning   125
Electrotherapy   437
Endodontic treatment   147

Entire cross-point region   426
Entropy coding   426
Epistemology system approach   444
Erectile dysfunction   281
Estimated glomerular filtration rate
    247
Excitation-conduction   117
Exosomes   137
Experienced   414
Eye movements   470

F

Fabrication   329
Facial   474
Fall detection   347
Fall Detection   184
Feature extraction   179, 198
Feature matching   417
Female   422
Femur   175
FES   454
Fibroblasts   137
Field Programmable Gate Array
    117, 203
Fingertip pulse wave   399
Finite element analysis   175
Finite-difference time-domain method
    289
FitzHugh-Nagumo   117
Flapless surgery   129, 133
Flickering light   486
Fluorescence   211, 474
fNIRS   389, 395, 486
fNIRS data   376
Force mapping   239
FPGA-in-the-loop simulation
    117, 203
Freehand ultrasound   163
Frictional respo   235
F-scan   503
F-Scan   495
Fucoidan   71
functional Near Infrared Spectroscopy
    389

G

Gait   495, 503
Galvanic   437
Gemcitabine   51
Gliclazide   302
Global health   15

Glomerular filtration rate (GFR)   274
GPRS   104
G-quadruplex DNA   207
G-quadruplex RNA   207
GSTP1   219
Guided waves   28
Guidewire   92
GYSVd-1   461

H

HA   336
HbE   253
HCV RNA   257
HDL Coder   117   203
Healthcare   87
Heart activities   75
Heart disease screening   75
Heart rate   113
Heart rate variability   96
Heat pressing   47
Hemodynamic response   486
High-heel   414
HIV-1   207
Holographic interferometry   440
Holter ECG records   96
Hormone of men   278, 281
HPC/H2O/H3PO4 tertiary system
    223
HPMC 4000   298
HRV   96
HSV-1   207
HSVd   461
Human activity recognition   347
Human pose estimation   409
Humeral long bone transverse fracture
    499
Hyaluronan   329
Hydrogel   321
Hydrothermal   243
Hydroxyapatite   332, 336
Hydroxypropyl cellulose (HPC)   223
Hydroxypropyl-beta-cyclodextrin
(HPβCD)   306
Hypermethylation   355
Hyperthermia   55

I

ICRICM algorithm   430
Ideal cross-point region   426, 430
In situ crosslinking hydrogel   317
Inclination angle   92

Independent Component Analysis 194
Infectious disease 15
Inhibition zone 362
Insomnia 171
Inspired Sinewave Technique 83
Interpolation 239
Intracanal dressing 147
Intraoral scanning 478
Ionization potential 464
IOS 285
Itraconazole 306

J

Joint angle 422
Jugular venous pulse 75

K

Keratinocytes 137, 227
Keratinocytes cell 231
Khmer 253
Kinematic 414
$k\sigma$-classifier 198

L

Large-eddy simulations 7
Latin dance shoes 422
Library 211
Liquid crystal 227, 231
Liquid crystals 239
Localized Surface Plasmon Resonance (LSPR) 1
Lossless image compression 430
Lower critical solution temperature (LCST) 223
Lung cancer 194
Lung Function Test 83
Lung Inhomogeneity 83
Luo-Rudy Phase I model 203
Lyapunov exponent 399

M

MagMOON 62
Magnetic nanoparticles 55, 71
Magnetic resonance imaging 55
Maximum likelihood ensemble filter (MLEF) 385
Mean absolute difference 163

Measuring 381
Medical image 405
Melting method 298
Metal film over nanosphere 15
Methylation 219
Methylation specific PCR 355
MFCC-DTW 417
Micro and macro shape deviations 450
Micro structure 100
Microfabrication 47
Micronization 302
Microwave 332
mindfulness meditations 171
Minimum inhibitory concentration 359
MIPAV 430
MM-SCP 343
Modified method 339
Modified rice starch 311
Monitoring 87
Multiple sequence alignment 351
Multiplex DNA detection 15
Multiplex PCR 339, 367, 371
Multiplex RT-PCR 461
Musculoskeletal 454

N

Nano powder 336
Nano suspension 293
Nanoflowers 243
Nanowave 15
Nerve conduction study (NCS) 270
Network design 87
Networks 113
Neurodegeneration 155
NIR imaging 190
Noninvasive 83
Non-invasive fixation 499
Nonlinear chaos analysis 399
Non-OR region 261
Numerical simulation 289

O

Oblique Fracture 491
Octapeptide repeats (OR) 261
Ocular artifact 179
Optical resonator 21
Oral health status 142
Ordinary Differential Equations 203

Orthopedic Implant 121, 491
Osseo-integration 129, 133
Osseointegration 43
Osteoarthritis 235
Osteoporosis 25 129
Ovary 198
OxyHb 395

P

$p^{16INK4\alpha}$ 219, 355
Pacemaker 79
Paediatric 175
Papillometre 486
Paracetamol 311
Parkinson 104
Passive mixing 11
PEG 6000 298
Periapical lesions 147
Permanent incisors 147
Phased array 32
Phenolic compounds 464
Photoluminescence 21
Phylogenetic tree 351
pK_a prediction 343
Plasmonic Nanosensors 1
Point set registration 409
Polycaprolactone 125
Polymers 293
Polysomnography 100
Polyvinyl Alcohol 321
Poroelastic 36
Porous cement 507
Post menopause 133
Posterior maxilla 129, 133
Predetermination 444
prefrontal cortex 389
Prefrontal-occipital fNIRS 171
Primary gender 278
Primary stability 43
Prion 261
Prion conversion 261
Probability of bits 426
Probe 211
Prosthetic 495
Prosthetic foot 503
Protein aggregation 155
Proteinuria 247
Psychophysiological condition 399
PVPA 317 329

Q

Quantitative ultrasound   43

R

Radiographs   175
Radiology   194
Radius   25
Radon transform   28, 32
Rapid prototyping   478
rapid prototyping technology   402
RARβ 355
RASSF1A   219
Rayleigh Light Scattering (RLS)   1
Realtime RT-PCR   257
Red blood cells   7
Rehabilitation   179
Remote monitoring   79
Renal transplant recipients   247
Resolution   28
RFID system   437
Rhinacanthus nasutus   362
Ribs suppression   194

S

Sawtooth wave   437
SCFV   215
SCI   454
Scoliosis   163
Screw Angle   491
Screw Fixation   491
Sebum   474
Secondary gender   278
SEM   336
Semi-analytical finite element   36
Semi-automatic program   159
Senna alata (L.)   362
Sensor   211
Sequencing   257
Sequencing techniques   253
Sericin   325
SERS   Molecular Sentinel-on-Chip
    15
Serum Cystatin C   247
Setting reaction   507
Signal analysis   179
Silk fiber   325
Simulation   87

Size of Tumor   381
sleep EEG state II   171
Small scale   458
Smartphone   481
Socket   104
Software   239, 434
Solid dispersion   293, 306
Solid Tumor   381
Solid Volume Fraction   39
solid works   402
Solubility   302
Speckle filters   198
Spectroscopic imaging ellipsometry   21
Spirometry   285
Splitting and recombination   11
Springs   503
Stabilizing agent   302
Stress   450
Subtypes, viral load   257
Supervise patients   113
Surface-enhanced Raman scattering
    15
Sustained release   311
Synergistic interaction   223

T

T. insectorum   359
TAR   121
Telemedicine   104, 108
TEM   336
Temperature sensing   113
TENS   437
Testosterone   278, 281
Tetraplegic   454
The types   257
Thickness estimation   92
Threshold algorithm   347, 376
Threshold algorithms   470
Tissue   62
Titanium dioxide   243
Total Ankle Replacement   121
Total hip joint endoprosthesis   450
Trabert   437
Training   434
Transit Time Spectrum   39
Trends   495
Tumor   71, 381
Type 2 diabetes   281
Typing   395

U

Ultrasonic crosstalk   32
Ultrasonic imaging   167
Ultrasound   25, 28, 36, 39, 198, 332
Ultrasound backscattered waves   289
Ultrasound scan   402   447
Ultrasound shear wave imaging   385
Uni-cortical non-locking bone screws
    499
Unilaterality   444
Urinary albumin excretion (UAE)   270

V

Vein detection   190
Vein infrared   190
Ventilators   434
Vertebral rotation   159
Vietnam   142
Viroid   461
visual cortex   389
Visual cortex   486
Voxel-based method   163

W

Wall displacement   447
Waveguide   36
Wavelet transform   179
WGM biosensing   21
Wheelchair control   417
Wireless   437
Wireless networks   96
Wireless sensor   113
Wireless sensor system   184
working memory   389
Wound dressing   321
Writing   395
WSN   87

X

Xanthan gum (XG)   223
X-ray imaging   92

Z

Zazen fasting therapy   399
ZnO MS   21

Druck: KN Digital Printforce GmbH · Schockenriedstraße 37 · 70565 Stuttgart